Lecture Notes in Computer Science 9816

Commenced Publication in 1973
Founding and Former Series Editors:
Gerhard Goos, Juris Hartmanis, and Jan van Leeuwen

Editorial Board

David Hutchison
  Lancaster University, Lancaster, UK
Takeo Kanade
  Carnegie Mellon University, Pittsburgh, PA, USA
Josef Kittler
  University of Surrey, Guildford, UK
Jon M. Kleinberg
  Cornell University, Ithaca, NY, USA
Friedemann Mattern
  ETH Zurich, Zürich, Switzerland
John C. Mitchell
  Stanford University, Stanford, CA, USA
Moni Naor
  Weizmann Institute of Science, Rehovot, Israel
C. Pandu Rangan
  Indian Institute of Technology, Madras, India
Bernhard Steffen
  TU Dortmund University, Dortmund, Germany
Demetri Terzopoulos
  University of California, Los Angeles, CA, USA
Doug Tygar
  University of California, Berkeley, CA, USA
Gerhard Weikum
  Max Planck Institute for Informatics, Saarbrücken, Germany

More information about this series at http://www.springer.com/series/7410

Matthew Robshaw · Jonathan Katz (Eds.)

Advances in Cryptology – CRYPTO 2016

36th Annual International Cryptology Conference
Santa Barbara, CA, USA, August 14–18, 2016
Proceedings, Part III

 Springer

Editors
Matthew Robshaw
Impinj, Inc.
Seattle, WA
USA

Jonathan Katz
University of Maryland
College Park, MD
USA

ISSN 0302-9743   ISSN 1611-3349   (electronic)
Lecture Notes in Computer Science
ISBN 978-3-662-53014-6   ISBN 978-3-662-53015-3   (eBook)
DOI 10.1007/978-3-662-53015-3

Library of Congress Control Number: 2016945783

LNCS Sublibrary: SL4 – Security and Cryptology

Printed on acid-free paper

This Springer imprint is published by Springer Nature
The registered company is Springer-Verlag GmbH Berlin Heidelberg

Preface

The 36th International Cryptology Conference (Crypto 2016) was held at UCSB, Santa Barbara, CA, USA, during August 14–18, 2016. The workshop was sponsored by the International Association for Cryptologic Research.

Crypto continues to grow. This year the Program Committee evaluated a record 274 submissions out of which 70 were chosen for inclusion in the program. Each paper was reviewed by at least three independent reviewers, with papers from Program Committee members receiving at least five reviews. Reviewers with potential conflicts of interest for specific papers were excluded from all discussions about those papers, and this policy was extended to the program chairs as well.

The 44 members of the Program Committee were aided in this complex and time-consuming task by many external reviewers. We would like to thank them all for their service, their expert opinions, and their spirited contributions to the review process. It was a tremendously difficult task to choose the program for this conference, as the quality of the submissions was very high. It was even harder to identify a single best paper, but our congratulations go to Elette Boyle, Niv Gilboa, and Yuval Ishai from IDC Herzliya, Ben Gurion University, and the Technion, respectively, whose paper "Breaking the Circuit Size Barrier for Secure Computation Under DDH" was awarded Best Paper. Our congratulations also go to Mark Zhandry of MIT and Princeton University who won the award for the Best Student Paper "The Magic of ELFs."

The invited speakers at Crypto 2016 were Brian Sniffen, Chief Security Architect at Akamai Technologies, Inc., and Paul Kocher, founder of Cryptography Research. Brian's presentation cast a fascinating light on the issues of real-world cryptographic deployment while Paul's presentation, a joint invitation from the program co-chairs of both Crypto 2016 and CHES 2016, marked 20 years since his publication of the first paper on side-channel attacks at Crypto 1996.

We are, of course, indebted to Brian LaMacchia, the general chair, as well as the local Organizing Committee, who together proved ideal liaisons for establishing the layout of the program and for supporting the speakers. Our job as program co-chairs was made much easier by the excellent tools developed by Shai Halevi; both Shai and Brian were always available at short notice to answer our queries. Finally, we would like to thank all the authors who submitted their work to Crypto 2016. Without you the conference would not exist.

August 2016

Matthew Robshaw
Jonathan Katz

Crypto 2016

The 36th IACR International Cryptology Conference

University of California, Santa Barbara, CA, USA
August 14–18, 2016

Sponsored by the *International Association for Cryptologic Research*

General Chair

Brian LaMacchia — Microsoft

Program Chairs

Matthew Robshaw — Impinj, USA
Jonathan Katz — University of Maryland, USA

Program Committee

Alex Biryukov — University of Luxembourg, Luxembourg
Anne Canteaut — Inria, France
Dario Catalano — Università di Catania, Italy
Nishanth Chandran — Microsoft Research, India
Melissa Chase — Microsoft Research, USA
Joan Daemen — STMicroelectronics, Belgium and Radboud University, The Netherlands
Martin Van Dijk — University of Connecticut, USA
Itai Dinur — Ben-Gurion University, Israel
Pierre-Alain Fouque — Université Rennes 1, France
Steven Galbraith — Auckland University, New Zealand
Sanjam Garg — University of California, Berkeley, USA
S. Dov Gordon — George Mason University, USA
Jens Groth — University College London, UK
Sorina Ionica — Université de Picardie, France
Tetsu Iwata — Nagoya University, Japan
Aggelos Kiayias — National and Kapodistrian University of Athens, Greece
Gregor Leander — Ruhr Universität Bochum, Germany
Shengli Liu — Shanghai Jiao Tong University, China
Alexander May — Ruhr Universität Bochum, Germany
Willi Meier — FHNW, Switzerland
Payman Mohassel — Visa Research, USA

Elke De Mulder	Cryptographic Research, France
Steven Myers	Indiana University, USA
Phong Nguyen	Inria, France and CNRS/JFLI and University of Tokyo, Japan
Kaisa Nyberg	Aalto University, Finland
Kenny Paterson	Royal Holloway University of London, UK
Thomas Peyrin	Nanyang Technological University, Singapore
Benny Pinkas	Bar-Ilan University, Israel
David Pointcheval	École Normale Supérieure, France
Manoj Prabhakaran	University of Illinois, USA
Bart Preneel	KU Leuven, Belgium
Mariana Raykova	Yale University, USA
Christian Rechberger	TU-Graz, Austria and DTU, Denmark
Mike Rosulek	Oregon State University, USA
Rei Safavi-Naini	University of Calgary, Canada
Alessandra Scafuro	Boston University and Northeastern University, USA
Patrick Schaumont	Virginia Tech, USA
Dominique Schröder	Saarland University, Germany
Jae Hong Seo	Myongji University, Korea
Yannick Seurin	ANSSI, France
Abhi Shelat	University of Virginia, USA
Nigel Smart	University of Bristol, UK
Ron Steinfeld	Monash University, Australia
Mehdi Tibouchi	NTT Secure Platform Laboratories, Japan

Additional Reviewers

Michel Abdalla	Foteini Baldimtsi	Dan Boneh
Masayuki Abe	Paulo Barreto	Jonathan Bootle
Arash Afshar	Gilles Barthe	Raphael Bost
Shashank Agrawal	Lejla Batina	Christina Boura
Shweta Agrawal	Christof Beierle	Florian Bourse
Ayo Akinyele	Mihir Bellare	Cyril Bouvier
Martin Albrecht	Fabrice Benhamouda	Elette Boyle
Gergely Alpar	Sanjay Bhattacherjee	Zvika Brakerski
Jacob Alperin-Sheriff	Jean-Francois Biasse	Lus Brandão
Elena Andreeva	Begul Bilgin	Anne Broadbent
Daniel Apon	Gaetan Bisson	Christina Brzuska
Gilad Asharov	Nir Bitansky	Christian Cachin
Gilles Van Assche	Simon Blackburn	Ran Canetti
Nuttapong Attrapadung	Olivier Blazy	Angelo De Caro
Saikrishna Badrinarayanan	Matthieu Bloch	Guilhem Castagnos
Josep Balasch	Céline Blondeau	Andrea Cerulli
	Andrej Bogdanov	Pyrros Chaidos

André Chailloux
Jie Chen
Céline Chevalier
Chongwon Cho
Seung Geol Choi
Ashish Choudhury
Sherman Chow
Kai-Min Chung
Michele Ciampi
Michael Clear
Ran Cohen
Geoffroy Couteau
Dana Dachman-Soled
Deepesh Data
Jean Paul Degabriele
David Derler
Daniel Dinu
Christoph Dobraunig
Yevgeniy Dodis
Nico Döttling
Natnatee Dokmai
Leo Ducas
Tuyet Duong
Keita Emura
Frederic Ezerman
Pooya Farshim
Sebastian Faust
Dario Fiore
Marc Fischlin
Joe Fitzsimons
Nils Fleischhacker
Emmanuel Fouotsa
Georg Fuchsbauer
Eiichiro Fujisaki
Martin Gagne
François Le Gall
Chaya Ganesh
Juan Garay
Christina Garman
Romain Gay
Essam Ghadafi
Benedikt Gierlichs
Niv Gilboa
Vipul Goyal
Frédéric Grosshans
Aurore Guillevic

Divya Gupta
Felix Günther
Shai Halevi
Mike Hamburg
Shuai Han
Helena Handschuh
Christian Hanser
Carmit Hazay
Ethan Heilman
Ryan Henry
Gottfried Herold
Felix Heuer
Viet Tung Hoang
Dennis Hofheinz
Ziyuan Hu
Yan Huang
Michael Hutter
Malika Izabachene
Håkon Jacobsen
Mahavir Jhawar
Dingding Jia
Keting Jia
Thomas Johansson
Aaron Johnson
Kimmo Järvinen
Yael Tauman Kalai
Bhavana Kanukurthi
Petteri Kaski
Marcel Keller
Nathan Keller
Carmen Kempka
Iordanis Kerenidis
Dmitry Khovratovich
Dakshita Khurana
Eike Kiltz
Jinsu Kim
Taechan Kim
Paul Kirchner
Elena Kirshanova
Susumu Kiyoshima
Simon Knellwolf
Stefan Koelbl
Vlad Kolesnikov
Takeshi Koshiba
Luke Kowalczyk
Thorsten Kranz

Daniel Kraschewski
Anna Krasnova
Hugo Krawczyk
Fernando Krell
Stephan Krenn
Ranjit Kumaresan
Alptekin Kupcu
Fabien Laguillaumie
Virginie Lallemand
Enrique Larraia
Changmin Lee
Hyung Tae Lee
Kwangsu Lee
Nikos Leonardos
Tancrède Lepoint
Anthony Leverrier
Benoit Libert
Fuchun Lin
Rachel Lin
Yehuda Lindell
Feng-Hao Liu
Yi-Kai Liu
Patrick Longa
Steve Lu
Stefan Lucks
Atul Luykx
Anna Lysyanskaya
Lin Lyu
Vadim Lyubashevsky
Mohammad Mahmoody
Hemanta Maji
Giulio Malavolta
Tal Malkin
Alex Malozemoff
Mark Marson
Daniel Masny
Takahiro Matsuda
Florian Mendel
Bart Mennink
Thyla van der Merwe
Peihan Miao
Christof Michel
Ian Miers
Andrew Miller
Brice Minaud
Kazuhiko Minematsu

Contents – Part III

IBE, ABE, and Functional Encryption

Automated Tools and Synthesis

Zero Knowledge

Theory

Quantum Techniques

Quantum Homomorphic Encryption
for Polynomial-Sized Circuits

Yfke Dulek[1,2,3(✉)], Christian Schaffner[1,2,3(✉)], and Florian Speelman[2,3(✉)]

[1] University of Amsterdam, Amsterdam, The Netherlands
C.Schaffner@uva.nl
[2] CWI, Amsterdam, The Netherlands
[3] QuSoft, Amsterdam, The Netherlands
{Y.M.Dulek,F.Speelman}@cwi.nl

Abstract. We present a new scheme for quantum homomorphic encryption which is compact and allows for efficient evaluation of arbitrary polynomial-sized quantum circuits. Building on the framework of Broadbent and Jeffery [BJ15] and recent results in the area of instantaneous non-local quantum computation [Spe15], we show how to construct quantum gadgets that allow perfect correction of the errors which occur during the homomorphic evaluation of T gates on encrypted quantum data. Our scheme can be based on any classical (leveled) fully homomorphic encryption (FHE) scheme and requires no computational assumptions besides those already used by the classical scheme. The size of our quantum gadget depends on the space complexity of the classical decryption function – which aligns well with the current efforts to minimize the complexity of the decryption function.

Our scheme (or slight variants of it) offers a number of additional advantages such as ideal compactness, the ability to supply gadgets "on demand", and circuit privacy for the evaluator against passive adversaries.

Keywords: Homomorphic encryption · Quantum cryptography · Quantum teleportation · Garden-hose model

1 Introduction

Fully homomorphic encryption (FHE) is the holy grail of modern cryptography. Rivest et al. were the first to observe the possibility of manipulating encrypted data in a meaningful way, rather than just storing and retrieving it [RAD78]. After some partial progress [GM84, Pai99, BGN05, IP07] over the years, a breakthrough happened in 2009 when Gentry presented a fully-homomorphic encryption (FHE) scheme [Gen09]. Since then, FHE schemes have been simplified [VDGHV10] and based on more standard assumptions [BV11]. The exciting developments around FHE have sparked a large amount of research in other areas such as functional encryption [GKP+13a, GVW13, GKP+13b, SW14] and obfuscation [GGH+13].

Developing quantum computers is a formidable technical challenge, so it currently seems likely that quantum computing will not be available

© International Association for Cryptologic Research 2016
M. Robshaw and J. Katz (Eds.): CRYPTO 2016, Part III, LNCS 9816, pp. 3–32, 2016.
DOI: 10.1007/978-3-662-53015-3_1

immediately to everyone and hence quantum computations have to be outsourced. Given the importance of classical[1] FHE for "computing in the cloud", it is natural to wonder about the existence of encryption schemes which can encrypt *quantum data* in such a way that a server can carry out arbitrary *quantum computations* on the encrypted data (without interacting with the encrypting party[2]). While previous work on *quantum homomorphic encryption* has mostly focused on information-theoretic security (see Sect. 1.2 below for details), schemes that are based on computational assumptions have only recently been thoroughly investigated by Broadbent and Jeffery. In [BJ15], they give formal definitions of quantum fully homomorphic encryption (QFHE) and its security and they propose three schemes for quantum homomorphic encryption assuming the existence of classical FHE.

A natural idea is to encrypt a message qubit with the quantum one-time pad (i.e. by applying a random Pauli operation), and send the classical keys for the quantum one-time pad along as classical information, encrypted by the classical FHE scheme. This basic scheme is called CL in [BJ15]. It is easy to see that CL allows an evaluator to compute arbitrary Clifford operations on encrypted qubits, simply by performing the actual Clifford circuit, followed by homomorphically updating the quantum one-time pad keys according to the commutation rules between the performed Clifford gates and the Pauli encryptions. The CL scheme can be regarded as analogous to additively homomorphic encryption schemes in the classical setting. The challenge, like multiplication in the classical case, is to perform non-Clifford gates such as the T gate. Broadbent and Jeffery propose two different approaches for doing so, accomplishing homomorphic encryption for circuits with a limited number of T gates. These results lead to the following main open problem:

Is it possible to construct a quantum homomorphic scheme that allows evaluation of polynomial-sized quantum circuits?

1.1 Our Contributions

We answer the above question in the affirmative by presenting a new scheme TP (as abbreviation for teleportation) for quantum homomorphic encryption which is both compact and efficient for circuits with polynomially many T gates. The scheme is secure against chosen plaintext attacks from quantum adversaries, as formalized by the security notion *q-IND-CPA security* defined by Broadbent and Jeffery [BJ15].

Like the schemes proposed in [BJ15], our scheme is an extension of the Clifford scheme CL . We add auxiliary quantum states to the evaluation key which we call quantum *gadgets* and which aid in the evaluation of the T gates. The size of a gadget depends only on (a certain form of) the space complexity of the

---

[1] Here and throughout the article, we use "classical" to mean "non-quantum".

[2] In contrast to *blind* or *delegated quantum computation* where some interaction between client and server is usually required, see Sect. 1.2 for references.

decryption function of the classical FHE scheme. This relation turns out to be very convenient, as classical FHE schemes are often optimized with respect to the complexity of the decryption operation (in order to make them bootstrappable). As a concrete example, if we instantiate our scheme with the classical FHE scheme by Brakerski and Vaikuntanathan [BV11], each evaluation gadget of our scheme consists of a number of qubits which is polynomial in the security parameter.

In TP, we require exactly one evaluation gadget for every T gate that we would like to evaluate homomorphically. Intuitively, after a T gate is performed on a one-time-pad encrypted qubit $X^a Z^b |\psi\rangle$, the result might contain an unwanted phase P^a depending on the key a with which the qubit was encrypted, since $T X^a Z^b |\psi\rangle = P^a X^a Z^b T |\psi\rangle$. Obviously, the evaluator is not allowed to know the key a. Instead, he holds an encryption $\tilde{a}$ of the key, produced by a classical FHE scheme. The evaluator can teleport the encrypted qubit "through the gadget" [GC99] in a way that depends on $\tilde{a}$, in order to remove the unwanted phase. In more detail, the quantum part of the gadget consists of a number of EPR pairs which are prepared in a way that depends on the secret key of the classical FHE scheme. Some classical information is provided with the gadget that allows the evaluator to homomorphically update the encryption keys after the teleportation steps. On a high level, the use of an evaluation gadget corresponds to a *instantaneous non-local quantum computation*[3] where one party holds the secret key of the classical FHE scheme, and the other party holds the input qubit and a classical encryption of the key to the quantum one-time pad. Together, this information determines whether an inverse phase gate $P^\dagger$ needs to be performed on the qubit or not. Very recent results by Speelman [Spe15] show how to perform such computations with a bounded amount of entanglement. These techniques are the crucial ingredients for our construction and are the reason why the *garden-hose complexity* [BFSS13] of the decryption procedure of the classical FHE is related to the size of our gadgets.

The quantum part of our evaluation gadget is strikingly simple, which provides a number of advantages. To start with, the evaluation of a T gate requires only one gadget, and does not cause errors to accumulate on the quantum state. The scheme is very compact in the sense that the state of the system after the evaluation of a T gate has the same form as after the initial encryption, except for any classical changes caused by the classical FHE evaluation. This kind of compactness also implies that individual evaluation gadgets can be supplied "on demand" by the holder of the secret key. Once an evaluator runs out of gadgets, the secret key holder can simply supply more of them.

Furthermore, TP does not depend on a specific classical FHE scheme, hence any advances in classical FHE can directly improve our scheme. Our requirements for the classical FHE scheme are quite modest: we only require the classical scheme to have a space-efficient decryption procedure and to be secure against quantum adversaries. In particular, no circular-security assumption is required.

---

[3] This term is not related to the term 'instantaneous quantum computation' [SB08], and instead first was used as a specific form of non-local quantum computation, one where all parties have to act simultaneously.

Since we supply at most a polynomial number of evaluation gadgets, our scheme TP is leveled homomorphic by construction, and we can simply switch to a new classical key after every evaluation gadget. In fact, the Clifford gates in the quantum evaluation circuit only require additive operations from the classical homomorphic scheme, while each T gate needs a fixed (polynomial) number of multiplications. Hence, we do not actually require fully homomorphic classical encryption, but leveled fully homomorphic schemes suffice.

Finally, circuit privacy in the passive setting almost comes for free. When wanting to hide which circuit was evaluated on the data, the evaluating party can add an extra randomization layer to the output state by applying his own one-time pad. We show that if the classical FHE scheme has the circuit-privacy property, then this extra randomization completely hides the circuit from the decrypting party. This is not unique to our specific scheme: the same is true for CL.

In terms of applications, our construction can be appreciated as a constant-round scheme for *blind delegated quantum computation*, using computational assumptions. The server can evaluate a universal quantum circuit on the encrypted input, consisting of the client's quantum input and a (classical) description of the client's circuit. In this context, it is desirable to minimize the quantum resources needed by the client. We argue that our scheme can still be used for constant-round blind delegated quantum computation if we limit either the client's quantum memory or the types of quantum operations the client can perform.

As another application, we can instantiate our construction with a classical FHE scheme that allows for *distributed* key generation and decryption amongst different parties that all hold a share of the secret key [AJLA+12]. In that case, it is likely that our techniques can be adapted to perform *multiparty quantum computation* [BCG+06] in the semi-honest case. However, the focus of this article lies on the description and security proof of the new construction, and more concrete applications are the subject of upcoming work.

1.2   Related Work

Early classical FHE schemes were limited in the sense that they could not facilitate arbitrary operations on the encrypted data: some early schemes only implemented a single operation (addition or multiplication) [RSA78, GM84, Pai99]; later on it became possible to combine several operations in a limited way [BGN05, GHV10, SYY99]. Gentry's first fully homomorphic encryption scheme [Gen09] relied on several non-standard computational assumptions. Subsequent work [BGV12, BV11] has relaxed these assumptions or replaced them with more conventional assumptions such as the hardness of learning with errors (LWE), which is believed to be hard also for quantum attackers. It is impossible to completely get rid of computational assumptions for a classical FHE scheme, since the existence of such a scheme would imply the existence of an information-theoretically secure protocol for private information retrieval (PIR) [KO97] that breaks the lower bound on the amount of communication required for that task [CKGS98, Fil12].

While quantum fully homomorphic encryption (QFHE) is closely related to the task of blind or delegated quantum computation [Chi05, BFK09, ABOE10, VFPR14, FBS+14, Bro15a, Lia15], QFHE does not allow interaction between the client and the server during the computation. Additionally, in QFHE, the server is allowed to choose which unitary it wants to apply to the (encrypted) data.

Yu et al. [YPDF14] showed that perfectly information-theoretically secure QFHE is not possible unless the size of the encryption grows exponentially in the input size. Thus, any scheme that attempts to achieve information-theoretically secure QFHE has to leak some proportion of the input to the server [AS06, RFG12] or can only be used to evaluate a subset of all unitary transformations on the input [RFG12, Lia13, TKO+14]. Like the multiplication operation is hard in the classical case, the hurdle in the quantum case seems to be the evaluation of non-Clifford gates. A recent result by Ouyang et al. provides information-theoretic security for circuits with at most a constant number of non-Clifford operations [OTF15].

Broadbent and Jeffery [BJ15] proposed two schemes that achieve homomorphic encryption for nontrivial sets of quantum circuits. Instead of trying to achieve information-theoretic security, they built their schemes based on a classical FHE scheme and hence any computational assumptions on the classical scheme are also required for the quantum schemes. Computational assumptions allow bypassing the impossibility result from [YPDF14] and work toward a (quantum) fully homomorphic encryption scheme.

Both of the schemes presented in [BJ15] are extensions of the scheme CL described in Sect. 1.1. These two schemes use different methods to implement the evaluation of a T gate, which we briefly discuss here. In the EPR scheme, some entanglement is accumulated in a special register during every evaluation of a T gate, and stored there until it can be resolved in the decryption phase. Because of this accumulation, the complexity of the decryption function scales (quadratically) with the number of T gates in the evaluated circuit, thereby violating the compactness requirement of QFHE. The scheme AUX also extends CL, but handles T gates in a different manner. The evaluator is supplied with auxiliary quantum states, stored in the evaluation key, that allow him to evaluate T gates and immediately remove any error that may have occurred. In this way, the decryption procedure remains very efficient and the scheme is compact. Unfortunately, the required auxiliary states grow doubly exponential in size with respect to the T depth of the circuit, rendering AUX useful only for circuits with constant T depth. Our scheme TP is related to AUX in that extra resources for removing errors are stored in the evaluation key. In sharp contrast to AUX, the size of the evaluation key in TP only grows linearly in the number of T gates in the circuit (and polynomially in the security parameter), allowing the scheme to be leveled fully homomorphic. Since the evaluation of the other gates causes no errors on the quantum state, no gadgets are needed for those; any circuit containing polynomially many T gates can be efficiently evaluated.

1.3   Structure of the Paper

We start by introducing some notation in Sect. 2 and presenting the necessary preliminaries on quantum computation, (classical and quantum) homomorphic encryption, and the garden-hose model which is essential to the most-general construction of the gadgets. In Sect. 3, we describe the scheme TP and show that it is compact. The security proof of TP is somewhat more involved, and is presented in several steps in Sect. 4, along with an informal description of a circuit-private variant of the scheme. In Sect. 5, the rationale behind the quantum gadgets is explained, and some examples are discussed to clarify the construction. We conclude our work in Sect. 6 and propose directions for future research.

2   Preliminaries

2.1   Quantum Computation

We assume the reader is familiar with the standard notions in the field of quantum computation (for an introduction, see [NC00]). In this subsection, we only mention the concepts that are essential to our construction.

The single-qubit *Pauli group* is, up to global phase, generated by the bit flip and phase flip operations,

$$X = \begin{bmatrix} 0 & 1 \\ 1 & 0 \end{bmatrix}, \quad Z = \begin{bmatrix} 1 & 0 \\ 0 & -1 \end{bmatrix}.$$

A *Pauli operator* on n qubits is simply any tensor product of n independent single-qubit Pauli operators. All four single-qubit Pauli operators are of the form $X^a Z^b$ with $a, b \in \{0, 1\}$. Here, and in the rest of the paper, we ignore the global phase of a quantum state, as it is not observable by measurement.

The *Clifford group* on n qubits consists of all unitaries C that commute with the Pauli group, i.e. the Clifford group is the normalizer of the Pauli group. Since all Pauli operators are of the form $X^{a_1} Z^{b_1} \otimes \cdots \otimes X^{a_n} Z^{b_n}$, this means that C is a Clifford operator if for any $a_1, b_1, \ldots, a_n, b_n \in \{0, 1\}$ there exist $a'_1, b'_1, \ldots, a'_n, b'_n \in \{0, 1\}$ such that (ignoring global phase):

$$C(X^{a_1} Z^{b_1} \otimes \cdots \otimes X^{a_n} Z^{b_n}) = (X^{a'_1} Z^{b'_1} \otimes \cdots \otimes X^{a'_n} Z^{b'_n})C.$$

All Pauli operators are easily verified to be elements of the Clifford group, and the entire Clifford group is generated by

$$P = \begin{bmatrix} 1 & 0 \\ 0 & i \end{bmatrix}, \quad H = \frac{1}{\sqrt{2}} \begin{bmatrix} 1 & 1 \\ 1 & -1 \end{bmatrix}, \quad \text{and} \quad CNOT = \begin{bmatrix} 1 & 0 & 0 & 0 \\ 0 & 1 & 0 & 0 \\ 0 & 0 & 0 & 1 \\ 0 & 0 & 1 & 0 \end{bmatrix}.$$

(See for example [Got98].) The Clifford group does not suffice to simulate arbitrary quantum circuits, but by adding any single non-Clifford gate, any quantum circuit can be efficiently simulated with only a small error. As in [BJ15], we choose this non-Clifford gate to be the T gate,

$$\mathsf{T} = \begin{bmatrix} 1 & 0 \\ 0 & e^{i\pi/4} \end{bmatrix}.$$

Note that the T gate, because it is non-Clifford, does not commute with the Pauli group. More specifically, we have $\mathsf{T}\mathsf{X}^a\mathsf{Z}^b = \mathsf{P}^a\mathsf{X}^a\mathsf{Z}^b\mathsf{T}$. It is exactly the formation of this P gate that has proven to be an obstacle to the design of an efficient quantum homomorphic encryption scheme.

We use $|\psi\rangle$ or $|\varphi\rangle$ to denote pure quantum states. Mixed states are denoted with ρ or σ. Let $\mathbb{I}_d$ denote the identity matrix of dimension d: this allows us to write the *completely mixed state* as $\mathbb{I}_d/d$.

Define $|\Phi^+\rangle := \frac{1}{\sqrt{2}}(|00\rangle + |11\rangle)$ to be an EPR pair.

If X is a random variable ranging over the possible basis states B for a quantum system, then let $\rho(X)$ be the density matrix corresponding to X, i.e. $\rho(X) := \sum_{b \in B} \Pr[X = b]|b\rangle\langle b|$.

Applying a Pauli operator that is chosen uniformly at random results in a single-qubit completely mixed state, since

$$\forall \rho : \sum_{a,b \in \{0,1\}} \left(\frac{1}{4}\mathsf{X}^a\mathsf{Z}^b\rho(\mathsf{X}^a\mathsf{Z}^b)^\dagger\right) = \frac{\mathbb{I}_2}{2}$$

This property is used in the construction of the *quantum one-time pad*: applying a random Pauli $\mathsf{X}^a\mathsf{Z}^b$ to a qubit completely hides the content of that qubit to anyone who does not know the key (a, b) to the pad. Anyone in possession of the key can decrypt simply by applying $\mathsf{X}^a\mathsf{Z}^b$ again.

2.2 Homomorphic Encryption

This subsection provides the definitions of (classical and quantum) homomorphic encryption schemes, and the security conditions for such schemes. In the current work, we only consider homomorphic encryption in the public-key setting. For a more thorough treatment of these concepts, and how they can be transferred to the symmetric-key setting, see [BJ15].

The Classical Setting. A classical homomorphic encryption scheme HE consists of four algorithms: key generation, encryption, evaluation, and decryption. The key generator produces three keys: a public key and evaluation key, both of which are publicly available to everyone, and a secret key which is only revealed to the decrypting party. Anyone in possession of the public key can encrypt the inputs $x_1, \ldots, x_\ell$, and send the resulting ciphertexts $c_1, \ldots, c_\ell$ to an evaluator who evaluates some circuit C on them. The evaluator sends the result to a party that possesses the secret key, who should be able to decrypt it to $\mathsf{C}(x_1, \ldots, x_\ell)$.

More formally, HE consists of the following four algorithms which run in probabilistic polynomial time in terms of their input and parameters [BV11]:

$(pk, evk, sk) \leftarrow$ HE.KeyGen$(1^\kappa)]$ where $\kappa \in \mathbb{N}$ is the *security parameter*. Three keys are generated: a public key pk, which can be used for the encryption of messages; a secret key sk used for decryption; and an evaluation key evk that may aid in evaluating the circuit on the encrypted state. The keys pk and evk are announced publicly, while sk is kept secret.

$c \leftarrow \mathsf{HE.Enc}_{pk}(x)$ for some one-bit message $x \in \{0,1\}$. This probabilistic procedure outputs a ciphertext c, using the public key pk.

$c' \leftarrow \mathsf{HE.Eval}^{\mathsf{C}}_{evk}(c_1, \ldots, c_\ell)$ uses the evaluation key to output some ciphertext c' which decrypts to the evaluation of circuit C on the decryptions of $c_1, \ldots, c_\ell$. We will often think of Eval as an evaluation of a function f instead of some canonical circuit for f, and write $\mathsf{HE.Eval}^{f}_{evk}(c_1, \ldots, c_\ell)$ in this case.

$x' \leftarrow \mathsf{HE.Dec}_{sk}(c)$ outputs a message $x' \in \{0,1\}$, using the secret key sk.

In principle, $\mathsf{HE.Enc}_{pk}$ can only encrypt single bits. When encrypting an n-bit message $x \in \{0,1\}^n$, we encrypt the message bit-by-bit, applying the encryption procedure n times. We sometimes abuse the notation $\mathsf{HE.Enc}_{pk}(x)$ to denote this bitwise encryption of the string x.

For HE to be a homomorphic encryption scheme, we require *correctness* in the sense that for any circuit C, there exists a negligible[4] function η such that, for any input x,

$$\Pr[\mathsf{HE.Dec}_{sk}(\mathsf{HE.Eval}^{\mathsf{C}}_{evk}(\mathsf{HE.Enc}_{pk}(x))) \neq \mathsf{C}(x)] \leq \eta(\kappa).$$

In this article, we assume for clarity of exposition that classical schemes HE are perfectly correct, and that it is possible to immediately decrypt after encrypting (without doing any evaluation).

Another desirable property is *compactness*, which states that the complexity of the decryption function should not depend on the size of the circuit: a scheme is compact if there exists a polynomial $p(\kappa)$ such that for any circuit C and any ciphertext c, the complexity of applying $\mathsf{HE.Dec}$ to the result of $\mathsf{HE.Eval}^{C}(c)$ is at most $p(\kappa)$.

A scheme that is both correct for all circuits and compact, is called *fully* homomorphic. If it is only correct for a subset of all possible circuits (e.g. all circuits with no multiplication gates) or if it is not compact, it is considered to be a *somewhat* homomorphic scheme. Finally, a *leveled* fully homomorphic scheme is (compact and) homomorphic for all circuits up to a variable depth L, which is supplied as an argument to the key generation function [Vai11].

We will use the notation $\widetilde{x}$ to denote the result of running $\mathsf{HE.Enc}_{pk}(x)$: that is, $\mathsf{Dec}_{sk}(\widetilde{x}) = x$ with overwhelming probability. In our construction, we will often deal with multiple classical key sets $(pk_i, sk_i, evk_i)_{i \in I}$ indexed by some set I. In that case, we use the notation $\widetilde{x}^{[i]}$ to denote the result of $\mathsf{HE.Enc}_{pk_i}(x)$, in order to avoid confusion. Here, pk_i does *not* refer to the ith bit of the public key: in case we want to refer to the ith bit of some string s, we will use the notation $s[i]$.

When working with multiple key sets, it will often be necessary to transform an already encrypted message $\widetilde{x}^{[i]}$ into an encryption $\widetilde{x}^{[j]}$ using a different key

---

[4] A *negligible function* η is a function such that for every positive integer d, $\eta(n) < 1/n^d$ for big enough n.

set $j \neq i$. To achieve this transformation, we define the procedure $\mathsf{HE.Rec}_{i \to j}$ that can always be used for this *recryption* task as long as we have access to an encrypted version $\widetilde{sk}_i^{[j]}$ of the old secret key sk_i. Effectively, $\mathsf{HE.Rec}_{i \to j}$ homomorphically evaluates the decryption of $\widetilde{x}^{[i]}$:

$$\mathsf{HE.Rec}_{i \to j}(\widetilde{x}^{[i]}) := \mathsf{HE.Eval}_{evk_j}^{\mathsf{HE.Dec}}\left(\widetilde{sk}_i^{[j]}, \mathsf{HE.Enc}_{pk_j}(\widetilde{x}^{[i]})\right).$$

The Quantum Setting. A quantum homomorphic encryption scheme QHE, as defined in [BJ15], is a natural extension of the classical case, and differs from it in only a few aspects. The secret and public keys are still classical, but the evaluation key is allowed to be a quantum state. This means that the evaluation key is not necessarily reusable, and can be consumed during the evaluation procedure. The messages to be encrypted are qubits instead of bits, and the evaluator should be able to evaluate quantum circuits on them.

All definitions given above carry over quite naturally to the quantum setting (see also [BJ15]):

$(pk, \rho_{evk}, sk) \leftarrow \mathsf{QHE.KeyGen}(1^\kappa)$ where $\kappa \in \mathbb{N}$ is the security parameter. In contrast to the classical case, the evaluation key is a quantum state.

$\sigma \leftarrow \mathsf{QHE.Enc}_{pk}(\rho)$ produces, for every valid public key pk and input state ρ from some message space, to a quantum cipherstate σ in some cipherspace.

$\sigma' \leftarrow \mathsf{QHE.Eval}_{\rho_{evk}}^{\mathsf{C}}(\sigma)$ represents the evaluation of a circuit C. If C requires n input qubits, then σ should be a product of n cipherstates. The evaluation function maps it to a product of n' states in some output space, where n' is the number of qubits that C would output. The evaluation key ρ_{evk} is consumed in the process.

$\rho' \leftarrow \mathsf{QHE.Dec}_{sk}(\sigma')$ maps a single state σ' from the output space to a quantum state ρ' in the message space. Note that if the evaluation procedure QHE.Eval outputs a product of n' states, then QHE.Dec needs to be run n' times.

The decryption procedure differs from the classical definition in that we require the decryption to happen subsystem-by-subsystem: this is fundamentally different from the more relaxed notion of *indivisible schemes* [BJ15] where an auxiliary quantum register may be built up for the entire state, and the state can only be decrypted as a whole. In this work, we only consider the divisible definition.

Quantum Security. The notion of security that we aim for is that of *indistinguishability under chosen-plaintext attacks*, where the attacker may have quantum computational powers (q-IND-CPA). This security notion was introduced in [BJ15, Definition 3.3] (see [GHS15] for a similar notion of the security of classical schemes against quantum attackers) and ensures semantic security [ABF+16]. We restate it here for completeness.

Definition 1 [BJ15]. *The quantum CPA indistinguishability experiment with respect to a scheme* QHE *and a quantum polynomial-time adversary* $\mathscr{A} = (\mathscr{A}_1, \mathscr{A}_2)$, *denoted by* $\mathsf{PubK}_{\mathscr{A},\mathsf{QHE}}^{\mathsf{cpa}}(\kappa)$, *is defined by the following procedure:*

1. KeyGen(1^κ) *is run to obtain keys* (pk, sk, ρ_{evk}).
2. *Adversary* $\mathscr{A}_1$ *is given* (pk, ρ_{evk}) *and outputs a quantum state on* $\mathcal{M} \otimes \mathcal{E}$.
3. *For* $r \in \{0, 1\}$, *let* $\Xi_{QHE}^{cpa,r} : D(\mathcal{M}) \to D(\mathcal{C})$ *be:* $\Xi_{QHE}^{cpa,0}(\rho) = \mathsf{QHE.Enc}_{pk}(|0\rangle\langle 0|)$ *and* $\Xi_{QHE}^{cpa,1}(\rho) = \mathsf{QHE.Enc}_{pk}(\rho)$. *A random bit* $r \in \{0, 1\}$ *is chosen and* $\Xi_{QHE}^{cpa,r}$ *is applied to the state in* $\mathcal{M}$ *(the output being a state in* $\mathcal{C}$*)*.
4. *Adversary* $\mathscr{A}_2$ *obtains the system in* $\mathcal{C} \otimes \mathcal{E}$ *and outputs a bit* r'.
5. *The output of the experiment is defined to be 1 if* $r' = r$ *and 0 otherwise. In case* $r = r'$, *we say that* $\mathscr{A}$ *wins the experiment.*

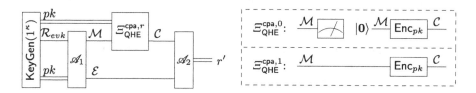

Fig. 1. The quantum CPA indistinguishability experiment $\mathsf{PubK}_{\mathscr{A},QHE}^{cpa}(\kappa)$. Double lines represent classical information flow, and single lines represent quantum information flow. The adversary $\mathscr{A}$ is split up into two separate algorithms $\mathscr{A}_1$ and $\mathscr{A}_2$, which share a working memory represented by the quantum state in register $\mathcal{E}$. [BJ15, reproduced with permission of the authors]

The game $\mathsf{PubK}_{\mathscr{A},QHE}^{cpa}(\kappa)$ is depicted in Fig. 1. Informally, the challenger randomly chooses whether to encrypt some message, chosen by the adversary, or instead to encrypt the state $|0\rangle\langle 0|$. The adversary has to guess which of the two happened. If he cannot do so with more than negligible advantage, the encryption procedure is considered to be q-IND-CPA secure:

Definition 2 [BJ15, Definition 3.3]. *A (classical or quantum) homomorphic encryption scheme* S *is q-IND-CPA secure if for any quantum polynomial-time adversary* $\mathscr{A} = (\mathscr{A}_1, \mathscr{A}_2)$ *there exists a negligible function* η *such that:*

$$\Pr[\mathsf{PubK}_{\mathscr{A},S}^{cpa}(\kappa) = 1] \leq \frac{1}{2} + \eta(\kappa).$$

Analogously to $\mathsf{PubK}_{\mathscr{A},S}^{cpa}(\kappa)$, in the game $\mathsf{PubK}_{\mathscr{A},S}^{cpa-mult}(\kappa)$, the adversary can give multiple messages to the challenger, which are either all encrypted, or all replaced by zeros. Broadbent and Jeffery [BJ15] show that these notions of security are equivalent.

2.3 Garden-Hose Complexity

The *garden-hose model* is a model of communication complexity which was introduced by Buhrman et al. [BFSS13] to study a protocol for position-based quantum cryptography. The model recently saw new use, when Speelman [Spe15] used it to construct new protocols for the task of instantaneous non-local quantum computation, thereby breaking a wider class of schemes for position-based quantum cryptography. (Besides the garden-hose model, this construction used tools from secure delegated computation. These techniques were first used in the setting of instantaneous non-local quantum computation by Broadbent [Bro15b].)

We will not explain the garden-hose model thoroughly, but instead give a short overview. The garden-hose model involves two parties, Alice with input x and Bob with input y, that jointly want to compute a function f. To do this computation, they are allowed to use garden hoses to link up pipes that run between them, one-to-one, in a way which depends on their local inputs. Alice also has a water tap, which she connects to one of the pipes. Whenever $f(x, y) = 0$, the water has to exit at an open pipe on Alice's side, and whenever $f(x, y) = 1$ the water should exit on Bob's side.

The applicability of the garden-hose model to our setting stems from a close correspondence between protocols in the garden-hose model and teleporting a qubit back-and-forth; the 'pipes' correspond to EPR pairs and the 'garden hoses' can be translated into Bell measurements. Our construction of the gadgets in Sect. 5.2 will depend on the number of pipes needed to compute the decryption function HE.Dec of a classical fully homomorphic encryption scheme. It will turn out that any log-space computable decryption function allows for efficiently constructable polynomial-size gadgets.

3   The TP Scheme

Our scheme TP (for teleportation) is an extension of the scheme CL presented in [BJ15]: the quantum state is encrypted using a quantum one-time pad, and Clifford gates are evaluated simply by performing the gate on the encrypted state and then homomorphically updating the encrypted keys to the pad. The new scheme TP, like AUX, includes additional resource states (gadgets) in the evaluation key. These gadgets can be used to immediately correct any P errors that might be present after the application of a T gate. The size of the evaluation key thus grows linearly with the upper bound to the number of T gates in the circuit: for every T gate the evaluation key contains one gadget, along with some classical information on how to use that gadget.

3.1   Gadget

In this section we only give the general form of the gadget, which suffices to prove security. The explanation on how to construct these gadgets, which depend on the decryption function of the classical homomorphic scheme HE.Dec, is deferred to Sect. 5.

Recall that when a T gate is applied to the state $X^a Z^b |\psi\rangle$, an unwanted P error may occur since $TX^a Z^b = P^a X^a Z^b T$. If a is known, this error can easily be corrected by applying $P^\dagger$ whenever $a = 1$. However, as we will see, the evaluating party only has access to some encrypted version $\tilde{a}$ of the key a, and hence is not able to decide whether or not to correct the state.

We show how the key generator can create a gadget ahead of time that corrects the state, conditioned on a, when the qubit $\mathsf{P}^a\mathsf{X}^a\mathsf{Z}^b\mathsf{T}|\psi\rangle$ is teleported through it. The gadget will not reveal any information about whether or not a P gate was present before the correction. Note that the value of a is completely unknown to the key generator, so the gadget cannot depend on it. Instead, the gadget will depend on the secret key sk, and the evaluator will use it in a way that depends on $\widetilde{a}$.

The intuition behind our construction is as follows. A gadget consists of a set of fully entangled pairs that are crosswise linked up in a way that depends on the secret key sk and the decryption function of the classical homomorphic scheme HE. If the decryption function $\mathsf{HE.Dec}$ is simple enough, i.e. computable in logarithmic space or by low-depth binary circuit, the size of this state is polynomial in the security parameter.

Some of these entangled pairs have an extra inverse phase gate applied to them. Note that teleporting any qubit $\mathsf{X}^a\mathsf{Z}^b|\psi\rangle$ through, for example, $(\mathsf{P}^\dagger \otimes \mathbb{I})|\varPhi^+\rangle$, effectively applies an inverse phase gate to the qubit, which ends up in the state $\mathsf{X}^{a'}\mathsf{Z}^{b'}\mathsf{P}^\dagger|\psi\rangle$, where the new Pauli corrections a',b' depend on a,b and the outcome of the Bell measurement.

When wanting to remove an unwanted phase gate, the evaluator of the circuit teleports a qubit through this gadget state in a way which is specified by $\widetilde{a}$. The gadget state is constructed so that the qubit follows a path through this gadget which passes an inverse phase gate if and only if $\mathsf{HE.Dec}_{sk}(\widetilde{a})$ equals 1. The Pauli corrections can then be updated using the homomorphically-encrypted classical information and the measurement outcomes.

Specification of Gadget. Assume $\mathsf{HE.Dec}$ is computable in space logarithmic in the security parameter κ. In Sect. 5 we will show that there exists an efficient algorithm $\mathsf{TP.GenGadget}_{pk'}(sk)$ which produces a gadget: a quantum state $\varGamma_{pk'}(sk)$ of the form as specified in this section.

The gadget will able to remove a single phase gate P^a, using only knowledge of $\widetilde{a}$, where $\widetilde{a}$ decrypts to a under the secret key sk. The public key pk' is used to encrypt all classical information which is part of the gadget.

The quantum part of the gadget consists of $2m$ qubits, with m some number which is polynomial in the security parameter κ. Let $\{(s_1,t_1),(s_2,t_2),\ldots,(s_m,t_m)\}$ be disjoint pairs in $\{1,2,\ldots,2m\}$, and let $p \in \{0,1\}^m$ be a string of m bits. Let $g(sk)$ be a shorthand for the tuple of both of these, together with the secret key sk;

$$g(sk) := (\{(s_1,t_1),(s_2,t_2),\ldots,(s_m,t_m)\},p,sk).$$

The tuple $g(sk)$ is the classical information that determines the structure of the gadget as a function of the secret key sk. The length of $g(sk)$ is not dependent on the secret key: the number of qubits m and the size of sk itself are completely determined by the choice of protocol HE and the security parameter κ.

For any bitstring $x,z \in \{0,1\}^m$, define the quantum state

$$\gamma_{x,z}\big(g(sk)\big) := \prod_{i=1}^m \mathsf{X}^{x[i]}\mathsf{Z}^{z[i]}\big(\mathsf{P}^\dagger\big)^{p[i]}|\varPhi^+\rangle\langle\varPhi^+|_{s_i t_i}\mathsf{P}^{p[i]}\mathsf{Z}^{z[i]}\mathsf{X}^{x[i]}.$$

(Here the single-qubit gates are applied to s_i, the first qubit of the entangled pair.) This quantum state is a collection of maximally-entangled pairs of qubits, some with an extra inverse phase gate applied, where the pairs are determined by the disjoint pairs $\{(s_1, t_1), (s_2, t_2), \ldots, (s_m, t_m)\}$ chosen earlier. The entangled pairs have arbitrary Pauli operators applied to them, described by the bitstrings x and z.

Note that, no matter the choice of gadget structure, averaging over all possible x, z gives the completely mixed state on $2m$ qubits,

$$\frac{1}{2^{2m}} \sum_{x,z \in \{0,1\}^m} \gamma_{x,z}\big(g(sk)\big) = \frac{\mathbb{I}_{2^{2m}}}{2^{2m}}.$$

This property will be important in the security proof; intuitively it shows that these gadgets do not reveal any information about sk whenever x and z are encrypted with a secure classical encryption scheme.

The entire gadget then is given by

$$\Gamma_{pk'}(sk) = \rho(\mathsf{HE.Enc}_{pk'}\big(g(sk)\big)) \otimes \frac{1}{2^{2m}} \sum_{x,z \in \{0,1\}^m} \rho(\mathsf{HE.Enc}_{pk'}(x, z)) \otimes \gamma_{x,z}\big(g(sk)\big).$$

To summarize, the gadget consists of a quantum state $\gamma_{x,z}\big(g(sk)\big)$, instantiated with randomly chosen x, z, the classical information denoting the random choice of x, z, and the other classical information $g(sk)$ which specifies the gadget. All classical information is homomorphically encrypted with a public key pk'.

Since this gadget depends on the secret key sk, simply encrypting this information using the public key corresponding to sk would not be secure, unless we assume that $\mathsf{HE.Dec}$ is circularly secure. In order to avoid the requirement of circular security, we will always use a fresh, independent key pk' to encrypt this information. The evaluator will have to do some recrypting before he is able to use this information, but otherwise using independent keys does not complicate the construction much. More details on how the evaluation procedure deals with the different keys is provided in Sect. 3.4.

Usage of Gadget. The gadget is used by performing Bell measurements between pairs of its qubits, together with an input qubit that needs a correction, without having knowledge of the structure of the gadget.

The choice of measurements can be generated by an efficient (classical) algorithm $\mathsf{TP.GenMeasurement}(\widetilde{a})$ which produces a list M containing m disjoint pairs of elements in $\{0, 1, 2, \ldots, 2m\}$. Here the labels 1 to $2m$ refer to the qubits that make up a gadget and 0 is the label of the qubit with the possible P error. The pairs represent which qubits will be connected through Bell measurements; note that all but a single qubit will be measured according to M.

Consider an input qubit, in some arbitrary state $\mathsf{P}^a|\psi\rangle$, i.e. the qubit has an extra phase gate if $a = 1$. Let $\widetilde{a}$ be an encrypted version of a, such that $a = \mathsf{HE.Dec}_{sk}(\widetilde{a})$. Then the evaluator performs Bell measurements on $\Gamma_{pk'}(sk)$ and the input qubit, according to $M \leftarrow \mathsf{TP.GenMeasurement}(\widetilde{a})$. By construction, one out the $2m + 1$ qubits is still unmeasured. This qubit will be in the state

$X^{a'}Z^{b'}|\psi\rangle$, for some a' and b', both of which are functions of the specification of the gadget, the measurement choices which depend on $\widetilde{a}$, and the outcomes of the teleportation measurements. Also see Sect. 3.4 (and the full version of this paper) for a more in-depth explanation of how the accompanying classical information is updated.

Intuitively, the 'path' the qubit takes through the gadget state, goes through one of the fully entangled pairs with an inverse phase gate whenever $\mathsf{HE.Dec}_{sk}(\widetilde{a}) = 1$, and avoids all such pairs whenever $\mathsf{HE.Dec}_{sk}(\widetilde{a}) = 0$.

3.2   Key Generation

Using the classical $\mathsf{HE.KeyGen}$ as a subroutine to create multiple classical homomorphic keysets, we generate a classical secret and public key, and a classical-quantum evaluation key that contains L gadgets, allowing evaluation of a circuit containing up to L T gates. Every gadget depends on a different secret key, and its classical information is always encrypted using the next public key. The key generation procedure $\mathsf{TP.KeyGen}(1^\kappa, 1^L)$ is defined as follows:

1. For $i = 0$ to L: execute $(pk_i, sk_i, evk_i) \leftarrow \mathsf{HE.KeyGen}(1^\kappa)$ to obtain $L + 1$ independent classical homomorphic key sets.
2. Set the public key to be the tuple $(pk_i)_{i=0}^L$.
3. Set the secret key to be the tuple $(sk_i)_{i=0}^L$.
4. For $i = 0$ to $L - 1$: Run the procedure $\mathsf{TP.GenGadget}_{pk_{i+1}}(sk_i)$ to create the gadget $\Gamma_{pk_{i+1}}(sk_i)$ as described in Sect. 3.1.
5. Set the evaluation key to be the set of all gadgets created in the previous step (including their encrypted classical information), plus the tuple $(evk_i)_{i=0}^L$. The resulting evaluation key is the quantum state

$$\bigotimes_{i=0}^{L-1}\left(\Gamma_{pk_{i+1}}(sk_i) \otimes |evk_i\rangle\langle evk_i|\right).$$

3.3   Encryption

The encryption procedure $\mathsf{TP.Enc}$ is identical to $\mathsf{CL.Enc}$, using the first public key pk_0 for the encryption of the one-time-pad keys. We restate it here for completeness.

Every single-qubit state σ is encrypted separately with a quantum one-time pad, and the pad key is (classically) encrypted and appended to the quantum encryption of σ, resulting in the classical-quantum state:

$$\sum_{a,b \in \{0,1\}} \frac{1}{4}\rho(\mathsf{HE.Enc}_{pk_0}(a), \mathsf{HE.Enc}_{pk_0}(b)) \otimes X^a Z^b \sigma Z^b X^a.$$

3.4   Circuit Evaluation

Consider a circuit with n wires. The evaluation of the circuit on the encrypted data is carried out one gate at a time.

Recall that our quantum circuit is written using a gate set that consists of the Clifford group generators $\{\mathsf{H}, \mathsf{P}, \mathsf{CNOT}\}$ and the T gate. A Clifford gate may affect multiple wires at the same time, while T gates can only affect a single qubit. Before the evaluation of a single gate U, the encryption of an n-qubit state ρ is of the form

$$\left(\mathsf{X}^{a_1}\mathsf{Z}^{b_1} \otimes \cdots \otimes \mathsf{X}^{a_n}\mathsf{Z}^{b_n}\right)\rho\left(\mathsf{X}^{a_1}\mathsf{Z}^{b_1} \otimes \cdots \otimes \mathsf{X}^{a_n}\mathsf{Z}^{b_n}\right).$$

The evaluating party holds the encrypted versions $\widetilde{a_1}^{[i]}, \ldots, \widetilde{a_n}^{[i]}$ and $\widetilde{b_1}^{[i]}, \ldots, \widetilde{b_n}^{[i]}$, with respect to the ith key set for some i (initially, $i = 0$). The goal is to obtain a quantum encryption of the state $U\rho U^\dagger$, such that the evaluator can homomorphically compute the encryptions of the new keys to the quantum one-time pad. If U is a Clifford gate, these encryptions will still be in the ith key. If U is a T gate, then all encryptions are transferred to the $(i + 1)$th key during the process.

- If U is a Clifford gate, we proceed exactly as in CL.Eval. The gate U is simply applied to the encrypted qubit, and since U commutes with the Pauli group, the evaluator only needs to update the encrypted keys in a straightforward way. For a detailed description of this computation, also see the full version of this paper, or e.g. [BJ15, Appendix C].
- If $U = \mathsf{T}$, the evaluator should start out by applying a T gate to the appropriate wire w. Afterwards, the qubit at wire w is in the state

$$\left(\mathsf{P}^{a_w}\mathsf{X}^{a_w}\mathsf{Z}^{b_w}\mathsf{T}\right)\rho_w\left(\mathsf{T}^\dagger\mathsf{X}^{a_w}\mathsf{Z}^{b_w}(\mathsf{P}^\dagger)^{a_w}\right).$$

In order to remove the P error, the evaluator uses one gadget $\Gamma_{pk_{i+1}}(sk_i)$ from the evaluation key; he possesses the classical information $\widetilde{a_w}^{[i]}$ encrypted with the correct key to be able to compute measurements $M \leftarrow$ TP.GenMeasurement$(\widetilde{a_w}^{[i]})$ and performs the measurements on the pairs given by M. Afterwards, using his own measurement outcomes, the classical information accompanying the gadget (encrypted using pk_{i+1}), and the recryptions of $\widetilde{a_w}^{[i]}$ and $\widetilde{b_w}^{[i]}$ into $\widetilde{a_w}^{[i+1]}$ and $\widetilde{b_w}^{[i+1]}$, the evaluator homomorphically computes the new keys $\widetilde{a_w'}^{[i+1]}$ and $\widetilde{b_w'}^{[i+1]}$. See also Fig. 2 and see the full version of this paper for a detailed description of the update algorithm. After these computations, the evaluator also recrypts the keys of all other wires into the $(i + 1)$th key set.

At the end of the evaluation of some circuit C containing k T gates, the evaluator holds a one-time-pad encryption of the state $C|\psi\rangle$, together with the keys to the pad, classically encrypted in the kth key. The last step is to recrypt (in $L - k$ steps) this classical information into the Lth (final) key. Afterwards, the quantum state and the key encryptions are sent to the decrypting party.

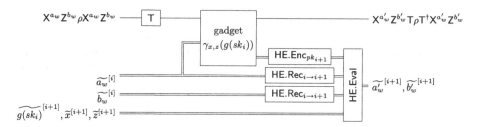

Fig. 2. The homomorphic evaluation of the $(i+1)$th T gate of the circuit. The gadget is consumed during the process. After the use of the gadget, the evaluator encrypts his own classical information (including measurement outcomes) in order to use it in the homomorphic computation of the new keys. HE.Eval evaluates this fairly straightforward computation that consists mainly of looking up values in a list and adding them modulo 2. Note that $\widetilde{sk_i}^{[i+1]}$, needed for the recryption procedures, is contained in the evaluation key.

3.5   Decryption

The decryption procedure is identical to CL.Dec. For each qubit, $\mathsf{HE.Dec}_{sk_L}$ is run twice in order to retrieve the keys to the quantum pad. The correct Pauli operator can then be applied to the quantum state in order to obtain the desired state $\mathsf{C}|\psi\rangle$.

The decryption procedure is fairly straightforward, and its complexity does not depend on the circuit that was evaluated. This is formalized in a compactness theorem for the TP scheme:

Theorem 1. *If* HE *is compact, then* TP *is compact.*

Proof. Note that because the decryption only involves removing a one-time pad from the quantum ciphertext produced by the circuit evaluation, this decryption can be carried out a single qubit at a time. By compactness of HE, there exists a polynomial $p(\kappa)$ such that for any function f, the complexity of applying HE.Dec to the output of $\mathsf{HE.Eval}^f$ is at most $p(\kappa)$. Since the keys to the quantum one-time pad of any wire w are two single bits encrypted with the classical HE scheme, decrypting the keys for one wire requires at most $2p(\kappa)$ steps. Obtaining the qubit then takes at most two steps more for (conditionally) applying X^{a_w} and Z^{b_w}. The total number of steps is polynomial in κ and independent of C, so we conclude that TP is compact.   □

4   Security of TP

In order to guarantee the privacy of the input data, we need to argue that an adversary that does not possess the secret key cannot learn anything about the data with more than negligible probability. To this end, we show that TP is q-IND-CPA secure, i.e. no polynomial-time quantum adversary can tell the difference between an encryption of a real message and an encryption of $|0\rangle\langle0|$, even if he gets to choose the message himself (recall the definition of q-IND-CPA security from Sect. 2.2). Like in the security proofs in [BJ15], we use a reduction argument to relate the probability of being able to distinguish between the two encryptions to the probability of winning an indistinguishability experiment for the classical HE, which we already know to be small. The aim of this section is to prove the following theorem:

Theorem 2. *If* HE *is q-IND-CPA secure, then* TP *is q-IND-CPA secure for circuits containing up to polynomially (in κ) many* T *gates.*

In order to prove Theorem 2, we first prove that an efficient adversary's performance in the indistinguishability game is only negligibly different whether or not he receives a real evaluation key with real gadgets, or just a completely mixed quantum state with encryptions of 0's accompanying them (Corollary 1). Then we argue that without the evaluation key, an adversary does not receive more information than in the indistinguishability game for the scheme CL, which has already been shown to be q-IND-CPA secure whenever HE is.

We start with defining a sequence of variations on the TP scheme. For $\ell \in \{0, \ldots, L\}$, let $\mathsf{TP}^{(\ell)}$ be identical to TP, except for the key generation procedure: $\mathsf{TP}^{(\ell)}.\mathsf{KeyGen}$ replaces, for every $i \geq \ell$, all classical information accompanying the ith gadget with the all-zero string before encrypting it. For any number i, define the shorthand

$$g_i := g(sk_i).$$

As seen in Sect. 3.1, the length of the classical information does not depend on sk_i itself, so a potential adversary cannot gain any information about sk_i just from this encrypted string. In summary,

$$\mathsf{TP}^{(\ell)}.\mathsf{KeyGen}(1^\kappa, 1^L) := \bigotimes_{i=0}^{L-1} |evk_i\rangle\langle evk_i| \otimes \bigotimes_{i=0}^{\ell-1} \Gamma_{pk_{i+1}}(sk_i) \otimes$$

$$\bigotimes_{i=\ell}^{L-1} \left(\rho(\mathsf{HE.Enc}_{pk_{i+1}}(0^{|g_i|})) \otimes \right.$$

$$\left. \frac{1}{2^{2m}} \sum_{x,z \in \{0,1\}^m} \rho(\mathsf{HE.Enc}_{pk_{i+1}}(0^m, 0^m)) \otimes \gamma_{x,z}(g_i) \right).$$

Intuitively, one can view $\mathsf{TP}^{(\ell)}$ as the scheme that provides only ℓ usable gadgets in the evaluation key. Note that $\mathsf{TP}^{(L)} = \mathsf{TP}$, and that in $\mathsf{TP}^{(0)}$, only the classical evaluation keys remain, since without the encryptions of the classical x and z, the quantum part of the gadget is just the completely mixed state. That is, we can rewrite the final line of the previous equation as

$$\frac{1}{2^{2m}} \sum_{x,z \in \{0,1\}^m} \rho(\mathsf{HE.Enc}_{pk_{i+1}}(0^m, 0^m)) \otimes \gamma_{x,z}(g_i)$$

$$= \rho(\mathsf{HE.Enc}_{pk_{i+1}}(0^m, 0^m)) \otimes \frac{\mathbb{I}_{2^{2m}}}{2^{2m}}. \tag{1}$$

With the definitions of the new schemes, we can lay out the steps to prove Theorem 2 in more detail. First, we show that in the quantum CPA indistinguishability experiment, any efficient adversary interacting with $\mathsf{TP}^{(\ell)}$ only has negligible advantage over an adversary interacting with $\mathsf{TP}^{(\ell-1)}$, i.e. the scheme where the classical information $g_{\ell-1}$ is removed (Lemma 1). By iteratively applying this argument, we are able to argue that the advantage of an adversary who interacts with $\mathsf{TP}^{(L)}$ over one who interacts with $\mathsf{TP}^{(0)}$ is also negligible (Corollary 1). Finally, we conclude the proof by arguing that $\mathsf{TP}^{(0)}$ is q-IND-CPA secure by comparison to the CL scheme.

Lemma 1. *Let $0 < \ell \leq L$. If HE is q-IND-CPA secure, then for any quantum polynomial-time adversary $\mathscr{A} = (\mathscr{A}_1, \mathscr{A}_2)$, there exists a negligible function η such that*

$$\Pr[\mathsf{PubK}^{\mathsf{cpa}}_{\mathscr{A},\mathsf{TP}^{(\ell)}}(\kappa) = 1] - \Pr[\mathsf{PubK}^{\mathsf{cpa}}_{\mathscr{A},\mathsf{TP}^{(\ell-1)}}(\kappa) = 1] \leq \eta(\kappa).$$

Proof. The difference between schemes $\mathsf{TP}^{(\ell)}$ and $\mathsf{TP}^{(\ell-1)}$ lies in whether the gadget state $\gamma_{x_{\ell-1}, z_{\ell-1}}(g_{\ell-1})$ is supplemented with its classical information $\widetilde{g_{\ell-1}}, \widetilde{x_{\ell-1}}, \widetilde{z_{\ell-1}}$, or just with an encryption of $0^{|g_{\ell-1}|+2m}$.

Let $\mathscr{A} = (\mathscr{A}_1, \mathscr{A}_2)$ be an adversary for the game $\mathsf{PubK}^{\mathsf{cpa}}_{\mathscr{A},\mathsf{TP}^{(\ell)}}(\kappa)$. We will define an adversary $\mathscr{A}' = (\mathscr{A}'_1, \mathscr{A}'_2)$ for $\mathsf{PubK}^{\mathsf{cpa-mult}}_{\mathscr{A}',\mathsf{HE}}(\kappa)$ that will either simulate the game $\mathsf{PubK}^{\mathsf{cpa}}_{\mathscr{A},\mathsf{TP}^{(\ell)}}(\kappa)$ or $\mathsf{PubK}^{\mathsf{cpa}}_{\mathscr{A},\mathsf{TP}^{(\ell-1)}}(\kappa)$. Which game is simulated will depend on some $s \in_R \{0, 1\}$ that is unknown to $\mathscr{A}'$ himself. Using the assumption that HE is q-IND-CPA secure, we are able to argue that $\mathscr{A}'$ is unable to recognize which of the two schemes was simulated, and this fact allows us to bound the difference in success probabilities between the security games of $\mathsf{TP}^{(\ell)}$ and $\mathsf{TP}^{(\ell-1)}$. The structure of this proof is very similar to e.g. Lemma 5.3 in [BJ15]. The adversary $\mathscr{A}'$ acts as follows (see also Fig. 3):

$\mathscr{A}'_1$ takes care of most of the key generation procedure: he generates the classical key sets 0 through $\ell - 1$ himself, generates the random strings $x_0, z_0, \ldots, x_{\ell-1}, z_{\ell-1}$, and constructs the gadgets $\gamma_{x_0, z_0}(g_0), \ldots, \gamma_{x_{\ell-1}, z_{\ell-1}}(g_{\ell-1})$ and their classical information $g_0, \ldots, g_{\ell-1}$. He encrypts the classical information using the appropriate public keys. Only $g_{\ell-1}, x_{\ell-1}$ and $z_{\ell-1}$ are left unencrypted: instead of encrypting these strings himself using pk_ℓ, $\mathscr{A}'_1$ sends the strings for encryption to the challenger. Whether the challenger really encrypts $g_{\ell-1}, x_{\ell-1}$ and $z_{\ell-1}$ or replaces the strings with a string of zeros, determines which of the two schemes is simulated. $\mathscr{A}'$ is unaware of the random choice of the challenger.

The adversary $\mathscr{A}'_1$ also generates the extra padding inputs that correspond to the already-removed gadgets ℓ up to $L - 1$. Since these gadgets consist of all-zero strings encrypted with independently chosen public keys that are not used anywhere else, together with a completely mixed quantum state (as shown in Eq. 1), the adversary can generate them without needing any extra information.

$\mathscr{A}_2'$ feeds the evaluation key and public key, just generated by $\mathscr{A}_1'$, to $\mathscr{A}_1$ in order to obtain a chosen message $\mathcal{M}$ (plus the auxiliary state $\mathcal{E}$). He then picks a random $r \in_R \{0, 1\}$ and erases $\mathcal{M}$ if and only if $r = 0$. He encrypts the result according to the TP.Enc procedure (using the public key $(pk_i)_{i=0}^L$ received from $\mathscr{A}_1'$), and gives the encrypted state, plus $\mathcal{E}$, to $\mathscr{A}_2$, who outputs r' in an attempt to guess r. $\mathscr{A}_2'$ now outputs 1 if and only if the guess by $\mathscr{A}$ was correct, i.e. $r \equiv r'$.

Because HE is q-IND-CPA secure, the probability that $\mathscr{A}'$ wins $\mathsf{PubK}_{\mathscr{A}', \mathsf{HE}}^{\mathsf{cpa-mult}}(\kappa)$, i.e. that $s' \equiv s$, is at most $\frac{1}{2} + \eta'(\kappa)$ for some negligible function η'. There are two scenarios in which $\mathscr{A}'$ wins the game:

- $s = 1$ and $\mathscr{A}$ guesses r correctly: If $s = 1$, the game that is being simulated is $\mathsf{PubK}_{\mathscr{A}, \mathsf{TP}^{(\ell)}}^{\mathsf{cpa}}(\kappa)$. If $\mathscr{A}$ wins the simulated game ($r \equiv r'$), then $\mathscr{A}'$ will correctly output $s' = 1$. (If $\mathscr{A}$ loses, then $\mathscr{A}'$ outputs 0, and loses as well).
- $s = 0$ and $\mathscr{A}$ does not guess r correctly: If $s = 0$, the game that is being simulated is $\mathsf{PubK}_{\mathscr{A}, \mathsf{TP}^{(\ell-1)}}^{\mathsf{cpa}}(\kappa)$. If $\mathscr{A}$ loses the game ($r \not\equiv r'$), then $\mathscr{A}'$ will correctly output $s' = 0$. (If $\mathscr{A}$ wins, then $\mathscr{A}'$ outputs 1 and loses).

From the above, we conclude that

$$\Pr[s=1] \cdot \Pr[\mathsf{PubK}_{\mathscr{A}, \mathsf{TP}^{(\ell)}}^{\mathsf{cpa}}(\kappa) = 1] + \Pr[s=0] \cdot \Pr[\mathsf{PubK}_{\mathscr{A}, \mathsf{TP}^{(\ell-1)}}^{\mathsf{cpa}}(\kappa) = 0] \leq \frac{1}{2} + \eta'(\kappa)$$

$$\Leftrightarrow \quad \frac{1}{2}\Pr[\mathsf{PubK}_{\mathscr{A}, \mathsf{TP}^{(\ell)}}^{\mathsf{cpa}}(\kappa) = 1] + \frac{1}{2}\left(1 - \Pr[\mathsf{PubK}_{\mathscr{A}, \mathsf{TP}^{(\ell-1)}}^{\mathsf{cpa}}(\kappa) = 1]\right) \leq \frac{1}{2} + \eta'(\kappa)$$

$$\Leftrightarrow \quad \Pr[\mathsf{PubK}_{\mathscr{A}, \mathsf{TP}^{(\ell)}}^{\mathsf{cpa}}(\kappa) = 1] - \Pr[\mathsf{PubK}_{\mathscr{A}, \mathsf{TP}^{(\ell-1)}}^{\mathsf{cpa}}(\kappa) = 1] \leq 2\eta'(\kappa)$$

Set $\eta(\kappa) := 2\eta'(\kappa)$, and the proof is complete. $\square$

By applying Lemma 1 iteratively, L times in total, we can conclude that the difference between $\mathsf{TP}^{(L)}$ and $\mathsf{TP}^{(0)}$ is negligible, because the sum of polynomially many negligible functions is still negligible:

Corollary 1. *If L is polynomial in κ, then for any quantum polynomial-time adversary $\mathscr{A} = (\mathscr{A}_1, \mathscr{A}_2)$, there exists a negligible function η such that*

$$\Pr[\mathsf{PubK}_{\mathscr{A}, \mathsf{TP}^{(L)}}^{\mathsf{cpa}}(\kappa) = 1] - \Pr[\mathsf{PubK}_{\mathscr{A}, \mathsf{TP}^{(0)}}^{\mathsf{cpa}}(\kappa) = 1] \leq \eta(\kappa).$$

Using Corollary 1, we can finally prove the q-IND-CPA security of our scheme $\mathsf{TP} = \mathsf{TP}^{(L)}$.

Proof of Theorem 2. The scheme $\mathsf{TP}^{(0)}$ is very similar to CL in terms of its key generation and encryption steps. The evaluation key consists of several classical evaluation keys, plus some completely mixed states and encryptions of 0 which we can safely ignore because they do not contain any information about the encrypted message. In both schemes, the encryption of a qubit is a quantum one-time pad together with the encrypted keys. The only difference is that in $\mathsf{TP}^{(0)}$, the public key and evaluation key form a tuple containing, in addition to pk_0 and evk_0 which are used for the encryption of the quantum one-time

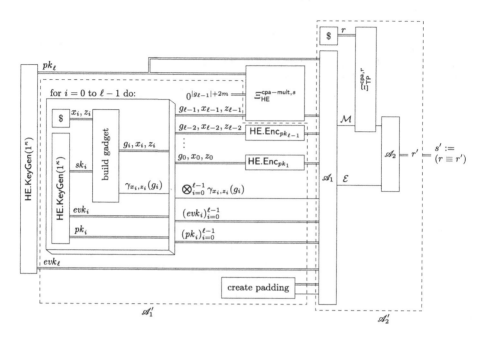

Fig. 3. A strategy for the game $\mathsf{PubK}^{\mathsf{cpa-mult}}_{\mathscr{A}',\mathsf{HE}}(\kappa)$, using an adversary $\mathscr{A}$ for $\mathsf{PubK}^{\mathsf{cpa}}_{\mathscr{A},\mathsf{TP}^{(\ell)}}(\kappa)$ as a subroutine. All the wires that form an input to $\mathscr{A}_1$ together form the evaluation key and public key for $\mathsf{TP}^{(\ell)}$ or $\mathsf{TP}^{(\ell-1)}$, depending on s. Note that $\Xi^{\mathsf{cpa},r}_{\mathsf{TP}} = \Xi^{\mathsf{cpa},r}_{\mathsf{TP}^{(\ell)}} = \Xi^{\mathsf{cpa},r}_{\mathsf{TP}^{(\ell-1)}}$, so $\mathscr{A}'_2$ can run either one of these independently of s (i.e. without having to query the challenger). The 'create padding' subroutine generates dummy gadgets for ℓ up to $L-1$, as described in the definition of $\mathscr{A}_1$.

pad, a list of public/evaluation keys that are independent of the encryption. These keys do not provide any advantage (in fact, the adversary could have generated them himself by repeatedly running $\mathsf{HE.KeyGen}(1^\kappa, 1^L)$). Therefore, we can safely ignore these keys as well.

In [BJ15, Lemma 5.3], it is shown that CL is q-IND-CPA secure. Because of the similarity between CL and $\mathsf{TP}^{(0)}$, the exact same proof shows that $\mathsf{TP}^{(0)}$ is q-IND-CPA secure as well, that is, for any $\mathscr{A}$ there exists a negligible function η' such that

$$\Pr[\mathsf{PubK}^{\mathsf{cpa}}_{\mathscr{A},\mathsf{TP}^{(0)}}(\kappa) = 1] \leq \frac{1}{2} + \eta'(\kappa).$$

Combining this result with Corollary 1, it follows that

$$\Pr[\mathsf{PubK}^{\mathsf{cpa}}_{\mathscr{A},\mathsf{TP}}(\kappa) = 1] \leq \Pr[\mathsf{PubK}^{\mathsf{cpa}}_{\mathscr{A},\mathsf{TP}^{(0)}}(\kappa) = 1] + \eta(\kappa)$$

$$\leq \frac{1}{2} + \eta'(\kappa) + \eta(\kappa).$$

Since the sum of two negligible functions is itself negligible, we have proved Theorem 2.

□

4.1 Circuit Privacy

The scheme TP as presented above ensures the privacy of the input data. It does not guarantee, however, that whoever generates the keys, encrypts, and decrypts cannot gain information about the circuit C that was applied to some input ρ by the evaluator. Obviously, the output value $C\rho C^\dagger$ often reveals something about the circuit C, but apart from this necessary leakage of information, one may require a (quantum) homomorphic encryption scheme to ensure *circuit privacy* in the sense that an adversary cannot statistically gain any information about C from the output of the evaluation procedure that it could not already gain from $C\rho C^\dagger$ itself.

We claim that circuit privacy for TP in the semi-honest setting (i.e. against passive adversaries[5]) can be obtained by modifying the scheme only slightly, given that the classical encryption scheme has the circuit privacy property.

Theorem 3. *If* HE *has circuit privacy in the semi-honest setting, then* TP *can be adapted to a quantum homomorphic encryption scheme with circuit privacy.*

Proof Sketch. If the evaluator randomizes the encryption of the output data by applying a quantum one-time pad to the (already encrypted) result of the evaluation, the keys themselves are uniformly random and therefore do not reveal any information about what circuit was evaluated. The evaluator can then proceed to update the classical encryptions of those keys accordingly, and by the circuit privacy of the classical scheme, the resulting encrypted keys will also contain no information about the computations performed. Because of space constraints, the full detailed proof is given in the full version of this paper.   □

5 Constructing the Gadgets

In this section we will first show how to construct gadgets that have polynomial size whenever the scheme HE has a decryption circuit with logarithmic depth (i.e., the decryption function is in NC^1). This construction will already be powerful enough to instantiate TP with current classical schemes for homomorphic encryption, since these commonly have low-depth decryption circuits. Afterwards, in Sect. 5.2, we will present a larger toolkit to construct gadgets, which is efficient for a larger class of possible decryption functions. To illustrate these techniques, we apply these tools to create gadgets for schemes that are based on Learning With Errors (LWE). Finally, we will reflect on the possibility of constructing these gadgets in scenarios where quantum power is limited.

5.1 For Log-Depth Decryption Circuits

The main tool for creating gadgets that encode log-depth decryption circuits comes from Barrington's theorem: a classic result in complexity theory,

---

[5] Note that there various ways to define passive adversaries in the quantum setting [DNS10, BB14]. Here, we are considering adversaries that follow all protocol instructions exactly.

which states that all boolean circuits of logarithmic depth can be encoded as polynomial-sized width-5 permutation branching programs. Every instruction of such a branching program will be encoded as connections between five Bell pairs.

Definition 3. *A width-k permutation branching program of length L on an input $x \in \{0,1\}^n$ is a list of L instructions of the form $\langle i_\ell, \sigma_\ell^1, \sigma_\ell^0 \rangle$, for $1 \leq \ell \leq L$, such that $i_\ell \in [n]$, and σ_ℓ^1 and σ_ℓ^0 are elements of S_k, i.e., permutations of $[k]$. The program is executed by composing the permutations given by the instructions 1 through L, selecting σ_ℓ^1 if $x_{i_\ell} = 1$ and selecting σ_ℓ^0 if $x_{i_\ell} = 0$. The program rejects if this product equals the identity permutation and accepts if it equals a fixed k-cycle.*

Since these programs have a very simple form, it came as a surprise when they were proven to be quite powerful [Bar89].

Theorem 4 (Barrington [Bar89]). *Every fan-in 2 boolean circuit C of depth d can be simulated by a width-5 permutation branching program of length at most 4^d.*

Our gadget construction will consist of first transforming the decryption function HE.Dec into a permutation branching program, and then encoding this permutation branching program as a specification of a gadget, as produced by TP.GenGadget$_{pk'}(sk)$, and usage instructions TP.GenMeasurement($\tilde{a}$).

Theorem 5. *Let HE.Dec$_{sk}(\tilde{a})$ be the decryption function of the classical homomorphic encryption scheme HE. If HE.Dec is computable by a boolean fan-in 2 circuit of depth $O(\log(\kappa))$, where κ is the security parameter, then there exist gadgets for TP of size polynomial in κ.*

Proof. Our description will consist of three steps. First, we write HE.Dec as a width-5 permutation branching program, of which the instructions alternately depend on the secret key sk and on the ciphertext $\tilde{a}$. Secondly, we specify how to transform these instructions into a gadget which almost works correctly, but for which the qubit ends up at an unknown location. Finally, we complete the construction by executing the inverse program, so that the qubit ends up at a known location.

The first part follows directly from Barrington's theorem. The effective input of HE.Dec can be seen as the concatenation of the secret key sk and the ciphertext $\tilde{a}$. Since by assumption the circuit is of depth $O(\log \kappa)$, there exists width-5 permutation branching program $\mathcal{P}$ of length $L = \kappa^{O(1)}$, with the following properties. We write

$$\mathcal{P} = \left(\langle i_1, \sigma_1^1, \sigma_1^0 \rangle, \langle i_2, \sigma_2^1, \sigma_2^0 \rangle, \ldots, \langle i_L, \sigma_L^1, \sigma_L^0 \rangle \right)$$

as the list of instructions of the width-5 permutation branching program. Without loss of generality[6], we can assume that the instructions alternately depend on bits

---

[6] This can be seen by inserting dummy instructions that always perform the identity permutation between any two consecutive instructions that depend on the same variable. Alternatively, it would be possible to improve the construction by 'multiplying out' consecutive instructions whenever they depend on the same variable.

of $\tilde{a}$ and bits of sk. That is, the index i_ℓ refers to a bit of $\tilde{a}$ if ℓ is odd, and to a bit of sk if ℓ is even. There are L instructions in total, of which $L/2$ are odd-numbered and $L/2$ are even.

The output of $\mathsf{TP.GenGadget}_{pk'}(sk)$, i.e., the list of pairs that defines the structure of the gadget, will be created from the even-numbered instructions, evaluated using the secret key sk. For every even-numbered $\ell \leq L$, we connect ten qubits in the following way. Suppose the ℓth instruction evaluates to some permutation $\sigma_\ell := \sigma_\ell^{sk_{i_\ell}}$. Label the 10 qubits of this part of the gadget by $1_{\ell,\text{in}}, 2_{\ell,\text{in}}, \ldots, 5_{\ell,\text{in}}$ and $1_{\ell,\text{out}}, 2_{\ell,\text{out}}, \ldots, 5_{\ell,\text{out}}$. These will correspond to 5 EPR pairs, connected according to the permutation: $(1_{\ell,\text{in}}, \sigma_\ell(1)_{\ell,\text{out}})$, $(2_{\ell,\text{in}}, \sigma_\ell(2)_{\ell,\text{out}})$, etc., up to $(5_{\ell,\text{in}}, \sigma_\ell(5)_{\ell,\text{out}})$.

After the final instruction of the branching program, σ_L, also perform an inverse phase gate $\mathsf{P}^\dagger$ on the qubits labeled as $2_{L,\text{out}}, 3_{L,\text{out}}, 4_{L,\text{out}}, 5_{L,\text{out}}$. Execution of the gadget will teleport the qubit through one of these whenever $\tilde{a} = 1$.

For this construction, $\mathsf{TP.GenMeasurement}(\tilde{a})$ will be given by the odd instructions, which depend on the bits of $\tilde{a}$. Again, for all odd $\ell \leq L$, let $\sigma_\ell := \sigma_\ell^{\tilde{a}_{i_\ell}}$ be the permutation given by the evaluation of instruction ℓ on $\tilde{a}$. For all ℓ strictly greater than one, the measurement instructions will be: perform a Bell measurement according to the permutation σ_ℓ between the 'out' qubits of the previous set, and the 'in' qubits of the next. The measurement pairs will then be $(1_{\ell-1,\text{out}}, \sigma(1)_{\ell,\text{in}})$, $(2_{\ell-1,\text{out}}, \sigma(2)_{\ell,\text{in}})$, up to $(5_{\ell-1,\text{out}}, \sigma(5)_{\ell,\text{in}})$.

For $\ell = 1$, there is no previous layer to connect to, only the input qubit. For that, we add the measurement instruction $(0, \sigma(1)_{1,\text{in}})$, where 0 is the label of the input qubit.

By Barrington's theorem, if $\mathsf{HE.Dec}_{sk}(\tilde{a}) = 0$ then the product, say τ, of the permutations coming from the evaluated instructions equals the identity. In that case, consecutively applying these permutations on '1', results in the unchanged starting value, '1'. If instead the decryption would output 1, the consecutive application results in another value in $\{2, 3, 4, 5\}$, because in that case, τ is a k-cycle. After teleporting a qubit through these EPR pairs, with teleportation measurements chosen accordingly, the input qubit will be present at $\tau(1)_{L,\text{out}}$, with an inverse phase gate if $\tau(1)$ is unequal to 1.

The gadget constructed so far would correctly apply the phase gate, conditioned on $\mathsf{HE.Dec}_{sk}(\tilde{a})$, with one problem: afterward, the qubit is at a location unknown to the user of the gadget, because the user cannot compute τ.

We fix this problem in the following way: execute the inverse branching program afterwards. The entire construction is continued in the same way, but the instructions of the inverse program are used. The inverse program can be made from the original program by reversing the order of instructions, and then for each permutation using its inverse permutation instead. The first inverse instruction is $\langle i_L, (\sigma_L^1)^{-1}, (\sigma_L^0)^{-1} \rangle$, then $\langle i_{L-1}, (\sigma_{L-1}^1)^{-1}, (\sigma_{L-1}^0)^{-1} \rangle$, with final instruction $\langle i_1, (\sigma_1^1)^{-1}, (\sigma_1^0)^{-1} \rangle$. One small detail is that i_L is used twice in a row, breaking the alternation; the user of the gadget can simply perform the

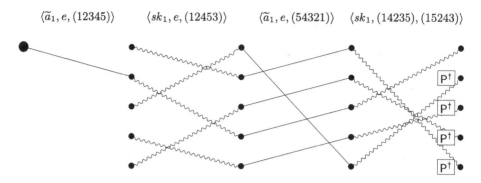

$\langle \widetilde{a}_1, e, (12345) \rangle$ $\langle sk_1, e, (12453) \rangle$ $\langle \widetilde{a}_1, e, (54321) \rangle$ $\langle sk_1, (14235), (15243) \rangle$

Fig. 4. Structure of the (first half of the) gadget, with measurements, coming from the 5-permutation branching program for the OR function on the input $(0,0)$. The example program's instructions are displayed above the permutations. The solid lines correspond to Bell measurements, while the wavy lines represent EPR pairs.

measurements that correspond to the identity permutation e in between, since $(\sigma_L^0)(\sigma_L^0)^{-1} = (\sigma_L^1)(\sigma_L^1)^{-1} = e$.

After having repeated the construction with the inverse permutation branching program, the qubit is guaranteed to be at the location where it originally started: $\sigma_1(1)$ of the final layer of five qubits – that will then be the corrected qubit which is the output of the gadget.

The total number of qubits which form the gadget, created from a width-5 permutation branching program of length L, of which the instructions alternate between depending on $\widetilde{a}$ and depending on sk, is $2 \cdot (5L) = 10L$. □

Example. The OR function on two bits can be computed using a width-5 permutation branching program of length 4, consisting of the following list of instructions:

1. $\langle 1, e, (12345) \rangle$
2. $\langle 2, e, (12453) \rangle$
3. $\langle 1, e, (54321) \rangle$
4. $\langle 2, (14235), (15243) \rangle$

As a simplified example, suppose the decryption function $\mathsf{HE.Dec}_{sk}(\widetilde{a})$ is sk_1 OR $\widetilde{a}_1$. Then, for one possible example set of values of $\widetilde{a}$ and sk, half of the gadget and measurements will be as given in Fig. 4. To complete this gadget, the same construction is appended, reflected horizontally.

5.2   For Log-Space Computable Decryption Functions

Even though the construction based on Barrington's theorem has enough power for current classical homomorphic schemes, it is possible to improve on this construction in two directions. Firstly, we extend our result to be able to handle a larger class of decryption functions: those that can be computed in logarithmic space, instead of only NC^1. Secondly, for some specific decryption functions, executing the construction of Sect. 5.1 might produce significantly larger gadgets than necessary. For instance, even for very simple circuits of depth $\log \kappa$, Barrington's theorem produces programs of length κ^2 — a direct approach can often easily improve on the exponent of the polynomial. See also the garden-hose protocols for equality [Mar14, CSWX14] and the majority function [KP14] for examples of non-trivial protocols that are much more efficient than applying Barrington's theorem as a black box.

Theorem 6. *Let* $\mathsf{HE.Dec}_{sk}(\widetilde{a})$ *be the decryption function of the classical homomorphic encryption scheme* HE. *If* $\mathsf{HE.Dec}$ *is computable by a Turing machine that uses space* $O(\log \kappa)$, *where* κ *is the security parameter, then there exist gadgets for* TP *of size polynomial in* κ.

A detailed explanation of how to construct a gadget for a log-space computation is given in the full version of this paper. All more-complicated constructions use a different language than the direct encoding of the previous section: there is a natural way of writing the requirements on the gadgets as a two-player task, and then writing strategies for this task in the *garden-hose model*. Writing these gadgets in terms of the garden-hose model, even though it adds a layer of complexity to the construction, gives more insight into the structure of the gadgets, and forms its original inspiration. We therefore sketch the link between log-space computation and gadget construction within this framework.

Besides clarifying the log-space construction, viewing the gadget construction as an instantiation of the garden-hose mode also makes it easier to construct gadgets for specific cases. Earlier work developed protocols in the garden-hose model for several functions, see for instance [Spe11, BFSS13, KP14], and connections to other models of computation. These results on the garden-hose model might serve as building blocks to create more efficient gadgets for specific decoding functions of classical homomorphic schemes, that are potentially much smaller than those created as a result of following the general constructions of Theorem 5 or 6.

The scheme by Brakerski and Vaikuntanathan [BV11] is well-suited for our construction, and its decryption function is representative for a much wider class of schemes which are based on the hardness of Learning With Errors (LWE). As an example, we construct gadgets that enable quantum homomorphic encryption based on the BV11 scheme in the full version of our paper.

5.3 Constructing Gadgets Using Limited Quantum Resources

In a setting where a less powerful client wants to delegate some quantum computation to a more powerful server, it is important to minimize the amount of effort required from the client. In delegated quantum computation, the complexity of a protocol can be measured by, among other things, the total amount of communication between client and server, the number of rounds of communication, and the quantum resources available to the client, such as possible quantum operations and memory size.

We claim that TP gives rise to a three-round delegated quantum computation protocol in a setting where the client can perform only Pauli and swap operations. TP.Enc and TP.Dec only require local application of Pauli operators to a quantum state, but TP.KeyGen is more involved because of the gadget construction. However, when supplied with a set of EPR pairs from the server (or any other untrusted source), the client can generate the quantum evaluation key for TP using only Pauli and swap operations. Even if the server produces some other state than the claimed list of EPR pairs, the client can prevent the leakage of information about her input by encrypting the input with random Pauli operations. More details are supplied in the appendix of the full version of this paper.

Alternatively, TP can be regarded as a two-round delegated quantum computation protocol in a setting where the client can perform arbitrary Clifford operations, but is limited to a constant-sized quantum memory, given that HE.Dec is in NC^1. In that case, the gadgets can be constructed ten qubits at a time, by constructing the sets of five EPR pairs as specified in Sect. 5.1. By decomposing the 5-cycles into products of 2-cycles, the quantum memory can even be reduced to only four qubits. The client sends these small parts of the gadgets to the server as they are completed. Because communication remains one-way until all gadgets have been sent, this can be regarded as a single round of communication.

6   Conclusion

We have presented the first quantum homomorphic encryption scheme TP that is compact and allows evaluation of circuits with polynomially many T gates in the security parameter, i.e. arbitrary polynomial-sized circuits. Assuming that the number of wires involved in the evaluation circuit is also polynomially related to the security parameter, we may consider TP to be leveled fully homomorphic. The scheme is based on an arbitrary classical FHE scheme, and any computational assumptions needed for the classical scheme are also required for security of TP. However, since TP uses the classical FHE scheme as a black box, any FHE scheme can be plugged in to change the set of computational assumptions.

Our constructions are based on a new and interesting connection between the area of instantaneous non-local quantum computation and quantum homomorphic encryption. Recent techniques developed by Speelman [Spe15], based on the garden-hose model [BFSS13], have turned out to be crucial for our construction of quantum gadgets which allow homomorphic evaluation of T gates on encrypted quantum data.

6.1   Future Work

Since Yu et al. [YPDF14] showed that information-theoretically secure QFHE is impossible (at least in the exact case), it is natural to wonder whether it is possible to construct a non-leveled QFHE scheme based on computational assumptions. If such a scheme is not possible, can one find lower bounds on the size of the evaluation key of a compact scheme? Other than the development of more efficient QFHE schemes, one can consider the construction of QFHE schemes with extra properties, such as circuit privacy against active adversaries. It is also interesting to look at other cryptographic tasks that might be executed using QFHE. In the classical world for example, multiparty computation protocols can be constructed from fully homomorphic encryption [CDN01]. We consider it likely that our new techniques will also be useful in other contexts such as quantum indistinguishability obfuscation [AF16].

Acknowledgements. We acknowledge useful discussions with Anne Broadbent, Harry Buhrman, and Leo Ducas. We thank Stacey Jeffery for providing the inspiration for a crucial step in the security proof, and Gorjan Alagic and Anne Broadbent for helpful comments on a draft of this article. This work was supported by the 7th framework EU SIQS and QALGO, and a NWO VIDI grant.

References

[ABF+16] Alagic, G., Broadbent, A., Fefferman, B., Gagliardoni, T., Schaffner, C., St. Jules, M.: Computational security of quantum encryption (2016). arXiv preprint arXiv:1602.01441

[ABOE10] Aharonov, D., Ben-Or, M., Eban, E.: Interactive proofs for quantum computations. In: Proceeding of Innovations in Computer Science (ICS 2010), pp. 453–469 (2010)

[AF16] Alagic, G., Fefferman, B.: On quantum obfuscation (2016). arXiv preprint arXiv:1602.01771

[AJLA+12] Asharov, G., Jain, A., López-Alt, A., Tromer, E., Vaikuntanathan, V., Wichs, D.: Multiparty computation with low communication, computation and interaction via threshold FHE. In: Pointcheval, D., Johansson, T. (eds.) EUROCRYPT 2012. LNCS, vol. 7237, pp. 483–501. Springer, Heidelberg (2012)

[AS06] Arrighi, P., Salvail, L.: Blind quantum computation. Int. J. Quantum Inf. **4**(05), 883–898 (2006)

[Bar89] Barrington, D.A.: Bounded-width polynomial-size branching programs recognize exactly those languages in NC1. J. Comput. Syst. Sci. **164**, 150–164 (1989)

[BB14] Baumeler, Ä., Broadbent, A.: Quantum private information retrieval has linear communication complexity. J. Cryptol. **28**(1), 161–175 (2014)

[BCG+06] Ben-Or, M., Crépeau, C., Gottesman, D., Hassidim, A., Smith, A.: Secure multiparty quantum computation with (only) a strict honest majority. In: 47th Annual IEEE Symposium on Foundations of Computer Science (FOCS 2006), pp. 249–260 (2006)

[BFK09] Broadbent, A., Fitzsimons, J., Kashefi, E.: Universal blind quantum computation. In: 50th Annual IEEE Symposium on Foundations of Computer Science, FOCS 2009, pp. 517–526. IEEE (2009)

[BFSS13] Buhrman, H., Fehr, S., Schaffner, C., Speelman, F.: The garden-hose model. In: Proceedings of the 4th Innovations in Theoretical Computer Science Conference, pp. 145–158. ACM (2013)

[BGN05] Boneh, D., Goh, E.-J., Nissim, K.: Evaluating 2-DNF formulas on ciphertexts. In: Kilian, J. (ed.) TCC 2005. LNCS, vol. 3378, pp. 325–341. Springer, Heidelberg (2005)

[BGV12] Brakerski, Z., Gentry, C., Vaikuntanathan, V.: (Leveled) fully homomorphic encryption without bootstrapping. In: Proceedings of the 3rd Innovations in Theoretical Computer Science Conference, pp. 309–325. ACM (2012)

[BJ15] Broadbent, A., Jeffery, S.: Quantum homomorphic encryption for circuits of low T-gate complexity. In: Gennaro, R., Robshaw, M. (eds.) CRYPTO 2015. LNCS, vol. 9216, pp. 609–629. Springer, Heidelberg (2015)

[Bro15a] Broadbent, A.: Delegating private quantum computations. Can. J. Phys. 93(9), 941–946 (2015)

[Bro15b] Broadbent, A.: Popescu-Rohrlich correlations imply efficient instantaneous nonlocal quantum computation (2015). arXiv preprint arXiv:1512.04930

[BV11] Brakerski, Z., Vaikuntanathan, V.: Efficient fully homomorphic encryption from (standard) LWE. In: 2011 IEEE 52nd Annual Symposium on Foundations of Computer Science (FOCS), pp. 97–106, October 2011

[CDN01] Cramer, R., Damgård, I.B., Nielsen, J.B.: Multiparty computation from threshold homomorphic encryption. In: Pfitzmann, B. (ed.) EUROCRYPT 2001. LNCS, vol. 2045, pp. 280–300. Springer, Heidelberg (2001)

[Chi05] Childs, A.M.: Secure assisted quantum computation. Quantum Inf. Comput. 5(6), 456–466 (2005)

[CKGS98] Chor, B., Kushilevitz, E., Goldreich, O., Sudan, M.: Private information retrieval. J. ACM (JACM) 45(6), 965–981 (1998)

[CSWX14] Chiu, W.Y., Szegedy, M., Wang, C., Xu, Y.: The garden hose complexity for the equality function. In: Gu, Q., Hell, P., Yang, B. (eds.) AAIM 2014. LNCS, vol. 8546, pp. 112–123. Springer, Heidelberg (2014)

[DNS10] Dupuis, F., Nielsen, J.B., Salvail, L.: Secure two-party quantum evaluation of unitaries against specious adversaries. In: Rabin, T. (ed.) CRYPTO 2010. LNCS, vol. 6223, pp. 685–706. Springer, Heidelberg (2010)

[FBS+14] Fisher, K.A.G., Broadbent, A., Shalm, L.K., Yan, Z., Lavoie, J., Prevedel, R., Jennewein, T., Resch, K.J.: Quantum computing on encrypted data. Nat. Commun. 5 (2014). Article number: 3074

[Fil12] Fillinger, M.: Lattice based cryptography and fully homomorphic encryption. Master of Logic Project (2012). http://homepages.cwi.nl/schaffne/courses/reports/MaxFillinger_FHE_2012.pdf

[GC99] Gottesman, D., Chuang, I.L.: Quantum teleportation is a universal computational primitive. Nature 402, 390–393 (1999)

[Gen09] Gentry, C.: Fully homomorphic encryption using ideal lattices. In: STOC, vol. 9, pp. 169–178 (2009)

[GGH+13] Garg, S., Gentry, C., Halevi, S., Raykova, M., Sahai, A., Waters, B.: Candidate indistinguishability obfuscation and functional encryption for all circuits. In: 2013 IEEE 54th Annual Symposium on Foundations of Computer Science (FOCS), pp. 40–49. IEEE (2013)

[GHS15] Gagliardoni, T., Hülsing, A., Schaffner, C.: Semantic security, indistinguishability in the quantum world (2015). arXiv preprint arXiv:1504.05255

[GHV10]   Gentry, C., Halevi, S., Vaikuntanathan, V.: A simple BGN-type cryptosystem from LWE. In: Gilbert, H. (ed.) EUROCRYPT 2010. LNCS, vol. 6110, pp. 506–522. Springer, Heidelberg (2010)

[GKP+13a] Goldwasser, S., Kalai, Y.T., Popa, R.A., Vaikuntanathan, V., Zeldovich, N.: How to run turing machines on encrypted data. In: Canetti, R., Garay, J.A. (eds.) CRYPTO 2013, Part II. LNCS, vol. 8043, pp. 536–553. Springer, Heidelberg (2013)

[GKP+13b] Goldwasser, S., Kalai, Y., Popa, R.A., Vaikuntanathan, V., Zeldovich, N.: Reusable garbled circuits and succinct functional encryption. In: Proceedings of the 45th Annual ACM Symposium on Theory of Computing, STOC 2013, pp. 555–564 (2013)

[GM84]   Goldwasser, S., Micali, S.: Probabilistic encryption. J. Comput. Syst. Sci. 28(2), 270–299 (1984)

[Got98]   Gottesman, D.: Theory of fault-tolerant quantum computation. Phys. Rev. A 57, 127–137 (1998)

[GVW13]   Gorbunov, S., Vaikuntanathan, V., Wee, H.: Attribute-based encryption for circuits. In: Proceedings of the 45th Annual ACM Symposium on Theory of Computing, STOC 2013, pp. 545–554 (2013)

[IP07]   Ishai, Y., Paskin, A.: Evaluating branching programs on encrypted data. In: Vadhan, S.P. (ed.) TCC 2007. LNCS, vol. 4392, pp. 575–594. Springer, Heidelberg (2007)

[KO97]   Kushilevitz, E., Ostrovsky, R.: Replication is not needed: single database, computationally-private information retrieval. In: FOCS, p. 364. IEEE (1997)

[KP14]   Klauck, H., Podder, S.: New bounds for the garden-hose model. In: 34th International Conference on Foundation of Software Technology and Theoretical Computer Science, pp. 481–492 (2014)

[Lia13]   Liang, M.: Symmetric quantum fully homomorphic encryption with perfect security. Quantum Inf. Process. 12(12), 3675–3687 (2013)

[Lia15]   Liang, M.: Quantum fully homomorphic encryption scheme based on universal quantum circuit. Quantum Inf. Process. 14(8), 2749–2759 (2015)

[Mar14]   Margalit, O.: On the riddle of coding equality function in the garden hose model. In: Information Theory and Applications Workshop (ITA), pp. 1–5. IEEE (2014)

[NC00]   Nielsen, M., Chuang, I.: Quantum Computation and Quantum Information. Cambridge University Press, Cambridge (2000)

[OTF15]   Ouyang, Y., Tan, S.-H., Fitzsimons, J.: Quantum homomorphic encryption from quantum codes (2015). arXiv preprint arXiv:1508.00938

[Pai99]   Paillier, P.: Public-key cryptosystems based on composite degree residuosity classes. In: Stern, J. (ed.) EUROCRYPT 1999. LNCS, vol. 1592, pp. 223–238. Springer, Heidelberg (1999)

[RAD78]   Rivest, R.L., Adleman, L., Dertouzos, M.L.: On data banks, privacy homomorphisms. Found. Secur. Comput. 4(11), 169–180 (1978)

[RFG12]   Rohde, P.P., Fitzsimons, J.F., Gilchrist, A.: Quantum walks with encrypted data. Phys. Rev. Lett. 109(15), 150501 (2012)

[RSA78]   Rivest, R.L., Shamir, A., Adleman, L.: A method for obtaining digital signatures and public-key cryptosystems. Commun. ACM 21(2), 120–126 (1978)

[SB08]   Shepherd, D., Bremner, M.J.: Instantaneous quantum computation (2008). arXiv preprint arXiv:0809:0847

[Spe11] Speelman, F.: Position-based quantum cryptography, the garden-hose game. Master's thesis, University of Amsterdam. arXiv:1210.4353

[Spe15] Speelman, F.: Instantaneous non-local computation of low T-depth quantum circuits (2015). arXiv preprint arXiv:1505.02695

[SW14] Sahai, A., Waters, B.: How to use indistinguishability obfuscation: deniable encryption, and more. In: Proceedings of the 46th Annual ACM Symposium on Theory of Computing, STOC 2014, pp. 475–484 (2014)

[SYY99] Sander, T., Young, A., Yung, M.: Non-interactive cryptocomputing for NC1. In: 40th Annual Symposium on Foundations of Computer Science, pp. 554–566. IEEE (1999)

[TKO+14] Tan, S.-H., Kettlewell, J.A., Ouyang, Y., Chen, L., Fitzsimons, J.F.: A quantum approach to fully homomorphic encryption (2014). arXiv preprint arXiv:1411.5254

[Vai11] Vaikuntanathan, V.: Computing blindfolded: new developments in fully homomorphic encryption. In: 2011 IEEE 52nd Annual Symposium on Foundations of Computer Science (FOCS), pp. 5–16. IEEE (2011)

[VDGHV10] van Dijk, M., Gentry, C., Halevi, S., Vaikuntanathan, V.: Fully homomorphic encryption over the integers. In: Gilbert, H. (ed.) EUROCRYPT 2010. LNCS, vol. 6110, pp. 24–43. Springer, Heidelberg (2010)

[VFPR14] Dunjko, V., Fitzsimons, J.F., Portmann, C., Renner, R.: Composable security of delegated quantum computation. In: Sarkar, P., Iwata, T. (eds.) ASIACRYPT 2014, Part II. LNCS, vol. 8874, pp. 406–425. Springer, Heidelberg (2014)

[YPDF14] Li, Y., Pérez-Delgado, C.A., Fitzsimons, J.F.: Limitations on information-theoretically-secure quantum homomorphic encryption. Phys. Rev. A **90**, 050303 (2014)

Adaptive Versus Non-Adaptive Strategies in the Quantum Setting with Applications

Frédéric Dupuis[2], Serge Fehr[1], Philippe Lamontagne[3(✉)], and Louis Salvail[3]

[1] CWI, Amsterdam, The Netherlands
[2] Faculty of Informatics, Masaryk University, Brno, Czech Republic
[3] Université de Montréal (DIRO), Montréal, Canada
lamontph@iro.umontreal.ca

Abstract. We prove a general relation between *adaptive* and *non-adaptive* strategies in the quantum setting, i.e., between strategies where the adversary can or cannot adaptively base its action on some auxiliary quantum side information. Our relation holds in a very general setting, and is applicable as long as we can control the bit-size of the side information, or, more generally, its "information content". Since adaptivity is notoriously difficult to handle in the analysis of (quantum) cryptographic protocols, this gives us a very powerful tool: as long as we have enough control over the side information, it is sufficient to restrict ourselves to non-adaptive attacks.

We demonstrate the usefulness of this methodology with two examples. The first is a quantum bit commitment scheme based on *1-bit cut-and-choose*. Since bit commitment implies oblivious transfer (in the quantum setting), and oblivious transfer is universal for two-party computation, this implies the universality of 1-bit cut-and-choose, and thus solves the main open problem of [9]. The second example is a quantum bit commitment scheme proposed in 1993 by Brassard *et al.* It was originally suggested as an unconditionally secure scheme, back when this was thought to be possible. We partly restore the scheme by proving it secure in (a variant of) the bounded quantum storage model.

In both examples, the fact that the adversary holds quantum side information obstructs a direct analysis of the scheme, and we circumvent it by analyzing a non-adaptive version, which can be done by means of known techniques, and applying our main result.

1 Introduction

Adaptive Versus Non-Adaptive Attacks. We consider attacks on cryptographic schemes, and we compare adaptive versus non-adaptive strategies for the adversary. In our context, a strategy is *adaptive* if the adversary's action can depend on some auxiliary side information, and it is *non-adaptive* if the adversary has no access to any such side information. Non-adaptive strategies are typically much easier to analyze than adaptive ones.

Adaptive strategies are clearly more powerful than non-adaptive ones, but this advantage is limited by the amount and quality of the side-information available to the attacker. In the classical case, this can be made precise by the following

© International Association for Cryptologic Research 2016
M. Robshaw and J. Katz (Eds.): CRYPTO 2016, Part III, LNCS 9816, pp. 33–59, 2016.
DOI: 10.1007/978-3-662-53015-3_2

simple argument. If the side information consists of a classical n-bit string, then adaptivity increases the adversary's success probability in breaking the scheme by at most a factor of 2^n. Indeed, a particular non-adaptive strategy is to try to guess the n-bit side information and then apply the best adaptive strategy. Since the guess will be correct with probability at least 2^{-n}, it follows that $P_{\text{succ}}^{\text{NA}} \geq 2^{-n} P_{\text{succ}}^{\text{A}}$, and thus $P_{\text{succ}}^{\text{A}} \leq 2^n P_{\text{succ}}^{\text{NA}}$, where $P_{\text{succ}}^{\text{A}}$ and $P_{\text{succ}}^{\text{NA}}$ respectively denote the optimal adaptive and non-adaptive success probabilities for the adversary to break the scheme. Even though there is an exponential loss, this is a very powerful relation between adaptive and non-adaptive strategies as it applies very generally, and it provides a non-trivial bound as long as we can control the size of the side information, and the non-adaptive success probability is small enough.

Our Technical Result. In this work, we consider the case where the side information (and the cryptographic scheme as a whole) may be *quantum*. A natural question is whether the same (or a similar) relation holds between adaptive and non-adaptive quantum strategies. The quantum equivalent to guessing the side information would be to emulate the n-qubit quantum side information by the completely mixed state $\frac{\mathbb{I}_A}{2^n}$. Since it always holds that $\rho_{AB} \leq 2^{2n} \frac{\mathbb{I}_A}{2^n} \otimes \rho_B$, we immediately obtain a similar relation $P_{\text{succ}}^{\text{A}} \leq 2^{2n} P_{\text{succ}}^{\text{NA}}$, but with an additional factor of 2 in the exponent. The bound is tight for certain choices of ρ_{AB}, and thus this additional loss is unavoidable in general; this seems to mostly answer the above question.

In this work, we show that this is actually not yet the end of the story. Our main technical result consists of a more refined treatment — and analysis — of the relation between adaptive and non-adaptive quantum strategies. We show that in a well-defined and rather general context, we can actually bound $P_{\text{succ}}^{\text{A}}$ as

$$P_{\text{succ}}^{\text{A}} \leq 2^{I_{\max}^{\text{acc}}(B;A)} P_{\text{succ}}^{\text{NA}},$$

where $I_{\max}^{\text{acc}}(B;A)$ is a new (quantum) information measure that is upper bounded by the number of qubits of A. As such, we not only recover the classical relation $P_{\text{succ}}^{\text{A}} \leq 2^n P_{\text{succ}}^{\text{NA}}$ in the considered context, but we actually improve on it.

In more detail, we consider an abstract "game", specified by an arbitrary bipartite quantum state ρ_{AB}, of which the adversary Alice and a challenger Bob hold the respective registers A and B, and by an arbitrary family $\{E^j\}_{j \in \mathcal{J}}$ of binary-outcome POVMs acting on register B. The game is played as follows: Alice chooses an index j, communicates it to Bob, and Bob measures his state B using the POVM $E^j = \{E_0^j, E_1^j\}$ specified by Alice. Alice wins the game if Bob's measurement outcome is 1. In the adaptive version of the game, Alice can choose the index j by performing a measurement on A; in the non-adaptive version, she has to decide upon j without resorting to A. As we will see, this game covers a large class of quantum cryptographic schemes, where Bob's binary measurement outcome specifies whether Alice succeeded in breaking the scheme.

Our main result shows that in any such game it holds that $P_{\text{succ}}^{\text{A}} \leq 2^n P_{\text{succ}}^{\text{NA}}$ where $n = H_0(A)$, i.e., the number of qubits of A. Actually, as already mentioned, we show a more general and stronger bound $P_{\text{succ}}^{\text{A}} \leq 2^{I_{\max}^{\text{acc}}(B;A)} P_{\text{succ}}^{\text{NA}}$ that also applies if we have no bound on the number of qubits of A, but we have some control over its "information content" $I_{\max}^{\text{acc}}(B; A)$, which is a new information measure that we introduce and show to be upper bounded by $H_0(A)$.

To give a first indication of the usefulness of our result, we observe that it easily provides a lower-bound on the quantity, *or quality*, of entanglement (as measured by $I_{\max}^{\text{acc}}(B; A)$) that a dishonest committer needs in order to carry out the standard attack [18] on a quantum bit commitment scheme. Let Alice be the committer and Bob the receiver in a bit commitment scheme in which the opening phase consists of Alice announcing a classical string j and Bob applying a verification described by POVM $\{E_{\text{accept}}^j, E_{\text{reject}}^j\}$. In the standard attack, Alice always commits to 0 while purifying her actions and applies an operation on her register if she wants to change her commitment to 1. If we let ρ_{AB} be the state of Bob's register B that corresponds to a commitment to 0, then the probability that a memoryless Alice successfully changes her commitment to 1 is $P_{\text{succ}}^{\text{NA}} = \max_j \text{tr}(E_{\text{accept}}^j \rho_{AB})$ where the maximum is over all j that open 1. If Alice holds a register A entangled with B, our main result implies that $I_{\max}^{\text{acc}}(B; A)$ must be proportional to $-\log P_{\text{succ}}^{\text{NA}}$ for Alice to have a constant probability of changing her commitment.

But the real potential lies in the observation that adaptivity is notoriously difficult to handle in the analysis of cryptographic protocols, and as such our result provides a very powerful tool: as long as we have enough control over the side information, it is sufficient to restrict ourselves to non-adaptive attacks.

Applications. We demonstrate the usefulness of this methodology by proving the security of two commitment schemes. In both examples, the fact that the adversary holds quantum side information obstructs a direct analysis of the scheme, and we circumvent it by analyzing a non-adaptive version and applying our general result.

One-Bit Cut-and-Choose is Universal for Two-Party Computation. As a first example, we propose and prove secure a quantum bit commitment scheme that uses an ideal *1-bit cut-and-choose* primitive 1CC (see Fig. 1 in Sect. 4) as a black box. Since bit commitment (BC) implies oblivious transfer (OT) in the quantum setting [2,7,20], and oblivious transfer is universal for two-party computation, this implies the universality of 1CC and thus completes the zero/xor/one law proposed in [9]. Indeed, it was shown in [9] that in the information-theoretic quantum setting, every primitive is either trivial (zero), universal (one), or can be used to implement an XOR — *except* that there was one missing piece in their characterization: it excluded 1CC (and any primitive that implies 1CC but not 2CC). How 1CC fits into the landscape was left as an open problem in [9]; we resolve it here.

The BCJL Bit Commitment Scheme in (A Variant of) The Bounded Quantum Storage Model. As a second application, we consider a general class of non-interactive commitment schemes and we show that for any such scheme, security against an adversary with no quantum memory *at all* implies security in a slightly strengthened version of the standard bounded quantum storage model[1], with a corresponding loss in the error parameter.[2]

As a concrete example scheme, we consider the classic BCJL scheme that was proposed in 1993 by Brassard *et al.* [6] as a candidate for an unconditionally-secure scheme — back when this was thought to be possible — but until now has resisted any rigorous *positive* security analysis. Our methodology of relating adaptive to non-adaptive security allows us to prove it secure in (a variant of) the bounded quantum storage model.

2 Preliminaries

2.1 Basic Notation

For any string $x = (x_1, \ldots, x_n) \in \{0,1\}^n$ and any subset $t = \{t_1, \ldots t_k\} \subseteq [n]$, we write x_t for the substring $x_t = (x_{t_1}, \ldots, x_{t_k}) \in \{0,1\}^{|t|}$. The n-bit all-zero string is denoted as 0^n. The Hamming distance between two strings $x, y \in \{0,1\}^n$ is defined as $d(x,y) = \sum_{i=1}^{n} x_i \oplus y_i$. For $\delta > 0$ and $x \in \{0,1\}^n$, $B^\delta(x)$ denotes the set of all n bit strings at Hamming distance at most δn from x. We denote by $\lg(\cdot)$ the logarithm with respect to base 2. It is well known that the set $B^\delta(x)$ contains at most $2^{nh(\delta)}$ strings where $h(\delta) = -\delta \lg(\delta) - (1-\delta) \lg(1-\delta)$ is the binary entropy function.

Ideal cryptographic *functionalities* (or *primitives*) are referenced by their name written in sans-serif font. They are fully described by their input/output behaviour (see, e.g., functionality 1CC described in Fig. 1 in Sect. 4). Cryptographic *protocols* have their names written in small capitals with a primitive name in superscript if the protocol has black-box access to this primitive (e.g. protocol BC^{1CC} in Sect. 4).

2.2 Quantum States and More

We assume familiarity with the basic concepts of quantum information; we merely fix notation and terminology here. We label quantum registers by capital letters A, B etc. and their corresponding Hilbert spaces are respectively denoted by $\mathcal{H}_A, \mathcal{H}_B$ etc. We say that a quantum register A is "empty" if $\dim(\mathcal{H}_A) = 1$. The state of a quantum register is specified by a density operator ρ, a positive semidefinite trace-1 operator. We typically write ρ_A for the state of A, etc. The set of density operators for register A is denoted $\mathcal{D}(\mathcal{H}_A)$. We write $X \geq 0$ to express that the operator X is positive semidefinite, and $Y \geq X$ to express that $Y - X$ is positive semidefinite.

---

[1] Beyond bounding the adversary's quantum memory, we also restrict its measurements to be projective; this can be justified by the fact that to actually implepro-jections onto thement a non-projective measurement, additional quantum memory is needed.

[2] We have already shown above how to argue for the standard attack [18] against quantum bit commitment schemes; taking care of *arbitrary* attacks is more involved.

We measure the distance between two states ρ and σ in terms of their *trace distance* $D(\rho, \sigma) := \frac{1}{2}\|\rho - \sigma\|_1$, where $\|X\|_1 := \text{tr}(\sqrt{X^\dagger X})$ is the *trace norm*. We say that ρ and σ are ϵ-close if $D(\rho, \sigma) \leq \epsilon$, and we call them *indistinguishable* if their trace distance is negligible (in the security parameter).

The *computational* (or rectilinear) basis for a single qubit quantum register is denoted by $\{|0\rangle_+, |1\rangle_+\}$, and the *diagonal* basis by $\{|0\rangle_\times, |1\rangle_\times\}$. Recall that $|0\rangle_\times = \frac{1}{\sqrt{2}}(|0\rangle_+ + |1\rangle_+)$ and $|1\rangle_\times = \frac{1}{\sqrt{2}}(|0\rangle_+ - |1\rangle_+)$. For any $x \in \{0,1\}^n$ and $\theta \in \{+, \times\}^n$, we set $|x\rangle_\theta := \bigotimes_{i=1}^n |x_i\rangle_{\theta_i}$. In the following, we will view and represent any sequence of diagonal and computational bases by a bit string $\theta \in \{0,1\}^n$, where $\theta_i = 0$ represents the computational basis and $\theta_i = 1$ the diagonal basis. In other words, for $b \in \{0,1\}$, $|b\rangle_0 := |b\rangle_+$ and $|b\rangle_1 := |b\rangle_\times$. And for $\theta, x \in \{0,1\}^n$, we define $|x\rangle_\theta := \bigotimes_{i=1}^n |x_i\rangle_{\theta_i}$.

Operations on quantum registers are modeled as completely-positive trace-preserving (CPTP) maps. To indicate that a CPTP map $\mathcal{E}$ takes inputs in A and outputs to B, we use subscript $A \to B$. If $\mathcal{E}_{A \to B}$ is a CPTP map acting on register A, we slightly abuse notation and write $\mathcal{E}(\rho_{AC})$ instead of $\mathcal{E} \otimes \mathbb{I}_C(\rho_{AC})$ where $\mathbb{I}_C$ is the CPTP map that leaves register C unchanged. A *measurement* on a quantum register A, producing a measurement outcome X, is a CPTP map $\mathcal{E}_{A \to X}$ of the form

$$\mathcal{E}(\rho_A) = \sum_{x \in \mathcal{X}} \text{tr}(E_x \rho_A)|x\rangle\langle x|_X,$$

where $\{|x\rangle\}$ a basis of $\mathcal{H}_X$ and $E = \{E_x\}_{x \in \mathcal{X}}$ is a POVM, i.e., a collection of positive semidefinite operators satisfying $\sum_{x \in \mathcal{X}} E_x = \mathbb{I}$.

The *spectral norm* of an operator X is defined as $\|X\| := \max_{|u\rangle} \|X|u\rangle\|$, where the maximum is over all normalized vectors $|u\rangle$, and an operator is called an *orthogonal projector* if $X^\dagger = X$ and $X^2 = X$. The following was shown in [8].

Lemma 1. *For any two orthogonal projectors X and Y: $\|X + Y\| \leq 1 + \|XY\|$.*

2.3 Entropy and Privacy Amplification

In the following, the two notions of entropy that we will be dealing with are the min-entropy and the zero-entropy of a quantum register. They are defined as follows:

Definition 1. *The* min-entropy *of a bipartite quantum state ρ_{AB} relative to register B is the largest number $\text{H}_\infty(A|B)_\rho$ such that there exists a $\sigma_B \in \mathcal{D}(\mathcal{H}_B)$,*

$$2^{-\text{H}_\infty(A|B)_\rho} \cdot \mathbb{I}_A \otimes \sigma_B \geq \rho_{AB}.$$

The zero-entropy *of a state ρ_A is defined as*

$$\text{H}_0(A)_\rho = \lg(\text{rank}(\rho_A)).$$

We write $\text{H}_\infty(A|B)$ and $\text{H}_0(A)$ when the state of the registers is clear from the context.

The min-entropy has the following operational interpretation [13]. Let ρ_{XB} be a so-called cq-state, i.e., of the from $\rho_{XB} = \sum_x P_X(x)|x\rangle\langle x|_X \otimes \rho_B^x$. Then $P_{\text{guess}}(X|B) = 2^{-H_\infty(X|B)_\rho}$ where $P_{\text{guess}}(X|B)$ is the probability of guessing the value of the classical random variable X, maximized over all POVMs on B.

Let $\mathcal{G}_n$ be a family of hash functions $g : \{0,1\}^n \to \{0,1\}$ with a binary output. The family $\mathcal{G}_n$ is said to be *two-universal* if for any $x, y \in \{0,1\}^n$ with $x \neq y$ and $G \in_R \mathcal{G}_n$,

$$\Pr\left(G(x) = G(y)\right) \leq \frac{1}{2}.$$

Privacy amplification against quantum side information, in case of hash functions with a binary-output, can be stated as follows:

Theorem 1 (Privacy Amplification [19]). *Let $\mathcal{G}_n$ be a two-universal family of hash functions $g : \{0,1\}^n \to \{0,1\}$ with a binary output. Furthermore, let $\rho_{XE} = \sum_{x\in\{0,1\}^n} P_X(x)|x\rangle\langle x|_X \otimes \rho_E^x$ be an arbitrary cq-state, and let*

$$\rho_{YGXE} := \frac{1}{|\mathcal{G}_n|} \sum_{g\in\mathcal{G}_n} \sum_{x\in\{0,1\}^n} P_X(x)|g(x)\rangle\langle g(x)|_Y \otimes |g\rangle\langle g|_G \otimes |x\rangle\langle x|_X \otimes \rho_E^x$$

be the state obtained by choosing a random g in $\mathcal{G}_n$, applying g to the value stored in X, and storing the result in register Y. Then,

$$D\left(\rho_{YGE}, \frac{\mathbb{I}_Y}{2} \otimes \rho_{GE}\right) \leq \frac{1}{2} \cdot 2^{-\frac{1}{2}(H_\infty(X|E)-1)}.$$

3   Main Result

We consider an abstract game between two parties Alice and Bob. The game is specified by a joint state ρ_{AB}, shared between Alice and Bob who hold respective registers A and B, and by a non-empty finite family $\mathbf{E} = \{E^j\}_{j\in\mathcal{J}}$ of binary-outcome POVMs $E^j = \{E_0^j, E_1^j\}$ acting on B. An execution of the game works as follows: Alice announces an index $j \in \mathcal{J}$ to Bob, and Bob measures register B of the state ρ_{AB} using the POVM E^j specified by Alice's choice of j. Alice *wins* the game if the measurement outcome is 1. We distinguish between an *adaptive* and a *non-adaptive* Alice. An *adaptive* Alice can obtain j by performing a measurement on her register A of ρ_{AB}; on the other hand, an *non-adaptive* Alice has to produce j from scratch, i.e., without accessing A. This motivates the following formal definitions.

Definition 2. *Let ρ_{AB} be a bipartite quantum state, and let $\mathbf{E} = \{E^j\}_{j\in\mathcal{J}}$ be a non-empty finite family of binary-outcome POVMs $E^j = \{E_0^j, E_1^j\}$ acting on B. Then, we define*

$$P_{\text{succ}}(\rho_{AB}, \mathbf{E}) := \max_{\{F_j\}_j} \sum_{j\in\mathcal{J}} \text{tr}\left((F_j \otimes E_1^j)\rho_{AB}\right),$$

where the maximum is over all POVMs $\{F_j\}_{j\in\mathcal{J}}$ *acting on* A. *We call* $P_{\text{succ}}(\rho_{AB}, \mathbf{E})$ *the adaptive success probability, and we call* $P_{\text{succ}}(\rho_B, \mathbf{E})$ *the non-adaptive success probability, where the latter is naturally understood by considering an "empty"* A, *and it equals*

$$P_{\text{succ}}(\rho_B, \mathbf{E}) = \max_{j\in\mathcal{J}} \text{tr}\big(E_1^j \rho_B\big).$$

If ρ_{AB} *and* $\mathbf{E}$ *are clear from the context, we write* $P_{\text{succ}}^{\text{A}}$ *and* $P_{\text{succ}}^{\text{NA}}$ *instead of* $P_{\text{succ}}(\rho_{AB}, \mathbf{E})$ *and* $P_{\text{succ}}(\rho_B, \mathbf{E})$.

As a matter of fact, for the sake of generality, we consider a setting with an additional quantum register A' to which both the adaptive and the non-adaptive Alice have access to, but, as above only the adaptive Alice has access to A. In that sense, we will compare an adaptive with a *semi-adaptive* Alice. Formally, we will consider a tripartite state $\rho_{AA'B}$ and relate $P_{\text{succ}}(\rho_{AA'B}, \mathbf{E})$ to $P_{\text{succ}}(\rho_{A'B}, \mathbf{E})$. Obviously, the special case of an "empty" A' will then provide a relation between $P_{\text{succ}}^{\text{A}}$ and $P_{\text{succ}}^{\text{NA}}$.

We now introduce a new measure of (quantum) information $I_{\max}^{\text{acc}}(B; A|A')_\rho$, which will relate the adaptive to the non- or semi-adaptive success probability in our main theorem. In its unconditional form $I_{\max}^{\text{acc}}(B; A)_\rho$, it is the accessible version of the max-information $I_{\max}(B; A)_\rho$ introduced in [3]; this means that it is the amount of max-information that can be accessed via measurements on Alice's share.

Definition 3. *Let* $\rho_{AA'B}$ *be a tripartite quantum state. Then, we define* $I_{\max}^{\text{acc}}(B; A|A')_\rho$ *as the smallest real number such that, for every measurement* $\mathcal{M}_{AA'\to X}$ *there exists a measurement* $\mathcal{N}_{A'\to X}$ *such that*

$$\mathcal{M}(\rho_{AA'B}) \leq 2^{I_{\max}^{\text{acc}}(B;A|A')_\rho} \mathcal{N}(\rho_{A'B}).$$

The unconditional version $I_{\max}^{\text{acc}}(B; A)_\rho$ *is naturally defined by considering* A' *to be "empty"; the above condition then coincides with*

$$\mathcal{M}(\rho_{AB}) \leq 2^{I_{\max}^{\text{acc}}(B;A)_\rho} \sigma_X \otimes \rho_B,$$

for some normalized density matrix $\sigma_X \in \mathcal{D}(\mathcal{H}_X)$, *which can be interpreted as the outcome of a measurement* $\mathcal{N}_{\mathbb{C}\to X}$ *on an "empty" register.*

We are now ready to state and prove our main result.

Theorem 2. *Let* $\rho_{AA'B}$ *be a tripartite quantum state, and let* $\mathbf{E} = \{E^j\}_{j\in\mathcal{J}}$ *be a non-empty finite family of binary-outcome POVMs* E^j *acting on* B. *Then, we have that*

$$P_{\text{succ}}(\rho_{AA'B}, \mathbf{E}) \leq 2^{I_{\max}^{\text{acc}}(B;A|A')_\rho} P_{\text{succ}}(\rho_{A'B}, \mathbf{E}).$$

By considering an "empty" A', we immediately obtain the following.

Corollary 1. *Let* ρ_{AB} *be a bipartite quantum state, and let* $\mathbf{E} = \{E^j\}_{j\in\mathcal{J}}$ *be as above. Then,*

$$P_{\text{succ}}^{\text{A}} \leq 2^{I_{\max}^{\text{acc}}(B;A)_\rho} P_{\text{succ}}^{\text{NA}}.$$

Proof (of Theorem 2*).* Let $\{F_j\}_{j\in\mathcal{J}}$ be an arbitrary POVM acting on AA', and let $\mathcal{M}_{AA'\to J}$ be the corresponding measurement $\mathcal{M}(\sigma_{AA'}) = \sum_j \mathrm{tr}(F_j\sigma)|j\rangle\langle j|$. We define the map

$$\mathcal{E}_{JB\to\mathbb{C}}(\sigma_{JB}) := \sum_j \mathrm{tr}((|j\rangle\langle j| \otimes E_1^j)\sigma_{JB}),$$

which is completely positive (but not trace-preserving in general). From the definition of $I_{\max}^{\mathrm{acc}}$, we know that there exists a measurement $\mathcal{N}_{A'\to J}$, i.e., a CPTP map of the form $\mathcal{N}(\sigma_{A'}) = \sum_j \mathrm{tr}(F'_j\sigma)|j\rangle\langle j|$ for a POVM $\{F'_j\}_{j\in\mathcal{J}}$ acting on A', such that

$$\mathcal{M}(\rho_{AA'B}) \leq 2^{I_{\max}^{\mathrm{acc}}(B;A|A')_\rho}\mathcal{N}(\rho_{A'B}).$$

Applying $\mathcal{E}$ on both sides gives

$$(\mathcal{E}\circ\mathcal{M})(\rho_{AA'B}) \leq 2^{I_{\max}^{\mathrm{acc}}(B;A|A')_\rho}(\mathcal{E}\circ\mathcal{N})(\rho_{A'B}),$$

and expanding both sides using the definitions of $\mathcal{E}$, $\mathcal{M}$ and $\mathcal{N}$ gives

$$\sum_j \mathrm{tr}((F_j\otimes E_1^j)\rho_{AA'B}) \leq 2^{I_{\max}^{\mathrm{acc}}(B;A|A')_\rho}\sum_j \mathrm{tr}((F'_j\otimes E_1^j)\rho_{A'B})$$

$$\leq 2^{I_{\max}^{\mathrm{acc}}(B;A|A')_\rho}P_{\mathrm{succ}}(\rho_{A'B}, \mathbf{E}).$$

This yields the theorem statement, since the left-hand side equals to $P_{\mathrm{succ}}(\rho_{AA'B}, \mathbf{E})$ when maximized over the choice of the POVM $\{F_j\}_{j\in\mathcal{J}}$. $\square$

By the following proposition, we see that Corollary 1 implies a direct generalization of the classical bound, which ensures that giving access to n bits increases the success probability by at most 2^n, to qubits.

Proposition 1. *For any ρ_{AB}, we have that $I_{\max}^{\mathrm{acc}}(B;A)_\rho \leq H_0(A)_\rho$.*

Proof. Let $|\psi\rangle_{ABR}$ be a purification of ρ_{AB} and let $\mathcal{M}_{A\to X}$ be a measurement on A. Since $|\psi\rangle$ is also a purification of ρ_A, there exists a linear operator $V_{\bar{A}\to BR}$ from a register $\bar{A}$ of the same dimension as A into BR such that $|\psi\rangle_{ABR} = (\mathbb{I}_A \otimes V)|\Phi\rangle_{A\bar{A}}$, with $|\Phi\rangle = \sum_i |i\rangle_A \otimes |i\rangle_{\bar{A}}$. Now, first note that

$$2^{-H_0(A)}(\mathcal{M}\otimes\mathbb{I})(\Phi_{A\bar{A}}) = \sum_x \lambda_x |x\rangle\langle x|_X \otimes \omega_{\bar{A}}^x \leq \sum_x \lambda_x |x\rangle\langle x|_X \otimes \mathbb{I}_{\bar{A}},$$

where $\{\lambda_x\}$ is a probability distribution, and each $\omega_{\bar{A}}^x$ is normalized because $\mathrm{tr}(\Phi) = 2^{H_0(A)}$. Multiplying both sides of the inequality by $2^{H_0(A)}$ and conjugating by V, we get

$$(\mathcal{M}\otimes\mathbb{I})(|\psi\rangle\langle\psi|) \leq 2^{H_0(A)}\sum_x \lambda_x |x\rangle\langle x| \otimes VV^\dagger.$$

Using the fact that $VV^\dagger = \psi_{BR} := \mathrm{tr}_A(|\psi\rangle\langle\psi|)$, this yields

$$(\mathcal{M}\otimes\mathbb{I})(|\psi\rangle\langle\psi|) \leq 2^{H_0(A)}\sum_x \lambda_x |x\rangle\langle x| \otimes \psi_{BR}.$$

Tracing out R on both sides and defining $\sigma_X = \sum_x \lambda_x |x\rangle\langle x|$ then yields

$$(\mathcal{M} \otimes \mathbb{I})(\rho_{AB}) \leq 2^{H_0(A)} \sigma_X \otimes \rho_B,$$

which proves the claim.   □

One might naively expect that also the conditional version $I_{\max}^{\mathrm{acc}}(B; A|A')_\rho$ is upper bounded by $H_0(A)_\rho$, implying a corresponding statement for a *semi-adaptive* Alice: giving access to n *additional* qubits increases the success probability by at most 2^n. However, this is not true, as the following example illustrates. Let register B contain two random classical bits, and let A and A' be two qubit registers, containing one of the four Bell states, and which one it is, is determined by the two classical bits. Alice's goal is to guess the two bits. Clearly, A' alone is useless, and thus a semi-adaptive Alice having access to A' has a guessing probability of at most $\frac{1}{4}$. On the other hand, adaptive Alice can guess them with certainty by doing a Bell measurement on AA'.

However, Proposition 1 does generalize to the conditional version in case of a *classical* A'.

Proposition 2. *For any state ρ_{ZAB} with classical Z:*

$$I_{\max}^{\mathrm{acc}}(B; A|Z)_\rho \leq \max_z I_{\max}^{\mathrm{acc}}(B; A)_{\rho^z} \leq H_0(A)_\rho.$$

An additional property of $I_{\max}^{\mathrm{acc}}$ is that quantum operations that are in tensor product form on registers A and B cannot increase the max-accessible-information.

Proposition 3. *Let $\mathcal{E}_{AB \to A'B'}$ be a CPTP map of the form $\mathcal{E} = \mathcal{E}^A \otimes \mathcal{E}^B$. Then*

$$I_{\max}^{\mathrm{acc}}(B'; A')_{\mathcal{E}(\rho)} \leq I_{\max}^{\mathrm{acc}}(B; A)_\rho.$$

The proofs the two previous results can be found in Appendix A.

4   Application 1: 1CC Is Universal

4.1   Background

It is a well-known fact that information-theoretically secure two-party computation is impossible without assumptions. As a result, one of the natural questions that arises is: what are the minimal assumptions required to achieve it? One way to attack this question is to try to identify the simplest cryptographic primitives which, when made available in a black-box way to the two parties, allow them to perform arbitrary two-party computations. We then say that such a primitive is "universal". Perhaps the best known such primitive is one-out-of-two oblivious transfer (OT), which has been shown to be universal by Kilian [10]. Since then, the power of various primitives for two-party computation has been studied in much more detail [11,12,14–17]. Recently, it has been shown in [16] that every non-trivial two-party primitive (i.e. any primitive that cannot be done

$$x \longrightarrow \boxed{\text{CC}} \longleftarrow c \in \{0,1\}$$

$$c \longleftarrow \quad \longrightarrow w = \begin{cases} \bot & \text{if } c = 0 \\ x & \text{if } c = 1. \end{cases}$$

Fig. 1. The cut-and-choose functionality. The one-bit and two-bit versions of the functionality refer to the length of x. One player chooses x, and the other player chooses whether he wants to see x or not. The first player then learns the choice that was made.

from scratch without assumptions) can be used as a black-box to implement one of four basic primitives: oblivious transfer (OT), bit commitment (BC), an XOR between Alice's and Bob's inputs, or a primitive called *cut-and-choose* (CC) as depicted in Fig. 1.

Interestingly, this picture becomes considerably simpler when we consider quantum protocols. First, BC can be used to implement OT [2,7,20] and is therefore universal. Furthermore, as was shown in [9], even a 2-bit cut-and-choose (2CC) is universal in the quantum setting, giving rise to what they call a zero/xor/one law: every primitive is either trivial (zero), universal (one), or can be used to implement an XOR. However, there was one missing piece in this characterization: it applies to all functionalities except those that are sufficient to implement 1-bit cut-and-choose (1CC), but not 2CC. In this section, we resolve this issue by showing that 1CC is universal. We do this by presenting a quantum protocol for bit commitment that uses 1CC as a black box, and we prove its security using our adaptive to non-adaptive reduction.

4.2   The Protocol

The protocol is given in Fig. 2, where Alice is the committer and Bob the receiver. The protocol is parameterized by $N \in \mathbb{N}$, which acts as security parameter, and by constants q, τ and r, where $q, \tau > 0$ are small and $r < 1$ is close to 1. Intuitively, our bit commitment protocol uses the 1CC primitive to ensure that the state Alice sends to Bob is close to what it is supposed to be: $|0^N\rangle_\theta$ for some randomly chosen but fixed basis θ. Indeed, the 1CC primitive allows Bob to sample a small random subset of the qubits and check for correctness on that subset; if the state looks correct on this subset, we expect that it cannot be too far off on the unchecked part.

Note that our protocol uses the B92 [1] encoding ($\{|0\rangle_+, |0\rangle_\times\}$), rather than the more common BB84 encoding. This allows us to get away with a *one*-bit cut-and-choose functionality; with the BB84 encoding, Alice would have to "commit" to *two* bits: the basis and the measurement outcome.

We use the quantum sampling framework of Bouman and Fehr [4] to analyze the checking procedure of the protocol. Actually, we use the *adaptive* version of [9], which deals with an Alice that can decide on the next basis adaptively depending on what Bob has asked to see so far. On the other hand, to deal with Bob choosing his sample subset adaptively depending on what he has seen so

---

$$\text{COMMIT}^{\text{1CC}}_{N,q,\tau,r}(b):$$

1. Alice chooses random $\theta \in \{0,1\}^N$ and sends the N qubit state $|0^N\rangle_\theta$ to Bob.
2. For $i = 1 \ldots N$, do
   (a) Alice and Bob invoke an instance of 1CC: Alice inputs the bit θ_i, and Bob inputs bit 1 with probability q and bit 0 with probability $1 - q$.
   (b) If Bob's input was 1, he measures the ith qubit of the state he received in basis θ_i and checks that the result is 0. If it is not, he aborts.
   Let $t \subset [N]$ the set of positions that Bob checked, $\bar{t} = [N] \setminus t$, and $n = |\bar{t}|$.
3. Alice aborts the protocol if Bob checked more than $2qN$ positions.
4. Bob chooses a generator matrix G of a $[n, k, d]$-code with rate $k/n \geq r$ and $d/n \geq \tau$, and he sends G to Alice, who checks that indeed $k/n \geq r$.
5. Alice picks a random member $g \in \mathcal{G}_n$ from a family $\mathcal{G}_n$ of two-universal hash functions, and she computes the syndrome s of $\theta_{\bar{t}}$ for the linear code. Then, she sends g, s and $w = g(\theta_{\bar{t}}) \oplus b$ to Bob.

$$\text{REVEAL}^{\text{1CC}}:$$

1. Alice sends $\theta_{\bar{t}}$ and b to Bob.
2. Bob checks that he obtains 0 by measuring the ith qubit in basis θ_i for all $i \in \bar{t}$. He checks that $\theta_{\bar{t}}$ has syndrome s, and that $g(\theta_{\bar{t}}) \oplus w = b$. If one of the above checks failed, he aborts.
3. If Bob is honest, this eventuality only occurs with probability less than $2 \exp(-2q^2 N)$ according to Hoeffding's inequality.

Fig. 2. Bit commitment protocol BC^{1CC} based on the 1-bit cut-and-choose primitive.

far, we require the sample subset to be rather small, so that we can then apply union bound over all possible choices.

4.3   Security Proofs

We use the standard notion of hiding for a (quantum) bit commitment scheme.

Definition 4 (Hiding). *A bit-commitment scheme is ϵ-hiding if, for any dishonest receiver Bob, his state ρ_0 corresponding to a commitment to $b = 0$ and his state ρ_1 corresponding to a commitment to $b = 1$ satisfy $D(\rho_0, \rho_1) \leq \epsilon$.*

Since the proof that our protocol is hiding uses a standard approach, we only briefly sketch it.

Theorem 3. *Protocol* $\text{COMMIT}^{\text{1CC}}_{N,q,\tau,r}$ *is* $2^{-\frac{1}{2}N(\lg(1/\gamma)-2q-(1-r))}$*-hiding, where* $\gamma = \cos^2(\pi/8) \approx 0.85$ *(and hence* $\lg(1/\gamma) \approx 0.23$*).*

Proof (sketch). We need to argue that there is sufficient min-entropy in $\theta_{\bar{t}}$ for Bob; then, privacy amplification does the job. This means that we have to show that Bob has small success probability in guessing $\theta_{\bar{t}}$. What makes the argument slightly non-trivial is that Bob can choose t depending on the qubits $|0^N\rangle_\theta$. Note that since Alice aborts in case $|t| > 2qN$, we may assume that $|t| \le 2qN$.

It is a straightforward calculation to show that Bob's success probability in guessing θ right after step 1 of the protocol, i.e., when given the qubits $|0^N\rangle_\theta$, is γ^N, where $\gamma = \cos^2(\pi/8) \approx 0.85$. From this it then follows that right after step 2, Bob's success probability in guessing $\theta_{\bar{t}}$ is at most $\gamma^N \cdot 2^{2qN}$: if it was larger, then he could guess θ right after step 1 with probability larger than γ^N by simulating the sampling and guessing the $|t| \le 2qN$ bits θ_i that Alice provides. It follows that right after step 2, Bob's min-entropy in $\theta_{\bar{t}}$ is $N(\lg(1/\gamma) - 2q)$. Finally, by the chain rule for min-entropy, Bob's min-entropy in $\theta_{\bar{t}}$ when additionally given the syndrome s is $N\big(\lg(1/\gamma) - 2q\big) - (n - k) = N\big(\lg(1/\gamma) - 2q\big) - n(1 - k/n) \ge N\big(\lg(1/\gamma) - 2q - (1 - r)\big)$. The statement then directly follows from privacy amplification (Theorem 1) and the triangle inequality. $\square$

As for the binding property of our commitment scheme, as we will show, we achieve a strong notion of security that not only guarantees the existence of a bit to which Alice is bound in that she cannot reveal the other bit, but this bit is actually *universally extractable* from the classical information held by Bob together with the inputs to the 1CC:

Definition 5 (Universally Extractable). *A bit-commitment scheme (in the 1CC-hybrid model) is ϵ-universally extractable if there exists a function c that acts on the classical information $view_{Bob,1CC}$ held by Bob and 1CC after the commit phase, so that for any pure commit and open strategy for dishonest Alice, she has probability at most ϵ of successfully unveiling the bit $1 - c(view_{Bob,1CC})$.*

Our strategy for proving the binding property for our protocol is as follows. First, we show that due to the checking part, the (joint) state after the commit phase is of a restricted form. Then, we show that, based on this restriction on the (joint) state, a *non-adaptive* Alice who has no access to her quantum state, cannot open to the "wrong" bit. And finally, we apply our main result to conclude security against a general (adaptive) Alice.

The following lemma follows immediately from (the adaptive version of) Bouman and Fehr's quantum sampling framework [4,9]. Informally, it states that if Bob did not abort during sampling, then the post-sampling state of Bob's register is close to the correct state, up to a few errors. In other words, after the commit phase, Bob's state is a superposition of strings close to 0^n in the basis specified by $\theta_{\bar{t}}$.

Lemma 2. *Consider an arbitrary pure strategy for Alice in protocol* $\text{COMMIT}_{N,q,\tau,r}^{1CC}$. *Let ρ_{AB} be the joint quantum state at the end of the commit phase, conditioned (and thus dependent) on t, θ, g, w and s. Then, for any $\delta > 0$, on average over the choices of t, θ, g, w and s, the state ρ_{AB} is ϵ-close to an "ideal state" $\tilde{\rho}_{AB}$ (which is also dependent on t, θ etc.) with the property that the conditional state of $\tilde{\rho}_{AB}$ conditioned on Bob not aborting is pure and of the form*

$$|\phi_{AB}\rangle = \sum_{y \in B^\delta(0^n)} \alpha_y |\xi^y\rangle_A |y\rangle_{\theta_{\bar{t}}} \tag{1}$$

where $|\xi^y\rangle$ are arbitrary states on Alice's register and $\epsilon \leq \sqrt{4\exp(-q^2\delta^2 N/8)}$.

The following lemma implies that after the commit phase, if Alice and Bob share a state of the form of (1), then a non-adaptive Alice is bound to a fixed bit which is defined by some string θ'.

Lemma 3. For any t, θ and s there exists θ' with syndrome s such that for every $\theta'' \neq \theta'$ with syndrome s, and for every state $|\phi_{AB}\rangle$ of the form of (1),

$$\mathrm{tr}\big((\mathbb{I} \otimes |0\rangle\langle 0|_{\theta''})\phi_{AB}\big) \leq 2^{-\frac{d}{2}+nh(\delta)}.$$

Proof. Let $\theta' \in \{0,1\}^n$ be the string with syndrome s closest to $\theta_{\bar{t}}$ (in Hamming distance). Then, since the set of strings with a fixed syndrome form an error correcting code of distance d, every other $\theta'' \in \{0,1\}^n$ of syndrome s is at distance at least $d/2$ from $\theta_{\bar{t}}$. Bob's reduced density operator of state (1) is $\phi_B = \sum_{y,y' \in B^\delta(0^n)} \alpha_y \alpha_{y'}^* \langle \xi_{y'}|\xi_y\rangle |y\rangle\langle y'|_{\theta_{\bar{t}}}$. Using the fact that $d(\theta_{\bar{t}}, \theta'') \geq d/2$ for every $\theta'' \neq \theta'$ (and hence $|\mathrm{tr}(|0\rangle\langle 0|_{\theta''}|y\rangle\langle y'|_{\theta_{\bar{t}}})| \leq 2^{-\frac{d}{2}}$) and the triangle inequality, we get:

$$\mathrm{tr}(|0\rangle\langle 0|_{\theta''}\phi_B) \leq 2^{-\frac{d}{2}} \sum_{y,y' \in B^\delta(0^n)} |\alpha_y \alpha_{y'}^* \langle \xi_{y'}|\xi_y\rangle|$$

$$\leq 2^{-\frac{d}{2}} \sum_{y,y' \in B^\delta(0^n)} |\alpha_y||\alpha_{y'}^*|$$

$$= 2^{-\frac{d}{2}} \left(\sum_y |\alpha_y|\right)^2$$

$$\leq 2^{-\frac{d}{2}+nh(\delta)},$$

where the last inequality is argued by viewing $\sum_y |\alpha_y|$ as inner product of the vectors $\sum_y |\alpha_y||y\rangle$ and $\sum_y |y\rangle$, and applying the Cauchy-Schwarz inequality. $\quad\square$

We are now ready to prove that the scheme is universally extractable:

Theorem 4. For any $\delta > 0$, $\mathrm{COMMIT}^{\mathrm{1CC}}_{N,q,\tau,r}$ is ϵ-universally extractable with

$$\epsilon \leq 2^{-N(1-2q)(\tau/2-2h(\delta))} + \sqrt{4\exp(-q^2\delta^2 N/8)}.$$

Proof. We need to show the existence of a binary-valued function $c(\theta, t, g, w, s)$ as required by Definition 5, i.e., such that for any commit strategy, there is no opening strategy that allows Alice to unveil $\bar{c}$, except with small probability. We define this function as $c(t, \theta, g, s, w) := g(\theta') \oplus w$ where θ' is as in Lemma 3, depending on t, θ and s only.

Now, consider an arbitrary pure strategy for Alice in protocol $\mathrm{COMMIT}^{\mathrm{1CC}}$. Let θ, g, w and s be the values chosen by Alice during the commit phase and let

ρ_{AB} be the joint state of Alice and Bob after the commit phase. Fix $\delta > 0$ and consider the states $\tilde{\rho}_{AB}$ and $|\phi_{AB}\rangle$ as promised by Lemma 2. Recall that ρ_{AB} is ϵ-close to $\tilde{\rho}_{AB}$ (on average over θ, g, w and s, and for $\epsilon \leq \sqrt{4\exp(-q^2\delta^2 N/8)}$), and $\tilde{\rho}_{AB}$ is a mixture of Bob aborting in the commit phase and of $|\phi_{AB}\rangle$; therefore, we may assume that Alice and Bob share the pure state $\phi_{AB} = |\phi_{AB}\rangle\langle\phi_{AB}|$ instead of ρ_{AB} by taking into account the probability at most ϵ that the two states behave differently.

Let $\mathcal{B}$ be the set of strings θ'' with syndrome s such that $g(\theta'') \oplus w = \bar{c}$ and let $\mathbf{E} = \{\{E_0^{\theta''}, E_1^{\theta''}\}\}_{\theta'' \in \mathcal{B}}$ be the family of POVMs that correspond to Bob's verification measurement when Alice announces θ'', i.e. where $E_1^{\theta''} = |0\rangle\langle 0|_{\theta''}$ and $E_0^{\theta''} = \mathbb{I} - |0\rangle\langle 0|_{\theta''}$. Then, Alice's probability of successfully unveiling bit $\bar{c}$ equals $P_{\text{succ}}(\phi_{AB}, \mathbf{E})$ as defined in Sect. 3. In order to apply Corollary 1, we must first control the size of the side-information that Alice holds. By looking at the definition of $|\phi_{AB}\rangle$ in (1), we notice that it is a superposition of at most $|\mathcal{B}^\delta(0^n)| \leq 2^{nh(\delta)}$ terms. Therefore, the rank of ϕ_A is at most $2^{nh(\delta)}$ and $\mathrm{H}_0(A) \leq nh(\delta)$. We can now bound Alice's probability of opening $\bar{c}$:

$$P_{\text{succ}}(\phi_{AB}, \mathbf{E}) \leq 2^{\mathrm{H}_0(A)} P_{\text{succ}}(\phi_B, \mathbf{E}) \leq 2^{-\frac{d}{2}+2nh(\delta)} \leq 2^{-n(\tau/2 - 2h(\delta))}$$

where the first inequality follows from Corollary 1 and Proposition 1, and the second from the bound on $\mathrm{H}_0(A)$ and from Lemma 3.   □

Regarding the choice of parameters q, τ and r, and the choice of the code, we note that the Gilbert-Varshamov bound guarantees that the code defined by a random binary $n \times (n - rn)$ generator matrix G has minimal distance $d \geq \tau n$, except with negligible probability, as long as $r < 1 - h(\tau)$. On the other hand, for the hiding property, we need that $r > 1 - 0.23 + 2q$. As such, as long as $h(\tau) < 0.23 - 2q$, there exists a suitable rate r and a suitable generator matrix G, so that our scheme offers statistical security against both parties.

4.4   Universality of 1CC

By using our 1CC-based bit commitment scheme BC^{1CC} in the standard construction for obtaining OT from BC in the quantum setting [2,7], we can conclude that 1CC implies OT in the quantum setting, and since OT is universal we thus immediately obtain the universality of 1CC. However, strictly speaking, this does not solve the open problem of [9] yet. The caveat is that [9] asks about the universality of 1CC in the *UC security model* [20], in other words, whether 1CC is "universally-composable universal". So, to truly solve the open problem of [9] we still need to argue *UC security* of the resulting OT scheme, for instance by arguing that our scheme BC^{1CC} is UC secure.

UC-security of BC^{1CC} against malicious Alice follows immediately from our binding criterion (Definition 5); after the commit phase, Alice is bound to a bit that can be extracted in a black-box way from the classical information held by Bob and the 1CC functionality. Thus, a simulator can extract that bit from malicious Alice and input it into the ideal commitment functionality, and since Alice is bound to this bit, this ideal-world attack is indistinguishable from the real-world attack.

However, it is not clear if BC^{1CC} is UC-secure against malicious Bob. The problem is that it is unclear whether it is *universally equivocable*, which is a stronger notion than the standard hiding property (Definition 4).

Nevertheless, we *can* still obtain a UC-secure OT scheme in the 1CC-hybrid model, and so solve the open problem of [9]. For that, we slightly modify the standard BC-based OT scheme [2,7] with BC instantiated by BC^{1CC} as follows: for every BB84 qubit that the receiver is meant to measure, he commits to the basis using BC^{1CC}, but he uses the 1CC-functionality *directly* to "commit" to the measurement outcome, i.e., he inputs the measurement outcome into 1CC — and if the sender asks 1CC to reveal it, the receiver also unveils the accompanying basis by opening the corresponding commitment.

Definition 5 ensures universal extractability of the committed bases and thus of the receiver's input. This implies UC-security against dishonest receiver. In order to argue UC-security against dishonest sender, we consider a simulator that acts like the honest receiver, i.e., chooses random bases and commits to them, but only measures those positions that the sender wants to see — because the simulator controls the 1CC-functionality he can do that. Then, once he has learned the sender's choices for the bases, he can measure all (remaining) qubits in the correct basis, and thus reconstruct *both* messages and send them to the ideal OT functionality. The full details of the proof are in Appendix B.

5   Application 2: On the Security of BCJL   Commitment Scheme

In this section, we show that for a wide class of bit-commitment schemes, the binding property of the scheme in (a slightly strengthened version of) the *bounded-quantum-storage model* reduces to its binding property against a dishonest committer that has *no quantum memory at all*. We then demonstrate the usefulness of this on the example of the BCJL commitment scheme [6].

5.1   Setting up the Stage

The class of schemes to which our reduction applies consists of the schemes that are non-interactive: all communication goes from Alice, the committer, to Bob, the verifier. Furthermore, we require that Bob's verification be "projective" in the following sense.

Definition 6. *We say that a bit-commitment scheme is* non-interactive and with projective verification, *if it is of the following form.*

Commit: *Alice sends a classical message x and a quantum register B to Bob.*
Opening to b: *Alice sends a classical opening y_b to Bob, and Bob applies a binary-outcome projective measurement $\{\mathbb{V}_{x,y_b}, \mathbb{I} - \mathbb{V}_{x,y_b}\}$ to register B.*

Since x is fixed after the commit phase, we tend to leave the dependency of $\mathbb{V}_{x,y_b}$ from x implicit and write $\mathbb{V}_{y_b}$ instead. Also, to keep language simple, we will just speak of a *non-interactive* bit-commitment scheme and drop the *projective verification* part in the terminology.

We consider the security — more precisely: the binding property — of such bit-commitment schemes in a slightly strengthened version of the bounded-quantum-storage model [8], where we bound the quantum memory of Alice, but we also restrict her measurement (for producing y_b in the opening phase) to be *projective*. This restriction on Alice's measurement is well justified since a general non-projective measurement requires additional quantum storage in the form of an ancilla to be performed coherently. From a technical perspective, this restriction (as well as the restriction on Bob's verification) is a byproduct of our proof technique, which requires the measurement operator describing the (joint) opening procedure to be repeatable; avoiding it is an open question.[3]

Formally, we capture the binding property as follows in this variation of the bounded-quantum-storage model.

Definition 7 (Binding). *A non-interactive bit commitment scheme is called ϵ-binding against q-quantum-memory-bounded (or q-QMB for short) projective adversaries if, for all states $\rho_{AB} \in \mathcal{D}(\mathcal{H}_A \otimes \mathcal{H}_B)$ with $\dim(\mathcal{H}_A) \leq 2^q$ and for all classical messages x,*

$$P_0^A(\rho_{AB}) + P_1^A(\rho_{AB}) \leq 1 + \epsilon$$

where

$$P_b^A(\rho_{AB}) := \max_{\{\mathbb{F}_{y_b}\}_{y_b}} \sum_{y_b} \operatorname{tr}((\mathbb{F}_{y_b} \otimes \mathbb{V}_{x,y_b})\rho_{AB})$$

is the probability of successfully opening bit b, maximized over all projective measurements $\{\mathbb{F}_{y_b}\}_{y_b}$.

In case $q = 0$, where the above requirement reduces to

$$P_0^{NA}(\rho_{AB}) + P_1^{NA}(\rho_{AB}) \leq 1 + \epsilon \quad \text{with} \quad P_b^{NA}(\rho_{AB}) := \max_{y_b} \operatorname{tr}(\mathbb{V}_{x,y_b}\rho_B)$$

and $\rho_B = \operatorname{tr}_A(\rho_{AB})$, we also speak of ϵ-binding against *non-adaptive* adversaries.

On the Binding Criterion for Non-interactive Commitment Schemes.
Binding criteria analogous to the one specified in Definition 7 have traditionally been weak notions of security against dishonest committers for quantum commitment schemes, as opposed to criteria that are more in the spirit of a bit that cannot be opened by the adversary. While more convenient for proving security of commitment schemes, a notable flaw of the $p_0 + p_1 \leq 1 + \epsilon$ definition is that it does not rule out the following situation. An adversary might, by some complex measurement, either completely ruin its capacity to open the commitment, or be

---

[3] The standard technique (using Naimark's dilation theorem) does not work here.

able to open the bit of its choice. Then the total probability of opening 0 and 1 sum to 1, but, conditioned on the second outcome of this measurement, they sum to 2. This is obviously an undesirable property of a quantum bit-commitment scheme.

Non-interactive schemes that are secure according to Definition 7 are binding in a stronger sense. For instance, the above problem of the $p_0 + p_1 \leq 1 + \epsilon$ definition does not hold for non-interactive schemes. If a scheme is ϵ-binding, then any state ρ obtained by conditioning on some measurement outcome must satisfy $P_0^A(\rho) + P_1^A(\rho) \leq 1 + \epsilon$. If the total probability of opening 0 and 1 was any higher, then the adversary could have prepared the state ρ in the first place, contradicting the fact that the protocol is ϵ-binding. It remains an open question how to accurately describe the security of non-interactive commitment schemes that satisfy Definition 7.

5.2   The General Reduction

We want to reduce security against a q-QMB projective adversary to the security against a non-adaptive adversary (which should be much easier to show) by means of applying our general adaptive-to-non-adaptive reduction. However, Corollary 1 does not apply directly; we need some additional gadget, which is in the form of the following lemma. It establishes that if there is a commit strategy for Alice so that the cumulative probability of opening 0 and 1 exceeds 1 by a non-negligible amount, then there is also a commit strategy for her so that she can open 0 *with certainty* and 1 with still a non-negligible probability.

Lemma 4. *Let $\rho \in \mathcal{D}(\mathcal{H}_A \otimes \mathcal{H}_B)$ and $\epsilon > 0$ be such that $P_0^A(\rho) + P_1^A(\rho) \geq 1 + \epsilon$. Then, there exists $\rho^0 \in \mathcal{D}(\mathcal{H}_A \otimes \mathcal{H}_B)$ such that $P_0^A(\rho^0) = 1$ and $P_1^A(\rho^0) \geq \epsilon^2$.*

Proof. Let $\{\mathbb{F}_{y_0}\}_{y_0}$ and $\{\mathbb{G}_{y_1}\}_{y_1}$ be the projective measurements maximizing $P_0^A(\rho)$ and $P_1^A(\rho)$, respectively. Define the projections onto the 0/1-accepting subspaces as

$$\mathbb{P}_0 := \sum_{y_0} \mathbb{F}_{y_0} \otimes \mathbb{V}_{y_0} \text{ and } \mathbb{P}_1 := \sum_{y_1} \mathbb{G}_{y_1} \otimes \mathbb{V}_{y_1}.$$

Since $\operatorname{tr}((\mathbb{P}_0 + \mathbb{P}_1)\rho) = P_0^A(\rho) + P_1^A(\rho) \geq 1 + \epsilon$, it follows that $\|\mathbb{P}_0 + \mathbb{P}_1\| \geq 1 + \epsilon$. From Lemma 1, we have that

$$1 + \|\mathbb{P}_1 \mathbb{P}_0\| \geq \|\mathbb{P}_0 + \mathbb{P}_1\| \geq 1 + \epsilon.$$

Therefore there exists $|\phi\rangle$ such that $\|\mathbb{P}_1 \mathbb{P}_0 |\phi\rangle\| \geq \epsilon$. Define $|\phi_0\rangle := \mathbb{P}_0 |\phi\rangle / \|\mathbb{P}_0 |\phi\rangle\|$, which we claim has the required properties. The probability to open 0 from $|\phi_0\rangle$ is $\|\mathbb{P}_0 |\phi_0\rangle\|^2 = 1$, and the probability to open 1 from $|\phi_0\rangle$ is $\|\mathbb{P}_1 \mathbb{P}_0 |\phi_0\rangle\|^2 = \|\mathbb{P}_1 \mathbb{P}_0 |\phi\rangle\|^2 / \|\mathbb{P}_0 |\phi\rangle\|^2 \geq \epsilon^2$. $\qquad\square$

Now, we are ready to state and prove the general reduction.

Theorem 5. *If a non-interactive bit-commitment scheme is ϵ-binding against non-adaptive adversaries, then it is $(2^{\frac{1}{2}q}\sqrt{\epsilon})$-binding against q-QMB projective adversaries.*

Proof. Let $\rho_{AB} \in \mathcal{D}(\mathcal{H}_A \otimes \mathcal{H}_B)$ be the joint state of Alice and Bob where $\dim(\mathcal{H}_A) \leq 2^q$ and let $\alpha > 0$ be such that the opening probabilities satisfy $P_0^A(\rho) + P_1^A(\rho) = 1 + \alpha$. From Lemma 4, we know that there exists $\rho_{AB}^0 \in \mathcal{D}(\mathcal{H}_A \otimes \mathcal{H}_B)$ constructed from ρ such that

$$P_0^A(\rho^0) = 1 \text{ and } P_1^A(\rho^0) \geq \alpha^2.$$

We use Corollary 1 and the assumption that the protocol is ϵ-binding against non-adaptive adversaries to show that α cannot be too large. Let $\{\mathbb{F}_{y_0}\}_{y_0}$ be the measurement that maximizes $P_0^A(\rho^0)$. Let us consider Bob's reduced density operator of ρ^0:

$$\rho_B^0 = \mathrm{tr}_A(\rho_{AB}^0) = \sum_{y_0} \mathrm{tr}_A((\mathbb{F}_{y_0} \otimes \mathbb{I})\rho_{AB}^0) = \sum_{y_0} \lambda_{y_0} \sigma_{y_0}$$

where for each y_0, it holds that $\mathrm{tr}(\mathbb{V}_{y_0}\sigma_{y_0}) = 1$. This implies $\mathrm{tr}(\mathbb{V}_{y_1}\sigma_{y_0}) \leq \epsilon$ for every y_1 that opens 1 from our assumption of the non-adaptive security of the commitment scheme. Then

$$P_1^{NA}(\rho_{AB}^0) = \max_{y_1} \mathrm{tr}(\mathbb{V}_{y_1}\rho_B^0) = \max_{y_1} \sum_{y_0} \lambda_{y_0} \mathrm{tr}(\mathbb{V}_{y_1}\sigma_{y_0}) \leq \epsilon.$$

Applying Corollary 1 completes the proof:

$$\alpha^2 \leq P_1^A(\rho^0) \leq 2^{I_{\max}^{\mathrm{acc}}(B;A)_{\rho_0}} P_1^{NA}(\rho^0) \leq 2^{H_0(A)_{\rho_0}} \epsilon \leq 2^q \epsilon.$$

$\square$

5.3   Special Case: The BCJL Bit-Commitment Scheme

In this subsection, we use the results of the previous section to prove the security of the BCJL scheme in the bounded storage model against projective measurement attacks.

The BCJL bit-commitment scheme was proposed in 1993 by Brassard et al. [6]. They proposed to hide the committed bit using a two-universal family of hash functions applied on the codeword of an error correcting code and then send this codeword through BB84 qubits. The idea behind this protocol is that privacy amplification hides the committed bit while the error correcting code makes it hard to change the value of this bit without being detected. While their intuition was correct, their proof ultimately was not, as shown by Mayers' impossibility result for bit commitment [18].

The following scheme (Fig. 3) differs only slightly from the original [6], this allows us to recycle some of the analysis from Sect. 4.

$$\textsc{commit}(b):$$

1. Let $\mathcal{C}$ be a linear $[n, k, d]$-error correcting code with $k > 0.78n$.
2. Alice chooses a random string $x \in_R \{0, 1\}^n$ and a random basis $\theta \in_R \{+, \times\}^n$. She sends $|x\rangle_\theta$ to Bob.
3. Upon reception of the qubits, Bob chooses a random basis $\hat{\theta}$ and measures the qubits in basis $\hat{\theta}$. Let $\hat{x}$ denote his measurement result.
4. Let $\mathcal{G}_n$ be a family of two-universal hash functions. Alice chooses $g \in_R \mathcal{G}_n$ and computes the syndrome s of x for the linear code $\mathcal{C}$. She sends $w = g(x) \oplus b$, g and s to Bob.

REVEAL:

1. Alice sends x, θ and b to Bob.
2. Bob checks that x has syndrome s. He then checks that $\hat{x}_i = x_i$ for every i such that $\hat{\theta}_i = \theta_i$. He accepts if $g(x) \oplus w = b$ and none of the above tests failed.

Fig. 3. The BCJL bit-commitment scheme

Theorem 6. BCJL *is statistically hiding as long as* $0.22 - (1 - k/n) \in \Omega(1)$.

The proof of Theorem 6 is straightforward. It follows the same approach as that of Theorem 3 by noticing that Bob has the same uncertainty about each x_i as he had about θ_i in protocol $\textsc{commit}^{1CC}$.

Instead of proving that BCJL is binding, we prove that an equivalent scheme BCJL$_\delta$ (see Fig. 4) is binding. The BCJL$_\delta$ scheme is a modified version of BCJL in which Bob has unlimited quantum memory and stores the qubits sent by Alice during the commit phase instead of measuring them. The opening phase of BCJL$_\delta$ is characterized by a parameter δ which determines how close it is to the opening phase of BCJL. The following lemma shows that the two protocols are equivalent from Alice's point of view; if Alice can cheat an honest Bob then she can cheat a Bob with unbounded quantum computing capabilities.

Lemma 5. *Let* $\delta > 0$. *If* BCJL$_\delta$ *is* ϵ-binding *then* BCJL *is* $(\epsilon + 2 \cdot 2^{-\delta n})$-binding.

Proof. Let (x, θ) be an opening to 0. First notice that Bob's actions in BCJL are equivalent to holding onto his state until the opening procedure, measuring in basis θ and verifying $x_T = \hat{x}_T$ for a randomly chosen sample $T \subseteq [n]$. From this point of view, Bob's measurement result is identically distributed in both protocols and we can speak of $\hat{x}$ without ambiguity. If $d(x, \hat{x}) > \delta n$, then the probability that $x_i = \hat{x}_i$ for all $i \in T$ is at most $2^{-\delta n}$. Therefore, if Bob rejects in REVEAL$_\delta$ with measurement outcome $\hat{x}$, then the probability that he rejects in REVEAL with the same outcome is at least $1 - 2^{-\delta n}$. If we let p_0 denote Bob's accepting probability in the original protocol and p_0^δ in the modified protocol, we have $p_0 \le p_0^\delta + 2^{-\delta n}$. Since the same holds for openings to 1, we have

---

COMMIT$_\delta(b)$:

1. Let $\mathcal{C}$ be a linear $[n, k, d]$-error correcting code with $k > 0.78n$.
2. Alice chooses a random string $x \in_R \{0, 1\}^n$ and a random basis $\theta \in_R \{+, \times\}^n$. She sends $|x\rangle_\theta$ to Bob.
3. Bob stores all the qubits he received to measure them later.
4. Let $\mathcal{G}_n$ be a family of two-universal hash functions. Alice chooses $g \in_R \mathcal{G}_n$ and computes the syndrome s of x for the linear code $\mathcal{C}$. She sends $w = g(x) \oplus b$, g and s to Bob.

REVEAL$_\delta$:

1. Alice sends x, θ and b to Bob.
2. Bob checks that x has syndrome s. He then measures his stored qubits in basis θ to obtain $\hat{x}$. He accepts if $d(x, \hat{x}) \leq \delta n$ and $g(x) \oplus w = b$ and none of the above tests failed.

Fig. 4. The BCJL$_\delta$ bit-commitment scheme.

$$p_0 + p_1 \leq p_0^\delta + p_1^\delta + 2 \cdot 2^{-\delta n} \leq 1 + \epsilon + 2 \cdot 2^{-\delta n}.$$

$\square$

The following proposition establishes the security of BCJL$_\delta$ in the non-adaptive setting. Its proof is straightforward and can be found in Appendix A.

Proposition 4. BCJL$_\delta$ is $2^{-d/2+\delta n+h(\delta)n}$-binding against non-adaptive adversaries.

Since the bit-commitment scheme BCJL$_\delta$ is non-interactive, it directly follows from Theorem 5 and Proposition 4 that BCJL$_\delta$ is $2^{\frac{1}{2}(q-d/2+\delta n+h(\delta)n)}$-binding against q-QMB projective adversaries. Combining the above with Lemma 5, we have the following statement for the BCJL scheme.

Theorem 7. *The* BCJL *bit-commitment scheme is* $(2^{\frac{1}{2}(q-d/2+\delta n+h(\delta)n)} + 2 \cdot 2^{-\delta n})$-*binding against* q-QMB *projective adversaries.*

Acknowledgments. FD acknowledges the support of the Czech Science Foundation (GAČR) project no. GA16-22211S and of the EU FP7 under grant agreement no. 323970 (RAQUEL). LS is supported by Canada's NSERC discovery grant.

A   Additional proofs

Proposition 2. *For any state* ρ_{ZAB} *with classical* Z:

$$I_{\max}^{\text{acc}}(B; A|Z)_\rho \leq \max_z I_{\max}^{\text{acc}}(B; A)_{\rho^z} \leq H_0(A)_\rho.$$

Proof. By assumption, ρ_{ZAB} is of the form $\rho_{ZAB} = \sum_z P_Z(z)|z\rangle\langle z| \otimes \rho_{AB}^z$. Let $\mathcal{M}_{ZA \to X}$ be a measurement on Z and A. By linearity, and by definition of $I_{\max}^{\mathrm{acc}}$, we have that

$$\mathcal{M}(\rho_{ZAB}) = \sum_z P_Z(z)\mathcal{M}\big(|z\rangle\langle z| \otimes \rho_{AB}^z\big)$$

$$\leq \sum_z P_Z(z) \cdot 2^{I_{\max}^{\mathrm{acc}}(B;A|Z)_{|z\rangle\langle z|\otimes\rho^z}} \mathcal{N}^z\big(|z\rangle\langle z| \otimes \rho_B^z\big)$$

for suitably chosen measurements $\mathcal{N}_{Z \to X}^z$. Now, noting that $I_{\max}^{\mathrm{acc}}(B; A|Z)_{|z\rangle\langle z|\otimes\rho^z} = I_{\max}^{\mathrm{acc}}(B; A)_{\rho^z}$, and that there exists a fixed measurement $\mathcal{N}_{Z \to X}$ so that $\mathcal{N}^z(|z\rangle\langle z|) = \mathcal{N}(|z\rangle\langle z|)$ for all z, it follows that

$$\mathcal{M}(\rho_{ZAB}) \leq 2^{\max_z I_{\max}^{\mathrm{acc}}(B;A)_{\rho^z}} \mathcal{N}(\rho_{ZB}),$$

which implies the first claimed inequality. The second inequality follows immediately by observing that $I_{\max}^{\mathrm{acc}}(B; A)_{\rho^z} \leq H_0(A)_{\rho^z} \leq H_0(A)_\rho$. $\quad\square$

Proposition 3. *Let $\mathcal{E}_{AB \to A'B'}$ be a CPTP map of the form $\mathcal{E} = \mathcal{E}^A \otimes \mathcal{E}^B$. Then*

$$I_{\max}^{\mathrm{acc}}(B'; A')_{\mathcal{E}(\rho)} \leq I_{\max}^{\mathrm{acc}}(B; A)_\rho.$$

Proof. Since the CPTP map $\mathcal{E}^B$ commutes with any measurement applied on Alice's register, it cannot increase the maximal accessible information.

To show that the CPTP map $\mathcal{E}^A$ cannot increase $I_{\max}^{\mathrm{acc}}$, it suffices to show that for every measurement $\mathcal{M}$ on register A, the CPTP map $\mathcal{M} \circ \mathcal{E}^A$ is also a measurement. Let $\{E_k\}_k$ be the Kraus operators associated with $\mathcal{E}^A$ and let $\{F_x\}_x$ be the POVM operators describing the measurement $\mathcal{M}$. Then, the positive operators $F_x' := \sum_k E_k^\dagger F_x E_k$ describe a POVM $\mathcal{M}'$, and

$$\mathcal{M} \circ \mathcal{E}^A(\rho) = \mathcal{M}'(\rho) \leq 2^{I_{\max}^{\mathrm{acc}}(B;A)_\rho} \sigma_X \otimes \rho_B$$

by the definition of $I_{\max}^{\mathrm{acc}}(B; A)_\rho$ for some normalized σ_X. $\quad\square$

Proposition 4. *Protocol* BCJL$_\delta$ *is* $2^{-d/2+\delta n+h(\delta)n}$*-binding against non-adaptive adversaries.*

Proof. Let $\rho_{AB} \in \mathcal{D}(\mathcal{H}_A \otimes \mathcal{H}_B)$ be the joint state of Alice and Bob and let $\mathbb{V}_{x,\theta}^\delta := \sum_{z \in B^\delta(x)} |z\rangle\langle z|_\theta$ be the projective measurement corresponding to Bob's verification procedure in protocol BCJL$_\delta$ if Alice announced (x, θ). Using Lemma 1, we have that for any two distinct openings (x, θ) and (x', θ'),

$$\mathrm{tr}(\mathbb{V}_{x,\theta}^\delta \rho_B) + \mathrm{tr}(\mathbb{V}_{x',\theta'}^\delta \rho_B) = \mathrm{tr}((\mathbb{V}_{x,\theta}^\delta + \mathbb{V}_{x',\theta'}^\delta)\rho_B)$$

$$\leq ||\mathbb{V}_{x,\theta}^\delta + \mathbb{V}_{x',\theta'}^\delta||$$

$$\leq 1 + ||\mathbb{V}_{x,\theta}^\delta \mathbb{V}_{x',\theta'}^\delta||.$$

Using techniques from [5], we can show that

$$||\mathbb{V}_{x,\theta}^\delta \mathbb{V}_{x',\theta'}^\delta|| \leq \max_{\substack{z \in B^\delta(x) \\ z' \in B^\delta(x')}} |\langle z|_\theta|z'\rangle_{\theta'}| \sqrt{|B^\delta(x)||B^\delta(x')|}.$$

Using the fact that $d(z, z') \geq d - 2\delta n$ for $z \in B^\delta(x)$ and $z' \in B^\delta(x')$ for any two strings x and x' with the same syndrome, and the fact that $|B^\delta(x)| \leq 2^{h(\delta)n}$, it follows that when maximizing over openings to 0 and 1, we obtain

$$P_0^{NA}(\rho_{AB}) + P_1^{NA}(\rho_{AB}) \leq 1 + 2^{-d/2 + \delta n + h(\delta)n}.$$

$\square$

B   UC-Completeness of 1CC

B.1   The UC Model

In order to show that a scheme securely implements a given functionality F in the universally composable (UC) model, one has to show that for any *adversary* that attacks the scheme by corrupting participants, there exists a *simulator* S that instead attacks the functionality, but is indistinguishable from the adversary from an outside observer's perspective. More precisely, one considers an *environment* Z that interacts with the adversary in the *real* model where the scheme is executed, or with S in the *ideal* model where the functionality F is executed, and it provides input to and obtains output from the uncorrupt players (see Fig. 5). The scheme is said to *statistically quantum-UC-emulate* the functionality if the environment cannot distinguish the real from the ideal model with non-negligible probability. For a more detailed description of the quantum UC framework, we refer to [9,20].

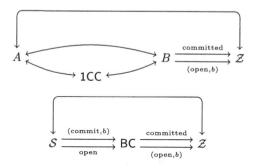

Fig. 5. The real model (top) and the ideal model (bottom) for protocol BC^{1CC} and functionality BC, respectively, with a dishonest Alice. BC^{1CC} statistically quantum-UC-emulates BC (against dishonest Alice) if the two models are indistinguishable for Z.

Most UC security proofs follow a similar mold. S internally runs a copy of the adversary, and it simulates the actions and interactions of the honest party, and of functionalities that are possibly used as subroutines in the scheme. S must look like the real model adversary to the environment Z, so it forwards any message it receives from Z to (its internal execution of) the adversary and vice versa. Furthermore, from the interaction with the adversary, it extracts the input(s) it has to provide to F (see Fig. 6).

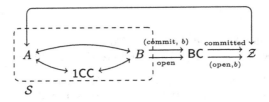

Fig. 6. The standard way for constructing $\mathcal{S}$: run dishonest Alice internally and simulate honest Bob and the calls to the functionality 1CC, and extract the input to BC.

In all our proofs below, the honest party is simulated by $\mathcal{S}$ by running it *honestly*, up to possible small modifications that are unnoticeable to the adversary, and that do not affect the (simulated) honest party's output. As such, in our proofs below, for showing indistinguishability of the real and the ideal model, it is sufficient to argue that, in the ideal model, the output of the simulated honest party equals what F outputs to $\mathcal{Z}$ upon the input that is provided by $\mathcal{S}$.

B.2 UC Security of OT from 1CC

As explained in Sect. 4.4, our protocol $\text{BC}^{1\text{CC}}$ does not seem to satisfy the UC security definition in case of a corrupted verifier Bob. As such, we cannot conclude UC security of the standard BC-based OT scheme [2,7] with BC instantiated by $\text{BC}^{1\text{CC}}$. Instead, we show UC security of OT from 1CC by means of the following strategy.

First, we show UC security of $\text{BC}^{1\text{CC}}$ against a corrupted committer Alice (Proposition 5). Then, we show that BC and 1CC together imply 2CC (actually, a variation of 2CC that gives Alice the option to abort) by means of a straightforward protocol (Proposition 6), and we recall that 2CC implies OT by means of the protocol $\text{OT}^{2\text{CC}}$ from [9]. Instantiating the underlying functionality BC by $\text{BC}^{1\text{CC}}$ then gives us a protocol $\text{OT}^{1\text{CC}}$ (Fig. 8) with UC security against a corrupted receiver (Lemma 6). Finally, it is rather straightforward to prove UC security of $\text{OT}^{1\text{CC}}$ against a corrupted sender directly (Lemma 7).

Proposition 5. *Protocol* $\text{BC}^{1\text{CC}}$ *statistically quantum-UC-emulates* BC *against corrupted committer Alice.*

Proof. The construction of $\mathcal{S}$ follows the paradigm outlined above. $\mathcal{S}$ runs dishonest Alice internally, and it simulates honest Bob and 1CC by running them honestly. Note that $\mathcal{S}$ gets to see Alice's inputs to 1CC. Once Alice announces g, w and s at the end of the commit phase, $\mathcal{S}$ computes $b = g(\theta') \oplus w$, where θ' is the string of syndrome s closest to the stored $\theta_{\bar{t}}$, and inputs "(commit, b)" into the BC functionality. Finally, when corrupted Alice opens her commitment, $\mathcal{S}$ inputs "open" into BC if Bob accepted the opening, and inputs "abort" if Bob aborted.

It now follows immediately from Lemma 3 that the bit b' output by the simulated Bob equals the bit b computed by $\mathcal{S}$ and input to BC, except with negligible probability. As such, real and ideal model are statistically indistinguishable. $\square$

Parties: The sender Alice and the receiver Bob.
Inputs: Alice receives $s_0, s_1 \in \{0, 1\}$ and Bob receives $c \in \{0, 1\}$.

1. Alice inputs "(commit, s_0)" in BC, Bob receives "committed".
2. Alice and Bob invoke 1CC with respective inputs s_1 and c.
3. If Alice receives $c = 1$ from 1CC, she sends "open" to BC. Bob receives s_0 from 1CC and s_1 from BC.
4. Alice outputs c, Bob outputs (s_0, s_1) if $c = 1$ and $\perp$ if $c = 0$. Bob outputs "abort" if $c = 1$ but Alice refuses to open her commitment.

Fig. 7. Protocol $2\text{CC}^{\text{BC},\text{1CC}}$.

Consider the candidate 2-bit cut-and-choose protocol $2\text{CC}^{\text{BC},\text{1CC}}$ from Fig. 7. This protocol does not implement the full-fledged 2CC functionality, but a variation 2CC' that gives the sender the option to abort after it sees the receiver's input c. This is because in the protocol the sender can refuse to open its commitments (or try to cheat when opening them so that the receiver will reject). In that case, the receiver will only learn one of the receiver's two inputs. This will not influence the security of the resulting OT scheme since aborting in any instance of 2CC' will stop the protocol.

Formally, 2CC' is described as follows: it first waits for inputs (s_0, s_1) from Alice and c from Bob. Upon reception of both inputs, it sends c to Alice. If $c = 0$, it sends $\perp$ to Bob. If $c = 1$, it waits for response "abort" or "continue" from Alice. On input "continue", 2CC' outputs (s_0, s_1) to Bob and on input "abort", it outputs "abort".

Proposition 6. *Protocol $2\text{CC}^{\text{BC},\text{1CC}}$ statistically quantum-UC-emulates* 2CC'.

Proof. We first consider a corrupted sender Alice. $\mathcal{S}$ simulates Bob, BC and 1CC by running them honestly. After step 2, when $\mathcal{S}$ has learned Alice's respective inputs s_0 and s_1 to BC and 1CC, it inputs (s_0, s_1) into the functionality 2CC'. After receiving c from the 2CC', $\mathcal{S}$ makes Bob input c into the 1CC. If $c = 0$ then the simulated Bob and 2CC' both output $\perp$. If $c = 1$ then Alice is supposed to open her commitment. If she refuses then $\mathcal{S}$ inputs "abort" into 2CC', and the simulated Bob and 2CC' both output "abort". Otherwise, i.e., if Alice opens the commitment (to s_0), $\mathcal{S}$ inputs "continue", and the simulated Bob and 2CC' both output (s_0, s_1). This proves the claim for a corrupted sender Alice. Security against a corrupted receiver Bob is similarly straightforward. □

Corollary 2. *Protocol 2CC^{1CC}, obtained by replacing each instance of BC by BC^{1CC}, statistically quantum-UC-emulates* 2CC' *against corrupted sender.*

Proof. Since BC^{1CC} statistically quantum UC-emulates BC against malicious committer, and since the sender in $2\text{CC}^{\text{BC},\text{1CC}}$ is the committer of BC, we can replace BC with BC^{1CC} in protocol $2\text{CC}^{\text{BC},\text{1CC}}$ and still maintain UC-security against corrupted sender. □

Parameters: A family $\mathcal{F} = \{\{0,1\}^n \rightarrow \{0,1\}^\ell\}$ of universal hash functions.
Parties: The sender Alice and the receiver Bob.
Inputs: Alice receives $s_0, s_1 \in \{0,1\}^\ell$ and Bob receives $c \in \{0,1\}$.

1. Alice chooses $x^A \in_R \{0,1\}^n$ and $\theta^A \in_R \{+, \times\}^n$ and sends the state $|x^A\rangle_{\theta^A}$ to Bob.
2. Upon reception, Bob chooses $\theta^B \in_R \{+, \times\}^n$ and measures the received state in basis θ^B. Lets x^B denote the measurement outcome.
3. For $i = 1 \ldots n$, do
   (a) Alice and Bob perform protocol COMMIT^{1CC} with Bob as the sender and input θ_i^B.
   (b) Alice chooses a selection bit $t_i \in_R \{0,1\}$ and they invoke an instance of 1CC with Bob as the sender and inputs t_i and x_i^B.
   (c) Whenever $t_i = 1$, Bob opens the ith commitment using protocol REVEAL^{1CC}.
4. If for some i s.t. $t_i = 1$, $\theta_i^A = \theta_i^B$, but $x_i^B \neq x_i^A$, Alice aborts. Bob aborts if $t_i = 1$ for more than $3n/5$ positions. Let $\hat{x}^A$ (resp. $\hat{\theta}^A, \hat{x}^B, \hat{\theta}^B$) be the restriction of x^A (resp. θ^A, x^B, θ^B) to the indices i for which $t_i = 0$.
5. Alice sends $\hat{\theta}^A$ to Bob. Bob constructs sets $I_c = \{i \mid \hat{\theta}_i^A = \hat{\theta}_i^B\}$ and $I_{1-c} = \{i \mid \hat{\theta}_i^A \neq \hat{\theta}_i^B\}$ then sends (I_0, I_1) to Alice.
6. Alice chooses $f \in_R \mathcal{F}$, computes $m_i = s_i \oplus f(\hat{x}_{I_i}^A)$ for $i = 0, 1$ and sends (f, m_0, m_1) to Bob.
7. Bob outputs $s = m_c \oplus f(\hat{x}_{I_c}^B)$.

Fig. 8. Protocol OT^{1CC}.

Lemma 6. *Protocol OT^{1CC} statistically quantum UC-emulates* OT *for corrupted receiver.*

Proof. Note that steps 3a through 3c of protocol OT^{1CC} are identical to protocol 2CC^{1CC} defined above with Bob as the sender and Alice as the receiver. Since 2CC^{1CC} statistically quantum-UC-emulates $2\text{CC}'$ against corrupted sender, we may replace steps 3a–3c by a single call to $2\text{CC}'$ with Bob as the sender and Alice as the receiver, and analyze the security of this protocol instead. The only difference between this protocol and the 2CC-based oblivious-transfer protocol from [9] is that the former uses $2\text{CC}'$ instead. However, this change does not affect UC-security since any adversary that aborts during one of the 2CC^{1CC} subroutines is indistinguishable from an adversary that aborts right after the same subroutine. It directly follows from the analysis of [9], that protocol OT^{1CC} statistically quantum-UC-emulates OT against corrupted receiver. $\square$

Lemma 7. *Protocol OT^{1CC} statistically quantum UC-emulates* OT *for corrupted sender.*

Proof. Let Alice be the corrupted sender and Bob the honest receiver. $\mathcal{S}$ simulates Bob and 1CC by running them honestly, *except* that Bob does *not* measure the received state in step 2 but stores it, and in step 3b, whenever Alice inputs $t_i = 1$ into 1CC, $\mathcal{S}$ "rushes" and measures the ith qubit in basis θ_i^B and inputs the outcome x_i^B in the 1CC. Furthermore, in step 5, $\mathcal{S}$ replies to Alice with a random partition (I_0, I_1). At the end of the protocol, $\mathcal{S}$ measures the remaining qubits in Alice's basis $\hat{\theta}^A$ to obtain $\hat{x}^B$, computes $s_i = m_i \oplus f(\hat{x}_{I_i}^B)$ for $i = 0, 1$, and sends (s_0, s_1) to the ideal OT functionality.

The output of OT, i.e., s_c, coincides with the string that a fully honest Bob would have output; hence, we have indistinguishability between the real and the ideal model. □

Theorem 8. 1CC *is statistically quantum UC-complete.*

Proof. We have shown that OT^{1CC} statistically quantum-UC-emulates OT. Since OT is quantum-UC-complete, we conclude that 1CC is also quantum-UC-complete. □

References

1. Bennett, C.H.: Quantum cryptography using any two nonorthogonal states. Phys. Rev. Lett. **68**, 3121–3124 (1992)
2. Bennett, C.H., Brassard, G., Crépeau, C., Skubiszewska, M.-H.: Practical quantum oblivious transfer. In: Feigenbaum, J. (ed.) CRYPTO 1991. LNCS, vol. 576, pp. 351–366. Springer, Heidelberg (1992)
3. Berta, M., Christandl, M., Renner, R.: The quantum reverse Shannon theorem based on one-shot information theory. Commun. Math. Phys. **306**(3), 579–615 (2011)
4. Bouman, N.J., Fehr, S.: Sampling in a quantum population, and applications. In: Rabin, T. (ed.) CRYPTO 2010. LNCS, vol. 6223, pp. 724–741. Springer, Heidelberg (2010)
5. Bouman, N.J., Fehr, S., González-Guillén, C., Schaffner, C.: An all-but-one entropic uncertainty relation, and application to password-based identification. In: Kawano, Y. (ed.) TQC 2012. LNCS, vol. 7582, pp. 29–44. Springer, Heidelberg (2012)
6. Brassard, G., Crépeau, C., Jozsa, R., Langlois, D.: A quantum bit commitment scheme provably unbreakable by both parties. In: Proceedings of the 34th Annual IEEE Symposium on the Foundation of Computer Science, pp. 362–371 (1993)
7. Crépeau, C.: Quantum oblivious transfer. J. Mod. Opt. **41**(12), 2445–2454 (1994)
8. Damgård, I., Fehr, S., Salvail, L., Schaffner, C.: Cryptography in the bounded-quantum-storage model. SIAM J. Comput. **37**(6), 1865–1890 (2008)
9. Fehr, S., Katz, J., Song, F., Zhou, H.-S., Zikas, V.: Feasibility and completeness of cryptographic tasks in the quantum world. In: Sahai, A. (ed.) TCC 2013. LNCS, vol. 7785, pp. 281–296. Springer, Heidelberg (2013)
10. Kilian, J.: Founding cryptography on oblivious transfer. In: Proceedings of the ACM Symposium on Theory of Computing, STOC 1988, pp. 20–31. ACM, New York (1988)
11. Kilian, J.: A general completeness theorem for two party games. In: Proceedings of the Twenty-Third Annual ACM Symposium on Theory of Computing, STOC 1991, pp. 553–560 (1991)

12. Kilian, J.: More general completeness theorems for secure two-party computation. In: Proceedings of the Thirty-Second Annual ACM Symposium on Theory of Computing, STOC 2000, pp. 316–324 (2000)
13. König, R., Renner, R., Schaffner, C.: The operational meaning of min- and max-entropy. IEEE Trans. Inf. Theor. **55**(9), 4337–4347 (2009)
14. Kraschewski, F.: Complete primitives for information-theoretically secure two-party computation. Ph.D. thesis, Karlsruhe Institute of Technology (2013)
15. Kraschewski, D., Müller-Quade, J.: Completeness theorems with constructive proofs for finite deterministic 2-party functions. In: Ishai, Y. (ed.) TCC 2011. LNCS, vol. 6597, pp. 364–381. Springer, Heidelberg (2011)
16. Maji, H.K., Prabhakaran, M., Rosulek, M.: A zero-one law for cryptographic complexity with respect to computational UC security. In: Rabin, T. (ed.) CRYPTO 2010. LNCS, vol. 6223, pp. 595–612. Springer, Heidelberg (2010)
17. Maji, H.K., Prabhakaran, M., Rosulek, M.: A unified characterization of completeness and triviality for secure function evaluation. In: Galbraith, S., Nandi, M. (eds.) INDOCRYPT 2012. LNCS, vol. 7668, pp. 40–59. Springer, Heidelberg (2012)
18. Mayers, D.: Unconditionally secure quantum bit commitment is impossible. Phys. Rev. Lett. **78**, 3414–3417 (1997)
19. Renner, R.S., König, R.: Universally composable privacy amplification against quantum adversaries. In: Kilian, J. (ed.) TCC 2005. LNCS, vol. 3378, pp. 407–425. Springer, Heidelberg (2005)
20. Unruh, D.: Universally composable quantum multi-party computation. In: Gilbert, H. (ed.) EUROCRYPT 2010. LNCS, vol. 6110, pp. 486–505. Springer, Heidelberg (2010)

Semantic Security and Indistinguishability in the Quantum World

Tommaso Gagliardoni[1]([✉]), Andreas Hülsing[2]([✉]),
and Christian Schaffner[3,4,5]([✉])

[1] Technische Universität Darmstadt, Darmstadt, Germany
tommaso@gagliardoni.net
[2] TU Eindhoven, Eindhoven, The Netherlands
andreas@huelsing.net
[3] Institute for Logic, Language and Compuation (ILLC), University of Amsterdam, Amsterdam, The Netherlands
c.schaffner@uva.nl
[4] Centrum Wiskunde & Informatica (CWI), Amsterdam, The Netherlands
[5] QuSoft, Amsterdam, The Netherlands

Abstract. At CRYPTO 2013, Boneh and Zhandry initiated the study of quantum-secure encryption. They proposed first indistinguishability definitions for the quantum world where the actual indistinguishability only holds for classical messages, and they provide arguments why it might be hard to achieve a stronger notion. In this work, we show that stronger notions are achievable, where the indistinguishability holds for quantum superpositions of messages. We investigate exhaustively the possibilities and subtle differences in defining such a quantum indistinguishability notion for symmetric-key encryption schemes. We justify our stronger definition by showing its equivalence to novel quantum semantic-security notions that we introduce. Furthermore, we show that our new security definitions cannot be achieved by a large class of ciphers – those which are quasi-preserving the message length. On the other hand, we provide a secure construction based on quantum-resistant pseudorandom permutations; this construction can be used as a generic transformation for turning a large class of encryption schemes into quantum indistinguishable and hence quantum semantically secure ones. Moreover, our construction is the first completely classical encryption scheme shown to be secure against an even stronger notion of indistinguishability, which was previously known to be achievable only by using quantum messages and arbitrary quantum encryption circuits.

1 Introduction

Quantum computers [20] threaten many cryptographic schemes. By using Shor's algorithm [22] and its variants [25], an adversary in possession of a quantum computer can break the security of every scheme based on factorization and discrete logarithms, including RSA, ElGamal, elliptic-curve primitives and many others. Moreover, longer keys and output lengths are required in order to maintain

© International Association for Cryptologic Research 2016
M. Robshaw and J. Katz (Eds.): CRYPTO 2016, Part III, LNCS 9816, pp. 60–89, 2016.
DOI: 10.1007/978-3-662-53015-3_3

the security of block ciphers and hash functions [5,12]. These difficulties led to the development of *post-quantum cryptography* [2], i.e., classical cryptography resistant against quantum adversaries.

When modeling the security of cryptographic schemes, care must be taken in defining exactly what property one wants to achieve. In classical security models, all parties and communications are classical. When these notions are used to prove *post-quantum* security, one must consider adversaries having access to a quantum computer. This means that, while the communication between the adversary and the user is still classical, the adversary might carry out computations on a quantum computer.

Such post-quantum notions of security turn out to be unsatisfying in certain scenarios. For instance, consider quantum adversaries able to use *quantum superpositions* of messages $\sum_x \alpha_x |x\rangle$ instead of classical messages when communicating with the user, even though the cryptographic primitive is still classical. This kind of scenario is considered, e.g., in [4,8,23,26,28]. Such a setting might for example occur in a situation where one party using a quantum computer encrypts messages for another party that uses a classical computer and an adversary is able to observe the outcome of the quantum computation before measurement. Other examples are an attacker which is able to trick a classical device into showing quantum behavior, or a classical scheme which is used as subprotocol in a larger quantum protocol. Another possibility occurs when using obfuscation. There are applications where one might want to distribute the obfuscated code of a symmetric-key encryption scheme (with the secret key hardcoded) in order to allow a third party to generate ciphertexts without being able to retrieve the key - think of this as building public-key encryption from symmetric-key encryption using Indistinguishability Obfuscation. Because in these cases an adversary receives the classical code for producing encryptions, he could implement the code on his local quantum computer and query the resulting quantum circuit on a superposition of inputs. Moreover, even in quantum reductions for classical schemes situations could arise where superposition access is needed. A typical example are impossibility results (such as meta-reductions [7]), where giving the adversary additional power often rules out a broader range of secure reductions. Notions covering such settings are often called *quantum-security* notions. In this work we propose new quantum-security notions for encryption schemes.

For encryption, the notion of *semantic security* [10,11] has been traditionally used. This notion models in abstract terms the fact that, without the corresponding decryption key, it is impossible not only to correctly decrypt a ciphertext, but even to recover any non-trivial information about the underlying plaintext. The exact definition of semantic security is cumbersome to work with in security proofs as it is simulation-based. Therefore, the simpler notion of *ciphertext indistinguishability* has been introduced. This notion is given in terms of an interactive game where an adversary has to distinguish the encryptions of two messages of his choice. The advantage of this definition is that it is easier to work with than (but equivalent to) semantic security.

To the best of our knowledge, no quantum semantic-security notions for classical encryption schemes have been proposed so far. For indistinguishability,

Boneh and Zhandry introduced indistinguishability notions for quantum-secure encryption under chosen-plaintext attacks in a recent work [4]. They consider a model (IND-qCPA) where a quantum adversary can query the encrypting device in superposition during a learning phase but is limited to classical communication during the actual challenge phase. However, in the symmetric-key scenario, this approach has the following shortcoming: If we assume that an adversary can get quantum access in a learning phase, it seems unreasonable to assume that he cannot get such access when the actual message of interest is encrypted. Boneh and Zhandry showed that a seemingly natural notion of quantum indistinguishability is unachievable. In order to restore a meaningful definition, they resorted to the compromise of IND-qCPA.

Our Contributions. In this paper we achieve two main results. On the one hand, we initiate the study of semantic security in the quantum world, providing new definitions and a thorough discussion about the motivations and difficulties of modeling these notions correctly. This study is concluded by a suitable definition of *quantum semantic security under chosen plaintext attacks (qSEM-qCPA)*. On the other hand, we extend the fundamental work initiated in [4] in finding suitable notions of indistinguishability in the quantum world. We show that the compromise that had to be reached there in order to define an achievable notion instead of a more natural one (i.e., IND-qCPA vs. fqIND-qCPA) can be overcome – although not trivially. We show how various other possible notions of quantum indistinguishability can be defined. All these security notions span a tree of possibilities which we analyze exhaustively in order to find the most suitable definition of *quantum indistinguishability under chosen plaintext attacks (qIND-qCPA)*. We prove this notion to be achievable, strictly stronger than IND-qCPA, and equivalent to qSEM-qCPA, thereby completing an elegant framework of security notions in the quantum world, see Fig. 2 below for an overview.

Furthermore, we formally define the notion of a *core function* and *quasi-length-preserving ciphers* – encryption schemes which essentially do not increase the plaintext size, such as stream ciphers and many block ciphers including AES – and we show the impossibility of achieving our new security notion for this kind of schemes. While this impossibility might look worrying from an application perspective, we also present a transformation that turns a block cipher into an encryption scheme fulfilling our notion. This transformation also works in respect to an even stronger notion of indistinguishability in the quantum world, which was introduced in [6], and previously only known to be achievable in the setting of *computational quantum encryption*, that is, the scenario where all the parties have quantum computing capabilities, and encryption is performed through arbitrary quantum circuits operating on quantum data. Even if this scenario goes in a very different direction from the scope of our work, it is interesting to note that our construction is the first fully classical scheme secure even in respect to such a purely quantum notion of security.

Related Work. The idea of considering scenarios where a quantum adversary can force other parties into quantum behaviour has been first considered in [8]. Attacks exploiting classical encryptions in quantum superposition have been described in [13,16,17,21]. In [4] the authors also consider the security of signature schemes where the adversary can have quantum access to a signing oracle. Quantum superposition queries have also been investigated relatively to the random oracle model [3]. Another quantum indistinguishability notion has been suggested (but not further analyzed) by Velema in [24]. Prior work has considered the security of quantum methods to encrypt classical data in the computational setting [15,27]. In concurrent and independent work, Broadbent and Jeffery [6] introduce indistinguishability notions for the public- and secret-key encryption of quantum messages in the context of fully homomorphic quantum computation. We refer to Page 15 for a more detailed description of how their definitions relate to our framework. A more complete overview for these notions, including semantic security for quantum encryption schemes, can be found in another concurrent work [1].

2   Preliminaries

In this section, we briefly recall the classical security notions for encryption schemes secure against chosen plaintext attacks (CPA). In addition, we revisit the two existing indistinguishability notions for the quantum world. We start by introducing notation we will use throughout the paper.

We say that a function $f : \mathbb{N} \to \mathbb{R}$ is *polynomially bounded* iff there exists a polynomial p and a value $\bar{n} \in \mathbb{N}$ such that: for every $n \geq \bar{n}$ we have that $f(n) \leq p(n)$; in this case we will just write $f = \text{poly}\,(n)$. We say that a function $\varepsilon : \mathbb{N} \to \mathbb{R}$ is *negligible*, if and only if for every polynomial p, there exists an $n_p \in \mathbb{N}$ such that $\varepsilon(n) \leq \frac{1}{p(n)}$ for every $n \geq n_p$; in this case we will just write $\varepsilon = \text{negl}\,(n)$. In this work, we focus on secret-key encryption schemes. In all that follows we use $n \in \mathbb{N}$ as the security parameter.

Definition 2.1 (Secret-Key Encryption Scheme [10]). *A secret-key encryption scheme is a triple of probabilistic polynomial-time algorithms* (Gen, Enc, Dec) *operating on a message space* $\mathcal{M} = \{0,1\}^m$ *(where* $m = \text{poly}\,(n) \in \mathbb{N}$*) that fulfills the following two conditions:*

1. *The key generation algorithm* Gen(1^n) *on input of security parameter n in unary outputs a bitstring k.*
2. *For all k in the range of* Gen(1^n) *and any message $x \in \mathcal{M}$, the algorithms* Enc *(encryption) and* Dec *(decryption) satisfy* $\Pr[\text{Dec}(k, \text{Enc}(k, x)) = x] = 1$, *where the probability is taken over the internal coin tosses of* Enc *and* Dec.

We write $\mathcal{K}$ for the range of Gen(1^n) (the key space) and Enc$_k(x)$ for Enc(k, x).

2.1   Classical Security Notions: IND-CPA and SEM-CPA

We turn to security notions for encryption schemes. In this work, we will only look at the notions of indistinguishability of ciphertexts under adaptively chosen plaintext attack (IND-CPA), and semantic security under adaptively chosen plaintext attack (SEM-CPA), which are known to be equivalent (e.g., [10]).

Game-Based Definitions. In general these notions can be defined as a game between a challenger $\mathcal{C}$ and an adversary $\mathcal{A}$. First, $\mathcal{C}$ generates a legitimate key running $k \longleftarrow \mathsf{Gen}(1^n)$ which he uses throughout the game. The game starts with a first learning phase. A challenge phase follows where $\mathcal{A}$ receives a challenge. Afterwards, a second learning phase follows, and finally $\mathcal{A}$ has to output a solution. The learning phases define the type of attack, and the challenge phase the notion captured by the game. We give all our definitions by referring to this game framework and by defining a learning and a challenge phase.

The CPA Learning Phase: $\mathcal{A}$ is allowed to adaptively ask $\mathcal{C}$ for encryptions of messages of his choice. $\mathcal{C}$ answers the queries using key k. Note that this is equivalent to saying that $\mathcal{A}$ gets oracle access to an encryption oracle that was initialized with key k.

The IND Challenge Phase: $\mathcal{A}$ defines a challenge template consisting of two equal-length messages x_0, x_1, and sends it to $\mathcal{C}$. The challenger $\mathcal{C}$ samples a random bit $b \xleftarrow{\$} \{0,1\}$ uniformly at random, and replies with the encryption $\mathsf{Enc}_k(x_b)$. $\mathcal{A}$'s goal is to guess b.

Definition 2.2 (IND-CPA). *A secret-key encryption scheme is called IND-CPA secure if the success probability of any probabilistic polynomial-time adversary winning the game defined by CPA learning phases and an IND challenge phase is at most negligibly (in n) close to 1/2.*

The SEM Challenge Phase: $\mathcal{A}$ sends $\mathcal{C}$ a challenge template (S_m, h_m, f_m) consisting of a poly-sized circuit S_m specifying a distribution over m-bit long plaintexts, an advise function $h_m : \{0,1\}^m \to \{0,1\}^*$, and a target function $f_m : \{0,1\}^m \to \{0,1\}^*$. The challenger $\mathcal{C}$ replies with the pair $(\mathsf{Enc}_k(x), h_m(x))$ where x is sampled according to S_m. $\mathcal{A}$'s challenge is to output $f_m(x)$.

In the definition of semantic security it is not required that $\mathcal{A}$'s probability of winning the game is always negligible. Instead, $\mathcal{A}$'s success probability is compared to that of a simulator $\mathcal{S}$ that plays in a *reduced game*: On one hand, $\mathcal{S}$ gets no learning phases. On the other hand, during the challenge phase, $\mathcal{S}$ does not receive the ciphertext but only the output of the advice function. This use of a simulator is what makes the notion hard to work with in proofs as one has to construct a simulator for every possible $\mathcal{A}$ to prove a scheme secure.

Definition 2.3 (SEM-CPA). *A secret-key encryption scheme is called SEM-CPA secure if for any probabilistic polynomial-time adversary $\mathcal{A}$ there exists a probabilistic polynomial-time simulator $\mathcal{S}$ such that the challenge templates produced by $\mathcal{S}$ and $\mathcal{A}$ are identically distributed and the success probability of $\mathcal{A}$ winning the game defined by CPA learning phases and a SEM challenge phase (computed over the coins of $\mathcal{A}$, Gen, and S_m) is negligibly close (in n) to the success probability of $\mathcal{S}$ winning the reduced game.*

Semantic security models what we want an encryption scheme to achieve: An adversary given a ciphertext can learn nothing about the encrypted message which he could not also learn from his knowledge of the message distribution and possibly existing side-information (modeled by h_m). Indistinguishability of ciphertexts is an equivalent technical notion introduced to simplify proofs.

2.2 Previous Notions of Security in the Quantum World

We briefly recall the results from [4] about quantum indistinguishability notions. We refer to [20] for commonly used notation and quantum information-theoretic concepts. Given security parameter n, let $\{\mathcal{H}_n\}_n$ be a family of complex Hilbert spaces such that $\dim \mathcal{H}_n = 2^{\mathrm{poly}\,(n)}$. We assume that $\mathcal{H}_n$ contains all the subspaces where the message states, the ciphertext states and any auxiliary state live. For the sake of simplicity we will not make a distinction when writing that a state $|\varphi\rangle$ belongs to one particular subspace, and we will omit the index n when the security parameter is implicit, therefore writing just $|\varphi\rangle \in \mathcal{H}$. We will denote pure states with ket notation, e.g., $|\varphi\rangle$, while mixed states will be denoted by lowercase Greek letters, e.g. ρ. We start by defining what we call a *classical description of a quantum state*:

Definition 2.4 (Classical Description). *A* classical description *of a quantum state ρ is a (classical) bitstring describing a quantum circuit S which (takes no input but starts from a fixed initial state $|0\rangle$ and) outputs ρ.*

This definition will be used later in our new notions of security. We deviate here from the traditional meaning of 'classical description' referring to individual numerical entries of the density matrix. The reason is that our definition also covers the cases where those numerical entries are not easily computable, as long as we can give an explicit constructive procedure for that state. Clearly, every pure quantum state $|\varphi\rangle$ has a classical description given by a description of the quantum circuit which implements the unitary that maps $|0\rangle$ to $|\varphi\rangle$. The classical description of a mixed state ρ_A is given by the circuit which first creates a purification $|\varphi\rangle_{AR}$ of ρ_A and then only outputs the A register. Note that a state admitting a classical description cannot be entangled with any other system.

For encryption, following the approach in [4] and many other works, we define the following:

Definition 2.5 (Quantum Encryption Oracle [4]). *Let* Enc *be the encryption algorithm of a secret-key encryption scheme $\mathcal{E}$. We define the* quantum encryption oracle U_{Enc_k} *associated with $\mathcal{E}$ and initialized with key k as (a family of) unitary operators defined by:*

$$U_{\mathsf{Enc}_k} : \sum_{x,y} \alpha_{x,y} |x\rangle |y\rangle \mapsto \sum_{x,y} \alpha_{x,y} |x\rangle |y \oplus \mathsf{Enc}_k(x)\rangle \tag{1}$$

where the same randomness r is used in superposition in all the executions of $\mathsf{Enc}_k(x)$ within one query[1] – for each new query, a fresh independent r is used.

---

[1] As shown in [4], this is not restrictive.

The first indistinguishability notion proposed in [4] replaces all classical communication between $\mathcal{A}$ and $\mathcal{C}$ by quantum communication. $\mathcal{A}$ and $\mathcal{C}$ are now quantum circuits operating on quantum states, and sharing a certain number of qubits (the quantum communication register). The definition for the new security game is obtained from Definition 2.2 by changing the learning and challenge phases as follows:

Quantum CPA Learning Phase (qCPA): $\mathcal{A}$ gets oracle access to U_{Enc_k}.

Fully Quantum IND Challenge Phase (fqIND): $\mathcal{A}$ prepares the communication register in the state $\sum_{x_0,x_1,y} \alpha_{x_0,x_1,y} |x_0\rangle |x_1\rangle |y\rangle$, consisting of two m-qubit states (the two input-message superpositions) and an ancilla state to store the ciphertext. $\mathcal{C}$ samples a bit $b \xleftarrow{\$} \{0,1\}$ and applies the transformation:

$$\sum_{x_0,x_1,y} \alpha_{x_0,x_1,y} |x_0\rangle |x_1\rangle |y\rangle \mapsto \sum_{x_0,x_1,y} \alpha_{x_0,x_1,y} |x_0\rangle |x_1\rangle |y \oplus \mathsf{Enc}_k(x_b)\rangle .$$

$\mathcal{A}$'s goal is to output b.

The resulting security notion in [4] is called *indistinguishability under fully quantum chosen-message attacks (IND-fqCPA)*. We decided to rename it to *fully quantum indistinguishability under quantum chosen-message attacks (fqIND-qCPA)* in order to fit into our naming scheme: It consists of a quantum CPA learning phase and a fully quantum IND challenge phase.

Definition 2.6 (fqIND-qCPA). *A secret-key encryption scheme is said to be fqIND-qCPA secure if the success probability of any quantum probabilistic polynomial-time adversary winning the game defined by qCPA learning phases and a fqIND challenge phase is at most negligibly close (in n) to $1/2$.*

As already observed in [4], this notion is unachievable. The separation by Boneh and Zhandry exploits the entanglement of quantum states, namely the fact that entanglement can be created between plaintext and ciphertext.

Theorem 2.7 (BZ Attack [4, Theorem 4.2]). *No symmetric-key encryption scheme can achieve fqIND-qCPA security.*

Proof. The attack works as follows: The adversary $\mathcal{A}$ chooses as challenge messages the states $|0^m\rangle$ and $H|0^m\rangle$ (where H denotes the m-fold tensor Hadamard transform), i.e. he prepares the register in the state $\sum_x \frac{1}{2^{m/2}} |0^m, x, 0^m\rangle$. When the challenger $\mathcal{C}$ performs the encryption, we can have two cases:

– if $b = 0$, i.e. the first message state is chosen, the state is transformed into

$$\sum_x \frac{1}{2^{m/2}} |0^m, x, \mathsf{Enc}_k(0^m)\rangle = |0^m\rangle \otimes H|0^m\rangle \otimes |\mathsf{Enc}_k(0^m)\rangle ;$$

– if $b = 1$, i.e. the second message state is chosen, the state is transformed into

$$\sum_x \frac{1}{2^{m/2}} |0^m, x, \mathsf{Enc}_k(x)\rangle = |0^m\rangle \otimes \sum_x \frac{1}{2^{m/2}} |x, \mathsf{Enc}_k(x)\rangle .$$

Notice that in the second case we have a fully entangled state between the second and the third register. At this point, $\mathcal{A}$ does the following:

1. measures (traces out) the third register;
2. applies again H to the second register;
3. measures the second register;
4. outputs $b' = 1$ iff the outcome of this last measurement is 0^m, else outputs 0.

In fact, if $b = 0$, then the second register is left untouched: By applying again the Hadamard transformation it will be reset to the state $|0^m\rangle$, and a measurement on this state will yield 0^m with probability 1. If $b = 1$ instead, tracing out one half of a fully entangled state results in a complete mixture in the second register. Applying a Hadamard transform and measuring in the computational basis necessarily gives a fully random outcome, and hence outcome 0^m only with probability $\frac{1}{2^m}$, which is negligible in n, because $m = \text{poly}(n)$.   □

Theorem 2.7 implies that the fqIND-qCPA notion is too strong. In order to weaken it, the following notion of indistinguishability under adaptively chosen quantum plaintext attacks was introduced:

Definition 2.8 (IND-qCPA [4]). *A secret-key encryption scheme is said to be* IND-qCPA *secure if the success probability of any quantum probabilistic polynomial-time adversary winning the game defined by qCPA learning phases and a classical IND challenge phase is at most negligibly close (in n) to 1/2.*

In this definition, the CPA queries are allowed to be quantum, but the challenge query is required to be classical. It has been shown that, under standard computational assumptions, IND-qCPA is strictly stronger than IND-CPA:

Theorem 2.9 (IND-CPA $\neq$ IND-qCPA [4, Theorem 4.8]). *If classically secure PRFs exist and order-finding in prime groups is classically hard, then there exists an encryption scheme $\mathcal{E}$ which is IND-CPA secure, but not IND-qCPA secure.*

3 New Notions of Quantum Indistinguishability

IND-qCPA might be viewed as classical indistinguishability (IND) under a quantum chosen plaintext attack (qCPA). The authors in [4] resorted to this definition in order to overcome their impossibility result on one seemingly natural notion of quantum indistinguishability (fqIND-qCPA) which turned out to be too strong. This raises the question whether IND-qCPA is the only possible quantum indistinguishability notion (and hence no classical encryption scheme can achieve indistinguishability of ciphertext superpositions) or if there exists a stronger notion which can be achieved.

In this section we show that by defining fqIND-qCPA, there are many choices which are made implicitly, and that on the other hand there exist other possible quantum indistinguishability notions. We discuss these choices spanning a binary 'security tree' of possible notions. Afterwards, we obtain a small set of candidate notions, eliminating those that are either ill-posed or unachievable because of the BZ attack from Theorem 2.7. In all these notions, we implicitly assume 'quantum CPA learning phases', as in the case of IND-qCPA. However, we limit the discussion in this section to the design of a quantum challenge phase. In the end, we select a suitable 'qIND-'notion amongst all the possible candidate ones.

3.1   The 'Security Tree'

To define a general notion of indistinguishability in the quantum world, we have to consider many different distinctions for possible candidate models. For example, can we rule out certain forms of entanglement? How? Does the adversary have complete control over the challenger device? Each of these distinctions leads to a fork in a 'security-model binary tree'. We analyze every 'leaf' of the tree[2]. Some of them lead to unreasonable or ill-posed models, some of them yield unachievable security notions, and others are analyzed in more detail.

Game Model: Oracle ($\mathcal{O}$) vs. Challenger ($\mathcal{C}$). This distinction decides how the game, and especially the challenge phase, is implemented. In the classical world, the following two cases are equivalent but in the quantum world they differ. In the *oracle* model, the adversary $\mathcal{A}$ gets oracle access to encryption and challenge oracles, i.e., he plays the game by performing calls to unitary gates $\mathcal{O}_1, \ldots, \mathcal{O}_q$. In this case $\mathcal{A}$ is modeled as a quantum circuit which implements a sequence of unitary gates $U_0, \ldots, U_q$, intertwined by calls to the $\mathcal{O}_i$'s. Given an input state $|\varphi\rangle$, the adversary therefore computes the state:

$$U_q \mathcal{O}_q \ldots U_1 \mathcal{O}_1 U_0 \, |\varphi\rangle \, .$$

The *structure* of the *oracle* gates $\mathcal{O}_i$ itself is unknown to $\mathcal{A}$, who is only allowed to apply them in a black-box way. The fqIND notion uses this model.

In what we call the *challenger* model instead, the game is played against an *external (quantum) challenger*. Here, $\mathcal{A}$ is a quantum circuit which shares a quantum register (the communication channel) with another quantum circuit $\mathcal{C}$. The main difference is that in this case we can also consider what happens if $\mathcal{C}$ has additional input or output lines out of $\mathcal{A}$'s control. Moreover, $\mathcal{A}$ does not automatically gain access to the inverse (adjoint) of quantum operations performed by $\mathcal{C}$, and $\mathcal{C}$ cannot be 'rewound' by the adversary, which would be far too powerful possibilities. This scenario also covers the case of 'unidirectional' state transmission, i.e., when qubits are sent over a quantum channel to another party, and they are not available afterwards until that party sends them back. Regardless, in security proofs in the ($\mathcal{C}$) model, it is still allowed for an external entity (e.g. a simulator, or a reduction) to rewind the joint circuit composed by adversary and challenger together, if need be. However, we are not aware of any known reduction involving rewinding in this form for encryption schemes in the quantum world.

---

[2] We do not rule out that some of them might eventually lead to the same model.

In order to keep consistency with this choice of the model, when also considering qCPA queries, we implicitly assume the same access mode to the Enc_k oracle as in the qIND game. That is, if we are in the $(\mathcal{O})$ scenario, during the qCPA phase $\mathcal{A}$ has quantum oracle access to Enc_k. In the $(\mathcal{C})$ case, instead, superposition access to Enc_k is provided to $\mathcal{A}$ by an external challenger.

At first glance, the $(\mathcal{O})$ model intuitively represents the scenario where $\mathcal{A}$ has almost complete control of some encryption device, whereas the $(\mathcal{C})$ model is more suited to a 'network' scenario where $\mathcal{A}$ wants to compromise the security of some external target.

Plaintexts: Quantum States (Q) vs. Classical Description (c). In the (Q) model, the two m-qubit plaintexts chosen by $\mathcal{A}$ for the challenge template can be arbitrary (BQP-producible) quantum states and can be entangled with each other and other states. In the (c) model, instead, $\mathcal{A}$ is only allowed to choose *classical descriptions* of two m-qubit quantum states according to Definition 2.4, thus being only allowed to send classical information to $\mathcal{C}$: the challenger $\mathcal{C}$ will read the states' descriptions and will build one of the two states depending on his challenge bit b.

In classical models, there is no difference between sending a description of a message or the message itself. In the quantum world, there is a big difference between these two cases, as the latter allows $\mathcal{A}$ to establish entanglement of the message(s) with other registers. This is not possible when using classical descriptions. It might intuitively appear that the (Q) model (considered for the fqIND-qCPA notion) is more natural. However, the (c) scenario models the case where $\mathcal{A}$ is well aware of the message that is encrypted, but the message is not constructed by $\mathcal{A}$ himself. Giving $\mathcal{A}$ the ability to choose the challenge messages for the IND game models the worst case that might happen: $\mathcal{A}$ knows that the ciphertext he receives is the encryption of one out of the two messages that he can distinguish best. This closely reflects the intuition behind the classical IND notions: in that game, the adversary is allowed to send the two messages not because in the real world he would be allowed to do so, but because we want to achieve security even for the best possible choice of messages from the adversary's perspective. Hence, the (c) model is a valid alternative. Will further discuss the difference between these two models later.

Relaying of Plaintext States: Yes (Y) vs. No (n). If $\mathcal{C}$ is *not relaying* (n), this means that the two plaintext states chosen by $\mathcal{A}$ will not be 'sent back' to $\mathcal{A}$ (in other words: their registers will not be available anymore to $\mathcal{A}$ after the challenge encryption). In circuit terms, this means that at the beginning of the game, $\mathcal{C}$ will have (one or two) ancilla registers in his internal (private) memory. During the encryption phase, $\mathcal{C}$ will swap these register(s) with the content of the original plaintext register(s), hence transferring their original content outside of $\mathcal{A}$'s control.

If the challenger is relaying (Y) instead, this means that the two plaintext states will be left in the original register (or channel), and may be accessed by $\mathcal{A}$ at any moment. This is the model considered for fqIND.

Again, the (Y) case is more fitting to those cases where $\mathcal{A}$ 'implements locally' the encryption device and has almost full control of it, whereas the (n) case is more appropriate when the game is played against some external entity which is not under $\mathcal{A}$'s control. This is a rather natural assumption, for example, when states are sent over some quantum channel and not returned. We stress that this distinction in relaying is not trivial: it is not possible for $\mathcal{A}$, in general, to simulate relaying by keeping internal states entangled with the plaintexts. As an example, consider the attack in Theorem 2.7: it is easy to see that this cannot be performed without relaying.

Type of Unitary Transformation: (1) vs. (2). In quantum computing, the 'canonical' way of evaluating a function $f(x)$ in superposition is by using an auxiliary register:

$$\sum_{x,y} \alpha_{x,y} |x,y\rangle \mapsto \sum_{x,y} \alpha_{x,y} |x, y \oplus f(x)\rangle \,.$$

This way ensures that the resulting operator is invertible, even if f is not. We call this *type-(1) transformations*: if Enc_k is an encryption mapping m-bit plaintexts to ℓ-bit ciphertexts, the resulting operator in this case will act on $m + \ell$ qubits in the following way:

$$\sum_{x,y} \alpha_{x,y} |x,y\rangle \mapsto \sum_{x,y} \alpha_{x,y} |x, y \oplus \mathsf{Enc}_k(x)\rangle \,,$$

where the y's are ancillary values. This approach is also used for fqIND.

In our case, though, we do not consider arbitrary functions, but encryptions, which act as *bijections* on some bit-string spaces (assuming that the randomness is treated as an input). Therefore, provided that the encryption does not change the size of a message, the following transformation is also invertible:

$$\sum_{x} \alpha_x |x\rangle \mapsto \sum_{x} \alpha_x |\mathsf{Enc}_k(x)\rangle \,. \tag{2}$$

For the more general case of arbitrary message expansion factors, we will consider transformations of the form:

$$\sum_{x,y} \alpha_{x,y} |x,y\rangle \mapsto \sum_{x,y} \alpha_{x,y} |\varphi_{x,y}\rangle \,,$$

where the length of the ancilla register is $|y| = |\mathsf{Enc}_k(x)| - |x|$ and $\varphi_{x,0} = \mathsf{Enc}_k(x)$ for every x – i.e., initializing the ancilla y register in the $|0\rangle$ state produces a correct encryption, which is what we expect from an honest quantum executor. One might ask what happens if the ancilla is not initialized to 0, and we leave the general case of arbitrary ancillas manipulation as an interesting open problem, but we stress the fact that this behavior is not considered in the case of honest parties. We call these *type-(2) transformations*[3].

---

[3] These are called *minimal quantum oracles* in [14].

Notice that, in general, type-(1) and type-(2) transformations are very different: having quantum oracle access to a type-(2) unitary $U_{\mathsf{Enc}}^{(2)}$ and its adjoint also gives access to the related type-(2) *decryption oracle* $U_{\mathsf{Dec}}^{(2)} : \sum_x \alpha_x |\mathsf{Enc}_k(x)\rangle \mapsto \sum_x \alpha_x |x\rangle$. In fact, notice that $(U_{\mathsf{Enc}}^{(2)})^\dagger = U_{\mathsf{Dec}}^{(2)}$, while the adjoint of a type-(1) encryption operator, $(U_{\mathsf{Enc}}^{(1)})^\dagger$, is generally *not* a type-(1) decryption operator. In particular, type-(2) operators are 'more powerful' in the sense that knowledge of the secret key is required in order to build any efficient quantum circuit implementing them. However, we stress the fact that whenever access to a decryption oracle is allowed, the two models are completely equivalent, because then we can simulate a type-(2) operator by using ancilla qubits and 'uncomputing' the resulting garbage lines (see Fig. 1) (as we will see, this will be the case for the challenger in our qIND notion).

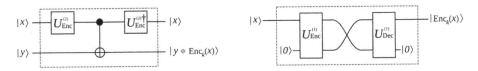

Fig. 1. Equivalence between type-(1) and type-(2) in the case of 1-qubit messages. Left: building a type-(1) encryption oracle by using a type-(2) encryption oracle (and its inverse) as a black-box. Right: building a type-(2) encryption oracle by using type-(1) encryption and decryption oracles as black-boxes.

3.2 Analysis of the Models

By considering these 4 distinctions in the security tree we have $2^4 = 16$ possible candidate models to analyze. We label each of these candidate models by appending each one of the 4 labels of every tree branch in brackets. Clearly, 16 different definitions of quantum indistinguishability is too much, but luckily most of these are unreasonable or unachievable. To start with, we can ignore the following:

Leaves of the Form ($\mathcal{O}c\dots$). In the $\mathcal{O}$ scenario, the oracle is actually a quantum gate inside $\mathcal{A}$'s quantum circuitry. Therefore $\mathcal{A}$ has the capability of querying the oracle on states which are possibly entangled with other registers kept by $\mathcal{A}$ itself.

Leaves of the Form ($\mathcal{O}Qn\dots$). Again, the oracle is a gate which has no internal memory to store and keep the plaintext states sent by $\mathcal{A}$.

Leaves of the Form ($\dots Y2$). Relaying is not taken into account in type-(2) transformations. In these transformations, to some extent, one of the two plaintext registers is *always* relayed (after having been 'transformed' into a ciphertext). If the other plaintext was to be relayed as well, this would immediately compromise indistinguishability (because one of the two states would be modified and the other not, and both of them would be handed over to $\mathcal{A}$).

Excluding these options leaves us with 7 models, but it is easy to see that 3 of them are unachievable because of the attack from Theorem 2.7. This is the case for $(\mathcal{O}QY1)$ (which is exactly fqIND-qCPA), $(\mathcal{C}QY1)$, and $(\mathcal{C}cY1)$. Of the remaining 4, notice that $(\mathcal{C}Qn1)$ and $(\mathcal{C}cn1)$ are equivalent to the IND-qCPA notion from [4]. The reason is that from $\mathcal{A}$'s perspective, a non-relaying $\mathcal{C}$ is indistinguishable from a $\mathcal{C}$ tracing out (measuring) the plaintext register (otherwise $\mathcal{A}$ and $\mathcal{C}$ could communicate faster than light). This measuring operation would make the ciphertext collapse into a single (classical) ciphertext. And since tracing out the challenge register and applying the type-(1) operator $U_{\text{Enc}}^{(1)}$ commute, one can consider (without loss of generality) the case that $\mathcal{A}$ himself first measures the plaintext register, and then initiates a classical IND query with $\mathcal{C}$, therefore recovering a classical definition of IND challenge query[4]. Therefore, using any of $(\mathcal{C}Qn1)$ or $(\mathcal{C}cn1)$ would lead to a weaker notion of quantum indistinguishability. Since we are interested in achieving stronger notions, we will hence consider the more challenging scenarios $(\mathcal{C}Qn2)$ and $(\mathcal{C}cn2)$.

This argument also leads to the following interesting observation. Ultimately, whether a challenger (or encryption device) performs type-(1) or type-(2) operations depends on its *architecture* which we cannot say anything about - we will focus on the $(\dots 2)$ models in order to be on the 'safe side', as they lead to security notions which are harder to achieve. In order to design a secure encryption device, it is good advice to avoid the possibility that it can be accessed in type-(2) mode. For such a device, it would be sufficient to provide IND-qCPA security, which is weaker and therefore easier to achieve. Clearly, providing guidelines on how to construct encryption devices resilient to type-(2) access lies outside the scope of this work.

3.3   qIND

At this point we are left with only two candidate notions: $(\mathcal{C}cn2)$ and $(\mathcal{C}Qn2)$. From now on we will denote them as *'quantum indistinguishability of ciphertexts'* (qIND) and *'general quantum indistinguishability of ciphertexts'* (gqIND) resp., and we summarize the resulting challenge phases as follows.

Quantum IND Challenge Phase (qIND): $\mathcal{A}$ chooses two quantum states ρ_0, ρ_1 having efficient (poly-sized) classical descriptions, and sends to $\mathcal{C}$ a challenge template consisting of these two classical descriptions according to Definition 2.4. $\mathcal{C}$ samples a bit b and replies to $\mathcal{A}$ with the state obtained by applying the type-(2) operator $U_{\text{Enc}_k}^{(2)}$ as defined in (2) to ρ_b. $\mathcal{A}$'s goal is to output b.

General Quantum IND Challenge Phase (gqIND): $\mathcal{A}$ chooses two quantum states ρ_0, ρ_1, and sends them to $\mathcal{C}$. $\mathcal{C}$ samples a bit b, discards (traces out)

---

[4] However, we stress that this interpretation is not entirely correct. In fact, one might consider composition scenarios where the IND query is just an intermediate step, and the plaintext and ciphertext registers are reunited at some later step. In such scenarios, not relaying would not be equivalent to measuring. We ignore such considerations in this work, and leave the general case of composable security as an interesting open question.

ρ_{1-b}, and replies to $\mathcal{A}$ with the state obtained by applying the type-(2) operator $U_{\mathsf{Enc}_k}^{(2)}$ as defined in (2) to ρ_b. $\mathcal{A}$'s goal is to output b.

Using these challenge phases and the notion of a qCPA learning phase, we define qIND-qCPA and gqIND-qCPA as follows.

Definition 3.1 (qIND-qCPA). *A secret-key encryption scheme is said to be qIND-qCPA secure if the success probability of any quantum probabilistic polynomial time adversary winning the game defined by qCPA learning phases and the qIND challenge phase above is at most negligibly close (in n) to 1/2.*

Definition 3.2 (gqIND-qCPA). *A secret-key encryption scheme is said to be gqIND-qCPA secure if the success probability of any quantum probabilistic polynomial time adversary winning the game defined by qCPA learning phases and the gqIND challenge phase above is at most negligibly close (in n) to 1/2.*

Since we mainly consider type-(2) transformations from now on, we will overload notation and also use U_{Enc_k} to denote the type-(2) encryption operator.

Theorem 3.3 (gqIND-qCPA $\Rightarrow$ qIND-qCPA). *Let $\mathcal{E}$ be a symmetric-key encryption scheme. If $\mathcal{E}$ is gqIND-qCPA secure, then $\mathcal{E}$ is also qIND-qCPA secure.*

The reason is that quantum states admitting an efficient classical description (used in qIND) are just a special case of arbitrary quantum plaintext states (used in gqIND). Despite this implication, we will mainly focus on the qIND notion in the following, and we will use the gqIND notion only as a comparison to other existing notions. The main reason for this choice is that in the context of classical encryption schemes resistant to superposition quantum access, we believe that it is important to not lose focus of what the capabilities of a 'reasonable' adversary should be. Namely, recall the following classical IND argument: *allowing the adversary to send plaintexts to the challenger is equivalent to the fact that indistinguishability must hold even for the most favorable case from the adversary's perspective.* Such an argument does *not* hold anymore quantumly. In fact, the (Q) model considered in gqIND presents the following issues:

- it allows entanglement between the adversary and the challenger: $\mathcal{A}$ could prepare a state of the form $\rho_{AB} = \frac{1}{\sqrt{2}}|00\rangle + \frac{1}{\sqrt{2}}|11\rangle$, sending ρ_A as a plaintext but keeping ρ_B;
- it allows the adversary to create certain non-reproduceable states. For example, consider the state $|\psi\rangle = \sum_{x \in X} \frac{1}{\sqrt{|X|}}|x, h(x)\rangle$, where h is a collision-resistant hash function. $\mathcal{A}$ could measure the second register, obtaining a random outcome y, and knowing therefore that the remaining state is the superposition of the preimages of y, $|\psi_y\rangle = \sum_{x \in X : h(x) = y} \frac{1}{\sqrt{|\{x \in X : h(x) = y\}|}}|x\rangle$. $\mathcal{A}$ could then use $|\psi_y\rangle$ as a plaintext in the challenge phase, but note that $\mathcal{A}$ cannot reproduce $|\psi_y\rangle$ for a given value y.

Both of the above examples are not reasonable in our scenario. Entanglement between $\mathcal{A}$ and $\mathcal{C}$ represents a sort of 'quantum watermarking' of messages, which goes beyond what a meaningful notion of indistinguishability should achieve. Knowledge of intermediate, unpredictable measurements also renders $\mathcal{A}$ too powerful, because it gives $\mathcal{A}$ access to information not available to $\mathcal{C}$ itself - e.g., in the example above $\mathcal{C}$ would not even know the value of y. As it is $\mathcal{C}$ who prepares the state to be encrypted, it is reasonable to assume that it is $\mathcal{C}$ who should know these intermediate measurements, not $\mathcal{A}$. In the example above, what $\mathcal{A}$ could see instead (provided he knows the circuit generating the state, as we assume in qIND) is that the plaintext is a mixture $\Psi = \sum_y \psi_y$ for all possible values of y.

The possibility offered by gqIND of allowing the adversary to play the IND game with arbitrary states is certainly elegant from a theoretical point of view, but from the perspective of the quantum security of the kind of schemes we are considering, it is too broad in scope. The (c) model used in qIND, on the other hand, inherently provides guidelines and reasonable limitations on what a quantum adversary can or cannot do. Also, qIND is often easier to deal with: notice that in the (c) model, unlike in the (Q) model, $\mathcal{A}$ always receives back an unentangled state from a challenge query. In security reductions, this means that we can more easily simulate the challenger, and that we do not have to take care of measures of entanglement when analyzing the properties of quantum states - for example, indistinguishability of states can be shown by only resorting to the *trace norm* instead of the more general *diamond norm*.

Furthermore, it is important to notice that all our new results in Sect. 6 are unaffected by the choice of either qIND or gqIND. Our impossibility result from Theorem 6.3 holds for qIND, and hence also for gqIND because of Theorem 3.3. On the other hand, the security proof of Construction 6.6 (Theorem 6.9) is given for gqIND, and holds therefore also for qIND. In fact, it remains unclear whether a separation between qIND and gqIND can be found at all in the realm of classical encryption schemes. We leave this as an interesting open question.

Finally, we note that the q-IND-CPA-2 indistinguishability notion for secret-key encryption of quantum messages introduced by Broadbent and Jeffery [6, Appendix B] resembles our gqIND notion, and it is in fact equivalent to it in the case that the encryption operation is a symmetric-key classical functionality operating in type-(2) mode.

Theorem 3.4 (gqIND-qCPA $\Leftrightarrow$ q-IND-CPA-2). *Let $\mathcal{E}$ be a symmetric-key encryption scheme. Then $\mathcal{E}$ is gqIND-qCPA secure if and only if $\mathcal{E}$ is q-IND-CPA-2 secure.*

A proof of the above theorem can be found in the full version [9]. A generalization of q-IND-CPA-2 to arbitrary quantum encryption schemes, together with equivalent notions of quantum semantic security, was given and analyzed in [1]. All these security notions are given in the context of 'fully quantum encryption', in the sense that the encryption schemes considered in [6] and [1] are arbitrary quantum circuits acting natively on quantum data, while in this work we consider the quantum security of classical encryption schemes. The fully quantum homomorphic schemes which are shown to be secure in [6], and the other quantum encryption schemes shown to be secure in [1], do not fall into the category of *classical* encryption schemes which we are studying here. On the other hand, as Theorem 6.9 shows, our Construction 6.6 is the first known example of a classical symmetric-key encryption scheme which is secure even against these kinds of 'fully quantum' security notions.

4   New Notions of Quantum Semantic Security

In this section, we initiate the study of suitable definitions of semantic security in the quantum world. As in the classical case, we are particularly interested in notions that can be proven equivalent to some version of quantum indistinguishability. So these definitions actually describe the *semantics* of the equivalent IND notions. As in the classical case, we present these notions in the non-uniform model of computation.

Working towards a quantum SEM notion, we restrict our analysis to the SEM challenge phase. For the learning phase, we stick to the 'qCPA learning phase', as in Definition 2.5, where the adversary has access to a quantum encryption oracle. In the end, we give a definition for quantum semantic security under quantum chosen-plaintext attacks (qSEM-qCPA) which we later prove equivalent to qIND-qCPA, thereby adding semantics to our qIND-qCPA notion.

4.1   Classical Semantic Security Under Quantum CPA

As a first notion of semantic security in the quantum world, we consider what happens if, like in the IND-qCPA notion, we stick to the classical definition but we allow for a quantum chosen-plaintext-attack phase. The definition uses a SEM-qCPA game that is obtained by combining qCPA learning phases with a classical SEM challenge phase as defined in Sect. 2. As in the classical case, $\mathcal{A}$'s success probability is compared to that of a simulator $\mathcal{S}$ that plays in a reduced game: $\mathcal{S}$ gets no learning phase and during the challenge phase it only receives the advice $h_m(x)$, not the ciphertext.

Definition 4.1 (SEM-qCPA). *A secret-key encryption scheme is called SEM-qCPA-secure if for every quantum polynomial-time machine $\mathcal{A}$, there exists a quantum polynomial-time machine $\mathcal{S}$ such that the challenge templates produced by $\mathcal{S}$ and $\mathcal{A}$ are identically distributed and the success probability of $\mathcal{A}$ winning the game defined by qCPA learning phases and a SEM challenge phase is negligibly close (in n) to the success probability of $\mathcal{S}$ winning the reduced game.*

Spoiler. It is easy to see that the SEM-qCPA notion of semantic security is equivalent to IND-qCPA, see Theorem 5.1.

In the full version [9] we discuss what happens if one also allows quantum advice states in this scenario, and why this option would not add anything meaningful.

4.2 Quantum Semantic Security

We now define *quantum semantic security under chosen-plaintext attacks* (qSEM-qCPA). As in the classical case, we want the definition of semantic security to formally capture what we intuitively understand as a strong security notion. In the quantum case, there are several choices to be made. We start by giving our formal definition of quantum semantic security, and justify our choices afterwards.

Quantum SEM (qSEM) Challenge Phase: $\mathcal{A}$ sends to $\mathcal{C}$ a challenge template consisting of classical descriptions of

– a quantum circuit G_m taking poly (n)-bit classical input and outputting m-qubit plaintext states,
– a quantum circuit h_m taking m-qubit plaintexts as input and outputting poly (n)-qubit advice states,
– a quantum circuit f_m taking m-qubit plaintexts as input and outputting poly (n)-qubit target states.

The challenger $\mathcal{C}$ samples $y \xleftarrow{\$} \{0,1\}^{\text{poly}(n)}$ and computes two copies of the plaintext $\rho_y = G_m(y)$. One is used to compute auxiliary information $h_m(\rho_y)$ and one to compute the ciphertext $U_{\mathsf{Enc}_k} \rho_y U_{\mathsf{Enc}_k}^\dagger$. $\mathcal{C}$ then replies with the pair $\left(U_{\mathsf{Enc}_k} \rho_y U_{\mathsf{Enc}_k}^\dagger, h_m(\rho_y) \right)$. $\mathcal{A}$'s goal is to output $f_m(\rho_y)$. We say that $\mathcal{A}$ *wins the qSEM-qCPA game* if no quantum polynomial-time distinguisher can distinguish $\mathcal{A}$'s output from the target state $f_m(\rho_y)$ with non-negligible advantage.

In the reduced game, $\mathcal{S}$ receives no encryption, but only the auxiliary information $h_m(\rho_y)$ from $\mathcal{C}$. Analogously to the above case, $\mathcal{S}$ *wins the qSEM-qCPA game* if no quantum polynomial-time distinguisher can distinguish $\mathcal{S}$'s output from the target state $f_m(\rho_y)$ with non-negligible advantage.

Definition 4.2 (qSEM-qCPA). *A secret-key encryption scheme is called qSEM-qCPA-secure if for every quantum polynomial-time machine $\mathcal{A}$, there exists a quantum polynomial-time machine $\mathcal{S}$ such that the challenge templates produced by $\mathcal{S}$ and $\mathcal{A}$ are identically distributed and the success probability of $\mathcal{A}$ winning the game defined by qCPA learning phases and a qSEM challenge phase is negligibly close (in n) to the success probability of $\mathcal{S}$ winning the reduced game.*

When defining quantum semantic security, we have to deal with several issues: First, we have to define how the plaintext distribution is described. In the classical definition, the distribution is produced by a (classical) circuit G_m running on uniform input bits. We take the same approach here, but let G_m output m-qubit plaintexts.

The second question is how to define the advice function. While the input should be the plaintext quantum state ρ_y, the output could be either quantum or classical. We decided to allow quantum advice as it leads to a more general model and it includes classical outputs as a special case. In order for the challenger to compute both the encryption of the plaintext state ρ_y and the advice state $h_m(\rho_y)$ without violation of the no-cloning theorem, we exploit how we generate the message state. We simply run S_m twice on the same classical randomness y to generate two copies of the plaintext state ρ_y. Another option would have been to allow for entanglement between the plaintext message ρ_y and the advice state $h_m(\rho_y)$. Allowing such entanglement would model side-channel information the attacker could obtain, for instance by learning the content of some internal register of the attacked device. However, the resulting notion would not be equivalent with qIND-qCPA anymore, because in qIND-qCPA, the challenge plaintexts are provided by their classical descriptions and can therefore not be entangled with the attacker.

Third, we have chosen to model the target function f_m in the same way as the advice function h_m, i.e. we allow arbitrary quantum circuits that might output quantum states. The reasoning behind allowing quantum output is again to use the strongest possible, most general model. Allowing quantum output however leads to the problem that, in general, we cannot physically test anymore if an adversary $\mathcal{A}$ outputs exactly the result of the target function $f_m(\rho_y)$. One option would be to require $\mathcal{A}$'s output to be close to $f_m(\rho_y)$ in terms of their trace distance. But two quantum states can be quantum-polynomial-time indistinguishable even if their trace distance is large[5]. Since we are only interested in computational security notions, we solve this problem by requiring QPT indistinguishability as success condition for winning the SEM game.

Spoiler. Our qSEM-qCPA notion of semantic security is equivalent to qIND-qCPA, and unachievable for those schemes which leave the size of the message unchanged (like most block ciphers), see Sect. 6.1.

5   Relations

In this section we show relations between our new notions of indistinguishability and semantic security in the quantum world. It is already known [10,11] that classically, IND-CPA and semantic security are equivalent. Our goal is to show a similar equivalence for our new notions, plus to show a hierarchy of equivalent security notions. Our results are summarized in Fig. 2.

Theorem 5.1 (IND-qCPA $\Leftrightarrow$ SEM-qCPA). *Let $\mathcal{E}$ be a symmetric-key encryption scheme. Then $\mathcal{E}$ is IND-qCPA secure if and only if $\mathcal{E}$ is SEM-qCPA secure.*

We split the proof of Theorem 5.1 into two propositions – one per direction. They closely follow the proofs for the classical case (see [10, Proof of Th. 5.4.11]), we recall them as they work as guidelines for the following proofs.

---

[5] Think of two different classical ciphertexts which are encrypted using a quantum-computationally secure encryption scheme. Then, the ciphertext states are orthogonal (and hence their trace distance is maximal), but they are computationally indistinguishable.

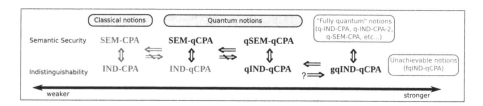

Fig. 2. The relations between notions of indistinguishability and semantic security in the quantum world (previously known results in gray).

Proposition 5.2 (IND-qCPA $\Rightarrow$ SEM-qCPA).

Proposition 5.3 (SEM-qCPA $\Rightarrow$ IND-qCPA).

Proof (of Proposition 5.2 – Sketch). The idea of the proof is to hand $\mathcal{A}$'s circuit as non-uniform advice to the simulator $\mathcal{S}$. $\mathcal{S}$ runs $\mathcal{A}$'s circuit and impersonates the challenger $\mathcal{C}$ by generating a new key and answering all of $\mathcal{A}$'s queries using this key. When it comes to the challenge query, $\mathcal{S}$ encrypts the $1\ldots1$ string of the same length as the original message. It follows from the indistinguishability of encryptions that the adversary's success probability in this game must be negligibly close to its success probability in the real semantic-security game, which concludes the proof. The only difference in the -qCPA case is that $\mathcal{A}$ and $\mathcal{S}$ are quantum circuits, and that $\mathcal{S}$ has to emulate the quantum encryption oracle instead of a classical one. $\square$

Proof (of Proposition 5.3). We recall here the full proof as it is short. Assume there exists an efficient distinguisher $\mathcal{A}$ against the IND-qCPA security of $\mathcal{E}$. Then we show how to construct an oracle machine $\mathcal{M}^{\mathcal{A}}$ that has access to $\mathcal{A}$ and breaks the SEM-qCPA security of the scheme. $\mathcal{M}^{\mathcal{A}}$ runs $\mathcal{A}$, emulating the quantum encryption oracle by simply forwarding all the qCPA queries to its own oracle. As $\mathcal{A}$ executes an IND challenge query on m-bit messages (x_0, x_1), $\mathcal{M}^{\mathcal{A}}$ produces the SEM template (G_m, h_m, f_m) with G_m describing the uniform distribution over $\{x_0, x_1\}$, $h_m = 1^n$ (or any other function such that $h_m(x_0) = h_m(x_1)$), and f_m a function that fulfills $f_m(x_0) = 0$ and $f_m(x_1) = 1$ (i.e., the distinguishing function). Then $\mathcal{M}^{\mathcal{A}}$ performs a SEM challenge query with this template, and given challenge ciphertext c, uses it to answer $\mathcal{A}$'s query. If, at that point, $\mathcal{A}$ performs more qCPA queries, $\mathcal{M}^{\mathcal{A}}$ answers again by forwarding all these queries to its own oracle. Finally, $\mathcal{M}^{\mathcal{A}}$ outputs $\mathcal{A}$'s output. As $\mathcal{A}$ distinguishes encryptions of x_0 and x_1 with non-negligible success probability, $\mathcal{A}$ will return the correct value of f_m with recognizably higher probability than guessing. As h_m is independent of the encrypted message, no simulator can do better than guessing. Hence, $\mathcal{M}^{\mathcal{A}}$ has a non-negligible advantage to output the right value of f_m. $\square$

Theorem 5.4 (qIND-qCPA $\Leftrightarrow$ qSEM-qCPA). *Let $\mathcal{E}$ be a symmetric-key encryption scheme. Then $\mathcal{E}$ is qIND-qCPA secure if and only if $\mathcal{E}$ is qSEM-qCPA secure.*

Again, we split the proof of Theorem 5.4 into two propositions.

Proposition 5.5 (qIND-qCPA $\Rightarrow$ qSEM-qCPA).

Proposition 5.6 (qSEM-qCPA $\Rightarrow$ qIND-qCPA).

Proof (of Proposition 5.5 – Sketch). The proof follows that of Proposition 5.2, with some careful observations. Since $\mathcal{A}$ is a QPT adversary against the qSEM-qCPA game, $\mathcal{A}$'s circuit has a short classical representation ξ. So $\mathcal{S}$ gets ξ as non-uniform advice and hence can implement and run $\mathcal{A}$. The simulator $\mathcal{S}$ simulates $\mathcal{C}$ for $\mathcal{A}$ by generating a new key and answering all of $\mathcal{A}$'s qCPA queries. When it comes to the challenge query, $\mathcal{A}$ produces a qSEM template, which $\mathcal{S}$ forwards to the real $\mathcal{C}$. Then $\mathcal{S}$ forwards $\mathcal{C}$'s reply, plus a bogus encrypted state (e.g., $U_{\mathsf{Enc}_k} |1 \dots 1\rangle$), to $\mathcal{A}$. If at this point $\mathcal{A}$ outputs a state φ which can be efficiently distinguished from the correct $f_m(\rho_y)$ computed by the real $\mathcal{C}$, we would have an efficient distinguisher against the qIND-qCPA security of the scheme. Hence, $\mathcal{A}$'s (and therefore also $\mathcal{S}$'s) output must be indistinguishable from $f_m(\rho_y)$ for any QPT distinguisher, which concludes the proof. $\qquad\square$

Proof (of Proposition 5.6). This is also similar to the proof of Proposition 5.3. Given an efficient distinguisher $\mathcal{A}$ for the qIND-qCPA game, our adversary for the qSEM-qCPA game is an oracle machine $\mathcal{M}^{\mathcal{A}}$ running $\mathcal{A}$ and acting as follows. Concerning $\mathcal{A}$'s qCPA queries, as usual $\mathcal{M}^{\mathcal{A}}$ just forwards everything to the qSEM-qCPA challenger $\mathcal{C}$. When $\mathcal{A}$ performs a challenge qIND query by sending the classical descriptions of two states φ_0 and φ_1, $\mathcal{M}^{\mathcal{A}}$ prepares the qSEM template (G_m, h_m, f_m), with G_m outputing φ_0 for half of the possible y values and φ_1 for the other half, $h_m(\rho_y) = 1^n$, and f_m the identity map $f_m(\rho_y) = \rho_y$. Then $\mathcal{M}^{\mathcal{A}}$ performs a qSEM challenge query with this template. Given challenge ciphertext state $U_{\mathsf{Enc}_k} \varphi_b U_{\mathsf{Enc}_k}^{\dagger}$ (for $b \in \{0,1\}$), he forwards it as an answer to $\mathcal{A}$'s challenge query. As $\mathcal{A}$ distinguishes $U_{\mathsf{Enc}_k} \varphi_0 U_{\mathsf{Enc}_k}^{\dagger}$ from $U_{\mathsf{Enc}_k} \varphi_1 U_{\mathsf{Enc}_k}^{\dagger}$ with non-negligible success probability, $\mathcal{A}$ returns the correct value of b with non-negligible advantage over guessing. Then $\mathcal{M}^{\mathcal{A}}$, having recorded a copy of the classical descriptions of φ_0 and φ_1, is able to compute the state $f_m(\varphi_b)$ exactly, and consequently win the qSEM-qCPA game with non-negligible advantage. As h_m generates the same advice state $h_m(\rho_y) = 1^n$ independently of the encrypted message, no simulator can do better than guessing the plaintext. This concludes the proof. $\qquad\square$

Finally, we show the separation result between the two classes of security we have identified (we show it between IND-qCPA and qIND-qCPA). This shows that qIND-qCPA (and equivalently qSEM-qCPA) is a strictly stronger notion than IND-qCPA (which is equivalent to SEM-qCPA).

Theorem 5.7 (IND-qCPA $\not\Rightarrow$ qIND-qCPA). *There exists a symmetric-key encryption scheme $\mathcal{E}$ which is IND-qCPA secure but not qIND-qCPA secure.*

Proof (of Theorem 5.7). The scheme we use as a counterexample is the one from [10] (Construction 5.3.9). It has been proven in [4] that this scheme is IND-qCPA secure if the used PRF is post-quantum secure. We exhibit a distinguisher $\mathcal{A}$ which breaks the qIND-qCPA security of this scheme with high probability. For ease of notation we restrict to the case of single-bit messages 0 and 1. $\mathcal{A}$ will simply choose as challenge states: $|\varphi_0\rangle = H|0\rangle = \frac{1}{\sqrt{2}}|0\rangle + \frac{1}{\sqrt{2}}|1\rangle$, and $|\varphi_1\rangle = H|1\rangle = \frac{1}{\sqrt{2}}|0\rangle - \frac{1}{\sqrt{2}}|1\rangle$. When the challenger $\mathcal{C}$ applies the type-2 transformation to either of these two states, it is easy to see that in any case the state is left unchanged. This is because U_{Enc_k} just applies a permutation in the space of the basis elements, but $|\varphi_0\rangle$ and $|\varphi_1\rangle$ have the same amplitudes on all their components, except for the sign. As these two states are orthogonal, they can be reliably distinguished by the adversary $\mathcal{A}$ who can then win the qIND-qCPA game with probability 1.   □

The above proof can be generalized to message states of arbitrary length, as our impossibility result in Sect. 6.1 shows.

6   Impossibility and Achievability Results

In this section we show that qIND-qCPA (and equivalently qSEM-qCPA) is impossible to achieve for encryption schemes which do not expand the message (such as stream ciphers and many block ciphers, without considering the randomness part in the ciphertext). Therefore, for a scheme to be secure according to this new definition, it is necessary (but not sufficient) to increase the message size during the encryption. Interestingly, such an increase happens in most public-key post-quantum encryption schemes, like for example LWE based schemes [18] or the McEliece scheme [19].

Then we propose a construction of a qIND-qCPA–secure symmetric-key encryption scheme. Our construction works for any (quantum-secure) pseudo-random permutation (PRP). Given that block ciphers are usually modelled as PRPs, it seems reasonable to assume that we can obtain a secure scheme when using block ciphers with sufficiently large key and block size. Hence, our construction can be used to patch existing schemes, or as a guideline in the design of quantum-secure encryption schemes from block ciphers.

6.1   Impossibility Result

First we formally define what it means for a cipher to expand or keep constant the message size by defining the *core function* of a (secret-key) encryption scheme. Intuitively, the definition splits the ciphertext into the randomness and a part carrying the message-dependent information. This definition covers most encryption schemes in the literature.

Definition 6.1 (Core Function). *Let* (Gen, Enc, Dec) *be a secret-key encryption scheme. We call the function* $f : \mathcal{K} \times \{0,1\}^\tau \times \mathcal{M} \to \mathcal{Y}$ *the* core function *of the encryption scheme if, for some* $\tau \in \mathbb{N}$:

- *for all $k \in \mathcal{K}$ and $x \in \mathcal{M}$, $\mathsf{Enc}_k(x)$ can be written as $(r, f(k, r, x))$, where $r \in \{0, 1\}^\tau$ is independent of the message; and*
- *there exists a function f' such that for all $k \in \mathcal{K}, r \in \{0, 1\}^\tau, x \in \mathcal{M}$, we have: $f'(k, r, f(k, x, r)) = x$.*

For example, in case of Construction 5.3.9 from [10] (where $Enc_k(x)$ is defined as $(r, F_k(r) \oplus x)$ for a PRF F) the core function is $f(k, r, x) = F_k(r) \oplus x$, with $f'(k, r, z) = z \oplus F_k(r)$.

Definition 6.2 (Quasi–Length-Preserving Encryption). *We call a secret-key encryption scheme with core function f quasi–length-preserving if*

$$\forall x \in \mathcal{M}, r \in \{0, 1\}^\tau, k \in \mathcal{K} \Rightarrow |f(k, x, r)| = |x|,$$

i.e., if the output of the core function has the same bit length as the message.

Continuing the above example, Construction 5.3.9 from [10] is quasi–length-preserving.

The crucial observation is the following: For a quasi–length-preserving encryption scheme, the space of possible input and (core function) output bit-strings (with respect to plaintext and ciphertext) coincide, therefore these ciphers act as permutations on this space. This means that if we start with an input state which is a superposition of *all* the possible basis states, all of them with the *same* amplitude, this state will be unchanged by the unitary type-2 encryption operation (because it will just 'shuffle' in the basis-state space amplitudes which are exactly the same).

Theorem 6.3 (Impossibility Result). *No quasi–length-preserving secret-key encryption scheme can be qIND secure.*

Proof. Let $(\mathsf{Gen}, \mathsf{Enc}, \mathsf{Dec})$ be a quasi–length-preserving scheme. We show an attack that is a generalization of the distinguishing attack in Theorem 5.7.

1. for m-bit message strings, the distinguisher $\mathcal{D}$ sets the two plaintext states for the qIND- game to be: $|\varphi_0\rangle = H |0^m\rangle, |\varphi_1\rangle = H |1^m\rangle$, where H is the m-fold tensor Hadamard transformation. Notice that both these states admit efficient classical representations, and are thus allowed in the qIND game.
2. The challenger flips a random bit b and returns $|\psi\rangle = U_{\mathsf{Enc}_k} |\varphi_b\rangle$.
3. $\mathcal{D}$ applies H to the core-function part of the ciphertext $|\psi\rangle$ and measures it in the computational basis. $\mathcal{D}$ outputs 0 if and only if the outcome is 0^m, and outputs 1 otherwise.

As already observed, applying U_{Enc_k} to $H |0^m\rangle$ leaves the state untouched: since the encryption oracle merely performs a permutation in the basis space, and since $|\varphi_0\rangle$ is a superposition of every basis element with the same amplitude, it follows that whenever b is equal to 0, the ciphertext state will be unchanged. In this case, after applying the self-inverse transformation H again, $\mathcal{D}$ obtains measurement outcome 0^m with probability 1. On the other hand, if $b = 1$, $|\varphi_1\rangle =$

$\frac{1}{2^{m/2}} \sum_y (-1)^{y \cdot 1^m} |y\rangle$ where $a \cdot b$ denotes the bitwise inner product between a and b. Hence, $|\varphi_1\rangle$ is a superposition of every basis element where (depending on the parity of y) half of the elements have a positive amplitude and the other half have a negative one, but all of them will be equal in absolute value. Applying $U_{\mathsf{Enc},k}$ to this state, results in $\frac{1}{2^{m/2}} \sum_y (-1)^{y \cdot 1^m} |\mathsf{Enc}_k(y)\rangle$. After re-applying H, the amplitude of the basis state $|0^m\rangle$ becomes $\sum_y (-1)^{y \cdot 1^m + \mathsf{Enc}_k(y) \cdot 0^m}$ which is easily calculated to be 0. Hence, the above attack gives $\mathcal{D}$ a way of perfectly distinguishing between encryptions of the two plaintext states. $\square$

Notice that the above attack also works if $\mathcal{A}$ is allowed to send quantum states to $\mathcal{C}$ directly. Therefore, it also holds for the gqIND notion of quantum indistinguishability described in Sect. 3. In particular, the above theorem shows that [10, Construction 5.3.9], which in [4] was shown to be IND-qCPA if the used PRF is quantum secure, does not fulfill qIND, nor gqIND.

This attack is a consequence of the well-known fact that, in order to perfectly (information-theoretically) encrypt a single quantum bit, *two* bits of classical information are needed: one to hide the basis bit, and one to hide the phase (i.e. the signs of the amplitudes). The fact that we are restricted to quantum operations of the form U_{Enc_k} - that is, quantum instantiations of classical encryptions - means that we cannot afford to hide the phase as well, and this restriction allows for an easy distinguishing procedure.

6.2 Secure Construction

Here we propose a construction of a qIND-qCPA secure symmetric-key encryption scheme from any family of quantum-secure pseudorandom permutations (see the full version [9] for formal definitions).

Construction 6.4. *For security parameter n, let $m = poly(n)$ and $\tau = poly(n)$. Consider an efficient family of permutations $\Pi_{m+\tau} = (\mathcal{I}, \Pi, \Pi^{-1})$ with key space $\mathcal{K}_\Pi$ that operates on bit strings of length $m + \tau$, and consider a plaintext message space $\mathcal{M} = \{0,1\}^m$, key space $\mathcal{K} = \mathcal{K}_\Pi$, and ciphertext space $\mathcal{C} = \{0,1\}^{m+\tau}$. The construction is given by the following algorithms:*

Key generation algorithm $k \longleftarrow \mathsf{Gen}(1^n)$: *on input of security parameter n, the key generation algorithm runs $k \longleftarrow \mathcal{I}(1^{m+\tau})$ and returns secret key k.*

Encryption algorithm $y \longleftarrow \mathsf{Enc}_k(x)$: *on input of message $x \in \mathcal{M}$ and key $k \in \mathcal{K}$, the encryption algorithm samples a τ-bit string $r \xleftarrow{\$} \{0,1\}^\tau$ uniformly at random, and outputs $y = \pi_k(x\|r)$ ($\|$ denotes string concatenation).*

Decryption algorithm $x \longleftarrow \mathsf{Dec}_k(y)$: *on input of ciphertext $y \in \mathcal{C}$ and key $k \in \mathcal{K}$, the decryption algorithm first runs $x' = \pi_k^{-1}(y)$, and then returns the first m bits of x'.*

The soundness of the construction can be easily checked. The security is stated in the following theorem.

Theorem 6.5 (qIND-qCPA Security of Construction 6.4). *If $\Pi_{m+\tau}$ is a family of quantum-secure pseudorandom permutations (qPRP), then the encryption scheme* (Gen, Enc, Dec) *defined in Construction 6.4 is qIND-qCPA secure.*

In the next section, we prove the security of a more powerful scheme which includes the above theorem as special case of a single message block.

6.3 Length Extension

Construction 6.4 has the drawback that the message length is upper bounded by the input length of the qPRP (minus the bit length of the randomness). However, like in the case of block ciphers, we can overcome this issue with a *mode of operation*. More specifically, we can handle arbitrary message lengths by splitting the message into m-bit blocks and applying the encryption algorithm of Construction 6.4 independently to each message block (using the same key but new randomness for each block). This procedure is akin to a 'randomized ECB mode', in the sense that each message block is processed separately, like in the ECB (Electronic Code Book) mode, but in our case the underlying cipher is inherently randomized (since we use fresh randomness for each block), so we can still achieve qCPA security. For simplicity we consider only message lengths which are multiples of m. The construction can be generalized to arbitrary message lengths using standard padding techniques. Moreover, the randomness for every block can be generated efficiently using a random seed and a post-quantum secure PRNG.

Construction 6.6. *For security parameter n, let $m = poly(n)$ and $\tau = poly(n)$. Consider an efficient family of permutations $\Pi_{m+\tau} = (\mathcal{I}, \Pi, \Pi^{-1})$ with key space $\mathcal{K}_\Pi$ that operates on bit strings of length $m + \tau$, and consider a plaintext message space $\mathcal{M} = \{0,1\}^{\mu m}$ for $\mu \in \mathbb{N}, \mu = poly(n)$, key space $\mathcal{K} = \mathcal{K}_\Pi$, and ciphertext space $\mathcal{C} = \{0,1\}^{\mu(m+\tau)}$. The construction is given by the following algorithms:*

Key generation algorithm $k \longleftarrow$ Gen(1^n): *on input of security parameter n, the key generation algorithm runs $k \longleftarrow \mathcal{I}(1^{m+\tau})$ and returns secret key k.*

Encryption algorithm $y \longleftarrow$ Enc$_k(x)$: *on input of message $x \in \mathcal{M}$ and key $k \in \mathcal{K}$, the encryption algorithm splits x into μ m-bit blocks $x_1, \ldots, x_\mu$. For each block x_i, the encryption algorithm samples a new τ-bit string $r_i \xleftarrow{\$} \{0,1\}^\tau$ uniformly at random, and outputs $y_i = \pi_k(x_i \| r_i)$ ($\|$ denotes string concatenation). The ciphertext is $y = y_1 \| \ldots \| y_\mu$.*

Decryption algorithm $x \longleftarrow$ Dec$_k(y)$: *on input of ciphertext $y \in \mathcal{C}$ and key $k \in \mathcal{K}$, the decryption algorithm first splits y into μ $m+\tau$-bit blocks $y_1, \ldots, y_\mu$. Then, it runs $x_i' = (\pi_k^{-1}(y_i))_m$ for each block (where $(s)_m$ refers to taking the first m bits of bit string s). It returns the plaintext $x' = x_1', \ldots, x_\mu'$.*

The soundness of the construction can be checked easily. For the security, we observe that splitting a μm-qubit plaintext state into μ blocks of m-qubits can introduce entanglement between the blocks. We will address this issue through the following technical lemma.

Lemma 6.7. *Let $\mathcal{E}$ be the quantum channel that takes as input an arbitrary m-qubit state, attaches another τ qubits in state $|0\rangle$, and then applies a permutation picked uniformly at random from $S_{2^{m+\tau}}$ to the computational basis space. Let $\mathcal{T}$ be the constant channel which maps any m-qubit state to the totally mixed state on $m + \tau$ qubits. Then, $\|\mathcal{E} - \mathcal{T}\|_\diamond \leq 2^{-\tau+2}$.*

Proof. In order to consider the fact that the m-qubit input state might be entangled with something else, we have to start with a purification of such a state. Formally, this is a bipartite pure $2m$-qubit state $|\phi\rangle_{XY} = \sum_{x,y} \alpha_{x,y} |x\rangle_X |y\rangle_Y$ whose m-qubit Y register is input into the channel and gets transformed into $id_X \otimes \mathcal{E}(|\phi\rangle\langle\phi|) = \mathrm{tr}_\Pi |\psi\rangle\langle\psi|$ where

$$|\psi\rangle = \sum_{x\in\{0,1\}^m, y\in\{0,1\}^m, \pi\in S_{2^{m+\tau}}} \alpha_{x,y} |x\rangle_X |\pi(y\|0)\rangle_C |\pi\rangle_\Pi .$$

By definition of the diamond-norm, we have to show that for any $2m$-qubit state ρ, we have that $\|(id \otimes \mathcal{E})(\rho) - (id \otimes \mathcal{T})(\rho)\|_{\mathrm{tr}} \leq 2^{-\tau+2}$. Due to the convexity of the trace distance, we may assume that $\rho = |\phi\rangle\langle\phi|$ is pure with $|\phi\rangle_{XY} = \sum_{x,y} \alpha_{x,y} |x\rangle_X |y\rangle_Y$. Hence, we obtain

$$(id_X \otimes \mathcal{E})(|\phi\rangle\langle\phi|) = \mathrm{tr}_\Pi |\psi\rangle\langle\psi|$$

$$= \frac{1}{2^{m+\tau}!} \sum_{x,x',y,y',\pi} \alpha_{x,y}\overline{\alpha_{x',y'}} |x\rangle\langle x'|_X \otimes |\pi(y\|0)\rangle \langle\pi(y'\|0)|_C$$

$$= \frac{1}{2^{m+\tau}!} \sum_{x,x',y} \alpha_{x,y}\overline{\alpha_{x',y}} |x\rangle\langle x'|_X \otimes \sum_\pi |\pi(y\|0)\rangle \langle\pi(y\|0)|_C$$

$$+ \frac{1}{2^{m+\tau}!} \sum_{x,x',y\neq y'} \alpha_{x,y}\overline{\alpha_{x',y'}} |x\rangle\langle x'|_X \otimes \sum_\pi |\pi(y\|0)\rangle \langle\pi(y'\|0)|_C$$

$$= \sum_{x,x',y} \alpha_{x,y}\overline{\alpha_{x',y}} |x\rangle\langle x'|_X \otimes \frac{1}{2^{m+\tau}} \sum_z |z\rangle\langle z|_C$$

$$+ \sum_{x,x',y\neq y'} \alpha_{x,y}\overline{\alpha_{x',y'}} |x\rangle\langle x'|_X \otimes \frac{1}{2^{m+\tau}(2^{m+\tau}-1)} \sum_{z\neq z'} |z\rangle\langle z'|_C$$

$$= \mathrm{tr}_Y |\phi\rangle\langle\phi| \otimes \tau_C + \chi_{XC}$$

$$= (id_X \otimes \mathcal{T})(|\phi\rangle\langle\phi|) + \chi_{XC},$$

where we defined the "difference state"

$$\chi_{XC} := \sum_{x,x',y\neq y'} \alpha_{x,y}\overline{\alpha_{x',y'}} |x\rangle\langle x'|_X \otimes \frac{1}{2^{m+\tau}(2^{m+\tau}-1)} \sum_{z\neq z'} |z\rangle\langle z'|_C.$$

In order to conclude, it remains to show that $\|\chi_{XC}\|_{\mathrm{tr}} \leq 2^{-\tau+2}$. For the C-register $\chi_C = \frac{1}{2^{m+\tau}(2^{m+\tau}-1)} \sum_{z\neq z'} |z\rangle\langle z'|_C$, one can verify that the $2^{m+\tau}$ eigenvalues are $(c \cdot (2^{m+\tau} - 1), -c, -c, \ldots, -c)$ where $c := \frac{1}{2^{m+\tau}(2^{m+\tau}-1)}$. Hence, the trace norm (which is the sum of the absolute eigenvalues) is exactly $c \cdot 2(2^{m+\tau} - 1) = 2^{-m-\tau+1}$.

For the X-register, we split χ_X into two parts $\chi_X = \xi_X - \xi'_X$ where

$$\xi_X := \sum_{x,x'} |x\rangle\langle x'| \sum_{y,y'} \alpha_{x,y} \overline{\alpha_{x',y'}},$$

$$\xi'_X := \sum_{x,x'} |x\rangle\langle x'| \sum_{y} \alpha_{x,y} \overline{\alpha_{x',y}},$$

and use the triangle inequality for the trace norm $\|\chi_X\|_{\mathrm{tr}} = \|\xi_X - \xi'_X\|_{\mathrm{tr}} \leq \|\xi_X\|_{\mathrm{tr}} + \|\xi'_X\|_{\mathrm{tr}}$. Observe that $\|\xi_X\|_{\mathrm{tr}} = \|\sum_{x,y} \alpha_{x,y} |x\rangle \sum_{x',y'} \overline{\alpha_{x',y'}} \langle x'| \|_{\mathrm{tr}} = \||s\rangle\langle s|\|_{\mathrm{tr}}$ for the (non-normalized) vector $|s\rangle := \sum_{x,y} \alpha_{x,y} |x\rangle$. Hence, the trace-norm $\|\xi_X\|_{\mathrm{tr}} = |\langle s \mid s \rangle| = \sum_x |\sum_y \alpha_{x,y}|^2 \leq \sum_x \sum_y |\alpha_{x,y}|^2 \cdot 2^m = 2^m$ by the Cauchy-Schwarz inequality and the normalization of the $\alpha_{x,y}$'s. Furthermore, we note that ξ'_X is exactly the reduced density matrix of $|\phi\rangle_{XY}$ after tracing out the Y register. Hence, ξ'_X is positive semi-definite and its trace norm is equal to its trace which is 1. In summary, we have shown that

$$\|\chi_{XC}\|_{\mathrm{tr}} = \|\chi_X\|_{\mathrm{tr}} \cdot \|\chi_C\|_{\mathrm{tr}} \leq (\|\xi_X - \xi'_X\|_{\mathrm{tr}}) \cdot 2^{-m-\tau+1}$$
$$\leq (\|\xi_X\|_{\mathrm{tr}} + \|\xi'_X\|_{\mathrm{tr}}) \cdot 2^{-m-\tau+1} \leq (2^m + 1) \cdot 2^{-m-\tau+1} \leq 2^{-\tau+2}.$$

$\square$

If we consider a slightly different encryption channel $\mathcal{E}^T$ which still maps m qubits to $m + \tau$ qubits but where the permutation π is not picked uniformly from $S_{2^{m+\tau}}$, but instead we are guaranteed that a certain set $T \subset \{0,1\}^{m+\tau}$ of outputs never occurs, we can consider such permutations w.l.o.g. as picked uniformly at random from a smaller set $S_{2^{m+\tau}-|T|}$. In this setting, we are interested in the distance of the encryption operation $\mathcal{E}^T$ from the slightly different constant channel $\mathcal{T}^T$ which maps all inputs to the $(m + \tau)$-qubit state which is completely mixed on the smaller set $\{0,1\}^{m+\tau} \setminus T$. By modifying slightly the proof of Lemma 6.7 we get the following.

Corollary 6.8. *Let $\mathcal{E}^T$ and $\mathcal{T}^T$ be the channels defined above. Then,*

$$\|\mathcal{E}^T - \mathcal{T}^T\|_\diamond \leq \frac{4}{2^\tau - |T|/2^m}. \tag{3}$$

We can now prove the security of Construction 6.6. We give the proof for gqIND-qCPA, and then qIND-qCPA follows immediately from Theorem 3.3.

Theorem 6.9 (gqIND-qCPA Security of Construction 6.6**).** *If $\Pi_{m+\tau}$ is a family of quantum-secure pseudorandom permutations (qPRP), then the encryption scheme* (Gen, Enc, Dec) *defined in Construction 6.6 is gqIND-qCPA secure.*

Proof. We want to show that no QPT distinguisher $\mathcal{D}$ can win the gqIND-qCPA game with probability substantially better than guessing. We first transform the game through a short game-hopping sequence into an indistinguishable game for which we can bound the success probability of any such $\mathcal{D}$.

Game 0. This is the original gqIND-qCPA game.

Game 1. This is like Game 0, but instead of using a permutation drawn from the qPRP family $\Pi_{m+\tau}$, a random permutation $\pi \in S_{2^{m+\tau}}$ is chosen from the set of all permutations over $\{0,1\}^{m+\tau}$. The difference in the success probability of $\mathcal{D}$ winning one or the other of these two games is negligible. Otherwise, we could use $\mathcal{D}$ to distinguish a random permutation drawn from $\Pi_{m+\tau}$ from one drawn from $S_{2^{m+\tau}}$. This would contradict the assumption that $\Pi_{m+\tau}$ is a qPRP.

Game 2. This is like Game 1, but $\mathcal{D}$ is guaranteed that the randomness used for each encryption query are μ new random τ-bit strings that were not used before. In other words, the challenger keeps track of all random values used so far and excludes those when sampling a new randomness. Since in Game 1 the same randomness is sampled twice only with negligible probability, the probability of winning these two games differs by at most a negligible amount.

Game 3. This is like Game 2 except that the answer to each query asked by $\mathcal{D}$ also contains the randomness $r_1, \ldots, r_\mu$ used by the challenger for answering that query. Clearly, $\mathcal{D}$'s probability of winning this game is at least the probability of winning Game 2.

When the modified gqIND game 3 starts, $\mathcal{D}$ chooses two different plaintext states and sends them to the challenger, who will then choose one of them and send it back encrypted with fresh randomness $\hat{r}_1, \ldots, \hat{r}_\mu$. Let Q denote the set of $q \cdot \mu = poly(n)$ query values used during the previous qCPA-phase. We have to consider that from this phase, $\mathcal{D}$ knows a set $T \subset \{0,1\}^{m+\tau}$ of 'taken' outputs, i.e. he knows that any $\pi(x \| \hat{r}_i)$ will not take one of these values as $\hat{r}_i$ has not been used before. So, from the adversary's point of view, π is a permutation randomly chosen from S', the set of those permutations over $\{0,1\}^{m+\tau}$ that fix these $|T|$ values. In order to simplify the proof, we will consider a very conservative bound where $|T| = q \cdot \mu \cdot 2^m$, and the size of S' is $|S'| = (2^{m+\tau} - |T|)!$ (notice that this bound is very conservative because it assumes that the adversary learns 2^m different (classical) ciphertexts for every of the $q \cdot \mu$ 'taken' randomnesses, but as we will see, this knowledge will be still insufficient to win the game).

By construction, the encryption of a μm-qubit (possibly mixed) state σ is performed in μ separate blocks of m qubits each. We are guaranteed that fresh randomness is used in each block, hence it follows from Corollary 6.8 that $\mathsf{Enc}_k(\sigma)$ is negligibly close to the ciphertext state where the first $m+\tau$ qubits are replaced with the completely mixed state (by noting that $|T|/2^m = q \cdot \mu$ is polynomial in n in our case, and hence the right-hand side of (3) is negligible). Another application of Corollary 6.8 gives negligible closeness to the ciphertext state where the first $2(m+\tau)$ qubits are replaced with the completely mixed state etc. After μ applications of Corollary 6.8, we have shown that $\mathsf{Enc}_k(\sigma)$ is negligibly close to the totally mixed state on $\mu(m+\tau)$ qubits. As this argument can be made for any cleartext state σ, we have shown that from $\mathcal{D}$'s point of view, all encrypted states are negligibly close to the totally mixed state and therefore cannot be distinguished. $\qquad\square$

Corollary 6.10 (qIND-qCPA Security of Construction 6.6). *If $\Pi_{m+\tau}$ is a family of quantum-secure pseudorandom permutations (qPRP), then the encryption scheme* (Gen, Enc, Dec) *defined in Construction 6.6 is qIND-qCPA secure.*

7   Conclusions and Further Directions

We believe that many of the current security notions used in different areas of cryptography are unsatisfying in case quantum computers become reality. In this respect, our work contributes to a better understanding of which properties are important for the long-term security of modern cryptographic primitives. Our work leads to many interesting follow-up questions.

There are many other directions to investigate, once the basic framework of 'indistinguishability versus semantic security' presented in this work is completed. A natural direction is to look at quantum CCA1 security in this framework. This topic was also initiated in [4] relative to the IND-qCPA model; it would be interesting to extend the definition of CCA1 security to stronger notions obtained by starting from our qIND-qCPA model.

In Sect. 3.3 we left open the interesting question on whether it is possible at all to find a separating example between the notions of qIND and gqIND. That is, find a symmetric-key encryption scheme $\mathcal{E}$ which is qIND-secure, but not gqIND-secure. Finding such an example (or provable lack of) would shed further light on the security model we consider.

We have so far not taken into account models where the adversary is allowed to initialize the ancilla qubits used in the encryption operation used by the challenger (i.e. the $|y\rangle$ in $|x, y\rangle \mapsto |x, y \oplus \mathsf{Enc}_k(x)\rangle$). These models lead to the study of *quantum fault attacks*, because they model cases where the adversary is able to 'watermark' or tamper with part of the challenger's internal memory. Moreover, we have not considered superpositions of keys or randomness: these lead to a quantum study of *weak-key* and *bad-randomness* models. The authors of this paper are not aware of any results in these directions.

One outstanding open problem is to define CCA2 (adaptive chosen ciphertext attack) security in the quantum world. The problem is that in the CCA2 game the challenger has to ensure that the attacker does not ask for a decryption of the actual challenge ciphertext leading to a trivial break. While this is easily implemented in the classical world, it raises several issues in the quantum world. What does it mean for a ciphertext to be different from the challenge ciphertext? And, more importantly: *How can the challenger check?* There might be several reasonable ways to solve the first issue but, as long as the queries are not classical, we are not aware of any possibility to solve the second issue without disturbing the challenge ciphertext and the query states.

Our secure construction shows how to turn block ciphers into qIND-qCPA secure schemes. An interesting research question is whether there exists a general patch transforming an IND-qCPA secure scheme into a qIND-qCPA secure one. It is also important to study how our transformation can be applied to modes of operation different from Construction 6.6.

Acknowledgements. The authors would like to thank Ronald de Wolf and Boris Škorić for helpful discussions, the anonymous reviewers for useful comments, and the organizers of Dagstuhl Seminar 15371 "Quantum Cryptanalysis" for networking, useful interactions, and support. T.G. was supported by the German Federal Ministry of Education and Research (BMBF) within EC SPRIDE and CROSSING. A.H. was supported by the Netherlands Organisation for Scientific Research (NWO) under grant 639.073.005 and the Commission of the European Communities through the Horizon 2020 program under project number 645622 PQCRYPTO. C.S. was supported by a 7th framework EU SIQS grant and a NWO VIDI grant. Part of this work was supported by the COST Action IC1306.

References

1. Alagic, G., Broadbent, A., Fefferman, B., Gagliardoni, T., Schaffner, C., St. Jules, M.: Computational security of quantum encryption (2016). http://arXiv.org/abs/1602.01441
2. Bernstein, D.J., Buchmann, J., Dahmen, E. (eds.): Post-Quantum Cryptography. Springer, Heidelberg (2009)
3. Boneh, D., Dagdelen, Ö., Fischlin, M., Lehmann, A., Schaffner, C., Zhandry, M.: Random Oracles in a quantum world. In: Lee, D.H., Wang, X. (eds.) ASIACRYPT 2011. LNCS, vol. 7073, pp. 41–69. Springer, Heidelberg (2011)
4. Boneh, D., Zhandry, M.: Secure signatures and chosen ciphertext security in a quantum computing world. In: Canetti, R., Garay, J.A. (eds.) CRYPTO 2013, Part II. LNCS, vol. 8043, pp. 361–379. Springer, Heidelberg (2013)
5. Brassard, G., Hoyer, P., Tapp, A.: Quantum algorithm for the collision problem (1997). arXiv:quant-ph/9705002
6. Broadbent, A., Jeffery, S.: Quantum homomorphic encryption for circuits of low T-gate complexity. In: Gennaro, R., Robshaw, M. (eds.) CRYPTO 2015. LNCS, vol. 9216, pp. 609–629. Springer, Heidelberg (2015)
7. Dagdelen, Ö., Fischlin, M., Gagliardoni, T.: The Fiat-Shamir transformation in a quantum world. In: ASIACRYPT 2013, Part II, pp. 62–81 (2013)
8. Damgård, I., Funder, J., Nielsen, J.B., Salvail, L.: Superposition attacks on cryptographic protocols. In: Padró, C. (ed.) ICITS 2013. LNCS, vol. 8317, pp. 146–165. Springer, Heidelberg (2014)
9. Gagliardoni, T., Hülsing, A., Schaffner, C.: Semantic security and indistinguishability in the quantum world. Cryptology ePrint Archive, Report 2015/355 (2015)
10. Goldreich, O.: Foundations of Cryptography. Basic Applications, vol. 2. Cambridge University Press, Cambridge (2004)
11. Goldwasser, S., Micali, S.: Probabilistic encryption. J. Comput. Syst. Sci. **28**(2), 270–299 (1984)
12. Grover, L.K.: A fast quantum mechanical algorithm for database search. In: STOC 1996, pp. 212–219. ACM Press (1996)
13. Kaplan, M., Leurent, G., Leverrier, A., Naya-Plasencia, M.: Breaking symmetric cryptosystems using quantum period finding. In: CRYPTO 2016 (2016, to appear)
14. Kashefi, E., Kent, A., Vedral, V., Banaszek, K.: Comparison of quantum oracles. Phys. Rev. A **65**(5), 050304 (2002)
15. Koshiba, T.: Security notions for quantum public-key cryptography. IEICE Trans. Fundam. Electron. Commun. Comput. Sci. **J90-A**(5), 367–375 (2007)
16. Kuwakado, H., Morii, M.: Quantum distinguisher between the 3-round Feistel cipher and the random permutation. In: ISIT 2010, pp. 2682–2685 (2010)

17. Kuwakado, H., Morii, M.: Security on the quantum-type Even-Mansour cipher. In: ISITA 2012, pp. 312–316 (2012)
18. Lindner, R., Peikert, C.: Better key sizes (and attacks) for LWE-based encryption. In: Kiayias, A. (ed.) CT-RSA 2011. LNCS, vol. 6558, pp. 319–339. Springer, Heidelberg (2011)
19. McEliece, R.J.: A public-key cryptosystem based on algebraic coding theory. DSN Prog. Rep. **42**(44), 114–116 (1978)
20. Nielsen, M., Chuang, I.: Quantum Computation and Quantum Information. Cambridge University Press, Cambridge (2000)
21. Santoli, T., Schaffner, C.: Using Simon's algorithm to attack symmetric-key cryptographic primitives (2016)
22. Shor, P.W.: Algorithms for quantum computation: discrete logarithms and factoring. In: FOCS 1994, pp. 124–134. IEEE Computer Society Press (1994)
23. Unruh, D.: Quantum proofs of knowledge. In: Pointcheval, D., Johansson, T. (eds.) EUROCRYPT 2012. LNCS, vol. 7237, pp. 135–152. Springer, Heidelberg (2012)
24. Velema, M.: Classical encryption and authentication under quantum attacks. Master's thesis, Master of Logic, University of Amsterdam (2013)
25. Watrous, J.: Quantum algorithms for solvable groups. In: STOC 2001, pp. 60–67. ACM Press (2001)
26. Watrous, J.: Zero-knowledge against quantum attacks. SIAM J. Comput. **39**(1), 25–58 (2009)
27. Xiang, C., Yang, L.: Indistinguishability and semantic security for quantum encryption scheme. In: Proceeding of the SPIE, vol. 8554, pp. 85540G–85540G-8 (2012)
28. Zhandry, M.: How to construct quantum random functions. In: FOCS 2012, pp. 679–687. IEEE (2012)

Spooky Encryption

Spooky Encryption and Its Applications

Yevgeniy Dodis[1]([⊠]), Shai Halevi[2]([⊠]), Ron D. Rothblum[3]([⊠]),
and Daniel Wichs[4]([⊠])

[1] NYU, New York, NY, USA
dodis@cs.nyu.edu
[2] IBM Research, Yorktown Heights, NY, USA
shaih@alum.mit.edu
[3] MIT, Cambridge, MA, USA
rothblum@gmail.com
[4] Northeastern University, Boston, MA, USA
danwichs@gmail.com

Abstract. Consider encrypting n inputs under n independent public keys. Given the ciphertexts $\{c_i = \mathsf{Enc}_{\mathsf{pk}_i}(x_i)\}_i$, Alice outputs ciphertexts $c'_1, \ldots, c'_n$ that decrypt to $y_1, \ldots, y_n$ respectively. What relationships between the x_i's and y_i's can Alice induce?

Motivated by applications to delegating computations, Dwork et al. [11] showed that a semantically secure scheme disallows *signaling* in this setting, meaning that y_i cannot depend on x_j for $j \neq i$. On the other hand if the scheme is homomorphic then any *local* (component-wise) relationship is achievable, meaning that each y_i can be an arbitrary function of x_i. However, there are also relationships which are neither signaling nor local. Dwork et al. asked if it is possible to have encryption schemes that support such "spooky" relationships. Answering this question is the focus of our work.

Our first result shows that, under the LWE assumption, there exist encryption schemes supporting a large class of "spooky" relationships, which we call *additive function sharing* (AFS) spooky. In particular, for any polynomial-time function f, Alice can ensure that $y_1, \ldots, y_n$ are random subject to $\sum_{i=1}^n y_i = f(x_1, \ldots, x_n)$. For this result, the public keys all depend on common public randomness. Our second result shows that, assuming sub-exponentially hard indistinguishability obfuscation (iO) (and additional more standard assumptions), we can remove the common randomness and choose the public keys completely independently. Furthermore, in the case of $n = 2$ inputs, we get a scheme that supports an even larger class of spooky relationships.

We discuss several implications of AFS-spooky encryption. Firstly, it gives a strong counter-example to a method proposed by Aiello et al. [1] for building arguments for NP from homomorphic encryption. Secondly, it gives a simple 2-round multi-party computation protocol where, at the end of the first round, the parties can locally compute an additive secret sharing of the output. Lastly, it immediately yields a function secret sharing (FSS) scheme for all functions.

We also define a notion of *spooky-free* encryption, which ensures that no spooky relationship is achievable. We show that any non-malleable encryption scheme is spooky-free. Furthermore, we can construct spooky-free *homomorphic* encryption schemes from SNARKs, and

M. Robshaw and J. Katz (Eds.): CRYPTO 2016, Part III, LNCS 9816, pp. 93–122, 2016.
DOI: 10.1007/978-3-662-53015-3_4

it remains an open problem whether it is possible to do so from falsifiable
assumptions.

1 Introduction

Imagine Alice and Bob, standing on different planets light years apart. They
are "simultaneously" given some input bits x_1 and x_2 respectively, and must
answer by outputting bits y_1 and y_2 respectively. Classical physics allows them to
implement *local* (component-wise) strategies where y_1 is an arbitrary function of
x_1 and y_2 is a function of x_2. On the other hand, the impossibility of faster-than-
light communication disallows *signaling* strategies, meaning that the distribution
of y_1 cannot depend on the value of x_2 and vice versa.

However, there are strategies that are neither local nor signaling. For example, perhaps Alice and Bob want to ensure that y_1, y_2 are random bits subject
to $y_1 \oplus y_2 = x_1 \wedge x_2$. In this case, the distribution of y_1 does not depend on x_2
(and vice versa) so the strategy is not signaling, but it's also not local. Surprisingly some such strategies which are neither signaling nor local are achievable
using quantum mechanics, if Alice and Bob share an entangled quantum state.
Einstein referred to this phenomenon as "spooky action at a distance".

In this work, we consider an analogous scenario, first considered by Dwork
et al. [11], where the separation between x_1, x_2 is enforced not via physical
distance but by encrypting these bits under two independent public keys.[1] Here
Alice gets the two ciphertexts $c_1 \leftarrow \mathsf{Enc}_{\mathsf{pk}_1}(x_1), c_2 \leftarrow \mathsf{Enc}_{\mathsf{pk}_2}(x_2)$, and outputs
two other ciphertexts c_1', c_2' which are decrypted as $y_i \leftarrow \mathsf{Dec}_{\mathsf{sk}_i}(c_i'), i = 1, 2$.
As in the physical analogy, here too we can rule out signaling strategies (if the
encryption is semantically secure), and can implement local strategies (if the
encryption is homomorphic). But can we replace the entangled state from above
by a special "spooky encryption scheme" that would allow Alice to implement
spooky strategies? Answering this question is the focus of this work, and we
obtain the following results:

- Assuming the hardness of learning with errors (LWE), there exists a secure
  encryption scheme in which Alice can implement a wide class of spooky strategies that we call *additive function sharing* (AFS) spooky. Namely, for any
  two-argument function $f : (\{0,1\}^*)^2 \rightarrow \{0,1\}$, Alice can convert encryption
  of inputs $c_i \leftarrow \mathsf{Enc}_{\mathsf{pk}_i}(x_i)$ to encryption of outputs $y_i \leftarrow \mathsf{Dec}_{\mathsf{sk}_i}(c_i')$, ensuring
  that $y_1 \oplus y_2 = f(x_1, x_2)$, except for a small error probability.
  This construction, described in Sect. 3, uses the LWE-based multi-key FHE
  schemes from [7,22,26], and it inherits from these multi-key scheme their
  dependence on a common random string.
- In Sect. 4 we describe a spooky scheme that supports arbitrary two-input
  spooky relations on short inputs, as well as a very wide class of two-input
  spooky relations on long inputs. This construction uses probabilistic indistinguishability obfuscation (piO), which is an extension of iO to probabilistic

---

[1] Dwork *et al.* considered PIR rather than encryption, but the translation is immediate.

circuits recently introduced by Canetti *et al.* [6], in conjunction with lossy encryption schemes which are homomorphic and ensure circuit privacy against malicious adversaries. This construction works in the plain model without common-random string and has no error, and it can be realized based on exponentially strong iO, exponentially strong PRFs, and DDH.

– In Sect. 5 we describe a transformation from a scheme that supports only two-input spooky relations on one-bit inputs to one that supports AFS spooky relations on arbitrary number of inputs (of arbitrarily length each). This transformation can be applied to both our LWE-based and piO-based constructions from above.

– We show several implications of (AFS-)spooky encryption. On a negative, it gives a strong counter-example to a method proposed by Aiello et al. [1] for building succinct arguments for NP from homomorphic encryption[2], resolving a question posted by [11]. On a positive, it immediately yields a function secret sharing (FSS) scheme for all functions [4,15], and also gives a simple 2-round multi-party computation protocol where, at the end of the first round, the parties can locally compute an additive secret sharing of the output. These application are discussed in Sect. 6.

– We also study in Sect. 7 the concept of *spooky free* encryption, i.e., an encryption scheme where we can prove that no spooky strategy is feasible. We show that any non-malleable encryption scheme is spooky-free, and also build spooky-free *homomorphic* encryption schemes from SNARKs. It remains an open problem to construct spooky-free homomorphic encryption under more standard assumptions. Spooky-free homomorphic encryption can be used to instantiate the approach of Aiello et al. to get succinct arguments for NP.

1.1 Technical Overview

LWE-Based Construction. Our LWE-based construction builds on the multi-key FHE schemes from [7,22,26]. In these schemes (after some syntactic massaging) secret keys and single-key ciphertexts are vectors in $\mathbb{Z}_q^n$, and decryption consists of computing $w = \langle s, c \rangle \bmod q$, then rounding to the nearest multiple of $q/2$, outputting zero if w is closer to 0 or one if w is closer to $q/2$.

These schemes, however, also support homomorphic computation across ciphertexts relative to different keys. Roughly, they feature a "lifting procedure" where a dimension-n ciphertext vector relative to one key s_i is "lifted" to a dimension ℓn vector $c' = (c'_1, \ldots, c'_\ell)$ relative to the concatenated key $s' = (s_1, \ldots, s_\ell)$ of dimension ℓn. These lifted ciphertexts can still be computed on, and the decryption procedure proceeds just as before, except using the higher-dimension vectors. Namely, to decrypt c' using s', one first computes the inner product $w' = \langle s', c' \rangle$ modulo q, then rounds to the nearest multiple of $q/2$. In other words, we compute the individual inner products $w_i = \langle s_i, c'_i \rangle$, then add them all up and round to the nearest multiple of $q/2$.

---

[2] Although included in the ICALP conference proceedings, the article [1] was withdrawn before the conference and was not presented there.

We observe (cf. Lemma 1) that for the special case of two keys, $\ell = 2$, instead of adding the w_i's and then rounding, we can first round each w_i to the nearest multiple of $q/2$ and then add, and this yields the same result with high probability. Specifically, the error probability is proportional to the rounding error for the overall sum w'. This observation immediately yields additive function sharing (AFS) spooky encryption for two-argument functions: We use one of the schemes from [7,22,26] to encrypt the two arguments x_1, x_2 under two keys, then use the multi-key evaluation procedure to compute a multi-key ciphertext $c' = (c'_1, c'_2)$ encrypting the value $f(x_1, x_2)$. Viewing each c'_i as a single-key ciphertext, we apply the usual decryption procedure to each of them, and the resulting two bits are an additive secret sharing of $f(x_1, x_2)$, except with a small error probability. The error probability can be made negligible by relying on LWE with a super-polynomial approximation factor.

piO-Based Construction. In Sect. 4 we show that using iO we can construct an AFS encryption scheme without CRS and without errors, and moreover we can support *arbitrary spooky relations* on two bits, not just additive sharing. For this overview, let us focus on the simpler task of constructing AFS spooky scheme for the multiplication function $\mathsf{MULT}(b_1, b_2) = b_1 \cdot b_2$.

The starting point of the construction takes a homomorphic encryption scheme $(\mathsf{Gen}, \mathsf{Enc}, \mathsf{Dec}, \mathsf{Eval})$ and adds to the public key an obfuscation of the randomized functionality that decrypts, computes the functions f, and re-encrypts secret-sharing of the result. Specifically, let us denote for any $x_1, y_1 \in \{0, 1\}$ the function $f_{x_1,y_1}(x_2) = x_1 \cdot x_2 \oplus y_1$, and consider the following randomized program:

Program $P_{\mathsf{sk}_1,\mathsf{pk}_1}(c_1, \mathsf{pk}_2, c_2)$	
1. $y_1 \leftarrow \{0, 1\}$.	4. $c'_2 = \mathsf{Eval}(\mathsf{pk}_2, f_{x_1,y_1}, c_2)$.
2. $c'_1 \leftarrow \mathsf{Enc}_{\mathsf{pk}_1}(y_1)$.	5. Output (c'_1, c'_2).
3. $x_1 = \mathsf{Dec}_{\mathsf{sk}_1}(c_1)$.	

Given the two pairs $(\mathsf{pk}_1, \mathsf{Enc}_{\mathsf{pk}_1}(x_1))$, $(\mathsf{pk}_2, \mathsf{Enc}_{\mathsf{pk}_2}(x_2))$, and access to the program $P_{\mathsf{sk}_1,\mathsf{pk}_1}$, we can run $P_{\mathsf{sk}_1,\mathsf{pk}_1}(c_1, \mathsf{pk}_2, c_2)$ to get two ciphertexts c'_1 and c'_2, encrypting y_1, y_2, respectively, such that $y_1 \oplus y_2 = x_1 \cdot x_2$. We would like, therefore, to add an obfuscation of $P_{\mathsf{sk}_1,\mathsf{pk}_1}$ to the public key, thereby obtaining AFS spooky multiplication.

As described, however, this construction is not even secure when $P_{\mathsf{sk}_1,\mathsf{pk}_1}(c_1, \mathsf{pk}_2, c_2)$ is only accessed by a perfect black box. The reason is that if the underlying homomorphic encryption is not *circuit private*, then the evaluated ciphertext c'_2 could leak information about x_1. To fix this issue, we require the use of circuit-private homomorphic encryption in this construction. In fact, since the adversary could run the program $P_{\mathsf{sk}_1,\mathsf{pk}_1}(c_1, \mathsf{pk}_2, c_2)$ on arbitrary inputs of its choice, we need a stronger notion of *circuit privacy against malicious adversaries* [24], that guarantees privacy even if the public-key and ciphertext given to the evaluation algorithm are generated adversarially.

Using a malicious circuit private homomorphic encryption scheme, the construction above would be secure if the program $P_{\mathsf{sk}_1,\mathsf{pk}_1}(c_1, \mathsf{pk}_2, c_2)$ is accessed as a perfect black box (e.g., using VBB obfuscation). However, we would like to rely on the weaker notion of indistinguishability obfuscation (iO), or rather probabilistic iO [6] (since we are dealing with a randomized program). We need to somehow argue that the secret key sk_1 that is encoded within the program $P_{\mathsf{sk}_1, \mathsf{pk}_1}$ is hidden by the weaker obfuscation, and we do it using a technique from the work of Canetti *et al.* [6], employing a *lossy* encryption scheme.

We note that the construction above only uses homomorphic computations for single-bit functions (in addition to probabilistic iO), and there are only four such function (identity, negation, constant 0 and constant 1). A secure and malicious-circuit-private encryption scheme that supports these operations was constructed by Naor and Pinkas [23] based on the DDH assumption.

From 2-Spooky to n-Spooky. Both the LWE and piO based constructions above only support two-argument spooky relations. Specifically the LWE-based scheme only supports AFS-spooky relations for two-argument functions, and the piO-based scheme supports a large class of spooky relations but again, only on two inputs. We extend the supported spooky relations by showing how to transform a scheme that supports (multiple hops of) AFS-spooky two-input multiplication and single-key additive homomorphism, into a leveled AFS spooky scheme for any number of inputs of any length.

The transformation is inspired by the Goldreich-Micali-Wigderson MPC protocol [16]: Suppose that we are given n public keys $\mathsf{pk}_1, \ldots, \mathsf{pk}_n$, bit-by-bit encryptions of the input values $\mathsf{Enc}_{\mathsf{pk}_i}(x_i)$, and an arithmetic circuit C : $(\{0,1\}^*)^n \to \{0,1\}$ that we want to evaluate (i.e., to produce encrypted shares of $C(x_1, \ldots, x_n)$). We process the circuit gate by gate, while maintaining the invariant that for every wire w we produce ciphertexts $\mathsf{Enc}_{\mathsf{pk}_1}(w_1), \ldots, \mathsf{Enc}_{\mathsf{pk}_n}(w_n)$ such that $\oplus_{i \in [n]} w_i$ is equal to the wire w's value. The wires are processed inductively:

1. For an *input* wire holding a bit b, which is part of the j'th input x_j, we take the ciphertext c that encrypts b relative to pk_j, and append to it the ciphertexts $c_i \leftarrow \mathsf{Enc}_{\mathsf{pk}_i}(0)$ for all $i \neq j$. Clearly the ciphertexts $(c_1, \ldots, c_{j-1}, c, c_{j+1}, \ldots, c_n)$ are encryptions of an additive sharing of the wire's value b.

2. For an addition gate with input wires u, v and output wire w, by induction we already have $\mathsf{Enc}_{\mathsf{pk}_1}(u_1), \ldots, \mathsf{Enc}_{\mathsf{pk}_n}(u_n)$ and $\mathsf{Enc}_{\mathsf{pk}_1}(v_1), \ldots, \mathsf{Enc}_{\mathsf{pk}_n}(v_n)$. Using just an additive homomorphism on each key individually, we can produce $\mathsf{Enc}_{\mathsf{pk}_1}(u_1 \oplus v_1), \ldots, \mathsf{Enc}_{\mathsf{pk}_n}(u_n \oplus v_n)$ which is the desired secret sharing.

3. For a multiplication gate with input wires u, v and output wire w, again by induction we already have $\mathsf{Enc}_{\mathsf{pk}_1}(u_1), \ldots, \mathsf{Enc}_{\mathsf{pk}_n}(u_n)$ and $\mathsf{Enc}_{\mathsf{pk}_1}(v_1), \ldots,$ $\mathsf{Enc}_{\mathsf{pk}_n}(v_n)$. Using the AFS spooky multiplication we compute an encrypted *tensor product* of the $\boldsymbol{u}$ and $\boldsymbol{v}$ vectors. Namely, for every i, j we use spooky multiplication to compute

$$\left(\mathsf{Enc}_{\mathsf{pk}_i}(x_{i,j}), \mathsf{Enc}_{\mathsf{pk}_j}(y_{i,j}) \right) \leftarrow \mathsf{SpookyMult}\left(\mathsf{Enc}_{\mathsf{pk}_i}(u_i), \mathsf{Enc}_{\mathsf{pk}_j}(u_j) \right),$$

such that $x_{i,j} \oplus y_{i,j} = u_i \cdot v_j$. Then we collapse this tensor product back into an n-vector using the additive homomorphism relative to each key separately.

That is, for every $i \in [n]$ we can compute a ciphertext $\mathsf{Enc}_{\mathsf{pk}_i}(w_i)$ such that $w_i = \bigoplus_{j \in [n]} x_{i,j} \oplus \bigoplus_{j \in [n]} y_{j,i}$. We observe that these ciphertexts form a secret sharing of $u \cdot v$. Indeed, adding up the plaintexts we get:

$$\bigoplus_{i \in [n]} \left(\bigoplus_{j \in [n]} x_{i,j} \oplus \bigoplus_{j \in [n]} z_{j,i} \right) = \bigoplus_{i,j \in [n]} (x_{i,j} \oplus y_{i,j}) = \bigoplus_{i,j \in [n]} u_i \cdot v_j = \left(\bigoplus_i u_i \right) \cdot \left(\bigoplus_j v_j \right) \tag{1}$$

Thus, if the scheme can support $2d$ interleaved hops of (two-key) spooky multiplication and (single-key) additive homomorphism then it is an AFS-spooky scheme for the class of all depth d arithmetic circuits. We note that the resulting scheme does not depend on the number of inputs or their length, and it only depends on the complexity of C inasmuch as the underlying scheme depends on the depth of the evaluated circuit.

Applications of Spooky Encryption. In Sect. 6 we describe both positive and negative applications of spooky encryption. On the positive, it immediately yields a function secret sharing (FSS) scheme for all functions [4,15]. Previously such a general function secret sharing scheme was only known to follow from sub-exponentially hard indistinguishability obfuscation [4] whereas we can base it on LWE (using our LWE based spooky encryption).

Spooky encryption also gives a simple 2-round multi-party computation protocol. Roughly, AFS-spooky encryption lets each party broadcast an encryption of its input under its own key, then everyone individually performs the AFS-spooky evaluation locally, each party can locally decrypt and recover a share of the output, and the output is recover using another round of communication. There are some technicalities that should be addressed for this idea to work, and perhaps the easiest way of addressing them is to use AFS-spooky encryption to construct *multi-key FHE with threshold decryption (TMFHE)*, which can then be used to get a two-round protocol as done in [22]. Using our obfuscation based construction (which does not require a CRS), this gives the first 2-round semi-honest secure MPC protocol in the plain model.[3]

On the negative side, AFS-spooky encryption yields a counter-example for the transformation of Aiello et al. [1] from multi-prover (MIP) to single-prover protocols. Their idea was to send all of the MIP queries to a single prover, but encrypted under independents keys of a homomorphic encryption scheme. The single prover can homomorphically implement the actions of the MIP provers on the individual encrypted queries, and hopefully the fact that the queries are encrypted under independent keys means that no cross-influence is possible. It is easy to see that spooky encryption violates this hope (by its very nature). Moreover, we show that this transformation can lead to a total break of soundness - in Sect. 6.1 we show how using AFS-spooky encryption can lead to an

---

[3] In contrast, [12] and [22] construct 2-round protocols in the *CRS* model. As for security against a *malicious* adversary, [20] show that 5 rounds are necessary in the plain model (with respect to black-box proofs of security).

unsound single-prover protocol, when the transformation is applied to a simple two-prover protocol for graph 3-colorability.

Spooky-Free Encryption. Finally, in Sect. 7 we discuss the notion of spooky-free (SF) encryption, which provably ensures that any correlation that an attacker can induce between the original messages $(m_1, \ldots, m_n)$ and "tampered messages" $(m'_1, \ldots, m'_n)$, can be simulated by a "local simulator" that produces m'_i only as a function of m_i (and some shared randomness), see Definition 6. To validate this definition, we show that a spooky-free FHE suffices to prove the security of the natural approach of Aiello *et al.* [1], which was discussed above, of converting a succinct MIP into a succinct one-round argument discussed above. Indeed, spooky-freeness ensures that the attacker cannot cause more damage from seeing all n ciphertexts than what it could have done by seeing each plaintext independently.

We then turn to the systematic study of spooky-free encryption. First, we show that spooky-freeness implies semantic security. On the other hand, a very weak form of non-malleability (called 1-non-malleability here, or 1-bounded CCA security in [8]) implies spooky-freeness. However, since the scheme is non-malleable, it is inherently not homomorphic and so we cannot use it to obtain a delegation scheme via the foregoing approach.

Indeed, to instantiate the approach of Aiello *et al.* constructing succinct arguments for NP, we need a *homomorphic* encryption scheme which is spooky free. As a proof of concept, in the full paper [9] we show how to built such a homomorphic spooky-free encryption using succinct non-interactive arguments of knowledge (SNARKs [3,14]), true-simulation-extractable NIZKs [10] and regular FHE. While the use of SNARKs makes this construction uninteresting in the application to succinct arguments, the clean definition of SF-encryption, coupled with our "proof of concept" implementation, might open the door for more useful future constructions.

1.2 Related Work

The starting point for this line of work is the natural approach, suggested by Aiello *et al.* [1], for constructing a secure delegation scheme by combining a multi-prover interactive proof-system (MIP) with a homomorphic encryption scheme as described above. This intuition was questioned by Dwork *et al.* [11] and our work confirms that the approach of [1] is not always secure.

An approach to overcoming this barrier was taken by Kalai *et al.* [18,19]. They designed a specific MIP (for $\mathcal{P}$) that is sound even against arbitrary no-signaling adversaries. Since semantic-security rules out signaling strategies, they obtain a secure delegation protocol for any language in $\mathcal{P}$.

Spooky Free vs. Homomorphism Extraction. Bitansky and Chiesa defined in [2] a security notion called *homomorphism extraction*, that they show can be used to securely instantiate the construction of Aiello *et al.* and get succinct arguments for NP. Intuitively, this notion says that to produce a valid encryption of m'

from an encryption of m, you must know a function f such that $m' = f(m)$. Compared to our notion of spooky-free (which is also sufficient for the Aiello *et al.* transformation), the main difference is that of "extraction vs. soundness", so homomorphism extraction seems a stronger requirement. For example, homomorphism extraction implies some form of "plaintext awareness" and therefore is non-trivial even for schemes that aren't homomorphic, whereas we show that any non-malleable encryption scheme is spooky-free.

Multi-key FHE. A notion that is related to spooky-encryption, introduced by López-Alt *et al.* [21] is that of multi-key FHE. In a multi-key FHE, similarly to a spooky encryption scheme, the homomorphic evaluation procedure gets as input n ciphertexts encrypted under different keys. The difference is that the output of the evaluation in a multikey FHE is a single ciphertext that can only be decrypted by combining all the n keys. In contrast, in a spooky encryption scheme the result of the spooky evaluation is n ciphertexts, $c_1, \ldots, c_n$ where each c_i is encrypted under the ith original. Thus, spooky encryption can be thought of as a specific type of multi-key FHE.

2   Definitions

2.1   Local, No-Signaling, and Spooky Relations

We say that two distributions D_1, D_2 over a (finite) universe $\mathcal{U}$ are ε-close if their statistical distance $\frac{1}{2}||D_1 - D_2||_1$ is at most ε, and denote it by $D_1 \overset{\varepsilon}{\approx} D_2$. We write $D_1 \equiv D_2$ to denote that the distributions are identical. We say that D_1, D_2 are δ-far if their statistical distance is *at least δ*.

Definition 1. *Let $f : \{0,1\}^{\ell_1} \times \cdots \{0,1\}^{\ell_n} \to \{0,1\}^{\ell'_1} \times \cdots \{0,1\}^{\ell'_n}$ be a randomized mapping from n input to n outputs. For input $\boldsymbol{x} = (x_1, \ldots, x_n)$ to f, we denote the i'th component of the output by $f(\boldsymbol{x})_i$, and more generally for a subset $I \subset [n]$ we denote the projected input by $\boldsymbol{x}_I = (x_i : i \in I)$ and the projected output by $f(\boldsymbol{x})_I = (f(\boldsymbol{x})_i : i \in I)$.*

- *f is local if there exist n randomized "component mappings" $f_i : \{0,1\}^{\ell_i} \to \{0,1\}^{\ell'_i}$ such that for all $(x_1, \ldots, x_n) \in \{0,1\}^{\ell_1} \times \cdots \{0,1\}^{\ell_n}$, the distribution $f(x_1, \ldots, x_n)$ is a product distribution $f(x_1, \ldots, x_n) \equiv f_1(x_1) \times \cdots \times f_n(x_n)$.*
- *f is no-signaling if for every subset $I \in [n]$ and every two inputs $\boldsymbol{x}, \boldsymbol{x}'$ with the same I projection, $\boldsymbol{x}_I = \boldsymbol{x}'_I$, the corresponding projected distributions are equal, $f(\boldsymbol{x})_I \equiv f(\boldsymbol{x}')_I$.*
- *We say that f is ε-spooky for some $\varepsilon > 0$ if it is no-signaling, but for every local f' there exists some input $\boldsymbol{x}$ such that $f(\boldsymbol{x})$ and $f'(\boldsymbol{x})$ are at least ε-far.*

These definitions extends to an ensemble of mappings $F = \{f_k : k \in \mathbb{N}\}$, with the mapping parameters n, ℓ_i, ℓ'_i and the distance bound ε possibly depending on the ensemble parameter k. In this case we say that F is spooky if the f_k's are ε-spooky for a non-negligible $\varepsilon = \varepsilon(k)$.

As an example, consider the randomized function $f(x_1, x_2) = (y_1, y_2)$ where y_1, y_2 are uniformly random subject to $y_1 \oplus y_2 = x_1 \wedge x_2$. This function is no-signaling since the distributions $f(x)_1$ and $f(x)_2$ are individually uniform, no matter what x is. However, it's easy to show that for any local function $f' = (f'_1, f'_2)$ there is an input $x = (x_1, x_2)$ such that $\Pr[f'_1(x_1) \oplus f'_2(x_2) = x_1 \wedge x_2] \leq 1/2$. Therefore the function f is ε-spooky for $\varepsilon = 1/2$.

2.2   Spooky Encryption

A public-key encryption scheme consists of a tuple (Gen, Enc, Dec) of polynomial-time algorithms. The key-generation algorithm Gen gets as input a security parameter $\kappa \in \mathbb{N}$ and outputs a pair of public/private keys (pk, sk). The encryption algorithm Enc gets as input the public-key pk and a bit $m \in \{0, 1\}^{\mathsf{poly}(\kappa)}$ and outputs a ciphertext c, whereas the decryption algorithm Dec gets as input the private-key sk and the ciphertext c and outputs the plaintext bit m. The basic correctness guarantee is that $\Pr[\mathsf{Dec}_{\mathsf{sk}}(\mathsf{Enc}_{\mathsf{pk}}(m)) = m] > 1 - \mathsf{negl}(k)$, where the probability is over the randomness of all these algorithms. The security requirement is that for every pair of polynomial-sized adversaries (A_1, A_2) it holds that

$$\Pr_{\substack{(\mathsf{pk},\mathsf{sk}) \leftarrow \mathsf{Gen}(1^\kappa) \\ b \leftarrow \{0,1\}}} \left[\begin{array}{c} (m_0, m_1) \leftarrow A_1(\mathsf{pk}) \text{ s.t. } |m_0| = |m_1| \\ A_2 \left(\mathsf{pk}, \mathsf{Enc}_{\mathsf{pk}}(m_b) \right) = b \end{array} \right] \leq \frac{1}{2} + \mathsf{negl}(\kappa).$$

If the message space consists of just a single bit then we say that the scheme is a bit encryption scheme.

Definition 2 (Spooky Encryption). *Let* (Gen, Enc, Dec) *be a public-key bit-encryption scheme and* Spooky-Eval *be a polynomial-time algorithm that takes as input a (possibly randomized) circuit with* $n = n(\kappa)$ *inputs and n outputs,* $C : (\{0,1\}^*)^n \to (\{0,1\}^*)^n$, *and also n pairs of (public-key, ciphertext), and outputs n ciphertexts.*

Let $\mathcal{C}$ be a class of such circuits, we say that (Gen, Enc, Dec, Spooky-Eval) *is a* $\mathcal{C}$-*spooky encryption scheme if for every security parameter κ, every randomized circuit $C \in \mathcal{C}$, and every input* $\boldsymbol{x} = (x_1, \dots, x_n)$ *for C, the distributions*

$$SPOOK[C, x_1, \dots, x_n] \stackrel{\text{def}}{=}$$

$$\left\{ (\mathsf{Dec}(\mathsf{sk}_1, c'_1), \dots, \mathsf{Dec}(\mathsf{sk}_n, c'_n)) : \begin{array}{c} \forall i \in [n] \ (\mathsf{pk}_i, \mathsf{sk}_i) \leftarrow \mathsf{Gen}(1^\kappa), \\ c_i \leftarrow \mathsf{Enc}(\mathsf{pk}_i, x_i), \\ (c'_1, \dots, c'_n) \leftarrow \mathsf{Spooky\text{-}Eval}(C, (\mathsf{pk}_i, c_i)_i) \end{array} \right\}$$

and $C(x_1, \dots, x_n)$ are close upto a negligible distance in κ.

We note that the name *spooky encryption* stems from the application of Definition 2 to circuits C that compute spooky mappings. Indeed, as shown by Dwork *et al.* [11], the semantic security of (Gen, Enc, Dec) implies that only (almost) no-signaling C's can be realized, and every homomorphic scheme can realize C's that compute product mappings.

Spooky Encryption with CRS. We say that $(\mathsf{Gen}, \mathsf{Enc}, \mathsf{Dec}, \mathsf{Spooky\text{-}Eval})$ is a $\mathcal{C}$-spooky encryption scheme with CRS if Definition 2 is satisfied except that we allow all algorithms (and the adversary) to get as input also a public uniformly distributed common random string.

2.3   Additive-Function-Sharing Spooky Encryption

An important special case of spooky encryption allow us to take encryptions $c_i \leftarrow \mathsf{Enc}_{\mathsf{pk}_i}(x_i)$ under n independent keys of inputs $x_1, \ldots, x_n$ to an n-argument function f, and produce new ciphertexts under the same n keys that decrypt to additive secret-shares of $y = f(x_1, \ldots, x_n)$. An encryption scheme that supports such "non-interactive sharing" is called *additive-function-sharing spooky encryption* (or AFS-spooky). Several variants of this concept are defined below:

- We can either insist on getting a *random* secret sharing of y, or contend ourselves with *any* secret sharing. Below we call the latter variant *weak* AFS-spooky, and the former is strong AFS-spooky (or just AFS-spooky).
- Similarly to homomorphic encryption schemes, we can have either a *leveled* variant where key-generation receives an additional depth parameter d and the result supports only circuits of depth upto d, or a *fully* AFS-spooky scheme that supports any circuit with a fixed parameter setting.
- We can either allow non-negligible error probability (i.e., the probability that the computation fails to produce a secret-sharing of the right output y), or insist on a negligible error probability. Below we denote by ε-AFS-spooky the variant where the error probability is bounded by some ε (that need not be negligible), and the variant with negligible error probability is just AFS-spooky.
- Sometimes we want to consider only two-argument functions $f(x_1, x_2)$, a scheme that only supports two-argument functions is called AFS-2-spooky.

The formal definition itself is provided in the full version [9], where we also show that the weak and strong variants are essentially equivalent.

3   LWE-Based Spooky Encryption

3.1   Learning with Errors (LWE) and Multi-key FHE

The LWE assumption roughly says that adding just a little noise to a set of linear equations makes them hard to solve. In our context, we consider equations modulo some integer q and the noise consists of numbers whose magnitude is much smaller than q, as expressed via a noise distribution χ that yields such "small numbers" with high probability. Below we identify $\mathbb{Z}_q$ with the symmetric interval $[-q/2, q/2)$ and let $[x]_q$ denote the reduction of x modulo q into this interval.

Definition 3 (Learning with Errors [28]). *Let $n = n(\kappa), q = q(\kappa) \in \mathbb{Z}$ be functions of the security parameter κ and $\chi = \{\chi(\kappa)\}_\kappa$ be a distribution ensemble over $\mathbb{Z}$. The decision-LWE assumption with parameters (n, q, χ) says that for any polynomial $m = m(\kappa) \in \mathbb{Z}$, the following two distribution ensembles are computationally indistinguishable*

$$\mathcal{LWE}[n, m, q, \chi] \overset{\text{def}}{=} \{(A, b) : A \leftarrow \mathbb{Z}_q^{n \times m}, \ s \leftarrow \mathbb{Z}_q^n, \ e \leftarrow \chi^m, \ b := [sA + e]_q\},$$

and $\mathcal{U}[n, m, q] \overset{\text{def}}{=} \{(A, b) : A \leftarrow \mathbb{Z}_q^{n \times m}, \ b \leftarrow \mathbb{Z}_q^m\}$ *(i.e., uniform over $\mathbb{Z}_q^{(n+1) \times m}$).*

For $\alpha = \alpha(\kappa) \in (0, 1)$, the α-DLWE assumption asserts the existence of parameters n, q, χ as above with n polynomial in κ, such that $e \leftarrow \chi$ yields $|e| < \alpha q$ with overwhelming probability.

Note that the α-DLWE assumption becomes stronger as α gets smaller, and it is known to be false in the extreme case where $\alpha = 2^{-\Omega(n)}$ using lattice-reduction techniques. On the other hand, we have ample evidence to belive the α-DLWE assumption with $\alpha = 1/\text{poly}(n)$ [5,25,28], and it is commonly belived to hold also for super-polynomially (and perhaps even sub-exponentially) small α's.

We show that assuming hardness of the learning-with-errors problem, there exists a function-secret sharing (in the common-random-string model) for any n-argument function f. Our construction can be built on the multi-key fully homomorphic encryption construction of Mukherjee and Wichs [22] or the one of Peikert and Shiehian [26], which are variations of the Clear-McGoldrick scheme from [7]. We summarize the relevant properties of these constructions:

Theorem 1 [7,22,26]. *Assuming the hardness of α-DLWE (for some $\alpha(\kappa)$), there exists a multi-key homomorphic encryption with the following properties:*

- *The construction works in the common-random-string model. For parameters $n, m, q = \text{poly}(\kappa)$, all instances have access to a uniformly random matrix $A \in \mathbb{Z}_q^{(n-1) \times m}$.*
- *For any depth parameter d, the scheme supports multi-key evaluation of depth-d circuits using public keys of size $d \cdot \text{poly}(\kappa)$, while secret keys are vectors $s \in \mathbb{Z}_q^n$, regardless of the depth parameter.*
  Specifically, there is an efficient procedure Eval that is given as input:
  - *Parameters $d, \ell \in \mathbb{N}$;*
  - *A depth-d circuit computing an ℓ-argument function $f : (\{0,1\}^*)^\ell \to \{0,1\}$;*
  - *Public keys $(\text{pk}_1, \ldots, \text{pk}_n)$ and fresh encryptions (bit-by-bit) of each argument $x_i \in \{0,1\}^*$ under key pk_i, denoted $c_i \leftarrow \text{Enc}_{\text{pk}_i}(x_i)$.*

On such input, the Eval procedure outputs a dimension $n\ell$-vector, $c' = (c'_1 \ldots c'_\ell)$ (with each $c'_i \in \mathbb{Z}_q^n$),[4] such that for the secret keys s_i corresponding to pk_i it holds that

$$\sum_{i=1}^{\ell} \langle s_i, c'_i \rangle = \lfloor q/2 \rfloor \cdot f(x_1, \ldots, x_n) + e \pmod{q}$$

for some error $e \in \mathbb{Z}_q$ with $|e| < \alpha q \cdot \text{poly}(\kappa)$.

---

[4] Referring to [22, Sect. 5.4], the vector c'_i is the result of the product $\hat{C}^{(i)} \times \hat{G}^{-1}(\hat{w}^T)$, without the added noise term e_i^{sm}.

By further making a circular-security assumption, there exists a scheme that supports evaluation of circuits of any depth without growing the public keys.

3.2 LWE-Based AFS Spooky Encryption

Below we show that under the decision-LWE assumption we can construct AFS-spooky encryption schemes (in the common-random-string model). Namely, for every n-argument function $f(x_1, \ldots, x_n)$, given encryption of the arguments under n independent public keys, we can compute an encryption of shares under the same keys of an additive secret-sharing of the output $y = f(x_1, \ldots, x_n)$.

Theorem 2. *Assuming the hardness of α-DLWE, there exists a leveled ε-AFS-2-Spooky encryption scheme for $\varepsilon = \alpha \cdot \mathsf{poly}(\kappa)$. Further making a circular-security assumption, we get a (non-leveled) ε-AFS-2-spooky encryption scheme.*

Proof. We show that the encryption scheme from Theorem 1 is already essentially a leveled weak AFS-2-spooky encryption scheme. Specifically, Theorem 1 tells us that given the description of a depth-d circuit C, computing a 2-argument function $f : (\{0,1\}^*)^2 \to \{0,1\}$, together with two public-key and corresponding bit-by-bit encryptions, $c_i \leftarrow \mathsf{Enc}_{\mathsf{pk}_i}(x_i)$, the Eval procedure yields $(c'_1, c'_2) \leftarrow \mathsf{Eval}(C, (\mathsf{pk}_1, c_1), (\mathsf{pk}_2, x_2))$ such that $\langle \mathsf{sk}_1, c'_1 \rangle + \langle \mathsf{sk}_2, c'_2 \rangle = y \cdot q/2 + e \pmod{q}$, where the sk_i's are the secret keys corresponding to the pk_i's, $y = f(x_1, x_2)$, and $|e| < \alpha q \cdot \mathsf{poly}(\kappa) = \varepsilon q$.

Denote $v_i = [\langle \mathsf{sk}_i, c'_i \rangle]_q$ for $i = 1, 2$ and $v = [v_1 + v_2]_q$. Lemma 1 below says that instead of first adding the v_i's and then rounding to the nearest multiple of $q/2$, we can first round and then add, and this will yield the same result except with error probability of at most 2ε. The only catch is that Lemma 1 assumes that v_1, v_2 are chosen at random subject to their sum modulo q being v, whereas in our case we do not have this guarantee. To account for this, we modify our $\mathsf{Spooky\text{-}Eval}$ procedure, letting it choose a random shift amount $\delta \in \mathbb{Z}_q$ and adding/subtracting it from v_1, v_2, respectively. More detail is provided in the full version [9].

Lemma 1. *Fix some modulus $q \in \mathbb{Z}$, bit $b \in \{0,1\}$, and a value $v \in \mathbb{Z}_q$ such that $v = b \cdot q/2 + e \pmod{q}$ for some bounded error $|e| < q/4$. Consider choosing v_1, v_2 uniformly at random in $\mathbb{Z}_q$ subject to $v_1 + v_2 = v \pmod{q}$, and denote $v_i = b_i \cdot q/2 + e_i \pmod{q}$ with $b_i = [[v_i \cdot 2/q]]_2 \in \{0,1\}$ and $|e_i| \leq q/4$. Then $\Pr_{v_1, v_2}[b_1 \oplus b_2 = b] > 1 - 2(|e| + 1)/q$.*

Proof. We break the proof into four cases, namely $b = 0$ vs. $b = 1$ and $e \geq 0$ vs. $e < 0$. Below we prove only the first the case $b = 0$ and $v = e \geq 0$, the other three cases are similar. For the first case consider choosing at random $v_1 \in \mathbb{Z}_q$ and setting $v_2 = [v - v_1]_q = [e - v_1]_q$. It is straightforward (but tedious) to check that the condition $b_1 \oplus b_2 = b = 0$ is satisfied whenever we have

$$\text{either } v_1, v_2 \in \left(\tfrac{-q}{4} + e, \tfrac{q}{4}\right), \text{ or } v_1, v_2 \in \left[\tfrac{-q}{2}, \tfrac{-q}{4}\right) \cup \left(\tfrac{q}{4} + e, \tfrac{q}{2}\right).$$

For example when $v_1 \in \left(\frac{q}{4} + e, \frac{q}{2}\right)$ then we have $b_1 = 1$ and

$$v_2 = e - v_1 \in \left(e - \frac{q}{2}, \; e - (\frac{q}{4} + e)\right) \subseteq \left(-\frac{q}{2}, \; -\frac{q}{4}\right),$$

so we get also $b_2 = 1$ and therefore $b_1 \oplus b_2 = 0 = b$.

The only error regions are $v_1, v_2 \in (\frac{-q}{4}, \frac{-q}{4} + e)$, $v_1, v_2 \in (\frac{q}{4}, \frac{q}{4} + v)$, and (depending on rounding) also upto two of the four points $v_1 \in \{\frac{\pm q}{4}, \frac{\pm q}{4} + e\} \cap \mathbb{Z}$.

3.3   Beyond AFS-2-Spooky Encryption

The construction from Theorem 2 does not directly extend to functions with more than two arguments, since Lemma 1 no longer holds for more than two v_i's (even for the no-error case of $e = 0$). Instead, we can use the GMW-like transformation that was sketched in the introduction and is described in detail in Sect. 5 to get a general AFS-spooky scheme.

To support this transformation, we need an AFS-2-spooky scheme which is *multi-hop* (in the sense of [13]), i.e. we need to apply the spooky evaluation procedure not just to fresh ciphertexts, but also to evaluated ciphertexts that resulted from previous applications of spooky evaluation. The AFS-2-spooky scheme in Theorem 2 may or may not have this property, depending on the underlying multi-key FHE scheme. In particular the Peikert-Shiehian scheme in [26] is "natively multi-hop," so we can base our construction on that scheme and get directly a multi-hop AFS-2-spooky scheme which is suitable for our transformation.

On the other hand, the schemes from [7,22] support only one hop, since only fresh cipehrtexts can be processed in a multi-key fashion. We can stil use them for our purposes by applying the same bootstrapping-based transformation as in [13, Theorem 4], which transforms any compact fully-homomorphic scheme to a multi-hop one:[5] More details are provided in the full version [9].

Theorem 3. *Assuming the hardness of α-DLWE, there exists a leveled FHE scheme that supports d interleaved levels of AFS-2-spooky multiplications and single-key addition, with total error probability $\varepsilon = \alpha \cdot d \cdot \mathsf{poly}(\kappa)$.*

Corollary 1. *Assuming the hardness of α-DLWE, there exists a leveled ε-AFS-spooky encryption scheme for $\varepsilon = \alpha \cdot d \cdot \mathsf{poly}(\kappa)$. Further making a circular-security assumption, we get a (non-leveled) ε-AFS-spooky encryption scheme.*   □

4   piO Based Spooky Encryption

In this section we show a construction based on probabilistic iO, in conjunction with lossy homomorphic encryption, that can support many 2-key spooky relations, even beyond AFS-spooky. Compared to our LWE-based construction from

---

[5] The transformation in [13] is described for single-key FHE schemes, but it applies also to multi-key schemes.

Sect. 3, the construction here does not need a CRS and has zero error probability, and it supports more spooky distributions. On the other hand, we are making a much stronger assumption here, and also we need a different scheme for different spooky relations.[6]

The construction in this section supports in particular the functionality that we need for our generic transformation from Sect. 5, that turns an AFS-2-spooky scheme to an AFS-n-spooky one. The resulting AFS-n-spooky also does not use a CRS and has no error probability. Moreover, applying this transformation yields a single scheme supporting all AFS-spooky relations.

Organization of this Section. In Sect. 4.1 we introduce our tools, defining probabilistic indistinguishability obfuscation (using a slightly weaker variant of the definition of Canetti et al. [6]) and lossy homomorphic encryption with malicious circuit privacy. In Sect. 4.2 we describe and prove our construction for 2-input spooky encryption scheme, and finally in Sect. 4.3 we show how to obtain a multi-input AFS-spooky encryption.

4.1 Tools

Probabilistic Indistinguishability Obfuscation. Our construction uses probabilistic iO, a notion that was recently introduced by Canetti et al. [6]. Loosely speaking, this is an obfuscator for probabilistic circuits with the guarantee that the obfuscations of any two "equivalent" circuits are computationally indistinguishable.

Canetti et al. define several variants of piO, where the main distinction is the precise formulation of what it means for circuits to be equivalent. Our definition corresponds to a (weakened variant) of their X-Ind piO (which can be realized assuming sub-exponentially secure iO and sub-exponentially secure OWF, see Theorem 4 below). Roughly, our variant only considers pairs of circuits with the property that for *every* input, their output distributions are *identical*, while the definition in [6] allows a small statistical gap.

To formally define piO, we consider a (possibly randomized) PPT sampling algorithm S that given as input a security parameter 1^κ, outputs a triple (C_0, C_1, z), where C_0 and C_1 are randomized circuits (to be obfuscated) and z is some auxiliary input. We say that a sampler S is an *equivalent-circuit-sampler* if with probability 1 it outputs circuits C_0 and C_1 such that for every x the circuits $C_0(x)$ and $C_1(x)$ generate identical distributions.

Definition 4 (Probabilistic Indistinguishable Obfuscation (piO), [6]). *A* probabilistic indistinguishability obfuscator *is a probabilistic polynomial-time algorithm* piO *that, given as input a security parameter 1^κ and a* probabilistic *circuit C, outputs a circuit $C' = $ piO$(1^\kappa, C)$ (which may be deterministic) of size at most $|C'| = $ poly$(\kappa, |C|)$ such that the following two properties hold:*

---

[6] We can extend the construction so that a single scheme can handle an entire class of spooky relations, as long as we can describe relations in that class and verify that a given relation is no-signaling.

1. *For every individual input x, the distribution $C(x)$ and $\big(\mathsf{piO}(1^\kappa, C)\big)(x)$ are identical.*[7]
2. *For every equivalent-circuit-sampler S, drawing $(C_0, C_1, z) \leftarrow S(1^\kappa)$ we get computationally indistinguishable distributions:*

$$\{(C_0, C_1, z, \mathsf{piO}(1^\kappa, C_0))\} \stackrel{c}{=} \{(C_0, C_1, z, \mathsf{piO}(1^\kappa, C_1))\}$$

We note that our correctness guarantee is incomparable to that given by [6]. Indeed, motivated by their PRF based construction, the definition in [6] basically requires that no PPT adversary can distinguish between oracle access to C and to $\mathsf{piO}(1^\kappa, C)$ (so long as the adversary is not allowed to repeat its queries). On the one hand our definition is weaker in that it only considers each input individually, but on the other hand it is stronger in that it requires that for each such individual input the distributions are *identical*. Our correctness guarantee can be easily obtained from the construction in [6], by using an underlying PRF $\{f_s\}_s$ with the property that $f_s(x)$ is individually uniformly random for every x. The latter can be easily obtained by taking any PRF and xor-ing its output with a fixed random string.

Theorem 4 [6]. *Assuming the existence of a sub-exponentially indistinguishable indistinguishability obfuscator for circuits and a sub-exponentially secure puncturable PRF, there exists a probabilistic indistinguishability obfuscator.*

Lossy Encryption. Loosely speaking, a lossy encryption scheme has a procedure $\widetilde{\mathsf{Gen}}$ for generating "lossy public keys." These keys are indistinguishable from normal public keys, but have the property that ciphertexts generated using such lossy keys contain no information about their plaintext. We defer the formal definition to the full version [9].

Malicious Circuit-Private Encryption. A public-key encryption scheme (Gen, Enc, Dec), with message space $\{0,1\}^\ell$, is a homomorphic encryption scheme for a class of Boolean circuits $\mathcal{C}$ on ℓ-bit inputs if there exists a PPT algorithm Eval, such that for every key-pair (pk, sk), circuit $C \in \mathcal{C}$ and ciphertext $c = \mathsf{Enc}_{\mathsf{pk}}(x)$, where $x \in \{0,1\}^\ell$, on input (C, c) the algorithm $\mathsf{Eval}_{\mathsf{pk}}$ outputs c^* such that $\mathsf{Dec}_{\mathsf{sk}}(c^*) = C(x)$. If the length of c^* does not depend on C then we say that the scheme is compact.

As noted in the introduction, our construction requires a homomorphic encryption scheme that has *malicious circuit privacy*, which means that the ciphertext c^* does not reveal any non-trivial information about the circuit C which was used to generate it, even for an adversarially chosen public-key pk and ciphertext c. We defer the formal definition to the full version [9].

Malicious circuit privacy for evaluating NC_1 circuits can be achieved by a "folklore" combination of an information theoretic variant of Yao's garbled circuit [17] with an oblivious transfer protocol that has perfect security against a

---

[7] The latter distribution is defined also over the randomnees of piO. Note that this does not imply that the joint distribution for multiple inputs will be the same in the two cases.

malicious receiver. The latter can be constructed based on DDH [23]. Moreover, these schemes can be made lossy using standard techniques.

Moreover, we can apply the techniques of Ostrovsky et al. [24] to bootstrap this result to any poly-circuit, assuming the existence of (leveled) fully homomorphic encryption with NC_1 decryption. The latter scheme can be instantiated based on LWE, see more details in the full version [9]. Hence we obtain:

Theorem 5. *Assuming the hardness of* LWE *and* DDH, *there exists a lossy leveled fully-homomorphic encryption scheme with malicious circuit privacy.*

4.2   Two-Key Spooky Encryption from piO

Our construction relies on a property of two-input relations that we call *re-sampleability*. Roughly, it should be possible to sample efficiently from the distribution of the second coordinate conditioned on a particular fixed value for the first coordinate.

Definition 5 (Efficiently Re-sampleable). *A randomized polynomial-size circuit* $C : \{0,1\}^{\ell_1} \times \{0,1\}^{\ell_2} \to \{0,1\}^{\ell_1'} \times \{0,1\}^{\ell_2'}$ *is* efficiently re-sampleable *if there exists a polynomial-size randomized "resampling circuit" RS_C, such that for any input (x_1, x_2) to C, the distribution $C(x_1, x_2)$ is identical to the "resampled distribution" $\{(y_1, y_2') : (y_1, y_2) \leftarrow C(x_1, x_2), \ y_2' \leftarrow RS_C(x_1, x_2, y_1)\}$.*

We construct a 2-key spooky scheme that supports any 2 input/output circuit that is both *efficiently re-sampleable* and *no-signaling*.

Theorem 6 (2-Key Spooky Encryption from piO). *Let* $C : \{0,1\}^{\ell_1} \times \{0,1\}^{\ell_2} \to \{0,1\}^{\ell_1'} \times \{0,1\}^{\ell_2'}$ *be an efficiently re-sampleable no-signaling circuit, with re-sampling circuit RS_C. If there exist (1)* piO, *and (2) a perfectly-lossy homomorphic encryption scheme that can evaluate C and RS_C, and is perfectly malicious circuit private, then there exists a C-spooky encryption scheme, which is also perfectly lossy (and hence semantically secure).*

We stress that the encryption scheme that we need for Theorem 6 must be able to evaluate C and RS_C and be perfectly malicious circuit private, but *it need not be compact*. In the full paper we describe such a scheme for NC_1 circuits based on DDH. Hence, under DDH and piO, we get a C-spooky scheme for every re-sampleable and no-signaling C in NC_1. Moreover, we can use the techniques of Ostrovsky et al. [24] to supports any poly-size circuit, assuming both DDH and FHE. Since [6] show that full-fledged FHE can be built based on piO, we get a construction under DDH and piO for C-spooky scheme for every re-sampleable and no-signaling polynomial-size circuit C.

Remark 1 (Almost No-Signaling). A natural relaxation of no-signaling circuits, considered in previous works (e.g., [11,18,19]), allows the distributions $C(x, y)_1$ and $C(x, y')_1$ to be indistinguishable (rather than identical). Such circuit is called *almost no-signaling.*

It is clear that for a secure C-spooky encryption scheme to exist, C must be (at least) almost no-signaling (cf. [11]). However our construction does not extend to the "almost" case, Theorem 6 requires that C to be perfectly no-signaling, i.e. $C(x,y)_1$ and $C(x,y')_1$ must be identically distributed for all x, y, y'. Supporting almost no-signaling circuits is left to future work.

Proof of Theorem 6. Let piO be a probabilistic indistinguishability obfuscator and let (Gen, Enc, Dec) be the encryption scheme from the theorem statement, with $\widetilde{\text{Gen}}$ the corresponding lossy key generation algorithm and Eval the homomorphic evaluation algorithm with malicious circuit privacy.

Each instance of our construction uses two public/secret keys pairs, where only the first pair is used for "normal encryption and decryption," and the other pair is only used for spooky evaluation. In addition to the two pairs, the public key also contains an obfuscated program that implements spooky evaluation using the secret key. That obfuscated program has a secret key hard-wired, and given two ciphertexts c_1, c_2 it decrypt the first one, then evaluates the re-sampling circuit RS_C homomorphically on the other. A complete description of the resulting scheme is found in Fig. 1.

We first show that the scheme supports spooky evaluation of C and then show that it is a lossy encryption scheme (and in particular is semantically secure).

Lemma 2. *The scheme* (Gen-Spooky, Enc-Spooky, Dec-Spooky, Spooky-Eval) *is C-spooky.*

Proof. The spooky evaluation procedure gets as input two public-keys pk-spooky$_1$ = $(\text{pk1}_1, \text{pk2}_1, \tilde{P}_1)$, pk-spooky$_2$ = $(\text{pk1}_2, \text{pk2}_2, \tilde{P}_2)$, and matching ciphertexts c_1 = Enc-Spooky(pk-spooky$_1, x_1$) and c_2 = Enc-Spooky (pk-spooky$_2, x_2$) (for some inputs x_1, x_2 to C). It simply runs the obfuscated program $\tilde{P}_1 = \text{piO}(1^\kappa, P[\text{sk1}_1, \text{pk2}_1])$ on input $(c_1, \text{pk1}_2, c_2)$ and returns its output.

By construction and using the correctness of piO, this procedure outputs c'_1 and c'_2 such that $c'_1 \leftarrow \text{Enc}(\text{pk2}_1, y_1)$, where $y_1 \leftarrow (C(x_1, 0^{\ell_2}))_1$, and $c'_2 \leftarrow \text{Eval}_{\text{pk1}_2}(RS[x_1, y_1, r], c_2)$, where $RS[x_1, y_1, r](x_2) \equiv RS_C(x_1, x_2, y_1; r)$. By the no-signaling property y_1 is distributed identically to $y'_1 \leftarrow (C(x_1, x_2))_1$ and so c'_2 is distributed as $\text{Eval}_{\text{pk1}_2}(RS[x_1, y'_1, r], c_2)$. Hence

$$\text{Dec-Spooky}(\text{sk-spooky}_1, c'_1) = \text{Dec}_{\text{sk1}_1}(\text{Enc}(\text{pk2}_1, y'_1)) = y'_1$$
$$\text{and Dec-Spooky}(\text{sk-spooky}_2, c'_2) = RS[x_1, y'_1, r](x_2) = RS_C(x_1, x_2, y'_1; r)_2.$$

By the definition of re-sampling, the joint distribution $\Big(\text{Dec-Spooky}(\text{sk-spooky}_1, c'_1), \text{Dec-Spooky}(\text{sk-spooky}_2, c'_2)\Big)$ is identical to $C(x_1, x_2)$, as required.

Lemma 3. *The scheme* (Gen-Spooky, Enc-Spooky, Dec-Spooky) *is a perfectly lossy encryption scheme.*

The probabilistic circuit $P[\mathsf{sk1}, \mathsf{pk2}](c_1, \mathsf{pk}, c)$:

Hardwired: a private-key $\mathsf{sk1}$ and a public-key $\mathsf{pk2}$.
Input: a ciphertext c_1 (presumably under $\mathsf{pk1}$),
   and additional (presumably matching) public-key pk and ciphertext c.

1. Decrypt $x_1 \leftarrow \mathsf{Dec}_{\mathsf{sk1}}(c_1)$;[a]
2. Choose randomness $r, r' \leftarrow \{0,1\}^*$ for C and RS_C, respectively;
3. Set $y_1 \leftarrow C(x_1, 0^{\ell_2}; r)_1$ and encrypt $c_1' \leftarrow \mathsf{Enc}_{\mathsf{pk2}}(y_1)$; $\overbrace{\phantom{C(x_1,0^{\ell_2};r)_1}}^{=y_1}$
4. Define the circuit $\mathsf{RS}[x_1, r, r'](x_2) \equiv RS_C(x_1, x_2, C(x_1, 0^{\ell_2}; r)_1; r')$;
5. Compute homomorphically $c_2' \leftarrow \mathsf{Eval}_{\mathsf{pk}}(\mathsf{RS}[x_1, r, r'], c)$.
6. Output $\big((2, c_1'), (1, c_2')\big)$.[b]

piO based Spooky Encryption

- $\underline{\mathsf{Gen\text{-}Spooky}(1^\kappa)}$:
   1. Select $(\mathsf{pk1}, \mathsf{sk1}), (\mathsf{pk2}, \mathsf{sk2}) \leftarrow \mathsf{Gen}(1^\kappa)$, and set $\tilde{P} \leftarrow \mathsf{piO}(1^\kappa, P[\mathsf{sk1}, \mathsf{pk2}])$.
   2. Output the secret key $\mathsf{sk\text{-}spooky} = (\mathsf{sk1}, \mathsf{sk2})$ and public key $\mathsf{pk\text{-}spooky} = \big(\mathsf{pk1}, \mathsf{pk2}, \tilde{P}\big)$.
- $\underline{\mathsf{Enc\text{-}Spooky}\big((\mathsf{pk1}, \mathsf{pk2}, \tilde{P}), x\big)}$: Output $\big(1, \mathsf{Enc}_{\mathsf{pk1}}(x)\big)$.
- $\underline{\mathsf{Dec\text{-}Spooky}\big((\mathsf{sk1}, \mathsf{sk2}), (tag, c)\big)}$: If $tag = 1$ output $\mathsf{Dec}_{\mathsf{sk1}}(c)$, else output $\mathsf{Dec}_{\mathsf{sk2}}(c)$.
- $\underline{\mathsf{Spooky\text{-}Eval}\big((\mathsf{pk1}_1, \mathsf{pk2}_1, \tilde{P}_1), c_1, (\mathsf{pk1}_2, \mathsf{pk2}_2, \tilde{P}_2), c_2,\big)}$: Output $\tilde{P}_1(c_1, \mathsf{pk1}_2, c_2)$.

---

[a] We assume that Dec always returns some value, even if c_1 is not a valid ciphertext.
[b] The tags "2", "1" signal to the decryption algorithm which secret key to use.

Fig. 1. piO based spooky encryption

The probabilistic circuit $P'[\mathsf{sk1}, \mathsf{pk2}](c_1, \mathsf{pk}, c)$:

The same as $P[\mathsf{sk1}, \mathsf{pk2}](c_1, \mathsf{pk}, c)$, but setting $c_1 \leftarrow \mathsf{Enc}_{\mathsf{pk}_2}(0^{\ell_1})$ in Step 3 rather than $c_1 \leftarrow \mathsf{Enc}_{\mathsf{pk}_2}(y_1)$.

The probabilistic circuit $P''[\mathsf{pk2}]$:

Hardwired: a public-key $\mathsf{pk2}$.
Input: a ciphertext c_1, a public-key pk and a ciphertext c (presumably under pk).

1. Encrypt $c_1' \leftarrow \mathsf{Enc}_{\mathsf{pk2}}(0^{\ell_1})$.
2. Choose randomness $r \leftarrow \{0,1\}^*$ for C, and define $f[r](\cdot) \equiv C(0^{\ell_1}, \cdot\,; r)_2$.
3. Compute homomorphically $c_2' \leftarrow \mathsf{Eval}_{\mathsf{pk}}(f[r], c)$.
4. Output $\big((2, c_1), (1, c_2')\big)$.

Fig. 2. The probabilistic circuits $P'[\mathsf{sk1}, \mathsf{pk2}]$ and $P''[\mathsf{pk2}]$

Proof. We need to show that there is an alternative key-generation procedure Gen-Spooky, producing public keys that are indistinguishable from the real ones, but such that ciphertexts encrypted relative to these keys contain no information about the encrypted plaintext.

The main challenge in establishing the lossiness of the scheme is in showing that the public-keys are indistinguishable from lossy keys despite the obfuscated programs in the public-key (which depend on the corresponding secret keys). Toward that end, we will (gradually) show that these obfuscated programs are computationally indistinguishable from programs that do not depend on the secret keys.

Below we state and prove a few claims, where we consider the distributions $(\mathsf{pk1}, \mathsf{sk1}), (\mathsf{pk2}, \mathsf{sk2}) \leftarrow \mathsf{Gen}(1^\kappa)$ and $\widetilde{\mathsf{pk1}}, \widetilde{\mathsf{pk2}} \leftarrow \widetilde{\mathsf{Gen}}(1^\kappa)$, where $\widetilde{\mathsf{Gen}}$ is the lossy key-generation of the underlying encryption scheme.

Claim 4.1. $\Big(\mathsf{pk1}, \mathsf{pk2}, \mathsf{piO}(1^\kappa, P[\mathsf{sk1}, \mathsf{pk2}])\Big) \overset{c}{\equiv} \Big(\mathsf{pk1}, \widetilde{\mathsf{pk2}}, \mathsf{piO}(1^\kappa, P[\mathsf{sk1}, \widetilde{\mathsf{pk2}}])\Big).$

Proof. Follows from the indistinguishability between standard and lossy public-keys of the underlying scheme.

Claim 4.2. $\Big(\mathsf{pk1}, \widetilde{\mathsf{pk2}}, \mathsf{piO}(1^\kappa, P[\mathsf{sk1}, \widetilde{\mathsf{pk2}}])\Big) \overset{c}{\equiv} \Big(\mathsf{pk1}, \widetilde{\mathsf{pk2}}, \mathsf{piO}(1^\kappa, P'[\mathsf{sk1}, \widetilde{\mathsf{pk2}}])\Big),$ where $P'[\mathsf{sk1}, \widetilde{\mathsf{pk2}}]$ is similar to $P[\mathsf{sk1}, \widetilde{\mathsf{pk2}}]$ except that it encrypts 0^{ℓ_1} rather than y_1 in Step 3, see Fig. 2.

Proof. Since $\widetilde{\mathsf{pk2}}$ is a *lossy* public-key, $\mathsf{Enc}_{\widetilde{\mathsf{pk2}}}(0^{\ell_1})$ and $\mathsf{Enc}_{\widetilde{\mathsf{pk2}}}(y_1)$ are *identically* distributed. Hence P and P' have identical output distribution for every input, and so their piO-obfuscations are indistinguishable.

We proceed to the main claim:

Claim 4.3. $\Big(\mathsf{pk1}, \widetilde{\mathsf{pk2}}, \mathsf{piO}(1^\kappa, P'[\mathsf{sk1}, \widetilde{\mathsf{pk2}}])\Big) \overset{c}{\equiv} \Big(\mathsf{pk1}, \widetilde{\mathsf{pk2}}, \mathsf{piO}(1^\kappa, P''[\widetilde{\mathsf{pk2}}])\Big),$ where the program $P''[\widetilde{\mathsf{pk2}}]$, defined in Fig. 2, does not have the secret key $\mathsf{sk1}$ (hence it cannot recover x_1 or compute y_1), so on $c = \mathsf{Enc}_{\mathsf{pk}}(x_2)$ it evaluates homomorphically $f''(x_2) = C(0^{\ell_1}, x_2)_2$ rather than $f'(x_2) = RS_C(x_1, x_2, y_1)$.

Proof. We will show that for every valid secret key $\mathsf{sk1}$ and arbitrary public key $\widetilde{\mathsf{pk2}}$, the randomized programs $P'[\mathsf{sk1}, \widetilde{\mathsf{pk2}}]$ and $P''[\widetilde{\mathsf{pk2}}]$ are functionally identical, in the sense that their outputs are identically distributed for every input. The claim will then follow from the fact that piO is a probabilistic indistinguishability obfuscator (see Definition 4).

Note that the first output $c_1' = \mathsf{Enc}_{\widetilde{\mathsf{pk2}}}(0^{\ell_1})$ is generated identically by the two programs, and is independent of everything else that happens in these programs, so we only need to show that the second output c_2' is identically distributed. To show this, we first establish that c_2' is an encryption under pk of a value y_2 that is distributed identically in the two programs, and then we appeal to the malicious circuit-privacy of the underlying scheme to conclude that also c_2' itself is identically distributed.

For starters, fix some arbitrary $x_1 \in \{0,1\}^{\ell_1}$ and $x \in \{0,1\}^{\ell_2}$, and consider the following distributions

$$\mathcal{D}_1[x_1, x] = \{y_1 \leftarrow C(x_1, 0^{\ell_2})_1, \text{ output } y_2 \leftarrow RS_C(x_1, x, y_1)\}, \text{ // Output distribution of } P'$$
$$\mathcal{D}_2[x_1, x] = \{y_1 \leftarrow C(x_1, x)_1, \quad \text{output } y_2 \leftarrow RS_C(x_1, x, y_1)\},$$
$$\mathcal{D}_3[x_1, x] = \{\text{output } y_2 \leftarrow C(x_1, x)_2\},$$
$$\mathcal{D}_4[x] = \{\text{output } y_2 \leftarrow C(0^{\ell_1}, x)_2\}. \qquad \text{ // Output distribution of } P''$$

Since C is a no-signaling circuit then $\mathcal{D}_1[x_1, x] = \mathcal{D}_2[x_1, x]$ and $\mathcal{D}_3[x_1, x] = \mathcal{D}_4[x]$, and since R_C is the re-sampling circuit for C then we also have $\mathcal{D}_2[x_1, x] = \mathcal{D}_3[x_1, x]$. We therefore conclude that the two distributions $\mathcal{D}_1[x_1, x]$ and $\mathcal{D}_4[x]$ are identical for every x_1, x.

Now consider $x_1 = \mathsf{Dec}_{\mathsf{sk1}}(c_1)$, and x which is the "effective plaintext" for pk, c (such x must exist since the underlying scheme is malicious circuit-private). Recall that the second output of $P'[\mathsf{sk1}, \widetilde{\mathsf{pk2}}]$ consists of a homomorphic evaluation of $\mathcal{D}_1[x_1, x]$, while the second output of $P''[\widetilde{\mathsf{pk2}}]$ consists of homomorphic evaluation of $\mathcal{D}_4[x]$. Using perfect malicious circuit privacy, we conclude that these outputs are identically distributed.

Having established that the output distributions of $P'[\mathsf{sk1}, \widetilde{\mathsf{pk2}}]$ and $P''[\widetilde{\mathsf{pk2}}]$ are identical (for every input), the claim follows because piO is a probabilistic indistinguishability obfuscator.

Claim 4.4. $\left(\mathsf{pk1}, \widetilde{\mathsf{pk2}}, \mathsf{piO}(P''_{\widetilde{\mathsf{pk2}}})\right) \stackrel{c}{=} \left(\widetilde{\mathsf{pk1}}, \widetilde{\mathsf{pk2}}, \mathsf{piO}(P''_{\widetilde{\mathsf{pk2}}})\right).$

Proof. This claim follows from the indistinguishability between standard and lossy public-keys of the underlying scheme.

Combining Claims 4.1-4.3, the two distributions $\left(\mathsf{pk1}, \mathsf{pk2}, \mathsf{piO}(P_{\mathsf{sk1}, \mathsf{pk2}})\right)$ and $\left(\widetilde{\mathsf{pk1}}, \widetilde{\mathsf{pk2}}, \mathsf{piO}(P''_{\widetilde{\mathsf{pk2}}})\right)$ are computationally indistinguishable. We complete the proof of Lemma 3 by observing that keys drawn from the latter distribution are lossy, since the key $\widetilde{\mathsf{pk1}}$ is lossy, the Enc-Spooky procedure just uses the underlying encryption procedure with $\widetilde{\mathsf{pk1}}$, and the program $P''[\widetilde{\mathsf{pk2}}]$ that we obfuscate is independent of $\widetilde{\mathsf{pk1}}$.

4.3 piO Based Multi-key Spooky Encryption

To obtain spooky encryption for more than two inputs, we would like to invoke our general transformation from 2-key spooky encryption to n-key spooky encryption (see Theorem 8). The scheme in Theorem 6 supports spooky multiplication, but we need it to support *multiple alternating hops* of (single-key) additive homomorphism and spooky multiplication. This is obtained by the following lemma:

Lemma 4. *Assume the existence of (1) piO and (2) a lossy encryption scheme that is homomorphic for all one-bit to one-bit functions with perfect malicious* *circuit privacy. Then, for every $d = d(\kappa)$, there exists an encryption scheme that supports d interleaved levels of AFS-2-spooky multiplications and single-key additions.*

Proof (Proof Sketch). To obtain an additive homomorphism, we use a construction of Canetti *et al.* [6] which, assuming piO, transforms any lossy encryption into a d-leveled FHE. This is done by taking d copies of keys of the original lossy scheme and publishing $d - 1$ obfuscated programs where the ith obfuscated program takes as input two ciphertexts encrypted under the ith key, decrypts them (using the ith private-key which is hard-wired) applies one operation (AND, XOR, NAND, etc.) and encrypts the result under the $(i + 1)^{\text{th}}$ key. Using the fact that the scheme is lossy, Canetti *et al.* show that the piO obfuscation hides the hard-wired private keys and semantic security is maintained.

For our application, we need to compute multiple spooky multiplications, and then sum them up with single-key addition. To get n-input AFS-spooky we need to sum up n ciphertexts, which can be done using an addition tree of depth $d = \log n$.

Looking more closely at the construction from [6], we observe that by setting $d = i \log n$ we can already support i interleaving hops of (single-key) additive homomorphism and 2-input spooky multiplications. This follows since the transformation in [6] has the property that after every additive homomorphic operation, we obtain a fresh ciphertext (under a new-key).

Using the scheme from Lemma 4 and applying Theorem 8, we get:

Theorem 7 (n-Key Spooky from piO). *Assume existence of (1) piO and (2) a lossy encryption scheme that is homomorphic for all single-bit to single-bit functions with perfect malicious circuit privacy. Then there exists a leveled AFS-spooky encryption scheme.*

5   From 2-Input to n-Input AFS-Spooky

Theorem 8 (2-Spooky to n-Spooky). *Let $d = d(\kappa)$ and assume that there exists a public-key bit-encryption scheme that supports $2d$ (interleaving) hops of (1) single-key compact additive homomorphism and (2) two-key spooky multiplication. Then, that same scheme is a d-level AFS-spooky encryption.*

Proof. Let (Gen, Enc, Dec) be the encryption scheme in the theorem statement, let Spooky-Mult be the spooky multiplication PPT algorithm and let Eval be the single-key homomorphic evaluation algorithm (that supports compact additive homomorphism). We show a procedure that given as input:

1. A depth-d, fan-in-2, n-input arithmetic circuit over $\mathsf{GF}(2)$, $C : (\{0,1\}^*)^n \to \{0,1\}$;
2. n public-keys $\mathsf{pk}_1, \ldots, \mathsf{pk}_n$; and
3. n ciphertexts $c_1, \ldots, c_n$, where $c_j = \mathsf{Enc}(\mathsf{pk}_j, x_j)$,

outputs a sequence of ciphertexts $c_1', \ldots, c_n'$ such that $\sum_{j \in [n]} \mathsf{Dec}_{\mathsf{sk}_j}(c_j') = C(x_1, \ldots, x_n)$ (where addition is over $\mathsf{GF}(2)$).

The procedure processes the circuit wire by wire. We maintain the invariant that whenever a wire w is processed, the procedure generates ciphertexts $c_1^{(w)}, \ldots, c_n^{(w)}$ such that $\sum_{j \in [n]} \mathsf{Dec}_{\mathsf{sk}_j}(c_j^{(w)})$ is the correct value of the wire w when the circuit C is evaluated on input $(x_1, \ldots, x_n)$. Furthermore, if the wire w is at distance i from the input then $c_1^{(w)}, \ldots, c_n^{(w)}$ have passed at most $2i$ hops of homomorphic operations. In particular, at the end of the process the procedure will have generated the sequence of ciphertexts $c_1^{\mathsf{out}}, \ldots, c_n^{\mathsf{out}}$ such that $\sum_{j \in [n]} \mathsf{Dec}_{\mathsf{sk}_j}(c_j^{\mathsf{out}})$ is equal to the output value of the circuit, as required. We proceed to describe how the wires are (inductively) processed.

Consider an *input* wire w, corresponding to an input bit b which is part of the ith input x_i, and for which we are given the input ciphertext $c = \mathsf{Enc}_{\mathsf{pk}_i}(b)$. For that wire we set $c_i^{(w)} = c$ and $c_j^{(w)} = \mathsf{Enc}_{\mathsf{pk}_{j'}}(0)$ for all $j \neq i$. Hence, $\sum_{j \in [n]} \mathsf{Dec}_{\mathsf{sk}_j}(c_j^{(w)}) = \mathsf{Dec}_{\mathsf{sk}_i}(c) = b$, which is the correct value for the wire w.

Consider a gate g with input wires u, v and output wire w. Let b_u (resp., b_v) be the value on the wire u (resp., v) when C is evaluated on input $(x_1, \ldots, x_n)$. By induction, we have already generated ciphertexts $c_1^{(u)}, \ldots, c_n^{(u)}$ and $c_1^{(v)}, \ldots, c_n^{(v)}$ such that $\sum_{j \in [n]} \mathsf{Dec}_{\mathsf{sk}_j}(c_j^{(u)}) = b_u$ and $\sum_{j \in [n]} \mathsf{Dec}_{\mathsf{sk}_j}(c_j^{(v)}) = b_v$.

For the case that g is an addition gate, we set $c_j^{(w)} = \mathsf{Eval}\left(\mathsf{pk}_j, \oplus, c_j^{(u)}, c_j^{(v)}\right)$ and we get:

$$\sum_{j \in [n]} \mathsf{Dec}_{\mathsf{sk}_j}(c_j^{(w)}) = \sum_{j \in [n]} \mathsf{Dec}_{\mathsf{sk}_j}(\mathsf{Eval}_{\mathsf{pk}_j}(\oplus, c_j^{(u)}, c_j^{(v)})) = \sum_{j \in [n]} \mathsf{Dec}_{\mathsf{sk}_j}(c_j^{(u)}) \oplus \mathsf{Dec}_{\mathsf{sk}_j}(c_j^{(v)}) = b_u \oplus b_v,$$

which is the correct value for the wire w. Furthermore, each new ciphertext was obtained by just a single homomorphic operation.

Now consider the case that g is a multiplication gate. We first compute auxiliary ciphertexts $(f_{j,j'}, g_{j,j'}) = \mathsf{Spooky\text{-}Mult}(\mathsf{pk}_j, \mathsf{pk}_{j'}, c_j^{(u)}, c_{j'}^{(v)})$, for every $j, j' \in [n]$. We then set

$$c_j^{(w)} = \mathsf{Eval}_{\mathsf{pk}_j}(\oplus, f_{j,1}, \ldots, f_{j,n}, g_{1,j}, \ldots, g_{n,j}).$$

We obtain that:

$$\sum_{j \in [n]} \mathsf{Dec}_{\mathsf{sk}_j}\left(c_j^{(w)}\right) = \sum_{j \in [n]} \mathsf{Dec}_{\mathsf{sk}_j}\left(\mathsf{Eval}_{\mathsf{pk}_j}(\oplus, x_{j,1}, \ldots, x_{j,n}, y_{1,j}, \ldots, y_{n,j})\right)$$

$$= \sum_{j \in [n]} \sum_{j' \in [n]} \mathsf{Dec}_{\mathsf{sk}_j}(f_{j,j'}) \oplus \mathsf{Dec}_{\mathsf{sk}_j}(g_{j',j})$$

$$= \sum_{j \in [n]} \sum_{j' \in [n]} \mathsf{Dec}_{\mathsf{sk}_j}(c_j^{(u)}) \cdot \mathsf{Dec}_{\mathsf{sk}_j}(c_{j'}^{(v)})$$

$$= \left(\sum_{j \in [n]} \mathsf{Dec}_{\mathsf{sk}_j}(c_j^{(u)})\right) \cdot \left(\sum_{j' \in [n]} \mathsf{Dec}_{\mathsf{sk}_j}(c_{j'}^{(v)})\right) = b_u \cdot b_v,$$

which is the correct value for the wire w (where the fourth equality is due to the Spooky-Mult guarantee). Furthermore, each new ciphertext was obtained by applying two hops of homomorphic operations.

6 Applications of Spooky Encryption

6.1 Counter Example for the [1] Heuristic

Building on [11], we show that AFS-2-spooky encryption gives a counter-example to a natural method proposed by Aiello et al. [1] for building succinct arguments for NP, resolving a question posed by [11]. The suggestion of Aiello *et al.* [1] was to take any multi-prover interactive proof-system (MIP) and to use that proof-system using only a *single* prover by sending all of the MIP queries encrypted under independents keys of a homomorphic encryption scheme.[8] The fact that the scheme is homomorphic allows the honest prover to answer the different queries (homomorphically) and the intuition was that the use of different keys means that only local homomorphisms are possible. Dwork *et al.* [11] questioned this intuition and raised the question of whether there exist spooky encryption schemes that allow for other kinds of attacks which can break the soundness of the [1] protocol. We show that this is indeed the case: there exists an MIP (suggested by [11]) which, when combined with any AFS-2-spooky encryption scheme via the [1] transformation, yields an insecure protocol. The MIP that we use is based on a PCP for 3-coloring due to Petrank [27]:

Theorem 9 [27]. *There exists a universal constant $\varepsilon > 0$ such that distinguishing between the following two types of graphs is NP complete:*

- *G is 3-colorable.*
- *Every 3-coloring of G has at least ε fraction of monochromatic edges.*

This PCP leads to the following natural MIP protocol between a verifier V and two non-communicating provers P_1 and P_2 (who, in case G is 3-colorable, also have access to the same 3-coloring of G).

1. V chooses a random edge $(u, v) \in E$, then with probability $1/3$ it sets $q_1 = u$ and $q_2 = v$, with probability $1/3$ it sets $q_1 = u$ and $q_2 = u$, and with probability $1/3$ it sets $q_1 = v$ and $q_2 = v$. V sends q_1 to P_1 and q_2 to P_2.
2. Each P_i sends the color $a_i \in \{0, 1, 2\}$ of the vertex q_i (encoded as two bits).
3. V accepts if $q_1 = q_2$ and $a_1 = a_2$, or if $q_1 \neq q_2$ and $a_1 \neq a_2$.

Completeness and soundness are easy to see, some details are given in the full version [9].

Insecurity of the 3-Coloring MIP. Composed the foregoing MIP with any AFS-2-spooky encryption scheme yields an insecure protocol. More specifically, the

---

[8] Actually, the original suggestion in [1] was to use a PCP (rather than an MIP). Dwork *et al.* [11] show that using PCPs is not sound and raise the question of whether soundness can be obtained by replacing the PCP with an MIP.

cheating prover is given ciphertexts $c_1 = \mathsf{Enc}_{\mathsf{pk}_1}(q_1)$ and $c_2 = \mathsf{Enc}_{\mathsf{pk}_2}(q_2)$. Loosely speaking, using the spooky evaluation algorithm it can produce ciphertexts $\mathsf{Enc}_{\mathsf{pk}_1}(a_1)$ and $\mathsf{Enc}_{\mathsf{pk}_2}(a_2)$ for bits $a_1, a_2 \in \{0,1\}$ such that $a_1 = a_2$ if and only if $u = v$. It sends as its answers to V the ciphertext $\big(\mathsf{Enc}_{\mathsf{pk}_1}(0), \mathsf{Enc}_{\mathsf{pk}_1}(a_1)\big)$ as its answer to the first query and $\big(\mathsf{Enc}_{\mathsf{pk}_1}(0), \mathsf{Enc}_{\mathsf{pk}_1}(a_2)\big)$ as its answer to the second query (the extra encryption of 0 is used simply because the verifier expects an answer with 2 bits).

Now, if the verifier choose $q_1 = u$ and $q_2 = v$ (corresponding to the first of the three possibilities) then $q_1 \neq q_2$ and so $a_1 \neq a_2$ and the verifier accepts. Otherwise, (i.e. if $q_1 = q_2$) then we have that $a_1 = a_2$ and again the verifier accepts. Hence, we have shown a strategy that breaks the soundness of the scheme with probability 1.

6.2   2-Round MPC from AFS-Spooky Encryption

AFS-spooky encryption seems to be a useful tool for minimally-interactive multi-party protocols: it lets each party broadcast an encryption of its input under its own key, then everyone individually performs the AFS-spooky evaluation locally, and each party can locally decrypt and recover a share of the output (relative to an additive n-out-of-n secret-sharing scheme). Finally another round of communication can be used to recover the secret from all the shares. Implementing this the approach requires attention to some details, such as ensuring that the spooky evaluation is deterministic (so that all the parties arrive at the same sharing) and making the shares simulatable. The latter can be done by having each party distribute a random additive sharing of 0 in the first round, and then adding all their received shares to their spooky generated share before broadcasting it in the second round.

A different (but similar) avenue for implementing 2-round MPC, is by reducing AFS-spooky encryption to *multi-key FHE with threshold decryption (TMFHE)*. This primitive was recently formalized by Mukherjee and Wichs [22], who showed how to use it to generically construct 2-round MPC. Just like spooky encryption, a TMFHE scheme can homomorphically process n ciphertexts $c_1, \ldots, c_n$, encrypting values $x_1, \ldots, x_n$ under independent public keys $\mathsf{pk}_1, \ldots, \mathsf{pk}_n$, producing for any function f a ciphertext $c^* = \mathsf{Eval}(f, (\mathsf{pk}_1, c_1), \ldots, (\mathsf{pk}_n, c_n))$. The ciphertexts c^* cannot be decrypted by any single secret keys sk_i individually, but each party can compute a partial decryption $y_i = \mathsf{PartDec}_{\mathsf{sk}_i}(c^*)$ and these y's can be combined to get $y = \mathsf{FinDec}(y_1, \ldots, y_n) = f(x_1, \ldots, x_n)$. For security, Mukherjee and Wichs required that for each individual i, the partial decryption y_i can be simulated given the evaluated ciphertext c^*, the final output y and the secret key sk_j for $j \neq i$ (see [22] for formal definitions).

We observe that an AFS-spooky encryption with perfect correctness immediately yields a TMFHE scheme. The homomorphic evaluation procedure Eval of the TMFHE runs the $\mathsf{Spooky\text{-}Eval}$ procedure of the AFS-spooky encryption and sets $c^* = (c_1', \ldots, c_n')$ to be the resulting ciphertexts. The partial decryption procedure $\mathsf{PartDec}_{\mathsf{sk}_i}(c^*)$ outputs $y_i = \mathsf{Dec}_{\mathsf{sk}_i}(c_i')$ and the combination procedure

$\mathsf{FinDec}(y_1, \ldots, y_n)$ outputs $y = \bigoplus_{i=1}^{n} y_i$. For security, we observe that each partial decryption y_i can be simulated given $c^* = (c'_1, \ldots, c'_n)$, y and sk_j for $j \neq i$ by computing $y_j = \mathsf{Dec}_{\mathsf{sk}_j}(c'_j)$ and setting $y_i = y \oplus (\bigoplus_{j \neq i} y_j)$.[9] This proves the following theorem.

Theorem 10. *An AFS-spooky encryption scheme with perfect correctness implies a multi-key FHE with threshold decryption (TMFHE).*

Using the above theorem and the results of [22] which constructs a 2-round MPC from TMFHE, we get the following corollaries.

Corollary 2. *Assuming the existence of a weak AFS-spooky encryption scheme:*

- *There exists a 2-round* MPC *protocol with semi-honest security. If the encryption scheme is in the plain model then so is the* MPC *protocol and if the encryption scheme requires a CRS then so does the* MPC *protocol.*
- *Furthermore, assuming the existence of NIZKs in the CRS model, there exists a 2-round* MPC *protocol with malicious security in the CRS model.*

Combining this with our construction of AFS-spooky encryption without a CRS from iO, we get the first construction of a 2-round semi-honest MPC protocol in the plain model.

Corollary 3. *Assume existence of (1)* piO *and (2) a lossy encryption scheme that is homomorphic for all single-bit to single-bit functions with perfect malicious circuit privacy. Then, there exists a 2-round* MPC *protocol with semi-honest security in the plain model.*

6.3   Function Secret Sharing

Function secret sharing (FSS), recently introduced by Boyle, Gilboa and Ishai [4], allows a dealer to split a function f into k succinctly described functions $\hat{f}_1, \ldots, \hat{f}_k$ such that (1) any strict subset of the $\hat{f}_i$'s reveals nothing about f and (2) for any x it holds that the values $\hat{f}_1(x), \ldots, \hat{f}_k(x)$ are an additive secret sharing of $f(x)$. Boyle *et al.* gave constructions under standard assumptions for certain restricted families of functions and a general construction for *any* poly-size circuit, based on piO. We show how to construct such a general FSS scheme given any AFS-spooky encryption scheme. In particular, we obtain a leveled FSS scheme assuming only LWE.

To construct such an FSS scheme, the dealer first generates a k-out-of-k secret sharing $f_1, \ldots, f_k$ of the *description* of the function f. The dealer also generates k key pairs $(\mathsf{pk}_i, \mathsf{sk}_i)_{i \in [k]}$ for the AFS spooky scheme and publishes $\hat{f}_i \overset{\mathrm{def}}{=} (\mathsf{sk}_i, \mathsf{pk}_1, \ldots, \mathsf{pk}_k, \mathsf{Enc}_{\mathsf{pk}_1}(f_1), \ldots, \mathsf{Enc}_{\mathsf{pk}_k}(f_k))$ as the ith share. Assuming

---

[9] We note that imperfect correctness of the AFS-spooky scheme will translate into a security problem for the TMFHE scheme, as the simulated y_i will have a different distribution than the real ones.

the scheme is semantically secure, any strict subset of the $\hat{f}_i$'s hides the original function f (upto its description length).

For the FSS functionality, given an input x we can consider the circuit C_x that takes as input k shares of a function f, adds them up and applies the resulting function to the input x (which, say, is hardwired). To evaluate $\hat{f}_i$ on x, we run the spooky evaluation algorithm, which we assume wlog is deterministic, on $\mathsf{Enc}_{\mathsf{pk}_1}(f_1), \ldots, \mathsf{Enc}_{\mathsf{pk}_k}(f_k)$ with respect to the circuit C_x. Thus, given each $\hat{f}_i$ separately, we can generate the same ciphertexts $c_1, \ldots, c_k$ which are encryptions of an additive secret sharing of $f(x)$. Each function $\hat{f}_i$ can then be used to decrypt c_i and publish its share of $f(x)$.

A De-centralized View. We remark that the above construction can be viewed as a de-centralized FSS. More specifically, we may have some k (not necessarily secret or functional) shares $f_1, \ldots, f_k$ of a function f, where each share is owned by a different player. Player i can generate a key pair $(\mathsf{pk}_i, \mathsf{sk}_i)$ and broadcast $(\mathsf{pk}_i, \mathsf{Enc}_{\mathsf{pk}_i}(f_i))$ to all other players. Using our scheme, after learning the input x, the players can (non-interactively) generate an additive secret sharing of $f(x)$.

7 Spooky-Free Encryption

We turn now to study *spooky-free* encryption, i.e. an encryption scheme that ensures that no spooky relations can be realized by an adversary. The formal definition roughly states that any correlation that an attacker can induce between the original messages $(m_1, \ldots, m_n)$ and "tampered messages" $(m'_1, \ldots, m'_n)$, can be simulated by a "local simulator" that produces m'_i only as a function of m_i (and some shared randomness).

Definition 6 (Spooky-Free Encryption). *An encryption scheme* $(\mathsf{Gen}, \mathsf{Enc}, \mathsf{Dec})$ *is* spooky-free *if for every PPT adversary* $\mathcal{A}$ *there exists a PPT simulator* $\mathcal{S}$ *such that for all PPT message distributions* $\mathcal{D}$, *the two distributions* $\mathbf{REAL}_{\mathcal{D},\mathcal{A}}(\kappa)$ *and* $\mathbf{SIM}_{\mathcal{D},\mathcal{S}}(\kappa)$ *specified below are computationally indistinguishable:*

$\underline{\mathbf{REAL}_{\mathcal{D},\mathcal{A}}(\kappa)}$: *1. Sample* $(m_1, \ldots, m_n, \alpha) \leftarrow \mathcal{D}(1^\kappa)$; // α *is auxiliary information*
  2. Choose $(\mathsf{pk}_i, \mathsf{sk}_i) \leftarrow \mathsf{Gen}(1^\kappa)$, *set* $c_i \leftarrow \mathsf{Enc}_{\mathsf{pk}_i}(m_i)$ *for* $i = 1, \ldots, n$;
  3. Let $(c'_1, \ldots, c'_n) \leftarrow \mathcal{A}(\mathsf{pk}_1, \ldots, \mathsf{pk}_n, c_1, \ldots, c_n)$;
  4. Set $m'_i = \mathsf{Dec}_{\mathsf{sk}_i}(c'_i)$ *for* $i = 1, \ldots, n$;
  5. Output $(m_1, \ldots, m_n, m'_1, \ldots, m'_n, \alpha)$.

$\underline{\mathbf{SIM}_{\mathcal{D},\mathcal{S}}(\kappa)}$: *1. Sample* $(m_1, \ldots, m_n, \alpha) \leftarrow \mathcal{D}(1^\kappa)$; // α *is auxiliary information*
  2. Sample a random r, *let* $m'_i = \mathcal{S}(1^\kappa, 1^n, i, m_i; r)$ *for* $i = 1, \ldots, n$;
  3. Output $(m_1, \ldots, m_n, m'_1, \ldots, m'_n, \alpha)$.

It is not hard to see that spooky-freeness for $n \geq 2$ implies semantic security. As a small subtlety, here the attacker must choose the messages it claims to distinguish before seeing the public-key, since the message sampler $\mathcal{D}$ does not know anything public keys used in the real experiment. (We defined it this way since stronger security was not needed for our delegation application.) Of course, this minor difference from standard semantic security is without loss of generality when the message space is polynomial small (e.g., for bit encryption).

Lemma 5. *A spooky-free scheme for $n \geq 2$ is semantically secure (in the "selective" sense discussed above).*

Proof. Suppose that a scheme (Enc, Dec, Gen) is *not semantically secure*, and let B be an attacker that can distinguish $\mathsf{Enc}_{\mathsf{pk}}(x_0)$ from $\mathsf{Enc}_{\mathsf{pk}}(x_1)$. We use B to construct a sampler $\mathcal{D}$ and attacker $\mathcal{A}$ that can fool any simulator $\mathcal{S}$ with non-negligible probability. We assume that $\mathcal{D}$ and $\mathcal{A}$ (and $\mathcal{S}$) know the messages x_0 and x_1 whose encryption B can distinguish.

$\mathcal{D}$ draws at random $m_1 \leftarrow \{x_0, x_1\}$ and sets $m_i := 0$ for $i > 1$. Upon seeing n ciphertexts $c_1, \ldots, c_n$, $\mathcal{A}$ gives c_1 to B, asking him to guess whether it encrypts x_0 or x_1. Let σ be the guess that B makes, then we know that $m_1 = x_\sigma$ with probability $\geq 1/2 + \varepsilon$. $\mathcal{A}$ then sets $c_i' = c_i$ for all $i \neq 2$, and sets c_2' to be a fresh encryption of x_σ under pk_2.

As we can see, the output of the real experiment has the tuple (m_1, m_2') distributed as (x_b, x_σ), where b is a random bit and $\sigma = b$ with probability $\geq 1/2 + \varepsilon$. On the other hand, the simulator for the second message m_2' is only given $m_2 = 0$ as the input, and has to guess σ' s.t., $\Pr[b = \sigma'] \geq 1/2 + \varepsilon$, which is impossible information-theoretically.

In the full version of this work [9] we show that spooky-free homomorphic encryption is exactly the ingredient needed to instantiate the idea of Aiello et al. [1] for converting general multi-prover (MIP) systems into single-prover arguments.[10] We also show there that non-malleable encryption is always spooky-free (albeit without any homomorphic capabilities), and we construct a spooky-free FHE scheme using a strong security component called succinct non-interactive argument of knowledge (SNARK).[11]

Spooky-Free Encryption with CRS. Definition 6 can be naturally extended to the common-reference-string model. We use this relaxation in the full version to gain somewhat better efficiency (at the price of a slightly harder proof of security). We note that, unlike the setting of spooky encryption from Sect. 3, we do not need the CRS to get the desired functionality, but rather use it only to improve efficiency. Our construction remains spooky-free (but slower) if all the public keys are chosen completely independently.

---

[10] An alternate route for instantiating the [1] idea due to [18,19] is to use special types of MIP, which satisfy a stronger soundness condition, together with any (possibly spooky) homomorphic encryption scheme.

[11] Of course, this construction does not give any new one-round delegation schemes, since SNARKs trivially imply the existence of such a scheme directly (i.e., without building spooky-free encryption). Still, if better constructions of spooky-free FHE are found, they would immediately imply new delegation schemes for NP.

Acknowledgments. This work was done in part while the authors were visiting the Simons Institute for the Theory of Computing, supported by the Simons Foundation and by the DIMACS/Simons Collaboration in Cryptography through NSF grant #CNS-1523467.

The first author was partially supported by gifts from VMware Labs and Google, and NSF grants 1319051, 1314568, 1065288, 1017471.

The second author was supported in part by the Defense Advanced Research Projects Agency (DARPA) and Army Research Office(ARO) under Contract No. W911NF-15-C-0236.

The third author was supported by NSF MACS - CNS-1413920, DARPA IBM - W911NF-15-C-0236, SIMONS Investigator award Agreement Dated 6-5-12 and DARPA NJIT - W911NF-15-C-0226.

The last author was supported in part by NSF grants CNS-1347350, CNS-1314722, CNS-1413964.

References

1. Aiello, W., Bhatt, S., Ostrovsky, R., Rajagopalan, S.R.: Fast verification of any remote procedure call: short witness-indistinguishable one-round proofs for NP. In: Welzl, E., Montanari, U., Rolim, J.D.P. (eds.) ICALP 2000. LNCS, vol. 1853, pp. 463–474. Springer, Heidelberg (2000)
2. Bitansky, N., Chiesa, A.: Succinct arguments from multi-prover interactive proofs and their efficiency benefits. In: Safavi-Naini, R., Canetti, R. (eds.) CRYPTO 2012. LNCS, vol. 7417, pp. 255–272. Springer, Heidelberg (2012)
3. Boneh, D., Segev, G., Waters, B.: Targeted malleability: homomorphic encryption for restricted computations. In: Goldwasser, S. (ed.) Innovations in Theoretical Computer Science, Cambridge, MA, USA, 8–10 January 2012, pp. 350–366. ACM (2012)
4. Boyle, E., Gilboa, N., Ishai, Y.: Function secret sharing. In: Oswald, E., Fischlin, M. (eds.) EUROCRYPT 2015. LNCS, vol. 9057, pp. 337–367. Springer, Heidelberg (2015)
5. Brakerski, Z., Langlois, A., Peikert, C., Regev, O., Stehlé, D.: Classical hardness of learning with errors. In: Boneh, D., Roughgarden, T., Feigenbaum, J. (eds.) Symposium on Theory of Computing Conference, STOC 2013, Palo Alto, CA, USA, 1–4 June 2013, pp. 575–584. ACM (2013)
6. Canetti, R., Lin, H., Tessaro, S., Vaikuntanathan, V.: Obfuscation of probabilistic circuits and applications. In: Dodis, Y., Nielsen, J.B. (eds.) TCC 2015, Part II. LNCS, vol. 9015, pp. 468–497. Springer, Heidelberg (2015)
7. Clear, M., McGoldrick, C.: Multi-identity and multi-key leveled FHE from learning with errors. In: Gennaro, R., Robshaw, M. (eds.) CRYPTO 2015. LNCS, vol. 9216, pp. 630–656. Springer, Heidelberg (2015)
8. Cramer, R., Hanaoka, G., Hofheinz, D., Imai, H., Kiltz, E., Pass, R., Shelat, A., Vaikuntanathan, V.: Bounded CCA2-secure encryption. In: Kurosawa, K. (ed.) ASIACRYPT 2007. LNCS, vol. 4833, pp. 502–518. Springer, Heidelberg (2007)
9. Dodis, Y., Halevi, S., Rothblum, R.D., Wichs, D.: Spooky encryption and its applications. IACR Cryptology ePrint Archive 2016:272 (2016)
10. Dodis, Y., Haralambiev, K., López-Alt, A., Wichs, D.: Efficient public-key cryptography in the presence of key leakage. In: Abe, M. (ed.) ASIACRYPT 2010. LNCS, vol. 6477, pp. 613–631. Springer, Heidelberg (2010)

11. Dwork, C., Langberg, M., Naor, M., Nissim, K., Reingold, O.: Succinct proofs for NP and spooky interactions (2004, Unpublished manuscript). http://www.cs.bgu. ac.il/~kobbi/papers/spooky_sub_crypto.pdf

12. Garg, S., Gentry, C., Halevi, S., Raykova, M.: Two-round secure MPC from indistinguishability obfuscation. In: Lindell, Y. (ed.) TCC 2014. LNCS, vol. 8349, pp. 74–94. Springer, Heidelberg (2014)

13. Gentry, C., Halevi, S., Vaikuntanathan, V.: i-hop homomorphic encryption and rerandomizable Yao circuits. In: Rabin, T. (ed.) CRYPTO 2010. LNCS, vol. 6223, pp. 155–172. Springer, Heidelberg (2010). http://eprint.iacr.org/2010/145

14. Gentry, C., Wichs, D.: Separating succinct non-interactive arguments from all falsifiable assumptions. In: STOC, pp. 99–108 (2011)

15. Gilboa, N., Ishai, Y.: Distributed point functions and their applications. In: Nguyen, P.Q., Oswald, E. (eds.) EUROCRYPT 2014. LNCS, vol. 8441, pp. 640–658. Springer, Heidelberg (2014)

16. Goldreich, O., Micali, S., Wigderson, A.: How to play any mental game or a completeness theorem for protocols with honest majority. In: Proceedings of the 19th Annual ACM Symposium on Theory of Computing, New York, USA, pp. 218–229 (1987)

17. Ishai, Y., Kushilevitz, E.: Randomizing polynomials: a new representation with applications to round-efficient secure computation. In: 41st Annual Symposium on Foundations of Computer Science, FOCS 2000, 12–14 November 2000, Redondo Beach, California, USA, pp. 294–304 (2000)

18. Kalai, Y.T., Raz, R., Rothblum, R.D.: Delegation for bounded space. In: STOC, pp. 565–574 (2013)

19. Kalai, Y.T., Raz, R., Rothblum, R.D.: How to delegate computations: the power of no-signaling proofs. In: Symposium on Theory of Computing, STOC 2014, New York, NY, USA, May 31–June 03 2014, pp. 485–494 (2014)

20. Katz, J., Ostrovsky, R.: Round-optimal secure two-party computation. In: Franklin, M. (ed.) CRYPTO 2004. LNCS, vol. 3152, pp. 335–354. Springer, Heidelberg (2004)

21. López-Alt, A., Tromer, E., Vaikuntanathan, V.: On-the-fly multiparty computation on the cloud via multikey fully homomorphic encryption. In: Proceedings of the 44th Symposium on Theory of Computing Conference, STOC 2012, New York, NY, USA, 19–22 May 2012, pp. 1219–1234 (2012)

22. Mukherjee, P., Wichs, D.: Two round multiparty computation via multi-key FHE. In: Fischlin, M., Coron, J.-S. (eds.) EUROCRYPT 2016. LNCS, vol. 9666, pp. 735–763. Springer, Heidelberg (2016). doi:10.1007/978-3-662-49896-5_26. http://eprint.iacr.org/2015/345

23. Naor, M., Pinkas, B.: Efficient oblivious transfer protocols. In: Proceedings of the Twelfth Annual Symposium on Discrete Algorithms, 7–9 January 2001, Washington, DC, USA, pp. 448–457 (2001)

24. Ostrovsky, R., Paskin-Cherniavsky, A., Paskin-Cherniavsky, B.: Maliciously circuit-private FHE. In: Garay, J.A., Gennaro, R. (eds.) CRYPTO 2014, Part I. LNCS, vol. 8616, pp. 536–553. Springer, Heidelberg (2014). https://eprint.iacr.org/2013/307

25. Peikert, C.: Public-key cryptosystems from the worst-case shortest vector problem: extended abstract. In: Mitzenmacher, M. (ed.) Proceedings of the 41st Annual ACM Symposium on Theory of Computing, STOC 2009, Bethesda, MD, USA, May 31–June 2 2009, pp. 333–342. ACM (2009)

26. Peikert, C., Shiehian, S.: Multi-key FHE from LWE, revisited. Cryptology ePrint Archive, report 2016/196 (2016). http://eprint.iacr.org/
27. Petrank, E.: The hardness of approximation: gap location. Comput. Complex. **4**, 133–157 (1994)
28. Regev, O.: On lattices, learning with errors, random linear codes, and cryptography. J. ACM **56**(6), 34:1–34:40 (2009)

Spooky Interaction and Its Discontents: Compilers for Succinct Two-Message Argument Systems

Cynthia Dwork[1], Moni Naor[2], and Guy N. Rothblum[3(✉)]

[1] Microsoft Research, Mountain View, USA
[2] Department of Computer Science and Applied Math,
Weizmann Institute of Science, Rehovot, Israel
[3] Samsung Research America, Mountain View, USA
rothblum@alum.mit.edu

Abstract. We are interested in constructing short two-message arguments for various languages, where the complexity of the verifier is small (e.g. linear in the input size, or even sublinear if the input is coded appropriately).

In 2000 Aiello et al. suggested the tantalizing possibility of obtaining such arguments for all of NP. These have proved elusive, despite extensive efforts. Our work builds on the compiler of Kalai and Raz, which takes as input an interactive proof system consisting of several rounds and produces a two-message argument system. The proof of soundness of their compiler relies on superpolynomial hardness assumptions.

In this work we obtain a succinct two-message argument system for any language in NC, where the verifier's work is linear (or even polylogarithmic). Soundness relies on any standard (polynomially hard) private information retrieval scheme or fully homomorphic encryption scheme. This is the first non trivial two-message succinct argument system that is based on a standard polynomial-time hardness assumption. We obtain this result by proving that the compiler is sound (under standard polynomial hardness assumptions) if the verifier in the original protocol runs in logarithmic space and public coins. We obtain our two-message argument by applying the compiler to an interactive proof protocol of Goldwasser, Kalai and Rothblum. On the other hand, we prove that under standard assumptions there is a sound interactive proof protocol that, when run through the compiler, results in a protocol that is not sound.

M. Naor—Incumbent of the Judith Kleeman Professorial Chair. Research supported in part by grants from the Israel Science Foundation, BSF and Israeli Ministry of Science and Technology and from the I-CORE Program of the Planning and Budgeting Committee and the Israel Science Foundation (grant No. 4/11). Part of this work was done while visiting Microsoft Research.

G.N. Rothblum—Part of this work was done while the author was at Microsoft Research Silicon Valley.

© International Association for Cryptologic Research 2016
M. Robshaw and J. Katz (Eds.): CRYPTO 2016, Part III, LNCS 9816, pp. 123–145, 2016.
DOI: 10.1007/978-3-662-53015-3_5

1   Introduction

Imagine going on vacation and upon your return you find that not only has your home computer ordered the garden robot to mow the lawn but it has also commissioned from some vendor a lengthy computation that you have been postponing for a while. While verifying that the lawn has been properly mowed is simple, you are suspicious about the computation and would like to receive a confirmation that it was performed correctly. Ideally such a proof would be a short certificate attached to the result of the program. Hence we are interested in proofs or arguments[1] that are either non-interactive or two-message.

The problem of constructing short two-message arguments for various languages where the verifier is very efficient (e.g. linear in the input size, or even sublinear if it is coded properly) has received quite a lot of attention over the last twenty years (see Sect. 1.1). Suppose that we have a low communication public coins *interactive* (multi-round) protocol for proving (or arguing) membership in the language. A possible approach for obtaining short arguments is using a "compiler" that takes any protocol consisting of several rounds and removes the need for interaction, producing a two-message argument system.

One approach to constructing such a compiler is having the verifier encrypt and send all of its (public coin) challenges, where encryption is performed using a very malleable[2] scheme, such as Private Information Retrieval (PIR) or Fully Homomorphic Encryption (FHE)[3]. The prover uses the malleability to simulate his actions had the queries been given in plaintext, generating and sending back the appropriate ciphertexts (which should correspond to encryption of the answers the prover would give in the plain protocol). This compiler was studied by Kalai and Raz [KR09], as well as Kalai et al. [KRR14].

We investigate whether this compiler can be proved secure under *standard* cryptographic assumptions. That is, we do not want to base security on assumptions such as "having access to random oracles" or on so-called "knowledge assumptions", i.e. that one can extract from any machine that computes a certain function a value that is *seemingly* essential to that computation. Furthermore, we prefer not to rely on "super-polynomial hardness assumptions", i.e. that a certain cryptographic primitive is so hard it remains secure even if given enough time to break another primitive. In particular, such assumptions are not *falsifiable* in the sense of Naor [Nao03],[4] and they assume a strict hierarchy beyond $P \neq NP$. Also we want the underlying assumptions to be simple to state such as "*Learning With Errors is hard*". We prove positive and negative results about the above compiler:

---

[1] An argument is a "proof" that is sound (under cryptographic assumptions) so long as its creator is computationally bounded.

[2] Malleable in the cryptographic sense means that it is possible to manipulate a given ciphertext to generate related ciphertexts.

[3] As we shall see, the latter is needed if the prover's messages in the multi-round protocol depend on super-logarithmically-many bits sent by the verifier.

[4] A "falsifiable" cryptographic assumption is one that can be refuted efficiently. Falsifiability is a basic "litmus test" for cryptographic assumptions.

– Assume FHE or PIR exist. Then there exists a sound interactive proof protocol, and there exists an FHE or PIR (respectively) scheme E, such that when the compiler is applied to the proof system using E, *the resulting two-message argument is insecure*. In fact, the compiler (when applied to this protocol) is insecure using all known FHE schemes. See Theorem 2.

– For any FHE or PIR, if the verifier in the original protocol is log-space and uses only public coins, then the compiled argument is sound (See Theorem 4). Combining this with the work of Goldwasser et al. [GKR15], we obtain a succinct two-message argument system for any language in NC, where the verifier's work is linear (or even polylogarithmic if the input is coded appropriately). See Theorem 5. This is the first succinct two-message argument based on standard polynomial-time hardness assumptions.

1.1 Background

Obtaining succinct (e.g. sub-linear) two-message proof (or argument) systems has been a long-standing goal in the literature. Two primary approaches have been explored, both using cryptography to transform information-theoretic proof systems (interactive proofs, PCPs, or multi-prover interactive proofs) into non-interactive or two-message computationally-sound arguments.

Two-Message Arguments from PIR or FHE. The compiler studied in this work is rooted in a tantalizing suggestion of Aiello et al. [ABOR00a] in 2000, who proposed combining two powerful tools: The PCP (probabilistically checkable proofs) Theorem[5] and Computational PIR schemes[6] in order to obtain a succinct two-message argument system. In particular, leveraging the full strength of the PCP theorem, one could hope to obtain such arguments for all of NP. However, shortly thereafter Dwork et al. [DLN+] pointed out problems in the proof and showed various counter examples for techniques of proving such a statement (see [ABOR00b]). No direct refutation was shown.[7]

Kalai and Raz [KR09] modified the Aiello et al. method, and suggested using it as a general compiler for turning public-coin *interactive proofs* (rather than PCPs) into two argument systems (without increasing the communication significantly, see below). They showed that, for any interactive proof system, one can tailor the compiler to that proof system by taking a large enough security parameter (polynomial in the communication of the proof system), and obtain

---

[5] That states that for every language $L \in$ NP there exist a polynomial size witness/proof that may be verified, with constant error probability, by probing only a constant number of locations of the proof.

[6] Enabling a two-party protocol where one party holds a long string S and the other party is interested the value of the string at a particular location i; the second party does not want to reveal i and the goal is to have a low communication (much shorter than S) protocol; See Sect. 2.

[7] The original Aiello et al. [ABOR00a] protocol had an additional oversight, having to do with verifying the consistency of query answers. As Dwork et al. [DLN+] showed, this can be corrected using a probabilistic consistency check.

a secure two-message argument. This requires subexponential hardness assumptions about the PIR or FHE. Applying the compiler to the interactive proofs of Goldwasser et al. [GKR08,GKR15] (see below), they obtain two-message arguments for bounded-depth computations.

Kalai et al. [KRR14] study *no-signalling* proof systems, a restricted type of multi-prover interactive proof. They showed that, fixing any no-signalling proof, the compiler can also be tailored to that proof system, again giving a secure two-message argument (and also using sub-exponential hardness assumptions). Since no-signalling proof systems are more powerful than interactive proofs, they obtain two-message arguments for a larger class of computations (going from bounded-depth in [KR09] to P in [KRR14]).

Kilian, Micali, et Sequelae. In 1992 Kilian [Kil92] suggested a short argument system for any language in NP. The protocol required 4 messages and the total amount of bits sent was polylog(n), where n is the input length, times a security parameter. The cryptographic assumption needed was fairly conservative, namely the existence of collision-resistant hash functions. Provided the prover has a witness, the work done by the prover is polynomial in the instance plus the witness sizes. In this protocol, the prover first committed to a PCP proof using a hash function provided by the verifier via a Merkle Tree (first two messages). The verifier then issued queries to the PCP and the prover needed to open its commitment in the specified locations in a way that would make the PCP verifier accept[8] (requiring two additional messages).

Some time later, Micali [Mic00] suggested using the Fiat-Shamir methodology [FS86] of removing interaction from public-coin protocols using an idealized hash function (random oracle) to obtain a two-message (or even non-interactive) succinct argument for any language in NP. Micali's work raised the issue of whether it is possible to obtain such argument systems in the "real world" (rather than in an idealized model). Barak and Goldreich [BG08] showed that security for the 4-message protocol could be based on standard (polynomial-time) hardness assumptions, but no secure instantiation of non-interactive arguments for NP is known under standard cryptographic assumptions.

Negative Results and Perspective. On the negative side, Gentry and Wichs [GW11] have shown that constructing two-message adaptively sound arguments for NP is going to be tricky: take any short two-message (even designated-verifier) proof system for NP, and assume that there are exponentially hard one-way functions. Then, paradoxically, any black-box reduction from a cheating prover to a falsifiable assumption can actually be used to break the assumption. One can interpret this result in several ways: (*i*) We need to find non black-box techniques in this realm. (*ii*) We should explore the boundaries of the Gentry-Wichs proof, i.e. when can we obtain black-box reductions and in particular what happens to computation in P (as opposed to NP). (*iii*) Use

---

[8] This is not a precise representation of Kilian's work, for instance the PCP Theorem did not exist in its 'final' form when he proved his result.

a non-falsifiable assumption. We prefer the first two interpretations, but there are quite a few works taking approach (*iii*). Thus, Kalai and Raz [KR09] and Kalai et al. [KRR14] used super-polynomial hardness assumptions and obtained two-message succinct protocols for all languages computable in bounded depth and in P (respectively). Several works, Mie [Mie08], Groth [Gro10], Bitansky et al. [BCCT12] and Goldwasser et al. [GLR11] used a knowledge assumption (where one assumes that in order to perform a certain computation another piece of information is necessary and extractable).

Proofs for Muggles. Goldwasser et al. [GKR08, GKR15] were able to obtain a succinct interactive *proof* system (with many rounds) for any language that can be computed using small-depth circuits of polynomial size (NC, or, more generally, bounded-depth circuits). The prover in their system runs in polynomial time. The verifier runs in nearly-linear time and logarithmic space, and uses only public-coin. The communication and round complexities are related to the circuit depth (using bounded fan-in circuits).

Other Related Works. Paneth and Rothblum [PR14] construct non-interactive arguments in a common reference string model for any computation in P. Their constructions are based on efficiently falsifiable assumptions over multilinear maps. Candidates for multilinear maps have been suggested recently, starting with the work of Garg et al. [GGH13a], but the security of these objects is not yet well understood, and is an active area of research. Looking ahead, we note that our construction of two-message arguments for bounded-depth computations is currently the only other construction based on efficiently falsifiable assumptions; we assume only PIR or FHE, rather than assumptions over multilinear maps. In a different vein, Bitansky et al. [BGL+15] construct non-interactive arguments using Indistinguishability Obfuscation (IO). This can be instantiated using the candidate of Garg et al. [GGH+13b] (which itself builds on multilinear maps).

Gennaro et al. [GGP10] have suggested a combination of garbled circuits and FHE in order to obtain non-interactive verification of outsourced work. In their setting a long setup message is sent by the verifier (whose length is proportional to the *total* amount of work) and for each subsequent input the verifier only needs to send a message proportional in length to the input size. The prover sends a short message and verification is quick.

1.2   Our Results

We investigate the compiler for converting public-coin interactive protocols into two-message protocols and show positive and negative results. On the positive side, we show that if the verifier uses only public coins and logarithmic space (and in particular, it has no secret memory), then *the compiler is secure*. This result can then be used to show that *any language in* NC *has a succinct two-message protocol based on any FHE*. More generally, if the computation involves a circuit of depth $D(n)$, then it can be proven by sending a message whose length is polynomial in $D(n)$ times the length of FHE ciphertexts. This is because not only does NC have log-space public-coin interactive proofs [FL93], but these can be made succinct, and moreover, such interactive proofs exist for any bounded-depth computation [GKR15]. These results are described in Sect. 5.

An application of the positive results could be for cases where exhaustive search is involved, and the entity performing the search wishes to show that it was unsuccessful or that the given result is the best possible. (A recent instance of such cases occurs in pools for mining Bitcoins: the goal is to search for a "nonce" that when added to the current block and hashed yields a certain number of ending 0's.) Such an entity (in the Bitcoin case, the participant in the pool) can provide a two-message argument that the computation was properly performed but alas, the search was not successful; the length of the argument is poly-logarithmic in the space searched (in case of Bitcoin the argument length would be polynomial in the length of the nonce). See details in Sect. 5.3.

On the negative side, we show that if FHE schemes exist, then there exists a simple three-message interactive proof (i.e. with unconditional soundness) that, when compiled, yields an unsound argument. In particular, this example means that to instantiate the compiler one must consider the protocol compiled and take into account the communication and runtimes of the parties. This is described in Sect. 4.

The Compiler is described in detail in Sect. 3 and general definitions are given in Sect. 2.

2   Definitions and Basic Properties

A function $\mu \colon \mathbb{N} \to [0,1]$ is *negligible*, denoted by $\mu = negl(n)$, if for every polynomial p, there exists $n_0 \in \mathbb{N}$ such that for every $n \geq n_0$, $\mu(n) \leq \frac{1}{p(n)}$.

2.1   Interactive Protocols

In this work, an interactive protocol consists of a pair $(\mathcal{P}, \mathcal{V})$ of interactive Turing machines that are run on a common input x, whose length we denote by $n = |x|$. The first machine is called *the prover* and is denoted by $\mathcal{P}$, and the second machine, which is probabilistic, is called *the verifier* and is denoted by $\mathcal{V}$. At the end of the protocol, the verifier accepts or rejects (this is the protocol's output).

Public-Coin Protocols. An interactive protocol is *public coins* if each bit sent from the verifier to the prover is uniformly random and independent of the rest of the communication transcript.

Definition 1 (Interactive Proof [GMR89]). *An interactive protocol $(\mathcal{P}, \mathcal{V})$ (as above) is an* Interactive Proof *for a language L if it satisfies the following two properties:*

- **completeness:** *For every $x \in L$, if $\mathcal{V}$ interacts with $\mathcal{P}$ on common input x, then $\mathcal{V}$ accepts with probability 1.*[9]

---

[9] More generally, there could be a small completeness error.

- s_{IP}-**soundness:** *For every $x \notin L$ and every (computationally unbounded) cheating prover strategy $\mathcal{P}^*$, the probability that the verifier $\mathcal{V}$ accepts when interacting with $\mathcal{P}^*$ is at most $s_{IP} = s_{IP}(n)$, where s_{IP} is called the* soundness error *of the proof-system. The probability is over the verifier's coin tosses.*

A verifier is log-space and public-coin *if it is public coin (as above), and uses only a $O(\log n)$-size memory tape (on top of one-way access to the communication and randomness tapes).*

Definition 2 (λ-History-Aware Interactive Proof). *An interactive proof is $\lambda = \lambda(n)$-history-aware if on top of the requirements of Definition 1, it is also the case that each message sent by the (honest) prover P is only a function of the last λ bits sent by the verifier.*

Note that in the above definition we make no assumptions on the strategies that can be employed by cheating provers. Note also that we do not use the related "history ignorant" terminology of [KR09], as we prefer the convenience of the "history-aware" definition.

We add explicit timing and probability parameters to the usual definition of argument systems.

Definition 3 (Argument System). *An interactive protocol $(\mathcal{P}, \mathcal{V})$ (as above) is an* Argument System *for L if it is complete, as per Definition 1, and satisfies computational soundness:*

- s_{arg}-**soundness against** T_{arg}-**time cheating provers:** *For every $x \notin L$ and every cheating prover $\mathcal{P}^*$ whose strategy can be implemented by a $T_{arg} = T_{arg}(n)$-time Turing Machine, the probability that the verifier $\mathcal{V}$ accepts when interacting with $\mathcal{P}^*$ is at most $s_{arg} = s_{arg}(n)$. The probability is over the verifier's coin tosses.*

2.2 FHE

Both FHE and PIR schemes allow one party to send to another party a relatively short string that is an encryption of a query. The second party then computes a ciphertext of a message that is supposed to be a function of the original message and information that the second party possesses. In the case of FHE, the query is a vector y in (say) $\{0,1\}^m$, the second party possesses a function $f: \{0,1\}^m \to \{0,1\}$, and the answer-ciphertext is an encryption of $f(y)$.

Definition 4 (Fully Homomorphic Encryption). *An FHE scheme is defined by algorithms:* KeyGen, Enc, Dec, Eval, *who all get as part of their input the security parameter 1^κ and an input length parameter m (we omit these two inputs when they are clear from the context). The KeyGen algorithm outputs a pair of public and secret keys (pk, sk). The encryption algorithm* Enc *takes the public key and a vector $y \in \{0,1\}^m$, and outputs an encryption of y. The* Eval *algorithm takes as input the public key pk, an encryption of y (under pk) and* a function $f\colon \{0,1\}^m \to \{0,1\}$, and outputs an encryption of $f(y)$. Finally, the decryption algorithm takes as input the secret key sk and the encryption of $f(y)$ produced by Eval and outputs the plaintext $f(y)$. We require:

- **Completeness:** $\forall \kappa, m \in \mathcal{N}, y \in \{0,1\}^m$, for any function f of circuit-size $\mathrm{poly}(m)$ and (pk, sk) generated by KeyGen, we have:

$$\mathsf{Dec}(sk, \mathsf{Eval}(pk, f, \mathsf{Enc}(pk, y))) = f(y).$$

- **Semantic Security:** $\forall \kappa, m \in \mathcal{N}, y, y' \in \{0,1\}^m$, the distributions $\mathsf{Enc}(pk, y)$ and $\mathsf{Enc}(pk, y')$ (where pk is generated by KeyGen) are $\mathrm{negl}(\kappa)$-indistinguishable.
- **Complexity:** The algorithm KeyGen runs in time $\mathrm{poly}(\kappa)$. The algorithms Enc, Dec run in time $\mathrm{poly}(\kappa, m)$. The algorithm Eval runs in time $\mathrm{poly}(\kappa, m, |f|)$. The outputs of Enc and Eval are of length $\mathrm{poly}(\kappa, m)$.

The possible existence of FHE scheme was an open question for many years until Gentry's work [Gen09] and we know now that FHE schemes can be constructed under standard lattice assumptions such as LWE [BV14].

2.3   PIR

Here the query is an index $y \in [\lambda]$, the second party possesses a database $Z \in \{0,1\}^\lambda$, and the answer-ciphertext is an encryption of Z_y.

Definition 5 (Private Information Retrieval (PIR) Scheme). *A PIR scheme is defined by three algorithms:* Enc, Dec, Eval, *who all get as part of their input the security parameter 1^κ and database length λ (we omit these two inputs when they are clear from the context).* Enc *also takes as input an index $y \in [\lambda]$, and outputs an "encryption" c of y, and a "secret key" sk for decryption. The* Eval *algorithm takes as input an encryption of y and a database $Z \in \{0,1\}^\lambda$, and outputs an "encryption" of Z_y (the y-th bit of Z). Finally,* Dec *takes as input the secret key sk and a ciphertext generated by* Eval *and outputs Z_y. We make the following requirements:*

- **Completeness:** $\forall \kappa, \lambda \in \mathcal{N}, y \in [\lambda], Z \in \{0,1\}^\lambda$, and $(c, sk) \leftarrow \mathsf{Enc}(y)$ we have that:

$$\mathsf{Dec}(sk, \mathsf{Eval}(Z, c)) = Z_y.$$

- **Semantic Security:** $\forall \kappa, \lambda \in \mathcal{N}, y, y' \in [\lambda]$, taking $(c, sk) \leftarrow \mathsf{Enc}(y)$ and $(c', sk') \leftarrow \mathsf{Enc}(y')$, the distributions of c and of c' are $\mathrm{negl}(\kappa)$-indistinguishable.
- **Complexity:** The algorithms Enc, Dec run in time $\mathrm{poly}(\kappa, \log \lambda)$. The algorithm Eval runs in time $\mathrm{poly}(\kappa, \lambda)$. In particular, the outputs of Enc and Eval are of length $\mathrm{poly}(\kappa, \log \lambda)$.

. PIR Schemes exist under a variety of assumptions such as quadratic residuosity [KO97], Φ-hiding [CMS99] and Learning with Errors [BV14].

3   Detailed Description of the Compiler

3.1   The Compiler: FHE Variant

We now describe the compiler in detail, focusing first on the FHE variant and in Sect. 3.2 the PIR variant. The compiler starts with a many-round public-coin interactive proof $(\mathcal{P}_{IP}, \mathcal{V}_{IP})$, and produces a two-message argument system $(\mathcal{P}_{arg}, \mathcal{V}_{arg})$. It is based on that of Kalai and Raz [KR09]. However, we leave the security parameter free, rather than tailoring it to the Interactive Proof $(\mathcal{P}_{IP}, \mathcal{V}_{IP})$ to be compiled.[10]

We denote by $\mathsf{Enc}_{pk}(y)$ the encryption of $y \in \{0,1\}^m$ under public key pk. Note that this is really a *distribution* on ciphertexts. Assume that $(\mathcal{P}_{IP}, \mathcal{V}_{IP})$ consists of k rounds (and $2k$ messages). For each round i, in which the verifier should send a random value α_i, the (compiled) verifier chooses an independent key pk_i from the underlying encryption system. In a single message, it sends the public keys $\{pk_i\}_{i=1}^k$ and the ciphertexts $\{a_i = \mathsf{Enc}_{pk_i}(\alpha_1, \alpha_2, \ldots \alpha_i)\}_{i=1}^k$. That is, the ciphertext $a_i = \mathsf{Enc}_{pk_i}(\alpha_1, \alpha_2, \ldots \alpha_i)$ is the encryption of the messages sent in rounds $1, \ldots, i$ of the simulated protocol. The prover uses the ciphertext a_i to homomorphically compute an encryption of the answer that it would have sent given the queries $\alpha_1, \alpha_2, \ldots \alpha_i$, i.e., what it "would have done" at round i.

Let the (efficiently computable) function that computes the i-th prover message in the interactive protocol be $\mathcal{P}_i(\alpha_1, \alpha_2, \ldots \alpha_i)$. So the prover computes and sends $b_i = \mathsf{Enc}_{pk_i}(\mathcal{P}_i(\alpha_1, \alpha_2, \ldots \alpha_i))$. This is done simultaneously for all the rounds. The verifier then decrypts the messages it receives, where β_i is the decryption of the ciphertext b_i, and accepts if and only if the simulated verifier accepts the transcript $(\alpha_1, \beta_1, \ldots, \alpha_k, \beta_k)$.

The resulting protocol is given in Fig. 1. By construction, it is a two-message protocol. Completeness rests on the completeness of the FHE scheme, i.e. if the scheme is complete, then so is the resulting protocol. The communication complexity of the new protocol increases: the verifier sends $k^2/2$ bit-encryptions and k public keys. The prover responds with k ciphertexts. Letting γ bound the length of ciphertexts and public keys, the total communication complexity is $O(k^2 \cdot \gamma)$. *The soundness of the resulting protocol is the main issue addressed in this work.*

Historical Note: Multi-prover Proof Systems. A related compiler starts with a multi-prover two-message scheme instead of a single-prover protocol (this is closer to the original idea of [ABOR00a]). As in the above compiler, the idea is for the verifier to encrypt the queries using independent keys and then ask that the prover perform the computation it would have done to answer the queries in the original protocol.

---

[10] The compiler can be based on FHE or PIR, see Sect. 3.2 for the PIR-based variant.

Protocol $(\mathcal{P}_{arg}, \mathcal{V}_{arg})(x, 1^\kappa)$ **for Language** L

The compiler uses an FHE scheme, with security parameter κ. Without loss of generality, we assume that each message in $\Pi = (\mathcal{P}_{IP}, \mathcal{V}_{IP})$ is only a single bit.

$\mathcal{V}_{arg} \to \mathcal{P}_{arg}$: The verifier $\mathcal{V}_{arg}$ simulates the interactive verifier $\mathcal{V}_{IP}$ to generate its k challenges $\alpha_1, \ldots, \alpha_k \in \{0, 1\}$. For each $i \in [k]$, $\mathcal{V}_{arg}$ chooses keys (pk_i, sk_i) for the PIR or FHE.
$\mathcal{V}_{arg}$ then sends the keys and ciphertexts $\{pk_i, \mathsf{Enc}_{pk_i}(\alpha_1, \ldots, \alpha_i)\}_{i \in [k]}$ (as a single message).

$\mathcal{P}_{arg} \to \mathcal{V}_{arg}$: The prover $\mathcal{P}_{arg}$ simulates the interactive prover $\mathcal{P}_{IP}$, where the function $\mathcal{P}_i$ that computes $\mathcal{P}_{IP}$'s i-th message is applied to the encrypted challenges sent under pk_i. I.e., $\mathcal{P}_{arg}$ homomorphically computes $b_i = \mathsf{Enc}_{pk_i}(\mathcal{P}_i(\alpha_1, \ldots, \alpha_i))$.
$\mathcal{P}_{arg}$ then sends th ciphertexts $\{b_i\}_{i \in [k]}$ to $\mathcal{V}_{arg}$ (as a single message)

Verification: The verifier $\mathcal{V}_{arg}$ decrypts each ciphertext b_i to retrieve the message β_i. It accepts if and only if the interactive verifier $\mathcal{V}_{IP}$ accepts the transcript $(\alpha_1, \beta_1, \ldots, \alpha_k, \beta_k)$.

Fig. 1. Compiler (FHE variant): compiling k-round interactive proof $(\mathcal{P}_{IP}, \mathcal{V}_{IP})$ to 2-msg argument $(\mathcal{P}_{arg}, \mathcal{V}_{arg})$. Based on [KR09].

3.2   The Compiler: PIR Variant

The PIR-based variant of the compiler is given in Fig. 2. We assume that the interactive proof to be compiled $(\mathcal{P}_{IP}, \mathcal{V}_{IP})$ is only λ-history-aware, so the prover's i-th message only depends on the last λ bits sent by the verifier.

The verifier $\mathcal{V}_{arg}$ simulates the interactive verifier $\mathcal{V}_{IP}$ to generate its k messages $\alpha_1, \ldots, \alpha_k \in \{0, 1\}$. For each $i \in [k]$, $\mathcal{V}_{arg}$ uses the PIR scheme to compute: $(c_i, sk_i) \leftarrow \mathsf{Enc}(\alpha_{i-\lambda+1}, \ldots, \alpha_i)$, where we interpret α_j as α_1 if $j < 1$ (this will occur when $i < \lambda$). For each $i \in [k]$, the prover $\mathcal{P}_{arg}$ interprets c_i, which encrypts a string of $\lambda' \leq \lambda$ bits, as a PIR query into a database of size at most $2^{\lambda'}$. For each i, $\mathcal{P}_{arg}$ computes a database $Z^{(i)}$ containing all $2^{\lambda'}$ answers, one for each possible λ'-bit history that $\mathcal{P}_{IP}$ might have encountered in its i-th round in the underlying interactive proof $(\mathcal{P}_{IP}, \mathcal{V}_{IP})$. $\mathcal{P}_{arg}$ responds with $b_i \leftarrow \mathsf{Eval}(Z^{(i)}, c_i)$, which contains the answer to the λ'-bit history encrypted in the i-th verifier query.

In the compiled protocol, the *honest* prover $\mathcal{P}_{arg}$ runs in time 2^λ. When λ is at most logarithmic in the input length, the running time of $\mathcal{P}_{arg}$ remains polynomial. Indeed, in our positive result we apply the compiler to the interactive proof of [GKR15], which has a logarithmic λ (see Sect. 5). The communication complexity of the new protocol is as follows: the verifier sends k PIR queries into a database of size 2^λ, and the prover responds with k answers to the PIR queries. Letting γ be the communication complexity of the PIR scheme (the combined length of the PIR query and response for databases of size 2^λ), the total communication complexity is $k\gamma$. Note that (for logarithmic λ) this is an improvement over the $O(k^2\gamma)$ obtained using FHE.

Protocol $(\mathcal{P}_{arg}, \mathcal{V}_{arg})(x, 1^\kappa)$ for Language L

The compiler uses a PIR scheme with security parameter κ. Without loss of generality, we assume that each message in $\Pi = (\mathcal{P}_{IP}, \mathcal{V}_{IP})$ is only a single bit.

$\mathcal{V}_{arg} \rightarrow \mathcal{P}_{arg}$: The verifier $\mathcal{V}_{arg}$ simulates the interactive verifier $\mathcal{V}_{IP}$ to generate its k messages $\alpha_1, \ldots, \alpha_k \in \{0,1\}$. For each $i \in [k]$, $\mathcal{V}_{arg}$ uses the PIR scheme to compute:

$$(c_i, sk_i) \leftarrow \mathsf{Enc}(\alpha_{i-\lambda+1}, \ldots, \alpha_i).$$

$\mathcal{V}_{arg}$ then sends the PIR queries $\{c_i\}_{i \in [k]}$ (as a single message).

$\mathcal{P}_{arg} \rightarrow \mathcal{V}_{arg}$: For each $i \in [k]$, the prover $\mathcal{P}_{arg}$ interprets c_i, which encrypts a λ-bit string, as a PIR query into a database of size 2^λ. For each i, $\mathcal{P}_{arg}$ computes a database $Z^{(i)}$ containing all 2^λ answers, one for each possible λ-bit history that $\mathcal{P}_{IP}$ might have encountered in its i-th round in the underlying interactive proof $(\mathcal{P}_{IP}, \mathcal{V}_{IP})$.

For each $i \in [k]$, $\mathcal{P}_{arg}$ computes $b_i \leftarrow \mathsf{Eval}(Z^{(i)}, c_i)$, and sends the strings $\{b_i\}_{i \in [k]}$ to $\mathcal{V}_{arg}$ (as a single message).

Verification: The verifier $\mathcal{V}_{arg}$ decrypts each value b_i using the key sk_i and retrieves a message β_i. It accepts if and only if the interactive verifier $\mathcal{V}_{IP}$ accepts the transcript $(\alpha_1, \beta_1, \ldots, \alpha_k, \beta_k)$.

Fig. 2. Compiler (PIR variant). Compiling k-round λ-history-aware interactive proof $(\mathcal{P}_{IP}, \mathcal{V}_{IP})$ to 2-message argument $(\mathcal{P}_{arg}, \mathcal{V}_{arg})$

4   The Negative Result: A Protocol that Does Not Compile Well

We now present a protocol $(\mathcal{P}_{IP}, \mathcal{V}_{IP})$ that does not compile well under the Compiler of Sect. 3. We work with the FHE variant of the compiler. The results hold *mutatis mutandis* for the PIR variant (see Remark 1 below).

This is what we would like to be able to say: "For any possible instantiation of the compiler with an FHE, there exists an (unconditionally sound) interactive protocol that, when compiled, yields an unsound two-message argument system." What we can actually say is: "For any possible instantiation of the compiler with an FHE, there exist another instantiation of the compiler with a (different) FHE and an interactive protocol that, when compiled, yields an unsound two-message argument". That is, given the FHE we will need to modify it a bit (still getting an FHE), so that the compiler will fail. We suggest two alternate modifications to the underlying FHE to undermine the compiler. We stress that the compiler fails under all known implementations of FHE (without any modification).

The rough idea of the interactive protocol $(\mathcal{P}_{IP}, \mathcal{V}_{IP})$ is for the prover to commit to a string $x_p \in \{0,1\}^n$ in the first round. The verifier then sends a "guess" for this commitment string in the form of $x_v \in \{0,1\}^n$ chosen uniformly at random. Finally the prover opens his commitment to the string x_p. The prover wins if the opening of x_p is legitimate (i.e. accepted by the receiver in the commitment protocol) and $x_v = x_p$. Obviously, since x_v is chosen after x_p, if the commitment is perfect (there is only one way to open any commitment), then the probability that the prover succeeds is $1/2^n$. This as an interactive proof protocol for the empty language.

Perfect and Weak Commitments. The protocol $(\mathcal{P}_{IP}, \mathcal{V}_{IP})$ uses a public-key encryption scheme to commit to the string x_p. To commit, the prover sends a public key pk and an encryption of x_p. To open the commitment, he sends the randomness used in the encryption. The resulting protocol is sound so long as the encryption scheme is *committing*: for every public key, and every ciphertext, there is only one message-randomness pair that yields that ciphertext.

We cannot base the above protocol on "non-committing" encryption, where some ciphertexts can be opened to *all* possible values, as is the case in deniable encryption [CDNO97, SW14]. If we use such an encryption scheme in the above protocol, then the resulting protocol will not be sound (the prover can win). Simply put, the prover can open the encryption as the string that the verifier sent.

Nevertheless, we can relax the commitment property a bit. Instead of a perfect commitment, we can use a *weak commitment* scheme, where we modify the requirement that there is a unique opening, to one where there are few openings. If there are at most w different values that can be opened, then the probability of the prover winning the above game (guessing at most w strings that include the one chosen by the verifier) is at most $w/2^n$. We can then use any semantically secure public-key encryption scheme (even a non-committing one) to get a weak commitment as follows. The commitment is as above (i.e. consists of a public key and a ciphertext). To open the commitment, the committer sends a decryption key sk corresponding to pk. Assuming the decryption algorithm is deterministic (which is w.l.o.g, since we can fix the coins), there is a unique plaintext corresponding to the ciphertext given a candidate for sk.

For this to make sense we need to make sure that the length of the decryption key is much shorter than $n = |x|$, and we get a weak commitment as above (the number of possible openings is bounded from above by the number of decryption keys, $w = 2^{|sk|}$, much smaller than $2^{|x|}$).

4.1 The Protocol $(\mathcal{P}_{IP}, \mathcal{V}_{IP})$

$(\mathcal{P}_{IP}, \mathcal{V}_{IP})$ is an interactive proof for the empty language, i.e. the verifier should reject any input w.h.p. The proposed protocol consists of 4 messages and uses public coins, where the first message, sent by the verifier, is empty. We can base it on any committing encryption scheme as above (and in parenthesis describe how to deal with non committing encryption). The notation $\mathsf{Enc}_{pk}(x, r)$ indicates that the message x is encrypted under public key pk using randomness r.

The protocol $(\mathcal{P}_{IP}, \mathcal{V}_{IP})$ is:

1. $\mathcal{V} \mapsto \mathcal{P}$: Empty message.
2. $\mathcal{P} \mapsto \mathcal{V}$: The prover uses a committing encryption scheme. It picks public key pk_p, string $x_p \in \{0,1\}^n$, randomness r_p and sends $(pk_p, c_p = \mathsf{Enc}_{pk_p}(x_p, r_p))$. (Same is done in the non-committing case.)
3. $\mathcal{V} \mapsto \mathcal{P}$: The verifier picks random $x_v \in \{0,1\}^n$ and sends it.
4. $\mathcal{P} \mapsto \mathcal{V}$: The prover sends $m = (x_p, r_p)$.

Verification. The verifier checks whether $x_p = x_v$ and $c_p = \mathsf{Enc}_{pk_p}(x_p, r_p)$ and accepts if they are both satisfied. Note that to perform this check, the verifier needs to "remember" pk_p, c_p and x_v (we refer to this fact in Sect. 5).

In the non-committing encryption variant of the protocol, in Step 4 the prover sends the decryption key sk_p corresponding to pk_p. In the verification step, the verifier decrypts c_p using sk_p and accepts if the answer equals x_v.

Theorem 1. *For any perfectly committing encryption scheme (respectively, non-committing scheme), the above protocol is complete and sound, with soundness error at most $1/2^n$ (respectively, $2^{|sk|}/2^n$ for the non-commiting variant).*

4.2 The Compiled Protocol

Let $(\mathcal{P}_{arg}, \mathcal{V}_{arg})$, described next, be the argument system obtained by applying the compiler of Fig. 1 to the interactive proof $(\mathcal{P}_{IP}, \mathcal{V}_{IP})$ for the empty language, described in Sect. 4.1.

$\mathcal{V}_{arg} \mapsto \mathcal{P}_{arg}$: Verifier picks and sends $pk_{v,1}$ (for the first round's empty message), and $pk_{v,2}, c_v = \mathsf{Enc}_{pk_{v,2}}(x_v, r_v)$ (for the verifier's second message in $(\mathcal{P}_{IP}, \mathcal{V}_{IP})$).
$\mathcal{P}_{arg} \mapsto \mathcal{V}_{arg}$: The (honest) prover $\mathcal{P}_{arg}$ sends $\mathsf{Enc}_{pk_{v,1}}((pk_p, c_p = \mathsf{Enc}_{pk_p}(x_p, r_p)), r')$ (for the first prover message), and $\mathsf{Enc}_{pk_{v,2}}(m = (x_p, r_p), r'')$ (for the second message).
The verifier decrypts the first prover message using $sk_{v,1}$ to retrieve (pk_p, c_p), decrypts the second message using $sk_{v,2}$ to retrieve $m = (x_p, r_p)$, and accepts if the original protocol's verifier accepts.

Speaking intuitively, the compiler will fail because a cheating prover can use the encryption c_v of the message x_v that the compiled verifier sends him to come up with a commitment to $x_p = x_v$. The challenge to the cheating prover then is how to obtain an encryption of the randomness r_p sent in Step 3 of $(\mathcal{P}_{IP}, \mathcal{V}_{IP})$. This seems like quite an obstacle for an arbitrary FHE scheme.

Suppose, however, that FHE scheme E "makes the cheating prover's life easy". That is, every encryption also includes an encryption of the randomness r (using freshly chosen randomness r_{add}) (the decryption algorithm simply ignores the second part of the ciphertext). The cheating prover for $(\mathcal{P}_{arg}, \mathcal{V}_{arg})$ can use this encryption of the randomness r_p to break soundness. Any FHE can be tweaked in this way, without harming its homomorphic or security properties. The case of a non-committing encryption is handled similarly.

Breaking the Compiled Protocol. Given $pk_{v,1}, pk_{v,2}$ and $c_v = \mathsf{Enc}_{pk_{v,2}}(x_v, r_v)$, the cheating prover P^* sends $\mathsf{Enc}_{pk_{v,1}}(pk_{v,2}, c_v)$ as its first message, and c_v as its second message. Recall that $c_v = \mathsf{Enc}_{pk_{v,2}}(x_v, r_v)$, and this includes both an encryption of x_v and of r_v, since we assumed that the cryptosystem Enc is such that it also gives an encryption of the random string. Thus, by following this strategy, P^* makes $\mathcal{V}_{arg}$ accept with probability 1 (based on perfect completeness of the encryption scheme).

Alternatively, for non-committing encryption, we assume that the public key includes an encryption of the secret key. Thus, the public key $pk_{v,2}$ includes $\mathsf{Enc}_{pk_{v,2}}(sk_{v,2})$, which can be sent by the cheating prover P^* as its second message to "de-commit" and break security. Thus, we can use a "circular-secure" FHE scheme, where semantic security holds even when the public key includes an encryption of the secret key, to show that the compiler fails. We note that it is not known in general whether "circular security" holds, namely whether including an encryption of the secret key in the public key *always* preserves semantic security (see e.g. Rothblum [Rot13]). However, for known FHE schemes, an encryption of the secret key is already included in the public key to enable "bootstrapping" (see e.g. Gentry [Gen09]). Thus, *the compiler is insecure when instantiated with all known concrete FHE candidates.* Moreover, even if a new (and non-committing) FHE candidate is discovered, we can modify its public key to include an encryption of the secret key. If the modified scheme remains semantically secure, then (as above) the compiler fails on $(\mathcal{P}_{IP}, \mathcal{V}_{IP})$. Thus, proving that the compiler is secure with the modified scheme would require proving that the original FHE scheme is *not* circular secure.

We can see that in both cases the cheating prover P^* succeeds and the verifier accepts. We therefor have:

Theorem 2. *If an FHE scheme Enc exists, then there exists an instantiation of the compiler of Fig. 1 with a (possibly) modified FHE scheme and a sound protocol in the public coins model such that the resulting compiled protocol is not sound.*

Remark 1. The same protocol "misbehaves" under the PIR-based compiler in Fig. 2. The compiled protocol is unsound when the compiler uses a PIR scheme that is directly based on the same FHE: where ciphertexts are modified to include encryptions of the randomness (in the perfectly committing case), or of the secret key (in the weakly committing case). The PIR scheme operates by sending the (modified) encryption of the index being queried. The Eval algorithm responds with an answer-ciphertext, and the Dec algorithm simply decrypts this ciphertext using the FHE decryption.

5   Positive Results

We show that the Compiler of Fig. 1 is secure when applied to interactive proofs with a public-coin log-space verifier. More generally, the compiler is secure for interactive proofs where (for any fixed partial transcript) the *optimal* continuation strategy, i.e. the strategy that maximizes the verifier's success probability, can be computed in polynomial time. We only assume the existence of standard (polynomially hard) PIR or FHE. This result is in Theorem 4 below, which we prove using a careful analysis of the Kalai-Raz compiler [KR09]. Recall that the negative example of Sect. 4 shows that the compiler is *insecure* for general interactive proofs. In particular, recall that (as noted above) the verifier needs enough space to "remember" pk_p, c_p, x_v. Thus, we need to leverage additional structure in order to prove security, and we do so via the space complexity of the verifier (or the optimal-continuation strategy). Kalai and Raz, in contrast, showed that security could be obtained by making super-polynomial hardness assumptions and simultaneously tailoring the compiler's security parameter to a given interactive proof system (that is, they choose the security parameter after seeing the interactive proof to be compiled).

Recall that for any language computable by depth $D(n)$ (log-space uniform) circuits, there exists an interactive proof where the verifier uses logarithmic space and public coins [GKR15]. By applying the compiler to these interactive proofs, we obtain a succinct two-message argument using any (polynomially-hard) PIR scheme. The communication is $\tilde{O}(D(n)\cdot\gamma)$, where γ is the communication required by the PIR scheme (for a database of length poly(n)). Similarly to [KR09], we use the fact that every prover message in the succinct interactive proof only depends on a logarithmic number of bits sent by the verifier. This result is in Theorem 5. In Sect. 5.3 we discuss applications to proving the results of an exhaustive search.

5.1   Security of the Compiler

As described above, our main insight is that if there is a polynomial time algorithm that computes an optimal prover-strategy for any interactive proof, then we can compile the protocol with no significant soundness error. For a log-space public-coin verifier this is possible and arguments of this type were used by Condon [Con91] and Fortnow and Sipser (see [For89]) to show that these proof systems can only compute languages in P. In fact, for any fixed partial transcript between the prover and verifier, we show how to *efficiently* compute an optimal prover strategy for continuing the interaction. The main cause for the super-polynomial security gap in the Kalai-Raz security reduction was the running time required to compute an optimal prover strategy (for a fixed partial transcript). For general interactive proofs, this requires time that is exponential in the communication. We leverage the *polynomial-time* transcript-completion algorithm to obtain a tighter security reduction and our main result.

We proceed by first defining the optimal transcript completion task for an interactive proof. We then show that (*i*) for log-space public-coin interactive proofs, there is an *efficient* transcript completion algorithm (Theorem 3), and (*ii*) we can strengthen the Kalai-Raz security reduction from the security of the argument to the security of the PIR scheme to reduce the loss in security: rather than *exponential* in the communication complexity of the interactive proof, as in [KR09], it is *polynomial* in the time required for optimal transcript completion and in the security parameter (Theorem 4).

Definition 6 (Optimal Transcript Completion). *We say that a public-coin interactive proof* $(\mathcal{P}_{IP}, \mathcal{V}_{IP})$ *with communication complexity* ℓ_{IP} *supports* optimal transcript completion *in time* $T_{IP}(n)$, *if there exists an algorithm* $\mathcal{P}'$, *running in time* $T_{IP}(n)$, *that, on input* any *partial transcript of communication with the verifier* $\mathcal{V}_{IP}$ *and ending with a message sent by* $\mathcal{V}_{IP}$, *produces a next message from the prover to the verifier, satisfying the following guarantee: For any algorithm* $\mathcal{P}^*$ *(with unbounded running time), the probability that* $\mathcal{V}_{IP}$ *accepts when the transcript is completed using* $\mathcal{P}^*$ *is no greater than the probability that* $\mathcal{V}_{IP}$ *accepts when the transcript is completed using* $\mathcal{P}'$.

Remark 2. We note that even if the interactive proof has an efficient (i.e. poly(n)-time) honest prover, there may not be an efficient optimal transcript completion algorithm $\mathcal{P}'$. Indeed, we can build explicit examples under (mild) cryptographic assumptions—e.g. using the Interactive Proof described in Sect. 4.

We now show that any interactive proof with a log-space public-coin verifier supports (polynomial time) optimal transcript completion:

Theorem 3 (See also [Con91, For89]). *Let* $(\mathcal{P}_{IP}, \mathcal{V}_{IP})$ *be a log-space public-coin interactive proof. Then* $(\mathcal{P}_{IP}, \mathcal{V}_{IP})$ *supports optimal transcript completion in* poly(n) *time.*

Proof (Proof Sketch). The transcript completion algorithm $\mathcal{P}'$ constructs the verifier's directed layered state graph, where every node is a possible memory configuration for the verifier. We assume w.l.o.g. that the verifier keeps track of the current round, and that each message in the protocol is a single bit. Each round consists of a pair of messages: a single random bit sent from the verifier to the prover, and a single bit sent in response from the prover to the verifier. For round i and memory configuration u, and round $(i + 1)$ and memory configuration v, there is an edge from (i, u) to $(i + 1, v)$ iff starting with configuration u just after round i, the verifier can reach configuration v just after round $(i + 1)$. Since each round consists of two single-bit messages, each node in the graph has at most 4 successors. This completes the description of the graph, and we note that it is of poly(n) size (because the verifier runs in $O(\log n)$ space).

Now, for any verifier state (i, u) we compute the optimal prover strategy as follows. We start from terminal nodes and work our way "backwards" in the state graph to nodes corresponding to states in earlier rounds. For each node/state, we compute the probability that the verifier accepts once it reaches that state, and the optimal prover response to the next verifier challenge. For node (i, u) we denote this by $a(i, u)$. For terminal nodes, the acceptance probability is either 0 or 1 (depending on whether this is a rejecting or accepting node). Once the accepting probabilities have been computed for all round-$(i + 1)$-states we can compute the accepting probabilities (and best prover responses) for round-i-states. For a non-terminal node (i, u), there are 4 possible transcripts for round i: $00, 01, 10, 11$, where transcript (α, β) leads to state $(i + 1, w_{\alpha, \beta})$ (here $\alpha, \beta \in \{0, 1\}$). For each verifier challenge α, the "best response" is the message leading to the state that maximizes the acceptance probability:

$$b(\alpha) = argmax_{\beta \in \{0,1\}} (a(i+1, w_{\alpha,\beta}))$$

and the (maximal) probability of acceptance is

$$a(i, u) = \frac{a(i+1, w_{0,b(0)}) + a(i+1, w_{1,b(1)})}{2}.$$

By construction, this procedure computes an optimal prover strategy for any verifier state.

Given a partial transcript, $\mathcal{P}'$ can compute the current verifier state and use this procedure to complete a best-response strategy.

Next, we give a strengthened security analysis for the Kalai-Raz compiler, where the reduction is parameterized in terms of the time required for optimal transcript completion (see in comparison Lemma 4.2 of [KR09] where "brute-force" completion is used):

Theorem 4. *Let $\Pi = (\mathcal{P}_{IP}, \mathcal{V}_{IP})$ be a public-coin interactive proof for a language L that is $\lambda(n)$-history-aware (as in Definition 2), with completeness $c_{IP}(n)$, soundness $s_{IP}(n)$ and communication complexity $\ell_{IP}(n)$. Let PIR be a PIR scheme with communication $\ell_{PIR}(m, \kappa)$.*

Then $(\mathcal{P}_{arg}, \mathcal{V}_{arg})$, the argument system of Fig. 2 instantiated with $(\mathcal{P}_{IP}, \mathcal{V}_{IP})$ and PIR, is a 2-message argument system for L, with completeness $c_{arg} = c_{IP}$, communication complexity $\ell_{arg}(n, \kappa) = \ell_{IP}(n) \cdot \ell_{PIR}(2^{\lambda(n)}, \kappa)$, and the following properties:

1. **Computational Soundness:** *If the interactive proof supports optimal transcript completion, as in Definition 6, in time $T_{IP}(n)$, and the PIR system is secure against adversaries running in time $T_{PIR}(\kappa)$, then the argument system has soundness $s_{arg}(n) = (\ell_{IP}(n) \cdot (s_{IP}(n) + negl(\kappa)))$ against adversaries running in time $T_{arg} = T_{PIR}(\kappa)/poly(n, \ell_{IP}(n), T_{IP}(n))$.*
2. **Honest Prover Complexity.** *$\mathcal{P}_{arg}$ runs in time $poly(T_{IP}(n), \kappa, 2^{\lambda(n)})$.*

Security of the Compiler Using FHE. An analogous theorem statement holds for the FHE version of the compiler. The advantage over the PIR version is that there is no need to assume that the interactive proof is λ-history aware. The disadvantage (other than the stronger assumption) is the quadratic blowup in the communication complexity (see Sect. 3).

Proof (Proof of Theorem 4). We assume w.l.o.g. that the each message sent by the verifier is 1 bit long, and take k to be the number of rounds. Suppose that there exists an input $x^* \notin L$ and a cheating prover $\mathcal{P}^*_{arg}$ that manages to convince $\mathcal{V}_{arg}$ with probability ε.

Definition of p_i. For $i \in \{0, 1, \ldots, k\}$, define p_i to be the success probability of the following process called Experiment$_i$: run the argument system with the cheating prover $\mathcal{P}^*_{arg}$. Let $\{(pk_i, sk_i)\}_{i \in [k]}$ be the keys and α_i be the bits encrypted in the challenges sent by $\mathcal{V}_{arg}$. Let $\{b_i\}_{i \in [k]}$ be the ciphertext answers returned by $\mathcal{P}^*_{arg}$, and let β_i be the plaintext value in b_i. Fixing the partial transcript $(\alpha_1, \beta_1, \ldots, \alpha_i, \beta_i)$ for the first i rounds, run the optimal-completion strategy $\mathcal{P}'_{IP}$ with the verifier $\mathcal{V}_{IP}$ (who simply generates random messages) to complete the transcript (i.e. for the last $(k - i)$ rounds), generating messages $(\alpha'_{i+1}, \beta'_{i+1}, \ldots, \alpha'_k, \beta'_k)$. Experiment$_i$ succeeds if and only if $\mathcal{V}_{IP}$ accepts the resulting transcript $(\alpha_1, \beta_1, \ldots, \alpha_i, \beta_i, \alpha'_{i+1}, \beta'_{i+1}, \ldots, \alpha'_k, \beta'_k)$.

Claim. There exists $i^* \in [k]$ s.t.:

$$p_{i^*} - p_{i^*-1} \geq \frac{\varepsilon - s_{IP}}{k}$$

Proof. The proof is by a hybrid argument. p_0 is bounded by the success probability of a cheating prover in the (sound) interactive proof, and thus it is at most s_{IP} (since $x^* \notin L$). p_k is exactly the success probability of the cheating prover $\mathcal{P}^*_{arg}$ in the two-message argument system, and thus by assumption it is at least ε.

Breaking the Encryption. We show that any $\mathcal{P}^*_{arg}$ that succeeds with probability at least ε, can be used to break semantic security of the encryption scheme with advantage at least $(\varepsilon - s_{IP})/k$. We do this by constructing a distinguisher for the following two distributions. In both distributions, generate keys $(pk_i, sk_i)_{i \in [k]}$, random bits $(\alpha_1, \ldots, \alpha_k)$, and the challenge ciphertexts $\{a_i\}_{i \in [k]}$. The distribution outputs all of the public keys and ciphertexts, the first $(i^* - 1)$ secret keys, and all plaintext values except the i^*-th, i.e. $\{\alpha_i\}_{i \neq i^*}$. So far the distributions are identical, the only difference is in a final output value α (see below). In particular, a sample from D_1 or D_2 is of the form:

$$\left(\{pk_i, a_i\}_{i \in [k]}, \{sk_i\}_{i < i^*}, (\alpha_1, \ldots, \alpha_{i^*-1}, \alpha, \alpha_{i^*+1}, \ldots, \alpha_k) \right),$$

where in D_1 we set $\alpha = \alpha_{i^*}$, and in D_2 we draw a uniformly random and independent bit α'_{i^*}, and set $\alpha = \alpha'_{i^*}$. By construction, if the distinguisher distinguishes D_1 and D_2 with non-negligible advantage, then semantic security is broken.

We now use the cheating prover $\mathcal{P}^*_{arg}$ to construct such a distinguisher. The distinguisher runs $\mathcal{P}^*_{arg}$ on the public keys and ciphertexts (these are distributed identically in D_1 and D_2). $\mathcal{P}^*_{arg}$ outputs its response ciphertexts $\{b_1, \ldots, b_k\}$, and the distinguisher uses the secret keys $\{sk_i\}_{i < i^*}$ to retrieve the plaintexts $(\beta_1, \ldots, \beta_{i^*-1})$. Starting with the partial transcript $(\alpha_1, \beta_1, \ldots, \alpha_{i^*-1}, \beta_{i^*-1}, \alpha)$, the distinguisher completes the transcript by simulating the interaction between the "optimal-completion prover" $\mathcal{P}'_{IP}$ and the verifier $\mathcal{V}_{IP}$. It outputs 1 if the verifier accepts and 0 otherwise. Observe that:

- On distribution D_2, the probability that the distinguisher outputs 1 is exactly p_{i^*-1}: the distribution of the partial transcript $(\alpha_1, \beta_1, \ldots, \alpha_{i^*-1}, \beta_{i^*-1})$ is identical to Experiment$_{i^*-1}$. When drawing according to D_2, the i^*-th verifier challenge is α'_{i^*}, which is uniformly random and independent of the preceding partial transcript, as it is in Experiment$_{i^*-1}$. The remainder of the transcript is also generated using the optimal-completion strategy, exactly as in $\mathsf{Experiment}_{i^*-1}$. The verifier accepts with probability p_{i^*-1}.

- On distribution D_1, the probability that the distinguisher outputs 1 is *at least* p_{i^*}: the distribution of the partial transcript $(\alpha_1, \beta_1, \dots, \alpha_{i^*-1}, \beta_{i^*-1})$ is identical to $\mathsf{Experiment}_{i^*}$. When drawing by D_1, the i^*-th challenge α_{i^*} equals the plaintext encrypted in the ciphertext a_{i^*}, exactly as in $\mathsf{Experiment}_{i^*}$. Here, however, the i^*-th prover message β'_{i^*} is drawn according to the optimal completion strategy, whereas in $\mathsf{Experiment}_{i^*}$ we use the plaintext β_{i^*} generated by $\mathcal{P}^*_{arg}$. Still, since the remainder of the transcript will be computed using random verifier queries, replacing the i^*-th prover message with the optimal-completion strategy cannot decrease the probability that the verifier accepts. The verifier accepts with probability p_{i^*}.

The above distinguisher runs in time $|\mathcal{P}^*_{arg}| + k \cdot T_{IP}(n)$, and has advantage at least $(\varepsilon - s_{IP})/k$ in distinguishing the distributions D_1 and D_2. The theorem follows.

5.2   Succinct Two-Message Arguments

Instantiating the secure compiler of Theorem 4 with the Interactive Proof of Theorem 6 below, we obtain succinct two-message arguments for bounded-depth computations:

Theorem 5. *Assume the existence of a PIR scheme with communication $\ell_{PIR}(m, \kappa)$ that is secure against time $T_{PIR}(\kappa)$, as per Definition 5. Then any language L that can be computed by logspace-uniform circuits of size $\mathrm{poly}(n)$ and depth $D(n) \geq \log n$ has a two-message argument system $(\mathcal{P}_{arg}, \mathcal{V}_{arg})$ with perfect completeness and negligible soundness error against adversaries that run in time $T_{PIR}(\kappa)/\mathrm{poly}(n)$. The communication complexity is $\ell_{PIR}(\mathrm{poly}(n), \kappa) \cdot \mathrm{poly}(D(n))$. $\mathcal{P}_{arg}$ runs in time $\mathrm{poly}(\kappa, n)$ and $\mathcal{V}_{arg}$ runs in time $n \cdot \mathrm{poly}(\kappa, D(n))$.*

This represents an exponential improvement in the security of the resulting argument system. Previously, Kalai and Raz [KR09] showed a similar result, but proved security of the argument system against adversaries running in time $T_{PIR}(\kappa)/2^{\mathrm{poly}(D(n))}$.[11] Interpreting their result, for a language L in NC, to obtain any argument system with $\mathrm{poly}(n)$ communication complexity and security against polynomial-time adversaries (i.e., a non-trivial argument system), quasi-polynomial hardness assumptions are needed (as one needs to have a PIR scheme that is secure against adversaries running in time $T_{PIR}(\kappa) \gg 2^{\mathrm{poly}(D(n))}$). In comparison, our results show that a PIR scheme secure against *polynomial*-time adversaries is sufficient for obtaining $\mathrm{poly}(n)$ communication.

Before proving Theorem 5, we review the main result of Goldwasser et al. [GKR08, GKR15]:

---

[11] The denominator in their result was super-polynomial in n; In particular, it was at least $n^{D(n)}$.

Theorem 6 (GKR Interactive Proof [GKR15]**).** *Any language L that can be computed by logspace-uniform circuits of size* $\mathrm{poly}(n)$ *and depth* $D(n) \geq \log n$ *has a multi-round interactive proof* $(\mathcal{P}_{IP}, \mathcal{V}_{IP})$ *with perfect completeness, negligible soundness error, and communication complexity* $D(n) \cdot \mathrm{polylog}(n)$. *Moreover,* $\mathcal{P}_{IP}$ *runs in time* $\mathrm{poly}(n)$ *and is* $O(\log n)$-*history-aware;* $\mathcal{V}_{IP}$ *is a public-coin logspace verifier, runs in time* $n \cdot \mathrm{poly}(D(n))$, *and sends messages of length* $O(\log(n))$.

Proof (Proof of Theorem 5). By Theorem 6, the GKR Interactive Proof [GKR15] for L is λ-history-aware, has a log-space public-coin verifier, and verifier messages of length $O(\log n)$. By Theorem 3, we conclude that it supports optimal transcript completion in time $\mathrm{poly}(n)$. Plugging this interactive proof into the transformation of Fig. 2, and using Theorem 4, we obtain a two-message argument with negligible soundness error against $\mathrm{poly}(n, \kappa)$-time adversaries. The communications complexity and the prover and verifier running times follow by the parameters of the interactive proof, the PIR scheme, and Theorem 4.

5.3   Application to Exhaustive Search

The methods of this section are appropriate for the verification of a type of computation that is often distributed among not completely trusted servers, that of exhaustive search. In this setting there is some space of solutions S that is partitioned into subspaces $\{S_i\}_i$ and processor i is assigned to search all possible solutions in S_i.

Usually it is easy to identify a successful search, say it satisfies some set of constraints. Therefore it is easy to verify the work of a processor that was successful. But how about verifying the work of an unsuccessful search? How can such an unlucky processor convince, say a central authority, that it performed the computation properly?

A good illustrating example to consider is the case of *Bitcoin mining* (but see caveat below), used to maintain the so called *block chain of transactions*. Here the processors (miners) are looking for a value called a 'nonce', such that when the current block content is (cryptographically) hashed together with the nonce, the result ends with a certain number of leading 0's (i.e. is numerically smaller than the current difficulty target).

A successful search is worth a certain number of Bitcoins (25 as of 2015). Now suppose there is a pool of miners who cooperate in order to reduce the variance in the reward. How can the pool manager verify that searches over the nonce space that were not successful were properly executed?

Given that exhaustive search is "embarrassingly parallel" (one for which no effort is required to separate the problem into a number of parallel tasks), it follows that we can apply the framework of Theorem 5.

We can express the search that a processor i should perform of the set S_i as a set of constraints over the set S_i, and for each element in S_i check whether it satisfies the constraints. The circuit will be of depth proportional to $\log |S_i|$ plus the depth of checking whether the set of constraints is satisfied. By Theorem 5, assuming the appropriate PIR scheme exists, we will have an argument system whose length is proportional to polynomial in $\log |S_i|$ plus the complexity of checking an instance. In the context of Bitcoin this means that any participant can provide a proof of search whose length is proportional to the nonce length plus the complexity of the hash functions.

Bitcoin pools reward members who come up with nearly-satisfying solutions, i.e. partition the prize according to the closeness. This makes perfect sense in the random oracle world. Our solution may be viewed as more equitable, and does not rely on random oracles for fairness: everybody gets rewarded for the work they perform.

Caveat: given that many things in the Bitcoin world are based on heuristics and on modeling functions as random oracles, the above ideas can probably be thought of as casting pearls before swine. So we prefer to think of it as an illustrating example rather than an actual application.

Acknowledgments. We thank Pavel Hubáček and Ilan Komargodski for helpful comments on the paper.

References

[ABOR00a] Aiello, W., Bhatt, S.N., Ostrovsky, R., Rajagopalan, S.: Fast verification of any remote procedure call: short witness-indistinguishable one-round proofs for NP. In: Welzl, E., Montanari, U., Rolim, J.D.P. (eds.) ICALP 2000. LNCS, vol. 1853, pp. 463–474. Springer, Heidelberg (2000)

[ABOR00b] Aiello, W., Bhatt, S.N., Ostrovsky, R., Rajagopalan, S.: Fast verification of any remote procedure call: short witness-indistinguishable one-round proofs for NP. IACR Cryptology ePrint Archive 2000:018 (2000)

[BCCT12] Bitansky, N., Canetti, R., Chiesa, A., Tromer, E.: From extractable collision resistance to succinct non-interactive arguments of knowledge, and back again. In: Goldwasser, S. (ed.) ITCS, pp. 326–349. ACM (2012)

[BG08] Barak, B., Goldreich, O.: Universal arguments and their applications. SIAM J. Comput. **38**(5), 1661–1694 (2008)

[BGL+15] Bitansky, N., Garg, S., Lin, H., Pass, R., Telang, S.: Succinct randomized encodings and their applications. In: Proceedings of the Forty-Seventh Annual ACM on Symposium on Theory of Computing, STOC 2015, Portland, OR, USA, 14–17 June 2015, pp. 439–448 (2015)

[BV14] Brakerski, Z., Vaikuntanathan, V.: Efficient fully homomorphic encryption from (standard) LWE. SIAM J. Comput. **43**(2), 831–871 (2014)

[CDNO97] Canetti, R., Dwork, C., Naor, M., Ostrovsky, R.: Deniable encryption. In: Kaliski Jr., B.S. (ed.) CRYPTO 1997. LNCS, vol. 1294, pp. 90–104. Springer, Heidelberg (1997)

[CMS99] Cachin, C., Micali, S., Stadler, M.A.: Computationally private information retrieval with polylogarithmic communication. In: Stern, J. (ed.) EUROCRYPT 1999. LNCS, vol. 1592, pp. 402–414. Springer, Heidelberg (1999)

[Con91] Condon, A.: Space-bounded probabilistic game automata. J. ACM **38**, 472–494 (1991)

[DLN+] Dwork, C., Langberg, M., Naor, M., Nissim, K., Reingold, O.: Succinct proofs for NP ander spooky interactions. http://www.wisdom.weizmann.ac.il/~naor/PAPERS/spooky.pdf

[FL93] Fortnow, L., Lund, C.: Interactive proof systems and alternating time-space complexity. Theor. Comput. Sci. **113**(1), 55–73 (1993)

[For89] Fortnow, L.: Complexity-theoretic aspects of interactive proof systems. Technical report, Ph.D. thesis, Laboratory for Computer Science, MIT (1989)

[FS86] Fiat, A., Shamir, A.: How to prove yourself: practical solutions to identification and signature problems. In: Odlyzko, A.M. (ed.) CRYPTO 1986. LNCS, vol. 263, pp. 186–194. Springer, Heidelberg (1987)

[Gen09] Gentry, C.: Fully homomorphic encryption using ideal lattices. In: Proceedings of the 41st Annual ACM Symposium on Theory of Computing, STOC 2009, Bethesda, MD, USA, May 31–June 2, 2009, pp. 169–178 (2009)

[GGH13a] Garg, S., Gentry, C., Halevi, S.: Candidate multilinear maps from ideal lattices. In: Johansson, T., Nguyen, P.Q. (eds.) EUROCRYPT 2013. LNCS, vol. 7881, pp. 1–17. Springer, Heidelberg (2013)

[GGH+13b] Garg, S., Gentry, C., Halevi, S., Raykova, M., Sahai, A., Waters, B.: Candidate indistinguishability obfuscation and functional encryption for all circuits. In: 54th Annual IEEE Symposium on Foundations of Computer Science, FOCS 2013, 26–29 October 2013, Berkeley, CA, USA, pp. 40–49 (2013)

[GGP10] Gennaro, R., Gentry, C., Parno, B.: Non-interactive verifiable computing: outsourcing computation to untrusted workers. In: Rabin, T. (ed.) CRYPTO 2010. LNCS, vol. 6223, pp. 465–482. Springer, Heidelberg (2010)

[GKR08] Goldwasser, S., Kalai, Y.T., Rothblum, G.N.: Delegating computation: interactive proofs for muggles. In: STOC, pp. 113–122 (2008)

[GKR15] Goldwasser, S., Kalai, Y.T., Rothblum, G.N.: Delegating computation: interactive proofs for muggles. J. ACM **62**(4), 27 (2015)

[GLR11] Goldwasser, S., Lin, H., Rubinstein, A.: Delegation of computation without rejection problem from designated verifier cs-proofs. IACR Cryptology ePrint Archive 2011:456 (2011)

[GMR89] Goldwasser, S., Micali, S., Rackoff, C.: The knowledge complexity of interactive proof systems. SIAM J. Comput. **18**(1), 186–208 (1989)

[Gro10] Groth, J.: Short pairing-based non-interactive zero-knowledge arguments. In: Abe, M. (ed.) ASIACRYPT 2010. LNCS, vol. 6477, pp. 321–340. Springer, Heidelberg (2010)

[GW11] Gentry, C., Wichs, D.: Separating succinct non-interactive arguments from all falsifiable assumptions. In: Fortnow, L., Vadhan, S.P. (eds.) STOC, pp. 99–108. ACM (2011)

[Kil92] Kilian, J.: A note on efficient zero-knowledge proofs and arguments (extended abstract). In: Kosaraju, S.R., Fellows, M., Wigderson, A., Ellis, J.A. (eds.) STOC, pp. 723–732. ACM (1992)

[KO97] Kushilevitz, E., Ostrovsky, R.: Replication is NOT needed: SINGLE database, computationally-private information retrieval. In: 38th Annual Symposium on Foundations of Computer Science, FOCS 1997, Miami Beach, Florida, USA, 19–22 October 1997, pp. 364–373 (1997)

[KR09] Kalai, Y.T., Raz, R.: Probabilistically checkable arguments. In: Halevi, S. (ed.) CRYPTO 2009. LNCS, vol. 5677, pp. 143–159. Springer, Heidelberg (2009)

[KRR14] Kalai, Y.T., Raz, R., Rothblum, R.D.: How to delegate computations: the power of no-signaling proofs. In: STOC 2014, pp. 485–494 (2014)

[Mic00] Micali, S.: Computationally sound proofs. SIAM J. Comput. **30**(4), 1253–1298 (2000)

[Mie08] Mie, T.: Polylogarithmic two-round argument systems. J. Math. Cryptol. **2**(4), 343–363 (2008)

[Nao03] Naor, M.: On cryptographic assumptions and challenges. In: Boneh, D. (ed.) CRYPTO 2003. LNCS, vol. 2729, pp. 96–109. Springer, Heidelberg (2003)

[PR14] Paneth, O., Rothblum, G.N.: Publicly verifiable non-interactive arguments for delegating computation. IACR Cryptology ePrint Archive 2014:981 (2014)

[Rot13] Rothblum, R.D.: On the circular security of bit-encryption. In: Sahai, A. (ed.) TCC 2013. LNCS, vol. 7785, pp. 579–598. Springer, Heidelberg (2013)

[SW14] Sahai, A., Waters, B.: How to use indistinguishability obfuscation: deniable encryption, and more. In: Proceedings of the 46th Annual ACM Symposium on Theory of Computing, STOC 2014, pp. 475–484. ACM, New York (2014)

Secure Computation and Protocols II

Adaptively Secure Garbled Circuits
from One-Way Functions

Brett Hemenway[1], Zahra Jafargholi[2], Rafail Ostrovsky[3],
Alessandra Scafuro[2,4(✉)], and Daniel Wichs[2]

[1] University of Pennsylvania, Philadelphia, USA
fbrett@cis.upenn.edu
[2] Northeastern University, Boston, USA
{zahra,wichs}@ccs.neu.edu
[3] University of California, Los Angeles, USA
rafail@cs.ucla.edu
[4] Boston University, Boston, USA
scafuro@bu.edu

Abstract. A garbling scheme is used to garble a circuit C and an input x in a way that reveals the output $C(x)$ but hides everything else. In many settings, the circuit can be garbled *off-line* without strict efficiency constraints, but the input must be garbled very efficiently *on-line*, with much lower complexity than evaluating the circuit. Yao's garbling scheme [31] has essentially optimal on-line complexity, but only achieves *selective security*, where the adversary must choose the input x prior to seeing the garbled circuit. It has remained an open problem to achieve *adaptive security*, where the adversary can choose x after seeing the garbled circuit, while preserving on-line efficiency.

In this work, we modify Yao's scheme in a way that allows us to prove adaptive security under one-way functions. In our main instantiation we achieve on-line complexity only proportional to the width w of the circuit. Alternatively we can also get an instantiation with on-line complexity only proportional to the depth d (and the output size) of the circuit, albeit incurring in a $2^{O(d)}$ security loss in our reduction. More broadly, we relate the on-line complexity of adaptively secure garbling schemes in our framework to a certain type of *pebble* complexity of the circuit. As our main tool, of independent interest, we develop a new

R. Ostrovsky—Supported in part by NSF grants 09165174, 1065276, 1118126 and 1136174, US-Israel BSF grant 2008411, OKAWA Foundation Research Award, IBM Faculty Research Award, Xerox Faculty Research Award, B. John Garrick Foundation Award, Teradata Research Award, Lockheed-Martin Corporation Research Award and by DARPA Safeware program. The views expressed are those of the author and do not reflect the official policy or position of the Department of Defense or the U.S. Government.

A. Scafuro—Supported by NSF grants 1012798, CNS-1414119.

D. Wichs—Supported by NSF grants CNS-1347350, CNS-1314722, CNS-1413964. This work was done in part while some of the authors were visiting the Simons Institute for the Theory of Computing, supported by the Simons Foundation and by the DIMACS/Simons Collaboration in Cryptography through NSF grant CNS-1523467.

© International Association for Cryptologic Research 2016
M. Robshaw and J. Katz (Eds.): CRYPTO 2016, Part III, LNCS 9816, pp. 149–178, 2016.
DOI: 10.1007/978-3-662-53015-3_6

notion of *somewhere equivocal* encryption, which allows us to efficiently equivocate on a small subset of the message bits.

Keywords: Adaptive security · Garbled circuits · Online/offline two-party computation

1   Introduction

Garbled Circuits. A *garbling scheme* (also referred to as a randomized encoding) can be used to garble a circuit C and an input x to derive a garbled circuit $\widetilde{C}$ and a garbled input $\widetilde{x}$. It's possible to evaluate $\widetilde{C}$ on $\widetilde{x}$ and get the correct output $C(x)$. However, the garbled values $\widetilde{C}, \widetilde{x}$ should not reveal anything else beyond this. In particular, there is a simulator that can simulate $\widetilde{C}, \widetilde{x}$ given only $C(x)$.

The notion of garbled circuits was introduced by Yao in (oral presentations of) [31,32], and can be instantiated based on one-way functions. Garbled circuits have since found countless applications in diverse areas of cryptography, most notably to secure function evaluation (SFE) starting with Yao's work, but also in parallel cryptography [5,6], verifiable computation [7,16], software protection [20,22], functional encryption [19,21,30], key-dependent message security [3,9], obfuscation [4] and many others. These applications rely on various efficiency/functionality properties of garbled circuits and a comprehensive study of this primitive is explored in the work of Bellare et al. [12].

On-line Complexity. In many applications, the garbled circuit $\widetilde{C}$ can be computed in an *off-line* pre-processing phase before the input is known, and therefore the efficiency of this procedure may not be of paramount importance. On the other hand, once the input x becomes available in the *on-line* phase, creating the garbled input $\widetilde{x}$ should be extremely efficient. Therefore, the main efficiency measure that we consider here is the *on-line complexity*, which is the time it takes to garble an input x, and hence also a bound on the size of $\widetilde{x}$. Ideally, the on-line complexity should only be linear in the input size $|x|$ and independent of the potentially much larger circuit size $|C|$.[1]

Yao's Scheme. Yao's garbling scheme already achieves essentially optimal online complexity, where the time to garble an input x and the size of $\widetilde{x}$ are only linear in the input size $|x|$, independent of the circuit size.[2] However, it only realizes a weak notion of security called *selective security*, which corresponds to a setting where adversary must choose the input x before seeing the garbled

---

[1] Note that, without any other restrictions on the structure of the garbling scheme, there is a trivial scheme where $\widetilde{C}$ is empty and $\widetilde{x} = C(x)$, whose on-line complexity is proportional to $|C|$.

[2] More precisely, in Yao's garbled circuits, the garbled input is of size $|x| \cdot \mathsf{poly}(\lambda)$ where λ is the security parameter. The work of [8] shows how to reduce this to $|x| + \mathsf{poly}(\lambda)$ assuming stronger assumptions such as DDH, RSA or LWE.

circuit $\widetilde{C}$. In particular, the adversary first chooses both C and x and then gets the garbled values $\widetilde{C}, \widetilde{x}$ which are either correctly computed using the "real" garbling scheme or "simulated" using only $C(x)$. The adversary should not be able to distinguish between the real world and the simulated world.

Selective vs. Adaptive Security. Selective security is often unsatisfactory in precisely the scenarios envisioned in the off-line/on-line setting, where the garbled circuit $\widetilde{C}$ is given out first and the garbled input $\widetilde{x}$ is only given out later. In such settings, the adversary may be able to (partially) influence the choice of the input x after seeing the garbled circuit $\widetilde{C}$. Therefore, we need a stronger notion called *adaptive security*, defined via the following two stage game:

1. The adversary chooses a circuit C and gets the garbled circuit $\widetilde{C}$.
2. After seeing $\widetilde{C}$ the adversary adaptively chooses an input x and gets the garbled input $\widetilde{x}$.

In the real world $\widetilde{C}, \widetilde{x}$ are computed correctly using the garbling scheme, while in the simulated world they are created by a simulator who only gets the output $C(x)$ in step (2) of the game but does not get the input x. The adversary should not be able to distinguish these two worlds.

The work of Bellare, Hoang and Rogaway [11] gave the first thorough treatment of adaptively secure garbling schemes and showed that this notion is crucial in many applications. They point out that it remains unknown whether Yao's garbling scheme or any of its many variants can satisfy adaptive security, and the proof techniques that work in the selective security setting do not extend to the adaptive setting. They left it as the main open problem to construct adaptively secure garbling schemes where the on-line complexity is smaller than the circuit size.[3]

Finally we emphasize that the problem of achieving adaptively secure garbled circuits is different from the problem of achieving adaptively secure two-party computation (with constant rounds) using an approach based on garbled circuits. The latter means that the adversary can *corrupt the players* adaptively. It is not known whether either problem can be reduced to the other.

1.1   Prior Approaches to Adaptive Security

Lower Bound and Yao's Scheme. The work of Applebaum et al. [8] (see also [24]) gives a lower bound on the on-line complexity of circuit garbling in the adaptive setting, showing that the size of the garbled input $\widetilde{x}$ must exceed the *output size* of the circuit. This is in contrast to the selective security setting, where

---

[3] The adaptive security notion we described, is denoted prv1 by [11]. They also consider a stronger variant called prv2, where the adversary adaptively chooses bits of the input x one at a time and gets the corresponding bits of the garbled input $\widetilde{x}$. They show that there is an efficiency preserving transformation from prv1 to prv2 following the ideas from [20]. Therefore, in this work we can focus solely on achieving prv1.

Yao's garbling scheme achieves on-line complexity that depends only on the input size and not the output size. In particular, this shows that Yao's garbling scheme cannot directly be adaptively secure.

Complexity Leveraging. It turns out that there is a simple and natural modification of Yao's garbling scheme (i.e., by withholding the mapping of output-wire keys to output bits until the on-line phase) that would match the above lower bound and could plausibly be conjectured to provide adaptive security. In fact, one can prove that the above variant of Yao's scheme is secure in the adaptive setting using *complexity leveraging*, but only at a 2^n security loss in the reduction, where n is the input size. There is no known proof of security that avoids this loss.[4]

One-Time Pad and Random-Oracles. An alternate approach, suggested by [11], is to use one-time pad encryption to encrypt a Yao garbled circuit in the off-line phase and then provide the decryption key with the garbled input in the on-line phase. Intuitively, since a one-time pad encryption is "non-committing" and the ciphertext can be *equivocated* to any possible message by providing a corresponding key, the adversary does not gain any advantage in seeing such a ciphertext in the off-line phase. Unfortunately, this solution blows up the on-line complexity to be at least as large as the circuit size.

The work of [11] also noted that one can replace the one-time pad encryption in the above solution with a random-oracle based encryption scheme, which can be equivocated by programming random oracle outputs. This gives an adaptively secure garbled circuit construction with optimal parameters in the random oracle model. In fact, this approach can even be used to prove security in parameter regimes that beat the lower bound of [8], and therefore we should be suspicious about it's implications in the standard model, when the random oracle is replaced by a hash function. In particular, the construction is using the random oracle for equivocation in ways that we know to be uninstantiable in the standard model [29].

UCE-Security. Bellare et al. [10] show that a variant of Yao garbled circuits (which does not violate the lower bound of [8]) can be proven secure when instantiated with a hash function that satisfies a security notion called *Universal Computational Extractor (UCE)* security. However, UCE is a strong, non-standard and non-falsifiable assumption.

Heavy Hammers. Lastly, we mention two approaches that get adaptively secure garbled circuits with good on-line complexity under significantly stronger assumptions than one-way functions. The work of Boneh et al. [13] implicitly provides such schemes where the on-line complexity is proportional to the input/output size and the depth d of the circuit, under the *learning with errors*

---

[4] Even if we're willing to assume exponentially secure primitives, the use of complexity leveraging blows up parameter sizes so that the garbled input must be of size at least $n^2 \cdot \mathsf{poly}(\lambda)$ where λ is the security parameter to get any meaningful security.

assumption with a modulus-to-noise ratio of $2^{\mathsf{poly}(d)}$. This translates to assuming the hardness of lattice problems with $2^{\mathsf{poly}(d)}$ approximation factors. The work of Ananth and Sahai [2] shows how to get an essentially optimal scheme, where the on-line complexity is only proportional to the input/output size of the circuit, assuming *indistinguishability obfuscation*. In terms of both assumptions and practical efficiency, these schemes are a far cry from Yao's original scheme.

1.2 Our Results

In this work, we construct the first adaptively secure garbling scheme whose on-line complexity is smaller than the circuit size and which only relies on the existence of one-way functions. Our construction is an adaptation of Yao' scheme that maintains essentially all of its desirable properties, such as having highly parallelizable circuit garbling and projective/decomposable input garbling.[5] In particular, our construction simply encrypts a Yao garbled circuit with a *somewhere equivocal* symmetric-key encryption scheme, which is a new primitive that we define and construct from one-way functions. The encrypted Yao garbled circuit is sent in the off-line phase, and the Yao garbled input along with the decryption key is sent in the on-line phase. We get various provably secure instantiations of the above approach depending on how we set the parameters of the encryption scheme.

As our main instantiation, we get a garbling scheme whose on-line complexity is $w \cdot \mathsf{poly}(\lambda)$ where w is the *width* of the circuit and λ is the security parameter, but is otherwise independent of the depth d of the circuit.[6] Note that, if we think of the circuit as representing a Turing Machine or RAM computation, then the width w of the circuit corresponds to the maximum of the input size n, output size m, and space complexity s of the computation, meaning that our on-line complexity is $(n + m + s) \cdot \mathsf{poly}(\lambda)$, but otherwise independent of the run-time of the computation.

Alternately, we also get a different instantiation where the on-line complexity is only $(n + m + d) \cdot \mathsf{poly}(\lambda)$, where n is the input size, m is the output size, and d is the depth of the circuit, but is otherwise independent of the circuit's width w. In this case, we also incur a $2^{O(d)}$ security loss in our reduction, but this can be a significant improvement over the naive complexity-leveraging approach which incurs a 2^n security loss, where n is the input size. In particular, in the case of NC^1 circuits where $d = O(\log n)$, we get a polynomial reduction and achieve optimal on-line complexity of $(n + m) \cdot \mathsf{poly}(\lambda)$.[7]

---

[5] Each bit of the garbled input only depends on one bit of the original input.

[6] We consider circuits made up of fan-in 2 gates with arbitrary fan-out. The circuit is composed of levels and wires can only connect gates in level i with those at the next level $i + 1$. The width of the circuit is the maximal number of gates in any level and the depth is the number of levels.

[7] For NC^1 circuits, there are perfectly (information theoretically) secure variants of Yao [25,26] which also achieve adaptive security. However, the on-line complexity in these schemes grows *exponentially* in the circuit depth d whereas ours is only linear in d. For example, for a boolean NC^1 circuit with depth $d = 100 \log n$, the on-line complexity of those schemes is $O(n^{100})$ whereas ours would be $O(n)$.

More broadly, we develop a connection between constructing adaptively secure schemes in our framework and a certain type of *pebble complexity* of the given circuit. The size of the garbled input is proportional to the maximal number of pebbles and the number of hybrids in our reduction is proportional to the number of moves needed to pebble the circuit.

1.3   Applications of Our Results

We briefly mention how our results can be used to get concrete improvements in several applications of garbled circuits in prior works.

On-line/Off-line Two-Party Computation. One of the main uses of garbled circuits is in two-party secure computation protocols. In this setting, Alice holds an input x_A, Bob holds an input x_B and they wish to compute $f(x_A, x_B)$. To do so, Alice creates a garbled circuit $\widetilde{C}_f$ for the function f and sends $\widetilde{C}_f$ along with her portion of the garbled input $\widetilde{x}_A$ to Bob. Bob runs an oblivious transfer (OT) protocol to get the garbled version of his input $\widetilde{x}_B$ without revealing x_B to Alice. This can be done if the garbling scheme is projective/decomposable (see footnote 9) so that each bit of the garbled input only depends on one bit of the original input. Security against fully malicious parties can be obtained via zero-knowledge proofs or cut-and-choose techniques. It is possible to instantiate the above construction with selectively secure garbled circuits, by having Bob commit to x_B before he gets the garbled circuit $\widetilde{C}_f$. This ensures that the choice of the input cannot depend on the garbled circuit.

However, in many cases, creating the garbled circuit $\widetilde{C}_f$ for the function f is expensive and we would like to do this off-line before the inputs x_A, x_B are known to Alice and Bob. Once the inputs become known, the on-line phase should be extremely efficient, and ideally much smaller than the size of the circuit of f. This setting was recently explored in the work of Lindell and Riva [28] who showed how to solve this problem very efficiently using cut-and-choose techniques, given an adaptively secure garbling scheme with low on-line complexity. To instantiate the latter primitive, they relied on the random oracle model. Using our construction of adaptively secure garbled circuit, we can instantiate the scheme of [28] in the standard model, where the on-line complexity of the two-party computation protocol would match that of our garbling schemes.

One-Time Programs and Verifiable Computation. As noted by [11], two prior works from the literature on one-time programs [20] and verifiable computation [16] implicitly require adaptively secure garbling.[8] In both cases, we can plug in our construction of adaptively secure garbling to these constructions.

---

[8] The work of [20] requires an even stronger notion of adaptivity called prv2 but this can be generically achieved given an adaptively secure scheme in our sense. See footnote 7.

In the case of one-time programs, the on-line complexity of the garbling scheme translates to the number of hardware tokens needed to create the one-time program. In the case of verifiable computation, the on-line complexity of the garbling scheme translates to the complexity of the verification protocol – it is essential that this is smaller than the circuit size to make the verification protocol non-trivial.

Compact Functional Encryption. The recent work of [1] shows how to convert any selectively secure functional encryption (FE) scheme into an adaptively secure FE. However, their transformation is not compact and the ciphertext size is as large as the maximum circuit size of the allowed functions. This is true even if the selectively secure FE that they start with is compact. Implicitly, the main bottleneck in the transformation is having adaptively secure garbled circuits with low on-line complexity. The work of [2] gives an alternate and modular transformation from a selectively secure compact FE to an adaptively secure one using adaptively secure garbled circuits (actually, their main construction is for Turing Machines and relies on garbling TMs which require heavier machinery – however, it can be scaled down to work for circuits to get the above result). This transformation applies to both bounded-collusion schemes and unbounded-collusion schemes. By plugging in our construction of adaptively secure garbled circuits into the above result we get a transformation from compact selectively secure FE to adaptive FE where the ciphertext size is only proportional to the on-line complexity of our garbling scheme.

1.4   Our Techniques

In order to explain our techniques, we must first explain the difficulties in proving the adaptive security of Yao's garbling schemes. Since these difficulties are subtle, we begin with a description of the scheme and the proof of selective security, following Lindell and Pinkas [27]. This allows us to fix a precise notation and terminology which will be needed to also explain our new construction and proof. We expect that the reader is already familiar with the basics of Yao circuits and refer to [27] for further details.

Yao's Scheme and the Challenge of Adaptive Security. *Yao's Scheme.* For each wire w in the circuit, we pick two keys k_w^0, k_w^1 for a symmetric-key encryption scheme. For each gate in the circuit computing a function $g : \{0,1\}^2 \to \{0,1\}$ and having input wires a, b and output wire c we create a *garbled gate* consisting of 4 randomly ordered ciphertexts created as:

$$
\begin{aligned}
c_{0,0} &= \mathsf{Enc}_{k_a^0}(\mathsf{Enc}_{k_b^0}(k_c^{g(0,0)})) & c_{1,0} &= \mathsf{Enc}_{k_a^1}(\mathsf{Enc}_{k_b^0}(k_c^{g(1,0)})), \\
c_{0,1} &= \mathsf{Enc}_{k_a^0}(\mathsf{Enc}_{k_b^1}(k_c^{g(0,1)})) & c_{1,1} &= \mathsf{Enc}_{k_a^1}(\mathsf{Enc}_{k_b^1}(k_c^{g(1,1)}))
\end{aligned}
\tag{1}
$$

where $(\mathsf{Enc}, \mathsf{Dec})$ is a CPA-secure encryption scheme. The garbled circuit $\widetilde{C}$ consists of all of the gabled gates, along with an *output mapping* $\{k_w^0 \to 0, k_w^1 \to 1\}$

which gives the correspondence between the keys and the bits they represent for each output wire w. To garble an n-bit value $x = x_1 x_2 \cdots x_n$, the garbled input $\widetilde{x}$ consists of the keys $k_{w_i}^{x_i}$ for the n input wires w_i.

To evaluate the garbled circuit on the garbled input, it's possible to decrypt (exactly) one ciphertext in each garbled gate and get the key $k_w^{v(w)}$ corresponding to the bit $v(w)$ going over the wire w during the computation $C(x)$. Once the keys for the output wires are computed, it's possible to recover the actual output bits by looking them up in the output mapping.

Selective Security Simulator. To prove the selective security of Yao's scheme, we need to define a simulator that gets the output $y = y_1 y_2 \cdots y_m = C(x)$ and must produce $\widetilde{C}$, $\widetilde{x}$. The simulator picks random keys k_1^0, k_w^1 for each wire w just like the real scheme, but it creates the garbled gates as follows:

$$
\begin{aligned}
c_{0,0} &= \mathsf{Enc}_{k_a^0}(\mathsf{Enc}_{k_b^0}(k_c^0)) & c_{1,0} &= \mathsf{Enc}_{k_a^1}(\mathsf{Enc}_{k_b^0}(k_c^0)), \\
c_{0,1} &= \mathsf{Enc}_{k_a^0}(\mathsf{Enc}_{k_b^1}(k_c^0)) & c_{1,1} &= \mathsf{Enc}_{k_a^1}(\mathsf{Enc}_{k_b^1}(k_c^0))
\end{aligned}
\tag{2}
$$

where all four ciphertext encrypt the same key k_c^0. It creates the output mapping $\{k_w^0 \to y_w, k_w^1 \to 1 - y_w\}$ by "programming it" so that the key k_w^0 corresponds to the correct output bit y_w for each output wire w. This defines the simulated garbled circuit $\widetilde{C}$. To create the simulated garbled input $\widetilde{x}$ the simulator simply gives out the keys k_w^0 for each input wire w. Note that, when evaluating the simulated garbled circuit on the simulated garbled input, the adversary only sees the keys k_w^0 for every wire w.

Selective Security Hybrids. To prove indistinguishability between the real world and the simulation, there is a series of carefully defined hybrid games that switch the distribution of one garbled gate at a time, starting with the input level and proceeding up the circuit level by level. In each step we switch the distribution of the ciphertexts in the targeted gate to:

$$
\begin{aligned}
c_{0,0} &= \mathsf{Enc}_{k_a^0}(\mathsf{Enc}_{k_b^0}(k_c^{v(c)})) & c_{1,0} &= \mathsf{Enc}_{k_a^1}(\mathsf{Enc}_{k_b^0}(k_c^{v(c)})), \\
c_{0,1} &= \mathsf{Enc}_{k_a^0}(\mathsf{Enc}_{k_b^1}(k_c^{v(c)})) & c_{1,1} &= \mathsf{Enc}_{k_a^1}(\mathsf{Enc}_{k_b^1}(k_c^{v(c)}))
\end{aligned}
\tag{3}
$$

where $v(c)$ is the correct *value* of the bit going over the wire c during the computation of $C(x)$.

Let us give names to the three modes for creating garbled gates that we defined above: (1) is called RealGate mode, (2) is called SimGate mode, and (3) is called InputDepSimGate mode, since the way that it is defined depends adaptively on the choice of the input x.

We can switch a gate from RealGate to InputDepSimGate mode if the gates in the previous level are in InputDepSimGate mode (or we are in the input level) by CPA security of encryption. In particular, we are *not* changing the value contained in ciphertext $c_{v(a),v(b)}$ encrypted under the keys $k_a^{v(a)}, k_b^{v(b)}$ that the adversary obtains during evaluation, but we *can* change the values contained in all of the other ciphertexts since the keys $k^{1-v(a)}, k^{1-v(b)}$ do not appear anywhere inside the garbled gates in the previous level.

At the end of the above sequence of hybrid games, all gates are switched from RealGate to InputDepSimGate mode and the output mapping is computed as in the real world. The resulting distribution is *statistically identical* to the simulation where all the gates are in SimGate mode and the output mapping is programmed. This is because, at any level that's not the output, the keys k_c^0, k_c^1 are used completely identically in the subsequent level so there is no difference between always encrypting $k_c^{v(c)}$ (InputDepSimGate) and k_c^0 (SimGate). At the output level there is no difference between encrypting $k_c^{v(c)}$ and giving the real mapping $k_c^{v(c)} \rightarrow y_c$ or encrypting k_c^0 and giving the programmed mapping $k_c^0 \rightarrow y_c$ where y_c is the output bit on wire c.

Challenges in Achieving Adaptive Security. There are two issues in using the above strategy in the adaptive setting: an immediate but easy to fix problem and a more subtle but difficult to overcome problem.

The first immediate issue is that the selective simulator needs to know the output $y = C(x)$ to create the garbled circuit $\widetilde{C}$ and in particular to program the output mapping $\{k_w^0 \rightarrow y_w, k_w^1 \rightarrow 1 - y_w\}$ for the output wires w. However, the adaptive simulator does not get the output y until *after* it creates the garbled circuit $\widetilde{C}$. Therefore, we cannot (even syntactically) use the selective security simulator in the adaptive setting. This issue turns out to be easy to fix by modifying the construction to send the output-mapping as part of the garbled input $\widetilde{x}$ in the on-line phase, rather than as part of the garbled circuit $\widetilde{C}$ in the off-line phase. This modification raises on-line complexity to also being linear in the output size of the circuit, which we know to be necessary by the lower bound of [8]. With this modification, the adaptive simulator can program the output mapping after it learns the output $y = C(x)$ in the on-line phase and therefore we get a syntactically meaningful simulation strategy in the adaptive setting.

The second problem is where the true difficulty lies. Although we have a syntactically meaningful simulation strategy, the previous proof of indistinguishability of the real world and the simulation completely breaks down in the adaptive setting. Recall that the proof consisted of a sequence of hybrids where we changed one garbled gate at a time (starting from the input level) from RealGate mode to the InputDepSimGate mode. In the latter mode, the gate is created in a way that depends on the input x, but in the adaptive setting the input x is chosen adaptively after the garbled circuit is created, leading to a circularity. In other words, the distribution of InputDepSimGate as specified in Eq. (3) doesn't even syntactically make sense in the adaptive setting. Therefore, *although we have a syntactically meaningful simulation strategy for the adaptive setting, we do not have any syntactically meaningful sequence of intermediate hybrids to prove indistinguishability between the real world and the simulated world.*

(One could hope to bypass InputDepSimGate mode altogether and define the hybrids by changing a gate directly from RealGate mode to SimGate mode. Unfortunately, this change is easily distinguishable already for the very first gate we would hope to change at the input level – the output value on the gate would no longer be $v(w)$ but 0 which may result in an overall incorrect output since we have not programmed the output map yet. On the other hand, we cannot immediately jump to a hybrid where we program the output map since all of the keys and their semantics are contained under encryption in prior levels of the circuit and we haven't argued about the security of the ciphertexts in these levels yet.)

Our Solution. *Outer Encryption Layer.* Our construction starts with the approach of [11] which is to encrypt the entire garbled circuit with an additional outer encryption layer in the off-line phase (this is unrelated to the encryption used to construct the garbled gates). Then, in the on-line phase, we give out the secret key for this outer encryption scheme. The approach of [11] required a symmetric-key, one-time encryption scheme which is *equivocal,* meaning that the ciphertext doesn't determine the message and it is possible to come up with a secret key that can open the ciphertext to any possible message. Unfortunately, any fully equivocal encryption scheme where a ciphertext can be opened to any message (e.g., the one-time pad) must necessarily have a secret key size which is as large as the message size. In our case, this is the entire garbled circuit and therefore this ruins the on-line efficiency of the scheme. Our main idea is to use a new type of *partially* equivocal encryption scheme, we call *somewhere equivocal.*

Somewhere Equivocal Encryption. Intuitively, a somewhere equivocal encryption scheme allows us to create a simulated ciphertext which contains "holes" in some small subset of the message bit positions I chosen by the simulator, but all other message bits are fixed. The simulator can later equivocate this ciphertext and "plug the holes" with any bits it wants by deriving a corresponding secret key. An adversary cannot distinguish between seeing a real encryption of some message $m = m_1 m_2 \cdots m_n$ and the real secret key, from seeing a simulated encryption created using only $(m_i)_{i \notin I}$ with "holes" in positions I and an equivocated secret key that later plugs the holes to the correct bits $(m_i)_{i \in I}$. We show how to construct somewhere equivocal encryption using one-way functions. The size of the secret key is only proportional to the maximum number of holes $t = |I|$ that we allow, which we call the "equivocation parameter", but can be much smaller than the message size.[9]

Our proof of security departs significantly from that of [11]. In particular, our simulator does *not* take advantage of the equivocation property of the encryption scheme at all, and in fact, our simulation strategy is identical to the adaptive simulator we outlined above for the variant of Yao's garbling where the output map is sent in the on-line phase. However, we crucially rely on the equivocation

---

[9] A different notion of partially equivocal encryption, called *somewhat non-committing* encryption, was introduced in [15]. The latter notion allows a ciphertext to be opened to some small, polynomial size, set of messages which can be chosen arbitrarily by the simulator at encryption time. The two notions are incomparable.

property to carefully define a meaningful sequence of hybrids that allows us to prove the indistinguishability of the real and simulated worlds.

Hybrids for Adaptive Security. We define hybrid distributions where various garbled gates will be created in one of three modes discussed above: RealGate, SimGate and InputDepSimGate. However, to make the last option meaningful (even syntactically) in the adaptive setting, we rely on the somewhere equivocal encryption scheme. For these hybrids, when we create the encrypted garbled circuit in the off-line phase, we will simulate the outer encryption layer with a ciphertext that contains "holes" in place of all gates that are in InputDepSimGate mode. Only when we open the outer encryption in the on-line phase after the input x is chosen, we will "plug the holes" by sampling these gates correctly in InputDepSimGate mode in a way that depends on the input x. Our equivocation parameter t for the somewhere equivocal encryption scheme therefore needs to be large enough to support the maximum number of gates in InputDepSimGate mode that we will have in any hybrid.

Sequence of Hybrids. For our main result, we use the following sequence of hybrids to prove indistinguishability of real and simulated worlds. We start by switching the first two levels of gates (starting with the input level) to InputDepSimGate mode. We then switch the first level of gates to SimGate mode and switch the third level InputDepSimGate mode. We continue this process, where in each step i we maintain level i in InputDepSimGate mode but switch the previous level $i - 1$ from InputDepSimGate to SimGate and then switch the next level $i+1$ from RealGate to InputDepSimGate. Eventually all gates will be in SimGate mode as we wanted. We can switch a level $i - 1$ from InputDepSimGate to SimGate mode when the subsequent level i is in InputDepSimGate mode since the keys k_c^0, k_c^1 for wires c crossing from level $i - 1$ to i are used identically in level i and therefore there is statistically no difference between encrypting the key $k_c^{v(c)}$ (InputDepSimGate) and k_c^0 (SimGate). We can also switch a level $i + 1$ from RealGate to InputDepSimGate when the previous level i is InputDepSimGate (or $i + 1$ is the input level) by CPA security following the same argument as in the selective setting. With this strategy, at any point in time we have at most two levels in InputDepSimGate mode and therefore our equivocation parameter only needs to be proportional to the circuit width w.

Connection to Pebbling. We can generalize the above idea and get other meaningful sequences of hybrids with different parameters and implications. We can think of the process of switching between RealGate, SimGate and InputDepSimGate modes as a new kind of *graph pebbling game*, where pebbles can be placed on the graph representing the circuit according to certain rules. Initially, all gates are in RealGate mode, which we associate with *not having any pebble* on them. We associate InputDepSimGate mode with having a *black pebble* and SimGate mode with having a *gray pebble*. The rules of the game go as follows:

– We can place or remove a black pebble on a gate as long as both predecessors of that gate have black pebbles on them (or the gate is an input gate).

- We can replace a black pebble with a gray pebble on a gate as long as all successors of that gate have black or gray pebbles on them (or the gate is an output gate).

The goal of the game is to end up with a gray pebble on every gate. Any such pebbling strategy leads to a sequence of hybrids that shows the indistinguishability between the real world and the simulation. The number of moves needed to complete the pebbling corresponds to the number of hybrids in our proof, and therefore the security loss of our reduction. The maximum number of black pebbles that are in play at any given time corresponds to the equivocation parameter needed for our somewhere equivocal encryption scheme.

For example, the sequence of hybrids discussed above corresponds to a pebbling strategy where the number of black pebbles used is linear in the circuit width w (but independent of the depth) and the number of moves is linear in the circuit size. We give an alternate recursive pebbling strategy where the number of black pebbles used is linear in the circuit depth d (but independent of the width) and the number of moves is $2^{O(d)}$ times the circuit size.

Constructing Somewhere Equivocal Encryption. Lastly, we discuss our construction of somewhere equivocal encryption from one-way functions, which may be of independent interest. Recall that a somewhere equivocal encryption provides a method for equivocating some small number (t out of n) of bits of the message.

Our construction is based on the techniques developed in recent constructions of *distributed point functions* [14,17]. These techniques give us a way to construct a pseudorandom function (PRF) family f_k with the following equivocation property: for any input x, we can create two PRF keys k_0, k_1 that each individually look uniformly random but such that $f_{k_0}(x') = f_{k_1}(x')$ for all $x' \neq x$ and $f_{k_0}(x) \neq f_{k_1}(x)$. The construction is based on a clever adaptation of the Goldreich-Goldwasser-Micali (GGM) PRF [18].

Using distributed point functions, we can immediately create a somewhere equivocal encryption with equivocation parameter $t = 1$. We rely on a PRF family f_k with the above equivocation property and with one-bit output. To encrypt a message $m = m_1 m_2 \cdots m_n \in \{0,1\}^n$ we create a ciphertext $c = f_k(1) \oplus m_1 || f_k(2) \oplus m_2 || \cdots || f_k(n) \oplus m_n$ using the PRF outputs as a one-time pad. To create a simulated encryption with a hole in position i, the simulator samples two PRF keys k_0, k_1 that only differ on input $x = i$. The simulator encrypts the n-bit message by setting the unknown value in position i to $m_i := 0$ and using k_0. If it later wants to open this value to 0, it sets the decryption key to k_0 else k_1.

We can extend the above approach to an arbitrarily large equivocation parameter t, by using the XOR of t independently chosen PRFs with the above equivocation property. The key size will be $t \cdot \mathsf{poly}(\lambda)$.

2   Preliminaries

General Notation. For a positive integer n, we define the set $[n] := \{1, \ldots, n\}$. We use the notation $x \leftarrow X$ for the process of sampling a value x according to the distribution X. For a vector $\overline{m} = (m_1, m_2, \cdots, m_n)$, and a subset $P \subset [n]$, we use $(m_i)_{i \in P}$ to denote a vector containing only the values m_i in positions $i \in P$ and $\perp$ symbols in all other positions. We use $(m_i)_{i \notin P}$ as shorthand for $(m_i)_{i \in [n] \backslash P}$.

Circuit Notation. A boolean circuit C consists of gates $\mathsf{gate}_1, \ldots, \mathsf{gate}_q$ and wires $w_1, w_2, \ldots, w_p$. A gate is defined by the tuple $\mathsf{gate}_i = (g, w_a, w_b, w_c)$ where $g : \{0, 1\}^2 \to \{0, 1\}$ is the function computed by the gate, w_a, w_b are the incoming wires, and w_c is the outgoing wire. Although each gate has a unique outgoing wire w_c, this wire can be used as an incoming wire to several different gates and therefore this models a circuit with fan-in 2 and unbounded fan-out. We let q denote the number of gates in the circuit, n denotes the number of input wires and m denote the number of output wires. The total number of wires is $p = n + q$ (since each wire can either be input wire or an outgoing wire of some gate). For convenience, we denote the n input wires by $\mathsf{in}_1, \ldots, \mathsf{in}_n$ and the m output wires by $\mathsf{out}_1, \ldots, \mathsf{out}_m$. For $x \in \{0, 1\}^n$ we write $C(x)$ to denote the output of evaluating the circuit C on input x.

We say C is leveled, if each gate has an associated level and any gate at level l has incoming wires only from gates at level $l - 1$ and outgoing wires only to gates at level $l + 1$. We let the *depth* d denote the number of levels and the *width* w denote the maximum number of gates in any level.

A circuit C is fully specified by a list of gate tuples $\mathsf{gate}_i = (g, w_a, w_b, w_c)$. We use $\Phi_{\mathsf{topo}}(C)$ to refer to the topology of a circuit– which indicates how gates are connected, without specifying the function implemented by each gate. In other words, $\Phi_{\mathsf{topo}}(C)$ is the list of *sanitized gate tuples* $\widehat{\mathsf{gate}}_i = (\perp, w_a, w_b, w_c)$ where the function g that the gate implements is removed from the tuple.

3   Garbling Scheme

We now give a formal definition of a garbling scheme. There are many variants of such definitions in the literature, and we refer the reader to [12] for a comprehensive treatment.

Definition 1. *A Garbling Scheme is a tuple of PPT algorithms* $\mathsf{GC} = (\mathsf{GCircuit}, \mathsf{GInput}, \mathsf{Eval})$ *such that:*

- $(\widetilde{C}, k) \xleftarrow{\$} \mathsf{GCircuit}(1^\lambda, C)$: *takes as input a security parameter* λ, *a circuit* $C : \{0, 1\}^n \to \{0, 1\}^m$, *and outputs the garbled circuit* $\widetilde{C}$, *and key* k.
- $\tilde{x} \leftarrow \mathsf{GInput}(k, x)$: *takes as input* $x \in \{0, 1\}^n$, *and key* k *and outputs* $\tilde{x}$.
- $y = \mathsf{Eval}(\widetilde{C}, \tilde{x})$: *given a garbled circuit* $\widetilde{C}$ *and a garbled input* $\tilde{x}$ *output* $y \in \{0, 1\}^m$.

Correctness. *There is a negligible function* ν *such that for any* $\lambda \in \mathbb{N}$, *any circuit* C *and input* x *it holds that* $\Pr[C(x) = \mathsf{Eval}(\widetilde{C}, \tilde{x})] = 1 - \nu(\lambda)$, *where* $(\widetilde{C}, k) \leftarrow \mathsf{GCircuit}(1^\lambda, C)$, $\tilde{x} \leftarrow \mathsf{GInput}(k, x)$.

Adaptive Security. *There exists a PPT simulator* Sim = (SimC, SimIn) *such that, for any PPT adversary* $\mathcal{A}$, *there exists a negligible function* ν *such that:*

$$\Pr[\mathsf{Exp}^{\mathsf{adaptive}}_{\mathcal{A},\mathsf{GC},\mathsf{Sim}}(1^\lambda,0) = 1] - \Pr[\mathsf{Exp}^{\mathsf{adaptive}}_{\mathcal{A},\mathsf{GC},\mathsf{Sim}}(1^\lambda,1) = 1] \leq \nu(\lambda)$$

where the experiment $\mathsf{Exp}^{\mathsf{adaptive}}_{\mathcal{A},\mathsf{GC},\mathsf{Sim}}(1^\lambda, b)$ *is defined as follows:*

1. *The adversary* $\mathcal{A}$ *specifies* C *and gets* $\widetilde{C}$ *where* $\widetilde{C}$ *is created as follows:*
   - *if* $b = 0$: $(\widetilde{C}, k) \leftarrow \mathsf{GCircuit}(1^\lambda, C)$,
   - *if* $b = 1$: $(\widetilde{C}, \mathsf{state}) \leftarrow \mathsf{SimC}(1^\lambda, \Phi_{\mathsf{topo}}(C))$, *where* $\Phi_{\mathsf{topo}}(C)$ *reveals the topology of* C.
2. *The adversary* $\mathcal{A}$ *specifies* x *and gets* $\tilde{x}$ *created as follows:*
   - *if* $b = 0$, $\tilde{x} \leftarrow \mathsf{GInput}(k, x)$,
   - *if* $b = 1$, $\tilde{x} \leftarrow \mathsf{SimIn}(C(x), \mathsf{state})$.
3. *Finally, the adversary outputs a bit* b', *which is the output of the experiment.*

On-line Complexity. The time it takes to garble an input x, (i.e., time complexity of $\mathsf{GInput}(\cdot, \cdot)$) is the *on-line complexity* of the scheme. Clearly the on-line complexity of the scheme gives a bound on the size of the garbled input $\tilde{x}$. Ideally, the on-line complexity should be much smaller than the circuit size $|C|$.

Projective Scheme. A garbling scheme is *projective* if each bit of the garbled input $\tilde{x}$ only depends on one bit of the actual input x. In other words, each bit of the input, is garbled independently of other bits of the input. Projective schemes are essential for two-party computation where the garbled input is transmitted using an oblivious transfer (OT) protocol. Our constructions will be projective.

Hiding Topology. A garbling scheme that satisfies the above security definition may reveal the topology of the circuit C. However, there is a way to transform any such garbling scheme into one that hides everything, including the topology of the circuit, without a significant asymptotic efficiency loss. More precisely, we rely on the fact that there is a function $\mathsf{HideTopo}(\cdot)$ that takes a circuit C as input and outputs a functionally equivalent circuit C', such that for any two circuits C_1, C_2 of equal size, if $C_1' = \mathsf{HideTopo}(C_1)$ and $C_2' = \mathsf{HideTopo}(C_2)$, then $\Phi_{\mathsf{topo}}(C_1') = \Phi_{\mathsf{topo}}(C_2')$. An easy way to construct such function $\mathsf{HideTopo}$ is by setting C' to be a universal circuit, with a hard-coded description of the actual circuit C. Therefore, to get a topology-hiding garbling scheme, we can simply use a topology-revealing scheme but instead of garbling the circuit C directly, we garble the circuit $\mathsf{HideTopo}(C)$.

4   Somewhere Equivocal Symmetric-Key Encryption

We introduce a new cryptographic primitive called *somewhere* equivocal encryption scheme. Intuitively, a somewhere equivocal encryption scheme allows one to create a simulated ciphertext which contain "holes" in some small subset of the messages in positions I chosen by the simulator, but all other messages are fixed. The simulator can later equivocate this ciphertext and "plug the holes" with any message it wants by deriving a corresponding secret key.

In more detail, encryptions can be computed in two modes: real mode and simulated mode. In the real mode, a key $\mathsf{key} \leftarrow \mathsf{KeyGen}(1^\lambda)$ is generated using the honest key generation procedure and a vector of n messages $\overline{m} = m_1, \ldots, m_n$ is encrypted using the honest encryption procedure $\overline{c} \leftarrow \mathsf{Enc}(\mathsf{key}, \overline{m})$.

In the simulated mode, there is an encryption procedure SimEnc that given a set I (set of holes) and only a subset of messages $(m_i)_{i \notin I}$, outputs simulated ciphertext $\overline{c}$ that is equivocal in positions I. In a later stage, upon learning the remaining messages $(m_i)_{i \in I}$, there exists a procedure SimKey that plugs the holes by generating a key key' that will decrypt $\overline{c}$ correctly according to $\overline{m}$.

The security property that we require is that the distributions of $\{\overline{c}, \mathsf{key}\}$ generated in the two modes are indistinguishable. To capture this property, one could envision a non-adaptive security game where and adversary $\mathcal{A}$ first selects the full vector $\overline{m}$ and the set I, then it receives the tuple $(\overline{c}, \mathsf{key})$ and needs to distinguish which distribution it belongs to. However, such security definition is not sufficient for our indistinguishability proof where instead we need an adversary to decide on the missing messages *after* she receives the ciphertext $\overline{c}$. Therefore, we consider an adaptive security definition where the security game is defined in two stages: in the first stage, the adversary chooses I, an *incomplete* vector of messages $(m_i)_{i \notin I}$, and a *challenge* index $j \notin I$ and receives the ciphertext $\overline{c}$. In the second stage, the adversary sends the remaining messages $(m_i)_{i \in I}$ and gets key. The adversary knows that all positions in I are equivocal and are plugged to the values $(m_i)_{i \in I}$ chosen in the second stage. The challenge is to distinguish whether the position j is also equivocal or not. Note that this two-stage (adaptive) security definition is stronger than the non-adaptive security definition sketched above. For completeness, we give the simpler non-adaptive definition and prove the above implication in the full version [23].

Definition 2. *A somewhere equivocal encryption scheme with* block-length s, message-length n *(in blocks), and* equivocation-parameter t *(all polynomials in the security parameter) is a tuple of probabilistic polynomial algorithms* $\Pi = (\mathsf{KeyGen}, \mathsf{Enc}, \mathsf{Dec}, \mathsf{SimEnc}, \mathsf{SimKey})$ *such that:*

- *The key generation algorithm* KeyGen *takes as input the security parameter* 1^λ *and outputs a key:* $\mathsf{key} \leftarrow \mathsf{KeyGen}(1^\lambda)$.
- *The encryption algorithm* Enc *takes as input a vector of n messages* $\overline{m} = m_1, \ldots, m_n$, *with* $m_i \in \{0,1\}^s$, *and a key* key, *and outputs ciphertext* $\overline{c} \leftarrow \mathsf{Enc}(\mathsf{key}, \overline{m})$.
- *The decryption algorithm* Dec *takes as input ciphertext* $\overline{c}$ *and a key* key *and outputs a vector of messages* $\overline{m} = m_1, \ldots, m_n$. *Namely,* $\overline{m} \leftarrow \mathsf{Dec}(\mathsf{key}, \overline{c})$.
- *The simulated encryption algorithm* SimEnc *takes as input a set of indexes* $I \subset [n]$, *such that* $|I| \leq t$, *and a vector of* $n - |I|$ *messages* $(m_i)_{i \notin I}$ *and outputs ciphertext* $\overline{c}$, *and a state* state. *Namely,* $(\mathsf{state}, \overline{c}) \leftarrow \mathsf{SimEnc}((m_i)_{i \notin I}, I)$.
- *The simulated key algorithm* SimKey, *takes in the variable* state *and messages* $(m_i)_{i \in I}$ *and outputs a key* key'. *Namely,* $\mathsf{key}' \leftarrow \mathsf{SimKey}(\mathsf{state}, (m_i)_{i \in I})$.

and satisfies the following properties:

Correctness. *For every* key $\leftarrow$ KeyGen(1^λ), $\overline{m} \in \{0,1\}^{s \times n}$ *it holds that:*

$$\mathsf{Dec}(\mathsf{key}, (\mathsf{Enc}(\mathsf{key}, \overline{m})) = \overline{m}$$

Simulation with No Holes. *We require that the distribution of* $(\overline{c}, \mathsf{key})$ *computed via* $(\overline{c}, \mathsf{state}) \leftarrow \mathsf{SimEnc}(\overline{m}, \emptyset)$ *and* key $\leftarrow \mathsf{SimKey}(\mathsf{state}, \emptyset)$ *to be identical to* key $\leftarrow \mathsf{KeyGen}(1^\lambda)$ *and* $\overline{c} \leftarrow \mathsf{Enc}(\mathsf{key}, \overline{m})$. *In other words, simulation when there are no holes (i.e.,* $I = \emptyset$) *is identical to honest key generation and encryption.*
Security. *For any PPT adversary* $\mathcal{A}$, *there is a negligible function* $\nu(\lambda)$ *s.t.:*

$$\Pr[\mathsf{Exp}^{\mathsf{simenc}}_{\mathcal{A},\Pi}(1^\lambda, 0) = 1] - \Pr[\mathsf{Exp}^{\mathsf{simenc}}_{\mathcal{A},\Pi}(1^\lambda, 1) = 1] \leq \nu(\lambda)$$

where the experiment $\mathsf{Exp}^{\mathsf{simenc}}_{\mathcal{A},\Pi}$ *is defined as follows:*
Experiment $\mathsf{Exp}^{\mathsf{simenc}}_{\mathcal{A},\Pi}(1^\lambda, b)$
1. *The adversary* $\mathcal{A}$ *on input* 1^λ *outputs a set* $I \subseteq [n]$ *s.t.* $|I| < t$, *vector* $(m_i)_{i \notin I}$, *and a challenge index* $j \in [n] \setminus I$. *Let* $I' = I \cup j$.
2.   – *If* $b = 0$, *compute* $\overline{c}$ *as follows:* $(\mathsf{state}, \overline{c}) \leftarrow \mathsf{SimEnc}((m_i)_{i \notin I}, I)$.
       – *If* $b = 1$, *compute* $\overline{c}$ *as follows:* $(\mathsf{state}, \overline{c}) \leftarrow \mathsf{SimEnc}((m_i)_{i \notin I'}, I')$.
3. *Send* $\overline{c}$ *to the adversary* $\mathcal{A}$.
4. *The adversary* $\mathcal{A}$ *outputs the set of remaining messages* $(m_i)_{i \in I}$.
       – *If* $b = 0$, *compute* key *as follows:* key $\leftarrow \mathsf{SimKey}(\mathsf{state}, (m_i)_{i \in I})$.
       – *If* $b = 1$, *compute* key *as follows:* key $\leftarrow \mathsf{SimKey}(\mathsf{state}, (m_i)_{i \in I'})$.
5. *Send* key *to the adversary* $\mathcal{A}$.
6. $\mathcal{A}$ *outputs* b' *which is the output of the experiment.*

In the full version of this paper, [23], we construct somewhere equivocal encryption from one-way functions, proving the following theorem.

Theorem 1. *Assuming the existence of one-way functions, there exists a somewhere equivocal encryption scheme for any polynomial message-length* n, *block-length* s, *and equivocation parameter* t, *having key size* $t \cdot s \cdot \mathsf{poly}(\lambda)$ *and ciphertext of size* $n \cdot s$ *bits.*

5 Adaptively Secure Garbling Scheme and Simulator

In this section we describe our garbling scheme and the simulation strategy.

5.1 Construction

Our adaptively-secure garbling scheme consists in two simple steps: (1) garble the circuit using Yao's garbling scheme; (2) hide the garbled circuit (without the output tables) under an **outer** layer of encryption instantiated with a *somewhere-equivocal* encryption scheme. In the on-line phase, the garbled input consists of Yao's garbled input plus the output tables. Next we provide the formal description of our scheme that contains the details of Yao's garbling scheme.

Let C be a leveled boolean circuit with fan-in 2 and unbounded fan-out, with inputs size n, output size m, depth d and width w. Let q denote the number of gates in C. Recall that wires are uniquely identified with labels $w_1, w_2, \ldots, w_p$, and a circuit C is specified by a list of gate tuples $\mathsf{gate} = (g, w_a, w_b, w_c)$. To simplify the description of our construction, we first describe the procedure for garbling a single gate, that we denote by $\mathsf{GarbleGate}$. Let $\Gamma = (G, E, D)$ be a CPA-secure symmetric-key encryption scheme satisfying the special correctness property, that is, the decryption procedure will abort if an incorrect key is used. $\mathsf{GarbleGate}(g, \{k_a^\sigma, k_b^\sigma, k_c^\sigma\}_{\sigma \in \{0,1\}})$ computes 4 ciphertexts $c_{\sigma_0, \sigma_1} : \sigma_0, \sigma_1 \in \{0, 1\}$ as defined below and outputs them in a random order as $\widetilde{g} = [c_1, c_2, c_3, c_4]$.

$$c_{0,0} \leftarrow E_{k_a^0}(E_{k_b^0}(k_c^{g(0,0)})) \quad c_{0,1} \leftarrow E_{k_a^0}(E_{k_b^1}(k_c^{g(0,1)}))$$
$$c_{1,0} \leftarrow E_{k_a^1}(E_{k_b^0}(k_c^{g(1,0)})) \quad c_{1,1} \leftarrow E_{k_a^1}(E_{k_b^0}(k_c^{g(1,1)}))$$

Let $\Pi = (\mathsf{KeyGen}, \mathsf{Enc}, \mathsf{Dec}, \mathsf{SimEnc}, \mathsf{SimKey})$ be a somewhere-equivocal symmetric-encryption scheme as defined in Sect. 4. Recall that in this primitive the plaintext is a vector of n blocks, each of which has s bits. In our construction we use the following parameters: the vector size $n = q$ is the number of gates and the block size $s = |\widetilde{g}|$ is the size of a single garbled gate. The equivocation parameter t is defined by the strategy used in the security proof and will be specified later. The garbling scheme is formally described in Fig. 1.

<u>GCircuit$(1^\lambda, C)$</u>

1. Garble Circuit (Yao's scheme)

- (Wires) $k_{w_i}^\sigma \leftarrow G(1^\lambda)$, $i \in [p]$, $\sigma \in \{0, 1\}$.
  (Input wires) $K = (k_{\mathsf{in}_i}^0, k_{\mathsf{in}_i}^1)_{i \in [n]}$.
- (Gates) For $\mathsf{gate}_i = (g, w_a, w_b, w_c)$ in C:
  $\widetilde{g}_i \leftarrow \mathsf{GarbleGate}(g, \{k_{w_a}^\sigma, k_{w_b}^\sigma, k_{w_c}^\sigma\}_{\sigma \in \{0,1\}})$
- (Output tables) For each output $j \in [m]$:
  $\widetilde{d}_j := [(k_{\mathsf{out}_j}^0 \to 0), (k_{\mathsf{out}_j}^1 \to 1)]$.

2. Outer Encryption

- key $\overset{\$}{\leftarrow} \mathsf{KeyGen}(1^\lambda)$.
- $\widetilde{C} \leftarrow \mathsf{Enc}(\mathsf{key}, (\widetilde{g}_1, \ldots, \widetilde{g}_q))$.

Output $\widetilde{C}, k = (K, \mathsf{key}, (\widetilde{d}_j)_{j \in [m]})$.

<u>GInput(x, k)</u>

- (Select input keys)
  $K^x = (k_{\mathsf{in}_1}^{x_1}, \ldots, k_{\mathsf{in}_n}^{x_n})$.
- **Output** $\widetilde{x} = (K^x, \mathsf{key}, (\widetilde{d}_j)_{j \in [m]})$.

<u>Eval$(\widetilde{C}, \widetilde{x})$</u>

1. Parse $\widetilde{x} = (K, \mathsf{key}, (\widetilde{d}_j)_{j \in [m]})$.
2. Decrypt Outer Encryption
   $(\widetilde{g}_i)_{i \in q} \leftarrow \mathsf{Dec}(\mathsf{key}, \widetilde{C})$.
3. Evaluate Circuit.
   Parse $K = (k_{\mathsf{in}_1}, \ldots, k_{\mathsf{in}_n})$.
   For each level $j = 1, \ldots, d$,
   For $\widehat{\mathsf{gate}}_i = (\perp, w_a, w_b, w_c)$ at level j:
   - Let $\widetilde{g}_i = [c_1, c_2, c_3, c_4]$;
   - For $\delta \in [4]$ let $k_{w_c}' \leftarrow D_{k_{w_a}}(D_{k_{w_b}}(c_\delta))$
     If $k_{w_c}' \neq \perp$ then set $k_{w_c} := k_{w_c}'$.
4. Decrypt output.
   For $j \in [m]$:
   - Parse $\widetilde{d}_j = [(k_{\mathsf{out}_j}^0 \to 0), (k_{\mathsf{out}_j}^1 \to 1)]$.
   - Set $y_j = b$ iff $k_{\mathsf{out}_j} = k_{\mathsf{out}_j}^b$.

Output $y_1, \ldots, y_m$.

Fig. 1. Adaptively-secure garbling scheme.

5.2   Adaptive Simulator

The adaptive security simulator for our garbling scheme is essentially the same as the static security simulator for Yao's scheme (as in [27]), with the only difference that the output table is sent in the on-line phase, and is computed adaptively to map to the correct output. Note that the garbled circuit simulator does not rely on the simulation properties of the somewhere equivocal encryption scheme - these are only used in the proof of indistinguishability.

More specifically, the adaptive simulator $(\mathsf{SimC}, \mathsf{SimIn})$ works as follows. In the off-line phase, SimC computes the garbled gates using procedure $\mathsf{GarbleSimGate}$, that generates 4 ciphertexts that encrypt the same output key. More precisely, $\mathsf{GarbleSimGate}(\{k_{w_a}^{\sigma}, k_{w_b}^{\sigma}\}_{\sigma \in \{0,1\}}, k_{w_c}')$ takes both keys for input wires w_a, w_b and a single key for the output wire w_c, that we denote by k_{w_c}'. It then outputs $\tilde{g}_c = [c_1, c_2, c_3, c_4]$ where the ciphertexts, arranged in random order, are computed as follows.

$$c_{0,0} \leftarrow E_{k_a^0}(E_{k_b^0}(k_c'))\quad c_{0,1} \leftarrow E_{k_a^0}(E_{k_b^1}(k_c'))$$
$$c_{1,0} \leftarrow E_{k_a^1}(E_{k_b^0}(k_c'))\quad c_{1,1} \leftarrow E_{k_a^1}(E_{k_b^0}(k_c'))$$

The simulator invokes $\mathsf{GarbleSimGate}$ on input $k_c' = k_c^0$. It then encrypts the garbled gates so obtained by using the honest procedure for the somewhere equivocal encryption.

In the on-line phase, SimIn, on input $y = C(x)$ adaptively computes the output tables so that the evaluator obtains the correct output. This is easily achieved by associating each bit of the output, y_j, to the only key encrypted in the output gate g_{out_j}, which is $k_{\mathsf{out}_j}^0$. For the input keys, SimIn just sends keys $k_{\mathsf{in}_i}^0$ for each $i \in [n]$. The detailed definition of $(\mathsf{SimC}, \mathsf{SimIn})$ is provided in Fig. 2.

Simulator

$\mathsf{SimC}(1^\lambda, \Phi_{\mathsf{topo}}(C))$

- (Wires) $k_{w_i}^{\sigma} \leftarrow G(1^\lambda)$ for $i \in [p], \sigma \in \{0,1\}$.
- (Garbled gates) For each gate $\widetilde{\mathsf{gate}}_i = (\perp, w_a, w_b, w_c)$ in $\Phi_{\mathsf{topo}}(C)$:
  $\tilde{g}_i \leftarrow \mathsf{GarbleSimGate}(\{k_{w_a}^{\sigma}, k_{w_b}^{\sigma}\}_{\sigma \in \{0,1\}}, k_{w_c}^0)$.
- (Outer Encryption): key $\overset{\$}{\leftarrow} \mathsf{KeyGen}(1^\lambda)$, $\widetilde{C} \leftarrow \mathsf{Enc}(\mathsf{key}, \tilde{g}_1, \ldots, \tilde{g}_q)$.
- **Output** $\widetilde{C}$, state $= (\{k_{w_i}^{\sigma}\}, \mathsf{key})$.

$\mathsf{SimIn}(y, \mathsf{state})$

- Generate output table: $\tilde{sd}_j \leftarrow [(k_{\mathsf{out}_j}^{y_j} \to 0), (k_{\mathsf{out}_j}^{1-y_j} \to 1)]_{j \in [m]}$. // ensures $k_{\mathsf{out}_j}^0 \to y_j$
- **Output** $\tilde{x} = ((k_{\mathsf{in}_i}^0)_{i \in [n]}, \mathsf{key}, (\tilde{sd}_j)_{j \in [m]})$.

Fig. 2. Simulator for adaptive security.

6   Hybrid Games

We now need to prove the indistinguishability of our garbling scheme and the simulation. We devise a modular approach for proving indistinguishability using different strategies that result in different parameters. We first provide a template for defining hybrid games, where each such hybrid game is parametrized by a *circuit configuration*, that is, a vector indicating the way the gates are garbled and encrypted. Then we define the rules that allow us to indistinguishably move from one configuration to another. With this framework in place, an indistinguishability proof consists of a strategy to move from the circuit configuration of the real game to the circuit configuration of the simulated game, using the allowed rules.

6.1   Template for Defining Hybrid Games

Gate/Circuit Configuration. We start by defining a *gate configuration*. A gate configuration is a pair (outer mode, garbling mode) indicating the way a gate is computed. The outer encryption mode can be {EquivEnc, BindEnc} depending on whether the outer encryption contains a "hole" in place of that gate or whether it is binding on that gate. The garbling mode can be {RealGate, SimGate, InputDepSimGate} which corresponds to the distributions outlined in Fig. 3. We stress that, if the garbling mode of a gate is InputDepSimGate then we require that the outer encryption mode is EquivEnc. This means that there are 5 valid gate configurations for each gate.

RealGate	SimGate	InputDepSimGate
$c_{0,0} \leftarrow E_{k_a^0}(E_{k_b^0}(k_c^{g(0,0)}))$	$c_{0,0} \leftarrow E_{k_a^0}(E_{k_b^0}(k_c^0))$	$c_{0,0} \leftarrow E_{k_a^0}(E_{k_b^0}(k_c^{v(c)}))$
$c_{0,1} \leftarrow E_{k_a^0}(E_{k_b^1}(k_c^{g(0,1)}))$	$c_{0,1} \leftarrow E_{k_a^0}(E_{k_b^1}(k_c^0))$	$c_{0,1} \leftarrow E_{k_a^0}(E_{k_b^1}(k_c^{v(c)}))$
$c_{1,0} \leftarrow E_{k_a^1}(E_{k_b^0}(k_c^{g(1,0)}))$	$c_{1,0} \leftarrow E_{k_a^1}(E_{k_b^0}(k_c^0))$	$c_{1,0} \leftarrow E_{k_a^1}(E_{k_b^0}(k_c^{v(c)}))$
$c_{1,1} \leftarrow E_{k_a^1}(E_{k_b^1}(k_c^{g(1,1)}))$	$c_{1,1} \leftarrow E_{k_a^1}(E_{k_b^1}(k_c^0))$	$c_{1,1} \leftarrow E_{k_a^1}(E_{k_b^1}(k_c^{v(c)}))$

Fig. 3. Garbling gate modes: RealGate (left), SimGate (center), InputDepSimGate (right). The value $v(c)$ depends on the input x and corresponds to the bit going over the wire c in the computation $C(x)$.

A *circuit configuration* simply consists of the gate configuration for each gate in the circuit. More specifically, we represent a circuit configuration by a tuple $(I, (\text{mode}_i)_{i \in [q]})$ where

– Set $I \subseteq [q]$ contains the indices of the gates i whose outer mode is EquivEnc.
– The value $\text{mode}_i \in \{\text{RealGate}, \text{SimGate}, \text{InputDepSimGate}\}$ describes the garbling mode of gate i.

A *valid circuit configuration* is one where all indexes i such that $\text{mode}_i = $ InputDepSimGate satisfy $i \in I$.

Game $\mathsf{Hyb}(I, (\mathsf{mode}_i)_{i \in [q]})$

<u>Garble Circuit $\underline{C}$:</u>

- **Garble Gates**
  (Wires) $k_{w_i}^\sigma \leftarrow G(1^\lambda)$ for $i \in [p]$, $\sigma \in \{0, 1\}$.
  (Gates) For each $\mathsf{gate}_i = (g, w_a, w_b, w_c)$ in C.
  - If $\mathsf{mode}_i = \mathsf{RealGate}$: run $\widetilde{g}_i \leftarrow \mathsf{GarbleGate}(g, \{k_{w_a}^\sigma, k_{w_b}^\sigma, k_{w_c}^\sigma\}_{\sigma \in \{0,1\}})$.
  - if $\mathsf{mode}_i = \mathsf{SimGate}$: run $\widetilde{g}_i \leftarrow \mathsf{GarbleSimGate}(\{k_{w_a}^\sigma, k_{w_b}^\sigma\}_{\sigma \in \{0,1\}}, k_{w_c}^0)$.
- **Outer Encryption.**
  1. $(\mathsf{state}, \widetilde{C}) \leftarrow \mathsf{SimEnc}((\widetilde{g}_i)_{i \notin I}, I)$.
  2. Output $\widetilde{C}$.

<u>Garble Input x:</u>

(Compute adaptive gates)
For each $i \in I$ s.t. $\mathsf{mode}_i = \mathsf{InputDepSimGate}$:

  Let $\mathsf{gate}_i = (g_i, w_a, w_b, w_c)$, and let $v(c)$
  be the bit on the wire w_c during the computation $C(x)$.
  Set $\widetilde{g}_i \leftarrow \mathsf{GarbleSimGate}((k_{w_a}^\sigma, k_{w_b}^\sigma)_{\sigma \in \{0,1\}}, k_{w_c}^{v(c)})$.

(Decryption key) $\mathsf{key}' \leftarrow \mathsf{SimKey}(\mathsf{state}, (\widetilde{g}_i)_{i \in I})$
(Output tables) Let $y = C(x)$. For $j = 1, \ldots, m$:
Let i be the index of the gate with output wire out_j.

  - If $\mathsf{mode}_i \neq \mathsf{SimGate}$, set $\widetilde{d}_j := [(k_{\mathsf{out}_j}^0 \to 0), (k_{\mathsf{out}_j}^1 \to 1)]$,
  - else, set $\widetilde{d}_j := [(k_{\mathsf{out}_j}^{y_j} \to 0), (k_{\mathsf{out}_j}^{1-y_j} \to 1)]$.

(Select input keys) For $j = 1, \ldots, n$:

  - If all gates i having in_j as an input wire satisfy $\mathsf{mode}_i = \mathsf{SimGate}$, then
    set $K[i] := k_{\mathsf{in}_i}^0$,
  - else set $K[i] := k_{\mathsf{in}_i}^{x_i}$.

Output $\widetilde{x} := (K, \mathsf{key}', \{\widetilde{d}_j\}_{j \in [m]})$.

Fig. 4. The hybrid game.

The Hybrid Game $\mathsf{Hyb}(I, (\mathsf{mode}_i)_{i \in [q]})$. Every valid circuit configuration I, $(\mathsf{mode}_i)_{i \in [q]}$ defines a hybrid game $\mathsf{Hyb}(I, (\mathsf{mode}_i)_{i \in [q]})$ as specified formally Fig. 4 and described informally below. The hybrid game consists of two procedures: GCircuit' for creating the garbled circuit $\widetilde{C}$ and GInput' for creating the garbled input $\widetilde{x}$ respectively. The garbled circuit is created by picking random keys $k_{w_j}^\sigma$ for each wire w_j. For each gate i, such that $\mathsf{mode}_i \in \{\mathsf{RealGate}, \mathsf{SimGate}\}$ it creates a garbled gate $\widetilde{g}_i$ using the corresponding distribution as described in Fig. 3. The garbled circuit $\widetilde{C}$ is then created by simulating the outer encryption using the values $\widetilde{g}_i$ in locations $i \notin I$ and "holes" in the locations I. The garbled input is created by first sampling the garbled gates $\widetilde{g}_i$ for each i such that $\mathsf{mode}_i = \mathsf{InputDepSimGate}$ using the corresponding distribution in Fig. 3 and using knowledge of the input x. Then the decryption key key is simulated by plugging in the holes in locations I with the correctly sampled garbled gates

$\widetilde{g}_i$. There is some subtlety about how the input labels $K[i]$ and the output label maps $\widetilde{d}_j$ are created when computing $\widetilde{x}$:

- If all of the gates having in_i as an input wire are in SimGate mode, then $K[i] := k_{\mathsf{in}_i}^0$ else $K[i] := k_{\mathsf{in}_i}^{x_i}$.
- If the unique gate having out_j as an output wire is in SimGate mode, then we give the simulated output map $\widetilde{d}_j := [(k_{\mathsf{out}_j}^{y_j} \to 0), (k_{\mathsf{out}_j}^{1-y_j} \to 1)]$ else the real one $\widetilde{d}_j := [(k_{\mathsf{out}_j}^0 \to 0), (k_{\mathsf{out}_j}^1 \to 1)]$.

Real game and Simulated Game. By definition of adaptively secure garbled circuits (Definition 1), the real game $\mathsf{Exp}_{\mathcal{A},\mathsf{GC},\mathsf{Sim}}^{\mathsf{adaptive}}(1^\lambda, 0)$ is equivalent to $\mathsf{Hyb}(I = \emptyset, (\mathsf{mode}_i = \mathsf{RealGate})_{i \in [q]})$ and the simulated game $\mathsf{Exp}_{\mathcal{A},\mathsf{GC},\mathsf{Sim}}^{\mathsf{adaptive}}(1^\lambda, 1)$ is equivalent to $\mathsf{Hyb}(I = \emptyset, (\mathsf{mode}_i = \mathsf{SimGate})_{i \in [q]})$. Therefore, the main aim is to show that these hybrids are indistinguishable.[10]

6.2   Rules for Indistinguishable Hybrids

Next, we provide rules that allow us to move from one configuration to another and prove that the corresponding hybrid games are indistinguishable. We define three rules that allow us to do this. We define $\mathsf{mode} \overset{\mathsf{def}}{=} (\mathsf{mode}_i)_{i \in [q]}$.

Indistinguishability Rule 1: Changing the Outer Encryption Mode BindEnc $\leftrightarrow$ EquivEnc. This rule allows to change the outer encryption of a single gate. It says that one can move from a valid circuit configuration (I, mode) to a circuit configuration (I', mode) where $I' = I \cup j$. Thus one more gate is now computed equivocally (and vice versa).

Lemma 1. *Let (I, mode) be any valid circuit configuration, let $j \in [q] \setminus I$ and let $I' = I \cup j$. Then $\mathsf{Hyb}(I, \mathsf{mode}) \overset{\mathsf{comp}}{\approx} \mathsf{Hyb}(I', \mathsf{mode})$ are computationally indistinguishable as long as $\Pi = (\mathsf{KeyGen}, \mathsf{Enc}, \mathsf{Dec}, \mathsf{SimEnc}, \mathsf{SimKey})$ is a somewhere equivocal encryption scheme with equivocation parameter t such that $|I'| \leq t$.*

Proof. Towards a contradiction, assume there exists a PPT distinguisher $\mathcal{A}$ that distinguishes the distributions $H_0 = \mathsf{Hyb}(I, \mathsf{mode})$ and $H_1 = \mathsf{Hyb}(I', \mathsf{mode})$ as defined in the Lemma.

We construct a distinguisher B for the security of somewhere equivocal encryption scheme as follows. Informally, adversary B is playing in experiment $\mathsf{Exp}_{B,\Pi}^{\mathsf{simenc}}(1^\lambda, b)$ and uses her oracle access to SimEnc to reproduce the distribution of H_b. B, on input I, j and $\mathsf{mode} = \mathsf{mode}_1, \ldots, \mathsf{mode}_q$ computes each garbled gate $\widetilde{g}_i$ on its own exactly as in H_0/H_1 accordingly to mode_i. B computes the outer encryptions of the gates by sending the gates, along with sets I, j to $\mathsf{Exp}^{\mathsf{simenc}}$.

---

[10] Note that, the games $\mathsf{Hyb}(\cdots)$ use the simulated encryption and key generation procedures of the somewhere equivocal encryption, while the games $\mathsf{Exp}_{\mathcal{A},\mathsf{GC},\mathsf{Sim}}^{\mathsf{adaptive}}(1^\lambda, b)$ only use the real key generation and encryption procedures. However, by definition, these are equivalent when $I = \emptyset$ (no "holes").

In the on-line phase, after obtaining x from $\mathcal{A}$, B computes the values for the missing gates $(\widetilde{g}_i)_{i \in I}$ and send them to $\mathsf{Exp}^{\mathsf{simenc}}$, and obtain a key key'. B uses such key to compute the garbled inputs $\widetilde{x}$.

Now, if B is playing the game $\mathsf{Exp}^{\mathsf{simenc}}_{B,\Pi}(1^\lambda, b)$ with a bit b, then the view generated by B is distributed identically to H_b. Thus, B distinguishes whether it is playing the game with $b = 0$ or $b = 1$ with the same probability that $\mathcal{A}$ distinguishes H_0 from H_1. A more detailed description of adversary B is provided in the full version [23].

Indistinguishability Rule 2. Changing the Garbling Mode, RealGate ↔ InputDepSimGate. This rule allows us to change the mode of a gate j from RealGate to InputDepSimGate as long as $j \in I$ and that $\mathsf{gate}_j = (g, w_a, w_b, w_c)$ has incoming wires w_a, w_b that are either input wires or are the outgoing wires of some predecessor gates both of which are in InputDepSimGate mode.

Double Encryption Security. For convenience, we use the notion of double encryption security, following [27]. This notion is implied by standard CPA security but is more convenient to use in our security proof of garbled circuit security. See the full version [23] for more details.

Definition 3 (Predecessor/Successor/Sibling Gates). *Given a circuit C and a gate $j \in [q]$ of the form $\mathsf{gate}_j = (g, w_a, w_b, w_c)$ with incoming wires w_a, w_b and outgoing wire w_c:*

- *We define the predecessors of j, denoted by $\mathsf{Pred}(j)$, to be the set of gates whose outgoing wires are either w_a or w_b. If w_a, w_b are input wires then $\mathsf{Pred}(j) = \emptyset$, else $|\mathsf{Pred}(j)| = 2$.*
- *We define the successors of j, denoted by $\mathsf{Succ}(j)$ to be the set of gates that contain w_c as an incoming wire. If w_c is an output wires then $\mathsf{Succ}(j) = \emptyset$.*
- *We define the siblings of j, denoted by $\mathsf{Siblings}(j)$ to be the set of gates that contain either w_a or w_b as an incoming wire.*

Lemma 2. *Let $(I, \mathsf{mode} = (\mathsf{mode}_i)_{i \in [q]})$ be a valid circuit configuration and let $j \in I$ be an index such that $\mathsf{mode}_j = \mathsf{RealGate}$ and for all $i \in \mathsf{Pred}(j)$: $\mathsf{mode}_i = \mathsf{InputDepSimGate}$. Let $\mathsf{mode}' = (\mathsf{mode}'_i)_{i \in [q]}$ be defined by $\mathsf{mode}'_i = \mathsf{mode}_i$ for all $i \neq j$ and $\mathsf{mode}'_j = \mathsf{InputDepSimGate}$. Then the games $\mathsf{Hyb}(I, \mathsf{mode}) \overset{comp}{\approx} \mathsf{Hyb}(I, \mathsf{mode}')$ are computationally indistinguishable as long as $\Gamma = (G, E, D)$ is an encryption scheme secure under chosen double encryption.*

Proof. Let I, mode, j and mode' be as in the statement of the Lemma. Towards a contradiction, assume that there exists a PPT adversary $\mathcal{A}$ distinguishing distributions generated in $H^0 := \mathsf{Hyb}(I, \mathsf{mode})$ and $H^1 := \mathsf{Hyb}(I, \mathsf{mode}')$.

We construct an adversary B that breaks the CPA-security of the inner encryption scheme $\Gamma = (G, E, D)$ which is used to garble gates. More specifically, we show that B wins the chosen double encryption security game which is implied by CPA security. Informally, B, on input mode, I and target gate j aims to use her CPA-oracle access in $\mathsf{Exp}^{\mathsf{double}}(1^\lambda, b)$ to generate a distribution H^b. Recall that the only difference between H^0 and H^1 is in the way that the garble gate $\tilde{g}_j$ is computed. On a high level, the reduction B will compute all garbled gates $\tilde{g}_i$ for $i \neq j$, according to experiment $\mathsf{Hyb}(I, \mathsf{mode})$, and will compute the garbled gate $\tilde{g}_j$ using the ciphertexts obtained as a challenge in the experiment $\mathsf{Exp}^{\mathsf{double}}(1^\lambda, b)$.

In more detail, let $\mathsf{gate}_j = (g, w_a, w_b, w_c)$ be the target gate. Recall $j \in I$ and therefore the value $\tilde{g}_j$ is only needed in the on-line phase. If the values going over the wires w_a, w_b during the computation $C(x)$ are α, β respectively, the reduction B will know all wire keys *except* for $k_{w_a}^{1-\alpha}, k_{w_b}^{1-\beta}$. To create the garbled gate $\tilde{g}_j$ it will create the ciphertext $c_{\alpha,\beta}$ as an encryption of $k_{w_c}^{g(\alpha,\beta)}$ on its own, but the remaining three ciphertexts $c_{\alpha',\beta'}$ will come from the experiment $\mathsf{Exp}^{\mathsf{double}}(1^\lambda, b)$ as either encryptions of different values $k_{w_c}^{g(\alpha',\beta')}$ (real) or of the same value $k_{w_c}^{g(\alpha,\beta)}$.

The one subtlety is that reduction needs to create encryptions under the keys $k_{w_a}^{1-\alpha}, k_{w_b}^{1-\beta}$ to create garbled gates $\tilde{g}_i$ for gates i that are siblings of gate j. It can do that by using the encryption oracles which are given to it as part of the experiment $\mathsf{Exp}^{\mathsf{double}}(1^\lambda, b)$. However, since some of the sibling gates i might be in RealGate or SimGate modes, the reduction needs to create these encryptions already in the offline phase and therefore needs to know the values of α, β in the offline phase before the input x is chosen. To deal with this, we simply have the reduction *guess* the bits α, β randomly in the offline phase. If in the online phase it finds out that the guess is incorrect it outputs a random bit and aborts, else continues. See the full version [23], for a detailed description of the reduction B.

Let *Correct* be the event that B guesses α and β correctly. Then

$$|\Pr[\mathsf{Exp}_B^{\mathsf{double}}(1^\lambda, 0) = 1] - \Pr[\mathsf{Exp}_B^{\mathsf{double}}(1^\lambda, 1) = 1]|$$

$$= \frac{1}{4}|\Pr[\mathsf{Exp}_B^{\mathsf{double}}(1^\lambda, 0) = 1|Correct] - \Pr[\mathsf{Exp}_B^{\mathsf{double}}(1^\lambda, 1) = 1|Correct]|$$

$$= \frac{1}{4}|\Pr[H_{\mathcal{A}}^0(1^\lambda) = 1] - \Pr[H_{\mathcal{A}}^1(1^\lambda)]|$$

$$\implies |\Pr[H_{\mathcal{A}}^0(1^\lambda) = 1] - \Pr[H_{\mathcal{A}}^1(1^\lambda)]|$$

$$\leq 4|\Pr[\mathsf{Exp}_B^{\mathsf{double}}(1^\lambda, 0) = 1] - \Pr[\mathsf{Exp}_B^{\mathsf{double}}(1^\lambda, 1) = 1]| \leq \mathsf{negl}(\lambda)$$

which proves the Lemma.

Indistinguishability Rule 3. Changing the Garbling Mode: InputDepSimGate $\leftrightarrow$ SimGate. This rule allows us to change the mode of a gate j from InputDepSimGate to SimGate under the condition that all successor gates $i \in \mathsf{Succ}(j)$ satisfy that $\mathsf{mode}_i \in \{\mathsf{InputDepSimGate}, \mathsf{SimGate}\}$.

Lemma 3. *Let $(I, \mathsf{mode} = (\mathsf{mode}_i)_{i \in [q]})$ be a valid circuit configuration and let $j \in I$ be an index such that $\mathsf{mode}_j = \mathsf{InputDepSimGate}$ and for all $i \in \mathsf{Succ}(j)$ we have $\mathsf{mode}_i \in \{\mathsf{SimGate}, \mathsf{InputDepSimGate}\}$. Let $\mathsf{mode}' = (\mathsf{mode}_i')_{i \in [q]}$ be defined by $\mathsf{mode}_i' = \mathsf{mode}_i$ for all $i \neq j$ and $\mathsf{mode}_j' = \mathsf{SimGate}$. Then the games $\mathsf{Hyb}(I, \mathsf{mode}) \equiv \mathsf{Hyb}(I, \mathsf{mode}')$ are identically distributed.*

Proof. Define $H_0 := \mathsf{Hyb}(I, \mathsf{mode})$ and $H_1 := \mathsf{Hyb}(I, \mathsf{mode}')$. Let $\mathsf{gate}_j = (g, w_a, w_b, w_c)$, and let $v(c)$ be the bit on the wire w_c during the computation $C(x)$, which is defined in the on-line phase.

The main difference between the hybrids is how the garbled gate $\widetilde{g}_j$ is created:

- In H_0, we set $\widetilde{g}_j \leftarrow \mathsf{GarbleSimGate}((k_{w_a}^\sigma, k_{w_b}^\sigma)_{\sigma \in \{0,1\}}, k_{w_c}^{v(c)})$.
- In H_1, we set $\widetilde{g}_j \leftarrow \mathsf{GarbleSimGate}((k_{w_a}^\sigma, k_{w_b}^\sigma)_{\sigma \in \{0,1\}}, k_{w_c}^0)$.

If j is not an output gate, and all successor gates $i \in \mathsf{Succ}(j)$ are in $\{\mathsf{SimGate}, \mathsf{InputDepSimGate}\}$ modes then the keys $k_{w_c}^0$ and $k_{w_c}^1$ are treated symmetrically everywhere in the game other than in $\widetilde{g}_j$. Therefore, by symmetry, there is no difference between using $k_{w_c}^0$ and $k_{w_c}^{v(c)}$ in $\widetilde{g}_j$

If j is an output gate then the keys $k_{w_c}^0$ and $k_{w_c}^1$ are only used in $\widetilde{g}_j$ and in the output map $\widetilde{d}_j$. Therefore, by symmetry, there is no difference between using $k_{w_c}^{y_j}$ in $\widetilde{g}_j$ and setting $\widetilde{d}_j := [(k_{\mathsf{out}_j}^0 \to 0), (k_{\mathsf{out}_j}^1 \to 1)]$ (in H_0) versus using $k_{w_c}^0$ in $\widetilde{g}_j$ and setting $\widetilde{d}_j := [(k_{\mathsf{out}_j}^{y_j} \to 0), (k_{\mathsf{out}_j}^{1-y_j} \to 1)]$ (in H_1).

One last difference between the hybrids occurs if some wire in_i becomes only connected to gates that are in $\mathsf{SimGate}$ in H_1. In this case, when we create the garbled input $\widetilde{x}$, then in H_0 we give $K[i] := k_{\mathsf{in}_i}^{x_i}$ but in H_1 we give $K[i] := k_{\mathsf{in}_i}^0$. Since the keys $k_{\mathsf{in}_i}^0, k_{\mathsf{in}_i}^1$ are treated symmetrically everywhere in the game (both in H_0 and H_1) other than in $K[i]$, there is no difference between setting $K[i] := k_{\mathsf{in}_i}^0$ versus $K[i] := k_{\mathsf{in}_i}^{x_i}$.

7   Pebbling and Sequences of Hybrid Games

In the last section we defined hybrid games parameterized by a configuration (I, mode). We also gave 3 rules, which describe ways that allow us to indistinguishably move from one configuration to another. Now our goal is to use the given rules so as to define a *sequence of indistinguishable hybrid games* that takes us from the *real game* $\mathsf{Hyb}(I = \emptyset, (\mathsf{mode}_i = \mathsf{RealGate})_{i \in [q]})$ to the simulation $\mathsf{Hyb}(I = \emptyset, (\mathsf{mode}_i = \mathsf{SimGate})_{i \in [q]})$.

Pebbling Game. We show that the problem of finding such sequences of hybrid games can be captured by a certain type of *pebbling game* on the circuit C. Each gate can either have *no pebble*, a *black pebble*, or a *gray pebble* on it (this will correspond to $\mathsf{RealGate}, \mathsf{InputDepSimGate}$ and $\mathsf{SimGate}$ modes respectively). Initially, the circuit starts out with no pebbles on any gate. The game consist of the following possible moves:

Rule A. We can place or remove a black pebble on a gate as long as both predecessors of that gate have black pebbles (or the gate is an input gate).

Rule B. We can replace a black pebble with a gray one, only if successors of that gate have black or gray pebbles on them (or the gate is an output gate).

A *pebbling* of a circuit C is a sequence of γ moves that follow rules A and B and that end up with a gray pebble on every gate. We say that a pebbling uses

t black pebbles if this is the maximal number of black pebbles on the circuit at any point in time during the game.

From Pebbling to Sequence of Hybrids. In our next theorem we prove that any pebbling of a circuit C results in a sequence of hybrids that shows indistinguishability of the real and simulated games. The number of hybrids is proportional to the number of moves in the pebbling and the equivocation parameter is proportional to the number of black pebbles it uses.

Theorem 2. *Assume that there is a pebbling of the circuit C in γ moves. Then there is a sequence of $2 \cdot \gamma + 1$ hybrid games, starting with the real game $\mathsf{Hyb}(I = \emptyset, (\mathsf{mode}_i = \mathsf{RealGate})_{i \in [q]})$ and ending with the simulated game $\mathsf{Hyb}(I = \emptyset, (\mathsf{mode}_i = \mathsf{SimGate})_{i \in [q]})$ such that any two adjacent hybrid games in the sequence are indistinguishable by rules 1, 2 or 3 from the previous section. Furthermore if pebbling uses t^* black pebbles then every hybrid $\mathsf{Hyb}(I, \mathsf{mode})$ in the sequence satisfies $|I| \leq t^*$. In particular, indistinguishability holds as long as the equivocation parameter is at least t^*.*

Proof. A *pebble configuration* specifies whether each gate contains no pebble, a black pebble, or a gray pebble. A pebbling in γ moves gives rise to a sequence of $\gamma + 1$ pebble configurations starting with no pebbles and ending with a gray pebble on each gate. Each pebble configuration follows from the preceding one by a move that satisfies pebbling rules A or B.

We let each pebble configuration define a hybrid $\mathsf{Hyb}(I, \mathsf{mode})$ where:

- For every gate $i \in [q]$, we set $\mathsf{mode}_i = \mathsf{RealGate}$ if gate i has no pebble, $\mathsf{mode}_i = \mathsf{InputDepSimGate}$ if gate i has a black pebble, and $\mathsf{mode}_i = \mathsf{SimGate}$ if gate i has a gray pebble.
- We set I to be the set of gates with black pebbles on them.

Therefore a pebbling defines a sequence of hybrids $\mathsf{Hyb}_\alpha = \mathsf{Hyb}(I^\alpha, \mathsf{mode}^\alpha)$ for $\alpha = 0, \ldots, \gamma$ where $\mathsf{Hyb}_0 = \mathsf{Hyb}(\emptyset, (\mathsf{mode}_i^0 = \mathsf{RealGate})_{i \in [q]})$ is the real game and $\mathsf{Hyb}_\gamma = \mathsf{Hyb}(\emptyset, (\mathsf{mode}_i^\gamma = \mathsf{SimGate})_{i \in [q]})$ is the simulated game, and each Hyb_α is induced by the pebbling configuration after α moves. We will need to add additional intermediate hybrids (which we call "half steps") to ensure that each pair of consecutive hybrids is indistinguishable by rules 1, 2 or 3. We do this as follows:

- Assume that move $\alpha + 1$ of the pebbling applies rule A to place a black pebble on gate j.
  Let $\mathsf{Hyb}_\alpha = \mathsf{Hyb}(I^\alpha, \mathsf{mode}^\alpha)$ and $\mathsf{Hyb}_{\alpha+1} = \mathsf{Hyb}(I^{\alpha+1}, \mathsf{mode}^{\alpha+1})$. Then $I^{\alpha+1} = I^\alpha \cup \{j\}$, $\mathsf{mode}_i^{\alpha+1} = \mathsf{mode}_i^\alpha$ for all $i \neq j$, and $\mathsf{mode}_j^\alpha = \mathsf{RealGate}$, $\mathsf{mode}_j^{\alpha+1} = \mathsf{InputDepSimGate}$.
  Define the intermediate "half-step" hybrid $\mathsf{Hyb}_{\alpha+\frac{1}{2}} := \mathsf{Hyb}(I^{\alpha+1}, \mathsf{mode}^\alpha)$.

  It holds that $\mathsf{Hyb}_\alpha \overset{\mathrm{comp}}{\approx} \mathsf{Hyb}_{\alpha+\frac{1}{2}}$ by rule 1, and $\mathsf{Hyb}_{\alpha+\frac{1}{2}} \overset{\mathrm{comp}}{\approx} \mathsf{Hyb}_{\alpha+1}$ by rule 2. The conditions needed to apply rule 2 are implied by pebbling rule A.

- Assume that move $\alpha + 1$ of the pebbling applies rule A to remove a black pebble from gate j.

  Let $\mathsf{Hyb}_\alpha = \mathsf{Hyb}(I^\alpha, \mathsf{mode}^\alpha)$ and $\mathsf{Hyb}_{\alpha+1} = \mathsf{Hyb}(I^{\alpha+1}, \mathsf{mode}^{\alpha+1})$. Then $I^{\alpha+1} = I^\alpha \setminus \{j\}$, $\mathsf{mode}_i^{\alpha+1} = \mathsf{mode}_i^\alpha$ for all $i \neq j$, and $\mathsf{mode}_j^\alpha = \mathsf{InputDepSimGate}$, $\mathsf{mode}_j^{\alpha+1} = \mathsf{RealGate}$.

  Define the intermediate "half-step" hybrid $\mathsf{Hyb}_{\alpha+\frac{1}{2}} := \mathsf{Hyb}(I^\alpha, \mathsf{mode}^{\alpha+1})$.

  It holds that $\mathsf{Hyb}_\alpha \overset{\mathsf{comp}}{\approx} \mathsf{Hyb}_{\alpha+\frac{1}{2}}$ by rule 2, and $\mathsf{Hyb}_{\alpha+\frac{1}{2}} \overset{\mathsf{comp}}{\approx} \mathsf{Hyb}_{\alpha+1}$ by rule 1. The conditions needed to apply rule 2 are implied by pebbling rule A.

- Assume that move $\alpha + 1$ of the pebbling applies rule B to replace a black pebble with a gray pebble on gate j.

  Let $\mathsf{Hyb}_\alpha = \mathsf{Hyb}(I^\alpha, \mathsf{mode}^\alpha)$ and $\mathsf{Hyb}_{\alpha+1} = \mathsf{Hyb}(I^{\alpha+1}, \mathsf{mode}^{\alpha+1})$. Then $I^{\alpha+1} = I^\alpha \setminus \{j\}$, $\mathsf{mode}_i^{\alpha+1} = \mathsf{mode}_i^\alpha$ for all $i \neq j$, and $\mathsf{mode}_j^\alpha = \mathsf{InputDepSimGate}$, $\mathsf{mode}_j^{\alpha+1} = \mathsf{SimGate}$.

  Define the intermediate "half-step" hybrid $\mathsf{Hyb}_{\alpha+\frac{1}{2}} := \mathsf{Hyb}(I^\alpha, \mathsf{mode}^{\alpha+1})$.

  It holds that $\mathsf{Hyb}_\alpha \overset{\mathsf{comp}}{\approx} \mathsf{Hyb}_{\alpha+\frac{1}{2}}$ by rule 3, and $\mathsf{Hyb}_{\alpha+\frac{1}{2}} \overset{\mathsf{comp}}{\approx} \mathsf{Hyb}_{\alpha+1}$ by rule 1. The conditions needed to apply rule 3 are implied by pebbling rule B.

Therefore the sequence $\mathsf{Hyb}_0, \mathsf{Hyb}_{\frac{1}{2}}, \mathsf{Hyb}_1, \mathsf{Hyb}_{1+\frac{1}{2}}, \mathsf{Hyb}_2, \ldots, \mathsf{Hyb}_\gamma$ consisting of $2\gamma + 1$ hybrids satisfies the conditions of the theorem.

Combining Theorems 2 and 1 we obtain the following corollary.

Corollary 1. *There exists an adaptively secure garbling scheme such that the following holds. Assuming the existence of one-way functions, there is an instantiation of the garbling scheme that has on-line complexity $(n + m + t^*)\mathsf{poly}(\lambda)$ for any circuit C that admits a pebbling with $\gamma = \mathsf{poly}(\lambda)$ moves and t^* black pebbles. Furthermore, assuming the existence of sub-exponentially secure one-way functions, there is an instantiation of the garbling scheme that has on-line complexity $(n + m + t^*)\mathsf{poly}(\lambda, \log \gamma)$ for any circuit C admits a pebbling strategy with $\gamma = 2^{\mathsf{poly}(\lambda)}$ moves and t^* black pebbles.*

Proof. We instantiate our construction from Sect. 5 with a CPA-secure "inner encryption" Γ having special correctness, and a somewhere-equivocal "outer encryption" Π from Sect. 4 using an equivocation parameter $t = t^*$. Both components can be instantiated from one-way functions. Assuming that $\gamma = \mathsf{poly}(\lambda)$, Theorem 2 tells us that the resulting garbling scheme is adaptively secure as long as Γ, Π are. The on-line complexity consists of $n + m$ keys for Γ along with the key of Π for a total of $(n + m)\mathsf{poly}(\lambda) + t^*\mathsf{poly}(\lambda)$ as claimed.

When $\gamma = 2^{\mathsf{poly}(\lambda)}$, then Theorem 2 tells us that the resulting garbling scheme is adaptively secure as long as the schemes Γ, Π provide a higher level of security so as to survive $2\gamma + 1$ hybrids, meaning that the distinguishing advantage for each of the schemes needs to be $2^{-(2\gamma+1)}\mathsf{negl}(\lambda)$. This can be accomplished assuming sub-exponentially secure one-way functions by setting the security parameter of Γ, Π to some $\lambda' = \mathsf{poly}(\lambda, \log \gamma)$ and results in on-line complexity $(n + m)\mathsf{poly}(\lambda, \log \gamma) + t^*\mathsf{poly}(\lambda, \log \gamma)$ as claimed.

7.1 Pebbling Strategies

In this section we give two pebbling strategies for arbitrary circuit with width w, depth d, and q gates. The first strategy uses $O(q)$ moves and $O(w)$ black pebbles. The second strategy uses $O(q2^d)$ moves and $O(d)$ black pebbles.

Strategy 1. To pebble the circuit proceed as follows:

Pebble(C):
1. Put a black pebble on each gate at the input level (level 1).
2. For $i = 1$ to $d - 1$, repeat:
   (a) Put a black pebble on each gate at level $i + 1$.
   (b) For each gate at level i, replace the black pebble with a gray pebble.
   (c) $i \leftarrow i + 1$
3. For each gate at level d, replace the black pebble with a gray pebble.

This strategy uses $\gamma = 2q$ moves and $t^* = 2w$ black pebbles. By instantiating Corollary 1 with this strategy, we obtain the following corollary.

Corollary 2. *Assuming the existence of one-way functions there exists an adaptively secure garbling scheme with on-line complexity $w \cdot \text{poly}(\lambda)$, where w is the width of the circuit.*

Strategy 2. This is a recursive strategy defined as follows.

- Pebble(C):
  • For each gate i in C starting with the gates at the top level moving to the bottom level:
    1. RecPutBlack(C, i)
    2. Replace the black pebble on gate i with a gray pebble.
- RecPutBlack(C, i): // Let LeftPred(C, i) and RightPred(C, i) are the two predecessors of gate i in C.
  1. If gate i is an input gate, put a black pebble on i and **return**.
  2. Run RecPutBlack($C, $ LeftPred(C, i)), RecPutBlack($C, $ RightPred(C, i))
  3. Put a black pebble on gate i.
  4. Run RecRemoveBlack($C, $ LeftPred(C, i))
     and RecRemoveBlack($C, $ RightPred(C, i))
- RecRemoveBlack(C, i): This is the same as RecPutBlack, except that instead of putting a black pebble on gate i, in steps 1 and 3, we remove it.

To analyze the correctness of this strategy, we note the following invariants: if the circuit C is in a configuration where it does not contain any pebbles at any level below that of gate i, then (1) the procedure RecPutBlack(C, i) results in a configuration where a single black pebble is added to gate i, but nothing else changes, (2) the procedure RecRemoveBlack(C, i) results in a configuration where a single black pebble is removed from gate i, but nothing else changes. Using these two invariants the correctness of of the entire strategy follows.

To calculate the number of black pebbles used and the number of moves that the above strategy takes to pebble C, we use the following simple recursive equations. Let #PebPut(d) and #PebRem(d) be the number of black pebbles on gate i and below it used to execute RecPutBlack and RecRemoveBlack on a gate at level d, respectively. We have,

$$\text{\#PebPut}(1) = 1, \quad \text{\#PebPut}(d) \leq \max(\text{\#PebPut}(d-1), \text{\#PebRem}(d-1)) + 2$$
$$\text{\#PebRem}(1) = 1, \quad \text{\#PebRem}(d) \leq \max(\text{\#PebPut}(d-1), \text{\#PebRem}(d-1)) + 2$$

Therefore the strategy requires at most $2d$ black pebbles to pebble the circuit.

To calculate the number of moves it takes run Pebble(C), we use the following recursive equations. Let #Moves(d) be the number of moves it takes to put a black pebble on, or remove a black pebble from, a gate at level d. Then

$$\text{\#Moves}(1) = 1, \quad \text{\#Moves}(d) = 4(\text{\#Moves}(d-1)) + 1$$

Hence, each call of RecPutBlack takes at most 4^d moves, and the total number of moves to pebble the circuit is at most $q4^d$.

In summary, the above gives us a strategy to pebble any circuit with at most $\gamma = q4^d$ moves and $t^* = 2d$ black pebbles. By instantiating Corollary 1 with the above strategy, we obtain the following corollary.

Corollary 3. *Assuming the existence of (standard) one-way functions, there exists an adaptively secure garbling schemes that has on-line complexity* $(n + m)\mathsf{poly}(\lambda)$ *for all circuits having depth* $d = O(\log \lambda)$.

Assuming the existence of sub-exponentially secure one-way functions, there exists an adaptively secure garbling scheme that has on-line complexity $(n + m)\mathsf{poly}(\lambda, d)$, *for arbitrary circuits of depth* $d = \mathsf{poly}(\lambda)$.

8   Conclusions

We have shown how to achieve adaptively secure garbling schemes under one-way functions by augmenting Yao's construction with an additional layer of somewhere-equivocal encryption. The on-line complexity in our constructions can be significantly smaller than the circuit size. In our main instantiation, the on-line complexity only scales with the width w of the circuit, which corresponds to the space complexity of the computation.

It remains as an open problem to get the optimal on-line complexity $(n + m)\mathsf{poly}(\lambda)$ which does not depend on the circuit depth or width. Currently, this is only known assuming the existence of indistinguishability obfuscation and therefore it remains open to achieve the above under one-way functions or even stronger assumptions such as DDH or LWE. It also remains open if Yao's scheme (or more precisely, a variant of it where the output map is sent in the on-line phase) can already achieve adaptive security without relying on somewhere-equivocal encryption. We have no proof nor a counter-example. It would be interesting to see if there is some simple-to-state standard-model security assumption that one could make on the encryption scheme used to create the garbled gates in Yao's construction (e.g., circular security, key-dependent message security, etc.), under which one could prove that the resulting garbling scheme is adaptively secure.

References

1. Ananth, P., Brakerski, Z., Segev, G., Vaikuntanathan, V.: From selective to adaptive security in functional encryption. In: Gennaro, R., Robshaw, M. (eds.) CRYPTO 2015. LNCS, vol. 9216, pp. 657–677. Springer, Heidelberg (2015)
2. Ananth, P., Sahai, A.: Functional encryption for turing machines. Cryptology ePrint Archive, Report 2015/776 (2015). http://eprint.iacr.org/
3. Applebaum, B.: Key-dependent message security: generic amplification and completeness. In: Paterson, K.G. (ed.) EUROCRYPT 2011. LNCS, vol. 6632, pp. 527–546. Springer, Heidelberg (2011)
4. Applebaum, B.: Bootstrapping obfuscators via fast pseudorandom functions. In: Sarkar, P., Iwata, T. (eds.) ASIACRYPT 2014, Part II. LNCS, vol. 8874, pp. 162–172. Springer, Heidelberg (2014)
5. Applebaum, B., Ishai, Y., Kushilevitz, E.: Cryptography in NC^0. In: 45th FOCS, pp. 166–175. IEEE Computer Society Press, October 2004
6. Applebaum, B., Ishai, Y., Kushilevitz, E.: Computationally private randomizing polynomials and their applications. In: 20th Annual IEEE Conference on Computational Complexity (CCC 2005), San Jose, CA, USA, 11–15 June 2005, pp. 260–274. IEEE Computer Society (2005)
7. Applebaum, B., Ishai, Y., Kushilevitz, E.: From secrecy to soundness: efficient verification via secure computation. In: Abramsky, S., Gavoille, C., Kirchner, C., Meyer auf der Heide, F., Spirakis, P.G. (eds.) ICALP 2010. LNCS, vol. 6198, pp. 152–163. Springer, Heidelberg (2010)
8. Applebaum, B., Ishai, Y., Kushilevitz, E., Waters, B.: Encoding functions with constant online rate or how to compress garbled circuits keys. In: Canetti, R., Garay, J.A. (eds.) CRYPTO 2013, Part II. LNCS, vol. 8043, pp. 166–184. Springer, Heidelberg (2013)
9. Barak, B., Haitner, I., Hofheinz, D., Ishai, Y.: Bounded key-dependent message security. In: Gilbert, H. (ed.) EUROCRYPT 2010. LNCS, vol. 6110, pp. 423–444. Springer, Heidelberg (2010)
10. Bellare, M., Hoang, V.T., Keelveedhi, S.: Instantiating random oracles via UCEs. In: Canetti, R., Garay, J.A. (eds.) CRYPTO 2013, Part II. LNCS, vol. 8043, pp. 398–415. Springer, Heidelberg (2013)
11. Bellare, M., Hoang, V.T., Rogaway, P.: Adaptively secure garbling with applications to one-time programs and secure outsourcing. In: Wang, X., Sako, K. (eds.) ASIACRYPT 2012. LNCS, vol. 7658, pp. 134–153. Springer, Heidelberg (2012)
12. Bellare, M., Hoang, V.T., Rogaway, P.: Foundations of garbled circuits. In: ACM CCS 2012, pp. 784–796. ACM Press, October 2012
13. Boneh, D., Gentry, C., Gorbunov, S., Halevi, S., Nikolaenko, V., Segev, G., Vaikuntanathan, V., Vinayagamurthy, D.: Fully key-homomorphic encryption, arithmetic circuit ABE and compact garbled circuits. In: Nguyen, P.Q., Oswald, E. (eds.) EUROCRYPT 2014. LNCS, vol. 8441, pp. 533–556. Springer, Heidelberg (2014)
14. Boyle, E., Gilboa, N., Ishai, Y.: Function secret sharing. In: Oswald, E., Fischlin, M. (eds.) EUROCRYPT 2015. LNCS, vol. 9057, pp. 337–367. Springer, Heidelberg (2015)

15. Garay, J.A., Wichs, D., Zhou, H.-S.: Somewhat non-committing encryption and efficient adaptively secure oblivious transfer. In: Halevi, S. (ed.) CRYPTO 2009. LNCS, vol. 5677, pp. 505–523. Springer, Heidelberg (2009)
16. Gennaro, R., Gentry, C., Parno, B.: Non-interactive verifiable computing: outsourcing computation to untrusted workers. In: Rabin, T. (ed.) CRYPTO 2010. LNCS, vol. 6223, pp. 465–482. Springer, Heidelberg (2010)
17. Gilboa, N., Ishai, Y.: Distributed point functions and their applications. In: Nguyen, P.Q., Oswald, E. (eds.) EUROCRYPT 2014. LNCS, vol. 8441, pp. 640–658. Springer, Heidelberg (2014)
18. Goldreich, O., Goldwasser, S., Micali, S.: On the cryptographic applications of random functions. In: Blakely, G.R., Chaum, D. (eds.) CRYPTO 1984. LNCS, vol. 196, pp. 276–288. Springer, Heidelberg (1985)
19. Goldwasser, S., Kalai, Y.T., Popa, R.A., Vaikuntanathan, V., Zeldovich, N.: Reusable garbled circuits and succinct functional encryption. In: Boneh, D., Roughgarden, T., Feigenbaum, J. (eds.) 45th ACM STOC, pp. 555–564. ACM Press, June 2013
20. Goldwasser, S., Kalai, Y.T., Rothblum, G.N.: One-time programs. In: Wagner, D. (ed.) CRYPTO 2008. LNCS, vol. 5157, pp. 39–56. Springer, Heidelberg (2008)
21. Gorbunov, S., Vaikuntanathan, V., Wee, H.: Functional encryption with bounded collusions via multi-party computation. In: Safavi-Naini, R., Canetti, R. (eds.) CRYPTO 2012. LNCS, vol. 7417, pp. 162–179. Springer, Heidelberg (2012)
22. Goyal, V., Ishai, Y., Sahai, A., Venkatesan, R., Wadia, A.: Founding cryptography on tamper-proof hardware tokens. In: Micciancio, D. (ed.) TCC 2010. LNCS, vol. 5978, pp. 308–326. Springer, Heidelberg (2010)
23. Hemenway, B., Jafargholi, Z., Ostrovsky, R., Scafuro, A., Wichs, D.: Adaptively secure garbled circuits from one-way functions. IACR Cryptology ePrint Archive 2015:1250 (2015)
24. Hubacek, P., Wichs, D.: On the communication complexity of secure function evaluation with long output. In: ITCS 2015, pp. 163–172. ACM, January 2015
25. Ishai, Y., Kushilevitz, E.: Randomizing polynomials: a new representation with applications to round-efficient secure computation. In: 41st Annual Symposium on Foundations of Computer Science, FOCS 2000, Redondo Beach, California, USA, 12–14 November 2000, pp. 294–304 (2000)
26. Ishai, Y., Kushilevitz, E.: Perfect constant-round secure computation via perfect randomizing polynomials. In: Widmayer, P., Triguero, F., Morales, R., Hennessy, M., Eidenbenz, S., Conejo, R. (eds.) ICALP 2002. LNCS, vol. 2380, pp. 244–256. Springer, Heidelberg (2002)
27. Lindell, Y., Pinkas, B.: A proof of security of Yao's protocol for two-party computation. J. Cryptology 22(2), 161–188 (2009)
28. Lindell, Y., Riva, B.: Cut-and-choose yao-based secure computation in the online/offline and batch settings. In: Garay, J.A., Gennaro, R. (eds.) CRYPTO 2014, Part II. LNCS, vol. 8617, pp. 476–494. Springer, Heidelberg (2014)
29. Nielsen, J.B.: Separating random oracle proofs from complexity theoretic proofs: the non-committing encryption case. In: Yung, M. (ed.) CRYPTO 2002. LNCS, vol. 2442, pp. 111–126. Springer, Heidelberg (2002)
30. Sahai, A., Seyalioglu, H.: Worry-free encryption: functional encryption with public keys. In: ACM CCS 2010, pp. 463–472. ACM Press, October 2010
31. Yao, A.C.-C.: Protocols for secure computations (extended abstract). In: 23rd FOCS, pp. 160–164. IEEE Computer Society Press, November 1982
32. Yao, A.C.-C: How to generate and exchange secrets (extended abstract). In: 27th FOCS, pp. 162–167. IEEE Computer Society Press, October 1986

Rate-1, Linear Time and Additively Homomorphic UC Commitments

Ignacio Cascudo[1], Ivan Damgård[2], Bernardo David[2(✉)], Nico Döttling[3],
and Jesper Buus Nielsen[2]

[1] Aalborg University, Aalborg, Denmark
[2] Aarhus University, Aarhus, Denmark
bernardo@cs.au.dk
[3] University of California, Berkeley, USA

Abstract. We construct the first UC commitment scheme for binary strings with the optimal properties of rate approaching 1 and linear time complexity (in the amortised sense, using a small number of seed OTs). On top of this, the scheme is additively homomorphic, which allows for applications to maliciously secure 2-party computation. As tools for obtaining this, we make three contributions of independent interest: we construct the first (binary) linear time encodable codes with non-trivial distance and rate approaching 1, we construct the first almost universal hash function with small seed that can be computed in linear time, and we introduce a new primitive called interactive proximity testing that can be used to verify whether a string is close to a given linear code.

1 Introduction

Commitment schemes are one of the fundamental building blocks of cryptographic protocols. In a nutshell, a commitment scheme is a two party protocol that allows a prover P to *commit* to a secret without revealing it to the verifier V. Later on, in an unveil phase P can convince V that the commitment contains a specific secret. Classically, two security properties are required of commitment schemes: The hiding property requires that the verifier V does not learn anything about the committed secret before the unveil and the binding property requires that the prover P cannot change the committed secret after the commit phase.

A stronger security requirement is (stand-alone) simulation-based security, where we require that any interaction with a commitment protocol is indistinguishable from a perfectly secure ideal commitment. Commitment schemes that satisfy these security notions can be realized stand-alone (i.e. no trusted setup required) from basic and highly efficient cryptographic primitives such as pseudorandom generators [Nao91].

However, commitment schemes are rarely used just by themselves; they are used as components in complex protocols. In such a situation, the stand-alone simulation-based security guarantee breaks down as several (nested) instances of a commitment protocol might be executed with correlated secrets.

M. Robshaw and J. Katz (Eds.): CRYPTO 2016, Part III, LNCS 9816, pp. 179–207, 2016.
DOI: 10.1007/978-3-662-53015-3_7

The most prominent security framework that captures this scenario of protocols running in a larger *context* is Canetti's UC framework [Can01]. UC security offers very strong composability guarantees; in particular UC secure protocols can be used in arbitrary contexts retaining their security properties. This however comes at a price, as UC commitments cannot be realized without trusted setup assumptions such as common reference strings [CF01]. On the positive side, it is well known that realizing UC secure commitments is sufficient for general UC secure two-party and multiparty computation [CLOS02].

Any commitment scheme that is UC secure must be both straight-line *extractable* and *equivocal*, meaning a simulator must have means to efficiently obtain the message in a commitment sent by a malicious prover and also change the contents of a commitment sent to a malicious verifier without having (non-black-box) access to these machines. To obtain these strong properties, earlier constructions of UC commitments (e.g. [CF01,Lin11,BCPV13]) relied on expensive public key primitives for every single instance of the protocol, which makes them considerably less efficient than stand-alone secure commitments (which, as mentioned above, can be realized from minimal cryptographic primitives). The most efficient UC commitment protocols based directly on public key assumptions [Lin11,BCPV13] require exponentiations in groups of larger order and have therefore a typical computational complexity of $\Omega(n^3)$ for commitments to n-bit messages.

A recent line of research [GIKW14,DDGN14,CDD+15,FJNT16,Bra16] is concerned with the construction of UC secure commitments schemes for which the use of public key primitives is confined to a once-and-for-all setup phase, the cost of which can be amortized over many sessions later on.

This gives us the possibility to build extremely efficient commitment schemes. Let us therefore consider what we can hope to achieve: Clearly, the best running time we can have is $O(n)$ for committing and opening n bits, since one must look at the entire committed string. As for communication, let us define the rate of a commitment scheme to be the size of the committed string divided by the size of a commitment. Now, a UC commitment must be of size at least the string committed to, because the simulator we need for the proof of UC security must be able to extract the committed string from the commitment. Therefore the rate of a UC commitment scheme must be at most 1.[1] If a commitment to n bits has size $n + o(n)$ bits, we will say it has rate approaching 1.

Another desirable property for commitment schemes is the additively homomorphic property: we interpret the committed strings as vectors over some finite field, and V can add any two commitments, to vectors $\boldsymbol{a}, \boldsymbol{b}$. The result will be a commitment that can be opened to (only) $\boldsymbol{a} + \boldsymbol{b}$ while revealing nothing else about $\boldsymbol{a}$ and $\boldsymbol{b}$. Note that this additive property is crucial in applications of string commitments to secure computation: In [FJN+13], it was shown how to do maliciously secure 2-party computation by doing cut-and-choose on garbled

---

[1] However, as we shall see, if one only needs to commit to *random* bit strings, one can hope to generate these pseudorandomly from a short seed, and have rate higher than 1 for commitment (but of course not for opening).

gates rather than on garbled circuits. This performs asymptotically better than conventional cut-and-choose but requires an additive commitment scheme to "glue" the garbled gates together to a circuit. In [AHMR15], additive commitments were used in a similar way for secure RAM computation. Any efficiency improvements for commitments are directly inherited by these applications.

1.1 Previous Work

In [GIKW14,Bra16] rate 1 was achieved. On the other hand, [DDGN14] achieved constant rate and additively homomorphic commitments. In follow-up work, linear time and additive homomorphism were achieved in [CDD+15], and shortly after, in [FJNT16], rate 1 and additively homomorphic commitments were achieved.

Now, the obvious question is of course: *can this line of research be closed, by constructing a commitment scheme with the optimal properties of rate 1 and linear time – and also with the additive property?*

To see why the answer is not clear from previous works in this line of research [GIKW14,DDGN14,CDD+15,FJNT16], let us briefly describe the basic ideas in those constructions:

P will encode the vector s to commit to using a linear error correcting code C, to get an encoding $C(s)$. Now he additively secret-shares each entry in $C(s)$ and a protocol is executed in which V learns one share of each entry while P does not know which shares are given to V. This phase uses a small number of seed OT's that are done in a once-and-for-all set-up phase analogous to the set-up of "watchlists" in the MPC-in-the-head and IPS compiler constructions [IPS09,IPS08]. To open, P reveals the codeword and both shares of each entry. V checks that the shares are consistent with those he already knew, reconstructs $C(s)$ and checks that it is indeed a codeword. This is clearly hiding because V has no information on $C(s)$ at commit time. Binding also seems to follow easily: if P wants to change his mind to another codeword, he has to change many entries and hence at least one share of each modified entry. We can expect that V will notice this with high probability since P does not know which share he can change without being caught. There is a problem, however: a corrupt P does not have to send shares of a codeword at commitment time, so he does not have to move all the way from a codeword to the next one, and it may not be clear (to the simulator) which string is being committed.

Three solutions to this have been proposed in earlier work: in [CDD+15] the minimum distance of C is chosen so large that P's only chance is to move to the closest codeword. This has a cost in efficiency and also means we cannot have the additive property: if we add codewords with errors, the errors may accumulate and binding no longer holds. In [DDGN14], a verifiable secret-sharing scheme was used on top of the coding, this allows V to do some consistency checks that forces P to use codewords, except with negligible probability. But it also introduces a constant factor overhead which means there is no hope to get rate 1. Finally, in [FJNT16], the idea was to force P to open some random linear combinations of the codewords. In the case of binary strings, k linear combinations must be opened, where k is the security parameter. This indeed forces P to use codewords and gives us the additive property. Also, a couple of tricks were proposed in [FJNT16] which gives commitments with rate 1, if the code C has rate 1. On the other hand, they could not get linear time this way, first because no linear time encodable codes with rate approaching 1 were known[2], and second because one needs to visit each of prover's codewords $\Omega(k)$ times to compute the linear combinations.

Table 1. Comparison between the UC commitment schemes presented in [GIKW14, DDGN14, CDD+15, FJNT16, Bra16] and the scheme presented in this paper (Π_{HCOM}).

Scheme	Rate 1	Linear time	Additively homomorphic
[GIKW14]	✓	✗	✗
[DDGN14]	✗	✗	✓
[CDD+15]	✗	✓	✓
[FJNT16]	✓	✗	✓
[Bra16]	✓	✗	✗
This work	✓	✓	✓

1.2   Our Contribution

In this paper, we show that we can indeed have UC commitments that have simultaneously rate approaching 1, linear time and additive homomorphism. A comparison between our results and previous works can be seen in Table 1. While we follow the same blue-print as in previous work, we overcome the obstacles outlined above via three technical contributions that are of independent interest:

1. We introduce a primitive we call interactive proximity testing that can be used to verify whether a given string s is in an *interleaved* linear code $C^{\odot m}$, or at least close to $C^{\odot m}$.[3] The idea is to choose a random almost universal and linear hash function h and test whether $h(s) \in h(C^{\odot m})$. We show that if s is "too far" away from $C^{\odot m}$, then this test will fail with high probability. Intuitively, this makes sense to use in a 2-party protocol because the party holding s can allow the other party to do the test while only revealing a small amount of information on s, namely $h(s)$. Of course, this assumes that the verifying party has a way to verify that the hash value is correct, more details on this are given later.

---

[2] Of course, rate 1 and linear time is trivial if there are no demands to the distance: just use the identity as encoding. What we mean here is that the code has length $n + o(n)$ and yet, as n grows, the distance remains larger than some parameter k.

[3] A codeword in an interleaved code is a matrix in which all m columns are in some underlying code C.

2. In order to be able to use interactive proximity testing efficiently in our protocol, we construct the first family of (linear) almost universal hash functions that can be computed in linear time, where for a fixed desired collision probability, the size of the seed only depends logarithmically on the input size. We note that the verification method from [FJNT16] is a special case of our proximity testing, where the hash function is a *random* linear function (which cannot be computed in linear time)[4].

3. We present the first *explicit* construction of linear time encodable (binary) codes with rate approaching 1. The construction is basically a family of iterated Sipser-Spielman codes [Spi96] and uses a family of explicit expander graphs constructed by Capalbo et al. [CRVW02]. Previous linear time encodable codes [Spi96, GI02, GI03, GI05, DI14] did not approach rate 1, which was a clear obstacle to our results.

2 Preliminaries

In the sections we establish notation and introduce notions that will be used throughout the paper. We borrow much of the notation from [CDD+15].

2.1 Notation

The set of the n first positive integers is denoted $[n] = \{1, 2, \ldots, n\}$. Given a finite set D, sampling a uniformly random element from D is denoted $r \xleftarrow{\$} D$ and sampling a uniformly random subset of n elements from D is denoted $\{r_1, \ldots, r_n\} \xleftarrow{\$} D$. Vectors of elements of some field are denoted by bold lower-case letters, while matrices are denoted by bold upper-case letters. Concatenation of vectors is represented by $\|$. For $z \in \mathbb{F}^k$, $z[i]$ denotes the i'th entry of the vector, where $z[1]$ is the first element of z. For a matrix $\mathbf{M} \in \mathbb{F}^{n \times k}$, we let $\mathbf{M}[\cdot, j]$ denote the j'th column of $\mathbf{M}$ and $\mathbf{M}[i, \cdot]$ denote the i'th row. The column span of $\mathbf{M}$, denoted by $\langle \mathbf{M} \rangle_{\text{col}}$ is the vector subspace of $\mathbb{F}^n$ spanned over $\mathbb{F}$ by the columns $\mathbf{M}[\cdot, 1], \ldots, \mathbf{M}[\cdot, k]$ of $\mathbf{M}$. The row support of $\mathbf{M}$ is the set of indices $I \subseteq \{1, \ldots, n\}$ such that $\mathbf{M}[i, \cdot] \neq \mathbf{0}$.

We say that a function ϵ is negligible in n if for every positive polynomial p there exists a constant c such that $\epsilon(n) < \frac{1}{p(n)}$ when $n > c$. Two ensembles $X = \{X_{\kappa,z}\}_{\kappa \in \mathbb{N}, z \in \{0,1\}^*}$ and $Y = \{Y_{\kappa,z}\}_{\kappa \in \mathbb{N}, z \in \{0,1\}^*}$ of binary random variables are said to be *statistically indistinguishable*, denoted by $X \approx_s Y$, if for all z it holds that $|\Pr[\mathcal{D}(X_{\kappa,z}) = 1] - \Pr[\mathcal{D}(Y_{\kappa,z}) = 1]|$ is negligible in κ for every probabilistic algorithm (distinguisher) $\mathcal{D}$. In case this only holds for computationally bounded (non-uniform probabilistic polynomial-time (*PPT*)) distinguishers we say that X and Y are *computationally indistinguishable* and denote it by $\approx_c$.

---

[4] On the other hand, we pay a small price for having a non-random function, namely the output size for the hash function needs to be $\Theta(s) + \log(m)$ rather than $\Theta(s)$, where s is the security parameter and m is the number of commitments.

2.2   Coding Theory

We denote finite fields by $\mathbb{F}$ and write $\mathbb{F}_q$ for the finite field of size q. For a vector $\boldsymbol{x} \in \mathbb{F}^n$, we denote the Hamming-weight of $\boldsymbol{x}$ by $\|\boldsymbol{x}\|_0 = |\{i \in [n] : \boldsymbol{x}[i] \neq 0\}|$. Let $\mathsf{C} \subset \mathbb{F}^n$ be a linear subspace of $\mathbb{F}^n$. We say that C is an $\mathbb{F}$-linear $[n, k, s]$ code, if C has dimension k and it holds for every nonzero $\boldsymbol{x} \in \mathsf{C}$ that $\|\boldsymbol{x}\|_0 \geq s$, i.e., the minimum distance of C is at least s. The distance $\mathrm{dist}(\mathsf{C}, \boldsymbol{x})$ between C and a vector $\boldsymbol{x} \in \mathbb{F}^n$ is the minimum of $\|\boldsymbol{c} - \boldsymbol{x}\|_0$ when $\boldsymbol{c} \in \mathsf{C}$. The rate of an $\mathbb{F}$-linear $[n, k, s]$ code is $\frac{k}{n}$ and its relative minimum distance is $\frac{s}{n}$.

A matrix $\mathbf{G} \in \mathbb{F}^{n \times k}$ is a generator matrix of C if $\mathsf{C} = \{\mathbf{G}\boldsymbol{x} : \boldsymbol{x} \in \mathbb{F}^k\}$. The code C is systematic if it has a generator matrix $\mathbf{G}$ such that the submatrix given by the top k rows of $\mathbf{G}$ is the identity matrix $\mathbf{I} \in \mathbb{F}^{k \times k}$. A matrix $\mathbf{P} \in \mathbb{F}^{(n-k) \times n}$ of maximal rank $n - k$ is a parity check matrix of C if $\mathbf{P}\boldsymbol{c} = 0$ for all $\boldsymbol{c} \in \mathsf{C}$. When we have fixed a parity check matrix $\mathbf{P}$ of C we say that the syndrome of an element $\boldsymbol{v} \in \mathbb{F}^n$ is $\mathbf{P}\boldsymbol{v}$.

For an $\mathbb{F}$-linear $[n, k, s]$ code C, we denote by $\mathsf{C}^{\odot m}$ the m-*interleaved product* of C, which is defined by

$$\mathsf{C}^{\odot m} = \{\mathbf{C} \in \mathbb{F}^{n \times m} : \forall i \in [m] : \mathbf{C}[\cdot, i] \in \mathsf{C}\} \ .$$

In other words, $\mathsf{C}^{\odot m}$ consists of all $\mathbb{F}^{n \times m}$ matrices for which all columns are in C. We can think of $\mathsf{C}^{\odot m}$ as a linear code with symbol alphabet $\mathbb{F}^m$, where we obtain codewords by taking m arbitrary codewords of C and bundling together the components of these codewords into symbols from $\mathbb{F}^m$. For a matrix $\mathbf{E} \in \mathbb{F}^{n \times m}$, $\|\mathbf{E}\|_0$ is the number of nonzero rows of $\mathbf{E}$, and the code $\mathsf{C}^{\odot m}$ has minimum distance at least s' if all nonzero $\mathbf{C} \in \mathsf{C}^{\odot m}$ satisfy $\|\mathbf{C}\|_0 \geq s'$. Furthermore, $\mathbf{P}$ is a parity-check matrix of C if and only if $\mathbf{P}\mathbf{C} = \mathbf{0}$ for all $\mathbf{C} \in \mathsf{C}^{\odot m}$.

2.3   Universal Composability

The results presented in this paper are proven secure in the Universal Composability (UC) framework introduced by Canetti in [Can01]. We refer the reader to Appendix A and [Can01] for further details.

Adversarial Model: Our protocols will be proved secure against static and active adversaries. This means that corruption is assumed to take place before the protocols starts execution and that the adversary may deviate from the protocol in any arbitrary way.

Setup Assumption: Since UC commitment protocols cannot be obtained in the plain model [CF01], they need a setup assumption, i.e., some resource available to all parties before the protocol starts. In the case of our protocol, our goal is to prove security in the $\mathcal{F}_{\mathrm{OT}}$-hybrid model [Can01, CLOS02], where the parties have access to an ideal 1-out-of-2 OT functionality. In order to attain this, we first prove our protocol secure in the $\mathcal{F}_{\mathrm{ROT}}$-hybrid model, where the resource

---

Functionality $\mathcal{F}_{\mathrm{HCOM}}$

$\mathcal{F}_{\mathrm{HCOM}}$ interacts with a sender P_s, a receiver P_r and an adversary $\mathcal{S}$ and it proceeds as follows:

- **Commit Phase:** The length of the committed messages λ is fixed and known to all parties.
  - If P_s is honest, upon receiving a message (commit, $sid, ssid, P_s, P_r$) from P_s, sample a random $\boldsymbol{m} \leftarrow \{0,1\}^\lambda$, record the tuple $(ssid, P_s, P_r, \boldsymbol{m})$, send the message (commit, $sid, ssid, P_s, P_r, \boldsymbol{m}$) to P_s and send the message (receipt, $sid, ssid, P_s, P_r$) to P_r and $\mathcal{S}$. Ignore any future commit messages with the same $ssid$ from P_s to P_r.
  - If P_s is corrupted, upon receiving a message (commit, $sid, ssid, P_s, P_r, \boldsymbol{m}$) from P_s, where $\boldsymbol{m} \in \{0,1\}^\lambda$, record the tuple $(ssid, P_s, P_r, \boldsymbol{m})$ and send the message (receipt, $sid, ssid, P_s, P_r$) to P_r and $\mathcal{S}$. Ignore any future commit messages with the same $ssid$ from P_s to P_r.
  - If a message (abort, $sid, ssid$) is received from $\mathcal{S}$, the functionality halts.
- **Open Phase:** Upon receiving a message (reveal, $sid, ssid$) from P_s: If a tuple $(ssid, P_s, P_r, \boldsymbol{m})$ was previously recorded, then send the message (reveal, $sid, ssid, P_s, P_r, \boldsymbol{m}$) to P_r and $\mathcal{S}$. Otherwise, ignore.
- **Addition:** Upon receiving a message (add, $sid, ssid_1, ssid_2, ssid_3, P_s, P_r$) from P_s: If tuples $(ssid_1, P_s, P_r, \boldsymbol{m}_1)$, $(ssid_2, P_s, P_r, \boldsymbol{m}_2)$ were previously recorded and $ssid_3$ is unused, record $(ssid_3, P_s, P_r, \boldsymbol{m}_1 + \boldsymbol{m}_2)$ and send the message (add, $sid, ssid_1, ssid_2, ssid_3, P_s, P_r,$ success) to P_s, P_r and $\mathcal{S}$.

Fig. 1. Functionality $\mathcal{F}_{\mathrm{HCOM}}$

that the parties have access to is an 1-out-of-2 random OT functionality, which we describe below. Since we can implement $\mathcal{F}_{\mathrm{ROT}}$ in the $\mathcal{F}_{\mathrm{OT}}$-hybrid model, as shown in Appendix B, the composability guarantees of the UC framework imply that we can achieve security for our commitment scheme in the $\mathcal{F}_{\mathrm{OT}}$-hybrid model too.

Ideal Functionalities: In Sect. 5, we construct an additively homomorphic string commitment protocol that UC-realizes the functionality $\mathcal{F}_{\mathrm{HCOM}}$, which is described in Fig. 1. This functionality basically augments the standard multiple commitments functionality $\mathcal{F}_{\mathrm{MCOM}}$ from [CLOS02] by introducing a command for adding two previously stored commitments and an abort command in the Commit Phase. $\mathcal{F}_{\mathrm{HCOM}}$ differs from a similar functionality of [CDD+15] in that it gives an honest sender commitments to random messages instead of letting it submit a message as input. In order to model corruptions, functionality $\mathcal{F}_{\mathrm{HCOM}}$ lets a corrupted sender choose the messages it wants to commit to. The abort is necessary to deal with inconsistent commitments that could be sent by a corrupted party.

In fact, our additively homomorphic commitment protocol is constructed in the $\mathcal{F}_{\mathrm{ROT}}$-hybrid model. Functionality $\mathcal{F}_{\mathrm{ROT}}$ models a random oblivious transfer of $n \times m$ matrices $\mathbf{R}_0, \mathbf{R}_1$ where the receiver learns a matrix $\mathbf{S}$ where each row

---

Functionality $\mathcal{F}_{\mathrm{ROT}}$

$\mathcal{F}_{\mathrm{ROT}}$ interacts with a sender P_s, a receiver P_r and an adversary $\mathcal{A}$, and it proceeds as follows:

- If both parties are honest, $\mathcal{F}_{\mathrm{ROT}}$ waits for messages (sender, sid, $ssid$) and (receiver, sid, $ssid$) from P_s and P_r, respectively. Then $\mathcal{F}_{\mathrm{ROT}}$ samples random bits $b_1, \ldots, b_n \xleftarrow{\$} \{0,1\}^n$ and two random matrices $\mathbf{R_0}, \mathbf{R_1} \xleftarrow{\$} \{0,1\}^{n \times m}$ with n rows and m columns. It computes a matrix $\mathbf{S}$ such that for $i = 1, \ldots, n$: $\mathbf{S}[i, \cdot] = \mathbf{R_{b_i}}[i, \cdot]$. It sends $(sid, ssid, \mathbf{R_0}, \mathbf{R_1})$ to P_s and $(sid, ssid, b_1, \ldots, b_n, \mathbf{S})$ to P_r. That is, for each row-position, P_r learns a row of $\mathbf{R_0}$ or of $\mathbf{R_1}$, but P_s does not know the selection.
- If P_s is corrupted, $\mathcal{F}_{\mathrm{ROT}}$ waits for messages (receiver, sid, $ssid$) from P_r and (adversary, sid, $ssid$, $\mathbf{R_0}, \mathbf{R_1}$) from $\mathcal{A}$. $\mathcal{F}_{\mathrm{ROT}}$ samples $(b_1, \ldots, b_n) \xleftarrow{\$} \{0,1\}^n$, sets $\mathbf{S}[i, \cdot] = \mathbf{R_{b_i}}[i, \cdot]$ for $i = 1, \ldots, n$ and sends $(sid, ssid, b_1, \ldots, b_n, \mathbf{S})$ to P_r.
- If P_r is corrupted, $\mathcal{F}_{\mathrm{ROT}}$ waits for messages (sender, sid, $ssid$) from P_s and (adversary, sid, $ssid$, $b_1, \ldots, b_n, \mathbf{S}$) from $\mathcal{A}$. $\mathcal{F}_{\mathrm{ROT}}$ samples random matrices $\mathbf{R_0}, \mathbf{R_1} \xleftarrow{\$} \{0,1\}^{n \times m}$, subject to $\mathbf{S}[i, \cdot] = \mathbf{R_{b_i}}[i, \cdot]$, for $i = 1, \ldots, n$. $\mathcal{F}_{\mathrm{ROT}}$ sends $(sid, ssid, \mathbf{R_0}, \mathbf{R_1})$ to P_s.

Notice that S can equivalently be specified as $\mathbf{S} = \boldsymbol{\Delta}\mathbf{R_1} + (\mathbf{I} - \boldsymbol{\Delta})\mathbf{R_0}$, where $\mathbf{I}$ is the identity matrix and $\boldsymbol{\Delta}$ is the diagonal matrix with $b_1, \ldots, b_n$ on the diagonal.

Fig. 2. Functionality $\mathcal{F}_{\mathrm{ROT}}$

is selected from either $\mathbf{R_0}$ or $\mathbf{R_1}$. Notice that this functionality can be trivially realized in the standard $\mathcal{F}_{\mathrm{OT}}$-hybrid model as shown in Appendix B. We define $\mathcal{F}_{\mathrm{OT}}$ in Appendix B and $\mathcal{F}_{\mathrm{ROT}}$ in Fig. 2 following the syntax of [CLOS02]. Notice that $\mathcal{F}_{\mathrm{OT}}$ can be efficiently UC-realized by the protocol in [PVW08], which can be used to instantiate the setup phase of our commitment protocols.

3   Interactive Proximity Testing

In this section, we will introduce our interactive proximity testing technique. It consists in the following argument: suppose we sample a function $\mathbf{H}$ from an almost universal family of linear hash functions (from $\mathbb{F}^m$ to $\mathbb{F}^\ell$), and we apply this to each of the rows of a matrix $\mathbf{X} \in \mathbb{F}^{n \times m}$, obtaining another matrix $\mathbf{X'} \in \mathbb{F}^{n \times \ell}$; because of linearity, if $\mathbf{X}$ belonged to an interleaved code $\mathsf{C}^{\odot m}$, then $\mathbf{X'}$ belongs to the interleaved code $\mathsf{C}^{\odot \ell}$. This suggests that we can test whether $\mathbf{X}$ is close to $\mathsf{C}^{\odot m}$ by testing instead if $\mathbf{X'}$ is close to $\mathsf{C}^{\odot \ell}$. Theorem 1 states that indeed the test gives such guarantee (with high probability over the choice of the hash function) and moreover, if these elements are close to the respective codes, the "error patterns" (the set of rows that have to be modified in each of the matrices in order to correct them to codewords) are the same.

Definition 1 (Almost Universal Linear Hashing). *We say that a family $\mathcal{H}$ of linear functions $\mathbb{F}^n \to \mathbb{F}^s$ is ϵ-almost universal, if it holds for every non-zero $\mathbf{x} \in \mathbb{F}^n$ that*

$$\Pr_{\mathbf{H} \xleftarrow{\$} \mathcal{H}} [\mathbf{H}(\mathbf{x}) = 0] \leq \epsilon,$$

where $\mathbf{H}$ is chosen uniformly at random from the family $\mathcal{H}$. We say that $\mathcal{H}$ is universal, if it is $|\mathbb{F}^{-s}|$-almost universal. We will identify functions $H \in \mathcal{H}$ with their transformation matrix and write $\mathbf{H}(\mathbf{x}) = \mathbf{H} \cdot \mathbf{x}$.

We will first establish a property of almost universal hash functions that can be summarized as follows. Applying a randomly chosen linear hash function $\mathbf{H}$ from a suitable family $\mathcal{H}$ to a matrix $\mathbf{M}$ will preserve its rank, unless the rank of $\mathbf{M}$ exceeds a certain threshold r. If the rank of $\mathbf{M}$ is bigger than r, we still have the guarantee that the rank of $\mathbf{H} \cdot \mathbf{M}$ does not drop below r.

Lemma 1. *Let $\mathcal{H} : \mathbb{F}^m \to \mathbb{F}^{r+s+t}$ be a family of $|\mathbb{F}|^{-(r+s)}$-almost universal linear functions. Fix a matrix $\mathbf{M} \in \mathbb{F}^{m \times n}$. Then it holds for $\mathbf{H} \xleftarrow{\$} \mathcal{H}$ that*

$$\Pr[\operatorname{rank}(\mathbf{H} \cdot \mathbf{M}) < \min(\operatorname{rank}(\mathbf{M}), r)] \leq |\mathbb{F}|^{-s}.$$

Remark 1. Since rank is preserved by transposition, we can state the consequence of the Lemma equivalently as

$$\Pr[\operatorname{rank}(\mathbf{M}^\top \mathbf{H}^\top) < \min(\operatorname{rank}(\mathbf{M}^\top), r)] \leq |\mathbb{F}|^{-s}.$$

Proof. If $\operatorname{rank}(\mathbf{M}) = 0$ the statement is trivial. Thus assume $\operatorname{rank}(\mathbf{M}) > 0$. Let $V = \langle \mathbf{M} \rangle_{\mathrm{col}}$ be the column-span of $\mathbf{M}$. We will first compute $\mathbb{E}[|\ker(\mathbf{H}) \cap V| - 1]$. By linearity of expectation we have that

$$\mathbb{E}[|\ker(\mathbf{H}) \cap V| - 1] = \mathbb{E}[|\{\mathbf{v} \in V \setminus \{0\} : \mathbf{H}(\mathbf{v}) = 0\}|]$$
$$= \sum_{\mathbf{v} \in V \setminus \{0\}} \Pr_{\mathbf{H}}[\mathbf{H}(\mathbf{v}) = 0]$$
$$\leq (|V| - 1)|\mathbb{F}|^{-(r+s)}$$
$$\leq |V| \cdot |\mathbb{F}|^{-(r+s)}.$$

As $|\ker(\mathbf{H}) \cap V| - 1$ is non-negative, it follows by the Markov inequality that

$$\Pr[|\ker(\mathbf{H}) \cap V| - 1 \geq |V| \cdot |\mathbb{F}|^{-r}] \leq \frac{\mathbb{E}[|\ker(\mathbf{H}) \cap V| - 1]}{|V| \cdot |\mathbb{F}|^{-r}}$$
$$\leq \frac{|V||\mathbb{F}|^{-(r+s)}}{|V| \cdot |\mathbb{F}|^{-r}}$$
$$= |\mathbb{F}|^{-s}.$$

Thus it follows that

$$\Pr[|\ker(\mathbf{H}) \cap V| > |V| \cdot |\mathbb{F}|^{-r}] \leq |\mathbb{F}|^{-s}. \tag{1}$$

– If $\mathsf{rank}(\mathbf{M}) = \dim(V) \leq r$, then it holds that

$$|V| \cdot |\mathbb{F}|^{-r} \leq 1$$

and (1) implies
$$\Pr[|\ker(\mathbf{H}) \cap V| > 1] \leq |\mathbb{F}|^{-s}.$$

But since $\dim(\ker(\mathbf{H}) \cap V) = \mathsf{rank}(\mathbf{M}) - \mathsf{rank}(\mathbf{HM})$, this means that

$$\Pr[\mathsf{rank}(\mathbf{HM}) < \mathsf{rank}(\mathbf{M})] \leq |\mathbb{F}|^{-s}.$$

– On the other hand, if $\mathsf{rank}(\mathbf{M}) = \dim(V) \geq r$, then we can restate (1) as

$$\Pr[\dim(\ker(\mathbf{H}) \cap V) > \mathsf{rank}(\mathbf{M}) - r] \leq |\mathbb{F}|^{-s}.$$

Again using $\dim(\ker(\mathbf{H}) \cap V) = \mathsf{rank}(\mathbf{M}) - \mathsf{rank}(\mathbf{HM})$ we obtain

$$\Pr[\mathsf{rank}(\mathbf{HM}) < r] \leq |\mathbb{F}|^{-s}.$$

All together, we obtain

$$\Pr[\mathsf{rank}(\mathbf{HM}) < \min(\mathsf{rank}(\mathbf{M}), r)] \leq |\mathbb{F}|^{-s},$$

which concludes the proof. $\qquad\square$

The next lemma states that we obtain a lower bound the distance of a matrix $\mathbf{X}$ from an interleaved code $\mathsf{C}^{\odot m}$ by the rank of $\mathbf{PX}$.

Lemma 2. *Let C be a $\mathbb{F}$-linear $[n, k, s]$ code with a parity check matrix $\mathbf{P}$. It holds for every $\mathbf{X} \in \mathbb{F}^{n \times m}$ that $\mathsf{dist}(\mathsf{C}^{\odot m}, \mathbf{X}) \geq \mathsf{rank}(\mathbf{PX})$.*

Proof. Let $\mathbf{E} \in \mathbb{F}^{n \times m}$ be a matrix of minimal row support such that $\mathbf{X} - \mathbf{E} \in \mathsf{C}^{\odot m}$, i.e., $\|\mathbf{E}\|_0 = \mathsf{dist}(\mathsf{C}^{\odot m}, \mathbf{X})$. Clearly $\mathbf{PX} = \mathbf{PE}$. It follows that

$$\mathsf{rank}(\mathbf{PX}) = \mathsf{rank}(\mathbf{PE}) \leq \mathsf{rank}(\mathbf{E}) \leq \|\mathbf{E}\|_0 = \mathsf{dist}(\mathsf{C}^{\odot m}, \mathbf{X}).$$

Lemma 3. *Let C be a $\mathbb{F}$-linear $[n, k, s]$ code with a parity check matrix $\mathbf{P}$. Let $\mathbf{X} \in \mathbb{F}^{n \times m}$ and $\mathbf{X}' \in \mathbb{F}^{n \times m'}$. If it holds that*

$$\langle \mathbf{PX} \rangle_{\mathrm{col}} \subseteq \langle \mathbf{PX}' \rangle_{\mathrm{col}},$$

then for any $\mathbf{C}' \in \mathsf{C}^{\odot m'}$ there exists a $\mathbf{C} \in \mathsf{C}^{\odot m}$ such that the row support of $\mathbf{X} - \mathbf{C}$ is contained in the row support of $\mathbf{X}' - \mathbf{C}'$. As a consequence, it also holds that

$$\mathsf{dist}(\mathsf{C}^{\odot m}, \mathbf{X}) \leq \mathsf{dist}(\mathsf{C}^{\odot m'}, \mathbf{X}').$$

Proof. As $\langle \mathbf{PX} \rangle_{\mathrm{col}} \subseteq \langle \mathbf{PX}' \rangle_{\mathrm{col}}$, we can express $\mathbf{PX}$ as

$$\mathbf{PX} = \mathbf{PX}'\mathbf{T},$$

for a matrix $\mathbf{T} \in \mathbb{F}^{m' \times m}$. This implies that $\mathbf{P}(\mathbf{X} - \mathbf{X}'\mathbf{T}) = 0$, from which it follows that $\mathbf{X} - \mathbf{X}'\mathbf{T} \in \mathsf{C}^{\odot m}$. Thus there exists a $\hat{\mathbf{C}} \in \mathsf{C}^{\odot m}$ with

$$\mathbf{X} - \mathbf{X}'\mathbf{T} = \hat{\mathbf{C}}. \tag{2}$$

Now fix an arbitrary $\mathbf{C}' \in \mathsf{C}^{\odot m'}$. Rearranging Eq. (2), we obtain

$$\mathbf{X} - (\hat{\mathbf{C}} + \mathbf{C}'\mathbf{T}) = (\mathbf{X}' - \mathbf{C}')\mathbf{T}.$$

Setting $\mathbf{C} = \hat{\mathbf{C}} + \mathbf{C}'\mathbf{T}$ it follows directly that the row support of $\mathbf{X} - \mathbf{C}$ is contained in the row support of $\mathbf{X}' - \mathbf{C}'$, as $\mathbf{X} - \mathbf{C} = (\mathbf{X}' - \mathbf{C}')\mathbf{T}$.

Theorem 1. *Let $\mathcal{H} : \mathbb{F}^m \rightarrow \mathbb{F}^{2s+t}$ be a family of $|\mathbb{F}|^{-2s}$-almost universal $\mathbb{F}$-linear hash functions. Further let C be an $\mathbb{F}$-linear $[n, k, s]$ code. Then for every $\mathbf{X} \in \mathbb{F}^{n \times m}$ at least one of the following statements holds, except with probability $|\mathbb{F}|^{-s}$ over the choice of $\mathbf{H} \xleftarrow{\$} \mathcal{H}$:*

1. $\mathbf{X}\mathbf{H}^\top$ has distance at least s from $\mathsf{C}^{\odot(2s+t)}$
2. For every $\mathbf{C}' \in \mathsf{C}^{\odot(2s+t)}$ there exists a $\mathbf{C} \in \mathsf{C}^{\odot m}$ such that $\mathbf{X}\mathbf{H}^\top - \mathbf{C}'$ and $\mathbf{X} - \mathbf{C}$ have the same row support

Remark 2. If the first item in the statement of the Theorem does not hold, the second one must hold. Then we can efficiently recover a codeword $\mathbf{C}$ with distance at most $s - 1$ from $\mathbf{X}$ using erasure correction, given a codeword $\mathbf{C}' \in \mathsf{C}^{\odot(2s+t)}$ with distance at most $s - 1$ from $\mathbf{X}\mathbf{H}^\top$. More specifically, we compute the row support of $\mathbf{X}\mathbf{H}^\top - \mathbf{C}'$, erase the corresponding rows of $\mathbf{X}$ and recover $\mathbf{C}$ from $\mathbf{X}$ using erasure correction[5]. The last step is possible as the distance between $\mathbf{X}$ and $\mathbf{C}$ is at most $s - 1$.

Proof. We will distinguish two cases, depending on whether $\mathrm{rank}(\mathbf{PX}) \geq s$ or $\mathrm{rank}(\mathbf{PX}) < s$.

- Case 1: $\mathrm{rank}(\mathbf{PX}) \geq s$. It follows by Lemma 1 that $\mathrm{rank}(\mathbf{PXH}^\top)$ is at least s, except with probability $|\mathbb{F}|^{-s}$ over the choice of $\mathbf{H} \xleftarrow{\$} \mathcal{H}$. Thus fix a $\mathbf{H} \in \mathcal{H}$ with $\mathrm{rank}(\mathbf{PXH}^\top) \geq s$. It follows by Lemma 2 that $\mathrm{dist}(\mathsf{C}^{\odot m}, \mathbf{XH}^\top) \geq s$, *i.e.*, the first item holds.
- Case 2: $\mathrm{rank}(\mathbf{PX}) < s$. It follows from Lemma 1 that $\mathrm{rank}(\mathbf{PXH}^\top) = \mathrm{rank}(\mathbf{PX})$, except with probability $|\mathbb{F}|^{-s}$ over the choice of $\mathbf{H} \xleftarrow{\$} \mathcal{H}$. Thus fix a $\mathbf{H} \in \mathcal{H}$ with $\mathrm{rank}(\mathbf{PXH}^\top) = \mathrm{rank}(\mathbf{PX})$. Since $\langle \mathbf{PXH}^\top \rangle_{\mathrm{col}} \subseteq \langle \mathbf{PX} \rangle_{\mathrm{col}}$ and $\mathrm{rank}(\mathbf{PXH}^\top) = \mathrm{rank}(\mathbf{PX})$, it holds that $\langle \mathbf{PXH}^\top \rangle_{\mathrm{col}} = \langle \mathbf{PX} \rangle_{\mathrm{col}}$. It follows from Lemma 3 that for every $\mathbf{C}' \in \mathsf{C}^{\odot(2s+t)}$ there exists a $\mathbf{C} \in \mathsf{C}^{\odot m}$ such that $\mathbf{XH}^\top - \mathbf{C}'$ and $\mathbf{X} - \mathbf{C}$ have the same row support, *i.e.*, the second item holds.

4   Linear Time Primitives

In this section, we will provide constructions of almost universal hash functions and rate-1 codes with linear time complexity.

---

[5] Recall that erasure correction for linear codes can be performed efficiently via gaussian elimination.

4.1 Linear Time Almost Universal Hashing with Short Seeds

Theorem 2 [IKOS08, DI14]. *Fix a finite field* $\mathbb{F}$ *of constant size. For all integers* n, m *with* $m \leq n$ *there exists a family of linear universal hash functions* $\mathcal{G} : \mathbb{F}^n \rightarrow \mathbb{F}^m$ *such that each function* $G \in \mathcal{G}$ *can be described by* $O(n)$ *bits and computed in time* $O(n)$.

It is well know that evaluating a polynomial of degree at most d over a field $\mathbb{F}$ is a $(d-1)/|\mathbb{F}|$-almost universal hash function. We will use the family provided in Theorem 2 to *pre-hash* the input in a block-wise manner, such that the computation time of the polynomial hash function becomes linear in the size of the original input. A similar *speed-up* trick was used in [IKOS08] to construct several cryptographic primitives, for instance pseudorandom functions, that can be computed in linear time.

Lemma 4. *Let* $d = d(s)$ *be a positive integer. Let* $\mathbb{F}$ *be a finite field of constant size and* $\mathbb{F}'$ *be an extension field of* $\mathbb{F}$ *of degree* $\lceil s + \log_{|\mathbb{F}|}(d) \rceil$. *Let* $n = n(s, d)$ *be such that a multiplication in* $\mathbb{F}'$ *can be performed in time* $O(n)$. *Let* $\mathcal{G} : \mathbb{F}^n \rightarrow \mathbb{F}'$ *be a family of* $\mathbb{F}$-*linear universal hash functions which can be computed in time* $O(n)$ *and has seed length* $O(n)$. *For a function* $G \in \mathcal{G}$ *and an element* $\alpha \in \mathbb{F}'$, *define the function* $H_{G,\alpha} : \mathbb{F}^{d \cdot n} \rightarrow \mathbb{F}' \cong \mathbb{F}^{s + \log_{|\mathbb{F}|}(d)}$ *by*

$$H_{G,\alpha}(\mathbf{x}) = \sum_{i=0}^{d-1} G(\mathbf{x}_i)\alpha^i,$$

where $\mathbf{x} = (\mathbf{x}_0, \ldots, \mathbf{x}_{d-1}) \in (\mathbb{F}^n)^d$. *Define the family* $\mathcal{H}$ *by* $\mathcal{H} = \{H_{G,\alpha} : G \in \mathcal{G}, \alpha \in \mathbb{F}'\}$. *Then the family* $\mathcal{H}$ *is* 2^{-s}-*almost universal, has sub-linear seed-length* $O(n)$ *and can be computed in linear time* $O(d \cdot n)$.

Remark 3. We can choose the function $n(s, d)$ as small as $O((s + \log_{|\mathbb{F}|}(d)) \cdot \text{polylog}(s + \log_{|\mathbb{F}|}(d)))$, if a fast multiplication algorithm for $\mathbb{F}'$ is used.

Proof. We will first show that $\mathcal{H}$ is 2^{-s} almost universal. Let $\mathbf{x} = (\mathbf{x}_0, \ldots, \mathbf{x}_{d-1}) \neq 0$. Thus there exists an $i \in \{0, \ldots, d-1\}$ such that $\mathbf{x}_i \neq 0$. Consequently, it holds for a randomly chosen $G \xleftarrow{\$} \mathcal{G}$ that $G(\mathbf{x}_i) \neq 0$, except with probability $1/|\mathbb{F}'|$. Suppose now that $0 \neq (G(\mathbf{x}_0), \ldots, G(\mathbf{x}_{d-1})) \in \mathbb{F}'^d$.

$$P(X) = \sum_{i=0}^{d-1} G(\mathbf{x}_i)X^i$$

is a non-zero polynomial of degree at most $d-1$, and consequently $P(X)$ has at most $d-1$ zeros. It follows that for a random $\alpha \xleftarrow{\$} \mathbb{F}'$ that

$$H_{G,\alpha}(\mathbf{x}) = \sum_{i=0}^{d-1} G(\mathbf{x}_i)\alpha^i = P(\alpha) \neq 0,$$

except with probability $(d-1)\,|\mathbb{F}'|$. All together, we can conclude that $H_{G,\alpha}(\mathbf{x}) \neq 0$, except with probability

$$1/|\mathbb{F}'| + (d-1)/|\mathbb{F}'| = d/|\mathbb{F}'| = |\mathbb{F}|^{-s}$$

over the choice of $G \overset{\$}{\leftarrow} \mathcal{G}$ and $\alpha \overset{\$}{\leftarrow} \mathbb{F}'$, as $|\mathbb{F}'| = |\mathbb{F}|^{s+\log_{|\mathbb{F}|}(d)}$.

Notice that the seed size of $H_{G,\alpha}$ is

$$|G| + \log(|\mathbb{F}'|) = O(n) + (s + \log_{|\mathbb{F}|}(d))\log(|\mathbb{F}|) = O(n).$$

We will finally show that for any choice of $G \in \mathcal{G}$ and $\alpha \in \mathbb{F}'$ the function $H_{G,\alpha}$ can be computed in linear time in the size of its input $\mathbf{x}$. Computing $G(\mathbf{x}_1), \ldots, G(\mathbf{x}_d)$ takes time $O(d \cdot n)$, as computing each $G(\mathbf{x}_i)$ takes time $O(n)$. Next, evaluating the polynomial $P(X) = \sum_{i=0}^{d-1} G(\mathbf{x}_i)X^i$ at α naively costs $d-1$ additions and $2(d-1)$ multiplications. Since both additions and multiplications in $\mathbb{F}'$ can be performed in time $O(n)$, the overall cost of evaluating $P(X)$ at α can be bounded by $O(d \cdot n)$. All together, we can compute $H_{G,\alpha}$ in time $O(d \cdot n)$, which is linear in the size of the input.

Instantiating the family $\mathcal{G}$ in Lemma 4 with the family provided in Theorem 2, we obtain the following theorem.

Theorem 3. *Fix a finite field $\mathbb{F}$ of constant size. There exists an explicit family $\mathcal{H} : \mathbb{F}^n \to \mathbb{F}^{s+O(\log(n))}$ of $|\mathbb{F}|^{-s}$-universal hash functions that can be represented by $O(s^2)$ bits and computed in time $O(n)$.*

4.2 Linear Time Rate-1 Codes

For the construction in this section we will need a certain kind of expander graph, called unique-neighbor expander.

Definition 2. *Let $\Gamma = (L, R, E)$ be a bipartite graph of left-degree d with $|L| = n$ and $|R| = m$. We say that Γ is a (n, m, d, w)-unique neighbor expander, if for every non-empty subset $S \subseteq L$ of size at most w, there exists at least one vertex $r \in R$ such that $|\Gamma(r) \cap S| = 1$, where $\Gamma(r) = \{l \in L : (l, r) \in E\}$ is the neighborhood of r.*

Lemma 5. *Fix a finite field $\mathbb{F}$ of constant size. Let C be an $\mathbb{F}$-linear $[m, k, s]$ code. Further let Γ be a (n, m, d, w)-unique-neighbor expander such that $w \cdot d < s$. Let $\mathbf{H}_\Gamma$ be the adjacency matrix of Γ. Then the code $\mathsf{C}' = \{\mathbf{c} \in \mathbb{F}^n | \mathbf{H}_\Gamma \cdot \mathbf{c} \in \mathsf{C}\}$ is an $\mathbb{F}$-linear $[n, n - m + k, w]$ code.*

Proof. Clearly, C' has length n. If $\mathbf{H}_\mathsf{C}$ is a parity check matrix of C, then $\mathbf{H}_\mathsf{C} \cdot \mathbf{H}_\Gamma \in \mathbb{F}^{(m-k)\times n}$ is a parity check matrix of C'. Thus, the dimension of C' is at least $n - m + k$. Now, let $\mathbf{e} \in \mathbb{F}^n$ be a non-zero vector of weight less than w. Then, by the unique neighbor expansion property of Γ, $\mathbf{H}_\Gamma \cdot \mathbf{e}$ is a non-zero vector of weight at most $d \cdot w$. But now it immediately holds that $\mathbf{H}_\Gamma \cdot \mathbf{e} \notin \mathsf{C}$, as C has distance at least $s > d \cdot w$. Thus, C' has minimum distance at least w.

Remark 4. The same arguments show that if Γ is a (n, m, d, w)-unique-neighbor expander (with no additional conditions on the parameters), the code $C'' = \{\mathbf{c} \in \mathbb{F}^n | \mathbf{H}_\Gamma \cdot \mathbf{c} = \mathbf{0}\}$ is an $\mathbb{F}$-linear $[n, n - m, w]$ code.

We will now use the statement of Lemma 5 on a suitable chain of expander graphs to obtain codes with rate 1 and a linear time parity check operation. We will use the following families of explicit expander graphs due to Capalbo et al. [CRVW02].

Theorem 4 [CRVW02]. *For all integers n, $m < n$ there exists an explicit (n, m, d, w)-unique-neighbor expander Γ with*

$$d = (\log(n) - \log(m))^{O(1)}$$

$$w = \Omega\left(\frac{m}{d}\right).$$

Moreover, if $n = O(m)$, then Γ can be constructed efficiently.

Lemma 6. *Fix a finite field $\mathbb{F}$ of constant size. There exists a constant $\gamma > 0$ and an explicit family $(C_s)_s$ of $\mathbb{F}$-linear codes, where C_s has length $O(s^2)$, minimum distance s and rate $1 - s^{-\gamma}$, i.e. the rate of C_s approaches 1. Moreover, the parity check operation of C can be performed in $O(s^2)$, which is linear in the codeword length.*

Proof. By Theorem 4, there exists a constant d and a constant α, such that for all choices of m there exists a $(2m, m, d, w)$-unique-neighbor expander Γ with $w \geq \alpha m/d$. Now let t be a constant such that $t \cdot \alpha \geq 1$ and let $\ell > 0$ be an integer. Choosing $m_i = t \cdot 2^{i-1} d^\ell s$, we obtain a chain of $(2m_i, m_i, d, w_i)$-unique-neighbor expanders Γ_i with $w_i \geq \alpha m_i/d$. For $1 \leq i \leq \ell$ we can get a lower bound for w_i by

$$w_i \geq \alpha m_i/d \geq \alpha t \cdot 2^{i-1} d^{\ell-1} s \geq d^{\ell-i} s,$$

as $\alpha t \geq 1$ and $i \geq 1$. We thus obtain that Γ_i is also a $(t \cdot 2^i d^\ell s, t \cdot 2^{i-1} d^\ell s, d, d^{\ell-i} s)$-unique neighbor expander.

We will choose $C_1 = \{\mathbf{c} \in \mathbb{F} : \mathbf{H}_{\Gamma_1}\mathbf{c} = 0\}$, which is a code of length $t2d^\ell s$ and distance at least $d^{\ell-1} s$ by Remark 4. Applying Lemma 5 on C_1 with the expander Γ_2, we obtain a code C_2 of length $t2^2 d^\ell s$ and distance $d^{\ell-2} s$. Iterating this procedure for $i \leq \ell$, applying Lemma 5 on C_i with expander Γ_{i+1}, we obtain codes C_i of length $t2^i d^\ell s$ and distance $d^{\ell-i} s$. Thus, C_ℓ is a code of length $t(2d)^\ell s$ and minimum distance s. By construction, the matrix

$$\mathbf{H}_\ell = \mathbf{H}_{\Gamma_1} \cdot \mathbf{H}_{\Gamma_2} \ldots \mathbf{H}_{\Gamma_\ell}$$

is its parity check matrix. Notice that multiplication $\mathbf{H}_\ell$ can be performed in linear time $O(t(2d)^\ell s)$ in the codeword length. This can be seen as multiplication with $\mathbf{H}_{\Gamma_i}$ can be performed it time $O(d \cdot t2^i d^\ell s)$ and thus multiplication with $\mathbf{H}_\ell$ can be performed in time

$$\sum_{i=1}^{\ell} O(d \cdot t2^i d^\ell s) = O(d^{\ell+1} ts \sum_{i=1}^{\ell} 2^i) = O(d^{\ell+1} ts 2^{\ell+1}) = O(t(2d)^\ell s)$$

We can also see from $\mathbf{H}_\ell$ that the dimension of C_ℓ is at least $t(2d)^\ell s - td^\ell s$, i.e. C_ℓ has rate $1 - 2^{-\ell}$. Now, choosing $\ell = \lceil \log(s)/\log(2d) \rceil$ we obtain a code C of length $O(s^2)$, minimum distance s and rate $1 - s^{-\gamma}$, where $\gamma \geq 1/\log(2d)$ is a constant. This concludes the proof.

We will now convert the codes constructed in Lemma 6 into codes with linear time encoding operation. The idea is simple: compute a syndrome of a message with respect to the parity check matrix promised in Lemma 6, encode this syndrome using a good code C_2 and append the encoded syndrome to the message. This systematic code has a linear time encoding operation, and the next Lemma shows that it has also good distance and rate.

Lemma 7. *Fix a finite field $\mathbb{F}$ of constant size. Let C_1 be an $\mathbb{F}$-linear $[n, n-m, d]$ code with linear time computable parity check operation with respect to a parity check matrix $\mathbf{H}_1$. Further let C_2 be an $\mathbb{F}$-linear $[l, m, d]$ code with a linear time encoding operation with respect to a generator matrix $\mathbf{G}_2$. Then the code C_3, defined via the encoding operation $\mathbf{x} \mapsto (\mathbf{x}, \mathbf{G}_2 \cdot \mathbf{H}_1 \cdot \mathbf{x})$ is an $\mathbb{F}$-linear $[n+l, n, d]$ code with linear time encoding operation.*

Proof. The fact that C_3 is linear time encodable follows immediately, as multiplication with both $\mathbf{H}_1$ and $\mathbf{G}_2$ are linear time computable. Moreover, it also follows directly from the definition of C_3 that C_3 has length $n + l$ and dimension n. We will now show that C_3 has minimum distance at least d. Let $\mathbf{e} = (\mathbf{e}_1, \mathbf{e}_2) \in \mathbb{F}^{n+l}$ be a non-zero vector of weight less than d. Clearly it holds that both $\mathbf{e}_1$ and $\mathbf{e}_2$ have weight less than d. If $\mathbf{e}_2$ is non-zero, then $\mathbf{e}_2 \notin C_2$, as C_2 has minimum distance d. On the other hand, if $\mathbf{e}_1$ is non-zero, then $\mathbf{H}_1 \cdot \mathbf{e}_1$ is non-zero as C_1 has distance d. But then, $\mathbf{G}_2 \cdot \mathbf{H}_1 \mathbf{e}_1$ has weight at least d, as C_2 has minimum distance d and $\mathbf{H}_1 \cdot \mathbf{e}_1$ is non-zero. Consequently, $\mathbf{G} \cdot \mathbf{H}_1 \cdot \mathbf{e}_1 \neq \mathbf{e}_2$, as $\mathbf{e}_2$ has weight less than d. We conclude that $(\mathbf{e}_1, \mathbf{e}_2) \notin C_3$.

To use Lemma 7, we need a family of linear time encodable codes with constant rate and constant relative minimum distance. Such codes were first constructed by Spielman [Spi96].

Theorem 5 [Spi96, GI05]. *Fix a finite field $\mathbb{F}$ of constant size. Then there exists a family $\{C_n\}$ of $\mathbb{F}$-linear codes with constant rate and constant relative minimum distance which supports linear time encoding.*

We can now bootstrap the statement of Lemma 6 into a linear time encodable code of rate 1 using Lemma 7 and Theorem 5.

Theorem 6. *Fix a finite field $\mathbb{F}$ of constant size. There exists a constant $\gamma > 0$ and an explicit family of $\mathbb{F}$-linear codes $(C_s)_s$ of length $O(s^2)$, minimum distance s and rate $1 - s^{-\gamma}$, which approaches 1. Moreover, C has an encoding algorithm Enc that runs in time $O(s^2)$, which is linear in the codeword length.*

5   Linear Time and Rate 1 Additive Commitments

In this section we construct a protocol for additively homomorphic commitments that UC realizes functionality $\mathcal{F}_{HCOM}$. This protocol achieves (amortized) linear computational complexity for both parties and rate 1, meaning that the ratio between the size of the committed messages and the size of the data exchanged by the parties in the protocol approaches one. We will show how to make commitments to random strings, which allows the protocol to achieve *sublinear* communication complexity in the commitment phase while keeping rate 1 in the opening phase, a property that finds applications in different scenarios of multiparty computation [FJN+13]. This protocol can be trivially extended to standard commitments by having the sender also send the difference between the random and the desired strings. The resulting protocol maintains rate 1 and linear computational complexity.

The construction in this section will be based on a systematic binary linear code C, an $[n, k, s]$ code, where s is the statistical security parameter and n is $k + O(s)$. It follows from the construction in Sect. 4.2 that for any desired value of s, we can make such a code for any k, that is, the rate tends to 1 as k grows, and furthermore that encoding in C takes linear time. We also need a family of linear time computable almost universal hash functions $\mathcal{H}$. Furthermore the functions in $\mathcal{H}$ must be linear. The functions will map m-bit strings to l-bit strings, where m is a parameter that can be chosen arbitrarily large (but polynomially related to n, k and l). We use the construction from Theorem 3, and hence, since we will need collision probability 2^{-2s}, we set $l = 2s + \log(m)$.

We will build commitments to k-bit random strings, and the protocol will produce $m - l$ such commitments. In Appendix C we show how our protocol can be used to commit to arbitrary messages achieving still preserving linear computational complexity and rate-1. In fact, in Sect. 5.1 we show that we can get even higher rate when committing to random messages. In the following, all vectors and matrices will be assumed to have binary entries. The construction can easily be generalized to other finite fields. The Commitment Phase is described in Fig. 3 and the Addition procedure and Opening Phase are described in Fig. 4. Notice that the Opening Phase presented in Fig. 4 does not achieve rate-1 but we show how to do so in Sect. 5.1. The security of our protocols is formally stated in Theorem 7.

As shown in Appendix B, we can implement $\mathcal{F}_{ROT}$ based on n one-out-of-two OT's on short strings (of length equal to a computational security parameter) using a pseudo-random generator and standard techniques. This will give the result mentioned in the introduction: we can amortize the cost of the OT's over many commitments.

Theorem 7. Π_{HCOM} *UC-realizes* $\mathcal{F}_{HCOM}$ *in the* $\mathcal{F}_{ROT}$-*hybrid model with statistical security against a static adversary. Formally, there exists a simulator* $\mathcal{S}$ *such that for every static adversary* $\mathcal{A}$, *and any environment* $\mathcal{Z}$, *the environment cannot distinguish* Π_{HCOM} *composed with* $\mathcal{F}_{ROT}$ *and* $\mathcal{A}$ *from* $\mathcal{S}$ *composed with* $\mathcal{F}_{HCOM}$. *That is, we have*

$$\text{IDEAL}_{\mathcal{F}_{HCOM}, \mathcal{S}, \mathcal{Z}} \approx_s \text{HYBRID}_{\Pi_{HCOM}, \mathcal{A}, \mathcal{Z}}^{\mathcal{F}_{ROT}}.$$

Protocol Π_{HCOM} (Commitment Phase)

Let C be a systematic binary linear $[n, k, s]$ code, where s is the statistical security parameter and n is $k + O(s)$. Let $\mathcal{H}$ be a family of linear almost universal hash functions $\mathbf{H} : \{0,1\}^m \rightarrow \{0,1\}^l$. Protocol Π_{HCOM} is run by a sender P and a receiver V and proceeds as follows:

Commitment Phase

1. The parties P and V invoke $\mathcal{F}_{\text{ROT}}$ with inputs (sender, $sid, ssid$) and (receiver, $sid, ssid$), respectively. P receives $(sid, ssid, \mathbf{R_0}, \mathbf{R_1})$ from $\mathcal{F}_{\text{ROT}}$ and sets $\mathbf{R} = \mathbf{R_0} + \mathbf{R_1}$. V receives $(sid, ssid, b_1, \ldots, b_n, \mathbf{S})$ from $\mathcal{F}_{\text{ROT}}$ and sets the diagonal matrix $\mathbf{\Delta}$ such that it contains $b_1, \ldots, b_n$ in the diagonal. $\mathbf{R}$ will contain in the top k rows the data to commit to. Note that $\mathbf{R_0}, \mathbf{R_1}$ forms an additive secret sharing of $\mathbf{R}$, and in each row V knows shares from either $\mathbf{R_0}$ or $\mathbf{R_1}$.

2. P now adjusts the bottom $n - k$ rows of $\mathbf{R}$ so that all columns are codewords in C, and V will adjust his shares accordingly, as follows: P constructs a matrix $\mathbf{W}$ with dimensions as $\mathbf{R}$ and 0s in the top k rows, such that $\mathbf{A} := \mathbf{R} + \mathbf{W} \in C^{\odot m}$ (recall that C is systematic). P sends $(sid, ssid, \mathbf{W})$ to V (of course, only the bottom $n - k = O(s)$ rows need to be sent).

3. P sets $\mathbf{A_0} = \mathbf{R_0}, \mathbf{A_1} = \mathbf{R_1} + \mathbf{W}$ and V sets $\mathbf{B} = \mathbf{\Delta W} + \mathbf{S}$. Note that now we have

$$\mathbf{A} = \mathbf{A_0} + \mathbf{A_1}, \quad \mathbf{B} = \mathbf{\Delta A_1} + (\mathbf{I} - \mathbf{\Delta})\mathbf{A_0}, \quad \mathbf{A} \in C^{\odot m} \; ,$$

i.e., $\mathbf{A}$ is additively shared and for each row index, V knows either a row from $\mathbf{A_0}$ or from $\mathbf{A_1}$.

4. V chooses a seed H' for a random function $\mathbf{H} \in \mathcal{H}$ and sends $(sid, ssid, H')$ to P, we identify the function with its matrix (recall that all functions in $\mathcal{H}$ are linear).

5. P computes $\mathbf{T_0} = \mathbf{A_0 H}, \mathbf{T_1} = \mathbf{A_1 H}$ and sends $(sid, ssid, \mathbf{T_0}, \mathbf{T_1})$ to V. Note that $\mathbf{AH} = \mathbf{A_0 H} + \mathbf{A_1 H} = \mathbf{T_0} + \mathbf{T_1}$, and $\mathbf{AH} \in C^{\odot l}$. So we can think of $\mathbf{T_0}, \mathbf{T_1}$ as an additive sharing of $\mathbf{AH}$, where again V knows some of the shares, namely the rows of $\mathbf{BH}$.

6. V checks that $\mathbf{\Delta T_0} + (\mathbf{I} - \mathbf{\Delta})\mathbf{T_1} = \mathbf{BH}$ and that $\mathbf{T_0} + \mathbf{T_1} \in C^{\odot l}$. If any check fails, he aborts.

7. We sacrifice some of the columns in $\mathbf{A}$ to protect P's privacy: Note that each column j in $\mathbf{AH}$ is a linear combination of some of the columns in $\mathbf{A}$, we let $\mathbf{A}(j)$ denote the index set for these columns. Now for each j the parties choose an index $a(j) \in \mathbf{A}(j)$ such that all $a(j)$'s are distinct. P and V now discard all columns in $\mathbf{A}, \mathbf{A_0}, \mathbf{A_1}$ and $\mathbf{B}$ indexed by some $a(j)$. For simplicity in the following, we renumber the remaining columns from 1.

8. P saves $\mathbf{A}, \mathbf{A_0}$ and $\mathbf{A_1}$, and V saves $\mathbf{B}$ and $\mathbf{\Delta}$ (all of which now have $m - l$ columns).

Fig. 3. Protocol Π_{HCOM} (commitment phase)

Protocol Π_{HCOM} (Addition and Opening Phase)

Assuming that the Commitment phase has been completed as specified in Figure 3, Protocol Π_{HCOM} is run by a sender P and a receiver V and proceeds as follows:

Addition of Commitments

1. To add commitments with index i and j, P appends the column $\mathbf{A}[\cdot, j] + \mathbf{A}[\cdot, i]$ to $\mathbf{A}$, likewise he appends to $\mathbf{A_0}$ and $\mathbf{A_1}$ the sum of their i'th and j'th columns. P sends (add, $sid, ssid, i, j$) to V.
2. Upon receiving (add, $sid, ssid, i, j$), V appends $\mathbf{B}[\cdot, j] + \mathbf{B}[\cdot, i]$ to $\mathbf{B}$. Note that this maintains the properties $\mathbf{A} = \mathbf{A_0} + \mathbf{A_1}$, $\mathbf{B} = \mathbf{\Delta A_1} + (\mathbf{I} - \mathbf{\Delta})\mathbf{A_0}$, and $\mathbf{A} \in \mathsf{C}^{\odot m'}$, where m' is the current number of columns.

Opening Phase

1. To open commitment number j, P sends $(sid, ssid, \mathbf{A_0}[\cdot, j], \mathbf{A_1}[\cdot, j])$ to V and halts.
2. V checks that $\mathbf{A_0}[\cdot, j] + \mathbf{A_1}[\cdot, j] \in \mathsf{C}$ and that for $i = 1, \ldots, n$, it holds that $\mathbf{B}[i, j] = \mathbf{A_{b_i}}[i, j]$ (recall that b_i is the i'th entry on the diagonal of $\mathbf{\Delta}$). If this is the case, he outputs the first k entries in $\mathbf{A_0}[\cdot, j] + \mathbf{A_1}[\cdot, j]$ as the opened string and halts, otherwise, he aborts outputting $(sid, ssid, \perp)$.

Fig. 4. Protocol Π_{HCOM} (addition and opening phase)

Proof. Simulation when both players are honest is trivial, so the theorem follows from the Lemmas 8 and 9 below, which establish security against a corrupt P and a corrupt V, respectively.

Lemma 8. *There exists a simulator $\mathcal{S}_P$ such that for every static adversary $\mathcal{A}$ who corrupts P, and any environment $\mathcal{Z}$, the environment cannot distinguish Π_{HCOM} composed with $\mathcal{F}_{ROT}$ and $\mathcal{A}$ from $\mathcal{S}_P$ composed with $\mathcal{F}_{HCOM}$. That is, we have*

$$\mathsf{IDEAL}_{\mathcal{F}_{HCOM}, \mathcal{S}_P, \mathcal{Z}} \approx_s \mathsf{HYBRID}_{\Pi_{HCOM}, \mathcal{A}, Z}^{\mathcal{F}_{ROT}}$$

Proof. Assume that the sender P is corrupted. We use $\hat{P}$ to denote the corrupted sender. In the UC framework, this is actually the adversary, which might in turn be controlled by the environment. We describe the simulator $\mathcal{S}_P$ in Fig. 5. The simulator $\mathcal{S}_P$ will run protocol Π_{HCOM} with an internal copy of $\hat{P}$ exactly as the honest V would have done. First, $\mathcal{S}_P$ runs the instance of $\mathcal{F}_{ROT}$ used by $\hat{P}$ and V exactly as in the real execution. In the commitment phase, if V aborts, then the simulator aborts. If V does not abort, then the simulator inspects $\mathcal{F}_{ROT}$ and reads off the matrices $\mathbf{R_0}$ and $\mathbf{R_1}$ that $\hat{P}$ gave as input. Now let $\mathbf{W}$ be the correction matrix sent by $\hat{P}$ and define $\mathbf{A_0} = \mathbf{R_0}$ and $\mathbf{A_1} = \mathbf{R_1} + \mathbf{W}$. Let $\mathbf{A} = \mathbf{A_0} + \mathbf{A_1}$. Notice that because $\hat{P}$ is malicious, it might be the case that $\mathbf{A} \notin \mathsf{C}^{\odot m}$.

Simulator $\mathcal{S}_P$

Simulator $\mathcal{S}_P$ interacts with environment $\mathcal{Z}$, functionality $\mathcal{F}_{\text{HCOM}}$ and an internal copy of adversary $\hat{P}$. Upon being activated by $\mathcal{Z}$, $\mathcal{S}_V$ proceeds as follows:

1. **Emulating $\mathcal{F}_{\text{ROT}}$:** Upon receiving (adversary, $sid, ssid, \mathbf{R_0}, \mathbf{R_1}$) from $\hat{P}$, $\mathcal{S}_P$, stores $(sid, ssid, \mathbf{R_0}, \mathbf{R_1})$ samples $(b_1, .., b_n) \overset{\$}{\leftarrow} \{0,1\}^n$, sets $\mathbf{S}[i, \cdot] = \mathbf{R}_{b_i}[i, \cdot]$ for $i = 1, \ldots, n$ and stores $(sid, ssid, b_1, \ldots, b_n, \mathbf{S})$.

2. **Commitment Phase:** Upon receiving $(sid, ssid, \mathbf{W})$ from $\hat{P}$, $\mathcal{S}_P$ runs the rest of the steps of the commitment phase of Π_{HCOM} exactly like an honest V would do. If an honest V would abort at any point then $\mathcal{S}_P$ also aborts. Otherwise, $\mathcal{S}_P$ uses its knowledge of $(sid, ssid, \mathbf{R_0}, \mathbf{R_1})$ to reconstruct $\mathbf{A}$. For $j = 1, \ldots, m - l$, $\mathcal{S}_P$ decodes column $\mathbf{A}[\cdot, j]$ obtaining message $\boldsymbol{m}_j$ and sends $(\text{commit}, sid, ssid_j, P_s, P_r, \boldsymbol{m}_j)$ to $\mathcal{F}_{\text{HCOM}}$. We will show that if $\mathcal{S}_P$ does not abort after executing V's steps in Π_{HCOM}, then the remaining $m - l$ columns of $\mathbf{A}$ can indeed be decoded to their corresponding committed messages except with negligible probability.

3. **Addition:** Upon receiving $(\text{add}, sid, ssid, i, j)$ from $\hat{P}$, $\mathcal{S}_P$ execute the steps of Π_{HCOM} for addition, chooses an unused ssid $ssid_a$ and sends $(\text{add}, sid, ssid_i, ssid_j, ssid_a, P_s, P_r)$ to $\mathcal{F}_{\text{HCOM}}$.

4. **Opening Phase:** Upon receiving $(sid, ssid, \mathbf{A_0}[\cdot, j], \mathbf{A_1}[\cdot, j])$ from $\hat{P}$, $\mathcal{S}_P$ runs the checks performed by V exactly as in Π_{HCOM}. If $\mathbf{A_0}[\cdot, j], \mathbf{A_1}[\cdot, j]$ is not a consistent opening, $\mathcal{S}_P$ outputs whatever $\hat{P}$ outputs and aborts. Otherwise $\mathcal{S}_P$ sends $(\text{reveal}, sid, ssid_j)$ to $\mathcal{F}_{\text{HCOM}}$, outputs whatever $\hat{P}$ outputs and halts.

Fig. 5. Simulator $\mathcal{S}_P$

We now describe how the simulator decodes the columns of $\mathbf{A}$. The simulator will identify $< s$ rows such that $\mathbf{A}$ is in $\mathsf{C}^{\odot m}$ except for the identified rows. As the code has distance s, this allows to erasure decode each column j of $\mathbf{A}$ to C and the corresponding decoded message will be the extracted message $\boldsymbol{m}_j$ that the simulator will input to $\mathcal{F}_{\text{HCOM}}$. We now give the details.

Let $R \subset [n]$ be a set of indices specifying rows of $\mathbf{A}$. For a column vector $\boldsymbol{c} \in \mathbb{F}^n$ we let $\pi_R(\boldsymbol{c}) = (\boldsymbol{c}[i])_{i \in [n] \setminus R}$ be the vector punctured at the indices $i \in R$. For a matrix $\mathbf{M}$ we let $\mathbf{M}_R = \pi_R(\mathbf{M})$ be the matrix with each column punctured using π_R and for a set S we let $S_R = \{\pi_R(s) | s \in S\}$. The simulator will need to find $R \subset [n]$ with $|R| < s$ such that

$$\mathbf{A}_R \in \mathsf{C}_R^{\odot m}. \tag{3}$$

It should furthermore hold that

$$\mathbb{H}_\infty((b_i)_{i \in R} | \hat{P}) = 0 \tag{4}$$

$$\mathbb{H}_\infty((b_i)_{i \in [n] \setminus R} | \hat{P}) = n - |R|, \tag{5}$$

where $\hat{P}$ here denotes the view of $\hat{P}$ in the simulator so far, *i.e.*, the adversary can guess R and each choice bit b_i for $i \in R$ with certainty at this point in the simulation and has no extra information on b_i for $i \notin R$.

Define $\mathbf{T} := \mathbf{A}\mathbf{H}$. Let $\hat{\mathbf{T}}_0$ and $\hat{\mathbf{T}}_1$ be the values sent by P and let $\hat{\mathbf{T}} = \hat{\mathbf{T}}_0 + \hat{\mathbf{T}}_1$. Let $\mathbf{T}_0 = \mathbf{R}_0\mathbf{H}$ and $\mathbf{T}_1 = (\mathbf{R}_1 + \mathbf{W})\mathbf{H}$ be the values that $\hat{P}$ should have sent. Let $\mathbf{T} = \mathbf{T}_0 + \mathbf{T}_1$. Let R be the smallest set such that $\hat{\mathbf{T}}_R = \mathbf{T}_R$. We claim that this set fulfills (3), (4) and (5).

We know that the receiver did not abort, which implies that $\mathbf{\Delta}\hat{\mathbf{T}}_0 + (\mathbf{I} - \mathbf{\Delta})\hat{\mathbf{T}}_1 = \mathbf{B}\mathbf{H}$. The i'th row of $\mathbf{\Delta}\hat{\mathbf{T}}_0 + (\mathbf{I} - \mathbf{\Delta})\hat{\mathbf{T}}_1$ can be seen to be $\hat{\mathbf{T}}_{b_i}[i, \cdot]$. The i'th row of $\mathbf{B}$ can be seen to be $b_i\mathbf{W}[i, \cdot] + \mathbf{R}_{b_i}$, so the i'th row of $\mathbf{B}\mathbf{H}$ is $\mathbf{T}_{b_i}[i, \cdot]$. We thus have for all i that

$$\hat{\mathbf{T}}_{b_i}[i, \cdot] = \mathbf{T}_{b_i}[i, \cdot].$$

For each $i \in R$ we have that $\hat{\mathbf{T}}[i, \cdot] \neq \mathbf{T}[i, \cdot]$, so we must therefore have for all $i \in R$ that

$$\hat{\mathbf{T}}_{1-b_i}[i, \cdot] \neq \mathbf{T}_{1-b_i}[i, \cdot].$$

It follows that if V for position i had chosen the choice bit $1 - b_i$ instead of b_i, then the protocol would have aborted. Since $\hat{P}$ can compute the correct values $\mathbf{T}_{b_i}[i, \cdot]$ and $\mathbf{T}_{1-b_i}[i, \cdot]$ it also knows which value of b_i will make the test pass. By assumption the protocol did not abort. This proves (4). It also proves that the probability of the protocol not aborting and R having size $|R|$ is at most $2^{-|R|}$ as $\hat{P}$ has no information on $b_1, \ldots, b_n$ prior to sending $\hat{\mathbf{T}}_0$ and $\hat{\mathbf{T}}_1$ so $\hat{P}$ can guess $(b_i)_{i \in R}$ with probability at most $2^{-|R|}$. It is easy to see that the value of the bits b_i for $i \notin R$ do not affect whether or not the test succeeds. Therefore these bits are still uniform in the view of $\hat{P}$ at this point.

In particular, we can therefore continue under the assumption that $|R| < s$. We can then apply Theorem 1 where we set $\mathbf{X} = \mathbf{A}$. From $|R| < s$ it follows that $\mathbf{X}\mathbf{H}$ has distance less than s to $\mathsf{C}^{\odot m}$, so we must be in case 2 in Theorem 1. Now, since the receiver checks that $\hat{\mathbf{T}} \in \mathsf{C}^{\odot l}$ and the protocol did not abort, we in particular have that $\hat{\mathbf{T}}_R \in \mathsf{C}_R^{\odot l}$ from which it follows that $\mathbf{T}_R \in \mathsf{C}_R^{\odot l}$, which in turn implies that $\mathbf{A}_R\mathbf{H} \in \mathsf{C}_R^{\odot l}$ and thus $\mathbf{X}_R\mathbf{H} \in \mathsf{C}_R^{\odot l}$. We can therefore pick a codeword $\mathbf{C}' \in \mathsf{C}^{\odot l}$ such that the row support of $\mathbf{X}\mathbf{H} - \mathbf{C}'$ is R. From Theorem 1 we then get that there exists $\mathbf{C} \in \mathsf{C}^{\odot m}$ such that the row support of $\mathbf{A} - \mathbf{C}$ is R. From this it follows that $\mathbf{A}_R = \mathbf{C}_R$, which implies (3).

Now notice that since C has distance s and $|R| < s$ the punctured code C_R will have distance at least 1. Therefore the simulator can from each column $\mathbf{A}[i, \cdot]_R \in \mathsf{C}_R$ decode the corresponding message $\boldsymbol{m}_j \in \{0, 1\}^k$. This the message that the simulator will input to $\mathcal{F}_{\mathrm{HCOM}}$ on behalf of $\hat{P}$.

In order to fool $\mathcal{S}_P$ and open a commitment to a different message than the one that has been extracted from $\mathbf{A}[\cdot, j]$, $\hat{P}$ would have to provide $\mathbf{A}'_0[\cdot, j], \mathbf{A}'_1[\cdot, j]$ such that $\mathbf{A}'[\cdot, j] = \mathbf{A}'_0[\cdot, j] + \mathbf{A}'_1[\cdot, j]$ is a valid codeword of C corresponding to a different message $\boldsymbol{m}'$. However, notice that since C_R has distance $s - |R|$, that would require $\hat{P}$ to modify an additional $s - |R|$ positions of $\mathbf{A}$ that are not contained in $\mathbf{B}$ so that it does not get caught in the checks performed by a honest V in the opening phase. That means that $\hat{P}$ would have to guess $s - |R|$ of the choice bits b_i for $i \notin R$. It follows from (5) that this will succeed with probability at most $2^{s-|R|}$.

Simulator $\mathcal{S}_V$

Simulator $\mathcal{S}_V$ interacts with environment $\mathcal{Z}$, functionality $\mathcal{F}_{\text{HCOM}}$ and an internal copy of adversary $\hat{V}$. Upon being activated by $\mathcal{Z}$, $\mathcal{S}_V$ proceeds as follows:

1. **Emulating $\mathcal{F}_{\text{ROT}}$:** Upon receiving (adversary, $sid, ssid, b_1, \ldots, b_n, \mathbf{S}$) from $\hat{V}$, $\mathcal{S}_V$ perfectly simulates $\mathcal{F}_{\text{ROT}}$ by sampling random matrices $\mathbf{R_0}, \mathbf{R_1} \xleftarrow{\$} \{0,1\}^{n \times m}$, subject to $\mathbf{S}[i, \cdot] = \mathbf{R}_{b_i}[i, \cdot]$, for $i = 1, \ldots, n$. Finally, it stores $(sid, ssid, \mathbf{R_0}, \mathbf{R_1})$.

2. **Commitment Phase:** Upon receiving (receipt, $sid, ssid, P_s, P_r$) from $\mathcal{F}_{\text{HCOM}}$, $\mathcal{S}_V$ runs the steps of P in the commitment phase exactly as in Π_{HCOM}.

3. **Addition:** Upon receiving (add, $sid, ssid_1, ssid_2, ssid_3, P_s, P_r, \text{success}$) from $\mathcal{F}_{\text{HCOM}}$, $\mathcal{S}_V$ runs the steps of P exactly as in Π_{HCOM} (setting i and j corresponding to $ssid_1, ssid_2$).

4. **Opening Phase:** Upon receiving (reveal, $sid, ssid, P_s, P_r, \mathsf{m}$) from $\mathcal{F}_{\text{HCOM}}$, $\mathcal{S}_V$ uses its knowledge of $b_1, \ldots, b_n$ to compute alternative columns $\mathbf{A_0}'[\cdot, j], \mathbf{A_1}'[\cdot, j]$ such that $\mathbf{A}'[\cdot, j] = \mathbf{A_0}'[\cdot, j] + \mathbf{A_1}'[\cdot, j]$ is a valid commitment to m that can opened without being caught by $\hat{V}$ even though m is different from the messages committed to in the commitment phase. Namely, let G be the generating matrix of C, $\mathcal{S}_V$ computes $\mathsf{c_m} = G\mathsf{m}$, initially sets $\mathbf{A_0}'[\cdot, j] = \mathbf{A_0}[\cdot, j], \mathbf{A_1}'[\cdot, j] = \mathbf{A_1}[\cdot, j]$ and then sets $\mathbf{A}'_{1-b_i}[\cdot, j] = \mathsf{c_m}[j] - \mathbf{A}_{b_i}[\cdot, j]$. Note that matrices $\mathbf{A_0}'[\cdot, j], \mathbf{A_1}'[\cdot, j]$ only differ from matrices $\mathbf{A_0}[\cdot, j], \mathbf{A_1}[\cdot, j]$ obtained in the commitment phase in positions that are not known by $\hat{V}$. Finally, $\mathcal{S}_V$ sends $(sid, ssid, \mathbf{A_0}'[\cdot, j], \mathbf{A_1}'[\cdot, j])$ to $\hat{V}$, outputs whatever $\hat{V}$ outputs and halts.

Fig. 6. Simulator $\mathcal{S}_V$

Lemma 9. *There exists a simulator $\mathcal{S}_V$ such that for every static adversary $\mathcal{A}$ who corrupts V, and any environment $\mathcal{Z}$, the environment cannot distinguish Π_{HCOM} composed with $\mathcal{F}_{\text{ROT}}$ and $\mathcal{A}$ from $\mathcal{S}_V$ composed with $\mathcal{F}_{\text{HCOM}}$. That is, we have*

$$\text{IDEAL}_{\mathcal{F}_{\text{HCOM}}, \mathcal{S}_V, \mathcal{Z}} \approx_s \text{HYBRID}_{\Pi_{\text{HCOM}}, \mathcal{A}, \mathcal{Z}}^{\mathcal{F}_{\text{ROT}}}$$

Proof. In case V is corrupted, the simulator $\mathcal{S}_V$ has to run Π_{HCOM} with an internal copy of $\hat{V}$, commit to a dummy string and then be able to *equivocate* this commitment (*i.e.* open it to an arbitrary message) when it gets the actual message from $\mathcal{F}_{\text{HCOM}}$. In order to achieve this, we can construct a $\mathcal{S}_V$ that executes the commitment phase exactly as in Π_{HCOM} only deviating in the opening phase. Note that after the commit phase $\hat{V}$ has no information at all about the committed strings. This holds because the additive shares in $\mathbf{S}$ trivially contain no information and furthermore because the columns in sacrificed positions $a(j)$ contain uniformly random data and are never opened. This completely randomizes the data seen by $\hat{V}$ in the verification stage $(\mathbf{T_0}, \mathbf{T_1})$.

Therefore, $\mathcal{S}_V$ can use its knowledge of $b_1, \ldots, b_n$ to open a commitment to an arbitrary message without being caught, by modifying position of the matrices that are unknown to $\hat{V}$ (*i.e.* unknown to V in the real world).

We describe $\mathcal{S}_V$ in Fig. 6. Note that $\mathcal{S}_V$ exactly follows all the steps of Π_{HCOM} (and $\mathcal{F}_{\text{ROT}}$) except for when it opens commitments. Instead, in the opening phase, $\mathcal{S}_V$ sends $\mathbf{A_0}'[\cdot, j], \mathbf{A_1}'[\cdot, j]$, which differ from $\mathbf{A_0}[\cdot, j], \mathbf{A_1}[\cdot, j]$ that was set in the commitment phase and that would be sent in a real execution of Π_{HCOM}. However, $\mathbf{A_0}'[\cdot, j], \mathbf{A_1}'[\cdot, j]$, only differ from $\mathbf{A_0}[\cdot, j], \mathbf{A_1}[\cdot, j]$ in positions that are unknown by V. Hence, the joint distribution of the ideal execution with simulator $\mathcal{S}_V$ is statistically indistinguishable from the real execution of Π_{HCOM} with a corrupted receiver.

5.1   Computational Complexity and Rate

It is straightforward to verify that the Commitment, Addition and Opening protocols run in linear time, or more precisely, that the computational cost per bit committed to is constant. Indeed, it follows easily from the fact that C is linear time encodable and that H can be computed in linear time. This holds, even if we consider the cost of implementing $\mathcal{F}_{\text{ROT}}$ and $\mathcal{F}_{\text{OT}}$: the first cost is linear if we use a PRG that costs only a constant number of operations per output bit. The cost of the OT operations is amortized away if we consider a sufficiently large number of commitments.

Furthermore, the commitment protocol achieves rate 1, i.e., the amortized communication overhead per committed bit is $o(1)$ as we increase the number of bits committed in one commitment. This follows from the fact that C is rate-1 and that the communication cost of the verification in the final steps of the protocol only depends on the security parameter, and hence is "amortized away". Note that in the case where the sender only wants to be committed to random messages, it is possible to achieve rate higher than 1 in the commitment phase. This happens if we plug in the implementation of $\mathcal{F}_{\text{ROT}}$ based on $\mathcal{F}_{\text{OT}}$, since then the random strings output from $\mathcal{F}_{\text{ROT}}$ are generated locally from a short seed using a PRG, and later the sender is only required to send the bottom $n-k$ rows of $\mathbf{W}$ and the matrices $\mathbf{T_0}, \mathbf{T_1}$, which are both of the order of $O(s)$.

The opening protocol does not achieve rate 1 as it stands because the communication is about twice the size of the committed string (both $\mathbf{A_0}[\cdot, j]$ and $\mathbf{A_1}[\cdot, j]$ are sent). However, for sufficiently long messages, we can get rate-1 by using the same verification method as used in the commitment protocol, namely we open many commitments at once. We can think of this as opening an entire matrix $\mathbf{A}$ instead of its columns one by one. The idea is then that V selects a hash function $\mathbf{H}$ and P sends $\mathbf{A}$ as well as $\mathbf{T_0} = \mathbf{A_0 H}$ and $\mathbf{T_1} = \mathbf{A_1 H}$. The receiver checks that $\mathbf{AH} = \mathbf{T_0} + \mathbf{T_1}$, that all columns in $\mathbf{A}$ are in C and that $\mathbf{\Delta T_0} + (\mathbf{I} - \mathbf{\Delta})\mathbf{T_1} = \mathbf{BH}$. This can be shown secure by essentially the same proof as we used to show the commitment protocol secure against a corrupt sender. Now the communication overhead for verification is insignificant for large enough matrices $\mathbf{A}$.

Acknowledgements. A major part of this work was done while Ignacio Cascudo and Nico Döttling were also with Aarhus University.

The authors acknowledge support from the Danish National Research Foundation and The National Science Foundation of China (under the grant 61361136003) for the Sino-Danish Center for the Theory of Interactive Computation and from the Center for Research in Foundations of Electronic Markets (CFEM), supported by the Danish Strategic Research Council.

In addition, Ignacio Cascudo acknowledges support from the Danish Council for Independent Research, grant no. DFF-4002-00367.

Nico Döttling gratefully acknowledges support by the DAAD (German Academic Exchange Service) under the postdoctoral program (57243032). While at Aarhus University, he was supported by European Research Council Starting Grant 279447. His research is also supported in part from a DARPA/ARL SAFEWARE award, AFOSR Award FA9550-15-1-0274, and NSF CRII Award 1464397. The views expressed are those of the author and do not reflect the official policy or position of the Department of Defense, the National Science Foundation, or the U.S. Government.

Jesper Buus Nielsen was supported by European Research Council Starting Grant 279447.

The authors thank the anonymous reviewers of CRYPTO 2016 for their comments, which contributed to improve the paper.

A   Universal Composability

We adopt description of the Universal Composability (UC) framework given in [CDD+15]. In this framework, protocol security is analyzed under the real-world/ideal-world paradigm, *i.e.* by comparing the real world execution of a protocol with an ideal world interaction with the primitive that it implements. The model has a *composition theorem*, that basically states that UC secure protocols can be arbitrarily composed with each other without any security compromises. This desirable property not only allows UC secure protocols to effectively serve as building blocks for complex applications but also guarantees security in practical environments where several protocols (or individual instances of protocols) are executed in parallel, such as the Internet.

In the UC framework, the entities involved in both the real and ideal world executions are modeled as probabilistic polynomial-time Interactive Turing Machines (ITM) that receive and deliver messages through their input and output tapes, respectively. In the ideal world execution, dummy parties (possibly controlled by an ideal adversary S referred to as the *simulator*) interact directly with the ideal functionality $\mathcal{F}$, which works as a trusted third party that computes the desired primitive. In the real world execution, several parties (possibly corrupted by a real world adversary $\mathcal{A}$) interact with each other by means of a protocol π that realizes the ideal functionality. The real and ideal executions are controlled by the *environment* $\mathcal{Z}$, an entity that delivers inputs and reads the outputs of the individual parties, the adversary $\mathcal{A}$ and the simulator S. After a real or ideal execution, $\mathcal{Z}$ outputs a bit, which is considered as the output of the execution. The rationale behind this framework lies in showing that the environment $\mathcal{Z}$ (that represents all the things that happen outside of the protocol execution) is not able to efficiently distinguish between the real and ideal executions, thus implying that the real world protocol is as secure as the ideal functionality.

We denote by $\mathsf{REAL}_{\pi,\mathcal{A},\mathcal{Z}}(\kappa, z, \bar{r})$ the output of the environment $\mathcal{Z}$ in the real-world execution of protocol π between n parties with an adversary $\mathcal{A}$ under security parameter κ, input z and randomness $\bar{r} = (r_{\mathcal{Z}}, r_{\mathcal{A}}, r_{P_1}, \ldots, r_{P_n})$, where $(z, r_{\mathcal{Z}})$, $r_{\mathcal{A}}$ and r_{P_i} are respectively related to $\mathcal{Z}$, $\mathcal{A}$ and party i. Analogously, we denote by $\mathsf{IDEAL}_{\mathcal{F},\mathcal{S},\mathcal{Z}}(\kappa, z, \bar{r})$ the output of the environment in the ideal interaction between the simulator $\mathcal{S}$ and the ideal functionality $\mathcal{F}$ under security parameter κ, input z and randomness $\bar{r} = (r_{\mathcal{Z}}, r_{\mathcal{S}}, r_{\mathcal{F}})$, where $(z, r_{\mathcal{Z}})$, $r_{\mathcal{S}}$ and $r_{\mathcal{F}}$ are respectively related to $\mathcal{Z}$, $\mathcal{S}$ and $\mathcal{F}$. The real world execution and the ideal executions are respectively represented by the ensembles $\mathsf{REAL}_{\pi,\mathcal{A},\mathcal{Z}} = \{\mathsf{REAL}_{\pi,\mathcal{A},\mathcal{Z}}(\kappa, z, \bar{r})\}_{\kappa \in \mathbb{N}}$ and $\mathsf{IDEAL}_{\mathcal{F},\mathcal{S},\mathcal{Z}} = \{\mathsf{IDEAL}_{\mathcal{F},\mathcal{S},\mathcal{Z}}(\kappa, z, \bar{r})\}_{\kappa \in \mathbb{N}}$ with $z \in \{0,1\}^*$ and a uniformly chosen $\bar{r}$.

In addition to these two models of computation, the UC framework also considers the $\mathcal{G}$-hybrid world, where the computation proceeds as in the real-world with the additional assumption that the parties have access to an auxiliary ideal functionality $\mathcal{G}$. In this model, honest parties do not communicate with the ideal functionality directly, but instead the adversary delivers all the messages to and from the ideal functionality. We consider the communication channels to be ideally authenticated, so that the adversary may read but not modify these messages. Unlike messages exchanged between parties, which can be read by the adversary, the messages exchanged between parties and the ideal functionality are divided into a *public header* and a *private header*. The public header can be read by the adversary and contains non-sensitive information (such as session identifiers, type of message, sender and receiver). On the other hand, the private header cannot be read by the adversary and contains information such as the parties' private inputs. We denote the ensemble of environment outputs that represents the execution of a protocol π in a $\mathcal{G}$-hybrid model as $\mathsf{HYBRID}^{\mathcal{G}}_{\pi,\mathcal{A},\mathcal{Z}}$ (defined analogously to $\mathsf{REAL}_{\pi,\mathcal{A},\mathcal{Z}}$). UC security is then formally defined as:

Definition 3. *A n-party ($n \in \mathbb{N}$) protocol π is said to UC-realize an ideal functionality $\mathcal{F}$ in the $\mathcal{G}$-hybrid model if, for every adversary $\mathcal{A}$, there exists a simulator $\mathcal{S}$ such that, for every environment $\mathcal{Z}$, the following relation holds:*

$$\mathsf{IDEAL}_{\mathcal{F},\mathcal{S},\mathcal{Z}} \approx \mathsf{HYBRID}^{\mathcal{G}}_{\pi,\mathcal{A},\mathcal{Z}}$$

We say that the protocol is *statistically secure* if the same holds for all $\mathcal{Z}$ with unbounded computing power.

Functionality $\mathcal{F}_{\text{OT}}$

$\mathcal{F}_{\text{OT}}$ interacts with a sender P_s, a receiver P_r and an adversary $\mathcal{S}$, and it proceeds as follows:

- Upon receiving a message (sender, $sid, ssid, \boldsymbol{x}_0, \boldsymbol{x}_1$) from P_s, where each $\boldsymbol{x}_i \in \{0,1\}^\lambda$, store the tuple $(ssid, \boldsymbol{x}_0, \boldsymbol{x}_1)$ (The lengths of the strings λ is fixed and known to all parties). Ignore further messages from P_s to P_r with the same $ssid$.
- Upon receiving a message (receiver, $sid, ssid, c$) from P_r, where $c \in \{0,1\}$, check if a tuple $(ssid, \boldsymbol{x}_0, \boldsymbol{x}_1)$ was recorded. If yes, send (received, $sid, ssid, \boldsymbol{x}_c$) to P_r and (received, $sid, ssid$) to P_s and halt. If not, send nothing to P_r (but continue running).

Fig. 7. Functionality $\mathcal{F}_{\text{OT}}$

B   Implementing $\mathcal{F}_{\text{ROT}}$

For the sake of simplicity we construct our commitment protocol in the $\mathcal{F}_{\text{ROT}}$-hybrid model. Here we show that $\mathcal{F}_{\text{ROT}}$ can be realized in the $\mathcal{F}_{\text{OT}}$-hybrid model in a straightforward manner. Intuitively, we have P_s sample two random matrices $\mathbf{R}_0, \mathbf{R}_1$ and do an OT for each row, where it inputs a row from each matrix (*i.e.* $\mathbf{R}_0[i, \cdot], \mathbf{R}_1[i, \cdot]$) and P_r inputs a random choice bit. However, this naïve construction has communication complexity that depends on the size of the matrices, since it needs n OTs of m-bit strings.

Using both a pseudorandom number generator prg and access to $\mathcal{F}_{\text{OT}}$ (Fig. 7), it is possible to realize $\mathcal{F}_{\text{ROT}}$ in a way that its communication complexity only depends on the number of rows of the matrices and a computational security parameter (Fig. 8). This is a key fact in achieving rate 1 for our commitment scheme, since the number of rows required in our protocol is independent from the number of commitments to be executed, allowing the communication cost to be amortized over many commitments. On the other hand, (amortized) linear time can be obtained by employing a pseudorandom number generator that only requires a constant number of operations per output bit (*e.g.* [VZ12]). As shown in the protocol description, expensive OT operations are only used n (the number of rows) times while the number of calls to the prg is a fraction of the number of commitments (the number of columns). This allows us to obtain an arbitrary number of commitments from a fixed number of OTs and a small number of calls to prg.

Let prg : $\{0,1\}^\kappa \rightarrow \{0,1\}^\ell$ be a pseudorandom number generator that stretches a seed $s \xleftarrow{\$} \{0,1\}^\kappa$ into a pseudorandom string $r \in \{0,1\}^\ell$. Intuitively, for $i = 1, \ldots, n$, we call $\mathcal{F}_{\text{OT}}$ with P_s's input equal to $\boldsymbol{r}_{0,i}, \boldsymbol{r}_{1,i} \xleftarrow{\$} \{0,1\}^\kappa$ and P_r's input equal to $b_i \xleftarrow{\$} \{0,1\}$. After all the OTs are done, P_s sets $\mathbf{R}_0[i, \cdot] = \text{prg}(\boldsymbol{r}_{0,i})$ and $\mathbf{R}_1[i, \cdot] = \text{prg}(\boldsymbol{r}_{1,i})$, while P_r sets $\mathbf{S}[i, \cdot] = \text{prg}(\boldsymbol{r}_{b_i,i})$. The output matrices have n rows and ℓ columns. However, an arbitrary number of columns m can be obtained by saving the last κ bits of every output of prg, repeatedly running prg using these bits as seeds and concatenating the outputs (minus the last κ bits) until m bits are obtained.

Protocol Π_{ROT}

1. **OT Phase:** For $i = 1, \ldots, n$, P_s samples random $r_{0,i}, r_{1,i} \xleftarrow{\$} \{0,1\}^\kappa$ and sends $(\text{sender}, sid, ssid_i, r_{0,i}, r_{1,i})$ to $\mathcal{F}_{\mathrm{OT}}$, while P_r samples $b_i \xleftarrow{\$} \{0,1\}$ and sends $(\text{receiver}, sid, ssid_i, b_i)$ to $\mathcal{F}_{\mathrm{OT}}$.
2. **Seed Expansion Phase:** For $i = 1, \ldots, n$, P_s sets $\mathbf{R}_0[i, \cdot] = \mathsf{prg}(r_{0,i})$ and $\mathbf{R}_1[i, \cdot] = \mathsf{prg}(r_{1,i})$, while P_r sets $\mathbf{S}[i, \cdot] = \mathsf{prg}(r_{b_i,i})$. P_s outputs $\mathbf{R}_0, \mathbf{R}_1$ and P_r outputs $b_1, \ldots, b_n, \mathbf{S}$.

Fig. 8. Protocol Π_{ROT}

Lemma 10. *Π_{ROT} UC-realizes $\mathcal{F}_{\mathrm{ROT}}$ in the $\mathcal{F}_{\mathrm{OT}}$-hybrid model with computational security against a static adversary. Formally, there exists a simulator $\mathcal{S}$ such that for every static adversary $\mathcal{A}$ and any environment $\mathcal{Z}$:*

$$\mathsf{IDEAL}_{\mathcal{F}_{\mathrm{ROT}}, \mathcal{S}, \mathcal{Z}} \approx_c \mathsf{HYBRID}^{\mathcal{F}_{\mathrm{OT}}}_{\Pi_{\mathrm{ROT}}, \mathcal{A}, \mathcal{Z}}$$

Proof (Sketch). The simulator $\mathcal{S}$ acts as $\mathcal{F}_{\mathrm{OT}}$ when running Π_{ROT} with an internal copy of $\mathcal{A}$. In case P_s is corrupted, $\mathcal{S}$ extracts the inputs $(r_{0,1}, r_{1,1}), \ldots, (r_{0,n}, r_{1,n})$ given by $\mathcal{A}$ to $\mathcal{F}_{\mathrm{OT}}$, constructs $\mathbf{R}_0, \mathbf{R}_1$ according to the protocol and sends them to $\mathcal{F}_{\mathrm{ROT}}$. In case P_r is corrupted, $\mathcal{S}$ extracts the inputs $b_1, \ldots, b_n$, samples random matrices $\mathbf{R}_0, \mathbf{R}_1 \xleftarrow{\$} \{0,1\}^{n \times \ell}$, constructs $\mathbf{S}$ according to the protocol and sends $(b_1, \ldots, b_n), \mathbf{S}$ to $\mathcal{F}_{\mathrm{ROT}}$. Basically, the ideal and the real distributions are computationally indistinguishable due to prg's pseudorandomness, *i.e.* an environment $\mathcal{Z}$ that distinguishes between the ideal and real distributions could be used to distinguish a pseudorandom string output by prg from a uniformly random string of same size.

C   Committing to Arbitrary Messages

Protocol Π_{HCOM} described in Sect. 5 realizes $\mathcal{F}_{\mathrm{HCOM}}$ and thus only allows the sender to commit to random messages. However, this can be trivially used to commit to arbitrary messages while preserving all properties of our scheme, namely, additive homomorphism, linear computational complexity and rate 1. This is achieved by having the sender also give the receiver the difference between the random string that it is committed to through Π_{HCOM} and the arbitrary string that he wishes to commit to. First, P runs the commitment phase of Π_{HCOM} and becomes committed to a string $\boldsymbol{m}$, then it computes $\boldsymbol{c} = \boldsymbol{m}' - \boldsymbol{m}$ and sends $\boldsymbol{c}$ to V (where $\boldsymbol{m}'$ is the message that P wishes to commit to). The addition of two commitments can proceed the same way as in Π_{HCOM} with an extra step of setting $\boldsymbol{c}_3 = \boldsymbol{c}_1 + \boldsymbol{c}_2 = \boldsymbol{m}'_1 + \boldsymbol{m}'_2 - \boldsymbol{m}_1 - \boldsymbol{m}_2$. In the opening phase, P proceeds exactly like in Π_{HCOM} and V obtains the intended message by computing $\boldsymbol{m}' = \boldsymbol{c} + \boldsymbol{m}$. We call this protocol Π_{AHCOM} and described it in the $\mathcal{F}_{\mathrm{HCOM}}$-hybrid model in Fig. 9.

Protocol Π_{AHCOM}

Protocol Π_{AHCOM} is run by a sender P with input $\boldsymbol{m}' \in \{0,1\}^k$ and a receiver V interacting with $\mathcal{F}_{\text{HCOM}}$, and proceeds as follows:

1. **Commitment Phase:**
   (a) P sends (commit, $sid, ssid, P_s, P_r$) to $\mathcal{F}_{\text{HCOM}}$. Upon receiving (commit, $sid, ssid, P_s, P_r, \boldsymbol{m}$) as answer, P sets $\boldsymbol{c} = \boldsymbol{m}' - \boldsymbol{m}$, and sends $(\boldsymbol{c}, sid, ssid,)$ to V.

2. **Addition:**
   (a) P sends (add, $sid, ssid_1, ssid_2, ssid_3, P_s, P_r$) to $\mathcal{F}_{\text{HCOM}}$ and sets $\boldsymbol{c}_3 = \boldsymbol{c}_1 + \boldsymbol{c}_2 = \boldsymbol{m}'_1 + \boldsymbol{m}'_2 - \boldsymbol{m}_1 - \boldsymbol{m}_2$.
   (b) Upon receiving (add, $sid, ssid_1, ssid_2, ssid_3, P_s, P_r, \text{success}$) from $\mathcal{F}_{\text{HCOM}}$, V also sets $\boldsymbol{c}_3 = \boldsymbol{c}_1 + \boldsymbol{c}_2 = \boldsymbol{m}'_1 + \boldsymbol{m}'_2 - \boldsymbol{m}_1 - \boldsymbol{m}_2$.

3. **Opening Phase:**
   (a) P sends (reveal, $sid, ssid$) to $\mathcal{F}_{\text{HCOM}}$ and halts.
   (b) Upon receiving (reveal, $sid, ssid, P_s, P_r, \boldsymbol{m}$) from $\mathcal{F}_{\text{HCOM}}$, V computes $\boldsymbol{m}' = \boldsymbol{c} + \boldsymbol{m}$ and outputs $\boldsymbol{m}'$. Note that, even if $\boldsymbol{c}$ is an addition of two commitments $\boldsymbol{c}_1$ and $\boldsymbol{c}_2$, this procedure is still valid since $\boldsymbol{c}_3 = \boldsymbol{c}_1 + \boldsymbol{c}_2 = \boldsymbol{m}'_1 + \boldsymbol{m}'_2 - \boldsymbol{m}_1 - \boldsymbol{m}_2$.

Fig. 9. Protocol Π_{AHCOM}: using Π_{HCOM} to commit to arbitrary messages.

The security of Π_{AHCOM} can be trivially observed since all we do is sending the difference between a random string that the sender is already committed to and the arbitrary string he wishes to commit to. The random string acts as a one-time pad hiding all information and binding is guaranteed by $\mathcal{F}_{\text{HCOM}}$, hence we obtain statistical security in the $\mathcal{F}_{\text{HCOM}}$-hybrid model (which is realized by Π_{HCOM}). Notice that the extra communication does not reduce the rate of the resuting commitment scheme, since in Π_{HCOM}'s commitment phase only the $n - k$ bottom rows of $\mathbf{W}$ are sent[6] and here we send the remaining k bits that define $\boldsymbol{m}'$. Moreover, it is possible to embed the difference $\boldsymbol{c}$ in $\mathbf{W}$ so that no extra rounds are required.

References

[AHMR15] Afshar, A., Hu, Z., Mohassel, P., Rosulek, M.: How to efficiently evaluate RAM programs with malicious security. In: Oswald, E., Fischlin, M. (eds.) EUROCRYPT 2015. LNCS, vol. 9056, pp. 702–729. Springer, Heidelberg (2015)

[BCPV13] Blazy, O., Chevalier, C., Pointcheval, D., Vergnaud, D.: Analysis and improvement of Lindell's UC-secure commitment schemes. In: Jacobson Jr., M.J., Locasto, M., Mohassel, P., Safavi-Naini, R. (eds.) ACNS 2013. LNCS, vol. 7954, pp. 534–551. Springer, Heidelberg (2013)

---

[6] Apart from $\mathbf{T}_0, \mathbf{T}_1$, which only depend on the security parameter and are amortized over many commitments.

[Bra16] Brandão, L.T.A.N.: Very-efficient simulatable flipping of many coins into a well. In: Cheng, C.M., et al. (eds.) PKC 2016. LNCS, vol. 9615, pp. 297–326. Springer, Heidelberg (2016). doi:10.1007/978-3-662-49387-8_12

[Can01] Canetti, R.: Universally composable security: a new paradigm for cryptographic protocols. In: FOCS, pp. 136–145. IEEE Computer Society (2001)

[CDD+15] Cascudo, I., Damgård, I., David, B.M., Giacomelli, I., Nielsen, J.B., Trifiletti, R.: Additively homomorphic UC commitments with optimal amortized overhead. In: Katz, J. (ed.) PKC 2015. LNCS, vol. 9020, pp. 495–515. Springer, Heidelberg (2015)

[CF01] Canetti, R., Fischlin, M.: Universally composable commitments. In: Kilian, J. (ed.) CRYPTO 2001. LNCS, vol. 2139, p. 19. Springer, Heidelberg (2001)

[CLOS02] Canetti, R., Lindell, Y., Ostrovsky, R., Sahai, A.: Universally composable two-party and multi-party secure computation. In: STOC, pp. 494–503 (2002)

[CRVW02] Capalbo, M.R., Reingold, O., Vadhan, S.P., Wigderson, A.: Randomness conductors and constant-degree lossless expanders. In: Proceedings on 34th Annual ACM Symposium on Theory of Computing, 19–21 May 2002, Montréal, Québec, Canada, pp. 659–668 (2002)

[DDGN14] Damgård, I., David, B., Giacomelli, I., Nielsen, J.B.: Compact VSS and efficient homomorphic UC commitments. In: Sarkar, P., Iwata, T. (eds.) ASIACRYPT 2014, Part II. LNCS, vol. 8874, pp. 213–232. Springer, Heidelberg (2014)

[DI14] Druk, E., Ishai, Y.: Linear-time encodable codes meeting the gilbert-varshamov bound and their cryptographic applications. In: Naor, M. (ed.) Innovations in Theoretical Computer Science, ITCS 2014, Princeton, NJ, USA, 12–14 January 2014, pp. 169–182. ACM (2014)

[FJN+13] Frederiksen, T.K., Jakobsen, T.P., Nielsen, J.B., Nordholt, P.S., Orlandi, C.: MiniLEGO: efficient secure two-party computation from general assumptions. In: Johansson, T., Nguyen, P.Q. (eds.) EUROCRYPT 2013. LNCS, vol. 7881, pp. 537–556. Springer, Heidelberg (2013)

[FJNT16] Frederiksen, T.K., Jakobsen, T.P., Nielsen, J.B., Trifiletti, R.: On the complexity of additively homomorphic uc commitments. In: Kushilevitz, E., et al. (eds.) TCC 2016-A. LNCS, vol. 9562, pp. 542–565. Springer, Heidelberg (2016). doi:10.1007/978-3-662-49096-9_23

[GI02] Guruswami, V., Indyk, P.: Near-optimal linear-time codes for unique decoding and new list-decodable codes over smaller alphabets. In: Reif, J.H. (ed.) Proceedings on 34th Annual ACM Symposium on Theory of Computing, 19–21 May 2002, Montréal, Québec, Canada, pp. 812–821. ACM (2002)

[GI03] Guruswami, V., Indyk, P.: Linear time encodable and list decodable codes. In: Larmore and Goemans [LG03], pp. 126–135

[GI05] Guruswami, V., Indyk, P.: Linear-time encodable/decodable codes with near-optimal rate. IEEE Trans. Inf. Theor. **51**(10), 3393–3400 (2005)

[GIKW14] Garay, J.A., Ishai, Y., Kumaresan, R., Wee, H.: On the complexity of UC commitments. In: Nguyen, P.Q., Oswald, E. (eds.) EUROCRYPT 2014. LNCS, vol. 8441, pp. 677–694. Springer, Heidelberg (2014)

[IKOS08] Ishai, Y., Kushilevitz, E., Ostrovsky, R., Sahai, A.: Cryptography with constant computational overhead. In: Dwork, C. (ed.) STOC, pp. 433–442. ACM (2008)

[IPS08] Ishai, Y., Prabhakaran, M., Sahai, A.: Founding cryptography on oblivious transfer – efficiently. In: Wagner, D. (ed.) CRYPTO 2008. LNCS, vol. 5157, pp. 572–591. Springer, Heidelberg (2008)

[IPS09] Ishai, Y., Prabhakaran, M., Sahai, A.: Secure arithmetic computation with no honest majority. In: Reingold, O. (ed.) TCC 2009. LNCS, vol. 5444, pp. 294–314. Springer, Heidelberg (2009)

[LG03] Larmore, L.L., Goemans, M.X. (eds.) Proceedings of the 35th Annual ACM Symposium on Theory of Computing, 9–11 June 2003, San Diego, CA, USA. ACM (2003)

[Lin11] Lindell, Y.: Highly-efficient universally-composable commitments based on the DDH assumption. In: Paterson, K.G. (ed.) EUROCRYPT 2011. LNCS, vol. 6632, pp. 446–466. Springer, Heidelberg (2011)

[Nao91] Naor, M.: Bit commitment using pseudorandomness. J. Cryptol. 4(2), 151–158 (1991)

[PVW08] Peikert, C., Vaikuntanathan, V., Waters, B.: A framework for efficient and composable oblivious transfer. In: Wagner, D. (ed.) CRYPTO 2008. LNCS, vol. 5157, pp. 554–571. Springer, Heidelberg (2008)

[Spi96] Spielman, D.A.: Linear-time encodable and decodable error-correcting codes. IEEE Trans. Inf. Theor. 42(6), 1723–1731 (1996)

[VZ12] Vadhan, S., Zheng, C.J.: Characterizing pseudoentropy and simplifying pseudorandom generator constructions. In: Proceedings of the 44th Symposium on Theory of Computing, pp. 817–836. ACM (2012)

UC Commitments for Modular Protocol Design and Applications to Revocation and Attribute Tokens

Jan Camenisch[1]($^{(\boxtimes)}$), Maria Dubovitskaya[1]($^{(\boxtimes)}$), and Alfredo Rial[2]($^{(\boxtimes)}$)

[1] IBM Reseach - Zurich, Rüschlikon, Switzerland
{jca,mdu}@zurich.ibm.com
[2] University of Luxembourg, Luxembourg, Luxembourg
alfredo.rial@uni.lu

Abstract. Complex cryptographic protocols are often designed from simple cryptographic primitives, such as signature schemes, encryption schemes, verifiable random functions, and zero-knowledge proofs, by bridging between them with commitments to some of their inputs and outputs. Unfortunately, the known universally composable (UC) functionalities for commitments and the cryptographic primitives mentioned above do not allow such constructions of higher-level protocols as hybrid protocols. Therefore, protocol designers typically resort to primitives with property-based definitions, often resulting in complex monolithic security proofs that are prone to mistakes and hard to verify.

We address this gap by presenting a UC functionality for non-interactive commitments that enables modular constructions of complex protocols within the UC framework. We also show how the new functionality can be used to construct hybrid protocols that combine different UC functionalities and use commitments to ensure that the same inputs are provided to different functionalities. We further provide UC functionalities for attribute tokens and revocation that can be used as building blocks together with our UC commitments. As an example of building a complex system from these new UC building blocks, we provide a construction (a hybrid protocol) of anonymous attribute tokens with revocation. Unlike existing accumulator-based schemes, our scheme allows one to accumulate several revocation lists into a single commitment value and to hide the revocation status of a user from other users and verifiers.

Keywords: Universal composability · Commitments · Attribute tokens · Revocation · Vector commitments

1 Introduction

Complex cryptographic protocols are often designed from simple cryptographic primitives, such as signature schemes, encryption schemes, verifiable random

---

This work was supported by the European Commission through the Seventh Framework Programme, under grant agreements #321310 for the PERCY grant.
A Rial—Work done while at IBM Research – Zurich.

© International Association for Cryptologic Research 2016
M. Robshaw and J. Katz (Eds.): CRYPTO 2016, Part III, LNCS 9816, pp. 208–239, 2016.
DOI: 10.1007/978-3-662-53015-3_8

functions, zero-knowledge proofs, and commitment schemes. Proving the security of such cryptographic protocols as well as verifying their security proofs are far from trivial and rather error-prone. Composability frameworks such as the Universal Composability (UC) framework [6] can help here. They guarantee that cryptographic primitives remain secure under arbitrary composition and thus enable a modular design and security analysis of cryptographic protocols constructed from such primitives. That is, they allow one to describe higher-level protocols as hybrid protocols that use the ideal functionalities of primitives rather than their realizations. Unfortunately, the known UC functionalities for cryptographic primitives allow only for very simple hybrid protocols, and thus protocols found in the literature foremost use basic ideal functionalities, such as the common reference string functionality $\mathcal{F}_{\mathrm{CRS}}$, registration functionality $\mathcal{F}_{\mathrm{REG}}$, and secure message transmission functionality $\mathcal{F}_{\mathrm{SMT}}$, and resort to constructions with property-based primitives, which typically results in complex monolithic security proofs that are prone to mistakes and hard to verify.

Consider for instance a two-party protocol where one party needs to compute a complex function $\mathcal{F}$ (that might include commitments, signatures, encryption, and zero-knowledge proofs) on the input and send the output to the second party. The original input of the first party might be hidden from the second party. The most common approach for building such a protocol is to describe the ideal functionality for that function $\mathcal{F}$, provide a monolithic realization, and prove that the latter securely implements the former. Following this approach, however, will result into a complex security proof.

A better approach would be a modular construction that breaks down the complex function into smaller building blocks each realized by a separate func-tionality. This will result in much simpler and structured protocols and proofs. However, this construction approach requires a mechanism to ensure that the input values to different subfunctionalities are the same. The most natural way to implement such a mechanism is to use cryptographic commitment functionality ($\mathcal{F}_{\mathrm{COM}}$) and build a realization in the $\mathcal{F}_{\mathrm{COM}}$-hybrid model.

In a nutshell, the hybrid protocol would work as follows. The ideal functional-ities of the building blocks are modified in such a way that they also accept com-mitments to the input values as input. When a party needs to guarantee that the inputs to two or more functionalities are equal, the party first sends that input value to the commitment functionality to obtain a commitment and the corre-sponding opening, and then sends the input to those functionalities along with the commitment and the opening. When the second party receives a commitment it can perform the verification without learning the original input value.

As a concrete example, consider a privacy-preserving attribute-based creden-tial system [3] that uses a commitment to a revocation handle to bridge a proof of knowledge of a signature that signs the revocation handle with a proof that the committed revocation handle is not revoked. The commitment guarantees that the same revocation handle is used in both proofs even if they are computed separately by different building blocks. This allows the composition of a protocol for proving possession of a signature with a protocol for proving non-revocation using commitments. The construction, definitions, and security proofs of such systems are all property-based and indeed rather complex [3]. Simplifying such a construction and its security proofs by using the UC model seems very attractive. However, that requires an ideal functionality for commitments that mirrors the way property-based primitives are combined with commitments. Unfortunately, none of the existing UC functionalities for commitments [6,8,9,11,12,15–17,19] fit this bill because they do not output cryptographic values or implement any other mechanism to ensure that a committed message is equal to the input of other functionalities. With the existing functionalities for commitments, the committer sends the committed message to the functionality, which informs the verifying party that a commitment has been sent. When the committer wishes to open the commitment, the functionality sends the message to the verifying party. Because no cryptographic value is ever output by the known functionalities, they cannot be used in our revocation example to guarantee the equality of the revocation handle or in any other similar case where one has to ensure that the message sent as input to the functionality for commitments equals the message sent as input to other functionalities. However, as we shall see, outputting just a cryptographic value for a commitment will not be sufficient.

1.1   UC Non-interactive Commitments for Hybrid Protocols

We provide a new ideal functionality $\mathcal{F}_{\text{NIC}}$ for commitments. The main differences between $\mathcal{F}_{\text{NIC}}$ and the existing commitment functionalities are that ours outputs cryptographic values and is non-interactive. In this respect it is similar to the signature functionality $\mathcal{F}_{\text{SIG}}$ [8].

Our functionality behaves as follows. When a party wishes to commit to a message, $\mathcal{F}_{\text{NIC}}$ computes a cryptographic commitment and an opening for that commitment (using algorithms provided by the simulator/environment upon initialization) and sends them as output to the calling party. When a party wishes to verify a commitment, it sends the commitment, the message and the opening to the functionality, which verifies the commitment and sends the verification result to the party. Therefore, our functionality does not involve interaction between a committer and a verifier. Furthermore, when a party requests a commitment to a message, the identity of the verifier is not sent to the functionality. Analogously, when a party verifies a commitment, the identity of the committer is not sent to the functionality.

$\mathcal{F}_{\text{NIC}}$ ensures that commitments are hiding and binding. We show that $\mathcal{F}_{\text{NIC}}$ can be realized by a standard commitment scheme that is binding and has a trapdoor (which implies it is hiding), such as the Pedersen commitment scheme [22]. All extra properties, such as non-malleability, simulation-sound trapdoor [18], etc., that are required to construct the standard UC functionalities are not necessary. We prove that the construction realizes $\mathcal{F}_{\text{NIC}}$ in the $\mathcal{F}_{\text{CRS}}$-hybrid model, which is also required for UC commitments in general [8].

There are protocols, however, that require extractable commitments. This is similar to requiring extractability in zero-knowledge proofs (ZKP). For some protocols, extractability is needed and thus a functionality for ZKP of knowledge must be used, whereas for other protocols sound ZKPs are sufficient and it is possible (and more efficient) to use a functionality for zero-knowledge that does not require extractability.

Therefore, we also propose an ideal functionality $\mathcal{F}_{\text{ENIC}}$ for extractable commitments and give a construction that realizes $\mathcal{F}_{\text{ENIC}}$. We compare both functionalities in Sect. 3.1 and explain why $\mathcal{F}_{\text{NIC}}$ suffices for some cases.

1.2 Modular Protocol Design in $\mathcal{F}_{\text{NIC}}$-Hybrid Model

Our ideal functionality for commitments can be used to construct higher-level protocols in a hybrid model because it allows one to bridge different ideal functionalities. To this end, the ideal functionalities of the building blocks can be modified so that their input values are accompanied by commitments and corresponding openings. These commitment and opening values are generated by the party providing the input using $\mathcal{F}_{\text{NIC}}$. Then, to convince a second party that the same inputs were provided to different functionalities, the first party sends the commitments to the second party, who will then also input the commitments to the different functionalities. For this to work, the building-block functionalities need to validate the commitments received and check whether the openings provided are correct. As functionalities cannot interact with each other, a verification algorithm COM.Verify needs to be provided as part of the commitment for local verification. The main challenge now is to ensure that a local verification implies a global binding property enforced by $\mathcal{F}_{\text{NIC}}$. We show how this challenge can be overcome.

We remark that our technique for modular protocol design based on $\mathcal{F}_{\text{NIC}}$ is very general. Any functionality that needs to be used in a protocol can be amended to receive committed inputs and to check those inputs by running COM.Verify. Therefore, our technique allows one to modularly describe a wide variety of hybrid protocols in the UC model. Moreover, we believe a similar approach could also be applied to functionalities that output cryptographic values that need to be verified inside other functionalities.

1.3 Example: Flexible Revocation for Attribute-Based Credentials

As a real-life example of building a complex system from our UC building blocks, we provide a construction for anonymous attribute tokens with revocation. We first provide the respective ideal functionalities for revocation and attribute tokens (signatures with the proofs of knowledge of signature possession). Then, we construct a protocol that uses those functionalities together with $\mathcal{F}_{\text{NIC}}$ to compose a protocol for proving possession of a non-revoked credential (signature). In fact, unlike existing accumulator-based schemes, our new scheme allows one to accumulate several revocation lists into a single commitment value and to hide the revocation status of a user from other users and verifiers.

In the literature, different privacy-preserving revocation mechanisms have been proposed for attribute-based credentials, such as signature lists [20], accumulators [1,5,21], and validity refreshing [2]. We provide a detailed overview of the related work on revocation in the full version of this paper. In some cases, credentials need to be revoked globally, e.g., when the related secret keys have been exposed, the attribute values have changed, or the user loses her right to use a credential. Often, credentials may be revoked only for specific contexts, i.e., when a user is not allowed to use her credential with a particular verifier, but can still use it elsewhere.

In such scenarios, the revocation authority needs to maintain multiple revocation lists. Because of their binary value limitation, the existing revocation systems require a separate application of a revocation mechanism for each list. This imposes an extra storage and computational overhead, not only to the users, but also to the revocation authority. Furthermore, in signature lists and accumulators, the revocation lists are disclosed to the other users and verifiers.

We propose a mechanism that allows one to commit several revocation lists into a single commitment value. Each user needs only one witness for all the revocation lists. Using this witness, a user can prove in a privacy-preserving manner the revocation status of her revocation handle in a particular revocation list.

We provide two ideal functionalities $\mathcal{F}_{REV}$ for revocation and propose two different constructions built from the vector commitments [10]. The first one hides the revocation status of a user from other users and from the verifiers, whereas in the second one, as for accumulators, revocation lists are public. Additionally, our schemes are flexible in the sense that revocation lists can be added (up to a maximum number) and removed without any cost, i.e., the cost is the same as for a revocation status update that does not change the number of lists, whereas accumulators would require one to set up a new accumulator and to issue witnesses to users, or delete them.

We note that aside from extending the standard revocation scenario with a central revocation authority and multiple revocation lists, our revocation schemes can be used to build an efficient dynamic attribute-based access control system in a very elegant way. Instead of issuing a list of credentials to each user, each certifying a certain attribute or role, in our revocation scheme a user can be issued just one base credential, which can be made valid or revoked for any context. The resulting solution saves the users, verifiers and the revocation authority a lot of storage and computational effort. That is, instead of having multiple credentials and corresponding revocation witnesses, a single credential and a single witness suffice to achieve the same goal.

1.4   Paper Organization

The remainder of this paper is organized as follows. In Sect. 2, we introduce the notation and conventions used to describe functionalities and their realizations in the UC model. In Sect. 3, we provide the ideal functionalities for non-interactive commitments and extractable commitments, and show the corresponding constructions that securely realize those functionalities. We also describe the generic approach of how to build modular constructions in the $\mathcal{F}_{NIC}$-hybrid model and to prove them secure. In Sect. 5, we describe an ideal functionality for attribute tokens with revocation, $\mathcal{F}_{TR}$, and provide a hybrid construction, Π_{TR}, that uses $\mathcal{F}_{NIC}$, $\mathcal{F}_{REV}$ and $\mathcal{F}_{AT}$ to realize $\mathcal{F}_{TR}$. We prove that the construction Π_{TR} realizes $\mathcal{F}_{TR}$ in that section.

2   Universally Composable Security

The universal composability framework [6] is a framework for defining and analyzing the security of cryptographic protocols so that security is retained under arbitrary composition with other protocols. The security of a protocol is defined by means of an ideal protocol that carries out the desired task. In the ideal protocol, all parties send their inputs to an ideal functionality $\mathcal{F}$ for the task. The ideal functionality locally computes the outputs of the parties and provides each party with its prescribed output.

The security of a protocol φ is analyzed by comparing the view of an environment $\mathcal{Z}$ in a real execution of φ against that of $\mathcal{Z}$ in the ideal protocol defined in $\mathcal{F}_\varphi$. The environment $\mathcal{Z}$ chooses the inputs of the parties and collects their outputs. In the real world, $\mathcal{Z}$ can communicate freely with an adversary $\mathcal{A}$ who controls both the network and any corrupt parties. In the ideal world, $\mathcal{Z}$ interacts with dummy parties, who simply relay inputs and outputs between $\mathcal{Z}$ and $\mathcal{F}_\varphi$, and a simulator $\mathcal{S}$. We say that a protocol φ securely realizes $\mathcal{F}_\varphi$ if $\mathcal{Z}$ cannot distinguish the real world from the ideal world, i.e., $\mathcal{Z}$ cannot distinguish whether it is interacting with $\mathcal{A}$ and parties running protocol φ or with $\mathcal{S}$ and dummy parties relaying to $\mathcal{F}_\varphi$.

2.1   Notation

Let $k \in \mathbb{N}$ denote the security parameter and $a \in \{0,1\}^*$ denote an input. Two binary distribution ensembles $X = \{X(k,a)\}_{k\in\mathbb{N},a\in\{0,1\}^*}$ and $Y = \{Y(k,a)\}_{k\in\mathbb{N},a\in\{0,1\}^*}$ are indistinguishable ($X \approx Y$) if for any $c,d \in \mathbb{N}$ there exists $k_0 \in \mathbb{N}$ such that for all $k > k_0$ and all $a \in \cup_{\kappa \leq k^d}\{0,1\}^\kappa$, $|\Pr[X(k,a) = 1] - \Pr[Y(k,a) = 1]| < k^{-c}$. Let $\mathrm{REAL}_{\varphi,\mathcal{A},\mathcal{Z}}(k,a)$ denote the distribution given by the output of $\mathcal{Z}$ when executed on input a with $\mathcal{A}$ and parties running φ, and let $\mathrm{IDEAL}_{\mathcal{F}_\varphi,\mathcal{S},\mathcal{Z}}(k,a)$ denote the output distribution of $\mathcal{Z}$ when executed on input a with $\mathcal{S}$ and dummy parties relaying to $\mathcal{F}_\varphi$. We say that protocol φ securely realizes $\mathcal{F}_\varphi$ if, for all polynomial-time $\mathcal{A}$, there exists a polynomial-time $\mathcal{S}$ such that, for all polynomial-time $\mathcal{Z}$, $\mathrm{REAL}_{\varphi,\mathcal{A},\mathcal{Z}} \approx \mathrm{IDEAL}_{\mathcal{F}_\varphi,\mathcal{S},\mathcal{Z}}$.

A protocol $\varphi^{\mathcal{G}}$ securely realizes $\mathcal{F}$ in the $\mathcal{G}$-hybrid model when φ is allowed to invoke the ideal functionality $\mathcal{G}$. Therefore, for any protocol ψ that securely realizes functionality $\mathcal{G}$, the composed protocol φ^ψ, which is obtained by replacing each invocation of an instance of $\mathcal{G}$ with an invocation of an instance of ψ, securely realizes $\mathcal{F}$.

2.2   Conventions

When describing ideal functionalities, we use the following conventions:

Interface Naming Convention. An ideal functionality can be invoked by using one or more interfaces. The name of a message in an interface consists of three fields separated by dots, e.g., com.setup.ini in the commitment functionality described in Sect. 3.1. The first field indicates the name of the functionality and is the same in all interfaces of the functionality. This field is useful for distinguishing between invocations of different functionalities in a hybrid protocol that uses two or more different functionalities. The second field indicates the kind of action performed by the functionality and is the same in all messages that the functionality exchanges within the same interface. The third field distinguishes between the messages that belong to the same interface, and can take six different values. A message $*.*$.ini is the incoming message received by the functionality, i.e., the message through which the interface is invoked. A message $*.*$.end is the outgoing message sent by the functionality, i.e., the message that ends the execution of the interface. The message $*.*$.sim is used by the functionality to send a message to the simulator, and the message $*.*$.rep is used to receive a message from the simulator. The message $*.*$.req is used by the functionality to send a message to the simulator to request the description of algorithms from the simulator, and the message $*.*$.alg is used by the simulator to send the description of those algorithms to the functionality.

Subsession Identifiers. Some interfaces in a functionality can be invoked more than once. When the functionality sends a message $*.*$.sim to the simulator in such an interface, a subsession identifier *ssid* is included in the message. The subsession identifier must also be included in the response $*.*$.rep sent by the simulator. The subsession identifier is used to identify the message $*.*$.sim to which the simulator replies with a message $*.*$.rep. We note that, typically, the simulator in the security proof may not be able to provide an immediate answer to the functionality after receiving a message $*.*$.sim. The reason is that the simulator typically needs to interact with the copy of the real adversary it runs in order to produce the message $*.*$.rep, but the real adversary may not provide the desired answer or may provide a delayed answer. In such cases, when the functionality sends more than one message $*.*$.sim to the simulator, the simulator may provide delayed replies, and the order of those replies may not follow the order of the messages received.

Aborts. When we say that an ideal functionality $\mathcal{F}$ aborts after being activated with a message $(*, \ldots)$, we mean that $\mathcal{F}$ stops the execution of the instruction and sends a special abortion message $(*, \perp)$ to the party that invoked the functionality.

Network vs. Local Communication. The identity of an interactive Turing machine (ITM) instance (ITI) consists of a party identifier *pid* and a session identifier *sid*. A set of parties in an execution of a system of ITMs is a protocol instance if they have the same session identifier *sid*. ITIs can pass direct inputs to and outputs from "local" ITIs that have the same *pid*. An ideal functionality $\mathcal{F}$ has $pid = \perp$ and is considered local to all parties. An instance of $\mathcal{F}$ with the session identifier *sid* only accepts inputs from and passes outputs to machines with the same session identifier *sid*. Some functionalities require the session identifier to have some structure. Those functionalities check whether the session identifier possesses the required structure in the first message that invokes the functionality. For the subsequent messages, the functionality implicitly checks that the session identifier equals the session identifier used in the first message. Communication between ITIs with different party identifiers must take place over the network. The network is controlled by the adversary, meaning that he can arbitrarily delay, modify, drop, or insert messages.

Delayed Outputs. We say that an ideal functionality $\mathcal{F}$ *sends a public delayed output* v to a party $\mathcal{P}$ if it engages in the following interaction. $\mathcal{F}$ sends to simulator $\mathcal{S}$ a note that it is ready to generate an output to $\mathcal{P}$. The note includes value v, identity $\mathcal{P}$, and a unique identifier for this output. When $\mathcal{S}$ replies to the note by echoing the unique identifier, $\mathcal{F}$ outputs the value v to $\mathcal{P}$. A *private delayed output* is similar, but value v is not included in the note.

3   UC Non-interactive Commitments

In existing commitment functionalities [8], the committer sends the committed message to the functionality, which informs the verifying party that a commitment has been sent. When the committer wishes to open the commitment, the functionality sends the message to the verifying party.

In contrast, our commitment functionalities do not involve any interaction between committer and verifier. In our commitment functionality, any party is allowed to request a commitment, and, when doing so, the identity of the verifier is not specified. Analogously, any party can verify a commitment, and the identity of the committer is not specified during verification.

In Sect. 3.1, we describe two ideal functionalities for non-interactive commitments $\mathcal{F}_{\mathrm{NIC}}$ and $\mathcal{F}_{\mathrm{ENIC}}$. Our commitment functionalities are similar to the functionalities of public key encryption and signatures [7,8,14]. For example, the signature functionality receives a message from the signer, computes a signature, and sends that signature to the signer. A verifying party sends a message and a signature to the functionality, which verifies the signature and sends the verification result. One of the reasons that the signature functionality has a "signature string" as part of its interface is to support the modularity of modeling complex protocols such as sending an encrypted signature [7].

Analogously, our ideal functionalities (unlike existing UC ideal functionalities for commitments) can be used in a hybrid protocol that also uses other functionalities that receive commitments as inputs. In a nutshell, a party would obtain a tuple (*ccom, cm, copen*), which consists of a commitment, a message and an opening, from $\mathcal{F}_{\mathrm{NIC}}$ or $\mathcal{F}_{\mathrm{ENIC}}$, and send (*ccom, cm, copen*) as input to the other functionalities. The use of commitments as input to those functionalities is useful when it is necessary to ensure that the inputs to those functionalities are equal.

For instance, our construction of anonymous attribute tokens with revocation in Sect. 5.2 uses an anonymous attribute token functionality, $\mathcal{F}_{\mathrm{AT}}$, and a revocation functionality, $\mathcal{F}_{\mathrm{REV}}$, that receive commitments output by $\mathcal{F}_{\mathrm{NIC}}$ as input. The commitments allow us to prove that the revocation handle used as input to $\mathcal{F}_{\mathrm{REV}}$ equals the one used as input to $\mathcal{F}_{\mathrm{AT}}$.

$\mathcal{F}_{\mathrm{ENIC}}$ requires commitments to be extractable, whereas $\mathcal{F}_{\mathrm{NIC}}$ does not. $\mathcal{F}_{\mathrm{NIC}}$ suffices for our construction of anonymous attribute tokens with revocation described in Sect. 5.2. The reason is that, in that construction, commitments are always sent along with their openings or along with proofs of knowledge of their openings, which provides the extraction property. In Sect. 3.4, we show that $\mathcal{F}_{\mathrm{NIC}}$ can be realized by any trapdoor and binding commitment scheme. We describe a construction for $\mathcal{F}_{\mathrm{ENIC}}$ and prove its security in the full version of this paper.

3.1 Ideal Functionalities $\mathcal{F}_{\mathrm{NIC}}$ and $\mathcal{F}_{\mathrm{ENIC}}$ for Non-interactive Commitments

$\mathcal{F}_{\mathrm{NIC}}$ and $\mathcal{F}_{\mathrm{ENIC}}$ are parameterized with the system parameters sp. This allows the parameters of the commitment scheme to depend on parameters generated externally, which could also be used in other schemes. For example, if a commitment scheme is used together with a non-interactive zero-knowledge proof of knowledge scheme, sp could include parameters shared by both the parameters of the commitment scheme and the parameters of the proof of knowledge scheme.

Functionality $\mathcal{F}_{\mathrm{NIC}}$ and $\mathcal{F}_{\mathrm{ENIC}}$

$\mathcal{F}_{\mathrm{NIC}}$ and $\mathcal{F}_{\mathrm{ENIC}}$ are parameterized by system parameters sp. The following COM.TrapCom, COM.TrapOpen, COM.Extract, and COM.Verify are ppt algorithms.

1. On input (com.setup.ini, sid) from a party $\mathcal{P}_i$:
   (a) If $(sid, cparcom, \mathrm{COM.TrapCom}, \mathrm{COM.TrapOpen}, \mathrm{COM.Extract}, \mathrm{COM.Verify}, ctdcom)$ is already stored, include $\mathcal{P}_i$ in the set $\mathbb{P}$, and send a delayed output (com.setup.end, sid, OK) to $\mathcal{P}_i$.
   (b) Otherwise proceed to generate a random $ssid$, store $(ssid, \mathcal{P}_i)$ and send (com.setup.req, sid, $ssid$) to $\mathcal{S}$.
S. On input (com.setup.alg, sid, $ssid$, m) from $\mathcal{S}$:
   (a) Abort if no pair $(ssid, \mathcal{P}_i)$ for some $\mathcal{P}_i$ is stored.
   (b) Delete record $(ssid, \mathcal{P}_i)$.
   (c) If $(sid, cparcom, \mathrm{COM.TrapCom}, \mathrm{COM.TrapOpen}, \mathrm{COM.Extract}, \mathrm{COM.Verify}, ctdcom)$ is already stored, include $\mathcal{P}_i$ in the set $\mathbb{P}$ and send (com.setup.end, sid, OK) to $\mathcal{P}_i$.
   (d) Otherwise proceed as follows.
       i. Parse m as $(cparcom, \mathrm{COM.TrapCom}, \mathrm{COM.TrapOpen}, \mathrm{COM.Extract}, \mathrm{COM.Verify}, ctdcom)$.
       ii. Store $(sid, cparcom, \mathrm{COM.TrapCom}, \mathrm{COM.TrapOpen}, \mathrm{COM.Extract}, \mathrm{COM.Verify}, ctdcom)$ and initialize both an empty table Tbl_{com} and an empty set $\mathbb{P}$.
       iii. Include $\mathcal{P}_i$ in the set $\mathbb{P}$ and send (com.setup.end, sid, OK) to $\mathcal{P}_i$.
2. On input (com.validate.ini, sid, $ccom$) from a party $\mathcal{P}_i$:
   (a) Abort if $\mathcal{P}_i \notin \mathbb{P}$.
   (b) Parse $ccom$ as $(ccom', cparcom', \mathrm{COM.Verify}')$.
   (c) Set $v \leftarrow 1$ if $cparcom' = cparcom$ and $\mathrm{COM.Verify}' = \mathrm{COM.Verify}$. Otherwise, set $v \leftarrow 0$.
   (d) Send (com.validate.end, sid, v) to $\mathcal{P}_i$.

---

3. On input (com.commit.ini, sid, cm) from any honest party P_i:
   (a) Abort if $P_i \notin \mathbb{P}$ or if $cm \notin \mathcal{M}$, where $\mathcal{M}$ is defined in $cparcom$.
   (b) Compute $(ccom, cinfo) \leftarrow$ COM.TrapCom(sid, $cparcom$, $ctdcom$).
   (c) Abort if there is an entry $[ccom, cm', copen', 1]$ in Tbl_{com} such that $cm \neq cm'$ in Tbl_{com}.
   (d) Run $copen \leftarrow$ COM.TrapOpen(sid, cm, $cinfo$).
   (e) Abort if $1 \neq$ COM.Verify(sid, $cparcom$, $ccom$, cm, $copen$).
   (f) Append $[ccom, cm, copen, 1]$ to Tbl_{com}.
   (g) Set $ccom \leftarrow (ccom, cparcom, \text{COM.Verify})$.
   (h) Send (com.commit.end, sid, $ccom$, $copen$) to P_i.
4. On input (com.verify.ini, sid, $ccom$, cm, $copen$) from any honest party P_i:
   (a) Abort if $P_i \notin \mathbb{P}$ or if $cm \notin \mathcal{M}$ or if $copen \notin \mathcal{R}$, where $\mathcal{M}$ and $\mathcal{R}$ are defined in $cparcom$.
   (b) Parse $ccom$ as $(ccom', cparcom', \text{COM.Verify}')$. Abort if $cparcom' \neq cparcom$ or COM.Verify$' \neq$ COM.Verify.
   (c) If there is an entry $[ccom', cm, copen, u]$ in Tbl_{com}, set $v \leftarrow u$.
   (d) Else, proceed as follows:
       i. If there is an entry $[ccom', cm', copen', 1]$ in Tbl_{com} such that $cm \neq cm'$, set $v \leftarrow 0$.
       ii. Else, proceed as follows:
           A. $\boxed{\text{If } cm \neq \text{COM.Extract}(sid, ctdcom, ccom'), \text{ set } v \leftarrow 0. \text{ Else:}}$
           B. Set $v \leftarrow$ COM.Verify(sid, $cparcom$, $ccom'$, cm, $copen$).
           C. Append $[ccom', cm, copen, v]$ to Tbl_{com}.
   (e) Send (com.verify.end, sid, v) to P_i.

$\mathcal{F}_{\text{NIC}}$ and $\mathcal{F}_{\text{ENIC}}$ interact with parties P_i that create the parameters of the commitment scheme and compute and verify commitments. The interaction between $\mathcal{F}_{\text{NIC}}$ (or $\mathcal{F}_{\text{ENIC}}$) and P_i takes place through the interfaces com.setup.*, com.validate.*, com.commit.*, and com.verify.*.

1. Any party P_i can call the interface com.setup.* to initialize the functionality. Only the first call will affect the functionality.
2. Any party P_i uses the interface com.validate.* to verify that $ccom$ contains the correct commitment parameters and verification algorithm.
3. Any party P_i uses the interface com.commit.* to send a message cm and then obtain a commitment $ccom$ and an opening $copen$.
4. Any party P_i uses the interface com.verify.* to verify that $ccom$ is a commitment to the message cm with the opening $copen$.

$\mathcal{F}_{\text{NIC}}$ and $\mathcal{F}_{\text{ENIC}}$ use a table Tbl_{com}. Tbl_{com} consists of entries of the form $[ccom, cm, copen, u]$, where $ccom$ is a commitment, cm is a message, $copen$ is an opening, and u is a bit whose value is 1 if the tuple $(ccom, cm, copen)$ is valid and 0 otherwise.

In the figure below, we depict $\mathcal{F}_{\text{NIC}}$ and $\mathcal{F}_{\text{ENIC}}$ and use a box to indicate those computations that take place only in $\mathcal{F}_{\text{ENIC}}$.

We now discuss the four interfaces of the ideal functionalities $\mathcal{F}_{\text{NIC}}$ and $\mathcal{F}_{\text{ENIC}}$. We mention $\mathcal{F}_{\text{ENIC}}$ only in those computations that are exclusive to $\mathcal{F}_{\text{ENIC}}$.

1. The com.setup.ini message is sent by any party P_i. If the functionality has not yet been initialized, it will trigger a com.setup.req message to ask the simulator S to send algorithms COM.TrapCom, COM.TrapOpen, $\boxed{\text{COM.Extract,}}$ and COM.Verify, the commitment parameters and the trapdoor. Once the simulator has provided the algorithms for the first time, $\mathcal{F}_{\text{NIC}}$ stores the algorithms, the commitment parameters $cparcom$ and the trapdoor $ctdcom$, and then notifies P_i that initialization was successful. If the functionality has already been set up, P_i is just told that initialization was successful.

2. The com.validate.ini message is sent by an honest party P_i. $\mathcal{F}_{\text{NIC}}$ checks if P_i has already run the setup. This is needed because otherwise in the real-world protocol the party would have to retrieve the parameters to validate the commitment, and this retrieval cannot be simulated because $\mathcal{F}_{\text{NIC}}$ enforces that the validation of a commitment must be local. The computation and verification of commitments are also local. $\mathcal{F}_{\text{NIC}}$ parses the commitment, and checks if the parameters and the verification algorithm from the commitment match with those stored by the functionality.

3. The com.commit.ini message is sent by any honest party P_i on input a message cm. $\mathcal{F}_{\text{NIC}}$ aborts if P_i did not run the setup. $\mathcal{F}_{\text{NIC}}$ runs the algorithm COM.TrapCom on input $cparcom$ and $ctdcom$ to get a simulated commitment $ccom$ and state information $cinfo$. COM.TrapCom does not receive the message cm to compute $ccom$, and therefore a commitment scheme that realizes this functionality must fulfill the hiding property. $\mathcal{F}_{\text{NIC}}$ also aborts if the table Tbl_{com} already stores an entry $[ccom, cm', copen', 1]$ such that $cm \neq cm'$ because this would violate the binding property. $\mathcal{F}_{\text{NIC}}$ runs the algorithm COM.TrapOpen on input cm and $cinfo$ to get an opening $copen$ and checks the validity of $(ccom, cm, copen)$ by running COM.Verify. If COM.Verify outputs 1, $\mathcal{F}_{\text{NIC}}$ stores $[ccom, cm, copen, 1]$ in Tbl_{com}, appends $(cparcom, \text{COM.Verify})$ to $ccom$, and sends $(ccom, copen)$ to P_i.

4. The com.verify.ini message is sent by any honest party P_i on input a commitment $ccom$, a message cm and an opening $copen$. $\mathcal{F}_{\text{NIC}}$ aborts if P_i did not run the setup. If there is an entry $[ccom, cm, copen, u]$ already stored in Tbl_{com}, then the functionality returns the bit u. Therefore, a commitment scheme that realizes this functionality must be consistent. If there is an entry $[ccom, cm', copen', 1]$ such that $cm \neq cm'$, the functionality returns 0. Therefore, a scheme that realizes the functionality must fulfill the binding property. Else, in $\mathcal{F}_{\text{ENIC}}$, the functionality checks whether the output of COM.Extract equals the message sent for verification and rejects the commitment if that is not the case. Then, the functionality runs the algorithm COM.Verify to verify $(ccom, cm, copen)$. The functionality records the result in Tbl_{com} and returns that result.

The functionality $\mathcal{F}_{\text{NIC}}$ does not allow the computation and verification of commitments using any parameters $cparcom$ that were not generated by the functionality. As can be seen, the interfaces com.commit.$*$ and com.verify.$*$ use the commitment parameters that are stored by the functionality to compute and verify commitments. Therefore, a construction that realizes this functionality must ensure that the honest parties use the same commitment parameters. In general, such a "CRS-based" setup is required to realize UC commitments [8].

We note that we introduce the com.validate.∗ interface so that the parties can ensure that the commitment contains the right parameters and verification algorithm. This is needed especially for the parties that only receive a commitment value, without the opening. Otherwise, the com.verify.∗ interface can be called directly. Another way of doing this is to introduce an interface in the commitment functionality that returns the parameters and verification algorithm and require parties to call it first and compare the received parameters with the ones from the commitment.

3.2   Binding and Hiding Properties of $\mathcal{F}_{\mathrm{NIC}}$ and $\mathcal{F}_{\mathrm{ENIC}}$

Let us analyse the security properties of our two commitment functionalities. While inspection readily shows that both functionalities satisfy the standard binding and hiding properties, this merits some discussion.

We first note that both functionalities are perfectly hiding (because the commitment is computed independently of the message to be committed) and perfectly binding (the functionalities will accept only one value per commitment as committed value). Both properties being perfect seems like a contradiction, but it is not because the functionalities will only be *computationally* indistinguishable from their realizations. This implies of course that only computationally binding and hiding are enforced onto realizations.

Having said this, the binding property of $\mathcal{F}_{\mathrm{NIC}}$ merits further discussion, because, although it is guaranteed that adversarially computed commitments (outside $\mathcal{F}_{\mathrm{NIC}}$) can only be opened in one way, it is conceivable that an adversary could produce a commitment that it could open in two ways, and then, depending on its choice, provide one or the other opening, which would be allowed by $\mathcal{F}_{\mathrm{NIC}}$. This seems like a weaker property than what a traditional commitment scheme offers. There, after computing a commitment on input a message, that commitment can only be opened to that message. In this respect, we first remark that for traditional, perfectly hiding commitments, this might also be possible (unless one can extract more than one opening from an adversary, for instance, via rewinding). Second, we can show the following proposition, stating that for all realizations of $\mathcal{F}_{\mathrm{NIC}}$, no adversary is actually able to provide two different openings for adversarially generated commitments (the proof is provided in the full version of this paper).

Proposition 1. *For any construction Π_{NIC} that realizes $\mathcal{F}_{\mathrm{NIC}}$, there is no algorithm* COM.Verify *input by the simulator $\mathcal{S}_{\mathrm{NIC}}$ to $\mathcal{F}_{\mathrm{NIC}}$ such that, for any tuples $(ccom, cm, copen)$ and $(ccom, cm', copen')$ such that $cm \neq cm'$, $1 =$* COM.Verify$(sid, cparcom, ccom, cm, copen)$ *and* $1 =$ COM.Verify$(sid, cparcom, ccom, cm', copen')$.

Let us finally note that the behaviour of $\mathcal{F}_{\mathrm{ENIC}}$ is different here, i.e., if the extraction algorithm is deterministic, it is guaranteed that there exists only one value to which a commitment can be opened.

3.3 Using $\mathcal{F}_{\mathrm{NIC}}$ in Conjunction with Other Functionalities

We turn to our main goal, namely how $\mathcal{F}_{\mathrm{NIC}}$ can be used to ensure that the same value is used as input to different functionalities or that an output from one functionality is used as an input to another functionality. We show the first case in detail with a toy example and then discuss the second case.

Ensuring Consistent Inputs. Let us consider the case where a construction requires that one party provides the same value to two (or more) different functionalities. To achieve this, the two functionalities need to get as input that value and also a commitment to that value and the corresponding opening value. It is further necessary that (1) also the other parties input the same commitment to the functionalities (or, alternatively, get the commitment from the functionalities and then check that they get the same commitment from them); (2) it is verified that the commitment is valid w.r.t. $\mathcal{F}_{\mathrm{NIC}}$, and that (3) the functionalities are able to somehow verify whether the value provided is indeed contained in the commitment. For the last item, it would seem natural that $\mathcal{F}_{\mathrm{NIC}}$ would be queried, but the UC framework does not allow that, and therefore we need to use a different mechanism: the commitments themselves contain a verification algorithm such that if the algorithm accepts an opening, then it is implied that $\mathcal{F}_{\mathrm{NIC}}$ would also accept the value and the opening for that commitment.

To enable this, let us start with two observations. In Proposition 1, we showed that COM.Verify will only accept one opening per *adversarially* computed commitment. However, this is not sufficient, because COM.Verify could accept different openings for commitments computed by $\mathcal{F}_{\mathrm{NIC}}$ because in that case $\mathcal{F}_{\mathrm{NIC}}$ does not invoke COM.Verify when processing requests to com.verify.ini and it is indeed conceivable that COM.Verify could behave differently.

However, for any secure realization Π_{NIC}, calls to the algorithm com.verify.ini of Π_{NIC} are indistinguishable from calls on the com.verify.ini interface to $\mathcal{F}_{\mathrm{NIC}}\|\mathcal{S}_{\Pi_{\mathrm{NIC}}}$, and com.verify.ini must be a non-interactive algorithm. Therefore, if $\mathcal{S}_{\Pi_{\mathrm{NIC}}}$ (i.e., the simulator such that $\mathcal{F}_{\mathrm{NIC}}\|\mathcal{S}_{\Pi_{\mathrm{NIC}}}$ is indistinguishable from Π_{NIC}) provides the real-world algorithm com.verify.ini of Π_{NIC} as COM.Verify() algorithm to $\mathcal{F}_{\mathrm{NIC}}$, then calling COM.Verify() in another functionality to verify an opening and committed message w.r.t. a commitment will necessarily produce the same result as a call to the com.verify.ini interface to $\mathcal{F}_{\mathrm{NIC}}\|\mathcal{S}_{\Pi_{\mathrm{NIC}}}$. We will use the latter in an essential way when composing different functionalities, as we will illustrate with an example in the following.

We note that the assumption that $\mathcal{S}_{\Pi_{\mathrm{NIC}}}$ provides the algorithms to $\mathcal{F}_{\mathrm{NIC}}\|\mathcal{S}_{\Pi_{\mathrm{NIC}}}$ that are used in the real world is natural and not a serious restriction. After all, the purpose of defining a functionality using cryptographic algorithms is that the functionality specifies the behavior of the real algorithms, especially those that are used to verify cryptographic values. Assuming that the calls com.commit.ini and com.verify.ini to Π_{NIC} are local to the calling parties is also natural as this is how traditional commitment schemes are realized.

Furthermore, we note that $\mathcal{F}_{\mathrm{NIC}}$ restricts $\mathcal{S}_{\Pi_{\mathrm{NIC}}}$ to send the real-world verification algorithm as COM.Verify. The reason is that $\mathcal{F}_{\mathrm{NIC}}$ outputs COM.Verify inside *ccom* through the (com.commit.end, *sid*, *ccom*, *copen*) message. In the real world, any construction for $\mathcal{F}_{\mathrm{NIC}}$ outputs the real-world verification algorithm through the (com.commit.end, *sid*, *ccom*, *copen*) message. Therefore, because the outputs in the real-word and in the ideal-world must be indistinguishable, any simulator must input the real-world verification algorithm as COM.Verify to $\mathcal{F}_{\mathrm{NIC}}$. Otherwise the message com.commit.end in the ideal world can be distinguished from that in the real world by the environment.

Functionality $\mathcal{F}_i$

1. On input (f_i.in.ini, *sid*, a, *ccom*$_1$, *copen*) from a party P_1, check if *sid* = (P_1, P_2, sid') for some P_2 and *sid'*, and no record is stored. If so, record (a, *ccom*$_1$, *copen*) and send (f_i.in.end, *sid*) to P_1, otherwise (f_i.in.end, *sid*, $\perp$) to P_1.
2. On input (f_i.eval.ini, *sid*, *ccom*$_2$) from P_2, check if *sid* = (P_1, P_2, sid') for some P_1 and *sid'*, if a record (a, *ccom*$_1$, *copen*) is stored, and if *ccom*$_1$ = *ccom*$_2$ and COM.Verify(*sid*, *cparcom*, *ccom*$_1$, a, *copen*) = 1 holds. If so, send delayed (f_i.eval.end, *sid*, $f_i(a)$) to P_2. Otherwise send delayed (f_i.eval.end, *sid*, $\perp$) to P_2.

Functionality $\mathcal{F}_{(1,2)}$

1. On input (f_{12}.eval.ini, *sid*, a) from a party P_1, check if *sid* = (P_1, P_2, sid'). If so, send delayed (f_{12}.eval.end, *sid*, ($f_1(a), f_2(a)$)) to P_2 and otherwise send delayed (f_{12}.eval.end, *sid*, $\perp$) to P_1.

Construction $\Pi_{(1,2)}$

1. On input (f_{12}.eval.ini, *sid*, a), P_1 proceeds as follows.
   (a) i. P_1 checks if *sid* = (P_1, P_2, sid').
        ii. P_1 calls $\mathcal{F}_{\mathrm{NIC}}$ with (com.setup.ini, *sid*) and receives (com.setup.end, *sid*, OK).
        iii. P_1 calls $\mathcal{F}_{\mathrm{NIC}}$ with (com.commit.ini, *sid*, a) to receive (com.commit.end, *sid*, *ccom*, *copen*).
        iv. P_1 calls $\mathcal{F}_1$ with (f_1.in.ini, *sid*, a, *ccom*, *copen*) and receives (f_1.in.end, *sid*).
        v. P_1 calls $\mathcal{F}_2$ with (f_2.in.ini, *sid*, a, *ccom*, *copen*) and receives (f_2.in.end, *sid*).
        vi. P_1 sends (smt.send.ini, *sid*, *ccom*) to P_2 using $\mathcal{F}_{\mathrm{SMT}}$.
   (b) Upon receiving (smt.send.end, *sid*, *ccom*) from P_1 via $\mathcal{F}_{\mathrm{SMT}}$, P_2 proceeds as follows.
        i. P_2 checks if *sid* = (P_1, P_2, sid').
        ii. P_2 calls $\mathcal{F}_{\mathrm{NIC}}$ with (com.setup.ini, *sid*) and receives (com.setup.end, *sid*, OK).
        iii. P_2 calls $\mathcal{F}_{\mathrm{NIC}}$ with (com.validate.ini, *sid*, *ccom*).
        iv. P_2 calls $\mathcal{F}_1$ with (f_1.eval.ini, *sid*, *ccom*) and receives (f_1.eval.end, *sid*, $f_1(a)$).
        v. P_2 calls $\mathcal{F}_2$ with (f_2.eval.ini, *sid*, *ccom*) and receives (f_2.eval.end, *sid*, $f_2(a)$).
        vi. P_2 outputs (f_{12}.eval.end, *sid*, ($f_1(a), f_2(a)$)).
   If at any step a party receives a wrong message from a functionality or some check fails, it outputs (f_{12}.eval.end, *sid*, $\perp$).

We are now ready to show how our goal can be achieved using a toy example. To this end, let us define three two party functionalities $\mathcal{F}_1$, $\mathcal{F}_2$, and $\mathcal{F}_{(1,2)}$. The first two $\mathcal{F}_1$ and $\mathcal{F}_2$ compute the function $f_1(\cdot)$ and $f_2(\cdot)$, respectively, on P_1's input and send the result to P_2. Analogously, $\mathcal{F}_{(1,2)}$ computes $(f_1(\cdot), f_2(\cdot))$ on P_1's input and sends the result to P_2. Our goal is now to realize $\mathcal{F}_{(1,2)}$ by a hybrid protocol $\Pi_{(1,2)}$ using $\mathcal{F}_1$ and $\mathcal{F}_2$ to compute $f_1(\cdot)$ and $f_2(\cdot)$, respectively, and $\mathcal{F}_{\mathrm{NIC}}$ to ensure that the inputs to both $\mathcal{F}_1$ and $\mathcal{F}_2$ are the same. To achieve this, $\mathcal{F}_1$ and $\mathcal{F}_2$ will take as inputs also commitments and do some basic checks on them. These functionalities and construction $\Pi_{(1,2)}$ are as follows.

We next show that $\Pi_{(1,2)}$ realizes $\mathcal{F}_{(1,2)}$ and thereby give an example of a security proof that uses $\mathcal{F}_{\mathrm{NIC}}$ and does not need to reduce to property-based security definitions of a commitment scheme. Note that although formally we consider a $\mathcal{F}_{\mathrm{NIC}}\|\mathcal{S}_{\Pi_{\mathrm{NIC}}}$-hybrid protocol, our example protocol $\Pi_{(1,2)}$ uses $\mathcal{F}_{\mathrm{NIC}}$ in the same way as any other functionality, i.e., without having to consider the simulator $\mathcal{S}_{\Pi_{\mathrm{NIC}}}$ for some realization Π_{NIC} of $\mathcal{F}_{\mathrm{NIC}}$.

Theorem 1. *Assume that $\mathcal{F}_{\mathrm{NIC}}\|\mathcal{S}_{\Pi_{\mathrm{NIC}}}$ is indistinguishable from Π_{NIC} and that $\mathcal{S}_{\Pi_{\mathrm{NIC}}}$ provides Π_{NIC}'s verification algorithm as $\mathsf{COM.Verify}()$ to $\mathcal{F}_{\mathrm{NIC}}$. Then $\Pi_{(1,2)}$ realizes $\mathcal{F}_{(1,2)}$ in the $(\mathcal{F}_{\mathrm{SMT}}, \mathcal{F}_1, \mathcal{F}_2, \mathcal{F}_{\mathrm{NIC}}\|\mathcal{S}_{\Pi_{\mathrm{NIC}}})$-hybrid model. $\mathcal{F}_{\mathrm{SMT}}$ [6] is described in the full version.*

Proof. We provide a simulator $\mathcal{S}_{\Pi_{(1,2)}}$ and prove that $\mathcal{F}_{(1,2)}\|\mathcal{S}_{\Pi_{(1,2)}}$ is indistinguishable from $\Pi_{(1,2)}$ if there exists a Π_{NIC} that realizes $\mathcal{F}_{\mathrm{NIC}}$.

We consider four cases, depending on which party is corrupt. In case both P_1 and P_2 are corrupt, there is nothing to simulate. In case both parties are honest, the simulator will be asked by $\mathcal{F}_{(1,2)}$ to send $(\mathsf{f}_{12}.\mathsf{eval.end}, sid, (f_1(a), f_2(a)))$ to P_2 and then proceed as follows. First it initializes $\mathcal{F}_{\mathrm{NIC}}$. It then picks a random value a' and executes $\Pi_{(1,2)}$ as P_1 and P_2 using a' as the input of P_1 and running $\mathcal{F}_{\mathrm{SMT}}$, $\mathcal{F}_1$, $\mathcal{F}_2$, and $\mathcal{F}_{\mathrm{NIC}}\|\mathcal{S}_{\Pi_{\mathrm{NIC}}}$ as they are specified, with exception that when $\mathcal{F}_1$ and $\mathcal{F}_2$ would output $f_1(a')$ and $f_2(a')$, respectively, to P_2, the simulator instead make these two functionalities output $f_1(a)$ and $f_2(a)$, respectively (which are the values $\mathcal{S}_{\Pi_{(1,2)}}$ had obtained earlier from $\mathcal{F}_{(1,2)}$). If this protocol execution is successful, $\mathcal{S}_{\Pi_{(1,2)}}$ will let the delayed output $(\mathsf{f}_{12}.\mathsf{eval.end}, sid, (f_1(a), f_2(a)))$ to P_2 pass. Otherwise it will drop it, as it will have already sent $(\mathsf{f}_{12}.\mathsf{eval.end}, sid, \bot)$ to P_2 or P_1 according to the protocol specification. It is not hard to see that this simulation will cause the same distribution on the values sent to the adversary as the real protocol. The only difference is that the simulator uses a different input value for P_1 and the only other value that depends in a' is *copen* (by the specification of $\mathcal{F}_{\mathrm{NIC}}$). As the environment/adversary never sees any of these two values or any value that depends on it (which is seen by inspection of all simulated functionalities and because $f_1(a')$ and $f_2(a')$ are replaced by $f_1(a)$ and $f_2(a)$ in the outputs of $\mathcal{F}_1$ and $\mathcal{F}_2$), the argument follows.

As next case, assume that P_1 is honest and P_2 is corrupt. This case is similar to the one where both are honest. The simulator proceeds the same way only that it will not execute the steps of P_2 and it will allow the delivery of the message $(\mathsf{f}_{12}.\mathsf{eval.end}, sid, (f_1(a), f_2(a)))$ to P_2. The argument that the simulation is successful remains essentially the same. Here, the environment will additionally see *ccom* which, as said before, does not depend on a'.

As last case, assume that P_2 is honest and P_1 is corrupt. Thus, $\mathcal{S}_{\Pi_{(1,2)}}$ interacts with the adversarial P_1 and the environment/adversary, simulating $\Pi_{1,2}$ towards P_1 and the functionalities $\mathcal{F}_{\mathrm{SMT}}$, $\mathcal{F}_1$, $\mathcal{F}_2$, and $\mathcal{F}_{\mathrm{NIC}} \| \mathcal{S}_{\Pi_{\mathrm{NIC}}}$ towards both the environment and P_1, and finally P_1 towards $\mathcal{F}_{1,2}$. Simulation is straightforward: the simulator just runs everything as specified, learning P_1's input a from P_1's input to $\mathcal{F}_{\mathrm{NIC}} \| \mathcal{S}_{\Pi_{\mathrm{NIC}}}$, $\mathcal{F}_1$, and $\mathcal{F}_2$. If this simulation reaches Step 1(b)vi, $\mathcal{S}_{\Pi_{(1,2)}}$ will input that a to $\mathcal{F}_{(1,2)}$ as P_1, causing it to send a delayed output $(\mathsf{f}_{12}.\mathsf{eval}.\mathsf{end}, sid, (f_1(a), f_2(a)))$ to P_2 for $\mathcal{S}_{\Pi_{(1,2)}}$ to deliver, which it will do. This simulation will be correct, as long as P_1 cannot cause $\mathcal{F}_1$ and $\mathcal{F}_2$ to send a result for a different input value. However, this cannot happen because if both functionalities accept P_1's input, the committed value must be identical thanks to the properties of COM.Verify (cf. discussion above). $\qquad\square$

Comparison with a Construction that Used a Standard Commitment Scheme. One could of course also realize $\mathcal{F}_{(1,2)}$ with a construction that uses a standard commitment scheme, i.e., one defined by property-based security definitions, instead of $\mathcal{F}_{\mathrm{NIC}}$. The resulting construction and the security proof would be less modular, comparable to a construction that uses a standard signature scheme instead of $\mathcal{F}_{SIG}$. For the security proof, the overall strategy would be rather similar, the main difference being that one would have to do reductions to the properties of the commitment scheme, i.e., additional game hops. That is, one would have to show that the binding property does not hold if an adversarial P_1 manages to send different inputs to $\mathcal{F}_1$ and $\mathcal{F}_2$. Also, one would have to show that the hiding property does not hold if an adversarial P_2 is able to distinguish between the real protocols and the simulator that interacts with the functionality $\mathcal{F}_{(1,2)}$ and thus has to send P_2 a commitment to a different value.

Ensuring an Output is Used as an Input. Let us consider a two-party construction that requires that an output from one functionality be used as an input to another functionality. This can be achieved in different ways, the simplest way seems to be that the first party, upon obtaining its output from the first functionality, calls $\mathcal{F}_{\mathrm{NIC}}$ to obtain a commitment on that value and an opening and sends the commitment and the opening to the first functionality. The first functionality will then check whether the commitment indeed contains the output and, if so, will send the commitment to the second party who can then use that commitment as input to the second functionality. We leave the details of this to the reader.

3.4   Construction of UC Non-interactive Commitments

We now provide our construction for UC non-interactive commitments. It uses a commitment scheme (CSetup, Com, VfCom) that fulfils the binding and trapdoor properties [13].

Our construction works in the $\mathcal{F}_{\mathrm{CRS}}^{\mathrm{CSetup}}$-hybrid model, where parties use the ideal functionality $\mathcal{F}_{\mathrm{CRS}}^{\mathrm{CSetup}}$ that is parameterized by the algorithm CSetup, which takes as input the system parameters sp.

Construction Π_{NIC}^{sp}

Construction Π_{NIC} is parameterized by system parameters sp, and uses the ideal functionality $\mathcal{F}_{\text{CRS}}^{\text{CSetup},sp}$ and a commitment scheme (CSetup, Com, VfCom).

1. On input (com.setup.ini, sid), a party $\mathcal{P}$ executes the following program:
   (a) Send (crs.setup.ini, sid) to $\mathcal{F}_{\text{CRS}}^{\text{CSetup},sp}$ to receive (crs.setup.end, sid, par_c).
   (b) Store (par_c, VfCom) and output (com.setup.end, sid, OK).
2. On input (com.validate.ini, sid, $ccom$), a party $\mathcal{P}$ executes the following program:
   (a) If (par_c, VfCom) is not stored, abort.
   (b) Parse $ccom$ as $(ccom', par_c', \text{VfCom}')$.
   (c) Set $v \leftarrow 1$ if $par_c' = par_c$ and $\text{VfCom}' = \text{VfCom}$. Otherwise, set $v \leftarrow 0$.
   (d) Output (com.validate.end, sid, v).
3. On input (com.commit.ini, sid, cm), a party $\mathcal{P}$ executes the following program:
   (a) If (par_c, VfCom) is not stored, abort.
   (b) Abort if $cm \notin \mathcal{M}$, where $\mathcal{M}$ is defined in par_c.
   (c) Run $(com, open) \leftarrow \text{Com}(par_c, cm)$.
   (d) Output (com.commit.end, sid, $ccom \leftarrow (com, par_c, \text{VfCom})$, $open$).
4. On input (com.verify.ini, sid, $ccom$, cm, $copen$), $\mathcal{P}$ executes the following program:
   (a) If (par_c, VfCom) is not stored, abort.
   (b) Abort if $cm \notin \mathcal{M}$ or if $copen \notin \mathcal{R}$, where $\mathcal{M}$ and $\mathcal{R}$ are defined in par_c.
   (c) Parse $ccom$ as $(ccom', par_c', \text{VfCom}')$.
   (d) If $par_c' = par_c$ and $\text{VfCom}' = \text{VfCom}$ then run $v \leftarrow \text{VfCom}(par_c, ccom', cm, copen)$. Otherwise, set $v \leftarrow 0$.
   (e) Output (com.verify.end, sid, v).

Theorem 2. *The construction Π_{NIC} realizes $\mathcal{F}_{\text{NIC}}$ in the $\mathcal{F}_{\text{CRS}}^{\text{CSetup}}$-hybrid model if the underlying commitment scheme (CSetup, Com, VfCom) is binding and trapdoor.*

We provide the proof in the full version of this paper.

4   The Ideal Functionalities $\mathcal{F}_{\text{REV}}$ and $\mathcal{F}_{\text{AT}}$

We describe our ideal functionality for non-hiding and hiding revocation, $\mathcal{F}_{\text{REV}}$, in Sect. 4.1. Our constructions for non-hiding and hiding revocation and their security analysis can be found in the full version of this paper. The construction for non-hiding revocation uses a non-hiding vector commitment scheme, whereas the hiding construction employs a trapdoor vector commitment scheme. In the full version of this paper we also define the trapdoor property for vector commitments and propose a construction for non-hiding and trapdoor vector commitments.

We describe our ideal functionality for attribute tokens, $\mathcal{F}_{\text{AT}}$, in Sect. 4.2. We provide the construction and prove it secure in the full version of this paper.

4.1   Ideal Functionality for Revocation $\mathcal{F}_{\text{REV}}$

Here we describe our ideal functionality $\mathcal{F}_{\text{REV}}$ for revocation. $\mathcal{F}_{\text{REV}}$ interacts with a revocation authority $\mathcal{RA}$, users $\mathcal{U}$ and any verifying parties $\mathcal{P}$. The revocation authority $\mathcal{RA}$ associates a revocation status $\mathbf{x}[rh]$ with every revocation handle rh. A revocation status consists of m bits, such that each bit $\mathbf{x}[rh, j]$ denotes the revocation status of the revocation handle rh with respect to the revocation list $j \in [1, m]$. The time is divided into epochs ep, and the revocation authority $\mathcal{RA}$ can change the revocation status of every revocation handle at the beginning of each epoch.

A user $\mathcal{U}$ can obtain a proof pr that the revocation status of the revocation handle rh committed in a commitment $ccom$ is $\mathbf{x}[rh, j]$ for the list j at the epoch ep. Given pr, $ccom$, $\mathbf{x}[rh, j]$, j, and ep, any party $\mathcal{P}$ can verify the proof pr.

Functionality $\mathcal{F}_{\mathrm{REV}}$

REV.SimProve and REV.Extract are ppt algorithms. $\mathcal{F}_{\mathrm{REV}}$ is parameterized by system parameters sp, by a maximum number of revocation lists m, and a maximum number of revocation handles n.

1. On input (rev.setup.ini, sid) from $\mathcal{RA}$:
   (a) Abort if $sid \neq (\mathcal{RA}, sid')$ or if (sid) is already stored.
   (b) Store (sid).
   (c) Send (rev.setup.req, sid) to $\mathcal{S}$.

S. On input (rev.setup.alg, sid, par_r, td_r, REV.SimProve, REV.Extract) from $\mathcal{S}$:
   (a) Abort if (sid) is not stored or if $(sid, par_r, td_r,$ REV.SimProve, REV.Extract) is already stored.
   (b) Store $(sid, par_r, td_r,$ REV.SimProve, REV.Extract).
   (c) Initialize an empty table Tbl_{pr} and an empty set $\mathbb{P}$.
   (d) Send (rev.setup.end, sid, par_r) to $\mathcal{RA}$.

2. On input (rev.get.ini, sid) from a party $\mathcal{P}$:
   (a) If $(sid, par_r, td_r,$ REV.SimProve, REV.Extract) is stored, set $par_r' \leftarrow par_r$; else $par_r' \leftarrow \perp$.
   (b) Create a fresh $ssid$ and store $(ssid, \mathcal{P}, par_r')$.
   (c) Send (rev.get.sim, sid, $ssid$, par_r') to $\mathcal{S}$.

S. On input (rev.get.rep, sid, $ssid$) from $\mathcal{S}$:
   (a) Abort if $(ssid, \mathcal{P}, par_r')$ is not stored.
   (b) If $par_r' \neq \perp$, include $\mathcal{P}$ in the set $\mathbb{P}$.
   (c) Replace $(ssid, \mathcal{P}, par_r')$ with $(\mathcal{P}, par_r')$.
   (d) Send (rev.get.end, sid, par_r') to $\mathcal{P}$.

3. On input (rev.epoch.ini, sid, ep, $\langle \boxed{\text{H: }\mathcal{U}_i,} rh_i, \mathbf{x}[rh_i] \rangle_{i=1}^{n'}$) from $\mathcal{RA}$:
   (a) Abort if record $(sid, par_r, td_r,$ REV.SimProve, REV.Extract) is not stored, or if $(sid, ep, \langle \boxed{\text{H: }\mathcal{U}_i',} rh_i', \mathbf{x}[rh_i]' \rangle_{i=1}^{n'})$ is already stored, or if $n' > n$, or if, for $i = 1$ to n', $rh_i \notin [1, n]$ or $\mathbf{x}[rh_i] \notin [0, 2^m)$ $\boxed{\text{H: or } \mathcal{U}_i \text{ is not a valid user identifier}}$.
   (b) Store $(sid, ep, \langle \boxed{\text{H: }\mathcal{U}_i,} rh_i, \mathbf{x}[rh_i] \rangle_{i=1}^{n'})$.
   (c) Send (rev.epoch.sim, sid, ep $\boxed{\text{NH:, } \langle rh_i, \mathbf{x}[rh_i] \rangle_{i=1}^{n'}}$) to $\mathcal{S}$.

S. On input (rev.epoch.rep, sid, ep, $info$) from $\mathcal{S}$:
   (a) Abort if record $(sid, ep, \langle \boxed{\text{H: }\mathcal{U}_i,} rh_i, \mathbf{x}[rh_i] \rangle_{i=1}^{n'})$ is not stored or if $(sid, ep, info',$ $\mathsf{Tbl}_{ep})$ is already stored.

(b) Set $\mathsf{Tbl}_{ep} \leftarrow \big\langle\, \boxed{\mathsf{H}\!:\mathcal{U}_i,}\, rh_i, \mathbf{x}[rh_i]\big\rangle_{i=1}^{n'}$ and initialize a set $\mathbb{E}_{ep}$.

(c) $\boxed{\mathsf{H}\!:\text{Initialize a set } \mathbb{S}_{ep}.\ \text{If } \mathcal{RA} \text{ is corrupt, initialize an empty table } \mathsf{Tbl}_{ep}.}$

(d) Store $(sid, ep, info, \mathsf{Tbl}_{ep})$.

(e) Send $(\mathsf{rev.epoch.end}, sid, ep, info)$ to $\mathcal{RA}$.

4. On input $(\mathsf{rev.getepoch.ini}, sid, ep)$ from any party $\mathcal{P}$:

   (a) Abort if $\mathcal{P} \not\subseteq \mathbb{P}$.

   (b) If $(sid, ep, info, \mathsf{Tbl}_{ep})$ is stored, set $info' \leftarrow info$ $\boxed{\mathsf{NH}\!:\text{ and } \mathsf{Tbl}'_{ep} \leftarrow \mathsf{Tbl}_{ep}}$; else set $info' \leftarrow \bot$ $\boxed{\mathsf{NH}\!:\text{ and } \mathsf{Tbl}'_{ep} \leftarrow \bot}$.

   (c) Create a fresh $ssid$ and store $(ssid, \mathcal{P}, ep, info'\,\boxed{\mathsf{NH}\!:, \mathsf{Tbl}'_{ep}})$.

   (d) Send $(\mathsf{rev.getepoch.sim}, sid, ssid, ep, info'\,\boxed{\mathsf{NH}\!:, \mathsf{Tbl}'_{ep}})$ to $\mathcal{S}$.

S. On input $(\mathsf{rev.getepoch.rep}, sid, ssid)$ from $\mathcal{S}$:

   (a) Abort if record $(ssid, \mathcal{P}, ep, info'\,\boxed{\mathsf{NH}\!:, \mathsf{Tbl}'_{ep}})$ is not stored.

   (b) If $info' \neq \bot$ $\boxed{\mathsf{NH}\!:\text{ and } \mathsf{Tbl}'_{ep} \neq \bot}$, include $\mathcal{P}$ in $\mathbb{E}_{ep}$.

   (c) Delete record $(ssid, \mathcal{P}, ep, info'\,\boxed{\mathsf{NH}\!:, \mathsf{Tbl}'_{ep}})$.

   (d) Send $(\mathsf{rev.getepoch.end}, sid, info'\,\boxed{\mathsf{NH}\!:, \mathsf{Tbl}'_{ep}})$ to $\mathcal{P}$.

---

5. H: On input $(\mathsf{rev.getstatus.ini}, sid, rh_i, ep)$ from a user $\mathcal{U}_i$:

   (a) Abort if $\mathcal{U}_i \not\subseteq \mathbb{E}_{ep}$.

   (b) If $\mathcal{RA}$ is honest, abort if there is no entry $[\mathcal{U}_i', rh_i', \mathbf{x}'[rh_i]]$ in Tbl_{ep} such that $rh_i' = rh_i$ and $\mathcal{U}_i' = \mathcal{U}_i$.

   (c) Create a fresh $ssid$ and store $(ssid, \mathcal{U}_i, rh_i, ep)$.

   (d) If $\mathcal{RA}$ is corrupt, send $(\mathsf{rev.getstatus.sim}, sid, ssid, \mathcal{U}_i, rh_i, ep)$ to $\mathcal{S}$; else send $(\mathsf{rev.getstatus.sim}, sid, ssid, \mathcal{U}_i, ep)$ to $\mathcal{S}$.

S. H: On input $(\mathsf{rev.getstatus.rep}, sid, ssid, \mathbf{x}[rh_i])$, if $\mathcal{RA}$ is corrupt, or $(\mathsf{rev.getstatus.rep}, sid, ssid)$, if $\mathcal{RA}$ is honest, from $\mathcal{S}$:

   (a) Abort if record $(ssid, \mathcal{U}_i, rh_i, ep)$ is not stored.

   (b) Delete record $(ssid, \mathcal{U}_i, rh_i, ep)$.

   (c) If $\mathcal{RA}$ is corrupt, $\mathbf{x}[rh_i] \in [0, 2^m)$ and there is no entry $[\bot, rh_i', \mathbf{x}'[rh_i]]$ in Tbl_{ep} such that $rh_i' = rh_i$, store $[\bot, rh_i, \mathbf{x}[rh_i]]$ in Tbl_{ep}.

   (d) If $\mathcal{RA}$ is corrupt and $\mathbf{x}[rh_i] \neq \mathbf{x}'[rh_i]$, where $\mathbf{x}'[rh_i]$ is in the entry $[\bot, rh_i, \mathbf{x}'[rh_i]]$ in Tbl_{ep}, send $(\mathsf{rev.getstatus.end}, sid, \bot)$ to $\mathcal{U}_i$; else include $\mathcal{U}_i$ in the set $\mathbb{S}_{ep}$ and send $(\mathsf{rev.getstatus.end}, sid, \mathbf{x}'[rh_i])$ to $\mathcal{U}_i$.

---

6. On input $(\mathsf{rev.prove.ini}, sid, ccom, rh, copen, ep, j)$ from an honest user $\mathcal{U}_i$:

   (a) Parse $ccom$ as $(ccom', cparcom, \mathsf{COM.Verify})$.

   (b) Abort if $\mathcal{U}_i \not\subseteq \mathbb{E}_{ep}$, or if $1 \neq \mathsf{COM.Verify}(cparcom, ccom', rh, copen)$, or if there is no entry $[\boxed{\mathsf{H}\!:\mathcal{U}_i',}\, rh_i', \mathbf{x}[rh_i]]$ in Tbl_{ep} such that $rh_i' = rh_i$.

   (c) H: Abort if $\mathcal{U}_i \not\subseteq \mathbb{S}_{ep}$, or if $\mathcal{RA}$ is honest and there is no entry $[\mathcal{U}_i', rh_i', \mathbf{x}[rh_i]]$ in Tbl_{ep} such that $\mathcal{U}_i = \mathcal{U}_i'$ and $rh_i' = rh_i$.

(d) Run $pr \leftarrow$ REV.SimProve$(sid, par_r, cparcom, ccom', ep, info, j, \mathbf{x}[rh, j], td_r)$.

(e) Append $[\langle ccom, ep, j, \mathbf{x}[rh, j]\rangle, pr, 1]$ to Table Tbl_{pr}.

(f) Send (rev.prove.end, sid, pr) to $\mathcal{U}_i$.

7. On input (rev.verify.ini, $sid, ccom, ep, j, b, pr$) from an honest party $\mathcal{P}$:

(a) Abort if $\mathcal{P} \nsubseteq \mathbb{E}_{ep}$.

(b) Parse $ccom$ as $(ccom', cparcom, \mathsf{COM.Verify})$.

(c) If there is an entry $[\langle ccom, ep, j, b\rangle, pr, u]$ in Tbl_{pr}, set $v \leftarrow u$.

(d) Else, do the following:

   i. Extract $(rh, open, \mathbf{x}[rh]) \leftarrow$ REV.Extract$(sid, par_r, cparcom, ccom', ep, info, j,$ $b, td_r, pr)$.

   ii. If $(rh, copen, \mathbf{x}[rh]) = \perp$ or $1 \neq$ COM.Verify$(cparcom, ccom', copen, rh)$, set $v \leftarrow 0$.

   iii. Else, do the following:

       A. $\boxed{\text{H: If } \mathcal{RA} \text{ is corrupt and there is no entry } [\perp, rh', \mathbf{x}[rh]] \text{ in } \mathsf{Tbl}_{ep} \text{ such that } rh' = rh, \text{ store } [\perp, rh, \mathbf{x}[rh]].}$

       A. If $b = \mathbf{x}[rh, j]$, where $\mathbf{x}[rh, j]$ is in the entry $[\boxed{\text{H: } \mathcal{U}_i,} rh, \mathbf{x}[rh]] \in \mathsf{Tbl}_{ep}$ $\boxed{\text{H:, where, if } \mathcal{RA} \text{ is honest, } \mathcal{U}_i \text{ must be corrupt}}$, set $v \leftarrow 1$; else set $v \leftarrow 0$.

   iv. Append $[\langle ccom, ep, j, \mathbf{x}[rh, j]\rangle, pr, v]$ to Table Tbl_{pr}.

(e) Send (rev.verify.end, sid, v) to $\mathcal{P}$.

$\mathcal{F}_{\text{REV}}$ describes two ideal functionalities: a *hiding* revocation functionality where, if the revocation authority is honest, the revocation status of a revocation handle is only revealed to the user associated with that revocation handle, and a *non-hiding* revocation functionality where the revocation statuses of all revocation handles are public. We provide a unified description of both ideal functionalities. The box $\boxed{\text{H: } \dots}$ is used to describe something that occurs only in the hiding revocation functionality, whereas the box $\boxed{\text{NH: } \dots}$ is used in the same way for the non-hiding revocation functionality.

$\mathcal{F}_{\text{REV}}$ interacts with the revocation authority $\mathcal{RA}$, the users $\mathcal{U}$ and any verifying parties $\mathcal{P}$ through the interfaces rev.setup.*, rev.get.*, rev.epoch.*, rev.getepoch.*, rev.getstatus.*, rev.prove.*, and rev.verify.*.

1. The revocation authority $\mathcal{RA}$ uses the rev.setup.* interface to receive the revocation parameters par_r.

2. Any party $\mathcal{P}$ invokes the rev.get.* interface to receive par_r.

3. The revocation authority $\mathcal{RA}$ uses the rev.epoch.* interface to send a list $\langle \boxed{\text{H: } \mathcal{U}_i,} rh_i, \mathbf{x}[rh_i]\rangle_{i=1}^{n'}$ of revocation handles and revocation statuses for the epoch ep and receive the epoch information $info$ for the epoch ep.

4. Any party $\mathcal{P}$ uses the rev.getepoch.* interface to get the epoch information $info$ for the epoch ep. In the non-hiding functionality, $\mathcal{P}$ also obtains the full list of revocation handles and revocation statuses $\langle rh_i, \mathbf{x}[rh_i]\rangle_{i=1}^{n'}$ for the epoch ep.

5. In the hiding revocation functionality, a user $\mathcal{U}$ with a revocation handle rh uses the rev.getstatus.* interface to receive the revocation status $\mathbf{x}[rh]$ at a given epoch ep.

6. An honest user $\mathcal{U}$ uses the rev.prove.* interface to obtain a proof pr that the revocation status of the revocation handle rh committed in a commitment $ccom$ is $\mathbf{x}[rh, j]$ for the list j at the epoch ep.
7. Any honest party $\mathcal{P}$ uses the rev.verify.* interface to verify a proof pr on input $ccom$, $\mathbf{x}[rh, j]$, j and ep.

$\mathcal{F}_{\text{REV}}$ uses the following tables:

Tbl$_{ep}$. For the epoch ep, Tbl$_{ep}$ stores entries of the form $[\boxed{\text{H: }\mathcal{U},}\ rh, \mathbf{x}[rh]]$ that associate the revocation handle rh with the revocation status $\mathbf{x}[rh]$. In the hiding functionality, if $\mathcal{RA}$ is honest, a user $\mathcal{U}$ is also associated with rh.

Tbl$_{pr}$. Tbl$_{pr}$ stores entries of the form $[\langle ccom, ep, j, \mathbf{x}[rh, j]\rangle, pr, u]$, where $\langle ccom, ep, j, \mathbf{x}[rh, j]\rangle$ is part of the instance of a proof, pr is the proof, and u is a bit that indicates the validity of the proof.

$\mathcal{F}_{\text{REV}}$ also uses a set $\mathbb{P}$. $\mathbb{P}$ contains the identifiers of the parties that retrieved the revocation parameters par_r. Additionally, $\mathcal{F}_{\text{REV}}$ uses a set $\mathbb{E}_{ep}$ that, for an epoch ep, stores the identifiers of the parties that retrieved the epoch information $info$ and, in the non-hiding functionality, the revocation statuses in Tbl$_{ep}$. The hiding functionality $\mathcal{F}_{\text{REV}}$ also uses a set $\mathbb{S}_{ep}$, which for an epoch ep stores the identifiers of the parties that retrieved the revocation statuses $\mathbf{x}[rh]$ of their revocation handles.

We now discuss the seven interfaces of the ideal functionality $\mathcal{F}_{\text{REV}}$.

1. The rev.setup.ini message is sent by the revocation authority. $\mathcal{F}_{\text{REV}}$ aborts if the rev.setup.ini message has already been sent. Otherwise $\mathcal{F}_{\text{REV}}$ asks the simulator $\mathcal{S}$ to provide the parameters par_r, the trapdoor td_r, and the algorithms REV.SimProve and REV.Extract. When $\mathcal{S}$ provides them, $\mathcal{F}_{\text{REV}}$ aborts if they have already been sent. Otherwise, $\mathcal{F}_{\text{REV}}$ initializes an empty set $\mathbb{P}$ and a table Tbl$_{pr}$ to store proofs, and sends the revocation parameters to the revocation authority.
2. The rev.get.ini message is sent by any party $\mathcal{P}$ to get the parameters par_r.
3. The rev.epoch.ini message is sent by the revocation authority on input an epoch identifier ep and a list of revocation handles and revocation statuses $\langle \boxed{\text{H: }\mathcal{U}_i,}\ rh_i, \mathbf{x}[rh_i]\rangle_{i=1}^{n'}$. In the hiding revocation functionality, the list also includes user identifiers $\mathcal{U}_i$. $\mathcal{F}_{\text{REV}}$ asks $\mathcal{S}$ to provide the epoch information $info$ for the epoch ep. In the hiding functionality, $info$ is computed without knowledge of $\langle \boxed{\text{H: }\mathcal{U}_i,}\ rh_i, \mathbf{x}[rh_i]\rangle_{i=1}^{n'}$, whereas in the non-hiding functionality $\mathcal{S}$ receives $\langle rh_i, \mathbf{x}[rh_i]\rangle_{i=1}^{n'}$. The epoch information $info$ is later given as input to the algorithms REV.SimProve and REV.Extract. When $\mathcal{S}$ sends $info$, $\mathcal{F}_{\text{REV}}$ aborts if the list $\langle \boxed{\text{H: }\mathcal{U}_i,}\ rh_i, \mathbf{x}[rh_i]\rangle_{i=1}^{n'}$ for ep was not received before or if $info$ for the epoch ep has already been received. Otherwise $\mathcal{F}_{\text{REV}}$ creates a table Tbl$_{ep}$ to store the list for the epoch ep and stores $(sid, ep, info, \text{Tbl}_{ep})$. In the hiding functionality, if the revocation authority is corrupt, Tbl$_{ep}$ is left empty and therefore the information $\langle \boxed{\text{H: }\mathcal{U}_i,}\ rh_i, \mathbf{x}[rh_i]\rangle_{i=1}^{n'}$ is not required. The reason is that, if the hiding functionality requires this information when

$\mathcal{RA}$ is corrupt, a construction that realizes this functionality would need to allow the extraction of $\langle\, \boxed{\mathsf{H}: \mathcal{U}_i,}\, rh_i, \mathbf{x}[rh_i]\rangle_{i=1}^{n'}$ in the security proof. In the construction, this would imply the use of extractable vector commitments, which, for the sake of efficiency, we chose to avoid. Therefore, the hiding $\mathcal{F}_{\mathrm{REV}}$, if $\mathcal{RA}$ is corrupt, learns the revocation statuses when they are disclosed through the rev.getstatus.* interface or the rev.verify.* interface. Finally, $\mathcal{F}_{\mathrm{REV}}$ sends the epoch ep and the epoch information $info$ to the revocation authority.

4. The rev.getepoch.ini message is sent by any party $\mathcal{P}$ on input an epoch ep. After the simulator prompts the response with a message (rev.getepoch.rep, $sid, ssid$), the functionality sends the epoch information $info$ to $\mathcal{P}$. In the non-hiding case, the functionality also sends the revocation statuses of all revocation handles to $\mathcal{P}$.

5. In the hiding functionality, the rev.getstatus.ini message is sent by a user $\mathcal{U}_i$ on input a revocation handle rh_i and an epoch ep. $\mathcal{F}_{\mathrm{REV}}$ works differently, depending on whether the revocation authority is corrupt or not:

   (a) If $\mathcal{RA}$ is honest, $\mathcal{F}_{\mathrm{REV}}$ aborts if there is no entry in Tbl_{ep} for $\mathcal{U}_i$ and rh_i. After the simulator prompts the response with a message (rev.getstatus.rep, $sid, ssid$), $\mathcal{F}_{\mathrm{REV}}$ sends the revocation status of rh_i to $\mathcal{U}_i$.

   (b) If $\mathcal{RA}$ is corrupt, $\mathcal{F}_{\mathrm{REV}}$ asks the simulator to provide the revocation status of rh_i. Then, if it was not stored in Tbl_{ep}, $\mathcal{F}_{\mathrm{REV}}$ stores the revocation status of rh_i in Tbl_{ep} and sends it to $\mathcal{U}_i$. If it is already stored, $\mathcal{F}_{\mathrm{REV}}$ only sends the revocation status to the user if the status sent by the functionality equals the one stored. Therefore, even if $\mathcal{RA}$ is corrupt, $\mathcal{F}_{\mathrm{REV}}$ ensures that $\mathcal{RA}$ associates a unique revocation status with a revocation handle during a given epoch.

6. The rev.prove.ini message is sent by an honest user $\mathcal{U}_i$ on input a commitment $ccom$ to a revocation handle rh with the opening $copen$. $\mathcal{U}_i$ also inputs the epoch ep and the revocation list j. The commitment $ccom$ consists of a commitment value $ccom$, commitment parameters $cparcom$, and a commitment verification algorithm COM.Verify. $\mathcal{F}_{\mathrm{REV}}$ aborts if rh and $copen$ are not a valid opening of $ccom$. It also aborts if the revocation handle rh_i is not in Tbl_{ep}. (In the hiding functionality, it also aborts if the revocation authority is honest and rh_i is not associated with $\mathcal{U}_i$.) $\mathcal{F}_{\mathrm{REV}}$ runs REV.SimProve($sid, par_r, cparcom, ccom, ep, info_{ep}, j, \mathbf{x}[rh, j], td_r$) to compute a proof pr that $\mathbf{x}[rh, j]$ is the revocation status of revocation handle rh with respect to the revocation list j at epoch ep. We note that pr does not reveal any information on the revocation handle rh, the opening $copen$, or the revocation status $\mathbf{x}[rh]$ with respect to revocation lists other than j. $\mathcal{F}_{\mathrm{REV}}$ stores the proof as valid in Tbl_{pr}.

7. The rev.verify.ini message is sent by any honest party $\mathcal{P}$ on input a proof pr that b is the revocation status with respect to the revocation list j at epoch ep and to the revocation handle committed to in $ccom$. $ccom$ consists of a commitment value $ccom$, commitment parameters $cparcom$, and a commitment verification algorithm COM.Verify. If the proof and the instance are stored in Tbl_{pr}, $\mathcal{F}_{\mathrm{REV}}$ outputs the verification result stored in Tbl_{pr} to ensure consistence. Otherwise $\mathcal{F}_{\mathrm{REV}}$ runs the algorithm REV.Extract to extract the revo-

cation handle rh, the opening *copen*, and the revocation status $\mathbf{x}[rh]$ from the proof pr. Any construction that realizes $\mathcal{F}_{\mathrm{REV}}$ must allow extractable proofs. If extraction fails or if *ccom* is not a commitment to rh and *copen*, $\mathcal{F}_{\mathrm{REV}}$ marks the proof as invalid. Otherwise, if the revocation authority is corrupt and the revocation status of rh is not stored in Tbl_{ep}, $\mathcal{F}_{\mathrm{REV}}$ stores it in Tbl_{ep}. After that, $\mathcal{F}_{\mathrm{REV}}$ also marks the proof as invalid if rh is not in Tbl_{ep} or if $b \neq \mathbf{x}[rh, j]$, where $\mathbf{x}[rh, j]$ is the revocation status stored in Tbl_{ep} for the revocation handle rh. In the hiding functionality, the proof is also marked as invalid when $b = \mathbf{x}[rh, j]$ but the revocation authority is honest and the user $\mathcal{U}$ associated with rh is honest. The reason is that the hiding functionality must prevent corrupt users from computing proofs about revocation handles associated with honest users because this constitutes a violation of the privacy of the revocation statuses. A construction that realizes $\mathcal{F}_{\mathrm{REV}}$ must use non-malleable proofs, i.e., it should not be possible to obtain a new proof from a valid proof without knowing the witness. We note that proofs for honest users are computed by $\mathcal{F}_{\mathrm{REV}}$ in the rev.prove.* interface and registered in Tbl_{pr} as valid, and are thus accepted by $\mathcal{F}_{\mathrm{REV}}$ in the verification interface without running the algorithm REV.Extract.

We note that $\mathcal{F}_{\mathrm{REV}}$ does not allow parties to send their own revocation parameters par_r through the rev.prove.* and rev.verify.* interfaces. This means that a construction that realizes this functionality must use some form of trusted registration that allows the revocation authority to register par_r and the other parties to retrieve par_r in order to ensure that all honest parties use the same parameters.

$\mathcal{F}_{\mathrm{REV}}$ asks the simulator $\mathcal{S}$ to provide prove and extract algorithms at setup. Alternatively, it would be possible that the functionality asks the simulator to compute proofs and extract from proofs when the rev.prove.* and rev.verify.* interfaces are invoked. We chose the first alternative because it hides the computation and verification of proofs by the parties from the simulator.

4.2   Ideal Functionality for Anonymous Attribute Tokens $\mathcal{F}_{\mathrm{AT}}$

Next, we describe the ideal functionality of anonymous attribute tokens, $\mathcal{F}_{\mathrm{AT}}$. $\mathcal{F}_{\mathrm{AT}}$ interacts with an issuer $\mathcal{I}$, users $\mathcal{U}_i$ and any verifying parties $\mathcal{P}$. The issuer $\mathcal{I}$ issues some attributes $\langle a_l \rangle_{l=1}^L$ to a user $\mathcal{U}_i$. A user $\mathcal{U}_i$ can obtain a proof that some commitments $\langle ccom_l \rangle_{l=1}^L$ commit to attributes that were issued by $\mathcal{I}$. Any party $\mathcal{P}$ can verify a proof.

The interaction between the functionality $\mathcal{F}_{\mathrm{AT}}$ and the issuer $\mathcal{I}$, the users $\mathcal{U}_i$ and the verifying parties $\mathcal{P}$ takes place through the interfaces at.setup.*, at.get.*, at.issue.*, at.prove.*, and at.verify.*.

1. The issuer $\mathcal{I}$ uses the at.setup.* interface to initialize the functionality and obtain the parameters par_{at} of the anonymous attribute token scheme.
2. Any party $\mathcal{P}$ invokes the at.get.* interface to obtain the parameters par_{at}.
3. The issuer $\mathcal{I}$ uses the at.issue.* interface to issue some attributes $\langle a_l \rangle_{l=1}^L$ to a user $\mathcal{U}_i$.

4. An honest user $\mathcal{U}_i$ uses the at.prove.* interface to get a proof pr that some commitments $\langle ccom_l \rangle_{l=1}^L$ commit to attributes that were issued by the issuer $\mathcal{I}$ to the user $\mathcal{U}_i$.
5. Any honest party $\mathcal{P}$ uses the at.verify.* interface to verify a proof pr.

As we described before, the commitment parameters and the verification algorithm are attached to the commitment itself, and the functionality parses the commitment value to obtain them to run a commitment verification. $\mathcal{F}_{AT}$ uses the following tables.

Tbl$_a$. Tbl$_a$ stores entries of the form $[\mathcal{U}_i, \langle a_l \rangle_{l=1}^L]$, which map a user $\mathcal{U}_i$ to a list of attributes $\langle a_l \rangle_{l=1}^L$ issued by $\mathcal{I}$.

Tbl$_{pr}$. Tbl$_{pr}$ stores entries of the form $[\langle ccom_l \rangle_{l=1}^L, pr, u]$, which consist of a proof instance $\langle ccom_l \rangle_{l=1}^L$, a proof pr, and a bit u that indicates whether the proof is valid.

$\mathcal{F}_{AT}$ also uses a set $\mathbb{P}$. $\mathbb{P}$ contains the identifiers of the parties $\mathcal{P}$ that retrieved the attribute tokens parameters par_{at}.

Functionality $\mathcal{F}_{AT}$

AT.SimProve and AT.Extract are ppt algorithms. $\mathcal{F}_{AT}$ is parameterized by the system parameters sp, a maximum number of attributes L_{max} and a universe of attributes Ψ.

1. On input (at.setup.ini, sid) from $\mathcal{I}$:
   (a) Abort if $sid \neq (\mathcal{I}, sid')$ or if the tuple (sid) is already stored. Store (sid).
   (b) Send (at.setup.req, sid) to $\mathcal{S}$.
S. On input (at.setup.alg, sid, par_{at}, td_{at}, AT.SimProve, AT.Extract) from $\mathcal{S}$:
   (a) Abort if (sid) is not stored or if $(sid, par_{at}, td_{at}, $ AT.SimProve, AT.Extract$)$ is already stored. Store $(sid, par_{at}, td_{at}, $ AT.SimProve, AT.Extract$)$.
   (b) Initialize an empty table Tbl$_a$, an empty table Tbl$_{pr}$ and an empty set $\mathbb{P}$, and parse sid as $(\mathcal{I}, sid')$.
   (c) Send (at.setup.end, sid, par_{at}) to $\mathcal{I}$.
2. On input (at.get.ini, sid) from any party $\mathcal{P}$:
   (a) If there is a tuple $(sid, par_{at}, td_{at}, $ AT.SimProve, AT.Extract$)$ stored, set $par_{at}' \leftarrow par_{at}$; else set $par_{at}' \leftarrow \bot$.
   (b) Create a fresh $ssid$ and store $(ssid, \mathcal{P}, par_{at}')$.
   (c) Send (at.get.sim, sid, $ssid$, par_{at}') to $\mathcal{S}$.
S. On input (at.get.rep, sid, $ssid$) from $\mathcal{S}$:
   (a) Abort if $(ssid, \mathcal{P}, par_{at}')$ is not stored.
   (b) If $par_{at}' \neq \bot$, include $\mathcal{P}$ in the set $\mathbb{P}$.
   (c) Delete record $(ssid, \mathcal{P}, par_{at}')$.
   (d) Send (at.get.end, sid, par_{at}') to $\mathcal{P}$.
3. On input (at.issue.ini, sid, $\mathcal{U}_i$, $\langle a_l \rangle_{l=1}^L$) from $\mathcal{I}$:
   (a) Abort if there is no tuple $(sid, par_{at}, td_{at}, $ AT.SimProve, AT.Extract$)$ stored, or if $\mathcal{U}_i$ is not a valid user identifier, or if $\langle a_l \rangle_{l=1}^L \not\subseteq \Psi$, or if $L > L_{max}$.
   (b) Create a fresh $ssid$ and store $(ssid, \mathcal{U}_i, \langle a_l \rangle_{l=1}^L)$.
   (c) Send (at.issue.sim, sid, $ssid$, $\mathcal{U}_i$) to $\mathcal{S}$.

S. On input (at.issue.rep, sid, $ssid$) from $\mathcal{S}$:
  (a) Abort if $(ssid, \mathcal{U}_i, \langle a_l \rangle_{l=1}^L)$ is not stored or if $\mathcal{U}_i \not\subseteq \mathbb{P}$.
  (b) If $\mathcal{U}_i$ is honest, then append $[\mathcal{U}_i, \langle a_l \rangle_{l=1}^L]$ to Tbl_a; else append $[\mathcal{S}, \langle a_l \rangle_{l=1}^L]$ to Tbl_a.
  (c) Delete record $(ssid, \mathcal{U}_i, \langle a_l \rangle_{l=1}^L)$.
  (d) Send (at.issue.end, sid, $\langle a_l \rangle_{l=1}^L$) to $\mathcal{U}_i$.

4. On input (at.prove.ini, sid, $\langle ccom_l, a_l, copen_l \rangle_{l=1}^L$) from an honest user $\mathcal{U}_i$:
  (a) Parse $ccom_l$ as $(ccom'_l, cparcom_l, \mathsf{COM.Verify}_l)$ and abort if $1 \neq \mathsf{COM.Verify}_l(cparcom_l, ccom'_l, a_l, copen_l)$ for any $l \in [1, L]$, or if there is no entry $[\mathcal{U}_i, \langle a_l \rangle_{l=1}^L]$ in Tbl_a .
  (b) Run $pr \leftarrow \mathsf{AT.SimProve}(sid, par_{at}, \langle cparcom_l, ccom'_l \rangle_{l=1}^L, td_{at})$.
  (c) Append $[\langle ccom_l \rangle_{l=1}^L, pr, 1]$ to Table Tbl_{pr}.
  (d) Send (at.prove.end, sid, pr) to $\mathcal{U}_i$.

5. On input (at.verify.ini, sid, $\langle ccom_l \rangle_{l=1}^L$, pr) from an honest party $\mathcal{P}$:
  (a) Abort if $\mathcal{P} \not\subseteq \mathbb{P}$.
  (b) Parse $ccom_l$ as $(ccom'_l, cparcom_l, \mathsf{COM.Verify}_l)$ for any $l \in [1, L]$.
  (c) If there is an entry $[\langle ccom_l \rangle_{l=1}^L, pr, u]$ in Tbl_{pr}, set $v \leftarrow u$.
  (d) Else, do the following:
     i. Run $\langle a_l, copen_l \rangle_{l=1}^L \leftarrow \mathsf{AT.Extract}(sid, par_{at}, \langle cparcom_l, ccom'_l \rangle_{l=1}^L, td_{at}, pr)$.
     ii. If $\mathcal{I}$ is corrupt, proceed as follows. If $\langle a_l, copen_l \rangle_{l=1}^L = 1$, set $v \leftarrow 1$; else set $v \leftarrow 0$.
     iii. Else, if $\langle a_l, copen_l \rangle_{l=1}^L = \bot$ or $1 \neq \mathsf{COM.Verify}_l(cparcom_l, ccom'_l, a_l, copen_l)$ for any $l \in [1, L]$, set $v \leftarrow 0$.
     iv. Else, if there exists an entry $[\mathcal{S}, \langle a_l \rangle_{l=1}^L] \in \mathsf{Tbl}_a$, set $v \leftarrow 1$; else set $v \leftarrow 0$.
     v. Append $[\langle ccom_l \rangle_{l=1}^L, pr, v]$ to Table Tbl_{pr}.
  (e) Send (at.verify.end, sid, v) to $\mathcal{P}$.

We now discuss the five interfaces of the ideal functionality $\mathcal{F}_{AT}$.

1. The at.setup.ini message is sent by the issuer $\mathcal{I}$. The restriction that the issuer's identity must be included in the session identifier $sid = (\mathcal{I}, sid')$ guarantees that each issuer can initialize its own instance of the functionality. $\mathcal{F}_{AT}$ requests the simulator $\mathcal{S}$ for the parameters and algorithms. $\mathcal{S}$ sends the parameters par_{at}, the trapdoor td_{at}, and the algorithms (AT.SimProve, AT.Extract) for proof computation and proof extraction. Finally, $\mathcal{F}_{AT}$ initializes two empty tables, Tbl_a and Tbl_{pr}, and sends the received parameters par_{at} to $\mathcal{I}$.

2. The at.get.ini message allows any party to request par_{at}.

3. The at.issue.ini message is sent by the issuer on input a user identity and a list of attributes. $\mathcal{F}_{AT}$ creates a subsession identifier $ssid$ and sends the user identity to the simulator. The simulator indicates when the issuance is to be finalized by sending a (at.issue.rep, sid, $ssid$) message. At this point, the issuance is recorded in Tbl_a. If the user is honest, the issuance is recorded under the correct user's identity, which in the real world requires any instantiating protocol to set up an authenticated channel to the user to ensure this. If the user is corrupt, the attributes are recorded as belonging to the simulator, modeling that corrupt users may pool their attribute tokens. Note that the simulator is not given the attribute values issued, so the real-world protocol must hide these from the adversary.

4. The at.prove.ini message lets an honest user $\mathcal{U}_i$ request a proof that the attributes $\langle a_l \rangle_{l=1}^L$ issued to her by $\mathcal{I}$ are committed in the commitments $\langle com_l \rangle_{l=1}^L$. The commitment parameters and the commitment verification algorithm attached to each commitment value allow $\mathcal{F}_{AT}$ to compute and verify proofs about commitments to attributes, such that the commitments are generated externally, e.g., by the functionality $\mathcal{F}_{NIC}$.

   $\mathcal{F}_{AT}$ computes a proof by running AT.SimProve, which does not receive the witness as input. Therefore, any construction that realizes $\mathcal{F}_{AT}$ must use zero-knowledge proofs. $\mathcal{F}_{AT}$ stores the proof in Tbl_{pr} and sends the proof to $\mathcal{U}_i$.

5. The at.verify.ini message allows any honest party to request the verification of a proof pr with respect to the instance $\langle ccom_l \rangle_{l=1}^L$. If the instance-proof pair is stored in the table Tbl_{pr}, then the functionality replies with the stored verification result to ensure consistency. If not, the functionality runs the algorithm AT.Extract. If the issuer is corrupt, the functionality interprets the output of AT.Extract as a bit b that indicates whether the proof is valid or not. If the issuer is honest, the functionality interprets the output as the witness $\langle a_l, copen_l \rangle_{l=1}^L$. The functionality only marks the proof as correct if extraction did not fail, if $\langle a_l, copen_l \rangle_{l=1}^L$ are correct openings of the commitments in the instance, and if any corrupt user was issued the attributes $\langle a_l \rangle_{l=1}^L$. A construction that realizes $\mathcal{F}_{AT}$ must use non-malleable proofs, i.e., $\mathcal{F}_{AT}$ enforces that it is not possible to obtain a new proof from a valid proof without knowing the witness. We note that proofs for honest users are computed by $\mathcal{F}_{AT}$ in the at.prove.* interface and registered in Tbl_{pr} as valid, and are thus accepted by $\mathcal{F}_{AT}$ in the verification interface without running the algorithm AT.Extract. Finally, the functionality stores the proof-instance pair and the verification result in Tbl_{pr} and sends the verification result to the party.

We note that $\mathcal{F}_{AT}$ does not allow parties to send their own revocation parameters par_{at} through the at.prove.* and at.verify.* interfaces. This means that a construction that realizes this functionality must use some form of trusted registration that allows the issuer to register par_{at} and the other parties to retrieve par_{at} to ensure that all honest parties use the same parameters.

$\mathcal{F}_{AT}$ asks the simulator $\mathcal{S}$ to provide prove and extract algorithms at setup. Alternatively, it would be possible that the functionality asks the simulator to compute proofs and extract from proofs when the at.prove.* and at.verify.* interfaces are invoked. We chose the first alternative because it hides the computation and verification of proofs by the parties from the simulator.

We refer to the full version of the paper for a construction that realizes $\mathcal{F}_{AT}$ and its security analysis.

5   Anonymous Attribute Tokens with Revocation

In this section we a high-level description of the ideal functionality $\mathcal{F}_{TR}$ of anonymous attribute tokens with revocation. For its formal description we defer to the full version of this paper. Similarly, we here consider only the version of functionality $\mathcal{F}_{TR}$ where the revocation statuses of every revocation handle are public. We then describe a construction for this version of $\mathcal{F}_{TR}$ that uses the functionalities $\mathcal{F}_{NIC}$, $\mathcal{F}_{REV}$, and $\mathcal{F}_{AT}$ and illustrates how to modularly design hybrid protocols in the UC framework.

5.1 Ideal Functionality $\mathcal{F}_{TR}$ of Anonymous Attribute Tokens with Revocation

$\mathcal{F}_{TR}$ interacts with an issuer $\mathcal{I}$, a revocation authority $\mathcal{RA}$, users $\mathcal{U}_i$, and any verifying party $\mathcal{P}$. The issuer $\mathcal{I}$ issues some attributes $\langle a_l \rangle_{l=1}^{L}$ and a revocation handle rh to a user $\mathcal{U}_i$. The revocation authority $\mathcal{RA}$ associates a revocation status $\mathbf{x}[rh]$ with each revocation handle rh. $\mathbf{x}[rh]$ is a vector of m bits, such that each bit $\mathbf{x}[rh, j]$ denotes the revocation status of the revocation handle rh with respect to the revocation list $j \in [1, m]$. A user $\mathcal{U}_i$ can prove to any party that a set of attributes $\langle a_l \rangle_{l=1}^{L}$ and a revocation handle were issued by $\mathcal{I}$. $\mathcal{U}_i$ also proves that b is the revocation status that the revocation authority $\mathcal{RA}$ associated with rh for the revocation list j and the epoch ep.

The interaction between the functionality $\mathcal{F}_{TR}$ and the issuer $\mathcal{I}$, the revocation authority $\mathcal{RA}$, the users $\mathcal{U}_i$, and any verifying party $\mathcal{P}$ takes place through the interfaces tr.setupi.*, tr.setupra.*, tr.setupp.*, tr.issue.*, tr.epoch.*, tr.getepoch.*, and tr.prove.*.

1. The issuer $\mathcal{I}$ invokes the tr.setupi.* interface for initialization.
2. The revocation authority $\mathcal{RA}$ invokes the tr.setupra.* interface for initialization.
3. Any user or verifying party $\mathcal{P}$ invokes the tr.setupp.* interface for initialization.
4. The issuer $\mathcal{I}$ uses the tr.issue.* interface to issue the attributes $\langle a_l \rangle_{l=1}^{L}$ and the revocation handle rh to a user $\mathcal{U}_i$.
5. The revocation authority $\mathcal{RA}$ uses the tr.epoch.* interface to send a list of revocation handles rh along with their respective revocation statuses $\mathbf{x}[rh]$ for the epoch ep.
6. Any party $\mathcal{P}$ uses the rev.getepoch.* interface to get the full list of revocation handles and revocation statuses $\langle rh_i, \mathbf{x}[rh_i] \rangle_{i=1}^{n'}$ for the epoch ep.
7. A user $\mathcal{U}_i$ uses the tr.prove.* interface to prove that some attributes $\langle a_l \rangle_{l=1}^{L}$ and a revocation handle rh were issued by $\mathcal{I}$. $\mathcal{U}_i$ also proves that $\mathbf{x}[rh, j]$ is the revocation status associated with rh for the revocation list j and the epoch ep.

5.2 Construction of Anonymous Attribute Tokens with Revocation

We now describe the construction of anonymous attribute tokens with revocation Π_{TR}. The construction Π_{TR} works in the $(\mathcal{F}_{SMT}, \mathcal{F}_{ASMT}, \mathcal{F}_{NIC}, \mathcal{F}_{REV}, \mathcal{F}_{AT})$-hybrid model, where the parties make use of the ideal functionalities for secure message transmission $\mathcal{F}_{SMT}$ and anonymous secure message transmission $\mathcal{F}_{ASMT}$ in [4]. The parties also use the ideal functionality for commitments

$\mathcal{F}_{\mathrm{NIC}}$ described in Sect. 3.1, the ideal functionality for revocation $\mathcal{F}_{\mathrm{REV}}$ described in Sect. 4.1, and the ideal functionality for anonymous attribute tokens $\mathcal{F}_{\mathrm{AT}}$ described in Sect. 4.2.

This construction illustrates our mechanism for a modular design of hybrid protocols in the UC framework. In the issuing phase, the users receive attributes and a revocation handle from the issuer through $\mathcal{F}_{\mathrm{AT}}$. To compute an attribute token and show that it has not been revoked, a user first obtains a commitment to the revocation handle and an opening from $\mathcal{F}_{\mathrm{NIC}}$. Then the revocation handle, the commitment and the opening are sent to $\mathcal{F}_{\mathrm{REV}}$ to get a non-interactive proof of non-revocation. Similarly, the revocation handle, the commitment and the opening, along with commitments and openings to some attributes issued along with the revocation handle, are sent to $\mathcal{F}_{\mathrm{AT}}$ to obtain an attribute token. Thanks to the fact that $\mathcal{F}_{\mathrm{AT}}$ and $\mathcal{F}_{\mathrm{REV}}$ run the commitment verification algorithm, it is ensured that $\mathcal{F}_{\mathrm{AT}}$ and $\mathcal{F}_{\mathrm{REV}}$ receive the same revocation handle.

The construction Π_{TR} is executed by an issuer $\mathcal{I}$, a revocation authority $\mathcal{RA}$, users $\mathcal{U}_i$, and any verifying party $\mathcal{P}$. Those parties are activated through the tr.setupi.*, tr.setupra.*, tr.setupp.*, tr.issue.*, tr.epoch.*, tr.getepoch.*, and tr.prove.* interfaces. Briefly, the construction Π_{TR} works as follows.

1. The issuer $\mathcal{I}$ receives (tr.setupi.ini, sid) as input. If the functionality $\mathcal{F}_{\mathrm{AT}}$ was not set up, $\mathcal{I}$ invokes the at.setup.* interface of $\mathcal{F}_{\mathrm{AT}}$; else $\mathcal{I}$ aborts.
2. The revocation authority receives (tr.setupra.ini, sid) as input. $\mathcal{RA}$ aborts if setup has already been run. Otherwise, $\mathcal{RA}$ invokes the at.get.* interface of $\mathcal{F}_{\mathrm{AT}}$ and aborts if $\mathcal{F}_{\mathrm{AT}}$ does not return the attribute token parameters. In addition, $\mathcal{RA}$ invokes the com.setup.* interface of $\mathcal{F}_{\mathrm{NIC}}$ to setup the commitment functionality. Finally, $\mathcal{RA}$ invokes the rev.setup.* interface of $\mathcal{F}_{\mathrm{REV}}$.
3. A user or a verifying party $\mathcal{P}$ receives (tr.setupp.ini, sid) as input. $\mathcal{P}$ aborts if the setup has already been run. Otherwise $\mathcal{P}$ invokes the at.get.* interface of $\mathcal{F}_{\mathrm{AT}}$ and aborts if $\mathcal{F}_{\mathrm{AT}}$ does not return the attribute token parameters. Then $\mathcal{P}$ invokes the (rev.get.ini, sid) interface of $\mathcal{F}_{\mathrm{REV}}$ and aborts if $\mathcal{F}_{\mathrm{REV}}$ does not return the revocation parameters. Finally $\mathcal{P}$ invokes the com.setup.* interface of $\mathcal{F}_{\mathrm{NIC}}$.
4. The issuer $\mathcal{I}$ receives (tr.issue.ini, $sid, \mathcal{U}_i, \langle a_l \rangle_{l=1}^L, rh$) as input. If the issuer setup has not been run, $\mathcal{I}$ aborts. Otherwise $\mathcal{I}$ invokes the at.issue.* interface of $\mathcal{F}_{\mathrm{AT}}$ to issue the attributes $\langle a_l \rangle_{l=1}^L$ and the revocation handle rh to the user $\mathcal{U}_i$. $\mathcal{U}_i$ aborts if the user setup has not been run by $\mathcal{U}_i$.
5. The revocation authority $\mathcal{RA}$ receives (tr.epoch.ini, $sid, \langle rh_i, \mathbf{x}[rh_i] \rangle_{i=1}^{n'}$) as input. $\mathcal{RA}$ aborts if the $\mathcal{RA}$ setup has not been run or if the input values are invalid. Otherwise $\mathcal{RA}$ uses the rev.epoch.* interface of $\mathcal{F}_{\mathrm{REV}}$ to send the revocation information and obtain the epoch information $info$ for the current epoch ep.
6. A party $\mathcal{P}$ receives (tr.getepoch.ini, sid, ep) as input. $\mathcal{P}$ aborts if it did not run the setup. Otherwise $\mathcal{P}$ invokes the interface rev.getepoch.* of $\mathcal{F}_{\mathrm{REV}}$ to get the revocation information Tbl_{ep} for ep.

Construction Π_{TR}

Π_{TR} is parameterized by the system parameters sp and uses the ideal functionalities $\mathcal{F}_{\text{SMT}}$, $\mathcal{F}_{\text{ASMT}}$, $\mathcal{F}_{\text{NIC}}$, $\mathcal{F}_{\text{REV}}$, and $\mathcal{F}_{\text{AT}}$. The constants used are the maximum number of attributes $L_{\max}$, the universe of attributes Ψ, the maximum number of revocation lists m, and the maximum number of revocation handles n.

1. On input $(\text{tr.setupi.ini}, sid)$, $\mathcal{I}$ does the following:
   (a) $\mathcal{I}$ aborts if $sid \neq (\mathcal{I}, \mathcal{RA}, sid')$, or if $(sid, \text{tr.setupi})$ is already stored.
   (b) $\mathcal{I}$ sends the message $(\text{at.setup.ini}, sid)$ to $\mathcal{F}_{\text{AT}}$ and receives the message $(\text{at.setup.end}, sid, par_{at})$ from $\mathcal{F}_{\text{AT}}$.
   (c) Store $(sid, \text{tr.setupi})$.
   (d) Output $(\text{tr.setupi.end}, sid)$.

2. On input $(\text{tr.setupra.ini}, sid)$, $\mathcal{RA}$ does the following:
   (a) $\mathcal{RA}$ aborts if $(sid, \text{tr.setupra})$ is already stored.
   (b) $\mathcal{RA}$ sends $(\text{at.get.ini}, sid)$ to $\mathcal{F}_{\text{AT}}$ and receives $(\text{at.get.end}, sid, par_{at})$ from $\mathcal{F}_{\text{AT}}$. If $par_{at} = \bot$, $\mathcal{RA}$ aborts.
   (c) $\mathcal{RA}$ sends $(\text{com.setup.ini}, sid)$ to $\mathcal{F}_{\text{NIC}}$ and receives $(\text{com.setup.end}, sid, OK)$ from $\mathcal{F}_{\text{NIC}}$.
   (d) $\mathcal{RA}$ parses sid as $(\mathcal{I}, \mathcal{RA}, sid')$, sets $sid_{\text{REV}} \leftarrow (\mathcal{RA}, sid')$, sends $(\text{rev.setup.ini}, sid)$ to $\mathcal{F}_{\text{REV}}$ and receives $(\text{rev.setup.end}, sid, par_r)$ from $\mathcal{F}_{\text{REV}}$.
   (e) Store $(sid, \text{tr.setupra})$.
   (f) Output $(\text{tr.setupra.end}, sid)$.

3. On input $(\text{tr.setupp.ini}, sid)$, a user or verifying party $\mathcal{P}$ does the following:
   (a) $\mathcal{P}$ aborts if $(sid, \text{tr.setupp})$ is already stored.
   (b) $\mathcal{P}$ sends $(\text{at.get.ini}, sid)$ to $\mathcal{F}_{\text{AT}}$ and receives $(\text{at.get.end}, sid, par_{at})$ from $\mathcal{F}_{\text{AT}}$. If $par_{at} = \bot$, $\mathcal{P}$ aborts.
   (c) $\mathcal{P}$ parses sid as $(\mathcal{I}, \mathcal{RA}, sid')$, sets $sid_{\text{REV}} \leftarrow (\mathcal{RA}, sid')$, sends $(\text{rev.get.ini}, sid)$ to $\mathcal{F}_{\text{REV}}$ and receives $(\text{rev.get.end}, sid, par_r)$ from $\mathcal{F}_{\text{REV}}$. If $par_r = \bot$, $\mathcal{P}$ aborts.
   (d) $\mathcal{P}$ sends $(\text{com.setup.ini}, sid)$ to $\mathcal{F}_{\text{NIC}}$ and receives $(\text{com.setup.end}, sid, OK)$ from $\mathcal{F}_{\text{NIC}}$.
   (e) Store $(sid, \text{tr.setupp})$.
   (f) Output $(\text{tr.setupp.end}, sid)$.

4. On input $(\text{tr.issue.ini}, sid, \mathcal{U}_i, \langle a_l \rangle_{l=1}^{L}, rh)$, $\mathcal{I}$ and $\mathcal{U}_i$ do the following:
   (a) $\mathcal{I}$ aborts if $(sid, \text{tr.setupi})$ is not stored, or if $\langle a_l \rangle_{l=1}^{L} \not\subseteq \Psi$, or if $rh \notin [1, n]$, or if $L > L_{\max}$.
   (b) $\mathcal{I}$ sets $a_{L+1} \leftarrow rh$ and sends $(\text{at.issue.ini}, sid, \mathcal{U}_i, \langle a_l \rangle_{l=1}^{L+1})$ to $\mathcal{F}_{\text{AT}}$.
   (c) $\mathcal{U}_i$ receives $(\text{at.issue.end}, sid, \langle a_l \rangle_{l=1}^{L+1})$ from $\mathcal{F}_{\text{AT}}$.
   (d) $\mathcal{U}_i$ aborts if $(sid, \text{tr.setupp})$ is not stored.
   (e) $\mathcal{U}_i$ sets $rh \leftarrow a_{L+1}$, stores $[\langle a_l \rangle_{l=1}^{L}, rh]$, and outputs $(\text{tr.issue.end}, sid, \langle a_l \rangle_{l=1}^{L}, rh)$.

5. On input $(\text{tr.epoch.ini}, sid, ep, \langle rh_i, \mathbf{x}[rh_i] \rangle_{i=1}^{n'})$, $\mathcal{RA}$ does the following:
   (a) $\mathcal{RA}$ aborts if $(sid, \text{tr.setupra})$ is not stored, or if $n' > n$, or if, for $i = 1$ to n', $rh_i \notin [1, n]$ or $\mathbf{x}[rh_i] \notin [0, 2^m)$, or if $(ep, info)$ is already stored.
   (b) $\mathcal{RA}$ parses sid as $(\mathcal{I}, \mathcal{RA}, sid')$, sets $sid_{\text{REV}} \leftarrow (\mathcal{RA}, sid')$, sends $(\text{rev.epoch.ini}, sid_{\text{REV}}, ep, \langle rh_i, \mathbf{x}[rh_i] \rangle_{i=1}^{n'})$ to $\mathcal{F}_{\text{REV}}$, receives $(\text{rev.epoch.end}, sid_{\text{REV}}, ep, info)$ from $\mathcal{F}_{\text{REV}}$, and stores $(ep, info)$.
   (c) $\mathcal{RA}$ outputs $(\text{tr.epoch.end}, sid, ep)$.

6. On input $(\text{tr.getepoch.ini}, sid, ep)$, a party $\mathcal{P}$ does the following:

(a) $\mathcal{P}$ aborts if $(sid, \mathsf{tr.setupp})$ is not stored.

(b) $\mathcal{P}$ parses sid as $(\mathcal{I}, \mathcal{RA}, sid')$, sets $sid_{\mathrm{REV}} \leftarrow (\mathcal{RA}, sid')$, sends the message $(\mathsf{rev.getepoch.ini}, sid_{\mathrm{REV}}, ep)$ to $\mathcal{F}_{\mathrm{REV}}$, and receives the message $(\mathsf{rev.getepoch.end}, sid_{\mathrm{REV}}, info, \mathsf{Tbl}_{ep})$ from $\mathcal{F}_{\mathrm{REV}}$.

(c) If $info \neq \bot$, $\mathcal{P}$ sets $b \leftarrow 1$ and stores $[ep, info, \mathsf{Tbl}_{ep}]$; else $b \leftarrow 0$.

(d) $\mathcal{P}$ outputs $(\mathsf{tr.getepoch.end}, sid, b, \mathsf{Tbl}_{ep})$.

7. On input $(\mathsf{tr.prove.ini}, sid, \mathcal{P}, \langle a_l \rangle_{l=1}^L, rh, ep, j, b)$, a user $\mathcal{U}_i$ and a verifying party $\mathcal{P}$ do the following:

(a) $\mathcal{U}_i$ aborts if an entry $[\langle a_l' \rangle_{l=1}^L, rh']$ such that $rh = rh'$ and $\langle a_l \rangle_{l=1}^L = \langle a_l' \rangle_{l=1}^L$ is not stored.

(b) If $[ep, info, \mathsf{Tbl}_{ep}]$ is not stored, $\mathcal{U}_i$ aborts.

(c) $\mathcal{U}_i$ aborts if $b \neq \mathsf{x}[rh, j]$, where $\mathsf{x}[rh, j]$ is the stored revocation status of rh for list j at epoch ep.

(d) $\mathcal{U}_i$ sets $a_{L+1} \leftarrow rh$, and, for $l = 1$ to $L+1$, $\mathcal{U}_i$ sends $(\mathsf{com.commit.ini}, sid, a_l)$ to $\mathcal{F}_{\mathrm{NIC}}$, and receives $(\mathsf{com.commit.end}, sid, ccom_l, copen_l)$ from $\mathcal{F}_{\mathrm{NIC}}$.

(e) $\mathcal{U}_i$ parses sid as $(\mathcal{I}, \mathcal{RA}, sid')$, sets $sid_{\mathrm{REV}} \leftarrow (\mathcal{RA}, sid')$, sends $(\mathsf{rev.prove.ini}, sid_{\mathrm{REV}}, ccom_{L+1}, rh, copen_{L+1}, ep, j)$ to $\mathcal{F}_{\mathrm{REV}}$, and receives $(\mathsf{rev.prove.end}, sid_{\mathrm{REV}}, pr)$ from $\mathcal{F}_{\mathrm{REV}}$.

(f) $\mathcal{U}_i$ sends $(\mathsf{at.prove.ini}, sid, \langle ccom_l, a_l, copen_l \rangle_{l=1}^{L+1})$ to $\mathcal{F}_{\mathrm{AT}}$ and receives $(\mathsf{at.prove.end}, sid, pr')$ from $\mathcal{F}_{\mathrm{AT}}$.

(g) $\mathcal{U}_i$ picks a fresh sid_{ASMT} and sends $(\mathsf{asmt.send.ini}, sid_{\mathrm{ASMT}}, \langle sid, j, ep, \mathsf{x}[rh, j], (a_l, copen_l)_{l=1}^L, (ccom_l)_{l=1}^{L+1}, pr, pr' \rangle, \mathcal{P})$ to $\mathcal{F}_{\mathrm{ASMT}}$.

(h) $\mathcal{P}$ receives the message $(\mathsf{asmt.send.end}, sid_{\mathrm{ASMT}}, \langle sid, j, ep, \mathsf{x}[rh, j], (a_l, copen_l)_{l=1}^L, (ccom_l)_{l=1}^{L+1}, pr, pr' \rangle)$ from $\mathcal{F}_{\mathrm{ASMT}}$.

(i) $\mathcal{P}$ aborts if $[ep, info, \mathsf{Tbl}_{ep}]$ is not stored.

(j) For $l = 1$ to L, $\mathcal{P}$ sends $(\mathsf{com.verify.ini}, sid, ccom_l, a_l, copen_l)$ to $\mathcal{F}_{\mathrm{NIC}}$ and receives $(\mathsf{com.verify.end}, sid, v_l)$ from $\mathcal{F}_{\mathrm{NIC}}$.

(k) $\mathcal{P}$ sends $(\mathsf{com.validate.ini}, sid, ccom_{L+1})$ to $\mathcal{F}_{\mathrm{NIC}}$, receives $(\mathsf{com.validate.end}, sid, v'')$ from $\mathcal{F}_{\mathrm{NIC}}$.

(l) $\mathcal{P}$ parses sid as $(\mathcal{I}, \mathcal{RA}, sid')$, sets $sid_{\mathrm{REV}} \leftarrow (\mathcal{RA}, sid')$, sends $(\mathsf{rev.verify.ini}, sid_{\mathrm{REV}}, ccom_{L+1}, ep, j, \mathsf{x}[rh, j], pr)$ to $\mathcal{F}_{\mathrm{REV}}$, and receives $(\mathsf{rev.verify.end}, sid_{\mathrm{REV}}, v)$ from $\mathcal{F}_{\mathrm{REV}}$.

(m) $\mathcal{P}$ sends $(\mathsf{at.verify.ini}, sid, \langle ccom_l \rangle_{l=1}^{L+1}, pr')$ to $\mathcal{F}_{\mathrm{AT}}$ and receives $(\mathsf{at.verify.end}, sid, v')$ from $\mathcal{F}_{\mathrm{AT}}$.

(n) If $v = v' = v'' = 1$ and $v_l = 1$ for $l = 1$ to L, then $\mathcal{P}$ outputs $(\mathsf{tr.prove.end}, sid, \langle a_l \rangle_{l=1}^L, \mathsf{x}[rh, j], ep, j)$; else $\mathcal{P}$ aborts.

7. The user $\mathcal{U}_i$ receives $(\mathsf{tr.prove.ini}, sid, \mathcal{P}, \langle a_l \rangle_{l=1}^L, rh, ep, j, b)$ as input. $\mathcal{U}_i$ aborts if the revocation handle and the attributes were not issued to her, if $\mathcal{U}_i$ did not get the epoch ep or if the revocation status given by $\mathcal{RA}$ is not b for list j and rh at epoch ep. Otherwise $\mathcal{U}_i$ invokes the com.commit.* interface of functionality $\mathcal{F}_{\mathrm{NIC}}$ to obtain commitments and openings for the attributes and for the revocation handle that were issued by $\mathcal{I}$. Note that, for simplicity, we reveal all attributes of the token, except the revocation handle. Then $\mathcal{U}_i$ invokes the at.prove.* interface of $\mathcal{F}_{\mathrm{AT}}$ to get a proof that the committed attributes were issued by $\mathcal{I}$. $\mathcal{U}_i$ also invokes the rev.prove.* interface of $\mathcal{F}_{\mathrm{REV}}$ to get a proof that the revocation status of the committed revocation handle for list j at epoch ep is b. The user $\mathcal{U}_i$ sends the commitments, the proofs,

and the openings of the commitments to the attributes through an instance of $\mathcal{F}_{\text{ASMT}}$ to the verifying party $\mathcal{P}$. $\mathcal{P}$ aborts if it did not get the epoch ep. Otherwise $\mathcal{P}$ validates the commitment parameters for the commitment to the revocation handle by calling the com.validate.∗ interface of $\mathcal{F}_{\text{NIC}}$, because it cannot verify the full commitment itself without the opening, and invokes the com.verify.∗ interface of $\mathcal{F}_{\text{NIC}}$ to verify the openings of the commitments to the attributes revealed. The at.verify.∗ interface of $\mathcal{F}_{\text{AT}}$ and the rev.verify.∗ interface of $\mathcal{F}_{\text{REV}}$ are used to verify the respective proofs.

6   Conclusion and Future Work

We have proposed a method for the modular design of cryptographic protocols in the UC framework. Our method allows one to compose two or more functionalities and to ensure that some inputs to those functionalities are equal or an output of one functionality is used as input to another functionality. For this purpose, the functionalities are amended to receive commitments as inputs and to verify them. In addition, we propose new ideal functionalities for commitments that, unlike existing ones, output cryptographic commitments. To illustrate our framework, we have shown a protocol for attribute tokens with revocation that uses our commitment functionality and ideal functionalities for attribute tokens and for revocation that receive committed inputs. As future work, we consider the modular design of other cryptographic protocols with our method as well as investigating the relations between our UC-based definitions and game-based definitions for attribute-based tokens and revocation.

References

1. Camenisch, J., Kohlweiss, M., Soriente, C.: An accumulator based on bilinear maps and efficient revocation for anonymous credentials. In: PKC, pp. 481–500 (2009)
2. Camenisch, J., Kohlweiss, M., Soriente, C.: Solving revocation with efficient update of anonymous credentials. In: Garay, J.A., De Prisco, R. (eds.) SCN 2010. LNCS, vol. 6280, pp. 454–471. Springer, Heidelberg (2010)
3. Camenisch, J., Krenn, S., Lehmann, A., Mikkelsen, G.L., Neven, G., Pedersen, M.Ø.: Formal treatment of privacy-enhancing credential systems. In: Dunkelman, O., et al. (eds.) SAC 2015. LNCS, vol. 9566, pp. 3–24. Springer, Heidelberg (2016). doi:10.1007/978-3-319-31301-6_1
4. Camenisch, J., Lehmann, A., Neven, G., Rial, A.: Privacy-preserving auditing for attribute-based credentials. In: ESORICS, pp. 109–127 (2014)
5. Camenisch, J.L., Lysyanskaya, A.: Dynamic accumulators and application to efficient revocation of anonymous credentials. In: Yung, M. (ed.) CRYPTO 2002. LNCS, vol. 2442, p. 61. Springer, Heidelberg (2002)
6. Canetti, R.: Universally composable security: a new paradigm for cryptographic protocols. In: FOCS, pp. 136–145 (2001)
7. Canetti, R.: Universally composable signature, certification, and authentication. In: CSFW, p. 219 (2004)
8. Canetti, R., Fischlin, M.: Universally composable commitments. In: Kilian, J. (ed.) CRYPTO 2001. LNCS, vol. 2139, p. 19. Springer, Heidelberg (2001)

9. Canetti, R., Lindell, Y., Ostrovsky, R., Sahai, A.: Universally composable two-party and multi-party secure computation. In: STOC, pp. 494–503 (2002)
10. Catalano, D., Fiore, D.: Vector commitments and their applications. In: PKC, pp. 55–72 (2013)
11. Damgård, I.B., Nielsen, J.B.: Perfect hiding and perfect binding universally composable commitment schemes with constant expansion factor. In: Yung, M. (ed.) CRYPTO 2002. LNCS, vol. 2442, p. 581. Springer, Heidelberg (2002)
12. Frederiksen, T.K., Jakobsen, T.P., Nielsen, J.B., Trifiletti, R.: On the complexity of additively homomorphic UC commitments. ePrint, Report 2015/694
13. Groth, J.: Homomorphic trapdoor commitments to group elements. ePrint, 2009/007
14. Hofheinz, D., Backes, M.: How to break and repair a universally composable signature functionality. In: ICS, pp. 61–72 (2004)
15. Hofheinz, D., Müller-Quade, J.: Universally composable commitments using random oracles. In: Naor, M. (ed.) TCC 2004. LNCS, vol. 2951, pp. 58–76. Springer, Heidelberg (2004)
16. Katz, J.: Universally composable multi-party computation using tamper-proof hardware. In: EUROCRYPT, pp. 115–128 (2007)
17. Lindell, Y.: Highly-efficient universally-composable commitments based on the DDH assumption. In: EUROCRYPT, pp. 446–466 (2011)
18. MacKenzie, P., Yang, K.: On simulation-sound trapdoor commitments. In: EUROCRYPT, pp. 382–400 (2004)
19. Moran, T., Segev, G.: David, goliath commitments: UC computation for asymmetric parties using tamper-proof hardware. In: EUROCRYPT, pp. 527–544 (2008)
20. Nakanishi, T., Fujii, H., Yuta, H., Funabiki, N.: Revocable group signature schemes with constant costs for signing and verifying. In: IEICE Transactions on Fundamentals of Electronics, Communications and Computer Sciences, pp. 50–62 (2010)
21. Nguyen, L.: Accumulators from bilinear pairings and applications. In: Menezes, A. (ed.) CT-RSA 2005. LNCS, vol. 3376, pp. 275–292. Springer, Heidelberg (2005)
22. Pedersen, T.P.: Non-interactive and information-theoretic secure verifiable secret sharing. In: CRYPTO, pp. 129–140 (1992)

Probabilistic Termination and Composability of Cryptographic Protocols

Ran Cohen[1]([✉]), Sandro Coretti[2], Juan Garay[3], and Vassilis Zikas[4]

[1] Department of Computer Science, Bar-Ilan University, Ramat Gan, Israel
cohenrb@cs.biu.ac.il
[2] Department of Computer Science, ETH Zurich, Zürich, Switzerland
corettis@inf.ethz.ch
[3] Yahoo Research, Sunnyvale, USA
garay@yahoo-inc.com
[4] Department of Computer Science, RPI, Troy, USA
vzikas@cs.rpi.edu

Abstract. When analyzing the round complexity of multi-party computation (MPC), one often overlooks the fact that underlying resources, such as a broadcast channel, can by themselves be expensive to implement. For example, it is impossible to implement a broadcast channel by a (deterministic) protocol in a sub-linear (in the number of corrupted parties) number of rounds. The seminal works of Rabin and Ben-Or from the early 80's demonstrated that limitations as the above can be overcome by allowing parties to terminate in different rounds, igniting the study of protocols with probabilistic termination. However, absent a rigorous simulation-based definition, the suggested protocols are proven secure in a property-based manner, guaranteeing limited composability. In this work, we define MPC with probabilistic termination in the UC framework. We further prove a special universal composition theorem for probabilistic-termination protocols, which allows to compile a protocol using deterministic-termination hybrids into a protocol that uses expected-constant-round protocols for emulating these hybrids, preserving the expected round complexity of the calling protocol.

We showcase our definitions and compiler by providing the first composable protocols (with simulation-based security proofs) for the following primitives, relying on point-to-point channels: (1) expected-constant-round perfect Byzantine agreement, (2) expected-constant-round perfect parallel broadcast, and (3) perfectly secure MPC with round complexity independent of the number of parties.

The full version of this paper can be found at the *IACR Cryptology ePrint Archive* [16].

R. Cohen—Work supported by a grant from the Israel Ministry of Science, Technology and Space (grant 3-10883) and by the National Cyber Bureau of Israel.

S. Coretti—Work supported by the Swiss NSF project no. 200020-132794.

J. Garay and V. Zikas—Work done in part while the author was visiting the Simons Institute for the Theory of Computing, supported by the Simons Foundation and by the DIMACS/Simons Collaboration in Cryptography through NSF grant #CNS-1523467.

V. Zikas—Work supported in part by the Swiss NSF Ambizione grant PZ00P2_142549.

© International Association for Cryptologic Research 2016
M. Robshaw and J. Katz (Eds.): CRYPTO 2016, Part III, LNCS 9816, pp. 240–269, 2016.
DOI: 10.1007/978-3-662-53015-3_9

1 Introduction

In secure multi-party computation (MPC) [27,49] n parties $P_1, \ldots, P_n$ wish to jointly perform a computation on their private inputs in a secure way, so that no coalition of cheating parties can learn more information than their outputs (privacy) or affect the outputs of the computation any more than by choosing their own inputs (correctness).

While the original security definitions had the above property-based flavor (i.e., the protocols were required to satisfy correctness and privacy—potentially along with other security properties, such as fairness and input independence), it is by now widely accepted that security of multi-party cryptographic protocols should be argued in a simulation-based manner. Informally, in the simulation paradigm for security, the protocol execution is compared to an ideal world where the parties have access to a trusted third party (TTP, aka the "ideal functionality") that captures the security properties the protocol is required to achieve. The TTP takes the parties' inputs and performs the computation on their behalf. A protocol is regarded as secure if for any adversary attacking it, there exists an ideal adversary (the simulator) attacking the execution in the ideal world, such that no external distinguisher (environment) can tell the real and the ideal executions apart.

There are several advantages in proving a protocol secure in this way. For starters, the definition of the functionality captures all security properties the protocol is supposed to have, and therefore its design process along with the security proof often exposes potential design flaws or issues that have been overlooked in the protocol design. A very important feature of many simulation-based security definitions is composability, which ensures that a protocol can be composed with other protocols without compromising its security. Intuitively, composability ensures that if a protocol $\pi^{\mathcal{G}}$ which uses a "hybrid" $\mathcal{G}$ (a broadcast channel, for example) securely realizes functionality $\mathcal{F}$, and protocol ρ securely realizes the functionality $\mathcal{G}$, then the protocol $\pi^{\rho/\mathcal{G}}$, which results by replacing in π calls to $\mathcal{G}$ by invocations of ρ, securely realizes $\mathcal{F}$. In fact, simulation-based security is the one and only way known to ensure that a protocol can be generically used to implement its specification within an arbitrary environment.

Round Complexity. The prevalent model for the design of MPC protocols is the synchronous model, where the protocol proceeds in rounds and all messages sent in any given round are received by the beginning of the next round. In fact, most if not all implemented and highly optimized MPC protocols (e.g., [15,18,20,37, 43]) are in this model. When executing such synchronous protocols over large networks, one needs to impose a long round duration in order to account for potential delay at the network level, since if the duration of the rounds is too short, then it is likely that some of the messages that arrive late will be ignored or, worse, assigned to a later round. Thus, the round complexity, i.e., the number of rounds it takes for a protocol to deliver outputs, is an important efficiency metric for such protocols and, depending on the network parameters, can play a dominant role in the protocol's running time.

An issue often overlooked in the analysis of the round complexity of protocols is that the relation between a protocol's round complexity and its actual running time is sensitive to the "hybrids" (e.g., network primitives) that the protocol is assumed to have access to. For example, starting with the seminal MPC works [6,14,27,47,49], a common assumption is that the parties have access to a broadcast channel, which they invoke in every round. In reality, however, such a broadcast channel might not be available and would have to be implemented by a broadcast protocol designed for a point-to-point network. Using a standard (deterministic) broadcast protocol for this purpose incurs a linear (in n, the number of parties[1]) blow-up on the round complexity of the MPC protocol, as no deterministic broadcast protocol can tolerate a linear number of corruptions and terminate in a sublinear number of rounds [22,24]. Thus, even though the round complexity of these protocols is usually considered to be linear in the multiplicative depth d of the computed circuit, in reality their running time could become linear in nd (which can be improved to $O(n + d)$ [34]) when executed over point-to-point channels.[2]

In fact, all so-called constant-round multi-party protocols (e.g., [1,3,17,25, 30,32,38,44]) rely on broadcast rounds—rounds in which parties make calls to a broadcast channel—and therefore their running time when broadcast is implemented by a standard protocol would explode to be linear in n instead of constant.[3] As the results from [22,24] imply, this is not a consequence of the specific choice of protocol but a limitation of any protocol in which there is a round such that all parties are guaranteed to have received their output; consistently with the literature on fault-tolerant distributed computing, we shall refer to protocols satisfying this property as *deterministic-termination* protocols. In fact, to the best of our knowledge, even if we allow a negligible chance for the broadcast to fail, the fastest known solutions tolerating a constant fraction of corruptions follow the paradigm from [23] (see below), which requires a poly-logarithmic (in n) number of rounds.[4]

Protocols with Probabilistic Termination. A major breakthrough in fault-tolerant distributed algorithms (recently honored with the 2015 Dijkstra Prize in Distributed Computing), was the introduction of randomization to the field by Ben-Or [4] and Rabin [46], which, effectively, showed how to circumvent the above limitation by using randomization. Most relevant to this submission, Rabin [46] showed that

---

[1] More precisely, in the number of corruptions a protocol can tolerate, which is a constant fraction of n.

[2] Throughout this work we will consider protocols in which all parties receive their output. If one relaxes this requirement (i.e., allows that some parties may not receive their output and give up on fairness) then the techniques of Goldwasser and Lindell [29] allow for replacing broadcast with a constant-round multi-cast primitive.

[3] We remark that even though those protocols are for the computational setting, the lower bound on broadcast round complexity also applies.

[4] Note that this includes even FHE-based protocols, as they also assume a broadcast channel and their security fails if multi-cast over point-to-point channels is used instead.

linearly resilient *Byzantine agreement* protocols [40, 45] (BA, related to broadcast, possibility- and impossibility-wise) in expected *constant* rounds were possible, provided that all parties have access to a "common coin" (i.e., a common source of randomness).[5] This line of research culminated with the work of Feldman and Micali [23], who showed how to obtain a shared random coin with constant probability from "scratch," yielding a probabilistic BA protocol tolerating the maximum number of misbehaving parties ($t < n/3$) that runs in expected constant number of rounds. The randomized BA protocol in [23] works in the information-theoretic setting; these results were later extended to the computational setting by Katz and Koo [33], who showed that assuming digital signatures there exists an (expected-)constant-round protocol for BA tolerating $t < n/2$ corruptions. The speed-up on the running time in all these protocols, however, comes at the cost of uncertainty, as now they need to give up on guaranteed (eventual) termination (no fixed upper bound on their running time[6]) as well as on *simultaneous* termination (a party that terminates cannot be sure that other parties have also terminated[7]) [21]. These issues make the simulation-based proof of these protocols a very delicate task, which is the motivation for the current work.

What made the simulation-based approach a more accessible technique in security proofs was the introduction simulation-based security frameworks. The ones that stand out in this development—and are most often used in the literature—are Canetti's modular composition (aka stand-alone security) [9] and the universal composition (UC) frameworks [10, 11]. The former defines security of synchronous protocols executed in isolation (i.e., only a single protocol is run at a time, and whenever a subroutine-protocol is called, it is run until its completion); the latter allows protocols to be executed alongside arbitrary (other) protocols and be interleaved in an arbitrary manner. We remark that although the UC framework is inherently asynchronous, several mechanisms have been proposed to allow for a synchronous execution within it (e.g., [11, 12, 36, 39]).

Despite the wide-spread use of the simulation-based paradigm to prove security of protocols with deterministic termination, the situation has been quite different when probabilistic-termination protocols are considered. Here, despite the existence of round-efficient BA protocols as mentioned above [23, 33], to our knowledge, no formal treatment of the problem in a simulation-based model exists, which would allow us to apply the ingenious ideas of Rabin and Ben-Or in order to speed up cryptographic protocols. We note that Katz and Koo [33] even provided an expected-constant-round MPC protocol using their fast BA protocol as a subroutine, employing several techniques to ensure proper use of randomized BA. In lack, however, of a formal treatment, existing constructions

---

[5] Essentially, the value of the coin can be adopted by the honest parties in case disagreement at any given round is detected, a process that is repeated multiple times.

[6] Throughout this paper we use running time and round complexity interchangeably.

[7] It should be noted however that in many of these protocols there is a known (constant) "slack" of c rounds, such that if a party terminates in round r, then it can be sure that every honest party will have terminated by round $r + c$.

are usually proved secure in a property-based manner or rely on *ad hoc*, less studied security frameworks [42].[8]

A   simulation-based   and   composable   treatment   of   such probabilistic-termination (PT for short) protocols would naturally allow, for example, to replace the commonly used broadcast channel with a broadcast protocol, so that the expected running time of the resulting protocol is the same as the one of the original (broadcast-hybrid) protocol. A closer look at this replacement, however, exposes several issues that have to do not only with the lack of simulation-based security but also with other inherent limitations. Concretely, it is usually the case in an MPC protocol that the broadcast channel is accessed by several (in many cases by all) parties in the same (broadcast) round in parallel. Ben-Or and El-Yaniv [5] observed that if one naïvely replaces each such invocation by a PT broadcast protocol with expected constant running-time, then the expected number of rounds until *all* broadcasts terminate is no longer constant; in fact, it is not hard to see that in the case of [23], the expected round complexity would be logarithmic in the number of instances (and therefore also in the player-set size). Nevertheless, in [5] a mechanism was proposed for implementing such parallel calls to broadcast so that the total number of rounds remains constant.

The difficulties arising with generic parallel composition are not the only issue with PT protocols. As observed by Lindell et al. [42], composing such protocols in sequence is also problematic. The main issue here is that, as already mentioned, PT protocols do not have simultaneous termination and therefore a party cannot be sure how long after he receives his output from a call to such a PT protocol he can safely carry on with the execution of the calling protocol. Although PT protocols usually guarantee a constant "slack" of rounds (say, c) in the output of any two honest parties, the naïve approach of using this property to synchronize the parties—i.e., wait c rounds after the first call, $2c$ rounds after the second call, and so on—imposes an exponential blow-up on the round complexity of the calling protocol. To resolve this, [42] proposed using fixed points in time at which a re-synchronization subroutine is executed, allowing the parties to ensure that they never get too far out-of-sync. Alternative approaches for solving this issue was also proposed in [8,33] but, again, with a restricted (property-based) proof.

Despite their novel aspects, the aforementioned results on composition of PT protocols do not use simulation-based security, and therefore it is unclear how (or if) they could be used to, for example, instantiate broadcast within a higher-level cryptographic protocol. In addition, they do not deal with other important features of modern security definitions, such as adaptive security and strict polynomial time execution. In fact, this lack of a formal cryptographic treatment places some of their claims at odds with the state-of-the-art cryptographic definitions—somewhat pointedly, [5] claims adaptive security, which, although

---

[8] As we discuss below, the protocol of Katz and Koo has an additional issue with adaptive security in the rushing adversary model, as defined in the UC framework, similar to the issue exploited in [31].

it can be shown to hold in a property-based definition, is not achieved by the specified construction when simulation-based security is considered (cf. Sect. 5).

Our Contributions. In this paper we provide the first formal simulation-based (and composable) treatment of MPC with probabilistic termination. Our treatment builds on Canetti's universal composition (UC) framework [10,11]. In order to take advantage of the fast termination of PT protocols, parties typically proceed at different paces and therefore protocols might need to be run in an interleaved manner—e.g., in an MPC protocol a party might initiate the protocol for broadcasting his r-round message before other parties have received output from the broadcasting of messages for round $r - 1$. This inherent concurrency along with its support for synchrony makes the UC framework the natural candidate for our treatment.

Our motivating goal, which we achieve, is to provide a generic compiler that allows us to transform any UC protocol π making calls to deterministic-termination UC protocols ρ_i in a "stand-alone fashion" (similar to [9], i.e., the protocols ρ_i are invoked sequentially and in each round exactly one protocol is being executed by all the parties) into a protocol in which each ρ_i is replaced by a (faster) PT protocol ρ_i'. The compiled protocol achieves the same security as π and has (expected) round complexity proportional to $\sum_i d_i r_i$, where d_i is the expected number of calls π makes to ρ_i and r_i is the expected round complexity of ρ_i.

Towards this goal, the first step is to define what it means for a protocol to (UC-)securely realize a functionality with probabilistic termination in a simulation-based manner, by proposing an explicit formulation of the functionality that captures this important protocol aspect. The high-level idea is to parameterize the functionality with an efficiently sampleable distribution D that provides an upper bound on the protocol's running time (i.e., number of rounds), so that the adversary cannot delay outputs beyond this point (but is allowed to deliver the output to honest parties earlier, and even in different rounds).

Next, we prove our universal composability result. Informally, our result provides a generic compiler that takes as input a "stand-alone" protocol ρ, realizing a probabilistic-termination functionality $\mathcal{F}^D$ (for a given distribution D) while making sequential calls to (deterministic-termination) secure function evaluation (SFE)-like functionalities, and compiles it into a new protocol ρ' in which the calls to the SFEs are replaced by probabilistic-termination protocols realizing them. The important feature of our compiler is that in the compiled protocol, the parties do not need to wait for every party to terminate their emulation of each SFE to proceed to the emulation of the next SFE. Rather, shortly after a party (locally) receives its output from one emulation, it proceeds to the next one. This yields an (at most) multiplicative blow-up on the expected round complexity as discussed above. In particular, if the protocols used to emulate the SFE's are expected constant round, then the expected round complexity of ρ' is the same (asymptotically) as that of ρ.

We then showcase our definition and composition theorem by providing simulation-based (therefore composable) probabilistic-termination protocols and security proofs for several primitives relying on point-to-point channels: expected-constant-round perfect Byzantine agreement, expected-constant-round perfect parallel broadcast, and perfectly secure MPC with round complexity independent of the number of parties. Not surprisingly, the simulation-based treatment reveals several issues, both at the formal and at the intuitive levels, that are not present in a property-based analysis, and which we discuss along the way. We now elaborate on each application in turn. Regarding Byzantine agreement, we present a protocol that perfectly securely UC-implements the probabilistic-termination Byzantine agreement functionality for $t < n/3$ in an expected-constant number of rounds. (We will use RBA to denote probabilistic-termination BA, as it is often referred to as "randomized BA."[9]) Our protocol follows the structure of the protocol in [23], with a modification inspired by Goldreich and Petrank [28] to make it strict polynomial time (see the discussion below), and in a sense it can be viewed as the analogue for RBA of the well-known "CLOS" protocol for MPC [13]. Indeed, similarly to how [13] converted (and proved) the "GMW" protocol [26] from statically secure in the stand-alone setting into an adaptively secure UC version, our work transforms the broadcast and BA protocols from [23] into adaptively UC-secure randomized broadcast and RBA protocols.[10]

Our first construction above serves as a good showcase of the power of our composition theorem, demonstrating how UC-secure RBA is built in a modular manner: First, we de-compose the sub-routines that are invoked in [23] and describe simple(r) (SFE-like) functionalities corresponding to these sub-routines; this provides us with a simple "backbone" of the protocol in [23] making calls to these hybrids, which can be easily proved to implement expected-constant-round RBA. Next, we feed this simplified protocol to our compiler which outputs a protocol that implements RBA from point-to-point secure channels; our composition theorem ensures that the resulting protocol is also expected constant round.

There is a sticky issue here that we need to resolve for the above to work: the protocol in [23] does not have guaranteed termination and therefore the distribution of the terminating round is not sampleable by a strict probabilistic polynomial-time (PPT) machine.[11] A way around this issue would be to modify the UC model of execution so that the corresponding ITMs are expected PPTs. Such a modification, however, would impact the UC model of computation, and would therefore require a new proof of the composition theorem—a trickier task than one might expect, as the shift to expected polynomial-time simulation is known to introduce additional conceptual and technical difficulties (cf. [35]),

---

[9] BA is a deterministic output primitive and it should be clear that the term "randomized" can only refer to the actual number of rounds; however, to avoid confusion we will abstain from using this term for functionalities other than BA whose output might also be probabilistic.

[10] As we show, the protocol in [23] does not satisfy input independence, and therefore is not adaptively secure in a simulation-based manner (cf. [31]).

[11] All entities in UC, and in particular ideal functionalities, are strict interactive PPT Turing machines, and the UC composition theorem is proved for such PPT ITMs.

whose resolution is beyond the scope of this work. Instead, here we take a different approach which preserves full compatibility with the UC framework: We adapt the protocol from [23] using ideas from [28] so that it implements a functionality which samples the terminating round with almost the same probability distribution as in [23], but from a finite (linear-size) domain; as we show, this distribution is sampleable in strict polynomial time and can therefore be used by a standard UC functionality.

Next, we use our composition theorem to derive the first simulation-based and adaptively (UC) secure parallel broadcast protocol, which guarantees that all broadcast values are received within an expected constant number of rounds. This extends the results from [5,33] in several ways: first, our protocol is perfectly UC-secure which means that we can now use it within a UC-secure SFE protocol to implement secure channels, and second, it is adaptively secure against a rushing adversary.[12]

Finally, by applying once again our compiler to replace calls to the broadcast channel in the SFE protocol by Ben-Or, Goldwasser, and Wigderson [6] (which, recall, is perfectly secure against $t < n/3$ corruptions in the broadcast-hybrid model [2]) by invocations to our adaptively secure UC parallel broadcast protocol, we obtain the first UC-secure PT MPC protocol in the point-to-point secure channels model with (expected) round complexity $O(d)$, independently of the number of parties, where d is the multiplicative depth of the circuit being computed. As with RBA, this result can be seen as the first analogue of the UC compiler by Canetti et al. [13] for SFE protocols with probabilistic termination.

We stress that the use of perfect security to showcase our composition theorem is just our choice and not a restriction of our composition theorem. In fact, our theorem can be also applied to statistically or computationally secure protocols. Moreover, if one is interested in achieving better constants in the (expected) round complexity then one can use SFE protocols that attempt to minimize the use of the broadcast channel (e.g., [34]). Our composition theorem will give a direct methodology for this replacement and will, as before, eliminate the dependency of the round complexity from the number of parties.[13]

2   Model

We consider n parties $P_1, \ldots, P_n$ and an adaptive t-adversary, i.e., the adversary corrupts up to t parties during the protocol execution.[14] We work in the UC model and assume the reader has some familiarity with its basics. To capture synchronous protocols in UC we use the framework of Katz et al. [36]. Concretely,

---

[12] Although security against a "dynamic" adversary is also claimed in [5], the protocol does not implement the natural parallel broadcast functionality in the presence of an adaptive adversary (see Sect. 5).

[13] Note that even a single round of broadcast is enough to create the issues with parallel composition and non-simultaneous termination discussed above.

[14] In contrast, a *static* adversary chooses the set of corrupted parties at the onset of the computation.

the assumption that parties are synchronized is captured by assuming that the protocol has access to a "clock" functionality $\mathcal{F}_{\text{CLOCK}}$. The functionality $\mathcal{F}_{\text{CLOCK}}$ maintains an indicator bit which is switched once *all honest parties* request the functionality to do it. At any given round, a party asks $\mathcal{F}_{\text{CLOCK}}$ to turn the bit on only after having finished with all operations for the current round. Thus, this bit's value can be used to detect when every party has completed his round, in which case they can proceed to the next round. As a result, this mechanism ensures that no party sends his messages for round $r + 1$ before every party has completed round r. For clarity, we retain from writing this clock functionality in our theorem statement; however, all our results assume access to such a clock functionality.

In the communication network of [36], parties have access to bounded-delay secure channels. These channels work in a so-called "fetch" mode, i.e., in order to receive his output the receiver issues a `fetch-output` command. This allows to capture the property of a channel between a sender P_s and a receiver P_r, delaying the delivery of a message by an amount δ: as soon as the sender P_s submits an input y (message to be sent to the receiver) the channel functionality starts counting how many times the receiver requests it.[15] The first $\delta - 1$ such `fetch-output` requests (plus all such requests that are sent before the sender submits input) are ignored (and the adversary is notified about them); the δth `fetch-output` request following a submitted input y from the sender results in the channel sending (`output`, y) to P_r. In this work we take an alternative approach and model secure channels as special simple SFE functionalities. These SFEs also work in a fetch mode[16] and provide the same guarantee as the bounded-delay channels.

There are two important considerations in proving the security of a synchronous UC protocol: (1) The simulator needs to keep track of the protocol's current round, and (2) because parties proceed at the same pace, they can synchronize their reaction to the environment; most fully synchronous protocols, for example, deliver output exactly after a given number of rounds. In [36] this property is captured as follows: The functionality keeps track of which round the protocol would be in by counting the number of activations it receives from honest parties. Thus, if the protocol has a regular structure, where every party advances the round after receiving a fixed number μ of activations from its environment (all protocols described herein will be in this form), the functionality can easily simulate how rounds in the protocol advance by incrementing its round index whenever it receives μ messages from all honest parties; we shall refer to such a functionality as a *synchronous functionality*. Without loss of generality, in this work we will describe all functionalities for $\mu = 1$, i.e., once a functionality receives a message from every party it proceeds to the simulation of the next protocol round. We stress that this is done to simplify the description, and the

---

[15] Following the simplifying approach of [36], we assume that communication channels are single use, thus each message transmission uses an independent instance of the channel.

[16] In fact, for simplicity we assume that they deliver output on the first "fetch".

in an actual evaluation, as in the synchronous setting of [36], in order to give the simulator sufficiently many activations to perform its simulation, functionalities typically have to wait for $\mu > 1$ messages from each party where the last $\mu - 1$ of these messages are typically "dummy" activations (usually of the type fetch-output).

To further simplify the description of our functionalities, we introduce the following terminology. We say that *a synchronous functionality $\mathcal{F}$ is in round ρ* if the current value of the above internal round counter in $\mathcal{F}$ is $r = \rho$. All synchronous functionalities considered in this work have the following format: They treat the first message they receive from any party P_i as P_i's input[17]—if this message is not of the right form (input, ·) then a default value is taken as P_i input; as soon as an honest party sends its first message, any future message by this party is treated as a fetch-output message.

3 Secure Computation with Probabilistic Termination

The work of Katz et al. [36] addresses (synchronous) cryptographic protocols that terminate in a fixed number of rounds for all honest parties. However, as mentioned in Sect. 1, Ben-Or [4] and Rabin [46] showed that in some cases, great asymptotic improvements on the *expected* termination of protocols can be achieved through the use of randomization. Recall, for example, that in the case of BA, even though a lower bound of $O(n)$ on the round complexity of any deterministic BA protocol tolerating $t = \Omega(n)$ corruptions exists [22,24], Rabin's global-coin technique (fully realized later on in [23]) yields an expected-constant-round protocol. This speed-up, however, comes at a price, namely, of relinquishing both *fixed* and *simultaneous* termination [21]: the round complexity of the corresponding protocols may depend on random choices made during the execution, and parties may obtain output from the protocol in different rounds.

In this section we show how to capture protocols with such *probabilistic termination (PT)*, i.e., without fixed and without simultaneous termination, within the UC framework. To capture probabilistic termination, we first introduce a functionality template $\mathcal{F}_{\text{CSF}}$ called a *canonical synchronous functionality (CSF)*. $\mathcal{F}_{\text{CSF}}$ is a simple two-round functionality with explicit (one-round) input and (one-round) output phases. Computation with probabilistic termination is then defined by wrapping $\mathcal{F}_{\text{CSF}}$ with an appropriate functionality *wrapper* that enables non-fixed, non-simultaneous termination.

3.1 Canonical Synchronous Functionalities

At a high level, $\mathcal{F}_{\text{CSF}}$ corresponds to a generalization of the UC secure function evaluation (SFE) functionality to allow for potential leakage on the inputs to the

---

[17] Note that this implies that also protocol machines treats its first message as their input.

adversary and potential adversarial influence on the outputs.[18] In more detail, $\mathcal{F}_{\mathrm{CSF}}$ has two parameters: (1) a (possibly) randomized function f that receives $n+1$ inputs (n inputs from the parties and one additional input from the adversary) and (2) a leakage function l that leaks some information about the input values to the adversary.

$\mathcal{F}_{\mathrm{CSF}}$ proceeds in two rounds: in the first round all the parties hand $\mathcal{F}_{\mathrm{CSF}}$ their input values, and in the second round each party receives its output. This is very similar to the standard (UC) SFE functionality; the difference here is that whenever some input is submitted to $\mathcal{F}_{\mathrm{CSF}}$, the adversary is handed some leakage function of this input—similarly, for example, to how UC secure channels leak the message length to the adversary. The adversary can use this leakage when deciding the inputs of corrupted parties. Additionally, he is allowed to input an extra message, which—depending on the function f—might affect the output(s). The detailed description of $\mathcal{F}_{\mathrm{CSF}}$ is given in Fig. 1.

Functionality $\mathcal{F}_{\mathrm{CSF}}^{f,l}(\mathcal{P})$

$\mathcal{F}_{\mathrm{CSF}}$ proceeds as follows, parametrized by a function $f\colon (\{0,1\}^* \cup \{\bot\})^{n+1} \to (\{0,1\}^*)^n$ and a leakage function $l\colon (\{0,1\}^* \cup \{\bot\})^n \to \{0,1\}^*$, and running with parties $\mathcal{P} = \{P_1, \ldots, P_n\}$ and an adversary $\mathcal{S}$.

- Initially, set the input values $x_1, \ldots, x_n$, the output values $y_1, \ldots, y_n$, and the adversary's value a to $\bot$.
- In round $\rho = 1$:
  - Upon receiving $(\mathtt{adv\text{-}input}, \mathtt{sid}, v)$ from the adversary, set $a \leftarrow v$.
  - Upon receiving a message $(\mathtt{input}, \mathtt{sid}, v)$ from some party $P_i \in \mathcal{P}$, set $x_i \leftarrow v$ and send $(\mathtt{leakage}, \mathtt{sid}, P_i, l(x_1, \ldots, x_n))$ to the adversary.
- In round $\rho = 2$:
  - Upon receiving $(\mathtt{adv\text{-}input}, \mathtt{sid}, v)$ from the adversary, if $y_1 = \ldots = y_n = \bot$, set $a \leftarrow v$. Otherwise, discard the message.
  - Upon receiving $(\mathtt{fetch\text{-}output}, \mathtt{sid})$ from some party $P_i \in \mathcal{P}$, if $y_1 = \ldots = y_n = \bot$ compute $(y_1, \ldots, y_n) = f(x_1, \ldots, x_n, a)$. Next, send $(\mathtt{output}, \mathtt{sid}, y_i)$ to P_i and $(\mathtt{fetch\text{-}output}, \mathtt{sid}, P_i)$ to the adversary.

Fig. 1. The canonical synchronous functionality

Next, we point out a few technical issues about the description of $\mathcal{F}_{\mathrm{CSF}}$. Following the simplifications from Sect. 2, $\mathcal{F}_{\mathrm{CSF}}$ advances its round as soon as it receives $\mu = 1$ message from each honest party. This ensures that the adversary cannot make the functionality stall indefinitely. Thus, formally speaking, the functionality $\mathcal{F}_{\mathrm{CSF}}$ is not well-formed (cf. [13]), as its behavior depends on the identities of the corrupted parties.[19] We emphasize that the non-well-formedness

---

[18] Looking ahead, this adversarial influence will allow us to describe BA-like functionalities as simple and intuitive CSFs.

[19] This is, in fact, also the case for the standard UC SFE functionality.

relates only to advancing the rounds, and is unavoidable if we want to restrict the adversary not to block the evaluation indefinitely (cf. [36]).

We point out that as a generalization of the SFE functionality, CSFs are powerful enough to capture any deterministic well-formed functionality. In fact, all the basic (unwrapped) functionalities considered in this work will be CSFs. We now describe how standard functionalities from the MPC literature can be cast as CSFs:

- SECURE MESSAGE TRANSMISSION (AKA SECURE CHANNEL). In the *secure message transmission* (SMT) functionality, a sender P_i with input x_i sends its input to P_j. Since $\mathcal{F}_{\mathrm{CSF}}$ is an n-party functionality and involves receiving input messages from all n parties, we define the two-party task using an n-party function. The function to compute is $f_{\mathrm{SMT}}^{i,j}(x_1, \ldots, x_n, a) = (\lambda, \ldots, x_i, \ldots, \lambda)$ (where x_i is the value of the j'th coordinate) and the leakage function is $l_{\mathrm{SMT}}^{i,j}(x_1, \ldots, x_n) = y$, where $y = |x_i|$ in case P_j is honest and $y = x_i$ in case P_j is corrupted. We denote by $\mathcal{F}_{\mathrm{SMT}}^{i,j}$ the functionality $\mathcal{F}_{\mathrm{CSF}}$ when parametrized with the above functions $f_{\mathrm{SMT}}^{i,j}$ and $l_{\mathrm{SMT}}^{i,j}$, for sender P_i and receiver P_j.

- BROADCAST. In the (standard) *broadcast* functionality, a sender P_i with input x_i distributes its input to all the parties, i.e., the function to compute is $f_{\mathrm{BC}}^i(x_1, \ldots, x_n, a) = (x_i, \ldots, x_i)$. The adversary only learns the length of the message x_i before its distribution, i.e., the leakage function is $l_{\mathrm{BC}}^i(x_1, \ldots, x_n) = |x_i|$. This means that the adversary does not gain new information about the input of an honest sender before the output value for all the parties is determined, and in particular, the adversary *cannot* corrupt an honest sender and change its input *after* learning the input message. We denote by $\mathcal{F}_{\mathrm{BC}}^i$ the functionality $\mathcal{F}_{\mathrm{CSF}}$ when parametrized with the above functions f_{BC}^i and l_{BC}^i, for sender P_i.

- SECURE FUNCTION EVALUATION. In the *secure function evaluation* functionality, the parties compute a randomized function $g(x_1, \ldots, x_n)$, i.e., the function to compute is $f_{\mathrm{SFE}}^g(x_1, \ldots, x_n, a) = g(x_1, \ldots, x_n)$. The adversary learns the length of the input values via the leakage function, i.e., the leakage function is $l_{\mathrm{SFE}}(x_1, \ldots, x_n) = (|x_1|, \ldots, |x_n|)$. We denote by $\mathcal{F}_{\mathrm{SFE}}^g$ the functionality $\mathcal{F}_{\mathrm{CSF}}$ when parametrized with the above functions f_{SFE}^g and l_{SFE}, for computing the n-party function g.

- BYZANTINE AGREEMENT (AKA CONSENSUS). In the *Byzantine agreement* functionality, defined for the set V, each party P_i has input $x_i \in V$. The common output is computed such that if $n - t$ of the input values are the same, this will be the output; otherwise the adversary gets to decide on the output. The adversary is allowed to learn the content of each input value from the leakage (and so it can corrupt parties and change their inputs based on this information). The function to compute is $f_{\mathrm{BA}}(x_1, \ldots, x_n, a) = (y, \ldots, y)$ such that $y = x$ if there exists a value x such that $x = x_i$ for at least $n - t$ input values x_i; otherwise $y = a$. The leakage function is $l_{\mathrm{BA}}(x_1, \ldots, x_n) = (x_1, \ldots, x_n)$. We denote by $\mathcal{F}_{\mathrm{BA}}^V$ the functionality $\mathcal{F}_{\mathrm{CSF}}$ when parametrized with the above functions f_{BA} and l_{BA}, defined for the set V.

3.2 Probabilistic Termination in UC

Having defined CSFs, we turn to the notion of (non-reactive) computation with probabilistic termination. This is achieved by defining the notion of an *output-round randomizing wrapper*. Such a wrapper is parametrized by an efficient probabilistic algorithm D, termed the *round sampler*, that may depend on a specific protocol implementing the functionality. The round sampler D samples a round number ρ_{term} by which all parties are guaranteed to receive their outputs no matter what the adversary strategy is. Moreover, since there are protocols in which all parties terminate in the same round and protocols in which they do not, we consider two wrappers: the first, denoted $\mathcal{W}_{\text{strict}}$, ensures in a strict manner that all (honest) parties terminate in the same round, whereas the second, denoted $\mathcal{W}_{\text{flex}}$, is more flexible and allows the adversary to deliver outputs to individual parties at any time before round ρ_{term}.

A delicate issue that needs to be addressed is the following: While an ideal functionality can be used to abstractly describe a protocol's task, it cannot hide the protocol's round complexity. This phenomenon is inherent in the synchronous communication model: any environment can observe how many rounds the execution of a protocol takes, and, therefore, the execution of the corresponding ideal functionality must take the same number of rounds.[20]

As an illustration of this issue, let $\mathcal{F}$ be an arbitrary functionality realized by some protocol π. If $\mathcal{F}$ is to provide guaranteed termination (whether probabilistic or not), it must enforce an upper bound on the number of rounds that elapse until all parties receive their output. If the termination round of π is not fixed (but may depend on random choices made during its execution), this upper bound must be chosen according to the distribution induced by π.

Thus, in order to simulate correctly, the functionality $\mathcal{F}$ and π's simulator $\mathcal{S}$ must coordinate the termination round, and therefore $\mathcal{F}$ must pass the upper bound it samples to $\mathcal{S}$. However, it is not sufficient to simply inform the simulator about the guaranteed-termination upper bound ρ_{term}. Intuitively, the reason is that protocol π may make probabilistic choices as to the order in which it calls its hybrids (and, even worse, these hybrids may even have probabilistic termination themselves). Thus, $\mathcal{F}$ needs to sample the upper bound based on π and the protocols realizing the hybrids called by π. As $\mathcal{S}$ needs to emulate the entire protocol execution, it is now left with the task of trying to sample the protocol's choices conditioned on the upper bound it receives from $\mathcal{F}$. In general, however, it is unclear whether such a reverse sampling can be performed in (strict) polynomial time.

To avoid this issue and allow for an efficient simulation, we have $\mathcal{F}$ output all the coins that were used for sampling round ρ_{term} to $\mathcal{S}$. Because $\mathcal{S}$ knows the round sampler algorithm, it can reproduce the entire computation of the sampler and use it in its simulation. In fact, as we discuss below, it suffices for our proofs to have $\mathcal{F}$ output a *trace* of its choices to the simulator instead of all the coins that were used

---

[20] In particular, this means that most CSFs are not realizable, since they always guarantee output after two rounds.

to sample this trace. In the remainder of this section, we motivate and formally describe our formulation of such traces. The formal description of the wrappers, which in particular sample traces, can then be found at the end of this section.

Execution Traces. As mentioned above, in the synchronous communication model, the execution of the ideal functionality must take the same number of rounds as the protocol. For example, suppose that the functionality $\mathcal{F}$ in our illustration above is used as a hybrid by a higher-level protocol π'. The functionality $\mathcal{G}$ realized by π' must, similarly to $\mathcal{F}$, choose an upper bound on the number of rounds that elapse before parties obtain their output. However, this upper bound now not only depends on π' itself but also on π (in particular, when π is a probabilistic-termination protocol).

Given the above, the round sampler of a functionality needs to keep track of how the functionality was realized. This can be achieved via the notion of *trace*. A trace basically records which hybrids were called by a protocol, and in a recursive way, for each hybrid, which hybrids would have been called by a protocol realizing that hybrid. The recursion ends with the hybrids that are "assumed" by the model, called *atomic* functionalities.[21]

Building on our running illustration above, suppose protocol π' (realizing $\mathcal{G}$) makes ideal hybrid calls to $\mathcal{F}$ and to some atomic functionality $\mathcal{H}$. Assume that in an example execution, π' happens to make (sequential) calls to instances of $\mathcal{H}$ and $\mathcal{F}$ in the following order: $\mathcal{F}$, then $\mathcal{H}$, and finally $\mathcal{F}$ again. Moreover, assume that $\mathcal{F}$ is replaced by protocol π (realizing $\mathcal{F}$) and that π happens to make two (sequential) calls to $\mathcal{H}$ upon the first invocation by π', and three (sequential) calls to $\mathcal{H}$ the second time. Then, this would result in the trace depicted in Fig. 2.

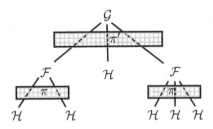

Fig. 2. Example of an execution trace

Assume that π is a probabilistic-termination protocol and π' a deterministic-termination protocol. Consequently, this means that $\mathcal{F}$ is in fact a flexibly wrapped functionality of some CSF $\mathcal{F}'$, i.e., $\mathcal{F} = \mathcal{W}_{\text{flex}}^{D_{\mathcal{F}}}(\mathcal{F}')$, where the distribution $D_{\mathcal{F}}$ samples (from a distribution induced by π) depth-1 traces with root $\mathcal{W}_{\text{flex}}^{D_{\mathcal{F}}}(\mathcal{F}')$ and leaves $\mathcal{H}$.[22] Similarly, $\mathcal{G}$ is a strictly wrapped functionality

---

[21] In this work, atomic functionalities are always $\mathcal{F}_{\text{PSMT}}$ CSFs.

[22] Note that the root node of the trace sampled from $D_{\mathcal{F}}$ is merely *labeled* by $\mathcal{W}_{\text{flex}}^{D_{\mathcal{F}}}(\mathcal{F}')$, i.e., this is not a circular definition.

of some CSF $\mathcal{G}'$, i.e., $\mathcal{G} = \mathcal{W}^{D_\mathcal{G}}_{\text{strict}}(\mathcal{G}')$, where the distribution $D_\mathcal{G}$ first samples (from a distribution induced by π') a depth-1 trace with root $\mathcal{W}^{D_\mathcal{G}}_{\text{strict}}(\mathcal{G}')$ and leaves $\mathcal{W}^{D_\mathcal{F}}_{\text{flex}}(\mathcal{F}')$ as well as $\mathcal{H}$. Then, each leaf node $\mathcal{W}^{D_\mathcal{F}}_{\text{flex}}(\mathcal{F}')$ is replaced by a trace (independently) sampled from $D_\mathcal{F}$. Thus, the example trace from Fig. 2 would look as in Fig. 3.

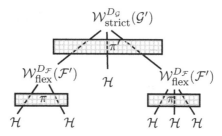

Fig. 3. An execution trace with probabilistic-termination and deterministic-termination protocols

Formally, a trace is defined as follows:

Definition 1 (Traces). *A trace is a rooted tree of depth at least 1, in which all nodes are labeled by functionalities and where every node's children are ordered. The root and all internal nodes are labeled by wrapped CSFs (by either of the two wrappers), and the leaves are labeled by unwrapped CSFs. The* trace complexity *of a trace T, denoted $c_{\text{tr}}(T)$, is the number of leaves in T. Moreover, denote by* $\text{flex}_{\text{tr}}(T)$ *the number of flexibly wrapped CSFs in T.*

Remark. The actual trace of a protocol may depend on the input values and the behavior of the adversary. For example, in the setting of Byzantine agreement, the honest parties may get the output faster in case they all have the same input, which results in a different trace. However, the wrappers defined below sample traces independently of the inputs. All protocols considered in this work can be shown to realize useful ideal functionalities in spite of this restriction.

Strict Wrapper Functionality. We now proceed to give the formal descriptions of the wrappers. The *strict wrapper functionality*, defined in Fig. 4, is parametrized by (a sampler that induces) a distribution D over traces, and internally runs a copy of a CSF functionality $\mathcal{F}$. Initially, a trace T is sampled from D; this trace is given to the adversary once the first honest party provides its input. The trace T is used by the wrapper to define the termination round $\rho_{\text{term}} \leftarrow c_{\text{tr}}(T)$. In the first round, the wrapper forwards all the messages from the parties and the adversary to (and from) the functionality $\mathcal{F}$. Next, the wrapper essentially waits until round ρ_{term}, with the exception that the adversary is allowed to send $(\texttt{adv-input}, \texttt{sid}, \cdot)$ messages and change its input to the function computed by the CSF. Finally, when round ρ_{term} arrives, the wrapper provides the output generated by $\mathcal{F}$ to all parties.

Wrapper Functionality $\mathcal{W}^D_{\text{strict}}(\mathcal{F})$

$\mathcal{W}_{\text{strict}}$, parametrized by an efficiently sampleable distribution D, internally runs a copy of $\mathcal{F}$ and proceeds as follows:

- Initially, sample a trace $T \leftarrow D$ and compute the output round $\rho_{\text{term}} \leftarrow c_{\text{tr}}(T)$. Send $(\text{trace}, \text{sid}, T)$ to the adversary.[a]
- At all times, forward $(\text{adv-input}, \text{sid}, \cdot)$ messages from the adversary to $\mathcal{F}$.
- In round $\rho = 1$: Forward $(\text{input}, \text{sid}, \cdot)$ messages from each party $P_i \in \mathcal{P}$ to $\mathcal{F}$. In addition, forward $(\text{leakage}, \text{sid}, \cdot)$ messages from $\mathcal{F}$ to the adversary.
- In rounds $\rho > 1$: Upon receiving a message $(\text{fetch-output}, \text{sid})$ from some party $P_i \in \mathcal{P}$, proceed as follows:
  - If $\rho = \rho_{\text{term}}$, forward the message to $\mathcal{F}$, and the response $(\text{output}, \text{sid}, y_i)$ to P_i.
  - Else, send $(\text{fetch-output}, \text{sid}, P_i)$ to the adversary.

---

[a]Technically, the trace is sent to the adversary at the first activation of the functionality along with the first message.

Fig. 4. The strict-wrapper functionality

Wrapper Functionality $\mathcal{W}^D_{\text{flex}}(\mathcal{F})$

$\mathcal{W}_{\text{flex}}$, parametrized by an efficiently sampleable distribution D, internally runs a copy of $\mathcal{F}$ and proceeds as follows:

- Initially, sample a trace $T \leftarrow D$ and compute the output round $\rho_{\text{term}} \leftarrow c_{\text{tr}}(T)$. Send $(\text{trace}, \text{sid}, T)$ to the adversary.[a] In addition, initialize termination indicators $\text{term}_1, \ldots, \text{term}_n \leftarrow 0$.
- At all times, forward $(\text{adv-input}, \text{sid}, \cdot)$ messages from the adversary to $\mathcal{F}$.
- In round $\rho = 1$: Forward $(\text{input}, \text{sid}, \cdot)$ messages from each party $P_i \in \mathcal{P}$ to $\mathcal{F}$. In addition, forward $(\text{leakage}, \text{sid}, \cdot)$ messages from $\mathcal{F}$ to the adversary.
- In rounds $\rho > 1$:
  - Upon receiving $(\text{fetch-output}, \text{sid})$ from some party $P_i \in \mathcal{P}$, proceed as follows:
    - If $\text{term}_i = 1$ or $\rho = \rho_{\text{term}}$ (and P_i did not receive output yet), forward the message to $\mathcal{F}$, and the output $(\text{output}, \text{sid}, y_i)$ to P_i.
    - Else, send $(\text{fetch-output}, \text{sid}, P_i)$ to the adversary.
  - Upon receiving $(\text{early-output}, \text{sid}, P_i)$ from the adversary, set $\text{term}_i \leftarrow 1$.

---

[a]Technically, the trace is sent to the adversary at the first activation of the functionality along with the first message.

Fig. 5. The flexible-wrapper functionality

Flexible-Wrapper Functionality. The *flexible-wrapper functionality*, defined in Fig. 5, follows in similar lines to the strict wrapper. The difference is that the adversary is allowed to instruct the wrapper to deliver the output to each party at any round. In order to accomplish this, the wrapper assigns a termination indicator term_i, initially set to 0, to each party. Once the wrapper receives an $\mathtt{early\text{-}output}$ request from the adversary for P_i, it sets $\mathsf{term}_i \leftarrow 1$. Now, when a party P_i sends a $\mathtt{fetch\text{-}output}$ request, the wrapper checks if $\mathsf{term}_i = 1$, and lets the party receive its output in this case (by forwarding the $\mathtt{fetch\text{-}output}$ request to $\mathcal{F}$). When the guaranteed-termination round ρ_{term} arrives, the wrapper provides the output to all parties that didn't receive it yet.

4   (Fast) Composition of PT Protocols

Canonical synchronous functionalities that are wrapped using the flexible wrapper (cf. Sect. 3.2), i.e., functionalities that correspond to protocols with non-simultaneous termination, are cumbersome to be used as hybrid functionalities for protocols. The reason is that the adversary can cause parties to finish in different rounds, and, as a result, after the execution of the first such functionality, the parties might be *out of sync*.

This "slack" can be reduced, however, only to a difference of one round, unless one is willing to pay a linear blow-up in round complexity [22,24]. Hence, all protocols must be modified to deal with a non-simultaneous start of (at least) one round, and protocols that introduce slack must be followed by a slack-reduction procedure. Since this is a tedious, yet systematic task, in this section we provide a generic compiler that transforms protocols designed in a simpler "stand-alone" setting, where all parties remain synchronized throughout the protocol (and no slack and round-complexity issues arise) into UC protocols that deal with these issues while maintaining their security.

Out starting point are protocols that are defined in the "stand-alone" setting. In such protocols all the hybrids are CSFs and are called in a strictly sequential manner.

Definition 2 (SNF). *Let* $\mathcal{F}_1, \ldots, \mathcal{F}_m$ *be canonical synchronous functionalities. A synchronous protocol* π *in the* $(\mathcal{F}_1, \ldots, \mathcal{F}_m)$*-hybrid model is in synchronous normal form (SNF) if in every round exactly one ideal functionality* $\mathcal{F}_i$ *is invoked by all honest parties, and in addition, no honest party hands inputs to other CSFs before this instance halts.*

Clearly, designing and proving the security of SNF protocols, which only make calls to simple two-round CSFs is a much simpler task than dealing with protocols that invoke more complicated hybrids, potentially with probabilistic termination (see Sect. 5 for concrete examples).

SNF protocols are designed as an intermediate step, since the hybrid functionalities $\mathcal{F}_i$ are two-round CSFs, and can, in general, not be realized by real-world protocols. To that end, we define a protocol compiler that transforms SNF protocols into (non-SNF) protocols making calls to wrapped hybrids that *can* be realized in the real world, while maintaining their security and asymptotic (expected) round complexity. At the same time, the compiler takes care of any potential slack that is introduced by the protocol and ensures that the protocol can be executed even if the parties do not start the protocol simultaneously.

In Sect. 4.1 we apply this approach to deterministic-termination protocols that use deterministic-termination hybrids, and in Sect. 4.2 generalize it to the probabilistic-termination setting. Section 4.3 covers the base case of realizing the wrapped $\mathcal{F}_{\text{PSMT}}$ using only $\mathcal{F}_{\text{SMT}}$ functionalities.

4.1   Composition with Deterministic Termination

We start by defining a slack-tolerant variant of the strict wrapper (cf. Sect. 3.2), which can be used even when parties operate with a (known) slack. Then, we show how to compile an SNF protocol π realizing a strictly-wrapped CSF $\mathcal{F}$ into a (non-SNF) protocol π' realizing a version of $\mathcal{F}$ wrapped with the slack-tolerant strict wrapper and making calls to wrapped hybrids.

Slack-Tolerant Strict Wrapper. The *slack-tolerant strict wrapper* $\mathcal{W}_{\text{sl-strict}}^{D,c}$, formally defined in Fig. 6, is parametrized by an integer $c \geq 0$, which denotes the

---

Wrapper Functionality $\mathcal{W}_{\text{sl-strict}}^{D,c}(\mathcal{F})$

$\mathcal{W}_{\text{sl-strict}}^{D,c}$, parametrized by an efficiently sampleable distribution D and a non-negative integer c, internally runs a copy of $\mathcal{F}$ and proceeds as follows:

- Initially, sample a trace $T \leftarrow D$ and compute the output round $\rho_{\text{term}} \leftarrow B_c \cdot c_{\text{tr}}(T)$, where $B_c := 3c + 1$. Send $(\texttt{trace}, \text{sid}, T)$ to the adversary.[a] Initialize slack indicators $c_1, \ldots, c_n \leftarrow 0$.
- At all times, forward $(\texttt{adv-input}, \text{sid}, \cdot)$ messages from the adversary to $\mathcal{F}$.
- In rounds $\rho = 1, \ldots, 2c + 1$: Upon receiving a message from some party $P_i \in \mathcal{P}$, proceed as follows:
  - If the message is $(\texttt{input}, \text{sid}, \cdot)$, forward it to $\mathcal{F}$, forward the $(\texttt{leakage}, \text{sid}, \cdot)$ message $\mathcal{F}$ subsequently outputs to the adversary, and set P_i's local slack $c_i \leftarrow \rho - 1$.
  - Else, send $(\texttt{fetch-output}, \text{sid}, P_i)$ to the adversary.
- In rounds $\rho > 2c + 1$: Upon receiving a message $(\texttt{fetch-output}, \text{sid})$ from some party $P_i \in \mathcal{P}$, proceed as follows:
  - If $\rho = \rho_{\text{term}} + c_i$, send the message to $\mathcal{F}$, and the output $(\texttt{output}, \text{sid}, y_i)$ to P_i.
  - Else, send $(\texttt{fetch-output}, \text{sid}, P_i)$ to the adversary.

---

[a]Technically, the trace is sent to the adversary at the first activation of the functionality along with the first message.

Fig. 6. The slack-tolerant strict wrapper functionality

amount of slack tolerance that is added, and a distribution D over traces. The wrapper $\mathcal{W}_{\text{sl-strict}}$ is similar to $\mathcal{W}_{\text{strict}}$ but allows parties to provide input within a window of $2c + 1$ rounds and ensures that they obtain output with the same slack they started with. The wrapper essentially increases the termination round by a factor of $B_c = 3c + 1$, which is due to the slack-tolerance technique used to implement the wrapped version of the atomic parallel SMT functionality (cf. Sect. 4.3).

Deterministic-Termination Compiler. Let $\mathcal{F}, \mathcal{F}_1, \ldots, \mathcal{F}_m$ be canonical synchronous functionalities, and let π an SNF protocol that UC-realizes the strictly wrapped functionality $\mathcal{W}_{\text{strict}}^D(\mathcal{F})$, for some distribution D, in the $(\mathcal{F}_1, \ldots, \mathcal{F}_m)$-hybrid model, assuming that all honest parties receive their inputs at the same round. We define a compiler $\text{Comp}_{\text{DT}}^c$, parametrized with a slack parameter $c \geq 0$, that receives as input the protocol π and distributions $D_1, \ldots, D_m$ over traces and replaces every call to a CSF $\mathcal{F}_i$ with a call to the wrapped CSF $\mathcal{W}_{\text{sl-strict}}^{D_i, c}(\mathcal{F}_i)$. We denote the output of the compiler by $\pi' = \text{Comp}_{\text{DT}}^c(\pi, D_1, \ldots, D_m)$.[23]

As shown below, π' realizes $\mathcal{W}_{\text{sl-strict}}^{D^{\text{full}}, c}(\mathcal{F})$, for a suitably adapted distribution D^{full}, assuming all parties start within $c + 1$ consecutive rounds. Consequently, the compiled protocol π' can handle a slack of up to c rounds while using hybrids that are realizable themselves.

Calling the wrapped CSFs instead of the CSFs $(\mathcal{F}_1, \ldots, \mathcal{F}_m)$ affects the trace corresponding to $\mathcal{F}$. The new trace $D^{\text{full}} = \text{full-trace}(D, D_1, \ldots, D_m)$ is obtained as follows:

1. Sample a trace $T \leftarrow D$, which is a depth-1 tree with root label $\mathcal{W}_{\text{strict}}^D(\mathcal{F})$ and leaves from the set $\{\mathcal{F}_1, \ldots, \mathcal{F}_m\}$.
2. For each leaf node $\mathcal{F}' = \mathcal{F}_i$, for some $i \in [m]$, sample a trace $T_i \leftarrow D_i$ and replace node $\mathcal{F}'$ by the trace T_i.
3. Output the resulting trace T'.

The following theorem states that the compiled protocol π' UC-realizes the wrapped functionality $\mathcal{W}_{\text{sl-strict}}^{D^{\text{full}}, c}(\mathcal{F})$.

Theorem 1. *Let $\mathcal{F}, \mathcal{F}_1, \ldots, \mathcal{F}_m$ be canonical synchronous functionalities, and let π an SNF protocol that UC-realizes $\mathcal{W}_{\text{strict}}^D(\mathcal{F})$ in the $(\mathcal{F}_1, \ldots, \mathcal{F}_m)$-hybrid model, for some distribution D, assuming that all honest parties receive their inputs at the same round. Let $D_1, \ldots, D_m$ be arbitrary distributions over traces, $D^{\text{full}} = \text{full-trace}(D, D_1, \ldots, D_m)$, and $c \geq 0$. Then, protocol $\pi' = \text{Comp}_{\text{DT}}^c(\pi, D_1, \ldots, D_m)$ UC-realizes $\mathcal{W}_{\text{sl-strict}}^{D^{\text{full}}, c}(\mathcal{F})$ in the $(\mathcal{W}_{\text{sl-strict}}^{D_1, c}(\mathcal{F}_1), \ldots, \mathcal{W}_{\text{sl-strict}}^{D_m, c}(\mathcal{F}_m))$-hybrid model, assuming that all honest parties receive their inputs within $c + 1$ consecutive rounds.*

---

[23] The distributions D_i depend on the protocols realizing the strictly wrapped functionalities $\mathcal{W}_{\text{sl-strict}}^{D_i, c}(\mathcal{F}_i)$. Note, however, that the composition theorems in Sects. 4.1 and 4.2 actually work for arbitrary distributions D_i.

The expected round complexity of the compiled protocol π' is

$$B_c \cdot \sum_{i \in [m]} d_i \cdot E[c_{\mathsf{tr}}(T_i)],$$

where d_i is the expected number of calls in π to hybrid $\mathcal{F}_i$, T_i is a trace sampled from D_i, and $B_c = 3c + 1$ is the blow-up factor in $\mathcal{W}_{\mathsf{sl\text{-}strict}}^{D^{\mathsf{full}}, c}$.

The proof of Theorem 1 can be found in the full version [16].

4.2   Composition with Probabilistic Termination

The composition theorem in Sect. 4.1 does not work if the protocol π itself introduces slack (e.g., the fast broadcast protocol by Feldman and Micali [23]) or if one of the hybrids needs to be replaced by a slack-introducing protocol (e.g., instantiating the broadcast hybrids with fast broadcast protocols in BGW [6]).

As in Sect. 4.1, we start by adjusting the flexible wrapper (cf. Sect. 3.2) to be slack-tolerant. In addition, the slack-tolerant flexible wrapper ensures that all parties will obtain their outputs within two consecutive rounds. Then, we show how to compile an SNF protocol π realizing a CSF $\mathcal{F}$, wrapped with the flexible wrapper, into a (non-SNF) protocol π' realizing a version of $\mathcal{F}$ wrapped with slack-tolerant flexible wrapper. The case where π implements a strictly wrapped CSF, but some of the hybrids are wrapped with the slack-tolerant flexible wrapper follows along similar lines.

Slack-Tolerant Flexible Wrapper. The *slack-tolerant flexible wrapper* $\mathcal{W}_{\mathsf{sl\text{-}flex}}^{D,c}$, formally defined in Fig. 7, is parametrized by an integer $c \geq 0$, which denotes the amount of slack tolerance that is added, and a distribution D over traces. The wrapper $\mathcal{W}_{\mathsf{sl\text{-}flex}}$ is similar to $\mathcal{W}_{\mathsf{flex}}$ but allows parties to provide input within a window of $2c + 1$ rounds and ensures that all honest parties will receive their output within two consecutive rounds. The wrapper essentially increases the termination round to

$$\rho_{\mathsf{term}} = B_c \cdot c_{\mathsf{tr}}(T) + 2 \cdot \mathsf{flex}_{\mathsf{tr}}(T) + c,$$

where the blow-up factor B_c is as explained in Sect. 4.1, and the additional factor of 2 results from the termination protocol described below for every flexibly wrapped CSF, which increases the round complexity by at most two additional rounds (recall that $\mathsf{flex}_{\mathsf{tr}}(T)$ denotes the number of such CSFs), and c is due to the potential slack. $\mathcal{W}_{\mathsf{sl\text{-}flex}}$ allows the adversary to deliver output at any round prior to ρ_{term} but ensures that all parties obtain output with a slack of at most one round. Moreover, it allows the adversary to obtain the output using the (get-output, sid) command, which is necessary in order to simulate the above termination protocol.

Wrapper Functionality $\mathcal{W}_{\text{sl-flex}}^{D,c}(\mathcal{F})$

$\mathcal{W}_{\text{sl-flex}}^{D,c}$, parametrized by an efficiently sampleable distribution D and a non-negative integer c, internally runs a copy of $\mathcal{F}$ and proceeds as follows:

- Initially, sample a trace $T \leftarrow D$ and compute the output round $\rho_{\text{term}} \leftarrow B_c \cdot c_{\text{tr}}(T) + B' \cdot \text{flex}_{\text{tr}}(T) + c$, where $B_c := 3c + 1$ and $B' = 2$. Send $(\text{trace}, \text{sid}, T)$ to the adversary.[a] Initialize termination indicators $\text{term}_1, \ldots, \text{term}_n \leftarrow 0$.
- At all times, forward $(\text{adv-input}, \cdot)$ messages from the adversary to $\mathcal{F}$.
- In rounds $\rho = 1, \ldots, 2c + 1$: Upon receiving a message from some party $P_i \in \mathcal{P}$, proceed as follows:
  - If the message is $(\text{input}, \text{sid}, \cdot)$, forward it to $\mathcal{F}$, forward the $(\text{leakage}, \text{sid}, \cdot)$ message $\mathcal{F}$ subsequently outputs to the adversary.
  - Else, send $(\text{fetch-output}, \text{sid}, P_i)$ to the adversary.
- In rounds $\rho > 2c + 1$:
  - Upon receiving a message $(\text{fetch-output}, \text{sid})$ from some party $P_i \in \mathcal{P}$, proceed as follows:
    * If $\text{term}_i = 1$ or $\rho = \rho_{\text{term}}$, forward the message to $\mathcal{F}$, and the output $(\text{output}, \text{sid}, y)$ to P_i.
    * Else, output $(\text{fetch-output}, \text{sid}, P_i)$ to the adversary.
  - Upon receiving $(\text{get-output}, \text{sid})$ from the adversary, if the output value y was not computed yet, send $(\text{fetch-output}, \text{sid})$ to $\mathcal{F}$ on behalf of some party P_i. Next, send $(\text{output}, \text{sid}, y)$ to the adversary.
  - Upon receiving $(\text{early-output}, \text{sid}, P_i)$ from the adversary, set $\text{term}_i \leftarrow 1$ and $\rho_{\text{term}} \leftarrow \min\{\rho_{\text{term}}, \rho + 1\}$.

---

[a] Technically, the trace is sent to the adversary at the first activation of the functionality along with the first message.

Fig. 7. The slack-tolerant flexible wrapper functionality

Probabilistic-Termination Compilers. Let $\mathcal{F}, \mathcal{F}_1, \ldots, \mathcal{F}_m$ be canonical synchronous functionalities, and let π be an SNF protocol that UC-realizes the flexibly wrapped functionality $\mathcal{W}_{\text{flex}}^D(\mathcal{F})$, for some distribution D, in the $(\mathcal{F}_1, \ldots, \mathcal{F}_m)$-hybrid model, assuming all parties start at the same round. Define the following compiler Comp_{PTR}, parametrized by a slack parameter $c \geq 0$. It receives as input the protocol π, distributions $D_1, \ldots, D_m$ over traces, and a subset $I \subseteq [m]$ indexing which CSFs $\mathcal{F}_i$ are to be wrapped with $\mathcal{W}_{\text{sl-flex}}$ and which with $\mathcal{W}_{\text{sl-strict}}$; it replaces every call to a CSF $\mathcal{F}_i$ with a call to the wrapped CSF $\mathcal{W}_{\text{sl-flex}}^{D_i,c}(\mathcal{F}_i)$ if $i \in I$ or to $\mathcal{W}_{\text{sl-strict}}^{D_i,c}(\mathcal{F}_i)$ if $i \notin I$.

In addition, the compiler adds the following termination procedure, based on an approach originally suggested by Bracha [7], which ensures all honest parties will terminate within two consecutive rounds:

- As soon as a party is ready to output a value y (according to the prescribed protocol) or upon receiving at least $t + 1$ messages (end, y) for the same value y (whichever happens first), it sends $(\text{end}, \text{sid}, y)$ to all parties.

- Upon receiving $n - t$ messages (end, sid, y) for a single value y, a party outputs y as the result of the computation and halts.

Observe that this technique only works for public-output functionalities, and, therefore, only CSFs with public output can be wrapped by $\mathcal{W}_{\text{sl-flex}}$. We denote the output of the compiler by $\pi' = \mathsf{Comp}^c_{\text{PTR}}(\pi, D_1, \ldots, D_m, I)$.

The following theorem states that the compiled protocol π' UC-realizes the wrapped functionality $\mathcal{W}^{D^{\text{full}}, c}_{\text{sl-flex}}(\mathcal{F})$, again for an adapted distribution D^{full}. Consequently, the compiled protocol π' can handle a slack of up to c rounds, while using hybrids that are realizable themselves, and ensuring that the output slack is at most one round (as opposed to π). Calling the wrapped hybrids instead of the CSFs affects the trace corresponding to $\mathcal{F}$ in exactly the same way as in the case with deterministic termination (cf. Sect. 4.1).[24]

Theorem 2. *Let* $\mathcal{F}, \mathcal{F}_1, \ldots, \mathcal{F}_m$ *be canonical synchronous functionalities, and let* π *an SNF protocol that UC-realizes* $\mathcal{W}^D_{\text{flex}}(\mathcal{F})$, *for some distribution* D, *in the* $(\mathcal{F}_1, \ldots, \mathcal{F}_m)$-*hybrid model, assuming that all honest parties receive their inputs at the same round. Let* $I \subseteq [m]$ *be the subset (of indices) of functionalities to be wrapped using the flexible wrapper, let* $D_1, \ldots, D_m$ *be arbitrary distributions over traces, denote* $D^{\text{full}} = \mathsf{full\text{-}trace}(D, D_1, \ldots, D_m)$ *and let* $c \geq 0$. *Assume that* $\mathcal{F}$ *and* $\mathcal{F}_i$ *for every* $i \in I$ *are public-output functionalities.*

Then, protocol $\mathsf{Comp}^c_{\text{PTR}}(\pi, D_1, \ldots, D_m, I)$ *UC-realizes* $\mathcal{W}^{D^{\text{full}}, c}_{\text{sl-flex}}(\mathcal{F})$ *in the* $(\mathcal{W}(\mathcal{F}_1), \ldots, \mathcal{W}(\mathcal{F}_m))$-*hybrid model, assuming that all honest parties receive their inputs within* $c + 1$ *consecutive rounds, where* $\mathcal{W}(\mathcal{F}_i) = \mathcal{W}^{D_i, c}_{\text{sl-flex}}(\mathcal{F}_i)$ *if* $i \in I$ *and* $\mathcal{W}(\mathcal{F}_i) = \mathcal{W}^{D_i, c}_{\text{sl-strict}}(\mathcal{F}_i)$ *if* $i \notin I$.

The expected round complexity of the compiled protocol π; *is*

$$B_c \cdot \sum_{i \in [m]} d_i \cdot E[c_{\text{tr}}(T_i)] + 2 \cdot \sum_{i \in [m]} d_i \cdot E[\text{flex}_{\text{tr}}(T_i)] + 2$$

where d_i *is the expected number of calls in* π *to hybrid* $\mathcal{F}_i$, T_i *is a trace sampled from* D_i, *and* $B_c = 3c + 1$ *is the blow-up factor.*

The proof of Theorem 2 can be found in the full version [16].

Consider now the scenario where SNF protocol π realizes a strictly wrapped functionality, yet soem of the CSF hybrids are to be wrapped by flexible wrappers. The corresponding compiler $\mathsf{Comp}_{\text{PT}}$ works as $\mathsf{Comp}_{\text{PTR}}$ except that it does not perform the slack-reduction protocol in the end. The proof of the following theorem follows that of Theorem 2.

Theorem 3. *Let* $\mathcal{F}, \mathcal{F}_1, \ldots, \mathcal{F}_m$ *be canonical synchronous functionalities, and let* π *an SNF protocol that UC-realizes* $\mathcal{W}^D_{\text{strict}}(\mathcal{F})$ *for some distribution* D, *in the* $(\mathcal{F}_1, \ldots, \mathcal{F}_m)$-*hybrid model, assuming that all honest parties receive their inputs at the same round. Let* $I \subseteq [m]$ *be the subset (of indices) of functionalities to*

---

[24] Of course, the root of the trace T sampled from D is a flexibly wrapped functionality $\mathcal{W}^D_{\text{flex}}(\mathcal{F})$ in the probabilistic-termination case.

be wrapped using the flexible wrapper, let $D_1, \ldots, D_m$ be arbitrary distributions over traces, denote $D^{\mathsf{full}} = \mathsf{full\text{-}trace}(D, D_1, \ldots, D_m)$ and let $c \geq 0$. Assume that $\mathcal{F}_i$ for every $i \in I$ is a public-output functionalities.

Then, protocol $\pi'\mathsf{Comp}_{\mathrm{PT}}^c(\pi, D_1, \ldots, D_m, I)$ UC-realizes $\mathcal{W}_{\mathrm{sl\text{-}flex}}^{D^{\mathsf{full}}, c}(\mathcal{F})$ in the $(\mathcal{W}(\mathcal{F}_1), \ldots, \mathcal{W}(\mathcal{F}_m))$-hybrid model, where $\mathcal{W}(\mathcal{F}_i) = \mathcal{W}_{\mathrm{sl\text{-}flex}}^{D_i, c}(\mathcal{F}_i)$ if $i \in I$ and $\mathcal{W}(\mathcal{F}_i) = \mathcal{W}_{\mathrm{sl\text{-}strict}}^{D_i, c}(\mathcal{F}_i)$ if $i \notin I$, assuming that all honest parties receive their inputs within $c + 1$ consecutive rounds.

The expected round complexity of the compiled protocol π' is

$$B_c \cdot \sum_{i \in [m]} d_i \cdot E[c_{\mathsf{tr}}(T_i)] + 2 \cdot \sum_{i \in [m]} d_i \cdot E[\mathsf{flex}_{\mathsf{tr}}(T_i)]$$

where d_i is the expected number of calls in π to hybrid $\mathcal{F}_i$, T_i is a trace sampled from D_i, and $B_c = 3c + 1$ is the blow-up factor.

4.3   Wrapping Secure Channels

The basis of the top-down, inductive approach taken in this work consists of providing protocols realizing wrapped atomic functionalities, using merely secure channels $\mathcal{F}_{\mathrm{SMT}}$. Due to the restriction to SNF protocols, which may only call a single CSF hybrid in any given round, a parallel variant $\mathcal{F}_{\mathrm{PSMT}}$ of $\mathcal{F}_{\mathrm{SMT}}$ (defined below) is used as an atomic functionality. This ensures that in SNF protocols parties can securely send messages to each other simultaneously.

Parallel SMT. The *parallel secure message transmission functionality* $\mathcal{F}_{\mathrm{PSMT}}$ is a CSF for the following functions f_{PSMT} and l_{PSMT}. Each party P_i has a vector of input values $(x_1^i, \ldots, x_n^i)$ such that x_j^i is sent from P_i to P_j. That is, the function to compute is $f_{\mathrm{PSMT}}((x_1^1, \ldots, x_n^1), \ldots, (x_1^n, \ldots, x_n^n), a) = ((x_1^1, \ldots, x_1^n), \ldots, (x_n^1, \ldots, x_n^n))$. As we consider rushing adversaries, that can determine the messages sent by the corrupted parties *after* receiving the messages sent by the honest parties, the leakage function should leak the messages that are to be delivered from honest parties to corrupted parties. Therefore, the leakage function is $l_{\mathrm{PSMT}}((x_1^1, \ldots, x_n^1), \ldots, (x_1^n, \ldots, x_n^n)) = (y_1^1, y_2^1, \ldots, y_{n-1}^n, y_n^n)$, where $y_j^i = |x_j^i|$ in case P_j is honest and $y_j^i = x_j^i$ in case P_j is corrupted.

Realizing Wrapped Parallel SMT. The remainder of this section deals with securely realizing $\mathcal{W}_{\mathrm{sl\text{-}strict}}^{D_{\mathrm{PSMT}}, c}(\mathcal{F}_{\mathrm{PSMT}})$ in the $\mathcal{F}_{\mathrm{SMT}}$-hybrid model, for a particular distribution D_{PSMT} and an arbitrary non-negative integer c. Note that the corresponding protocol π_{PSMT} is *not* an SNF protocol; this is of no concern since it directly realizes a wrapped functionality and therefore need not be compiled. There is a straight-forward (non-SNF) protocol realizing $\mathcal{F}_{\mathrm{PSMT}}$ in the $\mathcal{F}_{\mathrm{SMT}}$-hybrid model, and therefore (due to the UC composition theorem) it suffices to describe protocol π_{PSMT} in the $\mathcal{F}_{\mathrm{PSMT}}$-hybrid model.

A standard solution to overcome asynchrony by a constant number of rounds $c \geq 0$, introduced by Lindell et al. [41] and used by Katz and Koo [33], is to expand each communication round to $2c + 1$ rounds. Each party listens for messages throughout all $2c + 1$ rounds, and sends its own messages in round $c + 1$. It is straight-forward to verify that if the slack is c, i.e., the parties start within $c + 1$ rounds from each other, round r-messages (in the original protocol, without round-expansion) are sent, and delivered, before round $(r + 1)$-messages and after round $(r - 1)$-messages.

The solution described above does not immediately apply to our case, due to the nature of canonical synchronous functionalities. Recall that in a CSF the adversary can send an `adv-input` message (and affect the output) only before any honest party has received an output from the functionality. If only $2c + 1$ rounds are used a subtle problem arises: Assume for simplicity that $c = 1$ and say that P_1 is a fast party and P_2 is a slow party. Initially, P_1 listens for one round. In the second round P_2 listens and P_1 send its messages to all the parties. In the third round P_2 sends its messages and P_1 receives its message, produces output and completes the round. Now, P_2 listens for an additional round, and the adversary can send it messages on behalf of corrupted parties. In other words, the adversary can choose the value for P_2's output *after* P_1 has received its output – such a phenomena cannot be modeled using CSFs. For this reason we add an additional round where each party is idle; if P_1 waits one more round (without listening) before it produces its output, then P_2 will receive all the messages that determine its output, and so once P_1 produces output and completes, the adversary cannot affect the output of P_2.

As a result, in protocol π_{PSMT}, each round is expanded to $3c + 1$ rounds, where during the final c rounds, parties are simply idle and ignore any messages they receive. Denote by D_{PSMT} the deterministic distribution that outputs a depth-1 trace consisting of a single leaf $\mathcal{F}_{\text{PSMT}}$. In the full version [16] of this paper, we prove the following lemma.

Lemma 1. *Let $c \geq 0$. Protocol π_{PSMT} UC-realizes $\mathcal{W}_{\text{sl-strict}}^{D_{\text{PSMT}}^{\text{full}}, c}(\mathcal{F}_{\text{PSMT}})$ in the $\mathcal{F}_{\text{SMT}}$-hybrid model, assuming that all honest parties receive their inputs within $c + 1$ consecutive rounds.*

5 Applications of Our Fast Composition Theorem

In this section we demonstrate the power of our framework by providing some concrete applications. All of the protocols we present in this section enjoy perfect security facing adaptive adversaries corrupting less than a third of the parties.

5.1 Fast and Perfectly Secure Byzantine Agreement

We start by describing the binary and multi-valued randomized Byzantine agreement protocols (the definition of $\mathcal{F}_{\text{BA}}$ appears in Sect. 3.1). These protocols are based on techniques due to Feldman and Micali [23] and Turpin and Coan [48], with modifications to work in the UC framework. We provide simulation-based proofs for these protocols.

At a high level, protocol π_{RBA} proceeds as follows. Initially, each party sends its input to all other parties over a point-to-pint channel using $\mathcal{F}_{\text{PSMT}}$, and sets its vote to be its input bit. Next, the parties proceed in phases, where each phase consists of invoking the oblivious coin functionality $\mathcal{F}_{\mathrm{OC}}$ (see the full version) followed by a voting process consisting of three rounds of sending messages via $\mathcal{F}_{\mathrm{PSMT}}$. The voting ensures that (1) if all honest parties agree on their votes at the beginning of the phase, they will terminate at the end of the phase, (2) in each phase, all honest parties will agree on their votes at the end of each phase with probability at least p, and (3) if an honest party terminates in some phase then all honest parties will terminate with the same value by the end of the next phase. In the negligible event that the parties do not terminate after $\tau = \log^{1.5}(k) + 1$ phases, the parties use the Byzantine agreement functionality $\mathcal{F}_{\mathrm{BA}}$ in order to ensure termination.

In the full version [16] we prove the following theorem.

Theorem 4. *Let $c \geq 0$ and $t < n/3$. There exists an efficiently sampleable distribution D such that the functionality $\mathcal{W}_{\mathrm{sl\text{-}flex}}^{D,c}(\mathcal{F}_{\mathrm{BA}}^{\{0,1\}})$ has an expected constant round complexity, and can be UC-realized in the $\mathcal{F}_{\mathrm{SMT}}$-hybrid model, with perfect security, in the presence of an adaptive malicious t-adversary, assuming that all honest parties receive their inputs within $c + 1$ consecutive rounds.*

5.2 Fast and Perfectly Secure Parallel Broadcast

As discussed in Sect. 1 composing protocols with probabilistic termination naïvely does not retain expected round complexity. Ben-Or and El-Yaniv [5] constructed an elegant protocol for probabilistic-termination parallel broadcast[25] with a constant round complexity in expectation, albeit under a property-based security definition. In this section we adapt the [5] protocol to the UC framework and show that it does *not* realize the parallel broadcast functionality, but rather a weaker variant which we call *unfair parallel broadcast*. Next, we show how to use unfair parallel broadcast in order to compute (fair) parallel broadcast in constant excepted number of rounds.

In a standard broadcast functionality (cf. Sect. 3.1), the sender provides a message to the functionality which delivers it to the parties. Hirt and Zikas [31] defined the *unfair* version of the broadcast functionality, in which the functionality informs the adversary which message it received, and allows the adversary, based on this information, to corrupt the sender and replace the message. Following the spirit of [31], we now define the unfair parallel broadcast functionality, using the language of CSF.

– UNFAIR PARALLEL BROADCAST. In the *unfair parallel broadcast* functionality, each party P_i with input x_i distributes its input to all the parties. The adversary is allowed to learn the content of each input value from the leakage function (and so it can corrupt parties and change their messages prior to their distribution, based on this information). The function to compute is $f_{\mathrm{UPBC}}(x_1, \ldots, x_n, a) = ((x_1, \ldots, x_n), \ldots, (x_1, \ldots, x_n))$ and the leakage function

---

[25] In [5] the problem is referred to as "interactive consistency."

is $l_{\text{UPBC}}(x_1,\ldots,x_n) = (x_1,\ldots,x_n)$. We denote by $\mathcal{F}_{\text{UPBC}}$ the functionality $\mathcal{F}_{\text{CSF}}$ when parametrized with the above functions f_{UPBC} and l_{UPBC}.

In the full version [16] we present an adaptation of the [5] protocol, show that it perfectly UC-realizes (a wrapped version of) $\mathcal{F}_{\text{UPBC}}$, and prove the following result.

Theorem 5. *Let $c \geq 0$ and $t < n/3$. There exists an efficiently sampleable distribution D such that the functionality $\mathcal{W}^{D,c}_{\text{sl-flex}}(\mathcal{F}_{\text{UPBC}})$ has an expected constant round complexity, and can be UC-realized in the $\mathcal{F}_{\text{SMT}}$-hybrid model, with perfect security, in the presence of an adaptive malicious t-adversary, assuming that all honest parties receive their inputs within $c + 1$ consecutive rounds.*

We now turn to define the (fair) parallel broadcast functionality.

- PARALLEL BROADCAST. In the *parallel broadcast* functionality, each party P_i with input x_i distributes its input to all the parties. Unlike the unfair version, the adversary only learns the length of the honest parties' messages before their distribution, i.e., the leakage function is $l_{\text{PBC}}(x_1,\ldots,x_n) = (|x_1|,\ldots,|x_n|)$. It follows that the adversary cannot use the leaked information in a meaningful way when deciding which parties to corrupt. The function to compute is identical to the unfair version, i.e., $f_{\text{PBC}}(x_1,\ldots,x_n,a) = ((x_1,\ldots,x_n),\ldots,(x_1,\ldots,x_n))$. We denote by $\mathcal{F}_{\text{PBC}}$ the functionality $\mathcal{F}_{\text{CSF}}$ when parametrized with the above functions f_{PBC} and l_{PBC}.

Unfortunately, the unfair parallel broadcast protocol π_{UPBC} (see the full version [16]) fails to realize (a wrapped version of) the standard parallel broadcast functionality $\mathcal{F}_{\text{PBC}}$. The reason is similar to the argument presented in [31]: in the first round of the protocol, each party distributes its input, and since we consider a rushing adversary, the adversary learns the messages *before* the honest parties do. It follows that the adversary can corrupt a party *before* the honest parties receive the message and replace the message to be delivered. This attack cannot be simulated in the ideal world where the parties interact with $\mathcal{F}_{\text{PBC}}$, since by the time the simulator learns the broadcast message in the ideal world, the functionality does not allow to change it.

Although protocol π_{UPBC} does not realize $\mathcal{F}_{\text{PBC}}$, it can be used in order to construct a protocol that does. Each party commits to its input value before any party learns any new information, as follows. Each party, in parallel, first secret shares its input using a t-out-of-n secret-sharing protocol.[26] In the second step, every party, in parallel, broadcast a vector with all the shares he received, by use of the above unfair parallel broadcast functionality $\mathcal{F}_{\text{UPBC}}$, and each share is reconstructed based on the announced values. The reason this modification achieves fair broadcast is the following: If a sender P_i is not corrupted until he distributes his shares, then a t-adversary has no way of modifying the reconstructed output of P_i's input, since he can at most affect $t < n/3$ shares. Thus, the only

---

[26] In [31] verifiable secret sharing (VSS) is used; however, as we argue, this is not necessary.

way the adversary can affect any of the broadcast messages is by corrupting the sender independently of his input, an attack which is easily simulated. We describe this protocol, denoted π_{PBC}, in Fig. 8.

Protocol π_{PBC}

1. In the first round, upon receiving (input, sid, x_i) with $x_i \in V$ from the environment, P_i secret shares x_i using a t-out-of-n secret sharing scheme, denoted by $(x_i^1, \ldots, x_i^n)$. Next, P_i sends for every party P_j its share (sid, x_i^j) (via $\mathcal{F}_{\mathrm{PSMT}}$). Denote by x_j^i the value received from P_j.
2. In the second round, P_i broadcasts the values $\boldsymbol{x}_i = (x_1^i, \ldots, x_n^i)$ using the unfair parallel broadcast functionality, i.e., P_i sends (input, sid, $\boldsymbol{x}_i$) to $\mathcal{F}_{\mathrm{UPBC}}$. Denote by $\boldsymbol{y}_j = (y_1^j, \ldots, y_n^j)$ the value received from P_j. Now, P_i reconstructs all the input values, i.e., for every $j \in [n]$ reconstructs y_j from the shares $(y_j^1, \ldots, y_j^n)$, and outputs (output, sid, $(y_1, \ldots, y_n)$) .

Fig. 8. The parallel broadcast protocol, in the $(\mathcal{F}_{\mathrm{PSMT}}, \mathcal{F}_{\mathrm{UPBC}})$-hybrid model

We conclude with the following theorem, see the full version [16] for the proof.

Theorem 6. *Let $c \geq 0$ and $t < n/3$. There exists an efficiently sampleable distribution D such that the functionality $\mathcal{W}_{\mathrm{sl\text{-}flex}}^{D,c}(\mathcal{F}_{\mathrm{PBC}})$ has an expected constant round complexity, and can be UC-realized in the $\mathcal{F}_{\mathrm{SMT}}$-hybrid model, with perfect security, in the presence of an adaptive malicious t-adversary, assuming that all honest parties receive their inputs within $c + 1$ consecutive rounds.*

5.3   Fast and Perfectly Secure SFE

We conclude this section by showing how to construct a perfectly UC-secure SFE protocol which computes a given circuit in expected $O(d)$ rounds, independently of the number of parties, in the point-to-point channels model. The protocol is obtained by taking the protocol from [6],[27] denoted π_{BGW}. This protocol relies on (parallel) broadcast and (parallel) point-to-point channels, and therefore it can be described in the $(\mathcal{F}_{\mathrm{PSMT}}, \mathcal{F}_{\mathrm{PBC}})$-hybrid model. It follows from Theorem 3, that the compiled protocol $\mathsf{Comp}_{\mathrm{PT}}^c(\pi_{\mathrm{BGW}})$ UC-realizes the corresponding wrapped functionality $\mathcal{W}_{\mathrm{sl\text{-}flex}}^{D,c}(\mathcal{F}_{\mathrm{SFE}})$ (for an appropriate distribution D), in the $(\mathcal{W}_{\mathrm{sl\text{-}strict}}^{D_{\mathrm{PSMT}}^{\mathrm{full}},c}(\mathcal{F}_{\mathrm{PSMT}}), \mathcal{W}_{\mathrm{sl\text{-}flex}}^{D_{\mathrm{PBC}}^{\mathrm{full}},c}(\mathcal{F}_{\mathrm{PBC}}))$-hybrid model, resulting in the following.

Theorem 7. *Let f be an n-party function, C an arithmetic circuit with multiplicative depth d computing f, and $t < n/3$. Then there exists an efficiently*

---

[27] A full simulation proof of the protocol with a black-box straight-line simulation was recently given by [2] and [19].

sampleable distribution D such that the functionality $\mathcal{W}^{D,c}_{\text{sl-flex}}(\mathcal{F}^f_{\text{SFE}})$ has round complexity $O(d)$ in expectation, and can be UC-realized in the $\mathcal{F}_{\text{SMT}}$-hybrid model, with perfect security, in the presence of an adaptive malicious t-adversary, assuming that all honest parties receive their inputs within $c + 1$ consecutive rounds.

References

1. Asharov, G., Jain, A., López-Alt, A., Tromer, E., Vaikuntanathan, V., Wichs, D.: Multiparty computation with low communication, computation and interaction via threshold FHE. In: Pointcheval, D., Johansson, T. (eds.) EUROCRYPT 2012. LNCS, vol. 7237, pp. 483–501. Springer, Heidelberg (2012)
2. Asharov, G., Lindell, Y.: A full proof of the BGW protocol for perfectly-secure multiparty computation. Electron. Colloquium Comput. Complex. (ECCC) **18**, 36 (2011)
3. Beaver, D., Micali, S., Rogaway, P.: The round complexity of secure protocols (extended abstract). In: 22nd ACM STOC, pp. 503–513. ACM Press, May 1990
4. Ben-Or, M.: Another advantage of free choice: completely asynchronous agreement protocols (extended abstract). In: Probert, R.L., Lynch, N.A., Santoro, N. (eds.) 2nd ACM PODC, pp. 27–30. ACM Press, August 1983
5. Ben-O, M., El-Yaniv, R.: Resilient-optimal interactive consistency in constant time. Distrib. Comput. **16**(4), 249–262 (2003)
6. Ben-Or, M., Goldwasser, S., Wigderson, A.: Completeness theorems for non-cryptographic fault-tolerant distributed computation (extended abstract). In: 20th ACM STOC, pp. 1–10. ACM Press, May 1988
7. Bracha, G.: An asynchronou [(n − 1)/3]-resilient consensus protocol. In: Probert, R.L., Lynch, N.A., Santoro, N. (eds.) 3rd ACM PODC, pp. 154–162. ACM Press, August 1984
8. Cachin, C., Kursawe, K., Petzold, F., Shoup, V.: Secure and efficient asynchronous broadcast protocols. In: Kilian, J. (ed.) CRYPTO 2001. LNCS, vol. 2139, pp. 524–541. Springer, Heidelberg (2001)
9. Canetti, R.: Security and composition of multiparty cryptographic protocols. J. Cryptology **13**(1), 143–202 (2000)
10. Canetti, R.: Universally composable security: a new paradigm for cryptographic protocols. In: 42nd FOCS, pp. 136–145. IEEE Computer Society Press, October 2001
11. Canetti, R.: Universally composable signatures, certification and authentication. Cryptology ePrint Archive, Report 2003/239 (2003). http://eprint.iacr.org/2003/239
12. Canetti, R., Cohen, A., Lindell, Y.: A simpler variant of universally composable security for standard multiparty computation. In: Gennaro, R., Robshaw, M. (eds.) CRYPTO 2015. LNCS, vol. 9216, pp. 3–22. Springer, Heidelberg (2015)
13. Canetti, R., Lindell, Y., Ostrovsky, R., Sahai, A.: Universally composable two-party and multi-party secure computation. In: 34th ACM STOC, pp. 494–503. ACM Press, May 2002
14. Chaum, D., Crépeau, C., Damgård, I.: Multiparty unconditionally secure protocols (extended abstract). In: 20th ACM STOC, pp. 11–19. ACM Press, May 1988
15. Choi, S.G., Katz, J., Malozemoff, A.J., Zikas, V.: Efficient three-party computation from cut-and-choose. In: Garay, J.A., Gennaro, R. (eds.) CRYPTO 2014, Part II. LNCS, vol. 8617, pp. 513–530. Springer, Heidelberg (2014)

16. Cohen, R., Coretti, S., Garay, J.A., Zikas, V.: Probabilistic termination and composability of cryptographic protocols. Cryptology ePrint Archive, Report 2016/350 (2016). http://eprint.iacr.org/
17. Damgård, I.B., Ishai, Y.: Constant-round multiparty computation using a blackbox pseudorandom generator. In: Shoup, V. (ed.) CRYPTO 2005. LNCS, vol. 3621, pp. 378–394. Springer, Heidelberg (2005)
18. Damgård, I., Keller, M., Larraia, E., Pastro, V., Scholl, P., Smart, N.P.: Practical covertly secure MPC for dishonest majority – or: breaking the SPDZ limits. In: Crampton, J., Jajodia, S., Mayes, K. (eds.) ESORICS 2013. LNCS, vol. 8134, pp. 1–18. Springer, Heidelberg (2013)
19. Damgård, I., Nielsen, J.B.: Adaptive versus static security in the UC model. In: Chow, S.S.M., Liu, J.K., Hui, L.C.K., Yiu, S.M. (eds.) ProvSec 2014. LNCS, vol. 8782, pp. 10–28. Springer, Heidelberg (2014)
20. Damgård, I., Pastro, V., Smart, N., Zakarias, S.: Multiparty computation from somewhat homomorphic encryption. In: Safavi-Naini, R., Canetti, R. (eds.) CRYPTO 2012. LNCS, vol. 7417, pp. 643–662. Springer, Heidelberg (2012)
21. Dolev, D., Reischuk, R., Raymond Strong, H.: Early stopping in Byzantine agreement. J. ACM **37**(4), 720–741 (1990)
22. Dolev, D., Raymond Strong, H.: Authenticated algorithms for Byzantine agreement. SIAM J. Comput. **12**(4), 656–666 (1983)
23. Feldman, P., Micali, S.: An optimal probabilistic protocol for synchronous Byzantine agreement. SIAM J. Comput. **26**(4), 873–933 (1997)
24. Fischer, M.J., Lynch, N.A.: A lower bound for the time to assure interactive consistency. Inf. Process. Lett. **14**(4), 183–186 (1982)
25. Garg, S., Gentry, C., Halevi, S., Raykova, M.: Two-round secure MPC from indistinguishability obfuscation. In: TCC, pp. 74–94 (2014)
26. Goldreich, O., Micali, S., Wigderson, A.: Proofs that yield nothing but their validity and a methodology of cryptographic protocol design (extended abstract). In: 27th FOCS, pp. 174–187. IEEE Computer Society Press, October 1986
27. Goldreich, O., Micali, S., Wigderson, A.: How to play any mental game or a completeness theorem for protocols with honest majority. In: Aho, A. (ed.) 19th ACM STOC, pp. 218–229. ACM Press, May 1987
28. Goldreich, O., Petrank, E.: The best of both worlds: guaranteeing termination in fast randomized Byzantine agreement protocols. Inf. Process. Lett. **36**(1), 45–49 (1990)
29. Goldwasser, S., Lindell, Y.: Secure multi-party computation without agreement. J. Cryptology **18**(3), 247–287 (2005)
30. Dov Gordon, S., Liu, F.-H., Shi, E.: Constant-round MPC with fairness and guarantee of output delivery. In: Gennaro, R., Robshaw, M. (eds.) CRYPTO 2015. LNCS, vol. 9216, pp. 63–82. Springer, Heidelberg (2015)
31. Hirt, M., Zikas, V.: Adaptively secure broadcast. In: Gilbert, H. (ed.) EUROCRYPT 2010. LNCS, vol. 6110, pp. 466–485. Springer, Heidelberg (2010)
32. Ishai, Y., Prabhakaran, M., Sahai, A.: Founding cryptography on oblivious transfer– efficiently. In: Wagner, D. (ed.) CRYPTO 2008. LNCS, vol. 5157, pp. 572–591. Springer, Heidelberg (2008)
33. Katz, J., Koo, C.-Y.: On expected constant-round protocols for Byzantine agreement. In: Dwork, C. (ed.) CRYPTO 2006. LNCS, vol. 4117, pp. 445–462. Springer, Heidelberg (2006)
34. Katz, J., Koo, C.-Y.: Round-efficient secure computation in point-to-point networks. In: Naor, M. (ed.) EUROCRYPT 2007. LNCS, vol. 4515, pp. 311–328. Springer, Heidelberg (2007)

35. Katz, J., Lindell, Y.: Handling expected polynomial-time strategies in simulation-based security proofs. In: Kilian, J. (ed.) TCC 2005. LNCS, vol. 3378, pp. 128–149. Springer, Heidelberg (2005)

36. Katz, J., Maurer, U., Tackmann, B., Zikas, V.: Universally composable synchronous computation. In: Sahai, A. (ed.) TCC 2013. LNCS, vol. 7785, pp. 477–498. Springer, Heidelberg (2013)

37. Keller, M., Scholl, P., Smart, N.P.: An architecture for practical actively secure MPC with dishonest majority. In: Sadeghi, A.-R., Gligor, V.D., Yung, M. (eds.) ACM CCS 2013, pp. 549–560. ACM Press, November 2013

38. Kilian, J.: Founding cryptography on oblivious transfer. In: 20th ACM STOC, pp. 20–31. ACM Press, May 1988

39. Kushilevitz, E., Lindell, Y., Rabin, T.: Information-theoretically secure protocols and security under composition. In: Kleinberg, J.M. (ed.) 38th ACM STOC, pp. 109–118. ACM Press, May 2006

40. Lamport, L., Shostak, R.E., Pease, M.C.: The Byzantine generals problem. ACM Trans. Program. Lang. Syst. 4(3), 382–401 (1982)

41. Lindell, Y., Lysyanskaya, A., Rabin, T.: On the composition of authenticated Byzantine agreement. In: 34th ACM STOC, pp. 514–523. ACM Press, May 2002

42. Lindell, Y., Lysyanskaya, A., Rabin, T.: Sequential composition of protocols without simultaneous termination. In: Ricciardi, A. (ed.) 21st ACM PODC, pp. 203–212. ACM Press, July 2002

43. Lindell, Y., Pinkas, B., Smart, N.P., Yanai, A.: Efficient constant round multiparty computation combining BMR and SPDZ. In: Gennaro, R., Robshaw, M. (eds.) CRYPTO 2015. LNCS, vol. 9216, pp. 319–338. Springer, Heidelberg (2015)

44. Mukherjee, P., Wichs, D.: Two round multiparty computation via multi-key FHE. In: Fischlin, M., Coron, J.-S. (eds.) EUROCRYPT 2016. LNCS, vol. 9666, pp. 735–763. Springer, Heidelberg (2016). doi:10.1007/978-3-662-49896-5_26

45. Pease, M.C., Shostak, R.E., Lamport, L.: Reaching agreement in the presence of faults. J. ACM 27(2), 228–234 (1980)

46. Rabin, M.O.: Randomized Byzantine generals. In: 24th Annual Symposium on Foundations of Computer Science, Tucson, Arizona, USA, 7–9 November 1983, pp. 403–409. IEEE Computer Society (1983)

47. Rabin, T., Ben-Or, M.: Verifiable secret sharing and multiparty protocols with honest majority (extended abstract). In: 21st ACM STOC, pp. 73–85. ACM Press, May 1989

48. Turpin, R., Coan, B.A.: Extending binary Byzantine agreement to multivalued Byzantine agreement. Inf. Process. Lett. 18(2), 73–76 (1984)

49. Yao, A.C.-C.: Protocols for secure computations (extended abstract). In: 23rd FOCS, pp. 160–164. IEEE Computer Society Press, November 1982

Concurrent Non-Malleable Commitments (and More) in 3 Rounds

Michele Ciampi[1], Rafail Ostrovsky[2], Luisa Siniscalchi[1], and Ivan Visconti[1(⊠)]

[1] DIEM, University of Salerno, Fisciano, Italy
{mciampi,lsiniscalchi,visconti}@unisa.it
[2] UCLA, Los Angeles, USA
rafail@cs.ucla.edu

Abstract. The round complexity of commitment schemes secure against man-in-the-middle attacks has been the focus of extensive research for about 25 years. The recent breakthrough of Goyal et al. [22] showed that 3 rounds are sufficient for (one-left, one-right) non-malleable commitments. This result matches a lower bound of [41]. The state of affairs leaves still open the intriguing problem of constructing 3-round concurrent non-malleable commitment schemes.

In this paper we solve the above open problem by showing how to transform any 3-round (one-left one-right) non-malleable commitment scheme (with some extractability property) in a 3-round concurrent non-malleable commitment scheme. Our transform makes use of complexity leveraging and when instantiated with the construction of [22] gives a 3-round concurrent non-malleable commitment scheme from one-way permutations secure w.r.t. subexponential-time adversaries.

We also show a 3-round arguments of knowledge and a 3-round identification scheme secure against concurrent man-in-the-middle attacks.

Keywords: Non-malleability · Commitments · Identification schemes

1 Introduction

Commitment schemes are fundamental in Cryptography. They require a sender to fix a message that can not be changed anymore, but that will remain hidden to a receiver until the sender decides to reveal it.

In order to model modern real-world adversaries, commitment schemes have been proposed with additional security properties. Here we consider the intriguing question of constructing a scheme that remains secure against man-in-the-middle (MiM) attacks: a non-malleable (NM) commitment scheme [15].

Pass proved that NM commitments[1] require at least 3 rounds [41] when security is proved through a black-box reduction to a falsifiable (polynomial or subexponential time) hardness assumption. Instead by weakening the security definition admitting an inefficient challenger we know constructions of non-interactive NM commitments [38].

---

[1] We consider the notion of NM commitment w.r.t. commitment.

© International Association for Cryptologic Research 2016
M. Robshaw and J. Katz (Eds.): CRYPTO 2016, Part III, LNCS 9816, pp. 270–299, 2016.
DOI: 10.1007/978-3-662-53015-3_10

The round complexity of NM commitment schemes in the standard model has puzzled researchers for long time. Starting from the construction of [15] that required a logarithmic number of rounds, various constant-round schemes were proposed [1,19,20,28,29,42–44,46] reducing the round complexity to 4 rounds [5,11,23] with respect to concurrent MiM attacks, a setting that corresponds to what can actually happen when sender and receiver are connected through a communication network like the Internet. In such a much more interesting setting a MiM adversary receives multiple commitments from senders and sends his commitments to multiple receivers.

1.1   Towards 3-Round (Concurrent) NM Commitments

The existence of 3-round NM commitment schemes is an important question first because 3 is the best possible constant (in light of the lower bound of [41]), and second because 3 is the smallest number of rounds for a primitive that often makes use of commitment schemes: proofs of knowledge.

The importance of obtaining 3-round (and not just any constant-round) NM commitments motivated the very recent and innovative work of [22] that, by just relying on any non-interactive commitment scheme and exploiting the power of non-malleable codes in the split-state model, shows a 3-round NM commitment scheme. Interestingly, such construction is not claimed to be secure against concurrent man-in-the-middle attacks. Therefore the following natural and important question remains open.

Main Open Question: *Can we construct a 3-round concurrent non-malleable commitment scheme matching the lower bound of [41]?*

Other 3-Round Challenges. We list here 3 other interesting settings where no 3-round construction is known against concurrent MiM adversaries.

- Proofs[2] of knowledge are very useful in Cryptography. Despite their importance, there is no construction for 3-round proofs of knowledge (PoK) that is sufficiently secure under concurrent MiM attacks. This is due to the fact that such attacks are in general extremely difficult to deal with. Even though there exist constructions with a constant number of rounds, the case of just 3 rounds so far has remained unsolved.
- In [27][3] Lapidot and Shamir proposed a 3-round public-coin witness indistinguishable PoK for NP (the LS protocol) where the input (except its size) is needed only when playing the 3rd round. This special completeness property named "delayed input" in [12,13] has been used in many applications (e.g., [14,24,26,48,49] in particular recently [11,18,24,33]), and in [12,13] it was considered for the OR composition of Σ-protocols instead of relying on LS.

---

[2] For simplicity in the informal part of the paper we will not make a strict distinction between proofs and arguments. In the formal part we will use appropriate terms.

[3] See [37] for a detailed description of [27].

When a PoK is used as sub-protocol the delayed-input feature is instrumental to give a better round complexity to the external protocol. An additional features of delayed-input protocols is that they allow to shift large part of the computation to an off-line phase. Unfortunately the LS protocol and the PoKs of [12,12] are not secure against concurrent MiM attacks and this penalizes those applications where both round complexity and security against concurrent MiM attacks are important.

– We notice that identification schemes have been often proposed (e.g., [17]) through the paradigm of proving "knowledge" of a secret[4]. Under this formulation there are constant-round constructions that are proven secure against concurrent MiM attacks [2]. However no 3-round scheme known in literature is proven secure in presence of a concurrent MiM adversary.

1.2   Results of This Work

In this work we study 3-round commitment scheme in presence of concurrent MiM attacks and solve in the positive the above open problems.

3-Round Concurrent NM Commitment Schemes. In the main result of this submission, we show a transform that on input any 3-round NM commitment scheme[5] gives a 3-round concurrent NM commitment scheme. The construction of [22] can be used to instantiate our transform, therefore obtaining a 3-round concurrent NM commitment scheme based on any one-way permutation secure against subexponential-time adversaries. Moreover our scheme (still when instantiated with the one of [22] and using a proper one-way permutation) is public coin and (if desired[6]) has the delayed-input property.

Our transform extends the security of the underlying commitment scheme to multiple receivers. It is known that this implies security also with multiple senders [30]. The crucial idea of our transform is to combine the underlying NM commitment scheme along with a one-time pad, to produce a commitment of a message that by itself, in case of a malleability attack, will have sufficient structure to be recognized by a distinguisher in the session in which it appears. Therefore a successful concurrent MiM even playing multiple commitments with multiple receivers will have to maul the underlying commitment scheme in at least one session. Since the message has sufficient structure with respect to that

---

[4] Other notions based on signature or decryption capabilities are considered weaker since in some applications the verifier wants to make sure that the prover is the actual entity matching the announced identity. Indeed without a PoK a prover could give some partial information about his secret to others that can still succeed in convincing the verifier, even though they do not know the full secret.

[5] We also require the scheme to be extractable. Extractability often comes for free since it is commonly used in the non-malleability proof.

[6] Our transform can be instantiated in two ways. In the former the message to commit is required already when playing the first round, while in the latter the message to commit is required when playing the third round only.

single session, we are able to translate the concurrent MiM attack into a non-concurrent MiM that violates the security of the underlying (non-concurrent) NM commitment scheme. We will implement the idea of committing to a message with structure by forcing a successful concurrent MiM to commit to the solution of a puzzle in at least one session. We will use complexity leveraging to show that the attack of the concurrent MiM is indistinguishable from the attack of a polynomial-time simulator that plays with receivers only.

Just for completeness, we also show an explicit concurrent MiM adversary $\mathcal{A}$ for the scheme of [21]. The crucial point here, following a technique of [16] is that the scheme of [21] allows $\mathcal{A}$ to spread the message committed by the honest sender over several commitments that the adversary sends to multiple receivers. The scheme presented in [22] is slightly different and became available after our work was already submitted, therefore when describing $\mathcal{A}$ we stick with [21].

3-Round Arguments of Knowledge and ID Schemes Against Concurrent MiM Attacks. Our 3-round concurrent NM commitment scheme is a commit-and-prove argument of knowledge (AoK). This means that one can see our scheme as a commitment followed by an AoK about the committed value. By applying a simple change to the statement of the underlying AoK we obtain a 3-round concurrent NM witness-indistinguishable AoK (concurrent NMWIAoK) a notion introduced in [34] and later on extended in [31]. We stress that the delayed-input and public-coin properties of our commitment scheme are preserved by our concurrent NMWIAoK.

In [34] it is shown how to get concurrent NM zero knowledge (NMZK) in the bare public-key (BPK) model [6] with just two executions of a concurrent NMWIAoK. Therefore we directly obtain a round-efficient concurrent NMZKAoK in the BPK model. By making use of delayed-input completeness the simulator can extend a main thread avoiding issues due to aborting adversaries as discussed in [36,47].

Finally, we notice that one can get an identification scheme secure in the PoK sense in the concurrent[7] setting of [2] as well as under the stronger definition based on matching conversations of [3,25] naturally extended to concurrent sessions. Following [9,34], the key idea consists in using an identity that has two possible secrets such that knowledge of one witness does not allow to compute the other one in polynomial time. By using our implementation of a concurrent NMWIAoK for proving knowledge of a secret associated to such identity we obtain a 3-round identification scheme secure against concurrent MiM attacks.

Challenges for Future Work. The existence of OWPs is a standard falsifiable hardness assumption. Our scheme relies on a strengthening of this standard assumption w.r.t. subexponential-time adversaries. Notice that the lower bound of [41] still applies in case of subexponential-time hardness, therefore our 3-round

---

[7] In [2] the notion CR2 is proposed to deal with concurrent MiM attacks and reset attacks. Reset attack were also considered in the notion CR+ of [4]. Since reset attacks are out of the scope of this work, we focus on concurrent MiM attacks only.

concurrent non-malleable scheme is round optimal. Various natural and fascinating questions on commitments and proofs of knowledge remain open after our work and as such we think our results will motivate further research. Examples of open questions about concurrent NM commitments are the following: (1) the existence of 3-round schemes based on standard falsifiable hardness assumptions w.r.t. polynomial-time adversaries only; (2) the existence of 3-round schemes with black-box use of primitives; (3) the existence of practical schemes.

2   Notation, Definitions and Tools

We denote the security parameter by λ and use "$|$" as concatenation operator (i.e., if a and b are two strings then by $a|b$ we denote the concatenation of a and b). For a finite set Q, $x \leftarrow Q$ denotes the algorithm that chooses x from Q with uniform distribution. Usually we use the abbreviation PPT that stays for probabilistic polynomial-time. We use $\mathsf{poly}(\cdot)$ to indicate a generic polynomial function of the input.

A *polynomial-time relation* Rel (or *polynomial relation*, in short) is a subset of $\{0,1\}^* \times \{0,1\}^*$ such that membership of (x, w) in Rel can be decided in time polynomial in $|x|$. For $(x, w) \in$ Rel, we call x the *instance* and w a *witness* for x. For a polynomial-time relation Rel, we define the NP-language L_{Rel} as $L_{\mathsf{Rel}} = \{x | \exists w : (x, w) \in \mathsf{Rel}\}$. Analogously, unless otherwise specified, for an NP-language L we denote by Rel_L the corresponding polynomial-time relation (that is, Rel_L is such that $L = L_{\mathsf{Rel}_L}$).

Let A and B be two interactive probabilistic algorithms A and B. We denote by $\langle A(\alpha), B(\beta) \rangle(\gamma)$ the distribution of B's output after running on private input β with A using private input α, both running on common input γ. Typically, one of the two algorithms receives 1^λ as input. A *transcript* of $\langle A(\alpha), B(\beta) \rangle(\gamma)$ consists of the messages exchanged during an execution where A receives a private input α, B receives a private input β and both A and B receive a common input γ. Moreover, we will refer to the *view* of A as the messages it received during the execution of $\langle A(\alpha), B(\beta) \rangle(\gamma)$, along with its randomness and its input. We denote by A_r an algorithm A that receives as randomness r. We say that a protocol (A, B) is public coin if B sends to A random bits only.

A function $\nu(\cdot)$ from non-negative integers to reals is called negligible, if for every constant $c > 0$ and all sufficiently large $\lambda \in \mathbb{N}$ we have $\nu(\lambda) < \lambda^{-c}$. Standard definitions of one-way permutations (OWPs), proof/argument systems, witness indistinguishability (WI) and proofs of knowledge along with their strengthened versions secure again subexponential-time adversaries and adaptive-input selection can be found in the full version of this work [10].

2.1   Commitment Schemes

Definition 1 (Commitment Scheme). *Given a security parameter 1^λ, a commitment scheme* (Sen, Rec) *is a two-phase protocol between two PPT interactive algorithms, a sender* Sen *and a receiver* Rec. *In the commitment phase* Sen *on input a message m interacts with* Rec *to produce a commitment* com. *In the decommitment phase,* Sen *sends to* Rec *a decommitment information* d *such that* Rec *accepts m as the commitment of* com.

Formally, we say that CS $=$ (Sen, Rec) *is a perfectly binding commitment scheme if the following properties hold:*

Correctness:
- *Commitment phase. Let* com *be the commitment of the message m (i.e., com is the transcript of an execution of* CS $=$ (Sen, Rec) *where* Sen *runs on input a message m). Let* d *be the private output of* Sen *in this phase.*
- *Decommitment phase*[8]. Rec *on input m and* d *accepts m as decommitment of* com.

Hiding [32]: *for a* PPT *adversary* $\mathcal{A}$ *and a randomly chosen bit* $b \in \{0,1\}$, *consider the following hiding experiment* $\mathsf{ExpHiding}^b_{\mathcal{A},\mathsf{CS}}(\lambda)$:
- *Upon input* 1^λ, *the adversary* $\mathcal{A}$ *outputs a pair of messages* m_0, m_1 *that are of the same length.*
- Sen *on input the message* m_b *interacts with* $\mathcal{A}$ *to produce a commitment of* m_b.
- $\mathcal{A}$ *outputs a bit* b' *and this is the output of the experiment.*

For any PPT *adversary* $\mathcal{A}$, *there exist a negligible function* ν, *such that:*

$$\left| \mathrm{Prob}\left[\mathsf{ExpHiding}^0_{\mathcal{A},\mathsf{CS}}(\lambda) = 1 \right] - \mathrm{Prob}\left[\mathsf{ExpHiding}^1_{\mathcal{A},\mathsf{CS}}(\lambda) = 1 \right] \right| < \nu(\lambda).$$

Binding: *for every commitment* com *generated during the commitment phase by a possibly malicious unbounded sender* Sen* *interacting with an honest receiver* Rec, *there exists at most one message m that* Rec *accepts as decommitment of* com.

We also consider the definition of a commitment scheme where the hiding property still holds against an adversary $\mathcal{A}$ running in time bounded by $T = 2^{\lambda^\alpha}$ for some positive constant $\alpha < 1$. In this case we will say that a commitment scheme is T-hiding. We will also say that a commitment scheme is $\tilde{T}$-breakable to specify that an algorithm running in time $\tilde{T} = 2^{\lambda^\beta}$, for some positive constant $\beta < 1$, recovers the (if any) only message that can be successfully decommitment.

In the rest of the paper we also use a non-interactive commitment schemes, with secure parameter λ. In this case we consider a commitment scheme as a pair of PPT algorithms (NISen, NIRec) where:

- NISen takes as input $(m; \sigma)$, where $m \in \{0,1\}^{\mathsf{poly}(\lambda)}$ is the message to be committed and $\sigma \leftarrow \{0,1\}^\lambda$ is randomness, and outputs the commitment com and the decommitment dec;
- NIRec takes as input (dec, com, m) and outputs 1 if it accepts m as a decommitment of com and 0 otherwise.

---

[8] In this paper we consider a non-interactive decommitment phase only.

3-Round Extractable Commitment Schemes. Informally, a 3-round commitment scheme is extractable if there exists an efficient extractor that having black-box access to any efficient malicious sender ExCom* that successfully performs the commitment phase, outputs the only committed string that can be successfully decommitted.

Definition 2 (3-Round Extractable Commitment Scheme [45]). *A 3-round perfectly binding commitment scheme* ExCS = (ExCom, ExRec) *is an extractable commitment scheme if given oracle access to any malicious sender* ExCom*, *there exists an expected* PPT *extractor* Ext *that outputs a pair* (τ, σ^*) *such that the following properties hold:*

- **Simulatability:** *the simulated view* τ *is identically distributed to the view of* ExCom* *(when interacting with an honest* ExRec*) in the commitment phase.*
- **Extractability:** *there exists no decommitment of* τ *to* σ, *where* $\sigma \neq \sigma^*$.

2.2 Non-Malleable Commitment Schemes

Here we follow [30][9]. Let $\Pi = (\mathsf{Sen}, \mathsf{Rec})$ be a statistically binding commitment scheme. Consider MiM adversaries that are participating in left and right sessions in which $\mathsf{poly}(\lambda)$ commitments take place. We compare between a MiM and a simulated execution. In the MiM execution the adversary $\mathcal{A}$, with auxiliary information z, is simultaneously participating in $\mathsf{poly}(\lambda)$ left and right sessions. In the left sessions the MiM adversary $\mathcal{A}$ interacts with Sen receiving commitments to values $m_1, \ldots, m_{\mathsf{poly}(\lambda)}$ using identities $\mathsf{id}_1, \ldots, \mathsf{id}_{\mathsf{poly}(\lambda)}$ of its choice. In the right session $\mathcal{A}$ interacts with Rec attempting to commit to a sequence of related values $\tilde{m}_1, \ldots, \tilde{m}_{\mathsf{poly}(\lambda)}$ again using identities of its choice $\tilde{\mathsf{id}}_1, \ldots, \tilde{\mathsf{id}}_{\mathsf{poly}}(\lambda)$. If any of the right commitments is invalid, or undefined, its value is set to $\perp$. For any i such that $\tilde{\mathsf{id}}_i = \mathsf{id}_j$ for some j, set $\tilde{m}_i = \perp$ (i.e., any commitment where the adversary uses the same identity of one of the honest senders is considered invalid). Let $\mathsf{mim}_\Pi^{\mathcal{A}, m_1, \ldots, m_{\mathsf{poly}(\lambda)}}(z)$ denote a random variable that describes the values $\tilde{m}_1, \ldots, \tilde{m}_{\mathsf{poly}(\lambda)}$ and the view of $\mathcal{A}$, in the above experiment. In the simulated execution, an efficient simulator S directly interacts with Rec. Let $\mathsf{sim}_\Pi^S(1^\lambda, z)$ denote the random variable describing the values $\tilde{m}_1, \ldots, \tilde{m}_{\mathsf{poly}(\lambda)}$ committed by S, and the output view of S; whenever the view contains in the i-th right session the same identity of any of the identities of the left session, then $\tilde{m}_i$ is set to $\perp$.

We denote by $\tilde{\delta}$ a value associated with the right session (where the adversary $\mathcal{A}$ plays with a receiver MMRec) where δ is the corresponding value in the left session. For example, the sender commits to v in the left session while $\mathcal{A}$ commits to $\tilde{v}$ in the right session.

Definition 3 (Concurrent NM Commitment Scheme [30]). *A commitment scheme is* concurrent NM *with respect to commitment (or a many-many NM commitment scheme) if, for every* PPT *concurrent MiM adversary*

---

[9] In this paper we will consider only NM commitments w.r.t. commitments. Difficulties on achieving also the notion of NM w.r.t. decommitments were discussed in [7,35].

$\mathcal{A}$, there exists a PPT simulator S such that for all $m_i \in \{0,1\}^{\mathsf{poly}(\lambda)}$ for $i = \{1,\ldots,\mathsf{poly}(\lambda)\}$ the following ensembles are computationally indistinguishable:

$$\{\mathsf{mim}_\Pi^{\mathcal{A},m_1,\ldots,m_{\mathsf{poly}(\lambda)}}(z)\}_{z\in\{0,1\}^\star} \approx \{\mathsf{sim}_\Pi^S(1^\lambda,z)\}_{z\in\{0,1\}^\star}.$$

As in [30] we also consider relaxed notions of concurrent non-malleability: one-many and one-one NM commitment schemes. In a one-many NM commitment scheme, $\mathcal{A}$ participates in one left and polynomially many right sessions. In a one-one (i.e., a stand-alone secure) NM commitment scheme, we consider only adversaries $\mathcal{A}$ that participate in one left and one right session. We will make use of the following proposition of [30].

Proposition 1. Let $(\mathsf{Sen}, \mathsf{Rec})$ be a one-many NM commitment scheme. Then, $(\mathsf{Sen}, \mathsf{Rec})$ is also a concurrent (i.e., many-many) NM commitment scheme.

We also consider the definition of a NM commitment scheme secure against a MIM $\mathcal{A}$ running in time bounded by $T = 2^{\lambda^\alpha}$ for some positive constant $\alpha < 1$. In this case we will say that a commitment scheme is T-non-malleable.

When the identity is selected by the sender then the above id-based definitions guarantee non-malleability without ids as long as the MiM does not behave like a proxy (an unavoidable attack). Indeed the sender can pick as id the public key of a strong signature scheme signing the transcript. The MiM will have to use a different id or to break the signature scheme.

2.3   3-Round One-One NM Commitment Scheme

As main tool we need a 3-round one-one NM commitment scheme (NMCS) that enjoys the extractability property. In the rest of the paper we will refer to such a commitment scheme as $\Pi_{\mathsf{NM}} = (\mathsf{Sen}_{\mathsf{NM}}, \mathsf{Rec}_{\mathsf{NM}})$.

In [22] the authors provide the first 3-round one-one NM commitment scheme. Their scheme enjoys also the extractability property[10] and public coin.

By $\Pi_{\mathsf{NM}} = ((\mathsf{Sen}^1_{\mathsf{NM}}, \mathsf{Sen}^2_{\mathsf{NM}}), \mathsf{Rec}_{\mathsf{NM}})$ we denote a 3-round one-one NM commitment scheme such that:

- the algorithm $\mathsf{Sen}^1_{\mathsf{NM}}$ takes as input $(\mathsf{id}, m; \rho)$, where $\mathsf{id} \in \{0,1\}^\lambda$ is the identity, m is the message to be committed and $\rho \leftarrow \{0,1\}^\lambda$ is a randomness, and outputs a that is the first round of the commitment scheme to be sent to the receiver;
- the algorithm $\mathsf{Sen}^2_{\mathsf{NM}}$ takes as input $(\mathsf{id}, \mathsf{c}, m; \rho)$, where c is the second round received by Rec, m is the message to be committed, id is the same identity received as input by $\mathsf{Sen}^1_{\mathsf{NM}}$, ρ is the randomness, and outputs $(\mathsf{z}, \mathsf{dec})$ where z is the last round of the commitment, and dec is the decommitment value.

The reveal phase consists in sending dec and m to the receiver. The receiver $\mathsf{Rec}_{\mathsf{NM}}$, on input the randomness it used during the commitment phase, the transcript $\mathsf{com} = (\mathsf{a}, \mathsf{c}, \mathsf{z}, \mathsf{id})$, m and dec outputs 1 if dec is valid w.r.t. com and m and outputs 0 otherwise.

---

[10] Extractability is informally stated in Claim 12 of [21].

2.4   The LS Proof of Knowledge and NMWI Argument Systems

In this paper we use the 3-round public-coin WI adaptive proof of knowledge proposed by Lapidot and Shamir [27], that we denote by LS. LS is delayed-input since the inputs for the prover and the verifier are needed only to play the last round, while only the size of the common input is needed earlier. For this reason we will refer to a prover $\mathcal{P}$ as a pair $(\mathsf{P}^1, \mathsf{P}^2)$. More formally, LS for a relation Rel is a pair $\varPi = (\mathcal{P} = (\mathsf{P}^1, \mathsf{P}^2), \mathcal{V})$, with security parameter λ, where $\mathcal{P}$ executes the algorithms P^1 and P^2 defined as follows. The algorithm P^1, takes as input $(\ell; \alpha)$, ℓ is the instance length and $\alpha \leftarrow \{0,1\}^\lambda$ is the randomness, and outputs the 1st round of the LS protocol. The algorithm P^2 takes as input $(x, w, c; \alpha)$, where x, w are such that $(x, w) \in \mathsf{Rel}$, c is the challenge sent by $\mathcal{V}$ and α is the randomness[11] and outputs the 3rd round of the LS protocol.

In this paper we also consider a definition where the WI property of LS still holds against a distinguisher with running time bounded by $T = 2^{\lambda^\alpha}$ for some constant positive constant $\alpha < 1$. In this case we say that the instantiation of LS is T-witness indistinguishable (T-WI).

Witness Indistinguishability and MiM Attacks. The definition of non-malleable witness indistinguishability (NMWI) given in [34] requires that the witness *encoded in the proof* given by the MiM $\mathcal{A}$ be independent of the witness used by the honest prover in his proof. For details see [10].

3   3-Round Concurrent Non-Malleable Commitments

In this section we show our transform that takes as input a 3-round extractable one-one NM commitment scheme $\varPi_{\mathsf{NM}}$, a OWP f, a non-interactive perfectly binding commitment scheme NI, the 3-round delayed-input adaptive WI/PoK LS and outputs a 3-round fully concurrent (i.e., many-many) NM commitment scheme $\varPi_{\mathsf{MMCom}} = (\mathsf{MMSen}, \mathsf{MMRec})$.

Let m be the message that MMSen wants to commit. The high-level idea of our transform is depicted in Fig. 1. The sender MMSen, on input the session-id id and the message m, computes the 1st round of the protocol by running LS and sending the 1st round of NM to commit to a random message s_0 using id as session-id. In the 2nd round the receiver MMRec sends the challenges of NM and LS, also sends a random value Y in the range of the OWP f[12]. In the last round MMSen commits to message m using NI, therefore obtaining com, then computes the last round of NM, completes the transcript of LS, and finally sends a random string s_1. The protocol LS is used by MMSen to prove to MMRec that either she knows message m and the randomness used to compute com, or she knows the values (s_0, dec), such that $f(s_0 \oplus s_1) = Y$ and dec is a valid decommitment to s_0 w.r.t. the commitment computed using $\varPi_{\mathsf{NM}}$. We observe that MMSen

---

[11] The same α is passed to P^1 and P^2 so that P^2 can reconstruct the state of P^1.

[12] When sampling from the range of f corresponds to picking a random string, we have that our commitment scheme is public coin.

needs m only when computing the 3rd round, therefore our construction enjoys delayed-input correctness.

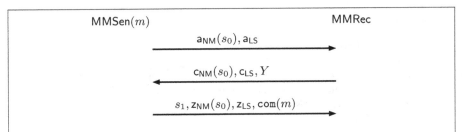

- Y is an element taken from the range of the OWP f.
- $\mathsf{com}(m)$ is the perfectly binding commitment of m computed using NI.
- $(\mathsf{a_{NM}}(s_0), \mathsf{c_{NM}}(s_0), \mathsf{z_{NM}}(s_0)) = \tau$ is the transcript of the execution of the NM commitment scheme Π_{NM} when the sender commits to s_0.
- $(\mathsf{a_{LS}}, \mathsf{c_{LS}}, \mathsf{z_{LS}}) = \pi$ is the transcript of LS proving knowledge of either m and the randomness used to compute com, or of (s_0, dec), s.t. $f(s_0 \oplus s_1) = Y$ and dec is a valid decommitment of s_0 w.r.t. τ.

Fig. 1. Informal description of our 3-round concurrent NM commitment scheme.

Our transform needs the following tools:

1. a OWP f that is secure against PPT adversaries and $\tilde{T}_f$-breakable;
2. a non interactive perfectly binding commitment scheme $\mathsf{NI} = (\mathsf{NISen}, \mathsf{NIRec})$ that is T_{NI}-hiding and $\tilde{T}_{\mathsf{NI}}$-breakable;
3. a 3-round extractable *one-one* NM commitment scheme $\Pi_{\mathsf{NM}} = (\mathsf{Sen_{NM}} = (\mathsf{Sen^1_{NM}}, \mathsf{Sen^2_{NM}}), \mathsf{Rec_{NM}})$ that is T_{NM}-hiding/non-malleable, and $\tilde{T}_{\mathsf{NM}}$-breakable;
4. the LS proof system $\mathsf{LS} = (\mathcal{P} = (\mathsf{P}^1, \mathsf{P}^2), \mathcal{V})$ for the language

$$L = \left\{ ((a,c,z), Y, s_1, \mathsf{com}, \mathsf{id}) : \exists\, (m, \sigma) \text{ s.t. } \mathsf{com} = \mathsf{NISen}(m; \sigma) \text{ OR} (\exists (\rho, s_0) \right.$$
$$\left. \text{s.t. } a = \mathsf{Sen^1_{NM}}(\mathsf{id}, s_0; \rho) \text{ AND } z = \mathsf{Sen^2_{NM}}(\mathsf{id}, c, s_0; \rho) \text{ AND } Y = f(s_0 \oplus s_1)) \right\}$$

that is T_{LS}-WI for the corresponding relation $\mathsf{Rel_L}$.

Let λ be the security parameter of our scheme. We will use wlog λ also as security parameter for the hardness to invert f with respect to polynomial time adversaries. Then we consider the following hierarchy of security levels for the above tools: $T_f << T_{\mathsf{NI}} << \sqrt{T_{\mathsf{NM}}} << T_{\mathsf{NM}} << \sqrt{T_{\mathsf{LS}}} << T_{\mathsf{LS}}$ where by "$T <<$ T'" we mean that "$T \cdot \mathrm{poly}(\lambda) < T'$". We also require that: (1) NI is T_{NI}-hiding, but is also $\tilde{T}_{\mathsf{NI}} = \sqrt{T_{\mathsf{NM}}}$-breakable; (2) Π_{NM} is T_{NM} hiding/non-malleable, but the hiding is also $\tilde{T}_{\mathsf{NM}} = \sqrt{T_{\mathsf{LS}}}$-breakable. Now we need to define different security parameters, one for each tool involved in the security proof to be consistent with the hierarchy of security levels defined above (a similar use of security parameters has been proposed in [46]). Given the security parameter λ of our scheme, we will make use of the following security parameters (all polynomially related to

---

Common input: Security parameters: λ, $(\lambda_{NI}, \lambda_{NM}, \lambda_{LS}, \ell) = \mathsf{Params}(\lambda)$.
MMSen's identity: $\mathtt{id} \in \{0,1\}^\lambda$.
Input to MMSen: $m \in \{0,1\}^{\mathsf{poly}\{\lambda\}}$.

Commitment Phase:

1. MMSen $\rightarrow$ MMRec
   1.1. Pick $s_0 \in \{0,1\}^\lambda$.
   1.2. Pick a randomness $\rho \in \{0,1\}^{\lambda_{NM}}$ and compute $\mathsf{a}_{NM} = \mathsf{Sen}^1_{NM}(\mathtt{id}, s_0; \rho)$.
   1.3. Pick a randomness $\alpha \in \{0,1\}^{\lambda_{LS}}$ and compute $\mathsf{a}_{LS} = \mathsf{P}^1(\ell; \alpha)$.
   1.4. Send $(\mathsf{a}_{NM}, \mathsf{a}_{LS})$ to MMRec.
2. MMRec $\rightarrow$ MMSen
   2.1. Pick a randomness γ and run Rec_{NM} on input $(\mathtt{id}, \mathsf{a}_{NM}; \gamma)$ to obtain c_{NM}.
   2.2. Pick a randomness β and run $\mathcal{V}$ to obtain c_{LS}.
   2.3. Pick a random $y \in \{0,1\}^\lambda$ and compute $Y = f(y)$.
   2.4. Send $(\mathsf{c}_{NM}, \mathsf{c}_{LS}, Y)$ to MMSen.
3. MMSen $\rightarrow$ MMRec
   3.1. Pick a randomness $\sigma \in \{0,1\}^{\lambda_{NI}}$ and compute $(\mathtt{com}, \mathtt{dec}) = \mathsf{NISen}(m; \sigma)$.
   3.2. Pick $s_1 \leftarrow \{0,1\}^\lambda$.
   3.3. Compute $(\mathsf{z}_{NM}, \mathtt{dec}_{NM}) = \mathsf{Sen}^2_{NM}(\mathtt{id}, \mathsf{c}_{NM}, s_0; \rho)$.
   3.4. Set $x = \big((\mathsf{a}_{NM}, \mathsf{c}_{NM}, \mathsf{z}_{NM}), Y, s_1, \mathtt{com}, \mathtt{id}\big)$ and $w = (m, \sigma, \bot, \bot)$ with $(|x| = \ell)$.
         Run $\mathsf{z}_{LS} = \mathsf{P}^2(x, w, \mathsf{c}_{LS}; \alpha)$ where x is the theorem to be proven and w is the witness.
   3.5. Send $(\mathsf{z}_{NM}, \mathtt{com}, \mathsf{z}_{LS}, s_1)$ to MMRec.
4. MMRec: Set $x = \big((\mathsf{a}_{NM}, \mathsf{c}_{NM}, \mathsf{z}_{NM}), Y, s_1, \mathtt{com}, \mathtt{id}\big)$ and abort iff $(\mathsf{a}_{LS}, \mathsf{c}_{LS}, \mathsf{z}_{LS})$ is not accepting for $\mathcal{V}$ with respect to x.

Decommitment Phase:

1. MMSen $\rightarrow$ MMRec: Send $(\mathtt{dec}, m, \mathtt{dec}_{NM}, s_0)$ to MMRec.
2. MMRec: Accept m as the committed message iff
   2.1. $\mathsf{NIRec}(\mathtt{dec}, \mathtt{com}, m) = 1$ and
   2.2. Rec_{NM} on input γ, $(\mathsf{a}_{NM}, \mathsf{c}_{NM}, \mathsf{z}_{NM}, \mathtt{id})$, s_0 and $\mathtt{dec}_{NM}$ outputs 1.

Fig. 2. Our 3-round concurrent NM commitment scheme.

λ and such that the above hierarchy of security levels holds): λ for f, λ_{NI} for NI, λ_{NM} for Π_{NM}, λ_{LS} for LS.

We denote by Params the function that on input λ outputs $(\lambda_{NI}, \lambda_{NM}, \lambda_{LS}, \ell)$ where ℓ is the size of the theorem to be proved using LS[13]. Our concurrent NM commitment scheme $\Pi_{MMCom} = (\mathsf{MMSen}, \mathsf{MMRec})$ is fully described in Fig. 2.

Theorem 1. *Suppose there exist OWPs secure against subexponential-time adversaries, then Π_{MMCom} is a perfectly binding delayed-input commitment scheme.*

---

[13] To compute 1st and 2nd round of LS only the length ℓ of the instance is required.

Proof. The delayed-input correctness of Π_{MMCom} follows by inspection from the delayed-input completeness of LS, and the correctness of Π_{NM} and NI.

Observe that the message given in output in the decommitment phase of Π_{MMCom} is the message committed using NI. Moreover the decommitment phase of Π_{MMCom} coincides with the decommitment of NI and Π_{NM}. Since NI and Π_{NM} is perfectly binding we have that Π_{MMCom} is perfectly binding too.

The hiding property follows from the non-malleability property proved in Theorem 2. Indeed the proof of Theorem 2 does not rely on the hiding of Π_{MMCom}.

Theorem 2. *Suppose there exist OWPs secure against subexponential-time adversaries, then Π_{MMCom} is concurrent (i.e., many-many) non-malleable.*

Proof. Since we can use Proposition 1, we only need to prove that our commitment enjoys one-many non-malleability. More formally with respect to a one-many adversary $\mathcal{A}$, we need to show that for all $m \in \{0,1\}^{\mathsf{poly}(\lambda)}$ it holds that: $\{\mathsf{mim}^{\mathcal{A},m}_{\Pi_{\mathsf{MMCom}}}(z)\}_{z\in\{0,1\}^*} \approx \{\mathsf{sim}^{S}_{\Pi_{\mathsf{MMCom}}}(1^\lambda, z)\}_{z\in\{0,1\}^*}$ where S is the simulator depicted in Fig. 3. This means that the real execution in which the sender runs MMSen to commit to a message m must be indistinguishable with respect to an execution in which a simulator S runs internally the MiM adversarial $\mathcal{A}$ sending a commitment of 0^λ, and then forwards the messages that $\mathcal{A}$ sends in the right sessions to receivers $\mathsf{MMRec}_1, \dots, \mathsf{MMRec}_{\mathsf{poly}(\lambda)}$.

In the security proof we denote by $\tilde{\delta}_i$ a value associated with the i-th right session (where the adversary $\mathcal{A}$ plays with a receiver MMRec_i with $i \in \{1, \dots, \mathsf{poly}(\lambda)\}$) where δ is the corresponding value in the left session. For example, the sender commits to v in the left session while $\mathcal{A}$ commits to $\tilde{v}_i$ in the i-th right session.

To prove the indistinguishability of the above two experiments we show 3 hybrid experiments[14] $\mathcal{H}^m_i(z)$ with $i = 1, 2, 3$, where m is the message committed in the left session. Following [28] we denote by $\{\mathsf{mim}^{\mathcal{A}}_{\mathcal{H}^m_i}(z)\}_{z\in\{0,1\}^*}$ the random variable describing the view of the MiM $\mathcal{A}$ combined with the value it commits in the right interaction in hybrid $\mathcal{H}^m_i(z)$ (as usual, the committed value is replaced by $\perp$ if the right interaction does not correspond to a commitment that can be successfully opened or if $\mathcal{A}$ has copied the identity of the left interaction).

The 1st hybrid is the experiment $\mathcal{H}^m_1(z)$ in which in the left session MMSen commits to m, while in the right session we run $\mathsf{MMRec}_1, \dots, \mathsf{MMRec}_{\mathsf{poly}(\lambda)}$ for the rights sessions played by $\mathcal{A}$.

$\mathcal{H}^m_1(z)$.

Left session:
1. First round.
   1.1. Pick $s_0 \leftarrow \{0,1\}^\lambda$.
   1.2. Compute $\mathsf{a}_{\mathsf{NM}} = \mathsf{Sen}^1_{\mathsf{NM}}(\mathsf{id}, s_0; \rho)$.
   1.3. Compute $\mathsf{a}_{\mathsf{LS}} = \mathsf{P}^1(1^{\lambda_{\mathsf{LS}}}, \ell; \alpha)$.
   1.4. Send $(\mathsf{a}_{\mathsf{NM}}, \mathsf{a}_{\mathsf{LS}})$ to $\mathcal{A}$.

---

[14] We will describe the hybrid experiments in a succinct way focusing on the key steps (e.g., omitting sampling of randomness, generation of parameters $\lambda_{\mathsf{NI}}, \lambda_{\mathsf{NM}}, \lambda_{\mathsf{LS}}, \ell$).

2. Third round, upon receiving (c_{NM}, c_{LS}, Y) from $\mathcal{A}$.
  2.1. Compute $(com, dec) = NISen(m; \sigma)$.
  2.2. Pick $s_1 \leftarrow \{0,1\}^\lambda$.
  2.3. Compute $(z_{NM}, dec_{NM}) = Sen_{NM}^2(id, c_{NM}, s_0; \rho)$.
  2.4. Set $x = ((a_{NM}, c_{NM}, z_{NM}), Y, s_1, com, id)$ and $w = (m, \sigma, \perp, \perp)$ with $(|x| = \ell)$. Run $z_{LS} = P^2(x, w, c_{LS}; \alpha)$.
  2.5. Send $(z_{NM}, com, z_{LS}, s_1)$ to $\mathcal{A}$.

Right sessions: act as a proxy between $\mathcal{A}$ and $MMRec_1, \ldots, MMRec_{poly(\lambda)}$.

We have that for all $m \in \{0,1\}^{poly(\lambda)}$ $\{mim_{\mathcal{H}_1^m}^{\mathcal{A}}(z)\}_{z \in \{0,1\}^*}$ corresponds to $\{mim_{\Pi_{MMCom}}^{\mathcal{A},m}(z)\}_{z \in \{0,1\}^*}$. We now prove that, for all $i \in \{1, \ldots, poly(\lambda)\}$ $\mathcal{A}$ does not manage to invert any values $\tilde{Y}_i$ in the right sessions by sending a value $\tilde{s}_{1i}$ such that $f(\tilde{s}_{0i} \oplus \tilde{s}_{1i}) = \tilde{Y}_i$ where $\tilde{s}_{0i}$ is the message committed in the i-th right session through NM.

Lemma 1. *Let p_i be the probability that in the i-th right session, for $i \in \{1, \ldots, poly(\lambda)\}$, $\mathcal{A}$ sends $\tilde{s}_{1i}$ such that $f(\tilde{s}_{1i} \oplus \tilde{s}_{0i}) = \tilde{Y}_i$ where $\tilde{s}_{0i}$ is the value committed using NM. Then $p_i < \nu(\lambda)$ for some negligible function ν.*

Proof. Suppose by contradiction that for a right session i the claim does not hold. We can construct an adversary $\mathcal{A}_f$ that inverts the OWP f in polynomial time. We consider a challenger $\mathcal{C}_f$ of f that chooses a random Y in the range of f and sends it to $\mathcal{A}_f$. $\mathcal{A}_f$ wins if it gives as output y such that $Y = f(y)$. Before describing the adversary we need to consider the augmented machine $\mathcal{S}_{n \to 1}$ that will be used by $\mathcal{A}_f$. $\mathcal{S}_{n \to 1}$ internally executes $\mathcal{A}$, and interacts with an external receiver Rec_{ext} of the protocol Π_{NM} acting as the sender.

$\mathcal{S}_{n \to 1}(Y, \varphi, z)$

1. Act in the left session with $\mathcal{A}$ (that runs using randomness φ) as in $\mathcal{H}_1^m(z)$.
2. For all $j \neq i \in \{1, \ldots poly(\lambda)\}$ run $MMRec_j$ as in $\mathcal{H}_1^m(z)$. Instead run $MMRec_i$ as described in steps 3, 4 and 5.
3. Upon receiving the 1st round of the i-th right session $(\tilde{a}_{NM_i}, \tilde{a}_{LS_i})$ from $\mathcal{A}$, send $\tilde{a}_{NM_i}$ to Rec_{ext}.
4. Upon receiving $\tilde{c}_{NM_i}$ from Rec_{ext}, run as follows:
  4.1. Run $\mathcal{V}$ to obtain $\tilde{c}_{LS_i}$.
  4.2. Set $\tilde{Y}_i = Y$.
  4.3. Send $(\tilde{c}_{NM_i}, \tilde{c}_{LS_i}, \tilde{Y}_i)$ to $\mathcal{A}$.
5. Upon receiving the 3rd round of the i-th right session $(\tilde{z}_{NM_i}, c\tilde{o}m_i, \tilde{z}_{LS_i}, \tilde{s}_{1i})$, set $\tilde{x} = ((\tilde{a}_{NM_i}, \tilde{c}_{NM_i}, \tilde{z}_{NM_i}), \tilde{Y}, \tilde{s}_{1i}, c\tilde{o}m_i, id)$ and abort iff $(\tilde{a}_{LS_i}, \tilde{c}_{LS_i}, \tilde{z}_{LS_i})$ is not accepting for $\mathcal{V}$ with respect to $\tilde{x}$.
6. Send $\tilde{z}_{NM_i}$ to Rec_{ext}.

Notice that the above execution of $\mathcal{S}_{n \to 1}$ is distributed identically to $\mathcal{H}_1^m(z)$ when Rec_{ext} plays identically as honest receiver. Now we can conclude the proof of this lemma by describing how $\mathcal{A}_f$ works. $\mathcal{A}_f$ runs the extractor of Π_{NM} using $\mathcal{S}_{n \to 1}$ as sender (recall that an extractor of Π_{NM} plays only having access to a sender of Π_{NM}). We have that the extractor with non-negligible probability outputs the committed message of an execution that inverts f. By using the randomness φ, $\mathcal{A}_f$ can reconstruct the view of $\mathcal{A}$ and retrive the value $\tilde{s}_{1i}$. Therefore $\mathcal{A}$ running in polynomial time[15] outputs with non-negligible probability the value $y = \tilde{s}_{0i} \oplus \tilde{s}_{1i}$ such that $f(y) = Y$.

We now consider the 2nd hybrid experiment $\mathcal{H}_2^m(z)$ where in the left session, after receiving Y from $\mathcal{A}$, the sender in time T_f finds a value y such that $Y = f(y)$. Then the sender sets and sends $s_1 = y \oplus s_0$, where s_0 is the value committed using Π_{NM}. The only difference between this hybrid experiment and $\mathcal{H}_1^m(z)$ is that $\mathcal{H}_2^m(z)$ runs in time sub-exponential in λ, and the value s_1 is equal to $y \oplus s_0$ where $Y = f(y)$.

$\mathcal{H}_2^m(z)$.

Left session:
1. First round.
   1.1. Pick $s_0 \leftarrow \{0,1\}^\lambda$.
   1.2. Compute $\mathsf{a}_{\mathsf{NM}} = \mathsf{Sen}_{\mathsf{NM}}^1(\mathsf{id}, s_0; \rho)$.
   1.3. Compute $\mathsf{a}_{\mathsf{LS}} = \mathsf{P}^1(1^{\lambda_{\mathsf{LS}}}, \ell; \alpha)$.
   1.4. Send $(\mathsf{a}_{\mathsf{NM}}, \mathsf{a}_{\mathsf{LS}})$ to $\mathcal{A}$.
2. Third round. Upon receiving $(\mathsf{c}_{\mathsf{NM}}, \mathsf{c}_{\mathsf{LS}}, Y)$ from $\mathcal{A}$.
   2.1. Compute $(\mathsf{com}, \mathsf{dec}) = \mathsf{NISen}(m; \sigma)$.
   2.2. Run in time T_f to compute y such that $Y = f(y)$.
   2.3. Set $s_1 = y \oplus s_0$.
   2.4. Compute $(\mathsf{z}_{\mathsf{NM}}, \mathsf{dec}_{\mathsf{NM}}) = \mathsf{Sen}_{\mathsf{NM}}^2(\mathsf{id}, \mathsf{c}_{\mathsf{NM}}, s_0; \rho)$.
   2.5. Set $x = ((\mathsf{a}_{\mathsf{NM}}, \mathsf{c}_{\mathsf{NM}}, \mathsf{z}_{\mathsf{NM}}), Y, s_1, \mathsf{com}, \mathsf{id})$ and $w = (m, \sigma, \perp, \perp)$ with $(|x| = \ell)$. Run $\mathsf{z}_{\mathsf{LS}} = \mathsf{P}^2(x, w, \mathsf{c}_{\mathsf{LS}}; \alpha)$.
   2.6. Send $(\mathsf{z}_{\mathsf{NM}}, \mathsf{com}, \mathsf{z}_{\mathsf{LS}}, s_1)$ to MMRec.
Right sessions: Act as a proxy between $\mathcal{A}$ and $\mathsf{MMRec}_1, \ldots, \mathsf{MMRec}_{\mathsf{poly}(\lambda)}$.

When switching from $\mathcal{H}_1^m(z)$ to $\mathcal{H}_2^m(z)$ we will make sure that the following two properties hold.

1. For all message $m \in \{0,1\}^{\mathsf{poly}(\lambda)}$ it holds that $\mathsf{mim}_{\mathcal{H}_1^m}^{\mathcal{A}}(z) \approx \mathsf{mim}_{\mathcal{H}_2^m}^{\mathcal{A}}(z)$.[16]
2. Let p_i be the probability that in the i-th right session of $\mathcal{H}_2$, for $i \in \{1, \ldots, \mathsf{poly}(\lambda)\}$, $\mathcal{A}$ sends $\tilde{s}_{1i}$ such that $f(\tilde{s}_{1i} \oplus \tilde{s}_{0i}) = \tilde{Y}_i$ where $\tilde{s}_{0i}$ is the value committed using NM. Then $p_i < \nu(\lambda)$ for some negligible function ν.

We now prove that the above two properties hold.

---

[15] The extractor is an expected polynomial-time algorithm while $\mathcal{A}_f$ must be a strict polynomial-time algorithm. Therefore $\mathcal{A}_f$ will run the extractor up to a given upperbounded number of steps that is higher than the expected running time of the extractor. Obviously with non-negligible probability the *truncated* extraction procedure will be completed successfully and this is sufficient for $\mathcal{A}_f$ to invert f. The same standard argument about truncating the execution of an expected polynomial-time algorithm will be needed later but for simplicity we will not repeat this discussion.

[16] To simplify the notation here, and in the rest of the proof, we will omit that the indistinguishability between two distributions must hold for every auxiliary input z.

Lemma 2. *For all message* $m \in \{0,1\}^{\mathsf{poly}(\lambda)}$ *it holds that* $\mathsf{mim}^{\mathcal{A}}_{\mathcal{H}^m_1}(z) \approx \mathsf{mim}^{\mathcal{A}}_{\mathcal{H}^m_2}(z)$.

Proof. Suppose by contradiction that the distribution of $\mathsf{mim}^{\mathcal{A}}_{\mathcal{H}^m_1}(z)$ is distinguishable from $\mathsf{mim}^{\mathcal{A}}_{\mathcal{H}^m_2}(z)$; this means that there exists a distinguisher $\mathcal{D}$ that can tell apart such two distributions. We now use $\mathcal{D}$ and $\mathcal{A}$ to construct an adversary $\mathcal{A}_{\mathsf{Hiding}}$ that breaks the hiding of Π_{NM} in time $\mathsf{poly}(\lambda) \cdot T_{\mathsf{NI}}$ therefore reaching a contradiction[17]. Let $\mathcal{C}_{\mathsf{Hiding}}$ be the challenger of the hiding game, we consider two randomly chosen challenge messages (m_0, m_1) sent to $\mathcal{C}_{\mathsf{Hiding}}$. We now provide a formal description of the adversary $\mathcal{A}_{\mathsf{Hiding}}$.

$\mathcal{A}_{\mathsf{Hiding}}(m_0, m_1, z)$

1. Upon receiving the 1st round a_{NM} from $\mathcal{C}_{\mathsf{Hiding}}$, run as follows:
   1.1. Compute $\mathsf{a}_{\mathsf{LS}} = \mathsf{P}^1(1^{\lambda_{\mathsf{LS}}}, \ell; \alpha)$.
   1.2. Send $(\mathsf{a}_{\mathsf{NM}}, \mathsf{a}_{\mathsf{LS}})$ to $\mathcal{A}$.
2. Upon receiving $(\mathsf{c}_{\mathsf{NM}}, \mathsf{c}_{\mathsf{LS}}, Y)$ from $\mathcal{A}$, send c_{NM} to $\mathcal{C}_{\mathsf{NM}}$.
3. Upon receiving the 3rd round z_{NM} from $\mathcal{C}_{\mathsf{Hiding}}$, run as follows:
   3.1. Compute y such that $Y = f(y)$, set $s_1 = m_0 \oplus y$.
   3.2. Compute $(\mathsf{com}, \mathsf{dec}) = \mathsf{NISen}(m; \sigma)$.
   3.3. Set $x = ((\mathsf{a}_{\mathsf{NM}}, \mathsf{c}_{\mathsf{NM}}, \mathsf{z}_{\mathsf{NM}}), Y, s_1, \mathsf{com}, \mathsf{id})$ and $w = (m, \sigma, \bot, \bot)$ with $(|x| = \ell)$. Run $\mathsf{z}_{\mathsf{LS}} = \mathsf{P}^2(x, w, \mathsf{c}_{\mathsf{LS}}; \alpha)$.
   3.4. Send $(\mathsf{z}_{\mathsf{NM}}, \mathsf{com}, \mathsf{z}_{\mathsf{LS}}, s_1)$ to $\mathcal{A}$.
4. Simulate $\mathsf{MMRec}_1, \ldots, \mathsf{MMRec}_{\mathsf{poly}(\lambda)}$ with $\mathcal{A}$ when $\mathcal{A}$ plays as a sender.
5. Let M be an empty tuple. For all $i \in \{1, \ldots, \mathsf{poly}(\lambda)\}$, consider $\tilde{\mathsf{com}}_i$, the non-interactive commitment received by MMRec_i, run in time T_{NI} to compute $\tilde{m}_i$ such that $\exists \; \tilde{\mathsf{dec}} : 1 = \mathsf{NIRec}(\tilde{\mathsf{com}}_i, \tilde{\mathsf{dec}}, \tilde{m}_i)$ and add $\tilde{m}_i$ to M.
6. Give M and the view of $\mathcal{A}$ to the distinguisher $\mathcal{D}$ and output what $\mathcal{D}$ outputs.

The proof ends with the observation that if $\mathcal{C}_{\mathsf{Hiding}}$ has committed to m_0 then the xor of the committed value with s_1 is equal to y such that $f(y) = Y$, like in $\mathcal{H}^m_2(z)$. If instead $\mathcal{C}_{\mathsf{Hiding}}$ has committed to m_1 then the xor of the committed value and s_1 is equal to a random value, like in $\mathcal{H}^m_1(z)$.

Lemma 3. *Let* p_i *be the probability that in the i-th right session of* $\mathcal{H}_2$, *for* $i \in \{1, \ldots, \mathsf{poly}(\lambda)\}$, $\mathcal{A}$ *sends* $\tilde{s}_{1i}$ *such that* $f(\tilde{s}_{1i} \oplus \tilde{s}_{0i}) = \tilde{Y}_i$ *where* $\tilde{s}_{0i}$ *is the value committed using* NM. *Then* $p_i < \nu(\lambda)$ *for some negligible function* ν.

Proof. Suppose by contradiction that for a right session i the claim does not hold. We can construct a distinguisher $\mathcal{D}_{\mathsf{NM}}$ and an adversary $\mathcal{A}_{\mathsf{NM}}$ that break the non-malleability of Π_{NM}. Let $\mathcal{C}_{\mathsf{NM}}$ be the challenger of the NM commitment and let (m_0, m_1) be two randomly chosen challenge messages given to $\mathcal{C}_{\mathsf{NM}}$.

---

[17] Recall that Π_{NM} is secure against adversaries running in time $\mathsf{poly}(\lambda) \cdot T_{\mathsf{NI}} < T_{\mathsf{NM}}$.

$\mathcal{A}_{\mathsf{NM}}\ (m_0, m_1, z)$

Left session:
1. Act as $\mathcal{A}_{\mathsf{Hiding}}$ acts in the left session.

Right sessions:
1. For all $j \neq i \in \{1, \ldots, \mathsf{poly}(\lambda)\}$ run MMRec_j as in $\mathcal{H}_2^m(z)$. Instead run MMRec_i as described in steps 1.1, 1.2 and 1.3.
   1.1. Forward $\tilde{\mathsf{a}}_{\mathsf{NM}_i}$ to $\mathsf{Rec}_{\mathsf{NM}}$.
   1.2. Upon receiving $\tilde{\mathsf{c}}_{\mathsf{NM}}$ from $\mathsf{Rec}_{\mathsf{NM}}$, pick a random $\tilde{\mathsf{c}}_{\mathsf{LS}_i}$, pick a random $\tilde{Y}_i$ and send $(\tilde{\mathsf{c}}_{\mathsf{NM}_i}, \tilde{\mathsf{c}}_{\mathsf{LS}_i}, \tilde{Y}_i)$ to $\mathcal{A}$.
   1.3. Upon receiving $\tilde{\mathsf{z}}_{\mathsf{NM}_i}$ from $\mathcal{A}$, send it to $\mathsf{Rec}_{\mathsf{NM}}$.

Let $\mathsf{mim}^{\mathcal{A}_{\mathsf{NM}}}(z)$ be the view of $\mathsf{mim}^{\mathcal{A}_{\mathsf{NM}}}(z)$ and the tuple of committed messages in the right session. The distinguisher $\mathcal{D}_{\mathsf{NM}}$ takes as input $\mathsf{mim}^{\mathcal{A}_{\mathsf{NM}}}(z)$ and acts as follows.

$\mathcal{D}_{\mathsf{NM}}(\mathsf{mim}^{\mathcal{A}_{\mathsf{NM}}}(z))$: Let $\tilde{s}_{0i}$ be the committed message sent in the i-right session by $\mathcal{A}_{\mathsf{NM}}$ to MMRec. Reconstruct the output messages of $\mathcal{A}$ (using the same randomness of $\mathsf{mim}^{\mathcal{A}_{\mathsf{NM}}}(z)$) to pick $\tilde{s}_{1i}$. If $f(\tilde{s}_{1i} \oplus \tilde{s}_{0i}) = \tilde{Y}_i$ output 1 and output 0 otherwise. The proof ends with the observation that if $\mathcal{C}_{\mathsf{NM}}$ has committed to m_0 then the xor of the committed value with s_{1i} is equal to y such that $f(y) = Y$ like in $\mathcal{H}_2^m$. If instead $\mathcal{C}_{\mathsf{Hiding}}$ has committed to m_1 then the xor of the committed value with s_{1i} is equal to a random string as in $\mathcal{H}_1^m$.

The 3rd hybrid experiment that we consider is equal to $\mathcal{H}_2^m(z)$ with the difference that the LS proof system is executed using s_0 and the randomness of the non-malleable commitment of s_0. Recall that $f(s_0 \oplus s_1) = Y$. We observe that in the left session of $\mathcal{H}_2^m(z)$ it already holds that $f(s_0 \oplus s_1) = Y$, therefore we can switch the witness used in LS and complete the execution of the proof system.

$\mathcal{H}_3^m(z)$.

Left sessions:
1. First round.
   1.1. Pick $s_0 \leftarrow \{0,1\}^\lambda$.
   1.2. Compute $\mathsf{a}_{\mathsf{NM}} = \mathsf{Sen}_{\mathsf{NM}}^1(\mathsf{id}, s_0; \rho)$.
   1.3. Compute $\mathsf{a}_{\mathsf{LS}} = \mathsf{P}^1(1^{\lambda_{\mathsf{LS}}}, \ell; \alpha)$.
   1.4. Send $(\mathsf{a}_{\mathsf{NM}}, \mathsf{a}_{\mathsf{LS}})$ to $\mathcal{A}$.
2. Third round. Upon receiving $(\mathsf{c}_{\mathsf{NM}}, \mathsf{c}_{\mathsf{LS}}, Y)$ from $\mathcal{A}$.
   2.1. Compute $(\mathsf{com}, \mathsf{dec}) = \mathsf{NISen}(m; \sigma)$.
   2.2. Run in time T_f to compute y such that $Y = f(y)$.
   2.3. Set $s_1 = s_0 \oplus y$.
   2.4. Compute $(\mathsf{z}_{\mathsf{NM}}, \mathsf{dec}_{\mathsf{NM}}) = \mathsf{Sen}_{\mathsf{NM}}^2(\mathsf{id}, \mathsf{c}_{\mathsf{NM}}, s_0; \rho)$.
   2.5. Compute $(\mathsf{com}, \mathsf{dec}) = \mathsf{NISen}(1^{\lambda_{\mathsf{NI}}}, m; \sigma)$.
   2.6. Set $x = ((\mathsf{a}_{\mathsf{NM}}, \mathsf{c}_{\mathsf{NM}}, \mathsf{z}_{\mathsf{NM}}), Y, s_1, \mathsf{com}, \mathsf{id})$ and $\underline{w = (\bot, \bot, s_0, \rho)}$ with $(|x| = \ell)$. Run $\mathsf{z}_{\mathsf{LS}} = \mathsf{P}^2(x, w, \mathsf{c}_{\mathsf{LS}}; \alpha)$.
   2.7. Send $(\mathsf{z}_{\mathsf{NM}}, \mathsf{com}, \mathsf{z}_{\mathsf{LS}}, s_1)$ to $\mathcal{A}$.

Right sessions: Act as a proxy between $\mathcal{A}$ and $\mathsf{MMRec}_1, \ldots, \mathsf{MMRec}_{\mathsf{poly}(\lambda)}$.

Even in this case we need to prove the following two properties.

1. For all message $m \in \{0,1\}^{\mathsf{poly}(\lambda)}$ it holds that $\mathsf{mim}^{\mathcal{A}}_{\mathcal{H}_2^m}(z) \approx \mathsf{mim}^{\mathcal{A}}_{\mathcal{H}_3^m}(z)$.
2. Let p_i be the probability that in the i-th right session of $\mathcal{H}_3$, for any $i \in \{1,\dots,\mathsf{poly}(\lambda)\}$, $\mathcal{A}$ sends $\tilde{s}_{1i}$ such that $f(\tilde{s}_{1i} \oplus \tilde{s}_{0i}) = \tilde{Y}_i$ where $\tilde{s}_{0i}$ is the value committed using NM. Then $p_i < \nu(\lambda)$ for some negligible function ν.

Lemma 4. *For any message* $m \in \{0,1\}^{\mathsf{poly}(\lambda)}$ *it holds that* $\mathsf{mim}^{\mathcal{A}}_{\mathcal{H}_2^m}(z) \approx \mathsf{mim}^{\mathcal{A}}_{\mathcal{H}_3^m}(z)$.

Proof. Suppose by contradiction that there exist a adversary $\mathcal{A}$ and a distinguisher $\mathcal{D}$ that can tell apart such two distributions. We can use this adversary and the associated distinguisher to construct an adversary $\mathcal{A}_{\mathsf{LS}}$ for the T_{LS}-witness-indistinguishable property of the LS protocol. Let $\mathcal{C}_{\mathsf{LS}}$ be the WI challenger, the adversary works as follows.

$\mathcal{A}_{\mathsf{LS}}(z)$

1. Pick $s_0 \leftarrow \{0,1\}^{\lambda}$.
2. Compute $\mathsf{a}_{\mathsf{NM}} = \mathsf{Sen}^1_{\mathsf{NM}}(\mathsf{id}, s_0; \rho)$.
3. Upon receiving a_{LS} from $\mathcal{C}_{\mathsf{LS}}$, send $(\mathsf{a}_{\mathsf{NM}}, \mathsf{a}_{\mathsf{LS}})$ to $\mathcal{A}$.
4. Upon receiving $(\mathsf{c}_{\mathsf{NM}}, \mathsf{c}_{\mathsf{LS}}, Y)$ from $\mathcal{A}$ run as follows:
   4.1. Run in time T_f to compute y such that $Y = f(y)$.
   4.2. Set $s_1 = s_0 \oplus y$.
   4.3. Compute $(\mathsf{z}_{\mathsf{NM}}, \mathsf{dec}_{\mathsf{NM}}) = \mathsf{Sen}^2_{\mathsf{NM}}(\mathsf{id}, \mathsf{c}_{\mathsf{NM}}, s_0; \rho)$.
   4.4. Compute $(\mathsf{com}, \mathsf{dec}) = \mathsf{NISen}(1^{\lambda_{\mathsf{NI}}}, m; \sigma)$.
   4.5. Set $x = ((\mathsf{a}_{\mathsf{NM}}, \mathsf{c}_{\mathsf{NM}}, \mathsf{z}_{\mathsf{NM}}), Y, s_1, \mathsf{com}, \mathsf{id})$, $w_0 = (\perp, \perp, s_0, \rho), w_1 = (m, \sigma, \perp, \perp)$ and send $(x, \mathsf{c}_{\mathsf{LS}}, w_0, w_1)$ to $\mathcal{C}_{\mathsf{LS}}$.
5. Upon receiving z_{LS} from $\mathcal{C}_{\mathsf{LS}}$, send $(\mathsf{z}_{\mathsf{NM}}, \mathsf{com}, \mathsf{z}_{\mathsf{LS}})$ to $\mathcal{A}$.
6. Simulate $\mathsf{MMRec}_1, \dots, \mathsf{MMRec}_{\mathsf{poly}(\lambda)}$ with $\mathcal{A}$, when $\mathcal{A}$ plays as a sender.
7. Let M be an empty tuple. For all $i \in \{1, \dots, \mathsf{poly}(\lambda)\}$, consider $\tilde{\mathsf{com}}_i$, the non-interactive commitment received by MMRec_i, and run in time $\tilde{T}_{\mathsf{NI}}$ to compute $\tilde{m}_i$ such that $\exists \, \tilde{\mathsf{dec}} : 1 = \mathsf{NIRec}(\tilde{\mathsf{com}}_i, \tilde{\mathsf{dec}}, \tilde{m}_i)$ and add $\tilde{m}_i$ to M.
8. Give M and the view of $\mathcal{A}$ to the distinguisher $\mathcal{D}$.
9. Output what $\mathcal{D}$ outputs.

The proof ends with the observation that if $\mathcal{C}_{\mathsf{LS}}$ has has used as witness the randomness of the non-malleable commitment of the value s_0 such that $f(s_0 \oplus s_1) = Y$ then we are in the hybrid experiment $\mathcal{H}_3^m(z)$. If instead $\mathcal{C}_{\mathsf{LS}}$ has used as a witness the randomness used to compute the non-interactive commitment NI then we are in the hybrid experiment $\mathcal{H}_2^m(z)$.

Lemma 5. *Let* p_i *be the probability that in the i-th right session of* $\mathcal{H}_3^m$, *for* $i \in \{1, \dots, \mathsf{poly}(\lambda)\}$, $\mathcal{A}$ *sends* $\tilde{s}_{1i}$ *such that* $f(\tilde{s}_{1i} \oplus \tilde{s}_{0i}) = \tilde{Y}_i$ *where* $\tilde{s}_{0i}$ *is the value committed using* NM. *Then* $p_i < \nu(\lambda)$ *for some negligible function* ν.

Proof. Suppose by contradiction that for a right session i the claim does not hold, then we can construct an adversary $\mathcal{A}'_{\mathsf{LS}}$ for the T_{LS} witness-indistinguishable property of the LS protocol. Let $\mathcal{C}_{\mathsf{LS}}$ be the WI challenger, the adversary works as follows.

$\mathcal{A}'_{\mathsf{LS}}(z)$

1. Pick $s_0 \leftarrow \{0,1\}^\lambda$.
2. Compute $\mathsf{a}_{\mathsf{NM}} = \mathsf{Sen}^1_{\mathsf{NM}}(\mathsf{id}, s_0; \rho)$.
3. Upon receiving a_{LS} from $\mathcal{C}_{\mathsf{LS}}$, send $(\mathsf{a}_{\mathsf{NM}}, \mathsf{a}_{\mathsf{LS}})$ to $\mathcal{A}$.
4. Upon receiving $(\mathsf{c}_{\mathsf{NM}}, \mathsf{c}_{\mathsf{LS}}, Y)$ from $\mathcal{A}$, run as follow:
   4.1. Run in time T_f to compute y such that $Y = f(y)$.
   4.2. Set $s_1 = s_0 \oplus y$.
   4.3. Compute $(\mathsf{z}_{\mathsf{NM}}, \mathsf{dec}_{\mathsf{NM}}) = \mathsf{Sen}^2_{\mathsf{NM}}(\mathsf{id}, \mathsf{c}_{\mathsf{NM}}, s_0; \rho)$.
   4.4. Compute $(\mathsf{com}, \mathsf{dec}) = \mathsf{NISen}(1^{\lambda_{\mathsf{NI}}}, m; \sigma)$.
   4.5. Set $x = ((\mathsf{a}_{\mathsf{NM}}, \mathsf{c}_{\mathsf{NM}}, \mathsf{z}_{\mathsf{NM}}), Y, s_1, \mathsf{com}, \mathsf{id})$, $w_0 = (\perp, \perp, s_0, \rho)$, $w_1 = (m, \sigma, \perp, \perp)$ and send $(x, \mathsf{c}_{\mathsf{LS}}, w_0, w_1)$ to $\mathcal{C}_{\mathsf{LS}}$.
5. Upon receiving z_{LS} from $\mathcal{C}_{\mathsf{LS}}$, send $(\mathsf{z}_{\mathsf{NM}}, \mathsf{com}, \mathsf{z}_{\mathsf{LS}})$ to $\mathcal{A}$.
6. Simulate $\mathsf{MMRec}_1, \ldots, \mathsf{MMRec}_{\mathsf{poly}(\lambda)}$ with $\mathcal{A}$, when $\mathcal{A}$ plays as a sender.
7. Run in time $\tilde{T}_{\mathsf{NM}}$ to extract the value $\tilde{s}_{0i}$ from the non-malleable commitment sent by $\mathcal{A}$ in the i-th session. Output 1 if $f(\tilde{s}_{0i} \oplus \tilde{s}_{1i}) = \tilde{Y}_i$ and output 0 otherwise.

The proof ends with the observation that if $\mathcal{C}_{\mathsf{LS}}$ has used $w_0 = (\perp, \perp, s_0, \rho)$ as a witness then $\mathcal{A}$ acts as in $\mathcal{H}_3^m(z)$, sending with non-negligible probability two shares such that the xor of them gives a puzzle solution. If $\mathcal{C}_{\mathsf{LS}}$ has used $w_1 = (m, \sigma, \perp, \perp)$ then the xor of the two shares is with overwhelming probability different from a puzzle solution as in $\mathcal{H}_2^m(z)$.

The next hybrid experiment that we consider is $\mathcal{H}_3^0(z)$. The only differences between this hybrid experiment and $\mathcal{H}_3^m(z)$ is that the sender, using NI, commits to a message 0^λ instead of m.

$\mathcal{H}_3^0(z)$.

Left session:
1. First round.
   1.1. Pick $s_0 \leftarrow \{0,1\}^\lambda$.
   1.2. Compute $\mathsf{a}_{\mathsf{NM}} = \mathsf{Sen}^1_{\mathsf{NM}}(\mathsf{id}, s_0; \rho)$.
   1.3. Compute $\mathsf{a}_{\mathsf{LS}} = \mathsf{P}^1(\ell; \alpha)$.
   1.4. Send $(\mathsf{a}_{\mathsf{NM}}, \mathsf{a}_{\mathsf{LS}})$ to $\mathcal{A}$.
2. Third round. Upon receiving $(\mathsf{c}_{\mathsf{NM}}, \mathsf{c}_{\mathsf{LS}}, Y)$ from $\mathcal{A}$, run as follows:
   2.1. Run in time T_f to compute y such that $Y = f(y)$.
   2.2. Set $s_1 = s_0 \oplus y$.
   2.3. Compute $(\mathsf{z}_{\mathsf{NM}}, \mathsf{dec}_{\mathsf{NM}}) = \mathsf{Sen}^2_{\mathsf{NM}}(\mathsf{id}, \mathsf{c}_{\mathsf{NM}}, s_0; \rho)$.
   2.4. Compute $(\mathsf{com}, \mathsf{dec}) = \underline{\mathsf{NISen}(0^\lambda; \sigma)}$.
   2.5. Set $x = ((\mathsf{a}_{\mathsf{NM}}, \mathsf{c}_{\mathsf{NM}}, \mathsf{z}_{\mathsf{NM}}), Y, s_1, \mathsf{com}, \mathsf{id})$ and $w = (\perp, \perp, s_0, \rho)$ with $(|x| = \ell)$. Run $\mathsf{z}_{\mathsf{LS}} = \mathsf{P}^2(x, w, \mathsf{c}_{\mathsf{LS}}; \alpha)$.
   2.6. Send $(\mathsf{z}_{\mathsf{NM}}, \mathsf{com}, \mathsf{z}_{\mathsf{LS}}, s_1)$ to $\mathcal{A}$.
Right sessions: Act as a proxy between $\mathcal{A}$ and $\mathsf{MMRec}_1, \ldots, \mathsf{MMRec}_{\mathsf{poly}(\lambda)}$.

We now prove the following properties.

1. Let p_i be the probability that in the i-th right session of $\mathcal{H}_3^0$, for any $i \in \{1, \ldots, \text{poly}(\lambda)\}$, $\mathcal{A}$ sends $\tilde{s}_{1i}$ such that $f(\tilde{s}_{1i} \oplus \tilde{s}_{0i}) = \tilde{Y}_i$ where $\tilde{s}_{0i}$ is the value committed using NM. Then $p_i < \nu(\lambda)$ for some negligible function ν.
2. For any message $m \in \{0,1\}^{\text{poly}(\lambda)}$ it holds that $\text{mim}_{\mathcal{H}_3^m}^{\mathcal{A}}(z) \approx \text{mim}_{\mathcal{H}_3^0}^{\mathcal{A}}(z)$.

Lemma 6. *Let p_i be the probability that in the i-th right session of $\mathcal{H}_3^0$, for $i \in \{1, \ldots, \text{poly}(\lambda)\}$, $\mathcal{A}$ sends $\tilde{s}_{1i}$ such that $f(\tilde{s}_{1i} \oplus \tilde{s}_{0i}) = \tilde{Y}_i$ where $\tilde{s}_{0i}$ is the value committed using NM. Then $p_i < \nu(\lambda)$ for some negligible function ν.*

Proof. Suppose by contradiction that there exists a right session $i \in \{1, \ldots, \text{poly}(\lambda)\}$ in which $\mathcal{A}$ commit to a string $\tilde{s}_0$ such that $f(\tilde{s}_{0i} \oplus \tilde{s}_{1i}) = \tilde{Y}_i$ using Π_{NM}. Then we can construct an adversary $\mathcal{A}_{\text{NI}}$ that breaks the hiding property of the non interactive commitment scheme NI. Let $\mathcal{C}_{\text{NI}}$ be the challenger that on input $m_0 = 0^\lambda$ and $m_1 = m$, picks a random bit b, computes $(\text{com}, \text{dec}) = \text{NISen}(1^{\lambda_{\text{NI}}}, m_b; \sigma)$ and sends com to $\mathcal{A}_{\text{NI}}$.

Before describing $\mathcal{A}_{\text{NI}}$ we need to consider, as in the proof of Lemma 1, a machine $\mathcal{S}_{n \to 1}$ that internally executes $\mathcal{A}$, and interacts with a receiver Rec_{ext} of the protocol Π_{NM} acting as the sender.

$\mathcal{S}_{n \to 1}(\text{com}, \varphi, z)$ Run $\mathcal{A}$ using randomness φ.

1. Pick $s_0 \leftarrow \{0,1\}^\lambda$.
2. Compute $a_{\text{NM}} = \text{Sen}_{\text{NM}}^1(\text{id}, s_0; \rho)$.
3. Compute $a_{\text{LS}} = \text{P}^1(1^{\lambda_{\text{LS}}}, \ell; \alpha)$.
4. Send $(a_{\text{NM}}, a_{\text{LS}})$ to $\mathcal{A}$.
5. Upon receiving $(c_{\text{NM}}, c_{\text{LS}}, Y)$ from $\mathcal{A}$, run as follows:
   5.1. Run in time T_f to compute y such that $Y = f(y)$.
   5.2. Set $s_1 = s_0 \oplus y$.
   5.3. Compute $(z_{\text{NM}}, \text{dec}_{\text{NM}}) = \text{Sen}_{\text{NM}}^2(\text{id}, c_{\text{NM}}, s_0; \rho)$.
   5.4. Set $x = ((a_{\text{NM}}, c_{\text{NM}}, z_{\text{NM}}), Y, s_1, \text{com}, \text{id})$ and $w = (\bot, \bot, s_0, \rho)$ with $(|x| = \ell)$. Run $z_{\text{LS}} = \text{P}^2(x, w, c_{\text{LS}}; \alpha)$.
   5.5. Send $(z_{\text{NM}}, \text{com}, z_{\text{LS}}, s_1)$ to $\mathcal{A}$.
6. Let $i \in \{1, \ldots, \text{poly}(\lambda)\}$ be the right session that contradicts the claim. For all $j \neq i \in \{1, \ldots \text{poly}(\lambda)\}$ run MMRec_j as in $\mathcal{H}_4(m, z)$. Run MMRec_i as follows.
   6.1. Upon receiving the 1rd round of the i-th right session $(\tilde{a}_{\text{NM}_i}, \tilde{a}_{\text{LS}_i})$ from $\mathcal{A}$, send $\tilde{a}_{\text{NM}_i}$ to the external receiver Rec_{ext}.
   6.2. Upon receiving $\tilde{c}_{\text{NM}_i}$ from Rec_{ext}, run as follows:
     i. Run $\mathcal{V}$ to obtain $\tilde{c}_{\text{LS}_i}$.
     ii. Pick a random $\tilde{Y}_i$.
     iii. Send $(\tilde{c}_{\text{NM}_i}, \tilde{c}_{\text{LS}_i}, \tilde{Y}_i)$ to $\mathcal{A}$.
   6.3. Upon receiving the 3rd round of the i-th right session $(\tilde{z}_{\text{NM}_i}, \tilde{\text{com}}_i, \tilde{z}_{\text{LS}_i}, \tilde{s}_{1i})$, set $\tilde{x} = ((\tilde{a}_{\text{NM}_i}, \tilde{c}_{\text{NM}_i}, \tilde{z}_{\text{NM}_i}), \tilde{Y}, \tilde{s}_{1i}, \tilde{\text{com}}_i, \tilde{\text{id}})$ and abort iff $(\tilde{a}_{\text{LS}_i}, \tilde{c}_{\text{LS}_i}, \tilde{z}_{\text{LS}_i})$ is not accepted by $\mathcal{V}$ with respect to $\tilde{x}$.
   6.4. Send $\tilde{z}_{\text{NM}_i}$ to Rec_{ext}.

Now we can conclude the proof of this lemma by describing how $\mathcal{A}_{\text{NI}}$ works. $\mathcal{A}_{\text{NI}}$ runs the extractor of the protocol Π_{NM} using $\mathcal{S}_{n \to 1}$ as sender (recall that an extractor of Π_{NM} plays only having access to a sender of Π_{NM}). Since the extractor with non-negligible probability outputs the committed message we have that $\mathcal{A}_{NI}$ retrives $\tilde{s}_{0i}$. Moreover $\mathcal{A}_{NI}$ gets $\tilde{s}_{1i}$ by reconstructing the view of $\mathcal{A}$ using the randomness φ. Since by contradiction $\mathcal{A}$ contradicts the claim of this lemma, we have that $\mathcal{A}_{NI}$ can break the hiding of NI because $f(\tilde{s}_{0i} \oplus \tilde{s}_{1i}) = \tilde{Y}$ with non-negligible probability in $\mathcal{H}_3^0(z)$ where $m_0 = 0^\lambda$ is committed in com, while the same happens with negligible probability only in $\mathcal{H}_3^m(z)$ where $m_1 = m$. Therefore if this happens, $\mathcal{A}_{NI}$ outputs 0, otherwise $\mathcal{A}_{NI}$ outputs a random bit.

Lemma 7. *For any message $m \in \{0,1\}^{\mathsf{poly}(\lambda)}$ it holds that $\mathsf{mim}_{\mathcal{H}_3^m}^{\mathcal{A}}(z) \approx \mathsf{mim}_{\mathcal{H}_3^0}^{\mathcal{A}}(z)$.*

Proof. Suppose by contradiction that there exists a distinguisher $\mathcal{D}$ and an adversary $\mathcal{A}$ such that $\mathsf{mim}_{\mathcal{H}_3^m}^{\mathcal{A}}(z)$ is distinguishable from $\mathsf{mim}_{\mathcal{H}_3^0}^{\mathcal{A}}(z)$ then we can construct an adversary $\mathcal{A}_{NI}$ that breaks the hiding property of the non-interactive commitment scheme NI. Let $\mathcal{C}_{NI}$ be the challenger that on input $m_0 = 0^\lambda$ and $m_1 = m$, picks a random bit b, computes $(\mathsf{com}, \mathsf{dec}) = \mathsf{NISen}(1^{\lambda_{NI}}, m_b; \sigma)$ and sends com to $\mathcal{A}_{NI}$. Before describing $\mathcal{A}_{NI}$, we consider the following experiment $\mathcal{E}_{m_b}(\varphi, \mathsf{com}, z)$.

$\mathcal{E}_{m_b}(\varphi, \mathsf{com}, z)$.
The randomness required from all next steps is take from φ.

Run $\mathcal{A}(z)$.
Left session:
1. First round.
   1.1. Pick $s_0 \leftarrow \{0,1\}^\lambda$.
   1.2. Compute $\mathsf{a}_{NM} = \mathsf{Sen}_{NM}^1(\mathsf{id}, s_0; \rho)$.
   1.3. Compute $\mathsf{a}_{LS} = \mathsf{P}^1(\ell; \alpha)$.
   1.4. Send $(\mathsf{a}_{NM}, \mathsf{a}_{LS})$ to $\mathcal{A}$.
2. Third round. Upon receiving $(\mathsf{c}_{NM}, \mathsf{c}_{LS}, Y)$ from $\mathcal{A}$, run as follows:
   2.1. Run in time T_f to compute y such that $Y = f(y)$.
   2.2. Set $s_1 = s_0 \oplus y$.
   2.3. Compute $(\mathsf{z}_{NM}, \mathsf{dec}_{NM}) = \mathsf{Sen}_{NM}^2(\mathsf{id}, \mathsf{c}_{NM}, s_0; \rho)$.
   2.4. Set $x = ((\mathsf{a}_{NM}, \mathsf{c}_{NM}, \mathsf{z}_{NM}), Y, s_1, \mathsf{com}, \mathsf{id})$ and $w = (\bot, \bot, s_0, \rho)$ with $(|x| = \ell)$. Run $\mathsf{z}_{LS} = \mathsf{P}^2(x, w, \mathsf{c}_{LS}; \alpha)$.
   2.5. Send $(\mathsf{z}_{NM}, \mathsf{com}, \mathsf{z}_{LS}, s_1)$ to $\mathcal{A}$.
Right sessions: Act as a proxy between $\mathcal{A}$ and $\mathsf{MMRec}_1, \ldots, \mathsf{MMRec}_{\mathsf{poly}(\lambda)}$.

Now we are ready to describe the adversary $\mathcal{A}_{NI}$ for the hiding of NI. $\mathcal{A}_{NI}$ executes the following steps.

1. Let M be an empty tuple. $\mathcal{A}_{NI}$ runs $\mathcal{E}_{m_b}(\varphi, \mathsf{com}, z)$.
2. For all $i \in \{1, \ldots, \mathsf{poly}(\lambda)\}$, $\mathcal{A}_{NI}$ runs the extractor of LS on the i-th right session of the execution of $\mathcal{E}_{m_b}(\varphi, \mathsf{com}, z)$ obtaining $\tilde{m}_i$ and adds it to M.
3. Using the randomness φ, $\mathcal{A}_{NI}$ reconstructs the view of $\mathcal{A}$ in the execution of $\mathcal{E}_{m_b}(\varphi, \mathsf{com}, z)$. Use such view and M as input to $\mathcal{D}$.
4. Output what $\mathcal{D}$ outputs.

Common input: Security parameters: λ, $(\lambda_{NI}, \lambda_{NM}, \lambda_{LS}, \ell) = \mathsf{Params}(\lambda)$. Identity: $\mathsf{id} \in \{0,1\}^\lambda$.

Internal simulation of the left session:

1. Pick $s_0 \leftarrow \{0,1\}^\lambda$.
2. Pick a randomness ρ, and compute $(\mathsf{dec}_{NM}, \mathsf{a}_{NM}) = \mathsf{Sen}_{NM}^1(\mathsf{id}, s_0; \rho)$.
3. Pick a randomness α and compute $\mathsf{a}_{LS} = \mathsf{P}^1(\ell; \alpha)$.
4. Send $(\mathsf{a}_{NM}, \mathsf{a}_{LS})$ to $\mathcal{A}$.
5. Upon receiving $(\mathsf{c}_{NM}, \mathsf{c}_{LS}, Y)$ from $\mathcal{A}$.
   5.1. Pick a randomness σ and compute $(\mathsf{com}, \mathsf{dec}) = \mathsf{NISen}(1^{\lambda_{NI}}, 0^\lambda; \sigma)$.
   5.2. Pick $s_1 \leftarrow \{0,1\}^\lambda$.
   5.3. Compute $\mathsf{z}_{NM} = \mathsf{Sen}_{NM}^2(\mathsf{id}, \mathsf{c}_{NM}, s_0; \rho)$.
   5.4. Set $x = \big((\mathsf{a}_{NM}, \mathsf{c}_{NM}, \mathsf{z}_{NM}), Y, s_1, \mathsf{com}, \mathsf{id}\big)$ and $w = (0^\lambda, \sigma, \perp, \perp)$ with $(|x| = \ell)$. Run $\mathsf{z}_{LS} = \mathsf{P}^2(x, w, \mathsf{c}_{LS}; \alpha)$ where x is the theorem to be proven and w is the witness.
   5.5. Send $(\mathsf{z}_{NM}, \mathsf{com}, \mathsf{z}_{LS}, s_1)$ to $\mathcal{A}$.

Stand-alone commitment:

1. S acts as a proxy between $\mathcal{A}$ and MMRec_i for $i = 1, \ldots, \mathsf{poly}(\lambda)$.

Fig. 3. The simulator S.

The proof ends with the observation that if $\mathcal{C}_{NI}$ has committed to 0^λ then the view of $\mathcal{A}$ and the distribution of the committed messages coincide with $\mathcal{H}_3^0(z)$, otherwise they coincide with $\mathcal{H}_3^m(z)$.

The entire security proof now is almost over because we have proved that for all $m \in \{0,1\}^{\mathsf{poly}(\lambda)}$ the following relation holds:

$$\{\mathsf{mim}_{\Pi_{\mathsf{MMCom}}}^{\mathcal{A}, m}(z)\}_{z \in \{0,1\}^\star} = \{\mathsf{mim}_{\mathcal{H}_1^m}^{\mathcal{A}}(z)\}_{z \in \{0,1\}^\star} \approx \{\mathsf{mim}_{\mathcal{H}_2^m}^{\mathcal{A}}(z)\}_{z \in \{0,1\}^\star} \approx$$
$$\{\mathsf{mim}_{\mathcal{H}_3^m}^{\mathcal{A}}(z)\}_{z \in \{0,1\}^\star} \approx \{\mathsf{mim}_{\mathcal{H}_3^0}^{\mathcal{A}}(z)\}_{z \in \{0,1\}^\star} \approx \{\mathsf{mim}_{\mathcal{H}_2^0}^{\mathcal{A}}(z)\}_{z \in \{0,1\}^\star} \approx$$
$$\{\mathsf{mim}_{\mathcal{H}_1^0}^{\mathcal{A}}(z)\}_{z \in \{0,1\}^\star} = \{\mathsf{sim}_{\Pi_{\mathsf{MMCom}}}^{S}(1^\lambda, z)\}_{z \in \{0,1\}^\star}.$$

We observe that in this proof we had to consider a delayed-input version of our commitment scheme. Indeed, the sender can choose the message m to be committed by sending the non-interactive commitment com of the message m in the 3rd round. It is easy to see that the same security proof still works when the non-interactive commitment is sent in the 1st round, but then clearly the delayed-input property is lost.

4 More Protocols Against Concurrent MiM Attacks

In this section we show 3-round arguments of knowledge and identification schemes that are secure against concurrent MiM attacks.

4.1   Non-Malleable WI Arguments of Knowledge

Our concurrent NM commitment scheme when instantiated without sessions ids, can be used to obtain almost directly a *commit-and-prove* AoK. Recall that in our scheme there is a non-interactive commitment com of m and then rest of the protocol is an AoK. This AoK is used by the sender to claim that either he knows the message committed in com, or he committed through Π_{NM} to a share s_0 that allows to compute the solution of the puzzle.

In order to be fully compliant with the notion of commit-and-prove AoK, we just need to make a trivial change to the statement of the LS subprotocol. Given an instance $x \in L$ and a witness w the prover of our commit-and-prove AoK uses the non-interactive commitment to commit to w, and uses the rest to prove that either he knows the committed message w that moreover is a witness for $x \in L$ or again, he committed through Π_{NM} to a share s_0 that allows to compute the solution of the puzzle.

More formally, we define a commit-and-prove AoK $\Pi_{CaP} = (\mathcal{P}_{CaP}, \mathcal{V}_{CaP})$ that corresponds to our concurrent NM commitment scheme with some minimal changes. First, $\mathcal{P}_{CaP}$ and $\mathcal{V}_{CaP}$ have as a common input an instance $x \in L$, where L is an NP-language. Second, $\mathcal{P}_{CaP}$ has as private input w such that $(x, w) \in \mathsf{Rel}_L$. Third, $\mathcal{P}_{CaP}$ runs MMSen on w, while $\mathcal{V}_{CaP}$ runs MMRec with the exception of running LS for the statement:

$$L_{CaP} = \left\{ (x, (a, c, z), Y, s_1, \mathsf{com}, \mathsf{id}) : (\exists\, (w, \sigma) \text{ s.t. } \mathsf{com} = \mathsf{NISen}(w; \sigma) \text{ AND } (x, w) \in \mathsf{Rel}_L) \right.$$

$$\left. \mathsf{OR}\ (\exists(\rho, s_0) \text{ s.t. } a = \mathsf{Sen}^1_{NM}(\mathsf{id}, s_0; \rho) \text{ AND } z = \mathsf{Sen}^2_{NM}(\mathsf{id}, c, s_0; \rho) \text{ AND } Y = f(s_0 \oplus s_1)) \right\}$$

that is WI for the corresponding NP relation $\mathsf{Rel}_{L_{CaP}}$.

Theorem 3. *Suppose there exist OWPs w.r.t. subexponential-time adversaries, then Π_{CaP} is a 3-round concurrent NMWI argument of knowledge.*

Proof. The proof of this theorem is pretty straightforward given the previous proof for the concurrent non-malleability of our commitment scheme, therefore here we just point out the main intuition.

First of all, Π_{CaP} is clearly a commit-and-prove AoK. Indeed, there exists a commitment of the witness and there is an AoK proving that the committed message is a witness. In order to see this, notice that for any PPT malicious prover succeeding with non-negligible probability in proving a statement $x \in L$, the extractor of LS (of course this needs to be run against an augmented machine) would return (in expected polynomial time and with overwhelming probability) the committed witness since otherwise it would return a share s_0 that combined with s_1 allows to invert the OWP in polynomial time.

We can now focus on the concurrent NMWI property, and we can assume (by contradiction) that the adversary succeeds in encoding in the right sessions witnesses that are related to the witnesses encoded in the left sessions. Notice that the proof is almost identical to the one of Theorem 2. We can indeed prove the case of one prover and multiple verifiers (i.e., one-many), and then we can apply the fact that any one-many NMWIAoK is also a concurrent NMWIAoK. Indeed this was used in [34] and follows similar arguments given in [30,42]. For the one-many case we can therefore follow the proof of Theorem 2 with the following trivial change. Instead of running hybrid experiments starting with a message m and ending with a message 0, in the proof of one-many concurrent NMWI we start with a witness w_0 and end with a witness w_1. Everything else remains untouched and all the reductions work directly.

Π_{CaP} can be instantiated to be public-coin and delayed-input, precisely as our concurrent NM commitment scheme. While what we discussed above applies to arguments only, techniques to obtain proofs can be found in [8].

Instances with Just One Witness and Non-Transferability. Recall that the definition of NMWI considers two experiments that differ only on the witness used by the prover. Therefore it is unclear which security is given by a NMWIAoK when the instance has only one witness. In order to understand the security guaranteed by Π_{CaP} in such a case, consider the proof of concurrent NMWI, and thus, in turn, consider the proof of concurrent non-malleability of our commitment scheme. Notice that while the sequence of hybrids goes from an experiment where the committed message is m to an experiment where the committed message is 0, there is an experiment $\mathcal{H}_3(\cdot, z)$ in which the committed message is irrelevant. Indeed, the entire execution is based on inverting the OWP, in encrypting it through the shares s_0 and s_1 and in using this witness in the execution of LS. This experiment can be seen as the execution of a quasi-polynomial time simulator that breaks the puzzle[18] following the approach of [39][19]. Therefore following the same observations of [39,40] on the security offered by quasi-polynomial time simulation, our concurrent NMWIAoK even for instances with just one witness would not help the adversary in proving a statement whose witness is much harder to compute than breaking the puzzle.

The above discussion explains also the non-transferability flavor of Π_{CaP}. Indeed, at first sight, a MiM attack of an adversary $\mathcal{A}$ to an AoK should be an attempt of $\mathcal{A}$ to transfer the proof that it gets from the prover to a verifier. As such, an AoK that is secure against concurrent MiM attacks should provide some non-transferability guarantee. Since the success of $\mathcal{A}$ during a MiM attack can be replicated without a MiM attack by a quasi-polynomial time simulator, we have that Π_{CaP} guarantees non-transferability whenever computing the witnesses for the considered instances is assumed to be harder than breaking the puzzle.

NMWI for NMZK in the Bare Public-Key (BPK) Model. In [34] it is shown that a concurrent NMWIAoK Π gives directly a concurrent NMZKAoK in the BPK model. The construction is straightforward as it just consists of running Π twice, first from the verifier to the prover (proving knowledge of one out of two secrets) and then from the prover to the verifier (proving knowledge of either a witness for $x \in L$ or of one out of the two secrets of the verifier). Our construction

---

[18] The puzzle can be implemented through a OWP that can be inverted in quasi-polynomial time.

[19] The work of Pass did not take into account MiM attacks.

from Theorem 3 when combined with the construction of [34] gives a candidate round-efficient concurrent NMZKAoK in the BPK model.

4.2 Identification Schemes

We show here a 3-round identification scheme secure against concurrent MiM attacks following the concept of proving knowledge of a secret.

Identification Schemes Based on Proving Knowledge of a Secret. The importance of this setting was for instance discussed in [9] mentioning the following example. Consider a verifier $\mathcal{V}$ that provides a service to restricted group of provers $\mathcal{P}$. A malicious prover $\mathcal{P}^\star$ could give to another party B that is not part of the group, some partial information about his secret that is sufficient for B to obtain the service from $\mathcal{V}$, while still B does not know $\mathcal{P}^\star$'s secret. The paradigm of proving knowledge of a secret in an identification scheme allows to prevent attacks like the one just described. When the identification scheme consists in proving knowledge of a secret the sole fact that B convinces $\mathcal{V}$ is sufficient to claim that one can extract the whole secret from B. This implies that B obtained $\mathcal{P}^\star$'s secret corresponding to his identity, and thus B is actually $\mathcal{P}^\star$[20].

We give a security definition that considers concurrent MiM attacks similarly to the definition CR2 (concurrent-reset on-line) of [2]. The definition of [2] also includes possible reset attacks in addition to allowing $\mathcal{A}$ to invoke multiple concurrent executions of the prover in the left sessions while $\mathcal{A}$ is interacting with the verifier. In the remaining part of this section we will ignore reset attacks since they are out of the purpose of our work. As described in [25] in most network-based settings reset attacks are not an issue. Following the notation of [25] we now give a formal security definitions for an identification scheme.

Definition 4. *Let $\Pi = (\mathcal{K}, \mathcal{P}, \mathcal{V})$ be a tuple of PPT algorithms. We say Π is an identification scheme secure against MiM attacks if the following two properties hold. (1) Correctness. For all $(\mathsf{pk}, \mathsf{sk}) \leftarrow \mathcal{K}(1^\lambda)$, $\mathrm{Prob}\left[\langle \mathcal{P}(\mathsf{sk}), \mathcal{V}\rangle(\mathsf{pk}) = 1\right] = 1$. (2) Security. For all PPT adversaries $\mathcal{A}$ there exists a negligible function ν such that $\mathrm{Prob}\left[(\mathsf{pk}, \mathsf{sk}) \leftarrow \mathcal{K}(1^\lambda) : \langle \mathcal{A}^{\mathcal{P}(\mathsf{sk})}, \mathcal{V}\rangle(\mathsf{pk}) = 1 \text{ AND } \tau \notin T\right] < \nu(\lambda)$, where $\mathcal{A}$ has oracle access to a stateful (i.e., non-resettable) $\mathcal{P}(\mathsf{sk})$, T is defined as the transcripts set of the interactions between $\mathcal{P}(\mathsf{sk})$ and $\mathcal{A}$, and τ is defined as the transcript of one of the interactions between $\mathcal{A}$ and $\mathcal{V}$. All interactions can be arbitrarily interleaved and $\mathcal{A}$ controls the scheduling of the messages.*

Identification Scheme from NMWI. Our construction $\Pi_{\mathsf{ID}} = (\mathcal{K}_{\mathsf{ID}}, \mathcal{P}_{\mathsf{ID}}, \mathcal{V}_{\mathsf{ID}})$ follows the approach of [9,34]. Let $f : \{0,1\}^\lambda \rightarrow \{0,1\}^\lambda$ be a OWP, let λ be the security parameter. The public key of $\mathcal{P}_{\mathsf{ID}}$ is the pair $(\mathsf{pk}_0, \mathsf{pk}_1)$, the secret key is sk_b for a randomly chosen bit b, such that $\mathsf{pk}_b = f(\mathsf{sk}_b)$. Therefore the algorithm $\mathcal{K}_{\mathsf{ID}}$ takes as input the security parameter and outputs $((\mathsf{pk}_0, \mathsf{pk}_1), \mathsf{sk}_b)$ as described above. The protocol simply consists in $\mathcal{P}_{\mathsf{ID}}$ running our 3-round

---

[20] This is instead not likely to happen in scenarios where the same secret key is used for other critical tasks such as signatures of any type of document.

concurrent NMWIAoK Π_{CaP} with $\mathcal{V}_{\mathsf{ID}}$ to prove that it *knows* the pre-image of either pk_0 or pk_1. Formally, let $L_{\mathtt{id}}$ be the following language $L_{\mathtt{id}} = \{(y_0, y_1) : \exists\, x \in \{0,1\}^\lambda$ such that $y_0 = f(x) \vee y_1 = f(x)\}$, then the identification scheme consists of $\mathcal{P}_{\mathsf{ID}}$ proving the statement $(\mathsf{pk}_0, \mathsf{pk}_1) \in L_{\mathtt{id}}$ using Π_{CaP}.

Theorem 4. *Assuming the existence of OWPs w.r.t. subexponential-time adversaries, there is an identification scheme secure against concurrent MiM attacks.*

The proof is again straight-forward. If a PPT $\mathcal{A}$ succeeds then concurrent NMWI of Π_{CaP} guarantees that the witness that he encoded in the proof is independent of the one encoded in the proofs given by $\mathcal{P}$. Therefore by using the AoK property of Π_{CaP} we can invert f with non-negligible probability.

5   Concurrent Malleability of [21]

Here we briefly explain the intuition behind the fact that the 3-round NM commitment scheme $\Pi_{\mathsf{NM}} = (\mathsf{Sen}_{\mathsf{NM}}, \mathsf{Rec}_{\mathsf{NM}})$ of [21] is malleable with respect to a concurrent MiM attack. We use ideas from [16]. We describe a succeeding concurrent MiM adversary $\mathcal{A}$ along with a distinguisher $\mathcal{D}$. We will refer to a NM commitment of the message m using the scheme Π_{NM} as $\mathsf{nmcom}(m)$. We stress that $\mathsf{nmcom}(m)$ is the result of a 3-round interaction between the sender $\mathsf{Sen}_{\mathsf{NM}}$ and the receiver $\mathsf{Rec}_{\mathsf{NM}}$. We start by describing the high-level idea of the protocol Π_{NM}. In the 1st round a left-state L is computed using a special split-state non-malleable code. Let $n = |\mathsf{L}|$. Then a non-interactive commitment $\mathsf{com}_{\mathsf{L}}$ of L is sent in the 1st round, while in the 3rd round the sender computes the right-state R corresponding to the message m and sends it in the clear. In parallel there is also a PoK of the message L committed in $\mathsf{com}_{\mathsf{L}}$. This PoK can be seen as a PoK of each bit of L. Therefore there are n PoKs where the j-th proof is used to prove knowledge of the bit L_j of L.

The actual scheme of [21] is more sophisticated than what we have just described, there are various other components but however they have no impact on the work done by our $\mathcal{A}$, so we will omit them from this short description. Essentially, we will show here that a simplified version of the scheme of [21] is concurrently malleable. However all our arguments apply to their full scheme.

The proposed adversary $\mathcal{A}$ interacts with one sender $\mathsf{Sen}_{\mathsf{NM}}$ in the left session and with many receiver $\mathsf{Rec}_{\mathsf{NM}1}, \ldots, \mathsf{Rec}_{\mathsf{NM}\mathsf{poly}(\lambda)}$ in the right sessions. The behavior of $\mathcal{A}$ in the left and right session can be summarized as following.

Left Session. $\mathsf{Sen}_{\mathsf{NM}}$ computes the 1st round of Π_{NM} as follows. First, he computes L, then he computes a perfectly binding commitment $\mathsf{com}_{\mathsf{L}}$ of L and computes n PoKs one for each bit of the message committed in $\mathsf{com}_{\mathsf{L}}$. In the last round of Π_{NM} $\mathsf{Sen}_{\mathsf{NM}}$ completes the n PoKs and sends R to $\mathcal{A}$ such that the pair (L, R) is a valid encoding of m according to the special non-malleable code. Hence in the left session $\mathcal{A}$ receives $\mathsf{com}_{\mathsf{L}}$, R and n PoKs one for each bit of the string committed in $\mathsf{com}_{\mathsf{L}}$, therefore a PoK for each bit L_j of L.

Right Sessions. In the right sessions $\mathcal{A}$ interacts with $\mathsf{Rec}_{\mathsf{NM}1}, \ldots,$ $\mathsf{Rec}_{\mathsf{NM}\mathsf{poly}(\lambda)}$ mauling the commitments received on the left. More specifically, it starts $2n$ right sessions where n of them should correspond to

$\mathsf{nmcom}(\mathsf{L}_1),\ldots,\mathsf{nmcom}(\mathsf{L}_n)$ such that $\mathsf{L} = \mathsf{L}_1\ldots\mathsf{L}_n$, and the other n sessions should correspond to invalid commitments (we refer to such commitments as $\mathsf{nmcom}(\bot)$).

More precisely, our adversary computes, for each bit L_j of L, two NM commitments $\mathsf{nmcom}(1^\lambda)$, $\mathsf{nmcom}(0^\lambda)$ such that if $\mathsf{L}_j = 1$ then $\mathsf{nmcom}(0^\lambda)$ is invalid, otherwise $\mathsf{nmcom}(1^\lambda)$ is invalid. In order to poison one out of $\mathsf{nmcom}(0^\lambda)$ and $\mathsf{nmcom}(1^\lambda)$, $\mathcal{A}$ will rely on the PoK of L_j received on the left. The PoK of L_j will be plugged in the PoKs of $\mathsf{nmcom}(0^\lambda)$ and in the PoKs of $\mathsf{nmcom}(1^\lambda)$. More precisely one of the n PoKs of $\mathsf{nmcom}(0^\lambda)$ that correspond to a PoK of the bit 0 will be replaced with the PoK of L_j. The same approach is applied when $\mathcal{A}$ computes $\mathsf{nmcom}(1^\lambda)$ with the only difference that the PoK that $\mathcal{A}$ will replace corresponds to a PoK of a bit 1. In this way only one out of $\mathsf{nmcom}(0^\lambda)$ and $\mathsf{nmcom}(1^\lambda)$ still remain a valid commitment. In particular $\mathsf{nmcom}(\mathsf{L}_j)$ will remain a valid commitment while $\mathsf{nmcom}(1 - \mathsf{L}_j)$ will be poisoned and thus will correspond to an invalid commitment (Fig. 4).

There is however a subtlety. Since the PoK played on the right is for one component copied from the PoK played on the left, it can be completed success-

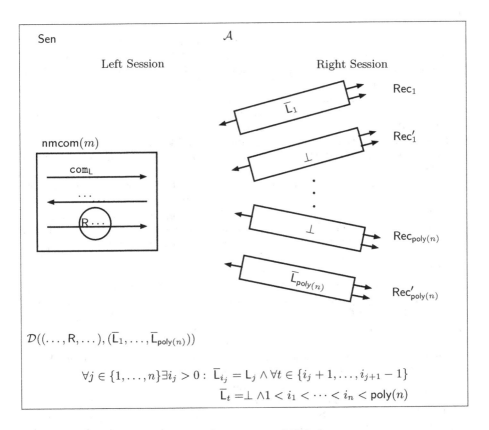

Fig. 4. The one-many MiM $\mathcal{A}$.

fully with constant probability and the adversary has to abort the session if it can not complete the PoK. Therefore each of the above $2n$ right sessions could be repeated multiple times, but however the total amount of right sessions will still be polynomial in the security parameter. Finally our distinguisher $\mathcal{D}$ given as input the committed bits $L_1, \ldots, L_n$ and R contained in the view of $\mathcal{A}$, can easily recover the message m committed in the left interaction.

Acknowledgments. We thank Vipul Goyal, and Silas Richelson for remarkable discussions on [22]. Research supported in part by "GNCS - INdAM", EU COST Action IC1306, NSF grants 1065276, 1118126 and 1136174, US-Israel BSF grant 2008411, OKAWA Foundation Research Award, IBM Faculty Research Award, Xerox Faculty Research Award, B. John Garrick Foundation Award, Teradata Research Award, and Lockheed-Martin Corporation Research Award. This material is based upon work supported in part by DARPA Safeware program. The views expressed are those of the authors and do not reflect the official policy or position of the Department of Defense or the U.S. Government. The work of the 1st, 3rd and 4th authors has been done in part while visiting UCLA.

References

1. Barak, B.: Constant-round coin-tossing with a man in the middle or realizing the shared random string model. In: Proceedings of 43rd Symposium on Foundations of Computer Science (FOCS 2002), Vancouver, BC, Canada, 16–19 November 2002, pp. 345–355 (2002)
2. Bellare, M., Fischlin, M., Goldwasser, S., Micali, S.: Identification protocols secure against reset attacks. In: Pfitzmann, B. (ed.) EUROCRYPT 2001. LNCS, vol. 2045, pp. 495–511. Springer, Heidelberg (2001)
3. Bellare, M., Rogaway, P.: Entity authentication and key distribution. In: Stinson, D.R. (ed.) CRYPTO 1993. LNCS, vol. 773, pp. 232–249. Springer, Heidelberg (1994)
4. Blundo, C., Persiano, G., Sadeghi, A.-R., Visconti, I.: Improved security notions and protocols for non-transferable identification. In: Jajodia, S., Lopez, J. (eds.) ESORICS 2008. LNCS, vol. 5283, pp. 364–378. Springer, Heidelberg (2008)
5. Brenner, H., Goyal, V., Richelson, S., Rosen, A., Vald, M.: Fast non-malleable commitments. In: Proceedings of the 22nd ACM SIGSAC Conference on Computer and Communications Security, Denver, CO, USA, 12–16 October 2015, pp. 1048–1057 (2015)
6. Canetti, R., Goldreich, O., Goldwasser, S., Micali, S.: Resettable zero-knowledge (extended abstract). In: Proceedings of the Thirty-Second Annual ACM Symposium on Theory of Computing, Portland, OR, USA, 21–23 May 2000, pp. 235–244 (2000). http://doi.acm.org/10.1145/335305.335334
7. Cao, Z., Visconti, I., Zhang, Z.: Constant-round concurrent non-malleable statistically binding commitments and decommitments. In: Nguyen, P.Q., Pointcheval, D. (eds.) PKC 2010. LNCS, vol. 6056, pp. 193–208. Springer, Heidelberg (2010)
8. Cao, Z., Visconti, I., Zhang, Z.: On constant-round concurrent non-malleable proof systems. Inf. Process. Lett. **111**(18), 883–890 (2011)
9. Cho, C., Ostrovsky, R., Scafuro, A., Visconti, I.: Simultaneously resettable arguments of knowledge. In: Cramer, R. (ed.) TCC 2012. LNCS, vol. 7194, pp. 530–547. Springer, Heidelberg (2012)

10. Ciampi, M., Ostrovsky, R., Siniscalchi, L., Visconti, I.: Concurrent non-malleable commitments (and more) in 3 rounds. In: Robshaw, M., Katz, J. (eds.) CRYPTO 2016. LNCS, vol. 9816, pp. 270–299. Springer, Heidelberg (2016). Cryptology ePrint Archive, Report 2016/566. http://eprint.iacr.org/

11. Ciampi, M., Ostrovsky, R., Siniscalchi, L., Visconti, I.: On round-efficient non-malleable protocols. Cryptology ePrint Archive, Report 2016/621 (2016). http://eprint.iacr.org/2016/621

12. Ciampi, M., Persiano, G., Scafuro, A., Siniscalchi, L., Visconti, I.: Improved OR-composition of sigma-protocols. In: Kushilevitz, E., et al. (eds.) TCC 2016-A. LNCS, vol. 9563, pp. 112–141. Springer, Heidelberg (2016). doi:10.1007/978-3-662-49099-0_5

13. Ciampi, M., Persiano, G., Scafuro, A., Siniscalchi, L., Visconti, I.: Online/offline OR composition of sigma protocols. In: Fischlin, M., Coron, J.-S. (eds.) EURO-CRYPT 2016. LNCS, vol. 9666, pp. 63–92. Springer, Heidelberg (2016). doi:10.1007/978-3-662-49896-5_3

14. Di Crescenzo, G., Persiano, G., Visconti, I.: Constant-round resettable zero knowledge with concurrent soundness in the bare public-key model. In: Franklin, M. (ed.) CRYPTO 2004. LNCS, vol. 3152, pp. 237–253. Springer, Heidelberg (2004)

15. Dolev, D., Dwork, C., Naor, M.: Non-malleable cryptography (extended abstract). In: Proceedings of the 23rd Annual ACM Symposium on Theory of Computing, New Orleans, Louisiana, USA, 5–8 May 1991, pp. 542–552 (1991)

16. Faust, S., Mukherjee, P., Nielsen, J.B., Venturi, D.: Continuous non-malleable codes. In: Lindell, Y. (ed.) TCC 2014. LNCS, vol. 8349, pp. 465–488. Springer, Heidelberg (2014)

17. Feige, U., Fiat, A., Shamir, A.: Zero knowledge proofs of identity. In: Proceedings of the 19th Annual ACM Symposium on Theory of Computing 1987, New York, USA, pp. 210–217 (1987)

18. Garg, S., Mukherjee, P., Pandey, O., Polychroniadou, A.: The exact round complexity of secure computation. In: Fischlin, M., Coron, J.-S. (eds.) EUROCRYPT 2016. LNCS, vol. 9666, pp. 448–476. Springer, Heidelberg (2016). doi:10.1007/978-3-662-49896-5_16

19. Goyal, V.: Constant round non-malleable protocols using one way functions. In: Proceedings of the 43rd ACM Symposium on Theory of Computing, STOC 2011, San Jose, CA, USA, 6–8 June 2011, pp. 695–704 (2011)

20. Goyal, V., Lee, C., Ostrovsky, R., Visconti, I.: Constructing non-malleable commitments: a black-box approach. In: 53rd Annual IEEE Symposium on Foundations of Computer Science, FOCS 2012, New Brunswick, NJ, USA, 20–23 October 2012, pp. 51–60 (2012)

21. Goyal, V., Pandey, O., Richelson, S.: Textbook non-malleable commitments. IACR Cryptology ePrint Archive 2015 (2015). Version 20151210: 144729 (posted10-Dec-2015 14: 47: 29 UTC). http://eprint.iacr.org/2015/1178

22. Goyal, V., Pandey, O., Richelson, S.: Textbook non-malleable commitments. In: Proceedings of the 48th Annual ACM Symposium on Theory of Computing, STOC 2016, Cambridge, MA, USA, 19–21 June 2016

23. Goyal, V., Richelson, S., Rosen, A., Vald, M.: An algebraic approach to non-malleability. In: 55th IEEE Annual Symposium on Foundations of Computer Science, FOCS 2014, Philadelphia, PA, USA, 18–21 October 2014, pp. 41–50 (2014)

24. Hazay, C., Venkitasubramaniam, M.: On the power of secure two-party computation. Cryptology ePrint Archive, Report 2016/074 (2016). http://eprint.iacr.org/

25. Katz, J.: Efficient cryptographic protocols preventing "Man-in-the-Middle" attacks. Ph.D. thesis, Columbia University (2002)

26. Katz, J., Ostrovsky, R.: Round-optimal secure two-party computation. In: Franklin, M. (ed.) CRYPTO 2004. LNCS, vol. 3152, pp. 335–354. Springer, Heidelberg (2004). http://dx.doi.org/10.1007/978-3-540-28628-8_21

27. Lapidot, D., Shamir, A.: Publicly verifiable non-interactive zero-knowledge proofs. In: Menezes, A., Vanstone, S.A. (eds.) CRYPTO 1990. LNCS, vol. 537, pp. 353–365. Springer, Heidelberg (1991)

28. Lin, H., Pass, R.: Constant-round non-malleable commitments from any one-way function. In: Proceedings of the 43rd ACM Symposium on Theory of Computing, STOC 2011, San Jose, CA, USA, 6–8 June 2011, pp. 705–714 (2011)

29. Lin, H., Pass, R.: Constant-round nonmalleable commitments from any one-way function. J. ACM **62**(1), 5:1–5:30 (2015)

30. Lin, H., Pass, R., Venkitasubramaniam, M.: Concurrent non-malleable commitments from any one-way function. In: Canetti, R. (ed.) TCC 2008. LNCS, vol. 4948, pp. 571–588. Springer, Heidelberg (2008)

31. Lin, H., Pass, R., Venkitasubramaniam, M.: A unified framework for concurrent security: universal composability from stand-alone non-malleability. In: Proceedings of the 41st Annual ACM Symposium on Theory of Computing, STOC 2009, Bethesda, MD, USA, May 31–June 2 2009, pp. 179–188 (2009)

32. Lindell, Y.: Foundations of cryptography 89–856 (2010). http://u.cs.biu.ac.il/~lindell/89-856/complete-89-856.pdf

33. Mittelbach, A., Venturi, D.: Fiat-shamir for highly sound protocols is instantiable. Cryptology ePrint Archive, Report 2016/313 (2016). http://eprint.iacr.org/

34. Ostrovsky, R., Persiano, G., Visconti, I.: Constant-round concurrent non-malleable zero knowledge in the bare public-key model. In: Aceto, L., Damgård, I., Goldberg, L.A., Halldórsson, M.M., Ingólfsdóttir, A., Walukiewicz, I. (eds.) ICALP 2008, Part II. LNCS, vol. 5126, pp. 548–559. Springer, Heidelberg (2008)

35. Ostrovsky, R., Persiano, G., Visconti, I.: Simulation-based concurrent nonmalleable commitments and decommitments. In: Reingold, O. (ed.) TCC 2009. LNCS, vol. 5444, pp. 91–108. Springer, Heidelberg (2009)

36. Ostrovsky, R., Rao, V., Scafuro, A., Visconti, I.: Revisiting lower and upper bounds for selective decommitments. In: Sahai, A. (ed.) TCC 2013. LNCS, vol. 7785, pp. 559–578. Springer, Heidelberg (2013)

37. Ostrovsky, R., Visconti, I.: Simultaneous resettability from collision resistance. Electronic Colloquium on Computational Complexity (ECCC) 19, 164 (2012)

38. Pandey, O., Pass, R., Vaikuntanathan, V.: Adaptive one-way functions and applications. In: Wagner, D. (ed.) CRYPTO 2008. LNCS, vol. 5157, pp. 57–74. Springer, Heidelberg (2008)

39. Pass, R.: Simulation in quasi-polynomial time, and its application to protocol composition. In: Biham, E. (ed.) EUROCRYPT 2003. LNCS, vol. 2656. Springer, Heidelberg (2003)

40. Pass, R.: Bounded-concurrent secure multi-party computation with a dishonest majority. In: Proceedings of the 36th Annual ACM Symposium on Theory of Computing, Chicago, IL, USA, 13–16 June 2004, pp. 232–241 (2004)

41. Pass, R.: Unprovable security of perfect NIZK and non-interactive non-malleable commitments. In: Sahai, A. (ed.) TCC 2013. LNCS, vol. 7785, pp. 334–354. Springer, Heidelberg (2013)

42. Pass, R., Rosen, A.: Concurrent non-malleable commitments. In: Proceedings of 46th Annual IEEE Symposium on Foundations of Computer Science (FOCS 2005), Pittsburgh, PA, USA, 23–25 October 2005, pp. 563–572 (2005)

43. Pass, R., Rosen, A.: New and improved constructions of non-malleable cryptographic protocols. In: Proceedings of the 37th Annual ACM Symposium on Theory of Computing, Baltimore, MD, USA, 22–24 May 2005, pp. 533–542 (2005)
44. Pass, R., Rosen, A.: Concurrent nonmalleable commitments. SIAM J. Comput. **37**(6), 1891–1925 (2008)
45. Pass, R., Wee, H.: Black-box constructions of two-party protocols from one-way functions. In: Reingold, O. (ed.) TCC 2009. LNCS, vol. 5444, pp. 403–418. Springer, Heidelberg (2009)
46. Pass, R., Wee, H.: Constant-round non-malleable commitments from subexponential one-way functions. In: Gilbert, H. (ed.) EUROCRYPT 2010. LNCS, vol. 6110, pp. 638–655. Springer, Heidelberg (2010)
47. Scafuro, A., Visconti, I.: On round-optimal zero knowledge in the bare public-key model. In: Pointcheval, D., Johansson, T. (eds.) EUROCRYPT 2012. LNCS, vol. 7237, pp. 153–171. Springer, Heidelberg (2012). http://dx.doi.org/10.1007/978-3-642-29011-4_11
48. Wee, H.: Black-box, round-efficient secure computation via non-malleability amplification. In: 51th Annual IEEE Symposium on Foundations of Computer Science, FOCS 2010, 23–26 October 2010, Las Vegas, Nevada, USA, pp. 531–540. IEEE Computer Society (2010)
49. Yung, M., Zhao, Y.: Generic and practical resettable zero-knowledge in the bare public-key model. In: Naor, M. (ed.) EUROCRYPT 2007. LNCS, vol. 4515, pp. 129–147. Springer, Heidelberg (2007)

IBE, ABE, and Functional Encryption

Programmable Hash Functions from Lattices: Short Signatures and IBEs with Small Key Sizes

Jiang Zhang[1(✉)], Yu Chen[2(✉)], and Zhenfeng Zhang[3(✉)]

[1] State Key Laboratory of Cryptology, P.O. Box 5159, Beijing 100878, China
jiangzhang09@gmail.com
[2] State Key Laboratory of Information Security,
Institute of Information Engineering, Chinese Academy of Sciences, Beijing, China
yuchen.prc@gmail.com
[3] Trusted Computing and Information Assurance Laboratory, Institute of Software,
Chinese Academy of Sciences, Beijing, China
zfzhang@tca.iscas.ac.cn

Abstract. Driven by the open problem raised by Hofheinz and Kiltz [34], we study the formalization of lattice-based programmable hash function (PHF), and give two types of constructions by using several techniques such as a novel combination of cover-free sets and lattice trapdoors. Under the Inhomogeneous Small Integer Solution (ISIS) assumption, we show that any (non-trivial) lattice-based PHF is collision-resistant, which gives a direct application of this new primitive. We further demonstrate the power of lattice-based PHF by giving generic constructions of signature and identity-based encryption (IBE) in the standard model, which not only provide a way to unify several previous lattice-based schemes using the partitioning proof techniques, but also allow us to obtain a new short signature scheme and a new fully secure IBE scheme with keys consisting of a logarithmic number of matrices/vectors in the security parameter κ. Besides, we also give a refined way of combining two concrete PHFs to construct an improved short signature scheme with short verification keys from weaker assumptions. In particular, our methods depart from the confined guessing technique of Böhl et al. [8] that was used to construct previous standard model short signature schemes with short verification keys by Ducas and Micciancio [24] and by Alperin-Sheriff [6], and allow us to achieve existential unforgeability against chosen message attacks (EUF-CMA) without resorting to chameleon hash functions.

1 Introduction

As a primitive capturing the partitioning proof techniques, programmable hash function introduced by Hofheinz and Kiltz [33] is a powerful tool to construct provably secure cryptographic schemes in the standard model. Informally, a PHF $\mathcal{H} = \{H_K\}$ is a keyed group hash function over some finite group $\mathbb{G}$, which can work in two (statistically) indistinguishable modes depending on how the key is generated: if the key K is generated in the normal mode, then the hash function behaves normally and maps an input X into a group element $H_K(X) \in \mathbb{G}$; while

M. Robshaw and J. Katz (Eds.): CRYPTO 2016, Part III, LNCS 9816, pp. 303–332, 2016.
DOI: 10.1007/978-3-662-53015-3_11

if the key K' is generated in the trapdoor mode, then (with the help of some trapdoor information td) it can additionally output a secret pair (a_X, b_X) such that $H_{K'}(X) = g^{a_X} h^{b_X}$ holds for some prior fixed group generators $g, h \in \mathbb{G}$. More formally, let $u, v \in \mathbb{Z}$ be some positive integers, $\mathcal{H}$ is said to be (u, v)-programmable if given any inputs $X_1, \ldots, X_u$ and $Y_1, \ldots, Y_v$ satisfying $X_i \neq Y_j$ for any i and j, the probability $\Pr[a_{X_1} = \cdots = a_{X_u} = 0 \wedge a_{Y_1}, \ldots, a_{Y_v} \neq 0] \geq 1/\mathrm{poly}(\kappa)$ for some polynomial $\mathrm{poly}(\kappa)$ in the security parameter κ, where the probability is over the random coins used in generating K' and td. This feature gives a partition of all inputs in terms of whether $a_X = 0$, and becomes very useful in security proofs when the discrete logarithm (DL) is hard in $\mathbb{G}$ [33].

Since its introduction, PHFs have attracted much attention from the research community [15,26,31,34,51], and had been used to construct many cryptographic schemes (such as short signature schemes [32]) in the standard model. However, both the definition and the constructions of traditional PHFs seem specific to hash functions defined over groups where the "DL problem" is hard. This might be the reason why almost all known PHFs were constructed from "DL groups". Actually, it was left as an open problem [34] to find instantiations of PHF from different assumptions, e.g., lattices.

Facing the rapid development of quantum computers, the past decade has witnessed remarkable advancement in lattice-based cryptography. Nevertheless, the silhouette of lattice-based PHFs is still not very clear. At Crypto 2013, Freire et al. [26] extended the notion of PHF to the multilinear maps setting. However, recent study shows that there is a long way to go before obtaining a practical and secure multilinear maps from lattices [16,18,19,27,35]. An intriguing question of great interest is to construct lattice-based PHFs or something similar based on standard hard lattice problems.

Lattice-Based Short Signatures. It is well-known that digital signature schemes [36] can be constructed from general assumptions, such as one-way functions. Nevertheless, these generic signature schemes suffer from either large signatures or large verification keys, thus a main open problem is to reduce the signature size as well as the verification key size. The first direct constructions of lattice-based signature schemes were given in [29,40]. Later, many works (e.g., [7,22,39]) significantly improved the efficiency of lattice-based signature schemes in the random oracle model. In comparison, the progress in constructing efficient lattice-based signature schemes in the standard model was relatively slow. At Eurocrypt 2010, Cash et al. [14] proposed a signature scheme with a linear number of vectors in the signatures. The first standard model short signature scheme with signatures consisting of a single lattice vector was due to Boyen [12], which was later improved by Micciancio and Peikert [43]. However, the verification keys of both schemes in [12,43] consist of a linear number of matrices.

In 2013, Böhl et al. [8] constructed a lattice-based signature scheme with constant verification keys by introducing the confined guessing proof technique. Later, Ducas and Micciancio [24] adapted the confined guessing proof technique to ideal lattices, and proposed a short signature scheme with logarithmic verification keys. Recently, Alperin-Sheriff [6] constructed a short signature with constant verification keys based on a stronger hardness assumption by using the idea of homomorphic trapdoor functions [30]. Due to the use of the confined guessing technique, the above three signature schemes [6, 8, 24] shared two undesired byproducts. First, the security can only be directly proven to be existentially unforgeable against non-adaptive chosen message attacks (EUF-naCMA). Even if an EUF-naCMA secure scheme can be transformed into an EUF-CMA secure one by using known techniques such as chameleon hash functions [37], in the lattice setting [24] this usually introduces an additional tag to each signature and roughly increases the signature size by twice. Second, a reduction loss about $(Q^2/\epsilon)^c$ for some parameter $c > 1$ seems unavoidable, where Q is the number of signing queries of the forger $\mathcal{F}$, and ϵ is the success probability of $\mathcal{F}$. Therefore, it is desirable to directly construct an EUF-CMA secure scheme that has short signatures, short verification keys, as well as a relatively tight security proof.

Identity-Based Encryption from Lattices. Shamir [48] introduced identity-based encryption (IBE) in 1984, but the first realizations were due to Boneh and Franklin from pairings [10] and Cocks from quadratic residues [17]. In the lattice setting, Gentry et al. [29] proposed the first IBE scheme based on the learning with errors (LWE) assumption in the random oracle model. Later, several works [2, 14, 23, 52] were dedicated to the study of lattice-based (hierarchical) IBE schemes also in the random oracle model. There were a few works focusing on designing standard model lattice-based IBE schemes [1, 2, 14]. Concretely, the scheme in [2] was only proven to be *selective-identity* secure in the standard model. By using standard complexity leverage technique [9], one can generally transform a selective-identity secure IBE scheme into a *fully secure* one. But the resulting scheme has to suffer from a reduction loss proportional to L, where L is the number of distinct identities for the IBE system and is independent from the number Q of the adversary's private key queries in the security proof. Since L is usually super-polynomial and much larger than Q, the above generic transformation is a very unsatisfying approach [28]. In [1, 14], the authors showed how to achieve *full security* against adaptive chosen-plaintext and chosen-identity attacks, but both standard model fully secure IBE schemes in [1, 14] had large master public keys consisting of a linear number of matrices. In fact, Agrawal, Boneh and Boyen left it as an open problem to find fully secure lattice-based IBE schemes with short master public keys in the standard model [1].

1.1 Our Contributions

Because of the (big) differences in the algebraic structures between lattices and DL groups, the traditional definition of PHFs does not seem to work on lattices. This makes it highly non-trivial to find instantiations of traditional PHFs on lattices. In this paper, we introduce the notion of lattice-based programmable hash function (PHF). Although our lattice-based PHF has gone beyond the realm of traditional PHFs, we prefer to still name it as PHF because it inherits the concept of traditional PHFs and aims at capturing the partitioning proof trick on lattices. By carefully exploiting the algebraic properties of lattices, we give several different constructions of lattice-based PHFs.

Under the Inhomogeneous Small Integer Solution (ISIS) assumption, we show that any (non-trivial) lattice-based PHF is collision-resistant. This gives a direct application of lattice-based PHFs. We further demonstrate the power of lattice-based PHFs by showing a generic way to construct short signature schemes. Under the ISIS assumption, our generic signature scheme is EUF-CMA secure in the standard model. We also give a generic IBE scheme from lattice-based PHFs with a property called high min-entropy. Under the LWE assumption, our generic IBE scheme is secure against adaptive chosen-plaintext and chosen-identity attacks in the standard model. Moreover, our IBE scheme can be extended to support hierarchical identities, and achieve chosen ciphertext security.

We find that lattice-based PHFs are implicitly used as the backbones in the signature schemes [12,43] and the IBE schemes [1]. Therefore, our results provide a way to unify and clarify those lattice-based cryptographic schemes using the partitioning proof strategy. Furthermore, by instantiating the generic schemes with our new PHF constructions, we obtain a new short signature scheme and a new IBE scheme. Compared to previous schemes, our instantiated schemes have several appealing advantages. Besides, we also construct an improved short signature scheme with short verification keys by carefully combining two concrete PHFs. Comparisons between our schemes and previous ones will be given in Sects. 1.3 and 1.4.

1.2   Techniques

We introduce the notion of lattice-based PHFs by carefully exploiting the specific algebraic structure of lattices. As the traditional PHFs, our lattice-based PHF $\mathcal{H} = \{H_K\}$ can work in two modes. Given a key K generated in either the normal mode or the trapdoor mode, the hash function H_K maps its input $X \in \mathcal{X}$ into a matrix $H_K(X) \in \mathbb{Z}_q^{n \times m}$ for some positive $n, m, q \in \mathbb{Z}$. In the trapdoor mode, there additionally exists a secret trapdoor td allowing to compute matrices $\mathbf{R}_X \in \mathbb{Z}_q^{\bar{m} \times m}$ and $\mathbf{S}_X \in \mathbb{Z}_q^{n \times n}$ for some integer $\bar{m} \in \mathbb{Z}$, such that $H_K(X) = \mathbf{A}\mathbf{R}_X + \mathbf{S}_X \mathbf{B} \in \mathbb{Z}_q^{n \times m}$ holds with respect to user-specified "generators" $\mathbf{A} \in \mathbb{Z}_q^{n \times \bar{m}}$ and $\mathbf{B} \in \mathbb{Z}_q^{n \times m}$. For non-triviality, we require that the keys generated in the two modes are statistically indistinguishable (even conditioned on the matrix $\mathbf{A}$ that was used to generate the trapdoor mode key), and that the two "generators" $\mathbf{A} \in \mathbb{Z}_q^{n \times \bar{m}}$ and $\mathbf{B} \in \mathbb{Z}_q^{n \times m}$ have essential differences for embedding hard lattice problems. More precisely, in our definition $\mathbf{A} \in \mathbb{Z}_q^{n \times \bar{m}}$ is required to be uniformly distributed (and thus can be used to embed the ISIS problem), while $\mathbf{B} \in \mathbb{Z}_q^{n \times m}$ is a trapdoor matrix that allows to efficiently sample short vector $\mathbf{e} \in \mathbb{Z}^m$ satisfying $\mathbf{B}\mathbf{e} = \mathbf{v}$ for any vector $\mathbf{v} \in \mathbb{Z}_q^n$.

In order to explore the differences between $\mathbf{A} \in \mathbb{Z}_q^{n \times \bar{m}}$ and $\mathbf{B} \in \mathbb{Z}_q^{n \times m}$ in the security reduction, we require that the largest singular value of $\mathbf{R}_X$ defined by $s_1(\mathbf{R}_X) = \max_{\mathbf{u}} \|\mathbf{R}_X \mathbf{u}\|$ is small where the maximum is taken over all unit vectors $\mathbf{u} \in \mathbb{R}^m$, and that $\mathbf{S}_X \in \mathcal{I}_n \cup \{\mathbf{0}\}$ where $\mathcal{I}_n$ is the set of invertible matrices in $\mathbb{Z}_q^{n \times n}$. More concretely, for any positive integer $u, v \in \mathbb{Z}$ and real $\beta \in \mathbb{R}$, a (u, v, β)-PHF $\mathcal{H}$ should satisfy the following two conditions: (1) $s_1(\mathbf{R}_X) \leq \beta$ holds for any input X; and (2) given any inputs $X_1, \ldots, X_u$ and $Y_1, \ldots, Y_v$

satisfying $X_i \neq Y_j$ for any i and j, the probability $\Pr[\mathbf{S}_{X_1} = \cdots = \mathbf{S}_{X_u} = \mathbf{0} \wedge \mathbf{S}_{Y_1}, \ldots, \mathbf{S}_{Y_v} \in \mathcal{I}_n]$ is at least $1/\mathrm{poly}(n)$, where the probability is taken over the random coins in producing td and K'. Besides, if the second condition only holds for some prior fixed $X_1, \ldots, X_u$ (chosen before generating the trapdoor mode key K'), we say that the hash function $\mathcal{H}$ is a weak (u, v, β)-PHF.

Looking ahead, if the trapdoor mode key K' is generated by using $\mathbf{A} \in \mathbb{Z}_q^{n \times \bar{m}}$ and trapdoor matrix $\mathbf{B} \in \mathbb{Z}_q^{n \times m}$, then for any input X the matrix $\mathbf{A}_X := (\mathbf{A} \| \mathbf{H}_{K'}(X)) = (\mathbf{A} \| \mathbf{A}\mathbf{R}_X + \mathbf{S}_X \mathbf{B}) \in \mathbb{Z}_q^{n \times (\bar{m}+m)}$ has a trapdoor $\mathbf{R}_X$ with respect to tag $\mathbf{S}_X$. The programmability comes from the fact that such a trapdoor enables us to sample short vector $\mathbf{e}$ satisfying $\mathbf{A}_X \mathbf{e} = \mathbf{v}$ for any vector $\mathbf{v} \in \mathbb{Z}_q^n$ when $\mathbf{S}_X$ is invertible, and loses this ability when $\mathbf{S}_X = \mathbf{0}$. This gives us the possibility to adaptively embed the ISIS problem depending on each particular input X. Since this feature is only useful when the key K' is used together with the "generator" $\mathbf{A} \in \mathbb{Z}_q^{n \times \bar{m}}$, we require the keys in both modes to be statistically indistinguishable even conditioned on the information of $\mathbf{A}$.

Our Type-I PHF construction is a high-level abstraction of the functions that were (implicitly) used in both signature schemes (e.g., [8,12,43]) and encryption schemes (e.g., [1,43]). Formally, let E be an encoding function from some domain $\mathcal{X}$ to $(\mathbb{Z}_q^{n \times n})^\ell$, where ℓ is an integer. Then, for any input $X \in \mathcal{X}$, the Type-I PHF construction $\mathcal{H} = \{\mathrm{H}_K\}$ from $\mathcal{X}$ to $\mathbb{Z}_q^{n \times m}$ is defined as $\mathrm{H}_K(X) = \mathbf{A}_0 + \sum_{i=1}^{\ell} \mathbf{C}_i \mathbf{A}_i$, where $K = (\mathbf{A}_0, \mathbf{A}_1, \ldots, \mathbf{A}_\ell)$ and $\mathrm{E}(X) = (\mathbf{C}_1, \ldots, \mathbf{C}_\ell)$. For appropriate choices of parameters and encoding function E, the literatures (implicitly) showed that the Type-I construction satisfies our definition of lattice-based PHFs. Concretely, if one sets $\mathcal{X} = \{0,1\}^\ell$, and $\mathrm{E}(X) = ((-1)^{X_1} \cdot \mathbf{I}_n, \ldots, (-1)^{X_\ell} \cdot \mathbf{I}_n)$ for any input $X = (X_1, \ldots, X_\ell)$, where $\mathbf{I}_n$ is the $n \times n$ identity matrix. Then, the instantiated PHF is exactly the hash functions that were used to construct the signature scheme in [12] and the IBE scheme in [1]. Since the Type-I PHF construction is independent from the particular choice of $\mathbf{B} \in \mathbb{Z}_q^{n \times m}$, it allows us to use any trapdoor matrix $\mathbf{B}$ when generating the trapdoor mode key. On the downside, such a construction has a large key size, i.e., the number of matrices in the key is linear in the input length ℓ.

Our Type-II PHF construction has keys only consisting of $O(\log \ell)$ matrices, which substantially reduces the key size by using a novel combination of the cover-free sets and the publicly known trapdoor matrix $\mathbf{B} = \mathbf{G}$ in [43], where $\mathbf{G} = \mathbf{I}_n \otimes \mathbf{g}^t \in \mathbb{Z}_q^{n \times nk}$, $k = \lceil \log_2 q \rceil$ and $\mathbf{g} = (1, 2, \ldots, 2^{k-1})^t \in \mathbb{Z}_q^k$. Concretely, for any positive $L \in \mathbb{Z}$, by $[L]$ we denote the set $\{0, 1, \ldots, L-1\}$. Recall that if $CF = \{CF_X\}_{X \in [L]}$ is a family of v-cover-free sets over domain $[N]$ for some integers $v, L, N \in \mathbb{Z}$, then for any subset $\mathcal{S} \subseteq [L]$ of size at most v and any $Y \notin \mathcal{S}$, there is at least one element $z^* \in CF_Y \subseteq [N]$ that is not included in the union set $\cup_{X \in \mathcal{S}} CF_X$. The property of cover-free sets naturally gives a partition of $[L]$, and was first used in constructing traditional PHFs in [32]. However, a direct application of the cover-free sets in constructing (lattice-based) PHFs will result in a very large key size (which is even worse than that of the Type-I PHF). Actually, for an input size $L = 2^\ell$, the key of the PHF in [32] should contain an associated element for each element in $[N]$, where N is as large as poly(ℓ). We solve this problem by using the nice property of $\mathbf{G}$ and the binary representation of the cover-free sets. Formally, let $\mathbf{G}^{-1}(\mathbf{C})$ be the binary decomposition of some matrix $\mathbf{C}$. By the definition of $\mathbf{G}$, we have $\mathbf{G} \cdot \mathbf{G}^{-1}(\mathbf{C}) = \mathbf{C}$. Now, we set the key K of the Type-II PHF as $K = (\mathbf{A}, \{\mathbf{A}_i\}_{i \in \{0,\dots,\mu-1\}})$, where $\mu = \lceil \log_2 N \rceil = O(\log \ell)$. For any input $X \in [L]$, we first map X into the corresponding set $CF_X \in CF$. Then, for each $z \in CF_X \subseteq [N]$, we "recover" an associated matrix $\mathbf{A}_z = \mathsf{Func}(K, z, 0)$ from K and the binary decomposition $(b_0, \dots, b_{\mu-1})$ of z, where Func is recursively defined as

$$\mathsf{Func}(K, z, i) = \begin{cases} \mathbf{A}_{\mu-1}, & \text{if } i = \mu - 1 \\ (\mathbf{A}_i - b_i\mathbf{G}) \cdot \mathbf{G}^{-1}(\mathsf{Func}(K, z, i+1)), & \text{otherwise} \end{cases}$$

Finally, we output the hash value $\mathrm{H}_K(X) = \mathbf{A} + \sum_{z \in CF_X} \mathbf{A}_z$.

In the trapdoor mode, we randomly choose a "target" element $z^* \in [N]$, and set $\mathbf{A} = \hat{\mathbf{A}}\mathbf{R} - (-1)^c \cdot \mathbf{G}$ and $\mathbf{A}_i = \hat{\mathbf{A}}\mathbf{R}_i + (1 - b_i^*) \cdot \mathbf{G}$ for all $i \in \{0,\dots,\mu-1\}$, where $(b_0^*,\dots,b_{\mu-1}^*)$ is the binary decomposition of z^* and c is the number of 1's in the vector $(b_0^*,\dots,b_{\mu-1}^*)$. By doing this, we have that $\mathbf{A}_z = \hat{\mathbf{A}}\hat{\mathbf{R}}_z + \hat{\mathbf{S}}_z\mathbf{G}$ holds for some matrices $\hat{\mathbf{R}}_z$ and $\hat{\mathbf{S}}_z = \prod_{i=0}^{\mu-1}(1 - b_i^* - b_i) \cdot \mathbf{I}_n$, where $(b_0, \dots, b_{\mu-1})$ is the binary decomposition of z. This means that $\hat{\mathbf{S}}_z = \mathbf{0}$ for any $z \neq z^*$, and $\hat{\mathbf{S}}_{z^*} = (-1)^c \cdot \mathbf{I}_n$. By the definition of $\mathrm{H}_K(X) = \mathbf{A} + \sum_{z \in CF_X} \mathbf{A}_z$, we have that $\mathrm{H}_K(X) = \hat{\mathbf{A}}\hat{\mathbf{R}}_X + \hat{\mathbf{S}}_X\mathbf{G}$ holds for some matrices $\hat{\mathbf{R}}_X = \mathbf{R} + \sum_{z \in CF_X} \hat{\mathbf{R}}_z$ and $\hat{\mathbf{S}}_X = -(-1)^c \cdot \mathbf{I}_n + \sum_{z \in CF_X} \hat{\mathbf{S}}_z$. Obviously, we have that $\hat{\mathbf{S}}_X = \mathbf{0}$ if and only if $z^* \in CF_X$, otherwise $\hat{\mathbf{S}}_X = -(-1)^c \cdot \mathbf{I}_n$. By the property of the cover-free sets, there is at least one element in $CF_Y \subseteq [N]$ that is not included in the union set $\cup_{X \in \mathcal{S}} CF_X$ for any $\mathcal{S} = \{X_1, \dots, X_v\}$ and $Y \notin \mathcal{S}$. Thus, if z^* is randomly chosen and is statistically hidden in the key $K = (\mathbf{A}, \{\mathbf{A}_i\}_{i \in \{0,\dots,\mu-1\}})$, we have the probability that $\mathrm{H}_K(X_i) = \hat{\mathbf{A}}\hat{\mathbf{R}}_{X_i} - (-1)^c \cdot \mathbf{G}$ for all $X_i \in \mathcal{S}$ and $\mathrm{H}_K(Y) = \hat{\mathbf{A}}\hat{\mathbf{R}}_Y$, is at least $1/N = 1/\mathrm{poly}(\ell)$.

1.3 Short Signatures

We now outline the idea on how to construct a generic signature scheme $\mathcal{SIG}$ from lattice-based PHFs in the standard model. Let $n, \bar{m}, m', \ell, q$ be some positive integers, and let $m = \bar{m} + m'$. Given a lattice-based PHF $\mathcal{H} = \{\mathrm{H}_K\}$ from $\{0,1\}^\ell$ to $\mathbb{Z}_q^{n \times m'}$, let $\mathbf{B} \in \mathbb{Z}_q^{n \times m'}$ be a trapdoor matrix that is compatible with $\mathcal{H}$. Then, the verification key of the generic signature scheme $\mathcal{SIG}$ consists of a uniformly distributed (trapdoor) matrix $\mathbf{A} \in \mathbb{Z}_q^{n \times \bar{m}}$, a uniformly random vector $\mathbf{u} \in \mathbb{Z}_q^n$, and a random key K for $\mathcal{H}$, i.e., $vk = (\mathbf{A}, \mathbf{u}, K)$. The signing key is a trapdoor $\mathbf{R}$ of $\mathbf{A}$ that allows to sample short vector $\mathbf{e}$ satisfying $\mathbf{A}\mathbf{e} = \mathbf{v}$ for any vector $\mathbf{v} \in \mathbb{Z}_q^n$. Given a message $M \in \{0,1\}^\ell$, the signing algorithm first computes $\mathbf{A}_M = (\mathbf{A}\|\mathrm{H}_K(M)) \in \mathbb{Z}_q^{n \times m}$, and then uses the trapdoor $\mathbf{R}$ to sample a short vector $\mathbf{e} \in \mathbb{Z}^m$ satisfying $\mathbf{A}_M\mathbf{e} = \mathbf{u}$ by employing the sampling algorithms in [14,29,43]. Finally, it returns $\sigma = \mathbf{e}$ as the signature on the message M. The verifier accepts $\sigma = \mathbf{e}$ as a valid signature on M if and only if $\mathbf{e}$ is short and

$\mathbf{A}_M\mathbf{e} = \mathbf{u}$. The correctness of the generic scheme $\mathcal{SIG}$ is guaranteed by the nice properties of the sampling algorithms in [29,43].

In addition, if $\mathcal{H} = \{H_K\}$ is a $(1, v, \beta)$-PHF for some integer v and real β, we can show that under the ISIS assumption, $\mathcal{SIG}$ is existentially unforgeable against adaptive chosen message attacks (EUF-CMA) in the standard model as long as the forger $\mathcal{F}$ makes at most $Q \le v$ signing queries. Intuitively, given an ISIS challenge instance $(\hat{\mathbf{A}}, \hat{\mathbf{u}})$ in the security reduction, the challenger first generates a trapdoor mode key K' for $\mathcal{H}$ by using $(\hat{\mathbf{A}}, \mathbf{B})$. Then, it defines $vk = (\hat{\mathbf{A}}, \hat{\mathbf{u}}, K')$ and keeps the trapdoor td of K' private. For message M_i in the i-th signing query, we have $\mathbf{A}_{M_i} = (\hat{\mathbf{A}}\|H_{K'}(M_i)) = (\hat{\mathbf{A}}\|\hat{\mathbf{A}}\mathbf{R}_{M_i} + \mathbf{S}_{M_i}\mathbf{B}) \in \mathbb{Z}_q^{n \times m}$. By the programmability of $\mathcal{H}$, with a certain probability we have that $\mathbf{S}_{M_i}$ is invertible for all the Q signing messages $\{M_i\}_{i \in \{1,...,Q\}}$, but $\mathbf{S}_{M^*} = \mathbf{0}$ for the forged message M^*. In this case, the challenger can use $\mathbf{R}_{M_i}$ to perfectly answer the signing queries, and use the forged message-signature pair (M^*, σ^*) to solve the ISIS problem by the equation $\mathbf{u} = \mathbf{A}_{M^*}\sigma^* = \hat{\mathbf{A}}(\mathbf{I}_{\bar{m}}\|\mathbf{R}_{M^*})\sigma^*$.

Each signature in the generic scheme $\mathcal{SIG}$ only has a single vector, which is as short as that in [12,43]. In fact, our generic scheme $\mathcal{SIG}$ encompasses the two signature schemes from [12,43] in the sense that both schemes can be seen as the instantiations of $\mathcal{SIG}$ using the Type-I PHF construction. Due to the inefficiency of the concrete PHFs, both schemes [12,43] had large verification keys consisting of a linear number of matrices. By instantiating $\mathcal{SIG}$ with our efficient Type-II PHF construction, we obtain a concrete scheme $\mathcal{SIG}_1$ with verification keys consisting of a logarithmic number of matrices. Unlike the prior schemes in [6,8,24], our methods do not use the confined guessing proof technique [8], and enable us to directly achieve EUF-CMA security without using chameleon hash functions. This also allows us to get a security proof of $\mathcal{SIG}_1$ with a reduction loss only about nv^2, which is independent from the forger's success probability ϵ. We remark that this improvement does not come for free: the underlying ISIS assumption should hold for parameter $\bar{\beta} = v^2 \cdot \tilde{O}(n^{5.5})$, where $v \ge Q$ is required.[1] By carefully combining our Type-II $(1, v, \beta)$-PHF with a simple weak Type-I PHF and introducing a very short tag to each signature, we further remove the condition $v \ge Q$ such that a much smaller $v = \omega(\log n)$ can be used to construct an improved short signature scheme $\mathcal{SIG}_2$ from (relatively) weaker ISIS assumption, which further removes a factor of Q^2 (resp. Q) from the ISIS parameter (resp. the reduction loss) of our generic signature scheme.

In Table 1, we give a (rough) comparison of lattice-based signature schemes in the standard model. For simplicity, the message length is set to be n. Let constant $c > 1$ and $d = O(\log_c n)$ be the parameters for the use of the confined guessing technique in [6,8,24]. We compare the size of verification keys and signatures in terms of the number of "basic" elements as in [6,24]. On general lattices, the "basic" element in the verification keys is a matrix over $\mathbb{Z}_q$ whose size is mainly determined by the underlying hard lattices, while the "basic" element in the signatures is a lattice vector. On ideal lattices, the "basic" element in the verification keys can be represented by a vector. Almost all schemes on general

---

[1] We write $f(n) = \tilde{O}(g(n))$ if $f(n) = O(g(n) \cdot \log^c(n))$ for some constant c.

Table 1. Rough comparison of lattice-based signatures in the standard model (Since all schemes only have a single "basic" element in the signing keys, we also omit the corresponding comparison in the size of signing keys for succinctness. The reduction loss is the ratio ϵ/ϵ' between the success probability ϵ of the forger and the success probability ϵ' of the reduction. Real $\bar{\beta}$ is the parameter for the (I)SIS problem, and "CMH?" denotes the necessity of chameleon hash functions to achieve EUF-CMA security. Constant $c > 1$ and $d = O(\log_c n)$ are the parameters in [6,8,24])

Schemes	Verification key	Signature	Reduction loss	(I)SIS param $\bar{\beta}$	CMH?
LM08 [40] *	1	$\log n$	Q	$\tilde{O}(n^2)$	No
CHKP10 [14]	n	$\log n$	Q	$\tilde{O}(n^{1.5})$	Yes
Boyen10 [12]	n	1	Q	$\tilde{O}(n^{3.5})$	No
MP12 [43] †	n	1	Q	$\tilde{O}(n^{2.5})$	Yes
BHJ$^+$14 [8]	1	d	$(Q^2/\epsilon)^c$	$\tilde{O}(n^{2.5})$	Yes
DM14 [24] *	d	1	$(Q^2/\epsilon)^c$	$\tilde{O}(n^{3.5})$	Yes
AS15 [6]	1	1	$(Q^2/\epsilon)^c$	$\tilde{O}(d^{2d} \cdot n^{5.5})$	Yes
Our $\mathcal{SIG}_1$	$\log n$	1	$n \cdot Q^2$	$Q^2 \cdot \tilde{O}(n^{5.5})$	No
Our $\mathcal{SIG}_2$	$\log n$	1	$Q \cdot \tilde{O}(n)$	$\tilde{O}(n^{5.5})$	No

lattices such as [6,8,12,14,43] and ours can be instantiated from ideal lattices, and thus roughly saves a factor of n in the verification key size. However, the two schemes [24,40] (marked with '*') from ideal lattices have no realizations over general lattices. We ignore the constant factors in the table to avoid clutter. Since all schemes only have a single "basic" element in the signing keys, we also omit the corresponding comparison in the size of signing keys for succinctness. Finally, we note that the signature scheme in [43] (marked with '†') is essentially identical to the one in [12] except that an improved security reduction under a weaker assumption was provided in the EUF–naCMA model. As shown in Table 1, the scheme in [6] only has a constant number of "basic" elements in the verification key. However, because a large (I)SIS parameter $\bar{\beta} = \tilde{O}(d^{2d} \cdot n^{5.5})$ is needed (which requires a super-polynomial modulus $q > \bar{\beta}$), the actual bit size to represent each "basic" element in [6] is at least $O(d) = O(\log n)$ times larger than that in [24] and our schemes. Even if we do not take account of the reduction loss, the bit size of the verification key in [6] is already as large as that in [24] and our schemes.

1.4   Identity-Based Encryptions

At STOC 2008, Gentry et al. [29] constructed a variant of the LWE-based public-key encryption (PKE) scheme [47]. Informally, the public key of their scheme [29] contained a matrix $\mathbf{A}$ and a vector $\mathbf{u}$, and the secret key was a short vector $\mathbf{e}$ satisfying $\mathbf{A}\mathbf{e} = \mathbf{u}$. Recall that in our generic signature scheme $\mathcal{SIG}$, any valid message-signature pair (M, σ) under the verification key $vk = (\mathbf{A}, \mathbf{u}, K)$ also satisfies an equation $\mathbf{A}_M\sigma = \mathbf{u}$, where $\mathbf{A}_M = (\mathbf{A}\|\mathrm{H}_K(M))$. A natural question is whether we can construct a generic IBE scheme from lattice-based PHFs by combining our generic signature scheme $\mathcal{SIG}$ with the PKE scheme in [29]. Concretely, let the master public key mpk and the master secret key msk of the IBE system be the verification key vk and the secret signing key sk of $\mathcal{SIG}$, respectively, i.e., $(mpk, msk) = (vk, sk)$. Then, for each identity id, we simply generate a "signature" $sk_{id} = \sigma$ on id under the master public key mpk as the user private key, i.e., $\mathbf{A}_{id}sk_{id} = \mathbf{u}$, where $\mathbf{A}_{id} = (\mathbf{A}\|\mathrm{H}_K(id))$. Finally, we run the encryption algorithm of [29] with "public key" $(\mathbf{A}_{id}, \mathbf{u})$ as a sub-routine to encrypt plaintexts under the identity id. The problem is that we do not know how to rely the security of the above "IBE" scheme on the LWE assumption.

Fortunately, the work [1] suggested a solution by adding an "artificial" noise in the ciphertext, which was later used in other advanced lattice-based encryption schemes such as functional encryptions [3]. To adapt their techniques to the above IBE construction, the challenge ciphertext $\mathbf{C}^*$ under identity id^* must contain a term $\mathbf{R}_{id^*}^t\mathbf{w}$ for some $\mathbf{w} \in \mathbb{Z}_q^{\bar{m}}$, where $\mathrm{H}_{K'}(id^*) = \mathbf{A}\mathbf{R}_{id^*}$ (i.e., $\mathbf{S}_{id^*} = \mathbf{0}$) for some trapdoor mode key K'. This means that $\mathbf{C}^*$ will leak some information of $\mathbf{R}_{id^*}$, which is not captured by our definition of lattice-based PHF, and might compromise the security of $\mathcal{H}$. An intuitive solution is directly resorting to an enhanced definition of PHF such that all the properties of $\mathcal{H}$ still hold even when the information of $\mathbf{R}_{id^*}^t\mathbf{w}$ (for any given $\mathbf{w}$) is leaked. For our particular generic construction of IBE, we can handle it more skillfully by introducing two seemingly relaxed conditions: (1) the PHF key K' in the trapdoor mode is still statistically close to the key K in the normal mode even conditioned on $(\mathbf{A}$ and) $\mathbf{R}_{id^*}^t\mathbf{w}$ for any given vector $\mathbf{w} \in \mathbb{Z}_q^{\bar{m}}$; (2) the hidden matrix $\mathbf{R}_{id^*}$ has high min-entropy in the sense that $\mathbf{R}_{id^*}^t\mathbf{w}$ (conditioned on $\mathbf{w}$) is statistically close to uniform over $\mathbb{Z}_q^m$ when $\mathbf{w} \in \mathbb{Z}_q^{\bar{m}}$ is uniformly random. Formally, we say that a PHF $\mathcal{H}$ has *high min-entropy* if it additionally satisfies the above two conditions. Intuitively, the high min-entropy property ensures that when $\mathbf{w}$ is uniformly random, $\mathbf{R}_{id^*}^t\mathbf{w}$ statistically leaks no information of $\mathbf{R}_{id^*}$, and thus will not affect the original PHF property of $\mathcal{H}$. In the security proof, we will make use of this fact by switching $\mathbf{w}$ to a uniformly random one under the LWE assumption. Interestingly, by choosing appropriate parameters, all our PHF constructions satisfy the high min-entropy property. In other words, such a property is obtained almost for free, which finally allows us to construct a generic IBE scheme $\mathcal{IBE}$ from lattice-based PHFs with high min-entropy. Similarly, our generic scheme $\mathcal{IBE}$ subsumes the concrete IBE schemes due to Agrawal et al. [1]. Besides, by instantiating $\mathcal{IBE}$ with our efficient Type-II PHF construction, we obtain the first standard model IBE scheme $\mathcal{IBE}_1$ with master public keys consisting of a logarithmic number of matrices. We also show how to extend our IBE scheme to a hierarchical IBE (HIBE) scheme and how to achieve CCA security, by using the trapdoor delegations [1,14,43] and the CHK transformation [13].

In Table 2, we give a (rough) comparison of lattice-based IBEs in the standard model. For simplicity, the identity length is set to be n. (Note that one can use a collision-resistant hash function with output length n to deal with identities with arbitrary length.) Similarly, we compare the size of master public keys and ciphertexts in terms of the number of "basic" elements. On general lattices, the

Table 2. Rough comparison of lattice-based IBEs in the standard model (Since all the schemes only have a single "basic" element in both the master secret key and the user private key, we omit them in the comparison for succinctness. The reduction loss is the ratio ϵ/ϵ' between the success probability ϵ of the attacker and the success probability ϵ' of the reduction. Real α is the parameter for the LWE problem, and "security" standards for the corresponding security model for security proofs)

Schemes	Master public key	Ciphertext	Reduction loss	LWE param $1/\alpha$	Security
ABB10a [2]	n^3	n^2	1	$\tilde{O}(n^{2n})$	Selective
ABB10b [1]	$1, n$	1	$1, Q$	$\tilde{O}(n^2)$	Selective, Full
CHKP10 [14]	n	n	Q^2	$\tilde{O}(n^{1.5})$	Full
Our $\mathcal{IBE}_1$	$\log n$	1	$n \cdot Q^2$	$Q^2 \cdot \tilde{O}(n^{6.5})$	Full

"basic" element in the master public keys is a matrix, while the "basic" element in the ciphertexts is a vector. If instantiated from ideal lattices, the "basic" element in the master public keys can be represented by a vector, and thus roughly saves a factor of n in the master public key size. We ignore the constant factor in the table to avoid clutter. Compared to the two fully secure IBEs [1,14] in the standard model, our concrete scheme $\mathcal{IBE}_1$ only has a logarithmic number of matrices in the master public key. However, such an improvement is not obtained without a penalty: the instantiated scheme $\mathcal{IBE}_1$ has a large security loss and requires a strong LWE assumption. Since both the improvement and the downside are inherited from the concrete Type-II PHF construction, this situation can be immediately changed if one can find a better lattice-based PHF.

1.5   Other Related Work

Hofheinz and Kiltz [33] first introduced the notion of PHF based on group hash functions, and gave a concrete $(2,1)$-PHF instantiation. Then, the work [32] constructed a $(u,1)$-PHF for any $u \geq 1$ by using cover-free sets. Later, Yamada et al. [51] reduced the key size from $O(u^2\ell)$ in [32] to $O(u\sqrt{\ell})$ by combining the two-dimensional representation of cover-free sets with the bilinear groups, where ℓ was the bit size of the inputs. At CRYPTO 2012, Hanaoka et al. [31] showed that it was impossible to construct *algebraic* $(u,1)$-PHF over prime order groups in a black-box way such that its key has less than u group elements.[2] Later, Freire et al. [26] got around the impossibility result of [31] and constructed a $(\text{poly},1)$-PHF by adapting PHFs to the multilinear maps setting. Despite its great theoretical interests, the current state of multilinear maps might be a big obstacle in any attempt to securely and efficiently instantiate the PHFs in [26]. More recently, Catalano et al. [15] introduced a variant of traditional PHF called

---

[2] Informally, an algorithm is algebraic if there is way to compute the representation of a group element output by the algorithm in terms of its input group elements [11].

asymmetric PHF over bilinear maps, and used it to construct (homomorphic) signature schemes with short verification keys.

All the above PHF constructions [15,26,32,33,51] seem specific to groups with nice properties, which might constitute a main barrier to instantiate them from lattices. Although several lattice-based schemes [1,14] had employed a similar partitioning proof trick as that was captured by the traditional PHFs, it was still an open problem to formalize and construct PHFs from lattices [34]. We put forward this study by introducing the lattice-based PHF and demonstrate its power in constructing lattice-based signatures and IBEs in the standard model. Our PHFs also provide a modular way to investigate several existing cryptographic constructions from lattices [1,12,43].

1.6 Roadmap

After some preliminaries in Sect. 2, we give the definition of lattice-based PHFs, and two types of constructions in Sect. 3. We construct signatures and IBEs from lattice-based PHFs in Sects. 4 and 5, respectively.

2 Preliminaries

2.1 Notation

Let κ be the natural security parameter, and all other quantities are implicitly dependent on κ. The function $\log_c$ denotes the logarithm with base c, and we use log to denote the natural logarithm. The standard notation O, ω are used to classify the growth of functions. If $f(n) = O(g(n) \cdot \log^c(n))$ for some constant c, we write $f(n) = \tilde{O}(g(n))$. By poly(n) we denote an arbitrary function $f(n) = O(n^c)$ for some constant c. A function $f(n)$ is negligible in n if for every positive c, we have $f(n) < n^{-c}$ for sufficiently large n. By negl(n) we denote an arbitrary negligible function. A probability is said to be overwhelming if it is $1 - \text{negl}(n)$. The notation $\leftarrow_r$ denotes randomly choosing elements from some distribution (or the uniform distribution over some finite set). If a random variable x follows some distribution D, we denote it by $x \backsim D$.

By $\mathbb{R}$ (resp. $\mathbb{Z}$) we denote the set of real numbers (resp. integers). For any positive $N \in \mathbb{Z}$, the notation $[N]$ denotes the set $\{0, 1, \ldots, N-1\}$. Vectors are used in the column form and denoted by bold lower-case letters (e.g., $\mathbf{x}$). Matrices are treated as the sets of column vectors and denoted by bold capital letters (e.g., $\mathbf{X}$). The concatenation of the columns of $\mathbf{X} \in \mathbb{R}^{n \times m}$ followed by the columns of $\mathbf{Y} \in \mathbb{R}^{n \times m'}$ is denoted as $(\mathbf{X} \| \mathbf{Y}) \in \mathbb{R}^{n \times (m+m')}$. For any element $0 \leq v \leq q$, we denote $\text{BitDecomp}_q(v) \in \{0,1\}^k$ as the k-dimensional bit-decomposition of v, where $k = \lceil \log_2 q \rceil$. By $\|\cdot\|$ and $\|\cdot\|_\infty$ we denote the l_2 and l_∞ norm, respectively. The norm of a matrix $\mathbf{X}$ is defined as the norm of its longest column (i.e., $\|\mathbf{X}\| = \max_i \|\mathbf{x}_i\|$). The largest singular value of a matrix $\mathbf{X}$ is $s_1(\mathbf{X}) = \max_{\mathbf{u}} \|\mathbf{X}\mathbf{u}\|$, where the maximum is taken over all unit vectors $\mathbf{u}$.

We say that a hash function $H : \mathbb{Z}_q^n \to \mathbb{Z}_q^{n \times n}$ is an encoding with full-rank differences (FRD) if the following two conditions hold: (1) for any $\mathbf{u} \neq \mathbf{v}$, the matrix $H(\mathbf{u}) - H(\mathbf{v}) \in \mathbb{Z}_q^{n \times n}$ is invertible over $\mathbb{Z}_q^{n \times n}$; and (2) H is computable in polynomial time in $n \log q$. As shown in [1,20], FRD encodings supporting the exponential size domain $\mathbb{Z}_q^n$ can be efficiently constructed.

2.2 Lattices and Gaussian Distributions

An m-dimensional full-rank lattice $\Lambda \subset \mathbb{R}^m$ is the set of all integral combinations of m linearly independent vectors $\mathbf{B} = (\mathbf{b}_1, \ldots, \mathbf{b}_m) \in \mathbb{R}^{m \times m}$, i.e., $\Lambda = \mathcal{L}(\mathbf{B}) = \{\sum_{i=1}^m x_i \mathbf{b}_i : x_i \in \mathbb{Z}\}$. For $\mathbf{x} \in \Lambda$, define the Gaussian function $\rho_{s,\mathbf{c}}(\mathbf{x})$ over $\Lambda \subseteq \mathbb{Z}^m$ centered at $\mathbf{c} \in \mathbb{R}^m$ with parameter $s > 0$ as $\rho_{s,\mathbf{c}}(\mathbf{x}) = \exp(-\pi \|\mathbf{x} - \mathbf{c}\|^2 / s^2)$. Let $\rho_{s,\mathbf{c}}(\Lambda) = \sum_{\mathbf{x} \in \Lambda} \rho_{s,\mathbf{c}}(\mathbf{x})$, and define the discrete Gaussian distribution over Λ as $D_{\Lambda,s,\mathbf{c}}(\mathbf{y}) = \frac{\rho_{s,\mathbf{c}}(\mathbf{y})}{\rho_{s,\mathbf{c}}(\Lambda)}$, where $\mathbf{y} \in \Lambda$. The subscripts s and $\mathbf{c}$ are taken to be 1 and $\mathbf{0}$ (resp.) when omitted. The following result was proved in [29,44,46].

Lemma 1. *For any positive integer $m \in \mathbb{Z}$, vector $\mathbf{y} \in \mathbb{Z}^m$ and large enough $s \geq \omega(\sqrt{\log m})$, we have that*

$$\Pr_{\mathbf{x} \leftarrow_r D_{\mathbb{Z}^m,s}} [\|\mathbf{x}\| > s\sqrt{m}] \leq 2^{-m} \quad \text{and} \quad \Pr_{\mathbf{x} \leftarrow_r D_{\mathbb{Z}^m,s}} [\mathbf{x} = \mathbf{y}] \leq 2^{1-m}.$$

Following [24,43], we say that a random variable X over $\mathbb{R}$ is subgaussian with parameter s if for all $t \in \mathbb{R}$, the (scaled) moment-generating function satisfies $\mathbb{E}(\exp(2\pi t X)) \leq \exp(\pi s^2 t^2)$. If X is subgaussian, then its tails are dominated by a Gaussian of parameter s, i.e., $\Pr[|X| \geq t] \leq 2\exp(-\pi t^2/s^2)$ for all $t \geq 0$. As a special case, any B-bounded symmetric random variable X (i.e., $|X| \leq B$ always) is subgaussian with parameter $B\sqrt{2\pi}$. Besides, we say that a random matrix $\mathbf{X}$ is subgaussian with parameter s if all its one-dimensional marginals $\mathbf{u}^t \mathbf{X} \mathbf{v}$ for unit vectors $\mathbf{u}, \mathbf{v}$ are subgaussian with parameter s. In such a definition, the concatenation of independent subgaussian vectors with parameter s, interpreted either as a vector or as a matrix, is subgaussian with parameter s. In particular, the distribution $D_{\Lambda,s}$ for any lattice $\Lambda \subset \mathbb{R}^n$ and $s > 0$ is subgaussian with parameter s. For random subgaussian matrix, we have the following result from the non-asymptotic theory of random matrices [49].

Lemma 2. *Let $\mathbf{X} \in \mathbb{R}^{n \times m}$ be a random subgaussian matrix with parameter s. There exists a universal constant $C \approx 1/\sqrt{2\pi}$ such that for any $t \geq 0$, we have $s_1(\mathbf{X}) \leq C \cdot s \cdot (\sqrt{m} + \sqrt{n} + t)$ except with probability at most $2\exp(-\pi t^2)$.*

Let $\mathbf{A} \in \mathbb{Z}_q^{n \times m}$ be a matrix for some positive $n, m, q \in \mathbb{Z}$, consider the following two lattices: $\Lambda_q^\perp(\mathbf{A}) = \{\mathbf{e} \in \mathbb{Z}^m \text{ s.t. } \mathbf{Ae} = \mathbf{0} \mod q\}$ and $\Lambda_q(\mathbf{A}) = \{\mathbf{y} \in \mathbb{Z}^m \text{ s.t. } \exists \mathbf{s} \in \mathbb{Z}^n, \mathbf{A}^t\mathbf{s} = \mathbf{y} \mod q\}$. By definition, we have $\Lambda_q^\perp(\mathbf{A}) = \Lambda_q^\perp(\mathbf{CA})$ for any invertible $\mathbf{C} \in \mathbb{Z}_q^{n \times n}$. In 1999, Ajtai [5] proposed the first trapdoor generation algorithm to output an essentially uniform trapdoor matrix $\mathbf{A}$ that allows to efficiently sample short vectors in $\Lambda_q^\perp(\mathbf{A})$. This trapdoor generation algorithm had been improved in [43]. Let $\mathbf{I}_n$ be the $n \times n$ identity matrix. We now recall the publicly known trapdoor matrix $\mathbf{G}$ in [43]. Formally, for any prime $q > 2$, integer $n \geq 1$ and $k = \lceil \log_2 q \rceil$, define $\mathbf{g} = (1, 2, \ldots, 2^{k-1})^t \in \mathbb{Z}_q^k$ and

$\mathbf{G} = \mathbf{I}_n \otimes \mathbf{g}^t \in \mathbb{Z}_q^{n \times nk}$, where '$\otimes$' represents the tensor product.[3] Then, the lattice $\Lambda_q^{\perp}(\mathbf{G})$ has a publicly known short basis $\mathbf{T} = \mathbf{I}_n \otimes \mathbf{T}_k \in \mathbb{Z}^{nk \times nk}$ with $\|\mathbf{T}\| \le \max\{\sqrt{5}, \sqrt{k}\}$. Let $(q_0, q_1, \dots, q_{k-1}) = \mathsf{BitDecomp}_q(q) \in \{0,1\}^k$, we have

$$
\mathbf{G} = \begin{pmatrix} \cdots \mathbf{g}^t \cdots & & & \\ & \cdots \mathbf{g}^t \cdots & & \\ & & \ddots & \\ & & & \cdots \mathbf{g}^t \cdots \end{pmatrix} \qquad \mathbf{T}_k = \begin{pmatrix} 2 & & & & q_0 \\ -1 & 2 & & & q_1 \\ & -1 & & & q_2 \\ & & \ddots & & \vdots \\ & & & 2 & q_{k-2} \\ & & & -1 & q_{k-1} \end{pmatrix}
$$

For any vector $\mathbf{u} \in \mathbb{Z}_q^n$, the basis $\mathbf{T} = \mathbf{I}_n \otimes \mathbf{T}_k \in \mathbb{Z}_q^{nk \times nk}$ can be used to sample short vector $\mathbf{e} \sim D_{\mathbb{Z}^{nk}, s}$ satisfying $\mathbf{Ge} = \mathbf{u}$ for any $s \ge \omega(\sqrt{\log n})$ in quasilinear time. Besides, one can deterministically compute a short vector $\mathbf{v} = \mathbf{G}^{-1}(\mathbf{u}) \in \{0,1\}^{nk}$ such that $\mathbf{Gv} = \mathbf{u}$. This fact will be frequently used in this paper.

Definition 1 (G-trapdoor [43]). *For any integers* $n, \bar{m}, q \in \mathbb{Z}, k = \lceil \log_2 q \rceil$, *and matrix* $\mathbf{A} \in \mathbb{Z}_q^{n \times \bar{m}}$, *the* **G-trapdoor** *for* $\mathbf{A}$ *is a matrix* $\mathbf{R} \in \mathbb{Z}^{(\bar{m}-nk) \times nk}$ *such that* $\mathbf{A} \begin{bmatrix} \mathbf{R} \\ \mathbf{I}_{nk} \end{bmatrix} = \mathbf{SG}$ *for some invertible tag* $\mathbf{S} \in \mathbb{Z}_q^{n \times n}$. *The quality of the trapdoor is measured by its largest singular value* $s_1(\mathbf{R})$.

If $\mathbf{R}$ is a G-trapdoor for $\mathbf{A}$, one can obtain a G-trapdoor $\mathbf{R}'$ for any extension $(\mathbf{A} \| \mathbf{B})$ by padding $\mathbf{R}$ with zero rows. In particular, we have $s_1(\mathbf{R}') = s_1(\mathbf{R})$. Besides, the rows of $\begin{bmatrix} \mathbf{R} \\ \mathbf{I}_{nk} \end{bmatrix}$ in Definition 1 can appear in any order, since this just induces a permutation of $\mathbf{A}$'s columns [43].

Proposition 1 [43]. *Given any integers* $n \ge 1$, $q > 2$, *sufficiently large* $\bar{m} = O(n \log q)$ *and a tag* $\mathbf{S} \in \mathbb{Z}_q^{n \times n}$, *there is an efficient randomized algorithm* $\mathsf{TrapGen}(1^n, 1^{\bar{m}}, q, \mathbf{S})$ *that outputs a matrix* $\mathbf{A} \in \mathbb{Z}_q^{n \times \bar{m}}$ *and a* **G-trapdoor** $\mathbf{R} \in \mathbb{Z}_q^{(\bar{m}-nk) \times nk}$ *with quality* $s_1(\mathbf{R}) \le \sqrt{\bar{m}} \cdot \omega(\sqrt{\log n})$ *such that the distribution of* $\mathbf{A}$ *is* negl(n)*-far from uniform and* $\mathbf{A} \begin{bmatrix} \mathbf{R} \\ \mathbf{I}_{nk} \end{bmatrix} = \mathbf{SG}$, *where* $k = \lceil \log_2 q \rceil$.

In addition, given a **G-trapdoor** $\mathbf{R}$ *of* $\mathbf{A} \in \mathbb{Z}_q^{n \times \bar{m}}$ *for some invertible tag* $\mathbf{S} \in \mathbb{Z}_q^{n \times n}$, *any* $\mathbf{U} \in \mathbb{Z}_q^{n \times n'}$ *for some integer* $n' \ge 1$ *and real* $s \ge s_1(\mathbf{R}) \cdot \omega(\sqrt{\log n})$, *there is an algorithm* $\mathsf{SampleD}(\mathbf{R}, \mathbf{A}, \mathbf{S}, \mathbf{U}, s)$ *that samples from a distribution within* negl(n) *statistical distance of* $\mathbf{E} \sim (D_{\mathbb{Z}^{\bar{m}}, s})^{n'}$ *satisfying* $\mathbf{AE} = \mathbf{U}$.

We also need the following useful facts from [29,43,46].

Lemma 3. *For any positive integer* n, *prime* $q > 2$, *sufficiently large* $m = O(n \log q)$ *and real* $s \ge \omega(\sqrt{\log m})$, *we have that for a uniformly random matrix* $\mathbf{A} \leftarrow_r \mathbb{Z}_q^{n \times m}$, *the following facts hold:*

---

[3] One can define $\mathbf{G}$ by using any base $b \ge 2$ and $\mathbf{g} = (1, b, \dots, b^{k-1})^t$ for $k = \lceil \log_b q \rceil$. In this paper, we fix $b = 2$ for simplicity.

- *for variable* $\mathbf{e} \sim D_{\mathbb{Z}^m, s}$, *the distribution of* $\mathbf{u} = \mathbf{A}\mathbf{e}$ mod q *is statistically close to uniform over* $\mathbb{Z}_q^n$;
- *for any* $\mathbf{c} \in \mathbb{R}^m$ *and every* $\mathbf{y} \in \Lambda_q^\perp(\mathbf{A})$, $\Pr_{\mathbf{x} \leftarrow_r D_{\Lambda_q^\perp(\mathbf{A}), s, \mathbf{c}}}[\mathbf{x} = \mathbf{y}] \leq 2^{1-m}$;
- *for any fixed* $\mathbf{u} \in \mathbb{Z}_q^n$ *and arbitrary* $\mathbf{v} \in \mathbb{R}^m$ *satisfying* $\mathbf{A}\mathbf{v} = \mathbf{u}$ mod q, *the conditional distribution of* $\mathbf{e} \sim D_{\mathbb{Z}^m, s}$ *given* $\mathbf{A}\mathbf{e} = \mathbf{u}$ mod q *is exactly* $\mathbf{v} + D_{\Lambda_q^\perp(\mathbf{A}), s, -\mathbf{v}}$.

2.3   Learning with Errors (LWE) and Small Integer Solutions (SIS)

For any positive integer n, q, real $\alpha > 0$, and any vector $\mathbf{s} \in \mathbb{Z}_q^n$, the distribution $A_{\mathbf{s}, \alpha}$ over $\mathbb{Z}_q^n \times \mathbb{Z}_q$ is defined as $A_{\mathbf{s}, \alpha} = \{(\mathbf{a}, \mathbf{a}^t \mathbf{s} + x \mod q) : \mathbf{a} \leftarrow_r \mathbb{Z}_q^n, x \leftarrow_r D_{\mathbb{Z}, \alpha q}\}$, where $D_{\mathbb{Z}, \alpha q}$ is the discrete Gaussian distribution over $\mathbb{Z}$ with parameter αq. For m independent samples $(\mathbf{a}_1, y_1), \ldots, (\mathbf{a}_m, y_m)$ from $A_{\mathbf{s}, \alpha}$, we denote it in matrix form $(\mathbf{A}, \mathbf{y}) \in \mathbb{Z}_q^{n \times m} \times \mathbb{Z}_q^m$, where $\mathbf{A} = (\mathbf{a}_1, \ldots, \mathbf{a}_m)$ and $\mathbf{y} = (y_1, \ldots, y_m)^t$. We say that an algorithm solves the $\mathrm{LWE}_{q, \alpha}$ problem if, for uniformly random $\mathbf{s} \leftarrow_r \mathbb{Z}_q^n$, given polynomial samples from $A_{\mathbf{s}, \alpha}$ it outputs $\mathbf{s}$ with noticeable probability. The decisional variant of LWE is that, for a uniformly random $\mathbf{s} \leftarrow_r \mathbb{Z}_q^n$, the solving algorithm is asked to distinguish $A_{\mathbf{s}, \alpha}$ from the uniform distribution over $\mathbb{Z}_q^n \times \mathbb{Z}_q$ (with only polynomial samples). For certain modulus q, the average-case decisional LWE problem is polynomially equivalent to its worst-case search version [47].

Proposition 2 [47]. *Let* $\alpha = \alpha(n) \in (0, 1)$ *and let* $q = q(n)$ *be a prime such that* $\alpha q > 2\sqrt{n}$. *If there exists an efficient (possibly quantum) algorithm that solves* $\mathrm{LWE}_{q, \alpha}$, *then there exists an efficient quantum algorithm for approximating* SIVP *(in the* l_2 *norm) on* n-*dimensional lattices, in the worst case, to within* $\tilde{O}(n/\alpha)$ *factors.*

The Small Integer Solution (SIS) problem was first introduced by Ajtai [4]. Formally, given positive $n, m, q \in \mathbb{Z}$, a real $\beta > 0$, and a uniformly random matrix $\mathbf{A} \in \mathbb{Z}_q^{n \times m}$, the $\mathrm{SIS}_{q, m, \beta}$ problem asks to find a non-zero vector $\mathbf{e} \in \mathbb{Z}^m$ such that $\mathbf{A}\mathbf{e} = \mathbf{0}$ mod q and $\|\mathbf{e}\| \leq \beta$. In [29], Gentry et al. introduced the ISIS problem, which was an inhomogeneous variant of SIS. Specifically, given an extra random syndrome $\mathbf{u} \in \mathbb{Z}_q^n$, the $\mathrm{ISIS}_{q, m, \beta}$ problem asks to find a vector $\mathbf{e} \in \mathbb{Z}^m$ such that $\mathbf{A}\mathbf{e} = \mathbf{u}$ mod q and $\|\mathbf{e}\| \leq \beta$. Both the two problems were shown to be as hard as certain worst-case lattice problems [29].

Proposition 3 [29]. *For any polynomially bounded* $m, \beta = poly(n)$ *and prime* $q \geq \beta \cdot \omega(\sqrt{n \log n})$, *the average-case problems* $\mathrm{SIS}_{q, m, \beta}$ *and* $\mathrm{ISIS}_{q, m, \beta}$ *are as hard as approximating* SIVP *on* n-*dimensional lattices, in the worst case, to within certain* $\gamma = \beta \cdot \tilde{O}(\sqrt{n})$ *factors.*

3   Programmable Hash Functions from Lattices

We now give the definition of lattice-based programmable hash function (PHF). Let $\ell, \bar{m}, m, n, q, u, v \in \mathbb{Z}$ be some polynomials in the security parameter κ. By $\mathcal{I}_n$

we denote the set of invertible matrices in $\mathbb{Z}_q^{n \times n}$. A hash function $\mathcal{H} : \mathcal{X} \to \mathbb{Z}_q^{n \times m}$ consists of two algorithms $(\mathcal{H}.\mathrm{Gen}, \mathcal{H}.\mathrm{Eval})$. Given the security parameter κ, the probabilistic polynomial time (PPT) key generation algorithm $\mathcal{H}.\mathrm{Gen}(1^\kappa)$ outputs a key K, i.e., $K \leftarrow \mathcal{H}.\mathrm{Gen}(1^\kappa)$. For any input $X \in \mathcal{X}$, the efficiently deterministic evaluation algorithm $\mathcal{H}.\mathrm{Eval}(K, X)$ outputs a hash value $\mathbf{Z} \in \mathbb{Z}_q^{n \times m}$, i.e., $\mathbf{Z} = \mathcal{H}.\mathrm{Eval}(K, X)$. For simplicity, we write $\mathrm{H}_K(X) = \mathcal{H}.\mathrm{Eval}(K, X)$.

Definition 2 (Lattice-Based Programmable Hash Function). *A hash function $\mathcal{H} : \mathcal{X} \to \mathbb{Z}_q^{n \times m}$ is a $(u, v, \beta, \gamma, \delta)$-PHF if there exist a PPT trapdoor key generation algorithm $\mathcal{H}.\mathrm{TrapGen}$ and an efficiently deterministic trapdoor evaluation algorithm $\mathcal{H}.\mathrm{TrapEval}$ such that given a uniformly random $\mathbf{A} \in \mathbb{Z}_q^{n \times \bar{m}}$ and a (public) trapdoor matrix $\mathbf{B} \in \mathbb{Z}_q^{n \times m}$,[4] the following properties hold:*

Syntax: *The PPT algorithm $(K', td) \leftarrow \mathcal{H}.\mathrm{TrapGen}(1^\kappa, \mathbf{A}, \mathbf{B})$ outputs a key K' together with a trapdoor td. Moreover, for any input $X \in \mathcal{X}$, the deterministic algorithm $(\mathbf{R}_X, \mathbf{S}_X) = \mathcal{H}.\mathrm{TrapEval}(td, K', X)$ returns $\mathbf{R}_X \in \mathbb{Z}_q^{\bar{m} \times m}$ and $\mathbf{S}_X \in \mathbb{Z}_q^{n \times n}$ such that $s_1(\mathbf{R}_X) \leq \beta$ and $\mathbf{S}_X \in \mathcal{I}_n \cup \{\mathbf{0}\}$ hold with overwhelming probability over the trapdoor td that is produced along with K'.*

Correctness: *For all possible $(K', td) \leftarrow \mathcal{H}.\mathrm{TrapGen}(1^\kappa, \mathbf{A}, \mathbf{B})$, all $X \in \mathcal{X}$ and its corresponding $(\mathbf{R}_X, \mathbf{S}_X) = \mathcal{H}.\mathrm{TrapEval}(td, K', X)$, we have $\mathrm{H}_{K'}(X) = \mathcal{H}.\mathrm{Eval}(K', X) = \mathbf{A}\mathbf{R}_X + \mathbf{S}_X\mathbf{B}$.*

Statistically Close Trapdoor Keys: *For all $(K', td) \leftarrow \mathcal{H}.\mathrm{TrapGen}(1^\kappa, \mathbf{A}, \mathbf{B})$ and $K \leftarrow \mathcal{H}.\mathrm{Gen}(1^\kappa)$, the statistical distance between $(\mathbf{A}, K')$ and $(\mathbf{A}, K)$ is at most γ.*

Well-distributed Hidden Matrices: *For all $(K', td) \leftarrow \mathcal{H}.\mathrm{TrapGen}(1^\kappa, \mathbf{A}, \mathbf{B})$, any inputs $X_1, \ldots, X_u, Y_1, \ldots, Y_v \in \mathcal{X}$ such that $X_i \neq Y_j$ for any i, j, let $(\mathbf{R}_{X_i}, \mathbf{S}_{X_i}) = \mathcal{H}.\mathrm{TrapEval}(td, K', X_i)$ and $(\mathbf{R}_{Y_i}, \mathbf{S}_{Y_i}) = \mathcal{H}.\mathrm{TrapEval}(td, K', Y_i)$. Then, we have that*

$$\Pr[\mathbf{S}_{X_1} = \cdots = \mathbf{S}_{X_u} = \mathbf{0} \wedge \mathbf{S}_{Y_1}, \ldots, \mathbf{S}_{Y_v} \in \mathcal{I}_n] \geq \delta,$$

where the probability is over the trapdoor td produced along with K'.

If γ is negligible and $\delta > 0$ is noticeable, we simply say that $\mathcal{H}$ is a (u, v, β)-PHF. Furthermore, if u (resp. v) is an arbitrary polynomial in κ, we say that $\mathcal{H}$ is a $(\mathrm{poly}, v, \beta)$-PHF (resp. $(u, \mathrm{poly}, \beta)$-PHF).

A **weak programmable hash function** is a relaxed version of PHF, where the $\mathcal{H}.\mathrm{TrapGen}$ algorithm additionally takes a list $X_1, \ldots, X_u \in \mathcal{X}$ as inputs such that the well-distributed hidden matrices property holds in the following sense: For all $(K', td) \leftarrow \mathcal{H}.\mathrm{TrapGen}(1^\kappa, \mathbf{A}, \mathbf{B}, \{X_1, \ldots, X_u\})$, any inputs $Y_1, \ldots, Y_v \in \mathcal{X}$ such that $Y_j \notin \{X_1, \ldots, X_u\}$ for all j, let $(\mathbf{R}_{X_i}, \mathbf{S}_{X_i}) = \mathcal{H}.\mathrm{TrapEval}(td, K', X_i)$ and $(\mathbf{R}_{Y_i}, \mathbf{S}_{Y_i}) = \mathcal{H}.\mathrm{TrapEval}(td, K', Y_i)$, we have that $\Pr[\mathbf{S}_{X_1} = \cdots = \mathbf{S}_{X_u} = \mathbf{0} \wedge \mathbf{S}_{Y_1}, \ldots, \mathbf{S}_{Y_v} \in \mathcal{I}_n] \geq \delta$, where the probability is over the trapdoor td produced along with K'.

---

[4] A general trapdoor matrix $\mathbf{B}$ is used for utmost generality, but the publicly known trapdoor matrix $\mathbf{B} = \mathbf{G}$ in [43] is recommended for both efficiency and simplicity.

Besides, a hash function $\mathcal{H} : \mathcal{X} \to \mathbb{Z}_q^{n \times m}$ can be a (weak) (u, v, β)-PHF for different parameters u and v, since there might exist different pairs of trapdoor key generation and trapdoor evaluation algorithms for $\mathcal{H}$. If this is the case, one can easily show that the keys output by these trapdoor key generation algorithms are statistically indistinguishable by definition.

3.1   Type-I Construction

We describe the Type-I construction of lattice-based PHFs in the following.

Definition 3. *Let $\ell, n, m, q \in \mathbb{Z}$ be some polynomials in the security parameter κ. Let E be a deterministic encoding from $\mathcal{X}$ to $(\mathbb{Z}_q^{n \times n})^\ell$, the hash function $\mathcal{H} = (\mathcal{H}.\mathrm{Gen}, \mathcal{H}.\mathrm{Eval})$ with key space $\mathcal{K} \subseteq (\mathbb{Z}_q^{n \times m})^{\ell+1}$ is defined as follows:*

- *$\mathcal{H}.\mathrm{Gen}(1^\kappa)$: Randomly choose $(\mathbf{A}_0, \dots, \mathbf{A}_\ell) \leftarrow_r \mathcal{K}$, return $K = \{\mathbf{A}_i\}_{i \in \{0, \dots, \ell\}}$.*
- *$\mathcal{H}.\mathrm{Eval}(K, X)$: Let $\mathrm{E}(X) = (\mathbf{C}_1, \dots, \mathbf{C}_\ell)$, return $\mathbf{Z} = \mathbf{A}_0 + \sum_{i=1}^{\ell} \mathbf{C}_i \mathbf{A}_i$.*

We note that the above hash function has actually been (implicitly) used to construct both signatures (e.g., [8,12,45]) and encryptions (e.g., [1,43]). Let $\mathbf{I}_n$ be the $n \times n$ identity matrix. In the following theorems, we summarize several known results which were implicitly proved in [1,12,43].

Theorem 1. *Let $\mathcal{K} = (\mathbb{Z}_q^{n \times m})^{\ell+1}$ and $\mathcal{X} = \{0,1\}^\ell$. In addition, given an input $X = (X_1, \dots, X_\ell) \in \mathcal{X}$, the encoding function $\mathrm{E}(X)$ returns $\mathbf{C}_i = (-1)^{X_i} \cdot \mathbf{I}_n$ for $i = \{1, \dots, \ell\}$. Then, for large enough integer $\bar{m} = O(n \log q)$ and any fixed polynomial $v = v(\kappa) \in \mathbb{Z}$, the instantiated hash function $\mathcal{H}$ of Definition 3 is a $(1, v, \beta, \gamma, \delta)$-PHF with $\beta \le \sqrt{\ell \bar{m}} \cdot \omega(\sqrt{\log n})$, $\gamma = \mathrm{negl}(\kappa)$ and $\delta = \frac{1}{q^t}(1 - \frac{v}{q^t})$, where t is the smallest integer satisfying $q^t > 2v$.*

Theorem 2. *For large enough $\bar{m} = O(n \log q)$, the hash function $\mathcal{H}$ given in Definition 3 is a weak $(1, \mathrm{poly}, \beta, \gamma, \delta)$-PHF with $\beta \le \sqrt{\ell \bar{m}} \cdot \omega(\sqrt{\log n})$, $\gamma = \mathrm{negl}(\kappa)$, and $\delta = 1$ when instantiated as follows:*

- *Let $\mathcal{K} = (\mathbb{Z}_q^{n \times m})^2$ (i.e., $\ell = 1$) and $\mathcal{X} = \mathbb{Z}_q^n$. Given an input $X \in \mathcal{X}$, the encoding $\mathrm{E}(X)$ returns $H(X)$ where $H : \mathbb{Z}_q^n \to \mathbb{Z}_q^{n \times n}$ is an FRD encoding.*
- *Let $\mathcal{K} = (\mathbb{Z}_q^{n \times m})^{\ell+1}$ and $\mathcal{X} = \{0,1\}^\ell$. Given an input $X = (X_1, \dots, X_\ell) \in \mathcal{X}$, the encoding $\mathrm{E}(X)$ returns $\mathbf{C}_i = X_i \cdot \mathbf{I}_n$ for all $i \in \{1, \dots, \ell\}$.*

Unlike the traditional PHFs [15,32,33] where a bigger u is usually better in constructing short signature schemes, our lattice-based PHFs seem more useful when the parameter v is bigger (e.g., a polynomial in κ). There is a simple explanation: although both notions aim at capturing some kind of partitioning proof trick, i.e., each programmed hash value contains a hidden element behaving as a trigger of some prior embedded trapdoors, for traditional PHFs the trapdoor is usually triggered when the hidden element is zero, while in the lattice setting the trapdoor is typically triggered when the hidden element is a non-zero invertible one. This also explains why previous known constructions on lattices (e.g., the instantiations in Theorems 1 and 2) are (weak) $(1, v, \beta)$-PHFs for some polynomial $v \in \mathbb{Z}$ and real $\beta \in \mathbb{R}$.

3.2 Type-II Construction

Let integers $\ell, \bar{m}, n, q, u, v, L, N$ be some polynomials in the security parameter κ, and let $k = \lceil \log_2 q \rceil$. We now exploit the nice property of the publicly known trapdoor matrix $\mathbf{B} = \mathbf{G} \in \mathbb{Z}_q^{n \times nk}$ to construct more efficient PHF from lattices for any $v = \mathrm{poly}(\kappa)$. We begin by first recalling the notion of cover-free sets. Formally, we say that set S does not cover set T if there exists at least one element $t \in T$ such that $t \notin S$. Let $CF = \{CF_X\}_{X \in [L]}$ be a family of subsets of $[N]$. The family CF is said to be v-cover-free over $[N]$ if for any subset $S \subseteq [L]$ of size at most v, then the union $\cup_{X \in S} CF_X$ does not cover CF_Y for all $Y \notin S$. Besides, we say that CF is η-uniform if every subset CF_X in the family $CF = \{CF_X\}_{X \in [L]}$ have size $\eta \in \mathbb{Z}$. Furthermore, there exists an efficient algorithm to generate cover-free sets [25,38]. Formally,

Lemma 4. *There is a deterministic polynomial time algorithm that on inputs integers $L = 2^\ell$ and $v \in \mathbb{Z}$, returns an η-uniform, v-cover-free sets $CF = \{CF_X\}_{X \in [L]}$ over $[N]$, where $N \leq 16v^2\ell$ and $\eta = N/4v$.*

In the following, we use the binary representation of $[N]$ to construct lattice-based PHFs with short keys.

Definition 4. *Let $n, q \in \mathbb{Z}$ be some polynomials in the security parameter κ. For any $\ell, v \in \mathbb{Z}$ and $L = 2^\ell$, let $N \leq 16v^2\ell, \eta \leq 4v\ell$ and $CF = \{CF_X\}_{X \in [L]}$ be defined as in Lemma 4. Let $\mu = \lceil \log_2 N \rceil$ and $k = \lceil \log_2 q \rceil$. Then, the hash function $\mathcal{H} = (\mathcal{H}.\mathrm{Gen}, \mathcal{H}.\mathrm{Eval})$ from $[L]$ to $\mathbb{Z}_q^{n \times nk}$ is defined as follows:*

- *$\mathcal{H}.\mathrm{Gen}(1^\kappa)$: Randomly choose $\hat{\mathbf{A}}, \mathbf{A}_i \leftarrow_r \mathbb{Z}_q^{n \times nk}$ for $i \in \{0, \ldots, \mu - 1\}$, return the key $K = (\hat{\mathbf{A}}, \{\mathbf{A}_i\}_{i \in \{0, \ldots, \mu-1\}})$.*
- *$\mathcal{H}.\mathrm{Eval}(K, X)$: Given $K = (\hat{\mathbf{A}}, \{\mathbf{A}_i\}_{i \in \{0, \ldots, \mu-1\}})$ and integer $X \in [L]$, the algorithm performs the **Procedure I** in Fig. 1 to compute $\mathbf{Z} = \mathrm{H}_K(X)$.*

We now show that for any prior fixed $v = \mathrm{poly}(\kappa)$, the hash function $\mathcal{H}$ given in Definition 4 is a $(1, v, \beta)$-PHF for some polynomially bounded $\beta \in \mathbb{R}$.

Theorem 3. *For any $\ell, v \in \mathbb{Z}$ and $L = 2^\ell$, let $N \leq 16v^2\ell, \eta \leq 4v\ell$ and $CF = \{CF_X\}_{X \in [L]}$ be defined as in Lemma 4. Then, for large enough $\bar{m} = O(n \log q)$, the hash function $\mathcal{H}$ in Definition 4 is a $(1, v, \beta, \gamma, \delta)$-PHF with $\beta \leq \mu v\ell\bar{m}^{1.5} \cdot \omega(\sqrt{\log \bar{m}})$, $\gamma = \mathrm{negl}(\kappa)$ and $\delta = 1/N$, where $\mu = \lceil \log_2 N \rceil$.*
In particular, if we set $\ell = n$ and $v = \omega(\log n)$, then $\beta = \tilde{O}(n^{2.5})$, and the key of $\mathcal{H}$ only consists of $\mu = O(\log n)$ matrices.

Proof. We now construct a pair of trapdoor algorithms for $\mathcal{H}$ as follows:

- *$\mathcal{H}.\mathrm{TrapGen}(1^\kappa, \mathbf{A}, \mathbf{G})$: Given a uniformly random $\mathbf{A} \in \mathbb{Z}_q^{n \times \bar{m}}$ and matrix $\mathbf{G} \in \mathbb{Z}_q^{n \times nk}$ for sufficiently large $\bar{m} = O(n \log q)$, let $s \geq \omega(\sqrt{\log \bar{m}}) \in \mathbb{R}$ satisfy the requirement in Lemma 3. Randomly choose $\hat{\mathbf{R}}, \mathbf{R}_i \leftarrow_r (D_{\mathbb{Z}^{\bar{m}}, s})^{nk}$ for $i \in \{0, \ldots, \mu-1\}$, and an integer $z^* \leftarrow_r [N]$. Let $(b_0^*, \ldots, b_{\mu-1}^*) = \mathrm{BitDecomp}_N(z^*)$, and let c be the number of 1's in the vector $(b_0^*, \ldots, b_{\mu-1}^*)$. Then, compute $\hat{\mathbf{A}} = \mathbf{A}\hat{\mathbf{R}} - (-1)^c \cdot \mathbf{G}$ and $\mathbf{A}_i = \mathbf{A}\mathbf{R}_i + (1 - b_i^*) \cdot \mathbf{G}$. Finally, return the key $K' = (\hat{\mathbf{A}}, \{\mathbf{A}_i\}_{i \in \{0, \ldots, \mu-1\}})$ and the trapdoor $td = (\hat{\mathbf{R}}, \{\mathbf{R}_i\}_{i \in \{0, \ldots, \mu-1\}}, z^*)$.*

Procedure I	Procedure II
$\mathbf{Z} := \hat{\mathbf{A}}$ For all $z \in CF_X$ $\quad (b_0, \ldots, b_{\mu-1}) := \mathsf{BitDecomp}_N(z)$ $\quad \mathbf{B}_z := \mathbf{A}_{\mu-1} - b_{\mu-1} \cdot \mathbf{G}$ $\quad$ For $i = \mu - 2, \ldots, 0$ $\quad\quad \mathbf{B}_z := (\mathbf{A}_i - b_i \cdot \mathbf{G}) \cdot \mathbf{G}^{-1}(\mathbf{B}_z)$ $\quad \mathbf{Z} := \mathbf{Z} + \mathbf{B}_z$ Return $\mathbf{Z}$	$\mathbf{R}_X := \hat{\mathbf{R}}, \mathbf{S}_X := -(-1)^c \cdot \mathbf{I}_n$ For all $z \in CF_X$ $\quad (b_0, \ldots, b_{\mu-1}) := \mathsf{BitDecomp}_N(z)$ $\quad \mathbf{B}_z := \mathbf{A}_{\mu-1} - b_{\mu-1} \cdot \mathbf{G}$ $\quad \mathbf{R}_z := \mathbf{R}_{\mu-1}$ $\quad \mathbf{S}_z := (1 - b_{\mu-1}^* - b_{\mu-1}) \cdot \mathbf{I}_n$ $\quad$ For $i = \mu - 2, \ldots, 0$ $\quad\quad \mathbf{B}_z := (\mathbf{A}_i - b_i \cdot \mathbf{G}) \cdot \mathbf{G}^{-1}(\mathbf{B}_z)$ $\quad\quad \mathbf{R}_z := \mathbf{R}_i \cdot \mathbf{G}^{-1}(\mathbf{B}_z) + (1 - b_i^* - b_i) \cdot \mathbf{R}_z$ $\quad\quad \mathbf{S}_z := (1 - b_i^* - b_i) \cdot \mathbf{S}_z$ $\quad \mathbf{R}_X := \mathbf{R}_X + \mathbf{R}_z, \mathbf{S}_X := \mathbf{S}_X + \mathbf{S}_z$ Return $(\mathbf{R}_X, \mathbf{S}_X)$

Fig. 1. The procedures used in Definition 4 and Theorem 3

- $\mathcal{H}.\mathsf{TrapEval}(td, K', X)$: Given td and an input $X \in [L]$, the algorithm first computes CF_X by Lemma 4. Then, let $(b_0^*, \ldots, b_{\mu-1}^*) = \mathsf{BitDecomp}_N(z^*)$, and perform the **Procedure II** in Fig. 1 to compute $(\mathbf{R}_X, \mathbf{S}_X)$.

Since $s \geq \omega(\sqrt{\log \bar{m}})$ and $\hat{\mathbf{R}}, \mathbf{R}_i \leftarrow_r (D_{\mathbb{Z}^{\bar{m}}, s})^{nk}$, each matrix in the key $K' = (\hat{\mathbf{A}}, \{\mathbf{A}_i\}_{i \in \{0, \ldots, \mu-1\}})$ is statistically close to uniform over $\mathbb{Z}_q^{n \times nk}$ by Lemma 3. Using a standard hybrid argument, it is easy to show that the statistical distance γ between $(\mathbf{A}, K')$ and $(\mathbf{A}, K)$ is negligible, where $K \leftarrow \mathcal{H}.\mathsf{Gen}(1^\kappa)$. In particular, this means that z^* is statistically hidden in K'.

For correctness, we first show that $\mathbf{B}_z = \mathbf{A}\mathbf{R}_z + \mathbf{S}_z\mathbf{G}$ always holds during the computation. By definition, we have that $\mathbf{B}_z = \mathbf{A}_{\mu-1} - b_{\mu-1} \cdot \mathbf{G} = \mathbf{A}\mathbf{R}_z + \mathbf{S}_z\mathbf{G}$ holds before entering the inner loop. Assume that $\mathbf{B}_z = \mathbf{A}\mathbf{R}_z + \mathbf{S}_z\mathbf{G}$ holds before entering the j-th (i.e., $i = j$) iteration of the inner loop, we now show that the equation $\mathbf{B}_z = \mathbf{A}\mathbf{R}_z + \mathbf{S}_z\mathbf{G}$ still holds after the j-th iteration. Since $\mathbf{A}_j - b_j \cdot \mathbf{G} = \mathbf{A}\mathbf{R}_j + (1 - b_j^* - b_j) \cdot \mathbf{G}$, we have that $\mathbf{B}_z := (\mathbf{A}_j - b_j \cdot \mathbf{G}) \cdot \mathbf{G}^{-1}(\mathbf{B}_z) = \mathbf{A}\mathbf{R}_j \cdot \mathbf{G}^{-1}(\mathbf{B}_z) + (1 - b_j^* - b_j) \cdot (\mathbf{A}\mathbf{R}_z + \mathbf{S}_z\mathbf{G})$. This means that if we set $\mathbf{R}_z := \mathbf{R}_j \cdot \mathbf{G}^{-1}(\mathbf{B}_z) + (1 - b_j^* - b_j) \cdot \mathbf{R}_z$ and $\mathbf{S}_z := (1 - b_j^* - b_j) \cdot \mathbf{S}_z$, the equation $\mathbf{B}_z = \mathbf{A}\mathbf{R}_z + \mathbf{S}_z\mathbf{G}$ still holds. In particular, we have that $\mathbf{S}_z = \prod_{i=0}^{\mu-1}(1 - b_i^* - b_i) \cdot \mathbf{I}_n$ holds at the end of the inner loop. It is easy to check that $\mathbf{S}_z = \mathbf{0}$ for any $z \neq z^*$, and $\mathbf{S}_z = (-1)^c \cdot \mathbf{I}_n$ for $z = z^*$, where c is the number of 1's in the binary vector $(b_0^*, \ldots, b_{\mu-1}^*) = \mathsf{BitDecomp}_N(z^*)$. The correctness of the trapdoor evaluation algorithm follows from that fact that $\mathbf{Z} = \mathcal{H}.\mathsf{Eval}(K', X) = \hat{\mathbf{A}} + \sum_{z \in CF_X} \mathbf{B}_z = \mathbf{A}\hat{\mathbf{R}} - (-1)^c \cdot \mathbf{G} + \sum_{z \in CF_X}(\mathbf{A}\mathbf{R}_z + \mathbf{S}_z\mathbf{G}) = \mathbf{A}\mathbf{R}_X + \mathbf{S}_X\mathbf{B}$. In particular, we have that $\mathbf{S}_X = -(-1)^c \cdot \mathbf{I}_n$ if $z^* \notin CF_X$, else $\mathbf{S}_X = \mathbf{0}$.

Since $s_1(\mathbf{G}^{-1}(\mathbf{B}_z)) \leq nk$ by the fact that $\mathbf{G}^{-1}(\mathbf{B}_z) \in \{0, 1\}^{nk \times nk}$, and $s_1(\hat{\mathbf{R}}), s_1(\mathbf{R}_i) \leq (\sqrt{\bar{m}} + \sqrt{nk}) \cdot \omega(\sqrt{\log \bar{m}})$ by Lemma 2, we have that $s_1(\mathbf{R}_z) \leq \mu\bar{m}^{1.5} \cdot \omega(\sqrt{\log \bar{m}})$ holds except with negligible probability for any $z \in CF_X$. Using $|CF_X| = \eta \leq 4v\ell$, the inequality $s_1(\mathbf{R}_X) \leq \mu v\ell\bar{m}^{1.5} \cdot \omega(\sqrt{\log \bar{m}})$

holds except with negligible probability for any $X \in [L]$. Besides, for any $X_1, Y_1, \ldots, Y_v \in [L]$ such that $X_1 \neq Y_j$ for all $j \in \{1, \ldots, v\}$, there is at least one element in $CF_{X_1} \subseteq [N]$ that does not belong to the union set $\cup_{j \in \{1, \ldots, v\}} CF_{Y_j}$. This is because the family $CF = \{CF_X\}_{X \in [L]}$ is v-cover-free. Since z^* is randomly chosen from $[N]$ and is statistically hidden in the key K', the probability $\Pr[z^* \in CF_{X_1} \wedge z^* \notin \cup_{j \in \{1, \ldots, v\}} CF_{Y_j}]$ is at least $1/N$. Thus, we have that $\Pr[\mathbf{S}_{X_1} = \mathbf{0} \wedge \mathbf{S}_{Y_1} = \cdots = \mathbf{S}_{Y_v} = -(-1)^c \cdot \mathbf{I}_n \in \mathcal{I}_n] \geq \frac{1}{N}$. $\square$

3.3 Collision-Resistance and High Min-Entropy

Collision-Resistance. Let $\mathcal{H} = \{\mathrm{H}_K : \mathcal{X} \to \mathcal{Y}\}_{K \in \mathcal{K}}$ be a family of hash functions with key space $\mathcal{K}$. We say that $\mathcal{H}$ is collision-resistant if for any PPT algorithm $\mathcal{C}$, its advantage

$$\mathrm{Adv}^{\mathrm{cr}}_{\mathcal{H}, \mathcal{C}}(\kappa) = \Pr[K \leftarrow_r \mathcal{K}; (X_1, X_2) \leftarrow_r \mathcal{C}(K, 1^\kappa) : X_1 \neq X_2 \wedge \mathrm{H}_K(X_1) = \mathrm{H}_K(X_2)]$$

is negligible in the security parameter κ.

Theorem 4. *Let $n, v, q \in \mathbb{Z}$ and $\bar{\beta}, \beta \in \mathbb{R}$ be polynomials in the security parameter κ. Let $\mathcal{H} = (\mathcal{H}.\mathrm{Gen}, \mathcal{H}.\mathrm{Eval})$ be a $(1, v, \beta, \gamma, \delta)$-PHF with $\gamma = \mathrm{negl}(\kappa)$ and noticeable $\delta > 0$. Then, for large enough $\bar{m}, m \in \mathbb{Z}$ and $v \geq 1$, if there exists an algorithm $\mathcal{C}$ breaking the collision-resistance of $\mathcal{H}$, there exists an algorithm $\mathcal{B}$ solving the $\mathrm{ISIS}_{q, \bar{m}, \bar{\beta}}$ problem for $\bar{\beta} = \beta\sqrt{m} \cdot \omega(\log n)$ with probability at least $\epsilon' \geq (\epsilon - \gamma)\delta$.*

For space reason, we defer the proof of Theorem 4 to the full version [53].

High Min-Entropy. Let $\mathcal{H} : \mathcal{X} \to \mathbb{Z}_q^{n \times m}$ be a $(1, v, \beta, \gamma, \delta)$-PHF with $\gamma = \mathrm{negl}(\kappa)$ and noticeable $\delta > 0$. Note that the well-distributed hidden matrices property of $\mathcal{H}$ holds even for an unbounded algorithm $\mathcal{A}$ that chooses $\{X_i\}$ and $\{Y_j\}$ after seeing K'. For any noticeable $\delta > 0$, this can only happen when the decomposition $\mathrm{H}_{K'}(X) = \mathbf{A}\mathbf{R}_X + \mathbf{S}_X\mathbf{B}$ is not unique (with respect to K') and the particular pair determined by td, i.e., $(\mathbf{R}_X, \mathbf{S}_X) = \mathcal{H}.\mathrm{TrapEval}(td, K', X)$, is information-theoretically hidden from $\mathcal{A}$. We now introduce a property called high min-entropy to formally capture this useful feature.

Definition 5 (PHF with High Min-Entropy). *Let $\mathcal{H} : \mathcal{X} \to \mathbb{Z}_q^{n \times m}$ be a $(1, v, \beta, \gamma, \delta)$-PHF with $\gamma = \mathrm{negl}(\kappa)$ and noticeable $\delta > 0$. Let $\mathcal{K}$ be the key space of $\mathcal{H}$, and let $\mathcal{H}.\mathrm{TrapGen}$ and $\mathcal{H}.\mathrm{TrapEval}$ be a pair of trapdoor generation and trapdoor evaluation algorithms for $\mathcal{H}$. We say that $\mathcal{H}$ is a PHF with high min-entropy if for uniformly random $\mathbf{A} \in \mathbb{Z}_q^{n \times \bar{m}}$ and (publicly known) trapdoor matrix $\mathbf{B} \in \mathbb{Z}_q^{n \times \bar{m}}$, the following conditions hold.*

1. *For any $(K', td) \leftarrow \mathcal{H}.\mathrm{TrapGen}(1^\kappa, \mathbf{A}, \mathbf{B})$, $K \leftarrow \mathcal{H}.\mathrm{Gen}(1^\kappa)$, any $X \in \mathcal{X}$ and any $\mathbf{w} \in \mathbb{Z}_q^{\bar{m}}$, the statistical distance between $(\mathbf{A}, K', \mathbf{R}_X^t\mathbf{w})$ and $(\mathbf{A}, K, \mathbf{R}_X^t\mathbf{w})$ is negligible in κ, where $(\mathbf{R}_X, \mathbf{S}_X) = \mathcal{H}.\mathrm{TrapEval}(td, K', X)$.*

2. *For any* $(K', td) \leftarrow \mathcal{H}.\text{TrapGen}(1^\kappa, \mathbf{A}, \mathbf{B})$, *any* $X \in \mathcal{X}$, *any uniformly random* $\mathbf{v} \in \mathbb{Z}_q^{\bar{m}}$, *and any uniformly random* $\mathbf{u} \leftarrow_r \mathbb{Z}_q^m$, *the statistical distance between* $(\mathbf{A}, K', \mathbf{v}, \mathbf{R}_X^t \mathbf{v})$ *and* $(\mathbf{A}, K', \mathbf{v}, \mathbf{u})$ *is negligible in* κ, *where* $(\mathbf{R}_X, \mathbf{S}_X) = \mathcal{H}.\text{TrapEval}(td, K', X)$.

Remark 1. Note that the well-distributed hidden matrices property of PHF only holds when the information (except that is already leaked via the key K') of the trapdoor td is hidden. This means that it provides no guarantee when some information of $\mathbf{R}_X$ for any $X \in \mathcal{X}$ (which is usually related to the trapdoor td) is given public. However, for a PHF with high min-entropy, this property still holds when the information of $\mathbf{R}_X^t \mathbf{v}$ for a uniformly random vector $\mathbf{v}$ is leaked.

For appropriate choices of parameters, the work [1] implicitly showed that the Type-I PHF construction satisfied the high min-entropy property. Now, we show that our Type-II PHF construction also has the high min-entropy property.

Theorem 5. *Let integers* $n, \bar{m}, q$ *be some polynomials in the security parameter* κ, *and let* $k = \lceil \log_2 q \rceil$. *For any* $\ell, v \in \mathbb{Z}$ *and* $L = 2^\ell$, *let* $N \leq 16v^2\ell, \eta \leq 4v\ell$ *and* $CF = \{CF_X\}_{X \in [L]}$ *be defined as in Lemma 4. Then, for large enough* $\bar{m} = O(n \log q)$, *the hash function* $\mathcal{H} : [L] \to \mathbb{Z}_q^{n \times nk}$ *given in Definition 4 (and proved in Theorem 3) is a PHF with high min-entropy.*

Proof. By Definition 4, the real key K of $\mathcal{H}$ is uniformly distributed over $(\mathbb{Z}_q^{n \times nk})^{2\mu+1}$. To prove that $\mathcal{H}$ satisfies the first condition of high min-entropy, we must show that for any $(K', td) \leftarrow \mathcal{H}.\text{TrapGen}(1^\kappa, \mathbf{A}, \mathbf{G})$, any $X \in \mathcal{X}$ and $(\mathbf{R}_X, \mathbf{S}_X) = \mathcal{H}.\text{TrapEval}(td, K', X)$, the key K' is statistically close to uniform over $(\mathbb{Z}_q^{n \times nk})^{2\mu+1}$ even conditioned on $\mathbf{R}_X^t \mathbf{w} \in \mathbb{Z}_q^{nk}$. Formally, for any $\mathbf{w} \in \mathbb{Z}_q^{\bar{m}}$, let $f_{\mathbf{w}} : \mathbb{Z}_q^{\bar{m} \times nk} \to \mathbb{Z}_q^{nk}$ be the function defined by $f_{\mathbf{w}}(\mathbf{X}) = \mathbf{X}^t \mathbf{w} \in \mathbb{Z}_q^{nk}$. Then, given $I = \{f_{\mathbf{w}}(\hat{\mathbf{R}}), \{f_{\mathbf{w}}(\mathbf{R}_i)\}_{i \in \{0,\dots,\mu-1\}})\}$ and (K', X, z^*), one can compute $\mathbf{R}_X^t \mathbf{w}$ by simulating the **Procedure II** in Theorem 3. Thus, it suffices to show that K' is statistically close to uniform over $(\mathbb{Z}_q^{n \times nk})^{2\mu+1}$ conditioned on I and z^*. Since each matrix in the key K' always has a form of $\mathbf{A}\tilde{\mathbf{R}} + b\mathbf{G}$ for some randomly chosen $\tilde{\mathbf{R}} \leftarrow_r (D_{\mathbb{Z}^{\bar{m}},s})^{nk}$, and a bit $b \in \{0,1\}$ depending on a random $z^* \leftarrow_r [N]$. Using a standard hybrid argument, it is enough to show that conditioned on $\mathbf{A}$ and $f_{\mathbf{w}}(\tilde{\mathbf{R}})$, $\mathbf{A}\tilde{\mathbf{R}}$ is statistically close to uniform over $\mathbb{Z}_q^{n \times nk}$.

Let $f'_{\mathbf{w}} : \mathbb{Z}_q^{\bar{m}} \to \mathbb{Z}_q$ be defined by $f'_{\mathbf{w}}(\mathbf{x}) = \mathbf{x}^t \mathbf{w}$, and let $\tilde{\mathbf{R}} = (\mathbf{r}_1, \dots, \mathbf{r}_{nk})$. Then, $f_{\mathbf{w}}(\tilde{\mathbf{R}}) = (f'_{\mathbf{w}}(\mathbf{r}_1), \dots, f'_{\mathbf{w}}(\mathbf{r}_{nk}))^t \in \mathbb{Z}_q^{nk}$. By Lemma 1, the guessing probability $\gamma(\mathbf{r}_i)$ is at most $2^{1-\bar{m}}$ for all $i \in \{1, \dots, nk\}$. By the generalized leftover hash lemma in [21], conditioned on $\mathbf{A}$ and $f'_{\mathbf{w}}(\mathbf{r}_i) \in \mathbb{Z}_q$, the statistical distance between $\mathbf{A}\mathbf{r}_i \in \mathbb{Z}_q^n$ and uniform over $\mathbb{Z}_q^n$ is at most $\frac{1}{2} \cdot \sqrt{2^{1-\bar{m}} \cdot q^n \cdot q}$, which is negligible if we set $\bar{m} = O(n \log q) > (n+1) \log q + \omega(\log n)$. Using a standard hybrid argument, we have that conditioned on $\mathbf{A}$ and $f_{\mathbf{w}}(\tilde{\mathbf{R}})$, the matrix $\mathbf{A}\tilde{\mathbf{R}} = (\mathbf{A}\mathbf{r}_1 \| \dots \| \mathbf{A}\mathbf{r}_{nk})$ is statistically close to uniform over $\mathbb{Z}_q^{n \times nk}$.

Now, we show that $\mathcal{H}$ satisfies the second condition in Definition 5. By Theorem 3 for any input X and $(\mathbf{R}_X, \mathbf{S}_X) = \mathcal{H}.\text{TrapEval}(td, K', X)$, we always have that $\mathbf{R}_X = \hat{\mathbf{R}} + \tilde{\mathbf{R}}$ for some $\tilde{\mathbf{R}}$ that is independent from $\hat{\mathbf{R}}$. Let $\mathbf{R}_X^t \mathbf{v} =$

$\hat{\mathbf{R}}^t\mathbf{v}+\tilde{\mathbf{R}}^t\mathbf{v} = \hat{\mathbf{u}}+\tilde{\mathbf{u}}$, it suffices to show that given K' and $\mathbf{v}$, the element $\hat{\mathbf{u}} = \hat{\mathbf{R}}^t\mathbf{v}$ is uniformly random. Since $\hat{\mathbf{R}} \leftarrow_r (D_{\mathbb{Z}^{\bar{m}},s})^{nk}$ for $s \geq \omega(\sqrt{\log \bar{m}})$ is only used to generate the matrix $\hat{\mathbf{A}} = \mathbf{A}\hat{\mathbf{R}} - (-1)^c \cdot \mathbf{G}$ in the key K', we have that for large enough $\bar{m} = O(n \log q)$, the pair $(\mathbf{A}\hat{\mathbf{R}}, \hat{\mathbf{u}}^t = \mathbf{v}^t\hat{\mathbf{R}})$ is statistically close to uniform over $\mathbb{Z}_q^{n \times nk} \times \mathbb{Z}_q^{nk}$ by the fact in Lemma 3.[5] Thus, $\mathbf{R}_X^t\mathbf{v} = \hat{\mathbf{R}}^t\mathbf{v} + \tilde{\mathbf{R}}^t\mathbf{v}$ is uniformly distributed over $\mathbb{Z}_q^{nk}$. This completes the proof of Theorem 5. $\square$

3.4 Programmable Hash Function from Ideal Lattices

As many cryptographic schemes over general lattices (e.g., [43]), we do not see any obstacle preventing us from adapting our definition and constructions of PHFs to ideal lattices defined over polynomial rings, e.g., $R = \mathbb{Z}[x]/(x^n + 1)$ or $R_q = \mathbb{Z}_q[x]/(x^n + 1)$ where n is a power of 2. In general, one can benefit from the rich algebraic structures of ideal lattices in many aspects. For example, compared to their counterparts over general lattices, the constructions over ideal lattices roughly save a factor of n in the key size (e.g., [41,42]).

At CRYPTO 2014, Ducas and Micciancio [24] proposed a short signature scheme by combining the confined guessing technique [8] with ideal lattices, which substantially reduced the verification key size from previous known $O(n)$ elements to $O(\log n)$ elements. We note that their construction implicitly used a weak $(1, \text{poly}, \beta)$-PHF for some $\beta = \text{poly}(\kappa) \in \mathbb{R}$ (we omit the details for not involving too many backgrounds on ideal lattices). But as noted by the authors, their methods used for constructing signatures with short verification keys (as well as the underlying PHF) seem specific to the ideal lattice setting, and thus cannot be instantiated from general lattices. In fact, it was left as an open problem [24] to construct a standard model short signature scheme with short verification keys from general lattices.

4 Short Signature Schemes from Lattice-Based PHFs

A digital signature scheme $\mathcal{SIG} = (\mathsf{KeyGen}, \mathsf{Sign}, \mathsf{Verify})$ consists of three PPT algorithms. Taking the security parameter κ as input, the key generation algorithm outputs a verification key vk and a secret signing key sk, i.e., $(vk, sk) \leftarrow \mathsf{KeyGen}(1^\kappa)$. The signing algorithm takes vk, sk and a message $M \in \{0,1\}^*$ as inputs, outputs a signature σ on M, briefly denoted as $\sigma \leftarrow \mathsf{Sign}(sk, M)$. The verification algorithm takes vk, message $M \in \{0,1\}^*$ and a string $\sigma \in \{0,1\}^*$ as inputs, outputs 1 if σ is a valid signature on M, else outputs 0, denoted as $1/0 \leftarrow \mathsf{Verify}(vk, M, \sigma)$. For correctness, we require that for any $(vk, sk) \leftarrow \mathsf{KeyGen}(1^\kappa)$, any message $M \in \{0,1\}^*$, and any $\sigma \leftarrow \mathsf{Sign}(sk, M)$, the equation $\mathsf{Verify}(vk, M, \sigma) = 1$ holds with overwhelming probability, where the probability is taken over the choices of the random coins used in KeyGen, Sign and Verify.

We defer the security definition of existential unforgeability against chosen message attacks (EUF-CMA) to the full version [53].

---

[5] This is because one can first construct a new uniformly random matrix $\mathbf{A}'$ by appending the row vector $\mathbf{v}^t$ to the rows of $\mathbf{A}$, and then apply the fact in Lemma 3.

4.1 A Short Signature Scheme with Short Verification Key

Let integers $\ell, n, m', v, q \in \mathbb{Z}, \beta \in \mathbb{R}$ be some polynomials in the security parameter κ, and let $k = \lceil \log_2 q \rceil$. Let $\mathcal{H} = (\mathcal{H}.\mathsf{Gen}, \mathcal{H}.\mathsf{Eval})$ be any $(1, v, \beta)$-PHF from $\{0,1\}^\ell$ to $\mathbb{Z}_q^{n \times m'}$. Let $\bar{m} = O(n \log q)$, $m = \bar{m} + m'$, and large enough $s > \max(\beta, \sqrt{m}) \cdot \omega(\sqrt{\log n}) \in \mathbb{R}$ be the system parameters. Our generic signature scheme $\mathcal{SIG} = (\mathsf{KeyGen}, \mathsf{Sign}, \mathsf{Verify})$ is defined as follows.

$\mathsf{KeyGen}(1^\kappa)$: Given a security parameter κ, compute $(\mathbf{A}, \mathbf{R}) \leftarrow \mathsf{TrapGen}(1^n, 1^{\bar{m}}, q, \mathbf{I}_n)$ such that $\mathbf{A} \in \mathbb{Z}_q^{n \times \bar{m}}$, $\mathbf{R} = \mathbb{Z}_q^{(\bar{m}-nk) \times nk}$, and randomly choose $\mathbf{u} \leftarrow_r \mathbb{Z}_q^n$. Then, compute $K \leftarrow \mathcal{H}.\mathsf{Gen}(1^\kappa)$, and return a pair of verification key and secret signing key $(vk, sk) = ((\mathbf{A}, \mathbf{u}, K), \mathbf{R})$.

$\mathsf{Sign}(sk, M \in \{0,1\}^\ell)$: Given $sk = \mathbf{R}$ and any message M, compute $\mathbf{A}_M = (\mathbf{A} \| \mathrm{H}_K(M)) \in \mathbb{Z}_q^{n \times m}$, where $\mathrm{H}_K(M) = \mathcal{H}.\mathsf{Eval}(K, M) \in \mathbb{Z}_q^{n \times m'}$. Then, compute $\mathbf{e} \leftarrow \mathsf{SampleD}(\mathbf{R}, \mathbf{A}_M, \mathbf{I}_n, \mathbf{u}, s)$, and return the signature $\sigma = \mathbf{e}$.

$\mathsf{Verify}(vk, M, \sigma)$: Given vk, a message M and a vector $\sigma = \mathbf{e}$, compute $\mathbf{A}_M = (\mathbf{A} \| \mathrm{H}_K(M)) \in \mathbb{Z}_q^{n \times m}$, where $\mathrm{H}_K(M) = \mathcal{H}.\mathsf{Eval}(K, M) \in \mathbb{Z}_q^{n \times m'}$. Return 1 if $\|\mathbf{e}\| \le s\sqrt{m}$ and $\mathbf{A}_M \mathbf{e} = \mathbf{u}$, else return 0.

The correctness of our scheme $\mathcal{SIG}$ can be easily checked. Besides, the schemes with linear verification keys in [12,43] can be seen as instantiations of $\mathcal{SIG}$ with the Type-I PHF construction in Theorem 1.[6] Since the size of the verification key is mainly determined by the key size of $\mathcal{H}$, one can instantiate $\mathcal{H}$ with our efficient Type-II PHF construction in Definition 4 to obtain a signature scheme with verification keys consisting of a logarithmic number of matrices. As for the security, we have the following theorem.

Theorem 6. *Let $\ell, n, \bar{m}, m', q \in \mathbb{Z}$ and $\bar{\beta}, \beta, s \in \mathbb{R}$ be some polynomials in the security parameter κ, and let $m = \bar{m} + m'$. Let $\mathcal{H} = (\mathcal{H}.\mathsf{Gen}, \mathcal{H}.\mathsf{Eval})$ be a $(1, v, \beta, \gamma, \delta)$-PHF from $\{0,1\}^\ell$ to $\mathbb{Z}_q^{n \times m'}$ with $\gamma = \mathsf{negl}(\kappa)$ and noticeable $\delta > 0$. Then, for large enough $\bar{m} = O(n \log q)$ and $s > \max(\beta, \sqrt{m}) \cdot \omega(\sqrt{\log n}) \in \mathbb{R}$, if there exists a PPT forger $\mathcal{F}$ breaking the EUF-CMA security of $\mathcal{SIG}$ with non-negligible probability $\epsilon > 0$ and making at most $Q \le v$ signing queries, there exists an algorithm $\mathcal{B}$ solving the $\mathsf{ISIS}_{q,\bar{m},\bar{\beta}}$ problem for $\bar{\beta} = \beta s \sqrt{m} \cdot \omega(\sqrt{\log n})$ with probability at least $\epsilon' \ge \epsilon\delta - \mathsf{negl}(\kappa)$.*

Since a proof sketch is given in Sect. 1.3, we omit the details of the proof. Let $\mathcal{SIG}_1$ denote the signature scheme obtained by instantiating $\mathcal{SIG}$ with our Type-II PHF construction in Definition 4. Then, the verification key of $\mathcal{SIG}_1$ has $O(\log n)$ matrices and each signature of $\mathcal{SIG}_1$ consists of a single lattice vector.

Corollary 1. *Let $n, q \in \mathbb{Z}$ be polynomials in the security parameter κ. Let $\bar{m} = O(n \log q), v = \mathsf{poly}(n)$ and $\ell = n$. If there exists a PPT forger $\mathcal{F}$ breaking the EUF-CMA security of $\mathcal{SIG}_1$ with non-negligible probability ϵ and making at most $Q \le v$ signing queries, then there exists an algorithm $\mathcal{B}$ solving the $\mathsf{ISIS}_{q,\bar{m},\bar{\beta}}$ problem for $\bar{\beta} = v^2 \cdot \tilde{O}(n^{5.5})$ with probability at least $\epsilon' \ge \frac{\epsilon}{16nv^2} - \mathsf{negl}(\kappa)$.*

---

[6] Note that the scheme in [12] used a syndrome $\mathbf{u} = \mathbf{0}$, we prefer to use a random chosen syndrome $\mathbf{u} \leftarrow_r \mathbb{Z}_q^n$ as that in [43] for simplifying the security analysis.

4.2 An Improved Short Signature Scheme from Weaker Assumption

Compared to prior constructions in [6,8,24], our $\mathcal{SIG}_1$ only has a reduction loss about $16nQ^2$, which does not depend on the forger's success probability ϵ. However, because of $v \geq Q$, our improvement requires the ISIS$_{q,\bar{m},\bar{\beta}}$ problem to be hard for $\bar{\beta} = Q^2 \cdot \tilde{O}(n^{5.5})$, which means that the modulus q should be bigger than $Q^2 \cdot \tilde{O}(n^{5.5})$. Even though q is still a polynomial of n in an asymptotic sense, it might be very large in practice. In this section, we further remove the direct dependency on Q from $\bar{\beta}$ by introducing a short tag about $O(\log Q)$ bits to each signature. For example, this only increases about 30 bits to each signature for a number $Q = 2^{30}$ of the forger's signing queries.

At a high level, our basic idea is to relax the requirement on a $(1, v, \beta)$-PHF $\mathcal{H} = \{\mathrm{H}_K\}$ so that a much smaller $v = \omega(\log n)$ can be used by employing a simple weak PHF $\mathcal{H}' = \{\mathrm{H}'_{K'}\}$ (recall that $v \geq Q$ is required in the scheme $\mathcal{SIG}$). Concretely, for each message M to be signed, instead of using $\mathrm{H}_K(M)$ in the signing algorithm of $\mathcal{SIG}$, we choose a short random tag $\mathbf{t}$, and compute $\mathrm{H}'_{K'}(\mathbf{t}) + \mathrm{H}_K(M)$ to generate the signature on M. Thus, if the trapdoor keys of both PHFs are generated by using the same "generators" $\mathbf{A}$ and $\mathbf{G}$, we have that $\mathrm{H}'_{K'}(\mathbf{t}) + \mathrm{H}_K(M) = \mathbf{A}(\mathbf{R}'_{\mathbf{t}} + \mathbf{R}_M) + (\mathbf{S}'_{\mathbf{t}} + \mathbf{S}_M)\mathbf{G}$, where $\mathrm{H}'_{K'}(\mathbf{t}) = \mathbf{A}\mathbf{R}'_{\mathbf{t}} + \mathbf{S}'_{\mathbf{t}}\mathbf{G}$ and $\mathrm{H}_K(M) = \mathbf{A}\mathbf{R}_M + \mathbf{S}_M\mathbf{G}$. Moreover, if we can ensure that $\mathbf{S}'_{\mathbf{t}} + \mathbf{S}_M \in \mathcal{I}_n$ when $\mathbf{S}'_{\mathbf{t}} \in \mathcal{I}_n$ or $\mathbf{S}_M \in \mathcal{I}_n$, then $\mathbf{S}_M$ is not required to be invertible for all the Q signing messages. In particular, $v = \omega(\log n)$ can be used as long as the probability that $\mathbf{S}'_{\mathbf{t}} + \mathbf{S}_M \in \mathcal{I}_n$ is invertible for all the Q signing messages, but $\mathbf{S}'_{\mathbf{t}^*} + \mathbf{S}_{M^*} = \mathbf{0}$ for the forged signature on the pair $(\mathbf{t}^*, M^*)$, is noticeable.

Actually, the weak PHF $\mathcal{H}'$ and the $(1, v, \beta)$-PHF $\mathcal{H} = (\mathcal{H}.\mathrm{Gen}, \mathcal{H}.\mathrm{Eval})$ are, respectively, the first instantiated Type-I PHF $\mathcal{H}'$ in Theorem 2 and the Type-II PHF $\mathcal{H} = (\mathcal{H}.\mathrm{Gen}, \mathcal{H}.\mathrm{Eval})$ given in Definition 4. Since $\mathcal{H}'$ is very simple, we directly plug its construction into our signature scheme $\mathcal{SIG}_2$. Specifically, let $n, q \in \mathbb{Z}$ be some polynomials in the security parameter κ, and let $k = \lceil \log_2 q \rceil$, $\bar{m} = O(n \log q)$, $m = \bar{m} + nk$ and $s = \tilde{O}(n^{2.5}) \in \mathbb{R}$. Let $H : \mathbb{Z}_q^n \to \mathbb{Z}_q^{n \times n}$ be the FRD encoding in [1] such that for any vector $\mathbf{v} = (v, 0 \ldots, 0)^t$, $\mathbf{v}_1, \mathbf{v}_2 \in \mathbb{Z}_q^n$, we have that $H(\mathbf{v}) = v\mathbf{I}_n$ and $H(\mathbf{v}_1) + H(\mathbf{v}_2) = H(\mathbf{v}_1 + \mathbf{v}_2)$ hold. For any $\mathbf{t} \in \{0, 1\}^\ell$ with $\ell < n$, we naturally treat it as a vector in $\mathbb{Z}_q^n$ by appending it $(n - \ell)$ zero coordinates. The weak PHF $\mathcal{H}'$ from $\{0, 1\}^\ell$ to $\mathbb{Z}_q^{n \times nk}$ has a form of $\mathrm{H}'_{K'}(\mathbf{t}) = \mathbf{A}_0 + H(\mathbf{t})\mathbf{G}$, where $K' = \mathbf{A}_0$. We restrict the domain of $\mathcal{H}'$ to be $\{0\} \times \{0, 1\}^\ell$ for $\ell \leq n - 1$ such that $\mathbf{S}'_{\mathbf{t}} + \mathbf{S}_M$ is invertible when $(\mathbf{S}'_{\mathbf{t}}, \mathbf{S}_M) \neq (\mathbf{0}, \mathbf{0})$. Our signature scheme $\mathcal{SIG}_2 = (\mathsf{KeyGen}, \mathsf{Sign}, \mathsf{Verify})$ is defined as follows.

$\mathsf{KeyGen}(1^\kappa)$: Given a security parameter κ, compute $(\mathbf{A}, \mathbf{R}) \leftarrow \mathsf{TrapGen}(1^n, 1^{\bar{m}}, q, \mathbf{I}_n)$ such that $\mathbf{A} \in \mathbb{Z}_q^{n \times \bar{m}}$, $\mathbf{R} = \mathbb{Z}_q^{(\bar{m}-nk) \times nk}$. Randomly choose $\mathbf{A}_0 \leftarrow_r \mathbb{Z}_q^{n \times nk}$ and $\mathbf{u} \leftarrow_r \mathbb{Z}_q^n$. Finally, compute $K \leftarrow \mathcal{H}.\mathrm{Gen}(1^\kappa)$, and return $(vk, sk) = ((\mathbf{A}, \mathbf{A}_0, \mathbf{u}, K), \mathbf{R})$.

$\mathsf{Sign}(sk, M \in \{0, 1\}^n)$: Given the secret key sk and a message M, randomly choose $\mathbf{t} \leftarrow_r \{0, 1\}^\ell$, and compute $\mathbf{A}_{M,\mathbf{t}} = (\mathbf{A}\|(\mathbf{A}_0 + H(0\|\mathbf{t})\mathbf{G}) + \mathrm{H}_K(M)) \in \mathbb{Z}_q^{n \times m}$, where $\mathrm{H}_K(M) = \mathcal{H}.\mathrm{Eval}(K, M) \in \mathbb{Z}_q^{n \times nk}$. Then, compute $\mathbf{e} \leftarrow \mathsf{SampleD}(\mathbf{R}, \mathbf{A}_{M,\mathbf{t}}, \mathbf{I}_n, \mathbf{u}, s)$, and return the signature $\sigma = (\mathbf{e}, \mathbf{t})$.

Verify(vk, M, σ): Given vk, message M and $\sigma = (\mathbf{e}, \mathbf{t})$, compute $\mathbf{A}_{M,\mathbf{t}} = (\mathbf{A} \| (\mathbf{A}_0 + H(0\|\mathbf{t})\mathbf{G}) + \mathrm{H}_K(M)) \in \mathbb{Z}_q^{n \times m}$, where $\mathrm{H}_K(M) = \mathcal{H}.\mathrm{Eval}(K, M) \in \mathbb{Z}_q^{n \times nk}$. Return 1 if $\|\mathbf{e}\| \leq s\sqrt{m}$ and $\mathbf{A}_{M,\mathbf{t}}\mathbf{e} = \mathbf{u}$. Otherwise, return 0.

Since $\mathbf{R}$ is a $\mathbf{G}$-trapdoor of $\mathbf{A}$, by padding with zero rows it can be extended to a $\mathbf{G}$-trapdoor for $\mathbf{A}_{M,\mathbf{t}}$ with the same quality $s_1(\mathbf{R}) \leq \sqrt{m} \cdot \omega(\sqrt{\log n})$. Since $s = \tilde{O}(n^{2.5}) > s_1(\mathbf{R}) \cdot \omega(\sqrt{\log n})$, the vector $\mathbf{e}$ output by SampleD follows the distribution $D_{\mathbb{Z}^m, s}$ satisfying $\mathbf{A}_{M,\mathbf{t}}\mathbf{e} = \mathbf{u}$. In other words, $\|\mathbf{e}\| \leq s\sqrt{m}$ holds with overwhelming probability by Lemma 1. This shows that $\mathcal{SIG}_2$ is correct.

Note that if we set $v = \omega(\log n)$, the key K only has $\mu = O(\log n)$ number of matrices and each signature consists of a vector plus a short ℓ-bit tag. We have the following theorem for security.

Theorem 7. *Let $\ell, \bar{m}, n, q, v \in \mathbb{Z}$ be polynomials in the security parameter κ. For appropriate choices of $\ell = O(\log n)$ and $v = \omega(\log n)$, if there exists a PPT forger $\mathcal{F}$ breaking the EUF-CMA security of $\mathcal{SIG}_2$ with non-negligible probability ϵ and making at most $Q = \mathrm{poly}(n)$ signing queries, there exists an algorithm $\mathcal{B}$ solving the $\mathrm{ISIS}_{q,\bar{m},\bar{\beta}}$ problem for $\bar{\beta} = \tilde{O}(n^{5.5})$ with probability at least $\epsilon' \geq \frac{\epsilon}{16 \cdot 2^\ell n v^2} - \mathrm{negl}(\kappa) = \frac{\epsilon}{Q \cdot \tilde{O}(n)}$.*

We defer the proof of Theorem 7 to the full version [53].

5   Identity-Based Encryptions from Lattice-Based PHFs

An identity-based encryption (IBE) scheme consists of four PPT algorithms $\mathcal{IBE} = (\mathsf{Setup}, \mathsf{Extract}, \mathsf{Enc}, \mathsf{Dec})$. Taking the security parameter κ as input, the randomized key generation algorithm Setup outputs a master public key mpk and a master secret key msk, denoted as $(mpk, msk) \leftarrow \mathsf{Setup}(1^\kappa)$. The (randomized) extract algorithm takes mpk, msk and an identity id as inputs, outputs a user private key sk_{id} for id, briefly denoted as $sk_{id} \leftarrow \mathsf{Extract}(msk, id)$. The randomized encryption algorithm Enc takes mpk, id and a plaintext M as inputs, outputs a ciphertext C, denoted as $C \leftarrow \mathsf{Enc}(mpk, id, M)$. The deterministic algorithm Dec takes sk_{id} and C as inputs, outputs a plaintext M, or a special symbol $\perp$, which is denoted as $M/\perp \leftarrow \mathsf{Dec}(sk_{id}, C)$. In addition, for all $(mpk, msk) \leftarrow \mathsf{Setup}(1^\kappa), sk_{id} \leftarrow \mathsf{Extract}(msk, id)$ and any plaintext M, we require that $\mathsf{Dec}(sk_{id}, C) = M$ holds for any $C \leftarrow \mathsf{Enc}(mpk, id, M)$.

5.1   An Identity-Based Encryption with Short Master Public Key

Let integers n, m', v, β, q be polynomials in the security parameter κ, and let $k = \lceil \log_2 q \rceil$. Let $\mathcal{H} = (\mathcal{H}.\mathrm{Gen}, \mathcal{H}.\mathrm{Eval})$ be any $(1, v, \beta)$-PHF with high min-entropy from $\{0,1\}^n$ to $\mathbb{Z}_q^{n \times m'}$. Let $\mathcal{H}.\mathrm{TrapGen}$ and $\mathcal{H}.\mathrm{TrapEval}$ be a pair of trapdoor generation and trapdoor evaluation algorithm of $\mathcal{H}$ that satisfies the conditions in Definition 5. For convenience, we set both the user identity space and the message space as $\{0,1\}^n$. Let integers $\bar{m} = O(n \log q), m = \bar{m} + m'$, $\alpha \in \mathbb{R}$, and large enough $s > \max(\beta, \sqrt{m}) \cdot \omega(\sqrt{\log n})$ be the system parameters. Our generic IBE scheme $\mathcal{IBE} = (\mathsf{Setup}, \mathsf{Extract}, \mathsf{Enc}, \mathsf{Dec})$ is defined as follows.

Setup(1^κ): Given a security parameter κ, compute $(\mathbf{A}, \mathbf{R}) \leftarrow$ TrapGen($1^n, 1^{\bar{m}}$, $q, \mathbf{I}_n$) such that $\mathbf{A} \in \mathbb{Z}_q^{n \times \bar{m}}$, $\mathbf{R} = \mathbb{Z}_q^{(\bar{m}-nk) \times nk}$. Randomly choose $\mathbf{U} \leftarrow_r \mathbb{Z}_q^{n \times n}$, and compute $K \leftarrow \mathcal{H}.\mathsf{Gen}(1^\kappa)$. Finally, return $(mpk, msk) = ((\mathbf{A}, K, \mathbf{U}), \mathbf{R})$.

Extract($msk, id \in \{0,1\}^n$): Given msk and a user identity id, compute $\mathbf{A}_{id} = (\mathbf{A}\|\mathbf{H}_K(id)) \in \mathbb{Z}_q^{n \times m}$, where $\mathbf{H}_K(id) = \mathcal{H}.\mathsf{Eval}(K, id) \in \mathbb{Z}_q^{n \times m'}$. Then, compute $\mathbf{E}_{id} \leftarrow \mathsf{SampleD}(\mathbf{R}, \mathbf{A}_{id}, \mathbf{I}_n, \mathbf{U}, s)$, and return $sk_{id} = \mathbf{E}_{id} \in \mathbb{Z}^{m \times n}$.

Enc($mpk, id \in \{0,1\}^n, M \in \{0,1\}^n$): Given mpk, id and plaintext M, compute $\mathbf{A}_{id} = (\mathbf{A}\|\mathbf{H}_K(id)) \in \mathbb{Z}_q^{n \times m}$, where $\mathbf{H}_K(id) = \mathcal{H}.\mathsf{Eval}(K, id) \in \mathbb{Z}_q^{n \times m'}$. Then, randomly choose $\mathbf{s} \leftarrow_r \mathbb{Z}_q^n$, $\mathbf{x}_0 \leftarrow_r D_{\mathbb{Z}^n, \alpha q}$, $\mathbf{x}_1 \leftarrow_r D_{\mathbb{Z}^{\bar{m}}, \alpha q}$, and compute $(K', td) \leftarrow \mathcal{H}.\mathsf{TrapGen}(1^\kappa, \mathbf{A}, \mathbf{B})$ for some trapdoor matrix $\mathbf{B} \in \mathbb{Z}_q^{n \times m'}$, $(\mathbf{R}_{id}, \mathbf{S}_{id}) = \mathcal{H}.\mathsf{TrapEval}(td, K', id)$. Finally, compute and return the ciphertext $\mathbf{C} = (\mathbf{c}_0, \mathbf{c}_1)$, where

$$\mathbf{c}_0 = \mathbf{U}^t\mathbf{s} + \mathbf{x}_0 + \frac{q}{2}M, \qquad \mathbf{c}_1 = \mathbf{A}_{id}^t\mathbf{s} + \begin{pmatrix} \mathbf{x}_1 \\ \mathbf{R}_{id}^t\mathbf{x}_1 \end{pmatrix} = \begin{pmatrix} \mathbf{A}^t\mathbf{s} + \mathbf{x}_1 \\ \mathbf{H}_K(id)^t\mathbf{s} + \mathbf{R}_{id}^t\mathbf{x}_1 \end{pmatrix}.$$

Dec($sk_{id}, \mathbf{C}$): Given $sk_{id} = \mathbf{E}_{id}$ and a ciphertext $\mathbf{C} = (\mathbf{c}_0, \mathbf{c}_1)$ under identity id, compute $\mathbf{b} = \mathbf{c}_0 - \mathbf{E}_{id}^t\mathbf{c}_1 \in \mathbb{Z}_q^n$. Then, treat each coordinate of $\mathbf{b} = (b_1, \ldots, b_n)^t$ as an integer in $\mathbb{Z}$, and set $M_i = 1$ if $|b_i - \lfloor \frac{q}{2} \rfloor| \leq \lfloor \frac{q}{4} \rfloor$, else $M_i = 0$, where $i \in \{1, \ldots, n\}$. Finally, return the plaintext $M = (M_0, \ldots, M_n)^t$.

By Proposition 1, we have that $s_1(\mathbf{R}) \leq O(\sqrt{\bar{m}}) \cdot \omega(\sqrt{\log n})$. For large enough $s \geq \sqrt{m} \cdot \omega(\sqrt{\log n})$, by the correctness of SampleD we know that $\mathbf{A}_{id}\mathbf{E}_{id} = \mathbf{U}$ and $\|\mathbf{E}_{id}\| \leq s\sqrt{m}$ hold with overwhelming probability. In this case, $\mathbf{c}_0 - \mathbf{E}_{id}^t\mathbf{c}_1 = \mathbf{c}_0 - \mathbf{E}_{id}^t(\mathbf{A}_{id}^t\mathbf{s} + \hat{\mathbf{x}}) = \mathbf{c}_0 - \mathbf{U}^t\mathbf{s} - \mathbf{E}_{id}^t\hat{\mathbf{x}} = \frac{q}{2}M + \mathbf{x}_0 - \mathbf{E}_{id}^t\hat{\mathbf{x}}$, where $\hat{\mathbf{x}} = \begin{pmatrix} \mathbf{x}_1 \\ \mathbf{R}_X^t\mathbf{x}_1 \end{pmatrix}$. Now, we estimate the size of $\|\mathbf{x}_0 - \mathbf{E}_{id}^t\hat{\mathbf{x}}\|_\infty$. Since $\mathbf{x}_0 \leftarrow_r D_{\mathbb{Z}^n, \alpha q}$, $\mathbf{x}_1 \leftarrow_r D_{\mathbb{Z}^{\bar{m}}, \alpha q}$, we have that $\|\mathbf{x}_0\|, \|\mathbf{x}_1\| \leq \alpha q\sqrt{m}$ holds with overwhelming probability by Lemma 1. In addition, using the fact that $s_1(\mathbf{R}_X) \leq \beta$, we have that $\|\hat{\mathbf{x}}\| \leq \alpha q\sqrt{m(\beta^2 + 1)}$. Thus, we have that $\|\mathbf{E}_{id}^t\hat{\mathbf{x}}\|_\infty \leq \alpha qms\sqrt{\beta^2 + 1}$, and $\|\mathbf{x}_0 - \mathbf{E}_{id}^t\hat{\mathbf{x}}\|_\infty \leq 2\alpha qms\sqrt{\beta^2 + 1}$. This means that the decryption algorithm is correct if we set parameters such that $2\alpha qms\sqrt{\beta^2 + 1} < \frac{q}{4}$ holds. For instance, we can set the parameters as follows: $m = 4n^{1+\psi}, s = \beta \cdot \omega(\sqrt{\log n}), q = \beta^2 m^2 \cdot \omega(\sqrt{\log n}), \alpha = (\beta^2 m^{1.5} \cdot \omega(\sqrt{\log n}))^{-1}$, where real $\psi \in \mathbb{R}$ satisfies $\log q < n^\psi$.

For security, we will use the notion called indistinguishable from random (known as INDr-ID-CPA) in [1], which captures both semantic security and recipient anonymity by requiring the challenge ciphertext to be indistinguishable from a uniformly random element in the ciphertext space. The formal definition of INDr-ID-CPA security is given in the full version [53]. Under the LWE assumption, our generic IBE scheme $\mathcal{IBE}$ is INDr-ID-CPA secure in the standard model.

Theorem 8. *Let $n, q, m' \in \mathbb{Z}$ and $\alpha, \beta \in \mathbb{R}$ be polynomials in the security parameter κ. For large enough $v = \mathrm{poly}(n)$, let $\mathcal{H} = (\mathcal{H}.\mathsf{Gen}, \mathcal{H}.\mathsf{Eval})$ be any $(1, v, \beta, \gamma, \delta)$-PHF with high min-entropy from $\{0,1\}^n$ to $\mathbb{Z}_q^{n \times m'}$, where $\gamma = \mathrm{negl}(\kappa)$ and $\delta > 0$ is noticeable. Then, if there exists a PPT adversary*

$\mathcal{A}$ breaking the INDr-ID-CPA security of $\mathcal{IBE}$ with non-negligible advantage ϵ and making at most $Q < v$ user private key queries, there exists an algorithm $\mathcal{B}$ solving the $\mathrm{LWE}_{q,\alpha}$ problem with advantage at least $\epsilon' \geq \epsilon\delta/3 - \mathrm{negl}(\kappa)$.

The proof is very similar to that in [1]. We defer it to the full version [53] for lack of space. Actually, by instantiating $\mathcal{H}$ in the generic scheme $\mathcal{IBE}$ with the Type-I PHF construction, we recover the fully secure IBE scheme due to Agrawal et al. [1]. Besides, if $\mathcal{H}$ is replaced by a weak $(1, v, \beta)$-PHF with high min-entropy, we can further show that the resulting scheme is INDr-sID-CPA secure, and subsumes the selectively secure IBE scheme in [1]. Formally,

Corollary 2. Let $n, m', q \in \mathbb{Z}$ and $\alpha, \beta \in \mathbb{R}$ be polynomials in the security parameter κ. For large enough $v = \mathrm{poly}(n)$, let $\mathcal{H} = (\mathcal{H}.\mathrm{Gen}, \mathcal{H}.\mathrm{Eval})$ be any weak $(1, v, \beta, \gamma, \delta)$-PHF with high min-entropy from $\{0,1\}^n$ to $\mathbb{Z}_q^{n \times m'}$, where $\gamma = \mathrm{negl}(\kappa)$ and $\delta > 0$ is noticeable. Then, under the $\mathrm{LWE}_{q,\alpha}$ assumption, the generic IBE scheme $\mathcal{IBE}$ is INDr-sID-CPA secure.

By instantiating the generic IBE scheme $\mathcal{IBE}$ with our efficient Type-II PHF in Definition 4, we can obtain a fully secure IBE scheme with master public key containing $O(\log n)$ number of matrices. Let $\mathcal{IBE}_1$ be the instantiated scheme.

Corollary 3. If there exists a PPT adversary $\mathcal{A}$ breaking the INDr-ID-CPA security of $\mathcal{IBE}_1$ with non-negligible advantage ϵ and making at most $Q = \mathrm{poly}(\kappa)$ user private key queries, then there exists an algorithm $\mathcal{B}$ solving the $\mathrm{LWE}_{q,\alpha}$ problem with advantage at least $\epsilon' \geq \frac{\epsilon}{48nQ^2} - \mathrm{negl}(\kappa)$.

Remark 2. Since our Type-II $(1, v, \beta)$-PHF depends on the parameter v in several aspects, the instantiated IBE scheme $\mathcal{IBE}_1$ relies on the particular number Q of user private key queries (because of $Q \leq v$) in terms of the master public key size and the reduction loss. On the first hand, the size of the master public key only depends on Q in a (somewhat) weak sense: for any polynomial Q it only affects the constant factor hidden in the number $O(\log n)$ of matrices in the master public key. When implementing the IBE scheme, one can either prior determine the target security level (or the maximum number Q of allowed user private key queries) before the setup phase, or set a super polynomial v to generate the master public keys. For example, for $v = n^{\log(\log n)}$, the master public key only contains $O(\log(\log n)\log n)$ matrices, which is still much smaller than the linear function $O(n)$ as that in [1,14]. On the other hand, the reduction loss of $\mathcal{IBE}_1$ also depends on Q (due to our proof of Theorem 3). Unlike the signature scheme $\mathcal{SIG}_2$, it is unclear if one can reduce the reduction loss with some modifications/improvements. Besides, it is also interesting to investigate the possibility of giving a proof of Theorem 3 with an improved $\delta > 0$.

5.2   Extensions

Hierarchical IBE. Using the trapdoor delegation techniques in [1,14,43], one can extend our generic IBE scheme $\mathcal{IBE}$ into a generic hierarchical IBE (HIBE)

scheme. We now give a sketch of the construction. For identity depth $d \geq 1$, we include d different PHF keys $\{K_i\}_{i \in \{1,...,d\}}$ in master public key, and the "public key" $\mathbf{A}_{id}$ for any identity $id = (id_1, \ldots, id_{d'})$ with depth $d' \leq d$ is defined as $\mathbf{A}_{id} = (\mathbf{A}\|\mathsf{H}_{K_1}(id_1)\|\cdots\|\mathsf{H}_{K_{d'}}(id_{d'}))$. Then, one can use $\mathbf{A}_{id}$ to encrypt plaintexts the same as in our generic IBE scheme. In order to enable the delegation of user private keys, the user private key should be replaced by a new trapdoor extended by the trapdoor of $\mathbf{A}$ using the algorithms in [1,14,43]. We note that as previous schemes using similar partitioning techniques [1,14], such a construction seems to inherently suffer from a reduction loss depending on the identity depth d in the exponent. It is still unclear whether one can adapt the dual system of Waters [50] to construct lattice-based (H)IBEs with tight security proofs.

Chosen Ciphertexts Security. Obviously, one can use the CHK technique in [13] to transform a CPA secure HIBE for identity depth d to a CCA secure HIBE for identity depth $d - 1$, by appending each identity in the encryption with the verification key of a one-time strongly EUF-CMA signature scheme. In our case, one can obtain an INDr-ID-CCA secure IBE scheme by using a two-level INDr-ID-CPA HIBE scheme. Since the CHK technique only requires "selective-security" to deal with the one-time signature's verification key, we can construct a more efficient CCA secure IBE scheme by directly combining a normal PHF with a weak one. Since a weak PHF is usually simpler and more efficient, the resulting IBE could be more efficient than the one obtained by directly applying the CHK technique to a two-level fully secure HIBE scheme. We now give the sketch of the improved construction. In addition to a normal PHF key K in the master public key of our generic IBE scheme $\mathcal{IBE}$, we also include it a weak PHF key K_1. When generating user private key for identity id, we compute a new trapdoor of $\mathbf{A}_{id} = (\mathbf{A}\|\mathsf{H}_K(id))$ as the user private key, by using the trapdoor delegation algorithms in [1,14,43]. In the encryption algorithm, we generate a one-time signature verification key vk (for simplicity we assume the length of vk is compatible with the weak PHF), and uses the matrix $\mathbf{A}_{id,vk} = (\mathbf{A}_{id}\|\mathsf{H}_{K_1}(vk)) = (\mathbf{A}\|\mathsf{H}_K(id)\|\mathsf{H}_{K_1}(vk))$ to encrypt the plaintext as $\mathcal{IBE}$.Enc. The decryption algorithm is the same as $\mathcal{IBE}$.Dec except that it first computes the "user private key" for $\mathbf{A}_{id,vk}$ from the user private key of $\mathbf{A}_{id}$.

Acknowledgments. We would like to thank Eike Kiltz and Xusheng Zhang for their helpful discussions. We also thank the anonymous reviewers of Crypto 2016 for their insightful advices. Jiang Zhang and Zhenfeng Zhang are supported by the National Grand Fundamental Research (973) Program of China under Grant No. 2013CB338003 and the National Natural Science Foundation of China under Grant No. U1536205. Yu Chen is supported by the National Natural Science Foundation of China under Grant Nos. 61303257 and 61379141, and by the State Key Laboratory of Cryptologys Open Project under Grant No. MMKFKT201511.

References

1. Agrawal, S., Boneh, D., Boyen, X.: Efficient lattice (H)IBE in the standard model. In: Gilbert, H. (ed.) EUROCRYPT 2010. LNCS, vol. 6110, pp. 553–572. Springer, Heidelberg (2010)

2. Agrawal, S., Boneh, D., Boyen, X.: Lattice basis delegation in fixed dimension and shorter-ciphertext hierarchical IBE. In: Rabin, T. (ed.) CRYPTO 2010. LNCS, vol. 6223, pp. 98–115. Springer, Heidelberg (2010)

3. Agrawal, S., Freeman, D.M., Vaikuntanathan, V.: Functional encryption for inner product predicates from learning with errors. In: Lee, D.H., Wang, X. (eds.) ASIACRYPT 2011. LNCS, vol. 7073, pp. 21–40. Springer, Heidelberg (2011)

4. Ajtai, M.: Generating hard instances of lattice problems (extended abstract). In: STOC 1996, pp. 99–108. ACM (1996)

5. Ajtai, M.: Generating hard instances of the short basis problem. In: Wiedermann, J., Emde Boas, P., Nielsen, M. (eds.) ICALP 1999. LNCS, vol. 1644, pp. 1–9. Springer, Heidelberg (1999)

6. Alperin-Sheriff, J.: Short signatures with short public keys from homomorphic trapdoor functions. In: Katz, J. (ed.) PKC 2015. LNCS, vol. 9020, pp. 236–255. Springer, Heidelberg (2015)

7. Bai, S., Galbraith, S.D.: An improved compression technique for signatures based on learning with errors. In: Benaloh, J. (ed.) CT-RSA 2014. LNCS, vol. 8366, pp. 28–47. Springer, Heidelberg (2014)

8. Böhl, F., Hofheinz, D., Jager, T., Koch, J., Seo, J.H., Striecks, C.: Practical signatures from standard assumptions. In: Johansson, T., Nguyen, P.Q. (eds.) EUROCRYPT 2013. LNCS, vol. 7881, pp. 461–485. Springer, Heidelberg (2013)

9. Boneh, D., Boyen, X.: Efficient selective-ID secure identity-based encryption without random oracles. In: Cachin, C., Camenisch, J.L. (eds.) EUROCRYPT 2004. LNCS, vol. 3027, pp. 223–238. Springer, Heidelberg (2004)

10. Boneh, D., Franklin, M.: Identity-based encryption from the weil pairing. In: Kilian, J. (ed.) CRYPTO 2001. LNCS, vol. 2139, pp. 213–229. Springer, Heidelberg (2001)

11. Boneh, D., Venkatesan, R.: Breaking RSA may not be equivalent to factoring. In: Nyberg, K. (ed.) EUROCRYPT 1998. LNCS, vol. 1403, pp. 59–71. Springer, Heidelberg (1998)

12. Boyen, X.: Lattice mixing and vanishing trapdoors: a framework for fully secure short signatures and more. In: Nguyen, P.Q., Pointcheval, D. (eds.) PKC 2010. LNCS, vol. 6056, pp. 499–517. Springer, Heidelberg (2010)

13. Canetti, R., Halevi, S., Katz, J.: Chosen-ciphertext security from identity-based encryption. In: Cachin, C., Camenisch, J.L. (eds.) EUROCRYPT 2004. LNCS, vol. 3027, pp. 207–222. Springer, Heidelberg (2004)

14. Cash, D., Hofheinz, D., Kiltz, E., Peikert, C.: Bonsai trees, or how to delegate a lattice basis. In: Gilbert, H. (ed.) EUROCRYPT 2010. LNCS, vol. 6110, pp. 523–552. Springer, Heidelberg (2010)

15. Catalano, D., Fiore, D., Nizzardo, L.: Programmable hash functions go private: constructions and applications to (homomorphic) signatures with shorter public keys. In: Gennaro, R., Robshaw, M. (eds.) CRYPTO 2015. LNCS, vol. 9216, pp. 254–274. Springer, Heidelberg (2015)

16. Cheon, J.H., Han, K., Lee, C., Ryu, H., Stehlé, D.: Cryptanalysis of the multilinear map over the integers. In: Oswald, E., Fischlin, M. (eds.) EUROCRYPT 2015. LNCS, vol. 9056, pp. 3–12. Springer, Heidelberg (2015)

17. Cocks, C.: An identity based encryption scheme based on quadratic residues. In: Honary, B. (ed.) Cryptography and Coding 2001. LNCS, vol. 2260, pp. 360–363. Springer, Heidelberg (2001)

18. Coron, J.S., et al.: Zeroizing without low-level zeroes: new MMAP attacks and their limitations. In: Gennaro, R., Robshaw, M. (eds.) CRYPTO 2015. LNCS, vol. 9215, pp. 247–266. Springer, Heidelberg (2015)

19. Coron, J.-S., Lepoint, T., Tibouchi, M.: Practical multilinear maps over the integers. In: Canetti, R., Garay, J.A. (eds.) CRYPTO 2013, Part I. LNCS, vol. 8042, pp. 476–493. Springer, Heidelberg (2013)

20. Cramer, R., Damgård, I.: On the amortized complexity of zero-knowledge protocols. In: Halevi, S. (ed.) CRYPTO 2009. LNCS, vol. 5677, pp. 177–191. Springer, Heidelberg (2009)

21. Dodis, Y., Rafail, O., Reyzin, L., Smith, A.: Fuzzy extractors: how to generate strong keys from biometrics and other noisy data. SIAM J. Comput. **38**, 97–139 (2008)

22. Ducas, L., Durmus, A., Lepoint, T., Lyubashevsky, V.: Lattice signatures and bimodal Gaussians. In: Canetti, R., Garay, J.A. (eds.) CRYPTO 2013, Part I. LNCS, vol. 8042, pp. 40–56. Springer, Heidelberg (2013)

23. Ducas, L., Lyubashevsky, V., Prest, T.: Efficient identity-based encryption over NTRU lattices. In: Sarkar, P., Iwata, T. (eds.) ASIACRYPT 2014, Part II. LNCS, vol. 8874, pp. 22–41. Springer, Heidelberg (2014)

24. Ducas, L., Micciancio, D.: Improved short lattice signatures in the standard model. In: Garay, J.A., Gennaro, R. (eds.) CRYPTO 2014, Part I. LNCS, vol. 8616, pp. 335–352. Springer, Heidelberg (2014)

25. Erdös, P., Frankl, P., Füredi, Z.: Families of finite sets in which no set is covered by the union of r others. Isr. J. Math. **51**(1–2), 79–89 (1985)

26. Freire, E.S.V., Hofheinz, D., Paterson, K.G., Striecks, C.: Programmable hash functions in the multilinear setting. In: Canetti, R., Garay, J.A. (eds.) CRYPTO 2013, Part I. LNCS, vol. 8042, pp. 513–530. Springer, Heidelberg (2013)

27. Garg, S., Gentry, C., Halevi, S.: Candidate multilinear maps from ideal lattices. In: Johansson, T., Nguyen, P.Q. (eds.) EUROCRYPT 2013. LNCS, vol. 7881, pp. 1–17. Springer, Heidelberg (2013)

28. Gentry, C.: Practical identity-based encryption without random oracles. In: Vaudenay, S. (ed.) EUROCRYPT 2006. LNCS, vol. 4004, pp. 445–464. Springer, Heidelberg (2006)

29. Gentry, C., Peikert, C., Vaikuntanathan, V.: Trapdoors for hard lattices and new cryptographic constructions. In: STOC 2008, pp. 197–206. ACM (2008)

30. Gorbunov, S., Vaikuntanathan, V., Wichs, D.: Leveled fully homomorphic signatures from standard lattices. In: STOC 2015, pp. 469–477. ACM (2015)

31. Hanaoka, G., Matsuda, T., Schuldt, J.C.N.: On the impossibility of constructing efficient key encapsulation and programmable hash functions in prime order groups. In: Safavi-Naini, R., Canetti, R. (eds.) CRYPTO 2012. LNCS, vol. 7417, pp. 812–831. Springer, Heidelberg (2012)

32. Hofheinz, D., Jager, T., Kiltz, E.: Short signatures from weaker assumptions. In: Lee, D.H., Wang, X. (eds.) ASIACRYPT 2011. LNCS, vol. 7073, pp. 647–666. Springer, Heidelberg (2011)

33. Hofheinz, D., Kiltz, E.: Programmable hash functions and their applications. In: Wagner, D. (ed.) CRYPTO 2008. LNCS, vol. 5157, pp. 21–38. Springer, Heidelberg (2008)

34. Hofheinz, D., Kiltz, E.: Programmable hash functions and their applications. J. Cryptol. **25**(3), 484–527 (2012)

35. Hu, Y., Jia, H.: Cryptanalysis of GGH map. In: Fischlin, M., Coron, J.-S. (eds.) EUROCRYPT 2016. LNCS, vol. 9665, pp. 537–565. Springer, Heidelberg (2016). doi:10.1007/978-3-662-49890-3_21
36. Katz, J.: Digital Signatures. Springer, Berlin (2010)
37. Krawczyk, H., Rabin, T.: Chameleon signatures. In: NDSS (2000)
38. Kumar, R., Rajagopalan, S., Sahai, A.: Coding constructions for blacklisting problems without computational assumptions. In: Wiener, M. (ed.) CRYPTO 1999. LNCS, vol. 1666, pp. 609–623. Springer, Heidelberg (1999)
39. Lyubashevsky, V.: Lattice signatures without trapdoors. In: Pointcheval, D., Johansson, T. (eds.) EUROCRYPT 2012. LNCS, vol. 7237, pp. 738–755. Springer, Heidelberg (2012)
40. Lyubashevsky, V., Micciancio, D.: Asymptotically efficient lattice-based digital signatures. In: Canetti, R. (ed.) TCC 2008. LNCS, vol. 4948, pp. 37–54. Springer, Heidelberg (2008)
41. Lyubashevsky, V., Peikert, C., Regev, O.: On ideal lattices and learning with errors over rings. In: Gilbert, H. (ed.) EUROCRYPT 2010. LNCS, vol. 6110, pp. 1–23. Springer, Heidelberg (2010)
42. Lyubashevsky, V., Peikert, C., Regev, O.: A toolkit for ring-LWE cryptography. In: Johansson, T., Nguyen, P.Q. (eds.) EUROCRYPT 2013. LNCS, vol. 7881, pp. 35–54. Springer, Heidelberg (2013)
43. Micciancio, D., Peikert, C.: Trapdoors for lattices: simpler, tighter, faster, smaller. In: Pointcheval, D., Johansson, T. (eds.) EUROCRYPT 2012. LNCS, vol. 7237, pp. 700–718. Springer, Heidelberg (2012)
44. Micciancio, D., Regev, O.: Worst-case to average-case reductions based on Gaussian measures. SIAM J. Comput. **37**, 267–302 (2007)
45. Nguyen, P.Q., Zhang, J., Zhang, Z.: Simpler efficient group signatures from lattices. In: Katz, J. (ed.) PKC 2015. LNCS, vol. 9020, pp. 401–426. Springer, Heidelberg (2015)
46. Peikert, C., Rosen, A.: Efficient collision-resistant hashing from worst-case assumptions on cyclic lattices. In: Halevi, S., Rabin, T. (eds.) TCC 2006. LNCS, vol. 3876, pp. 145–166. Springer, Heidelberg (2006)
47. Regev, O.: On lattices, learning with errors, random linear codes, and cryptography. In: STOC 2005, pp. 84–93. ACM (2005)
48. Shamir, A.: Identity-based cryptosystems and signature schemes. In: Blakely, G.R., Chaum, D. (eds.) CRYPTO 1984. LNCS, vol. 196, pp. 47–53. Springer, Heidelberg (1985)
49. Vershynin, R.: Introduction to the non-asymptotic analysis of random matrices (2010). arXiv preprint arXiv:1011.3027
50. Waters, B.: Dual system encryption: realizing fully secure IBE and HIBE under simple assumptions. In: Halevi, S. (ed.) CRYPTO 2009. LNCS, vol. 5677, pp. 619–636. Springer, Heidelberg (2009)
51. Yamada, S., Hanaoka, G., Kunihiro, N.: Two-dimensional representation of cover free families and its applications: short signatures and more. In: Dunkelman, O. (ed.) CT-RSA 2012. LNCS, vol. 7178, pp. 260–277. Springer, Heidelberg (2012)
52. Zhandry, M.: Secure identity-based encryption in the quantum random oracle model. In: Canetti, R., Safavi-Naini, R. (eds.) CRYPTO 2012. LNCS, vol. 7417, pp. 758–775. Springer, Heidelberg (2012)
53. Zhang, J., Chen, Y., Zhang, Z.: Programmable hash functions from lattices: short signatures and IBEs with small key sizes. Cryptology ePrint Archive, Report 2016/523 (2016)

Fully Secure Functional Encryption for Inner Products, from Standard Assumptions

Shweta Agrawal[1], Benoît Libert[2([✉])], and Damien Stehlé[2]

[1] IIT Delhi, New Delhi, India
[2] Laboratoire LIP (U. Lyon, CNRS, ENSL, INRIA, UCBL),
ENS de Lyon, Lyon, France
benoit.libert@ens-lyon.fr

Abstract. Functional encryption is a modern public-key paradigm where a master secret key can be used to derive sub-keys SK_F associated with certain functions F in such a way that the decryption operation reveals $F(M)$, if M is the encrypted message, and nothing else. Recently, Abdalla *et al.* gave simple and efficient realizations of the primitive for the computation of linear functions on encrypted data: given an encryption of a vector $\boldsymbol{y}$ over some specified base ring, a secret key $SK_{\boldsymbol{x}}$ for the vector $\boldsymbol{x}$ allows computing $\langle \boldsymbol{x}, \boldsymbol{y} \rangle$. Their technique surprisingly allows for instantiations under standard assumptions, like the hardness of the Decision Diffie-Hellman (DDH) and Learning-with-Errors (LWE) problems. Their constructions, however, are only proved secure against *selective* adversaries, which have to declare the challenge messages M_0 and M_1 at the outset of the game.

In this paper, we provide constructions that provably achieve security against more realistic *adaptive* attacks (where the messages M_0 and M_1 may be chosen in the challenge phase, based on the previously collected information) for the same inner product functionality. Our constructions are obtained from hash proof systems endowed with homomorphic properties over the key space. They are (almost) as efficient as those of Abdalla *et al.* and rely on the same hardness assumptions.

In addition, we obtain a solution based on Paillier's composite residuosity assumption, which was an open problem even in the case of selective adversaries. We also propose LWE-based schemes that allow evaluation of inner products modulo a prime p, as opposed to the schemes of Abdalla *et al.* that are restricted to evaluations of integer inner products of short integer vectors. We finally propose a solution based on Paillier's composite residuosity assumption that enables evaluation of inner products modulo an RSA integer $N = p \cdot q$.

We demonstrate that the functionality of inner products over a prime field is powerful and can be used to construct bounded collusion FE for all circuits.

Keywords: Functional encryption · Adaptive security · Standard assumptions · DDH · LWE · Extended LWE · Composite residuosity

© International Association for Cryptologic Research 2016
M. Robshaw and J. Katz (Eds.): CRYPTO 2016, Part III, LNCS 9816, pp. 333–362, 2016.
DOI: 10.1007/978-3-662-53015-3_12

1   Introduction

Functional encryption (FE) [19,56] is a generalization of public-key encryption, which overcomes the all-or-nothing, user-based access to data that is inherent to public key encryption and enables fine grained, role-based access that makes it very desirable for modern applications. A bit more formally, given an encryption $\mathsf{enc}(X)$ and a key corresponding to a function F, the key holder only learns $F(X)$ and nothing else. Apart from its theoretical appeal, the concept of FE also finds numerous applications. In cloud computing platforms, users can store encrypted data on a remote server and subsequently provide the server with a key SK_F which allows it to compute the function F of the underlying data without learning anything else.

In some cases, the message $X = (\mathsf{IND}, M)$ consists of an index IND (which can be thought of as a set of descriptive attributes) and a message M, which is sometimes called "payload". One distinguishes FE systems with public index, where IND is publicly revealed by the ciphertext but M is hidden, from those with private index, where IND and M are both hidden. Public index FE is popularly referred to as attribute based encryption.

A Brief History of FE. The birth of Functional Encryption can be traced back to Identity Based Encryption [17,57] which can be seen as the first non-trivial generalization of Public Key Encryption. However, it was the work of Sahai and Waters [56] that coined the term Attribute Based Encryption, and the subsequent, natural unification of all these primitives under the umbrella of Functional Encryption took place only relatively recently [19,49]. Constructions of public index FE have matured from specialized – equality testing [13,17,35], keyword search [1,16,44], Boolean formulae [42], inner product predicates [44], regular languages [58] – to general polynomial-size circuits [18,34,40] and even Turing machines [37]. The journey of private index FE has been significantly more difficult, with inner product predicate constructions [3,44] being the state of the art for a long time until the recent elegant generalization to polynomial-size circuits [41].

However, although private index FE comes closer than ever before to the goal of general FE, it falls frustratingly short. This is because all known constructions of private index FE only achieve *weak attribute hiding*, which severely restricts the function keys that the adversary can request in the security game – the adversary may request keys for functions f_i that do not decrypt the challenge ciphertext (IND^*, M^*), i.e., $f_i(\mathsf{IND}^*) \neq 0$ holds for all i. The most general notion of FE – private index, strongly attribute hiding – has been built for the restricted case of bounded collusions [38,39] or using the brilliant, but ill-understood[1] machinery of multi-linear maps [33] and indistinguishability obfuscation [33]. These constructions provide FE for general polynomial-size circuits and Turing machines [37], but, perhaps surprisingly, there has been little effort to build the general notion of FE ground-up, starting from smaller functionalities.

---

[1] Indeed, the two candidate multi-linear maps [24,32] put forth in 2013 were recently found to be insecure [23,43].

This appears as a gaping hole that begs to be filled. Often, from the practical standpoint, efficient constructions for a smaller range of functionalities, such as linear functions or polynomials, are extremely relevant, and such an endeavour will also help us understand the fundamental barriers that thwart our attempts for general FE. This motivates the question:

Can we build FE for restricted classes of functions, satisfying standard security definitions, under well-understood assumptions?

In 2015, Abdalla et al. [2] considered the question of building FE for linear functions. Here, a ciphertext C encrypts a vector $y \in \mathcal{D}^\ell$ over some ring $\mathcal{D}$, a secret key for the vector $x \in \mathcal{D}^\ell$ allows computing $\langle x, y \rangle$ and nothing else about y. Note that this is quite different from the inner product predicate functionality of [3,44]: the former computes the actual value of the inner product while the latter tests whether the inner product is zero or not, and reveals a hidden bit M if so. Abdalla *et al.* [2] showed, surprisingly, that this functionality allows for very simple and efficient realizations under standard assumptions like the Decision Diffie-Hellman (DDH) and Learning-with-Errors (LWE) assumptions [53]. The instantiation from DDH was especially unexpected since DDH is not known to easily lend itself to the design of such primitives.[2] What enables this surprising result is that the functionality itself is rather limited – note that with ℓ queries, the adversary can reconstruct the entire message vector. Due to this, the scheme need not provide *collusion resistance*, which posits that no collection of secret keys for functions $F_1, \ldots, F_q$ should make it possible to decrypt a ciphertext that no individual such key can decrypt. Collusion resistance is usually the chief obstacle in proving security of FE schemes. On the contrary, for linear FE constructions, if two adversaries combine their keys, they do get a valid new key, but this key gives them a plaintext which could anyway be computed by their individual plaintexts. Hence, collusion is permitted by the functionality itself, and constructions can be much simpler. As we shall see below, linear FE is already very useful and yields many interesting applications, as we discuss in the full version of the paper [4].

More recently, Bishop et al. [12] considered the same functionality as Abdalla *et al.* in the secret-key setting with the motivation of achieving function privacy.

While [12] considers adaptive adversaries, their construction requires bilinear maps and does not operate over standard DDH-hard groups. In the public-key setting, Abdalla *et al.* [2] only proved their schemes to be secure against *selective* adversaries, that have to declare the challenge messages M_0, M_1 of the semantic security game upfront, before seeing the master public key mpk. Selective security is usually too weak a notion for practical applications and is often seen as a stepping stone to proving full adaptive security. Historically, most flavors of functional encryption have been first realized for selective adversaries [13,33,42,44,56] before being upgraded to attain full security. Boneh and Boyen [14] observed that a standard complexity leveraging argument can be used

---

[2] And indeed, this unsuitability partially manifests itself in the limitation of message/function space of the aforementioned construction: message/function vectors must be short integer vectors, and the inner product is evaluated over the integers.

to argue that a selectively-secure system is also adaptively secure. However, this argument is not satisfactory in general as the reduction incurs an exponential security loss in the message length. Quite recently, Ananth et al. [8] described a generic method of building adaptively secure functional encryption systems from selectively secure ones. However their transformation is based on the existence of a *sufficiently expressive* selectively secure FE scheme, where sufficiently secure roughly means capable of evaluating a weak PRF. Since no such scheme from standard assumptions is known, their transformation does not apply to our case, and in any case would significantly increase the complexity of the construction, even if it did.

Our Results. In this paper, we describe fully secure functional encryption systems for the evaluation of inner products on encrypted data. We propose schemes that evaluate inner products of integer vectors, based on DDH, LWE and the Composite Residuosity hardness assumptions. Our DDH-based and LWE-based constructions for integer inner products are of efficiency comparable to those of Abdalla et al. [2] and rely on the same standard assumptions. Note that a system based on Paillier's composite residuosity assumption was an open problem even for the case of selective adversaries, which we resolve in this work.

Additionally, we propose schemes that evaluate inner products modulo a prime p or a composite $N = pq$, based on the LWE and Composite Residuosity hardness assumptions. In contrast, the constructions of [2] must restrict the ring $\mathcal{D}$ to the ring of integers, which is a significant drawback. Indeed, although their DDH-based realization allows evaluating $\langle x, y \rangle \mod p$ when the latter value is sufficiently small, their security proof restricts the functionality to the computation of $\langle x, y \rangle \in \mathbb{Z}$.

The functionality of inner products over a prime field is powerful: we show that it can be bootstrapped all the way to yield a conceptually simple construction for bounded collusion FE for all circuits. The only known construction for general FE handling bounded collusions is by Gorbunov et al. [39]. Our construction is conceptually simpler, albeit a bit more inefficient. Also, since it requires the inner product functionality over a prime field, it can only be instantiated with our LWE-based scheme for now.

1.1 Overview of Techniques

We briefly summarize our techniques below.

Fully Secure Linear FE: Hash Proof Systems. Our DDH-based construction and its security proof implicitly build on hash proof systems [26]. It involves public parameters comprised of group elements $\left(g, h, \{h_i = g^{s_i} \cdot h^{t_i}\}_{i=1}^{\ell}\right)$, where g, h generate a cyclic group $\mathbb{G}$ of prime order q, and the master secret key is $\mathsf{msk} = (s, t) \in \mathbb{Z}_q^{\ell} \times \mathbb{Z}_q^{\ell}$. On input of a vector $y = (y_1, \ldots, y_\ell) \in \mathbb{Z}_q^{\ell}$, the encryption algorithm computes $\left(g^r, h^r, \{g^{y_i} \cdot h_i^r\}_{i=1}^{\ell}\right)$ in such a way that a secret key of the form $SK_x = (\langle s, x \rangle, \langle t, x \rangle)$ allows computing $g^{\langle y, x \rangle}$ in the same way as in [2]. Despite its simplicity and its efficiency (only one more group element than in [2] is needed in the ciphertext), we show that the above system can be proved fully secure using arguments – akin to those of Cramer and Shoup [25] – which consider what the adversary knows about the master secret key $(\boldsymbol{s}, \boldsymbol{t}) \in \mathbb{Z}_q^\ell \times \mathbb{Z}_q^\ell$ in the information theoretic sense. The security proof is arguably simpler than its counterpart in the selective case [2]. As in all security proofs based on hash proof systems, it uses the fact that the secret key is known to the reduction at any time, which makes it simpler to handle secret key queries without knowing the adversary's target messages $\boldsymbol{y}_0, \boldsymbol{y}_1 \in \mathbb{Z}_q^\ell$ in advance.

While our DDH-based realization only enables efficient decryption when the inner product $\langle \boldsymbol{x}, \boldsymbol{y} \rangle$ is contained in a sufficiently small interval, we show how to eliminate this restriction using Paillier's cryptosystem in the same way as in [21,22]. We thus obtain the first solution based on the Composite Residuosity assumption, which was previously an open problem (even in the case of selective security).

LWE-Based Fully Secure Linear FE. Our LWE-based construction builds on the dual Regev encryption scheme from Gentry et al. [35]. Its security analysis requires more work. The master public key contains a random matrix $\mathbf{A} \in \mathbb{Z}_q^{m \times n}$. For simplicity, we restrict ourselves to plaintext vectors and secret key vectors with binary coordinates. Each vector coordinate $i \in \{1, \ldots, \ell\}$ requires a master public key component $\boldsymbol{u}_i^T = \boldsymbol{z}_i^T \cdot \mathbf{A} \in \mathbb{Z}_q^n$, for a small norm vector $\boldsymbol{z}_i \in \mathbb{Z}^m$ made of Gaussian entries which will be part of the master secret key $\mathsf{msk} = \{\boldsymbol{z}_i\}_{i=1}^\ell$. Each $\{\boldsymbol{u}_i\}_{i=1}^\ell$ can be seen as a syndrome in the GPV trapdoor function for which vector $\boldsymbol{z}_i$ is a pre-image. Our security analysis will rely on the fact that each GPV syndrome has a large number of pre-images and, conditionally on $\boldsymbol{u}_i \in \mathbb{Z}_q^n$, each $\boldsymbol{z}_i$ retains a large amount of entropy. In the security proof, this will allow us to apply arguments similar to those of hash proof systems [26] when we will generate the challenge ciphertext using $\{\boldsymbol{z}_i\}_{i=1}^\ell$. More precisely, when the first part $\mathbf{c}_0 \in \mathbb{Z}_q^m$ of the ciphertext is a random vector instead of an actual LWE sample $\mathbf{c}_0 = \mathbf{A} \cdot \mathbf{s} + \mathbf{e}_0$, the action of $\{\boldsymbol{z}_i\}_{i=1}^\ell$ on $\mathbf{c}_0 \in \mathbb{Z}_q^m$ produces vectors that appear statistically uniform to any legitimate adversary. In order to properly simulate the challenge ciphertext using the master secret key $\{\boldsymbol{z}_i\}_{i=1}^\ell$, we use a variant of the extended LWE assumption [50] (eLWE) so as to have the (hint) values $\{\langle \boldsymbol{z}_i, \mathbf{e}_0 \rangle\}_{i=1}^\ell$ at disposal. One difficulty is that the reductions from LWE to eLWE proved in [7,20] handle a single hint vector $\boldsymbol{z}$. Fortunately, we extend the techniques of Brakerski et al. [20] using the gadget matrix from [45] to obtain a reduction from LWE to the multi-hint variant of eLWE that we use in the security proof. More specifically, we prove that the multi-hint variant of eLWE remains at least as hard as LWE when the adversary obtains as many as $n/2$ hints, where n is the dimension of the LWE secret.

Evaluation Inner Products Modulo p. Our construction from the DDH assumption natively supports the computation of inner products modulo a prime p as long as the remainder $\langle \boldsymbol{x}, \boldsymbol{y} \rangle \bmod p$ falls in a polynomial-size interval. Under the Paillier and LWE assumptions, we first show how to compute integer inner products $\langle \boldsymbol{x}, \boldsymbol{y} \rangle \in \mathbb{Z}$. In a second step, we upgrade our Paillier and LWE-based systems so as to compute inner products modulo a composite $N = pq$ and a prime p, respectively, without leaking the actual value $\langle \boldsymbol{x}, \boldsymbol{y} \rangle$ over $\mathbb{Z}$.

Hiding anything but the remainder modulo N or p requires additional techniques. In the context of LWE-based FE, this is achieved by using an LWE modulus of the form $q = p \cdot p'$ and multiplying plaintexts by p', so that an inner product modulo q over the ciphertext space natively translates into an inner product modulo p for the underlying plaintexts.

The latter plaintext/ciphertext manipulations do not solve another difficulty which arises from the discrepancy between the base rings of the master key and the secret key vectors: indeed, the master key consists of integer vectors, whereas the secret keys are defined modulo an integer. When the adversary queries a secret key vector $\boldsymbol{x} \in \mathbb{Z}_p^\ell$ (or $\mathbb{Z}_N^\ell$), it gets the corresponding combination modulo p of the master key components. By making appropriate vector queries that are linearly dependent modulo p (and hence valid), an attacker could learn a combination of the master key components which is singular modulo p but invertible over the field of rational numbers: it would then obtain the whole master key! However, note that as long as the adversary only queries secret keys for $\ell - 1$ independent vectors over $\mathbb{Z}_p^\ell$ (or $\mathbb{Z}_N^\ell$), there is no reason not to reveal more than $\ell - 1$ secret keys overall. In order to make sure that the adversary only obtains redundant information by making more than $\ell - 1$ queries, we assume that a trusted authority keeps track of all vectors $\boldsymbol{x}$ for which secret keys were previously given out (more formally, the key generation algorithm is stateful).

Compiling Linear FE to Bounded Collusion General FE. We provide a conceptually simpler way to build q-query Functional Encryption for all circuits. The only known construction for this functionality was suggested by Gorbunov et al. in [39]. At a high level, the q-query construction by Gorbunov et al. is built in several layers, as follows:

1. They start with a single key FE scheme for all circuits, which was provided by [55].
2. The single FE scheme is compiled into a q-query scheme for NC_1 circuits. This is the most non-trivial part of the construction. They run N copies of the single key scheme, where $N = O(q^4)$. To encrypt, they encrypt the views of some N-party MPC protocol computing some functionality related to C, à la "MPC in the head". For the MPC protocol, they use the BGW [10] semi-honest MPC protocol without degree reduction and exploit the fact that this protocol is completely non-interactive when used to compute bounded degree functions. The key generator provides the decryptor with a subset of the single query FE keys, where the subsets are guaranteed to have small pairwise intersections. This subset of keys enables the decryptor to recover sufficiently many shares of $C(x)$ which enables her to compute $C(x)$ via polynomial interpolation. However, an attacker with q keys only learns a share x_i in the clear if two subsets of keys intersect, and due to small pairwise intersections, this does not occur often enough for him learn sufficiently many shares of x, hence, by the guarantees of secret sharing, input x remains hidden.

3. Finally, they bootstrap the q-query FE for NC_1 to a q-query FE for all circuits using computational randomized encodings [9]. They must additionally use cover-free sets to ensure that fresh randomness is used for each randomized encoding.

Our construction replaces steps 1 and 2 with a inner product modulo p FE scheme, and then uses step 3 as in [39]. Thus, the construction of single key FE in step 1 by Sahai and Seyalioglu, and the nontrivial "MPC in the head" of step 2 can both be replaced by the simple abstraction of an inner product FE scheme. For step 3, observe that the bootstrapping theorem of [39] provides a method to bootstrap an FE for NC_1 that handles q queries to an FE for all polynomial-size circuits that is also secure against q queries. The bootstrapping relies on the result of Applebaum et $al.$ [9, Theorem 4.11] which states that every polynomial time computable function f admits a perfectly correct computational randomized encoding of degree 3. In more detail, let $\mathcal{C}$ be a family of polynomial-size circuits. Let $C \in \mathcal{C}$ and let x be some input. Let $\widetilde{C}(x, R)$ be a randomized encoding of C that is computable by a constant depth circuit with respect to inputs x and R. Then consider a new family of circuits $\mathcal{G}$ defined by:

$$G_{C,\Delta}(x, R_1, \ldots, R_S) = \left\{ \widetilde{C}\left(x; \underset{a\in\Delta}{\oplus} R_a\right) : C \in \mathcal{C}, \ \Delta \subseteq [S]\right\},$$

for some sufficiently large S (quadratic in the number of queries q). As observed in [39], circuit $G_{C,\Delta}(\cdot, \cdot)$ is computable by a constant degree polynomial (one for each output bit). Given an FE scheme for $\mathcal{G}$, one may construct a scheme for $\mathcal{C}$ by having the decryptor first recover the output of $G_{C,\Delta}(x, R_1, \ldots, R_S)$ and then applying the decoder for the randomized encoding to recover $C(x)$.

However, to support q queries the decryptor must compute q randomized encodings, each of which needs fresh randomness. This is handled by hardcoding S random elements in the ciphertext and using random subsets $\Delta \subseteq [S]$ (which are cover-free with overwhelming probability) to compute fresh randomness $\underset{a\in\Delta}{\oplus} R_a$ for every query. The authors then conclude that bounded query FE for NC_1 suffices to construct a bounded query FE scheme for all circuits.

We observe that the ingredient required to bootstrap is *not FE for the entire circuit class* NC_1 but rather only the particular circuit class $\mathcal{G}$ as described above. This circuit class, being computable by degree 3 polynomials, may be supported by a linear FE scheme, via linearization of the degree 3 polynomials! To illustrate, let us consider FE secure only for a single key. Then, the functionality that the initial FE must support is exactly the randomized encoding of [9], which, indeed, is in NC_0. Now, to support q queries, we must ensure that each key uses a fresh piece of randomness, and this is provided using a cover-free set family S as in [39] – the key generator picks a random subset $\Delta \subseteq [S]$ and sums up its elements to obtain i.i.d. randomness for the key being requested. To obtain a random element in this manner, addition over the integers does not suffice, we must take addition modulo p. Here, our inner product modulo p construction comes to our rescue!

Putting it together, the encryptor encrypts all degree 3 monomials in the inputs $R_1, \ldots, R_S$ and $x_1, \ldots, x_\ell$. Note that this ciphertext is polynomial in size. Now, for a given circuit C, the keygen algorithm samples some $\Delta \subseteq [S]$ and computes the symbolic degree 3 polynomials which must be released to the decryptor. It then provides the linear FE keys to compute the same. By correctness and security of Linear FE as well as the randomizing polynomial construction, the decryptor learns $C(x)$ and nothing else. The final notion of security that we obtain is non-adaptive simulation based security NA-SIM [39,49], i.e. (poly, poly, 0) SIM security, where the adversary can request a polynomial number of pre-challenge keys, ask for polynomially sized challenge ciphertexts but may not request post-challenge keys. For more details, we refer the reader to Sect. 6. We note that the construction of [39] also achieves the stronger AD-SIM security, but for a scheme that supports only a *single* ciphertext and bounded number of keys. The bound on the number of ciphertexts is necessary due to a lower bound by [19]. The notion of single ciphertext, bounded key FE appears to be quite restrictive, hence we do not study AD-SIM security here.

We note that subsequent to our work, Agrawal and Rosen [6] used our adaptively secure mod p inner products FE scheme in a more sophisticated manner than we do here, to achieve ciphertext size that improves upon the construction of [39].

2   Background

In this section, we recall the hardness assumptions underlying the security of the schemes we will describe. The functionality and security definitions of functional and non-interactive controlled functional encryption schemes are given in the full version of the paper [4].

Our first scheme relies on the standard DDH assumption in ordinary (i.e., non-pairing-friendly) cyclic groups.

Definition 1. *In a cyclic group $\mathbb{G}$ of prime order q, the* **Decision Diffie-Hellman** *(DDH) problem is to distinguish the distributions* $D_0 = \{(g, g^a, g^b, g^{ab}) \mid g \hookleftarrow \mathbb{G}, a, b \hookleftarrow \mathbb{Z}_q\}, D_1 = \{(g, g^a, g^b, g^c) \mid g \hookleftarrow \mathbb{G}, a, b, c \hookleftarrow \mathbb{Z}_q\}.$

A variant of our first scheme relies on Paillier's composite residuosity assumption.

Definition 2 [51]. *Let $N = pq$, for prime numbers p, q. The* **Decision Composite Residuosity** *(DCR) problem is to distinguish the distributions $D_0 := \{z = z_0^N \bmod N^2 \mid z_0 \hookleftarrow \mathbb{Z}_N^*\}$ and $D_1 := \{z \hookleftarrow \mathbb{Z}_{N^2}^*\}.$*

Our third construction builds on the Learning-With-Errors (LWE) problem, which is known to be at least as hard as certain standard lattice problems in the worst case [20,54].

Definition 3. *Let q, α, m be functions of a parameter n. For a secret $\mathbf{s} \in \mathbb{Z}_q^n$, the distribution $A_{q,\alpha,\mathbf{s}}$ over $\mathbb{Z}_q^n \times \mathbb{Z}_q$ is obtained by sampling $\mathbf{a} \hookleftarrow \mathbb{Z}_q^n$ and an $e \hookleftarrow D_{\mathbb{Z},\alpha q}$, and returning $(\mathbf{a}, \langle \mathbf{a}, \mathbf{s} \rangle + e) \in \mathbb{Z}_q^{n+1}$. The Learning With Errors* *problem* $\mathsf{LWE}_{q,\alpha,m}$ *is as follows: For* $\mathbf{s} \hookleftarrow \mathbb{Z}_q^n$, *the goal is to distinguish between the distributions:*

$$D_0(\mathbf{s}) := U(\mathbb{Z}_q^{m \times (n+1)}) \quad \text{and} \quad D_1(\mathbf{s}) := (A_{q,\alpha,\mathbf{s}})^m.$$

We say that a PPT algorithm $\mathcal{A}$ *solves* $\mathsf{LWE}_{q,\alpha}$ *if it distinguishes* $D_0(\mathbf{s})$ *and* $D_1(\mathbf{s})$ *with non-negligible advantage (over the random coins of* $\mathcal{A}$ *and the randomness of the samples), with non-negligible probability over the randomness of* $\mathbf{s}$.

3   Fully Secure Functional Encryption for Inner Products from DDH

In this section, we show that an adaptation of the DDH-based construction of Abdalla *et al.* [2] provides full security under the standard DDH assumption. Like [2], the scheme computes inner products over $\mathbb{Z}$ as long as they land in a sufficiently small interval.

In comparison with the solution of Abdalla *et al.*, we only introduce one more group element in the ciphertext and all operations are just as efficient as in [2]. Our scheme is obtained by modifying [2] in the same way as Damgård's encryption scheme [27] was obtained from the Elgamal cryptosystem. The original DDH-based system of [2] encrypts a vector $\boldsymbol{y} = (y_1, \ldots, y_\ell) \in \mathbb{Z}_q^\ell$ by computing $(g^r, \{g^{y_i} \cdot h_i^r\}_{i=1}^\ell)$, where $\{h_i = g^{s_i}\}_{i=1}^\ell$ are part of the master public key and $\mathsf{sk}_{\boldsymbol{x}} = \sum_{i=1}^\ell s_i \cdot x_i \bmod q$ is the secret key associated with the vector $\boldsymbol{x} = (x_1, \ldots, x_\ell) \in \mathbb{Z}_q^\ell$. Here, we encrypt $\boldsymbol{y}$ in the fashion of Damgård's Elgamal, by computing $(g^r, h^r, \{g^{y_i} \cdot h_i^r\}_{i=1}^\ell)$. The decryption algorithm uses secret keys of the form $\mathsf{sk}_{\boldsymbol{x}} = (\sum_{i=1}^\ell s_i \cdot x_i, \sum_{i=1}^\ell t_i \cdot x_i)$, where $h_i = g^{s_i} \cdot h^{t_i}$ for each i and $\boldsymbol{s} = (s_1, \ldots, s_\ell) \in \mathbb{Z}_q^\ell$ and $\boldsymbol{t} = (t_1, \ldots, t_\ell) \in \mathbb{Z}_q^\ell$ are part of the master key msk.

The scheme and its security proof also build on ideas from the Cramer-Shoup cryptosystem [25,26]. Analogously to the bounded-collusion-resistant IBE schemes of Goldwasser *et al.* [36], the construction can be seen as an applying a hash proof system [26] with homomorphic properties over the key space. It also bears similarities with the broadcast encryption system of Dodis and Fazio [29] in the way to use hash proof systems to achieve adaptive security.

Setup$(1^\lambda, 1^\ell)$: Choose a cyclic group $\mathbb{G}$ of prime order $q > 2^\lambda$ with generators $g, h \hookleftarrow \mathbb{G}$. Then, for each $i \in \{1, \ldots, \ell\}$, sample $s_i, t_i \hookleftarrow \mathbb{Z}_q$ and compute $h_i = g^{s_i} \cdot h^{t_i}$. Define $\mathsf{msk} := \{(s_i, t_i)\}_{i=1}^\ell$ and

$$\mathsf{mpk} := \left(\mathbb{G}, g, h, \{h_i\}_{i=1}^\ell \right).$$

Keygen$(\mathsf{msk}, \boldsymbol{x})$: To generate a key for the vector $\boldsymbol{x} = (x_1, \ldots, x_\ell) \in \mathbb{Z}_q^\ell$, compute
$\mathsf{sk}_{\boldsymbol{x}} = (s_{\boldsymbol{x}}, t_{\boldsymbol{x}}) = (\sum_{i=1}^\ell s_i \cdot x_i, \sum_{i=1}^\ell t_i \cdot x_i) = (\langle \boldsymbol{s}, \boldsymbol{x} \rangle, \langle \boldsymbol{t}, \boldsymbol{x} \rangle)$.

Encrypt$(\mathsf{mpk}, \boldsymbol{y})$: To encrypt a vector $\boldsymbol{y} = (y_1, \ldots, y_\ell) \in \mathbb{Z}_q^\ell$, sample $r \hookleftarrow \mathbb{Z}_q$ and compute

$$C = g^r, \qquad D = h^r, \qquad \{E_i = g^{y_i} \cdot h_i^r\}_{i=1}^\ell.$$

Return $C_{\boldsymbol{y}} = (C, D, E_1, \ldots, E_\ell)$.

Decrypt(mpk, sk_x, C_y): Given $sk_x = (s_x, t_x)$, compute

$$E_x = (\prod_{i=1}^{\ell} E_i^{x_i})/(C^{s_x} \cdot D^{t_x}).$$

Then, compute and output $\log_g(E_x)$.

The decryption algorithm requires to compute a discrete logarithm. This is in general too expensive. Like in [2], this can be circumvented by imposing that the computed inner product lies in an interval $\{0, \ldots, L\}$, for some polynomially bounded integer L. Then, computing the required discrete logarithm may be performed in time $\widetilde{O}(L^{1/2})$ using Pollard's kangaroo method [52]. As reported in [11], this can be reduced to $\widetilde{O}(L^{1/3})$ operations by precomputing a table of size $\widetilde{O}(L^{1/3})$. Note that even though the functionality is limited (decryption may not be performed efficiently for all key vectors and for all message vectors), while proving security we will let the adversary query any key vector in $\mathbb{Z}_q^{\ell}$.

Before proceeding with the security proof, we would like to clarify that, although the scheme of [2] only decrypts values in a polynomial-size space, the usual complexity leveraging argument does not prove it fully secure via a polynomial reduction. Indeed, when ℓ is polynomial in λ, having the inner product $\langle y, x \rangle$ in a small interval does not mean that original vector $y \in \mathbb{Z}_q^{\ell}$ lives in a polynomial-size universe. In Sect. 5, we show how to eliminate the small-interval restriction using Paillier's cryptosystem [51].

The security analysis uses similar arguments to those of Cramer and Shoup [25, 26] in that it exploits the fact that mpk does not reveal too much information about the master secret key. At some step, the challenge ciphertext is generated using msk instead of the public key and, as long as msk retains a sufficient amount of entropy from the adversary's view, it will perfectly hide which vector among y_0, y_1 is actually encrypted. The reason why we can prove adaptive security is the fact that, as usual in security proofs relying on hash proof systems [25, 26], the reduction knows the master secret key at any time. It can thus correctly answer all secret key queries without knowing the challenge messages y_0, y_1 beforehand.

The DDH-based scheme can easily be generalized so as to rely on weaker variants of DDH, like the Decision Linear assumption [15] or the Matrix DDH assumption [31].

Theorem 1. *The scheme provides full security under the* DDH *assumption.* (The proof is given in the full version of the paper [4]).

4 Full Security Under the LWE Assumption

We describe two LWE-based schemes: the first one for integer inner products of short integer vectors, the second one for inner products over a prime field $\mathbb{Z}_p$.

In both cases, the security relies on the hardness of a variant of the extended-LWE problem. The extended-LWE problem introduced by O'Neill et al. [50] and further investigated in [7,20]. At a high level, the extended-LWE problem can be seen as $\mathsf{LWE}_{\alpha,q}$ with a fixed number m of samples, for which some extra information on the LWE noises is provided: the adversary is provided a given linear combination of the noise terms. More concretely, the problem is to distinguish between the distributions

$$\left(\mathbf{A},\ \mathbf{A}\cdot\mathbf{s}+\mathbf{e},\mathbf{z},\langle\mathbf{e},\mathbf{z}\rangle\right) \quad \text{and} \quad \left(\mathbf{A},\ \mathbf{u},\ \mathbf{z},\ \langle\mathbf{e},\mathbf{z}\rangle\right),$$

where $\mathbf{A}\hookleftarrow\mathbb{Z}_q^{m\times n}, \mathbf{s}\hookleftarrow\mathbb{Z}_q^n, \mathbf{u}\hookleftarrow\mathbb{Z}_q^m, \mathbf{e}\hookleftarrow D_{\mathbb{Z},\alpha q}^m$, and $\mathbf{z}$ is sampled from a specified distribution. Note that in [50], a noise was added to the term $\langle\mathbf{e},\mathbf{z}\rangle$. The LWE to extended-LWE reductions from [7,20] do not require such an extra noise term.

We will use a variant of extended-LWE for which multiple hints $(\mathbf{z}_i, \langle\mathbf{e},\mathbf{z}_i\rangle)$ are given, for the same noise vector $\mathbf{e}$.

Definition 4 (Multi-hint Extended-LWE). *Let q, m, t be integers, α be a real and τ be a distribution over $\mathbb{Z}^{t\times m}$, all of them functions of a parameter n. The multi-hint extended-LWE problem $\mathsf{mheLWE}_{q,\alpha,m,t,\tau}$ is to distinguish between the distributions of the tuples*

$$\left(\mathbf{A},\ \mathbf{A}\cdot\mathbf{s}+\mathbf{e},\mathbf{Z},\ \mathbf{Z}\cdot\mathbf{e}\right) \quad and \quad \left(\mathbf{A},\ \mathbf{u},\ \mathbf{Z},\ \mathbf{Z}\cdot\mathbf{e}\right),$$

where $\mathbf{A}\hookleftarrow\mathbb{Z}_q^{m\times n}, \mathbf{s}\hookleftarrow\mathbb{Z}_q^n, \mathbf{u}\hookleftarrow\mathbb{Z}_q^m, \mathbf{e}\hookleftarrow D_{\mathbb{Z},\alpha q}^m$, and $\mathbf{Z}\hookleftarrow\tau$.

A reduction from LWE to mheLWE is presented in Subsect. 4.3.

4.1 Integer Inner Products of Short Integer Vectors

In the description hereunder, we consider the message space $\mathcal{P}=\{0,\dots,P-1\}^\ell$, for some integer P and where $\ell\in\mathsf{poly}(n)$ denotes the dimension of vectors to encrypt. Secret keys are associated with vectors in $\mathcal{V}=\{0,\dots,V-1\}^\ell$ for some integer V. As in the DDH case, inner products are evaluated over $\mathbb{Z}$. However, unlike our DDH-based construction, we can efficiently decrypt without confining inner product values within a small interval: here the inner product between the plaintext and key vectors belongs to $\{0,\dots,K-1\}$ with $K=\ell PV$, and it is possible to set parameters so that the scheme is secure under standard hardness assumptions while K is more than polynomial in the security parameter. We compute ciphertexts using a prime modulus q, with q significantly larger than K.

Setup$(1^n, 1^\ell, P, V)$: Set integers $m, q \geq 2$, real $\alpha\in(0,1)$ and distribution τ over $\mathbb{Z}^{\ell\times m}$ as explained below. Set $K=\ell PV$. Sample $\mathbf{A}\hookleftarrow\mathbb{Z}_q^{m\times n}$ and $\mathbf{Z}\hookleftarrow\tau$. Compute $\mathbf{U}=\mathbf{Z}\cdot\mathbf{A}\in\mathbb{Z}_q^{\ell\times n}$. Define $\mathsf{mpk}:=(\mathbf{A},\mathbf{U},K,P,V)$ and $\mathsf{msk}:=\mathbf{Z}$.

Keygen$(\mathsf{msk},\mathbf{x})$: Given a vector $\mathbf{x}\in\mathcal{V}$, compute and return the secret key $\mathbf{z}_{\mathbf{x}}:=\mathbf{x}^T\cdot\mathbf{Z}\in\mathbb{Z}^m$.

Encrypt$(\mathsf{mpk},\mathbf{y})$: To encrypt a vector $\mathbf{y}\in\mathcal{P}$, sample $\mathbf{s}\hookleftarrow\mathbb{Z}_q^n, \mathbf{e}_0\hookleftarrow D_{\mathbb{Z},\alpha q}^m$ and $\mathbf{e}_1\hookleftarrow D_{\mathbb{Z},\alpha q}^\ell$ and compute

$$\mathbf{c}_0 = \mathbf{A}\cdot\mathbf{s}+\mathbf{e}_0\ \in\mathbb{Z}_q^m,$$

$$\mathbf{c}_1 = \mathbf{U}\cdot\mathbf{s}+\mathbf{e}_1+\left\lfloor\frac{q}{K}\right\rfloor\cdot\mathbf{y}\ \in\mathbb{Z}_q^\ell.$$

Then, return $C := (c_0, c_1)$.

Decrypt(mpk, x, z_x, C): Given a ciphertext $C := (c_0, c_1)$ and a secret key z_x for $x \in \mathcal{V}$, compute $\mu' = \langle x, c_1 \rangle - \langle z_x, c_0 \rangle \bmod q$ and output the value $\mu \in \{-K+1, \ldots, K-1\}$ that minimizes $||\lfloor \frac{q}{K} \rfloor \cdot \mu - \mu'|$.

Setting the Parameters. Let B_τ be such that with probability $\geq 1 - n^{-\omega(1)}$, each row of sample from τ has norm $\leq B_\tau$. As explained just below, correctness may be ensured by setting

$$\alpha^{-1} \geq K^2 B_\tau \omega(\sqrt{\log n}) \quad \text{and} \quad q \geq \alpha^{-1} \omega(\sqrt{\log n}).$$

The choice of τ is driven by the reduction from LWE to mheLWE (as summarized in Theorem 4), and more precisely from Lemma 4 (another constraint arises from the use of [35, Corollary 2.8] at the end of the security proof). We may choose $\tau = D_{\mathbb{Z}, \sigma_1}^{\ell \times m/2} \times (D_{\mathbb{Z}^{m/2}, \sigma_2, \delta_1} \times \ldots \times D_{\mathbb{Z}^{m/2}, \sigma_2, \delta_\ell})$, where $\delta_i \in \mathbb{Z}^\ell$ denotes the ith canonical vector, and the standard deviation parameters satisfy $\sigma_1 = \Theta(\sqrt{n \log m} \max(\sqrt{m}, K))$ and $\sigma_2 = \Theta(n^{7/2} m^{1/2} \max(m, K^2) \log^{5/2} m)$.

To ensure security based on $\mathsf{LWE}_{q, \alpha', m}$ in dimension $\geq c \cdot n$ for some $c \in (0, 1)$ via Theorems 2 and 4 below, one may further impose that $\ell \leq (1 - c) \cdot n$ and $m = \Theta(n \log q)$, to obtain $\alpha' = \Omega(\alpha/(n^6 K \log^2 q \log^{5/2} n))$. Note that $\mathsf{LWE}_{q, \alpha', m}$ enjoys reductions from lattice problems when $q \geq \Omega(\sqrt{n}/\alpha')$.

Combining the security and correctness requirements, we may choose $\alpha' = 1/((n \log q)^{O(1)} \cdot K^2)$ and $q = \Omega(\sqrt{n}/\alpha')$, resulting in LWE parameters that make LWE resist all known attacks running in time 2^λ, as long as $n \geq \widetilde{\Omega}(\lambda \log K)$.

Decryption Correctness. To show the correctness of the scheme, we first observe that, modulo q:

$$\mu' = \langle x, c_1 \rangle - \langle z_x, c_0 \rangle = \lfloor q/K \rfloor \cdot \langle x, y \rangle + \langle x, e_1 \rangle - \langle z_x, e_0 \rangle.$$

Below, we show that the magnitude of the term $\langle x, e_1 \rangle - \langle z_x, e_0 \rangle$ is $\leq \ell V B_\tau \alpha q \omega(\sqrt{\log n})$ with probability $\geq 1 - n^{-\omega(1)}$. Thanks to the choices of α and q, the latter upper bound is $\leq \lfloor q/K \rfloor/4$, which suffices to guarantee decryption correctness.

Note that e_1 is an integer Gaussian vector of dimension ℓ and standard deviation $\alpha q \geq \omega(\sqrt{\log n})$, and that $\|x\| \leq \sqrt{\ell} V$. As a result, we have that $|\langle x, e_1 \rangle| \leq \sqrt{\ell} V \alpha q \omega(\sqrt{\log n})$ holds with probability $1 - n^{-\omega(1)}$. Similarly, as $\|z_x\| \leq \ell V B_\tau$, we obtain that $|\langle z_x, e_0 \rangle| \leq \ell V B_\tau \alpha q \omega(\sqrt{\log n})$ holds with probability $1 - n^{-\omega(1)}$.

Full Security. In order to prove adaptive security of the scheme, we use the multi-hint extended-LWE from Definition 4. Before we provide the formal proof, we provide some intuition.

Intuition. Here we describe some challenges in proving adaptive security for our LWE construction. To begin we describe the approach used by Abdalla et al. [2] in showing selective security for a similar construction. In the selective game, the adversary must announce the challenge vectors y_0, y_1 at the outset of the game.

By definition of an admissible adversary, every query $\mathbf{x}^i$ made must satisfy the property that $\langle \mathbf{x}^i, (\mathbf{y}_0 - \mathbf{y}_1) \rangle = 0$ (over $\mathbb{Z}$) for all i. For ease of exposition, consider challenge messages $\mathbf{y}_0, \mathbf{y}_1$ that only differ in the last co-ordinate. Then, the simulator knows at the very beginning of the game, the subspace within which all queries must lie. Since the secret key is structured as $(\mathbf{x}^i)^T \mathbf{Z}$, it suffices for the simulator to pick all but the final column of $\mathbf{Z}$ in order to answer all legitimate key requests. It can set the public parameters by constructing all except one row of $\mathbf{U}$ using its choice of $\mathbf{Z}$, and receiving the final $\mathbf{u}_\ell$ from the LWE oracle. Now the challenge ciphertext can be embedded along this dimension to argue security.

In the adaptive game however, the simulator cannot know in advance which subspace the adversary's queries will lie in, hence it must pick the entire master secret key $\mathbf{Z}$ to answer key requests. Given that the simulator has no secrets, it is unclear how it may leverage the adversary. To handle this, our approach is to carefully analyze the entropy loss that occurs in the master secret $\mathbf{Z}$ via that keys seen by the adversary. We show that despite seeing linear relations involving $\mathbf{Z}$, there is enough residual entropy left in the master secret so that the challenge ciphertext created using this appears uniform to the adversary.

To the best of our knowledge, this proof technique has not been used in prior constructions of LWE based FE systems, which mostly rely on a "punctured trapdoor" approach. This approach roughly provides the simulator with a trapdoor that can be used to answer key requests but vanishes w.h.p for the challenge. Our simulator does not use trapdoors, but relies on an argument about entropy leakage as described above. We now proceed with the formal proof.

Theorem 2. *Assume that $\ell \leq n^{O(1)}$, $m \geq 4n \log_2 q$, $q > \ell K^2$ and τ is as described above. Then the functional encryption scheme above is fully secure, under the* $\mathsf{mheLWE}_{q,\alpha,m,\ell,\tau}$ *hardness assumption.*

Proof. The proof proceeds with a sequence of games that starts with the real game and ends with a game in which the adversary's advantage is negligible. For each i, we call S_i the event that the adversary wins in Game i.

Game 0: This is the genuine full security game. Namely: the adversary $\mathcal{A}$ is given the master public key mpk; in the challenge phase, adversary $\mathcal{A}$ comes up with two distinct vectors $\mathbf{y}_0, \mathbf{y}_1 \in \mathcal{P}$ and receives an encryption C of $\mathbf{y}_\beta$ for $\beta \hookleftarrow \{0, 1\}$ sampled by the challenger; when $\mathcal{A}$ halts, it outputs $\beta' \in \{0, 1\}$ and S_0 is the event that $\beta' = \beta$. Note that any vector $\mathbf{x} \in \mathcal{V}$ queried by $\mathcal{A}$ to the secret key extraction oracle must satisfy $\langle \mathbf{x}, \mathbf{y}_0 \rangle = \langle \mathbf{x}, \mathbf{y}_1 \rangle$ over $\mathbb{Z}$ if $\mathcal{A}$ is a legitimate adversary.

Game 1: We modify the generation of $C = (\mathbf{c}_0, \mathbf{c}_1)$ in the challenge phase. Namely, at the outset of the game, the challenger picks $\mathbf{s} \hookleftarrow \mathbb{Z}_q^n$, $\mathbf{e}_0 \hookleftarrow D_{\mathbb{Z},\alpha q}^m$ (which may be chosen ahead of time) as well as $\mathbf{Z} \hookleftarrow \tau$. The master public key mpk is computed by setting $\mathbf{U} = \mathbf{Z} \cdot \mathbf{A} \bmod q$. In the challenge phase, the challenger picks a random bit $\beta \hookleftarrow \{0, 1\}$ and encrypts $\mathbf{y}_\beta$ by computing (modulo q)

$$c_0 = \mathbf{A} \cdot \boldsymbol{s} + \boldsymbol{e}_0,$$
$$c_1 = \mathbf{Z} \cdot c_0 - \mathbf{Z} \cdot \boldsymbol{e}_0 + \boldsymbol{e}_1 + \lfloor q/K \rfloor \cdot \boldsymbol{y}_\beta,$$

with $\boldsymbol{e}_1 \hookleftarrow D_{\mathbb{Z},\alpha q}^\ell$. As the distribution of C is the same as in Game 0, we have $\Pr[S_1] = \Pr[S_0]$.

Game 2: We modify again the generation of $C = (c_0, c_1)$ in the challenge phase. Namely, the challenger picks $\boldsymbol{u} \hookleftarrow \mathbb{Z}_q^m$, sets $c_0 = \boldsymbol{u}$ and computes c_1 using $c_0, \mathbf{Z}$ and $\boldsymbol{e}_0$ as in Game 1.

Under the mheLWE hardness assumption with $t = \ell$, this modification has no noticeable effect on the behavior of $\mathcal{A}$. Below, we prove that $\Pr[S_2] \approx 1/2$, which completes the proof of the theorem.

Let $\boldsymbol{x}^i \in \mathcal{V}$ be the vectors corresponding to the secret key queries made by $\mathcal{A}$. As $\mathcal{A}$ is a legitimate adversary, we have $\langle \boldsymbol{x}^i, \boldsymbol{y}_0 \rangle = \langle \boldsymbol{x}^i, \boldsymbol{y}_1 \rangle$ over $\mathbb{Z}$ for each secret key query $\boldsymbol{x}^i$. Let $g \neq 0$ be the gcd of the coefficients of $\boldsymbol{y}_1 - \boldsymbol{y}_0$ and define $\boldsymbol{y} = (y_1, \ldots, y_\ell) = \frac{1}{g}(\boldsymbol{y}_1 - \boldsymbol{y}_0)$. We have that $\langle \boldsymbol{x}^i, \boldsymbol{y} \rangle = 0$ (over $\mathbb{Z}$) for all i. Consider the lattice $\{\boldsymbol{x} \in \mathbb{Z}^\ell : \langle \boldsymbol{x}, \boldsymbol{y} \rangle = 0\}$: all the queries $\boldsymbol{x}^i$ must belong to that lattice. Without loss of generality, we assume the n_0 first entries of $\boldsymbol{y}$ are zero (for some n_0), and all remaining entries are non-zero. Further, the rows of the following matrix form a basis of a full-dimensional sublattice:

$$\mathbf{X}_{top} = \begin{pmatrix} \mathbf{I}_{n_0} & & & & \\ \hline & -y_{n_0+2} & y_{n_0+1} & & \\ & & -y_{n_0+3} & y_{n_0+2} & \\ & & & \ddots & \ddots & \\ & & & & -y_\ell & y_{\ell-1} \end{pmatrix} \in \mathbb{Z}^{(\ell-1)\times\ell}.$$

We may assume that through the secret key queries, the adversary learns exactly $\mathbf{X}_{top}\mathbf{Z}$, as all the queried vectors $\boldsymbol{x}^i$ can be obtained as rational combinations of the rows of $\mathbf{X}_{top}$.

Let $\mathbf{X}_{bot} = \boldsymbol{y}^T \in \mathbb{Z}^{1\times\ell}$. Consider the matrix $\mathbf{X} \in \mathbb{Z}_q^{\ell\times\ell}$ obtained by putting $\mathbf{X}_{top}$ on top of $\mathbf{X}_{bot}$. We claim that $\mathbf{X}$ is invertible modulo q. To see this, observe that

$$\mathbf{X}\mathbf{X}^T = \begin{pmatrix} \mathbf{I}_{n_0} & & & & & \\ \hline & y_{n_0+1}^2 + y_{n_0+2}^2 & -y_{n_0+1}\cdot y_{n_0+3} & & & \\ & -y_{n_0+1}\cdot y_{n_0+3} & y_{n_0+2}^2 + y_{n_0+3}^2 & \ddots & & \\ & & \ddots & \ddots & \ddots & \\ & & & -y_{\ell-2}\cdot y_\ell & y_{\ell-1}^2 + y_\ell^2 & \\ \hline & & & & & \|\boldsymbol{y}\|^2 \end{pmatrix}$$

It can be proved by induction that its determinant is

$$\det(\mathbf{X}\mathbf{X}^T) = \left(\prod_{k=n_0+2}^{\ell-1} y_k^2 \right) \cdot \|\boldsymbol{y}\|^4.$$

As each of the y_k's is small and non-zero, they are all non-zero modulo prime q. Similarly, the integer $(\sum_{k=n_0+1}^{\ell} y_k^2)$ is non-zero and $< \ell P^2 < q$. This shows that $(\det \mathbf{X})^2 \neq 0 \bmod q$, which implies that $\mathbf{X}$ is invertible modulo q.

In Game 2, we have $\boldsymbol{c}_1 = \mathbf{Z}\boldsymbol{u} - \boldsymbol{f} + \lfloor q/K \rfloor \cdot \boldsymbol{y}_\beta$, with $\boldsymbol{f} := -\mathbf{Z}\boldsymbol{e}_0 + \boldsymbol{e}_1$. We write:

$$\boldsymbol{c}_1 = \mathbf{X}^{-1} \cdot \mathbf{X} \cdot (\mathbf{Z}\boldsymbol{u} - \boldsymbol{f} + \lfloor q/K \rfloor \cdot \boldsymbol{y}_\beta) \bmod q.$$

We will show that the distribution of $\mathbf{X} \cdot \boldsymbol{c}_1 \bmod q$ is (almost) independent of β. As $\mathbf{X}$ is (almost) independent of β and invertible over $\mathbb{Z}_q$, this implies that the distribution of $\boldsymbol{c}_1$ is (almost) independent of β and $\Pr[S_2] \approx 1/2$.

The first $\ell - 1$ entries of $\mathbf{X} \cdot \boldsymbol{c}_1$ do not depend on β because $\mathbf{X}_{top} \cdot \boldsymbol{y}_0 = \mathbf{X}_{top} \cdot \boldsymbol{y}_1 \bmod q$.

It remains to prove that the last entry of $\mathbf{X} \cdot \boldsymbol{c}_1 \bmod q$ is (almost) independent of β. For this, we show that the residual distribution of $\mathbf{X}_{bot}\mathbf{Z}$ given the tuple $(\mathbf{A}, \mathbf{Z}\mathbf{A}, \mathbf{X}_{top}\mathbf{Z})$ has high entropy. Using (a variant of) the leftover hash lemma with randomness $\mathbf{X}_{bot}\mathbf{Z}$ and seed $\boldsymbol{u}$, we will then conclude that given $(\mathbf{A}, \mathbf{Z}\mathbf{A}, \mathbf{X}_{top}\mathbf{Z})$, the pair $(\boldsymbol{u}, \mathbf{X}_{bot}\mathbf{Z}\boldsymbol{u})$ is close to uniform and hence statistically hides $\lfloor q/K \rfloor \cdot \boldsymbol{y}_\beta$ in $\boldsymbol{c}_1$.

Write $\mathbf{A} = (\mathbf{A}_1^T | \mathbf{A}_2^T)^T$ with $\mathbf{A}_1, \mathbf{A}_2 \in \mathbb{Z}_q^{(m/2) \times n}$. Similarly, write $\mathbf{Z} = (\mathbf{Z}_1 | \mathbf{Z}_2)$ with $\mathbf{Z}_1, \mathbf{Z}_2 \in \mathbb{Z}_q^{\ell \times (m/2)}$. Recall that by construction, every entry of $\mathbf{Z}_1$ is independently sampled from a zero-centered integer Gaussian of standard deviation parameter $\sigma_1 = \Theta(\sqrt{n \log m} \max(\sqrt{m}, K))$. Further, every entry of $\mathbf{Z}_2$ is independently sampled from a (not zero-centered) integer Gaussian of standard deviation parameter σ_2 that is larger than σ_1.

Lemma 1. *Conditioned on $(\mathbf{A}, \mathbf{Z}\mathbf{A}, \mathbf{X}_{top}\mathbf{Z}_1)$, the row vector $\mathbf{X}_{bot}\mathbf{Z}_1$ is distributed as $\boldsymbol{c} + D_{\|\boldsymbol{y}\|^2 \mathbb{Z}^{m/2}, \|\boldsymbol{y}\|\sigma_1, -\boldsymbol{c}}$ for some vector $\boldsymbol{c}$ that depends only on $\mathbf{X}_{top}\mathbf{Z}_1$.*

Proof. Thanks to [35, Corollary 2.8], we have that $\mathbf{Z}_2\mathbf{A}_2$ is within $2^{-\Omega(n)}$ statistical distance to uniform. It hence statistically hides the term $\mathbf{Z}_1\mathbf{A}_1$ in $\mathbf{Z}\mathbf{A} = \mathbf{Z}_1\mathbf{A}_1 + \mathbf{Z}_2\mathbf{A}_2$, and we obtain that given $(\mathbf{A}, \mathbf{Z}\mathbf{A})$, the distribution of each entry of $\mathbf{Z}_1$ is still $D_{\mathbb{Z},\sigma_1}$.

Note that in $\mathbf{X}_{top}\mathbf{Z}_1$ and $\mathbf{X}_{bot}\mathbf{Z}_1$, matrices $\mathbf{X}_{top}$ and $\mathbf{X}_{bot}$ act in parallel on the columns of $\mathbf{Z}_1$. To prove the claim, it suffices to consider the distribution of $\mathbf{X}_{bot}\boldsymbol{z}$ conditioned on $\mathbf{X}_{top}\boldsymbol{z}$, with $\boldsymbol{z}$ sampled from $D_{\mathbb{Z}^\ell,\sigma_1}$. Let $\boldsymbol{b} = \mathbf{X}_{top}\boldsymbol{z} \in \mathbb{Z}^{\ell-1}$ and fix $\boldsymbol{z}_0 \in \mathbb{Z}^\ell$ arbitrary such that $\boldsymbol{b} = \mathbf{X}_{top}\boldsymbol{z}_0$. The distribution of $\boldsymbol{z}$ given that $\mathbf{X}_{top}\boldsymbol{z} = \boldsymbol{b}$ is $\boldsymbol{z}_0 + D_{\Lambda,\sigma_1,-\boldsymbol{z}_0}$, with $\Lambda = \{\boldsymbol{x} \in \mathbb{Z}^\ell : \mathbf{X}_{top}\boldsymbol{x} = \mathbf{0}\}$. By construction of $\mathbf{X}$, we have that $\Lambda = \mathbb{Z}\boldsymbol{y}$. As a result, the conditional distribution of $\mathbf{X}_{bot}\boldsymbol{z}$ is $\boldsymbol{c} + D_{\|\boldsymbol{y}\|^2 \mathbb{Z}, \|\boldsymbol{y}\|\sigma_1, -\boldsymbol{c}}$ with $\boldsymbol{c} = \langle \boldsymbol{y}, \boldsymbol{z}_0 \rangle \in \mathbb{Z}$. $\square$

Now, let us write $\boldsymbol{u} = (\boldsymbol{u}_1^T | \boldsymbol{u}_2^T)^T$ for vectors $\boldsymbol{u}_1, \boldsymbol{u}_2 \in \mathbb{Z}_q^{m/2}$. We have $\mathbf{X}_{bot}\mathbf{Z}\boldsymbol{u} = \mathbf{X}_{bot}\mathbf{Z}_1\boldsymbol{u}_1 + \mathbf{X}_{bot}\mathbf{Z}_2\boldsymbol{u}_2$. Thanks to the claim above and the result of [35, Corollay 2.8], we obtain that the distribution of $(\boldsymbol{u}_1, \langle D_{\|\boldsymbol{y}\|^2 \mathbb{Z}^{m/2}, \|\boldsymbol{y}\|\sigma_1, -\boldsymbol{c}}, \boldsymbol{u}_1 \rangle)$ is within $2^{-\Omega(n)}$ statistical distance to uniform (note that $\|\boldsymbol{y}\|^2$ is invertible modulo q, that $D_{\|\boldsymbol{y}\|^2 \mathbb{Z}^{m/2}, \|\boldsymbol{y}\|\sigma_1, -\boldsymbol{c}} = \|\boldsymbol{y}\|^2 \cdot$

$D_{\mathbb{Z}^{m/2}, \sigma_1/\|y\|, -c/\|y\|^2}$, and that $\sigma_1/\|y\|$ satisfies the assumption of [35, Corollay 2.8]). This implies that given $(\mathbf{A}, \mathbf{ZA}, \mathbf{X}_{top}\mathbf{Z})$, the pair $(u, \mathbf{X}_{bot}\mathbf{Z}u)$ is close to uniform, which completes the security proof.   □

4.2   Inner Products Modulo a Prime p

We now modify the LWE-based scheme above so that it enables secure functional encryption for inner products modulo prime p. The plaintext and key vectors now belong to $\mathbb{Z}_p^\ell$.

Note that the prior scheme evaluates inner products over the integers and is insecure if ported as is to the modulo p setting. To see this, consider the following simple attack in which the adversary requests a single key $\mathbf{x}$ so that integer inner product with the challenge messages $\mathbf{y}_0$ and $\mathbf{y}_1$ are different by a multiple of p. Since the functionality posits that the inner product evaluations only agree modulo p, this is an admissible query. However, since decryption is performed over $\mathbb{Z}_q$ with q much larger than p, the adversary can easily distinguish. To prevent this attack, we scale the encrypted message by a factor of q/p (instead of $\lfloor q/K \rfloor$ as in the previous scheme): decryption modulo q forces arithmetic modulo p on the underlying plaintext.

A related difficulty in adapting the previous LWE-based scheme to modular inner products is the distribution of the noise component after inner product evaluation. Ciphertexts are manipulated modulo q, which internally manipulates plaintexts modulo p. If implemented naively, the carries of the plaintext computations may spill outside of the plaintext slots and bias the noise components of the ciphertexts. This may result in distinguishing attacks. To handle this, we take q a multiple of p. This adds some technical complications, as $\mathbb{Z}_q$ is hence not a field anymore.

A different attack is that the adversary may request keys for vectors that are linearly dependent modulo p but linearly independent over the integers. Note that with ℓ such queries, the attacker can recover the master secret key. To prevent this attack, we modify the scheme in that the authority is now stateful and keeps a record of all key queries made so far, so that it can make sure that key queries that are linearly dependent modulo p remain so modulo q. We also take q a power of p to simplify the implementation of this idea.

We note that for our application to bounded query FE for all circuits, all queries will be linearly independent modulo p, hence we will not require a stateful keygen. For details, see Sect. 6.

We now describe our scheme for inner products modulo p.

Setup$(1^n, 1^\ell, p)$: Set integers $m, q = p^k$ for some integer k, real $\alpha \in (0,1)$ and distribution τ over $\mathbb{Z}^{\ell \times m}$ as explained below. Sample $\mathbf{A} \hookleftarrow \mathbb{Z}_q^{m \times n}$ and $\mathbf{Z} \hookleftarrow \tau$. Compute $\mathbf{U} = \mathbf{Z} \cdot \mathbf{A} \in \mathbb{Z}_q^{\ell \times n}$. Define $\mathsf{mpk} := (\mathbf{A}, \mathbf{U})$ and $\mathsf{msk} := \mathbf{Z}$.

Keygen$(\mathsf{msk}, x, \mathsf{st})$: Given a vector $x \in \mathbb{Z}_p^\ell$, and an internal state st, compute the secret key z_x as follows. Recall that Keygen is a stateful algorithm with empty initial State st. At any point in the scheme execution, State st contains at most ℓ tuples $(x_i, \overline{x}_i, z_i)$ where the x_i's are (a subset of the) key queries that have been made so far, and the $(\overline{x}_i, z_i)$'s are the corresponding secret keys. If x is linearly independent from the x_i's modulo p, set $\overline{x} = x \in \mathbb{Z}^{\ell}$ (with coefficients in $[0, p)$), $z_x = \overline{x}^T \cdot Z \in \mathbb{Z}^m$ and add $(x, \overline{x}, z_x)$ to st. If $x = \sum_i k_i x_i \bmod p$ for some k_i's in $[0, p)$, then set $\overline{x} = \sum_i k_i \overline{x}_i \in \mathbb{Z}^{\ell}$ and $z_x = \sum_i k_i z_i \in \mathbb{Z}^m$. In both cases, return $(\overline{x}, z_x)$.

Encrypt(mpk, y): To encrypt a vector $y \in \mathbb{Z}_p^{\ell}$, sample $s \hookleftarrow \mathbb{Z}_q^n$, $e_0 \hookleftarrow D_{\mathbb{Z}, \alpha q}^m$ and $e_1 \hookleftarrow D_{\mathbb{Z}, \alpha q}^{\ell}$ and compute

$$c_0 = A \cdot s + e_0 \in \mathbb{Z}_q^m,$$
$$c_1 = U \cdot s + e_1 + p^{k-1} \cdot y \in \mathbb{Z}_q^{\ell}.$$

Then, return $C := (c_0, c_1)$.

Decrypt(mpk, $(\overline{x}, z_x), C$): Given a ciphertext $C := (c_0, c_1)$ and a secret key $(\overline{x}, z_x)$ for $x \in \mathbb{Z}_p^{\ell}$, compute $\mu' = \langle \overline{x}, c_1 \rangle - \langle z_x, c_0 \rangle \bmod q$ and output the value $\mu \in \mathbb{Z}_p$ that minimizes $|p^{k-1} \cdot \mu - \mu'|$.

Decryption Correctness. Correctness derives from the following observation:

$$\mu' = \langle \overline{x}, c_1 \rangle - \langle z_x, c_0 \rangle = p^{k-1} \cdot (\langle x, y \rangle \bmod p) + \langle \overline{x}, e_1 \rangle - \langle z_x, e_0 \rangle \bmod q.$$

By adapting the proof of the first LWE-based scheme, we can show that the magnitude of the term $\langle \overline{x}, e_1 \rangle - \langle z_x, e_0 \rangle$ is $\leq \ell^2 p^2 B_\tau \alpha q \omega(\sqrt{\log n})$ with probability $\geq 1 - n^{-\omega(1)}$. This follows from the bound $\|z_x\| \leq \ell \|\overline{x}\| \leq \ell^2 p^2 B_\tau$.

Setting the Parameters. The main difference with the previous LWE-based scheme with respect to parameter conditions is the choice of q of the form $q = p^k$ instead of q prime. As explained just above, correctness may be ensured by setting

$$\alpha^{-1} \geq \ell^2 p^3 B_\tau \omega(\sqrt{\log n}) \quad \text{and} \quad q \geq \alpha^{-1} \omega(\sqrt{\log n}).$$

The choice of τ is driven by Lemma 2 below (the proof requires that σ_1 is large) and the reduction from LWE to mheLWE (as summarized in Theorem 4), and more precisely from Lemma 4. We may choose $\tau = D_{\mathbb{Z}, \sigma_1}^{\ell \times m/2} \times (D_{\mathbb{Z}^{m/2}, \sigma_2, \delta_1} \times \ldots \times D_{\mathbb{Z}^{m/2}, \sigma_2, \delta_\ell})$, where $\delta_i \in \mathbb{Z}^{\ell}$ denotes the ith canonical vector, and the standard deviation parameters satisfy $\sigma_1 = \Theta(\sqrt{n \log m} \max(\sqrt{m}, K'))$ and $\sigma_2 = \Theta(n^{7/2} m^{1/2} \max(m, K'^2) \log^{5/2} m)$, with $K' = (\sqrt{\ell} p)^{\ell}$.

To ensure security based on $\mathsf{LWE}_{q, \alpha', m}$ in dimension $\geq c \cdot n$ for some $c \in (0, 1)$ via Theorems 2 and 4 below, one may further impose that $\ell \leq (1 - c) \cdot n$ and $m = \Theta(n \log q)$, to obtain $\alpha' = \Omega(\alpha/(n^6 K' \log^2 q \log^{5/2} n))$. Remember that $\mathsf{LWE}_{q, \alpha', m}$ enjoys reductions from lattice problems when $q \geq \Omega(\sqrt{n}/\alpha')$.

Note that the parameter conditions make the scheme efficiency degrade quickly when ℓ increases, as K' is exponential in ℓ. Assume that $p \leq n^{O(1)}$ and $\ell = \Omega(\log n)$. Then $\sigma_1, \sigma_2, 1/\alpha, 1/\alpha'$ and q can all be set as $2^{\tilde{O}(\ell)}$. To maintain security against all $2^{o(\lambda)}$ attacks, one may set $n = \tilde{\Theta}(\ell \lambda)$.

Theorem 3. *Assume that $\ell \leq n^{O(1)}$, $m \geq 4n \log_2 q$ and τ is as described above. Then the stateful functional encryption scheme above is fully secure, under the $\mathsf{mheLWE}_{q,\alpha,m,\ell,\tau}$ hardness assumption.*

Proof. The sequence of games in the proof of Theorem 2 can be adapted to the modified scheme. The main difficulty is to show that in the adapted version of the last game, the winning probability is close to $1/2$. Let us recall that game.

Game 2′: At the outset of the game, the challenger picks $s \hookleftarrow \mathbb{Z}_q^n$, $e_0 \hookleftarrow D_{\mathbb{Z},\alpha q}^m$ as well as $\mathbf{Z} \hookleftarrow \tau$. The master public key mpk is computed by setting $\mathbf{U} = \mathbf{Z} \cdot \mathbf{A} \bmod q$ and is provided to the adversary. In the challenge phase, adversary $\mathcal{A}$ comes up with two distinct vectors $\boldsymbol{y}_0, \boldsymbol{y}_1 \in \mathbb{Z}_p^\ell$. The challenger picks a random bit $\beta \hookleftarrow \{0,1\}$, $\boldsymbol{u} \hookleftarrow \mathbb{Z}_q^m$ and encrypts $\boldsymbol{y}_\beta$ by computing (modulo q)

$$c_0 = u,$$
$$c_1 = \mathbf{Z} \cdot c_0 - \mathbf{Z} \cdot e_0 + e_1 + p^{k-1} \cdot \boldsymbol{y}_\beta,$$

with $e_1 \hookleftarrow D_{\mathbb{Z},\alpha q}^\ell$. Note that any vector $\boldsymbol{x} \in \mathbb{Z}_p^\ell$ queried by $\mathcal{A}$ to the secret key extraction oracle must satisfy $\langle \boldsymbol{x}, \boldsymbol{y}_0 \rangle = \langle \boldsymbol{x}, \boldsymbol{y}_1 \rangle \bmod p$ if $\mathcal{A}$ is a legitimate adversary. Adversary $\mathcal{A}$ is then given a secret key $(\overline{\boldsymbol{x}}, \boldsymbol{z}_x)$ as in the real scheme. When $\mathcal{A}$ halts, it outputs $\beta' \in \{0,1\}$ and wins in the event that $\beta' = \beta$.

Define $\boldsymbol{y} = \boldsymbol{y}_1 - \boldsymbol{y}_0 \in \mathbb{Z}_p^\ell$. Let $\boldsymbol{x}_i \in \mathbb{Z}_p^\ell$ be the vectors corresponding to the secret key queries made by $\mathcal{A}$. As $\mathcal{A}$ is a legitimate adversary, we have $\langle \boldsymbol{x}_i, \boldsymbol{y} \rangle = 0 \bmod p$ for each secret key query $\boldsymbol{x}_i$.

We consider the view of the adversary after it has made exactly j key queries that are linearly independent modulo p, for each j from 0 up to $\ell - 1$. In fact, counter j may stop increasing before reaching $\ell-1$, but without loss of generality, we may assume that it eventually reaches $\ell - 1$. We are to show by induction that for any j, the view of the adversary is almost independent of β. In particular, for all $j < \ell - 1$, this implies that the $(j + 1)$th linearly independent key query is almost (statistically) independent of β. It also implies, for $j = \ell - 1$, that the adversary's view through Game 2′ is almost independent of β, which is exactly what we are aiming for. In what follows, we take $j \in \{0, \ldots, \ell - 1\}$, and assume that state st is independent from β. We also assume that the jth private key query occurs after the challenge phase since the adversary's view is trivially independent of β before the generation of the challenge ciphertext.

At this stage, the state st contains exactly j tuples $(\boldsymbol{x}_i, \overline{\boldsymbol{x}}_i, \boldsymbol{z}_i)$, where the vectors $\{\boldsymbol{x}_i\}_{i=1}^j$ form a $\mathbb{Z}_p$-basis of a subspace of the $(\ell - 1)$-dimensional vector space $\boldsymbol{y}^\perp := \{\boldsymbol{x} \in \mathbb{Z}_p^\ell : \langle \boldsymbol{x}, \boldsymbol{y} \rangle = 0 \bmod p\}$. From $\boldsymbol{y}$, we deterministically extend $\{\boldsymbol{x}_i\}_{i=1}^j$ into a basis of $\boldsymbol{y}^\perp$ that is statistically independent of β. A way to interpret this is to imagine that the challenger makes dummy private key queries $\{\boldsymbol{x}_i\}_{i=j+1}^{\ell-1}$ for itself so as to get a full basis of $\boldsymbol{y}^\perp$ and creates the corresponding $\{\overline{\boldsymbol{x}}_i\}_{i=j+1}^{\ell-1}$ in $\mathbb{Z}^\ell$. We define $\mathbf{X}_{top} \in \mathbb{Z}^{(\ell-1)\times\ell}$ as the matrix whose ith row is $\overline{\boldsymbol{x}}_i$ for all i, including the genuine and dummy keys. Through the secret key queries, the adversary learns at most $\mathbf{X}_{top}\mathbf{Z} \in \mathbb{Z}^{(\ell-1)\times m}$.

Let $\boldsymbol{x}' \in \mathbb{Z}_p^\ell$ be a vector that does not belong to $\boldsymbol{y}^\perp$, and $\mathbf{X}_{bot} \in \mathbb{Z}^{1\times\ell}$ be the canonical lift of $(\boldsymbol{x}')^T$ over the integers. Consider the matrix $\mathbf{X} \in \mathbb{Z}^{\ell\times\ell}$ obtained by putting $\mathbf{X}_{top}$ on top of $\mathbf{X}_{bot}$. By construction, the matrix $\mathbf{X}$ is invertible modulo p, and hence modulo $q = p^k$. Also, by induction and construction, $\mathbf{X} \in \mathbb{Z}^{\ell \times \ell}$ is statistically independent of $\beta \in \{0, 1\}$.

In Game $2'$, we have $\mathbf{c}_1 = \mathbf{Z}\mathbf{u} - \mathbf{f} + p^{k-1} \cdot \mathbf{y}_\beta$, with $\mathbf{f} := -\mathbf{Z}\mathbf{e}_0 + \mathbf{e}_1$. We write:

$$\mathbf{c}_1 = \mathbf{X}^{-1} \cdot \mathbf{X} \cdot (\mathbf{Z}\mathbf{u} - \mathbf{f} + p^{k-1} \cdot \mathbf{y}_\beta) \mod q.$$

We will show that the distribution of $\mathbf{X} \cdot \mathbf{c}_1 \mod q$ is (almost) independent of β. As the matrix $\mathbf{X}$ is independent of $\beta \in \{0, 1\}$ and invertible over $\mathbb{Z}_q$, this implies that the distribution of $\mathbf{c}_1$ is statistically independent of β (recall that $\mathbf{X}$ is information-theoretically known to $\mathcal{A}$, which means that, if $\mathbf{c}_1$ carries any noticeable information on β, so does $\mathbf{X} \cdot \mathbf{c}_1 \mod q$). This ensures that the winning probability in Game $2'$ is negligibly far from $1/2$.

First, the first $\ell - 1$ entries of $\mathbf{X} \cdot \mathbf{c}_1$ do not depend on β because we have the equality $p^{k-1} \cdot \mathbf{X}_{top} \cdot \mathbf{y}_0 = p^{k-1} \cdot \mathbf{X}_{top} \cdot \mathbf{y}_1 \mod q$ by construction of $\mathbf{X}_{top}$.

It remains to prove that the last entry of $\mathbf{X} \cdot \mathbf{c}_1 \mod q$ is (almost) independent of β. Let us write $\mathbf{A} = (\mathbf{A}_1^T | \mathbf{A}_2^T)^T$ with $\mathbf{A}_1, \mathbf{A}_2 \in \mathbb{Z}_q^{(m/2) \times n}$. Similarly, we also write $\mathbf{Z} = (\mathbf{Z}_1 | \mathbf{Z}_2)$ with $\mathbf{Z}_1, \mathbf{Z}_2 \in \mathbb{Z}^{\ell \times (m/2)}$. Recall that by construction, every entry of $\mathbf{Z}_1$ is independently sampled from a zero-centered integer Gaussian of standard deviation parameter $\sigma_1 = \Theta(\sqrt{n \log m} \max(\sqrt{m}, K'))$ with $K' = (\sqrt{\ell}p)^\ell$. Further, every entry of $\mathbf{Z}_2$ is independently sampled from a (not zero-centered) integer Gaussian of standard deviation parameter σ_2 that is larger than σ_1.

Lemma 2. *Conditioned on $(\mathbf{A}, \mathbf{Z}\mathbf{A}, \mathbf{X}_{top}\mathbf{Z}_1)$, the row vector $\mathbf{X}_{bot}\mathbf{Z}_1 \mod p$ is within negligible statistical distance from the uniform distribution over $\mathbb{Z}_p^{m/2}$.*

Proof. Thanks to [35, Corollary 2.8], we have that $\mathbf{Z}_2\mathbf{A}_2$ is within $2^{-\Omega(n)}$ statistical distance to uniform over $\mathbb{Z}_q^{(\ell-1) \times m}$. It hence statistically hides the term $\mathbf{Z}_1\mathbf{A}_1$ in $\mathbf{Z}\mathbf{A} = \mathbf{Z}_1\mathbf{A}_1 + \mathbf{Z}_2\mathbf{A}_2 \mod q$, and we obtain that given $(\mathbf{A}, \mathbf{Z}\mathbf{A})$, the distribution of each entry of $\mathbf{Z}_1$ is still $D_{\mathbb{Z},\sigma_1}$.

Note that in $\mathbf{X}_{top}\mathbf{Z}_1$ and $\mathbf{X}_{bot}\mathbf{Z}_1$, matrices $\mathbf{X}_{top}$ and $\mathbf{X}_{bot}$ act in parallel on the columns of $\mathbf{Z}_1$. To prove the claim, it suffices to consider the distribution of $\mathbf{X}_{bot}\mathbf{z}$ conditioned on $\mathbf{X}_{top}\mathbf{z}$, with $\mathbf{z}$ sampled from $D_{\mathbb{Z}^\ell,\sigma_1}$. Let $\mathbf{b} = \mathbf{X}_{top}\mathbf{z} \in \mathbb{Z}^{\ell-1}$ and fix $\mathbf{z}_0 \in \mathbb{Z}^\ell$ arbitrary such that $\mathbf{b} = \mathbf{X}_{top}\mathbf{z}_0$. The distribution of $\mathbf{z}$ given that $\mathbf{X}_{top}\mathbf{z} = \mathbf{b}$ is $\mathbf{z}_0 + D_{\Lambda,\sigma_1,-\mathbf{z}_0}$, with $\Lambda = \{\mathbf{x} \in \mathbb{Z}^\ell : \mathbf{X}_{top}\mathbf{x} = \mathbf{0}\}$ (where the equality holds over the integers). Note that Λ is a 1-dimensional lattice in $\mathbb{Z}^\ell$.

We can write $\Lambda = \mathbf{y}' \cdot \mathbb{Z}$, for some $\mathbf{y}' \in \mathbb{Z}^\ell$. Note that there exists $\alpha \in \mathbb{Z}_p \setminus \{0\}$ such that $\mathbf{y}' = \alpha \cdot \mathbf{y} \mod p$: otherwise, the vector $\mathbf{y}'/p$ would belong to $\Lambda \setminus \mathbf{y}' \cdot \mathbb{Z}$, contradicting the definition of $\mathbf{y}'$. Further, we have $\|\mathbf{y}'\| = \det \Lambda \leq \det \Lambda'$, where Λ' is the lattice spanned by the rows of $\mathbf{X}_{top}$ (see, e.g., [48], for properties on orthogonal lattices). Hadamard's bound implies that $\|\mathbf{y}'\| \leq (\sqrt{\ell}p)^{\ell-1}$.

By [35, Corollary 2.8], the fact that $\sigma_1 \geq \sqrt{n}(\sqrt{\ell}p)^\ell$ implies that the distribution $(D_{\Lambda,\sigma_1,-\mathbf{z}_0} \mod p\Lambda)$ is within $2^{-\Omega(n)}$ statistical distance from the uniform distribution over $\Lambda/p\Lambda \simeq \mathbf{y}\mathbb{Z}_p$. We conclude that the conditional distribution of $(\mathbf{X}_{bot}\mathbf{z} \mod p)$ is within exponentially small statistical distance from the uniform distribution over $\mathbb{Z}_p$ (here we use the facts that p is prime and that $\mathbf{X}_{bot} \mathbf{y} \neq 0 \bmod p$, by construction of $\mathbf{X}_{bot}$). $\qquad \square$

Now, write $\mathbf{u} = (\mathbf{u}_1^T | \mathbf{u}_2^T)^T$ with $\mathbf{u}_1, \mathbf{u}_2 \in \mathbb{Z}_q^{m/2}$. We have $\mathbf{X}_{bot} \mathbf{Z} \mathbf{u} = \mathbf{X}_{bot} \mathbf{Z}_1 \mathbf{u}_1 + \mathbf{X}_{bot} \mathbf{Z}_2 \mathbf{u}_2$. Thanks to Lemma 2 and a variant of the leftover hash lemma modulo $q = p^k$ (given in the full version of the paper [4]), we obtain that conditioned on $(\mathbf{A}, \mathbf{Z}\mathbf{A}, \mathbf{X}_{top} \mathbf{Z})$, the distribution of $(\mathbf{u}_1, \mathbf{X}_{bot} \mathbf{Z}_1 \mathbf{u}_1)$ is within $2^{-\Omega(n)}$ statistical distance to uniform modulo q (here we used the assumption that $m \geq k + n/(\log p)$). This implies that given $(\mathbf{A}, \mathbf{Z}\mathbf{A}, \mathbf{X}_{top} \mathbf{Z})$, the pair $(\mathbf{u}, \mathbf{X}_{bot} \mathbf{Z} \mathbf{u})$ is close to uniform, which completes the security proof. $\qquad \square$

4.3 Hardness of Multi-hint Extended-LWE

In this section, we prove the following theorem, which shows that for some parameters, the mheLWE problem is no easier than the LWE problem.

Theorem 4. *Let $n \geq 100$, $q \geq 2$, $t < n$ and m with $m = \Omega(n \log n)$ and $m \leq n^{O(1)}$. There exists $\xi \leq O(n^4 m^2 \log^{5/2} n)$ and a distribution τ over $\mathbb{Z}^{t \times m}$ such that the following statements hold:*

- *There is a reduction from $\mathsf{LWE}_{q,\alpha,m}$ in dimension $n - t$ to $\mathsf{mheLWE}_{q,\alpha\xi,m,t,\tau}$ that reduces the advantage by at most $2^{\Omega(t-n)}$,*
- *It is possible to sample from τ in time polynomial in n,*
- *Each entry of matrix τ is an independent discrete Gaussian $\tau_{i,j} = D_{\mathbb{Z},\sigma_{i,j},c_{i,j}}$ for some $c_{i,j}$ and $\sigma_{i,j} \geq \Omega(\sqrt{mn \log m})$,*
- *With probability $\geq 1 - n^{-\omega(1)}$, all rows from a sample from τ have norms $\leq \xi$.*

Our reduction from LWE to mheLWE proceeds as the reduction from LWE to extended-LWE from [20], using the matrix gadget from [45] to handle the multiple hints. We first reduce LWE to the following variant of LWE in which the first samples are noise-free. This problem generalizes the first-is-errorless LWE problem from [20].

Definition 5 (First-are-errorless LWE). *Let q, α, m, t be functions of a parameter n. The first-are-errorless LWE problem $\mathsf{faeLWE}_{q,\alpha,m,t}$ is defined as follows: For $\mathbf{s} \hookleftarrow \mathbb{Z}_q^n$, the goal is to distinguish between the following two scenarios. In the first, all m samples are uniform over $\mathbb{Z}_q^n \times \mathbb{Z}_q$. In the second, the first t samples are from $A_{q,\{0\},\mathbf{s}}$ (where $\{0\}$ denotes the distribution that is deterministically zero) and the rest are from $A_{q,\alpha,\mathbf{s}}$.*

Lemma 3. *For any $n > t$, $m, q \geq 2$, and $\alpha \in (0, 1)$, there is an efficient reduction from $\mathsf{LWE}_{q,\alpha,m}$ in dimension $n - t$ to $\mathsf{faeLWE}_{q,\alpha,m,t}$ in dimension n that reduces the advantage by at most 2^{-n+t+1}.*

The proof, postponed to the appendices, is a direct adaptation of the one of [20, Lemma 4.3].

In our reduction from faeLWE to mheLWE, we use the following gadget matrix from [45, Corollary 10]. It generalizes the matrix construction from [20, Claim 4.6].

Lemma 4. *Let n, m_1, m_2 with $100 \leq n \leq m_1 \leq m_2 \leq n^{O(1)}$. Let $\sigma_1, \sigma_2 > 0$ be standard deviation parameters such that $\sigma_1 \geq \Omega(\sqrt{m_1 n \log m_1})$, $m_1 \geq \Omega(n \log(\sigma_1 n))$ and $\sigma_2 \geq \Omega(n^{5/2}\sqrt{m_1}\sigma_1^2 \log^{3/2}(m_1\sigma_1))$. Let $m = m_1 + m_2$. There exists a probabilistic polynomial time algorithm that given n, m_1, m_2 (in unary) and σ_1, σ_2 as inputs, outputs $\mathbf{G} \in \mathbb{Z}^{m \times m}$ such that:*

- *The top $n \times m$ submatrix of $\mathbf{G}$ is within statistical distance $2^{-\Omega(n)}$ of $\tau = D_{\mathbb{Z},\sigma_1}^{n \times m_1} \times (D_{\mathbb{Z}^{m_2},\sigma_2,\boldsymbol{\delta}_1} \times \ldots \times D_{\mathbb{Z}^{m_2},\sigma_2,\boldsymbol{\delta}_n})^T$ with $\boldsymbol{\delta}_i$ denoting the ith canonical unit vector,*
- *We have $|\det(\mathbf{G})| = 1$ and $\|\mathbf{G}^{-1}\| \leq O(\sqrt{nm_2}\sigma_2)$, with probability $\geq 1 - 2^{-\Omega(n)}$.*

Lemma 5. *Let $n, m_1, m_2, m, \sigma_1, \sigma_2, \tau$ be as in Lemma 4, and $\xi \geq \Omega(\sqrt{nm_2}\sigma_2)$. Let $q \geq 2$, $t \leq n$, $\alpha \geq \Omega(\sqrt{n}/q)$. Let τ_t be the distribution obtained by keeping only the first t rows from a sample from τ. There is a (dimension-preserving) reduction from $\mathsf{faeLWE}_{q,\alpha,m,t}$ to $\mathsf{mheLWE}_{q,2\alpha\xi,m,t,\tau_t}$ that reduces the advantage by at most $2^{-\Omega(n)}$.*

Proof. Let us first describe the reduction. Let $(\mathbf{A}, \boldsymbol{b}) \in \mathbb{Z}_q^m \times \mathbb{Z}_q$ be the input, which is either sampled from the uniform distribution, or from distribution $A_{q,\{0\},\boldsymbol{s}}^t \times A_{q,\alpha,\boldsymbol{s}}^{m-t}$ for some fixed $\boldsymbol{s} \hookleftarrow \mathbb{Z}_q^n$. Our objective is to distinguish between the two scenarios, using an mheLWE oracle. We compute $\mathbf{G}$ as in Lemma 4 and let $\mathbf{U} = \mathbf{G}^{-1}$. We let $\mathbf{Z} \in \mathbb{Z}^{t \times m}$ denote the matrix formed by the top t rows of $\mathbf{G}$, and let $\mathbf{U}' \in \mathbb{Z}^{m \times (m-t)}$ denote the matrix formed by the right $m - t$ columns of $\mathbf{U}$. By construction, we have $\mathbf{Z}\mathbf{U}' = \mathbf{0}$. We define $\mathbf{A}' = \mathbf{U} \cdot \mathbf{A} \bmod q$. We sample $\boldsymbol{f} \hookleftarrow D_{\alpha q(\xi^2 \mathbf{I} - \mathbf{U}'\mathbf{U}'^T)^{1/2}}$ (thanks to Lemma 4 and the choice of ξ, the matrix $\xi^2 \mathbf{I} - \mathbf{U}'\mathbf{U}'^T$ is positive definite). We sample $\boldsymbol{e}'$ from $\{0\}^t \times D_{\alpha q}^{m-t}$ and define $\boldsymbol{b}' = \mathbf{U} \cdot (\boldsymbol{b} + \boldsymbol{e}') + \boldsymbol{f}$. We then sample $\boldsymbol{c} \hookleftarrow D_{\mathbb{Z}^m - \boldsymbol{b}', \sqrt{2}\alpha\xi q}$, and define $\boldsymbol{h} = \mathbf{Z}(\boldsymbol{f} + \boldsymbol{c})$.

Finally, the reduction calls the mheLWE oracle on input $(\mathbf{A}', \boldsymbol{b}' + \boldsymbol{c}, \mathbf{Z}, \boldsymbol{h})$, and outputs the reply.

Correctness is obtained by showing that distribution $A_{q,\{0\},\boldsymbol{s}}^t \times A_{q,\alpha,\boldsymbol{s}}^{m-t}$ is mapped to the mheLWE "LWE" distribution and that the uniform distribution is mapped to the mheLWE "uniform" distribution, up to $2^{-\Omega(n)}$ statistical distances (we do not discuss these tiny statistical discrepancies below). The proof is identical to the reduction analysis in the proof of [20, Lemma 4.7]. □

Theorem 4 is obtained by combining Lemmas 3, 4 and 5.

5 Constructions Based on Paillier

In this section, we show how to remove the main limitation of our DDH-based system which is its somewhat expensive decryption algorithm. To this end, we use Paillier's cryptosystem [51] and the property that, for an RSA modulus $N = pq$, the multiplicative group $\mathbb{Z}_{N^2}^*$ contains a subgroup of order N (generated by $(N + 1)$) in which the discrete logarithm problem is easy. We also rely on the observation [21,22] that combining the Paillier and Elgamal encryption schemes makes it possible to decrypt without knowing the factorization of $N = pq$.

5.1   Computing Inner Products over $\mathbb{Z}$

In the following scheme, key vectors $\boldsymbol{x}$ and message vectors $\boldsymbol{y}$ are assumed to be of bounded norm $\|\boldsymbol{x}\| \leq X$ and $\|\boldsymbol{y}\| \leq Y$, respectively. The bounds X and Y are chosen so that $X \cdot Y < N$, where N is the composite modulus of Paillier's cryptosystem. Decryption allows to recover $\langle \boldsymbol{x}, \boldsymbol{y} \rangle \bmod N$, which is exactly $\langle \boldsymbol{x}, \boldsymbol{y} \rangle$ over the integers, thanks to the norm bounds. The security proof further requires that $\ell Y^2 < N$ and we thus assume $X, Y < (N/\ell)^{1/2}$.

Setup$(1^\lambda, 1^\ell, X, Y)$: Choose safe prime numbers $p = 2p' + 1$, $q = 2q' + 1$ with sufficiently large primes $p', q' > 2^{l(\lambda)}$, for some polynomial l, and compute $N = pq > XY$. Then, sample $g' \hookleftarrow \mathbb{Z}_{N^2}^*$ and compute $g = g'^{2N} \bmod N^2$, which generates the subgroup of $(2N)$th residues in $\mathbb{Z}_{N^2}^*$ with overwhelming probability. Then, sample an integer vector $\boldsymbol{s} = (s_1, \ldots, s_\ell)^T \hookleftarrow D_{\mathbb{Z}^\ell, \sigma}$ with discrete Gaussian entries of standard deviation $\sigma > \sqrt{\lambda} \cdot N^{5/2}$ and compute $h_i = g^{s_i} \bmod N^2$. Define

$$\mathsf{mpk} := \left(N, g, \{h_i\}_{i=1}^\ell, Y \right)$$

and $\mathsf{msk} := (\{s_i\}_{i=1}^\ell, X)$. The prime numbers p, p', q, q' are no longer needed.

Keygen$(\mathsf{msk}, \boldsymbol{x})$: To generate a key for the vector $\boldsymbol{x} = (x_1, \ldots, x_\ell) \in \mathbb{Z}^\ell$ with $\|\boldsymbol{x}\| \leq X$, compute $\mathsf{sk}_{\boldsymbol{x}} = \sum_{i=1}^\ell s_i \cdot x_i$ over $\mathbb{Z}$.

Encrypt$(\mathsf{mpk}, \boldsymbol{y})$: To encrypt a vector $\boldsymbol{y} = (y_1, \ldots, y_\ell) \in \mathbb{Z}^\ell$ with $\|\boldsymbol{y}\| \leq Y$, sample $r \hookleftarrow \{0, \ldots, \lfloor N/4 \rfloor\}$ and compute

$$C_0 = g^r \bmod N^2,$$
$$C_i = (1 + y_i N) \cdot h_i^r \bmod N^2, \qquad \forall i \in \{1, \ldots, \ell\}.$$

Return $C_{\boldsymbol{y}} = (C_0, C_1, \ldots, C_\ell) \in \mathbb{Z}_{N^2}^{\ell+1}$.

Decrypt$(\mathsf{mpk}, \mathsf{sk}_{\boldsymbol{x}}, C_{\boldsymbol{y}})$: Given $\mathsf{sk}_{\boldsymbol{x}} \in \mathbb{Z}$, compute

$$C_{\boldsymbol{x}} = \left(\prod_{i=1}^\ell C_i^{x_i} \right) \cdot C_0^{-\mathsf{sk}_{\boldsymbol{x}}} \bmod N^2.$$

Then, compute and output $\log_{(1+N)}(C_{\boldsymbol{x}}) = \frac{C_{\boldsymbol{x}} - 1 \bmod N^2}{N}$.

As in previous constructions (including those of [2]), our security proof requires inner products to be evaluated over $\mathbb{Z}$, although the decryptor technically computes $\langle \boldsymbol{x}, \boldsymbol{y} \rangle \bmod N$. The reason is that, since secret keys are computed over the integers, our security proof only goes through if the adversary is restricted to only obtain secret keys for vectors $\boldsymbol{x}$ such that $\langle \boldsymbol{x}, \boldsymbol{y}_0 \rangle = \langle \boldsymbol{x}, \boldsymbol{y}_1 \rangle$ over $\mathbb{Z}$.

Theorem 5. *The scheme provides full security under the* DCR *assumption.* (The proof is available in the full version of the paper [4]).

5.2 A Construction for Inner Products over $\mathbb{Z}_N$

Here, we show that our first DCR-based scheme can be adapted in order to compute the inner product $\langle y, x \rangle \bmod N$ instead of computing it over $\mathbb{Z}$. To do this, a first difficulty is that, as in our LWE-based system, private keys are computed over the integers and the adversary may query private keys for vectors that are linearly dependent over $\mathbb{Z}_N^\ell$ but independent over $\mathbb{Z}^\ell$. This problem is addressed as previously, by having the authority keep track of all previously revealed private keys. As in our LWE-based construction over $\mathbb{Z}_p$, we also need to increase the size of private keys (by a factor $\approx \ell$) because we have to use a different information-theoretic argument in the last step of the security proof.

Setup$(1^\lambda, 1^\ell)$: Choose safe prime numbers $p = 2p'+1$, $q = 2q'+1$ with sufficiently large primes $p', q' > 2^{l(\lambda)}$, for some polynomial l, and compute $N = pq$. Then, sample $g' \hookleftarrow \mathbb{Z}_{N^2}^*$ and compute $g = g'^{2N} \bmod N^2$, which generates the subgroup of $(2N)$th residues in $\mathbb{Z}_{N^2}^*$ with overwhelming probability. Then, sample an integer vector $s = (s_1, \ldots, s_\ell)^T \hookleftarrow D_{\mathbb{Z}^\ell, \sigma}$ with discrete Gaussian entries of standard deviation $\sigma > \sqrt{\lambda}(\sqrt{\ell}N)^{\ell+1}$ and compute $h_i = g^{s_i} \bmod N^2$. Define msk $:= \{s_i\}_{i=1}^\ell$ and

$$\mathsf{mpk} := \left(N, g, \{h_i\}_{i=1}^\ell \right).$$

Keygen$(\mathsf{msk}, x, \mathsf{st})$: To generate the jth secret key sk_x for a vector $x \in \mathbb{Z}_N^\ell$ using the master secret key msk and an (initially empty) internal state st, a stateful algorithm is used. At any time, st contains at most ℓ tuples $(x_i, \overline{x}_i, z_{x_i})$ where the $(\overline{x}_i, z_{x_i})$'s are the previously revealed secret keys and the x_i are the corresponding vectors.
  - If x is linearly independent from the x_i's modulo N, set $\overline{x} = x \in \mathbb{Z}^\ell$ (with coefficients in $[0, N)$), $z_x = \langle s, x \rangle \in \mathbb{Z}$ and add $(x, \overline{x}, z_x)$ to st.
  - If $x = \sum_i k_i x_i \bmod N$ for some coefficients $\{k_i\}_{i \leq j-1}$ in $\mathbb{Z}_N$, then compute $\overline{x} = \sum_i k_i \cdot \overline{x}_i \in \mathbb{Z}^\ell$ and $z_x = \sum_i k_i \cdot z_{x_i} \in \mathbb{Z}^m$.
  In either case, return $\mathsf{sk}_x = (\overline{x}, z_x)$.

Encrypt(mpk, y): To encrypt a vector $y = (y_1, \ldots, y_\ell) \in \mathbb{Z}_N^\ell$, sample $r \hookleftarrow \{0, \ldots, \lfloor N/4 \rfloor\}$ and compute

$$C_0 = g^r \bmod N^2,$$
$$C_i = (1 + y_i N) \cdot h_i^r \bmod N^2, \qquad \forall i \in \{1, \ldots, \ell\}.$$

Return $C_y = (C_0, C_1, \ldots, C_\ell) \in \mathbb{Z}_{N^2}^{\ell+1}$.

Decrypt$(\mathsf{mpk}, \mathsf{sk}_x, C_y)$: Given $\mathsf{sk}_x = (\overline{x}, z_x) \in \mathbb{Z}^\ell \times \mathbb{Z}$ with $\overline{x} = (\overline{x}_1, \ldots, \overline{x}_\ell)$, compute

$$C_x = \left(\prod_{i=1}^\ell C_i^{\overline{x}_i} \right) \cdot C_0^{-z_x} \bmod N^2.$$

Then, compute and output $\log_{(1+N)}(C_x) = \frac{C_x - 1 \bmod N^2}{N}$.

From a security standpoint, the following result is proved in the full version of the paper [4].

Theorem 6. *The above stateful scheme provides full security under the* DCR *assumption.*

6 Bootstrapping Linear FE to Efficient Bounded FE for All Circuits

In this section, we describe how to compile our Linear FE scheme, denoted by LinFE which computes linear functions modulo p (for us $p = 2$), into a bounded collusion FE scheme for all circuits, denoted by BddFE. The underlying scheme LinFE is assumed to be AD-IND secure, which, by [49], is equivalent to non-adaptive simulation secure NA-SIM, since linear functions are "preimage sampleable". We refer the reader to [49] for more details.

Let $\mathcal{C}$ be a family of polynomial-size circuits. Let $C \in \mathcal{C}$ and let $\mathbf{x}$ be some input. Let $\widetilde{C}(\mathbf{x}, R)$ be a randomized encoding of C that is computable by a constant depth circuit with respect to inputs x and R (see [9]). Then consider a new family of circuits $\mathcal{G}$ defined by:

$$G_{C,\Delta}(x, R_1, \ldots, R_S) = \left\{ \widetilde{C}\left(x; \bigoplus_{a \in \Delta} R_a\right) : C \in \mathcal{C}, \ \Delta \subseteq [S] \right\},$$

for some S to be chosen below. As observed in [39, Sect. 6], circuit $G_{C,\Delta}(\cdot, \cdot)$ is computable by a constant degree polynomial (one for each output bit). Given an FE scheme for $\mathcal{G}$, one may construct a scheme for $\mathcal{C}$ by having the decryptor first recover the output of $G_{C,\Delta}(\mathbf{x}, R_1, \ldots, R_S)$ and then applying the decoder for the randomized encoding to recover $C(\mathbf{x})$.

Note that to support q queries the decryptor must compute q randomized encodings, each of which needs fresh randomness. As shown above, this is handled by hardcoding sufficiently many random elements in the ciphertext and taking a random subset sum of these to generate fresh random bits for each query. As in [39], the parameters are chosen so that the subsets form a cover-free system, so that every random subset yields fresh randomness (with overwhelming probability).

In more details, we let the set S, v, m be parameters to the construction. Let Δ_i for $i \in [q]$ be a uniformly random subset of S of size v. To support q queries, we identify the set $\Delta_i \subseteq S$ with query i. If $v = O(\lambda)$ and $S = O(\lambda \cdot q^2)$ then the sets Δ_i are cover-free with high probability. For details, we refer the reader to [39, Sect. 5]. We now proceed to describe our construction. Let $L \triangleq (\ell + S \cdot m)^3$, where $m \in \mathsf{poly}(\lambda)$ is the size of the random input in the randomized encoding and ℓ is the length of the messages to be encrypted.

BddFE.Setup$(1^\lambda, 1^\ell)$: Upon input the security parameter λ and the message space $\mathcal{M} = \{0, 1\}^\ell$, invoke $(\mathsf{mpk}, \mathsf{msk}) = \mathsf{LinFE.Setup}(1^\lambda, 1^L)$ and output it.

BddFE.KeyGen(msk, C): Upon input the master secret key and a circuit C, do:

1. Sample a uniformly random subset $\Delta \subseteq S$ of size v.
2. Express $C(\mathbf{x})$ by $G_{C,\Delta}(\mathbf{x}, R_1, \ldots, R_S)$, which in turn can be expressed as a sequence of degree 3 polynomials $P_1, \ldots, P_k$, where $k \in \mathsf{poly}(\lambda)$.
3. Linearize each polynomial P_i and let P_i' be its vector of coefficients. Note that the ordering of the coefficients can be aribitrary but should be public.
4. Output $\mathsf{BddFE.SK}_C = \{\mathsf{SK}_i = \mathsf{LinFE.KeyGen}(\mathsf{LinFE.msk}, P_i')\}_{i \in [k]}$.

BddFE.Enc(x, mpk): Upon input the public key and the plaintext $\mathbf{x}$, do:

1. Sample $R_1, \ldots, R_S \leftarrow \{0,1\}^m$.
2. Compute all symbolic monomials of degree 3 in the variables $x_1, \ldots, x_\ell$ and $R_{i,j}$ for $i \in [S]$, $j \in [m]$. The number of such monomials is $L = (\ell + S \cdot m)^3$. Arrange them according to the public ordering and denote the resulting vector by $\mathbf{y}$.
3. Output $\mathsf{CT_x} = \mathsf{LinFE.Enc}(\mathsf{LinFE.mpk}, \mathbf{y})$.

BddFE.Dec(mpk, CT$_\mathbf{x}$, SK$_C$): Upon input a ciphertext $\mathsf{CT_x}$ for vector $\mathbf{x}$, and a secret key $\mathsf{SK}_C = \{\mathsf{SK}_i\}_{i \in [k]}$ for circuit C, do the following:

1. Compute $G_{C,\Delta}(\mathbf{x}, R_1, \ldots, R_S) = \{P_i(\mathbf{Y})\}_{i \in [k]} = \{\mathsf{LinFE.Dec}(\mathsf{CT_x}, \mathsf{SK}_i)\}_{i \in [k]}$.
2. Run the decoder for the randomized encoding to recover $C(\mathbf{x})$ from $G_{C,\Delta}(\mathbf{x}, R_1, \ldots, R_S)$.

Correctness follows from the correctness of LinFE and the correctness of randomized encodings.

Security. The definition for q-NA-SIM security is provided in the full version of the paper [4]. We proceed to describe our simulator Bdd.Sim. Let RE.Sim be the simulator guaranteed by the security of randomized encodings and LinFE.Sim be the simulator guaranteed by the security of the LinFE scheme.

Simulator Bdd.Sim$\big(\{C_i, C_i(\mathbf{x}), \mathsf{SK}_i\}_{i \in [q^*]}\big)$: The simulator Bdd.Sim receives the secret key queries C_i, the corresponding (honestly generated) secret keys SK_i and the values $C_i(\mathbf{x})$ for $i \in [q^*]$ where $q^* \leq q$, and must simulate the ciphertext $\mathsf{CT_x}$. It proceeds as follows:

1. Sample $\Delta_1, \ldots, \Delta_q \subseteq S$, of size v each.
2. For each $i \in [q^*]$, invoke $\mathsf{RE.Sim}(C_i(x))$ to learn $G_{C_i}(\mathbf{x}, \hat{R}_i)$ for some $\hat{R}_i$ chosen by the simulator. Interpret

$$\hat{R}_i = \bigoplus_{a \in \Delta_i} R_a \quad \text{and} \quad G_{C_i, \Delta_i}(\mathbf{x}, R_1, \ldots, R_S) = G_{C_i}(\mathbf{x}, \hat{R}_i) = \big(P_1(\mathbf{Y}), \ldots, P_k(\mathbf{Y})\big).$$

3. Let $\mathsf{CT_x} = \mathsf{LinFE.Sim}\big(\{G_{C_i, \Delta_i}, G_{C_i, \Delta_i}(\mathbf{x}, R_1, \ldots, R_S), \mathsf{SK}_i\}_{i \in [q^*]}\big)$ and output it.

The correctness of Bdd.Sim follows from the correctness of RE.Sim and LinFE.Sim.

A last remaining technicality is that the most general version of our construction for FE for inner product modulo p is stateful. This is because a general adversary against LinFE may request keys that are linearly dependent modulo p

but linearly independent over the integers, thus learning new linear relations in the master secret. This forces the simulator (and hence the key generator) to maintain a state.

However, in our application, we can make do with a stateless variant, since all the queries will be linearly independent over $\mathbb{Z}_2$. To see this, note that in the above application of LinFE, each query is randomized by a unique random set Δ_i. Recall that by cover-freeness, the element $\underset{a \in \Delta_i}{\oplus} R_a$ must contain at least one fresh random element, say R^*, which is not contained by $\underset{j \neq i}{\cup} \Delta_j$. Stated a bit differently, if we consider the query vectors of size L, then cover-freeness implies that no query vector lies within the linear span of the remaining queries made by the adversary. For any query Q, there is at least one position $j \in [L]$ so that this position is nonzero in the L vector representing Q but zero for all other vectors. Hence the query vectors are linearly independent over $\mathbb{Z}_2$, for which case, our construction of Sect. 4.2 is stateless.

Acknowledgements. We thank Fabrice Benhamouda and Hoeteck Wee for helpful discussions. This work has been supported in part by ERC Starting Grant ERC-2013-StG-335086-LATTAC. Part of this work was also funded by the "Programme Avenir Lyon Saint-Etienne de l'Université de Lyon" in the framework of the programme "Investissements d'Avenir" (ANR-11-IDEX-0007).

A   Definitions for Functional Encryption

We now recall the syntax of Functional Encryption, as defined by Boneh et al. [19], and their indistinguishability-based security definition.

Definition 6 [19]. *A functionality F defined over $(\mathcal{K}, \mathcal{Y})$ is a function $F : \mathcal{K} \times \mathcal{Y} \to \Sigma \cup \{\bot\}$, where $\mathcal{K}$ is a key space, $\mathcal{Y}$ is a message space and Σ is an output space, which does not contain the special symbol $\bot$.*

Definition 7. *A functional encryption (FE) scheme for a functionality F is a tuple $\mathcal{FE} = $ (Setup, Keygen, Encrypt, Decrypt) of algorithms with the following specifications:*

Setup(1^λ): *Takes as input a security parameter 1^λ and outputs a master key pair (mpk, msk).*

Keygen(msk, K): *Given the master secret key msk and a key (i.e., a function) $K \in \mathcal{K}$, this algorithm outputs a key sk_K.*

Encrypt(mpk, Y): *On input of a message $Y \in \mathcal{Y}$ and the master public key mpk, this randomized algorithm outputs a ciphertext C.*

Decrypt(mpk, sk_K, C): *Given the master public key mpk, a ciphertext C and a key sk_K, this algorithm outputs $v \in \Sigma \cup \{\bot\}$.*

We require that, for all (mpk, msk) $\leftarrow$ Setup(1^λ), all keys $K \in \mathcal{K}$ and all messages $Y \in \mathcal{Y}$, if $\mathsf{sk}_K \leftarrow$ Keygen(msk, K) and $C \leftarrow$ Encrypt(mpk, Y), with overwhelming probability, we have Decrypt(mpk, sk_K, C) = $F(K, Y)$ whenever $F(K, Y) \neq \bot$.

In some cases, we will also give a state st as input to algorithm Keygen, so that a stateful authority may reply to key queries in a way that depends on the queries that have been made so far. In that situation, algorithm Keygen may additionally update state st.

INDISTINGUISHABILITY-BASED SECURITY. From a security standpoint, what we expect from a FE scheme is that, given $C \leftarrow$ Encrypt(mpk, Y), the only thing revealed by a secret key sk_K about the underlying Y is the function evaluation $F(K, Y)$. In the natural definition of indistinguishability-based security (see, e.g., [19]), one asks that no efficient adversary be able to differentiate encryptions of Y_0 and Y_1 without obtaining secret keys sk_K such that $F(K, Y_0) \neq F(K, Y_1)$.

A detailed definition of indistinguishability-based security (which initially comes from [19, Sect. 4]) is given in the full version of the paper. It captures *adaptive* security in that the adversary is allowed to choose the messages Y_0, Y_1 in the middle of the game, based on the information obtained so far. Abdalla *et al.* [2] considered a weaker notion, called *selective* security, where the adversary has to declare the messages Y_0, Y_1 before even seeing mpk. In this scenario, the adversary can receive the challenge ciphertext at the same time as the public key.

References

1. Abdalla, M., et al.: Searchable encryption revisited: consistency properties, relation to anonymous IBE, and extensions. In: Shoup, V. (ed.) CRYPTO 2005. LNCS, vol. 3621, pp. 205–222. Springer, Heidelberg (2005)
2. Abdalla, M., Bourse, F., De Caro, A., Pointcheval, D.: Simple functional encryption schemes for inner products. In: Katz, J. (ed.) PKC 2015. LNCS, vol. 9020, pp. 733–751. Springer, Heidelberg (2015)
3. Agrawal, S., Freeman, D.M., Vaikuntanathan, V.: Functional encryption for inner product predicates from learning with errors. In: Lee, D.H., Wang, X. (eds.) ASIACRYPT 2011. LNCS, vol. 7073, pp. 21–40. Springer, Heidelberg (2011)
4. Agrawal, S., Libert, B., Stehlé, D.: Fully secure functional encryption for inner products, from standard assumptions. Cryptology ePrint Archive: report 2015/608
5. Agrawal, S., Gorbunov, S., Vaikuntanathan, V., Wee, H.: Functional encryption: new perspectives and lower bounds. In: Canetti, R., Garay, J.A. (eds.) CRYPTO 2013, Part II. LNCS, vol. 8043, pp. 500–518. Springer, Heidelberg (2013)
6. Agrawal, S., Rosen, A.: Online-offline functional encryption for bounded collusions. Cryptology ePrint Archive, Report 2016/361 (2016). http://eprint.iacr.org/
7. Alperin-Sheriff, J., Peikert, C.: Circular and KDM security for identity-based encryption. In: Fischlin, M., Buchmann, J., Manulis, M. (eds.) PKC 2012. LNCS, vol. 7293, pp. 334–352. Springer, Heidelberg (2012)
8. Ananth, P., Brakerski, Z., Segev, G., Vaikuntanathan, V.: From selective to adaptive security in functional encryption. In: Gennaro, R., Robshaw, M. (eds.) CRYPTO 2015. LNCS, vol. 9216, pp. 657–677. Springer, Heidelberg (2015)
9. Applebaum, B., Ishai, Y., Kushilevitz, E.: Computationally private randomizing polynomials and their applications. Comput. Complex. 15(2), 115–162 (2006)

10. Ben-Or, M., Goldwasser, S., Wigderson, A.: Completeness theorems for non-cryptographic fault-tolerant distributed computation. In: Proceedings of STOC, pp. 1–10. ACM (1988)

11. Bernstein, D.J., Lange, T.: Computing small discrete logarithms faster. In: Galbraith, S., Nandi, M. (eds.) INDOCRYPT 2012. LNCS, vol. 7668, pp. 317–338. Springer, Heidelberg (2012)

12. Bishop, A., Jain, A., Kowalczyk, L.: Function-hiding inner product encryption. In: Iwata, T., et al. (eds.) ASIACRYPT 2015. LNCS, vol. 9452, pp. 470–491. Springer, Heidelberg (2015). doi:10.1007/978-3-662-48797-6_20

13. Boneh, D., Boyen, X.: Efficient selective-ID secure identity-based encryption without random oracles. In: Cachin, C., Camenisch, J.L. (eds.) EUROCRYPT 2004. LNCS, vol. 3027, pp. 223–238. Springer, Heidelberg (2004)

14. Boneh, D., Boyen, X.: Efficient selective identity-based encryption without random oracles. J. Cryptol. 24(4), 659–693 (2011)

15. Boneh, D., Boyen, X., Shacham, H.: Short group signatures. In: Franklin, M. (ed.) CRYPTO 2004. LNCS, vol. 3152, pp. 41–55. Springer, Heidelberg (2004)

16. Boneh, D., Di Crescenzo, G., Ostrovsky, R., Persiano, G.: Public key encryption with keyword search. In: Cachin, C., Camenisch, J.L. (eds.) EUROCRYPT 2004. LNCS, vol. 3027, pp. 506–522. Springer, Heidelberg (2004)

17. Boneh, D., Franklin, M.: Identity-based encryption from the Weil pairing. SIAM J. Comput. 32(3), 586–615 (2003). (electronic)

18. Boneh, D., Gentry, C., Gorbunov, S., Halevi, S., Nikolaenko, V., Segev, G., Vaikuntanathan, V., Vinayagamurthy, D.: Fully key-homomorphic encryption, arithmetic circuit ABE and compact garbled circuits. In: Nguyen, P.Q., Oswald, E. (eds.) EUROCRYPT 2014. LNCS, vol. 8441, pp. 533–556. Springer, Heidelberg (2014)

19. Boneh, D., Sahai, A., Waters, B.: Functional encryption: definitions and challenges. In: Ishai, Y. (ed.) TCC 2011. LNCS, vol. 6597, pp. 253–273. Springer, Heidelberg (2011)

20. Brakerski, Z., Langlois, A., Peikert, C., Regev, O., Stehlé, D.: On the classical hardness of learning with errors. In: Proceedings of STOC, pp. 575–584. ACM (2013)

21. Bresson, E., Catalano, D., Pointcheval, D.: A simple public-key cryptosystem with a double trapdoor decryption mechanism and its applications. In: Laih, C.-S. (ed.) ASIACRYPT 2003. LNCS, vol. 2894, pp. 37–54. Springer, Heidelberg (2003)

22. Camenisch, J.L., Shoup, V.: Practical verifiable encryption and decryption of discrete logarithms. In: Boneh, D. (ed.) CRYPTO 2003. LNCS, vol. 2729, pp. 126–144. Springer, Heidelberg (2003)

23. Cheon, J.H., Han, K., Lee, C., Ryu, H., Stehlé, D.: Cryptanalysis of the multilinear map over the integers. In: Oswald, E., Fischlin, M. (eds.) EUROCRYPT 2015. LNCS, vol. 9056, pp. 3–12. Springer, Heidelberg (2015)

24. Coron, J.-S., Lepoint, T., Tibouchi, M.: Practical multilinear maps over the integers. In: Canetti, R., Garay, J.A. (eds.) CRYPTO 2013, Part I. LNCS, vol. 8042, pp. 476–493. Springer, Heidelberg (2013)

25. Cramer, R., Shoup, V.: A practical public key cryptosystem provably secure against adaptive chosen ciphertext attack. In: Krawczyk, H. (ed.) CRYPTO 1998. LNCS, vol. 1462, pp. 13–25. Springer, Heidelberg (1998)

26. Cramer, R., Shoup, V.: Universal hash proofs and a paradigm for adaptive chosen ciphertext secure public-key encryption. In: Knudsen, L.R. (ed.) EUROCRYPT 2002. LNCS, vol. 2332, pp. 45–64. Springer, Heidelberg (2002)

27. Damgård, I.B.: Towards practical public key systems secure against chosen ciphertext attacks. In: Feigenbaum, J. (ed.) CRYPTO 1991. LNCS, vol. 576, pp. 445–456. Springer, Heidelberg (1992)
28. De Caro, A., Iovino, V., Jain, A., O'Neill, A., Paneth, O., Persiano, G.: On the achievability of simulation-based security for functional encryption. In: Canetti, R., Garay, J.A. (eds.) CRYPTO 2013, Part II. LNCS, vol. 8043, pp. 519–535. Springer, Heidelberg (2013)
29. Dodis, Y., Fazio, N.: Public key trace and revoke scheme secure against adaptive chosen ciphertext attack. In: Desmedt, Y.G. (ed.) PKC 2003. LNCS, vol. 2567, pp. 100–115. Springer, Heidelberg (2002)
30. Ducas, L., Durmus, A., Lepoint, T., Lyubashevsky, V.: Lattice signatures and bimodal Gaussians. In: Canetti, R., Garay, J.A. (eds.) CRYPTO 2013, Part I. LNCS, vol. 8042, pp. 40–56. Springer, Heidelberg (2013)
31. Escala, A., Herold, G., Kiltz, E., Ràfols, C., Villar, J.: An algebraic framework for Diffie-Hellman assumptions. In: Canetti, R., Garay, J.A. (eds.) CRYPTO 2013, Part II. LNCS, vol. 8043, pp. 129–147. Springer, Heidelberg (2013)
32. Garg, S., Gentry, C., Halevi, S.: Candidate multilinear maps from ideal lattices. In: Johansson, T., Nguyen, P.Q. (eds.) EUROCRYPT 2013. LNCS, vol. 7881, pp. 1–17. Springer, Heidelberg (2013)
33. Garg, S., Gentry, C., Halevi, S., Raykova, M., Sahai, A., Waters, B.: Candidate indistinguishability obfuscation and functional encryption for all circuits. In: Proceedings of FOCS, pp. 40–49 (2013)
34. Garg, S., Gentry, C., Halevi, S., Sahai, A., Waters, B.: Attribute-based encryption for circuits from multilinear maps. In: Canetti, R., Garay, J.A. (eds.) CRYPTO 2013, Part II. LNCS, vol. 8043, pp. 479–499. Springer, Heidelberg (2013)
35. Gentry, C., Peikert, C., Vaikuntanathan, V.: Trapdoors for hard lattices and new cryptographic constructions. In: Proceedings of STOC, pp. 197–206. ACM (2008)
36. Goldwasser, S., Lewko, A., Wilson, D.A.: Bounded-collusion IBE from key homomorphism. In: Cramer, R. (ed.) TCC 2012. LNCS, vol. 7194, pp. 564–581. Springer, Heidelberg (2012)
37. Goldwasser, S., Kalai, Y.T., Popa, R.A., Vaikuntanathan, V., Zeldovich, N.: How to run turing machines on encrypted data. In: Canetti, R., Garay, J.A. (eds.) CRYPTO 2013, Part II. LNCS, vol. 8043, pp. 536–553. Springer, Heidelberg (2013)
38. Goldwasser, S., Tauman Kalai, Y., Popa, R., Vaikuntanathan, V., Zeldovich, N.: Reusable garbled circuits and succinct functional encryption. In: Proceedings of STOC, pp. 555–564. ACM Press (2013)
39. Gorbunov, S., Vaikuntanathan, V., Wee, H.: Functional encryption with bounded collusions via multi-party computation. In: Safavi-Naini, R., Canetti, R. (eds.) CRYPTO 2012. LNCS, vol. 7417, pp. 162–179. Springer, Heidelberg (2012)
40. Gorbunov, S., Vaikuntanathan, V., Wee, H.: Attribute-based encryption for circuits. In: Proceedings of STOC, pp. 545–554. ACM Press (2013)
41. Gorbunov, S., Vaikuntanathan, V., Wee, H.: Predicate encryption for circuits from LWE. In: Gennaro, R., Robshaw, M. (eds.) CRYPTO 2015. LNCS, vol. 9216, pp. 503–523. Springer, Heidelberg (2015)
42. Goyal, V., Pandey, O., Sahai, A., Waters, B.: Attribute-based encryption for fine-grained access control of encrypted data. In: Proceedings of ACM-CCS 2006, pp. 89–98. ACM Press (2006)
43. Hu, Y., Jia, H.: Cryptanalysis of GGH map. In: Fischlin, M., Coron, J.-S. (eds.) EUROCRYPT 2016. LNCS, vol. 9665, pp. 537–565. Springer, Heidelberg (2016). doi:10.1007/978-3-662-49890-3_21

44. Katz, J., Sahai, A., Waters, B.: Predicate encryption supporting disjunctions, polynomial equations, and inner products. In: Smart, N.P. (ed.) EUROCRYPT 2008. LNCS, vol. 4965, pp. 146–162. Springer, Heidelberg (2008)

45. Ling, S., Phan, D.H., Stehlé, D., Steinfeld, R.: Hardness of k-LWE and applications in traitor tracing. In: Garay, J.A., Gennaro, R. (eds.) CRYPTO 2014, Part I. LNCS, vol. 8616, pp. 315–334. Springer, Heidelberg (2014)

46. Micciancio, D., Mol, P.: Pseudorandom knapsacks and the sample complexity of LWE search-to-decision reductions. In: Rogaway, P. (ed.) CRYPTO 2011. LNCS, vol. 6841, pp. 465–484. Springer, Heidelberg (2011)

47. Micciancio, D., Peikert, C.: Trapdoors for lattices: simpler, tighter, faster, smaller. In: Pointcheval, D., Johansson, T. (eds.) EUROCRYPT 2012. LNCS, vol. 7237, pp. 700–718. Springer, Heidelberg (2012)

48. Nguyên, P.Q., Stern, J.: Merkle-Hellman revisited: a cryptanalysis of the Qu-Vanstone cryptosystem based on group factorizations. In: Kaliski Jr., B.S. (ed.) CRYPTO 1997. LNCS, vol. 1294, pp. 198–212. Springer, Heidelberg (1997)

49. O'Neill, A.: Definitional issues in functional encryption. Cryptology ePrint Archive, Report 2010/556 (2010). http://eprint.iacr.org/

50. O'Neill, A., Peikert, C., Waters, B.: Bi-deniable public-key encryption. In: Rogaway, P. (ed.) CRYPTO 2011. LNCS, vol. 6841, pp. 525–542. Springer, Heidelberg (2011)

51. Paillier, P.: Public-key cryptosystems based on composite degree residuosity classes. In: Stern, J. (ed.) EUROCRYPT 1999. LNCS, vol. 1592, pp. 223–238. Springer, Heidelberg (1999)

52. Pollard, J.: Kangaroos, monopoly and discrete logarithms. J. Cryptol. **13**, 433–447 (2000)

53. Regev, O.: On lattices, learning with errors, random linear codes, cryptography. In: Proceedings of STOC, pp. 84–93. ACM (2005)

54. Regev, O.: On lattices, learning with errors, random linear codes, and cryptography. J. ACM **56**(6), 34 (2009)

55. Sahai, A., Seyalioglu, H.: Worry-free encryption: functional encryption with public keys. In: Proceedings of the 17th ACM Conference on Computer and Communications Security, CCS 2010 (2010)

56. Sahai, A., Waters, B.: Fuzzy identity-based encryption. In: Cramer, R. (ed.) EUROCRYPT 2005. LNCS, vol. 3494, pp. 457–473. Springer, Heidelberg (2005)

57. Shamir, A.: Identity-based cryptosystems and signature schemes. In: Blakely, G.R., Chaum, D. (eds.) CRYPTO 1984. LNCS, vol. 196, pp. 47–53. Springer, Heidelberg (1985)

58. Waters, B.: Functional encryption for regular languages. In: Safavi-Naini, R., Canetti, R. (eds.) CRYPTO 2012. LNCS, vol. 7417, pp. 218–235. Springer, Heidelberg (2012)

Circuit-ABE from LWE: Unbounded Attributes and Semi-adaptive Security

Zvika Brakerski[1,2]([✉]) and Vinod Vaikuntanathan[1,2]

[1] Weizmann Institute of Science, Rehovot, Israel
zvika.brakerski@weizmann.ac.il
[2] MIT, Cambridge, USA

Abstract. We construct an LWE-based key-policy attribute-based encryption (ABE) scheme that supports attributes of *unbounded polynomial length*. Namely, the size of the public parameters is a fixed polynomial in the security parameter and a depth bound, and with these fixed length parameters, one can encrypt attributes of arbitrary length. Similarly, any polynomial size circuit that adheres to the depth bound can be used as the policy circuit regardless of its input length (recall that a depth d circuit can have as many as 2^d inputs). This is in contrast to previous LWE-based schemes where the length of the public parameters has to grow linearly with the maximal attribute length.

We prove that our scheme is *semi-adaptively secure*, namely, the adversary can choose the challenge attribute after seeing the public parameters (but before any decryption keys). Previous LWE-based constructions were only able to achieve selective security. (We stress that the "complexity leveraging" technique is not applicable for unbounded attributes).

We believe that our techniques are of interest at least as much as our end result. Fundamentally, selective security and bounded attributes are both shortcomings that arise out of the current LWE proof techniques that *program the challenge attributes into the public parameters*. The LWE toolbox we develop in this work allows us to *delay this programming*. In a nutshell, the new tools include a way to generate an a-priori *unbounded* sequence of LWE matrices, and have fine-grained control over which trapdoor is embedded in each and every one of them, all with succinct representation.

1 Introduction

Key-policy attribute-based encryption [22,34] is a special type of public-key encryption scheme where a (master) public key mpk is used for encryption,

Z. Brakerski—Supported by the Israel Science Foundation (Grant No. 468/14), the Alon Young Faculty Fellowship, Binational Science Foundation (Grant No. 712307) and Google Faculty Research Award.

V. Vaikuntanathan—Research supported in part by DARPA Safeware Grant, NSF CAREER Award CNS-1350619, NSF Grant CNS-1413964 (MACS: A Modular Approach to Computer Security), US-Israel Binational Science Foundation Grant No. 712307, Alfred P. Sloan Research Fellowship, Microsoft Faculty Fellowship, NEC Corporation and a Steven and Renee Finn Career Development Chair from MIT.

© International Association for Cryptologic Research 2016
M. Robshaw and J. Katz (Eds.): CRYPTO 2016, Part III, LNCS 9816, pp. 363–384, 2016.
DOI: 10.1007/978-3-662-53015-3_13

and users are associated to secret keys sk_f corresponding to (policy) functions $f : \mathcal{X} \to \{0, 1\}$. The encryption of a message μ is labeled with a public attribute $x \in \mathcal{X}$, and can be decrypted using sk_f if and only if $f(x) = 0$.[1]

Intuitively, the security requirement is collusion resistance: a coalition of users learns nothing about the plaintext message μ if none of their individual keys are authorized to decrypt the ciphertext.

The past few years have seen much progress in constructing secure and efficient attribute-based encryption (ABE) schemes from different assumptions and for different settings. The first constructions [10,22,23,25,27,30,36] apply to predicates computable by Boolean formulas (which are equivalent to log-depth computations). More recently, important progress has been made on constructions for the set of all polynomial-size circuits (of a-priori bounded polynomial depth): Gorbunov et al. [19] gave a construction from the Learning With Errors (LWE) assumption, and Garg et al. [14] gave a construction using multilinear maps. In both constructions the policy functions are represented as Boolean circuits composed of fan-in 2 gates, and the secret key size is proportional to the *size* of the circuit. Boneh et al. [9] constructed an "arithmetic" ABE scheme where the secret key size is independent of the circuit-size of the function f, but rather depends only on the circuit-depth. This in turn gave the first construction of compact reusable garbled circuits [9], and led to constructions of predicate encryption [20], homomorphic signatures [21] and constrained pseudo-random functions [11].

However, despite all this progress, there are several deficiencies in these constructions. The first is that in all of them, the length of the attribute, represented as a binary string, has to be determined during the initial setup. This is a problem not just for ABE, but also for all downstream constructions (of succinct single-use functional encryption, homomorphic signatures, predicate encryption, and so on) where the size of the input to be encrypted (or signed) is limited by the initial setup.[2] We know of three exceptions to this: the first is the (selectively secure) ABE construction of Lewko et al. [26] that handles *Boolean formulas*, under assumptions on bilinear maps and the second is the (fully secure) inner product encryption and ABE construction of Okamoto and Takashima [31] that again only handles Boolean formulas. Finally, there is the recent work of Ananth and Sahai [7] who show a functional encryption scheme for Turing machines that can take arbitrarily long inputs. In particular, this gives rise to an ABE scheme with the same properties, however this construction uses the huge hammer of indistinguishability obfuscation (IO) unlike the ones in the previous paragraph.

Q1: Is there an ABE scheme for general circuits with unbounded attribute length under standard complexity assumptions?

---

[1] We follow, here and after, the convention that $f(x) = 0$ signifies the ability to decrypt. This is the opposite of the standard convention, and is done purely for our convenience in the technical sections.

[2] One can modify the circuit-ABE constructions of [9,20] to support unbounded attributes in the (programmable) random oracle model. Our focus in this paper is on constructions in the standard model.

The second shortcoming of the circuit-ABE constructions based on lattices and LWE is that they are only selectively secure. Selective security means that the attacker needs to decide which challenge attribute to attack before seeing the public parameters of the scheme or any of the keys. In adaptive security (also known as full security), the challenge attribute x^* can be chosen at any point, even depending on the public parameters and decryption keys obtained by the attacker.

While we do know of adaptively secure ABE for *formulas* [25] based on bilinear maps, and for circuits based on multilinear maps [15] and on indistinguishability obfuscation [37], achieving adaptive security in LWE-based constructions seems to require fundamentally new ideas. Recently, Ananth et al. [6] came up with a generic way to go from selective to adaptive security for (collusion-resistant) FE schemes, but their transformation does not work for ABE schemes.

A well known "hack" for getting around the selectiveness issue is to use "complexity leveraging". This technique is based on the observation that an adaptive adversary can be made selective at the cost of a factor 2^ℓ increase in the running time (or loss of $2^{-\ell}$ in the success probability), where ℓ is the maximum attribute length, just by guessing the challenge attribute ahead of time. Therefore, if we start with a selective scheme that is secure against $2^\ell \cdot \text{poly}(\lambda)$ adversaries, then it is also adaptively secure against $\text{poly}(\lambda)$ time adversaries. Since usually $\ell = \text{poly}(\lambda)$, this method leads to a considerable increase in security parameter. More importantly in our situation, if the attribute space is a-priori unbounded, then complexity leveraging cannot work at all.

An intermediate milestone to adaptively secure ABE is the weaker notion of semi-adaptive security, introduced by Chen and Wee [13]. Semi-adaptive security permits an adversary to choose the challenge attributes after it sees the public parameters, but before it sees the answers to any of its secret-key queries. Chen and Wee show a simpler construction of adaptively secure ABE for *formulas*. Note that for unbounded attributes, complexity leveraging is of no use for this notion as well.

Q2: *Is there an* adaptively (or even semi-adaptively) secure *ABE for general circuits under standard complexity assumptions?*

We resolve the first question and (semi-)resolve the second, as follows.

Theorem 1 (Informal). *Assuming the (polynomial) hardness of approximating worst-case lattice problems to within sub-exponential factor, there is a semi-adaptively secure ABE scheme for circuits of a-priori bounded (polynomial) depth which supports attributes of unbounded length.*

In particular, the setup procedure of our scheme does not require an upper bound on the length of the attributes that will be encrypted. Quite curiously, semi-adaptivity in our result seems to come *for free* from our techniques to achieve unbounded attribute ABE. We elaborate more on our techniques below.

1.1   Overview of Our Techniques

We start with an interpretation of the ABE scheme of Boneh et al. [9] (itself based on the homomorphic encryption scheme of Gentry et al. [18]) which will be instrumental for our presentation.

Given matrices $\mathbf{C}_1, \ldots, \mathbf{C}_\ell$ of appropriate dimension, and a function $f : \{0,1\}^\ell \to \{0,1\}$, represented as a Boolean circuit, one can compute a matrix $\mathbf{C}_f$ which is the "homomorphic evaluation" of f on $\{\mathbf{C}_i\}$. The property of $\mathbf{C}_f$ is that for all $x \in \{0,1\}^\ell$ there exists a low-norm matrix $\mathbf{H} = \mathbf{H}_{\vec{C},f,x}$ (that is, one with "fairly small" entries, the exact amplitude depends on the depth of f and does not matter for this high level description) for which

$$\mathbf{C}_f - f(x)\mathbf{G} = \left(\underbrace{[\mathbf{C}_1 \| \cdots \| \mathbf{C}_\ell]}_{\text{denote } \vec{C}} - \underbrace{[x_1\mathbf{G} \| \cdots \| x_\ell\mathbf{G}]}_{\text{denote } x\vec{G}} \right) \cdot \mathbf{H}.$$

The matrix $\mathbf{G}$ is a special "gadget matrix". This means that if $\mathbf{C}_i = \mathbf{A}\mathbf{R}_i + x_i\mathbf{G}$ for some low-norm matrix $\mathbf{R}_i$, then $\mathbf{C}_f$ can be expressed as $\mathbf{A}\mathbf{R}_f + f(x)\mathbf{G}$ for a somewhat low-norm matrix $\mathbf{R}_f$.

In the ABE scheme of Boneh et al. [9], the public parameters contain a matrix $\mathbf{A}$ and a set $\vec{C} = (\mathbf{C}_1, \ldots, \mathbf{C}_\ell)$ so that ℓ is the length of supported attributes. The parameters are chosen so that a secret trapdoor can always find a low norm solution $\mathbf{R}$ to any equation of the form $\mathbf{C} = \mathbf{A}\mathbf{R} + y\mathbf{G}$, for all $\mathbf{C}, y$. Encrypting a message to an attribute x is done (at a high level) by considering $[\mathbf{A}\|\vec{C} - x\vec{G}]$ as a public key to a dual-Regev encryption scheme [17] and encrypting relative to this key. An important feature of dual-Regev is that it is possible to modify a ciphertext which was encrypted with respect to a certain public key into one that is encrypted with respect to a related key, so long as the new key is obtained by multiplying the old key by a low-norm matrix. Therefore, given some function f, the ciphertext can be converted into one that corresponds to the public key $[\mathbf{A}\|\mathbf{C}_f - f(x)\mathbf{G}]$. Indeed, ABE secret-keys sk_f are generated as dual-Regev keys for the public key $[\mathbf{A}\|\mathbf{C}_f]$, and indeed they can decrypt whenever $f(x) = 0$.[3]

In the proof of security, $\mathbf{A}$ is generated without a trapdoor, but $\mathbf{C}_i$ are generated as $\mathbf{A}\mathbf{R}_i + x_i^*\mathbf{G}$ (which is indistinguishable from their honest distribution). This means that whenever $f(x^*) = 1$, the matrix $[\mathbf{A}\|\mathbf{C}_f]$ equals to $[\mathbf{A}\|\mathbf{A}\mathbf{R}_f + \mathbf{G}]$. It had been shown by [3,29] that if $\mathbf{R}_f$ is known, then dual-Regev keys can be generated *even without a trapdoor*. Finally, the challenge ciphertext is encrypted relative to $[\mathbf{A}\|\vec{C} - x^*\vec{G}] = \mathbf{A} \cdot [I\|\vec{R}]$, which can be shown to be LWE-hard to break if a trapdoor for $\mathbf{A}$ is not known (which indeed it isn't).

The absolutely vital technique that makes the proof of [9] work[4] is the ability to *embed the challenge attributes* into the public parameters. It is apparent from

---

[3] Note that this "negated policy" formulation is obviously equivalent to the standard formulation in the literature wherein decryption succeeds if $f(x) = 1$. From this point and on, purely for our convenience in the technical sections, we will assume that a ciphertext should be decryptable if $f(x) = 0$ and not decryptable otherwise.

[4] The proofs of the other circuit-ABE schemes from standard assumptions, namely [14,19], follow along similar lines.

this description that the [9] scheme is inherently selectively secure and attribute length bounded. It is important that in the security proof, the values of $\mathbf{C}_i$ are set ahead of time to the right values according to the challenge attributes x^*, making the proof inherently selectively secure. In fact, the entire paradigm of embedding the challenge ciphertext in the public parameters necessitates, for pure information-theoretic reasons, that the public parameters grow with the length of the challenge attribute.

The first thing that we should do if we want to stretch the [9] scheme to support unbounded length attributes, is to find a way to generate an unbounded number of $\mathbf{C}_i$ matrices out of a-priori bounded public parameters. Our first observation is that the scheme already exhibits a similar feature in a different context. Namely, the generation of many $\mathbf{C}_f$ out of a bounded number of $\mathbf{C}_i$. Indeed, in our scheme, the public parameters will contain $\mathbf{A}$ and a sequence of matrices $\vec{\mathbf{B}}$. We will consider a predefined and public sequence of functions ϕ_i, where $i = 1, 2, \ldots$, and let $\mathbf{C}_i$ be the output of homomorphic evaluation of ϕ_i on $\vec{\mathbf{B}}$. Thus, the scheme already allows us to generate exponentially many matrices out of a few.

This allows us to extend the functionality of the scheme to unbounded attribute length, but only syntactically, since the proof does not extend to this setting. In particular, if we try to program the matrices $\vec{\mathbf{B}}$ in the proof similarly to $\vec{\mathbf{C}}$ from previous works, we can set $\mathbf{B}_i = \mathbf{A}\mathbf{R}_i + \sigma_i\mathbf{G}$ for some string σ. If we do so, we will get that $\mathbf{C}_i = \mathbf{A}\mathbf{R}_{\phi_i} + \phi_i(\sigma)\mathbf{G}$, where $\mathbf{R}_{\phi_i}$ is low-norm and can be computed out of the $\mathbf{R}_i$ matrices. On the one hand, this is quite encouraging since it is not too far from what we need, if only there was a way to define ϕ_i and σ so that $\phi_i(\sigma) = x_i^*$ (the ith bit of the challenge attribute) we would be in business. On the other hand, this is of course impossible for mere information theoretic reasons, since the ϕ_i are public functions and σ has bounded length, so they cannot encode an x^* of arbitrary length.

Let us therefore take a step back and think, as an intermediate step, about a restricted security model where x^* is chosen randomly and not adversarially (except its length, which is still under the adversary's control). Indeed, a random x^* cannot be compressed, but in the proof of security we can swap x^* for a *pseudorandom* value that can be easily expressible as the output of a pseudorandom function. In particular, we define $\phi_i(\sigma) = \mathsf{PRF}_\sigma(i)$ for some pseudorandom function family. For a random seed σ, letting $x_i^* = \mathsf{PRF}_\sigma(i)$ will be indistinguishable from a random value, and will allow us to support random unbounded length attributes using the proof methods from above.

Indeed, we managed to hack the framework into producing an arbitrarily long sequence of $\mathbf{C}_i$ in such a way that each $\mathbf{C}_i$ encodes a trapdoor that corresponds to x_i^*. We view this as an interesting contribution by itself. However, we would like to support adversarially chosen attributes, and not just random ones. To do this, we will show how to "program" the challenge attribute into the PRF values *after the fact*. In particular, consider, as a mental experiment, an infinite string Δ which is defined such that $\Delta_i = x_i^* \oplus \mathsf{PRF}_\sigma(i)$. This string is pseudorandom to the adversary, but combining it with the PRF key σ, it contains the information about x^*. What we do in the proof, is generate decryption keys for functions

$f_\Delta(x) = f(x \oplus \Delta)$, instead of for f itself. This needs to be offset by changing the encryption algorithm to encrypt to $x \oplus \Delta$ rather than to x itself (which might seem impossible at this point, however see below). If we are able to offset our ciphertext, then the challenge ciphertext will now be encrypted respective to $x^* \oplus \Delta$ which is just our PRF value. All of this is done without the adversary noticing anything, because Δ just seems to him as a completely random string that does not depend on x^*.

We are left with two problems. The first and easier one is that Δ needs to be publicly known, but it has unlimited size and in the proof, we need to know x^* in order to generate it. This is easily managed by noticing that only the ℓ-prefix of Δ is needed in order to use a secret key for a function with ℓ-bit input. We will therefore append the appropriate prefix of Δ to any key that we release. This means that we only need to know the value of Δ when we answer key queries and not when we generate the public parameters. This very fact allows us to achieve *semi-adaptive security*, where x^* can be specified after the setup phase but before key generation. We note that of course setting Δ respective to x^* is only done in the proof. In the real scheme Δ is a random (or pseudorandom) string that is maintained by the key authority and whose prefixes are released as needed (it is important that the same Δ is used for all keys). A savvy reader would have noticed that this "delayed" definition of Δ is similar to non-committing proof techniques which, looking back, is not too surprising. It is also not hard to observe why this technique stops at *semi*-adaptive security: we managed to postpone defining Δ to the time when we generate the first secret key. Since Δ depends on x^* in the proof, we are restricted to the semi-adaptive world where all secret-key queries come after the challenge attributes have been declared.

The second and harder problem is how to encrypt in this brave new scheme. The encryption attribute needs to offset for the effect of Δ on the key, but Δ itself is not (and must not be) a part of the public parameters and is thus unknown to the encryption algorithm. This problem is solved by showing that we can encrypt for *all* possible values of Δ at the same time. Recall that in the encryption, we consider the matrices $\mathbf{C}_i - x_i \mathbf{G}$, for all i. In fact, the encryption process generates a piece of the ciphertext out of each of these matrices, and the collection of pieces constitutes the entire ciphertext. In order to allow for any possible value of Δ, we will generate a ciphertext piece $c_{i,0}$ for $\mathbf{C}_i - x_i \mathbf{G}$ (accounting for $\Delta_i = 0$) and a piece $c_{i,1}$ for $\mathbf{C}_i - (x_i \oplus 1)\mathbf{G}$ (accounting for $\Delta_i = 1$). This would allow us to take the relevant pieces and use them in the decryption process. Alas, the security of the [9] scheme shatters completely if the adversary is allowed to see encryption pieces relative to both $\mathbf{C}_i$ and $\mathbf{C}_i - \mathbf{G}$. It appears that we fixed functionality at the expense of security.

Our last technical contribution is to solve this problem by using ... attribute based encryption! (in fact, even identity based encryption would suffice, but with slightly worse parameters). As a part of our public parameters, we include parameters for a "small" ABE scheme that only needs to support bounded short attributes and low depth circuits. We will encrypt the ciphertext piece $c_{i,b}$ with respect to attribute (i, b) using the "small" scheme. Then, as a part of the functional key, we will also produce a "small" key that will allow to decrypt only attributes (i, b) for which $b = \Delta_i$. This means that an adversary can only see those ciphertext pieces that are needed for decryption. Furthermore, since the offset Δ is fixed, the adversary will only ever see $c_{i,0}$ or $c_{i,1}$ but not both, thus keeping security in tact. This completes the description of our scheme.

2   Preliminaries

2.1   Bounded Distributions and Swallowing

As in many previous works based on LWE, we will rely heavily on distributions that are supported over a bounded domain (with high probability). We will also rely on the fact that some distributions (e.g. sufficiently wide Gaussians) remain almost unchanged under small shifts. Formal definitions follow.

Definition 1. *A   distribution χ supported   over $\mathbb{Z}$ is (B, ϵ)-bounded   if* $\Pr_{x \xleftarrow{\$} \chi}[|x| > B] < \epsilon.$

Definition 2. *A distribution $\tilde{\chi}$ supported over $\mathbb{Z}$ is (B, ϵ)-swallowing if for all $y \in [-B, B] \cap \mathbb{Z}$ it holds that $\tilde{\chi}$ and $y + \tilde{\chi}$ are within ϵ statistical distance.*

The following is a straightforward application of the properties of rounded/discrete Gaussians.

Fact 1. *For every B, ϵ, δ there exists an efficiently sampleable distribution that is both (B, ϵ)-swallowing and $(B \cdot \sqrt{\log(1/\delta)}/\epsilon, O(\delta))$-bounded.*

Finally, we will define the notion of a distribution that is swallowing with respect to another.

Definition 3. *A distribution $\tilde{\chi}$ supported over $\mathbb{Z}$ is (χ, ϵ)-swallowing, for a distribution χ, if it holds that $\tilde{\chi}$ and $\chi + \tilde{\chi}$ are within ϵ statistical distance. We omit the ϵ when it indicates a negligible function in a security parameter that is clear from the context.*

The following corollary summarizes the swallowing properties required for our scheme.

Corollary 1. *Let $B(\lambda)$ be some function and let $\tilde{B}(\lambda) = B(\lambda) \cdot \lambda^{\omega(1)}$, then there exists an efficiently sampleable ensemble $\{\tilde{\chi}_\lambda\}_\lambda$ such that $\tilde{\chi}$ is χ-swallowing for any $B(\lambda)$-bounded $\{\chi_\lambda\}_\lambda$, and also $\tilde{B}(\lambda)$-bounded.*

2.2   Pseudorandom Functions

A pseudorandom function family is a pair of PPT algorithms PRF = (PRF.Gen, PRF.Eval), such that the key generation PRF.Gen(1^λ) takes as input the security parameter, and outputs a seed $\sigma \in \{0,1\}^\eta$ (where $\eta = \eta_\lambda$ is the key length). The evaluation algorithm PRF.Eval(σ, x) takes a seed $\sigma \in \{0,1\}^\eta$ and in input $x \in \{0,1\}^*$ and returns a bit $y \in \{0,1\}$.

Definition 4. *A family* PRF *as above is secure if for every polynomial time adversary* $\mathcal{A}$ *it holds that*

$$\left| \Pr[\mathcal{A}^{\mathsf{PRF.Eval}(\sigma,\cdot)}(1^\lambda) = 1] - \Pr[\mathcal{A}^{\mathcal{O}(\cdot)}(1^\lambda) = 1] \right| = \mathrm{negl}(\lambda),$$

where $\sigma = \mathsf{PRF.Gen}(1^\lambda)$ *and* $\mathcal{O}$ *is a random oracle. The probabilities are taken over all of the randomness of the experiment.*

2.3   KP-ABE with Unbounded Attribute Length

Let $\mathcal{F} = \{\mathcal{F}_\lambda\}_\lambda$ be an ensemble of function classes such that $\mathcal{F}_\lambda \subseteq \{0,1\}^* \to \{0,1\}$. We assume that the functions are represented as boolean circuits. A key-policy attribute based encryption (KP-ABE) scheme is defined by a tuple of PPT algorithms ABE = (ABE.Params, ABE.Enc, ABE.Keygen, ABE.Dec) such that:

- The setup algorithm ABE.Params(1^λ) takes the security parameter as input and outputs a master secret key msk and a set of public parameters pp.
- The encryption algorithm ABE.Enc$_{\mathsf{pp}}(\mu, x)$ uses the public parameters pp and takes as input a message μ from a message space $\mathcal{M} = \mathcal{M}_\lambda$ and an attribute $x \in \{0,1\}^*$. It outputs a ciphertext ct $\in \{0,1\}^*$.
- The key generation algorithm ABE.Keygen$_{\mathsf{msk}}(f)$ uses the master secret key msk and takes as input a function $f \in \mathcal{F}_\lambda$. It outputs a secret key sk$_f$.
- The decryption algorithm ABE.Dec$_{\mathsf{pp}}(\mathsf{sk}_f, x, \mathsf{ct})$ takes as input a function secret key sk$_f$, an attribute $x \in \{0,1\}^*$ and a ciphertext ct, and outputs a message $\mu' \in \mathcal{M}$.

Definition 5 (Correctness of KP-ABE). *A scheme* ABE *is* correct *if the following holds. Consider a sequence of functions* $\{f_\lambda \in \mathcal{F}_\lambda\}_\lambda$ *and a sequence of attributes* $\{x_\lambda \in \{0,1\}^*\}_\lambda$, *such that for all* λ, *the input size of* f *is exactly* $|x_\lambda|$ *and* $f_\lambda(x_\lambda) = 0$.[5] *For all such sequences and for any sequence* $\{m_\lambda \in \mathcal{M}_\lambda\}_\lambda$, *it holds that*

$$\Pr[\mathsf{ABE.Dec}_{\mathsf{pp}}(\mathsf{sk}_f, x, \mathsf{ct}) \neq \mu] = \mathrm{negl}(\lambda),$$

where $(\mathsf{msk}, \mathsf{pp}) = \mathsf{ABE.Params}(1^\lambda)$, $\mathsf{ct} = \mathsf{ABE.Enc}_{\mathsf{pp}}(\mu, x)$, $\mathsf{sk}_f = \mathsf{ABE.Keygen}_{\mathsf{msk}}(f)$.

Definition 6 (Security for KP-ABE). *Let* ABE *be a KP-ABE encryption scheme as above, and consider the following game between the challenger and adversary.*

1. *The challenger generates* $(\mathsf{msk}, \mathsf{pp}) = \mathsf{ABE.Params}(1^\lambda)$, *and sends* pp *to the adversary.*
2. *The adversary makes arbitrarily many key queries by sending functions* f_i *(represented as circuits) to the challenger. Upon receiving such function, the challenger creates* $\mathsf{sk}_i = \mathsf{ABE.Keygen}_{\mathsf{msk}}(f_i)$ *and sends* sk_i *to the adversary.*

---

[5] Recall our convention that $f(x) = 0$ is the event when decryption succeeds.

3. *The adversary sends an attribute x^* and a pair of messages m_0, m_1 to the challenger. The challenger samples $b \in \{0,1\}$ and computes the challenge ciphertext $ct^* = \mathsf{ABE.Enc_{pp}}(m_b, x^*)$. It sends ct^* to the adversary.*
4. *The adversary makes arbitrarily many key queries as in Step 2 above.*
5. *The adversary outputs $\tilde{b} \in \{0,1\}$.*
6. *Let* legal *denote the event where all key queries of the adversary are such that $f_i(x^*) = 1$. If* legal, *the output of the game is $b' = \tilde{b}$, otherwise the output b' is a uniformly random bit.*

The advantage *of an adversary $\mathcal{A}$ is $|\Pr[b' = b] - 1/2|$, where b, b' are generated in the game played between the challenger and the adversary $\mathcal{A}(1^\lambda)$. If x^* is too short or too long compared to the prescribed input size of f_i then it is truncated or padded with zeros appropriately (see discussion below).*

The game above is called the adaptive security game *for ABE, and it has relaxed variants. In the* selective security game, *the adversary sends x^* before Step 1. In the* semi-adaptive security game, *the adversary sends x^* before Step 2.*

The scheme ABE *is* adaptively/selectively/semi-adaptively *secure if any PPT adversary $\mathcal{A}$ only has negligible advantage in the adaptive/selective/semi-adaptive security game (respectively).*

Negated Policies. We allow decryption when $f(x) = 0$ and require that in the security game all queries are such that $f(x^*) = 1$. In LWE-based constructions it is often much more convenient to work with this negated version of the policy, which explains the apparent strangeness. This variant is obviously equivalent.

Discussion. Our definition does not place any restrictions on the attribute length so the only restriction comes from limiting the adversary to run in polynomial time (so it can only output x^* and f_i that are polynomially bounded). It is important to notice that in this regime, there are no known generic transformations from selective to semi-adaptive to adaptive security, even if we strengthen the hardness assumption. In particular, the complexity leveraging technique, in which the challenger "guesses" x^* in the beginning of the experiment, and a sub-exponential hardness assumption is made to account for the success probability of this guess, is no longer applicable. In this light, we view our semi-adaptive security improvement as *qualitative rather than quantitative.*

Lastly, we note that in the security definition (but not in the correctness requirement!) we chose to allow $f(x^*)$ to be well defined even if there is a mismatch between the input length of f and the length of x^* (by truncating x^* or padding with zeros). A different valid approach would be to consider an alternate, stronger, definition that if there is a mismatch then $f(x^*) = 1$ (and thus it is legal for the adversary to query any function that does not have the same input length as $|x^*|$). We notice that this notion of security is derived from ours by adding the length itself to the attribute. More explicitly, when you want to encrypt with attribute x of length ℓ, use the ABE scheme with attribute (ℓ, x), and in the key generation process, when you want to generate a key for function f, generate a key for $f'(\ell, x)$ that first checks that ℓ is indeed the intended input length. Therefore, using our definition does not limit generality in this aspect.

3   LWE, Trapdoors, Homomorphism

This section summarizes tools from previous works that are used in our construction. This includes the definition of the LWE problem and its relation to worst case lattice problems, the notion of trapdoors for lattices and operations on trapdoors, and homomorphic evaluation of matrices with special properties.

Learning with Errors (LWE). The Learning with Errors (LWE) problem was introduced by Regev [33] as a generalization of "learning parity with noise" [5,8]. We now define the decisional version of LWE. (Unless otherwise stated, we will treat all vectors as column vectors in this paper).

Definition 7 (Decisional LWE (DLWE) [33]). *Let λ be the security parameter, $n = n(\lambda)$, $m = m(\lambda)$, and $q = q(\lambda)$ be integers and $\chi = \chi(\lambda)$ be a probability distribution over $\mathbb{Z}$. The $\mathrm{DLWE}_{n,q,\chi}$ problem states that for all $m = \mathrm{poly}(n)$, letting $\mathbf{A} \leftarrow \mathbb{Z}_q^{n \times m}$, $\mathbf{s} \leftarrow \mathbb{Z}_q^n$, $\mathbf{e} \leftarrow \chi^m$, and $\mathbf{u} \leftarrow \mathbb{Z}_q^m$, the following distributions are computationally indistinguishable:*

$$(\mathbf{A}, \mathbf{s}^T \mathbf{A} + \mathbf{e}^T) \stackrel{c}{\approx} (\mathbf{A}, \mathbf{u}^T)$$

There are known quantum (Regev [33]) and classical (Peikert [32]) reductions between $\mathrm{DLWE}_{n,q,\chi}$ and approximating short vector problems in lattices. Specifically, these reductions take χ to be a discrete Gaussian distribution $D_{\mathbb{Z},\alpha q}$ for some $\alpha < 1$. We write $\mathrm{DLWE}_{n,q,\alpha}$ to indicate this instantiation. We now state a corollary of the results of [28,29,32,33]. These results also extend to additional forms of q (see [28,29]).

Corollary 2 [28,29,32,33]. *Let $q = q(n) \in \mathbb{N}$ be either a prime power $q = p^r$, or a product of co-prime numbers $q = \prod q_i$ such that for all i, $q_i = \mathrm{poly}(n)$, and let $\alpha \geq \sqrt{n}/q$. If there is an efficient algorithm that solves the (average-case) $\mathrm{DLWE}_{n,q,\alpha}$ problem, then:*

– *There is an efficient quantum algorithm that solves* $\mathsf{GapSVP}_{\widetilde{O}(n/\alpha)}$ *(and* $\mathsf{SIVP}_{\widetilde{O}(n/\alpha)}$*) on any n-dimensional lattice.*

– *If in addition $q \geq \tilde{O}(2^{n/2})$, there is an efficient classical algorithm for* $\mathsf{GapSVP}_{\widetilde{O}(n/\alpha)}$ *on any n-dimensional lattice.*

Recall that GapSVP_γ is the (promise) problem of distinguishing, given a basis for a lattice and a parameter d, between the case where the lattice has a vector shorter than d, and the case where the lattice doesn't have any vector shorter than $\gamma \cdot d$. SIVP is the search problem of finding a set of "short" vectors. The best known algorithms for GapSVP_γ [35] require at least $2^{\tilde{\Omega}(n/\log \gamma)}$ time. We refer the reader to [32,33] for more information.

In this work, we will only consider the case where $q \leq 2^n$. Furthermore, the underlying security parameter λ is assumed to be polynomially related to the dimension n.

Lastly, we derive the following corollary which will allow us to choose the LWE parameters for our scheme. The corollary follows immediately from the fact that the discrete Gaussian $D_{\mathbb{Z},\alpha q}$ is $(\alpha q \cdot t, 2^{-\Omega(t^2)})$-bounded for all t.

Corollary 3. *For all $\epsilon > 0$ there exist functions $q = q(n) \leq 2^n, \chi = \chi(n)$ such that χ is B-bounded for some $B = B(n)$, $q/B \geq 2^{n^\epsilon}$ and such that $\mathrm{DLWE}_{n,q,\chi}$ is at least as hard as the classical hardness of GapSVP_γ and the quantum hardness of SIVP_γ for $\gamma = 2^{\Omega(n^\epsilon)}$.*

The Gadget Matrix. Let $N = n \cdot \lceil \log q \rceil$ and define the "gadget matrix" $\mathbf{G} = \mathbf{g} \otimes \mathbf{I}_n \in \mathbb{Z}_q^{n \times N}$ where $\mathbf{g} = (1, 2, 4, \dots, 2^{\lceil \log q \rceil - 1}) \in \mathbb{Z}_q^{\lceil \log q \rceil}$. We will also refer to this gadget matrix as the "powers-of-two" matrix. We define the inverse function $\mathbf{G}^{-1} : \mathbb{Z}_q^{n \times m} \to \{0, 1\}^{N \times m}$ which expands each entry $a \in \mathbb{Z}_q$ of the input matrix into a column of size $\lceil \log q \rceil$ consisting of the bits of the binary representation of a. We have the property that for any matrix $\mathbf{A} \in \mathbb{Z}_q^{n \times m}$, it holds that $\mathbf{G} \cdot \mathbf{G}^{-1}(\mathbf{A}) = \mathbf{A}$.

Trapdoors. Let $n, m, q \in \mathbb{N}$ and consider a matrix $\mathbf{A} \in \mathbb{Z}_q^{n \times m}$. For all $\mathbf{V} \in \mathbb{Z}_q^{n \times m'}$, we let $\mathbf{A}_\tau^{-1}(\mathbf{V})$ denote the random variable whose distribution is a Gaussian $D_{\mathbb{Z}^m, \tau}^{m'}$ conditioned on $\mathbf{A} \cdot \mathbf{A}_\tau^{-1}(\mathbf{V}) = \mathbf{V}$. A τ-trapdoor for $\mathbf{A}$ is a procedure that can sample from the distribution $\mathbf{A}_\tau^{-1}(\mathbf{V})$ in time $\mathrm{poly}(n, m, m', \log q)$, for any $\mathbf{V}$. We slightly overload notation and denote a τ-trapdoor for $\mathbf{A}$ by $\mathbf{A}_\tau^{-1}$.

The following properties had been established in a long sequence of works.

Corollary 4 (Properties of Trapdoors [2–4,12,17,29]). *Lattice trapdoors exhibit the following properties.*

1. *Given $\mathbf{A}_\tau^{-1}$, one can obtain $\mathbf{A}_{\tau'}^{-1}$ for any $\tau' \geq \tau$.*
2. *Given $\mathbf{A}_\tau^{-1}$, one can obtain $[\mathbf{A} \| \mathbf{B}]_\tau^{-1}$ and $[\mathbf{B} \| \mathbf{A}]_\tau^{-1}$ for any $\mathbf{B}$.*
3. *For all $\mathbf{A} \in \mathbb{Z}_q^{n \times m}$ and $\mathbf{R} \in \mathbb{Z}^{m \times N}$, with $N = n \lceil \log q \rceil$, one can obtain $[\mathbf{AR} + \mathbf{G} \| \mathbf{A}]_\tau^{-1}$ for $\tau = O(m \cdot \|\mathbf{R}\|_\infty)$.*
4. *There exists an efficient procedure $\mathsf{TrapGen}(1^n, q)$ that outputs $(\mathbf{A}, \mathbf{A}_{\tau_0}^{-1})$ where $\mathbf{A} \in \mathbb{Z}_q^{n \times m}$ for some $m = O(n \log q)$ and is 2^{-n}-uniform, where $\tau_0 = O(\sqrt{n \log q \log n})$.*

Homomorphic Evaluation. Consider some $n, q \in \mathbb{N}$. Consider $\mathbf{C}_1, \dots, \mathbf{C}_\ell \in \mathbb{Z}_q^{n \times N}$ where $N = n \lceil \log q \rceil$, and denote $\vec{\mathbf{C}} = [\mathbf{C}_1 \| \cdots \| \mathbf{C}_\ell]$. Let f be a boolean circuit of depth d computing a function $\{0, 1\}^\ell \to \{0, 1\}$, and assume that f contains only NAND gates. We define $\mathbf{C}_f = \mathsf{Eval}(f, \vec{\mathbf{C}})$ recursively: associate $\mathbf{C}_1, \dots, \mathbf{C}_\ell$ with the input wires of the circuit. For every wire w in f, letting u, v be its predecessors and define $\mathbf{C}_w = \mathbf{G} - \mathbf{C}_u \cdot \mathbf{G}^{-1}(\mathbf{C}_v)$. Finally $\mathbf{C}_f$ is the matrix associated with the output wire.

Denoting $x\vec{\mathbf{G}} = [x_1\mathbf{G}\|\cdots\|x_\ell\mathbf{G}]$, it holds that if $\mathbf{C}_f = \mathsf{Eval}(f, \vec{\mathbf{C}})$, then $\mathbf{C}_f - f(x)\mathbf{G} = (\vec{\mathbf{C}} - x\vec{\mathbf{G}}) \cdot \mathbf{H}_{f,x,\vec{\mathbf{C}}}$, for a matrix $\mathbf{H}_{f,x,\vec{\mathbf{C}}}$ with $\left\|\mathbf{H}_{f,x,\vec{\mathbf{C}}}\right\|_\infty \leq (N+1)^d$.
In particular, if $\mathbf{C}_i = \mathbf{A}\mathbf{R}_i + x_i\mathbf{G}$, i.e. $\vec{\mathbf{C}} = \mathbf{A}\vec{\mathbf{R}} + x\vec{\mathbf{G}}$ for $\vec{\mathbf{R}} = [\mathbf{R}_1\|\cdots\|\mathbf{R}_\ell]$,
then $\mathbf{C}_f = \mathbf{A}\mathbf{R}_f + f(x)\mathbf{G}$ for $\mathbf{R}_f = \vec{\mathbf{R}} \cdot \mathbf{H}_{f,x,\vec{\mathbf{C}}}$ (where $\mathbf{H}$ is independent of $\vec{\mathbf{R}}$).

4   Our Scheme

We now present our scheme and prove its correctness and security. As in previous works on LWE-based ABE schemes [9,19], it would be easier for us to work with "negated policies", so that sk_f can decrypt ciphertexts with attribute x if $f(x) = 0$. We start by defining the class of depth bounded circuits, to which our construction is targeted.

Definition 8 (Depth-bounded Circuits). *The class of d-bounded circuits, denoted $\mathcal{P}_d$, for some function $d = d(\lambda)$ is the ensemble of functions $\{\mathcal{P}_{d,\lambda}\}_\lambda$ such that $\mathcal{P}_{d,\lambda}$ is the set of boolean circuits of depth at most $d(\lambda)$ and input length at most 2^ν for some $\nu(\lambda) = \omega(\log \lambda)$ which will be clear from the context.*

Next, we define another class of circuits. These are very simple circuits that contain a hardcoded string, and upon receiving an index and bit as input, they check whether the relevant location in the string is indeed the supplied value.

Definition 9. *Consider the family of circuits $\{\mathsf{BitCheck}_{\nu,x}\}$ s.t. for all $\nu \in \mathbb{N}$ and $x \in \{0,1\}^*$, $|x| \leq 2^\nu$, we define $\mathsf{BitCheck}_{\nu,x} : [2^\nu] \times [2^\nu] \times \{0,1\} \to \{0,1\}$ such that $\mathsf{BitCheck}_{\nu,x}(\ell, i, b) = 0$ if and only if $|x| = \ell$ and also $x_i = b$. Note that $\mathsf{BitCheck}_{\nu,x}$ can always be computed by a boolean circuit of depth $O(\log|x|) = O(\nu)$ (we assume that ℓ, i are in standard ν-bit binary representation).*

The Scheme. Let $\nu = \nu(\lambda)$ be any super-logarithmic function (so that 2^ν is super-polynomial). Let $\mathsf{oldABE} = (\mathsf{oldABE.Params}, \mathsf{oldABE.Enc}, \mathsf{oldABE.Keygen}, \mathsf{oldABE.Dec})$ be a selectively-secure key-policy ABE scheme for the function class $\{\{\mathsf{BitCheck}_{\nu(\lambda),x} : |x| \leq 2^\nu\}\}_\lambda$ where ν is as above (i.e. oldABE only need to support bounded length attributes, and furthermore this length can be any super-logarithmic function). Let PRF be a family of pseudorandom functions and let $\eta = \eta_\lambda$ be the seed length (for security parameter λ). Let d_{prf} be the depth of $\mathsf{PRF.Eval}(\sigma, x)$ for $|x| = \nu$ (by definition $d_{\mathsf{prf}} = \mathrm{poly}(\lambda)$).

We now present our ABE scheme for any class of circuits of a-priori polynomial depth bound. We note that as in previous works, we submit the depth bound as an additional parameter to the setup procedure. In order to support the class $\mathcal{P}_d$, the setup procedure is to be executed on input $(1^\lambda, 1^{d(\lambda)})$. Finally, the scheme is parameterized by a constant $\epsilon \in (0,1)$ that determines the trade-off between the lattice approximation factor on which security is based, and the efficiency of the scheme.

- ABE.Params$(1^\lambda, 1^d)$. We start by setting DLWE parameters based on Corollary 3. Let n be s.t. $(n^2 + 1)^{2(d_{\mathrm{prf}}+d)} \cdot 2^{3\nu} \leq 2^{n^\epsilon}$. The solution to the equation is of the form $n \leq (\lambda d)^{O(1/\epsilon)}$, which is polynomial in the security parameter for any constant ϵ. We choose q, χ, B accordingly based on Corollary 3, and note that by definition $q/B \geq (N+1)^{2(d_{\mathrm{prf}}+d)} \cdot 2^{3\nu}$ (recall that $N = n\lceil \log q\rceil$). We further let $\tilde{\chi}$ be a B'-swallowing and $\tilde{B}$-bounded distribution, for $B' = B \cdot m\eta N(N+1)^{d_{\mathrm{prf}}}$ and $\tilde{B} = 2^\nu \cdot B'$, whose existence is guaranteed by Corollary 1.

  Generate a matrix-trapdoor pair $(\mathbf{A}, \mathbf{A}_{\tau_0}^{-1}) = \mathsf{TrapGen}(1^n, q)$ (see Corollary 4), vector $\mathbf{v} \xleftarrow{\$} \mathbb{Z}_q^n$, and matrices $\mathbf{B}_1, \ldots, \mathbf{B}_\eta \xleftarrow{\$} \mathbb{Z}_q^{n \times N}$, and denote $\vec{\mathbf{B}} = [\mathbf{B}_1\| \ldots \|\mathbf{B}_\eta]$. We assume w.l.o.g that $m \geq n\lceil \log q\rceil + 2\lambda$ (otherwise random padding can be applied). Generate a key pair for oldABE: $(\mathsf{oldabemsk}, \mathsf{oldabepp}) = \mathsf{oldABE.Params}(1^\lambda)$. Generate a seed for a PRF $\sigma = \mathsf{PRF.Gen}(1^\lambda)$.

  We set $\mathsf{msk} = (\mathbf{A}_\tau^{-1}, \mathsf{oldabemsk}, \sigma)$ and $\mathsf{pp} = (\mathbf{A}, \vec{\mathbf{B}}, \mathsf{oldabepp})$.
- ABE.Enc$_{\mathsf{pp}}(\mu, x)$, where $\mathsf{pp} = (\mathbf{A}, \vec{\mathbf{B}}, \mathsf{oldabepp})$, $\mu \in \{0,1\}$ and $x \in \{0,1\}^*$. We let $\ell = |x|$ denote the length of the attribute string. For all $i \in [\ell]$, generate $\mathbf{C}_i = \mathsf{Eval}(\mathsf{PRF.Eval}(\cdot, i), \vec{\mathbf{B}})$. (Where $\mathsf{PRF.Eval}(\cdot, i)$ is the circuit that takes a seed σ and outputs $\mathsf{PRF.Eval}(\sigma, i)$.)

  Sample $\mathbf{s} \xleftarrow{\$} \mathbb{Z}_q^n$, $\mathbf{e} \xleftarrow{\$} \chi^m$, $e' \xleftarrow{\$} \chi$, let

$$\mathbf{c}_0^T = \mathbf{s}^T[\mathbf{A}\|\mathbf{v}] + [\mathbf{e}^T\|e'] + \mu\lfloor q/2\rfloor \cdot [\mathbf{0}^T\|1].$$

This is essentially a dual-Regev encryption of μ under public key $\mathbf{A}, \mathbf{v}$. The rest of the ciphertext will contain auxiliary information that will allow to decrypt given a proper functional secret key. Specifically, we sample for all $i \in [\ell]$ a noise vector $\tilde{\mathbf{e}}_i \xleftarrow{\$} \tilde{\chi}^N$, and compute

$$\mathbf{c}_{i,x_i \oplus \beta}^T = \mathbf{s}^T(\mathbf{C}_i - (x_i \oplus \beta)\mathbf{G}) + \tilde{\mathbf{e}}_i^T, \tag{1}$$

Finally, the vectors $\mathbf{c}_{i,x_i \oplus \beta}$ are encrypted again using the old ABE scheme:

$$\psi_{i,\beta} = \mathsf{oldABE.Enc}_{\mathsf{oldabepp}}(\mathbf{c}_{i,x_i \oplus \beta}, (\ell, i, \beta)).$$

The final ciphertext is

$$\mathsf{ct} = \left(\mathbf{c}_0, (\psi_{i,\beta})_{i \in [\ell], \beta \in \{0,1\}}\right).$$

- ABE.Keygen$_{\mathsf{msk}}(f)$. Given a circuit f computing a function $\{0,1\}^\ell \to \{0,1\}$, the key is generated as follows. We recall that we work with negated policies so sk_f should decrypt only when $f(x) = 0$.

  For all i, define $\Delta_i = \mathsf{PRF.Eval}(\sigma, i)$. Further let $\Delta_{\leq \ell} = \Delta_1 \cdots \Delta_\ell$ be the ℓ-prefix of the infinite string Δ (in fact, we can think of Δ as having length 2^ν, which is finite but super-polynomial).

  Generate a key for the old scheme $\mathsf{oldabesk}_\ell = \mathsf{oldABE.Keygen}_{\mathsf{oldabemsk}}$ $(\mathsf{BitCheck}_{\nu, \Delta_{\leq \ell}})$.

Note that $\Delta_{\leq \ell}$ and $\mathsf{oldabesk}_\ell$ depend only on msk and ℓ, and not on f, and therefore they can be generated and published once and for all for each value of ℓ (however, since ℓ is a-priori unbounded, it is impossible to publish this information for "all possible ℓ" at the same time). Define $f_\Delta : \{0,1\}^\ell \to \{0,1\}$ as $f_\Delta(x) = f(x \oplus \Delta_{\leq \ell})$.

For all $i \in [\ell]$, generate $\mathbf{C}_i = \mathsf{Eval}(\mathsf{PRF.Eval}(\cdot, i), \vec{\mathbf{B}})$ (as in the encryption algorithm). Let $\vec{\mathbf{C}} = [\mathbf{C}_1 \| \cdots \| \mathbf{C}_\ell]$ and set $\mathbf{C}_f = \mathsf{Eval}(f_\Delta, \vec{\mathbf{C}})$. Let

$$\mathbf{r}_f = [\mathbf{C}_f \| \mathbf{A}]_\tau^{-1}(\mathbf{v}),$$

where $\tau = 2^\nu \cdot mN^2(N+1)^{d+d_{\mathrm{prf}}} \geq \tau_0$ and $\mathbf{t}_f = [-\mathbf{r}_f^T \| 1]^T$. Note that $[\mathbf{C}_f \| \mathbf{A} \| \mathbf{v}] \cdot \mathbf{t}_f = 0$.

Output $\mathsf{sk}_f = (f, \Delta_{\leq \ell}, \mathsf{oldabesk}_\ell, \mathbf{t}_f)$.

- ABE.Dec($\mathsf{sk}_f, x, \mathsf{ct}$). Given $\mathsf{sk}_f = (f, \Delta_{\leq \ell}, \mathsf{oldabesk}_\ell, \mathbf{t}_f)$, $x \in \{0,1\}^\ell$ such that $f(x) = 0$, and $\mathsf{ct} = \left(\mathbf{c}_0, (\psi_{i,\beta})_{i \in [\ell], \beta \in \{0,1\}}\right)$, the decryption process runs as follows.

Use $\mathsf{oldabesk}_\ell$ to compute

$$\mathbf{c}_{i, x_i \oplus \Delta_i} = \mathsf{oldABE.Dec}(\mathsf{oldabesk}_\ell, \psi_{i, \Delta_i}, (\ell, i, \Delta_i)), \tag{2}$$

and recompose

$$\mathbf{c}_{x \oplus \Delta_{\leq \ell}}^T = [\mathbf{c}_{1, x_1 \oplus \Delta_1}^T \| \cdots \| \mathbf{c}_{\ell, x_\ell \oplus \Delta_\ell}^T].$$

We again compute $\mathbf{C}_i = \mathsf{Eval}(\mathsf{PRF.Eval}(\cdot, i), \vec{\mathbf{B}})$, $\vec{\mathbf{C}} = [\mathbf{C}_1 \| \cdots \| \mathbf{C}_\ell]$ and $\mathbf{C}_f = \mathsf{Eval}(f_\Delta, \vec{\mathbf{C}})$. We also compute $\mathbf{H} = \mathbf{H}_{f_\Delta, x \oplus \Delta_{\leq \ell}, \vec{\mathbf{C}}}$. Note that by the properties stated above, it holds that

$$(\vec{\mathbf{C}} - (x \oplus \Delta_{\leq \ell})\vec{\mathbf{G}}) \cdot \mathbf{H} = \mathbf{C}_f - f_\Delta(x \oplus \Delta_{\leq \ell})\mathbf{G} = \mathbf{C}_f,$$

since $f_\Delta(x \oplus \Delta_{\leq \ell}) = f(x) = 0$.

Recalling that $\mathbf{c}_{x \oplus \Delta_{\leq \ell}}^T$ is linear (up to noise) in $\vec{\mathbf{C}} - (x \oplus \Delta_{\leq \ell})\vec{\mathbf{G}}$, we will set $\mathbf{c}_f^T = \mathbf{c}_{x \oplus \Delta_{\leq \ell}}^T \cdot \mathbf{H}_{f_\Delta, x \oplus \Delta_{\leq \ell}, \vec{\mathbf{C}}}$, with intent to show that $\mathbf{c}_f^T$ is linear (up to noise) in $\mathbf{C}_f$.

Finally, we compute $\tilde{\mu} = [\mathbf{c}_f^T \| \mathbf{c}_0^T] \cdot \mathbf{t}_f$, and output $\mu' = 0$ if $|\tilde{\mu}| < q/4$ and $\mu' = 1$ if $|\tilde{\mu}| \geq q/4$.

4.1   Correctness

Let $\{(f_\lambda, x_\lambda)\}_\lambda$ be an arbitrary sequence of function-message pairs s.t. f_λ has size $\mathsf{poly}(\lambda)$, depth at most $d(\lambda)$, and $|x| = \ell(\lambda)$ for some $\ell(\lambda) \leq 2^{\nu(\lambda)}$. Consider properly generated $(\mathsf{pp}, \mathsf{msk}) = \mathsf{ABE.Params}(1^\lambda)$, a properly encrypted ciphertext $\mathsf{ct} = \mathsf{Enc}_{\mathsf{pp}}(\mu, x)$ for some value $\mu \in \{0,1\}$ and a properly generated functional key $\mathsf{sk}_f = \mathsf{ABE.Keygen}_{\mathsf{msk}}(f)$.

Consider the execution of ABE.Dec($\mathsf{sk}_f, x, \mathsf{ct}$). The correctness of oldABE implies that with all but negligible probability, the vectors $\mathbf{c}_{i, x_i \oplus \Delta_i}$ computed in Eq. (2) are indeed equal to the ones encrypted in Eq. (1). Namely, that

$$\mathbf{c}_{i, x_i \oplus \Delta_i}^T = \mathbf{s}^T(\mathbf{C}_i - (x_i \oplus \Delta_i)\mathbf{G}) + \mathbf{e}^T \mathbf{R}_i,$$

and therefore

$$\mathbf{c}_{x \oplus \Delta_{\leq \ell}}^T = \mathbf{s}^T (\vec{\mathbf{C}} - (x \oplus \Delta_{\leq \ell})\vec{\mathbf{G}}) + \tilde{\mathbf{e}}^T,$$

which, recalling that $f(x) = 0$ and denoting $\mathbf{H} = \mathbf{H}_{f_\Delta, x \oplus \Delta_{\leq \ell}, \vec{\mathbf{C}}}$, implies that

$$\mathbf{c}_f^T = \mathbf{c}_{x \oplus \Delta_{\leq \ell}}^T \cdot \mathbf{H} = \mathbf{s}^T \mathbf{C}_f + \tilde{\mathbf{e}}^T \mathbf{H}.$$

Finally, we get that

$$[\mathbf{c}_f^T \| \mathbf{c}_0^T] = \mathbf{s}^T [\mathbf{C}_f \| \mathbf{A} \| \mathbf{v}] + [\tilde{\mathbf{e}}^T \mathbf{H} \| \mathbf{e}^T \| e'] + \mu \lfloor q/2 \rfloor \cdot [\mathbf{0}^T \| 1],$$

and therefore that

$$[\mathbf{c}_f^T \| \mathbf{c}_0^T] \cdot \mathbf{t}_f = [\tilde{\mathbf{e}}^T \mathbf{H} \| \mathbf{e}^T \| e'] \cdot \mathbf{t}_f + \mu \lfloor q/2 \rfloor.$$

We conclude that we have correct decryption so long as $\left| [\tilde{\mathbf{e}}^T \mathbf{H} \| \mathbf{e}^T \| e'] \cdot \mathbf{t}_f \right|$ is bounded away from $q/4$. We will produce a fairly loose bound, since the asymptotic parameters will only be effected marginally. A precise analysis could be obtained using standard techniques. We recall that by the properties of discrete Gaussians, it holds that $\|\mathbf{t}_f\|_\infty \leq \tau \sqrt{m + N}$ with all but $2^{-(m+N)} = \mathrm{negl}(\lambda)$ probability, and also that asymptotically $\ell \leq 2^\nu$. Therefore, with all but negligible probability

$$\left[\tilde{\mathbf{e}}^T \mathbf{H} \| \mathbf{e}^T \| e'\right] \cdot \mathbf{t}_f \leq \left\| \left[\tilde{\mathbf{e}}^T \mathbf{H} \| \mathbf{e}^T \| e'\right] \right\|_\infty \cdot \|\mathbf{t}_f\|_\infty \cdot (N + m + 1)$$

$$\leq \left(\tilde{B} \cdot (N+1)^d \cdot (\ell N) + B \cdot (m+1) \right) \cdot \|\mathbf{t}_f\|_\infty \cdot (N + m + 1)$$

$$\leq \left(\tilde{B} \cdot (N+1)^d \cdot (\ell N) + B \cdot (m+1) \right) \tau \sqrt{m + N} \cdot (N + m + 1)$$

$$\leq B \cdot (N+1)^{2(d_{\mathrm{prf}}+d)} 2^{2\nu} \cdot \mathrm{poly}(n, \log q).$$

Since we set $q/B \geq (N+1)^{2(d_{\mathrm{prf}}+d)} 2^{3\nu}$, we get that correctness holds asymptotically for any such $\ell(\lambda), d(\lambda)$.

4.2 Security

We prove that our scheme is semi-adaptively secure as per Definition 6. Our proof heavily relies on the structure of the string Δ. Whereas Δ has a succinct representation as the output of a PRF, the view of the adversary does not depend on the seed of the PRF in any way except through the bits of Δ. Therefore, it follows from the pseudorandomness property that Δ is indistinguishable from a completely random string. It follows, therefore, that XORing x^* into Δ will go unnoticed by the adversary. However, this allows us to embed the challenge attribute in the public parameters in an indirect way, namely, now the XOR of the PRF's ith bit with Δ_i is exactly x_i^*. This means that $x_i^* \oplus \Delta_i = \mathrm{PRF}(i)$ and thus that $\mathbf{C}_i - (x_i^* \oplus \Delta_i)\mathbf{G}$ is independent of x^* itself and therefore can be known to the reduction ahead of time. This will allow us to apply similar techniques to those in [9] to prove security. A formal statement of the lemma together with a detailed sketch of the proof follows.

Lemma 1. *Let* PRF *be a family of secure pseudorandom functions as per Sect. 2.2, and let* oldABE *be a selectively secure ABE scheme for the function class* $\mathsf{BitCheck}_{\nu,x}$ *for some super-logarithmic* $\nu = \nu(\lambda)$. *Then under the* $\mathsf{DLWE}_{n,q,\chi}$ *assumption, the scheme* ABE *is a semi-adaptively secure ABE scheme for the function class* $\mathcal{P}_d$.

Proof (Extended sketch). We use ℓ^* to denote the length of the challenge attribute x^*. We also extend the notation x_i^* as follows: if $i \leq \ell^*$ then x_i^* denotes the ith bit of x^* as usual, however, for $i > \ell^*$ our convention is that $x_i^* = 0$.

The proof follows by a sequence of hybrids. We consider an adversary $\mathcal{A}$ for the semi-adaptive security game in Definition 6. Let $\mathrm{Adv}[\mathcal{A}]$ denote the advantage of $\mathcal{A}$ in the security game. We will denote by $\mathrm{Adv}_{\mathcal{H}}[\mathcal{A}]$ the advantage of $\mathcal{A}$ in the experiment described in hybrid $\mathcal{H}$.

Hybrid $\mathcal{H}_0$. This is the ABE semi-adaptive security game as per Definition 5. By definition $\mathrm{Adv}[\mathcal{A}] = \mathrm{Adv}_{\mathcal{H}_0}[\mathcal{A}]$.

Hybrid $\mathcal{H}_1$. In this hybrid, we change the way the (infinite) string Δ is defined. Recall that in the previous hybrid, $\Delta_i = \mathsf{PRF.Eval}(\sigma, i)$. However in this hybrid and throughout the proof we set

$$\Delta_i = \begin{cases} (\mathsf{PRF.Eval}(\sigma, i) \oplus x_i^*) & \text{if } i \leq \ell^*, \\ \mathsf{PRF.Eval}(\sigma, i) & \text{otherwise.} \end{cases} \tag{3}$$

Note that now x^* needs to be known in order to compute Δ. However, Δ is not used at all until the first key query is answered. Therefore, to execute this hybrid, the challenger only needs to know x^* before responding to the first key query, which is consistent with semi-adaptive security.

To see why the view of the adversary is indistinguishable in $\mathcal{H}_1$ and $\mathcal{H}_0$, consider replacing $\mathsf{PRF.Eval}(\sigma, i)$ with an oracle that returns a random bit for every i. In such case, the distributions in both hybrids are identical. Since σ itself is not used anywhere except to generate $\mathsf{PRF.Eval}(\sigma, i)$, the pseudorandomness of PRF guarantees that the views when using $\mathsf{PRF.Eval}(\sigma, i)$ are computationally indistinguishable. We conclude that

$$|\mathrm{Adv}_{\mathcal{H}_1}[\mathcal{A}] - \mathrm{Adv}_{\mathcal{H}_0}[\mathcal{A}]| = \mathrm{negl}(\lambda).$$

We remark that this is the only place where the pseudorandomness of the PRF is used, and from this hybrid and on one can think of σ as public.

Lastly, we notice that since we extended our notation so that $x_i^* = 0$ for $i > \ell^*$, we can say that from this hybrid and throughout the proof, it holds that $\Delta_i = \mathsf{PRF.Eval}(\sigma, i) \oplus x_i^*$ for all $i \in \mathbb{N}$.

Hybrid $\mathcal{H}_2$. We now change the way the matrices $\vec{\mathbf{B}}$ are generated. We will now generate $\mathbf{B}_i$ as follows: Sample $\mathbf{R}_i \xleftarrow{\$} \{0,1\}^{m \times N}$ and set $\mathbf{B}_i = \mathbf{A}\mathbf{R}_i + \sigma_i \mathbf{G}$. Indistinguishability will follow from the leftover hash lemma since $m \geq n\lceil \log q \rceil + 2\lambda$. We point out that one has to be careful when applying the leftover hash lemma since $\mathbf{A}$ is only statistically close to uniform, and it is generated together with $\mathbf{A}_{\tau_0}^{-1}$. We notice, however that $\mathbf{A}_{\tau_0}^{-1} - \mathbf{A} - \mathbf{AR}_i$ is a Markov chain, and therefore we can think about first sampling $\mathbf{A}$ and then sampling $\mathbf{A}_{\tau_0}^{-1}$ and $\mathbf{AR}_i$ independently from the marginals. Therefore, since $(\mathbf{A}, \mathbf{AR}_i)$ is statistically indistinguishable from uniform when $\mathbf{A}$ is uniform, it also holds true when $\mathbf{A}$ is only statistically close to uniform, and also holds true when $\mathbf{A}_{\tau_0}^{-1}$ is known as well.

$$|\mathrm{Adv}_{\mathcal{H}_2}[\mathcal{A}] - \mathrm{Adv}_{\mathcal{H}_1}[\mathcal{A}]| = \mathrm{negl}(\lambda).$$

We notice that in this hybrid, we now have that $\vec{\mathbf{B}} = \mathbf{A}\vec{\mathbf{R}} + \sigma\vec{\mathbf{G}}$, where $\vec{\mathbf{R}} = [\mathbf{R}_1 \| \cdots \| \mathbf{R}_\eta]$. Recalling that $\mathbf{C}_i = \mathrm{Eval}(\mathrm{PRF.Eval}(\cdot, i), \vec{\mathbf{B}})$, we can define $\mathbf{H}_i^* = \mathbf{H}_{\mathrm{PRF.Eval}(\cdot, i), \sigma, \vec{\mathbf{B}}}$, and it will hold that

$$\mathbf{C}_i = \mathbf{A}\vec{\mathbf{R}}\mathbf{H}_i^* + \mathrm{PRF.Eval}(\sigma, i) \cdot \mathbf{G} = \mathbf{A}\vec{\mathbf{R}}\mathbf{H}_i^* + (x_i^* \oplus \Delta_i)\mathbf{G}. \tag{4}$$

We recall that $\mathbf{H}_i^*$ is computable given σ, and furthermore $\|\mathbf{H}_i^*\|_\infty \le (N+1)^{d_{\mathrm{prf}}}$. If we denote $\vec{\mathbf{H}}^* = [\mathbf{H}_1^* \| \cdots \| \mathbf{H}_\ell^*]$, we conclude that

$$\vec{\mathbf{C}} - (x^* \oplus \Delta_{\le \ell})\vec{\mathbf{G}} = \mathbf{A}\vec{\mathbf{R}}\vec{\mathbf{H}}^*. \tag{5}$$

Hybrid $\mathcal{H}_3$. In this hybrid we will switch from generating sk_f using $\mathbf{A}_{\tau_0}^{-1}$ to generating them using $\vec{\mathbf{R}}$. We recall that we are only required to generate keys for f s.t. $f(x^*) = 1$, otherwise the adversary loses in the semi-adaptive security game.

We recall that by definition, in order to derive sk_f, we need to sample from $[\mathbf{C}_f \| \mathbf{A}]_\tau^{-1}$. We recall that we defined $\mathbf{C}_f = \mathrm{Eval}(f_\Delta, \vec{\mathbf{C}})$, and therefore, denoting $\mathbf{H} = \mathbf{H}_{f_\Delta, (x^* \oplus \Delta_{\le \ell}), \vec{\mathbf{C}}}$, it holds that

$$\mathbf{C}_f - f_\Delta(x^* \oplus \Delta_{\le \ell}) \cdot \mathbf{G} = \left(\vec{\mathbf{C}} - (x^* \oplus \Delta_{\le \ell})\vec{\mathbf{G}}\right) \cdot \mathbf{H}.$$

Plugging in Eq. (5), and since $f_\Delta(x^* \oplus \Delta_{\le \ell}) = f(x^*) = 1$, we get that

$$\mathbf{C}_f = \mathbf{A}\vec{\mathbf{R}}\vec{\mathbf{H}}^*\mathbf{H} + \mathbf{G}.$$

Therefore, $[\mathbf{C}_f \| \mathbf{A}] = [\mathbf{A} \cdot (\vec{\mathbf{R}}\vec{\mathbf{H}}^*\mathbf{H}) + \mathbf{G} \| \mathbf{A}]$. This means that given $\vec{\mathbf{R}}$ and the computable matrices $\vec{\mathbf{H}}^*, \mathbf{H}$, one can sample from $[\mathbf{C}_f \| \mathbf{A}]_\tau^{-1}$ for all values of $\tau \ge \tau'$ for $\tau' = O\left(m \cdot \left\|\vec{\mathbf{R}} \cdot \vec{\mathbf{H}}^* \cdot \mathbf{H}\right\|_\infty\right)$. Plugging in the known bounds, we get that

$$\tau' = O(m \cdot N\eta \cdot (N+1)^{d_{\mathrm{prf}}} \cdot N\ell \cdot (N+1)^d) = O(\ell) \cdot (N+1)^{d+d_{\mathrm{prf}}} \cdot mN^2,$$

Recall that we need to sample with $\tau = 2^\nu \cdot mN^2(N+1)^{d+d_{\mathrm{prf}}}$ which is asymptotically greater than τ', which is enabled by our parameter setting.

It follows that changing our method of sampling $\mathbf{r}_f$ does not change the resulting distribution, and therefore

$$\dot{\mathrm{A}}\mathrm{dv}_{\mathcal{H}_3}[\mathcal{A}] = \mathrm{Adv}_{\mathcal{H}_2}[\mathcal{A}].$$

We notice that in this hybrid, the challenger does not require $\mathbf{A}_{\tau_0}^{-1}$ at all.

Hybrid $\mathcal{H}_4$. In this hybrid, we change the distribution of $\mathbf{A}$ and sample it uniformly from $\mathbb{Z}_q^{n \times m}$ rather than via TrapGen. Since TrapGen samples $\mathbf{A}$ which is statistically indistinguishable from uniform, we conclude that the distribution produced in the two hybrids are statistically indistinguishable as well.

$$|\mathrm{Adv}_{\mathcal{H}_4}[\mathcal{A}] - \mathrm{Adv}_{\mathcal{H}_3}[\mathcal{A}]| = \mathrm{negl}(\lambda).$$

Hybrid $\mathcal{H}_5$. In this hybrid we change the way the challenge ciphertext is computed. Specifically we change the way we compute $\psi_{i,1-\Delta_i}$, for all i, and set

$$\psi_{i,1-\Delta_i} = \mathsf{oldABE.Enc}_{\mathsf{oldabepp}}(\mathbf{0}, (\ell^*, i, 1 - \Delta_i)),$$

where the zero vector has the same length as $\mathbf{c}_{i,x_i^* \oplus \Delta_i \oplus 1}$.

Since for all ℓ, i, $\mathsf{BitCheck}_{n,\Delta_{\leq \ell}}(\ell, i, 1 - \Delta_i) = 1$, and thus for all ℓ, the key $\mathsf{oldabesk}_\ell$ must not decrypt $\psi_{i,1-\Delta_i}$, we would like to use the security of oldABE to argue that $\mathcal{H}_5$ is computationally indistinguishable from $\mathcal{H}_4$. However, some care needs to be taken since we only assume that oldABE is *selectively* secure.

The formal proof will proceed via a hybrid argument going over all values of ℓ and β (note that we at this point we have an upper bound on ℓ given by the running time of $\mathcal{A}$). In the (i, β) hybrid, we change all ciphertexts $\psi_{i',\beta'}$ such that $(i', \beta') < (i, \beta)$ (lexicographically) to $\mathbf{0}$ if $\beta' \neq \Delta_{i'}$. To argue that two adjacent hybrids are indistinguishable, we rely on the selective hardness of oldABE for the fixed attribute (i, β) which can be provided in the beginning of the game as required for selective security.

We conclude that this hybrid is computationally indistinguishable from the previous one.

$$|\mathrm{Adv}_{\mathcal{H}_5}[\mathcal{A}] - \mathrm{Adv}_{\mathcal{H}_4}[\mathcal{A}]| = \mathrm{negl}(\lambda).$$

Hybrid $\mathcal{H}_6$. We again change the contents of the challenge ciphertext as follows. We generate $\mathbf{s}, \mathbf{e}, e'$ as before, and set $\mathbf{b}^T = \mathbf{s}^T \mathbf{A} + \mathbf{e}^T$, and $b' = \mathbf{s}^T \mathbf{v} + e'$. The vector $\mathbf{c}_0$ is generated identically to before, but we can express it in terms of $\mathbf{b}, b'$ as

$$\mathbf{c}_0^T = [\mathbf{b}^T \| b'] + \mu \lfloor q/2 \rfloor \cdot [\mathbf{0}^T \| 1].$$

We recall that as of the previous hybrid, the values $\mathbf{c}_{i,x_i^* \oplus \Delta_i \oplus 1}$ no longer appear in the challenge ciphertext, so they are not generated at all. The only change that we make is in the generation of $\mathbf{c}_{i,x_i^* \oplus \Delta_i}$. We recall that in the previous hybrid

$$\mathbf{c}_{i,x_i^* \oplus \Delta_i}^T = \mathbf{s}^T (\mathbf{C}_i - (x_i^* \oplus \Delta_i)\mathbf{G}) + \tilde{\mathbf{e}}_i^T.$$

and since at this point $(\mathbf{C}_i - (x_i^* \oplus \Delta_i)\mathbf{G}) = \mathbf{A}\vec{\mathbf{R}}\mathbf{H}_i^*$, as per Eq. (4), we had that

$$\mathbf{c}_{i,x_i^* \oplus \Delta_i}^T = \mathbf{s}^T \mathbf{A}\vec{\mathbf{R}}\mathbf{H}_i^* + \tilde{\mathbf{e}}_i^T.$$

In this hybrid, we change these values to

$$\mathbf{c}_{i,x_i^* \oplus \Delta_i}^T = \mathbf{b}^T \vec{\mathbf{R}}\mathbf{H}_i^* + \tilde{\mathbf{e}}_i^T = \mathbf{s}^T \mathbf{A}\vec{\mathbf{R}}\mathbf{H}_i^* + \mathbf{e}^T \vec{\mathbf{R}}\mathbf{H}_i^* + \tilde{\mathbf{e}}_i^T.$$

This distribution, however, is statistically close to the previous one, since the distribution $\mathbf{e}^T \vec{\mathbf{R}} \mathbf{H}_i^*$ is $(B \cdot m \cdot \eta N \cdot (N+1)^{d_{\mathrm{prf}}})$-bounded and since we selected $\tilde{\chi}$ to be $(Bm\eta N(N+1)^{d_{\mathrm{prf}}})$-swallowing, statistical indistinguishability follows by definition.

$$|\mathrm{Adv}_{\mathcal{H}_6}[\mathcal{A}] - \mathrm{Adv}_{\mathcal{H}_5}[\mathcal{A}]| = \mathrm{negl}(\lambda).$$

We note that in this hybrid, given $\mathbf{b}, b'$, the challenger does not need to know the values of $\mathbf{s}, \mathbf{e}, e'$ since they are not used directly.

Hybrid $\mathcal{H}_7$. In the final hybrid, we change the distribution of $\mathbf{b}, b'$ to be uniform in $\mathbb{Z}_q^m, \mathbb{Z}_q$, respectively. Indistinguishability follows by definition from the $\mathrm{DLWE}_{n,q,\chi}$ assumption. We have

$$|\mathrm{Adv}_{\mathcal{H}_7}[\mathcal{A}] - \mathrm{Adv}_{\mathcal{H}_6}[\mathcal{A}]| = \mathrm{negl}(\lambda).$$

Clearly, in this hybrid the adversary has no advantage since b' is uniform and completely masks the value of μ. It follows therefore that

$$\mathrm{Adv}_{\mathcal{H}_7}[\mathcal{A}] = 1/2,$$

and therefore

$$|\mathrm{Adv}[\mathcal{A}] - 1/2| = \mathrm{negl}(\lambda),$$

which completes the proof of security.

4.3 Conclusion

Finally we can put all the pieces together and state our result with all parameters.

Theorem 2. *Assume that* GapSVP *(respectively* SIVP*) is hard to approximate by a polynomial time classical (respectively quantum) algorithm to within a factor of 2^{n^ϵ}. Then for any polynomial $d = d(\lambda)$ there exists a correct and semi-adaptively secure ABE scheme for the policy class $\mathcal{P}_d$.*

Letting $k = (\lambda d)^{1/\epsilon}$, the public parameters of the scheme are of size $\mathrm{poly}(k)$, ciphertexts are of length $\ell \cdot \mathrm{poly}(k)$, where ℓ is the attribute length, and the key length is $\ell + \mathrm{poly}(k)$, where ℓ is the input length of the policy function (all $\mathrm{poly}(\cdot)$ notations indicate a specific polynomial function).

Proof. A secure family of pseudorandom functions can be instantiated based on the existence of any one-way function, and in particular on the hardness of lattice approximation to within $\mathrm{poly}(n) \ll 2^{n^\epsilon}$ factor.

We instantiate oldABE using the scheme from [9]. Recall that oldABE only needs to support attributes of length $O(\nu)$ and policies which can be represented by circuits of depth $O(\log(\nu))$. This means that such a scheme can be based on the hardness of DLWE with parameters that translate to the hardness of lattice approximation to within a factor of $2^{n^{o(1)}} \ll 2^{n^\epsilon}$. The keys and ciphertexts of oldABE will have overhead $\mathrm{poly}(\lambda)$ for a fixed polynomial.

Combining these primitives with the correctness analysis and with the security analysis in Lemma 1, the theorem follows.

References

1. Abdalla, M., Catalano, D., Dent, A.W., Malone-Lee, J., Neven, G., Smart, N.P.: Identity-based encryption gone wild. In: Bugliesi, M., Preneel, B., Sassone, V., Wegener, I. (eds.) ICALP 2006. LNCS, vol. 4052, pp. 300–311. Springer, Heidelberg (2006)
2. Agrawal, S., Boneh, D., Boyen, X.: Efficient lattice (H)IBE in the standard model. In: Gilbert, H. (ed.) EUROCRYPT 2010. LNCS, vol. 6110, pp. 553–572. Springer, Heidelberg (2010)
3. Agrawal, S., Boneh, D., Boyen, X.: Lattice basis delegation in fixed dimension and shorter-ciphertext hierarchical IBE. In: Rabin, T. (ed.) CRYPTO 2010. LNCS, vol. 6223, pp. 98–115. Springer, Heidelberg (2010)
4. Ajtai, M.: Generating hard instances of lattice problems (extended abstract). In: Miller, G.L. (ed.) Proceedings of the Twenty-Eighth Annual ACM Symposium on the Theory of Computing, Philadelphia, Pennsylvania, USA, 22–24 May 1996, pp. 99–108. ACM (1996)
5. Alekhnovich, M.: More on average case vs. approximation complexity. In: Proceedings of the 44th Symposium on Foundations of Computer Science (FOCS 2003), Cambridge, MA, USA, 11–14 October 2003, pp. 298–307. IEEE Computer Society (2003)
6. Ananth, P., Brakerski, Z., Segev, G., Vaikuntanathan, V.: From selective to adaptive security in functional encryption. In: Gennaro and Robshaw [16], pp. 657–677 (2015)
7. Ananth, P., Sahai, A.: Functional encryption for turing machines. IACR Cryptology ePrint Archive 2015:776 (2015)
8. Blum, A., Furst, M.L., Kearns, M., Lipton, R.J.: Cryptographic primitives based on hard learning problems. In: Stinson, D.R. (ed.) CRYPTO 1993. LNCS, vol. 773, pp. 278–291. Springer, Heidelberg (1994)
9. Boneh, D., Gentry, C., Gorbunov, S., Halevi, S., Nikolaenko, V., Segev, G., Vaikuntanathan, V., Vinayagamurthy, D.: Fully key-homomorphic encryption, arithmetic circuit ABE and compact garbled circuits. In: Nguyen, P.Q., Oswald, E. (eds.) EUROCRYPT 2014. LNCS, vol. 8441, pp. 533–556. Springer, Heidelberg (2014)
10. Boyen, X.: Attribute-based functional encryption on lattices. In: Sahai, A. (ed.) TCC 2013. LNCS, vol. 7785, pp. 122–142. Springer, Heidelberg (2013)
11. Brakerski, Z., Vaikuntanathan, V.: Constrained key-homomorphic PRFs from standard lattice assumptions. In: Dodis, Y., Nielsen, J.B. (eds.) TCC 2015, Part II. LNCS, vol. 9015, pp. 1–30. Springer, Heidelberg (2015)
12. Cash, D., Hofheinz, D., Kiltz, E., Peikert, C.: Bonsai trees, or how to delegate a lattice basis. J. Cryptol. 25(4), 601–639 (2012)
13. Chen, J., Wee, H.: Semi-adaptive attribute-based encryption and improved delegation for Boolean formula. In: Abdalla, M., De Prisco, R. (eds.) SCN 2014. LNCS, vol. 8642, pp. 277–297. Springer, Heidelberg (2014)
14. Garg, S., Gentry, C., Halevi, S., Sahai, A., Waters, B.: Attribute-based encryption for circuits from multilinear maps. In: Canetti, R., Garay, J.A. (eds.) CRYPTO 2013, Part II. LNCS, vol. 8043, pp. 479–499. Springer, Heidelberg (2013)
15. Garg, S., Gentry, C., Halevi, S., Zhandry, M.: Fully secure attribute based encryption from multilinear maps. IACR Cryptology ePrint Archive 2014:622 (2014)
16. Gennaro, R., Robshaw, M. (eds.): CRYPTO 2015. LNCS, vol. 9216. Springer, Heidelberg (2015)

17. Gentry, C., Peikert, C., Vaikuntanathan, V.: Trapdoors for hard lattices and new cryptographic constructions. In: STOC (2008)
18. Gentry, C., Sahai, A., Waters, B.: Homomorphic encryption from learning with errors: conceptually-simpler, asymptotically-faster, attribute-based. In: Canetti, R., Garay, J.A. (eds.) CRYPTO 2013, Part I. LNCS, vol. 8042, pp. 75–92. Springer, Heidelberg (2013)
19. Gorbunov, S., Vaikuntanathan, V., Wee, H.: Attribute-based encryption for circuits. In: Boneh, D., Roughgarden, T., Feigenbaum, J. (eds.) Symposium on Theory of Computing Conference, STOC 2013, Palo Alto, CA, USA, 1–4 June 2013, pp. 545–554. ACM (2013)
20. Gorbunov, S., Vaikuntanathan, V., Wee, H.: Predicate encryption for circuits from LWE. In: Gennaro and Robshaw [16], pp. 503–523 (2015)
21. Gorbunov, S., Vaikuntanathan, V., Wichs, D.: Leveled fully homomorphic signatures from standard lattices. In: Servedio, R.A., Rubinfeld, R. (eds.) Proceedings of the Forty-Seventh Annual ACM on Symposium on Theory of Computing, STOC 2015, Portland, OR, USA, 14–17 June 2015, pp. 469–477. ACM (2015)
22. Goyal, V., Pandey, O., Sahai, A., Waters, B.: Attribute-based encryption for fine-grained access control of encrypted data. In: Juels, A., Wright, R.N., di Vimercati, S.D.C. (eds.) Proceedings of the 13th ACM Conference on Computer and Communications Security, CCS 2006, Alexandria, VA, USA, 30 October–3 November 2006, pp. 89–98. ACM (2006)
23. Hohenberger, S., Waters, B.: Attribute-based encryption with fast decryption. In: Kurosawa, K., Hanaoka, G. (eds.) PKC 2013. LNCS, vol. 7778, pp. 162–179. Springer, Heidelberg (2013)
24. Katz, J., Wang, N.: Efficiency improvements for signature schemes with tight security reductions. In: Jajodia, S., Atluri, V., Jaeger, T. (eds.) Proceedings of the 10th ACM Conference on Computer and Communications Security, CCS 2003, Washington, DC, USA, 27–30 October 2003, pp. 155–164. ACM (2003)
25. Lewko, A., Okamoto, T., Sahai, A., Takashima, K., Waters, B.: Fully secure functional encryption: attribute-based encryption and (hierarchical) inner product encryption. In: Gilbert, H. (ed.) EUROCRYPT 2010. LNCS, vol. 6110, pp. 62–91. Springer, Heidelberg (2010)
26. Lewko, A., Waters, B.: Unbounded HIBE and attribute-based encryption. In: Paterson, K.G. (ed.) EUROCRYPT 2011. LNCS, vol. 6632, pp. 547–567. Springer, Heidelberg (2011)
27. Lewko, A., Waters, B.: New proof methods for attribute-based encryption: achieving full security through selective techniques. In: Safavi-Naini, R., Canetti, R. (eds.) CRYPTO 2012. LNCS, vol. 7417, pp. 180–198. Springer, Heidelberg (2012)
28. Micciancio, D., Mol, P.: Pseudorandom knapsacks and the sample complexity of LWE search-to-decision reductions. In: Rogaway, P. (ed.) CRYPTO 2011. LNCS, vol. 6841, pp. 465–484. Springer, Heidelberg (2011)
29. Micciancio, D., Peikert, C.: Trapdoors for lattices: simpler, tighter, faster, smaller. In: Pointcheval, D., Johansson, T. (eds.) EUROCRYPT 2012. LNCS, vol. 7237, pp. 700–718. Springer, Heidelberg (2012)
30. Okamoto, T., Takashima, K.: Fully secure functional encryption with general relations from the decisional linear assumption. In: Rabin, T. (ed.) CRYPTO 2010. LNCS, vol. 6223, pp. 191–208. Springer, Heidelberg (2010)
31. Okamoto, T., Takashima, K.: Fully secure unbounded inner-product and attribute-based encryption. In: Wang, X., Sako, K. (eds.) ASIACRYPT 2012. LNCS, vol. 7658, pp. 349–366. Springer, Heidelberg (2012)

32. Peikert, C.: Public-key cryptosystems from the worst-case shortest vector problem: extended abstract. In: Proceedings of the 41st Annual ACM Symposium on Theory of Computing, STOC 2009, Bethesda, MD, USA, 31 May–2 June 2009, pp. 333–342 (2009)
33. Regev, O.: On lattices, learning with errors, random linear codes, and cryptography. In: Proceedings of the 37th Annual ACM Symposium on Theory of Computing, Baltimore, MD, USA, 22–24 May 2005, pp. 84–93 (2005)
34. Sahai, A., Waters, B.: Fuzzy identity-based encryption. In: Cramer, R. (ed.) EUROCRYPT 2005. LNCS, vol. 3494, pp. 457–473. Springer, Heidelberg (2005)
35. Schnorr, C.: A hierarchy of polynomial time lattice basis reduction algorithms. Theor. Comput. Sci. **53**, 201–224 (1987)
36. Waters, B.: Functional encryption for regular languages. In: Safavi-Naini, R., Canetti, R. (eds.) CRYPTO 2012. LNCS, vol. 7417, pp. 218–235. Springer, Heidelberg (2012)
37. Waters, B.: A punctured programming approach to adaptively secure functional encryption. In: Gennaro and Robshaw [16], pp. 678–697 (2015)

Automated Tools and Synthesis

Design in Type-I, Run in Type-III: Fast and Scalable Bilinear-Type Conversion Using Integer Programming

Masayuki Abe[1,3]($\boxtimes$), Fumitaka Hoshino[1,4], and Miyako Ohkubo[2]

[1] Information Sharing Platform Laboratories, NTT Corporation, Tokyo, Japan
abe.masayuki@lab.ntt.co.jp
[2] Security Fundamentals Laboratory, CSRI, NICT, Tokyo, Japan
[3] Graduate School of Informatics, Kyoto University, Kyoto, Japan
[4] School of Computing, Department of Mathematical and Computing Science,
Tokyo Institute of Technology, Tokyo, Japan

Abstract. Bilinear type conversion is to convert cryptographic schemes designed over symmetric groups instantiated with imperilled curves into ones that run over more secure and efficient asymmetric groups. In this paper we introduce a novel type conversion method called *IPConv* using 0–1 Integer Programming. Instantiated with a widely available IP solver, it instantly converts existing intricate schemes, and can process large-scale schemes that involves more than a thousand variables and hundreds of pairings.

Such a quick and scalable method allows a new approach in designing cryptographic schemes over asymmetric bilinear groups. Namely, designers work without taking much care about asymmetry of computation but the converted scheme runs well in the asymmetric setting. We demonstrate the usefulness of conversion-aided design by presenting somewhat counterintuitive examples where converted DLIN-based Groth-Sahai proofs are more compact than manually built SXDH-based proofs.

Keywords: Conversion · Bilinear groups · Integer programming · Groth-Sahai proofs · Zero-knowledge

1 Introduction

1.1 Background

Prime-order bilinear groups consist of source groups $\mathbb{G}_0$ and $\mathbb{G}_1$, target group $\mathbb{G}_T$, and a pairing $e : \mathbb{G}_0 \times \mathbb{G}_1 \to \mathbb{G}_T$. In so called Type-I bilinear groups, $\mathbb{G}_0 = \mathbb{G}_1$, i.e., the pairing is symmetric. It has been a popular choice in early research and development. Recent progress in analyzing symmetric pairing groups instantiated with small characteristic curves [9, 26, 30, 31] motivates crypto designers to move to Type-III groups where $\mathbb{G}_0 \neq \mathbb{G}_1$, i.e., the pairing is asymmetric, and no efficient mapping is known between $\mathbb{G}_0$ and $\mathbb{G}_1$. For Type-III groups, no such

© International Association for Cryptologic Research 2016
M. Robshaw and J. Katz (Eds.): CRYPTO 2016, Part III, LNCS 9816, pp. 387–415, 2016.
DOI: 10.1007/978-3-662-53015-3_14

weakness has been observed until now and efficient instantiations have been developed. Yet Type-I setting is useful for presenting and understanding cryptographic schemes for their simplicity. Besides, number of schemes have been designed only for Type-I groups in the literature, e.g. [2, 8, 16, 36, 37, 40, 41].

Bilinear-type conversion is a method to translate schemes designed for Type-I groups into ones that work over Type-III groups. Cryptographic schemes designed in Type-I setting do not necessarily work in Type-III due to the presence of symmetric pairings, $e(X, X)$. A workaround is to convert the algorithm by *duplicating* the variables. That is, the variable is represented by a pair $(X, X') \in \mathbb{G}_0 \times \mathbb{G}_1$. Duplication however clearly slows down the performance since all relevant computations are 'duplicated' in $\mathbb{G}_0$ and $\mathbb{G}_1$ as well. Besides, duplication is not always possible due to mathematical constraints or external requirements. For instance, it is not known how to pick random and consistent pair X and X' while retaining the hardness of the discrete logarithm problem on X and X'. An automated conversion finds the best allocation of variables over $\mathbb{G}_0$ and $\mathbb{G}_1$ that makes all group operations doable with minimal overhead.

Besides saving existing schemes over Type-I groups, conversion plays the role in putting *"Design in Type-I and Run in Type-III"* paradigm into practice as suggested in the pioneering work by Akinyele et al. [7]. That is, let crypto designers focus on their high-level idea of construction without taking much care about asymmetry of computation by designing in Type-I setting, and then convert the results to obtain executable schemes over Type-III groups. For conversion tools to be useful, the processing speed and scalability are of importance on top of the performance of the final executables. Like compilers for high-level programming languages a conversion tool will be executed over and over again throughout the development. Quick response is strongly desired for productivity and stress-free developing environment. Its importance increases when large-scale systems that consist of several building blocks are targeted. Nevertheless, only small-scale monolithic schemes has been targeted so far. Hence the validity of the design paradigm has not been well substantiated yet.

1.2   Our Contribution

We propose a new efficient conversion algorithm, which we call 'IPConv', based on 0–1 Integer Programming (IP). A technical highlight that separates this work from previous ones [6, 7] is how to encode several kinds of constraints into a system of linear relations over binary variables, and how to implement ones metric into an objective function the 0–1 IP minimizes subject to the constraints. The idea of encoding computational constraints into an objective function follows from previous works. Our novelty is the encoding method that allows one to use Integer Programming that fits well to our optimization problem with various constraints. Besides, using such a tool is advantageous in the sense that there are publicly available (both commercial and non-commercial) software packages such as [5, 24, 28, 33–35].

Performance of IPConv is demonstrated by experiments over real cryptographic schemes in the literature. IPConv instantly completes the task even for complex schemes. To measure the scalability, large systems with thousands of variables and pairings are generated randomly subject to some reasonably looking structures. IPConv processed them in a few minutes to hours even with non-commercial IP solver SCIP [5] as an engine. The concrete figures of course become magnitude of better with a powerful commercial IP solver e.g. [28]. Scaling up to thousands of pairings may seem an overkill. However, for instance, schemes that include Groth-Sahai (GS) proof system [27] easily involve dozens or even hundreds of pairings when their security proofs are taken into account. Furthermore, tools such as [10–12] would allow automated synthesis that reach to or even exceed such a scale. Our method not only contributes to speedup the process of conversion but also opens the door to automated synthesis and optimization of large scale cryptographic applications over bilinear groups.

Next we, for the first time, prove the usefulness of the conversion-assisted design for middle-scale schemes. It is shown that schemes involving GS proofs based on decision linear assumption (DLIN) can be converted to ones based on XDLIN assumption [1] in Type-III so that they are more efficient than their direct instantiation based on the symmetric external Diffie-Hellman assumption (SXDH). The result may be counter-intuitive since the commitments and proofs of SXDH-based GS-proofs require less group elements than those based on DLIN. Key observations that explain our result are:

- Relations such as $e(X, A) = e(B, Y)$ for variables X and Y are considered as linear pairing product equations (PPEs) in Type-I whose proof consists of 3 elements whereas they are more costly two-sided PPEs in Type-III that costs 8 elements. Proving linear PPEs can be converted without duplicating the proofs and commitments in general.
- Commitments and proofs in the converted proof system are allocated mostly in $\mathbb{G}_0$ whereas they appear in both $\mathbb{G}_0$ and $\mathbb{G}_1$ in direct SXDH-based instantiation. Taking the fact that elements in $\mathbb{G}_1$ is typically twice as long as those in $\mathbb{G}_0$ in bits, the former can be shorter than the latter in some cases.

Our first example in Sect. 5.2 is a scheme for showing ones possession of a correct structure-preserving signature [3] on a public message in zero-knowledge. The scheme obtained by conversion yields proofs that are up to 50 % shorter (asymptotic in the message length) than those generated by direct constructions based on SXDH. It uses a novel fine-tuning for zero-knowledge GS-proofs (GSZK) presented in Sect. 5.1 that takes the above mentioned advantages.

Our second example in Sect. 5.3 is to demonstrate that our framework can be applied to schemes that is already designed in Type-III setting to seek for better instantiations. We pick an automorphic blind signature scheme [3] that involves GS-proofs and is secure under SXDH assumption in Type-III setting. We show that the proofs can be replaced with the DLIN-based ones and it can be converted to work in Type-III under XDLIN assumption. Though the GS-proofs are witness indistinguishable for this time, it still can take the above mentioned advantages and saves 28 % in the length of the signatures compared to the originally manufactured SXDH-based scheme.

Although our primary metric for optimization is the size of intended objects, we also compare their computational workload in the number of pairings in signature verification. Interestingly, the winner changes depending on the message size, acceptable duplication, and also the use of batch verification technique [13]. This unveils an open issue on optimization of schemes involving GS-proofs.

1.3 Related Works

There are some conversion systems in the literature. Early works on type conversion, e.g. [17–19,39], study and suggest heuristic guidelines for when a scheme allows or resists conversion. To our best knowledge, AutoGroup introduced by Akinyele et al. in [7] is the first automated conversion system that converts schemes from Type-I to Type-III. Given a target scheme described in their scheme description language, the system finds set of 'valid' solutions that satisfy constraints over pairings by using a satisfiability modulo theory solver [21]. It then search for the 'optimal' solution that conforms to other mathematical constraints and ones preferences. When there are number of possible solutions, the performance gets lower. In this pioneering work, the security of the resulting converted scheme was not guaranteed. In [4], Abe et al., established a theoretical ground for preserving security during conversion. Their framework, reviewed in Sect. 2, provides useful theorems for security guarantee. But their conversion algorithm is basically a brute-force search over all possible conversions and it requires exponential time in the number of pairings. Recently in [6], Akinyele, Garman, and Hohenberger introduced an upgraded system called AutoGroup+ that integrates the framework of [4] to AutoGroup. Though the system becomes more solid in terms of security, their approach for finding an optimal solution remains the same as before. They cover only small scale cryptographic schemes.

Regarding Groth-Sahai zero-knowledge proofs, the closest work is the one by Escala and Groth in [22]. They observe that commitment of $1_{\mathbb{Z}_p}$ can be seen as a commitment of the default generator G and uses the fact that a commitment of G can be equivocated to G^0 to construct more efficient zero-knowledge proofs for pairing product equations (PPEs) with constant pairings of the form $e(G, A)$ in Type-III setting. Our fine-tuning in Sect. 5.1 uses the same property for the commitment of G but use it in a different manner that is most effective in Type-I setting. Another close work is [25] that presents a DLIN-based variant of GS-proof system over asymmetric bilinear groups. Their scheme bases on SDLIN assumption where *independent* DLIN in $\mathbb{G}_0$ and $\mathbb{G}_1$ are assumed as hard, and uses independently generated CRSes for commitments in $\mathbb{G}_0$ and $\mathbb{G}_1$. Thus their proof system is inherently asymmetric, which cannot exploit nice properties of symmetric setting as done in this work. Besides, SDLIN-based instantiation is less efficient than SXDH-based one. We therefore use the original SXDH-based instantiation for comparison in this paper.

In [23,29], a more efficient instantiation of GS-proofs by using recently introduced Matrix assumptions. Although DLIN-based GS-proofs are used throughout this paper, matrix-based assumption might be an alternative to further gain efficiency if the Type-III analogue of the assumption is acceptable.

2   Conversion Based on Dependency Graphs

2.1   Overview

In this section we review the framework in [4]. To guarantee the security of the resulting scheme, it converts not only algorithms that form the target scheme but also all algorithms that appear in the security proof as well as underlying assumptions. Namely, it assumes that the security is proven by the existence of reduction algorithms from some assumptions in Type-I, and converts the algorithms and assumptions into Type-III. This way, the security proof is preserved under the converted assumption. It is proven in [4] that if the original assumptions are valid in Type-I generic bilinear group model [15], the converted assumptions are valid in Type-III generic bilinear group model. Most typically, the DLIN assumption is converted to XDLIN.

In their framework relations among variables in target algorithms are described by using a graph called a dependency graph, and the central task of conversion is reduced to find a 'split' of the graph so that each graph implies variables and computations in each source group in the Type-III setting.

We follow the framework of [4] that consists of the following four steps.

1. Verify that the target scheme in Type-I and its security proof follows the abstraction of bilinear groups.
2. Describe the generic bilinear group operations over source group $\mathbb{G}$ by using a dependency graph as we shall explain later.
3. Split the dependency graph into two that satisfy some conditions. The resulting graphs imply variables and group operations in $\mathbb{G}_0$ and $\mathbb{G}_1$ respectively.
4. Describe the resulting scheme in Type-III as suggested by the graphs.

As well as [4], we focus on step 3 and propose an efficient algorithm for the task of finding a split. Thus, when we conduct an experiment for demonstrating the performance, we start from a dependency graph as input and complete when a desirable split of the input graph is obtained.

2.2   Dependency Graph

A dependency graph is a directed graph that represents computational dependencies among variables storing source group elements in the target system. In Fig. 1, we show an example of a dependency graph for a program that computes some group operations over Type-I bilinear groups. In the right is a sample program that takes source group elements A, B, D as input and computes C and E via group operations (multiplication and exponentiation), and outputs a result of pairing $e(C, E)$. In the left is a dependency graph that corresponds to the algorithm. Nodes represent the source group elements and edges correspond to group operations. Each input to the pairing operation is represented by a connection to node $\mathsf{P}_{\mathsf{CE}}[b]$ called a pairing node. As the graph only describes relations between group elements via group operations, it does not show the structure of

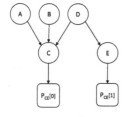

Sample(a, A, B, D):
$a \in \mathbb{Z}_p,\ A, B, C, D, E \in \mathbb{G}$

> if $a = 0$ then
> $\quad C := A \cdot B, \quad E := D$
> else
> $\quad C := D^a, \quad E := D^3$
> endif
> Output $e(C, E)$

Fig. 1. An example of a dependency graph for a program in Type-I bilinear groups.

the program like "if-then-else" directive or involve non source group elements like $a \in \mathbb{Z}_p$. Operations in the target group are irrelevant either.

There are several types of nodes in a dependency graph. Node types can be considered as attributes attached to the nodes or lists of nodes. We use either way according to the context.

- Pairing nodes ($\mathcal{P}$). They represent inputs to pairing operations. Every pairing node has only one incoming edge and no outgoing edges. Each pairing node is paired with another pairing node so that the pair constitutes an input to a pairing operation.
- Control nodes ($\mathcal{CT}$). These are the ones added to the graph to control the assignment to their parent nodes. A control node has one or more incoming edges but no outgoing edges. By specifying which group to assign to a control node, its parent nodes are also assigned to the same group. For instance, when two variables associated to nodes n and n' are to be compared, a control node is added with incoming edges from n and n'. This results in assigning n and n' to the same group the control node is assigned. The control nodes are used also to implement user specified preferences such as grouping as we shall explain later.
- Regular nodes ($\mathcal{R}$). All nodes other than pairing nodes and control nodes are regular nodes. Regular nodes may have other attributes named as follows.
  - Bottom nodes ($\mathcal{B}$). A regular node is a bottom node if it does not have outgoing edges. This includes a 'pseudo' bottom node that virtually works as a bottom node in a closure.
  - Prohibited nodes ($\mathcal{PH}$). These are nodes that must not be duplicated for some reasons. They are assigned to either of the source groups but the assignment is not fixed in advance. Nodes representing variables as an output of "hash-to-group" function that directly maps to group elements must be a prohibited node. Currently known technology does not allow us to hash an input onto two source group elements in a way that their exponents are unknown but remain in a preliminary fixed relation. Another example of the prohibiting nodes are inputs given to the target scheme from outside like messages in a signature scheme. They are subject to other building blocks and hence demanding duplicated messages

loses generality of the signature scheme. Thus it is generally desirable that messages are considered as prohibited nodes.

From the above classification, we have $V = \mathcal{P} \cup \mathcal{CT} \cup \mathcal{R}$. The nodes that will be assigned to either of the source groups exclusively are called constrained nodes. Precisely, we define constrained nodes $\mathcal{C}$ by $\mathcal{C} := \mathcal{P} \cup \mathcal{CT} \cup \mathcal{B} \cup \mathcal{PH}$.

2.3 Valid Split

It has been shown in [4] that if a dependency graph is split into two graphs that satisfy four conditions below then the converted scheme derived from the graphs works over Type-III bilinear groups and is secure in the same sense as the original scheme but based on converted assumptions. Such a pair of graphs is called a valid split. Let $\mathsf{Anc}(\Gamma, X)$ denote a subgraph of Γ that consists of X and all paths that reach to X. Let NoDup be a list of nodes representing variables as output of hash-to-group function.

Definition 1 (Valid Split). *Let $\Gamma = (V, E)$ be a dependency graph for $\tilde{\Pi}$. Let $P = (p_1[0], \ldots, p_{n_p}[1]) \subset V$ be pairing nodes. A pair of graphs $\Gamma_0 = (V_0, E_0)$ and $\Gamma_1 = (V_1, E_1)$ is a valid split of Γ with respect to $\mathsf{NoDup} \subseteq V$ if:*

1. *merging Γ_0 and Γ_1 recovers Γ,*
2. *for each $i \in \{0,1\}$ and every $X \in V_i \setminus P$, the subgraph $\mathsf{Anc}(\Gamma, X)$ is in Γ_i,*
3. *for each $i \in \{1, \ldots, n_p\}$, paring nodes $p_i[0]$ and $p_i[1]$ are separately included in V_0 and V_1, and*
4. *$V_0 \cap V_1 \cap \mathsf{NoDup} = \emptyset$.*

The first condition guarantees that all variables and computations are preserved during conversion. The second condition guarantees that all variables needed to compute a variable belong to the same source group. The third condition guarantees consistency of pairing operations by forcing that every pairing operation takes inputs from $\mathbb{G}_0$ and $\mathbb{G}_1$. The last condition is to conform with the constraint about the hash-to-group functions. In Fig. 2, we illustrate a valid split for the dependency graph shown in Fig. 1 and the resulting program in Type-III.

Note that a valid split as defined above only meets the mathematical constraint over the pairings and those given by NoDup. There could be large number of valid splits for a dependency graph and it is another issue how to pick the optimal one according the metric and constraints given by the user.

3 Finding Optimal Valid Split with IP

3.1 Users' Preferences

One may want to avoid duplication regarding specific set of variables as much as possible. Typical practical demands would be to look for the minimal duplication in the public key elements, or the smallest possible duplication in the instance of assumptions. We show in the following several types of preferences that can be handled in our conversion procedure.

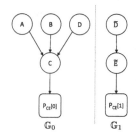

Sample(a, $A, B, D, \tilde{D}$):
$a \in \mathbb{Z}_p$, $A, B, C, D \in \mathbb{G}_0$, $\tilde{D}, \tilde{E} \in \mathbb{G}_1$
if $e(G, \tilde{D}) \neq e(D, \tilde{G})$ then err;

  if $a = 0$ then
    $C := A \cdot B$, $\tilde{E} := \tilde{D}$
  else
    $C := D^a$, $\tilde{E} := \tilde{D}^3$
  endif
output $e(C, \tilde{E})$

Fig. 2. A valid split for the dependency graph in Fig. 1, and a converted program.

1. **Priority.** We allow users to give a priority to some nodes so that they avoid duplication as much as possible than other nodes. Concretely, a priority is given by a list of sets of nodes. Let $(I_1, I_2, \cdots)$ be a sequence of non-empty sets of nodes where every set consists of arbitrary number of nodes and the sets are pairwise disjoint. It is considered that nodes in I_i are given more priority for non-duplication than those in I_{i+1}. For instance, suppose that I_1 includes nodes representing a public key and I_2 includes nodes representing a signature. By specifying (I_1, I_2) as a priority, a solution that includes less duplication in a public key is preferred. If only one node in a public key is duplicated in solution A, and all nodes in a signature are duplicated in solution B, then solution B will be taken. Unspecified nodes are given the least priority.
2. **Prohibiting Duplication.** By specifying a node as 'prohibited', the node will never be duplicated.
3. **Grouping.** By specifying a set of nodes, they are assigned to the same group. (But it does not solely mean no duplication for individual node.)
4. **Exclusive Assignment.** By specifying two nodes, different groups are assigned to each node. The specified nodes are implicitly specified as prohibited so that the exclusive assignment holds. This option, together with the prohibition, allows one to describe schemes designed in Type-III without concretely specifying groups to every variable.
5. **Specific Assignment.** By specifying a particular group to a particular node, the group is assigned to the node. (But the node may still be duplicated unless it is specified as 'prohibited' as well.)
6. **Magnification Factor.** Often a node represents multiple of variables treated in the same manner in the converting program. For instance, a message m consisting of several group elements $m = (m[0], \ldots, m[k])$ with constant k can be represented by a node referred to by m[i]. Such a node should have a magnification factor of k. It must be equal or larger than one.

In the next section, we explain how these preferences are incorporated to the objective function and constraints given to Integer Programming.

3.2 IPConv Procedure

We present a new method, which we call 'IPConv' for finding an optimal valid split. IPConv takes the task in the third step of the conversion procedure mentioned in Sect. 2.1. It takes as input a dependency graph Γ for source group $\mathbb{G}$ of Type-I scheme, and outputs two dependency graphs Γ_0 and Γ_1 for $\mathbb{G}_0$ and $\mathbb{G}_1$, respectively, of the converted Type-III scheme.

IPConv consists of the following stages. Details are given after the overview.

1. **Preprocessing on the Graph.** The input dependency graph is modified to implement some user-specified preferences. The output of this stage is the modified dependency graph and a list of constrained nodes.
2. **Establishing the Objective Function.** Binary variables that represent (non-)membership in each source group are placed on constraint nodes. They must satisfy relations for consistency and for user's preferences. Sanity checking is done to assure the existence of a solution that conforms to the constraints. Then the objective function over the variables is established.
3. **Running Integer Programming.** Run 0–1 Integer Programming for finding an assignment to the variables that minimizes the objective function subject to the constraints.
4. **Composing the Final Split.** The assignment decides which constraint nodes belong to which source group, and further decides on other nodes. Thus a valid split is composed from the assignment.

Preprocessing on the Graph. First of all, user preference in prohibiting duplication is dealt simply by including the specified nodes to the list of prohibited nodes $\mathcal{PH}$. A specific assignment to a specific node, say n, is handled by adding a new control node, c, and edge (n, c) to the graph. As the specific group is assigned to c, the same group must be assigned to n as well since n is an ancestor of c. Grouping of nodes $n_1, \ldots, n_k$ is handled in the same manner by adding a new control node c, and edges $(n_1, c), \ldots, (n_k, c)$ to the graph. This step outputs the updated graph with attributes that identifies the constraint nodes.

Establishing the Objective Function. By $|\mathbb{G}_b|$ we denote number of bits necessary to represent arbitrary element in $\mathbb{G}_b$. Let $\Gamma = (V, E)$ be a dependency graph. By $dec(n)$ for node $n \in V$, we denote all descendant nodes of n in Γ, i.e., all nodes that can be reached from n. For every node $n \in \mathcal{C}$ we associate a binary variable x_{nb} for $b = 0, 1$ that[1]:

$$x_{nb} = \begin{cases} 1 & (n \in \mathbb{G}_b) \\ 0 & (n \notin \mathbb{G}_b) \end{cases} \tag{1}$$

Let x denote the set of all those variables; $x := \{x_{nb} \mid n \in V, b \in \{0, 1\}\}$. Let $\Phi(x)$ be a collection of relations on variables in x needed for consistency of

---

[1] Instead, we can associate a single variable x_n set to b if the node is in $\mathbb{G}_b$ as done in our proof of concept implementation. It slightly reduces the number of relations, but here we choose x_{nb} for comprehensible explanation.

assignments. Since every constrained node should be exclusively assigned to either of the source groups, relation $x_{n0} + x_{n1} = 1$ for all $n \in \mathcal{C}$ are included in $\Phi(x)$. For every pair of pairing nodes, say n and n', they must get exclusive assignment to either of the source groups. Thus it must hold that $x_{n0} + x_{n'1} = 1$ and $x_{n1} + x_{n'0} = 1$. The same relation should hold for every pair of nodes specified to have exclusive assignment. For every pair of nodes n and n' in $\mathcal{C}$, if $n' \in dec(n)$, then $x_{n0} - x_{n'0} = 0$ and $x_{n1} - x_{n'1} = 0$ must be included in $\Phi(x)$ as they have to receive the same assignment. For a control node n for specifying assignment $\mathbb{G}_b$ to a regular node, relation $x_{nb} = 1$ is included in $\Phi(x)$. Control nodes for prohibiting duplication and grouping need no further treatment since they are already treated as a constrained nodes.

We apply a sanity checking that the constraints in $\Phi(x)$ are satisfiable. Observe that relations in $\Phi(x)$ can be seen as a system of equations over $GF(2)$. Then $\Phi(x)$ is satisfiable if and only if the system of equations is not overdetermined. Such a checking can be done in $O(|\mathcal{C}|^3)$ binary operations. Despite the asymptotic growth rate, the sanity check indeed finishes instantly even for large inputs and in fact negligible compared to the main workload shown in the next. By $\Phi(x) = 1$, we denote that constraints in $\Phi(x)$ are satisfiable. We denote $\Phi(x) = 0$, otherwise.

We then establish the objective function, $\mathcal{E}$, and constraints Ψ. Define a function $n \stackrel{?}{\in} \mathbb{G}_b$ for $n \in V$ and $b = 0, 1$ by

$$n \stackrel{?}{\in} \mathbb{G}_b = \begin{cases} 1 & (n \in \mathbb{G}_b) \\ 0 & (n \notin \mathbb{G}_b) \end{cases}. \tag{2}$$

For every node $n \in \mathcal{C}$, it is clear, by definition, that

$$(n \stackrel{?}{\in} \mathbb{G}_b) = x_{nb}. \tag{3}$$

For regular nodes (as defined in Sect. 2.2) other than those included in $\mathcal{C}$, i.e., $n \in V \setminus \mathcal{C} = \mathcal{R} \setminus (\mathcal{B} \cup \mathcal{PH})$, observe that $n \in \mathbb{G}_b$ holds if there is a constrained node in the descendant of n that is assigned to $\mathbb{G}_b$. Let $\mathcal{C}_n$ denote $\mathcal{C} \cap dec(n)$ that are the constrained nodes reached from node n. Then we have

$$(n \stackrel{?}{\in} \mathbb{G}_b) = \bigvee_{d \in \mathcal{C}_n} x_{db} = \neg \bigwedge_{d \in \mathcal{C}_n} \neg x_{db} = 1 - \prod_{d \in \mathcal{C}_n} (1 - x_{db}). \tag{4}$$

We now use a well known lemma [20] to remove the higher-order term in the above formula.

Lemma 2. *For binary variables* $x_1, \ldots, x_k$ *and* y, *relation*

$$\prod_{i=1}^{k} x_i = y \tag{5}$$

holds if and only if the following relations hold:

$$k - 1 - \sum_{i=1}^{k} x_i + y \geq 0 \quad and \quad x_i - y \geq 0 \text{ for all } i = 1, \ldots, k. \tag{6}$$

With this trick, we write (4) using a new variable, y_{nb}, as

$$(n \overset{?}{\in} \mathbb{G}_b) = 1 - \prod_{d \in C_n} (1 - x_{db}) = y_{nb} \tag{7}$$

and put constraints

$$\sum_{d \in C_n} x_{db} - y_{nb} \geq 0, \quad \text{and} \quad y_{nb} - x_{db} \geq 0 \quad \text{for all } d \in C_n. \tag{8}$$

Define function $eval(n)$ for every regular node $n \in \mathcal{R}$ by

$$eval(n) := \sum_{b \in \{0,1\}} w_{nb} \cdot (n \overset{?}{\in} \mathbb{G}_b) \tag{9}$$

where w_{nb} is a positive real number associated to node n. Also define

$$\begin{aligned}
eval\_max(n) &:= w_{n0} + w_{n1}, \\
eval\_2nd(n) &:= \begin{cases} w_{n0} + w_{n1} & (\text{if } w_{n0} = w_{n1}), \\ \max(w_{n0}, w_{n1}) & (\text{if } w_{n0} \neq w_{n1}), \end{cases} \\
eval\_min(n) &:= \min(w_{n0}, w_{n1}),
\end{aligned} \tag{10}$$

which means the maximum, second-minimum, and minimum value $eval(n)$ can take respectively.

Parameter w_{nb} represents the cost of having node n in $\mathbb{G}_b$ and the concrete value for the parameter is defined according to one's metrics. In this work, we set $w_{n0} := 1$ and $w_{n1} := 2$ according to the typical ratio of bit length of elements in $\mathbb{G}_0$ and $\mathbb{G}_1$. When a magnification factor k_n is defined, they are multiplied by k_n. The idea for the setting is that we seek for a conversion requiring minimum space for storing objects specified in the priority.

We then compose an objective function according to the given priority $(I_1, \ldots, I_k)$. Let I_{k+1} be regular nodes that do not appear in the priority, i.e., $I_{k+1} := \mathcal{R} \setminus (\bigcup_{i=1}^{k} I_i)$. For each node n, let

$$\Delta_n := eval\_max(n) - eval\_min(n) \tag{11}$$

which means the relative impact of duplicating n in the priority of n. And for each I_i, let

$$\Xi_i := \min_{n \in I_i} \{ eval\_2nd(n) - eval\_min(n) \}, \tag{12}$$

that is the relative minimum impact in the I_i of the assigning one single node to the larger group. For every I_i, we define priority factor ρ_i as

$$\rho_i \cdot \Xi_i > \sum_{j=i+1}^{k+1} \rho_j \sum_{n \in I_j} \Delta_n. \tag{13}$$

This means that assigning one single node to the larger group in any level of priority has more significant impact than duplicating all nodes in all lower levels of priority. For example, it is enough to let $\rho_{k+1} := 1$ and

$$\rho_i := 1 + \frac{1}{\Xi_i} \sum_{j=i+1}^{k+1} \rho_j \sum_{n \in I_j} \Delta_n \tag{14}$$

for $i = k$ down to 1. Let v denote all variables x_{nb} and y_{nb}. We define the target function $\mathcal{E}(v)$ by

$$\mathcal{E}(v) := \sum_{i=1}^{k+1} \rho_i \sum_{n \in I_i} eval(n) - eval\_min(n), \tag{15}$$

which is linear over variables in v. By $\Psi(v)$ we denote associated constraints that include all relations in $\Phi(x)$ and relations in (8). By $\Psi(v) = 1$ we denote that all constrains in $\Psi(v)$ are fulfilled. Otherwise $\Psi(v) = 0$.

Running 0–1 Integer Programming. Now we run 0–1 IP solver by giving $\mathcal{E}(v)$ and $\Psi(v)$ as input. The output is an assignment to v that minimizes $\mathcal{E}(v)$ subject to $\Psi(v) = 1$. Note that the IP solver, SCIP, used in our implementation recognizes unsolvable inputs by nature as a part of its functionality. It makes the sanity check in the previous stage redundant. Nevertheless, the sanity check in the earlier stage is useful for debugging.

Composing the Final Split. Given the assignment to v one can compute $(n \overset{?}{\in} \mathbb{G}_b)$ for all $n \in V$, and construct two dependency graphs for $\mathbb{G}_0$ and $\mathbb{G}_1$ in such a way that every edge (n, n') in the input dependency graph is included in at least one of the resulting graphs that include the destination n'. Since the assignment conforms to all given constraints, this yields a valid split. The split is optimal in the sense that it minimizes the target value $\mathcal{E}(v)$ that measures one's preferences. This completes the description of our IPConv method.

3.3   Optimality of the Output

According to our implementation of the objective function, IPConv outputs a solution whose variables given the top priority have minimal space to store. That is, those variables avoid duplication and are allocated in $\mathbb{G}_0$ as much as possible. Then, subject to the allocation in the top priority, variables in the second priority are allocated to have minimal space to store, and so forth. Concrete meaning of optimality is defined by the variables specified in the order of priority. If one's target is a public-key encryption scheme, for instance, and elements in a public-key are set as the top priority, the outcome is a scheme whose public-key has the shortest representation possible. (But it never reduces the number of group elements in the public-key, which is left for the designers' work.) To see the balance between several options in the order of priority, one may repeat

Table 1. Processing time of IPConv with SCIP. Figures in parenthesis are those of AutoGroup+ in the same environment. The upper half is small-scale monolithic schemes and the lower half is middle-scale schemes consisting of several building blocks. (# vertices) counts all nodes including the pairing nodes in the input graph. (# pairings) counts pairs of pairing nodes.

Target scheme	Graph size		Processing	Notes
	#vertices	#pairings	time	
Waters' DSE [41]	95	13	146 ms	(4639 ms)
BBS HIBE [14]	283	56	262 ms	(15667 ms)
BlindAutoSIG [3]	339	116	142 ms	-
AHO [3] + GSZK [27]	597	222	463 ms	-
Trace. Group Enc. [38]	1604	588	6306 ms	-

the conversion to the same scheme with different preferences. Each result of conversion is optimal with respect to the given preference.

In the context of bilinear-type conversion, optimizing the size of objects is a reasonable choice for better efficiency as avoiding duplication not only saves the space but also saves relevant computation. Yet extending the objective function to implement more elaborate metrics is a potential direction for further research. For instance, it is desirable to incorporate the cost of computation each variable is involved in. It requires the dependency graph to carry more information than the relations by group operations. We leave it for future development.

4   Performance

Throughout the paper, experiments are done on a standard PC: CPU: Intel Core i5-3570 3.40GHz, OS: Linux 3.16.0-34-generic #47-Ubuntu. For Integer Programming, we use SCIP [5] (non-commercial) and GUROBI [28] (commercial).

4.1   Processing Time for Real Schemes

Small-Scale Schemes. In the first two rows of Table 1, we show the processing time of IPConv for converting Boneh-Boyen HIBE [14] with $\ell = 9$ hierarchy, and Waters' Dual-system encryption [41]. Their dependency graphs are relatively small but have number of possible splits. A comparison to AutoGroup+ is done in the same environment. For fair comparison, we need to offset the overhead for processing high-level I/O format in AutoGroup+. According to [6], it takes about 500ms to handle the smallest case in their experiments. Even after offsetting similar amount as an overhead, the speedup with IPConv is obvious.

Middle-scale schemes. We also conduct experiments on middle scale schemes that involve GS-proofs and other building blocks. The results are summarized in Table 1.

AHO Signature + GSZK: Our first experiment is for a structure-preserving signature scheme in [3], a.k.a. AHO signature scheme, combined with zero-knowledge proof of a correct signature on a public message. We set the message length for AHO signatures to $n = 4$ and instantiate the zero-knowledge proof with the DLIN-based GS-proofs and convert the entire scheme to Type-III. More details appear in Sect. 5.

Blind Automorphic Signature Scheme: The second experiment is for the automorphic blind signature scheme from [3]. This experiment is to demonstrate that our framework can handle schemes that is already in Type-III. Overall structure of the target scheme is the same as the first one; a combination of a signature scheme and a NIWI GS-proof of a correct signature. Unlike the first one, however, the scheme is constructed under SXDH assumption that holds only in the Type-III setting. We describe a dependency graph for the scheme using exclusive assignment directive so that SXDH assumption is consistently incorporated to the framework. It may be interesting to see that assumptions are the only part that need to set constraints originated from the asymmetry of groups. Constraints in all upper layer algorithms are automatically taken from the assumptions. More details appear in Sect. 5.3.

Traceable Group Encryption: Our last experiment is for a traceable group encryption scheme from [38] that is more intricate involving several building blocks such as a tag-based encryption [32], AHO signatures, and one-time signatures, and GS-proofs. Taking reduction algorithms in the security proofs of each building block, the corresponding dependency graph becomes as large as consisting of 1604 nodes including 588×2 pairing nodes, which is beyond the scale that existing automated conversion can process within a practical time.

4.2   Scalability

Though the experiment in the previous section already demonstrates the scalability of IPConv to some extent, we would like to see overall behavior of IPConv against the size of inputs. Generally it is exponential due to the nature of IP. Yet it is worth to know the threshold for the practical use.

On Random Graphs. To measure the performance and the tolerance in the scale, it is necessary to sample dependency graphs from reasonable and scalable distribution. However, it is indeed impossible to consider the distribution over all constructable cryptographic schemes. It does not make sense to consider it over all possible graphs, either, since most of them do not correspond to meaningful cryptographic schemes. We therefore use some heuristics to define the distribution. Through the experiments in the previous section, we have observed that dependency graphs for real cryptographic schemes follow some structure.

We simulate it in a scalable manner in the following way: Let N be the number of regular nodes, P be the number of pairings, and k be the maximum fan-in to a regular node. Every regular node is indexed by $i \in \{1, \ldots, N\}$. Pairing nodes $p_{ij}[0]$ and $p_{ij}[1]$ represent a pairing with nodes i and j as input.

[Random Dependency Graph Generation]

1. Generate regular nodes $1, \ldots, N$.
2. For every regular node $i \in \{1, \ldots, N\}$, select $k' \leftarrow \{1, \ldots, k\}$ and repeat the following k' times:
   - Select $j \leftarrow \{1, \ldots, i - 1\}$.
   - Generate an edge (j, i).
3. Repeat the following P times:
   - Randomly select two regular nodes i and $j (\geq i)$ (discard and redo if the pair has been chosen before).
   - Generate pairing nodes $p_{ij}[0]$ and $p_{ij}[1]$ and edges $(i, p_{ij}[0])$ and $(j, p_{ij}[1])$.

Our preliminary experiment shows that large k results in so dense graphs that do not well simulate the graphs for real schemes in the previous section. Throughout our experiments, we set $k = 6$ and $N = P$ as they are close to the average for those in the real examples. With such a heuristic parameter setting we are not able to claim theoretical rigorousness to the result of our experiments. But they do show some tendency in the scalability.

We first examine the permissible scale of IPConv by measuring its processing time for random dependency graphs having up to 600 pairings and equal number of regular nodes. Figure 3 illustrates the results for 1200 inputs. IPConv finds an optimal solution in well affordable time up to around $N = P = 600$. But after that point, the processing time gets more dispersed depending on the input.

We next compare the performance with AutoGroup+. The result is illustrated in Fig. 4 that includes 250 samples for each AutoGroup+ and IPConv. Around 150 nodes, the SMT solver used in AutoGroup+ rarely fails for some

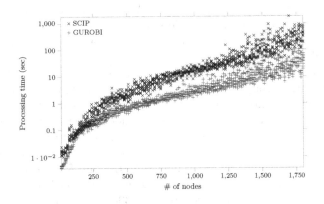

Fig. 3. Processing time in the semi-log scale for random dependency graphs.

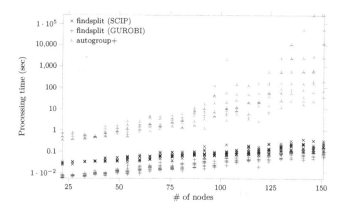

Fig. 4. Comparison between IPConv and AutoGroup+ regarding stability of processing time.

unidentified reason. With graphs containing 150 nodes, the processing time between two conversion methods differ 100 to 10^6 times. This result shows that middle to large scale conversion is out of the scope of AutoGroup+. Comparing the absolute processing time based on Fig. 4 is not perfectly fair as IPConv only takes the task of finding an optimal split whereas AutoGroup+ deals with higher-level inputs and outputs. But from the figure, one can see less dispersion in the processing time with IPConv, and its scalability is well observed.

On Cluster Graphs. We next evaluate the performance for more structured dependency graphs based on a prospect that large scale systems over bilinear groups are built in a modular fashion by combining several building blocks and GS-proofs. How would dependency graphs for such systems look like? Observe that, (1) only a small number of objects will be passed from one building block to others, (2) every building block would be used only through the legitimate interface during security proofs, and (3) the default generator is connected to a number of nodes in each building blocks. We thus foresee that a dependency graph for a modularly-built large-scale system would form sparsely connected clusters of dependency graphs with a single node that has relatively dense connection to nodes in every cluster.

We generate random cluster dependency graphs in a way that each cluster has similar volume and structure as that of AHO signature plus GS zero-knowledge proof in the previous experiment. Namely, a cluster consists of a randomly connected thirty six regular nodes and some of the nodes are involved in two random PPEs for GS zero-knowledge proofs whose dependency is automatically encoded to the graph. Then every two clusters are randomly connected each other with a fixed number of edges. The performance of IPConv for the random cluster graphs are measured up to $n = 19$ clusters. The experiment is repeated 10 times for each n. At $n = 19$, a graph consists of 13046 nodes and 5182 pairings in average. Comparing Fig. 5 with Fig. 3, there is a clear stretch in the handleable

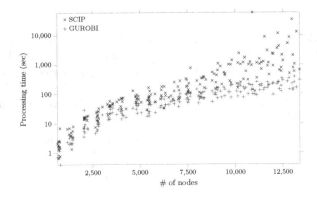

Fig. 5. Processing time in the semi-log scale for cluster dependency graphs.

number of vertices. If there are no connections between the clusters (except for those from the node representing the default generator), the processing time will be linear in the number of the clusters assuming that the processing time for each cluster is the same. We can thus see that the sparse connection among the clusters did not add much complexity.

5 Using Conversion in Cryptographic Design

In this section we show how conversion plays the role in designing cryptographic schemes. We begin by introducing a new fine-tuned construction of GS Zero-knowledge proofs in Type-I setting in Sect. 5.1. It is followed by an example that combines the GS ZK with the AHO signature scheme in Sect. 5.2. We then show another example in Sect. 5.3 that demonstrates conversion of an automorphic blind signature scheme designed originally in Type-III.

5.1 Fine-Tuned GS Proof of Correct Commitment via Conversion

In the Groth-Sahai NIZK for PPE relations, it is often needed to prove that $[X]$ is a correct commitment of a public constant A in such a way that the proof can be simulated with $X = 1_{\mathbb{G}}$. In the original paper [27], it is done by proving a relation represented by a general multi-scalar multiplication equation (MSE). We present a technique that does the job with a less costly linear pairing product equation (PPE).

THE ORIGINAL CONSTRUCTION. Recall that, in the symmetric setting under the DLIN assumption, committing to a scalar value $a \in \mathbb{Z}_p$ requires two random values, say r_1 and r_2, in $\mathbb{Z}_p$, and committing to a group element $A \in \mathbb{G}$ uses three random values, $s_1, s_2, s_3 \in \mathbb{Z}_p$. We denote the commitment by $[a; r_1, r_2]$, and $[A; s_1, s_2, s_3]$, respectively. The genuine prover algorithm computes a default commitment of $1_{\mathbb{Z}_p}$ as $[1_{\mathbb{Z}_p}; 0, 0]$, and a proof for multi-scalar multiplication equation

$$[X]^1 \cdot A^{-[1_{\mathbb{Z}_p}]} = 1_{\mathbb{G}}. \tag{16}$$

Zero-knowledge simulation with a hiding CRS is done as follows. The simulator opens the default commitment $[1_{\mathbb{Z}_p}; 0, 0]$ as $[0_{\mathbb{Z}_p}; r_1', r_2']$ by using the trapdoor. It then sets $X = 1_{\mathbb{G}}$ and computes $[X]$ which is perfectly indistinguishable from $[A]$. With respect to those commitments relation (16) is read as $[1_{\mathbb{G}}]^1 \cdot A^{-[0_{\mathbb{Z}_p}]} = 1_{\mathbb{G}}$, which is true. Thus the simulator can generate a proof following the legitimate procedure.

FINE-TUNING IN TYPE-I. Instead of using default $[1_{\mathbb{Z}_p}]$, the prover algorithm uses default commitment $[G^1; 0, 0, 0]$. Then prove a PPE

$$e([X], G)e(A^{-1}, [G^1]) = 1_{\mathbb{G}_T}. \tag{17}$$

instead of (16). Since we are considering the DLIN-based instantiation for now, (17) is a *linear* PPE that costs only 3 group elements whereas proof of (16) requires 9 elements.

Zero-knowledge simulation with a hiding CRS is done by first equivocating $[G^1; 0, 0, 0]$ into $[G^0; s_1, s_2, s_3]$ using the trapdoor. Then, by setting $X = 1_{\mathbb{G}}$, relation (17) is $e([1_{\mathbb{G}}], G)e(A^{-1}, [G^0]) = 1_{\mathbb{G}_T}$, which is true. Thus the zero-knowledge simulator can prove it using the witness.

CONVERTING TO TYPE-III. By converting the above proof system, we have an analogue proof system in the asymmetric setting based on the XDLIN assumption [1]. While the security is guaranteed by the conversion framework of [4], the quality of the resulting proof system must be examined.

Speaking from the conclusion, we have a clean split of its dependency graph without duplication except for the nodes representing the CRS. Thus, with duplicated CRS in $\mathbb{G}_0$ and $\mathbb{G}_1$, every group operation is done in either $\mathbb{G}_0$ or $\mathbb{G}_1$ and asymmetric pairing computation can be performed consistently. More importantly, the proof remains consisting of 3 group elements (and they are all in $\mathbb{G}_0$). Below, we present the resulting proof system in detail. It is particularly important to see that A and $[X]$ in (17) are in the same group without duplicating A. Full details are presented in the following.

To cope with the original description of the Groth-Sahai proof system, we switch to additive notation in the rest of this section. Let us define some notations used in the following. Let $(p, \mathbb{G}_0, \mathbb{G}_1, \mathbb{G}_T, e, G, \tilde{G})$ be an asymmetric bilinear group with $e : \mathbb{G}_0 \times \mathbb{G}_1 \to \mathbb{G}_T$. For $X, Y \in \mathbb{G}_b^n$, operation $X + Y$ denotes the result of element-wise group operations in $\mathbb{G}_b$. By $\mathrm{Mat}_{n \times m}$, we denote all matrices of size $n \times m$ over $\mathbb{Z}_p$. Let $\tilde{F}$ be a function that

$$\tilde{F}\left(\begin{pmatrix} X_1 \\ X_2 \\ X_3 \end{pmatrix}, \begin{pmatrix} Y_1 \\ Y_2 \\ Y_3 \end{pmatrix} \right) := \begin{pmatrix} \hat{e}(X_1, Y_1) & \hat{e}(X_1, Y_2) & \hat{e}(X_1, Y_3) \\ \hat{e}(X_2, Y_1) & \hat{e}(X_2, Y_2) & \hat{e}(X_2, Y_3) \\ \hat{e}(X_3, Y_1) & \hat{e}(X_3, Y_2) & \hat{e}(X_3, Y_3) \end{pmatrix} \tag{18}$$

where

$$\hat{e}(X, Y) = \begin{cases} e(X, Y) & (X \in \mathbb{G}_0 \wedge Y \in \mathbb{G}_1) \\ e(Y, X) & (Y \in \mathbb{G}_0 \wedge X \in \mathbb{G}_1) \, . \\ \bot & (\text{otherwise}) \end{cases} \tag{19}$$

By $X \tilde{\bullet} Y$, we denote $\tilde{F}(X,Y)$. For vectors $\boldsymbol{X} = (X^{(1)},\ldots,X^{(n)})$ and $\boldsymbol{Y} = (Y^{(1)},\ldots,Y^{(n)})$, we denote $\boldsymbol{X} \tilde{\bullet} \boldsymbol{Y}$ for shorthand of $\sum_{i=1}^n \left(X^{(i)} \tilde{\bullet} Y^{(i)}\right)$.

It is important to see that computation in $\tilde{F}$ and $\tilde{\bullet}$ can be carried out as long as X and Y are taken exclusively from $\mathbb{G}_0$ and $\mathbb{G}_1$. We use convention that large case letters like A represent elements in $\mathbb{G}_0$, and those with tilde like $\tilde{A}$ represent elements in $\mathbb{G}_1$.

Now we are ready to describe how to prove that $[X]$ is a correct commitment of $A \in \mathbb{G}_0$ with the GS proof system instantiated in Type-III setting based on XDLIN.

[CRS GENERATION]
Choose $\alpha, \beta, \xi_1, \xi_2 \leftarrow \mathbb{Z}_p$ and compute $G_1 := G^\alpha$, $G_2 := G^\beta$, $\boldsymbol{u}_1 := (G_1, \mathcal{O}, G)$, $\boldsymbol{u}_2 := (\mathcal{O}, G_2, G)$, and

$$\boldsymbol{u}_3 = (G_{31}, G_{32}, G_{33}) := \xi_1 \cdot \boldsymbol{u}_1 + \xi_2 \cdot \boldsymbol{u}_2 + (\mathcal{O}, \mathcal{O}, -\gamma \cdot G) \tag{20}$$
$$= (\xi_1 \cdot G_1, \xi_2 \cdot G_2, (\xi_1 + \xi_2 - \gamma) \cdot G) \tag{21}$$

where $\gamma = 0$ for binding and $\gamma = 1$ for hiding mode. Compute $\tilde{\boldsymbol{u}}_1$, $\tilde{\boldsymbol{u}}_2$, and $\tilde{\boldsymbol{u}}_3$ exactly in the same way using the same randomness $(\alpha, \beta, \xi_1, \xi_2)$ but with generator $\tilde{G}$ instead of G. Then CRS is $(\boldsymbol{u}, \tilde{\boldsymbol{u}})$ where

$$\boldsymbol{u} := \begin{pmatrix} \boldsymbol{u}_1 \\ \boldsymbol{u}_2 \\ \boldsymbol{u}_3 \end{pmatrix}, \quad \text{and} \quad \tilde{\boldsymbol{u}} := \begin{pmatrix} \tilde{\boldsymbol{u}}_1 \\ \tilde{\boldsymbol{u}}_2 \\ \tilde{\boldsymbol{u}}_3 \end{pmatrix}. \tag{22}$$

[PROVER ALGORITHM]
Given $A \in \mathbb{G}_0$ as a witness, first commit to $X := A$ using randomness $\mathcal{S}_X := (s_{1,X}, s_{2,X}, s_{3,X}) \leftarrow \text{Mat}_{1\times 3}$ as

$$[X] := (\mathcal{O}, \mathcal{O}, X) + \mathcal{S}_X \boldsymbol{u} = (C_{1,X}, C_{2,X}, C_{3,X}). \tag{23}$$

Set $(s_{1,\tilde{G}}, s_{2,\tilde{G}}, s_{3,\tilde{G}}) = (0,0,0) \in \mathbb{Z}_p^3$. Compute proof $\theta_{(17)}$ as

$$\theta_{(17)} := \begin{pmatrix} s_{1,X} & s_{1,\tilde{G}} \\ s_{2,X} & s_{2,\tilde{G}} \\ s_{3,X} & s_{3,\tilde{G}} \end{pmatrix} \begin{pmatrix} \mathcal{O} & \mathcal{O} & G \\ \mathcal{O} & \mathcal{O} & A^{-1} \end{pmatrix} = \begin{pmatrix} \mathcal{O} & \mathcal{O} & \theta_{1,(17)} \\ \mathcal{O} & \mathcal{O} & \theta_{2,(17)} \\ \mathcal{O} & \mathcal{O} & \theta_{3,(17)} \end{pmatrix}. \tag{24}$$

Output $[X]$ and $\theta_{(17)}$ as a proof. Dropping trivial elements, they consist of 6 group elements in $\mathbb{G}_0$.

[VERIFIER ALGORITHM]
Compute the default commitment of $\tilde{G}$ as

$$[\tilde{G}] := (\mathcal{O}, \mathcal{O}, \tilde{G}) + (0,0,0)\,\boldsymbol{u} = (\mathcal{O}, \mathcal{O}, \tilde{G}) = (\tilde{C}_{1,\tilde{G}}, \tilde{C}_{2,\tilde{G}}, \tilde{C}_{3,\tilde{G}}). \tag{25}$$

Then output 1 if the following holds. Output 0, otherwise.

$$\begin{pmatrix} C_{1,X} \\ C_{2,X} \\ C_{3,X} \end{pmatrix} \tilde{\bullet} \begin{pmatrix} \mathcal{O} \\ \mathcal{O} \\ \tilde{G} \end{pmatrix} + \begin{pmatrix} \tilde{C}_{1,\tilde{G}} \\ \tilde{C}_{2,\tilde{G}} \\ \tilde{C}_{3,\tilde{G}} \end{pmatrix} \tilde{\bullet} \begin{pmatrix} \mathcal{O} \\ \mathcal{O} \\ A^{-1} \end{pmatrix} = (\tilde{\boldsymbol{u}})^\top \tilde{\bullet} (\theta_{(17)})^\top \tag{26}$$

[Zero-Knowledge Simulation]
Generate CRS with $\gamma = 1$ (hiding mode). Given $A \in \mathbb{G}_0$ and trapdoor $(\alpha, \beta, \xi_1, \xi_2)$, set $(s_{1,\tilde{G}}, s_{2,\tilde{G}}, s_{3,\tilde{G}}) := (\xi_1, \xi_2, -1)$, which equivocate the default commitment $[\tilde{G}^1; 0, 0, 0]$ to $[\tilde{G}^0; \xi_1, \xi_2, -1]$. Also set $X := G^0$. Then follow the prover algorithm using these witnesses.

Direct Fine-Tuning in Type-III. The above idea can be applied to SXDH-based GS-proofs in Type-III as well. However, it is limited to the case where A is duplicated. The reason is that, relation (17) must be proved as one-side PPE in Type-III where involved commitments appear only in one side of the pairing operations. Namely, (17) has to be rewritten as

$$e([X], \tilde{G})e([G^1], \tilde{A}^{-1}) = 1_{\mathbb{G}_T}. \tag{27}$$

Thus we need $A \in \mathbb{G}_0$ to compute $[X]$ and additionally need $\tilde{A} \in \mathbb{G}_1$ to verify the proof.

If duplicating A is not acceptable, we have to get back to the original construction that proves MSE (16) instead. It costs 6 group elements. Note that it is also possible to prove (17) as a two-side PPE but it costs 8 group elements.

5.2   AHO Signature + GSZK

AHO signature scheme in Type-I setting is summarized as follows. Let $gk := (p, \mathbb{G}, \mathbb{G}_T, e, G)$ be a symmetric bilinear groups. A public-key is $(gk, A_0, A_1, A_2, A_3, B_0, B_1, B_2, B_3, G_z, G_r, H_z, H_u, G_1, \ldots, G_n, H_1, \ldots, H_n)$ for the message space of $\mathbb{G}^n$. A signature for message $(M_1, \ldots, M_n)$ is $\sigma = (Z, R, S, T, U, V, W) \in \mathbb{G}^7$. To prove possession of a correct signature for a message in the clear, a prover randomizes (S, T, V, W) into (S', T', V', W') in a way that $e(S, T) = e(S', T')$ and $e(V, W) = e(V', W')$ hold and then proves that pairing product equations

$$e(A_0, [A_1])\, e(A_2, [A_3]) = e(G_z, [Z])e(G_r, [R])e(S', [T'])\prod_{i=1}^{n} e(G_i, [M_i]) \tag{28}$$

$$e(B_0, [B_1])\, e(B_2, [B_3]) = e(H_z, [Z])e(H_u, [U])e(V', [W'])\prod_{i=1}^{n} e(H_i, [M_i]) \tag{29}$$

hold with respect to committed variables in the brackets. Additionally, relation (17) for every public value $X \in \{A_1, A_3, B_1, B_3, M_1, \ldots, M_n\}$ is proved by using the technique in Sect. 5.1 to show the correctness of the commitments.

We then consider four approaches to obtain Type-III counterpart of the above scheme. Table 2 summarizes the performance of the resulting schemes in Type-III in terms of the proof size and number of pairings in verification.

Conversion: By converting the above scheme we obtain a scheme in Type-III. Details for the proof part are presented in Appendix A. In the resulting scheme, CRS is entirely duplicated but elements in the proofs, public-keys,

Table 2. Comparison of proof size and number of pairings between conversion-aided and three direct constructions. The message is in $\mathbb{G}_0$. Proof size is for GS commitments and proofs. Column"naive" counts the number of pairings literally in the verification equations, and "batched" counts the number of pairings in batch verification.

Construction	Duplicated object	Proof size			# of pairings	
		$\mathbb{G}_0$	$\mathbb{G}_1$	In bits	Naive	Batched
Conversion	crs	$6n + 39$	6	$(6n + 51)\lambda$	$18n + 90$	$2n + 20$
Direct (1)	msg	$2n + 18$	$3n + 12$	$(8n + 42)\lambda$	$12n + 60$	$2n + 17$
Direct (2)	pk	$4n + 26$	$4n + 16$	$(12n + 58)\lambda$	$20n + 84$	$n + 23$
Direct (3)	-	$4n + 26$	$4n + 20$	$(12n + 66)\lambda$	$22n + 100$	$2n + 22$

and messages are assigned to either $\mathbb{G}_0$ or $\mathbb{G}_1$ without duplication. It is particularly important to point out that X and $[X]$ in (17) are assigned to the same group without duplicating X while proving (17) as a linear PPE. This approach is the most efficient in the proof size since most of commitments and proofs can be allocated in $\mathbb{G}_0$.

Direct Instantiation 1 (with Duplicated Messages): Next we consider instantiating the GS-proofs directly over Type-III groups based on the SXDH assumption. As observed in Sect. 5.1, the fine-tuned construction is only possible when public constants paired with committed variables are duplicated. Therefore, elements $\{A_1, A_3, B_1, B_3, M_1, \ldots, M_n\}$ have to be duplicated. Duplicated key elements, A_1, A_3, B_1, and B_3 will be a part of the public-key. On the other hand, duplicated message $M_1, \ldots, M_n$ must be sent to the verifier as a part of the proof.

Direct Instantiation 2 (with Duplicated Keys): When duplicating M_i is prohibiting, a workaround would be to commit to public-key elements G_i and H_i instead. Duplicated G_i and H_i can be included in the public-key (thus we do not count it in the proof size). Unfortunately, this approach is not efficient in terms of proof size since the proofs of correct commitment for both G_i and H_i doubles the proof length. On the other hand, it allows efficient batch verification. The reason is that pairings corresponding to $e([G_i], M_i)$ and $e([H_i], M_i)$ in the verification can be merged into one pairing associated to M_i while at least two pairings are needed to deal with $e(G_i, [M_i])$ and $e(H_i, [M_i])$ in the above approaches.

Direct Instantiation 3 (Without Duplication): Finally, we consider avoiding duplication at all in the direct instantiation of GS proofs in Type-III by following the original approach using MSE (16) as shown in the beginning of Sect. 5.1. As expected, both proof size and number of pairings increase due to the MSEs. Use of batch verification is not quite effective, either.

As we see from Table 2, there is no clear winner. The scheme obtained by conversion yields the most compact proofs for messages of $n > 5$. But for short and duplicable messages, direct construction produces more compact proofs.

Regarding the computational workload, when batch verification is taken into account, there is not much difference for small n no matter what approach is taken. But for large n, direct instantiation in Type-III with duplicated public-key is more advantageous.

5.3   Automorphic Blind Signature Scheme

Examples so far deals with schemes designed purely in Type-I. Now we show that schemes designed originally in Type-III are also incorporated into our framework for finding optimal deployment of source groups and perhaps finding more efficient GS-proofs used there.

In the automorphic blind signature scheme in [3], a blind signature is a GS-proof for one's possession of a correct (plain) automorphic signature on a clear message. A plain automorphic signature consists of five group elements $\sigma :=$ $(A, B, \tilde{D}, R, \tilde{S})$ verified by PPEs:

$$e(A, \tilde{Y} \cdot \tilde{D}) = e(K \cdot M, \tilde{G}) \, e(T, \tilde{S}), \tag{30}$$

$$e(B, \tilde{G}) = e(F, \tilde{D}), \quad e(R, \tilde{G}) = e(G, \tilde{S}). \tag{31}$$

An automorphic blind signature is a GS-proof of (30) and (31) with $(A, B, \tilde{D}, R, \tilde{S})$ as a witness. The security of the original construction bases on SXDH assumption and Asymmetric Double Hidden Strong DH Assumption (ADHDH) [3].

To incorporate the scheme into the conversion framework, we need to build a dependency graph in such a manner that the original scheme is included in a possible solution of conversion. First, a special treatment is needed to the nodes representing X and $\tilde{Y}$ that are already in the duplicated form since they should not be individually duplicated by conversion. We set dependency $Y \rightarrow X$, and prohibit duplication of X. In this way, Y will be duplicated so that X is assigned to $\mathbb{G}_b$ and $\tilde{Y}$ is assigned to $\mathbb{G}_{1-b}$. Such a treatment is applied to (M, N) and (R, S) as well. Second, we need to build a dependency graph for the assumptions. Since ADHDH is known to hold even in the Type-I generic bilinear group model, we simply ignore the distinction of $\mathbb{G}_0$ and $\mathbb{G}_1$. For SXDH, we prohibit duplication of any variable in its instance and use grouping of variables so that they are allocated to the same group. In this way, the assumption remains valid when converted back to Type-III. Finally, the GS-proof part is described by using the DLIN-based instantiation of GS-proofs. They are witness indistinguishable proofs and we do not rely on the fine-tuning as in the previous case.

After conversion, the resulting scheme in Type-III is secure based on SXDH, ADHDH with duplicated $\tilde{D}$, and XDLIN assumptions. We present details of the converted scheme for the part of generating and verifying a blind signature in Appendix B. Table 3 summarizes the performance in comparison with the original construction. The converted scheme saves 28 % of blind signature in bits and equal or slightly better in verification workload.

Table 3. Comparison of the signature size and number of pairings in verification between conversion-aided and direct instantiations of verifier's algorithm for the automorphic blind signature scheme [3]. The message is $(M, N) \in \mathbb{G}_0 \times \mathbb{G}_1$. Duplication of $\tilde{D}$ is needed for computing proofs but not for verification.

Construction	Duplicated objects	Size of blind sig. $\mathbb{G}_0$	$\mathbb{G}_1$	in bits	# of pairings Naive	Batched
Conversion	crs, $\tilde{D}$	24	6	$36\,\lambda$	64	13
Original [3]	-	18	16	$50\,\lambda$	68	13

6 Conclusion

We have proposed an efficient type conversion method based on 0–1 Integer Programming. It is shown how to represent several constraints into a system of linear binary equations so that a 0–1 IP solver can find an optimal solution that meets the constraints. The performance and scalability are demonstrated over real schemes and randomly generated samples.

Usefulness of the conversion-aided design approach is demonstrated by examples that outputs more compact GS-proofs than those manufactured directly in Type-III setting. A fine-tuning technique that improves the performance of converted GS-proofs is introduced.

Nevertheless, results in this paper can be seen as a step toward realizing automated modular design of cryptographic protocols. Depending on the target schemes, direct instantiation in Type-III based on SXDH can yield better results. It is in fact another optimization issue that machines can help to find a globally optimal solution. We include it as an interesting research and engineering target in our future plan.

Finally, a proof-of-concept implementation with source codes and data files for experiments in Sect. 5 are available from the authors for review. Open source development is certainly in our future plan.

Acknowledgements. The authors thank Susan Hohenberger Waters and co-authors of [6,7] for their help to understand AutoGroup. We also thank to Takeya Tango for an alternative sanity checking method. Special thanks to the developers of SCIP [5] for their quality software.

A Converted GSZK for AHO Signature

Let parameters for AHO signature scheme be asymmetric bilinear groups $gk := (p, \mathbb{G}_0, \mathbb{G}_1, \mathbb{G}_T, e, G, \tilde{G})$, verification-key $pk := (gk, \tilde{G}_z, \tilde{G}_r, \tilde{H}_z, \tilde{H}_u, \{\tilde{G}_i, \tilde{H}_i\}_{i=1}^n, \tilde{A}_0, A_1, \tilde{A}_1, A_2, \tilde{B}_0, B_1, \tilde{B}_1, B_2)$, message $msg := (M_1, \ldots, M_n)$, and signature $\sigma := (Z, R, U, \tilde{S}, T, \tilde{V}, W)$. CRS $\boldsymbol{u} \in \mathbb{G}_0^3$ and $\tilde{\boldsymbol{u}} \in \tilde{\mathbb{G}}_1^3$ are generated in exactly the same manner as described in Sect. 5.1. The relations to prove are PPEs (28), (29), and (17) re-numbered as follows.

$$\hat{e}(\tilde{A}_0, [A_1])\,\hat{e}(\tilde{A}_2, [A_3]) = \hat{e}(\tilde{G}_z, [Z])\,\hat{e}(\tilde{G}_r, [R])\,\hat{e}(\tilde{S}', [T'])\prod_{i=1}^{n}\hat{e}(\tilde{G}_i, [M_i]) \qquad (32)$$

$$\hat{e}(\tilde{B}_0, [B_1])\,\hat{e}(\tilde{B}_2, [B_3]) = \hat{e}(\tilde{H}_z, [Z])\,\hat{e}(\tilde{H}_u, [U])\,\hat{e}(\tilde{V}', [W'])\prod_{i=1}^{n}\hat{e}(\tilde{H}_i, [M_i]) \quad (33)$$

$$\hat{e}(\tilde{G}, [X])\,\hat{e}([\tilde{G}], X^{-1}) = 1_{\mathbb{G}_T} \text{ for each } X \in \{A_1, A_3, B_1, B_3, M_i\}. \qquad (34)$$

With pairing $\hat{e}$ defined as (19), the relations can be regarded as linear PPEs. In the rest of this section, we switch to additive notation for convenience of presenting GS-proofs.

[PROVER ALGORITHM]
Commit to $Y \in (Z, R, U, T', W', A_1, A_3, B_1, B_3, M_i)$ by computing

$$[Y] := (\mathcal{O}, \mathcal{O}, Y) + \mathcal{S}_Y\,\boldsymbol{u} = (C_{1,Y}, C_{2,Y}, C_{3,Y}) \in \mathbb{G}_0^3. \qquad (35)$$

with independently uniform $\mathcal{S}_Y \leftarrow \mathrm{Mat}_{1\times 3}$. Let $\mathcal{S}_{\tilde{G}} := (0,0,0) \in \mathbb{Z}_p^3$, and let

$$\mathcal{S}_{(32)} := \begin{pmatrix} \mathcal{S}_{A_1} \cdot \\ \mathcal{S}_{A_3} \\ \mathcal{S}_Z \\ \mathcal{S}_R \\ \mathcal{S}_{T'} \\ \mathcal{S}_{M_i} \end{pmatrix}, \quad \mathcal{S}_{(33)} := \begin{pmatrix} \mathcal{S}_{B_1} \\ \mathcal{S}_{B_3} \\ \mathcal{S}_Z \\ \mathcal{S}_U \\ \mathcal{S}_{W'} \\ \mathcal{S}_{M_i} \end{pmatrix}, \quad \text{and} \quad \mathcal{S}_{(34),X} := \begin{pmatrix} \mathcal{S}_{\tilde{G}} \\ \mathcal{S}_X \end{pmatrix}. \qquad (36)$$

Compute $\tilde{\theta}_{(32)}$, $\tilde{\theta}_{(33)}$ and $\theta_{(34),X}$ for $X \in \{A_1, A_3, B_1, B_3, M_1, \ldots, M_i\}$ where:

$$\tilde{\theta}_{(32)} := \mathcal{S}_{(32)}^{\top} \begin{pmatrix} \mathcal{O} & \mathcal{O} & \tilde{A}_0 \\ \mathcal{O} & \mathcal{O} & \tilde{A}_2 \\ \mathcal{O} & \mathcal{O} & \tilde{G}_z^{-1} \\ \mathcal{O} & \mathcal{O} & \tilde{G}_r^{-1} \\ \mathcal{O} & \mathcal{O} & \tilde{G}_t^{-1} \\ \mathcal{O} & \mathcal{O} & \tilde{G}_i^{-1} \end{pmatrix} = \begin{pmatrix} \mathcal{O} & \mathcal{O} & \tilde{\theta}_{1,(32)} \\ \mathcal{O} & \mathcal{O} & \tilde{\theta}_{2,(32)} \\ \mathcal{O} & \mathcal{O} & \tilde{\theta}_{3,(32)} \end{pmatrix} \in \tilde{\mathbb{G}}_1^{3\times 3}, \qquad (37)$$

$$\tilde{\theta}_{(33)} := \mathcal{S}_{(33)}^{\top} \begin{pmatrix} \mathcal{O} & \mathcal{O} & \tilde{B}_0 \\ \mathcal{O} & \mathcal{O} & \tilde{B}_2 \\ \mathcal{O} & \mathcal{O} & \tilde{H}_z^{-1} \\ \mathcal{O} & \mathcal{O} & \tilde{H}_u^{-1} \\ \mathcal{O} & \mathcal{O} & \tilde{H}_w^{-1} \\ \mathcal{O} & \mathcal{O} & \tilde{H}_i^{-1} \end{pmatrix} = \begin{pmatrix} \mathcal{O} & \mathcal{O} & \tilde{\theta}_{1,(33)} \\ \mathcal{O} & \mathcal{O} & \tilde{\theta}_{2,(33)} \\ \mathcal{O} & \mathcal{O} & \tilde{\theta}_{3,(33)} \end{pmatrix} \in \tilde{\mathbb{G}}_1^{3\times 3}, \qquad (38)$$

$$\theta_{(34),X} := \mathcal{S}_{(34),X}^{\top} \begin{pmatrix} \mathcal{O} & \mathcal{O} & G \\ \mathcal{O} & \mathcal{O} & X^{-1} \end{pmatrix} = \begin{pmatrix} \mathcal{O} & \mathcal{O} & \theta_{1,(34),X} \\ \mathcal{O} & \mathcal{O} & \theta_{2,(34),X} \\ \mathcal{O} & \mathcal{O} & \theta_{3,(34),X} \end{pmatrix} \in \mathbb{G}_0^{3\times 3}. \qquad (39)$$

Output all $[Y]$, $\tilde{\theta}_{(32)}$, $\tilde{\theta}_{(33)}$, and $\theta_{(34),X}$ dropping redundant $\mathcal{O}$.

[VERIFIER ALGORITHM]

Given the above proof and CRS as input, output 1 (as accept) if all the following equations hold. Output 0, otherwise.

$$\begin{pmatrix} \mathcal{O} \\ \mathcal{O} \\ \tilde{A}_0 \end{pmatrix} \tilde{\bullet} \begin{pmatrix} C_{1,A_1} \\ C_{2,A_1} \\ C_{3,A_1} \end{pmatrix} + \begin{pmatrix} \mathcal{O} \\ \mathcal{O} \\ \tilde{A}_2 \end{pmatrix} \tilde{\bullet} \begin{pmatrix} C_{1,A_3} \\ C_{2,A_3} \\ C_{3,A_3} \end{pmatrix} + \begin{pmatrix} \mathcal{O} \\ \mathcal{O} \\ \tilde{G}_z^{-1} \end{pmatrix} \tilde{\bullet} \begin{pmatrix} C_{1,Z} \\ C_{2,Z} \\ C_{3,Z} \end{pmatrix}$$

$$+ \begin{pmatrix} \mathcal{O} \\ \mathcal{O} \\ \tilde{G}_r^{-1} \end{pmatrix} \tilde{\bullet} \begin{pmatrix} C_{1,R} \\ C_{2,R} \\ C_{3,R} \end{pmatrix} + \begin{pmatrix} \mathcal{O} \\ \mathcal{O} \\ \tilde{S}'^{-1} \end{pmatrix} \tilde{\bullet} \begin{pmatrix} C_{1,T'} \\ C_{2,T'} \\ C_{3,T'} \end{pmatrix} + \sum_{i=1}^{n} \begin{pmatrix} \mathcal{O} \\ \mathcal{O} \\ \tilde{G}_i^{-1} \end{pmatrix} \tilde{\bullet} \begin{pmatrix} C_{1,M_i} \\ C_{2,M_i} \\ C_{3,M_i} \end{pmatrix}$$

$$= \left(\tilde{\theta}_{(32)} \right)^{\top} \tilde{\bullet} (u)^{\top} \tag{40}$$

$$\begin{pmatrix} \mathcal{O} \\ \mathcal{O} \\ \tilde{B}_0 \end{pmatrix} \tilde{\bullet} \begin{pmatrix} C_{1,B_1} \\ C_{2,B_1} \\ C_{3,B_1} \end{pmatrix} + \begin{pmatrix} \mathcal{O} \\ \mathcal{O} \\ \tilde{B}_2 \end{pmatrix} \tilde{\bullet} \begin{pmatrix} C_{1,B_3} \\ C_{2,B_3} \\ C_{3,B_3} \end{pmatrix} + \begin{pmatrix} \mathcal{O} \\ \mathcal{O} \\ \tilde{H}_z^{-1} \end{pmatrix} \tilde{\bullet} \begin{pmatrix} C_{1,Z} \\ C_{2,Z} \\ C_{3,Z} \end{pmatrix}$$

$$+ \begin{pmatrix} \mathcal{O} \\ \mathcal{O} \\ \tilde{H}_u^{-1} \end{pmatrix} \tilde{\bullet} \begin{pmatrix} C_{1,U} \\ C_{2,U} \\ C_{3,U} \end{pmatrix} + \begin{pmatrix} \mathcal{O} \\ \mathcal{O} \\ \tilde{V}'^{-1} \end{pmatrix} \tilde{\bullet} \begin{pmatrix} C_{1,W'} \\ C_{2,W'} \\ C_{3,W'} \end{pmatrix} + \sum_{i=1}^{n} \begin{pmatrix} \mathcal{O} \\ \mathcal{O} \\ \tilde{H}_i^{-1} \end{pmatrix} \tilde{\bullet} \begin{pmatrix} C_{1,M_i} \\ C_{2,M_i} \\ C_{3,M_i} \end{pmatrix}$$

$$= \left(\tilde{\theta}_{(33)} \right)^{\top} \tilde{\bullet} (u)^{\top} \tag{41}$$

$$\begin{pmatrix} C_{1,X} \\ C_{2,X} \\ C_{3,X} \end{pmatrix} \tilde{\bullet} \begin{pmatrix} \mathcal{O} \\ \mathcal{O} \\ \tilde{G} \end{pmatrix} + \begin{pmatrix} \tilde{C}_{1,\tilde{G}} \\ \tilde{C}_{2,\tilde{G}} \\ \tilde{C}_{3,\tilde{G}} \end{pmatrix} \tilde{\bullet} \begin{pmatrix} \mathcal{O} \\ \mathcal{O} \\ X^{-1} \end{pmatrix} = (\tilde{u})^{\top} \tilde{\bullet} \left(\theta_{(34),X} \right)^{\top} \tag{42}$$

for $X \in \{A_1, A_3, B_1, B_3, M_i\}$ where $(\tilde{C}_{1,\tilde{G}}, \tilde{C}_{2,\tilde{G}}, \tilde{C}_{3,\tilde{G}}) := (\mathcal{O}, \mathcal{O}, \tilde{G})$.

B Converted Automorphic Blind Signature Scheme

This section presents details of automorphic blind signature scheme obtained by conversion. A full description includes key generation, blinding, signing, unblinding, verification algorithms, and also security proofs. Here, we focus on presenting user's and verifier's algorithms in transferring a blind signature. They actually consist of prover and verifier algorithms like the previous case. CRS $u \in \mathbb{G}_0^3$ and $\tilde{u} \in \tilde{\mathbb{G}}_1^3$ are generated as described in Sect. 5.1. Let parameters be asymmetric bilinear groups $gk := (p, \mathbb{G}_0, \mathbb{G}_1, \mathbb{G}_T, e, G, \tilde{G})$, verification-key $pk := (gk, F, K, T, X(= G^x), \tilde{Y}(= \tilde{G}^x))$, message $(M(= G^m), \tilde{N}(= \tilde{G}^m))$. An automorphic blind signature is a witness indistinguishable GS-proof for relations (30) and (31) as re-numbered as follows.

$$\hat{e}([A], \tilde{Y}) \, \hat{e}([A], [\tilde{D}]) = \hat{e}(K, \tilde{G}) \, \hat{e}(M, \tilde{G}) \, \hat{e}(T, [S]) \tag{43}$$

$$\hat{e}([B], \tilde{G}) = \hat{e}(F, [\tilde{D}]) \tag{44}$$

$$\hat{e}([R], \tilde{G}) = \hat{e}(G, [S]) \tag{45}$$

With pairing $\hat{e}$ defined as (19), the second and third relations are regarded as linear PPEs. Again, we switch to additive notation while describing GS-proofs in the following.

[BLIND SIGNATURE ISSUING ALGORITHM]
Commit to $\delta \in (A, B, R)$ and $\tilde{\rho} \in (\tilde{D}, \tilde{S})$ by

$$[\delta] := (\mathcal{O}, \mathcal{O}, \delta) + \mathcal{S}_\delta \, \boldsymbol{u} = (C_{1,\delta}, C_{2,\delta}, C_{3,\delta}) \in \mathbb{G}_0^3, \text{ and} \tag{46}$$

$$[\tilde{\rho}] := (\mathcal{O}, \mathcal{O}, \tilde{\rho}) + \mathcal{S}_{\tilde{\rho}} \, \tilde{\boldsymbol{u}} = (C_{1,\tilde{\rho}}, C_{2,\tilde{\rho}}, C_{3,\tilde{\rho}}) \in \mathbb{G}_1^3 \tag{47}$$

where $\mathcal{S}_\delta \leftarrow \mathrm{Mat}_{1 \times 3}$ and $\mathcal{S}_{\tilde{\rho}} \leftarrow \mathrm{Mat}_{1 \times 3}$. Let T_p be a random 3×3 matrix over $\mathbb{Z}_p$. Compute $\theta_{(43)}$, $\theta_{(44)}$, and $\theta_{(45)}$ as:

$$\theta_{(43)} = \mathcal{S}_A^\top (\mathcal{O}, \mathcal{O}, X) + \mathcal{S}_A^\top (\mathcal{O}, \mathcal{O}, D) + \mathcal{S}_{\tilde{D}}^\top (\mathcal{O}, \mathcal{O}, A) + \mathcal{S}_A^\top \mathcal{S}_{\tilde{D}} \, \boldsymbol{u}$$
$$- \mathcal{S}_{\tilde{S}}^\top (\mathcal{O}, \mathcal{O}, T) + (T_p - T_p^\top) \, \boldsymbol{u} \tag{48}$$

$$\theta_{(44)} = \mathcal{S}_B^\top (\mathcal{O}, \mathcal{O}, G) - \mathcal{S}_{\tilde{D}}^\top (\mathcal{O}, \mathcal{O}, F) \tag{49}$$

$$\theta_{(45)} = \mathcal{S}_R^\top (\mathcal{O}, \mathcal{O}, G) - \mathcal{S}_{\tilde{S}}^\top (\mathcal{O}, \mathcal{O}, G) \tag{50}$$

Output all $[\delta]$, $[\tilde{\rho}]$, $\theta_{(43)}$, $\theta_{(44)}$, and $\theta_{(45)}$ without redundant $\mathcal{O}$ as a blind signature.

[VERIFIER ALGORITHM]
Given the above blind signature and message $msg := (M, \tilde{N})$, output 1 if all the following equations hold. Output 0, otherwise.

$$\begin{pmatrix} C_{1,A} \\ C_{2,A} \\ C_{3,A} \end{pmatrix} \tilde{\bullet} \begin{pmatrix} \mathcal{O} \\ \mathcal{O} \\ \tilde{Y} \end{pmatrix} + \begin{pmatrix} C_{1,A} \\ C_{2,A} \\ C_{3,A} \end{pmatrix} \tilde{\bullet} \begin{pmatrix} C_{1,\tilde{D}} \\ C_{2,\tilde{D}} \\ C_{3,\tilde{D}} \end{pmatrix} \tag{51}$$

$$= \begin{pmatrix} \mathcal{O} \\ \mathcal{O} \\ K \end{pmatrix} \tilde{\bullet} \begin{pmatrix} \mathcal{O} \\ \mathcal{O} \\ \tilde{G} \end{pmatrix} + \begin{pmatrix} \mathcal{O} \\ \mathcal{O} \\ M \end{pmatrix} \tilde{\bullet} \begin{pmatrix} \mathcal{O} \\ \mathcal{O} \\ \tilde{G} \end{pmatrix} + \begin{pmatrix} \mathcal{O} \\ \mathcal{O} \\ T \end{pmatrix} \tilde{\bullet} \begin{pmatrix} C_{1,\tilde{S}} \\ C_{2,\tilde{S}} \\ C_{3,\tilde{S}} \end{pmatrix} + \left(\theta_{(43)}\right)^\top \tilde{\bullet} \left(\tilde{\boldsymbol{u}}\right)^\top$$

$$\begin{pmatrix} \mathcal{O} \\ \mathcal{O} \\ \tilde{G} \end{pmatrix} \tilde{\bullet} \begin{pmatrix} C_{1,B} \\ C_{2,B} \\ C_{3,B} \end{pmatrix} = \begin{pmatrix} \mathcal{O} \\ \mathcal{O} \\ F \end{pmatrix} \tilde{\bullet} \begin{pmatrix} C_{1,\tilde{D}} \\ C_{2,\tilde{D}} \\ C_{3,\tilde{D}} \end{pmatrix} + \left(\theta_{(44)}\right)^\top \tilde{\bullet} \left(\tilde{\boldsymbol{u}}\right)^\top \tag{52}$$

$$\begin{pmatrix} \mathcal{O} \\ \mathcal{O} \\ \tilde{G} \end{pmatrix} \tilde{\bullet} \begin{pmatrix} C_{1,R} \\ C_{2,R} \\ C_{3,R} \end{pmatrix} = \begin{pmatrix} \mathcal{O} \\ \mathcal{O} \\ G \end{pmatrix} \tilde{\bullet} \begin{pmatrix} C_{1,\tilde{S}} \\ C_{2,\tilde{S}} \\ C_{3,\tilde{S}} \end{pmatrix} + \left(\theta_{(45)}\right)^\top \tilde{\bullet} \left(\tilde{\boldsymbol{u}}\right)^\top \tag{53}$$

References

1. Abe, M., Chase, M., David, B., Kohlweiss, M., Nishimaki, R., Ohkubo, M.: Constant-size structure-preserving signatures: generic constructions and simple assumptions. In: Wang, X., Sako, K. (eds.) ASIACRYPT 2012. LNCS, vol. 7658, pp. 4–24. Springer, Heidelberg (2012)

2. Abe, M., David, B., Kohlweiss, M., Nishimaki, R., Ohkubo, M.: Tagged one-time signatures: tight security and optimal tag size. In: Kurosawa, K., Hanaoka, G. (eds.) PKC 2013. LNCS, vol. 7778, pp. 312–331. Springer, Heidelberg (2013)

3. Abe, M., Fuchsbauer, G., Groth, J., Haralambiev, K., Ohkubo, M.: Structure-preserving signatures and commitments to group elements. J. Cryptol. **29**(2), 363–421 (2016)

4. Abe, M., Groth, J., Ohkubo, M., Tango, T.: Converting cryptographic schemes from symmetric to asymmetric bilinear groups. In: Garay, J.A., Gennaro, R. (eds.) CRYPTO 2014, Part I. LNCS, vol. 8616, pp. 241–260. Springer, Heidelberg (2014)

5. Achterberg, T.: CIP: solving constraint integer programs. Math. Program. Comput. **1**(1), 1–41 (2009). http://mpc.zib.de/index.php/MPC/article/view/4

6. Akinyele, J.A., Garman, C., Hohenberger, S.: Automating fast and secure translations from type-I to type-III pairing schemes. In: ACM CCS 2015, pp. 1370–1381. ACM (2015)

7. Akinyele, J.A., Green, M., Hohenberger, S.: Using SMT solvers to automate design tasks for encryption and signature schemes. In: ACM CCS 2013, pp. 399–410. ACM (2013)

8. Backes, M., Fiore, D., Reischuk, R.M.: Verifiable delegation of computation on outsourced data. In: ACM CCS 2013, pp. 863–874. ACM (2013)

9. Barbulescu, R., Gaudry, P., Joux, A., Thomé, E.: A quasi-polynomial algorithm for discrete logarithm in finite fields of small characteristic. IACR ePrint Archive, report 2013/400 (2013). http://eprint.iacr.org

10. Barthe, G., Fagerholm, E., Fiore, D., Mitchell, J., Scedrov, A., Schmidt, B.: Automated analysis of cryptographic assumptions in generic group models. In: Garay, J.A., Gennaro, R. (eds.) CRYPTO 2014, Part I. LNCS, vol. 8616, pp. 95–112. Springer, Heidelberg (2014)

11. Barthe, G., Fagerholm, E., Fiore, D., Scedrov, A., Schmidt, B., Tibouchi, M.: Strongly-optimal structure preserving signatures from type II pairings: synthesis and lower bounds. In: Katz, J. (ed.) PKC 2015. LNCS, vol. 9020, pp. 355–376. Springer, Heidelberg (2015)

12. Blanchet, B.: Cryptoverif: A computationally sound mechanized prover for cryptographic protocols. In: Dagstuhl seminar Formal Protocol Verification Applied, vol. 10 (2007)

13. Blazy, O., Fuchsbauer, G., Izabachène, M., Jambert, A., Sibert, H., Vergnaud, D.: Batch Groth–Sahai. In: Zhou, J., Yung, M. (eds.) ACNS 2010. LNCS, vol. 6123, pp. 218–235. Springer, Heidelberg (2010)

14. Boneh, D., Boyen, X.: Efficient selective-ID secure identity-based encryption without random oracles. In: Cachin, C., Camenisch, J.L. (eds.) EUROCRYPT 2004. LNCS, vol. 3027, pp. 223–238. Springer, Heidelberg (2004)

15. Boneh, D., Boyen, X., Shacham, H.: Short group signatures. In: Franklin, M. (ed.) CRYPTO 2004. LNCS, vol. 3152, pp. 41–55. Springer, Heidelberg (2004)

16. Boneh, D., Shacham, H.: Group signatures with verifier-local revocation. In: ACM CCS 2004, pp. 168–177. ACM (2004)

17. Chatterjee, S., Hankerson, D., Knapp, E., Menezes, A.: Comparing two pairing-based aggregate signature schemes. Des. Codes Crypt. **55**(2–3), 141–167 (2010)

18. Chatterjee, S., Menezes, A.: On cryptographic protocols employing asymmetric pairings - the role of psi revisited. IACR ePrint Archive, Report 2009/480 (2009). http://eprint.iacr.org

19. Chatterjee, S., Menezes, A.: On cryptographic protocols employing asymmetric pairings - the role of revisited. Discrete Appl. Math. **159**(13), 1311–1322 (2011)

20. Chen, D.-S., Batson, R.G., Dang, Y.: Applied Integer Programming: Modeling and Solution. Wiley, Hoboken (2009)
21. de Moura, L., Bjørner, N.S.: Z3: an efficient SMT solver. In: Ramakrishnan, C.R., Rehof, J. (eds.) TACAS 2008. LNCS, vol. 4963, pp. 337–340. Springer, Heidelberg (2008)
22. Escala, A., Groth, J.: Fine-tuning Groth-Sahai proofs. In: Krawczyk, H. (ed.) PKC 2014. LNCS, vol. 8383, pp. 630–649. Springer, Heidelberg (2014)
23. Escala, A., Herold, G., Kiltz, E., Ràfols, C., Villar, J.: An algebraic framework for Diffie-Hellman assumptions. In: Canetti, R., Garay, J.A. (eds.) CRYPTO 2013, Part II. LNCS, vol. 8043, pp. 129–147. Springer, Heidelberg (2013)
24. Gamrath, G., Lübbecke, M.E.: Experiments with a generic Dantzig-Wolfe decomposition for integer programs. In: Festa, P. (ed.) SEA 2010. LNCS, vol. 6049, pp. 239–252. Springer, Heidelberg (2010)
25. Ghadafi, E., Smart, N.P., Warinschi, B.: Groth–Sahai proofs revisited. In: Nguyen, P.Q., Pointcheval, D. (eds.) PKC 2010. LNCS, vol. 6056, pp. 177–192. Springer, Heidelberg (2010)
26. Göloglu, F., Granger, R., McGuire, G., Zumbrägel, J.: On the function field sieve and the impact of higher splitting probabilities: Application to discrete logarithms in f_2^{1971}. IACR ePrint Archive, Report 2013/074 (2013). http://eprint.iacr.org
27. Groth, J., Sahai, A.: Efficient noninteractive proof systems for bilinear groups. SIAM J. Comput. **41**(5), 1193–1232 (2012)
28. Gurobi Optimization, Inc., Gurobi optimizer reference manual. https://www.gurobi.com/documentation/6.5/refman.pdf. http://www.gurobi.com/
29. Herold, G., Hesse, J., Hofheinz, D., Ràfols, C., Rupp, A.: Polynomial spaces: a new framework for composite-to-prime-order transformations. In: Garay, J.A., Gennaro, R. (eds.) CRYPTO 2014, Part I. LNCS, vol. 8616, pp. 261–279. Springer, Heidelberg (2014)
30. Joux, A.: Faster index calculus for the medium prime case application to 1175-bit and 1425-bit finite fields. In: Johansson, T., Nguyen, P.Q. (eds.) EUROCRYPT 2013. LNCS, vol. 7881, pp. 177–193. Springer, Heidelberg (2013)
31. Joux, A.: A new index calculus algorithm with complexity l(1/4+o(1)) in very small characteristic. IACR ePrint Archive, Report 2013/095 (2013). http://eprint.iacr.org
32. Kiltz, E.: Chosen-ciphertext security from tag-based encryption. In: Halevi, S., Rabin, T. (eds.) TCC 2006. LNCS, vol. 3876, pp. 581–600. Springer, Heidelberg (2006)
33. Koch, T.: Rapid mathematical prototyping. Ph.D. thesis, Technische Universität Berlin (2004)
34. LINDO Systems. LINDO. http://www.lindo.com/
35. Melnick, M.: LiPS. http://lipside.sourceforge.net/
36. Libert, B., Joye, M.: Group signatures with message-dependent opening in the standard model. In: Benaloh, J. (ed.) CT-RSA 2014. LNCS, vol. 8366, pp. 286–306. Springer, Heidelberg (2014)
37. Libert, B., Joye, M., Yung, M., Peters, T.: Secure efficient history-hiding append-only signatures in the standard model. In: Katz, J. (ed.) PKC 2015. LNCS, vol. 9020, pp. 450–473. Springer, Heidelberg (2015)
38. Libert, B., Yung, M., Joye, M., Peters, T.: Traceable group encryption. In: Krawczyk, H. (ed.) PKC 2014. LNCS, vol. 8383, pp. 592–610. Springer, Heidelberg (2014)
39. Smart, N.P., Vercauteren, F.: On computable isomorphisms in efficient asymmetric pairing-based systems. Discrete Appl. Math. **155**(4), 538–547 (2007)

40. Waters, B.: Efficient identity-based encryption without random oracles. In: Cramer, R. (ed.) EUROCRYPT 2005. LNCS, vol. 3494, pp. 114–127. Springer, Heidelberg (2005)
41. Waters, B.: Dual system encryption: realizing fully secure IBE and HIBE under simple assumptions. In: Halevi, S. (ed.) CRYPTO 2009. LNCS, vol. 5677, pp. 619–636. Springer, Heidelberg (2009)

Linicrypt: A Model for Practical Cryptography

Brent Carmer$^{(\boxtimes)}$ and Mike Rosulek

Oregon State University, Corvallis, USA
{carmerb,rosulekm}@eecs.oregonstate.edu

Abstract. A wide variety of objectively practical cryptographic schemes can be constructed using only symmetric-key operations and linear operations. To formally study this restricted class of cryptographic algorithms, we present a new model called *Linicrypt*. A Linicrypt program has access to a random oracle whose inputs and outputs are field elements, and otherwise manipulates data only via fixed linear combinations.

Our main technical result is that it is possible to decide *in polynomial time* whether two given Linicrypt programs induce computationally indistinguishable distributions (against arbitrary PPT adversaries, in the random oracle model).

We show also that indistinguishability of Linicrypt programs can be expressed as an existential formula, making the model amenable to *automated program synthesis*. In other words, it is possible to use a SAT/SMT solver to automatically generate Linicrypt programs satisfying a given security constraint. Interestingly, the properties of Linicrypt imply that this synthesis approach is both sound and complete. We demonstrate this approach by synthesizing Linicrypt constructions of garbled circuits.

1   Introduction

Throughout cryptography, we find many examples of objectively practical constructions that share common features. In particular, they treat blocks of bits as atomic units, and manipulate these units by calling a symmetric-key primitive or by interpreting them as elements in a field and applying *strictly linear* operations to them. Below are just some examples:

- Standard block cipher modes like CBC, OFB, PCBC for privacy, and LRW modes [34] for tweakable block ciphers consist of calls to the underlying block cipher and XOR, the linear operation in $GF(2^n)$. (This ignores matters of padding/ciphertext stealing, where the input is not an exact multiple of field elements.)
- Constructions in other settings also consist of calls to an underlying symmetric primitive along with XOR operations: the Davies-Meyer construction & its variants [13,47] for collision-resistance; the Even-Mansour [18] and Feistel [35] constructions for PRPs; NMAC, HMAC [31], and VMAC [32] for authenticity; Naor's commitment scheme [41].

Authors supported by NSF award 1149647.

M. Robshaw and J. Katz (Eds.): CRYPTO 2016, Part III, LNCS 9816, pp. 416–445, 2016.
DOI: 10.1007/978-3-662-53015-3_15

- Some constructions use $GF(2^n)$-linear transformations with (fixed) coefficients other than 1 (i.e., these constructions use multiplication by fixed field elements). These include: OCB mode [50] for authenticated encryption, CMC mode [23] for disk encryption, XE/XEX modes [49] for tweakable block ciphers, PMAC [12] for authentication.
- Signing algorithms for lightweight one-time signature schemes like those of Lamport [33] and Winternitz [52] consist purely of calls to a one-way or [target] collision-resistant hash function. Variants like W-OTS+ [25] incorporate XOR operations. Few-time signature schemes like HORS and variants [45,48] also use only a random oracle. These simple signature schemes can be composed to give many-use signature schemes using Merkle trees [39] and derivatives thereof [11,14–16,21,40,44]. These extensions do not introduce any additional operations on the atomic field elements.
- Practical constructions of garbled circuits [22,29,30,42,53] simply use XOR and calls to an underlying hash function/KDF, while the construction of [46] uses polynomial interpolation (with fixed points of evaluation) over $GF(2^n)$, which is a linear operation.

1.1 Overview of Our Results

Inspired by the constructions above, we introduce a restricted model of computation called **Linicrypt**. Programs in the Linicrypt model have access to a random oracle (to model a symmetric-key primitive), whose inputs and outputs are elements of a field $\mathbb{F}$. The field $\mathbb{F}$ is public and its size should be exponential in the security parameter.

Beyond calling a random oracle, Linicrypt programs can manipulate field elements only by uniformly sampling them or by applying fixed linear combinations. More formally, a **(pure) Linicrypt program** is a fixed sequence of statements of the following form:

$$v_i \xleftarrow{\$} \mathbb{F}:\text{ sample a value uniformly from } \mathbb{F}.$$
$$v_i := \sum_j c_j v_j:\text{ apply a linear combination to existing variables, using } fixed \text{ coefficients.}$$
$$v_i := H(t\|v_{j_1}\|v_{j_2}\|\cdots\|v_{j_k}):\text{ call the random oracle on a set of existing variables, and optionally a string } t \text{ which is fixed with the program (useful for domain separation).}$$
$$\text{output } (v_{j_1},\ldots,v_{j_k}):\text{ output an ordered sequence of variables.}$$

Linicrypt is expressive enough to capture cryptographic construction of interest, but still restrictive enough that it provides several key benefits:

1. It is tractable to reason about cryptographic properties of Linicrypt programs. Our **main technical result** is that it is possible to decide, *in polynomial time*, whether two Linicrypt programs induce indistinguishable output distributions (in the random oracle model, against *arbitrary* PPT adversaries).

   We also point out that unforgeability properties (*e.g.*, given the output of a program $\mathcal{P}$, it is hard to predict an internal value v^*) can be easily transformed into indistinguishability properties, making many standard styles of security definition expressible (and efficiently decidable) in Linicrypt.

2. Unlike in other restricted models, Linicrypt programs manipulate data as atomic units. This makes it possible to prove fine-grained lower bounds *to the level of optimal constant factors* (*e.g.*, "this cryptographic task cannot be done in Linicrypt with keys smaller than 5λ bits"). Such lower bounds for Linicrypt hold in the random oracle model, and hence they also imply impossibility of a black-box construction from one-way functions.

3. The question of finding a Linicrypt program whose output is indistinguishable from some specification (*e.g.*, its output is pseudorandom) can be expressed as an existential formula. One can then use an SAT/SMT solver to find a witness — *i.e.*, automatically *synthesize* a secure Linicrypt construction. Additionally, if the formula is found to be unsatisfiable, it implies that no secure Linicrypt construction exists for the task — *i.e.*, this paradigm for program synthesis is both *sound* and *complete*.

In Sect. 2 we formally define Linicrypt, develop techniques to reason about its algorithms, and prove our main technical result. Later in Sect. 3 we give an example application of our approach to program synthesis. We show how to use an SMT solver to synthesize secure Linicrypt constructions of garbled circuits. Specifically, for a given boolean function $f : \{0,1\}^k \to \{0,1\}^\ell$ (*e.g.*, an adder, a multiplexer), we synthesize Linicrypt procedures to garble f (as an atomic unit) in a way that is compatible with the Free XOR optimization of [30].

1.2   Related Work and Inspiration

Minicrypt. Linicrypt is inspired in name by Impagliazzo's [26] Minicrypt, which refers to a hypothetical world in which one-way functions exist but no "fancier" cryptography is possible. Minicrypt is formalized (as in [27]) by having a random oracle and allowing adversaries to be computationally unbounded (but with only polynomially many queries to the oracle). In this way, the random oracle becomes the only available source of computational cryptography.

The main distinction therefore between Linicrypt & Minicrypt is the additional constraint of linearity. This restriction allows Linicrypt lower bounds to resolve optimal constant factors, whereas optimal constant factors are not typically well-defined in Minicrypt. For example, imagine instantiating a secure Minicrypt scheme with security parameter λ/c; as a function of λ, the resulting construction would typically have constants reduced by a factor of c but still be secure.

Generic Group Model. Linicrypt has many similarities to the generic group model (GGM) of Shoup [51]. In the GGM, adversaries are restricted to manipulating elements of a *cyclic group* in a black-box way using only the prescribed group operations. While the GGM was originally proposed as a heuristic model for *adversaries*, one can also use GGM *constructions* to prove lower bounds. Dodis *et al.* [17] show that full-domain hashing from RSA cannot be proven secure using techniques that treat the RSA group as a generic multiplicative group. Papakonstantinou *et al.* [43] show that identity-based encryption is impossible via a GGM construction (without a bilinear pairing).

GGM lower bounds can identify *optimal constant factors*, which is one of the goals of Linicrypt. A line of work by Abe *et al.* [1–3] considers the case of *structure-preserving* digital signatures. They prove (among other things) that 3 group elements are optimal for structure-preserving signatures implemented by GGM algorithms. More recently, synthesis has been effectively applied [7] to generate novel and optimal structure-preserving schemes.

Despite these similarities, we point out some important technical differences:

(1) In the GGM, group elements are represented via a random encoding into bits, and adversaries are allowed to "look at" these encodings. This is slightly less restricting than our compartmentalized approach in which encodings don't play a part (and hence Linicrypt programs cannot perform equality tests). In that regard, our model is similar to the generic-group variant of Maurer [38]. Since our goal is to place restrictions on constructions rather than adversaries, the distinction does not seem to be very significant.

(2) Linicrypt includes a random oracle, which has not yet been considered in GGM lower bound results to the best of our knowledge. The random oracle is indeed a source of technical complications in Linicrypt.

(3) Both Linicrypt and GGM allow only linear operations (*e.g.*, in the GGM, a value "in the exponent" can only be manipulated in linear ways). However, a Linicrypt program must apply linear operations with *fixed* (*i.e.*, known to the adversary) coefficients, while the GGM model allows constructions to choose random (secret) coefficients. This difference is what allows Diffie-Hellman-style constructions to be modeled in GGM but not in Linicrypt. Namely, a GGM algorithm can hide a random value "in the exponent" by performing the generic operation $g \mapsto g^x$, but the analogous operation in Linicrypt ($v \mapsto xv$) hides nothing since x would always be considered fixed.

Algebraic Cryptography Model. Applebaum *et al.* [6] define a model for *arithmetic cryptography*, building on earlier work by Ishai *et al.* [28]. Their model has some similarities to Linicrypt but also fundamental differences. Compared to Linicrypt, the arithmetic model allows for general field operations on its elements, not just linear combinations. More importantly, the defining feature of the arithmetic model is that the construction is *oblivious* to the underlying field/ring — the construction must work no matter what field/ring is used. In order to model cryptographic practice, Linicrypt allows the ring to be *specified* by the construction. Additionally, their model does not currently include random oracles, and hence it is only applicable to information-theoretic constructions or computational assumptions that can be obtained from the algebraic structure in a black-box way. The model is not equipped to consider standard assumptions like the existence of pseudorandom functions or collision-resistant hash functions.

Linear Garbling. In this work we study Linicrypt programs in the context of garbled circuit constructions. This is inspired in part by the lower bound of Zahur *et al.* [53]. They too observe that practical garbled circuit constructions consist of only linear operations and calls to a random oracle. They prove a lower bound, namely, that such "linear garbling schemes" require 2 field elements to garble a single AND gate.

In concurrent and independent work, Pastro *et al.* [36] extend the model of linear garbling and characterize security in terms of linear-algebraic properties like span. They generalize the garbling scheme of [53] to natively support low-degree polynomials (not just AND-gates).

Later in Sect. 3 we go into more detail about the ZRE lower bound in the context of Linicrypt. For now, we simply point out the main differences between our work and the two above: (1) in this work we present a full theory of Linicrypt, not constrained only to garbled circuits; (2) the above models of linear garbling only consider "Linicrypt programs" that make non-adaptive calls to the random oracle, whereas our general Linicrypt model has no such restriction (arguably, the ability to reason about arbitrary oracle queries is the most important feature of Linicrypt). The difference is important specifically in the context of garbled circuits since, in most schemes, adaptive oracle queries result when composing several gates together in a larger circuit.

Synthesis of Cryptographic Constructions. Synthesis has been effectively used in the generic group model to discover batching schemes for signature verification [5] and optimal structure-preserving signatures [7]. Both of these results synthesize constructions involving bilinear pairings.

Malozemoff *et al.* [37] synthesized IND-CPA secure block cipher modes by expressing the main loop of a mode as a directed graph. They defined typing rules for the vertices of this graph and showed that if a valid assignment of types exists, then the resulting scheme is secure. Using a SAT solver, they were able to check for valid type assignments for candidate modes and subsequently enumerate secure modes. In a followup work, Hoang *et al.* [24] extended the synthesis to authenticated encryption modes built from tweakable block ciphers.

Prior work of Gagné *et al.* [19,20] developed techniques for automated proofs of security for (CPA-secure) block cipher modes. Akinyele *et al.* [4] use an SMT solver to automate transformations of pairing-based signature schemes.

In all of the works involving block cipher modes [19,20,24,37] the techniques are developed for modes involving just XOR operations and [tweakable] block cipher calls. This corresponds to a natural special case of Linicrypt. We emphasize, however, that in these works the methods are sound but not complete.[1]

2 Linicrypt

2.1 Basic Model

A **pure Linicrypt program** over field $\mathbb{F}$ is a tuple $\mathcal{P} = (\text{in}, \text{out}, \text{cmds})$, where: in is a nonnegative integer, out is an ordered sequence of indices from

---

[1] In [37] the authors explicitly say, "we prevent a random value from both being output as ciphertext and input into a PRF . . . This does not mean there do not exist secure schemes which have this property; however, our tool does not allow such schemes". In [19,20] the techniques involve a logic that uses only local invariants.

$\{1, \ldots, |\mathsf{cmds}|\}$, and cmds is an ordered sequence of **Linicrypt commands**. The ith command in cmds must have one of the following forms:

- (INP, j), where $1 \le j \le \text{in}$ [retrieve a value from input]
- (SAMP) [sample an element of $\mathbb{F}$]
- $(\text{LIN}, c_1, \ldots, c_{i-1})$, where each $c_j \in \mathbb{F}$ [perform a linear combination of values]
- $(\text{HASH}, t, j_1, \ldots, j_k)$, where $t \in \{0,1\}^*$ and $j_1, \ldots, j_k < i$ [call the random oracle on a set of variables, and additional (fixed) string t]

Intuitively, the program $\mathcal{P}$ takes as input a vector from $\mathbb{F}^{\text{in}}$, then performs the operations specified by cmds. Each of the internal values of $\mathcal{P}$ is assigned to a variable $v[i]$. Finally, the program outputs the values whose indices are in the set out. More formally, we define the behavior of $\mathcal{P}$ as a process via:

$$
\begin{array}{|l|}
\hline
\mathcal{P}^H(\boldsymbol{x} \in \mathbb{F}^{\text{in}}): \\
\quad \text{for } i = 1 \text{ to } |\mathsf{cmds}|: \\
\qquad \text{if } \mathsf{cmds}[i] = (\text{INP}, j): \qquad\qquad\quad v[i] := \boldsymbol{x}[j] \\
\qquad \text{if } \mathsf{cmds}[i] = (\text{SAMP}): \qquad\qquad\quad\; v[i] \xleftarrow{\$} \mathbb{F} \\
\qquad \text{if } \mathsf{cmds}[i] = (\text{LIN}, c_1, \ldots, c_{i-1}): \;\; v[i] := \sum c_j v[j] \\
\qquad \text{if } \mathsf{cmds}[i] = (\text{HASH}, t, j_1, \ldots, j_k): \; v[i] := H(t; v[j_1], \ldots, v[j_k]) \\
\quad \text{return } \left(v[j]\right)_{j \in \text{out}} \\
\hline
\end{array}
$$

Note that H is an oracle with type $H : \{0,1\}^* \times \mathbb{F}^* \to \mathbb{F}$. In informal discussions, we often omit the first argument to H when it is an empty string.

2.2 Mixed Linicrypt Programs and Modelling Real-World Primitives

Most of the cryptographic primitives listed in the introduction cannot actually be implemented strictly as pure Linicrypt programs. For example, consider the one-time Winternitz signature of a single "digit" $x \in [m]$. The secret key $sk \leftarrow \mathbb{F}$ is chosen uniformly. The public key is then $pk := H^{(m)}(sk)$. To sign x, release $\sigma := H^{(x)}(sk)$. Then to verify, check $pk \overset{?}{=} H^{(m-x)}(\sigma)$.

The main operations in Winternitz are simply repeated calls to the hash/one-way function H, which are certainly allowed in Linicrypt. However, the signing algorithm *uses x in a non-linear way* — to choose how many Linicrypt commands to execute!

We therefore extend the scope of Linicrypt beyond *pure* Linicrypt programs. A **mixed Linicrypt program** is one in which we designate some inputs to be *non-linear* and the others to be linear. For instance, in the signing algorithm of Winternitz signatures there is a for-loop whose exit condition is non-linear in x.

We can associate any *mixed* Linicrypt program with a collection of *pure* Linicrypt programs. Think of any *mixed* Linicrypt program as a switch/case statement (based on its non-linear input) selecting which *pure* Linicrypt program to run. See Fig. 1 for the example of Winternitz signatures. Each $\mathsf{sign}(\cdot, x)$ is a pure Linicrypt program. Since x is public in the security definition for signatures,

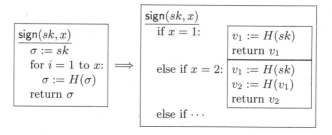

Fig. 1. The signing algorithm for one-time Winternitz signatures as a *mixed Linicrypt program*. Each inner box on the right-hand side is a *pure* Linicrypt programs, $\mathsf{sign}(\cdot, x)$, for fixed x.

we can express the security of the (mixed) signing algorithm in terms of the properties of each (pure) program $\mathsf{sign}(\cdot, x)$.

The way one decides to model some inputs as non-linear and other inputs as linear is highly *application-specific*. In general, it makes the most sense to let the length of non-linear inputs to be a *constant* c: First, the complexity of deciding security and synthesizing constructions grows exponentially with c. Second, this implies that all of the security properties are a result of the Linicrypt operations (the random oracle and linear operations over a field $\mathbb{F}$, whose size is exponential in the security parameter) and not the non-linear behavior. In other words, in a security game an adversary could guess with constant probability the non-linear input, leaving a residual *pure* Linicrypt program. So security is reduced to the security properties of the individual pure Linicrypt programs in the collection.

Throughout the rest of this section we develop a general theory of Linicrypt, and restrict our attention to *pure* Linicrypt programs. Later when discussing specific applications of Linicrypt to garbled circuits, we explicitly discuss *mixed* Linicrypt programs and non-linear inputs, etc.

2.3 Algebraic Representation

Let $\mathcal{P}$ be a (pure) Linicrypt program with notation as above. Say that $v[i]$ is a **derived** variable if $\mathsf{cmds}[i]$ is of the form $(\mathrm{LIN}, \cdots)$. Otherwise say that $v[i]$ is a **base** variable. That is, a base variable is the result of a command with one of SAMP, HASH, or INP. Let base denote the number of base variables. The main idea behind manipulating Linicrypt programs in an algebraic way is to observe that all values of importance can be expressed as linear functions of the *base* variables.

In more detail, fix an ordering of the base variables and denote them by the vector $\boldsymbol{v}_{\mathsf{base}}$. Then for the ith command in cmds, define $\mathsf{row}(i)$ to be the vector in $\mathbb{F}^{\mathsf{base}}$ such that $v[i] = \mathsf{row}(i) \cdot \boldsymbol{v}_{\mathsf{base}}$, where the $\cdot$ denotes dot product of vectors. More formally:

$$\mathsf{row}(i) \stackrel{\mathsf{def}}{=} \begin{cases} \overbrace{[0\ 0\ \cdots\ 0}^{j-1}\ 1\ 0\cdots0] & \text{if } v[i] \text{ is the } j\text{th base variable} \\ \sum_j c_j \mathsf{row}(j) & \text{if } \mathsf{cmds}[i] = (\mathrm{LIN}, c_1, \ldots, c_{i-1}) \end{cases}.$$

We create a matrix to represent the output of a Linicrypt program:

$$\mathcal{M} \stackrel{\text{def}}{=} \begin{bmatrix} -\ \mathsf{row}(o_1)\ - \\ \vdots \\ -\ \mathsf{row}(o_k)\ - \end{bmatrix}, \qquad \text{where out} = (o_1, \ldots, o_k).$$

$\mathcal{M}$ therefore characterizes the *direct* correlations among the program's output variables. Yet, it contains no information about how these variables may be *correlated via the random oracle!* So, our characterization of a Linicrypt program includes a set of **oracle constraints**. The idea behind an oracle constraint $\langle t, \mathcal{Q}, a \rangle$ is that if the random oracle is called on input $(t; \mathcal{Q} \times v_{\mathsf{base}})$ then the response will be $a \cdot v_{\mathsf{base}}$.

$$\mathcal{C} \stackrel{\text{def}}{=} \left\{ \left\langle t, \begin{bmatrix} -\ \mathsf{row}(j_1)\ - \\ \vdots \\ -\ \mathsf{row}(j_k)\ - \end{bmatrix}, \mathsf{row}(i) \right\rangle \ \middle|\ \mathsf{cmds}[i] = (\textsc{hash}, t, j_1, \ldots, j_k) \right\}$$

Without loss of generality, we can assume that no two constraints share $(t, \mathcal{Q})$ in common. Under that restriction, the set $\{a \mid \langle t, \mathcal{Q}, a \rangle \in \mathcal{C}\}$ is a linearly independent set — *i.e.*, the results of distinct random oracle queries are linearly independent.

Finally, we define the **algebraic representation** of a Linicrypt program $\mathcal{P}$ to be $(\mathcal{M}, \mathcal{C})$. We refer to $\mathcal{M}$ as the **output matrix** and $\mathcal{C}$ as the set of **oracle constraints**.

To demonstrate the different ways of viewing a Linicrypt program, consider the following example, with in $= 0$:

plain-language:	Linicrypt cmds:	var type:	matrix representation:				
$v_1 \leftarrow \mathbb{F}$	1: $(\textsc{samp})$	base	$\begin{bmatrix} v_1 \\ v_2 \\ v_3 \\ v_4 \\ v_5 \end{bmatrix}$	$=$	$\begin{bmatrix} 1 & 0 & 0 \\ 0 & 1 & 0 \\ 1 & -1 & 0 \\ 0 & 0 & 1 \\ 1 & 0 & 1 \end{bmatrix}$	$\begin{bmatrix} v_1 \\ v_2 \\ v_4 \end{bmatrix}$	
$v_2 \leftarrow \mathbb{F}$	2: $(\textsc{samp})$	base					
$v_3 := v_1 - v_2$	3: $(\textsc{lin}, 1, -1)$	derived					
$v_4 := H(\mathsf{foo}, v_3, v_2)$	4: $(\textsc{hash}, \mathsf{foo}, 3, 2)$	base					
$v_5 := v_4 + v_1$	5: $(\textsc{lin}, 1, 0, 0, 1)$	derived					
return (v_4, v_5)	// out $= (4, 5)$						

algebraic representation:

$$\mathcal{M} = \begin{bmatrix} 0 & 0 & 1 \\ 1 & 0 & 1 \end{bmatrix}; \quad \mathcal{C} = \left\{ \left\langle \mathsf{foo}, \begin{bmatrix} 1 & -1 & 0 \\ 0 & 1 & 0 \end{bmatrix}, [0\ 0\ 1] \right\rangle \right\}$$

There are three base variables. With v_4, v_5 being output variables, the output matrix $\mathcal{M}$ consists of $\mathsf{row}(4), \mathsf{row}(5)$. There is one HASH-command "$v_4 := H(\mathsf{foo}, v_3, v_2)$," leading to a single oracle constraint $\langle \mathsf{foo}, \begin{bmatrix} \mathsf{row}(3) \\ \mathsf{row}(2) \end{bmatrix}, \mathsf{row}(4) \rangle$.

In the rest of this paper, we specialize to input-less (*i.e.*, in $= 0$) Linicrypt programs. Restricting our domain to input-less programs simplifies the definitions & proofs. This is justified by our main application to garbled circuits. In the security definition for garbled circuits, the adversary chooses an input x to the function, but since we model x as non-linear input, what is left over is a collection of security experiments, one for each x, each involving an input-less (pure) Linicrypt program.

We hereafter overload notation and write $\mathcal{P} = (\mathcal{M}, \mathcal{C})$. We claim that $(\mathcal{M}, \mathcal{C})$ completely characterizes the behavior of $\mathcal{P}$. In more detail, let $\mathcal{P}$ be an input-less Linicrypt program, let $\mathcal{A}$ be an oracle machine, and consider the following **canonical simulation** of $\mathcal{P}$.

$\mathcal{S}^{\mathcal{A}}_{\mathcal{P}}()$:

1. $\boldsymbol{v}_{\mathsf{base}} \xleftarrow{\$} \mathbb{F}^{\mathsf{base}}$
2. $\boldsymbol{v}_{\mathsf{out}} := \mathcal{M}\, \boldsymbol{v}_{\mathsf{base}}$
3. $cache :=$ empty associative array
4. return $\mathcal{A}^{H}(\boldsymbol{v}_{\mathsf{out}})$, where H implemented as below:

$H(t; \boldsymbol{q} \in \mathbb{F}^{*})$:

// if the adversary found a collision among oracle constraints
5. if $\exists \langle t, \mathcal{Q}, \boldsymbol{a} \rangle, \langle t, \mathcal{Q}', \boldsymbol{a}' \rangle \in \mathcal{C}$ with $\boldsymbol{a} \neq \boldsymbol{a}'$ and $\mathcal{Q}\boldsymbol{v}_{\mathsf{base}} = \mathcal{Q}'\boldsymbol{v}_{\mathsf{base}} = \boldsymbol{q}$:   (1)
6.   abort
// if there is an oracle constraint for the query q
7. if $\exists \langle t, \mathcal{Q}, \boldsymbol{a} \rangle \in \mathcal{C}$ with $\mathcal{Q}\boldsymbol{v}_{\mathsf{base}} = \boldsymbol{q}$:
8.   return $\boldsymbol{a} \cdot \boldsymbol{v}_{\mathsf{base}}$
// honest simulation of a random oracle beyond this point
9. if $cache[t; \boldsymbol{q}]$ does not exist:
10. $cache[t; \boldsymbol{q}] \xleftarrow{\$} \mathbb{F}$
11. return $cache[t; \boldsymbol{q}]$

The idea is to simply sample *all* of the base variables upfront, instead of deriving some of them via calls to the random oracle. But then to make the simulation of the random oracle consistent, we "patch" the random oracle so that when queried on $(t, \mathcal{Q}\boldsymbol{v}_{\mathsf{base}})$, the consistent result $\boldsymbol{a} \cdot \boldsymbol{v}_{\mathsf{base}}$ is simulated (lines 7–8). The simulation aborts when two oracle constraints are in conflict (lines 5–6).

Lemma 1 (Canonical Simulation). *Let $\mathcal{P}$ be an input-less (i.e., in $= 0$) Linicrypt program that executes n HASH-commands. Then for all oracle machines $\mathcal{A}$:*

$$\Pr\left[\mathcal{S}^{\mathcal{A}}_{\mathcal{P}}() = 1\right] - \Pr_{H}\left[\mathcal{A}^{H}(\mathcal{P}^{H}()) = 1\right] \leq \frac{n(n+1)}{2|\mathbb{F}|}.$$

We emphasize that $\mathcal{A}$ here is an arbitrary program. It need not be linear, it may be computationally unbounded, and (at least for this lemma) it is even unrestricted in the number of oracle queries it makes.

Proof (Sketch). Conditioned on the simulation not aborting in line 6, the simulation is perfect. Essentially, each query to H answered in lines 7–8 is answered with a randomly chosen base variable (since each $\boldsymbol{a}$ is a canonical basis vector), exactly matching how queries are answered by an honest random oracle. Hence, the error in the simulation is the probability that the condition in line 5 is true. This happens if $\mathcal{Q}v_{\text{base}} = \mathcal{Q}'v_{\text{base}}$ for some distinct constraints $\langle t, \mathcal{Q}, a \rangle, \langle t, \mathcal{Q}', a' \rangle \in \mathcal{C}$. Since WLOG no two constraints share $(t, \mathcal{Q})$, we have that $\mathcal{Q} - \mathcal{Q}'$ is a nonzero matrix, and therefore that

$$\mathcal{Q}v_{\text{base}} = \mathcal{Q}'v_{\text{base}} \iff (\mathcal{Q} - \mathcal{Q}')v_{\text{base}} = 0 \iff v_{\text{base}} \in \text{kernel}(\mathcal{Q} - \mathcal{Q}').$$

Note that $\text{kernel}(\mathcal{Q} - \mathcal{Q}')$ is a proper subspace of $\mathbb{F}^{\text{base}}$ with maximum dimension (base $- 1$). Then, when v_{base} is chosen uniformly from $\mathbb{F}^{\text{base}}$, the probability that it is in a particular proper subspace is at most $|\mathbb{F}|^{\text{base}-1}/|\mathbb{F}|^{\text{base}} = 1/|\mathbb{F}|$. Recall that $\mathcal{P}$ executes n HASH-commands. Then there are $\binom{n}{2} = n(n+1)/2$ possible pairs of distinct oracle constraints. By the union bound, the probability that there exist some pair of oracle constraints with $\mathcal{Q}$ and $\mathcal{Q}'$ for which $v_{\text{base}} \in \text{kernel}(\mathcal{Q} - \mathcal{Q}')$ is at most $n(n+1)/2|\mathbb{F}|$.

2.4 Linear Transformations, Basis Changes and Composition

The algebraic representation for Linicrypt programs turns out to be convenient, as we can perform linear-algebraic manipulations to Linicrypt programs.

For instance, consider **applying a linear transformation** to a Linicrypt program. Let $\mathcal{P} = (\mathcal{M}, \mathcal{C})$ be a Linicrypt program. Recall that the width of the vectors in $\mathcal{M}$ and $\mathcal{C}$ is base. Now let B be a base $\times$ base matrix with entries in $\mathbb{F}$ and consider the Linicrypt representation $(\mathcal{M}B, \mathcal{C}B)$, where

$$\mathcal{C}B \overset{\text{def}}{=} \{\langle t, \mathcal{Q}B, aB \rangle \mid \langle t, \mathcal{Q}, a \rangle \in \mathcal{C}\}.$$

When B is an invertible matrix, we refer to $(\mathcal{M}B, \mathcal{C}B)$ as a **basis change** of B applied to $(\mathcal{M}, \mathcal{C})$. Such a basis change has no effect on the output distribution of the Linicrypt program. More precisely:

Proposition 2. *Let $\mathcal{P} = (\mathcal{M}, \mathcal{C})$ be an input-less Linicrypt program, and let $\mathcal{P}' = (\mathcal{M}B, \mathcal{C}B)$ for some invertible matrix B. Then for all oracle machines $\mathcal{A}$, we have:*

$$\Pr\left[\mathcal{S}_{\mathcal{P}}^{\mathcal{A}}() = 1\right] = \Pr\left[\mathcal{S}_{\mathcal{P}'}^{\mathcal{A}}() = 1\right].$$

Proof. A basis change by B is equivalent to adding a statement "$v_{\text{base}} := Bv_{\text{base}}$" between lines 1 and 2 in Eq. 1. Since B is invertible, this additional statement has no effect on the distribution of v_{base}.

Composition. We can use the idea of a linear transformation to reason algebraically about the composition of two Linicrypt programs. Let $\mathcal{P} = (\mathcal{M}, \mathcal{C})$ be a Linicrypt program with no input and out outputs, and let $\mathcal{P}' = (\mathcal{M}', \mathcal{C}')$ be a Linicrypt program with out inputs, so that it makes sense to feed the output of $\mathcal{P}$ as input to $\mathcal{P}'$. Without loss of generality, we make the following assumptions:

- Both programs have the same number of base variables (so that $\mathcal{M}, \mathcal{M}'$ have the same number of columns and so on).

– The first out base variables of $\mathcal{P}'$ are identified with its input variables.

The algebraic representation of $\mathcal{P}'$ implicitly treats all of its input variables as linearly independent. So the case when $\mathcal{M}$ has full rank is easiest. To compose the programs, one simply applies a basis change to either program to align $\mathcal{P}$'s output variables ($\mathcal{M}$) and $\mathcal{P}'$'s input variables (expressed as $[I \mid 0]$, where I is the out × out identity matrix), and similarly align the oracle constraints of the programs. If such a basis change has been applied, then the composed program's output is characterized by $\mathcal{M}'$ and its oracle constraints are simply $\mathcal{C} \cup \mathcal{C}'$.

However, in general the output of $\mathcal{P}$ may have linear correlations, and this can have a serious effect on the behavior of $\mathcal{P}'$. Take for example the case where $\mathcal{P}'$ takes two input variables (v_1, v_2) and outputs $H(v_1) - H(v_2)$. Then the behavior of $\mathcal{P}'$ is qualitatively different when v_1 and v_2 are linearly independent vs. when they are correlated as $v_1 = v_2$, for instance.

In general, we consider applying a linear transformation to $\mathcal{P}'$ that "collapses" the appropriate base variables (they become associated with the same vector in the algebraic representation). Collapsing input base variables may result in the collapse of oracle queries that use these variables. In the example above, $H(v_1)$ and $H(v_2)$ are themselves base variables which are linearly independent in general; yet they collapse to the same base variable when $v_1 = v_2$.

Hence, to compose $\mathcal{P}$ with $\mathcal{P}'$ we consider a linear transformation Γ applied to $\mathcal{P}'$, with the following properties:

1. Γ aligns the input variables of $\mathcal{P}'$ (the first out base variables) with the output $\mathcal{M}$ of $\mathcal{P}$. That is, $\mathcal{M} = [I \mid 0] \times \Gamma$ where I is the out × out identity matrix.
2. Γ consistently aligns the oracle queries of $\mathcal{P}'$ to those in $\mathcal{P}$. That is, if $\langle t, \mathcal{Q}, \boldsymbol{a} \rangle \in \mathcal{C}'\Gamma$, and $\langle t, \mathcal{Q}, \boldsymbol{a}' \rangle \in \mathcal{C}$, then $\boldsymbol{a} = \boldsymbol{a}'$.
3. Γ collapses appropriate oracle constraints in $\mathcal{P}'$: that is, if Γ causes (previously distinct) oracle constraints to now share the same t and $\mathcal{Q}$ components, then they must now also share the same $\boldsymbol{a}$ component. More formally, the constraints in $\mathcal{C}'\Gamma$ should all have distinct $t, \mathcal{Q}$ values. However, note that $\mathcal{C}'\Gamma$ may have fewer constraints than $\mathcal{C}'$ due to collapses induced by Γ.
4. Γ should only collapse base variables that are absolutely required by the above conditions. In other words, the rank of Γ should be as large as possible given the above constraints. Note that if $\mathcal{M}$ has full rank, then Γ will indeed be a basis change. However, in general Γ may not be a basis change — this is consistent with the fact that feeding linearly correlated values into $\mathcal{P}'$ may indeed fundamentally change its behavior. A basis change exactly preserves behavior.

Given such a transformation Γ, then $(\mathcal{M}'\Gamma, \mathcal{C} \cup \mathcal{C}'\Gamma)$ is an algebraic representation for the composition of programs $\mathcal{P}' \circ \mathcal{P}$.

2.5 Indistinguishability vs. Unpredictability

When we consider Linicrypt programs that implement cryptographic primitives, the most fundamental question is: when do two Linicrypt programs induce indistinguishable distributions (in the random oracle model)?

Definition 3. *Let $\mathcal{P}_1$ and $\mathcal{P}_2$ be two input-less Linicrypt programs over $\mathbb{F}$. Let $\lambda = \log |\mathbb{F}|$ be the security parameter. We say that $\mathcal{P}_1$ and $\mathcal{P}_2$ are **indistinguishable**, and write $\mathcal{P}_1 \cong \mathcal{P}_2$, if for every (possibly computationally unbounded) oracle machine $\mathcal{A}$ that queries its oracle a polynomial (in λ) number of times, we have*

$$\Pr[\mathcal{A}^H(\mathcal{P}_1^H()) = 1] - \Pr[\mathcal{A}^H(\mathcal{P}_2^H()) = 1] \text{ is negligible in } \lambda.$$

The probabilities are over the choice of random oracle H and the coins of $\mathcal{P}_1$, $\mathcal{P}_2$, and $\mathcal{A}$.

We point out that *indistinguishability can be used to reason about unforgeability properties* as well. Suppose $\mathcal{P}$ is a Linicrypt program that has some special internal variable v^*, and we wish to formalize the idea that "v^* is hard to predict (in the random oracle model) given the output of $\mathcal{P}$". Now define the following two related programs:

- $\mathcal{P}_1$ runs $\mathcal{P}$ and outputs whatever $\mathcal{P}$ outputs, along with an additional output $v_{\text{extra}} = H(t^*; v^*)$, where t^* is a "tweak" that is not used in $\mathcal{P}$.
- $\mathcal{P}_2$ runs $\mathcal{P}$ and outputs whatever $\mathcal{P}$ outputs, along with an additional output $v_{\text{extra}} \xleftarrow{\$} \mathbb{F}$.

Note that $\mathcal{P}_1$ and $\mathcal{P}_2$ are a Linicrypt programs if $\mathcal{P}$ is. Now observe that the following statements are equivalent:

1. Given the output of $\mathcal{P}$, the probability that an adversary (with access to the random oracle) outputs v^* is negligible.
2. Given the output of $\mathcal{P}$, the probability that an adversary queries the random oracle on $H(t^*; v^*)$ is negligible.
3. Given the output of $\mathcal{P}$, the value $H(t^*; v^*)$ is indistinguishable from uniform. This follows simply from the definition of the random oracle model, and the fact that $\mathcal{P}$ itself does not use any values of the form $H(t^*; \cdot)$.
4. $\mathcal{P}_1 \cong \mathcal{P}_2$.

Hence, standard unforgeability properties of a Linicrypt program can be expressed as the indistinguishability of two Linicrypt programs. From now on, we therefore focus on indistinguishability only. And indeed, our main characterization theorem will include reasoning like that above, regarding which oracle queries can be made by an adversary with non-negligible probability.

2.6   Normalization

We now describe a procedure for "normalizing" a Linicrypt program. Specifically, normalizing corresponds to removing "unnecessary" calls to the oracle. We illustrate the ideas with a brief example, below:

plain language:	Linicrypt cmds:	matrix representation:

$$v_1 \overset{\$}{\leftarrow} \mathbb{F}$$
$$v_2 := H(\texttt{foo}, v_1)$$
$$v_3 := v_1 - v_2$$
$$v_4 := H(\texttt{bar}, v_3)$$
$$v_5 := H(\texttt{baz}, v_3)$$
output (v_3, v_5)

1: (SAMP)
2: (HASH, foo, 1)
3: (LIN, 1, −1)
4: (HASH, bar, 3)
5: (HASH, baz, 3)

$$\begin{bmatrix} v_1 \\ v_2 \\ v_3 \\ v_4 \\ v_5 \end{bmatrix} = \begin{bmatrix} 1 & 0 & 0 & 0 \\ 0 & 1 & 0 & 0 \\ 1 & -1 & 0 & 0 \\ 0 & 0 & 1 & 0 \\ 0 & 0 & 0 & 1 \end{bmatrix} \begin{bmatrix} v_1 \\ v_2 \\ v_4 \\ v_5 \end{bmatrix}$$

This program has 3 oracle queries, two of which are "unnecessary" in some sense.

- It is instructive to consider what information the adversary can collect about the base variables v_{base}. From the output of $\mathcal{P}$, one obtains $v_3 = [1\ {-}1\ 0\ 0] \cdot v_{\mathsf{base}}$ and $v_5 = [0\ 0\ 0\ 1] \cdot v_{\mathsf{base}}$. Then one can call the oracle as $H(\texttt{bar}, v_3)$ to obtain $v_4 = [0\ 0\ 1\ 0] \cdot v_{\mathsf{base}}$. However, it is hard to predict $v_1 = [1\ 0\ 0\ 0] \cdot v_{\mathsf{base}}$ given just the output of $\mathcal{P}$. More specifically, $[1\ 0\ 0\ 0]$ is not in the span of $\{[1\ {-}1\ 0\ 0], [0\ 0\ 1\ 0], [0\ 0\ 0\ 1]\}$.

  In other words, the probability of an adversary querying H on v_1 is negligible, so we call this oracle query **unreachable**. Conditioned on the adversary not querying H on v_1, its output $v_2 = H(\texttt{foo}, v_1)$ looks uniformly random. Removing the corresponding oracle constraint therefore has negligible effect. Note that removing the oracle constraint corresponds to replacing "$v_2 := H(\texttt{foo}, v_1)$" with "$v_2 \overset{\$}{\leftarrow} \mathbb{F}$"; *i.e.*, changing cmds[2] from (HASH, foo, 1) to (SAMP).

- Oracle query $H(\texttt{bar}, v_3)$ is reachable, since the output of $\mathcal{P}$ includes v_3. However, its result is v_4 which is not used anywhere else in the program. This can be seen by observing that all other row vectors in the algebraic representation have a zero in the position corresponding to v_4. Hence this oracle call can be replaced with "$v_4 \overset{\$}{\leftarrow} \mathbb{F}$" with no effect on the adversary. We call this query *useless*.

- Oracle query $H(\texttt{baz}, v_3)$ is similarly reachable, but it is *useful*. The result of this query is $H(\texttt{baz}, v_3) = v_5$ which is included in the output of $\mathcal{P}$ and hence visible to the adversary. It cannot be removed because an adversary could query $H(\texttt{baz}, v_3)$ and check that it matches v_5 from the output.

More generally, we normalize a Linicrypt program by computing which oracle queries/constraints are *reachable* and which are *useless* in the above sense.

To compute which oracle queries are reachable, we perform the following procedure until it reaches a fixed point: Given Linicrypt program $\mathcal{P} = (\mathcal{M}, \mathcal{C})$, mark the rows of $\mathcal{M}$ as *reachable*. Then, if any oracle constraint $\langle t, \mathcal{Q}, a \rangle \in \mathcal{C}$ has every row of $\mathcal{Q}$ in the span of reachable vectors, then mark a as *reachable*.

Instead of computing which queries are useful, it is more straight-forward to compute which queries are *useless*, one by one. Intuitively, a constraint $\langle t, \mathcal{Q}, a \rangle$ is *useless* if a is linearly independent of all other vectors appearing in $\mathcal{M}$ and $\mathcal{C}'$ (either as rows of $\mathcal{M}$ or rows of some $\mathcal{Q}'$ or as an a'). After removing one useless constraint, other constraints might become useless. For instance, consider a Linicrypt program that outputs v but also internally computes $H(H(H(v)))$.

```
normalize(𝒫 = (ℳ, 𝒞)):
    Reachable := rows(ℳ)
    𝒞' := ∅
    until 𝒞' reaches a fixed point:
        for each ⟨t, 𝒬, a⟩ ∈ 𝒞 \ 𝒞':
            if rows(𝒬) ⊆ span(Reachable):
                add a to Reachable
                add ⟨t, 𝒬, a⟩ to 𝒞'

    Useless := ∅
    until Useless reaches a fixed point:
        V := (multiset of) all row vectors in ℳ and 𝒞' \ Useless
        for each ⟨t, 𝒬, a⟩ ∈ 𝒞' \ Useless:
            if a ∉ span(V \ {a}):
                add ⟨t, 𝒬, a⟩ to Useless

    𝒞'' := 𝒞' \ Useless

    return (ℳ, 𝒞'')
```

Fig. 2. Procedure to normalize a Linicrypt program. Since V is a multiset, we clarify that "$V \setminus \{a\}$" means to decrease the multiplicity of a in multiset V by only one. So $V \setminus \{a\}$ may yet include a. One reason for a to have high multiplicity in V is if a appears both in an oracle constraint and as a row of $\mathcal{M}$.

Only the outermost call to H is initially useless. After it is removed, the "new" outermost call is marked useless, and so on, until a fixed point is reached.

The details of the normalize procedure are given in Fig. 2. In the full version we prove the following:

Lemma 4. *If* $\mathcal{P}$ *is an input-less Linicrypt program, then* normalize($\mathcal{P}$) $\cong \mathcal{P}$ *(Fig. 2).*

2.7   Main Characterization

We can now present our main technical theorem about Linicrypt programs:

Theorem 5 (Linicrypt Characterization). *Let* $\mathcal{P}_1$ *and* $\mathcal{P}_2$ *be two input-less Linicrypt programs over* $\mathbb{F}$. *Then* $\mathcal{P}_1 \cong \mathcal{P}_2$ *if and only if* normalize($\mathcal{P}_1$) *and* normalize($\mathcal{P}_2$) *differ by a basis change.*

Proof (Proof Sketch). The nontrivial case is to show the $\Rightarrow$ direction. Without loss of generality assume that $\mathcal{P}_1$ and $\mathcal{P}_2$ are normalized, and suppose they do not differ by a basis change. The idea is to first construct a "profile" for $\mathcal{P}_1$ and for $\mathcal{P}_2$. In the code of normalize, we compute the reachable subspace of a program; the *profile* simply refers to the order in which reachable oracle constraints are activated during this process.

We use the profile to construct a family of *canonical distinguishers* for $\mathcal{P}_1$. It processes oracle constraints in the order determined by the profile. It maintains the invariant that at all stages of the computation, if $\mathcal{R}$ is the set of currently reachable vectors, the distinguisher holds $r = \mathcal{R} \times v_{\text{base}}$, where v_{base} refers to the base variables in the canonical simulation of $\mathcal{P}_1$.

A side-effect of normalization is that all oracle constraints are reachable and useful. Because of this, the set of reachable vectors will eventually contain non-trivial linear relations — as a matrix, the set of reachable vectors has a nontrivial kernel. A canonical distinguisher chooses some element z from this kernel and tests whether $z^\top r = 0$. By construction, $z^\top r = z^\top \mathcal{R} v_{\text{base}}$. Since $z \in \ker(\mathcal{R})$, the distinguisher always outputs true in the presence of $\mathcal{P}_1$.

Now the challenge is to show that, for some choice of $z \in \ker(\mathcal{R})$, the distinguisher outputs false with overwhelming probability in the presence of $\mathcal{P}_2$. To see why, we consider the first point at which the profiles of $\mathcal{P}_1$ and $\mathcal{P}_2$ disagree (if the profiles agree fully, then it is easy to obtain a basis change relating $\mathcal{P}_1$ to $\mathcal{P}_2$). The most nontrivial case is when $\mathcal{P}_1$ contains an oracle constraint that no basis change can bring into alignment with $\mathcal{P}_2$. This implies that when the distinguisher makes the query in the presence of $\mathcal{P}_2$, it will not trigger any oracle constraint and the result will be random and independent of everything else in the system. But because this oracle constraint was useful in $\mathcal{P}_1$, we can eventually choose a final kernel-test z that is "sensitive" to the result in the following way: While in $\mathcal{P}_1$, the kernel-test always results in zero, in $\mathcal{P}_2$ the kernel test will be independently random.

The actual proof is considerably more involved concerning the different cases for why the profiles of $\mathcal{P}_1$ and $\mathcal{P}_2$ disagree.

3   Synthesizing Linicrypt Garbled Circuits

In this section we describe how to express the security of garbled circuits in the language of Linicrypt, culminating in a method to leverage an SMT solver to automatically synthesize secure schemes. We assume some familiarity with the classical (textbook) Yao garbling scheme. Roughly speaking, each wire in the circuit is associated with two *labels* (bitstrings) W^0 and W^1, encoding FALSE and TRUE, respectively. The evaluator will learn exactly one of these two labels for each wire. Then, for each gate in the circuit, the evaluator uses the labels for the input wires, along with *garbled gate* information (classically, the garbled truth table), to compute the appropriate label on the output wire. We restrict our synthesis technique to the context of two basic garbled circuit techniques: *Free-XOR* and *Point-and-Permute*.

Free-XOR. In the Free-XOR garbling technique of Kolesnikov and Schneider [30], the garbler chooses a random Δ that is global, and arranges for $W^0 \oplus W^1 = \Delta$ on every wire. Hereafter, we typically write the FALSE label simply as W and the TRUE wirelabel as $W \oplus \Delta$; more generally, the wirelabel encoding b is $W \oplus b\Delta$.

Using Free-XOR, no ciphertexts are necessary to garble an XOR gate. For instance, let A and B be the FALSE input wirelabels. Set the FALSE output wirelabel to $C = A \oplus B$. Then when the evaluator holds wirelabels $A^* = A \oplus a\Delta$ and $B^* = B \oplus b\Delta$ (encoding a and b, respectively), she can compute $A^* \oplus B^* = A \oplus a\Delta \oplus B \oplus b\Delta = C \oplus (a \oplus b)\Delta$. That is, the result will be the wirelabel correctly encoding truth value $a \oplus b$. We note that no garbled gate information is required in the garbled circuit, nor must the evaluator perform any cryptographic operations to evaluate the gate — just an XOR of strings.

Free-XOR is ubiquitous in practical implementations of garbled circuits. For that reason (and because it conveniently reduces degrees of freedom over choice of wirelabels), we restrict our attention to garbling schemes that are compatible with Free-XOR.

Point-and-Permute and Non-linearity. The *point-and-permute* optimization of [8] is used in all practical garbling schemes. The idea is to append to each wirelabel a random bit χ (which we call the **"color bit"**). The two labels on each wire have opposite (but random) color bits.

Now consider the naive/classical garbling of an AND gate, in which the garbler generates 4 ciphertexts. Because color bits are independent of truth values, the garbler can arrange the ciphertexts in order of the color bits of the input wirelabels. The evaluator selects and decrypts the correct ciphertext indicated by the color bits of the input wirelabels she holds. Importantly, this makes the color bits *non-linear inputs* with respect to Linicrypt! The color bits determine which linear combination the evaluator will apply.

Similarly, the garbler's behavior is non-linear in a complementary way. We refer to σ as the **"select bit"** such that the wirelabel encoding truth value v has color $\chi = v \oplus \sigma$. Equivalently, σ is the (random) color bit of the FALSE wire. We emphasize that σ is known only to the garbler, and χ is known only to the evaluator, effectively hiding the truth value v. In typical garbling schemes, the garbler's behavior depends non-linearly on σ but is otherwise within the Linicrypt model.

We treat garbling schemes as mixed Linicrypt programs, as in Sect. 2.2. Then, a mixed Linicrypt garbling scheme is a collection of pure Linicrypt garbling programs indexed by color bits and select bits.

Restricting to Linicrypt with XOR *as the Linear Operation.* Technically speaking, a Linicrypt program is an infinite family of programs, one for each value of the security parameter. Unfortunately, we can only synthesize an object of finite size. Hence we restrict our focus to *single* Linicrypt programs that are compatible with an infinite family of fields/security parameters, in the following way.

Suppose a Linicrypt program uses field $GF(p)$ for prime p. Then that Linicrypt program is also compatible with field $GF(p^\lambda)$ for any λ, since $GF(p) \subseteq GF(p^\lambda)$ in a natural way. A very natural special case is $p = 2$, which corresponds to Linicrypt programs that use $GF(2^\lambda)$ and use only linear combinations with coefficients from $\{0, 1\}$ — in other words, Linicrypt programs that are restricted to using XOR as their only linear operation. Hereafter we restrict our attention to XOR-only Linicrypt programs.

3.1 Gate-Garbling

A garbling scheme for an entire circuit is a non-trivially large object — much too large to synthesize using a SAT/SMT solver. We instead focus on techniques for *garbling individual gates* in a way that allows them to be securely composed with other gates and the Free-XOR technique to yield a garbling scheme for arbitrary circuits.

Notation. A wirelabel that carries the truth-value FALSE is always signified W, a wirelabel that carries TRUE is always $W \oplus \Delta$, and a wirelabel carrying unknown truth-value is always W^*. We collect wirelabels into vectors notated as follows: $\boldsymbol{W} = W_1, \ldots, W_n$. Operations over vectors are computed componentwise. For instance, $\boldsymbol{A} \oplus \boldsymbol{B} = A_1 \oplus B_1, \ldots, A_n \oplus B_n$. When $\Delta \in GF(2^\lambda)$ and x is a string of n bits, we write $x\Delta$ to mean the vector $x_1\Delta, \ldots, x_n\Delta$. For example, if $\boldsymbol{W} = W_1, \ldots, W_n$ are a vector of FALSE wirelabels, then $\boldsymbol{W} \oplus x\Delta$ is a vector of wirelabels encoding truth values x.

Syntax. Let $\tau : \{0,1\}^m \to \{0,1\}^n$ be the functionality of an m-ary boolean gate that we wish to garble. Let $\sigma = \sigma_1 \,||\, \ldots \,||\, \sigma_m$ be a string of select bits and $\chi = \chi_1 \,||\, \ldots \,||\, \chi_m$ be a string of color bits. Then, a **free-XOR compatible garbled gate** consists of algorithms:

$$\mathsf{GateGb}(\sigma; A_1, \ldots, A_m, \Delta) \to (C_1, \ldots, C_n; G_1, \ldots, G_\ell)$$
$$\mathsf{GateEv}(\chi; A_1^*, \ldots, A_m^*, G_1, \ldots, G_\ell) \to (C_1^*, \ldots, C_n^*)$$

The semantics are as follows. GateGb takes m FALSE input wirelabels $\boldsymbol{A} = A_1, \ldots, A_m$, their select bits σ, and global constant Δ. It returns the n FALSE output wirelabels $\boldsymbol{C} = C_1, \ldots, C_m$, and garbled gate information $\boldsymbol{G} = G_1, \ldots, G_\ell$. The evaluator takes m input wirelabels with *unknown* truth values $\boldsymbol{A}^* = A_1^*, \ldots, A_m^*$, their color bits χ, and the garbled gate information $\boldsymbol{G}$. It returns output wirelabels with *unknown* truth values $\boldsymbol{C}^* = C_1^*, \ldots, C_n^*$.

We emphasize that when GateGb and GateEv are Linicrypt programs, all inputs and outputs besides σ and χ are field elements in $GF(2^\lambda)$.

Correctness. If a gate garbling scheme is correct, then the evaluator can always produce the correct output wirelabels according to τ. That is, when the evaluator holds wirelabels encoding x on the input wires, the result of evaluating the gate is the wirelabels encoding $\tau(x)$ on the output wires.

Definition 6. *A Free-XOR-compatible garbled gate ($\mathsf{GateGb}, \mathsf{GateEv}$) correctly computes functionality $\tau : \{0,1\}^m \to \{0,1\}^n$ if for all inputs $x \in \{0,1\}^m$, select bit strings $\sigma \in \{0,1\}^m$, and color bit string $\chi \in \{0,1\}^m$, with $x = \sigma \oplus \chi$, false input wirelabels $\boldsymbol{A} = A_1, \ldots, A_m$, global Free-XOR constant Δ:*

$$(\boldsymbol{C}, \boldsymbol{G}) \leftarrow \mathsf{GateGb}(\sigma; \boldsymbol{A}, \Delta) \implies \mathsf{GateEv}(\chi; \boldsymbol{A} \oplus x\Delta, \boldsymbol{G}) = \boldsymbol{C} \oplus \tau(x)\Delta$$

Security. One important consideration is that in the free-XOR setting, the labels of different wires can have linear correlations. The gate should be secure even for such correlated input wirelabels.[2]

We define security in terms of the evaluator's view in a typical garbling scenario. Then we define $\mathsf{View}_R^H(\chi, x)$ to encapsulate the information the evaluator sees for this gate, when the visible color bits are χ, the logical gate inputs are x, and the input wirelabels have correlations described by an $m \times m$ matrix R.

$$
\begin{array}{l}
\mathsf{View}_R^H(\chi, x): \\
\hline
\Delta, r_1, \ldots, r_m \leftarrow \{0,1\}^\lambda \\
A = (A_1, \ldots, A_m) := R \times [r_1, \ldots, r_m] \\
(C, G) \leftarrow \mathsf{GateGb}^H(\chi \oplus x; \, A, \Delta) \\
\text{return } (A \oplus x\Delta, \, G, \, C \oplus \tau(x)\Delta)
\end{array}
$$

We call R **non-degenerate** if no row of R is all-zeroes, as that would lead to a zero wirelabel (whose complementary wirelabel would immediately leak Δ). In particular, if $R = I$ then the wirelabels are independent.

Importantly, if GateGb^H is a Linicrypt program and parameters χ and x are fixed, then $\mathsf{View}_R^H(\chi, x)$ is a input-less Linicrypt program. We can therefore apply the results of Sect. 2 to reason about the indistinguishability and unforgeability properties required of View^H. The fact that these properties can be expressed algebraically is the core of our synthesis technique.

We define the following security property for a Free-XOR compatible garbled gate scheme:

Definition 7. *A Free-XOR compatible garbled gate is **secure** if:*

1. *for all χ, $x \in \{0,1\}^m$, all non-degenerate $R \in \{0,1\}^{m \times m}$, and all polynomial-time oracle algorithms A, the probability $\Pr[A^H(\mathsf{View}_R^H(\chi, x)) = \Delta]$ is negligible in λ,*
2. *for all $\chi, x, x' \in \{0,1\}^m$ and all non-degenerate $R \in \{0,1\}^{m \times m}$, we have $\mathsf{View}_R^H(\chi, x) \cong \mathsf{View}_R^H(\chi, x')$.*

In other words, the garbled gate should not leak Δ to the evaluator (this is important for arguing that such garbled gates compose to yield a garbling scheme for circuits), and the garbled gates should hide the truth value. Furthermore, this should hold for all ways that the input wire labels could be correlated.

Composition. We now discuss how (free-XOR-compatible) gate-level garbling procedures can be combined to yield a circuit garbling scheme. The details are given in Fig. 3. Roughly speaking, we follow the general approach of Free-XOR garbling, first choosing a global offset Δ. Recall that for each wire i we associate a wirelabel W_i encoding FALSE; $W_i \oplus \Delta$ will encode TRUE. These false wirelabels are chosen uniformly for input wires. Thereafter, we process gates in topological

---

[2] In fact, some natural garbled gate constructions are secure for independent input wirelabels but insecure when they are correlated, as illustrated strikingly in [9].

order. Each gate-garbling operation determines the garbled-gate information G as well as the FALSE wirelabels of the gate's output wires.

For each wire we choose a random select bit σ_i as described above. For each gate, the garbling scheme must provide a way for the evaluator to learn the correct color bits for the output wires. In many practical schemes, the random oracle calls used to evaluate the gate can serve double-duty and also be made to convey the color bits. However, in our case, we aim for complete generality so our scheme manually encrypts the color bits (the G' values in Fig. 3). In more detail, if the evaluator has color bits χ on the input wires, then she should obtain color bits $\sigma^{(out)} \oplus \tau(\sigma^{(in)} \oplus \chi)$ for the output wires, where $\sigma^{(in)}$ and $\sigma^{(out)}$ are the select bits for the input/output wires of this gate, respectively. We use the wirelabels encoding truth value $\sigma^{(in)} \oplus \chi^{(in)}$ as the key to a one-time encryption that encodes the output color bits.

We point out that these color-ciphertexts are of constant size — 2^m of them, each n bits long (e.g., for a traditional boolean gate with fan-in 2, the cost is 4 bits). As mentioned above, in specific cases it may be possible to eliminate the extra random oracle calls used for these color-bit encryptions.

One subtlety we point out is that each call to a gate-level garbling scheme is restriced to a disjoint set of possible random oracle calls — the gth gate is instructed to use $H(g; \cdot)$ as its random oracle. This domain separation is crucially important in arguing that the gate-level security properties are inherited by the circuit-level garbling scheme.

Lemma 8. *Let $\mathbb{B}$ be a set of boolean functions. Suppose for each $\tau \in \mathbb{B}$, $(\mathsf{GateGb}_\tau, \mathsf{GateEv}_\tau)$ is a correct and secure free-XOR-compatible gate garbling scheme for gate functionality τ (according to Definitions 6 and 7).*

Then the garbling scheme in Fig. 3 satisfies the prv, aut, and obv security definitions of [10] in the random oracle model, for circuits expressed in terms of $\mathbb{B}$-gates.

Proof (Proof Sketch). We sketch here the proof of prv-security; that is, if $f(x) = f(x')$ then (F, X, d) collectively hide whether they were generated with $X = \mathsf{En}(e, x)$ or $X = \mathsf{En}(e, x')$. The proofs of the other security properties obv & aut follow using standard modifications.

We show a sequence of hybrids, beginning with an interaction in which (F, X, d) are generated with $X = \mathsf{En}(e, x)$. In this initial hybrid, Gb is written in terms of what the garbler sees/knows. The only "persistent" values maintained throughout the main loop are the FALSE wirelabels W_i and select bits σ_i. We rearrange Gb to instead be in terms of what the evaluator sees: the "visible" wirelabels W^* and their color bits χ_i. We achieve this change by using x to compute the truth value v_i on each wire i. Then we replace all references to $W_i^{v_i}$ with W_i^*; references to $W_i^{\overline{v_i}}$ with $W_i^* \oplus \Delta$; references to σ_i with $\chi_i \oplus v_i$. The adversary's view in this modified hybrid is unchanged.

After this change, each main loop is a Linicrypt program that takes the previously-computed visible wirelabels, along with Δ, and computes the next garbled gate and output wirelabels (we ignore the encryptions of color bits for now).

$\mathsf{Gb}^H(1^\lambda, f)$:

$\Delta \leftarrow \{0,1\}^\lambda$

for each wire i of f:

$\quad \sigma_i \leftarrow \{0,1\}$

for each input wire i of f:

$\quad W_i \leftarrow \mathbb{F}$

$\quad e[i,0] := (W_i, \sigma_i); \quad e[i,1] := (W_i \oplus \Delta, \overline{\sigma_i})$

for each gate g in f, in topological order:

$\quad$ let g have input wires $i_1, \ldots, i_m$, output wires $j_1, \ldots, j_n$, functionality τ

$\quad \boldsymbol{W}^{(in)} := (W_{i_1}, \ldots, W_{i_m})$

$\quad \sigma^{(in)} := \sigma_{i_1} \| \cdots \| \sigma_{i_m}; \quad \sigma^{(out)} := \sigma_{j_1} \| \cdots \| \sigma_{j_n}$

$\quad (\boldsymbol{W}^{(out)}; \boldsymbol{G}) \leftarrow \mathsf{GateGb}_\tau^{H(g,\cdot)}(\sigma^{(in)}; \boldsymbol{W}^{(in)}, \Delta)$

$\quad (W_{j_1}, \ldots, W_{j_n}) := \boldsymbol{W}^{(out)}$

$\quad$ for χ in $\{0,1\}^m$:

$\quad\quad v := \sigma^{(in)} \oplus \chi$

$\quad\quad G'_\chi := H(\mathtt{color}\|g\|\chi; \boldsymbol{W}^{(in)} \oplus v\Delta) \oplus (\sigma^{(out)} \oplus \tau(v))$

$\quad F[g] := (\boldsymbol{G}; G'_{0^m}, \ldots, G'_{1^m})$

for each output wire i of f:

$\quad d[i,0] := H(\mathtt{out}\|i; W_i); \quad d[i,1] := H(\mathtt{out}\|i; W_i \oplus \Delta)$

return F, e, d

$\mathsf{En}(e,x)$:	$\mathsf{De}(d,Y)$:
for $i = 1$ to $|x|$:	for $i = 1$ to $|Y|$:
$\quad X_i = e[i, x_i]$	$\quad$ if $Y_i = d[i,0]$ then $y_i = 0$
return X	$\quad$ elsif $Y_i = d[i,1]$ then $y_i = 1$
	$\quad$ else return $\perp$
	return y

$\mathsf{Ev}^H(F, X)$:

for each input wire i of f:

$\quad (W_i^*, \chi_i) := X_i$

for each gate g in f, in topological order:

$\quad$ let g have input wires $i_1, \ldots, i_m$, output wires $j_1, \ldots, j_n$, functionality τ

$\quad \chi^{(in)} := \chi_{i_1} \| \cdots \| \chi_{i_m}$

$\quad (\boldsymbol{G}; G'_{0^m}, \ldots, G'_{1^m}) := F[g]$

$\quad (W_{j_1}^*, \ldots, W_{j_n}^*) \leftarrow \mathsf{GateEv}_\tau^{H(g,\cdot)}(\chi^{(in)}; W_{i_1}^*, \ldots, W_{i_m}^*, \boldsymbol{G})$

$\quad \chi_{j_1} \| \cdots \| \chi_{j_n} := H(\mathtt{color}\|g\|\chi^{(in)}; W_{i_1}^*, \ldots, W_{i_m}^*) \oplus G'_{\chi^{(in)}}$

for each output wire i of f:

$\quad Y_i := H(\mathtt{out}\|i; W_i^*)$

return Y

Fig. 3. Gate-level garbling composed into a circuit garbling scheme.

In fact, such a computation is precisely $\mathsf{View}_R(\chi, v)$ defined above, for some appropriate R that describes the correlations among previous input wirelabels.

The security of the GateGb components (Definition 6) says that $\mathsf{View}(\chi; v)$ and $\mathsf{View}(\chi; v')$ are indistinguishable. But this statement only applies when Δ

is a *local variable* to these views, whereas in the garbling scheme Δ is shared among all gates. So first we must argue that this shared state is not a problem. To do this, we prove a general composition lemma which shows that, if several programs *individually* satisfy Definition 6, and they use guaranteed disjoint calls to the random oracle, then their composition also satisfies Definition 6. It is in this composition lemma that we use the fact that the output of each View also hides Δ. We ensure disjointness of oracle queries by using random oracle $H(g; \cdot)$ when garbling gate g.

We use similar reasoning to handle the color bits, since they are not strictly within the scope of Linicrypt (they use distinct oracle calls and do not leak Δ). Collectively the entire output given to the adversary's view hides the truth values v_i which are used to select which View to run. The only other place where the v_i truth values are used is in the computation of the garbled decoding information d. And in this case, v_i are required only for the output wirelabels, which are the same when garbling either x or x'. Hence, we can replace x with x' with negligible effect on the adversary's view, and the proof is complete.

3.2   Synthesis Approach

One of our motivating goals for Linicrypt is the ability to synthesize secure cryptographic constructions. We do precisely that for free-XOR-compatible gate garbling schemes.

We have written a synthesis tool, **Linisynth** which takes as input the desired parameters of a garbled gate construction. These parameters include:

- The gate functionality $\tau : \{0,1\}^m \to \{0,1\}^n$
- The arity of the random oracle arity $\in \mathbb{N}$ (*e.g.*, whether the oracle is called with 1 or 2 field elements, etc.)
- The number of oracle queries made by GateGb and GateEv: $\mathsf{calls_{gb}}, \mathsf{calls_{ev}} \in \mathbb{N}$
- The size (in field elements) of the garbled gate information $\mathsf{size} \in \mathbb{N}$
- Whether adaptive queries to the oracle are allowed adaptive $\in \{0,1\}$ (see below).

Given such parameters, Linisynth constructs an appropriate SMT formula encoding the required security properties, invokes an SMT solver, and finally interprets the witness (if any) as a human-readable garbled gate construction.

High-Level Outline. Gate garbling schemes as defined in Definitions 6 and 7 are meant to be nonlinear in their use of inputs σ and χ. Hence, to synthesize a complete gate-garbling scheme, we must actually synthesize a *collection* of $\mathsf{GateGb}(\sigma; \cdots)$ and $\mathsf{GateEv}(\chi; \cdots)$ — one for each choice of σ and χ — each of which is a *pure* Linicrypt program.

We now describe roughly how the gate-garbling search problem is expressed as an existential SAT/SMT formula. Recall that pure Linicrypt programs can be represented algebraically as an output matrix $\mathcal{M}$ and a set of oracle constraints $\mathcal{C}$. When restricted to Free-XOR compatible garbling, the entries in these matrices are single bits. These bits comprise the existentially quantified variables of our SMT formula.

Not every bit in the oracle constraints $\mathcal{C}$ has to be an unconstrained variable. Specifically, if the Linicrypt program in question has k input variables, then we identify these with the first k base variables. This means that the first oracle query made by the program can be a linear combination *only* of these first k base variables. For the corresponding oracle constraint $\langle t, \mathcal{Q}, a \rangle$, this means that each row of $\mathcal{Q}$ must end in a certain number of zeroes — say, i zeroes. Then we can associate the output of this oracle query with the $(k+1)$th base variable, fixing a to be $[\underbrace{0 \;\cdots\; 0}_{k} \; 1 \; 0 \;\cdots\; 0]$. Then the next oracle query can be a linear combination of only the first $k+1$ variables, and so on. Overall, many of the existential variables comprising the oracle constraints can be fixed in this way. Furthermore, we can seamlessly enforce non-adaptive oracle queries by forcing all constraints $\langle t, \mathcal{Q}, a \rangle$ to have $\mathcal{Q}$ depending only on the input variables, and not on further base variables. This is what is referred to by the adaptive parameter.

We then express the requirements of Definitions 6 and 7 as clauses over the variables that comprise the programs themselves. The formula is satisfiable **if and only if** a secure gate-garbling scheme exists with the given parameters.

Correctness. Correctness (Definition 6) can be expressed in terms of composing $\mathsf{View}_R(\chi, x)$ (which generates input wirelabels along with the garbled gate information) with $\mathsf{GateEv}(\chi, \cdot)$ in a particular way. We can apply the concepts of Sect. 2.4 to reason about their composition.

We make some simplifying observations that lead us to synthesize only "minimal" gate garbling schemes:

- Correctness needs to hold only for independently distributed input wirelabels ($R = I$). In this setting, the wirelabel inputs to GateEv will have full rank.
- We can assume the garbled gate information has full rank. If any linear dependencies existed, then the same dependencies must exist in $\mathsf{GateGb}(\sigma, \cdot)$ for all σ, or else security is trivially violated (malicious evaluator can obtain information about σ by detecting a linear dependency among garbled gate info). Hence the correlations can be removed from *all* $\mathsf{GateGb}(\sigma, \cdot)$ and reconstructed if needed in *all* $\mathsf{GateGb}(\chi, \cdot)$. The result would be a smaller but equivalent & secure scheme.
- The *entire* input to GateEv (garbled gate information and input wirelabels *together*) has full rank. If there is a linear dependency between garbled gate information and input wirelabels, then the same dependency must exist regardless of σ, or else security will be trivially violated. Then again, the dependency could be removed from *all* $\mathsf{GateGb}(\sigma, \cdot)$ and reconstructed by *all* $\mathsf{GateGb}(\chi, \cdot)$, resulting in a smaller scheme.

We therefore consider a composition of $\mathsf{View}_R(\chi, x)$ and $\mathsf{GateEv}(\chi, \cdot)$ in which the input to GateEv is of full rank. This simplifies the task, since it now suffices to find a *basis change* to GateEv that aligns it with the corresponding output of $\mathsf{View}_R(\chi, x)$.

Let $\mathcal{M}_{R,\chi,x}$ denote the output matrix of $\mathsf{View}_R(\chi, x)$. We split this matrix into a top and bottom: $\mathcal{M}_{R,\chi,x}^{top}, \mathcal{M}_{R,\chi,x}^{bot}$, where the top matrix corresponds to the input wirelabels for x along with garbled gate information, while the bottom matrix corresponds to the output wirelabels for the result $\tau(x)$.

Following Sect. 2.4, we seek a basis change B such that $\mathcal{M}_{R,\chi,x}^{top} = [I \mid 0] \times B$, which represents the input base variables of $\mathsf{GateEv}(\chi, \cdot)$. The basis change must also bring all oracle constraints between the two programs into alignment. We assume that every oracle query made by GateEv is also made by GateGb. This is without loss of generality if we assume that GateEv is "minimal", since such oracle queries can be removed with no effect (if not, it is easy to see that correctness or security is violated). Hence, we check that for every oracle constraint in GateEv, the basis change brings one of the constraints of GateGb into agreement.

Having identified the correct basis change, we simply check that the output matrix of GateEv equals the output matrix $\mathcal{M}_{R,\chi,x}^{bot}$ (under the basis change). In other words, the wirelabels that GateEv outputs always coincide with the "correct" wirelabels specified by View_R.

We also must ensure that B is invertible. To do so we simply guess its inverse B^{-1} and check that $B \times B^{-1}$ is the identity matrix. We point out that multiplication of boolean matrices is straight-forward to express in an SMT formula.

Putting it all together, the clause is as follows. Recall that the input $x = \sigma \oplus \chi$, and that we have restricted $R = I$. We use $(\mathcal{M}_{R,\chi,x}, \mathcal{C}_{R,\chi,x})$ to refer to the algebraic representation of $\mathsf{View}_R(\chi, x)$, and use $(\mathcal{M}_{\mathsf{GateEv},\chi}, \mathcal{C}_{\mathsf{GateEv},\chi})$ to denote the algebraic representation of $\mathsf{GateEv}(\chi, \cdot)$.

$$\forall \sigma, \chi \in \{0,1\}^m : \exists B, B^{-1} : B \times B^{-1} = I$$
$$\wedge \left[\forall \langle t, \mathcal{Q}, \boldsymbol{a} \rangle \in \mathcal{C}_{\mathsf{GateEv},\chi} : \langle t, \mathcal{Q} \times B, \boldsymbol{a} \times B \rangle \in \mathcal{C}_{R,\chi,x} \right]$$
$$\wedge \, \mathcal{M}_{\mathsf{GateEv},\chi} \times B = \mathcal{M}_{R,\chi,x}^{bot} \wedge [I \mid 0] \times B = \mathcal{M}_{R,\chi,x}^{top}$$

We point out that the universal quantifiers are over a constant number of terms (2^{2m} choices of (σ, χ) and calls_{ev} constraints) and are explicitly expanded in the formula we pass to the SMT solver. Likewise, the test for $\langle t, \mathcal{Q} \times B, \boldsymbol{a} \times B \rangle \in \mathcal{C}_{R,\chi,x}$ is expressed as a logical-OR of calls_{gb} equality checks.

Security, Condition 1. The first condition of Definition 7 is that $\mathsf{row}(\Delta)$ is unreachable (in the sense of Fig. 2). If the SAT solver could discover the linear subspace $\mathcal{R}$ of reachable vectors, it could simply test whether this subspace includes $\mathsf{row}(\Delta)$. However, to do this iteratively as in Fig. 2 is impractical in a SAT formula, so we employ a trick.

Our idea is to *guess* a basis change B that maps the reachable space to some canonical form that is easily testable by the SAT solver. In particular, consider a basis change B under which the reachable vectors are exactly those that have zero in their rightmost several positions. The SAT formula can easily check for such a condition. To check that our guess for B indeed maps the reachable subspace to the desired canonical form, we observe that the reachable space is *characterized by* the following properties:

- Every row of the output matrix $\mathcal{M}$ is contained in the reachable space
- For every oracle constraint $\langle t, \mathcal{Q}, \boldsymbol{a} \rangle \in \mathcal{C}$, if every row of $\mathcal{Q}$ is in the reachable space, then so is $\boldsymbol{a}$.

For the reachable space after the basis change, the membership condition is simply that the vector ends in the correct number of zeroes.

We note that from the input parameters, we can compute the dimension of the reachable space (and from that derive the required number of trailing zeroes in the vectors) as $d = m + \mathsf{calls}_{\mathsf{ev}} + \mathsf{size}$, where m is the number of inputs, $\mathsf{calls}_{\mathsf{ev}}$ is the number of oracle queries allowed the evaluator, and size is the size of the garbled gate information. This assumes that each oracle query of GateEv increases the dimension of the reachable space — an assumption that is without loss of generality for "minimal" schemes since oracle queries not of this kind are superfluous.

Putting everything together, the formula is as follows. We write $(\mathcal{M}_{R,\chi,x}, \mathcal{C}_{R,\chi,x})$ to denote the algebraic representation of $\mathsf{View}_R(\chi, x)$, which can be obtained in a systematic way from the algebraic representation of $\mathsf{GateGb}(\chi; \cdot)$ (which comprise the existentially quantified variables of the SAT formula). We use $\mathrm{row}(\Delta)$ to refer to the appropriate vector in this representation.

$$\forall \sigma, \chi \in \{0,1\}^m, \textit{non-degenerate } R : \exists B, B^{-1} :$$
$$B \times B^{-1} = I \wedge \neg \mathsf{RightZeroes}(\mathrm{row}(\Delta) \times B) \wedge \mathsf{RightZeroes}(\mathcal{M}_{R,\chi,x} \times B)$$
$$\wedge \big[\forall \langle t, \mathcal{Q}, \boldsymbol{a} \rangle \in \mathcal{C}_{R,\chi,x} : \mathsf{RightZeroes}(\mathcal{Q} \times B) \Rightarrow \mathsf{RightZeroes}(\boldsymbol{a} \times B) \big]$$
$$(2)$$

Here $\mathsf{RightZeroes}$ simply means that the argument vector/matrix has the appropriate number of zeroes in its rightmost columns. The universal quantifiers are over a constant number of terms (2^{2m} choices of (σ, χ), 2^{m^2} choices of R, and $\mathsf{calls}_{\mathsf{gb}}$ constraints) and are explicitly expanded in the formula we pass to the SMT solver.

Security, Condition 2. The second condition of Definition 7 is that $\mathsf{View}_R(\chi, x)$ and $\mathsf{View}_R(\chi, x_0)$ are indistinguishable. Here we fix x_0 and show indistinguishability with respect to this fixed $\mathsf{View}_R(\chi, x_0)$. Since the programs involved are inputless Linicrypt programs, from Theorem 5 it suffices to show that they differ by a basis change after normalization (unreachable and useless oracle queries removed).

We make an assumption that all reachable oracle constraints in $\mathsf{View}_R(\chi, x)$ are in fact useful, and hence we can only synthesize gate-garbling schemes with this property. However, if a secure scheme has reachable and useless constraints in some $\mathsf{View}_R(\chi, x = \chi \oplus \sigma)$, then the same constraint must be also reachable and useless in all $\mathsf{View}_R(\chi, x' = \chi \oplus \sigma')$ by security. Hence it can be removed from every $\mathsf{GateGb}(\sigma; \cdot)$ resulting in an even less expensive yet equivalent and secure gate-garbling scheme.

To show that $\mathsf{View}_R(\chi, x)$ and $\mathsf{View}_R(\chi, x_0)$ are indistinguishable, we therefore only need to find a basis change aligning their output matrices and their *reachable* oracle constraints. Note that from the previous clause, the SAT solver has already obtained a basis B that maps the reachable subspace of $\mathsf{View}_R(\chi, x)$

to a canonical form (vectors ending in some number of zeroes). Hence we can easily check whether a given oracle constraint is reachable. Also note that B is not constrained in how it operates *within* the reachable subspace. Hence we can let this B basis serve double-duty and ask for it to also align the reachable subspace of $\mathsf{View}_R(\chi, x)$ to that of $\mathsf{View}_R(\chi, x_0)$.

In more detail, let $B_{R,\chi,x}$ be the basis matrix that is already quantified corresponding to $\mathsf{View}_R(\chi, x)$ from security condition 1. We want $\mathcal{M}_{R,\chi,x} \times B_{R,\chi,x}$ and $\mathcal{M}_{R,\chi,x_0} \times B_{R,\chi,x_0}$ to coincide, and we want $\mathcal{C}_{R,\chi,x} B_{R,\chi,x}$ and $\mathcal{C}_{R,\chi,x_0} B_{R,\chi,x_0}$ to coincide, but only for reachable constraints. Hence:

$$\mathcal{M}_{R,\chi,x} \times B_{R,\chi,x} = \mathcal{M}_{R,\chi,x_0} \times B_{R,\chi,x_0} \wedge$$

$$\Big[\forall \langle t, \mathcal{Q}, a \rangle \in \mathcal{C}_{R,\chi,x} : \mathsf{RightZeroes}(\mathcal{Q} \times B_{R,\chi,x})$$

$$\Rightarrow \langle t, \mathcal{Q} \times B_{R,\chi,x} \times B_{R,\chi,x_0}^{-1}, a \times B_{R,\chi,x} \times B_{R,\chi,x_0}^{-1} \rangle \in \mathcal{C}_{R,\chi,x_0}\Big]$$

Note that $\langle t, \mathcal{Q} \times B_{R,\chi,x} \times B_{R,\chi,x_0}^{-1}, a \times B_{R,\chi,x} \times B_{R,\chi,x_0}^{-1} \rangle \in \mathcal{C}_{R,\chi,x_0}$ is equivalent to saying $\langle t, \mathcal{Q} B_{R,\chi,x}, a B_{R,\chi,x} \rangle \in \mathcal{C}_{R,\chi,x_0} B_{R,\chi,x_0}$. Hence the bracketed expression captures the requirement that $\mathcal{C}_{R,\chi,x} B_{R,\chi,x}$ and $\mathcal{C}_{R,\chi,x_0} B_{R,\chi,x_0}$ coincide for reachable constraints.

As usual, the quantifications over constraints are expanded within the formula.

3.3   Implementation Results

We implemented Linisynth using Python and the SMT solver Z3[3]. Linisynth extracts the resulting witness and prints it as a human-readable garbling scheme. We used Linisynth to successfully synthesize variants of known gate garbling schemes as well as some of our own creations (*i.e.*, garbled LT gates and garbled EQ gates). Linicrypt can also enumerate constructions that satisfy given parameters. Our code is available at https://github.com/osu-crypto/linisynth.

Linisynth works as follows. For each value in the algebraic representation of GateGb and GateEv, it creates a boolean variable. After it has created all the variables, it makes a formula that constrains them in the following way. For each combination of σ and χ, the invertiblity, correctness, and security conditions from Sect. 3.2 hold (expressed as boolean formulas over the variables). This often results in rather large formulas (see Fig. 4). Linisynth then hands the formula over to Z3. If Z3 finds a solution, it maps the satisfying assignment back to the garbling scheme and prints it.

Synthesis Results. We rediscovered known constructions. For example, our tool was able to discover that XOR gates can be garbled for free. It also rediscovered many garbled AND-gate constructions that are equivalent to the half-gates construction of Zahur et al. [53] (costing 2 ciphertexts). An example of such a garbled AND-gate is given in Fig. 5. We synthesized garbling schemes for a number of different gates (garbled <, garbled =, garbled MUX, etc.), but they all had comparable performance to AND, explained below. A summary is presented in Fig. 4.

---

[3] https://github.com/Z3Prover/z3.

name	τ	size	arity	calls$_{gb}$	calls$_{ev}$	adaptive	vars	p-size	time	sat
free-xor	$\oplus : 2 \to 1$	0	1	0	0	0	224	5,102	1s	1
half-gate	$\wedge : 2 \to 1$	2	1	4	2	0	1,972	117,586	5s	1
half-gate-cheaper	$\wedge : 2 \to 1$	2	1	4	1	1	1,960	92,690	6.2h	0
half-gate-h2	$\wedge : 2 \to 1$	2	2	4	2	0	2,000	114,397	2h	0
one-third-gate	$\wedge : 2 \to 1$	1	1	4	2	1	4,104	716,454	74s	0
1-out-of-2-mux	MUX $: 3 \to 1$	2	1	4	2	1	9,416	654,433	29s	1
2-bit-eq	$= : 4 \to 1$	2	1	4	2	1	44,144	3,497,286	6m	1
2-bit-eq-small	$= : 4 \to 1$	1	1	4	2	1	39,248	3,535,942	6m	0
2-bit-leq	$\leq : 4 \to 1$	1	1	2	1	1	23,296	1,155,686	77s	0
2-bit-lt	$< : 4 \to 1$	2	1	4	2	1	44,144	3,502,425	3.5h	0

Fig. 4. Selection of our synthesis results on an Intel Xeon 3.4 GHz processor with 16 GB memory. Satisfiable schemes are listed in the full version. Notation: "$f : m \to n$" is shorthand for a function with m bits of input and n bits of output that performs the operation f on the input, "vars" and "p-size" refer to the number of variables and nodes in the security & correctness formula. "sat" refers to whether the formula was satisfiable.

---

half-gate   size $= 2$ calls$_{gb} = 4$ adaptive $= 0$

$\wedge : \{0,1\}^2 \to \{0,1\}$ arity $= 1$ calls$_{ev} = 2$ time $=$ 5s

$\mathsf{GateGb}^H(\sigma, A, B, \Delta)$: $\mathsf{GateEv}^H(\chi, A^*, B^*, G_0, G_1)$:

$\quad h_1 = H(A)$ $\quad$ return $[1,3]A^* + [0,2]B^* +$

$\quad h_2 = H(A + \Delta)$ $\qquad\qquad [0,1]G_0 + [1,3]G_1 +$

$\quad h_3 = H(A + B)$ $\qquad\qquad H(A^*) + H(A^* + B^*)$

$\quad h_4 = H(A + B + \Delta)$

$\quad G_0 = [0,2]\Delta + h_3 + h_4$

$\quad G_1 = A + B + [0,2]\Delta + h_1 + h_2 + h_3 + h_4$

$\quad C_0 = B + [0]\Delta + [0,2]h_1 + [1,3]h_2 + [1,2]h_3 + [0,3]h_4$

$\quad$ return G_0, G_1, C_0

Fig. 5. An example of one of our synthesized schemes. This scheme is an alternative to the half-gates AND gate of [53], with identical parameters (number of ciphertexts, and number of calls to H). The notation is as follows: GateGb: When S is a set of indices, "$[S]W$" refers to nonlinear behavior "if $\sigma \in S$ then W else 0^λ" GateEv: When S is a set of indices, "$[S]W$" refers to nonlinear behavior "if $\chi \in S$ then W else 0^λ"

We were not able to synthesize a garbling scheme better than 2 ciphertexts per AND gate. We suspect that this may be a hard limit (if compatibility with free-XOR is required), in support of the half-gates lower-bound presented in [53]. We formalize that hypothesis here. First, note that $\mathbb{B} = \{$ AND, NOT, XOR $\}$ is a universal basis for boolean circuits. Then take any boolean gate τ and decompose it into some combination of AND, NOT, and XOR. Let circ-min$_{\text{AND}}(\tau)$ be the minimum number of AND gates necessary to construct τ with basis $\mathbb{B}$. Our hypothesis is this: for all gates τ, the minimum number of ciphertexts to garble τ with full security and compatibility with free-XOR is $2 \times$ circ-min$_{\text{AND}}(\tau)$. Verification of this hypothesis is left as future work.

Enumeration of Solutions. Linisynth can also *enumerate* schemes. Let p be a formula generated according to Sect. 3.2 and let w be a satisfying assignment with $p(w) = 1$. When Linisynth gets w from the solver, it prints the corresponding scheme, sets $p := \neg w \wedge p$, and asks the solver to find a new solution. Since pysmt provides access to an active instance of Z3, we can use Z3's push/pop functionality to add an assertion without causing the solver to restart. Each new scheme is found in a fraction of the time it takes to find the first one. Using enumeration, we found thousands of schemes equivalent to half-gates (with parameters size = 4, arity = 1, calls$_{gb}$ = 4, calls$_{ev}$ = 2, and adaptive = 0).

Acknowledgement. We thank Viet Tung Hoang for pointing out to us some subtleties that arise when wires have correlated labels.

References

1. Abe, M., Groth, J., Haralambiev, K., Ohkubo, M.: Optimal structure-preserving signatures in asymmetric bilinear groups. In: Rogaway, P. (ed.) CRYPTO 2011. LNCS, vol. 6841, pp. 649–666. Springer, Heidelberg (2011)
2. Abe, M., et al.: Structure-preserving signatures from type II pairings. In: Garay, J.A., Gennaro, R. (eds.) CRYPTO 2014, Part I. LNCS, vol. 8616, pp. 390–407. Springer, Heidelberg (2014)
3. Abe, M., Groth, J., Ohkubo, M., Tibouchi, M.: Unified, minimal and selectively randomizable structure-preserving signatures. In: Lindell, Y. (ed.) TCC 2014. LNCS, vol. 8349, pp. 688–712. Springer, Heidelberg (2014)
4. Akinyele, J.A., Green, M., Hohenberger, S.: Using SMT solvers to automate design tasks for encryption and signature schemes. In: Sadeghi, A.-R., Gligor, V.D., Yung, M. (eds.) ACM CCS 2013, pp. 399–410. ACM Press, November 2013
5. Akinyele, J.A., Green, M., Hohenberger, S., Pagano, M.W.: Machine-generated algorithms, proofs and software for the batch verification of digital signature schemes. In: Yu, T., Danezis, G., Gligor, V.D. (eds.) ACM CCS 2012, pp. 474–487. ACM Press, October 2012
6. Applebaum, B., Avron, J., Brzuska, C.: Arithmetic cryptography: extended abstract. In: Roughgarden, T. (ed.) Proceedings of the Conference on Innovations in Theoretical Computer Science, ITCS 2015, Rehovot, Israel, 11–13 January 2015, pp. 143–151. ACM (2015)
7. Barthe, G., Fagerholm, E., Fiore, D., Scedrov, A., Schmidt, B., Tibouchi, M.: Strongly-optimal structure preserving signatures from type II pairings: synthesis and lower bounds. In: Katz, J. (ed.) PKC 2015. LNCS, vol. 9020, pp. 355–376. Springer, Heidelberg (2015)
8. Beaver, D., Micali, S., Rogaway, P.: The round complexity of secure protocols (extended abstract). In: 22nd ACM STOC, pp. 503–513. ACM Press, May 1990
9. Bellare, M., Hoang, V.T., Keelveedhi, S., Rogaway, P.: Efficient garbling from a fixed-key blockcipher. In: 2013 IEEE Symposium on Security and Privacy, pp. 478–492. IEEE Computer Society Press, May 2013
10. Bellare, M., Hoang, V.T., Rogaway, P.: Foundations of garbled circuits. In:Yu, T., Danezis, G., Gligor, V.D. (eds.) ACM CCS 2012, pp. 784–796. ACMPress, October 2012

11. Bernstein, D.J., et al.: SPHINCS: practical stateless hash-based signatures. In: Oswald, E., Fischlin, M. (eds.) EUROCRYPT 2015. LNCS, vol. 9056, pp. 368–397. Springer, Heidelberg (2015)
12. Black, J.A., Rogaway, P.: A block-cipher mode of operation for parallelizable message authentication. In: Knudsen, L.R. (ed.) EUROCRYPT 2002. LNCS, vol. 2332, pp. 384–397. Springer, Heidelberg (2002)
13. Black, J.A., Rogaway, P., Shrimpton, T.: Black-box analysis of the block-cipher-based hash-function constructions from PGV. In: Yung, M. (ed.) CRYPTO 2002. LNCS, vol. 2442, pp. 320–335. Springer, Heidelberg (2002)
14. Buchmann, J., Dahmen, E., Hülsing, A.: XMSS - a practical forward secure signature scheme based on minimal security assumptions. Cryptology ePrint Archive, Report 2011/484 (2011). http://eprint.iacr.org/2011/484
15. Buchmann, J., Dahmen, E., Klintsevich, E., Okeya, K., Vuillaume, C.: Merkle signatures with virtually unlimited signature capacity. In: Katz, J., Yung, M. (eds.) ACNS 2007. LNCS, vol. 4521, pp. 31–45. Springer, Heidelberg (2007)
16. Buchmann, J., García, L.C.C., Dahmen, E., Döring, M., Klintsevich, E.: CMSS – an improved merkle signature scheme. In: Barua, R., Lange, T. (eds.) INDOCRYPT 2006. LNCS, vol. 4329, pp. 349–363. Springer, Heidelberg (2006)
17. Dodis, Y., Haitner, I., Tentes, A.: On the instantiability of hash-and-sign RSA signatures. In: Cramer, R. (ed.) TCC 2012. LNCS, vol. 7194, pp. 112–132. Springer, Heidelberg (2012)
18. Even, S., Mansour, Y.: A construction of a cipher from a single pseudorandom permutation. In: Matsumoto, T., Imai, H., Rivest, R.L. (eds.) ASIACRYPT 1991. LNCS, vol. 739, pp. 210–224. Springer, Heidelberg (1993)
19. Gagné, M., Lafourcade, P., Lakhnech, Y., Safavi-Naini, R.: Automated security proof for symmetric encryption modes. In: Datta, A. (ed.) ASIAN 2009. LNCS, vol. 5913, pp. 39–53. Springer, Heidelberg (2009)
20. Gagné, M., Lafourcade, P., Lakhnech, Y., Safavi-Naini, R.: Automated verification of block cipher modes of operation, an improved method. In: Garcia-Alfaro, J., Lafourcade, P. (eds.) FPS 2011. LNCS, vol. 6888, pp. 23–31. Springer, Heidelberg (2012)
21. Goldreich, O.: Two remarks concerning the Goldwasser-Micali-Rivest signature scheme. In: Odlyzko, A.M. (ed.) CRYPTO 1986. LNCS, vol. 263, pp. 104–110. Springer, Heidelberg (1987)
22. Gueron, S., Lindell, Y., Nof, A., Pinkas, B.: Fast garbling of circuits under standard assumptions. In: Ray, I., Li, N., Kruegel, C. (eds.) ACM CCS 2015, pp. 567–578. ACM Press, October 2015
23. Halevi, S., Rogaway, P.: A tweakable enciphering mode. In: Boneh, D. (ed.) CRYPTO 2003. LNCS, vol. 2729, pp. 482–499. Springer, Heidelberg (2003)
24. Hoang, V.T., Katz, J., Malozemoff, A.J.: Automated analysis and synthesis of authenticated encryption schemes. In: Ray, I., Li, N., Kruegel, C. (eds.) ACM CCS 2015, pp. 84–95. ACM Press, October 2015
25. Hülsing, A.: W-OTS+ – shorter signatures for hash-based signature schemes. In: Youssef, A., Nitaj, A., Hassanien, A.E. (eds.) AFRICACRYPT 2013. LNCS, vol. 7918, pp. 173–188. Springer, Heidelberg (2013)
26. Impagliazzo, R.: A personal view of average-case complexity. In: Proceedings of the Tenth Annual Structure in Complexity Theory Conference, Minneapolis, Minnesota, USA, 19–22 June 1995, pp. 134–147. IEEE Computer Society (1995)
27. Impagliazzo, R., Rudich, S.: Limits on the provable consequences of one-way permutations. In: Goldwasser, S. (ed.) CRYPTO 1988. LNCS, vol. 403, pp. 8–26. Springer, Heidelberg (1990)

28. Ishai, Y., Prabhakaran, M., Sahai, A.: Secure arithmetic computation with no honest majority. In: Reingold, O. (ed.) TCC 2009. LNCS, vol. 5444, pp. 294–314. Springer, Heidelberg (2009)
29. Kolesnikov, V., Mohassel, P., Rosulek, M.: FleXOR: flexible garbling for XOR gates that beats free-XOR. In: Garay, J.A., Gennaro, R. (eds.) CRYPTO 2014, Part II. LNCS, vol. 8617, pp. 440–457. Springer, Heidelberg (2014)
30. Kolesnikov, V., Schneider, T.: Improved garbled circuit: free XOR gates and applications. In: Aceto, L., Damgård, I., Goldberg, L.A., Halldórsson, M.M., Ingólfsdóttir, A., Walukiewicz, I. (eds.) ICALP 2008, Part II. LNCS, vol. 5126, pp. 486–498. Springer, Heidelberg (2008)
31. Krawczyk, H., Bellare, M., Canetti, R.: HMAC: keyed-hashing for message authentication. In: IETF RFC 2104 (1997). https://www.ietf.org/rfc/rfc2104.txt
32. Krovetz, T., Dai, W.: VMAC: message authentication code using universal hashing. CFRG Working Group (2007). http://www.fastcrypto.org/vmac/draft-krovetz-vmac-01.txt
33. Lamport, L.: Constructing digital signatures from a one-way function. Technical report SRI-CSL-98, SRI International Computer Science Laboratory (1979)
34. Liskov, M., Rivest, R.L., Wagner, D.: Tweakable block ciphers. In: Yung, M. (ed.) CRYPTO 2002. LNCS, vol. 2442, pp. 31–46. Springer, Heidelberg (2002)
35. Luby, M., Rackoff, C.: How to construct pseudo-random permutations from pseudo-random functions. In: Williams, H.C. (ed.) CRYPTO 1985. LNCS, vol. 218, pp. 447–447. Springer, Heidelberg (1986)
36. Malkin, T., Pastro, V., Shelat, A.: An algebraic approach to garbling. Unpublished Manuscript (2016). Presented at Simons Institute workshop on securing computation: https://simons.berkeley.edu/talks/tal-malkin-2015-06-10
37. Malozemoff, A.J., Katz, J., Green, M.D.: Automated analysis and synthesis of block-cipher modes of operation. In: IEEE 27th Computer Security Foundations Symposium, CSF, pp. 140–152. IEEE (2014)
38. Maurer, U.M.: Abstract models of computation in cryptography. In: Smart, N.P. (ed.) Cryptography and Coding 2005. LNCS, vol. 3796, pp. 1–12. Springer, Heidelberg (2005)
39. Merkle, R.C.: A certified digital signature. In: Brassard, G. (ed.) CRYPTO 1989. LNCS, vol. 435, pp. 218–238. Springer, Heidelberg (1990)
40. Naor, D., Shenhav, A., Wool, A.: One-time signatures revisited: have they become practical? Cryptology ePrint Archive, Report 2005/442 (2005). http://eprint.iacr.org/2005/442
41. Naor, M.: Bit commitment using pseudorandomness. J. Cryptol. 4(2), 151–158 (1991)
42. Naor, M., Pinkas, B., Sumner, R.: Privacy preserving auctions and mechanism design. In: Proceedings of the 1st ACM Conference on Electronic Commerce, pp. 129–139. ACM, New York (1999)
43. Papakonstantinou, P.A., Rackoff, C.W., Vahlis, Y.: How powerful are the DDH hard groups? Cryptology ePrint Archive, Report 2012/653 (2012). http://eprint.iacr.org/2012/653
44. Pereira, G.C., Puodzius, C., Barreto, P.S.: Shorter hash-based signatures. J. Syst. Softw. 116, 95–100 (2016)
45. Pieprzyk, J., Wang, H., Xing, C.: Multiple-time signature schemes against adaptive chosen message attacks. In: Matsui, M., Zuccherato, R.J. (eds.) SAC 2003. LNCS, vol. 3006, pp. 88–100. Springer, Heidelberg (2004)

46. Pinkas, B., Schneider, T., Smart, N.P., Williams, S.C.: Secure two-party computation is practical. In: Matsui, M. (ed.) ASIACRYPT 2009. LNCS, vol. 5912, pp. 250–267. Springer, Heidelberg (2009)

47. Preneel, B., Govaerts, R., Vandewalle, J.: Hash functions based on block ciphers: a synthetic approach. In: Stinson, D.R. (ed.) CRYPTO 1993. LNCS, vol. 773, pp. 368–378. Springer, Heidelberg (1994)

48. Reyzin, L., Reyzin, N.: Better than BiBa: short one-time signatures with fast signing and verifying. In: Batten, L.M., Seberry, J. (eds.) ACISP 2002. LNCS, vol. 2384, pp. 144–153. Springer, Heidelberg (2002)

49. Rogaway, P.: Efficient instantiations of tweakable blockciphers and refinements to modes OCB and PMAC. In: Lee, P.J. (ed.) ASIACRYPT 2004. LNCS, vol. 3329, pp. 16–31. Springer, Heidelberg (2004)

50. Rogaway, P., Bellare, M., Black, J., Krovetz, T.: O.C.B: a block-cipher mode of operation for efficient authenticated encryption. In: ACM CCS 2001, pp. 196–205.ACM Press, November 2001

51. Shoup, V.: Lower bounds for discrete logarithms and related problems. In: Fumy, W. (ed.) EUROCRYPT 1997. LNCS, vol. 1233, pp. 256–266. Springer, Heidelberg (1997)

52. Winternitz, R.S.: Producing a one-way hash function from DES. In: Chaum, D. (ed.) CRYPTO 1983, pp. 203–207. Plenum Press, New York (1983)

53. Zahur, S., Rosulek, M., Evans, D.: Two halves make a whole. In: Oswald, E., Fischlin, M. (eds.) EUROCRYPT 2015. LNCS, vol. 9057, pp. 220–250. Springer, Heidelberg (2015)

Zero Knowledge

On the Relationship Between Statistical Zero-Knowledge and Statistical Randomized Encodings

Benny Applebaum and Pavel Raykov[⊠]

School of Electrical Engineering, Tel-Aviv University, Tel Aviv, Israel
{bennyap,pavelraykov}@post.tau.ac.il

Abstract. *Statistical Zero-knowledge proofs* (Goldwasser et al. [GMR89]) allow a computationally unbounded server to convince a computationally limited client that an input x is in a language Π without revealing any additional information about x that the client cannot compute by herself. *Randomized encoding* (RE) of functions (Ishai and Kushilevitz [IK00]) allows a computationally limited client to publish a single (randomized) message, $\mathsf{Enc}(x)$, from which the server learns whether x is in Π and nothing else.

It is known that $\mathcal{SRE}$, the class of problems that admit statistically private randomized encoding with polynomial-time client and computationally unbounded server, is contained in the class $\mathcal{SZK}$ of problems that have statistical zero-knowledge proof. However, the exact relation between these two classes, and, in particular, the possibility of equivalence was left as an open problem.

In this paper, we explore the relationship between $\mathcal{SRE}$ and $\mathcal{SZK}$, and derive the following results:

- In a non-uniform setting, statistical randomized encoding with one-side privacy ($\mathit{1RE}$) is equivalent to non-interactive statistical zero-knowledge ($\mathcal{NISZK}$). These variants were studied in the past as natural relaxation/strengthening of the original notions. Our theorem shows that proving $\mathcal{SRE} = \mathcal{SZK}$ is equivalent to showing that $\mathit{1RE} = \mathcal{SRE}$ and $\mathcal{SZK} = \mathcal{NISZK}$. The latter is a well-known open problem (Goldreich et al. [GSV99]).
- If $\mathcal{SRE}$ is non-trivial (not in $\mathcal{BPP}$), then infinitely-often one-way functions exist. The analog hypothesis for $\mathcal{SZK}$ yields only *auxiliary-input* one-way functions (Ostrovsky [Ost91]), which is believed to be a significantly weaker implication.
- If there exists an average-case hard language with *perfect randomized encoding*, then collision-resistance hash functions (CRH) exist. Again, a similar assumption for $\mathcal{SZK}$ implies only constant-round statistically-hiding commitments, a primitive which seems weaker than CRH.

We believe that our results sharpen the relationship between $\mathcal{SRE}$ and $\mathcal{SZK}$ and illuminates the core differences between these two classes.

© International Association for Cryptologic Research 2016
M. Robshaw and J. Katz (Eds.): CRYPTO 2016, Part III, LNCS 9816, pp. 449–477, 2016.
DOI: 10.1007/978-3-662-53015-3_16

1   Introduction

Consider a "computationally-weak" client, Alice, which holds an input $x \in \{0,1\}^n$ to a language, or promise problem, Π which is beyond her computational power. We will be interested in the following two related scenarios.

– Alice contacts a computationally-strong server Bob, and asks him to prove that x is a yes-instance of Π. The server wishes to do so without revealing any additional information about x that Alice cannot compute by herself. That is, we are interested in an interactive proof system in which, for every yes-instance, the client is able to simulate her view without any interaction with the server.
– Alice would like to send to the server Bob a single (randomized) message $\mathsf{Enc}(x)$ which allows Bob to tell whether x is a yes-instance or a no-instance but hides any other information about x. That is, the message $\mathsf{Enc}(x)$ should be *private* in the sense that all yes-instances (resp., no-instances) are mapped by $\mathsf{Enc}(x)$ to the same universal yes-distribution $\mathsf{Sim}_{\text{YES}}$ (resp., no-distribution Sim_{NO}); In addition, $\mathsf{Enc}(x)$ should be *correct* (i.e., it should be possible to decode membership in Π) and so the yes-distribution is required to be statistically-far from the no-distribution.

The first setting is captured by the notion of *zero-knowledge* (ZK) proofs introduced in [GMR89], while the second is captured by the notion of *randomized encoding* (RE) of functions [IK00, AIK04]. In this paper, we model the client as a polynomial-time machine, the server as a computationally-unbounded party, and ask for information-theoretic security.[1] Problems that admit such a statistical zero-knowledge proofs (resp., such statistical randomized encodings) give rise to the complexity class $\mathcal{SZK}$ (resp., $\mathcal{SRE}$).

The class $\mathcal{SZK}$ and its variants were extensively studied and we have relatively rich insights about its power and structure including non-trivial upper-bounds (e.g., $\mathcal{SZK} \subseteq \mathcal{AM} \cap \text{co-}\mathcal{AM}$ [AH87]), complete problems [SV03, GV99], and closure properties [Oka00, Vad99]. Unfortunately, the status of $\mathcal{SRE}$ is very different. Although randomized encoding are extensively used in cryptography (see the surveys [App11, Ish13]), the class $\mathcal{SRE}$ was left relatively unexplored. The main known result (observed in [App14]) is that

$$\mathcal{SRE} \subseteq \mathcal{SZK}.$$

That is, a statistical randomized encoding for a problem Π can be transformed into a statistical zero knowledge proof system for the same problem. The exact relation between $\mathcal{SRE}$ and $\mathcal{SZK}$, and, in particular, the intriguing possibility that these two classes are actually equivalent was left as an open problem. This question was recently addressed by Agrawal et al. [AIKP15] who provided an

---

[1] The literature contains many other natural choices for security (e.g., computational [AIK05]) and efficiency (e.g., client with low parallel complexity and polynomial-time server [AIK04]). Following Agrawal et al. [AIKP15], we view the current choice as a natural starting point for a complexity-theoretic treatment.

oracle separation between the two classes, in addition to candidates for problems in $\mathcal{SRE}$ that are not solvable in (non-uniform) polynomial-time. As usual, an oracle separation tells us that equivalence cannot be established via relativized techniques, and so it essentially addresses the *proof of equivalence* (or technical barriers against it). However, such separations tell us very little on the statement itself ($\mathcal{SRE} = \mathcal{SZK}$) and its potential implications on the landscape of computational complexity.[2]

1.1   Our Results

In this paper, we continue the complexity theoretic study of $\mathcal{SRE}$, as advocated by [AIKP15], and further explore the exact relationship between $\mathcal{SRE}$ and $\mathcal{SZK}$. We study variants of these classes, prove their equivalence, and sharpen the difference between $\mathcal{SRE}$ and $\mathcal{SZK}$. We also point out several interesting complexity-theoretic implications of an equivalence between $\mathcal{SRE}$ and $\mathcal{SZK}$. Overall, we believe that our results shed light on the causes for which $\mathcal{SZK}$ is (seemingly) more powerful than $\mathcal{SRE}$.

Non-interactive ZK is Equivalent to Semi-private RE. Zero-knowledge proofs differ from randomized-encoding in many aspects. Most notably, the flow of information is reversed (Server-to-Client for ZK-proofs vs. Client-to-Sever for encodings). Let us ignore this major difference and focus on two seemingly less important syntactic differences. First, recall that REs are non-interactive while zero-knowledge proofs are allowed to use interaction. Secondly, the privacy condition of REs should hold for both yes and no-instances, whereas the ZK condition is defined only with respect to yes-instances. In an attempt to make a "fair" comparison between these two notions, we consider *non-interactive* zero-knowledge proofs (NISZK) [BFM88] and statistical randomized encoding with *one-sided privacy* (1RE) [AIK04, AIK15].

The NISZK model, introduced by Blum et al. [BFM88], restricts the prover to send a single message to the verifier at the expense of allowing the parties to share a common reference string that was pre-sampled by a trusted (efficient) dealer.[3] The notion of statistical randomized encoding with one-sided privacy was introduced by Applebaum et al. [AIK04] (under the term semi-private encoding) as a relaxation of REs in which the privacy condition should hold only for yes-instances.

---

[2] Moreover, there are examples for classes which are separated relative to some oracle, but, without an oracle, are actually equal. (E.g., $\mathcal{IP}$ vs. $\mathcal{PSPACE}$; see the discussion in [CCG+94]).

[3] Our description corresponds to the *public-parameter model*, which is widely used in the literature (see [PS05] and references therein). This setting generalizes the original *common random string* (crs) model proposed by Blum et al. [BFM88], in which the reference string is simply a uniformly random string of polynomial length. Following [CCKV08], we use the superscripts PUB and CRS to distinguish between these two variants. Observe that $\mathcal{NISZK}^{\text{CRS}} \subseteq \mathcal{NISZK}^{\text{PUB}}$.

We show that the corresponding complexity classes $\mathcal{NISZK}^{\text{PUB}}$ and $1\mathcal{RE}$ are essentially equivalent.

Theorem 1. *It holds that* $\mathcal{NISZK}^{\text{PUB}} \subseteq 1\mathcal{RE}$ *and, in the non-uniform setting,* $1\mathcal{RE} \subseteq \mathcal{NISZK}^{\text{PUB}}$.

The "non-uniform" setting refers to the case where all efficient entities (the client, the dealer, and the RE/SZK simulators) are modeled by polynomial-size circuits. The theorem shows that, non-uniformly, the class $\mathcal{NISZK}^{\text{PUB}}$ is *equivalent* to the class $1\mathcal{RE}$. It is known that $\mathcal{NISZK}^{\text{PUB}} \subseteq \mathcal{SZK}$ [PS05] and, by definition, we have that $\mathcal{SRE} \subseteq 1\mathcal{RE}$. Hence, together with Theorem 1, we derive the following interesting picture (in the non-uniform setting):

$$\mathcal{SRE} \subseteq 1\mathcal{RE} = \mathcal{NISZK}^{\text{PUB}} \subseteq \mathcal{SZK}.$$

Note that if $\mathcal{SZK}$ collapses to $\mathcal{SRE}$ then all intermediate classes also collapse. This means that the question of putting $\mathcal{SZK}$ inside $\mathcal{SRE}$ boils down to two separate questions: "Can statistical zero-knowledge be made non-interactive?" ($\mathcal{NISZK}^{\text{PUB}} = \mathcal{SZK}$?) and "Can one-side privacy be upgraded to full privacy?" ($\mathcal{SRE} = 1\mathcal{RE}$?). Nicely, each of these well motivated questions is "pure" in the sense that it only addresses one object (either randomized encoding or zero-knowledge proofs). We further mention that the first question ($\mathcal{NISZK} = \mathcal{SZK}$?) is a well-known open problem that was studied before by [GSV99].[4]

Consequences of Randomized Encoding for Intractable Problems. Another way to compare $\mathcal{SZK}$ to $\mathcal{SRE}$ is by asking what are the consequences of the existence of computationally-intractable problems in the class. For example, the following theorem was proven by Ostrovsky.

Theorem 2 [Ost91]. *If* $\mathcal{SZK}$ *is not in* $\mathcal{BPP}$, *then Auxiliary-Input One-way functions exist.*[5]

Auxiliary-input one-way functions (ai-OWF) are keyed functions that achieve a very weak form of one-wayness. Roughly speaking, for each adversary there exists a set of hard keys on which the adversary fails to invert the function. (See [Gol01] for definition.) However, it may be the case that there is no universal set of keys which is simultaneously hard for all efficient adversaries.

For $\mathcal{SRE}$ we prove (Sect. 6) the following stronger implication:

Theorem 3. *If* $\mathcal{SRE}$ *is not in* $\mathcal{BPP}$, *then infinitely-often one-way functions exist.*

---

[4] More precisely, [GSV99] focused on the CRS model, and provided several necessary and sufficient conditions for the equality $\mathcal{NISZK}^{\text{CRS}} = \mathcal{SZK}$.

[5] This theorem, and all the other results in this section, is formulated in the uniform setting. If one considers a non-uniform variant of $\mathcal{SZK}$, then the theorem holds by changing $\mathcal{BPP}$ to $\mathcal{P}/\text{poly}$ and by relaxing the notion of AIOWFs to be computable by polynomial-size circuits. Similar modifications can be applied to the other theorems of this section.

Infinitely-often one-way functions (io-OWFs) are essentially standard one-way functions except that their hardness holds over a (universal) set of infinitely many input lengths. This notion is considered to be significantly stronger than ai-OWFs. For example, while it is possible to construct ai-OWFs based on the worst-case hardness of graph-isomorphism (GI), it is unknown how to obtain io-OWF from such an assumption. By Theorem 3, such a GI-based io-OWF would follow from the equivalence of $\mathcal{SZK}$ and $\mathcal{SRE}$. More generally, a proof of such an equivalence would allow us to base io-OWFs on worst-case hardness in $\mathcal{SZK}$ improving the 25-year old classical result of [Ost91].

Theorem 3 also explains why all the candidates of Agrawal et al. [AIKP15] for computationally-hard problems in $\mathcal{SRE}$ imply the existence of one-way functions – Such an assumption is inherently necessary to separate $\mathcal{SRE}$ from $\mathcal{BPP}$.

We can further ask what are the implications of an average-case hard problem in these complexity classes. Roughly speaking, a promise problem Π is average-case hard if it is equipped with a probability distribution D such that no efficient algorithm can classify correctly an instance x sampled from D with probability significantly better than $1/2$. Ostrovsky's result can be used to prove that the existence of an average-case hard language in $\mathcal{SZK}$ implies the existence of a one-way function. The following (stronger) theorem is implicit in the work of Ong and Vadhan [OV08].

Theorem 4 (implicit in [OV08]). *If there exists an average-case hard language in $\mathcal{SZK}$ then a constant-round statistically-hiding commitments (CRSC) exists.*

As a general primitive, CRCS implies the existence of one-way functions, and is believed to be strictly stronger due to the black-box separation of [HHRS15]. We derive a stronger implication if we have randomized encoding for an average-case hard problem. Specifically, we consider the class $\mathcal{PRE}$ of problems that admit *perfect randomized encoding* [AIK04] – a stronger variant of $\mathcal{SRE}$ which achieves perfect correctness (zero-decoding error), perfect privacy (the simulators perfectly simulate the encoding) and enjoys some additional syntactic properties. (See Sect. 4 for a formal definition.)

Theorem 5. *If there exists an average-case hard language in $\mathcal{PRE}$ then collision-resistance hash functions (CRH) exist.*

The proof of the theorem is sketched in Sect. 7. CRH imply CRSC but the converse is not known to be true. Hence, this implication is seemingly stronger than the one proven in [OV08]. Extending this theorem to the case of $\mathcal{SRE}$ is left as an interesting open problem.

2   Our Techniques

Let us outline the main ideas behind the proofs of Theorems 1, 3 and 5.

Proof of Theorem 1. We begin with the equivalence of $1\mathcal{RE}$ and $\mathcal{NISZK}^{\text{PUB}}$. It is instructive to note that all the complexity classes $\mathcal{SZK}, \mathcal{NISZK}, 1\mathcal{RE}$

and $\mathcal{SRE}$ essentially capture different variants of "statistical-distance" problems. Indeed, as we already saw, for a $\mathcal{SRE}$-problem Π, the membership of x boils down to determining whether the distribution $\mathsf{Enc}(x)$ is close to one of two distributions $\mathsf{Sim}_{\mathrm{YES}}$ and $\mathsf{Sim}_{\mathrm{NO}}$ which are statistically-far apart from each other. Notably, these distributions are *universal* and they depend only on the problem Π (and not on the input x). The work of [SV03] also shows that, for any $\mathcal{SZK}$-problem Π, there exists an efficient mapping from an instance x to a pair of distributions (A_x, B_x) which are statistically-close if x is a yes-instance and statistically-far otherwise. However, in contrast to the case of SREs, the distributions (A_x, B_x) are *instance dependent*. In particular, two different yes-instances x and x' may be mapped to completely different pairs of distributions (A_x, B_x) and $(A_{x'}, B_{x'})$.

In the intermediate notion of $\mathcal{NISZK}$, one of the distributions, say B, corresponds to the dealer's distribution and so it becomes universal [SCPY98, GSV99].[6] Correspondingly, all yes-instances x are mapped to this single universal distribution, i.e., $A_x \approx B$. (A_x essentially corresponds to the simulated version of the public-parameter). For no-instances, the distribution A_x may be instance-dependent. Similarly, for $\mathcal{1RE}$, only yes-instances are mapped by $\mathsf{Enc}(x)$ to some universal yes-distribution $\mathsf{Sim}_{\mathrm{YES}}$, whereas the encoding of a no-instance $\mathsf{Enc}(x)$ may be instance-dependent. Overall, the privacy properties of $\mathcal{1RE}$ and the zero-knowledge properties of $\mathcal{NISZK}$ match nicely. Still, there is one technical difference with respect to the requirements on the distributions of no-instances.

In $\mathcal{1RE}$, correctness requires the existence of a single decoder that distinguishes between the yes-distribution $\mathsf{Sim}_{\mathrm{YES}}$ and *all* possible no-distributions $\{\mathsf{Enc}(x)\}_{x \in \Pi_{\mathrm{NO}}}$. This means that $\mathsf{Sim}_{\mathrm{YES}}$ is "universally-far" from all the no-distributions. In contrast, the soundness property of $\mathcal{NISZK}$ requires from every no-distribution A_x to be "disjoint" from B in the following sense: A random sample from the universal distribution $b \xleftarrow{R} B$ should fall, with high probability, outside the support of A_x. To prove Theorem 1 we should be able to move from "universal-farness" to "disjointness" and vice versa. While it is relatively straightforward to convert disjointness to universal-farness (e.g., via parallel-repetition), the converse direction requires some work.

As a concrete (and somewhat simplified) example, imagine the case where we have a single pair of distributions X and Y, where X outputs, with probability $1 - \varepsilon$, a random n-bit string whose first bit is 1, and, with probability ε, a random n-bit string whose first bit is 0. Assume that Y does exactly the opposite. These distributions are $(1 - 2\varepsilon)$-far in statistical distance, but they do not satisfy the disjointness property as their supports are equal. The key observation is to note that a typical $y \xleftarrow{R} Y$ value, has much larger weight under Y compared to its weight under X. When these distributions are implemented by circuits that use m random bits as inputs, this means that the set of preimages $Y^{-1}(y)$ is likely to be significantly larger than the set $X^{-1}(y)$. In other words, the entropy e_1

---

[6] Interestingly, in the CRS model, this distribution is simply the uniform distribution and it is therefore also *problem-independent*.

of the conditional distribution $[r|Y(r) = y]$ is larger than the entropy e_2 of the conditional distribution $[r|X(r) = y]$. Following the approach of [GSV99], we can turn these distributions to be disjoint by hashing out about $e_1 \ll e \ll e_2$ random bits from r, and appending the result $h(r)$ to the output. That is, we define a pair of new distributions by $X' = (X(r), h, h(r))$ and $Y' = (Y(r), h, h(r))$ where h is sampled from a 2-universal family of hash functions.[7] One can now show that for a typical $y \xleftarrow{R} Y$ (and most h's), the conditional distribution $[h(r)|Y(r) = y]$ is almost uniform, whereas the conditional distribution $[h(r)|X(y) = y]$ has small support. This means that a random sample from Y' is likely to land out of the support of X', as required.

The actual construction introduces some additional technicalities. Most notably, it requires an estimation on the amount of entropy of the distribution which is sampled by $\mathsf{Sim}_{\mathrm{YES}}$, the simulator of the original encoding. We overcome this problem by treating this value as a non-uniform advice. We note that this advice is short (of logarithmic length) and so one may hope to simply try all possible values. The problem is that some of these values will violate the zero-knowledge property, while others would violate soundness. Unfortunately, we do not know how to "combine" together several faulty NISZK protocol into a single good protocol. The question of finding a way around this problem and achieving a fully uniform reduction is left for future research.

Proof of Theorem 3. Recall that Theorem 3 asserts that if infinitely-often one-way functions do not exist, then any language Π in $\mathcal{SRE}$ can be decided by some $\mathcal{BPP}$ algorithm A. The proof is based on the following observation: Given an instance x, one can probabilistically decide if $x \in \Pi$ by first sampling an encoding $y = \mathsf{Enc}(x)$, and then outputting "yes" if the weight of y under the distribution $\mathsf{Sim}_{\mathrm{YES}}$ is larger than its weight under $\mathsf{Sim}_{\mathrm{NO}}$. Note that the latter problem can be reduced to the following "distributional inversion" problem. Define the function

$$g(r, b) = \begin{cases} \mathsf{Sim}_{\mathrm{NO}}(r), & \text{if } b = 0, \\ \mathsf{Sim}_{\mathrm{YES}}(r), & \text{if } b = 1; \end{cases}$$

sample a random preimage (r, b) of y under g, and output the bit b. (I.e., when $b = 0$ the instance x is classified as a no-instance, and if $b = 1$ then x is classified as a yes-instance.) It can be shown, based on the privacy and the correctness guaranties of the encoding, that b is likely to classify x correctly. By the results of Impagliazzo and Luby [IL89], the distributional inversion problem can be efficiecntly solved (up to small, inverse-polynomial, deviation error), assuming that infinitely-often one-way functions do not exist.

It is instructive to compare the above to the SZK setting. The RE simulators give rise to a universal function g (independent of the instance x) whose inversion is as hard as deciding Π. In contrast, in the SZK setting, the corresponding distributions depend on x, and so deciding $x \in \Pi$ reduces to inverting an

---

[7] More generally, we could use any seeded randomness extractor that extracts e almost uniform bits from any e_2-bit source.

instance-dependent function g_x. Correspondingly, the intractability of Π yields only auxiliary-input one-way functions.

Proof of Theorem 5. In Theorem 5 we show that if an average-case hard language Π admits a prefect RE then CRH exist. The notion of perfect encoding guarantees that the image of the encoder Enc can be partitioned into two equal sets Y and N and that for any yes-instance (resp., no-instance) x, the mapping $\text{Enc}(x; r)$ is a bijection from the randomness space to Y (resp., N). Similarly both simulators, $\text{Sim}_{\text{YES}}(r)$ and $\text{Sim}_{\text{NO}}(r)$, form a bijective mapping from the randomness space to Y and N, respectively. Let us define a pair of functions, keyed by instances x, y,

$$h_x^0(r, b) = \begin{cases} g(x; r), & \text{if } b = 0, \\ \text{Sim}_{\text{NO}}(r), & \text{otherwise;} \end{cases} \qquad h_y^1(r, b) = \begin{cases} g(y; r), & \text{if } b = 0, \\ \text{Sim}_{\text{YES}}(r), & \text{otherwise;} \end{cases}$$

Since the encoding is perfect, h_x^0 and h_y^1 are permutations if x is a yes-instance and y is a no-instance; on the other hand, if x is a no-instance and y is a yes-instance the images of the functions are disjoint. Suppose that there exists an efficiently samplable distribution $\mathcal{Y}$ over yes-instances which is indistinguishable from some efficiently samplable distribution $\mathcal{N}$ over no-instances. Then, we can sample a pair of yes/no instances $(x, y) \xleftarrow{R} \mathcal{Y} \times \mathcal{N}$ which is indistinguishable from a pair of no/yes instances $(x', y') \xleftarrow{R} \mathcal{N} \times \mathcal{Y}$. This means that, although the functions h_x^0, h_y^1 are permutations with identical images, it is computationally hard to find a pair (u, v) which forms a "claw", i.e., $h_x^0(u) = h_y^1(u)$. (Indeed, a claw-finder can be used to distinguish (x, y) from (x', y').) Such *claw-free permutations* [Dam87, GMR88] imply the existence of CRH. The argument extends to the case where there exists only a single "hard" distribution over yes/no instances of Π (as opposed to a pair of "pure" distributions). In this case, we get claw-free *pseudo-permutations* [Rus95], whose existence still implies CRH.

2.1 A Broader Perspective

So far we emphasized the differences between $\mathcal{SRE}$ and $\mathcal{SZK}$, however, from a broader point of view, our results may be interpreted as saying that the two classes are actually close variants of each other. This is similar in spirit to a recent result [AR16] that reveals a close connection between *private simultaneous message protocols* (PSM) [FKN94] and *Zero-Information Arthur-Merlin* (ZAM) protocols [GPW15]. PSMs and ZAMs can be viewed as the communication-complexity analog of Randomized Encodings and Zero-Knowledge proofs, where instead of limiting the computational power of the client, we split it into two non-communicating (computationally-unbounded) parties Alice and Bob each holding different parts of the input $x = (x_A, x_B)$. It is shown in [AR16] that the communication complexity of ZAM protocols is closely related to the randomness complexity of (variants of) PSMs, and vice versa. This is conceptually similar to some of the current results (e.g., $1\mathcal{RE} = \mathcal{NISZK}^{\text{PUB}}$) though the computational setting introduces different technical challenges, and correspondingly it requires a significantly different approach.

Organization. We begin with some standard preliminaries in Sect. 3. In Sect. 4 we provide formal definitions of statistical zero knowledge proofs, statistical randomized encoding and their variants. Theorem 1 is proved in Sect. 5, Theorem 3 in Sect. 6 and Theorem 5 in Sect. 7.

3   Preliminaries

Basic Definitions. For a finite set S, let $s \xleftarrow{R} S$ denote an element that is sampled uniformly at random from S, and let $U(S)$ denote the corresponding random variable. The uniform distribution over n-bit strings is denoted by U_n. The *support* of a random variable X is the set $\text{supp}(X) := \{x \mid \Pr[X = x] > 0\}$. The Shannon *entropy* of X is $H(X) := -\sum_z \Pr[X = z] \log \Pr[X = z]$. For a distribution D, we let $\otimes^k D$ be the probability distribution over k-tuples where each element is sampled independently according to D. Similarly, for a randomized algorithm $F(x)$, we let $\otimes^k F(x)$ be a k-tuple of k independent samples of $F(x)$. We sometimes make the coins of a randomized algorithm F explicit by writing $F(x; r)$ where $r \xleftarrow{R} U_{s(x)}$ denotes the random coins used on an input x and $s(x)$ denotes the randomness complexity of F on an input x, which, by default, is assumed to solely depend on the length of x.

Statistical Distance. The *statistical distance* between a pair of random variables X and Y distributed over the set Z is defined as

$$\Delta(X; Y) := \frac{1}{2} \sum_{z \in Z} |\Pr[X = z] - \Pr[Y = z]|.$$

Equivalently, $\Delta(X; Y) = \max_A |\Pr[A(X) = 1] - \Pr[A(Y) = 1]|$ where the maximum ranges over all Boolean functions $A : Z \to \{0, 1\}$. We write

$$\Delta_{x_1 \xleftarrow{R} D_1, \ldots, x_k \xleftarrow{R} D_k} (F(x_1, \ldots, x_k); G(x_1, \ldots, x_k))$$

to denote the statistical distance between two random variables obtained as a result of sampling x_i's from D_i's and applying the functions F and G to $(x_1, \ldots, x_k)$, respectively. We will use the following properties of statistical distance and entropy.

Fact 1. *Let X and Y be a pair of random variables. Then the following holds:*

1. [Vad99, Fact 3.2.2] *For every (possibly randomized) function F, we have that* $\Delta(F(X); F(Y)) \leq \Delta(X; Y)$.
2. [Vad99, Fact 3.3.9] *Let D be the range of X and Y, then $|H(X) - H(Y)| \leq (\log |D|) \cdot \Delta(X; Y) + 1$.*
3. [Vad99, Lemma 3.1.15] *For any integer $q > 0$, we have that* $1 - 2\exp(-q(\Delta(X; Y))^2/2) \leq \Delta(\otimes^q X; \otimes^q Y) \leq q\Delta(X; Y)$.

4. [SV03, Fact 2.5] *Suppose that $X = (X_1, X_2)$ and $Y = (Y_1, Y_2)$ are distributed over a set $D \times E$ such that: (a) X_1 and Y_1 are identically distributed; and (b) with probability greater than $1 - \varepsilon$ over $x \xleftarrow{R} X_1$, we have $\Delta(X_2|_{X_1 = x}, Y_2|_{Y_1 = x}) \leq \delta$. Then $\Delta(X, Y) \leq \varepsilon + \delta$.*

5. (cf. Appendix A.1) *If $\Delta(X; Y) \geq 1 - \varepsilon$, then, for any $t > 1$, it holds that $\Pr_{x \xleftarrow{R} X}[\Pr[X = x] < t \cdot \Pr[Y = x]] \leq \varepsilon t$.*

Flattening. We will use the following notion of Δ-flat distributions from [GSV99].

Definition 1 (Flat Distributions). *Let X be a distribution. An element x of $\mathrm{supp}(X)$ is called ε-typical if $|\log(1/\Pr[X = x]) - H(X)| \leq \varepsilon$. We say that X is Δ-flat if for every $t > 0$ the probability that an element chosen from X is $(t \cdot \Delta)$-typical is at least $1 - 2^{-t^2 + 1}$.*

A 0-flat distribution is uniform on its support, and is simply referred to as a *flat* distribution. A natural way to flatten a distribution is via parallel repetition.

Lemma 1 (Flattening Lemma [Vad99, GSV99]). *Let D be a distribution such that for all x from $\mathrm{supp}(D)$ we have that $D(x) \geq 2^{-m}$. Then, for any $k \in \mathbb{N}$, the distribution $\otimes^k D$ is $(\sqrt{k} \cdot m)$-flat.*

Hashing. A family $\mathcal{H}$ of functions mapping a domain $\mathcal{D}$ to a range $\mathcal{R}$ is 2-*universal* [CW79] if for every two elements $x \neq y$ from $\mathcal{D}$ and a, b from $\mathcal{R}$ it holds that $\Pr_{h \xleftarrow{R} \mathcal{H}}[h(x) = a \wedge h(y) = b] = \frac{1}{|\mathcal{R}|^2}$. We write $\mathcal{H}_{n,m}$ to denote a 2-universal family from $\{0,1\}^n$ to $\{0,1\}^m$. There are efficient constructions of 2-universal families of hash functions $\mathcal{H}_{n,m}$ that can be evaluated and sampled in $\mathrm{poly}(n, m)$ time [CW79].

Lemma 2 (Leftover Hash Lemma [ILL89, GSV99]). *Let $\mathcal{H}$ be a 2-universal family of hash functions mapping a domain $\mathcal{D}$ to a range $\mathcal{R}$. Let X be a flat distribution on $\mathcal{D}$ such that for all $x \in \mathrm{supp}(X)$ we have that $\Pr[X = x] \leq \alpha/|\mathcal{R}|$. Then*

$$\Delta_{h \xleftarrow{R} \mathcal{H}} ((h, h(X)); (h, U(\mathcal{R}))) \leq O(\alpha^{1/3}).$$

Sampling Distributions via Circuits. Let X be a circuit with m input and n output gates. We will sometimes abuse notation and use X to denote the random variable $X(U_m)$ which corresponds to the output distribution of the circuit induced by "feeding" a uniformly chosen n-bit input. We let $X^{-1}(x)$ denote the set of preimages of x under X, i.e., $X^{-1}(x) := \{r \in \{0,1\}^m \mid X(r) = x\}$. Observe that $\Pr[X = x] = 2^{-m} \cdot |X^{-1}(x)|$.

4 $\mathcal{NISZK}$ and $\mathcal{SRE}$

A *promise problem* [ESY84] Π is a pair of two non-intersecting sets of strings $(\Pi_{\mathrm{YES}}, \Pi_{\mathrm{NO}})$. The strings in Π_{YES} are called *yes-instances* and the strings in Π_{NO} are called *no-instances*. Let $\chi_\Pi(x)$ be the characteristic function of Π which outputs 1 on yes-instances and 0 on no-instances. Note that a promise problem is a generalization of a language $L \subseteq \{0,1\}^*$, i.e., L is translated into a promise problem Π_L where L corresponds to the set of yes-instances and $\{0,1\}^* \setminus L$ corresponds to the set of no-instances. (See [Gol06] for a survey.)

Definition 2 (Statistical Randomized Encoding [IK00, AIK04]). *We say that an efficient randomized algorithm* Enc *is a ε-private and δ-correct statistical randomized encoding of a promise problem $\Pi = (\Pi_{\mathrm{YES}}, \Pi_{\mathrm{NO}})$ (abbreviated (ε, δ)-SRE), if the following holds:*

ε-PRIVACY FOR YES-INSTANCES: *There exists an efficient simulator* $\mathsf{Sim}_{\mathrm{YES}}$ *such that for every yes-instance x_{YES} of length n from Π,*

$$\Delta(\mathsf{Sim}_{\mathrm{YES}}(1^n); \mathsf{Enc}(x_{\mathrm{YES}})) \le \varepsilon(n).$$

ε-PRIVACY FOR NO-INSTANCES: *There exists an efficient simulator* $\mathsf{Sim}_{\mathrm{NO}}$, *such that for every no-instance x_{NO} of length n from Π,*

$$\Delta(\mathsf{Sim}_{\mathrm{NO}}(1^n); \mathsf{Enc}(x_{\mathrm{NO}})) \le \varepsilon(n).$$

δ-CORRECTNESS: *There exists a computationally-unbounded decoder* Dec, *such that for every instance $x \in (\Pi_{\mathrm{YES}} \cup \Pi_{\mathrm{NO}})$ of length n,*

$$\Pr[\mathsf{Dec}(\mathsf{Enc}(x)) \ne \chi_\Pi(x)] \le \delta(n).$$

By default, $\varepsilon(n)$ and $\delta(n)$ are required to be negligible functions.

Perfect Encoding [AIK04]. A randomized encoding which is 0-private (resp., 0-correct) is called *perfectly* private (resp., *perfectly correct*). For an input of length n, let $s(n)$ denote the length of the random strings used by Enc and let $t(n)$ be the output length of the encoding. A perfectly private and perfectly correct randomized encoding whose simulators $\mathsf{Sim}_{\mathrm{YES}}$ and $\mathsf{Sim}_{\mathrm{NO}}$ use $s(n)$ coins, $\mathrm{supp}(\mathsf{Sim}_{\mathrm{YES}}(1^n)) \cup \mathrm{supp}(\mathsf{Sim}_{\mathrm{NO}}(1^n)) = \{0,1\}^{t(n)}$, and $1 + s(n) = t(n)$ is called *perfect.* (See [AIK04] for an intuitive explanation of these requirements.)

One-Sided Encoding [AIK04, AIK15]. A randomized encoding which is ε-private on yes-instances and δ-correct is called *one-sided* (or *semi-private*) randomized encoding (denoted with (ε, δ)-1RE) [AIK04, AIK15]. Clearly, any (ε, δ)-SRE is also (ε, δ)-1RE, though the converse does not necessarily hold. A *disjoint one-sided* randomized encoding is an encoding which is ε-private on yes-instances and, instead of standard correctness, it satisfies the following ρ-*disjointness* property: For every no-instance x_{NO} of length n from Π, it holds that $\Pr[\mathsf{Sim}_{\mathrm{YES}}(1^n) \in \mathrm{supp}(\mathsf{Enc}(x_{\mathrm{NO}}))] \le \rho(n)$. We refer to such an encoding as (ε, ρ)-D1RE.

Definition 3 (Non-interactive Statistical Zero-Knowledge [BSMP91]). *A non-interactive statistical zero-knowledge proof system (NISZK) for a promise problem $\Pi = (\Pi_{\mathrm{YES}}, \Pi_{\mathrm{NO}})$ is defined by probabilistic algorithms* Prov *(prover),* Deal *(dealer),* Sim *(simulator), and a deterministic algorithm* Ver *(verifier), such that for every n-bit instance x the following holds*

α-COMPLETENESS: *If* $x \in \Pi_{\text{YES}}$ *then* $\Pr[\text{Ver}(x, \sigma, \text{Prov}(x, \sigma)) \neq 1] \leq \alpha(n)$, *where* $\sigma \xleftarrow{R} \text{Deal}(1^n)$.

β-SOUNDNESS: *If* $x \in \Pi_{\text{NO}}$ *then* $\Pr[\exists p = p(x, \sigma) : \text{Ver}(x, \sigma, p) = 1] \leq \beta(n)$, *where* $\sigma \xleftarrow{R} \text{Deal}(1^n)$.

γ-ZERO-KNOWLEDGE: *If* $x \in \Pi_{\text{YES}}$ *then the pair* (σ, p) *is* $\gamma(n)$-*close in statistical distance to the pair* (σ', p') *where* $\sigma \xleftarrow{R} \text{Deal}(1^n), p \xleftarrow{R} \text{Prov}(x, \sigma)$ *and* $(\sigma', p') \xleftarrow{R} \text{Sim}(x)$.

The algorithms Ver, Deal, *and* Sim *are required to be efficient, while the prover's algorithm* Prov *is allowed to be computationally unbounded. By default,* α, β *and* γ *are assumed to be negligible in* n.

Variants. In the special case where the dealer $\text{Deal}(1^n)$ samples σ uniformly from the set of all strings of length $r(n)$ (for some polynomial $r(\cdot)$), the proof system is called an interactive zero-knowledge proof system in the *common random string model* and is denoted by (α, β, γ)-NISZK^{CRS} [BFM88]. We will focus on the more general setting (defined above) where the dealer is allowed to use any arbitrary (polynomial-time samplable) distribution. This setting is referred to as the *public parameter model* and protocols in the model are denoted by (α, β, γ)-NISZK^{PUB}.[8]

Remark 1 (Efficiency: Uniformity vs. Non-Uniformity). Randomized encodings and non-interactive statistical-zero knowledge proof systems can be defined either in the uniform setting where all efficient entities (encoder, RE-simulator, verifier, dealer, and NISZK-simulator) are assumed to be probabilistic polynomial-time algorithms, or in the non-uniform setting where these entities are represented by probabilistic polynomial-time algorithms which take a non-uniform advice. We will emphasize this distinction only when it matters (Theorem 6), and otherwise, (when the results are insensitive to the difference) ignore it.

Definition 4 (Complexity classes). *The complexity class* $\mathcal{SRE}$ *(resp.,* $1\mathcal{RE}$, $\mathcal{NISZK}^{\text{PUB}}$*) is the set of all the promise problems that have an* SRE *(resp.,* 1RE, NISZK^{PUB}*).*

5 $\mathcal{NISZK}^{\text{pub}} = 1\mathcal{RE}$

In this section we will prove Theorem 1. We start by showing that the notions of 1RE and D1RE are equivalent in Sect. 5.1. Then, based on this equivalence we prove that $\mathcal{NISZK}^{\text{PUB}} = 1\mathcal{RE}$. In the first part of the proof we show that $\mathcal{NISZK}^{\text{PUB}} \subseteq 1\mathcal{RE}$ (cf. Sect. 5.2). In the second part of the proof we show that $1\mathcal{RE} \subseteq \mathcal{NISZK}^{\text{PUB}}$ (cf. Sect. 5.3).

---

[8] The class NISZK^{PUB} was implicitly considered in [BDLP88], and was later referred to as NISZK in the *auxiliary string model* [Dam00] and as *protocol-dependent* NISZK by [GB00]. Our terminology (NISZK in *public parameter model*) is taken from [PS05].

5.1   Equivalence of 1RE and D1RE

We start by showing how to convert a 1RE F for a promise problem Π into a D1RE G for the same problem. The construction is inspired by the techniques of [GSV99]. The encoding G consists of sufficiently many independent copies of F together with a hash of the randomness used to generate the copies. In order to achieve disjointness, while keeping privacy, the length of the hash is chosen such that for yes-instance the hash is close to uniform and in the case of no-instances the support of the hash output is relatively small.

We note that this construction is *non-uniform*. That is, the length of the hash is chosen using a non-uniform advice that depends on the entropy of the encoding distribution on yes-instances. It is an interesting open question whether one can give a uniform construction achieving disjointness.

Theorem 6. *If the promise problem Π has a (possibly non-uniform) 1RE F, then it also has a non-uniform D1RE G. Moreover, if F is uniform then G can be implemented based on F and an advice of $O(\log n)$ bits.*

Proof. Let Π be a promise problem that has an ε-private and δ-correct 1RE F, where ε and δ are negligible. Let Sim_F be the simulator showing the privacy of F on yes-instances. For an input length of n, let $m = m(n) = \mathrm{poly}(n)$ denote the maximum bit-length of the randomness used by Sim_F and F. We define a D1RE $G(x)$ for Π as follows:

---

1. Parameters: $q = 10^6 n m^2$, $m' = qm$.
2. Non-uniform advice $\ell := \lceil m' - H(S_n) - \sqrt{qn} \cdot m - 2n \rceil$.
3. Input: $x \in \{0,1\}^n$.
4. Sample randomness $r = (r_1, \ldots, r_q) \overset{R}{\leftarrow} \{0,1\}^{m'}$ (where $|r_i| = m$), and a pair-wise independent hash function $h \overset{R}{\leftarrow} \mathcal{H}_{m',\ell}$.
5. Output $((F(x;r_1), \ldots, F(x;r_q)), h, h(r))$.

To simplify notation, we let $J_x(r) = (F(x,r_1), \ldots, F(x,r_q))$ and write J_x to denote the distribution induced by a uniform choice of $r \overset{R}{\leftarrow} U_{m'}$. We let $S_n = \otimes^q \mathsf{Sim}_F(1^n)$, and let $\mathcal{H}$ denote the family $\mathcal{H}_{m',\ell}$.

We proceed with an analysis of the encoding G, starting with privacy. We define the simulator $\mathsf{Sim}_G(1^n)$ to generate the random variable $(S_n, U(\mathcal{H}), U_\ell)$. Fix some yes-instance x of length n from Π. Our goal is to show that the statistical distance $\varepsilon'(n)$ between $\mathsf{Sim}_G(1^n)$ and $G(x)$ is upper-bounded by some negligible function. First observe that, by the triangle inequality, ε' is upper-bounded by

$$\Delta(\mathsf{Sim}_G(1^n); \, (J_x, U(\mathcal{H}), U_\ell)) + \Delta((J_x, U(\mathcal{H}), U_\ell); \, G(x)). \tag{1}$$

By the ε-privacy of the original encoding and by Fact 1 item 3, the first summand satisfies

$$\Delta(\mathsf{Sim}_G(1^n); \, (J_x, U(\mathcal{H}), U_\ell)) = \Delta((S_n, U(\mathcal{H}), U_\ell); \, (J_x, U(\mathcal{H}), U_\ell))$$
$$\leq \Delta(S_n, J_x)$$

$$\leq q\varepsilon(n) = \text{neg}(n).$$

It is left to analyze the second summand in (1), i.e., to upper-bound the quantity

$$\underset{r \overset{R}{\leftarrow} \{0,1\}^{m'}, h \overset{R}{\leftarrow} \mathcal{H}}{\Delta} \left((J_x(r), h, U_\ell); \; (J_x(r), h, h(r)) \right). \tag{2}$$

Since the first entry is identically distributed in both distributions, it suffices to analyze the statistical distance between the two tuples conditioned on the outcome of the first entry J_x. Indeed, we prove the following claim.

Claim 1. With probability $1 - 2^{-\Omega(n)}$ over $z \overset{R}{\leftarrow} J_x$, it holds that

$$\underset{r \overset{R}{\leftarrow} \{0,1\}^{m'}, h \overset{R}{\leftarrow} \mathcal{H}}{\Delta} \left([J_x(r), h, U_\ell | J_x(r) = z]; \; [J_x(r), h, h(r) | J_x(r) = z] \right) < 2^{-\Omega(n)}. \tag{3}$$

It follows (by Fact 1 item 4) that (2) is upper-bounded by $2^{-\Omega(n)}$.

Proof. (Proof of Claim 1). Recall that on any input x the encoding F uses at most m random bits, and so any element in its support has weight at least 2^{-m}. Hence, due to the Flattening Lemma 1, the distribution J_x is Δ-flat for $\Delta = \sqrt{q}m$. Since $z \overset{R}{\leftarrow} J_x$ is $(\sqrt{n}\Delta)$-typical with probability at least $1 - O(2^{-n})$, it suffices to show that (3) holds for every $(\sqrt{n}\Delta)$-typical z.

Fix some $(\sqrt{n}\Delta)$-typical z from J_x and consider the distribution $(J_x(r), h, h(r))$ conditioned on $J_x(r) = z$. The conditional distribution of r is uniform over the set $J_x^{-1}(z)$. We will show below that

$$\log(|J_x^{-1}(z)|) \geq \ell + n \tag{4}$$

Therefore we can apply the Leftover Hash Lemma 2 to the distribution of $r \overset{R}{\leftarrow} J_x^{-1}(z)$ with $\mathcal{R} = \{0,1\}^\ell$ and $\alpha = 2^{-n}$, and conclude that the distribution of $(J_x(r), h, h(r))$ conditioned on $J_x(r) = z$ is $O(2^{-n/3})$-close to the distribution $(z, U(\mathcal{H}), U_\ell)$.

It remains to prove (4). First, we show that the entropies $H(J_x)$ and $H(S_n)$ are close. Indeed, by the privacy of F, we have that $\Delta(\text{Sim}_F(1^n); F(x)) \leq \varepsilon(n)$ and therefore (by Fact 1 item 3) $\Delta(J_x; S_n) \leq q\varepsilon(n)$. Hence, by Fact 1 item 2, we get that, for all sufficiently large n's,

$$|H(J_x) - H(S_n)| \leq m'q\varepsilon(n) + 1 \leq 2, \tag{5}$$

where the second inequality follows by noting that $\varepsilon(n)$ is negligible in n, and m', q are polynomials in n. Now, recall that z is $(\sqrt{n}\Delta)$-typical, and therefore $\log(|J_x^{-1}(z)|) \geq m' - H(J_x) - \sqrt{n}\Delta$. Plugging in (5) we conclude that

$$\log(|J_x^{-1}(z)|) \geq m' - H(S_n) - 2 - \sqrt{n}\Delta$$
$$\geq \underbrace{[m' - H(S_n) - \sqrt{n}\Delta - 2n]}_{=\ell} + (n - 3) + n$$
$$\geq \ell + n,$$

where the last inequality holds for $n \geq 3$. $\qquad\square$

We move on to prove the disjointness property. Fix some no-instance x. Our goal is to upper-bound

$$\Pr\left[\mathsf{Sim}_G(1^n) \in \mathrm{supp}(G(x))\right] = \Pr\left[(S_n, U(\mathcal{H}), U_\ell) \in \mathrm{supp}(G(x))\right] \qquad (6)$$

by some negligible function. For $z \overset{R}{\leftarrow} S_n$, let $\mathcal{E} = \mathcal{E}(z)$ be the event that $|J_x^{-1}(z)| \le 2^{\ell-n}$. By marginalizing the probability, we can upper-bound (6) by

$$\Pr_{z \overset{R}{\leftarrow} S_n, h \overset{R}{\leftarrow} \mathcal{H}, w \overset{R}{\leftarrow} \{0,1\}^\ell} [(z, h, w) \in \mathrm{supp}(G(x)) \mid \mathcal{E}(z)] + \Pr_{z \overset{R}{\leftarrow} S_n} [\neg\mathcal{E}(z)].$$

We will show that both the first and second summand are negligible in n.

Claim 2. $\Pr_{z \overset{R}{\leftarrow} S_n, h \overset{R}{\leftarrow} \mathcal{H}, w \overset{R}{\leftarrow} \{0,1\}^\ell} [(z, h, w) \in \mathrm{supp}(G(x)) \mid \mathcal{E}(z)] \le 2^{-n}.$

Proof. By definition $\mathrm{supp}(G(x)) = \{(J_x(r), h, h(r)) \mid r \in \{0,1\}^{m'}, h \in \mathcal{H}\}$. Therefore, for any fixed z and h the probability, over $w \overset{R}{\leftarrow} \{0,1\}^\ell$, that the triple (z, h, w) lands in $\mathrm{supp}(G(x))$ is exactly

$$\frac{|h(J_x^{-1}(z))|}{2^\ell} \le \frac{|J_x^{-1}(z)|}{2^\ell},$$

which is upper-bounded by $2^{\ell-n}/2^\ell = 2^{-n}$ when we condition on $\mathcal{E}(z)$.   □

We conclude the proof by showing that for $z \overset{R}{\leftarrow} S_n$ the event $\mathcal{E}(z)$ happens almost surely.

Claim 3. $\Pr_{z \overset{R}{\leftarrow} S_n} [\log |J_x^{-1}(z)| \le \ell - n] \ge 1 - 2^{-\Omega(n)}.$

Proof. Call z *good* if

$$z \text{ is } (\sqrt{n}\Delta)\text{-typical}, \qquad \text{where } \Delta = \sqrt{q}m, \qquad (7)$$

and

$$\Pr[S_n = z] \ge 2^{q/10} \Pr[J_x = z]. \qquad (8)$$

We begin by showing that, except with probability $2^{-\Omega(n)}$, a random $z \overset{R}{\leftarrow} S_n$ is good. First, recall that $\mathsf{Sim}_F(1^n)$ uses at most m random bits, and so any element in its support has weight at least 2^{-m}. Hence, due to the Flattening Lemma 1, the distribution S_n is Δ-flat for $\Delta = \sqrt{q}m$ which implies that a random $z \overset{R}{\leftarrow} S_n$ satisfies (7) with probability at least $1 - 2^{-\Omega(n)}$. Next, we show that, except with probability $2^{-\Omega(n)}$, a random $z \overset{R}{\leftarrow} S_n$ satisfies (8). Indeed, due to the correctness property of F, we have that $\Delta(\mathsf{Sim}_F(1^n); F(x)) \ge 1/2$ which implies (by Fact 1 item 3) that $\Delta(S_n, J_x) \ge 1 - 2\exp(-q/8)$. Applying Fact 1 item 5, we conclude that

$$\Pr_{z \overset{R}{\leftarrow} S_n} [\Pr[S_n = z] < t\Pr[J_x = z]] \le t \cdot 2\exp(-q/8),$$

for any $t \ge 1$. Taking $t := 2^{q/10}$, and noting that

$$t \cdot 2 \exp(-q/8) \le 2t \cdot 2^{-q/8} = 2 \cdot 2^{q/10} \cdot 2^{-q/8} = 2^{-q/40+1} = 2^{-\Omega(n)},$$

we conclude that (8) holds for all but $2^{-\Omega(n)}$-fraction of the $z \stackrel{R}{\leftarrow} S_n$. It follows, by a union-bound, that, except with probability $2^{-\Omega(n)}$, a random $z \stackrel{R}{\leftarrow} S_n$ is good.

Finally, we prove that for any good z it holds that $\log |J_x^{-1}(z)| \le \ell - n$. By definition

$$|J_x^{-1}(z)| = 2^{m'} \cdot \Pr[J_x = z]$$

and by (8) the latter is upper-bounded by

$$2^{m'-q/10} \cdot \Pr[S_n = z].$$

Recalling that $\Pr[S_n = z] \le 2^{-H(S_n)+\sqrt{n}\Delta}$ (since z is $\sqrt{n}\Delta$-typical) we conclude that

$$|J_x^{-1}(z)| \le 2^{m'-q/10-H(S_n)+\sqrt{n}\Delta}.$$

Hence, we get that

$$\log |J_x^{-1}(z)| \le m' - H(S_n) + \sqrt{n}\Delta - q/10$$
$$\le \underbrace{[(m' - H(S_n) - \sqrt{n}\Delta - 2n)]}_{=\ell} - n + \underbrace{(3n + 3\sqrt{n}\Delta - q/10)}_{T}.$$

Since $q = 10^6 n m^2$ the expression T is always negative, and the claim follows. $\square$

This completes the proof of Theorem 6.

Now we show that if we repeat a D1RE polynomially many times we preserve the privacy of the encoding on yes-instances and gain the correctness security property of 1RE.

Theorem 7. *Let Π be a promise problem that has an ε-private and ρ-disjoint D1RE F, where ε and ρ are negligible. Then, there exists G a 1RE for Π that is ε'-private and δ-correct, where ε' and δ are negligible.*

Proof. For an instance x of length n, we define a randomized encoding $G(x)$ to be $\otimes^n F(x)$. Since F is efficient, the encoding G is also efficient. We prove that G is a 1RE for Π.

PRIVACY FOR YES-INSTANCES: Let Sim_F be the simulator showing the privacy of F on yes-instances. Define $\mathsf{Sim}_G(1^n) := \otimes^n \mathsf{Sim}_F(1^n)$. Take any yes-instance x from Π. We have that

$$\Delta(\mathsf{Sim}_G(1^n); G(x)) = \Delta(\otimes^n \mathsf{Sim}_F(1^n); \otimes^n F(x)) \le n \cdot \varepsilon(n),$$

where the last inequality holds due to Fact 1 item 3. Since $\varepsilon(n)$ is negligible, we have that $\varepsilon'(n) := n \cdot \varepsilon(n)$ is also negligible.

CORRECTNESS: Let $Z = \bigcup_{x \in \Pi_{\text{NO}}} \text{supp}(G(x))$. The decoder Dec on input s outputs 0 if $s \in Z$; and outputs 1, otherwise. Clearly, a no-instance is always decoded correctly. For a yes-instance x, we upper-bound the decoding error by showing that $\Pr[G(x) \in Z]$ is negligible. Since G is ε'-private on yes-instances, we have that

$$\Pr[G(x) \in Z] \leq \Pr[\text{Sim}_G(1^n) \in Z] + \varepsilon'(n).$$

By ρ-disjointness, it holds that $\Pr[\text{Sim}_F(1^n) \in \text{supp}(F(x_{\text{NO}}))] \leq \rho(n)$, for any no-instance x_{NO}. This implies that if we repeat this experiment n times we get that $\Pr[\text{Sim}_G(1^n) \in \text{supp}(G(x_{\text{NO}}))] \leq \rho(n)^n$. By a union bound, we conclude that $\Pr[\text{Sim}_G(1^n) \in Z] \leq 2^n \rho(n)^n$, which implies that

$$\Pr[G(x) \in Z] \leq 2^n \rho(n)^n + \varepsilon'(n) \leq \text{neg}(n).$$

The theorem follows. $\qquad\qquad\qquad\qquad\qquad\qquad\qquad\qquad\qquad\qquad\qquad$ □

5.2 From NISZK^{pub} to 1RE

In this section we prove that $\mathcal{NISZK}^{\text{PUB}} \subseteq \mathcal{1RE}$.

Theorem 8. $\mathcal{NISZK}^{\text{PUB}} \subseteq \mathcal{1RE}$.

Proof. Let Π be a promise problem with (α, β, γ)-NISZK^{PUB} proof system consisting of $(\text{Prov}, \text{Ver}, \text{Deal}, \text{Sim}_{\text{ZK}})$, where α, β, γ are negligible. By Theorem 7, it suffices to show that Π has a (ε, ρ)-D1RE Enc for some negligible ε and ρ. For an n-bit string x, we define a randomized encoding $\text{Enc}(x)$ as follows[9]:

---
1. Compute $(\sigma, p) = \text{Sim}_{\text{ZK}}(x)$.
2. Compute the bit $b = \text{Ver}(x, \sigma, p)$.
3. If $b = 1$ output σ, otherwise output a fixed string $z_n \notin \text{supp}(\text{Deal}(1^n))$.[9]

Observe that Enc is efficient because Sim_{ZK} and Ver are efficient. We prove that Enc is a D1RE.

PRIVACY: We define $\text{Sim}_{\text{YES}}(1^n) = \text{Deal}(1^n)$ and prove that for any yes-instance x the distribution $\text{Sim}_{\text{YES}}(1^n)$ is $\varepsilon(n)$-close to $\text{Enc}(x)$ where $\varepsilon(n) = \alpha(n) + 2 \cdot \gamma(n) = \text{neg}(n)$. Fix some yes-instance x of length n. Due to the zero-knowledge property of NISZK, we have that

$$\underset{\sigma \xleftarrow{R} \text{Deal}(1^n)}{\Delta} (\text{Sim}_{\text{ZK}}(x), (\sigma, \text{Prov}(x, \sigma))) \leq \gamma(n).$$

By the definition of the statistical distance, this implies that

$$\left| \underset{\sigma \xleftarrow{R} \text{Deal}(n)}{\Pr} [\text{Ver}(\sigma, x, \text{Prov}(x, \sigma)) \neq 1] - \underset{(\sigma, p) \xleftarrow{R} \text{Sim}_{\text{ZK}}(x)}{\Pr} [\text{Ver}(\sigma, x, p) \neq 1] \right| \leq \gamma(n).$$

---

[9] For example, such a $z(n)$ can be efficiently constructed by appending a trailing 1 to the output of $\text{Deal}(1^n)$ and setting $z(n)$ to the all-zero string.

Because of the correctness property of NISZK, we have that

$$\Pr_{\sigma \xleftarrow{R} \mathsf{Deal}(n)} [\mathsf{Ver}(\sigma, x, \mathsf{Prov}(x, \sigma)) \neq 1] \leq \alpha(n).$$

This implies that

$$\Pr_{(\sigma, p) \xleftarrow{R} \mathsf{Sim}_{\mathsf{ZK}}(x)} [\mathsf{Ver}(\sigma, x, p) \neq 1] \leq \alpha(n) + \gamma(n).$$

The latter inequality means that in the execution of $\mathsf{Enc}(x)$ the bit b equals to 1 except with the probability $\alpha(n) + \gamma(n)$. Hence, $\Delta(\mathsf{Enc}(x); \mathsf{Sim}_{\mathsf{ZK}}(x)[1]) \leq \alpha(n) + \gamma(n)$, where $\mathsf{Sim}_{\mathsf{ZK}}(x)[1]$ denotes the first component of the tuple output by the simulator. Because of the zero-knowledge property of NISZK and due to Fact 1 item 1, we have that $\Delta(\mathsf{Sim}_{\mathsf{ZK}}(x)[1]; \mathsf{Deal}(1^n)) \leq \gamma(n)$. Finally, combining the last two inequalities, we get that

$$\Delta(\mathsf{Enc}(x); \mathsf{Deal}(1^n)) \leq \alpha(n) + 2 \cdot \gamma(n) = \mathrm{neg}(n).$$

DISJOINTNESS: Let x be a no-instance of Π. Let $E \subseteq \mathrm{supp}(\mathsf{Deal}(1^n))$ denote the set of the strings admitting a proof for the no-instance x, i.e., $E := \{\sigma \in \mathrm{supp}(\mathsf{Deal}(1^n)) \mid \exists p : \mathsf{Ver}(\sigma, x, p) = 1\}$. By Enc's construction we have that $\mathrm{supp}(\mathsf{Enc}(x)) \subseteq E \cup \{z_n\}$. This implies that

$$\Pr[\mathsf{Deal}(1^n) \in \mathrm{supp}(\mathsf{Enc}(x))] \leq \Pr[\mathsf{Deal}(1^n) \in E \cup \{z_n\}]$$
$$\overset{(\star)}{=} \Pr[\mathsf{Deal}(1^n) \in E]$$
$$\leq \beta(n),$$

where the last inequality follows from the soundness property of NISZK, and the equality $(\star)$ holds because $z_n \notin \mathrm{supp}(\mathsf{Deal}(1^n))$. $\qquad\square$

5.3   From 1RE to NISZK^{pub}

Theorem 9. *If the promise problem Π has a (possibly non-uniform) 1RE F, then it also has a non-uniform* NISZK^{PUB} *proof system. Moreover, if F is uniform then the* NISZK^{PUB} *proof system can be implemented based on F and an advice of $O(\log n)$ bits.*

Proof. Let $\Pi \in 1\mathcal{RE}$. Due to Theorem 6, there exists a non-uniform (ε, ρ)-D1RE Enc for Π such that ε and ρ are negligible. Let $s(n)$ denote the randomness complexity of the encoding Enc when it is applied to an n-bit input x, and let $\mathsf{Sim}_{\mathsf{RE}}$ be the simulator showing the privacy of Enc on yes-instances. We construct a proof system $(\mathsf{Prov}, \mathsf{Ver}, \mathsf{Deal}, \mathsf{Sim}_{\mathsf{ZK}})$ for Π as follows:

- Deal: Given 1^n, the dealer outputs $\mathsf{Sim}_{\mathsf{RE}}(1^n)$.
- Prov: Given an n-bit input x and a string σ from Deal, the prover samples a random $r \in \{0,1\}^{s(n)}$ subject to $\mathsf{Enc}(x, r) = \sigma$, and sends it to the verifier. If no such r exists the prover sends some arbitrary message.
- Ver: Given (x, σ, r), the verifier accepts if $\mathsf{Enc}(x, r) = \sigma$, and other rejects.
- Sim$_{\mathsf{ZK}}$: Given x, the simulator $\mathsf{Sim}_{\mathsf{ZK}}$ samples a random r and outputs the pair $(\mathsf{Enc}(x, r), r)$.

We show that $(\mathsf{Prov}, \mathsf{Ver}, \mathsf{Deal}, \mathsf{Sim}_{\mathrm{ZK}})$ forms a NISZK for Π.

COMPLETENESS: Consider some yes-instance x of length n. Recall that, by the privacy of D1RE, the simulator's distribution $\mathsf{Sim}_{\mathrm{RE}}(1^n)$ is $\varepsilon(n)$-close to $\mathsf{Enc}(x)$, which implies that

$$\Pr[\mathsf{Sim}_{\mathrm{RE}}(1^n) \in \mathrm{supp}(\mathsf{Enc}(x))] \geq 1 - \varepsilon(n).$$

Hence, except with probability $\varepsilon(n)$, for a string σ generated by $\mathsf{Sim}_{\mathrm{RE}}(1^n)$, the prover Prov can find r, such that $\mathsf{Enc}(x, r) = \sigma$.

SOUNDNESS: For all no-instances x of Π, we have that

$$\Pr_{\sigma \xleftarrow{R} \mathsf{Deal}(1^n)} [\exists p : V(x, \sigma, p) = 1] = \Pr_{\sigma \xleftarrow{R} \mathsf{Sim}_{\mathrm{RE}}(1^n)} [\sigma \in \mathrm{supp}(\mathsf{Enc}(x))] \leq \delta(n),$$

where the last inequality follows from the disjointness property of Enc.

ZERO KNOWLEDGE: For all yes-instances x of Π, we have that

$$\Delta_{\sigma \xleftarrow{R} \mathsf{Deal}(1^n)} (\mathsf{Sim}_{\mathrm{ZK}}(x); (\sigma, \mathsf{Prov}(x, \sigma))) =$$

$$\Delta_{\sigma \xleftarrow{R} \mathsf{Sim}_{\mathrm{RE}}(1^n), r \xleftarrow{R} \{0,1\}^{s(n)}} ((\mathsf{Enc}(x, r), r); (\sigma, \mathsf{Prov}(x, \sigma))) =$$

$$\Delta_{\sigma \xleftarrow{R} \mathsf{Sim}_{\mathrm{RE}}(1^n), r \xleftarrow{R} \{0,1\}^{s(n)}} ((\mathsf{Enc}(x, r), \mathsf{Prov}(x, \mathsf{Enc}(x, r))); (\sigma, \mathsf{Prov}(x, \sigma))) \leq$$

$$\Delta_{\sigma \xleftarrow{R} \mathsf{Sim}_{\mathrm{RE}}(1^n), r \xleftarrow{R} \{0,1\}^{s(n)}} (\mathsf{Enc}(x, r); \sigma) \leq$$

$$\varepsilon(n),$$

where the second equality follows by recalling that $\mathsf{Prov}(\sigma)$ samples a random r subject to $\mathsf{Enc}(x, r) = \sigma$ and so $(\mathsf{Enc}(x, r), r)$ is identically distributed to $(\mathsf{Enc}(x, r), \mathsf{Prov}(x, \mathsf{Enc}(x, r)))$, and the first inequality follows from Fact 1 item 1 $\square$.

6 If $\mathcal{SRE}$ Is Non-trivial Then One-Way Functions Exist

In this section we prove Theorem 3:

Theorem 3 (Restated). *If $\mathcal{SRE}$ is non-trivial (not in $\mathcal{BPP}$), then infinitely-often one-way functions exist.*

Proof. Assume that infinitely-often one-way functions do not exist. Impagliazzo and Luby [IL89] showed that in this case every efficiently computable function $g(x)$ can be "distributionally-inverted" in the following sense: For every inverse polynomial $\alpha(\cdot)$, there exists an efficient adversary A such that, for random $x \in \{0,1\}^n$, the pair $(x, g(x))$ is $\alpha(n)$-close to the pair $(A(g(x)), g(x))$. In other words, for most x's, A finds an almost uniform preimage of $g(x)$. We refer to α as the *deviation* of the inverter and set it to $1/10$.

We will show that such an inverter allows to put $\mathcal{SRE}$ in $\mathcal{BPP}$. Let Π be a promise problem in $\mathcal{SRE}$ with ε-private δ-correct statistical encoding Enc for some negligible ε and δ. Let $\mathsf{Sim}_{\mathrm{YES}}$ and $\mathsf{Sim}_{\mathrm{NO}}$ be the simulators of the encoding and define $\mathsf{Sim}(b, r)$ to be a "joint" simulator which takes as an input a single bit $b \in \{0, 1\}$ and random string r and outputs a sample from $\mathsf{Sim}_{\mathrm{YES}}(r)$ if $b = 1$ and from $\mathsf{Sim}_{\mathrm{NO}}(r)$ if $b = 0$.[10] We decide Π via the following $\mathcal{BPP}$ procedure B: Given a string $x \in \{0, 1\}^n$, sample an encoding $y \overset{R}{\leftarrow} \mathsf{Enc}(x)$ and α-distributionally invert the simulator Sim on the string y. Take the resulting preimage (b, r) (where r is the coins of the simulator) and output the bit b. We analyze the success probability of deciding Π with this procedure.

Claim 4. The procedure B decides Π with error probability of at most $1/6 + 5\delta + \varepsilon + \alpha$.

Proof. Let us focus on the case where $x \in \{0, 1\}^n$ is a yes-instance (the other case is symmetric). First consider an "ideal" version B' of the algorithm B in which (1) the string y is sampled from $\mathsf{Sim}_{\mathrm{YES}}(r)$ and (2) the distributional inversion algorithm is perfect and has zero deviation. Observe that the gap between the error probability of the real algorithm B to the error probability of the ideal algorithm B' is at most $\varepsilon + \alpha$ (this is due to ε-privacy and to α-deviation of the actual inverter). Hence, it suffices to show that the ideal version errs with probability of at most $1/6 + 5\delta$.

For a given encoding y, the ideal algorithm outputs the right answer $b = 1$ with probability $\frac{p_1(y)}{p_0(y) + p_1(y)}$ where $p_0(y)$ denotes the weight of y under the distribution sampled by $\mathsf{Sim}_{\mathrm{NO}}$ and $p_1(y)$ denotes the weight of y under $\mathsf{Sim}_{\mathrm{YES}}$. By the δ-correctness of the encoding and by Fact 1 item 5 (instantiated with $t = 5$), it holds that, except with probability at most 5δ over $y \overset{R}{\leftarrow} \mathsf{Sim}_{\mathrm{YES}}$, we have that $p_1(y) \geq 5p_0(y)$. It follows, by a union bound, that the ideal algorithm errs with probability of at most $5\delta + 1/6$, as required. $\square$

It remains to notice, that since δ and ε are negligible and α is an inverse polynomial, we have that Π can be decided with success probability at least $2/3$.

7   If $\mathcal{PRE}$ Is Hard on the Average Then CRH Exist

In this section we will study the consequences of the existence of an average-case hard problem $\Pi \in \mathcal{PRE}$.

---

[10] We omit the unary input 1^n of the simulators, and assume that the randomness complexity $m(n)$ of the simulators uniquely determines the instance length n. Similarly, we assume that the output of $\mathsf{Sim}(b, r)$ uniquely determines n. Both requirements can be achieved without loss of generality via standard padding conventions. (E.g., pad the randomness r and concatenate the input length 1^n to the encoding and to the output of Sim.).

Definition 5. *We say that a promise problem* $\Pi = (\Pi_{\text{YES}}, \Pi_{\text{NO}})$ *is hard on average if there exists an efficient sampler* S *that given* 1^n *outputs an* n-*bit instance of* Π *such that for every non-uniform efficient algorithm* A,

$$\left| \Pr_{x \xleftarrow{R} S(1^n)} [A(x) = \chi_\Pi(x)] - 1/2 \right| < \text{neg}(n).$$

We say that the problem has efficient Yes/No samplers *if it is possible to efficiently sample from the conditional Yes distribution* $Y_n = [S(1^n)|S(1^n) \in \Pi_{\text{YES}}]$ *and from the conditional No distribution* $N_n = [S(1^n)|S(1^n) \in \Pi_{\text{NO}}]$.

A collection of *claw-free pseudo-permutations* (CFPP) [Dam87, GMR88, Rus95] is a set of pairs of efficiently computable functions $f^0, f^1 : \{0,1\}^n \to \{0,1\}^n$ for which it is hard to find a pair (u, v) which forms a *claw*, i.e., $f^0(u) = f^1(v)$, or a *collapse*, i.e., $f^b(u) = f^b(v)$ and $u \neq v$ for some bit b. Collections of claw-free *permutations* (CFPs) correspond to the special case where f_0 and f_1 are permutations and so collapses simply do not exist.

Definition 6 (Claw-free Functions). *A collection of pairs of functions consists of an infinite set of indices, denoted* $\overline{I}$, *finite sets* D_i *for each* $i \in \overline{I}$, *and two functions* f_i^0 *and* f_i^1 *mapping* D_i *to* D_i, *respectively. Such a collection is called a* claw-free pseudo-permutations *if there exist three probabilistic polynomial-time algorithms* I, D, *and* F *such that the following conditions hold:*

EASY TO SAMPLE AND COMPUTE: *The random variable* $I(1^n)$ *is assigned values in the set* $\overline{I} \cap \{0,1\}^{p(n)}$ *for some polynomial* $p(\cdot)$. *For each* $i \in \overline{I}$, *the random variable* $D(i)$ *is distributed uniformly over* D_i. *For each* $i \in \overline{I}$, $b \in \{0,1\}$ *and* $x \in D_i$, $F(b, i, x) = f_i^b(x)$.

HARD TO FORM CLAWS: *A pair* (x, y) *satisfying* $f_i^0(x) = f_i^1(y)$ *is called a claw for index* i. *Let* C_i *denote the set of claws for index* i. *It is required that for every probabilistic polynomial-time algorithm* A,

$$\Pr_{i \xleftarrow{R} I(1^n)} [A(i) \in C_i] < \text{neg}(n).$$

HARD TO FORM COLLAPSES: *A pair* (x, y) *satisfying* $f_i^b(x) = f_i^b(y)$ *is called a collapse for an index* i *and a bit* b. *Let* $T_{i,b}$ *denote the set of collapses for* (i, b). *It is required that for every probabilistic polynomial-time algorithm* A *and every* $b \in \{0,1\}$,

$$\Pr_{i \xleftarrow{R} I(1^n)} [A(i) \in T_{i,b}] < \text{neg}(n).$$

If the last item holds for unbounded adversaries, i.e., f_i^0 *and* f_i^1 *are permutations over* D_i, *then the collection is called a collection of* claw-free permutations.

It is known that CFPP's imply Collision-Resistant Hash functions (CRH) [Rus95]. We will show that the existence of an average-case hard problem $\Pi \in \mathcal{PRE}$ implies the existence of CFPPs. We begin with the simpler case in which Π has an *efficient Yes/No samplers* and show that, in this case, we obtain a collection of claw-free permutations.

Theorem 10. *If there exists an average-case hard language in $\mathcal{PRE}$ with efficient Yes/No samplers then CFPs exist.*

We will need the following simple claim.

Claim 5. Let Π be a promise problem with perfect randomized encoding g whose simulators are $\mathrm{Sim_{YES}}$ and $\mathrm{Sim_{NO}}$. Define the functions h_x^0, h_y^1 which are indexed by a pair of instances (x, y) of Π as follows:

$$h_x^0(r, b) = \begin{cases} g(x; r), & \text{if } b = 0, \\ \mathrm{Sim_{NO}}(r), & \text{otherwise}; \end{cases} \qquad h_y^1(r, b) = \begin{cases} g(y; r), & \text{if } b = 0, \\ \mathrm{Sim_{YES}}(r), & \text{otherwise}; \end{cases} \qquad (9)$$

Then the following holds for any n-bit strings x and y:

1. If $x \in \Pi_{\mathrm{YES}}$, then h_x^0 is a permutation.
2. If $y \in \Pi_{\mathrm{NO}}$, then h_y^1 is a permutation.
3. If $(x, y) \in \Pi_{\mathrm{NO}} \times \Pi_{\mathrm{YES}}$ then $\mathrm{Im}\left(h_x^0\right) \cap \mathrm{Im}\left(h_y^1\right) = \emptyset$.

Proof. Let R_0 and R_1 denote $\mathrm{Im}(\mathrm{Sim_{NO}})$ and $\mathrm{Im}(\mathrm{Sim_{YES}})$, respectively. Let $s(n)$ denote the randomness complexity of g and let $t(n)$ denote the output length of g. Since g is a perfect randomized encoding, we have that $R_0 \cap R_1 = \emptyset$, $R_0 \cup R_1 = \{0, 1\}^{t(n)}$, and $t(n) = s(n) + 1$. Consider the case where $x \in \Pi_{\mathrm{YES}}$. Then $h_x^0(\cdot, 0) : \{0, 1\}^{s(n)} \to R_1$ is a bijection and $h_x^0(\cdot, 1) : \{0, 1\}^{s(n)} \to R_0$. Since $R_0 \cap R_1 = \emptyset$, the function $h_x^0(\cdot, \cdot)$ is a permutation on $R_0 \cup R_1 = \{0, 1\}^{t(n)}$. Similarly, if $y \in \Pi_{\mathrm{NO}}$, the function $h_y^1(\cdot, \cdot)$ is a permutation on $\{0, 1\}^{t(n)}$.

In order to prove the third item, we observe that if $x \in \Pi_{\mathrm{NO}}$, then $\mathrm{Im}\left(h_x^0\right) = R_0$; and if $y \in \Pi_{\mathrm{YES}}$, then $\mathrm{Im}\left(h_y^1\right) = R_1$. This implies that for all $(x, y) \in \Pi_{\mathrm{NO}} \times \Pi_{\mathrm{YES}}$ it holds that $\mathrm{Im}\left(h_x^0\right) \cap \mathrm{Im}\left(h_y^1\right) = R_0 \cap R_1 = \emptyset$. $\square$

We can now prove Theorem 10.

Proof (Proof of Theorem 10). Let Π be an average-case hard language with efficient Yes/No samplers (Y_n, N_n), and let g be a perfect randomized encoding for Π. For a pair of inputs (x, y) from Π, we say that (x, y) is a $(\mathrm{YES}, \mathrm{NO})$-instance (resp., $(\mathrm{NO}, \mathrm{YES})$), if x is a yes-instance and y is a no-instance (resp., if x is a no-instance and y is a yes-instance).

We construct a CFP family which is indexed by pairs $(x, y) \in \Pi_{\mathrm{YES}} \times \Pi_{\mathrm{NO}}$. Given a security parameter 1^n, an index (x, y) is chosen by sampling $x \xleftarrow{R} Y_n$ and $y \xleftarrow{R} N_n$. For each index (x, y) we let $f_{(x,y)}^0 \equiv h_x^0$ and $f_{(x,y)}^1 \equiv h_y^1$, where h_x^0 and h_x^1 are defined as in (9). Recall that the domain and range of $f_{x,y}^b$ are $\{0, 1\}^{t(n)}$ where $t(n)$ is the output length of g's output. Clearly this collection is efficiently samplable and efficiently computable. Moreover, since our sampler always samples a $(\mathrm{YES}, \mathrm{NO})$-instance (x, y), it holds, due to Claim 5, that $f_{(x,y)}^0 \equiv h_x^0$ and $f_{(x,y)}^1 \equiv h_y^1$ are permutations on $\{0, 1\}^{t(n)}$. We complete the proof by showing that claws are hard to find.

Recall that we assume that the distribution ensemble $\{Y_n\}$ is computationally indistinguishable from $\{N_n\}$. By a standard hybrid argument, it follows that the pair (Y_n, N_n) is computationally indistinguishable from the pair (Y_n, Y_n) which, in turn, is computationally indistinguishable from the pair (N_n, Y_n). Now assume, for the sake of contradiction, that there exists an efficient algorithm A that given $(x, y) \xleftarrow{R} (Y_n, N_n)$ can find claws with non-negligible probability ε. We can use A to distinguish (Y_n, N_n) from (N_n, Y_n) as follows: Given (x, y) call $A(x, y)$ and output 1 if A's output (u, v) forms a collision under h_x^0 and h_y^1. By assumption, the resulting distinguisher outputs 1 when $(x, y) \xleftarrow{R} (Y_n, N_n)$ with probability ε. In contrast, when $(x, y) \xleftarrow{R} (N_n, Y_n)$, the distinguisher never finds a claw since claws do not exist (due to Claim 5). Hence the distinguisher has a noticeable advantage of ε, in contradiction to our assumption.   □

We continue by considering the more general case where Π is hard on average but does not admit efficient Yes/No samplers, and obtain, in this case, claw-free *pseudo-permutations* (whose existence still implies collision-resistance hash functions).

Theorem 11. *If there exists an average-case hard language in $\mathcal{PRE}$ then claw-free pseudo-permutations (CFPP) exist.*

Proof. The construction is identical to the one presented in Theorem 10, except that the index $(x, y) \in \Pi \times \Pi$ is chosen by sampling both x and y independently from the distribution $S(1^n)$ over which Π is average-case hard. By definition, the collection $f_{(x,y)}^b = h_x^b$, where h is defined as in (9), is efficiently samplable and efficiently computable. We verify that it is CFPP.

We begin by showing that $f_{(x,y)}^0 = h_x^0$ is a pseudo-permutation (the case of $f_{(x,y)}^1$ is analogous). Assume for the sake of contradiction that there is an algorithm A that can find collapses for $f_{(x,y)}^0$ with a non-negligible probability ε. Using A we construct a new algorithm A' that has a non-negligible advantage in guessing $\chi_\Pi(x)$ for $x \xleftarrow{R} S(1^n)$. Given an input $x \xleftarrow{R} S(1^n)$, the algorithm A' samples $y \leftarrow S(1^n)$, and then invokes $A(x, y)$ to find a collapse (u, v) for $f_{(x,y)}^0 = h_x^0$. If A finds a valid collapse (i.e., $u \neq v$ and $h_x^0(u) = h_x^0(v)$), the algorithm A' classifies the input x as a no-instance and outputs 0; otherwise A' outputs a random bit. Recall that when x is a yes-instance the function h_x^0 is a permutation, and so it does not have collapses. Hence, A' outputs a correct answer whenever A finds a collapse. Also, when a collapse is not found, the success probability of A' is $1/2$. Hence, the overall success probability of A' is

$$\Pr_{x \xleftarrow{R} S(1^n)} [A'(x) = \chi_\Pi(x)] = 1/2 \cdot (1 - \varepsilon) + 1 \cdot \varepsilon = 1/2 + \varepsilon/2,$$

in contradiction to the average-case hardness of Π.

We move on to show that it is hard to find claws. Assume for the sake of contradiction that there exists an efficient algorithm A that finds claws with a non-negligible probability ε. We construct a new algorithm A' that has a non-negligible advantage in guessing $\chi_\Pi(x)$ for $x \xleftarrow{R} S(1^n)$. Let

$$p = \Pr_{x \xleftarrow{R} S(1^n), y \xleftarrow{R} S(1^n)} [A(x, y) \text{ finds a claw} \,|\, x \in \Pi_{\text{NO}}].$$

We distinguish between two cases based on the value of p.

First, consider the case where $p \geq \varepsilon/2$. Then, by an averaging argument, there exists some fixed no-instance x_0 for which

$$\Pr_y[A(x_0, y) \text{ finds a claw}] \geq \varepsilon/2.$$

Recall that when the index is a (NO, YES) pair there are no claws and so when A finds a claw, y must be a no-instance We can therefore construct a non-uniform algorithm that decides $y \xleftarrow{R} S(1^n)$ as follows: Call $A(x_0, y)$ and output zero ("no") if a collision is found and otherwise toss a random coin. The success probability is at least $\varepsilon/2 + (1 - \varepsilon/2)/2 = 1/2 + \varepsilon/4$.

Second, consider the case where $p < \varepsilon/2$. In this case, we determine whether $x \xleftarrow{R} S(1^n)$ is a yes-instance or a no-instance via the following procedure A'. Sample $y \xleftarrow{R} S(1^n)$, and call $A(x, y)$ if A returns a valid claw, outputs 1 (classify x as a yes-instance); otherwise, output a random bit. The success probability of A' can be marginalized as follows:

$$\Pr_x[A'(x) \text{ succeeds}] = \Pr_{x,y}[A'(x) \text{ succeeds} \,|\, A(x,y) \text{ finds a claw}] \cdot \varepsilon$$

$$+ \Pr_{x,y}[A'(x) \text{ succeeds} \,|\, A(x,y) \text{ doesn't find a claw}] \cdot (1 - \varepsilon)$$

$$= \Pr_x[x \in \Pi_{\text{YES}} \,|\, A(x,y) \text{ finds a claw}] \cdot \varepsilon + (1 - \varepsilon)/2,$$

Therefore, it suffices to show that

$$\Pr_x[x \in \Pi_{\text{YES}} \,|\, A(x, y) \text{ finds a claw}] \geq 2/3 \tag{10}$$

since this implies that A' succeeds with probability of at least $2/3 \cdot \varepsilon + (1 - \varepsilon)/2 = 1/2 + \varepsilon/6$. To prove (10), we upper-bound by $1/3$ the probability of the complementary event:

$$\Pr_x[x \in \Pi_{\text{NO}} \,|\, A(x, y) \text{ finds a claw}] =$$

$$\frac{\Pr_{x,y}[A(x,y) \text{ finds a claw} \,|\, x \in \Pi_{\text{NO}}] \cdot \Pr_x[x \in \Pi_{\text{NO}}]}{\Pr[A(x, y) \text{ finds a claw}]} \leq$$

$$\frac{(\varepsilon/2) \cdot (2/3)}{\varepsilon} =$$

$$\frac{1}{3},$$

where the inequality follows by our assumption ($p < \varepsilon/2$) and by the fact that $\Pr_x[x \in \Pi_{\text{NO}}] < 2/3$ (since otherwise the trivial adversary that always outputs 0 breaks the average-case hardness of Π over $S(1^n)$). The proof follows. $\square$

Acknowledgements. Research supported by the European Union's Horizon 2020 Programme (ERC-StG-2014-2020) under grant agreement no. 639813 ERC-CLC, ISF grant 1155/11, GIF grant 1152/2011, and the Check Point Institute for Information Security. This work was done in part while the first author was visiting the Simons Institute for the Theory of Computing, supported by the Simons Foundation and by the DIMACS/Simons Collaboration in Cryptography through NSF grant CNS-1523467.

A   Omitted Proofs

A.1   Proof of Item 5 of Fact 1

We prove that if $\Delta(X; Y) \geq 1 - \varepsilon$, then, for any $t > 1$, it holds that $\Pr_{x \xleftarrow{R} X}[\Pr[X = x] < t \cdot \Pr[Y = x]] \leq \varepsilon t$.

Proof. We start by proving an additional claim:

Claim 6. For any two distributions X, Y and a subset S of their domain, it holds that:
$$\Delta(X; Y) \leq 1 - \sum_{x \in S} \min(\Pr[X = x], \Pr[Y = x]).$$

Proof.

$$2\Delta(X; Y) = \sum_{x} |\Pr[X = x] - \Pr[Y = x]|$$

$$= \sum_{x \notin S} |\Pr[X = x] - \Pr[Y = x]| + \sum_{x \in S} |\Pr[X = x] - \Pr[Y = x]|$$

$$\leq \sum_{x \notin S} \Pr[X = x] + \sum_{x \notin S} \Pr[Y = x] + \sum_{x \in S} |\Pr[X = x] - \Pr[Y = x]|$$

$$= \sum_{x} \Pr[X = x] + \sum_{x} \Pr[Y = x] -$$

$$\sum_{x \in S} (\Pr[X = x] + \Pr[Y = y] - |\Pr[X = x] - \Pr[Y = x]|)$$

$$= 2 - 2 \sum_{x \in S} \min(\Pr[X = x], \Pr[Y = x]).$$

The last equality holds because $\sum_{x} \Pr[X = x] = 1 = \sum_{x} \Pr[Y = x]$, and for all a, b we have that $a + b - |a - b| = 2 \min(a, b)$. $\qquad\square$

Now we proceed to the proof of the lemma. Let $S := \{x \mid \Pr[X = x] < t \cdot \Pr[Y = x]\}$. Due to the claim, we have that

$$\Delta(X; Y) \leq 1 - \sum_{x \in S} \min(\Pr[X = x], \Pr[Y = x]) \tag{11}$$

We now give a lower bound for each summand $\min(\Pr[X = x], \Pr[Y = x])$. Namely, we show that

$$\forall x \in S \;\; \min(\Pr[X = x], \Pr[Y = x]) \geq \Pr[X = x]/t. \tag{12}$$

By the construction of S, we have that for any $x \in S$ $\Pr[Y = x] > \Pr[X = x]/t$. Hence, $\min(\Pr[X = x], \Pr[Y = x]) \geq \min(\Pr[X = x], \Pr[X = x]/t)$. Since $t > 1$, we have that $\min(\Pr[X = x], \Pr[X = x]/t) = \Pr[X = x]/t$. Combining inequalities 11 and 12, we get that

$$\Delta(X;Y) \leq 1 - \sum_{x \in S} \min(\Pr[X = x], \Pr[Y = x])$$

$$\leq 1 - \sum_{x \in S} \Pr[X = x]/t$$

$$= 1 - \Pr[X \in S]/t.$$

Recall that by assumption $1 - \varepsilon \leq \Delta(X;Y)$, and therefore, we conclude that $\varepsilon \geq \Pr[X \in S]/t$ implying that $\Pr[X \in S] \leq \varepsilon t$. $\qquad \square$

References

[AH87] Aiello, W., Håstad, J.: Perfect zero-knowledge languages can be recognized in two rounds. In: 28th Annual Symposium on Foundations of Computer Science, Los Angeles, California, USA, 27–29 October 1987, pp. 439–448. IEEE Computer Society (1987)

[AIK04] Applebaum, B., Ishai, Y., Kushilevitz, E.: Cryptography in NC^0. In: Proceedings of 45th Symposium on Foundations of Computer Science (FOCS 2004), 17–19 October 2004, Rome, Italy, pp. 166–175. IEEE Computer Society (2004)

[AIK05] Applebaum, B., Ishai, Y., Kushilevitz, E.: Computationally private randomizing polynomials and their applications. In: 20th Annual IEEE Conference on Computational Complexity (CCC 2005), 11–15 June 2005, San Jose, CA, USA, pp. 260–274. IEEE Computer Society (2005)

[AIK15] Applebaum, B., Ishai, Y., Kushilevitz, E., Minimizing locality of one-way functions via semi-private randomized encodings. Electronic Colloquium on Computational Complexity (ECCC), 22:45 (2015)

[AIKP15] Agrawal, S., Ishai, Y., Khurana, D., Paskin-Cherniavsky, A.: Statistical randomized encodings: a complexity theoretic view. In: Halldórsson, M.M., Iwama, K., Kobayashi, N., Speckmann, B. (eds.) ICALP 2015. LNCS, vol. 9134, pp. 1–13. Springer, Heidelberg (2015)

[App11] Applebaum, B.: Randomly encoding functions: a new cryptographic paradigm. In: Fehr, S. (ed.) ICITS 2011. LNCS, vol. 6673, pp. 25–31. Springer, Heidelberg (2011)

[App14] Applebaum, B.: Cryptography in Constant Parallel Time. Information Security and Cryptography. Springer, Heidelberg (2014)

[AR16] Applebaum, B., Raykov, P.: From private simultaneous messages to zero information Arthur-Merlin protocols and back. To appear in TCC 2016A, 2016. Available as eprint report 2015/1046. http://eprint.iacr.org/

[BDLP88] Brandt, J., Damgård, I.B., Landrock, P., Pedersen, T.P.: Zero-knowledge authentication scheme with secret key exchange (extended abstract). In: Goldwasser, S. (ed.) CRYPTO 1988. LNCS, vol. 403, pp. 583–588. Springer, Heidelberg (1990)

[BFM88] Blum, M., Feldman, P., Micali, S.: Non-interactive zero-knowledge and its applications (extended abstract). In: Simon, J. (ed.) Proceedings of the 20th Annual ACM Symposium on Theory of Computing, 2–4 May 1988, Chicago, Illinois, USA, pp. 103–112. ACM (1988)

[BSMP91] Blum, M., De Santis, A., Micali, S., Persiano, G.: Noninteractive zero-knowledge. SIAM J. Comput. 20(6), 1084–1118 (1991)

[Can08] Canetti, R. (ed.): Theory of Cryptography. LNCS, vol. 4948. Springer, Heidelberg (2008)

[CCG+94] Chang, R., Chor, B., Goldreich, O., Hartmanis, J., Håstad, J., Ranjan, D., Rohatgi, P.: The random Oracle hypothesis is false. J. Comput. Syst. Sci. 49(1), 24–39 (1994)

[CCKV08] Chailloux, A., Ciocan, D.F., Kerenidis, I., Vadhan, S.P.: Interactive and noninteractive zero knowledge are equivalent in the help model. In: Canetti [Can08], pp. 501–534

[CW79] Carter, L., Wegman, M.N.: Universal classes of hash functions. J. Comput. Syst. Sci. 18(2), 143–154 (1979). Preliminary version appeared in STOC 1977

[Dam87] Damgård, I.B.: Collision free hash functions and public key signature schemes. In: Price, W.L., Chaum, D. (eds.) EUROCRYPT 1987. LNCS, vol. 304, pp. 203–216. Springer, Heidelberg (1988)

[Dam00] Damgård, I.B.: Efficient concurrent zero-knowledge in the auxiliary string model. In: Preneel, B. (ed.) EUROCRYPT 2000. LNCS, vol. 1807, pp. 418–430. Springer, Heidelberg (2000)

[ESY84] Even, S., Selman, A.L., Yacobi, Y.: The complexity of promise problems with applications to public-key cryptography. Inf. Control 61(2), 159–173 (1984)

[FKN94] Feige, U., Kilian, J., Naor, M.: A minimal model for secure computation (extended abstract). In: Leighton, F.T., Goodrich, M.T. (eds.) Proceedings of the Twenty-Sixth Annual ACM Symposium on Theory of Computing, 23–25 May 1994, Montréal, Québec, Canada, pp. 554–563. ACM (1994)

[GB00] Gutfreund, D., Ben-Or, M.: Increasing the power of the dealer in non-interactive zero-knowledge proof systems. In: Okamoto, T. (ed.) ASIACRYPT 2000. LNCS, vol. 1976, pp. 429–442. Springer, Heidelberg (2000)

[GMR88] Goldwasser, S., Micali, S., Rivest, R.L.: A digital signature scheme secure against adaptive chosen-message attacks. SIAM J. Comput. 17(2), 281–308 (1988)

[GMR89] Goldwasser, S., Micali, S., Rackoff, C.: The knowledge complexity of interactive proof systems. SICOMP: SIAM J. Comput. 18, 291–304 (1989)

[Gol01] Goldreich, O.: The Foundations of Cryptography - Basic Techniques, vol. 1. Cambridge University Press, Cambridge (2001)

[Gol06] Goldreich, O.: On promise problems: a survey. In: Goldreich, O., Rosenberg, A.L., Selman, A.L. (eds.) Theoretical Computer Science. LNCS, vol. 3895, pp. 254–290. Springer, Heidelberg (2006)

[GPW15] Göös, M., Pitassi, T., Watson, T.: Zero-information protocols and unambiguity in Arthur-Merlin communication. In: Roughgarden, T. (ed.) Proceedings of the 2015 Conference on Innovations in Theoretical Computer Science, ITCS 2015, Rehovot, Israel, 11–13 January 2015, pp. 113–122. ACM (2015)

[GSV99] Goldreich, O., Sahai, A., Vadhan, S.P.: Can statistical zero knowledge be made non-interactive? Or on the relationship of SZK and NISZK. In: Wiener, M. (ed.) CRYPTO 1999. LNCS, vol. 1666, p. 467. Springer, Heidelberg (1999)

[GV99] Goldreich, O., Vadhan, S.P.: Comparing entropies in statistical zero knowledge with applications to the structure of SZK. In: Proceedings of the 14th Annual IEEE Conference on Computational Complexity, Atlanta, Georgia, USA, 4–6 May 1999, p. 54. IEEE Computer Society (1999)

[HHRS15] Haitner, I., Hoch, J.J., Reingold, O., Segev, G.: Finding collisions in interactive protocols - tight lower bounds on the round and communication complexities of statistically hiding commitments. SIAM J. Comput. 44(1), 193–242 (2015)

[IK00] Ishai, Y., Kushilevitz, E.: Randomizing polynomials: a new representation with applications to round-efficient secure computation. In: 41st Annual Symposium on Foundations of Computer Science, FOCS 2000, 12–14 November 2000, Redondo Beach, California, USA, pp. 294–304. IEEE Computer Society (2000)

[IL89] Impagliazzo, R., Luby, M.: One-way functions are essential for complexity based cryptography (extended abstract). In: 30th Annual Symposium on Foundations of Computer Science, Research Triangle Park, North Carolina, USA, 30 October–1 November 1989, pp. 230–235. IEEE Computer Society (1989)

[ILL89] Impagliazzo, R., Levin, L.A., Luby, M.: Pseudo-random generation from one-way functions (extended abstracts). In: Johnson, D.S. (ed.) Proceedings of the 21st Annual ACM Symposium on Theory of Computing, 14–17 May 1989, Seattle, Washigton, USA, pp. 12–24. ACM (1989)

[Ish13] Ishai, Y.: Randomization techniques for secure computation. In: Prabhakaran, M., Sahai, A. (eds.) Secure Multi-party Computation. Cryptology and Information Security Series, vol. 10, pp. 222–248. IOS Press, Amsterdam (2013)

[Oka00] Okamoto, T.: On relationships between statistical zero-knowledge proofs. J. Comput. Syst. Sci. 60(1), 47–108 (2000)

[Ost91] Ostrovsky, R.: One-way functions, hard on average problems, and statistical zero-knowledge proofs. In: Proceedings of the Sixth Annual Structure in Complexity Theory Conference, Chicago, Illinois, USA, 30 June–3 July 1991, pp. 133–138. IEEE Computer Society (1991)

[OV08] Ong, S.J., Vadhan, S.P.: An equivalence between zero knowledge and commitments. In: Canetti [Can08], pp. 482–500

[PS05] Pass, R., Shelat, A.: Unconditional characterizations of non-interactive zero-knowledge. In: Shoup, V. (ed.) CRYPTO 2005. LNCS, vol. 3621, pp. 118–134. Springer, Heidelberg (2005)

[Rus95] Russell, A.: Necessary and sufficient condtions for collision-free hashing. J. Cryptol. 8(2), 87–100 (1995)

[SCPY98] De Santis, A., Di Crescenzo, G., Persiano, G., Yung, M.: Image density is complete for non-interactive-SZK. In: Larsen, K.G., Skyum, S., Winskel, G. (eds.) ICALP 1998. LNCS, vol. 1443, pp. 784–795. Springer, Heidelberg (1998)

[SV03] Sahai, A., Vadhan, S.P.: A complete problem for statistical zero knowledge. J. ACM **50**(2), 196–249 (2003)

[Vad99] Vadhan, S.P.: A study of statistical zero-knowledge proofs. Ph.D. thesis (1999)

How to Prove Knowledge of Small Secrets

Carsten Baum[(⊠)], Ivan Damgård, Kasper Green Larsen, and Michael Nielsen

Department of Computer Science, Aarhus University, Aarhus, Denmark
{cbaum,ivan,larsen,mik}@cs.au.dk

Abstract. We propose a new zero-knowledge protocol applicable to additively homomorphic functions that map integer vectors to an Abelian group. The protocol demonstrates knowledge of a short preimage and achieves amortised efficiency comparable to the approach of Cramer and Damgård from Crypto 2010, but gives a much tighter bound on what we can extract from a dishonest prover. Towards achieving this result, we develop an analysis for bins-and-balls games that might be of independent interest. We also provide a general analysis of rewinding of a cut-and-choose protocol as well as a method to use Lyubachevsky's rejection sampling technique efficiently in an interactive protocol when many proofs are given simultaneously.

Our new protocol yields improved proofs of plaintext knowledge for (Ring-)LWE-based cryptosystems, where such general techniques were not known before. Moreover, they can be extended to prove preimages of homomorphic hash functions as well.

Keywords: Proofs of plaintext knowledge · Lattice-based encryption · Homomorphic hashing · Integer commitments

1   Introduction

Proofs of Knowledge. In a zero-knowledge protocol, a prover convinces a sceptical verifier that some claim is true (and in some cases that he knows a proof) while conveying no other knowledge than the fact that the claim is true. Zero-knowledge protocols are one of the most fundamental tools in cryptographic protocol design. In particular, one needs zero-knowledge proofs of knowledge in multiparty computation to have a player demonstrate that he knows the input he is providing. This is necessary to be able to show (UC-)security of a protocol.

C. Baum, I. Damgård and M. Nielsen—Supported by The Danish National Research Foundation and The National Science Foundation of China (under the grant 61061130540) for the Sino-Danish Center for the Theory of Interactive Computation, within which part of this work was performed; by the CFEM research center (supported by the Danish Strategic Research Council) within which part of this work was performed; and by the Advanced ERC grant MPCPRO.

K.G. Larsen—Supported by the Center for Massive Data Algorithmics, a Center of the Danish National Research Foundation, grant DNRF84, a Villum Young Investigator Grant and an AUFF Starting Grant.

M. Robshaw and J. Katz (Eds.): CRYPTO 2016, Part III, LNCS 9816, pp. 478–498, 2016.
DOI: 10.1007/978-3-662-53015-3_17

In this work, we will consider one-way functions $f : \mathbb{Z}^r \mapsto G$ where G is an Abelian group (written additively in the following), and where furthermore the function is additively homormorphic, i.e., $f(\boldsymbol{a}) + f(\boldsymbol{b}) = f(\boldsymbol{a} + \boldsymbol{b})$. We will call such functions *ivOWF*'s (for homomorphic One-Way Functions over Integer Vectors). This turns out to be a very general notion: the encryption function of several (Ring-)LWE-based cryptosystems can be seen an ivOWF (such as the one introduced in [BGV12] and used in the so-called SPDZ protocol [DPSZ12]). Even more generally, the encryption function of any semi-homomorphic cryptosystem as defined in [BDOZ11] is an ivOWF. Also, in commitment schemes for committing to integer values, the function one evaluates to commit is typically an ivOWF (see, e.g., [DF02]). Finally, hash functions based on lattice problems such as [GGH96, LMPR08], where it is hard to find a short preimage, are ivOWFs.

We will look at the scenario where a prover $\mathcal{P}$ and a verifier $\mathcal{V}$ are given $y \in G$ and $\mathcal{P}$ holds a short preimage $\boldsymbol{x}$ of y, i.e., such that $||\boldsymbol{x}|| \leq \beta$ for some β. $\mathcal{P}$ wants to prove in zero-knowledge that he knows such an $\boldsymbol{x}$. When f is an encryption function and y is a ciphertext, this can be used to demonstrate that the ciphertext decrypts and $\mathcal{P}$ knows the plaintext. When f is a commitment function this can be used to show that one has committed to a number in a certain interval.

An obvious but inefficient solution is the following 3-message protocol π:

(1) $\mathcal{P}$ chooses $\boldsymbol{r}$ at random such that $||\boldsymbol{r}|| \leq \tau \cdot \beta$ for some sufficiently large τ, the choice of which we return to below.
(2) $\mathcal{P}$ then sends $a = f(\boldsymbol{r})$ to $\mathcal{V}$.
(3) $\mathcal{V}$ sends a random challenge bit b.
(4) $\mathcal{P}$ responds with $\boldsymbol{z} = \boldsymbol{r} + b \cdot \boldsymbol{x}$.
(5) $\mathcal{V}$ checks that $f(\boldsymbol{z}) = a + b \cdot y$ and that $||\boldsymbol{z}|| \leq \tau \cdot \beta$.

If τ is sufficiently large, the distribution of $\boldsymbol{z}$ will be statistically independent of $\boldsymbol{x}$, and the protocol will be honest verifier statistical zero-knowledge[1]. On the other hand, we can extract a preimage of y from a cheating prover who can produce correct answers $\boldsymbol{z}_0, \boldsymbol{z}_1$ to $b = 0, b = 1$, namely $f(\boldsymbol{z}_1 - \boldsymbol{z}_0) = y$. Clearly, we have $||\boldsymbol{z}_1 - \boldsymbol{z}_0|| \leq 2 \cdot \tau \cdot \beta$. We will refer to the factor 2τ as the *soundness slack* of the protocol, because it measures the discrepancy between the interval used by the honest prover and what we can force a dishonest prover to do. The value of the soundness slack is important: if f is, e.g., an encryption function, then a large soundness slack will force us to use larger parameters for the underlying cryptosystem to ensure that the ciphertext decrypts even if the input is in the larger interval, and this will cost us in efficiency.

The naive protocol above requires an exponentially large slack to get zero-knowledge, but using Lyubachevsky's rejection sampling technique, the soundness slack can made polynomial or even constant (at least in the random oracle model).

---

[1] We will only be interested in honest verifier zero-knowledge here. In applications one would get security for malicious verifiers by generating the challenge in a trusted way, e.g., using a maliciously sure coin-flip protocol.

The obvious problem with the naive solution is that one needs to repeat the protocol k times where k is the statistical security parameter, to get soundness error probability 2^{-k}. This means that one needs to generate $\Omega(k)$ auxiliary f-values. We will refer to this as the *overhead* of the protocol and use it as a measure of efficiency.

One wants, of course as small overhead and soundness slack as possible, but as long as we only want to give a proof for a single f-value, we do not know how to reduce the overhead dramatically in general. But if instead we want to give a proof for k or more f-values, then we know how to reduce the *amortised* overhead: Cramer and Damgård [CD09] show how to get amortised overhead $O(1)$, but unfortunately the soundness slack is $2^{\Omega(k)}$, even if rejection sampling is used. In [DKL+13] two protocols were suggested, where one is only covertly secure, and we will not consider it here as our goal is full malicious security. The other one can achieve polynomial soundness slack with overhead $\Omega(\log(k)^2)$ and works only in the random oracle model[2].

1.1   Contributions and Techniques

In this work, we introduce a new paradigm for zero-knowledge proof of knowledge of preimage under an ivOWF, abbreviated ZKPoKP. For the first time, we are able to optimize both parameters, namely we obtain quasi-polynomial soundness slack (proportional to $(2k + 1)^{\log(k)/2}$) and $o(1)$ ciphertext overhead, all results hold in the standard model (no random oracles are needed).

For our zero-knowledge proof, we use the following high-level strategy:

(1) Use a cut-and-choose style protocol for the inputs $y_1, \ldots, y_n$.
(2) Repeat the following experiment several times:
  (2.1) Let the verifier randomly assign each y_i to one of several *buckets*.
  (2.2) For each bucket, add all elements that landed in the bucket and have the prover demonstrate that he knows a preimage of the sum.

The intuition behind the proof then goes as follows: the first step will ensure that we can extract *almost* all of the required n preimages, in fact all but k where k is the security parameter. In the second step, since we only have k elements left that were "bad" in the sense that we could not yet extract a preimage, then if we have more than k buckets, say ck for a constant $c > 1$, there is a significant probability that many of the bad elements will be alone in a bucket. If this happens, we can extract a preimage by linearity of f. Furthermore, the cost of doing such a step is at most n additions, plus work that only depends on the security parameter k and is insignificant if $n \gg k$. We can now repeat the experiment some number of times to extract the remaining bad elements,

---

[2] The protocol in [DKL+13] is actually stated as a proof of plaintext knowledge for random ciphertexts, but generalizes to a protocol for ivOWFs. It actually offers a tradeoff between soundness slack and overhead in the sense that the overhead is $M \cdot \log(k)$, where M has to be chosen such that $(1/s)^M$ is negligible. Thus one can choose s to be $\mathsf{poly}(k)$ and $M = \log(k)$, or s to be constant and $M = k$.

while adjusting the number of buckets carefully. We are then able to prove that we can extract all preimages quickly, namely after $\log(k)$ repetitions, and this is what give us the small soundness slack. In comparison, in [CD09], the extraction takes place in $\Omega(k)$ stages, which leads to an exponential soundness slack.

Along the way to our main result, we make two technical contributions: first, we show a general result on what you can extract by rewinding from a prover that successfully passes a cut-and-choose test. Second, we show a method for using rejection sampling efficiently in an interactive protocol. In comparison, the protocol from [DKL+13] also used rejection sampling to reduce the soundness slack, but in a more simplistic way that leads to a larger overhead. See Sect. 3.1 for more information on this.

Our protocol is honest verifier zero-knowledge and is sound in the sense of a standard proof of knowledge, i.e., we extract the prover's witness by rewinding. Nevertheless, the protocol can be readily used as a tool in a bigger protocol that is intended to be UC secure against malicious adversaries. Such a construction is already known from [DPSZ12]. See more details in Sect. 4. Here we also explain more concretely how to use our protocol when f is an encryption function.

1.2   Related Work

On a high level, our approach is related to Luby Transform (LT) codes [Lub02]: here, a sender encodes a codeword by splitting it into blocks of equal size and then sending random sums of these, until the receiver is able to reconstruct all such blocks (because all sums are formed independently, this yields a so-called *erasure code*). We could actually use the LT code approach to construct a protocol like ours, but it would not be a good solution: LT codes do not have to consider any noise expansion because they handle vectors over $\mathbb{Z}_2$, rather than integer vectors. This is a problem since in the worst case a block is reconstructed after n operations, where n is the number of blocks in total, which yields a noise bound that is exponential.

The same bound can be achieved using the technique due to Cramer and Damgård [CD09]. The main technique is to prove linear combinations of ciphertexts using regular 1 out of 2 zero-knowledge proofs. If enough equations are proven correctly, then one can use gaussian elimination to recompute the plaintexts. Unfortunately (as with LT codes) this leads to a blowup in the preimage size that can be exponential, which is not desireable for practical applications.

A different amortization technique was introduced in [DKL+13] and further improved in the full version of [BDTZ16]. The basic idea here is to produce a large number of *auxiliary* ciphertexts, open a part of them and open sums of the plaintexts to be proven and the plaintexts of the auxiliary ciphertexts. This experiment is repeated multiple times, and a combinatorial argument as in [NO09] can then be used to estimate the error probability. As already mentioned above, this proof technique needs $\Omega(\log(k))^2$ auxiliary ciphertexts per proven plaintext, which can be quite substantial for practical applications.

There has been other work conducted for specialized instances of ivOWFs, such as e.g. the proof of plaintext knowledge from [BCK+14] which only applies to Ring-LWE schemes[3]. Moreover the protocol of [LNSW13] can be applied to ivOWFs with a lattice structure, but the protocol comes with a large soundness gap per instance.

Notation. Throughout this work we will format vectors such as $\boldsymbol{b}$ in lower-case bold face letters, whereas matrices such as $\boldsymbol{B}$ will be in upper case. We refer to the ith position of vector $\boldsymbol{b}$ as $\boldsymbol{b}[i]$, let $[r] := \{1, ..., r\}$ and define for $\boldsymbol{b} \in \mathbb{Z}^r$ that $||\boldsymbol{b}|| = \max_{i \in [r]}\{|\boldsymbol{b}[i]|\}$. To sample a variable g uniformly at random from a set G we use $g \xleftarrow{\$} G$. Throughout this work we will let λ be a computational and k be a statistical security parameter. Moreover, we use the standard definition for polynomial and negligible functions and denote those as $\mathsf{poly}(\cdot), \mathsf{negl}(\cdot)$.

2   Homomorphic OWFs and Zero-Knowledge Proofs

In this section we will present an abstraction that covers as a special case proofs of plaintext knowledge for lattice-based cryptosystems, and many other cases as well, as explained in the introduction. We call the abstraction *homomorphic one-way functions over integer vectors*. It follows the standard definition of a OWF which can be found in [KL14].

Let $\lambda \in \mathbb{N}$ be the security parameter, G be an Abelian group, $\beta, r \in \mathbb{N}$, $f : \mathbb{Z}^r \to G$ be a function and $\mathcal{A}$ be any algorithm. Consider the following game:

$\mathsf{Invert}_{\mathcal{A}, f, \beta}(\lambda)$:

(1) Choose $\boldsymbol{x} \in \mathbb{Z}^r, ||\boldsymbol{x}|| \leq \beta$ and compute $y = f(\boldsymbol{x})$.
(2) On input $(1^\lambda, y)$ the algorithm $\mathcal{A}$ computes an $\boldsymbol{x}'$.
(3) Output 1 iff $f(\boldsymbol{x}') = y, ||\boldsymbol{x}'|| \leq \beta$, and 0 otherwise.

Definition 1 (Homomorphic OWF over Integer Vectors (ivOWF)). *A function $f : \mathbb{Z}^r \to G$ is called a homomorphic one-way function over the integers if the following conditions hold:*

(1) There exists a polynomial-time algorithm eval_f such that $\mathsf{eval}_f(\boldsymbol{x}) = f(\boldsymbol{x})$ for all $\boldsymbol{x} \in \mathbb{Z}^r$.
(2) For all $\boldsymbol{x}, \boldsymbol{x}' \in \mathbb{Z}^r$ it holds that $f(\boldsymbol{x}) + f(\boldsymbol{x}') = f(\boldsymbol{x} + \boldsymbol{x}')$.
(3) For every probabilistic polynomial-time algorithm $\mathcal{A}$ there exists a negligible function $\mathsf{negl}(\lambda)$ such that

$$\Pr[\mathsf{Invert}_{\mathcal{A}, f, \beta}(\lambda) = 1] \leq \mathsf{negl}(\lambda)$$

Our definition is rather broad and does capture, among other primitives, lattice-based encryption schemes such as [BGV12, GSW13, BV14] where the one-way property is implied by IND-CPA and β is as large as the plaintext space. Moreover it also captures hash functions such as [GGH96, LMPR08], where it is hard to find a preimage for all *sufficiently* short vectors that have norm smaller than β.

---

[3] Their approach only *almost* yields a proof a plaintext knowledge, due to technical limitations.

2.1 Proving Knowledge of Preimage

Consider a setting with two parties $\mathcal{P}$ and $\mathcal{V}$. $\mathcal{P}$ holds some values $\boldsymbol{x}_1, ..., \boldsymbol{x}_n \in \mathbb{Z}^r$, $\mathcal{V}$ has some $y_1, ..., y_n \in R$ and $\mathcal{P}$ wants to prove towards $\mathcal{V}$ that $y_i = f(\boldsymbol{x}_i)$ and that $\boldsymbol{x}_i$ is *short*, while not giving any knowledge about the $\boldsymbol{x}_i$ away. More formally, the relation that we want to give a zero-knowledge proof of knowledge for is

$$R_{\mathrm{KSP}} = \left\{ (v, w) \ \middle| \ \begin{array}{l} v = (y_1, ..., y_n) \wedge w = (\boldsymbol{x}_1, ..., \boldsymbol{x}_n) \wedge \\[2mm] \left[y_i = f(\boldsymbol{x}_i) \wedge ||\boldsymbol{x}_i|| \le \beta \right]_{i \in [n]} \end{array} \right\}$$

However, like all other protocols for this type of relation, we will have to live with a *soundness slack* τ as explained in the introduction. What this means more precisely is that there must exist a knowledge extractor with properties exactly as in the standard definition of knowledge soundness, but the extracted values only have to satisfy $[y_i = f(\boldsymbol{x}_i) \wedge ||\boldsymbol{x}_i|| \le \tau \cdot \beta]_{i \in [n]}$.

3 Proofs of Preimage

We start by constructing an *imperfect proof of knowledge*. That is, the protocol will allow to prove the above relation with a certain soundness slack, but the knowledge extractor is only required to extract almost all preimages. Furthermore, we will use this protocol as a subprotocol in our actual proof of knowledge. To show knowledge soundness, Goldreich and Bellare [BG93] have shown that it is sufficient to consider deterministic provers, therefore we only need to consider deterministic provers when proving the subprotocol.

On the Use of Rejection Sampling. Conceptually, the idea is to run the naive 3-message protocol π from the intro once for each of the n instances to prove. However, in order to have a small soundness slack, we want to make use of Lyubashevsky's rejection sampling technique [Lyu08, Lyu09]. The idea here is that the prover will sometimes abort the protocol after seeing the challenge if he notices that the random choices he made in the first message will lead him to reveal information about his witness if he were to send the final message. This is fine when used with the Fiat-Shamir heuristic because the prover only has to communicate the successful execution(s). But in our interactive situation, one has to allow for enough failed attempts so that the honest prover will succeed. The most straightforward idea is to have the prover start up one execution of π in parallel for each instance, complete those that are successful and try again for the rest (this was essentially the approach taken in [DKL+13]). The expected number of attempts needed is constant, so we get a protocol that is expected constant round, but may sometimes run for a longer time. Alternatively, the prover could start so many attempts in parallel for each instance that he is sure to finish one of them. This will be exact constant round but wasteful in terms of work needed.

Here, we obtain the best of both worlds. The idea is the following: we can make a large list L of T candidates for the prover's first message, and then do standard cut-and-choose where we open half of them to show that most of the remaining ones are correctly formed. Now, for every instance to prove, the prover will take the first unused one from L that leads to success and complete the protocol for that one. Again, since the *expected* number of attempts for one instance is very close to 1, and we run over many instances, L only needs to be of length $O(n)$, the prover will run out of candidates only with negligible probability. Further, since this can all be done in parallel, we get an exact constant round protocol.

On Extraction by Rewinding from Cut-and-Choose. When we need to extract knowledge from the prover in the imperfect proof, we need to exploit the fact that we do cut-and-choose on the list of candidates L as mentioned above, where each candidate is an image under f. If we just wanted to establish that most of the candidates are well formed in the sense that they are images of short enough inputs, it would be easy: if each candidate is opened with probability $1/2$, then if more than k candidates are not well formed, the prover clearly survives with probability at most 2^{-k}. However, we have to actually extract preimages of almost all candidates. Since we want to avoid using random oracles or other set-up assumptions, we can only resort to rewinding. Now it is not so clear what happens: it may be that all candidates are well formed, but the corrupt prover has some (unknown to us) strategy for which challenges he wants to respond to correctly. All we know is that he will answer a non-negligible fraction of them. We show that nevertheless, there is a rewinding strategy that will do almost as well as in the easy case, and we treat this in a separate general lemma, as we believe the solution to this is of independent interest.

To establish this general point of view, consider any polynomial time computable function $g : X \mapsto Y$ and a generic protocol between a prover $\mathcal{P}$ and a verifier $\mathcal{V}$ we call $\mathcal{P}_{\text{CutnChoose}}$ that works as follows:

(1) $\mathcal{P}$ chooses $x_1, ..., x_T \in X$ such that all x_i satisfy some predicate pre, we say x_i is good if it satisfies pre.
(2) $\mathcal{P}$ sets $y_i = g(x_i)$ for all i and sends $y_1, ..., y_T$ to $\mathcal{V}$.
(3) $\mathcal{V}$ chooses $s \in \{0,1\}^T$ uniformly at random and sends it $\mathcal{P}$.
(4) $\mathcal{P}$ returns $\{x_i \mid s[i] = 0\}$ and $\mathcal{V}$ accepts if $y_i = g(x_i)$ whenever $s[i] = 0$ and each such x_i is good.

Lemma 1 (Cut-and-Choose Rewinding Lemma). *There exists an extractor $\mathcal{E}$ such that the following holds: for any (deterministic) prover $\hat{\mathcal{P}}$ that makes the verifier in $\mathcal{P}_{\text{CutnChoose}}$ accept with probability $p > 2^{-k+1}$, where T is polynomial in k, $\mathcal{E}$ can extract from $\hat{\mathcal{P}}$ at least $T - k$ good x_i-values such that $g(x_i) = y_i$. $\mathcal{E}$ runs in expected time proportional to $O(\text{poly}(s) \cdot k^2/p)$, where s is the size of the inputs.*

Proof. Let $\hat{\mathcal{P}}$ be a deterministic prover that makes $\mathcal{V}$ accept in $\mathcal{P}_{\text{CutnChoose}}$ with probability $p > 2^{-k+1}$. Consider the following algorithm $\mathcal{E}$:

(1) Start $\hat{\mathcal{P}}$, who in turn outputs $y_1, ..., y_T$.

(2) Run T instances of $\hat{\mathcal{P}}$ in parallel, which we denote $\hat{\mathcal{P}}_1, ..., \hat{\mathcal{P}}_T$.

(3) Let $A = \emptyset$ and do the following until $|A| \geq T - k$:

   (3.1) For each $\hat{\mathcal{P}}_i$ sample a random challenge $s_i \xleftarrow{\$} \{0,1\}^T$, subject to $s_i[i] = 0$ and run each $\hat{\mathcal{P}}_i$ on challenge s_i.

   (3.2) For each instance $\hat{\mathcal{P}}_i$ that does not abort, check that the prover's response contains x_i such that $f(x_i) = y_i$. If so, then $A = A \cup \{x_i\}$.

(4) Output A.

We will now show that $\mathcal{E}$ runs in the required time. Denote the probability that $\hat{\mathcal{P}}_i$ outputs a good x_i in step (3) as p_i. We will say that p_i is bad if $p_i < p/k$, and good otherwise.

Let X_i be the event that $\hat{\mathcal{P}}_i$ eventually outputs a good x_i, where $X_i = 1$ if the event happened or $X_i = 0$ otherwise. If p_i is good then, after α iterations

$$\Pr[X_i = 0] = (1 - p/k)^\alpha \leq e^{-p/k \cdot \alpha}$$

so after at most $\alpha = k^2/p$ iterations we can expect that x_i was extracted except with probability negligible in k. This can then be generalized to the success of all $\hat{\mathcal{P}}_i$ (where p_i is good) by a union bound, and the probability of failing is still negligible because T is polynomial in k. Since the experiment of running k^2/p iterations produces success for all good p_i with probability essentially 1, the expected number of times we would need to repeat it to get success is certainly not more than 2, so the claimed expected run time follows, provided there are less than k bad p_i.

Hence, for the sake of contradiction, assume that there are k bad p_i which, for simplicity, are $p_1, ..., p_k$. In the protocol, the challenge s is chosen uniformly at random. The success probability of $\hat{\mathcal{P}}$ can be conditioned on the value of $s[1]$ as

$$p = \Pr[\hat{\mathcal{P}} \text{ succeeds}] = 1/2 \cdot p_1 + 1/2 \cdot \Pr[\hat{\mathcal{P}} \text{ succeeds} \mid s[1] = 1]$$

since p_1 is only of our concern if $s[1] = 0$. Conditioning additionally on $s[2]$ yields

$$p \leq 1/2 \cdot p_1 + 1/2 \cdot (1/2 \cdot 2 \cdot p_2 + 1/2 \cdot \Pr[\hat{\mathcal{P}} \text{ succeeds} \mid s[1] = 1 \wedge s[2] = 1])$$
$$= 1/2 \cdot (p_1 + p_2) + 1/4 \cdot \Pr[\hat{\mathcal{P}} \text{ succeeds} \mid s[1] = 1 \wedge s[2] = 1]$$

The reason the inequality holds is as follows: the probability that a random challenge asking to open a_2 will yield a preimage of a_2 is p_2. Now, conditioning on $s[1] = 1$, which occurs with probability $1/2$, will increase that probability from p_2 to at most $2p_2$.

Repeating the above argument generalizes to

$$p = \Pr[\hat{\mathcal{P}} \text{ succeeds}] \leq 1/2 \cdot (p_1 + ... + p_k) +$$
$$2^{-k} \cdot \Pr[\hat{\mathcal{P}} \text{ succeeds} \mid s[1] = 1 \wedge ... \wedge s[k] = 1]$$
$$< 1/2 \cdot p + 2^{-k}$$

which follows since the first k p_i were bad. But this last inequality implies $p < 2^{-k+1}$, and this contradicts the assumption we started from, that $p > 2^{-k+1}$. $\square$

3.1   The Imperfect Proof of Knowledge

We assume the existence of an auxiliary commitment scheme C_{aux} that is computationally hiding and perfectly binding, and which allows to commit to values from the group G that f maps into. The reason we need it is quite subtle and will show up in the proof of security of $\mathcal{P}_{\text{IMPERFECTPROOF}}$. We will denote a commitment using C_{aux} to a value $x \in G$ as $C_{aux}(x)$ (Fig. 1).

Procedure $\mathcal{P}_{\text{IMPERFECTPROOF}}$

Let f be an ivOWF and C_{aux} be a commitment scheme over G. $\mathcal{P}$ inputs w to the procedure and $\mathcal{V}$ inputs v.

innerProof(v, w, T, τ, β) :

(1) Let $v = (y_1, ..., y_n), w = (\boldsymbol{x}_1, ..., \boldsymbol{x}_n)$. $\mathcal{P}$ samples T values $\boldsymbol{g}_1, ..., \boldsymbol{g}_T \xleftarrow{\$} \mathbb{Z}^r$ such that $\|\boldsymbol{g}_i\| \leq \tau \cdot \beta$.

(2) $\mathcal{P}$ computes $a_i = f(\boldsymbol{g}_i)$, $d_i = C_{aux}(a_i)$ and sends $d = (d_1, ..., d_T)$ to $\mathcal{V}$.

(3) $\mathcal{V}$ samples $\boldsymbol{s} \xleftarrow{\$} \{0,1\}^T$ and sends $\boldsymbol{s}$ to $\mathcal{P}$.

(4) Both parties set $C = \{i \mid \boldsymbol{s}[i] = 1\}, O = [T] \setminus C$.

(5) $\mathcal{P}$ sends $(\boldsymbol{g}_i)_{i \in O}$ as well as the randomness used to generate d_i to $\mathcal{V}$ who checks that $\forall i \in O$:

   (5.1) $a_i = f(\boldsymbol{g}_i)$

   (5.2) $\|\boldsymbol{g}_i\| \leq \tau \cdot \beta$

   (5.3) $d_i = C_{aux}(a_i)$

   If any check fail he aborts.

(6) Let $Z = \emptyset$. For $i \in [n]$ do the following:

   (6.1) Find the first $j \in C \setminus Z$ such that $\|\boldsymbol{g}_j + \boldsymbol{x}_i\| \leq (\tau - 1) \cdot \beta$. If no such j exists then continue. Else set $\boldsymbol{z}_i = \boldsymbol{g}_j + \boldsymbol{x}_i$ and $Z = Z \cup \{1, ..., j\}$.

(7) If $|Z| < n$ then $\mathcal{P}$ aborts.

(8) Else $\mathcal{P}$ sends $(Z, \boldsymbol{z}_1, ..., \boldsymbol{z}_n, (a_j)_{j \in Z})$ as well as the randomness used to generate $d_j, j \in Z$ to $\mathcal{V}$.

(9) $\mathcal{V}$ checks that

   (9.1) $\|\boldsymbol{z}_i\| \leq (\tau - 1) \cdot \beta$

   (9.2) $a_j + y_i = f(\boldsymbol{z}_i)$

   (9.3) $d_j = C_{aux}(a_j)$

   If yes, then he outputs accept. Else, he outputs reject.

Fig. 1. Imperfect proof for the relation R_{KSP}

Theorem 1. *Let f be an ivOWF, k be a statistical security parameter, C_{aux} be a perfectly binding/computationally hiding commitment scheme over G, $\tau = 100 \cdot r$ and $T = 3 \cdot n, n \geq \max\{10, k\}$. Then $\mathcal{P}_{\text{IMPERFECTPROOF}}$ has the following properties:*

Correctness: *If $\mathcal{P}, \mathcal{V}$ are honest and run on an instance of R_{KSP}, then the protocol succeeds with probability at least $1 - \mathsf{negl}(k)$.*

Soundness: *For every deterministic prover $\hat{\mathcal{P}}$ that succeeds to run the protocol with probability $p > 2^{-k+1}$ one can extract at least $n - k$ values $\boldsymbol{x}'_i$ such that*

$f(x_i') = y_i$ and $||x_i'|| \leq 2 \cdot \tau \cdot \beta$, in expected time $O(\text{poly}(s) \cdot k^2/p)$ where s is the size of the input to the protocol.

Zero-Knowledge: The protocol is computational honest-verifier zero-knowledge.

Proof.

Completeness. By the homomorphic property of f, all the checked equations hold. The protocol can only abort if $\mathcal{P}$ aborts, which can only happen in step (7). We first show that $|C| \geq 1.1 \cdot n$ with all but negligible probability for large enough n. Using this, we show that $\Pr[\mathcal{P}_{\text{IMPERFECTPROOF}} \text{ aborts} \mid |C| \geq 1.1 \cdot n]$ is negligible in n.

Let $\#_1(s)$ denote the number of ones in s, then $\#_1(s) \sim \mathcal{BIN}_{1/2,T}$ where $\mathcal{BIN}$ is the Binomial distribution. Using the Chernoff bound we obtain

$$\Pr[\#_1(s) \leq 1.1 \cdot n \mid s \xleftarrow{\$} \{0,1\}^T] \leq \exp\left(-2\frac{(1/2 \cdot T - 1.1 \cdot n)^2}{T}\right)$$

$$= \exp\left(\frac{-32}{300} \cdot n\right)$$

Since $n \geq k$ this becomes negligible for large enough n and we can assume that $|C| \geq 1.1 \cdot n$.

Consider a single coordinate of a z_i. The chance that it fails the bound is $1/\tau$. Each vector has length r, so z_i exceeds the bound with probability $r/\tau = 1/100$. In such a case, $\mathcal{P}$ would take the next (independently chosen) g_j and try again. The ith attempt of $\mathcal{P}$ is denoted as X_i, where $X_i = 1$ if he fails and 0 otherwise. We allow $\mathcal{P}$ to do at most T of these attempts[4]. Then $X_i \sim \mathcal{B}_{1/100}$ and $X \sim \mathcal{BIN}_{1/100,T}, X = \sum X_i$. We set $\overline{X} = \frac{1}{T}X$ where $E[\overline{X}] = 1/100$. Using Hoeffding's inequality, one can show that the probability of failure is

$$\Pr[\overline{X} - E[\overline{X}] \geq 0.09] \leq \exp\left(-2.2 \cdot n \cdot 0.09^2\right)$$

which is negligible in k since we assume $n \geq k$.[5]

Soundness. Let $\hat{\mathcal{P}}$ be a deterministic prover that makes an honest $\mathcal{V}$ accept $\mathcal{P}_{\text{IMPERFECTPROOF}}$ with probability $p > 2^{-k+1}$. Consider the following algorithm $\mathcal{E}_{ImperfectProof}$:

(1) Start $\hat{\mathcal{P}}$, who in turn outputs $d = (d_1, ..., d_T)$.
(2) Observe that the first part of the protocol is an instance of $\mathcal{P}_{\text{CUTNCHOOSE}}$ with $g = C_{aux} \circ f$ and where a preimage g_i is good if $||g_i|| \leq \tau \cdot \beta$, We therefore run the extractor $\mathcal{E}$ guaranteed by Lemma 1 which gives us a set A with $T - k$ good g_i-values.

---

[4] The probability that $\mathcal{P}$ needs more auxiliary ciphertexts is $\approx 0.63^n$ and therefore negligible in n.

[5] In fact, setting $n = 40$ already makes $\mathcal{P}$ abort with probability 2^{-10}.

(3) Let $X = \emptyset$ and do the following until $|X| \geq n - k$:

(3.1) Run a regular instance with $\hat{\mathcal{P}}$.

(3.2) If the instance was accepting, then for each $\boldsymbol{z}_i$ with a corresponding $\boldsymbol{g}_j \in A, j \in C$ add the preimage to X, i.e. $X = X \cup \{\boldsymbol{z}_i - \boldsymbol{g}_j\}$.

(4) Output X.

We will now show that $\mathcal{E}_{ImperfectProof}$ runs in the required time. The run-time of $\mathcal{E}$ was established in Lemma 1. Using the set A it outputs, we can now argue that step (3) also terminates as required: $\mathcal{E}_{ImperfectProof}$ reaches step (3.2) after an expected number of $1/p$ rounds. At most k of the T preimages of a_j are not given in A and therefore step (3.2) is only executed once. From the bound on the a_i, the bound on the extracted $\boldsymbol{x}_i$ immediately follows.

Zero-Knowledge. Consider the following algorithm $\mathcal{S}_{\text{IMPERFECTPROOF}}$

(1) On input $(v = (y_1, ..., y_n), T, \tau, \beta)$ sample the string $\boldsymbol{s} \xleftarrow{\$} \{0,1\}^T$ as in the protocol.

(2) Compute the sets C, O as in $\mathcal{P}_{\text{IMPERFECTPROOF}}$. For each $i \in O$ sample $\boldsymbol{g}_i \in \mathbb{Z}^r, ||\boldsymbol{g}_i|| \leq \tau \cdot \beta$ and set $a_i = f(\boldsymbol{g}_i)$ as well as $d_i = C_{aux}(a_i)$.

(3) For $i \in C$ sample $\boldsymbol{z}_i \in \mathbb{Z}^r, ||\boldsymbol{z}_i|| \leq \tau \cdot \beta$ uniformly at random. Let $Z' := \{i \in C \mid ||\boldsymbol{z}_i|| \leq (\tau - 1) \cdot \beta\}$.

(4) If $|Z'| < n$ then for $i \in C$ set $\boldsymbol{g}_i = \boldsymbol{z}_i, a_i = f(\boldsymbol{z}_i), d_i = C_{aux}(a_i)$, output $(\boldsymbol{s}, d_1, ..., d_T, (a_i, \boldsymbol{g}_i)_{i \in O})$ and abort.

(5) If $|Z'| \geq n$ then let Z be the first n elements of Z'. For each $i \in C \setminus Z$ set $a_i = f(\boldsymbol{z}_i), d_i = C_{aux}(a_i)$.

(6) Denote Z as $Z = \{i_1, ..., i_n\}$. For all $i_j \in Z$ set $a_{i_j} = f(\boldsymbol{z}_{i_j}) - y_j, d_{i_j} = C_{aux}(a_{i_j})$.

(7) Output $(\boldsymbol{s}, d_1, ..., d_T, (a_i, \boldsymbol{g}_i)_{i \in O}, Z, (a_i, \boldsymbol{z}_i)_{i \in Z})$.

In the simulation, we can assume that there exists a witness w for v according to relation R_{KSP}. We first observe that if an a_i and its randomness when generating a d_i are ever revealed, then it holds that $d_i = C_{aux}(a_i)$. For those commitments that are not opened the computational hiding property implies that their distribution in the simulated case is indistinguishable from the real protocol.

What remains to study is the abort probability of the protocol, the sets C, O, Z and the $a_i, z_i, \boldsymbol{g}_i$. The choice of C, O is identical in $\mathcal{P}_{\text{IMPERFECTPROOF}}$, $\mathcal{S}_{\text{IMPERFECTPROOF}}$ for an honest verifier since they are computed the same way.

Abort Probability and Z. The probability of abort of $\mathcal{S}_{\text{IMPERFECTPROOF}}$ in step (4) is the same as in (7) in $\mathcal{P}_{\text{IMPERFECTPROOF}}$. This indeed is true if $\#_1(\boldsymbol{s}) < n$ and also if $\#_1(\boldsymbol{s}) \geq n, |Z'| < n$. The second is a little more subtle and can be seen by arguing what the chance is that a certain z_i ends up in Z': for the sake of simplicity, assume $n = r = 1$ since all these vectors and their entries are chosen i.i.d. In the above simulator, $z \in [-\tau \cdot \beta, \tau \cdot \beta]$ was chosen uniformly at random. Hence $z \notin Z'$ with probability $1/100$. In the case of $\mathcal{P}_{\text{IMPERFECTPROOF}}$ we have $z = x + g$ where $g \in [-\tau \cdot \beta, \tau \cdot \beta]$ was chosen uniformly at random and $x \in [-\beta, \beta]$, i.e. $z \in [-\tau \cdot \beta + x, \tau \cdot \beta + x]$ chosen uniformly at random from a *shifted* interval of equal length. But $[-(\tau-1)\cdot\beta, (\tau-1)\cdot\beta] \subset [-\tau\cdot\beta+x, \tau\cdot\beta+x]$ always holds due to the upper bound of x, hence the probability of abort is also $1/100$. By the same reasoning, Z has the same distribution in $\mathcal{S}_{\text{IMPERFECTPROOF}}$ and $\mathcal{P}_{\text{IMPERFECTPROOF}}$.

Distribution of g_j, a_j for $j \in O$. Due to the homomorphism of f, the checks from step (9) do also hold on the simulated output. For all $j \in O$ the distribution of the g_j, a_j is the same in both the protocol and $\mathcal{S}_{\text{IMPERFECTPROOF}}$ as the values are chosen exactly the same way.

Distribution of z_i, a_i for $i \in Z$. Consider the distribution of the z_{i_j}, a_{i_j} for $i_j \in Z$ when $\mathcal{S}_{\text{IMPERFECTPROOF}}$ runs successfully. By the above argument, the distribution of the z_{i_j} is independent of the x_j in $\mathcal{P}_{\text{IMPERFECTPROOF}}$. In $\mathcal{P}_{\text{IMPERFECTPROOF}}$ exactly those z_{i_j} will be sent to $\mathcal{V}$ where $z_{i_j} = g_{i_j} + x_j$ is in the correct interval. Since by our assumption there exists a witness $w = (x'_1, ..., x'_n)$ then due to the linearity of f there must exist a g'_{i_j} of the same bound as the g_i in the protocol, where $a_{i_j} = f(g'_{i_j})$ by linearity.

Why Using C_{aux}? It may not be directly obvious from the above proof why the commitment C_{aux} is necessary. But a problem can occur in step (7) of $\mathcal{P}_{\text{IMPERFECTPROOF}}$ with the elements with indices from $C \setminus Z$: although the simulator can simulate perfectly the choice of this set, we would have a problem if we had to reveal the corresponding $f(g_i)$-values. The issue is that the g_i's should be values that cause an abort and we cannot choose such values unless we know the corresponding secrets. One solution is to apply f to a random input and make a non-standard assumption that this cannot be distinguished from the real thing, but this is undesirable. Instead, sending $C_{aux}(a_i)$ allows to both hide the distribution, while soundness is still guaranteed because $C_{aux} \circ f$ is hard to invert due to the binding property of C_{aux}. $\qquad\qquad\square$

3.2 The Full Proof of Knowledge

We use the above imperfect protocol as a building block of the actual proof. After executing it with the (x_i, y_i) as input, we can assume that a preimage of most of the y_i's (in fact, all but k) can be extracted from the prover.

Our strategy for the last part of the protocol is to repeat the following procedure several times: we let the verifier randomly assign each y_i to one of several *buckets*. Then, for each bucket, we add all elements that landed in the bucket and have the prover demonstrate that he knows a preimage of the sum. The observation is that since we only have k elements left that were "bad" in the sense that we could not yet extract a preimage, then if we have more than k buckets, say ck for a constant $c > 1$, there is a significant probability that many of the bad elements will be alone in a bucket. If this happens, we can extract a preimage by linearity of f. Furthermore, the cost of doing such a step is at most n additions, plus work that only depends on the security parameter k and is insignificant if $n \gg k$. Now, by repeating this game some number of times with the right number of buckets, we shall see that we can extract all preimages quite quickly.

In the following, the experiment where we throw n values randomly into b buckets will be denoted $\text{Exp}(b, n)$. As is apparent from the above discussion, we will need to analyse the probability that the bad values will be "killed" by being alone in a bucket. That is, we need to consider $\text{Exp}(b, v)$, where $v \leq k$ can be thought of as the number of bad elements. We will say that an element *survives* if it is not alone in a bucket. We will write t independent repetitions of the experiment as $\text{Exp}^t(b, v)$ and we say that an element survives this if it survives in every repetition. The following lemma will be helpful:

Lemma 2. *Notation as above. Consider* $\text{Exp}^t(b, v)$ *and assume* $b \geq 4v$ *and* $t \geq 8$. *Then the probability* p *that at least* $v/4$ *elements survive satisfies* $p \leq \frac{3v}{4} \epsilon^{tv}$, *where* $\epsilon = \frac{e^{5/32}}{2^{5/16}} \approx 0.94$.

Proof. Consider the event of exactly s bad elements surviving $\text{Exp}^t(b, v)$ where $s \geq v/4$. The s surviving elements could be any of the v values, but must cover less than $s/2$ buckets in each repeated experiment, since surviving elements are not alone in a bucket. From this we get the bound

$$\Pr\left[s \text{ survive}\right] \leq \binom{v}{s}\left(\binom{b}{s/2}\left(\frac{s/2}{b}\right)^s\right)^t$$

on which we apply upper bounds on the binomial coefficients:

$$\leq \left(\frac{ve}{s}\right)^s\left(\left(\frac{be}{s/2}\right)^{s/2}\left(\frac{s/2}{b}\right)^s\right)^t$$

$$= \left(\frac{ve}{s}\right)^s\left(\frac{se}{2b}\right)^{ts/2}$$

$$= \left(\left(\frac{ve}{s}\right)^{1/t}\left(\frac{se}{2b}\right)^{1/2}\right)^{ts}$$

and finally maximize using $b \geq 4v, t \geq 8$ and $s \in [v/4, v]$:

$$\leq \left(\left(\frac{ve}{v/4}\right)^{1/8}\left(\frac{ve}{2 \cdot 4v}\right)^{1/2}\right)^{tv/4}$$

$$= \left((4e)^{1/8}\left(\frac{e}{8}\right)^{1/2}\right)^{tv/4}$$

$$= \left(\frac{e^{5/32}}{2^{5/16}}\right)^{tv}$$

Using this we can bound the probability p by union bound:

$$p = \Pr\left[\geq \frac{v}{4} \text{ survive}\right] \leq \sum_{s=v/4}^{v} \Pr\left[s \text{ survive}\right] \leq \frac{3v}{4}\left(\frac{e^{5/32}}{2^{5/16}}\right)^{tv}$$

$\square$

Before we continue, let us discuss the implications of the above Lemma. First, due to the first equation of the proof we yield that, except with probability p, in an instance of $\mathrm{Exp}^t(b, v)$ *at least one* of the t iterations contains at least $3v/4$ buckets with single elements. A second, somewhat surprising fact is that for fixed values of t, b the probability p is not monotone in v. This will be particularly difficult, as we only know upper bounds on the value v in the proof of the main protocol.

Our soundness argument implicitly defines an extraction algorithm that runs in $\log_2(k)$ rounds, where in each round the same total number of buckets is used (the number of buckets per iteration drops in half, but the total number of iterations per round doubles). What we then show (using the above Lemma) is that the upper bound on the number of unextracted preimages is reduced by a factor of 2 between each two successive rounds, while the error probability stays somewhat constant. This is due to the following thought experiment: assume as an invariant that, for $O(k)$ buckets and k balls, at least $k/2$ of these balls land in their own bucket except with probability $2^{-O(k)}$. By running the experiment again, we see that the error probability now increases because we now only use $k/2$ balls. But by independently running the experiment twice, one obtains that half of the $k/2$ balls are alone (in one of the two experiments) except with essentially the same probability as before. This now allows for a recursive extraction strategy.

Theorem 2. *Let f be an ivOWF, k be a statistical security parameter, β be a given upper bound and $n > k \cdot \log_2(k)$. Then $\mathcal{P}_{\mathrm{COMPLETEPROOF}}$ is an interactive honest-verifier zero-knowledge proof of the relation R_{KSP} with knowledge error 2^{-k+1}. More specifically, it has the following properties (Fig. 2):*

Correctness: *If $\mathcal{P}, \mathcal{V}$ are honest then the protocol succeeds with probability at least $1 - 2^{-O(k)}$.*

Soundness: *For every deterministic prover $\hat{\mathcal{P}}$ that succeeds to run the protocol with probability $p > 2^{-k+1}$ one can extract n values $\boldsymbol{x}_i'$ such that $f(\boldsymbol{x}_i') = y_i$ and $||\boldsymbol{x}_i'|| \leq O((2k+1)^{\log_2(k)/2} \cdot n \cdot r \cdot \beta)$ except with negligible probability, in expected time $\mathsf{poly}(s, k)/p$, where s is the size of the input to the protocol.*

Zero-Knowledge: *The protocol is computational honest-verifier zero-knowledge.*

Proof.

Correctness. The first call to $\mathcal{P}_{\mathrm{IMPERFECTPROOF}}$ in step (1) will succeed with all but negligible probability due to Theorem 1. For each i in step (2) the experiment is repeated 2^{i+4} times using $4k \cdot 2^{-i}$ buckets, hence the total number of sums for each such round is $64k$ which determines h. A set I_j as chosen in step (3) can have size at most n by definition, therefore $||\boldsymbol{\delta}_j|| \leq \beta \cdot n$. The call to $\mathcal{P}_{\mathrm{IMPERFECTPROOF}}$ in step (4) will then be successful according to Theorem 1 with overwhelming probability.

Soundness. We will first prove the existence of an efficient extractor, then give a bound on the extraction probability and only establish the bound on the norm of the preimage afterwards.

Procedure $\mathcal{P}_{\text{COMPLETEPROOF}}$

Let f be an ivOWF. $\mathcal{P}$ inputs w to the procedure and $\mathcal{V}$ inputs v. We assume for simplicity that the security parameter k is a 2-power.

proof(v, w, β) :

(1) Let $v = (y_1, ..., y_n)$, $w = (\boldsymbol{x}_1, ..., \boldsymbol{x}_n)$. Run innerProof$(v, w, 3n, 100r, \beta)$. If $\mathcal{V}$ in $\mathcal{P}_{\text{IMPERFECTPROOF}}$ aborts then abort, otherwise continue.

(2) For $i = 0, 1, \ldots, \log_2(k)$, execute (in parallel) $\text{Exp}^{t_i}(b_i, n)$, where $t_i = 2^{i+4}, b_i = 4k \cdot 2^{-i}$ and where $\mathcal{V}$ chooses the randomness for all the experiments, i.e., chooses how to distribute elements in buckets.

(3) Let $h = 64k \cdot (\log_2(k) + 1)$ be the total number of buckets used in all the experiments in the previous step and order all the buckets in some arbitrary order. Bucket j contains some subset of the n input values, let these be designated by index set I_j. Now, for $j = 1, \ldots, h$, both players compute $\gamma_j = \sum_{i \in I_j} v_i$ and $\mathcal{P}$ also computes $\boldsymbol{\delta}_j = \sum_{i \in I_j} \boldsymbol{x}_i$.

(4) Run innerProof$(\gamma, \boldsymbol{\delta}, 3h, 100r, n\beta)$. If $\mathcal{V}$ in $\mathcal{P}_{\text{IMPERFECTPROOF}}$ aborts then abort, otherwise accept.

Fig. 2. A protocol to prove the relation R_{KSP}

An Efficient Extractor. From the subprotocol used in step (1) and Theorem 1 all but k of the n ciphertexts can be extracted. The same holds for step (4) from which we can argue that at most k of the h sums are proven incorrectly. For each i, observe that of the $t_i = 2^{i+4}$ iterations, there must be at least 2^{i+3} of them that each contain at most $2^{-i} \cdot k/4$ bad buckets. For otherwise, we would have at least 2^{i+3} iterations that each have at least $2^{-i} \cdot k/4$ buckets which adds up to $2k$ bad buckets.

For $i = 0, 1, ..$ the number of bad values entering into the experiment $\text{Exp}^{t_i}(b_i, n)$ is $v_i \le k \cdot 2^{-i}$, except with negligible probability (we will consider the error probability later). This can be seen as follows: for $i = 0$, we have $v_0 \le k$ due to step (1). So let $v_i \le 2^{-i}k$, then by the proof of Lemma 2 at least one of the 2^{i+3} iterations, $3/4 \cdot v_i$ or more buckets contain only one of the not-yet extracted elements and can hence be extracted now. For this instance, we established that at most $2^{-i} \cdot k/4$ of the sums can be bad, hence

$$v_{i+1} \le v_i/4 + k/4 \cdot 2^{-i} \le k/4 \cdot 2^{-i} + k/4 \cdot 2^{-i} = k \cdot 2^{-i-1}$$

Hence after $\text{Exp}^{t_i}(b_i, v_i)$ we can extract at least $v_i/2$ of the ciphertexts. In the last round we have $v_{\log_2(k)} \le 2$ and must prove that after $\text{Exp}^{t_{\log_2(k)}}(b_{\log_2(k)}, v_{\log_2(k)})$, no unextracted preimages are left. Therefore consider the following two cases:

$v_{\log_2(k)} = 1$ In this case, for the experiment to fail the remaining unextracted preimage must be in the bad sum for all $8k$ instances. For each such instance, there are 4 buckets out of which at most 1 can be bad. The extraction will hence only fail with probability 2^{-16k}.

$v_{\log_2(k)} = 2$ To be able to extract, we want both unextracted preimages to be in different buckets and none of them in a bad bucket. The chance that the first preimage ends up in a bad bucket is $1/4$, and the second preimage can either fall into the bucket with the first preimage (with probability $1/4$) or in the bad bucket, so in total with probability at most $3/4$ one of the $8k$ iterations will fail, and all will be bad with probability at most $(3/4)^{8k} < 2^{-2k}$.

By a union bound, the last experiment will fail with probability at most $p_{log_2(k)} = 2^{-k}$.

For rounds $i = 0, ..., \log_2(k) - 1$, the extractor will only extract from the experiment i if $k \cdot 2^{-i-1} \leq v_i \leq k \cdot 2^{-i}$ and otherwise safely continue with round $i + 1$. By Lemma 2, extraction will fail in this round with probability at most

$$p_i \leq \max_{\tilde{v}_i \in [k \cdot 2^{-i-1}, k \cdot 2^{-i}]} \{3/4 \tilde{v}_i \cdot \epsilon^{t_i \cdot \tilde{v}_i}\}$$
$$< 3/4k \cdot 2^{-i} \cdot \epsilon^{2^{i+3} \cdot k \cdot 2^{-i-1}}$$
$$= 3k \cdot 2^{2-i} \cdot \epsilon^{4k}$$

because the actual value of v_i is unknown. The extraction process can fail if it fails in one of the experiments. By a union bound, we obtain

$$p_0 + ... + p_{\log_2(n)} < 3k \cdot 2^2 \cdot \epsilon^{4k} + ... + 3k \cdot 2^{2-\log_2(k)+1} \cdot \epsilon^{4k} + 2^{-k}$$
$$= 3k \cdot \epsilon^{4k} \cdot \sum_{j=0}^{\log_2(k)-1} 2^{2-j} + 2^{-k}$$
$$< 24k \cdot \epsilon^{4k} + 2^{-k}$$

which is in $2^{-O(k)}$ because $\epsilon < 1$ and constant. Since soundness for Theorem 1 fails with probability $2^{-O(k)}$ as well, this proves the claim.

Extraction Bound. Let $\tau = 100r$ be the slackness chosen for the instances of innerProof. Consider a value x_i' extracted in round 0, i.e. there exists a *good* $\delta_j, i \in I_j$ such that $x_i' = \delta_j - \sum_{o \in I_j \setminus \{x_i'\}} x_o'$ where all such x_o' were already extracted from $\mathcal{P}_{\text{IMPERFECTPROOF}}$ in step (1). Then

$$||x_i'|| \leq ||\delta_j - \sum_{o \in I_j \setminus \{i\}} x_o'||$$
$$\leq ||\delta_j|| + ||\sum_{o \in I_j \setminus \{i\}} x_o'||$$
$$\leq 2 \cdot \tau \cdot n \cdot \beta + (n-1) \cdot 2 \cdot \tau \cdot \beta$$
$$< 4 \cdot n \cdot \tau \cdot \beta$$
$$:= \beta_0$$

In round 1 each preimage that we extract will be a sum of preimage known from the cut-and-choose phase and those from round 0, where from the last round at most $k/2$ can be part of the sum. Calling this upper bound β_1 we obtain

$$\beta_1 = 2 \cdot \tau \cdot n \cdot \beta + \frac{k}{2}\beta_0$$

The above argument easily generalizes to an arbitrary round $i > 0$ where it then holds that

$$\beta_i = 2 \cdot \tau \cdot n \cdot \beta + \sum_{j=1}^{i-1} \frac{k}{2^j} \beta_{j-1}$$

because in round 0 we extracted at most $k/2$ preimages, in round 1 $k/4$ and so on. In particular, the above can be rewritten as

$$\beta_i = 2 \cdot \tau \cdot n \cdot \beta + \sum_{j=1}^{i} \frac{k}{2^j} \beta_{j-1}$$

$$= 2 \cdot \tau \cdot n \cdot \beta + \sum_{j=1}^{i-1} \frac{k}{2^j} \beta_{j-1} + \frac{k}{2^i} \beta_{i-1}$$

$$\leq \beta_{i-1} + \frac{k}{2^i} \beta_{i-1}$$

$$= \left(\frac{k}{2^i} + 1 \right) \beta_{i-1}$$

In particular, for the bound on the last preimages that are extracted in round $\log_2(k)$ one obtains

$$\beta_{\log_2(k)} = \prod_{i=1}^{\log_2(k)-1} \left(\frac{k}{2^i} + 1 \right) \beta_0$$

To compute a bound on the leading product, we consider the square of the above bound and reorder the terms as

$$\prod_{i=1}^{\log_2(k)-1} \left(\frac{k}{2^i} + 1 \right)^2 = \prod_{i=1}^{\log_2(k)-1} \left(\frac{k}{2^i} + 1 \right) \left(\frac{k}{2^{\log_2(k)-i}} + 1 \right)$$

$$= \prod_{i=1}^{\log_2(k)-1} \left(k + \frac{k}{2^i} + \frac{k}{2^{\log_2(k)-i}} + 1 \right)$$

$$< (2k+1)^{\log_2(k)}$$

and we can conclude that

$$\beta_{\log_2(k)} < (2k+1)^{\log_2(k)/2} \cdot 4 \cdot n \cdot \tau \cdot \beta$$

Zero-Knowledge. The simulation algorithm chooses the randomness for all the experiments like an honest $\mathcal{V}$ would do and then uses the simulator from Theorem 1 to simulate the calls to $\mathcal{P}_{\mathrm{IMPERFECTPROOF}}$. The computational HVZK property then follows directly from Theorem 1.   □

4   Applications

As a first general remark, we note that even though our protocol is only honest verifier zero-knowlegde and proved sound using extraction by rewinding, we can nevertheless use it as a tool in a bigger protocol that is intended to be UC secure against malicious adversaries. Such a construction is already known from [DPSZ12]. The idea is first to generate the verifier's challenge using a secure coin-flip protocol. Then honest verifier zero-knowledge suffices, and the cost of generating the challenge can be amortised over several proofs. Second, if the function f is an encryption function, the UC simulator can be given the secret key and can hence extract straight-line. Rewinding then only takes place in the reduction to show that the UC simulator works.

In the rest of this section we will first show how to rephrase lattice-based encryption as ivOWFs and then show how to generalize the result from the previous section such that it applies in this setting.

4.1   Encryption as ivOWFs

As an example, let us consider a variant of the homomorphic encryption scheme due to Brakerski et al. [BGV12]. Let $n, N, \lambda \in \mathbb{N}^+, p, q \in \mathbb{P}$ and $q \gg p$. Moreover, let χ be a distribution over $\mathbb{Z}$ such that, with overwhelming probability in k, $e \leftarrow \chi \Rightarrow |e| \leq q/2$. We consider λ to be the computational security parameter.

$\mathsf{KG}(1^\lambda)$: Sample $t \xleftarrow{\$} \chi^n, e \xleftarrow{\$} \chi^N$ and $\boldsymbol{B} \leftarrow \mathbb{Z}_q^{N \times n}$. Let $\boldsymbol{u}_1 = (1, 0, ..., 0) \in \mathbb{Z}_q^{n+1}$ be the unit vector for the first coordinate. Then compute

$$\boldsymbol{b} \leftarrow \boldsymbol{B}t + p \cdot \boldsymbol{e}$$

$$\boldsymbol{A} \leftarrow \left(\boldsymbol{u}_1^T \,\middle\|\, \begin{pmatrix} \boldsymbol{b}^T \\ -\boldsymbol{B}^T \end{pmatrix} \,\middle\|\, p \cdot \boldsymbol{I}_{n+1} \right)$$

where $\boldsymbol{I}_{n+1}$ is the identity matrix with $n + 1$ rows and columns. Output $\mathsf{pk} \leftarrow \boldsymbol{A}, \mathsf{sk} \leftarrow t$.

$\mathsf{Enc}_{\mathsf{pk}}\left(\begin{pmatrix} m \\ \boldsymbol{r} \end{pmatrix} \right)$: Check that $m \in \mathbb{Z}_p, \boldsymbol{r} \in \mathbb{Z}_q^{N+n+1}$ and output

$$\boldsymbol{c} \leftarrow \boldsymbol{A} \times \begin{pmatrix} m \\ \boldsymbol{r} \end{pmatrix}$$

$\mathsf{Dec}_{\mathsf{sk}}(\boldsymbol{s})$: Compute

$$m' \leftarrow \left(\left\langle \boldsymbol{c}, \begin{pmatrix} 1 \\ t \end{pmatrix} \right\rangle \bmod q \right) \bmod p$$

and output $m' \in \mathbb{Z}_p$.

For appropriately chosen parameters, the function $\mathsf{Enc}_{\mathsf{pk}}$ is an ivOWF (by the natural embedding of $\mathbb{Z}_q$ into the integers) assuming that the LWE problem is hard. It therefore seems natural to apply our proof framework in the above setting.

Unfortunately we have to show different bounds for different indices of the preimage, which is impossible for the existing proof.

Procedure $\mathcal{P}_{CompleteProof,f,\beta}$

$\mathcal{P}$ inputs w to the procedure and $\mathcal{V}$ inputs v. We assume for simplicity that the security parameter k is a 2-power.

pG(v, w, β) :

(1) Let $v = (y_1, ..., y_n), w = (x_1, ..., x_n)$. Run iPG$(v, w, 3n, 100r, \beta)$. If $\mathcal{V}$ in $\mathcal{P}_{ImperfectProof,f,\beta}$ aborts then abort, otherwise continue.

(2) For $i = 0, 1, ..., \log_2(k)$, execute (in parallel) Exp$^{t_i}(b_i, n)$, where $t_i = 2^{i+4}, b_i = 4k \cdot 2^{-i}$ and where $\mathcal{V}$ chooses the randomness for all the experiments, i.e., chooses how to distribute elements in buckets.

(3) Let $h = 64k \cdot (\log_2(k) + 1)$ be the total number of buckets used in all the experiments in the previous step and order all the buckets in some arbitrary order. Bucket j contains some subset of the n input values, let these be designated by index set I_j. Now, for $j = 1, ..., h$, both players compute $\gamma_j = \sum_{i \in I_j} v_i$ and $\mathcal{P}$ also computes $\delta_j = \sum_{i \in I_j} x_i$.

(4) Run iPG$(\gamma, \delta, 3h, 100r, n\beta)$. If $\mathcal{V}$ in $\mathcal{P}_{ImperfectProof,f,\beta}$ aborts then abort, otherwise accept.

Fig. 3. A protocol to prove the relation $R_{\text{KSP},f,\beta}$

4.2 Refining the Proof Technique

To gain more flexibility, we start out by defining a predicate $\mathsf{InfNorm}_\beta$, which we define as follows

$$\mathsf{InfNorm}_\beta(x) = \begin{cases} \top & \text{if } \beta \in \mathbb{N}^+ \wedge \forall i \in [r] : |x[i]| \leq \beta[i] \\ \bot & \text{else} \end{cases}$$

where β is supposed to be a *coordinatewise upper bound* on x.

We call a vector $x \in \mathbb{Z}^r$ to be β-*bounded* iff $\mathsf{InfNorm}_\beta(x) = \top$. For a function $f : \mathbb{Z}^r \to G$ and the set $\{c_1, ..., c_t\}$ one then tries to prove the following relation

$$R_{\text{KSP},f,\beta} = \Big\{ (v, w) \mid \quad v = (c_1, ..., c_t) \wedge w = (x_1, ..., x_t) \wedge$$

$$\big[c_i = f(x_i) \wedge \mathsf{InfNorm}_\beta(x_i)\big]_{i \in [t]} \Big\}$$

That is, a proof of plaintext knowledge for our defined cryptosystem would then set $\beta \in \mathbb{N}^{N+n+2}, f = \mathsf{Enc}_{\mathsf{pk}}$ with $\beta[1] = \beta_P, \beta[2] = ... = \beta[N + n + 2] = \beta_R$ where β_P is the bound on the plaintext and β_R on the randomness. One then uses a modified version of the proof $\mathcal{P}_{\text{IMPERFECTPROOF}}$, namely the protocol from Fig. 4 and moreover replaces $\mathcal{P}_{\text{COMPLETEPROOF}}$ with Fig. 3.

Theorems 1 and 2 directly generalize to the above setting due to the linearity of all operations (if the simulators for the rejection sampling just sample from the appropriate bound for each coordinate). This is possible because none of the success probabilities changes since these are independent of the bound β in the

---

Procedure $\mathcal{P}_{ImperfectProof,f,\boldsymbol{\beta}}$

Let f be an ivOWF and C_{aux} be a commitment scheme over G. $\mathcal{P}$ inputs w to the procedure and $\mathcal{V}$ inputs v.

$\mathsf{iPG}(v, w, T, \tau, \boldsymbol{\beta})$:

(1) Let $v = (y_1, ..., y_n), w = (\boldsymbol{x}_1, ..., \boldsymbol{x}_n)$. $\mathcal{P}$ samples T values $\boldsymbol{g}_1, ..., \boldsymbol{g}_T \xleftarrow{\$} \mathbb{Z}^r$ such that $\mathsf{InfNorm}_{\tau \cdot \boldsymbol{\beta}}(\boldsymbol{g}_i)$.

(2) $\mathcal{P}$ computes $a_i = f(\boldsymbol{g}_i)$, $d_i = C_{aux}(a_i)$ and sends $d = (d_1, ..., d_T)$ to $\mathcal{V}$.

(3) $\mathcal{V}$ samples $\boldsymbol{s} \xleftarrow{\$} \{0, 1\}^T$ and sends $\boldsymbol{s}$ to $\mathcal{P}$.

(4) Both parties set $C = \{i \mid \boldsymbol{s}[i] = 1\}$, $O = [T] \setminus C$.

(5) $\mathcal{P}$ sends $(\boldsymbol{g}_i)_{i \in O}$ as well as the randomness used to generate d_i to $\mathcal{V}$ who checks that $\forall i \in O$:

    (5.1) $a_i = f(\boldsymbol{g}_i)$

    (5.2) $\mathsf{InfNorm}_{\tau \cdot \boldsymbol{\beta}}(\boldsymbol{g}_i)$

    (5.3) $d_i = C_{aux}(a_i)$

    If any check fail he aborts.

(6) Let $Z = \emptyset$. For $i \in [n]$ do the following:

    (6.1) Find the first $j \in C \setminus Z$ such that $\|\boldsymbol{g}_j + \boldsymbol{x}_i\| \leq (\tau - 1) \cdot \boldsymbol{\beta}$. If no such j exists then continue. Else set $\boldsymbol{z}_i = \boldsymbol{g}_j + \boldsymbol{x}_i$ and $Z = Z \cup \{1, ..., j\}$.

(7) If $|Z| < n$ then $\mathcal{P}$ aborts.

(8) Else $\mathcal{P}$ sends $(Z, \boldsymbol{z}_1, ..., \boldsymbol{z}_n, (a_j)_{j \in Z})$ as well as the randomness used to generate $d_j, j \in Z$ to $\mathcal{V}$.

(9) $\mathcal{V}$ checks that

    (9.1) $\mathsf{InfNorm}_{(\tau - 1) \cdot \boldsymbol{\beta}}(\boldsymbol{z}_i)$

    (9.2) $a_j + y_i = f(\boldsymbol{z}_i)$

    (9.3) $d_j = C_{aux}(a_j)$

    If yes, then he outputs accept. Else, he outputs reject.

Fig. 4. Imperfect proof for the relation $R_{\mathrm{KSP},f,\boldsymbol{\beta}}$

first place. The above could be generalized to other predicates which e.g. enforce ℓ_2-norms. We leave this as future work.

References

[BCK+14] Benhamouda, F., Camenisch, J., Krenn, S., Lyubashevsky, V., Neven, G.: Better zero-knowledge proofs for lattice encryption and their application to group signatures. In: Sarkar, P., Iwata, T. (eds.) ASIACRYPT 2014. LNCS, vol. 8873, pp. 551–572. Springer, Heidelberg (2014)

[BDOZ11] Bendlin, R., Damgård, I., Orlandi, C., Zakarias, S.: Semi-homomorphic encryption and multiparty computation. In: Paterson, K.G. (ed.) EURO-CRYPT 2011. LNCS, vol. 6632, pp. 169–188. Springer, Heidelberg (2011)

[BDTZ16] Baum, C., Damgård, I., Toft, T., Zakarias, R.: Better preprocessing for secure multiparty computation. In: Manulis, M., Sadeghi, A.-R., Schneider, S. (eds.) ACNS 2016. LNCS, vol. 9696, pp. 327–345. Springer, Heidelberg (2016). doi:10.1007/978-3-319-39555-5_18

[BG93] Bellare, M., Goldreich, O.: On defining proofs of knowledge. In: Brickell, E.F. (ed.) CRYPTO 1992. LNCS, vol. 740, pp. 390–420. Springer, Heidelberg (1993)

[BGV12] Brakerski, Z., Gentry, C., Vaikuntanathan, V.: (Leveled) fully homomorphic encryption without bootstrapping. In: Proceedings of the 3rd Innovations in Theoretical Computer Science Conference, ITCS 2012, pp. 309–325. ACM, New York (2012)

[BV14] Brakerski, Z., Vaikuntanathan, V.: Efficient fully homomorphic encryption from (standard) LWE. SIAM J. Comput. **43**(2), 831–871 (2014)

[CD09] Cramer, R., Damgård, I.: On the amortized complexity of zero-knowledge protocols. In: Halevi, S. (ed.) CRYPTO 2009. LNCS, vol. 5677, pp. 177–191. Springer, Heidelberg (2009)

[DF02] Damgård, I.B., Fujisaki, E.: A statistically-hiding integer commitment scheme based on groups with hidden order. In: Zheng, Y. (ed.) ASIACRYPT 2002. LNCS, vol. 2501, pp. 125–142. Springer, Heidelberg (2002)

[DKL+13] Damgård, I., Keller, M., Larraia, E., Pastro, V., Scholl, P., Smart, N.P.: Practical covertly secure MPC for dishonest majority – or: breaking the SPDZ limits. In: Crampton, J., Jajodia, S., Mayes, K. (eds.) ESORICS 2013. LNCS, vol. 8134, pp. 1–18. Springer, Heidelberg (2013)

[DPSZ12] Damgård, I., Pastro, V., Smart, N., Zakarias, S.: Multiparty computation from somewhat homomorphic encryption. In: Safavi-Naini, R., Canetti, R. (eds.) CRYPTO 2012. LNCS, vol. 7417, pp. 643–662. Springer, Heidelberg (2012)

[GGH96] Goldreich, O., Goldwasser, S., Halevi, S.: Collision-free hashing from lattice problems. In: Electronic Colloquium on Computational Complexity (ECCC), vol. 3, pp. 236–241 (1996)

[GSW13] Gentry, C., Sahai, A., Waters, B.: Homomorphic encryption from learning with errors: conceptually-simpler, asymptotically-faster, attribute-based. In: Canetti, R., Garay, J.A. (eds.) CRYPTO 2013, Part I. LNCS, vol. 8042, pp. 75–92. Springer, Heidelberg (2013)

[KL14] Katz, J., Lindell, Y.: Introduction to Modern Cryptography. CRC Press, Boca Raton (2014)

[LMPR08] Lyubashevsky, V., Micciancio, D., Peikert, C., Rosen, A.: SWIFFT: a modest proposal for FFT hashing. In: Nyberg, K. (ed.) FSE 2008. LNCS, vol. 5086, pp. 54–72. Springer, Heidelberg (2008)

[LNSW13] Ling, S., Nguyen, K., Stehlé, D., Wang, H.: Improved zero-knowledge proofs of knowledge for the ISIS problem, and applications. In: Hanaoka, G., Kurosawa, K. (eds.) PKC 2013. LNCS, vol. 7778, pp. 107–124. Springer, Heidelberg (2013)

[Lub02] Luby, M.: Lt codes. In: Proceedings of the 43rd Symposium on Foundations of Computer Science, p. 271. IEEE Computer Society (2002)

[Lyu08] Lyubashevsky, V.: Lattice-based identification schemes secure under active attacks. In: Cramer, R. (ed.) PKC 2008. LNCS, vol. 4939, pp. 162–179. Springer, Heidelberg (2008)

[Lyu09] Lyubashevsky, V.: Fiat-Shamir with aborts: applications to lattice and factoring-based signatures. In: Matsui, M. (ed.) ASIACRYPT 2009. LNCS, vol. 5912, pp. 598–616. Springer, Heidelberg (2009)

[NO09] Nielsen, J.B., Orlandi, C.: LEGO for two-party secure computation. In: Reingold, O. (ed.) TCC 2009. LNCS, vol. 5444, pp. 368–386. Springer, Heidelberg (2009)

Efficient Zero-Knowledge Proof of Algebraic and Non-Algebraic Statements with Applications to Privacy Preserving Credentials

Melissa Chase[1], Chaya Ganesh[2(✉)], and Payman Mohassel[3]

[1] Microsoft Research, Redmond, USA
[2] Department of Computer Science, New York University, New York, USA
chaya.ganesh@gmail.com
[3] Visa Research, Foster City, USA

Abstract. Practical anonymous credential systems are generally built around sigma-protocol ZK proofs. This requires that credentials be based on specially formed signatures. Here we ask whether we can instead use a standard (say, RSA, or (EC)DSA) signature that includes formatting and hashing messages, as a credential, and still provide privacy. Existing techniques do not provide efficient solutions for proving knowledge of such a signature: On the one hand, ZK proofs based on garbled circuits (Jawurek et al. 2013) give efficient proofs for checking formatting of messages and evaluating hash functions. On the other hand they are expensive for checking algebraic relations such as RSA or discrete-log, which can be done efficiently with sigma protocols.

We design new constructions obtaining the best of both worlds: combining the efficiency of the garbled circuit approach for non-algebraic statements and that of sigma protocols for algebraic ones. We then discuss how to use these as building-blocks to construct privacy-preserving credential systems based on standard RSA and (EC)DSA signatures.

Other applications of our techniques include anonymous credentials with more complex policies, the ability to efficiently switch between commitments (and signatures) in different groups, and secure two-party computation on committed/signed inputs.

1 Introduction

Efficient Proofs. Zero knowledge proofs [GMR85] provide an extremely powerful tool, which allows a prover to convince a verifier that a statement is true without revealing any further information. Moreover, it has been shown that every NP language has a zero knowledge proof system [GMW87], opening up the possibility for a vast range of privacy preserving applications. However, while this is true in theory, designing proof systems that are efficient enough to be used is significantly more challenging. In reality, we only have a few techniques for efficient proofs, and those only apply to a restricted set of languages.

Almost exclusively, these proof systems focus on proving algebraic statements, i.e. statements about discrete logarithms, roots, or polynomial relationships between values [Sch90, GQ88, CS97b, GS08]. The most common and most

© International Association for Cryptologic Research 2016
M. Robshaw and J. Katz (Eds.): CRYPTO 2016, Part III, LNCS 9816, pp. 499–530, 2016.
DOI: 10.1007/978-3-662-53015-3_18

efficient of these systems fall into a class known as sigma protocols. Of course we could express any NP relation as a combination of algebraic statements, for example by expressing the relation as a circuit, and expressing each gate as an algebraic relation between input and output wires. But if we were to take this approach to prove a statement using sigma protocols we would need several exponentiations per gate in the circuit. This becomes prohibitively expensive for large circuits (for example a circuit computing a cryptographic hash function or block cipher).[1]

Recently, [JKO13] introduced a new approach for proving statements phrased as boolean circuits, based on garbled circuits. Their construction has the advantage that it only requires a few symmetric key operations per gate, making it dramatically more efficient than a sigma-protocol-based solution for non-algebraic statements. This means that it is finally practical to prove statements about complex operations such as hash functions or block ciphers. For instance, zero knowledge proofs for an AES circuit or a SHA256 circuit can be done in miliseconds on standard PCs using state of the art implementations for garbled circuits. On the other hand, expressing many public key operations as a circuit is still extremely expensive. (Consider for example a circuit computing modular exponentiation on a cryptographic group - the result would be much larger than the circuit computing a hash function, and computing a garbled circuit for such a computation would be too expensive to be practical.)

Now we have two very different techniques for achieving zero knowledge proofs for algebraic and non-algebraic statements. But in some applications, one is interested in proving statements that combine the two. For example, what if we want an efficient protocol for proving knowledge of a DSA or RSA signature, whose verification requires computing both a hash function and several exponentiations?

The state of the art fails to take advantage of the best of both worlds and has to forgo the efficiency of one approach to obtain the other's. One might consider directly combining both protocols, but a naive solution would allow a cheating prover to use a different witness for the algebraic and non-algebraic components of the computation and produce a convincing proof for a statement for which there is no single valid witness. Thus, one of the basic challenges is to bind the values committed to in the sigma protocols to the prover's inputs in the GC-based zero knowledge proof, without having to perform expensive group operations (e.g. exponentiation) inside the garbled circuit, and without proving large-circuit statements using sigma protocols.

Anonymous Credentials. Here, we primarily focus on the case of anonymous credentials, introduced by Chaum [Cha86], although we believe our results will be applicable to many other privacy protocols. A credential system allows a user to obtain credentials from an organization and at some later point prove

---

[1] SNARKs [Gro10, GGPR13] allow for very efficient verification and short proofs, but have similar shortcomings in prover efficiency as the prover performs public-key operations proportional to the size of the arithmetic circuit representing the statement.

to a verifier (either the same organization or some other party) that she has been given appropriate credentials. More specifically, the user's credentials will contain a set of attributes, and the verifier will require that the user prove that the attributes in his credential satisfy some policy. We say the system is anonymous if this proof does not reveal anything beyond this fact.

There have been several proposals for constructions of anonymous credential systems [CL01, CL04, BCKL08, Bra99, BL13]. In general, they all follow a similar approach: the credential is a signature from the organization on the user's attributes. To prove possession of valid credentials, the user will first commit to her attributes, then prove, in zero knowledge, knowledge of a signature on the committed attributes, and finally prove, again in zero knowledge, that the committed attributes satisfy the policy. To make these zero knowledge proofs efficient, most of the proposed credential systems are based on sigma protocols, which as described above give efficient proofs of knowledge for certain algebraic statements. This in turn means that the signatures used must be specially designed so that a sigma protocol can be used to prove knowledge of a signature on a committed message.[2]

But what if we want to base our credentials on a standard signature such as FDH-RSA or DSA which includes hashing the message? Or what if we want the user to be able to prove a statement about his attributes that is not easily expressible as an algebraic relation?

Our Results. We study the problem of combining proof systems for algebraic and non-algebraic statements, and obtain the following results.

- Given an algebraic commitment C, we propose two protocols for proving that C is a commitment to x such that $f(x) = 1$ where f is expressed as a boolean circuit. Both constructions have the desired property that the GC-based component is dominated by the cost of garbling f (i.e. not garbling expensive group operations), and the total number of public-key operations is independent of the size of f.
  More specifically, our first solution has public key operations proportional to the maximum bit length of the input ($|x|$), and symmetric-key operations proportional to the number of gates in f. The second has public-key operations proportional to the statistical security parameter s and symmetric-key operations proportional to the number of gates in $f + |x|s$.
  Existing solutions either require public-key operations proportional to the size of f, or need to garble circuits for expensive group operations such as exponentiations in large groups.

---

[2] Technically, [Bra99, BL13] work slightly differently in that the user and organization jointly compute the proof of knowledge of a signature as part of the credential issuance. However they still use a customized issuing protocol which would not be compatible with standardized signatures, and they use sigma protocols exactly as described here to prove that the committed attributes satisfy the policy.

- Building directly on these protocols, we show how to implement a proof that one committed message is the hash of another, and a proof that two commitments in different groups commit to the same value.
- Finally, we show how we can combine all of these protocols to obtain an efficient proof of knowledge of a signature on a committed message for RSA-FDH[3], DSA, and EC-DSA signatures.

Applications.

- **Anonymous Credentials Based on RSA, DSA, EC-DSA Signatures.** The most direct application in the context of anonymous credentials would be to use RSA, DSA, or EC-DSA signatures directly as credentials but still allow for privacy preserving presentation protocols. This would be slower than existing credential systems, but it would have the advantage that the issuer would not have to perform a complex protocol, but would only have to issue standardized signatures. It further enables interoperability with existing libraries and non-private credential applications. [4]

  Alternatively, we could construct a service which allows users to convert their non-private credentials (based on RSA/DSA/EC-DSA signatures) into traditional anonymous credentials (e.g. Idemix [ide10] or UProve [PZ13] tokens, or keyed-verification credentials [CMZ14]). Using our new protocol, the service could perform that conversion *without knowing the user's attributes*: the user would commit to his attributes, prove using our protocol that they have been signed, and then obtain from the service an anonymous credential encoding the same attributes. (All of these anonymous credential systems allow for issuing credentials on committed attributes.)
- **Anonymous Credentials with more General Policies.** Even if we consider a system based on traditional anonymous credentials, we might use the $\Pi_{\mathsf{Com},f}$ protocol (which we will describe in Sect. 3) to allow the user to prove that his attributes satisfy a more complicated policy. For example, he might want to release the hash of one of his attributes and prove that that has been done correctly, or prove that an attribute has been encrypted using a standard encryption scheme like RSA-OAEP.

  Our protocols could also be used to prove that a user's attributes fall in a given range, or to prove statements about comparisons between attributes. If the range of values possible for each attribute is small, we already have reasonably efficient solutions - the user can just commit to each bit of the value, and do a straightforward proof. However this becomes expensive when the range gets larger, in which case the most efficient known approach is based on integer

---

[3] This easily extends to standardized variants of RSA like RSA-PSS.

[4] Delignat-Lavaud et al. [DLFKP16] achieve a similar result using SNARKs, but with very different tradeoffs: their approach results in much shorter, non-interactive proofs, but much more expensive proof generation. They also explore several applications in more detail; in some of these applications, those which allow for interactive proofs, our protocols could be used to achieve these different tradeoffs.

commitments [FO97] and requires several exponentiations with an RSA modulus where the exponent is larger than the group order (e.g. a roughly 2000 bit exponentiation with a 2000 bit modulus for reasonable security parameters). Alternatively we can use our second scheme, which only requires a number of public-key operations linear in the security parameter (e.g. 60), and allows those operations to use much more efficient elliptic curve groups.

We note that the independent and concurrent work of [KKL+16] provides an alternative solution to the problem of anonymous credentials for general policies, using different techniques.

- **Converting Between Different Commitment Schemes.** There are many protocols based around commitments, and ideally we would be able to combine these protocols arbitrarily. For example, if we have an efficient protocol for proving that a committed tag matches one of the attributes in a user's credential, and another protocol for proving that a committed tag is not on a list of revoked values, then we would be able to combine the two protocols to prove that the user's credential has not been revoked. However, often the protocols will be based on different commitment schemes, or even worse, on schemes that operate in different sized groups. (For example UProve credentials can be instantiated in standardized elliptic curve groups like those used for ECDSA, while revocation systems like that in [Ngu05] require pairing groups; to combine the two we would need to find a pairing group whose group order matches one of the standardized curves. Finding a pairing group to match a specific group order often incurs a significant cost in efficiency.) With our protocol for converting between commitment schemes we could choose the most efficient groups for each, and then the user would merely prove that he has used the same attributes in each. Before our work, the only known approach to convert between groups of different sizes was to use integer commitments, which as described above can be quite expensive.

- **Other Privacy-Preserving Protocols.** We note that while anonymous credentials make a good motivating application, these problems (converting between commitments schemes, comparing committed values, or proving other non-algebraic statements) come up in many other privacy/anonymity scenarios.

- **2PC with Authenticated Input.** As input to a secure computation protocol, sometimes it is desirable to use previously committed [JS07] or signed [CZ09] inputs. In our constructions, we show how to commit to an input x and prove knowledge of x (or prove knowledge of a signature on x) and a non-algebraic statement $f(x) = 1$ using garbled circuits. As we discuss in Sect. 3.4, it is relatively easy to extend our construction to also allow secure two-party computation of $g(x, y)$ where x is the prover's input and y the verifier's, hence obtaining secure two-party computation on signed/committed inputs. The benefit of this approach is that checking the signature takes place outside the secure two-party computation and can be significantly more efficient.

2   Preliminaries

2.1   Simulation-Based Security

We use a simulation-based definition of security in the ideal/real world paradigm, which is formulated by specifying an ideal functionality. A protocol is secure if it "emulates" this ideal functionality in the presence of any adversary. Our definitions are in the stand-alone setting (as opposed to the UC framework). We formulate the simulation-based definitions by defining a functionality $\mathcal{F}$ in the ideal world. In the ideal world, all parties and the adversary $\mathcal{A}$ interact via $\mathcal{F}$. Let $IDEAL_{\mathcal{F},\mathcal{A}}(x_1, x_2)$ denote the output vector of the adversary and the honest party from the execution in the ideal world. In the real world, a protocol π is executed among the parties, and let $REAL_{\pi,\mathcal{A}}(x_1, x_2)$ denote the output of the adversary and the honest party from the execution of π. A two party protocol π securely realizes the functionality $\mathcal{F}$ if for any PPT adversary $\mathcal{A}$ in the real world, there exists a PPT adversary $\mathcal{S}$ in the ideal-world, such that

$$\{IDEAL_{\mathcal{F},\mathcal{S}}(x_1, x_2)\}_{x_1, x_2 s.t |x_1|=|x_2|} \overset{c}{\equiv} \{REAL_{\pi,\mathcal{A}}(x_1, x_2)\}_{x_1, x_2 s.t |x_1|=|x_2|}$$

that is, the two distributions are computationally indistinguishable.

2.2   Commitment Scheme

A commitment protocol involves two parties: the committer and the receiver. At a high level, it consists of two stages, a commitment phase and a de-commitment phase. In the commitment stage, the committer with a secret input m engages in a protocol with the receiver. At the end of this protocol, receiver does not know what m is (hiding property), and at the same time, the committer, can subsequently in the de-commitment phase, open only one possible value of m (binding property). Throughout the paper, we use algebraic commitment schemes that allow proving linear relationships among committed values. An example of such a scheme with computational binding and unconditional hiding properties based on the discrete logarithm problem is the one due to Pedersen [Ped91]. It works in a group G of prime order q. Given two random generators g and h such that $\log_g h$ is unknown, a value $x \in \mathbb{Z}_q$ is committed to by choosing r randomly from $\mathbb{Z}_q$, and computing $C_x = g^x h^r$. Protocols are known in literature to prove knowledge of a committed value, equality of two committed values, and so on, and the protocols can be combined in natural ways. In particular, Pedersen commitments allows proving linear relationships among committed values: Given C_x and C_y, prove that $y = ax + b$ for some public values a and b.

2.3   Committing OT

Similar to [JKO13] we need to need an OT protocol with a sender verifiability property- i.e. that at the end of the OTs, the sender is committed to its messages, and can be asked to reveal all its input messages to the receiver. This is closely related to the notion of committing OT [KS06], but can be achieved even more generally since we do not require individual commitments to sender's messages. In particular, as discussed in [JKO13] it can be satisfied by a protocol where the sender commits to a seed in the beginning of the protocol, and then runs any secure OT protocol using the output of a pseudorandom generator on the seed as its random tape. Then the open phase can be realized by letting the sender reveal the seed and all the input messages. The ideal functionality $\mathcal{F}_{COT}$ is defined in Fig. 1.

- The receiver inputs $(choose, b), b \in \{0, 1\}$, and the sender inputs (m_0, m_1).
- Output m_b to the receiver.
- On input open from the sender, send (m_0, m_1) to the receiver.

Fig. 1. The ideal functionality $\mathcal{F}_{COT}$

2.4 Garbled Circuits

We assume some familiarity with standard constructions of garbled circuits. We employ the abstraction of garbling schemes [BHR12] introduced by Bellare et al., but similar to [JKO13] we add a verification algorithm that can check correctness of the garbled circuit given all input labels to the circuit.

A garbling scheme is defined by a tuple of algorithms $\mathcal{G} = (\mathsf{Gb}, \mathsf{En}, \mathsf{De}, \mathsf{Eval}, \mathsf{Ve})$ such that:

- Gb is a randomized garbled circuit generation function that takes a security parameter, and the description of a boolean circuit f and outputs a garbled circuit GC and the encoding and decoding information e and d, respectively.
- The En algorithm takes the encoding information e output by Gb, and an input x to f, and outputs the garbled input corresponding to x.
- The Eval algorithm takes the garbled circuit GC and the garbled input, and outputs an encoded output.
- The De algorithm gets the encoded output and the decoding information d as input and returns a decoded output.
- The Ve algorithm gets as input a garbled circuit GC, the encoding information e, and a boolean function f, and outputs d or $\perp$.

In our constructions, we assume that the encoding information e is a vector of pairs of input labels, where the pair (K_i^0, K_i^1) denotes the input labels for 0 and 1 for input wire i in the circuit. Similarly, we assume that the decoding information d is a vector of pairs of output labels.

A garbling scheme may satisfy several properties such as *correctness, authenticity and privacy*. We review these notions next.

Definition 1. *A garbling scheme satisfies* **correctness** *if:*

- *for all boolean circuits f and all input x,*

$$\mathsf{De}(d, \mathsf{Eval}(GC, \mathsf{En}(e, x))) = f(x) \text{ whenever } (GC, e, d) \leftarrow \mathsf{Gb}(f, 1^\kappa)$$

- *for all boolean circuits f and all (possibly malicious) garbled circuits GC and encoding information e, decoding information d, and all input x,*

$$\text{if } d \leftarrow \mathsf{Ve}(GC, e, f) \text{ and } d \neq \bot \text{ then } \mathsf{De}(d, \mathsf{Eval}(GC, \mathsf{En}(e, x))) = f(x)$$

Definition 2. *A garbling scheme has* **authenticity** *if for every circuit f, input x, and PPT algorithm $\mathcal{A}$, the following probability*

$$\Pr[\exists y \neq f(x), y = \mathsf{De}(d, d') : (GC, e, d) \leftarrow \mathsf{Gb}(f, 1^\kappa), d' \leftarrow \mathcal{A}(GC, \mathsf{En}(e, x))]$$

is negligible in κ.

Definition 3. *A garbling scheme has* **privacy** *if there exists a PPT simulator $\mathcal{S}$ such that the following two distributions are indistinguishable:*

- *$Real(f, x)$: run $(GC, e, d) \leftarrow \mathsf{Gb}(f, 1^\kappa)$, and output $(GC, \mathsf{En}(e, x), d)$.*
- *$Ideal_{\mathcal{S}}(f, f(x))$: output $\mathcal{S}(f, f(x))$*

2.5   Zero-Knowledge Proofs

A Zero-knowledge (ZK) proof allows a prover to convince a verifier of the validity of a statement, without revealing any other information. Let $\mathcal{L}$ be the language associated with an NP relation R: $\mathcal{L} = \{x \mid \exists w : R(x, w) = 1\}$. A zero-knowledge proof for $\mathcal{L}$ lets the prover convince a verifier that $x \in \mathcal{L}$ for a common input x. A proof of knowledge captures not only the truth of a statement $x \in \mathcal{L}$, but also that the prover "possesses" a witness w to this fact. A proof of knowledge for a relation $R(\cdot, \cdot)$ is an interactive protocol where a prover P convinces a verifier V that P knows a w such that $R(x, y) = 1$, where x is a common input to P and V. The prover can always successfully convince the verifier if indeed P knows such a w. Conversely, if P can convince the verifier with reasonably high probability, then it "knows" such a w, that is, such a w can be efficiently computed given x and the code of P. The formal definition follows. In the following, $view_V$ is the "view" of the verifier in the interaction, consisting of its input x, its random coins, and the sequence of the prover's messages.

Definition 4 (ZK Proof of Knowledge). *An interactive protocol $\langle P, V \rangle$ is a zero-knowledge proof of knowledge for an NP relation R if the following properties are satisfied.*

1. *Completeness: $\forall x, y$ such that $R(x, y) = 1$,*

$$\Pr[\langle P(x, w), V(x) \rangle = 1] = 1$$

2. *Proof of Knowledge: For every polynomial time prover strategy P^*, $\exists$ an oracle PPT machine K called the extractor such that $K^{P^*}(x)$ outputs w' and*

$$\Pr[\langle P^*(x, w), V(x) \rangle = 1 \land R(x, w') = 0]$$

is negligible in κ.

3. *Zero-knowledge: For every polynomial time verifier V^*, there is a PPT algorithm S called the simulator such that for every $x \in L$, the following two distributions are indistinguishable:*
   - *$view_{V^*}(\langle P(x, w), V^*(x) \rangle)$*
   - *$S(x)$*

Honest-verifier zero-knowledge: An interactive proof system (P, V) for a language L is said to be honest-verifier zero knowledge if there exists a PPT algorithm S called the simulator such that for all $x \in L$, $view_V(\langle P(x, w), V(x) \rangle)$ and $S(x)$ are indistinguishable. This definition says that the verifier gains no knowledge from the interaction, as long as it runs the prescribed algorithm V. If the verifier tries to gain some knowledge from its interaction with the prover by deviating from the prescribed protocol, we should consider an arbitrary (but efficient) cheating verifier V^* as in the property 3 of the above definition which is full zero-knowledge. Efficient zero knowledge proofs are known which are based on sigma protocols. Sigma protocols are three round public-coin protocols and are honest-verifier zero-knowledge proof systems. There exist sigma protocols for various tasks like proving knowledge of discrete logarithm of a value, that a tuple is of the Diffie-Hellman type etc., and it is also possible to efficiently combine sigma protocols to prove compound statements. It is possible to efficiently compile a sigma protocol (which is honest-verifier ZK) into a zero-knowledge proof of knowledge. The Fiat-Shamir transform [FS86] converts any public-coin zero-knowledge proof into a zero-knowledge proof of knowledge and removes interaction, and is secure in the random oracle model [PS96]. Transformations in the common reference string model [Dam00, Lin15] are also known. The transformation of [Dam00] gives a 3-round concurrent zero-knowledge protocol in the CRS model, whereas [Lin15] is non-interactive.

In our constructions and protocols, we make use of interactive zero knowledge proofs of knowledge of discrete logarithms and relations between discrete logarithms. We use the following notation:

$$PK\{(x, y, \cdots) : \text{statements about } x, y, \cdots\}$$

In the above, $x, y, \cdots$ are secrets (discrete logarithms), the prover asserts knowledge of $x, y, \cdots$, and that they satisfy *statements*. The other values in the protocol are public.

2.6   ZK Proof Based on Garbled Circuits

Here, we review an important building block for our construction, i.e., the garbled-circuit-based ZK protocol of [JKO13]. To prove a statement $\exists w : R(x, w) = 1$ (for public R and x), the protocol proceeds as follows:

1. The verifier generates a garbled circuit computing $R(x, \cdot)$. Using a committing oblivious transfer, the prover obtains the wire labels corresponding to his private input w. Then the verifier sends the garbled circuit to the prover.
2. The prover evaluates the garbled circuit, obtaining a single garbled output (wire label). He commits to this garbled output.
3. The verifier opens his inputs to the committing oblivious transfer, giving the prover all garbled inputs. From this, the prover can check whether the garbled circuit was generated correctly. If so, the prover opens his commitment to the garbled output; if not, the prover aborts.
4. The verifier accepts the proof if the prover's commitment holds the output wire label corresponding to TRUE.

Security against a cheating prover follows from the properties of the circuit garbling scheme. Namely, the prover commits to the output wire label before the circuit is opened, so the *authenticity* property of the garbling scheme ensures that he cannot predict the TRUE output wire label unless he knows a w with $R(x, w) = $ TRUE. Security against a cheating verifier follows from correctness of the garbling scheme. The garbled output of a *correctly generated* garbled circuit reveals only the output of the (plain) circuit, and this garbled output is not revealed until the garbled circuit was shown to be correctly generated.

Note that in this protocol, the prover evaluates the garbled circuit on an input which is completely known to him. This is the main reason that the garbled circuit used for evaluation can also be later opened and checked for correctness, unlike in the setting of cut-and-choose for general 2PC. Along the same lines, it was further pointed out in [FNO15] that the circuit garbling scheme need not satisfy the *privacy* requirement of [BHR12], only the *authenticity* requirement. Removing the privacy requirement from the garbling scheme leads to a non-trivial reduction in garbled circuit size.

In one of our constructions (Sect. 3.2), the verifier does have a private input, but its input only needs to be kept private until the circuit is evaluated and the prover has committed to the output. In that scenario, we also invoke the *privacy* property of the garbling scheme as defined above.

Efficiency of Garbling Schemes. The state of the art garbling scheme uses the free-XOR technique [KS08] to garble XOR gates and the half-gate technique to garble AND gates [ZRE15]. For a circuit with g non-XOR gates, the total number of ciphertexts is $2g$, and the number of hash invocations is $4g$ for the garbler and $2g$ for the evaluator.

For *privacy-free* garbling, the costs are reduced by factor of two (see [FNO15, ZRE15]). In particular, for a circuit with g non-XOR gates, the total number of ciphertexts is g, and the number of hash invocations is $2g$ for the garbler and g for the evaluator.

We need to garble a few common building-block circuits in our constructions. It is helpful to review the size of these circuits based on the concrete constructions given in [KSS09]. The circuit for comparing ℓ bit integers requires 4ℓ non-XOR gates. The circuit for testing equality of ℓ bit integers also requires 4ℓ non-XOR gates. The circuit for adding two ℓ bit integers requires 4ℓ non-XOR gates, while the circuit for multiplying two ℓ bit integers requires $8\ell^2 - 4\ell$ non-XOR gates.

3   Proving Non-algebraic Statements on Algebraic Commitments

An important sub-protocol used in our constructions, is to commit to an input x using an algebraic commitment $\mathsf{Com}(x)$ (e.g. pedersen commitment), and perform a zero-knowledge proof of a non-algebraic statement about x, i.e. that $f(x) = 1$ for a boolean circuit f.

Such a protocol allows one to efficiently switch between proving algebraic statements on a committed input (e.g. proof of knowledge of a signature on a committed input) and non-algebraic statement (e.g. hashing, comparison, equality testing and more).

All our protocols are defined in terms of an ideal functionality, and are proven secure in the ideal/real world paradigm. We start by defining this task in terms of an ideal functionality in Fig. 2. We provide two instantiations for this functionality that provide different efficiency trade-offs depending on the input size and the algebraic commitment scheme used.

- The verifier inputs $\mathsf{Com}(x)$ and prover inputs the opening information x and the randomness.
- If $f(x) = 1$ and the opening to the commitment verifies, output accept to the verifier.

Fig. 2. The ideal functionality $\mathcal{F}_{\mathsf{Com},f}$

The starting point for both instantiations is the ZK-proof of non-algebraic statements based on garbled circuits [JKO13] (see Sect. 2.6). As the naive solution we could garble a circuit that takes x and the opening of $\mathsf{Com}(x)$ as prover's input and outputs 1 if $f(x) = 1$ and $\mathsf{Com}(x)$ correctly opens to x. The main drawback of this solution is that checking correctness of opening for an algebraic commitment requires performing expensive group operations (e.g. exponentiation) inside the garbled circuit which would dominate the computation/communication cost. Our two instantiations show how to avoid these costs and perform all algebraic operations outside the garbled circuit.

3.1   First Instantiation

In our first construction, we have the prover commit to each bit of x, i.e. $\mathsf{Com}(x_i)$ for all $i \in [n]$, and prove that when combined they yield x.

Then, following the GC-based approach, the verifier constructs a garbled circuit that computes $f(x)$, parties go through the steps of the GC-based ZK proof for the prover to prove knowledge of a value x' such that $f(x') = 1$. The main issue is that a malicious prover may use a different input $x' \neq x$ in the circuit than what he committed to.

But we observe that the input keys associated with x' in the GC (which are obtained through the COT), can function as one-time MACs on each bit of x' and can be used to enforce that $x' = x$ using efficient algebraic ZK proofs that take place outside the garbled circuit. In particular, immediately after the COTs, the prover commits to its input keys i.e. $K_i^{x'_i}$ for the ith bit of x'. When the GC is opened and both input keys K_i^0, K_i^1 are opened, the prover can provide ZK proofs that $K_i^{x'_i} = x_i K_i^1 + (1 - x_i) K_i^0$ if the commitment scheme provides efficient proofs of linear relations.

The complete protocol description in the COT-hybrid model is given in Fig. 3. We point out that steps 1, 6 and 13 are additions compared to the protocol of [JKO13].

Theorem 1. *Let $\mathcal{G}$ be a garbling scheme satisfying correctness and authenticity properties as defined in 2.4. Let Com be a secure commitment scheme, and let the proofs PK be implemented with a zero knowledge proof of knowledge. Then, the protocol $\Pi_{\mathsf{Com},f}$ in Fig. 3 securely implements $\mathcal{F}_{\mathsf{Com},f}$ in the presence of malicious adversaries in the $\mathcal{F}_{COT}$-hybrid model.*

Proof. **Corrupt Prover.**

The simulator works as follows: It uses the OT simulator to extract the prover's input x' to the OT. It then plays the role of the honest verifier in the rest of the simulation - it constructs the garbled circuit honestly and uses its input keys as verifier's inputs to the COT functionality. The simulator then extracts the value Z' committed to by the prover from the proofs of knowledge of opening in step 8. It also extracts prover's committed input x and the values K_i' that prover committed to in the protocol, using the extractor for the ZK proof of knowledge in step 13. The simulator finally outputs x and the opening extracted from the ZK proofs, iff all the following hold: $x = x'$, $f(x) = 1, Z$ is the one-key of the output wire, $K_i' = K_i^{x_i}$ for all i, the commitment in step 8 is opened to Z, and the ZK proofs of step 13 verifies. Note that in the ideal model the functionality will always output accept when the simulator sends this witness.

We now prove that a corrupt prover's view in the real protocol is indistinguishable from his view with the simulator via a series of intermediate games.

- Game Ideal: This is the interaction of the corrupt prover with the simulator and functionality as described above.
- Game G_0: This is the interaction of the corrupt prover with the simulator as described above, with the exception that instead of the simulator sending x and the opening to F, which outputs accept iff $f(x) = 1$, the game will output accept iff $f(x') = 1$ for the x' extracted from the OT (and all the other conditions listed hold). Since one of the conditions checks $x = x'$, this is identical.

Let $\mathcal{G} = (\mathsf{Gb}, \mathsf{En}, \mathsf{De}, \mathsf{Eval}, \mathsf{Ve})$ be a verifiable garbling scheme. Let F be the following functionality: it takes as input x, and outputs v such that $v = 1$ if $f(x) = 1$ and 0 otherwise. The prover has input x, the verifier is in possession of $C_x = \mathsf{Com}(x)$ and both parties have as input the security parameter κ.

1. The prover commits to the bits of x by sending bit-wise commitment to x: $C_i = \mathsf{Com}(x_i), \forall 1 \le i \le n$.
2. The verifier constructs a garbled circuit for F.

$$(GC, e, d) \leftarrow \mathsf{Gb}(1^\kappa, F)$$

3. The prover inputs his choice bits by sending (i, x_i) for all $i \in [n]$ to $\mathcal{F}_{COT}$.
4. The verifier inputs the wire labels corresponding to the prover's input by sending (i, K_i^0, K_i^1) for all $i \in [n]$ to $\mathcal{F}_{COT}$.
5. $\mathcal{F}_{COT}$ outputs K_i' for all $i \in [n]$ to the prover where $K_i' = K_i^{x_i}$.
6. The prover commits to the received input wire labels by sending $C_{K_i} = \mathsf{Com}(K_i')$ for all i.
7. The prover evaluates the garbled circuit

$$Z \leftarrow \mathsf{Eval}(GC, \{K_i'\}_{i \in [n]})$$

8. Prover commits to the garbled output Z by sending $\mathsf{Com}(Z)$ to the verifier and proves knowledge of opening.
9. Verifier sends open to $\mathcal{F}_{COT}$.
10. $\mathcal{F}_{COT}$ sends (K_i^0, K_i^1) to the prover for all $i \in [n]$.
11. Prover verifies that the correct circuit was garbled by running $\mathsf{Ve}(GC, \{K_i^0, K_i^1\}_{i \in [n]})$. If the output is not accept, the prover terminates. Otherwise if Ve outputs accept, he opens the commitment to the output Z by sending Z and the randomness used in $\mathsf{Com}(Z)$.
12. Verifier checks that the opening is correct and that $\mathsf{De}(d, Z) = 1$. If the opening is not correct or if $\mathsf{De}(d, Z) \ne 1$, the verifier outputs reject and terminates.
13. If the verifier did not terminate, the prover and the verifier engage in a Zero-knowledge protocol to prove the following:
    - $\mathsf{PK}\{(x_i, K_i', r, R) : C_i = \mathsf{Com}(x_i) \wedge C_{K_i} = \mathsf{Com}(K_i') \wedge K_i' = x_i K_i^1 + (1 - x_i)K_i^0\}, \forall 1 \le i \le n$.
    - $\mathsf{PK}\{(x, x_1, \cdots, x_n, r, r_1, \cdots r_n) : C_x = \mathsf{Com}(x) \wedge C_i = \mathsf{Com}(x_i) \wedge x = \sum 2^i x_i\}$
14. If the zero-knowledge proof verifies, the verifier outputs accept.

Fig. 3. The protocol $\Pi_{\mathsf{Com}, f}$

- Game G_1: This game, behaves exactly as in G_0 except for a slight change in the accept condition. It outputs accept if $f(x') = 1$ and $K_i' = K_i^{x_i}$ for all i and Z is the one-key of the output wire and the commitment in step 8 is correctly opened to Z, and all the ZK proofs verify (i.e. no $x = x'$ check). Indistinguishability:
Define the event Bad as the event that $x \ne x', f(x') = 1, Z$ is the one-key of the output wire, $K_i' = K_i^{x_i}$ for all i, and the opening to Z is correct and the ZK proofs of step 13 verify.

Observe that G_0 is identical to G_1 conditioned on $\overline{\text{Bad}}$. We now argue that $\Pr[\text{Bad}]$ is negligible, by observing that an adversary who makes us reject G_0 but accept in G_1, can only succeed with probability $1/2^s$ where s is a statistical security parameter, given the COT hybrid model. Without loss of generality lets assume the ith bit of x is 0 and ith bit of x' is 1. Then, the probability of the adversary guessing K_i^0 given only K_i^1 is less than $1/2^{|K_i^0|}$. Note that $|K_i^0|$ is the computational security parameter, which is 128 bits for an AES key. But without loss of security we can used a truncated K_i^0 (to its least significant s bits) in the ZK proofs of step 13.

Hence Games G_0 and G_1 are indistinguishable except with negligible probability in s.

- Game G_2: This game behaves as in G_1 except for another change in the accept condition. We accept if $f(x') = 1$ and ZK proofs of step 13 verifies and Z is the one-key of the output wire, and the commitment in step 8 is correctly opened to Z (i.e. no $K_i' = K_i^{x_i}$ check).

  If an adversary can distinguish between Games G_1 and G_2, we can break the *soundness of the ZK proof* of step 13. Therefore, G_1 and G_2 are indistinguishable.

- Game G_3: This game behaves as in G_2 except for a small change in accept condition. We accept if ZK proofs of step 13 verifies and Z is the one-key of the output wire, and the commitment in step 8 is correctly opened to Z (i.e. no $f(x') = 1$ check).

  Games G_2 and G_3 are identical, except when the following event occurs: $f(x') \neq 1$ and ZK proof of step 13 passes, and Z is the one-key of the output wire. When this event occurs, we accept in G_3 and rejects in G_2. We now argue that the probability of this event is negligible. For the sake of contradiction, assume the prover's input to OT is x' such that $f(x') \neq 1$, but the value committed to is the correct one-key Z for the output wire. We can use such a prover to break the authenticity of the garbling scheme (See Definition 2).

- Game G_4: This game behaves as in G_3 except for the accept condition. We accept if the ZK proofs of step 13 verifies and the commitment in step 8 opens correctly (i.e. no check that it is the same as extracted Z).

  An adversary who can distinguish between G_3 and G_4 can be used to violate the binding property of the commitment scheme.

  G_4 is identical to the real world game with an honest verifier.

Corrupt Verifier. The simulator commits to bits of a random value. It also uses a random value as prover's inputs to the COT, and receives the verifier's inputs to the COT functionality (K_i^0, K_i^1) for all i, i.e. the input keys to the verifier GC. The simulator then commits to the keys corresponding to the random input it used in the OTs.

It then runs $\mathsf{Ve}(GC, (K_i^0, K_i^1), f)$ to either obtain reject, or the decoding information d. If the output is reject it commits to a dummy value, else it commits to the one-key for the output wire, denoted by Z.

It then receives the "open" message from the verifier. If Ve had not output reject earlier, the simulator opens the commitment to Z and uses the simulator for the ZK proof to simulate the proofs of step 13. Otherwise, the simulator aborts.

- Game G_0: This is the interaction of the corrupt verifier with the simulator as described above.
- Game G_1: Is similar to game G_0 except that the real input x of prover is committed to.
  The two games are indistinguishable due to the hiding property of the commitment scheme.
- Game G_2: Is similar to G_1 except that instead of computing Z by running Ve, we run $\mathsf{Eval}(GC, K_i^{x_i})$ to compute and commit to Z.
  The two games are indistinguishable due to the second condition in the *correctness* property of the garbling scheme. Note that we are also implicitly using the committing OT property (the protocol described in the COT hybrid model) as the keys extracted in the OTs and what the functionality sends to the honest prover are the same.
- Game G_3: Is similar to G_2 except that the honest input x of the prover is used in the OTs.
  The two games are identical in the OT hybrid model.
- Game G_4: Is similar to G_3 except that the simulator commits to inputs keys associated with the real input x.
  The two games are identical due to the hiding property of the commitment scheme.
- Game G_5: Is similar to G_4 except that in step 13, the simulator performs the proofs, honestly.
  The two games are indistinguishable due to zero-knowledge property of the ZK proof.
  Note that G_5 is the real game with the honest prover.

3.2   Second Instantiation

We now give an alternative construction that implements the functionality in Fig. 2. In particular, we avoid the bit-wise commitments to each bit of x_i, and the associated bit-wise ZK proofs, and hence require fewer public-key operations (exponentiations) in the construction. On the other hand, the garbled circuit is augmented and hence a larger number of symmetric-key operations are needed.

The idea is as follows. In order to ensure that the prover uses the same input x in the GC, we have the circuit not only compute $f(x)$ but also a one-time MAC of x, i.e. $t = ax + b$ for random a and b of the verifier's choice. a and b are initially unknown to the prover, but are opened along with the GC after the prover has committed to t. Given a and b, the prover then provides a ZK proof that $\mathsf{Com}(t)$ is indeed the one-time MAC of x (using efficient proofs of linear relations). We note that the $t = ax + b$ operation performed in the circuit is on integers.

We note that our construction deviates from the standard construction of GC-based ZK where the verifier has no input, and privacy-free garbling is sufficient. In particular, we do invoke the privacy property of the garbling scheme in our construction to ensure that the prover does not learn a and b, until the opening stage.

The complete protocol description in the COT-hybrid model is given in Fig. 4.

Let $\mathcal{G} = (\mathsf{Gb}, \mathsf{En}, \mathsf{De}, \mathsf{Eval}, \mathsf{Ve})$ be a garbling scheme. Let F be the following functionality: it takes as inputs x, a, b and outputs v, t such that $v = 1$ if $f(x) = 1$ and 0 otherwise, and $t = ax + b$. The prover has input x, the verifier is in possession of $C_x = \mathsf{Com}(x)$. Both parties have as input the security parameter κ.

1. The verifier generates uniformly random integers a and b of length s and $n + s$ respectively. It commits to them by sending $C_a = \mathsf{Com}(a)$, $C_b = \mathsf{Com}(b)$ and proves knowledge of their opening.
2. The verifier constructs a garbled circuit for F.

$$(GC, e, d) \leftarrow \mathsf{Gb}(1^\kappa, F(x, a, b) = (f(x), ax + b))$$

3. The prover inputs his choice bits by sending (i, x_i) for all $i \in [n]$ to $\mathcal{F}_{COT}$.
4. The verifier inputs the wire keys corresponding to the prover's input by sending (i, K_i^0, K_i^1) for all $i \in [n]$ to $\mathcal{F}_{COT}$.
5. $\mathcal{F}_{COT}$ outputs K_i' for all $i \in [n]$ to the prover where $K_i' = K_i^{x_i}$.
6. The verifier sends the garbled circuit GC to the prover. Note that in what follows, for simplicity, we consider the input keys for a and b to be part of the GC itself, and hence not sent separately.
7. The prover evaluates the garbled circuit

$$(t', Z) \leftarrow \mathsf{Eval}(GC, \{K_i'\}_{i \in [n]})$$

8. Prover commits to the garbled output Z by sending $\mathsf{Com}(Z)$ to the verifier and proves knowledge of opening.
9. Verifier sends the decoding information d_t for t.
10. Prover decodes

$$t = \mathsf{De}(d_t, t')$$

and commits to the decoded output by sending $C_t = \mathsf{Com}(t)$, and proves knowledge of opening.
11. Verifier sends open to $\mathcal{F}_{COT}$.
12. $\mathcal{F}_{COT}$ sends (K_i^0, K_i^1) to the prover for all $i \in [n]$.
13. Verifier opens $\mathsf{Com}(a)$ and $\mathsf{Com}(b)$. Prover checks the openings and aborts if they fail.
14. Prover verifies that the correct circuit was garbled by running $\mathsf{Ve}(GC, \{K_i^0, K_i^1\}_{i \in [n]}, F)$. It also checks that garbled inputs for x, a, b are the correct one. If any of checks fail, the prover terminates. Otherwise, it receives the decoding vector d, and he opens the commitment to the output Z by sending Z and randomness.
15. Verifier checks that the opening is correct and that $\mathsf{De}(d, Z) = 1$. If the opening is not correct or if $\mathsf{De}(d, Z) \neq 1$, the verifier outputs reject and terminates.
16. If the verifier did not terminate, the prover and the verifier engage in a Zero-knowledge protocol to prove the following:

$$\mathsf{PK}\{(x, t, r, R) : C_x = \mathsf{Com}(x) \wedge C_t = \mathsf{Com}(t) \wedge t = ax + b\}$$

17. If the zero-knowledge proof verifies, the verifier outputs accept.

Fig. 4. The Protocol $\Pi_{\mathsf{MAC}, f}$

Theorem 2. *Let $\mathcal{G}$ be a garbling scheme satisfying correctness, authenticity, and privacy properties as defined in Sect. 2.4. Let* Com *be a secure commitment scheme, and let the proofs PK be implemented with a zero knowledge proof of knowledge. Then, the protocol $\Pi_{\mathsf{MAC},f}$ in Fig. 4 securely implements $\mathcal{F}_{\mathsf{Com},f}$ in the presence of malicious adversaries in the $\mathcal{F}_{COT}$-hybrid model.*

Proof. **Corrupt Prover.**

The simulator works as follows: It uses the OT simulator to extract the prover's input x' to the OT. It then plays the role of the honest verifier in the rest of the simulation - it chooses a, b randomly as the honest verifier would, constructs the garbled circuit honestly and uses its input keys as verifier's inputs to the COT functionality. The simulator then extracts the value Z' committed to by the prover from the proofs of knowledge of opening in step 8. It also extracts prover's committed input x and the tag t' that the prover committed to in the protocol, using the extractor for the ZK proof of knowledge in step 16. The simulator finally outputs x and the opening extracted from the ZK proofs, iff all the following hold: $x = x'$, $f(x) = 1$, Z is the one-key of the output wire, $t' = ax + b$, the commitment in step 8 is opened to Z, and the ZK proof of step 16 verifies. Note that in the ideal model the functionality will always output accept when the simulator sends this witness.

We now prove that a corrupt prover's view in the real protocol is indistinguishable from his view with the simulator by a series of intermediate games.

- Game Ideal: This is the interaction of the corrupt prover with the simulator and functionality as described above.
- Game G_0: This is the interaction of the corrupt prover with the simulator as described above, with the exception that instead of the simulator sending x and the opening to F, which outputs accept iff f(x)=1, the game will output accept iff $f(x') = 1$ for the x' extracted from the OT (and all the other conditions listed hold). Since one of the conditions checks $x = x'$, this is identical.
- Game G_1: In this game, the simulator behaves exactly as in G_0 except that it does not check the $x = x'$ condition.
  Define the event Bad as the event that $x \neq x'$ but $t = ax + b$. Observe that G_0 is identical to G_1 conditioned on $\overline{\mathsf{Bad}}$. We argue that $\Pr[\mathsf{Bad}]$ is negligible due to the unforgeability property of the one-time MAC, the hiding property of the commitment scheme, and the privacy of the garbled circuit.
  Consider a game where we run as in G_1 but stop after step 10, and look at the probability that in this gane $t' = ax + b$ but $x \neq x'$; if $\Pr[\mathsf{Bad}]$ is nonnegligible, this will be nonnegligible as well. Now, by the privacy of the garbled circuit, this is indistinguishable from a game where the verifier computes a tag t on x', and then constructs (GC, e, d) using the privacy simulator: $\mathcal{S}(F, (t, 1))$. Similarly, by the hiding of the commitment scheme this is still indistinguishable from a game where the verifier commits to random values instead of a, b. But if in this final game we get $t' = ax + b$ and $x \neq x'$ with non-negligible probability, then we can break the unforgeability of the MAC. The probability of forgery is bounded by $1/2^{|a|}$, and hence exponentially small in the statistical security parameter $s = |a|$.

- Game G_2: In this game, the simulator behaves as in G_1 except that it does not check the condition $t = ax + b$.

  If an adversary can distinguish between Games G_2 and G_1, we can break the soundness of the ZK proof of step 16.

- Game G_3: In this game, the simulator behaves as in G_2 except that we do not check the condition $f(x') = 1$.

  Games G_2 and G_3 are identical, except when the following event occurs: $f(x') \neq 1$ and ZK proof of tag verifies and Z is the one-key of the output wire. We now argue that the probability of this event is negligible. For the sake of contradiction, assume the prover's input to OT is x' such that $f(x') \neq 1$, but the value committed to is the correct one-key Z for the output wire. We can use such a prover to break the authenticity of the garbling scheme (See definition 2).

- Game G_4: In this game, the simulator behaves as in G_3 except for the accept condition. The simulator accepts if the ZK proofs of step 16 verifies and the commitment in step 8 opens correctly (i.e. no check that it is the same as extracted Z).

  An adversary who can distinguish between G_4 and G_3 can be used to violate the binding property of the commitment scheme.

  G_4 is identical to the real world game with an honest verifier.

Corrupt Verifier. The simulator extracts a and b from the proofs of knowledge of their openings by verifier. It uses a random value as prover's inputs to the COT, and receives the verifier's inputs to the COT functionality (K_i^0, K_i^1) for all i, i.e. the input keys to the verifier GC.

It then runs $\mathsf{Ve}(GC, (K_i^0, K_i^1), F)$ (and checks against the extracted a, b) to either obtain reject, or the decoding information d. If the output is reject it commits to dummy values for Z and t, else it commits to the one-key for the output wire denoted by Z, and dummy t.

The simulator receives the openings of $\mathsf{Com}(a)$ and $\mathsf{Com}(b)$. If the openings are not what it extracted earlier, or if Ve had output reject earlier, it aborts. Else, it opens the commitment to Z and uses the simulator for the ZK proof to simulate the proofs of step 16.

- Game G_0: This is the interaction of the corrupt verifier with the simulator as described above.
- Game G_1: Is similar to game G_0 except that $t = ax + b$ for the real input x of prover is committed to.

  The two games are indistinguishable due to the hiding property of the commitment scheme.

- Game G_2: Is similar to G_1 except that instead of computing Z and t by running Ve, we run $\mathsf{Eval}(GC, K_i^{x_i})$ to compute and commit to Z and t.

  The two games are indistinguishable due to the second condition in the *correctness* property of the garbling scheme, and binding property of commitments $\mathsf{Com}(a)$ and $\mathsf{Com}(b)$. Note that we are also implicitly using the committing

OT property (the protocol described in the COT hybrid model) as the keys extracted in the OTs and what the functionality sends to the honest prover are the same.

– Game G_3: Is similar to G_2 except that the honest input x of the prover is used in the OTs.
  The two games are identical in the OT hybrid model.

– Game G_4: Is similar to G_3 except that in step 13, the simulator performs the proofs honestly.
  The two games are indistinguishable due to zero-knowledge property of the ZK proof.
  Note that G_4 is the real game with the honest prover.

3.3   Efficiency Comparison and Optimizations

Efficiency Comparison. In our first instantiation, in addition to the cost associated with the GC-based ZK, i.e. the oblivious transfer for x and the cost of garbling f, $O(n)$ exponentiations are necessary to commit to each bit of input x and to perform the bitwise ZK proofs associated with them in the last step.

In our second instantiation, the bitwise commitments/proofs are eliminated (i.e. only a constant number of exponentiations) but instead the circuit for $ax+b$ needs to be garbled which requires $O(ns + s^2)$ additional symmetric-key operations when using textbook multiplication (we discuss range of values for s shortly). Using Karatsuba's multiplication algorithm [Knu69], this can potentially be further reduced.

The round complexity of both protocols is essentially the same as the GC-based ZK proof of [JKO13] (5 rounds), as the extra messages can be sent within the same rounds. (To simplify presentation, we used a separate step for each operation in our protocol description, but many of these can be combined.) A more round-efficient GC-based ZK proof would make our constructions more round efficient as well.

The first instantiation requires more exponentiations which are significantly costlier than their symmetric-key counterpart, but the total number of symmetric-key operations in the second instantiation is higher. Hence, when n is small, the first instantiation is likely more efficient, while when n is larger, the second instantiation will be the better option. Furthermore, if bit-wise commitment to the input is already necessary as part of the bigger protocol (as is the case in some of our constructions), the first instantiation may be the better choice. In the case where a comparison circuit $x < q$ is also computed, an additional $O(n)$ symmetric-key operations suffices.

Optimizations. Next we review a few other optimizations that improve efficiency of our instantiations.

– *Reducing exponentiations.* We consider the following optimization for the protocol $\Pi_{\text{Com},f}$ in Fig. 3 which reduces the number of exponentiations necessary for the ZK proofs significantly. In step 6, the prover commits to the sum of the keys received instead of individually to each wire key. The prover sends $\mathsf{Com}(S) = \mathsf{Com}\left(\sum_{i=1}^{n} K_i'\right)$ in step 6. We assume that the bit commitment scheme Com is homomorphic, and each wire key K_i is truncated to s bits and interpreted as a group element. Now, in the zero knowledge proofs of step 13, the prover proves the following statements which can be performed with fewer exponentiations:

- $\mathsf{PK}\{(x_i, S, r, R) : \mathsf{Com}(x_i) = g^{x_i} h^r \wedge \mathsf{Com}(S) = g^S h^R \wedge S = \sum_{i=1}^{n} \left(x_i K_i^1 + (1 - x_i) K_i^0\right)\}$
- $\mathsf{PK}\{(x, x_1, \cdots, x_n, r, r_1, \cdots r_n) : \mathsf{Com}(x) = g^x h^r \wedge \mathsf{Com}(x_i) = g^{x_i} h^{r_i} \wedge x = g^{\sum 2^i x_i} h^r\}$

We can show that if the sum extracted by the simulator from the commitment in step 6 is not equal to the sum of keys corresponding to the input x' extracted from COT, but the ZK proofs verify, then for some i, the prover must have correctly guessed K_i^b such that $b \neq x_i'$. The probability of this is negligible by the security of the COT protocol.

- *Privacy-free garbling.* As discussed earlier, in [FNO15] it is observed that privacy-free garbling is sufficient for GC-based ZK proofs of non-algebraic statements. This improves the communication/computation cost of garbled circuits in our first instantiation by a factor of two. But as mentioned earlier, the same cannot be said about our second construction since the privacy property of garbling is required to hide a and b in the earlier stage of the construction.

  But we can think of bigger circuit as consisting of two smaller circuits: one computing the function f and the other computing $ax+b$. If we split the computation into two garbled circuits with shared OT, then we can use the privacy free garbling scheme of [FNO15, ZRE15] for the first circuit as the verifier has no input, and use a standard garbling scheme for the $ax + b$ circuit.

- *Smaller multiplication circuit.* For the one-time MAC in the second protocol, a small a is sufficient for security - if the security (unforgeability) desired is 2^{-s}, it suffices for a to be s bits long. Hence, for a 512-bit input, a 40–80-bit a is sufficient to compute $ax + b$ which reduces the size of the multiplication circuit significantly.

3.4   Secure Computation on Committed/Signed Inputs

In the protocols described above, we have shown how to commit to a value $\mathsf{Com}(x)$ and then use a GC-based ZK proof to prove non-algebraic statements about x.

It is not hard to show that one can extend this approach, to a full-fledged secure two-party computation (2PC) of any function $g(x, y)$ where x is the committed input of the prover. In particular, note that in the ZK proof, the prover feeds its input x into the COTs in order to obtain its inputs keys to the GC of the ZK proof. In order to extend this to a secure 2PC based on garbled circuits, we let the prover play the role of the evaluator in a cut-and-choose 2PC based on garbled circuits, and use the same COT as above for the prover to obtain the garbled inputs for x in the 2PC. This would ensure that the same x that was used in the ZK proof is also used in the 2PC, and the ZK proof already ensures that this is the same input committed to in $\mathsf{Com}(x)$.

A subtle point here is that we need to open the sender's input to the COTs for the GC for the ZK but not for the GCs for the 2PC. This is supported by the committing OT of [S+11] (also see the discussion on COTs in [MR13]). It is interesting to explore the use of OT extension in such COTs where some sender inputs are opened while others are not.

We emphasize that the GCs for the 2PC only garble the desired function g, and hence the GC for the ZK proof is not part of any cut-and-choose. However, we note that the above technique is currently limited to the evaluator's input since the OTs for evaluator's input enable an almost-free check of equality of inputs in the 2PC and the ZK. Extending the ideas to both party's inputs is an interesting future direction.

This approach can be easily extended to prove other statements about x, such as proof of knowledge of a signature on x (hence signed-input 2PC) either using the techniques we give below in the case of RSA/DSA signatures, or using previous techniques to give a proof of knowledge of a CL signature [CL01].

4 Building Blocks for Privacy-Preserving Signature Verification

We introduce three important building blocks for our privacy-preserving signature verification protocols. Two of them can be directly instantiated using our $\mathcal{F}_{\mathsf{Com},f}$ functionality introduced in Sect. 3, while for the third one we provide a customized construction.

4.1 Proving that a Committed Value Is the Hash of Another Committed Value

Here, the goal is to commit to a message m and its hash $\mathcal{H}(m)$ and prove in zero-knowledge that one committed value is the hash of the other. We define the task in terms of an ideal functionality in Fig. 5.

- The verifier inputs $\mathsf{Com}(m), \mathsf{Com}(M)$ and the prover inputs the opening information (m, M) and the randomness.
- If $\mathcal{H}(m) = M$ and the openings to the commitments verify, output accept to the verifier.

Fig. 5. The ideal functionality $\mathcal{F}_{Hash}$

We now use the abstract functionality $\mathcal{F}_{\mathsf{Com},f}$ from Fig. 2 with a commitment scheme Com_h to instantiate a protocol that implements $\mathcal{F}_{Hash}$. Here, the input is $x = (m, M = \mathcal{H}(m))$ and the Com_h is defined as $\mathsf{Com}_h(x = (m, M)) = (\mathsf{Com}(m), \mathsf{Com}(M))$. To commit to bits of x, one can commit to bits of m and M individually. Com_h inherits efficient proofs of linear relations from Com as long as the proofs on m and M are performed separately. Given these, we show in Fig. 6 how to implement $\mathcal{F}_{Hash}$ by defining the right function f for the ideal functionality $\mathcal{F}_{\mathsf{Com},f}$.

---

1. The prover commits to $x = (m, M)$ by sending $\mathsf{Com}_h(x) = \mathsf{Com}(m), \mathsf{Com}(M)$ to the verifier.
2. The prover and the verifier run $\Pi_{\mathsf{Com},f}$ where f is the following functionality: f takes m and M as inputs and outputs v such that $v = 1$ if $\mathcal{H}(m) = M$ and 0 otherwise.

Fig. 6. The protocol Π_{Hash}

Theorem 3. *The protocol Π_{Hash} in Fig. 6 securely implements $\mathcal{F}_{Hash}$, given the ideal functionality $\mathcal{F}_{\mathsf{Com},f}$, in the presence of malicious adversaries.*

4.2   Proof of Equality of Committed Values in Different Groups

The goal is to prove that the value committed to in different prime groups of size p and q are the same. We define the task in terms of an ideal functionality, defined in Fig. 7. This can be achieved using standard techniques which involve using the integer commitment scheme by Damgard and Fujisaki [DF02] to prove properties about the discrete logarithms in $\mathbb{Z}$ (instead of modulo the order of the group). This requires that the verifier choose an RSA modulus $\tilde{N}$ such that the factorization is unknown to the prover, and prove that it is chosen correctly in an initial set-up phase. The prover also has to compute exponentiations in an RSA group where the exponents are $|\tilde{N}| + \kappa$ bits long. Since the group order is hidden, chinese remaindering cannot be used to speed up the exponentiations, and therefore the approach is fairly expensive. We give a protocol that avoids the integer commitment technique.

- The verifier inputs $\mathsf{Com}_p(x), \mathsf{Com}_q(y)$ and the prover inputs (x, y) and the opening information. p and q are public primes and $q < p$.
- If $0 \le x < p, 0 \le y < p, x \equiv y \mod q$, and the openings to the commitments verify, output accept to the verifier.

Fig. 7. The ideal functionality $\mathcal{F}_{Eq}$

In Fig. 8, we use the ideal functionality $\mathcal{F}_{\mathsf{Com},f}$ from Fig. 2 with a commitment scheme Com_{pq} to instantiate a protocol that implements $\mathcal{F}_{Eq}$. The scheme is defined as $\mathsf{Com}_{pq}(x) = (\mathsf{Com}_p(x), \mathsf{Com}_q(x))$, where it is assumed that Com_p and Com_q allow for proving linear relationships among committed values.

---

1. The prover commits to x and y by sending $\mathsf{Com}_p(x), \mathsf{Com}_q(y)$ to the verifier.
2. The prover and the verifier run $\Pi_{\mathsf{Com},f}$ where f is the following functionality: f takes x and checks that it is upper bounded by p and outputs v such that $v = 1$ if $x \leq p$ and 0 otherwise.

Fig. 8. The protocol Π_{Eq}

4.3 Proof of Equality of Discrete Logarithm of a Committed Value and Another Committed Value

Let $\mathbb{G}_1 = \langle G_1 \rangle$ and $\mathbb{G}_2 = \langle G_2 \rangle$ be two groups of order p and q respectively with $q|p-1$ and let $g \in \mathbb{G}_2$ be an element of order q . Given $y_1 = G_1^{g^x} H_1^{R_1}$ and $y_2 = G_2^x H_2^{R_2}$, we want to prove that the discrete logarithm w.r.t to base g of the value committed to in y_1 is equal to the value committed to in y_2. Let k be a security parameter. Following standard notation, we denote the protocol by $\mathsf{PK}\{(x, R_1, R_2) : y_1 = G_1^{g^x} H_1^{R_1} \wedge y_2 = G_2^x H_2^{R_2}\}$. The technique of our protocol is similar to [Sta96, CS97a], and is a variant of [MGGR13]. Our protocol is only honest verifier zero-knowledge. This HVZK protocol can be compiled into a full zero-knowledge proof of knowledge in the auxiliary string model using the technique of [Dam00].

Given $y_1 = G_1^{g^x} H_1^{R_1}$ and $y_2 = G_2^x H_2^{R_2}$

1. The prover computes the following $2k$ values: $u_i = G_1^{g^{\alpha_i}} H_1^{\beta_i}$ and $v_i = G_2^{\alpha_i} H_2^{\gamma_i}$ for $1 \leq i \leq k$, for randomly chosen $\alpha_i, \gamma_i \in \mathbb{Z}_q$ and $\beta_i \in \mathbb{Z}_p$, and sends u_i, v_i to the verifier.
2. The verifier chooses a random string c of length k as the challenge, and sends it to the prover.
3. For a challenge string $c = c_1 \ldots c_k$, compute and send the tuple (r_i, s_i, t_i)
   If $c_i = 0$,
   $$r_i = \alpha_i, s_i = \beta_i, t_i = \gamma_i$$
   If $c_i = 1$,
   $$r_i = \alpha_i - x \pmod{q}, s_i = \beta_i - R_1 g^{r_i} \pmod{p}, t_i = \gamma_i - R_2 \pmod{q}$$

4. Verification:
   If $c_i = 0$, check whether $u_i = G_1^{g^{r_i}} H_1^{s_i}$ and $v_i = G_2^{r_i} H_2^{t_i}$
   If $c_i = 1$, check if $u_i = y_1^{g^{r_i}} H_1^{s_i}$ and $v_i = y_2 G_2^{r_i} H_2^{t_i}$. The verifier accepts if Verification succeeds for all i.

Fig. 9. $\mathsf{PK}\{(x, R_1, R_2) : y_1 = G_1^{g^x} H_1^{R_1} \wedge y_2 = G_2^x H_2^{R_2}\}$

We will show that the protocol in Fig. 9 is correct, has a soundness error of $1/2^k$, and is honest verifier zero knowledge.

Proof.

- **Completeness:** If the prover and the verifier behave honestly, it is easy to see that verification conditions hold.
  If $c_i = 0$:

$$G_1^{g_i^r} H_1^{s_i} = G_1^{g^{\alpha_i}} H_1^{\beta_i} = u_i \text{ and } G_2^{r_i} H_2^{t_i} = G_2^{\alpha_i} H_2^{\gamma_i} = v_i$$

  If $c_i = 1$:

$$y_1^{g^{r_i}} H_1^{s_i} = (G_1^{g^x})^{g^{r_i}} (H_1^{R_1})^{g^{r_i}} H_1^{s_i} = G_1^{g^{\alpha_i}} H_1^{\beta_i} = u_i \text{ and}$$

$$y_2 G_2^{r_i} H_2^{t_i} = G_2^x H_2^{R_2} G_2^{r_i} H_2^{t_i} = v_i$$

- **Soundness:** We show an extractor that computes x, R_1, R_2 given two different accepting views with same commitments but different challenge strings. Say, we have two accepting views for challenges c and $\hat{c} \neq c$. Without loss of generality, let us assume that they differ in the jth position, and $c_j = 0$. We have,

$$u_j = G_1^{g^{r_j}} H_1^{s_j} = y_1^{g^{\hat{r}_j}} H_1^{\hat{s}_j}$$

$$G_1^{g^{r_j}} H_1^{s_j} = G_1^{g^x g^{\hat{r}_j}} H_1^{Rg^{\hat{r}_j} + \hat{s}_j}$$

$$g^x = g^{r_j - \hat{r}_j}$$

  We can compute (in $\mathbb{Z}_q$),

$$x = r_j - \hat{r}_j$$

  We have,

$$s_j = R_1 g^{\hat{r}_j} + \hat{s}_j$$

  and thus,

$$R_1 = \frac{s_j - \hat{s}_j}{g^{\hat{r}_j}}$$

  We also have

$$v_j = G_2^{r_j} H_2^{t_j} = y_2 G_2^{\hat{r}_j} H_2^{\hat{t}_j}$$

$$G_2^{r_j} H_2^{t_j} = G_2^{x + \hat{r}_j} H_2^{\hat{t}_j + R_2}$$

  and thus,

$$R_2 = t_j - \hat{t}_j$$

- **Honest Verifier Zero Knowledge:** We show a simulator such that the output of the simulator is statistically indistinguishable from the transcript of the protocol with a prover. The simulator on input c, randomly chooses $\alpha_i = r_i \in \mathbb{Z}_q, \beta_i = s_i \in \mathbb{Z}_p, \gamma_i = t_i \in \mathbb{Z}_q$ and computes for $1 \leq i \leq k$:
  If $c_i = 0$,

$$u_i = G_1^{g^{r_i}} H_1^{s_i} \text{ and } v_i = G_2^{r_i} H_2^{t_i}$$

  if $c_i = 1$,

$$u_i = y_1^{g^{r_i}} H_1^{s_i} \text{ and } v_i = y_2 G_2^{r_i} H_2^{t_i}$$

5   Privacy-Preserving FDH-RSA Signature Verification

The FDH-RSA Scheme. The Full Domain Hash RSA signature scheme FDH = (KeyGen, Sign, Verify) is defined as follows [BR93]. The KeyGen algorithm on input the security parameter k, selects two $k/2$-bit primes p and q and computes the modulus $N = pq$. It then chooses an exponent $e \in \mathbb{Z}^*_{\phi(N)}$, and computes d such that $ed = 1 \mod \phi(N)$. Return (pk, sk), where $pk = (N, e)$ and $sk = (N, d)$. The signature generation and verification are as follows and use a hash function $\mathcal{H} : \{0, 1\} \rightarrow \mathbb{Z}^*_N$.

$\mathsf{Sign}_{N,d}(M)$
$x = \mathcal{H}(M)$
$\sigma = x^d \mod N$
return σ

$\mathsf{Verify}_{N,e}(M, \sigma)$
$y = \sigma^e \mod N$
$y' = \mathcal{H}(M)$
if $(y = y')$ **then return** 1;
else return 0;

5.1   Proof of Knowledge of RSA Signatures

Given $\mathsf{Com}_N(m)$, a commitment to m in a group of order N, the following protocol is a zero knowledge proof of knowledge of a valid RSA signature on m.

1. The prover has input (m, σ) and the verifier is in possession of $\mathsf{Com}_N(m) = C_1 = g^m h^{r_1}$
2. The prover commits to $M = \mathcal{H}(m)$, that is, $M \in \mathbb{Z}_N$, compute $\mathsf{Com}_N(M) = C_2 = g^M h^{r_2}$, for randomly chosen $r_2 \in Z^*_N$. Send C_2 to the verifier and prove knowledge of opening.
3. The prover and verifier engage in the protocol Π_{Hash} with inputs (m, M) and (C_1, C_2) respectively.
4. The prover proves knowledge of e-th root of a committed value [CS97a]. Given $y = C_2 = g^M h^r$, prover proves knowledge of σ, such that, $y = g^{\sigma^e} h^r$.
   (a) The prover computes the following tuple:

$$(y_1, \cdots, y_{e-1}) \text{ where } y_i = g^{\sigma^i} h^{r_i}$$

   for randomly chosen $r_i \in \mathbb{Z}_N$, for $i = 1$ to $e - 1$.
   (b) The prover and the verifier run the following proof of knowledge:

$$\mathsf{PK}\{(\alpha, (\beta_1, \cdots, \beta_e)) : y_1 = g^\alpha h^{\beta_1} \wedge y_2 = y_1^\alpha h^{\beta_2} \wedge \cdots \wedge y = y_{e-1}^\alpha h^{\beta_e}\}$$

When e is one greater than a power of 2, we can employ optimizations like repeated squaring to prove knowledge of e-th root. Given $y = g^{\sigma^e} h^r$, for $e = 2^k + 1$, step 4 in the verification protocol can be now be realized as follows:

1. The prover computes the following tuple:

$$(y_0, y_1, \cdots, y_k) \text{ where } y_i = g^{\sigma^{2^i}} h^{r_i}$$

for randomly chosen $r_i \in \mathbb{Z}_N$, for $i = 1$ to k.

2. The prover and the verifier run the following proof of knowledge:

$$\mathsf{PK}\{(\alpha, \alpha_1, \cdots, \alpha_k, \beta, \beta_0, \cdots, \beta_k, R_0, \cdots, R_k) :$$
$$y_0 = g^\alpha h^\beta \wedge y_1 = y_0^\alpha h^{\beta_0} \wedge y_1 = g^{\alpha_1} h^{R_0} \wedge y_2 = y_1^{\alpha_1} h^{\beta_1}$$
$$\wedge\, y_2 = g^{\alpha_2} h^{R_1} \cdots \wedge y_k = y_{k-1}^{\alpha_{k-1}} h^{\beta_{k-1}} \wedge y_k = g^{\alpha_k} h^{R_{k-1}} \wedge y = y_k^\alpha h^{\beta_k}\}$$

It might be possible to improve the efficiency for some e's by using addition chains for the integer e. An addition chain for integer e is an ascending sequence $1 = e_0 < e_1 < \cdots e_r = e$ such that for $1 \leq i \leq r$, we have $e_i = e_j + e_k$. The prover, now, would have to provide only the y_i's for which i is an element of the addition chain for e. The relations among the y_i's will be sightly different, but can be proved in a similar way.

The above verification protocol can also be adapted to support variants of RSA-based signatures, like the probabilistic signature scheme (PSS) from [BR96]. PSS is a probabilistic generalization of FDH which uses two hash functions and more complicated padding. We can instantiate protocol $\Pi_{\mathsf{Com},f}$ with an f that verifies the additional checks of PSS to achieve privacy preserving verification of a PSS signature.

5.2 Proof of Security

We sketch a proof that the above protocol is a zero-knowledge proof of knowledge of an RSA signature on a committed message. The completeness follows easily from the security of protocol Π_{Hash}, and from the observation that

$$y = \left(y_{e-1}^\alpha\right) h^{\beta_e} = \left(\left(\cdots \left(g^\alpha h^{\beta_1}\right)^\alpha h^{\beta_2} \cdots\right)^\alpha h^{\beta_{e-1}}\right)^\alpha h^{\beta_e}$$

$$= g^{\alpha^e} h^{\beta_e + \alpha\beta_{e-1} + \cdots + \alpha^{e-1}\beta_1}$$

in step 4.

- Soundness: We show an extractor, that, given access to the prover, extracts (m, σ) such that $\mathsf{Verify}_{N,e}(m, \sigma) = 1$. The extractor invokes the simulator for the corrupt prover of protocol Π_{Hash} to extract m and M. It then runs the extractor corresponding to the proof in step 4b to extract α. By the security of Π_{Hash} and the binding property of Com, it follows that $\alpha^e \mod N = M = \mathcal{H}(m)$.
- Zero-knowledge: We sketch a simulator that simulates the verifier's view in the protocol. The simulator commits to a random value on behalf of the prover in step 2 by computing $C_2' = \mathsf{Com}(M')$. It sends C_2' to the verifier, proves knowledge of opening and invokes the simulator for the corrupt verifier of protocol Π_{Hash}. It then chooses $y_1, \cdots, y_{e-1} \in Z_N$ at random, and runs the simulator corresponding to the proof in step 4b. We can show that the view of the verifier in the protocol is indistinguishable from the view with the simulator.

6 Privacy-Preserving (EC)DSA Signature Verification

The DSA Scheme. The Digital Signature Algorithm (DSA) is a variant of the Elgamal signature scheme. The key generation, signature generation and verification algorithms are given next. The KeyGen algorithm chooses two primes p and q such that $q \mid p - 1$. Let g be an element of order q in $\mathbb{Z}_p^*$. It then chooses x randomly from $\{1, \cdots, q - 1\}$. The private key is set to be x and the public key is $(g, p, q, y), y = g^x \mod p$.

Sign(m)
$M \leftarrow \mathcal{H}(m)$
Pick a random $k, 1 \le k < q$
$r = (g^k \mod p) \mod q$
$s = k^{-1}(M + rx) \mod q$
return (r, s)

Verify($m, (r, s)$)
$M \leftarrow \mathcal{H}(m)$
$w = s^{-1} \mod q$
$u_1 = Mw \mod q$
$u_2 = rw \mod q$
if $r = (g^{u_1} y^{u_2} \mod p) \mod q$
then return ;
1 **else return** 0;

The ECDSA Scheme. ECDSA is the elliptic curve analogue of DSA. It works in an elliptic curve group $E(\mathbb{Z}_p)$. The ECDSA Key generation, signature and verification algorithms are given below. The KeyGen algorithm chooses an elliptic curve E defined over $\mathbb{Z}_p$ such that the number of points in $E(\mathbb{Z}_p)$ is divisible by a large prime n. Pick a point $P \in E(\mathbb{Z}_p)$ of order n. Let $d \in [1, n - 1]$ be a randomly chosen integer. Set $Q = dP$. The public key is (E, P, Q, n) and the private key is d.

Sign(m)
$M \leftarrow \mathcal{H}(m)$
Pick a random $k \in [1, n - 1]$
$kP = (x_0, y_0)$
$r = x_0 \mod n$
$s = k^{-1}(M + rd) \mod n$
return (r, s)

Verify($m, (r, s)$)
$M \leftarrow \mathcal{H}(m)$
if $r, s \notin [1, n - 1]$ **then return ;**
0
$w = s^{-1} \mod n$
$u_1 = Mw \mod n$
$u_2 = rw \mod n$
$(x_1, y_1) = u_1 P + u_2 Q$
$v = x_1 \mod n$
if $r = v$ **then return** 1;
else return 0;

6.1 Proof of Knowledge of DSA Signatures

Let (r, s) be the DSA signature on m. Let $\mathbb{G}_1 = \langle G_1 \rangle$ and $\mathbb{G}_2 = \langle G_2 \rangle$ be two distinct groups of order p and q respectively where p and q are the parameters of the DSA signature algorithm. One technical difficulty is that we have to show r in G_1 and G_2 is equal modulo q. For that purpose, we use our protocol Π_{Eq} from Fig. 8 to prove equality across groups. We also employ our protocol from Fig. 9 to prove equality of discrete logarithm of a committed value and another committed value. We now describe the DSA verification protocol in detail. Given a commitment to m, the following protocol is a zero-knowledge proof of knowledge of a valid DSA signature on m.

1. The verifier is in possession of $C_1 = \mathsf{Com}_q(m)$, and the prover has as input message $(m, (r, s))$ and the opening information of C_1 to m.
2. The prover commits to $M = \mathcal{H}(m)$, that is, $M \in \mathbb{Z}_q$, compute $C_2 = \mathsf{Com}_q(M)$ Send C_2 to the verifier and prove knowledge of opening.
3. Now the prover and verifier engage in the protocol Π_{Hash} to prove that $M = \mathcal{H}(m)$.
4. The prover commits to the signature (r, s) by sending $\mathsf{Com}_{pq}(r) = (\mathsf{Com}_p(r), \mathsf{Com}_q(r))$ and $\mathsf{Com}_q(s)$. The prover also commits to the following values: $u_1 = \mathcal{H}(m)s^{-1}, u_2 = rs^{-1}, \alpha = g^{u_1}, \beta = y^{u_2}$, where g is the generator of a cyclic group of order q in $\mathbb{Z}_p^*$ used in DSA signing, and y is the DSA public key. Prover sends $\mathsf{Com}_q(u_1), \mathsf{Com}_q(u_2), \mathsf{Com}_p(\alpha), \mathsf{Com}_p(\beta)$.
5. The prover and the verifier carry out the following Σ-protocol zero-knowledge proofs of knowledge:
   (a) $\mathsf{PK}\{(u_1, R_1, R_2) : \mathsf{Com}_p(\alpha) = G_1^{g^{u_1}} H_1^{R_1} \wedge \mathsf{Com}_q(u_1) = G_2^{u_1} H_1^{R_2}\}$
   (b) $\mathsf{PK}\{(u_2, R_1, R_2) : \mathsf{Com}_p(\beta) = G_1^{y^{u_2}} H_1^{R_1} \wedge \mathsf{Com}_q(u_2) = G_2^{u_2} H_1^{R_2}\}$
   (c) $\mathsf{PK}\{(r, \alpha, \beta, R_1, R_2, R_3) : \mathsf{Com}_p(\beta) = G_1^\beta H_1^{R_1} \wedge \mathsf{Com}_p(\alpha) = G_1^\alpha H_1^{R_2} \wedge \mathsf{Com}_p(r) = G_1^r H_1^{R_3} \wedge r = \alpha\beta\}$
   (d) $\mathsf{PK}\{(M, u_1, s, R_1, R_2, R_3) : \mathsf{Com}_q(M) = G_2^M H_2^{R_1} \wedge \mathsf{Com}_q(u_1) = G_2^{u_1} H_2^{R_2} \wedge \mathsf{Com}_q(s) = G_2^s H_2^{R_3} \wedge M = u_1 s\}$
   (e) $\mathsf{PK}\{(r, u_2, s, R_1, R_2, R_3) : \mathsf{Com}_q(r) = G_2^r H_2^{R_1} \wedge \mathsf{Com}_q(u_2) = G_2^{u_2} H_2^{R_2} \wedge \mathsf{Com}_q(s) = G_2^s H_2^{R_3} \wedge r = u_2 s\}$
6. The prover and verifier engage in Π_{Eq} with input $\mathsf{Com}_{pq}(r)$.

6.2   Proof of Security

We sketch a proof of the soundness and zero-knowledge properties of the above protocol. The completeness follows from security of Π_{Hash} and completeness of the proofs of knowledge in step 5.

– Proof of Knowledge: We show an extractor, that, given access to the prover, extracts $(m, (r, s))$ such that $\mathsf{Verify}(m, (r, s)) = 1$. The extractor invokes the simulator for the corrupt prover of protocol Π_{Hash} to extract m and M and the opening information for C_1.
   It then runs the extractor guaranteed by the proof of knowledge property of the proofs in step 5 to extract $u_1, u_2, \alpha, \beta, s, r$. Finally it returns $(m, (r, s))$ and the opening information. By security of Π_{Hash}, Π_{Eq} and the binding property of the commitment scheme Com, it follows that $r = g^{Ms^{-1}} y^{rs^{-1}}$ and $M = \mathcal{H}(m)$.
– Zero-knowledge: We sketch a simulator that simulates the verifier's view in the protocol. The simulator commits to a random value on behalf of the prover in step 2 by computing $C_2' = \mathsf{Com}(M')$. It sends C_2' to the verifier, proves knowledge of the opening and invokes the simulator for the corrupt verifier of protocol Π_{Hash}. It then commits to random values in step 4, and runs the

simulator corresponding to the proofs of knowledge in step 5. Finally in step 6, the simulator invokes the simulator for protocol Π_{Eq}. We can show that the view of the verifier in the protocol is indistinguishable from the view with the simulator.

6.3 Proof of Knowledge of ECDSA Signatures

Let (r, s) be the ECDSA signature on m. Let $\mathbb{G}_1 = \langle G_1 \rangle$ and $\mathbb{G}_2 = \langle G_2 \rangle$ be two distinct groups of order p and n respectively where p is the order of the field of the curve and n is the order of point P. Addition of elliptic curve points which is the group operation requires arithmetic operations in the underlying finite field $\mathbb{Z}_p$ of the curve E. We use a straight forward variant of the protocol in Fig. 9 to prove statements about multiples of an elliptic curve point (elliptic curve analogue of exponentiation) inside commitments.

1. The verifier is in possession of $C_1 = \mathsf{Com}_p(m)$ and the prover has as input (m, σ) and the opening of C_1 to m.
2. The prover commits to $M = \mathcal{H}(m)$, by computing $C_2 = \mathsf{Com}_p(M)$. Send C_2 to the verifier and prove knowledge of opening.
3. The prover and verifier engage in the protocol Π_{Hash} with inputs (m, M) and (C_1, C_2) respectively.
4. The prover commits to the signature (r, s) and proves knowledge of an opening. The prover sends $\mathsf{Com}_{pn}(r) = (\mathsf{Com}_p(r), \mathsf{Com}_n(r))$ and $\mathsf{Com}_n(s)$. The prover also commits to the following values: $u_1 = \mathcal{H}(m)s^{-1}, u_2 = rs^{-1}$, and the co-ordinates of the points $u_1 P = (\alpha_x, \alpha_y), u_2 Q = (\beta_x, \beta_y)$, where P is the point of order n in $E(\mathbb{Z}_p)$ used in ECDSA signing, and Q is the ECDSA public key. The prover sends $\mathsf{Com}_n(u_1)$, $\mathsf{Com}_n(u_2)$, $\mathsf{Com}_p(\alpha_x)$, $\mathsf{Com}_p(\alpha_y)$, $\mathsf{Com}_p(\beta_x)$, $\mathsf{Com}_p(\beta_y)$.
5. The prover and the verifier carry out the following Σ-protocol zero-knowledge proofs of knowledge:
   (a) $\mathsf{PK}\{(u_1, \alpha_x, \alpha_y, R_1, R_2, R_3) \; : \; \mathsf{Com}_p(\alpha_x) = G_1^{\alpha_x} H_1^{R_1} \wedge \mathsf{Com}_p(\alpha_y) = G_1^{\alpha_y} H_1^{R_2} \wedge \mathsf{Com}_n(u_1) = G_2^{u_1} H_1^{R_3} \wedge (\alpha_x, \alpha_y) = u_1 P\}$
   (b) $\mathsf{PK}\{(u_2, \beta_x, \beta_y, R_1, R_2, R_3) \; : \; \mathsf{Com}_p(\beta_x) = G_1^{\beta_x} H_1^{R_1} \wedge \mathsf{Com}_p(\beta_y) = G_1^{\beta_y} H_1^{R_2} \wedge \mathsf{Com}_n(u_2) = G_2^{u_2} H_1^{R_3} \wedge (\beta_x, \beta_y) = u_2 Q\}$
   (c) $\mathsf{PK}\{(r, \alpha_x, \alpha_y, \beta_x, \beta_y, R_1, R_2, R_3, R_4, R_5) \; : \; \mathsf{Com}_p(\beta_x) = G_1^{\beta_x} H_1^{R_1} \wedge \mathsf{Com}_p(\beta_y) = G_1^{\beta_y} H_1^{R_2} \wedge \mathsf{Com}_p(\alpha_x) = G_1^{\alpha_x} H_1^{R_3} \wedge \mathsf{Com}_p(\alpha_y) = G_1^{\alpha_y} H_1^{R_4} \wedge \mathsf{Com}_p(r) = G_1^r H_1^{R_5} \wedge r = ((\alpha_x, \alpha_y) + (\beta_x, \beta_y))_x\}$
   (d) $\mathsf{PK}\{(M, u_1, s, R_1, R_2, R_3) : \mathsf{Com}_n(M) = G_2^M H_2^{R_1} \wedge \mathsf{Com}_n(u_1) = G_2^{u_1} H_2^{R_2} \wedge \mathsf{Com}_n(s) = G_2^s H_2^{R_3} \wedge M = u_1 s\}$
   (e) $\mathsf{PK}\{(r, u_2, s, R_1, R_2, R_3) : \mathsf{Com}_n(r) = G_2^r H_2^{R_1} \wedge \mathsf{Com}_n(u_2) = G_2^{u_2} H_2^{R_2} \wedge \mathsf{Com}_n(s) = G_2^s H_2^{R_3} \wedge r = u_2 s\}$
6. The prover and verifier engage in Π_{Eq} with input $\mathsf{Com}_{pn}(r)$.

The above protocol can be proven to be a zero knowledge proof of knowledge of ECDSA signature. The proofs for correctness, soundness and zero-knowledge are similar to the proofs of the protocol for the DSA signature.

References

[BCKL08] Belenkiy, M., Chase, M., Kohlweiss, M., Lysyanskaya, A.: P-signatures and noninteractive anonymous credentials. In: Canetti, R. (ed.) TCC 2008. LNCS, vol. 4948, pp. 356–374. Springer, Heidelberg (2008)

[BHR12] Bellare, M., Hoang, V.T., Rogaway, P.: Foundations of garbled circuits. In: Proceedings of the 2012 ACM Conference on Computer and Communications Security, pp. 784–796. ACM (2012)

[BL13] Baldimtsi, F., Lysyanskaya, A.: Anonymous credentials light. In: Sadeghi, A.-R., Gligor, V.D., Yung, M. (eds.) ACM CCS 2013, pp. 1087–1098. ACM Press, November 2013

[BR93] Bellare, M., Rogaway, P.: Random oracles are practical: a paradigm for designing efficient protocols. In: Proceedings of the 1st ACM Conference on Computer and Communications Security, pp. 62–73. ACM (1993)

[BR96] Bellare, M., Rogaway, P.: The exact security of digital signatures - how to sign with RSA and Rabin. In: Maurer, U.M. (ed.) EUROCRYPT 1996. LNCS, vol. 1070, pp. 399–416. Springer, Heidelberg (1996)

[Bra99] Brands, S.: Rethinking public key infrastructure and digital certificates-building in privacy. Ph.D. thesis, Eindhoven Institute of Technology, Eindhoven, The Netherlands (1999)

[Cha86] Chaum, D.: Showing credentials without identification. In: Pichler, F. (ed.) EUROCRYPT 1985. LNCS, vol. 219, pp. 241–244. Springer, Heidelberg (1986)

[CL01] Camenisch, J.L., Lysyanskaya, A.: An efficient system for non-transferable anonymous credentials with optional anonymity revocation. In: Pfitzmann, B. (ed.) EUROCRYPT 2001. LNCS, vol. 2045, pp. 93–118. Springer, Heidelberg (2001)

[CL04] Camenisch, J.L., Lysyanskaya, A.: Signature schemes and anonymous credentials from bilinear maps. In: Franklin, M. (ed.) CRYPTO 2004. LNCS, vol. 3152, pp. 56–72. Springer, Heidelberg (2004)

[CMZ14] Chase, M., Meiklejohn, S., Zaverucha, G.: Algebraic MACs and keyed-verification anonymous credentials. In: Ahn, G.-J., Yung, M., Li, N. (eds.) ACM CCS 2014, pp. 1205–1216. ACM Press, November 2014

[CS97a] Camenisch, J.L., Stadler, M.A.: Efficient group signature schemes for large groups. In: Kaliski Jr., B.S. (ed.) CRYPTO 1997. LNCS, vol. 1294, pp. 410–424. Springer, Heidelberg (1997)

[CS97b] Camenisch, J.L., Stadler, M.A.: Efficient group signature schemes for large groups. In: Kaliski Jr., B.S. (ed.) CRYPTO 1997. LNCS, vol. 1294, pp. 410–424. Springer, Heidelberg (1997)

[CZ09] Camenisch, J., Zaverucha, G.M.: Private intersection of certified sets. In: Dingledine, R., Golle, P. (eds.) FC 2009. LNCS, vol. 5628, pp. 108–127. Springer, Heidelberg (2009)

[Dam00] Damgård, I.B.: Efficient concurrent zero-knowledge in the auxiliary string model. In: Preneel, B. (ed.) EUROCRYPT 2000. LNCS, vol. 1807, pp. 418–430. Springer, Heidelberg (2000)

[DF02] Damgård, I.B., Fujisaki, E.: A statistically-hiding integer commitment scheme based on groups with hidden order. In: Zheng, Y. (ed.) ASIACRYPT 2002. LNCS, vol. 2501, pp. 125–142. Springer, Heidelberg (2002)

[DLFKP16] Delignat-Lavaud, A., Fournet, C., Kohlweiss, M., Parno, B.: Cinderella: turning shabby X.509 certificates into elegant anonymous credentials with the magic of verifiable computation. In: IEEE Symposium on Security & Privacy 2016 (Oakland 2016). IEEE (2016)

[FNO15] Frederiksen, T.K., Nielsen, J.B., Orlandi, C.: Privacy-free garbled circuits with applications to efficient zero-knowledge. In: Oswald, E., Fischlin, M. (eds.) EUROCRYPT 2015. LNCS, vol. 9057, pp. 191–219. Springer, Heidelberg (2015)

[FO97] Fujisaki, E., Okamoto, T.: Statistical zero knowledge protocols to prove modular polynomial relations. In: Kaliski Jr., B.S. (ed.) CRYPTO 1997. LNCS, vol. 1294, pp. 16–30. Springer, Heidelberg (1997)

[FS86] Fiat, A., Shamir, A.: How to prove yourself: practical solutions to identification and signature problems. In: Odlyzko, A.M. (ed.) CRYPTO 1986. LNCS, vol. 263, pp. 186–194. Springer, Heidelberg (1987)

[GGPR13] Gennaro, R., Gentry, C., Parno, B., Raykova, M.: Quadratic span programs and succinct NIZKs without PCPs. In: Johansson, T., Nguyen, P.Q. (eds.) EUROCRYPT 2013. LNCS, vol. 7881, pp. 626–645. Springer, Heidelberg (2013)

[GMR85] Goldwasser, S., Micali, S., Rackoff, C.: The knowledge complexity of interactive proof-systems (extended abstract). In: Proceedings of the 17th Annual ACM Symposium on Theory of Computing, 6–8 May 1985, Providence, Rhode Island, USA, pp. 291–304 (1985)

[GMW87] Goldreich, O., Micali, S., Wigderson, A.: How to prove all NP-statements in zero-knowledge and a methodology of cryptographic protocol design. In: Odlyzko, A.M. (ed.) CRYPTO 1986. LNCS, vol. 263, pp. 171–185. Springer, Heidelberg (1987)

[GQ88] Guillou, L.C., Quisquater, J.-J.: A practical zero-knowledge protocol fitted to security microprocessor minimizing both transmission and memory. In: Günther, C.G. (ed.) EUROCRYPT 1988. LNCS, vol. 330, pp. 123–128. Springer, Heidelberg (1988)

[Gro10] Groth, J.: Short pairing-based non-interactive zero-knowledge arguments. In: Abe, M. (ed.) ASIACRYPT 2010. LNCS, vol. 6477, pp. 321–340. Springer, Heidelberg (2010)

[GS08] Groth, J., Sahai, A.: Efficient non-interactive proof systems for bilinear groups. In: Smart, N.P. (ed.) EUROCRYPT 2008. LNCS, vol. 4965, pp. 415–432. Springer, Heidelberg (2008)

[ide10] Specification of the identity mixer cryptographic library (revised version 2.3.0). Technical report RZ 3730, IBM Research, April 2010

[JKO13] Jawurek, M., Kerschbaum, F., Orlandi, C.: Zero-knowledge using garbled circuits: how to prove non-algebraic statements efficiently. In: Sadeghi, A.-R., Gligor, V.D., Yung, M. (eds.) ACM CCS 2013, pp. 955–966. ACM Press, November 2013

[JS07] Jarecki, S.: Efficient two-party secure computation on committed inputs. In: Naor, M. (ed.) EUROCRYPT 2007. LNCS, vol. 4515, pp. 97–114. Springer, Heidelberg (2007)

[KKL+16] Kolesnikov, V., Krawczyk, H., Lindell, Y., Malozemoff, A.J., Rabin, T.: Attribute-based key exchange with general policies. Cryptology ePrint Archive, Report 2016/518 (2016). http://eprint.iacr.org/

[Knu69] Knuth, D.E.: The Art of Computer Programming Vol. 2: Seminumerical Algorithms, pp. 229–279. Addison Wesley, Reading (1969)

[KS06] Kiraz, M., Schoenmakers, B.: A protocol issue for the malicious case of yaos garbled circuit construction. In: 27th Symposium on Information Theory in the Benelux, pp. 283–290 (2006)

[KS08] Kolesnikov, V., Schneider, T.: Improved garbled circuit: free XOR gates and applications. In: Aceto, L., Damgård, I., Goldberg, L.A., Halldórsson, M.M., Ingólfsdóttir, A., Walukiewicz, I. (eds.) ICALP 2008, Part II. LNCS, vol. 5126, pp. 486–498. Springer, Heidelberg (2008)

[KSS09] Kolesnikov, V., Sadeghi, A.-R., Schneider, T.: Improved garbled circuit building blocks and applications to auctions and computing minima. In: Garay, J.A., Miyaji, A., Otsuka, A. (eds.) CANS 2009. LNCS, vol. 5888, pp. 1–20. Springer, Heidelberg (2009)

[Lin15] Lindell, Y.: An efficient transform from sigma protocols to NIZK with a CRS and non-programmable random oracle. In: Dodis, Y., Nielsen, J.B. (eds.) TCC 2015, Part I. LNCS, vol. 9014, pp. 93–109. Springer, Heidelberg (2015)

[MGGR13] Miers, I., Garman, C., Green, M., Rubin, A.D.: Zerocoin: Anonymous distributed e-cash from bitcoin. In: IEEE Symposium on Security and Privacy (SP), pp. 397–411. IEEE (2013)

[MR13] Mohassel, P., Riva, B.: Garbled circuits checking garbled circuits: more efficient and secure two-party computation. In: Canetti, R., Garay, J.A. (eds.) CRYPTO 2013, Part II. LNCS, vol. 8043, pp. 36–53. Springer, Heidelberg (2013)

[Ngu05] Nguyen, L.: Accumulators from bilinear pairings and applications. In: Menezes, A. (ed.) CT-RSA 2005. LNCS, vol. 3376, pp. 275–292. Springer, Heidelberg (2005)

[Ped91] Pedersen, T.P.: Non-interactive and information-theoretic secure verifiable secret sharing. In: Feigenbaum, J. (ed.) CRYPTO 1991. LNCS, vol. 576, pp. 129–140. Springer, Heidelberg (1992)

[PS96] Pointcheval, D., Stern, J.: Security proofs for signature schemes. In: Maurer, U.M. (ed.) EUROCRYPT 1996. LNCS, vol. 1070, pp. 387–398. Springer, Heidelberg (1996)

[PZ13] Paquin, C., Zaverucha, G.: U-prove cryptographic specification v1.1 (revision 2) (2013). www.microsoft.com/uprove

[S+11] shelat, A., Shen, C.: Two-output secure computation with malicious adversaries. In: Paterson, K.G. (ed.) EUROCRYPT 2011. LNCS, vol. 6632, pp. 386–405. Springer, Heidelberg (2011)

[Sch90] Schnorr, C.-P.: Efficient identification and signatures for smart cards. In: Brassard, G. (ed.) CRYPTO 1989. LNCS, vol. 435, pp. 239–252. Springer, Heidelberg (1990)

[Sta96] Stadler, M.A.: Publicly verifiable secret sharing. In: Maurer, U.M. (ed.) EUROCRYPT 1996. LNCS, vol. 1070, pp. 190–199. Springer, Heidelberg (1996)

[ZRE15] Zahur, S., Rosulek, M., Evans, D.: Two halves make a whole. In: Oswald, E., Fischlin, M. (eds.) EUROCRYPT 2015. LNCS, vol. 9057, pp. 220–250. Springer, Heidelberg (2015)

Theory

Fine-Grained Cryptography

Akshay Degwekar$^{(\boxtimes)}$, Vinod Vaikuntanathan, and Prashant Nalini Vasudevan

MIT, CSAIL, Cambridge, USA
{akshayd,vinodv,prashvas}@mit.edu

Abstract. *Fine-grained cryptographic primitives* are ones that are secure against adversaries with an a-priori bounded polynomial amount of resources (time, space or parallel-time), where the honest algorithms use less resources than the adversaries they are designed to fool. Such primitives were previously studied in the context of time-bounded adversaries (Merkle, CACM 1978), space-bounded adversaries (Cachin and Maurer, CRYPTO 1997) and parallel-time-bounded adversaries (Håstad, IPL 1987). Our goal is come up with fine-grained primitives (in the setting of parallel-time-bounded adversaries) and to show unconditional security of these constructions when possible, or base security on widely believed separation of worst-case complexity classes. We show:

1. NC^1-cryptography: Under the assumption that $\mathsf{NC}^1 \neq \oplus\mathsf{L}/\mathsf{poly}$, we construct one-way functions, pseudo-random generators (with sublinear stretch), collision-resistant hash functions and most importantly, *public-key encryption schemes*, all computable in NC^1 and secure against all NC^1 circuits. Our results rely heavily on the notion of randomized encodings pioneered by Applebaum, Ishai and Kushilevitz, and crucially, make *non-black-box* use of randomized encodings for logspace classes.

2. AC^0-cryptography: We construct (unconditionally secure) pseudo-random generators with arbitrary polynomial stretch, weak pseudo-random functions, secret-key encryption and perhaps most interestingly, *collision-resistant hash functions*, computable in AC^0 and secure against all AC^0 circuits. Previously, one-way permutations and pseudo-random generators (with linear stretch) computable in AC^0 and secure against AC^0 circuits were known from the works of Håstad and Braverman.

1 Introduction

The last four decades of research in the theory of cryptography has produced a host of fantastic notions, from public-key encryption [DH76, RSA78, GM82]

Research supported in part by NSF Grants CNS-1350619 and CNS-1414119, Alfred P. Sloan Research Fellowship, Microsoft Faculty Fellowship, the NEC Corporation, a Steven and Renee Finn Career Development Chair from MIT. This work was also sponsored in part by the Defense Advanced Research Projects Agency (DARPA) and the U.S. Army Research Office under contracts W911NF-15-C-0226.

M. Robshaw and J. Katz (Eds.): CRYPTO 2016, Part III, LNCS 9816, pp. 533–562, 2016.
DOI: 10.1007/978-3-662-53015-3_19

and zero-knowledge proofs [GMR85] in the 1980s, to fully homomorphic encryption [RAD78, Gen09, BV11] and program obfuscation [BGI+01, GGH+13, SW14] in the modern day. Complexity theory is at the heart of these developments, playing a key role in coming up with precise mathematical definitions as well as constructions whose security can be reduced to precisely stated computational hardness assumptions.

However, the uncomfortable fact remains that a vast majority of cryptographic constructions rely on *unproven assumptions*. At the very least, one requires that NP $\nsubseteq$ BPP [IL89], but that is hardly ever enough — when designing advanced cryptographic objects, cryptographers assume the existence of one-way functions as a given, move up a notch to assuming the hardness of specific problems such as factoring, discrete logarithms and the approximate shortest vector problem for lattices, and, more recently, even more exotic assumptions. While there are some generic transformations between primitives, such as from one-way functions to pseudo-random generators and symmetric encryption (e.g., [HILL99]), there are large gaps in our understanding of relationships between most others. In particular, it is a wide open question whether NP $\nsubseteq$ BPP suffices to construct even the most basic cryptographic objects such as one-way functions, or whether it is possible to construct public-key encryption assuming only the existence of one-way functions (for some partial negative results, see [BT03, AGGM06, BB15, IR88]).

In this work, we ask if a weaker version of these cryptographic primitives can be constructed, with security against a *bounded* class of adversaries, based on either *mild complexity-theoretic assumptions* or *no assumptions* at all. Indeed, this question has been asked by several researchers in the past.

1. Merkle [Mer78] constructed a non-interactive key exchange protocol (and thus, a public-key encryption scheme) where the honest parties run in linear time $O(n)$ and security is shown against adversaries that run in time $o(n^2)$. His assumption was the existence of a random function that both the honest parties and the adversary can access (essentially, the random oracle model [BR93]). Later, the assumption was improved to exponentially strong one-way functions [BGI08]. This work is timeless, not only because it jump-started public-key cryptography, but also because it showed how to obtain a primitive with much structure (trapdoors) from one that apparently has none (namely, random oracles and exponentially strong one-way functions).

2. Maurer [Mau92] introduced the bounded storage model, which considers adversaries that have an a-priori bounded amount of space but unbounded computation time. Cachin and Maurer constructed symmetric-key encryption and key-exchange protocols that are *unconditionally secure* in this model [CM97] assuming that the honest parties have storage $O(s)$ and the adversary has storage $o(s^2)$ for some parameter s. There has been a rich line of work on this model [CM97, AR99, DM04] following [Mau92].

3. Implicit in the work of Håstad [Has87] is a beautiful construction of a one-way permutation that can be computed in NC^0 (constant-depth circuits with

AND and OR gates of unbounded fan-in and NOT gates), but inverting which is hard for any AC^0 circuit. Here is the function:

$$f(x_1, x_2, \ldots, x_n) = \left(x_1, x_1 \oplus x_2, x_2 \oplus x_3, \ldots, x_{n-1} \oplus x_n\right)$$

Clearly, each output bit of this function depends on at most two input bits. Inverting the function implies in particular the ability to compute x_n, which is the parity of all the output bits. This is hard for AC^0 circuits as per [FSS84, Ajt83, Hås86].

All these works share two common features. First, security is achieved against a class of adversaries with bounded resources (time, space and parallel time, respectively, in the three works above). Secondly, the honest algorithms use less resources than the class of adversaries they are trying to fool. We propose to call the broad study of such cryptographic constructions *fine-grained cryptography*, and construct several fine-grained cryptographic primitives secure against parallel-time-bounded adversaries.

We study two classes of low-depth circuits (as adversaries). The first is AC^0, which is the class of functions computable by *constant-depth* polynomial-sized circuits consisting of AND, OR, and NOT gates of *unbounded fan-in*, and the second is NC^1, the class of functions computable by *logarithmic-depth* polynomial-sized circuits consisting of AND, OR, and NOT gates of *fan-in 2*. In both cases, we mean the non-uniform versions of these classes. Note that this also covers the case of adversaries that are randomized circuits with these respective restrictions. This is because for any such randomized adversary $\mathcal{A}$ there is a non-uniform adversary $\mathcal{A}'$ that performs as well as $\mathcal{A}$ – $\mathcal{A}'$ is simply $\mathcal{A}$ hard-coded with the randomness that worked best for it.

Early developments in circuit lower bounds [FSS84, Ajt83, Hås86] showed progressively better and *average-case* and *exponential* lower for the PARITY function against AC^0 circuits. This has recently been sharpened to an average-case depth hierarchy theorem [RST15]. We already saw how these lower bounds translate to meaningful cryptography, namely one-way permutations against AC^0 adversaries. Extending this a little further, a reader familiar with Braverman's breakthrough result [Bra10] (regarding the pseudorandomness of n^ϵ-wise independent distributions against AC^0) will notice that his result can be used to construct large-stretch pseudo-random generators that are computable by fixed-depth AC^0 circuits and are pseudo-random against arbitrary constant-depth AC^0 circuits. Can we do more? *Can we construct secret-key encryption, collision-resistant hash functions, and even trapdoor functions, starting from known lower bounds against* AC^0 *circuits?* Our first contribution is a positive answer to some of these questions.

Our second contribution is to study adversaries that live in NC^1. In this setting, as we do not know any lower bounds against NC^1, we are forced to rely on an unproven complexity-theoretic assumption; however, we aim to limit this to a worst-case, widely believed, separation of complexity classes. Here, we construct several cryptographic primitives from the *worst-case* hardness assumption that $\oplus L/poly \not\subseteq NC^1$, the most notable being an additively-homomorphic public-key encryption scheme where the key generation, encryption and decryption algorithms are all computable in $\mathsf{AC}^0[2]$ (constant-depth circuits with MOD2 gates; note that $\mathsf{AC}^0[2] \subsetneq \mathsf{AC}^0$ [Raz87, Smo87]), and the scheme is semantically secure against NC^1 adversaries. ($\oplus\mathsf{L}/\mathsf{poly}$ can be thought of as the class of languages with polynomial-sized branching programs. Note that by Barrington's Theorem [Bar86], all languages in NC^1 have polynomial-sized branching programs of constant width.)

Apart from theoretical interest stemming from the fact that these are rather natural objects, one possible application of such constructions (that was suggested to us independently by Ron Rothblum and Yuval Ishai) is in using them in conjunction with other constructions that are secure against polynomial-time adversaries under stronger assumptions. This could be done to get hybrids that are secure against polynomial-time adversaries under these stronger assumptions while also being secure against bounded adversaries unconditionally (or under minimal assumptions). For instance, consider an encryption scheme where the message is first encrypted using the AC^0-encryption scheme from Sect. 5.3, and the resultant ciphertext is then encrypted using a scheme that works in AC^0 and is secure against polynomial-time adversaries under some standard assumptions (see [AIK04] for such schemes). This resultant scheme can be shown to be secure against polynomial-time adversaries under the same assumptions while being unconditionally secure against AC^0 adversaries.

We now briefly describe the relation between our results and the related work on randomized encodings [IK00, AIK04], and move on to describing the results in detail.

Relation to Randomized Encodings and Cryptography in NC^0. Randomized encodings of Ishai and Kushilevitz [IK00, AIK04] are a key tool in our results against NC^1 adversaries. Using randomized encodings, Applebaum, Ishai and Kushilevitz [AIK04] showed how to convert several cryptographic primitives computable in logspace classes into ones that are computable in NC^0. The difference between their work and ours is two-fold: (1) Their starting points are cryptographic schemes secure against arbitrary polynomial-time adversaries, which rely on average-case hardness assumptions, whereas in our work, the focus is on achieving security either with no unproven assumptions or only worst-case assumptions; of course, our schemes are not secure against polynomial-time adversaries, but rather, limited adversarial classes; (2) In the case of public-key encryption, they manage to construct key generation and encryption algorithms that run in NC^0, but the decryption algorithm retains its high complexity. In contrast, in this work, we can construct public key encryption (against NC^1 adversaries) where the encryption algorithm can be computed in NC^0 and the key generation and decryption in $\mathsf{AC}^0[2]$.

A Remark on Cryptographic vs. Non-cryptographic Constructions. An important *desideratum* for us is that the (honest) algorithms in our constructions can be implemented with fewer resources than the adversary that they are trying to fool. We call such constructions *cryptographic* in contrast to *non-cryptographic* constructions where this is not necessarily the case. Perhaps the clearest and the most well-known example of this distinction is the case of pseudo-random generators (PRGs) [BM84, Yao82, NW94]. Cryptographic PRGs, pioneered in the works of Blum, Micali and Yao [BM84, Yao82] are functions computable in a *fixed* polynomial time that produce outputs that are indistinguishable from random against *any* polynomial-time machine. The designer of the PRG does not know the precise power of the adversary: he knows that the adversary is polynomial-time, but not *which* polynomial. On the other hand, non-cryptographic ("Nisan-Wigderson type") PRGs [NW94] take more time to compute than the adversaries they are designed to fool.

Our constructions will be exclusively in the *cryptographic* regime. For example, our one-way functions, pseudo-random generators and collision-resistant hash functions against AC^0 are computable by circuits of fixed polynomial size $q(\lambda)$ and fixed (constant) depth d, and maintain security (in the appropriate sense) against adversarial circuits of size $p'(\lambda)$ and depth d' for any polynomial function p' and any constant d'.

1.1 Our Results and Techniques

Our results are grouped into two classes — primitives secure against NC^1 circuits based on minimal worst-case assumptions, and those that are unconditionally secure against AC^0 circuits. In the description below and throughout the rest of the paper, all algebra is over $\mathbb{F}_2$.

Constructions Against NC^1 Adversaries. We construct one-way functions (OWFs), pseudo-random generators (PRGs), additively homomorphic public-key encryption (PKE), and collision-resistant hash functions (CRHFs) that are computable in NC^1 and are secure against NC^1 adversaries, based on the worst-case assumption that $\oplus L/poly \not\subseteq NC^1$. An important tool we use for these constructions is the notion of *randomized encodings* of functions introduced in [IK00].

A randomized encoding of a function f is a randomized function $\hat{f}$ that is such that for any input x, the distribution of $\hat{f}(x)$ reveals $f(x)$, but nothing more about x. We know through the work of [IK00, AIK04] that there are randomized encodings for the class $\oplus L/poly$ that can be computed in (randomized, uniform) NC^0. Randomized encodings naturally offer a flavor of worst-to-average case reductions as they reduce the problem of evaluating a function on a given input to deciding some property of the distribution of its encoding. Our starting point is the observation, implicit in [AIK04, AR15], that they can be used to generically construct infinitely-often one-way functions and pseudo-random generators with additive stretch, computable in NC^0 and secure against NC^1 adversaries (assuming, again, that $\oplus L/poly \not\subseteq NC^1$). We start with the following general theorem.

Theorem 1.1 (Informal). *Let $\mathcal{C}_1$ and $\mathcal{C}_2$ be two classes such that $\mathcal{C}_2 \not\subseteq \mathcal{C}_1$ and $\mathcal{C}_2$ has perfect randomized encodings computable in $\mathcal{C}_1$. Then, there are OWFs and PRGs that are computable in $\mathcal{C}_1$ and are secure against arbitrary adversarial functions in $\mathcal{C}_1$.*

Informally, the argument for Theorem 1.1 is the following: Let L be the language in $\mathcal{C}_2$ but not $\mathcal{C}_1$. The PRG is a function that takes an input r and outputs the randomized encoding of the indicator function for membership in L on the input 0^λ, using r as the randomness (where λ is a security parameter). Any adversary that can distinguish the output of this function from random can be used to decide if a given x is in the language L by computing the randomized encoding of x and feeding it to the adversary. This gives us a PRG with a non-zero additive stretch (and also a OWF) if the randomized encoding has certain properties (they need to be *perfect*) — see Sect. 3 for details.

While we have one way functions and pseudorandom generators, a black-box construction of public key cryptosystems from randomized encodings seems elusive. Our first contribution in this work is to use the algebraic structure of the randomized encodings for $\oplus L/poly$ to construct an additively homomorphic public key encryption scheme secure against NC^1 circuits (assuming that $\oplus L/poly \not\subseteq NC^1$).

Additively Homomorphic Public-Key Encryption. The key attribute of the randomized encodings of [IK00, AIK04] for $\oplus L/poly$ is that the encoding is not a structureless string. Rather, the randomized encodings of computations are matrices whose rank corresponds to the result of the computation. Our public-key encryption construction uses two observations:

- Under the assumption $\oplus L/poly \not\subseteq NC^1$, there exist, for an infinite number of values of n, distributions D_0^n and D_1^n over $n \times n$ matrices of rank $(n-1)$ and n, respectively, that are indistinguishable to NC^1 circuits.
- It is possible to sample a matrix $\mathbf{M}$ from D_0^n along with the non-zero vector $\mathbf{k}$ in its kernel. The sampling can be accomplished in NC^1 or even $AC^0[2]$.

The public key in our scheme is such a matrix $\mathbf{M}$, and the secret key is the corresponding $\mathbf{k}$. Encryption of a bit b is a vector $\mathbf{r}^T\mathbf{M} + b\mathbf{t}^T$, where $\mathbf{r}$ is a random vector[1] and $\mathbf{t}$ is a vector such that $\langle \mathbf{t}, \mathbf{k} \rangle = 1$. In effect, the encryption of 0 is a random vector in the row-span of $\mathbf{M}$ while the encryption of 1 is a random vector outside. Decryption of a ciphertext $\mathbf{c}$ is simply the inner product $\langle \mathbf{c}, \mathbf{k} \rangle$. Semantic security against NC^1 adversaries follows from the fact that D_0^n and D_1^n are indistinguishable to NC^1 circuits. In particular, (1) We can indistinguishably replace the public key by a random full rank matrix $\mathbf{M}'$ chosen from D_n^1; and (2) with $\mathbf{M}'$ as the public key, encryptions of 0 are identically distributed to the encryptions of 1. The following is an informal restatement of Theorem 4.1.

Theorem 1.2 (Informal). *If $\oplus L/poly \neq NC^1$, there is a public-key encryption scheme secure against NC^1 adversaries where key generation, encryption and decryption are all computable in (randomized) $AC^0[2]$.*

The scheme above is additively homomorphic, and thus, collision-resistant hash functions (CRHF) against NC^1 follow immediately from the known generic transformations [IKO05] which work in NC^1.

---

[1] We maintain the convention that all vectors are by default column vectors. For a vector $\mathbf{r}$, the notation $\mathbf{r}^T$ denotes the row vector that is the transpose of $\mathbf{r}$.

Theorem 1.3 (Informal). *If* $\oplus L/poly \neq NC^1$, *there is a family of collision-resistant hash functions that is secure against* NC^1 *adversaries where both sampling hash functions and evaluating them can be performed in (randomized)* $AC^0[2]$.

We remark that in a recent work, Applebaum and Raykov [AR15] construct CRHFs against polynomial-time adversaries under the assumption that there are average-case hard functions with perfect randomized encodings. Their techniques also carry over to our setting and imply, for instance, the existence of CRHFs against NC^1 under the assumption that there is a language that is average-case hard for NC^1 that has perfect randomized encodings that can be computed in NC^1. This does not require any additional structure on the encodings apart from perfectness, but does require average-case hardness in place of our worst-case assumptions.

Constructions Against AC^0 Adversaries. We construct one-way functions (OWFs), pseudo-random generators (PRGs), weak pseudo-random functions (weak PRFs), symmetric-key encryption (SKE) and collision-resistant hash functions (CRHFs) that are computable in AC^0 and are unconditionally secure against arbitrary AC^0 circuits. While some constructions for OWFs and PRGs against AC^0 were already known [Hås86, Bra10], and the existence of weak PRFs and SKE, being minicrypt primitives, is not that surprising, the possibility of unconditionally secure CRHFs against AC^0 is somewhat surprising, and we consider this to be our primary contribution in this section. We also present a candidate construction for public-key encryption, but we are unable to prove its unconditional security against AC^0 circuits.

As we saw earlier, Håstad [Has87] constructed one-way permutations secure against AC^0 circuits based on the hardness of computing PARITY. When allowed polynomial running time, we have black-box constructions of pseudorandom generators [HILL99] and pseudorandom functions [GGM86] from one-way functions. But because these reductions are not implementable in AC^0, getting primitives computable in AC^0 requires more effort.

Our constructions against AC^0 adversaries are primarily based on the theorem of Braverman [Bra10] (and its recent sharpening by Tal [Tal14]) regarding the pseudo-randomness of polylog-wise independent distributions against constant depth circuits. We use this to show that AC^0 circuits cannot distinguish between the distribution $(\mathbf{A}, \mathbf{Ak})$, where $\mathbf{A}$ is a random "sparse" matrix of dimension $poly(n) \times n$ and $\mathbf{k}$ is a uniformly random secret vector, from the distribution $(\mathbf{A}, \mathbf{r})$, where $\mathbf{r}$ is a uniformly random vector. Sparse here means that each row of $\mathbf{A}$ has at most $polylog(n)$ many ones.

(This is shown as follows. It turns out that with high probability, a matrix chosen in this manner is such that any set of $polylog(n)$ rows is linearly independent (Lemma 2.5). Note that when a set of rows of $\mathbf{A}$ is linearly independent, the corresponding set of bits in $\mathbf{Ak}$ are uniformly distributed. This implies that if all $polylog(n)$-sized sets of rows of $\mathbf{A}$ are linearly independent, then $\mathbf{Ak}$ is $polylog(n)$-wise independent. This fact, along with the theorems regarding pseudo-randomness mentioned above prove the indistinguishability by AC^0.)

We also crucially use the fact, from [AB84], that the inner product of an arbitrary vector with a sparse vector can be computed in constant depth.

OWFs and PRGs. This enables us to construct PRGs in NC^0 with constant multiplicative stretch and in AC^0 with polynomial multiplicative stretch. The construction is to fix a sparse matrix $\mathbf{A}$ with the linear independence properties mentioned above, and the PRG output on seed $\mathbf{k}$ is $\mathbf{Ak}$. Pseudo-randomness follows from the indistinguishability arguments above. This is stated in the following informal restatement of Theorem 5.1. We need to show that there exist such matrices $\mathbf{A}$ in which any polylog-sized set of rows are linearly independent, and yet are sparse. As we show in Sect. 2.3, there are indeed matrices that have these properties.

Theorem 1.4 (Informal). *For any constant c, there is a family of circuits $\left\{ C_n : \{0,1\}^n \to \{0,1\}^{n^c} \right\}$ such that for any n, each output bit of C_n depends on at most $O(c)$ input bits. Further, for large enough n, AC^0 circuits cannot distinguish the output distribution $C_n(U_n)$ from U_{n^c}.*

We note that similar techniques have been used in the past to construct PRGs that fool circuit families of a fixed constant depth - see, for instance, [Vio12].

Weak PRFs Against AC^0. A Pseudo-Random Function family (PRF) is a collection of functions such that a function chosen at random from this collection is indistinguishable from a function chosen at random from the set of all functions (with the appropriate domain and range), based on just a polynomial number of evaluations of the respective functions. In a *Strong* PRF, the distinguisher is allowed to specify (even adaptively) the input points at which it wants the function to be evaluated. In a *Weak* PRF, the distinguisher is given function evaluations at input points chosen uniformly at random.

We construct Weak PRFs against AC^0 that are unconditionally secure. In our construction, each function in the family is described by a vector $\mathbf{k}$. The computation of the pseudo-random function proceeds by mapping its input $\mathbf{x}$ to a sparse vector $\mathbf{a}$ and computing the inner product $\langle \mathbf{a}, \mathbf{k} \rangle$ over $\mathbb{F}_2$. Given polynomially many samples of the form $(\mathbf{a}, \langle \mathbf{a}, \mathbf{k} \rangle)$, one can write this as $(\mathbf{A}, \mathbf{Ak})$, where $\mathbf{A}$ is a matrix with random sparse rows. Our mapping of $\mathbf{x}$'s to $\mathbf{a}$'s is such that $(\mathbf{A}, \mathbf{Ak})$ is in some sense the only useful information contained in a set of random function evaluations. This is now indistinguishable from $(\mathbf{A}, \mathbf{r})$ where $\mathbf{r}$ is uniformly random, via the arguments mentioned earlier in this section. The following is an informal restatement of Theorem 5.2.

Theorem 1.5 (Informal). *There is a Weak Pseudo-Random Function family secure against AC^0 adversaries and is such that both sampling a function at random and evaluating it can be performed in AC^0.*

We note that while our construction only gives us quasi-polynomial security (that is, an adversary might be able to achieve an inverse quasi-polynomial advantage in telling our functions from random) as opposed to exponential security, we show that this is an inherent limitation of weak PRFs computable in AC^0. Roughly speaking, due to the work of [LMN93], we know that a constant fraction of the Fourier mass of any function on n inputs computable in AC^0 is concentrated on Fourier coefficients upto some $\mathsf{polylog}(n)$. So there is at least one coefficient in the case of such a function that is at least $\Omega\left(\frac{1}{2^{\mathsf{polylog}(n)}}\right)$ in absolute value, whereas in a random function any coefficient would be exponentially small. So, by guessing and estimating a Fourier coefficient of degree at most $\mathsf{polylog}(n)$ (which can be done in AC^0), one can distinguish functions computed in AC^0 from a random function with some $\Omega\left(\frac{1}{2^{\mathsf{polylog}(n)}}\right)$ advantage.

Symmetric Key Encryption Against AC^0. In the case of polynomial-time adversaries and constructions, weak PRFs generically yield symmetric key encryption schemes, and this continues to hold in our setting. However, we present an alternative construction that has certain properties that make it useful in the construction of collision-resistant hash functions later on. The key in our scheme is again a random vector $\mathbf{k}$. The encryption of a bit b is a random sparse vector $\mathbf{c}$ such that $\langle \mathbf{c}, \mathbf{k} \rangle = b$ over $\mathbb{F}_2$. (Similar schemes, albeit without the sparsity, have been employed in the past in the leakage-resilience literature — see [GR12] and references therein.)

Encryption is performed by rejection sampling to find such a $\mathbf{c}$, and decryption is performed by computing $\langle \mathbf{c}, \mathbf{k} \rangle$, which can be done in constant depth owing to the sparsity of $\mathbf{c}$. We reduce the semantic security of this construction to the indistinguishability of the distributions $(\mathbf{A}, \mathbf{Ak})$ and $(\mathbf{A}, \mathbf{r})$ mentioned earlier. Note that this scheme is additively homomorphic, a property that will be useful later. The following is an informal restatement of Theorem 5.3.

Theorem 1.6 (Informal). *There is a Symmetric Key Encryption scheme that is secure against* AC^0 *adversaries and is such that key generation, encryption and decryption are all computable in (randomized)* AC^0.

Collision Resistance Against AC^0. Our most important construction against AC^0, which is what our encryption scheme was designed for, is that of Collision Resistant Hash Functions. Note that while there are generic transformations from additively homomorphic encryption schemes to CRHFs [IKO05], these transformations do not work in AC^0 and hence do not yield the construction we desire.

Our hash functions are described by matrices where each column is the encryption of a random bit under the above symmetric encryption scheme. Given such a matrix $\mathbf{M}$ that consists of encryptions of the bits of a string m, the evaluation of the function on input $\mathbf{x}$ is $\mathbf{Mx}$. Note that we wish to perform this computation in constant depth, and this turns out to be possible to do correctly for most keys because of the sparsity of our ciphertexts.

Finding a collision for a given key $\mathbf{M}$ is equivalent to finding a vector $\mathbf{u}$ such that $\mathbf{Mu} = 0$. By the additive homomorphism of the encryption scheme, and the fact that $\mathbf{0}$ is a valid encryption of 0, this implies that $\langle m, \mathbf{u} \rangle = 0$. But this is non-trivial information about m, and so should violate semantic security. However showing that this is indeed the case turns out to be somewhat non-trivial.

In order to do so, given an AC^0 adversary A that finds collisions for the hash function with some non-negligible probability, we will need to construct another AC^0 adversary, B, that breaks semantic security of the encryption scheme. The most straightforward attempt at this would be as follows. B selects messages $\mathbf{m}_0$ and $\mathbf{m}_1$ at random and sends them to the challenger who responds with $\mathbf{M} = \mathsf{Enc}(\mathbf{m}_b)$. B then forwards this to A who would then return, with non-negligible probability, a vector $\mathbf{u}$ such that $\langle \mathbf{u}, \mathbf{m}_b \rangle = 0$. If B could compute $\langle \mathbf{u}, \mathbf{m}_0 \rangle$ and $\langle \mathbf{u}, \mathbf{m}_1 \rangle$, B would then be able to guess b correctly with non-negligible advantage. The problem with this approach is that $\mathbf{u}$, $\mathbf{m}_0$ and $\mathbf{m}_1$ might all be of high Hamming weight, and this being the case, B would not be able to compute the above inner products.

The solution to this is to choose $\mathbf{m}_0$ to be a sparse vector and $\mathbf{m}_1$ to be a random vector and repeat the same procedure. This way, B can compute $\langle \mathbf{u}, \mathbf{m}_0 \rangle$, and while it still cannot check whether $\langle \mathbf{u}, \mathbf{m}_1 \rangle = 0$, it can instead check whether $\mathbf{Mu} = \mathbf{0}$ and use this information. If it turns out that $\mathbf{Mu} = \mathbf{0}$ and $\langle \mathbf{u}, \mathbf{m}_0 \rangle \neq 0$, then B knows that $b = 1$, due to the additive homomorphism of the encryption scheme. Also, when $b = 1$, since $\mathbf{m}_0$ is independent of $\mathbf{m}_1$, the probability that A outputs $\mathbf{u}$ such that $\langle \mathbf{u}, \mathbf{m}_0 \rangle \neq 0$ is non-negligible. Hence, by guessing $b = 1$ when this happens and by guessing b at random otherwise, B can gain non-negligible advantage against semantic security. This achieves the desired contradiction and demonstrates the collision resistance of our construction. The following is an informal restatement of Theorem 5.4.

Theorem 1.7 (Informal). *There is a family of Collision Resistant Hash Functions that is secure against* AC^0 *adversaries and is such that both sampling a hash function at random and evaluating it can be performed in (randomized)* AC^0.

Candidate Public Key Encryption Against AC^0. We also propose a candidate Public Key Encryption scheme whose security we cannot show. It is similar to the LPN-based cryptosystem in [Ale03]. The public key is a matrix of the form $\mathbf{M} = (\mathbf{A}, \mathbf{Ak})$ where $\mathbf{A}$ is a random $n \times n$ matrix and $\mathbf{k}$, which is also the secret key, is a random sparse vector of length n. To encrypt 0, we choose a random sparse vector $\mathbf{r}$ and output $\mathbf{c}^T = \mathbf{r}^T \mathbf{M}$, and to encrypt 1 we just output a random vector $\mathbf{c}^T$ of length $(n+1)$. Decryption is simply the inner product of $\mathbf{c}$ and the vector $(\mathbf{k}\ 1)^T$, and can be done in AC^0 because $\mathbf{k}$ is sparse.

1.2   Other Related Work: Cryptography Against Bounded Adversaries

The big bang of public-key cryptography was the result of Merkle [Mer78] who constructed a key exchange protocol where the honest parties run in linear time $O(n)$ and security is obtained against adversaries that run in time $o(n^2)$. His assumption was the existence of a random function that both the honest parties and the adversary can access. Later, the assumption was improved to strong one-way functions [BGI08]. This is, indeed, a fine-grained cryptographic protocol in our sense.

The study of ϵ-biased generators [AGHP93, MST06] is related to this work. In particular, ϵ-biased generators with exponentially small ϵ give us almost k-wise independent generators for large k, which in turn fool AC^0 circuits by a result of Braverman [Bra10]. This and other techniques have been used in the past to construct PRGs that fool circuits of a fixed constant depth, with the focus generally being on optimising the seed length [Vio12, TX13].

The notion of precise cryptography introduced by Micali and Pass [MP06] studies reductions between cryptographic primitives that can be computed in linear time. That is, they show constructions of primitive B from primitive A such that if there is a $\mathsf{TIME}(f(n))$ algorithm that breaks primitive B, there is a $\mathsf{TIME}(O(f(n)))$ algorithm that breaks A.

Maurer [Mau92] introduced the bounded storage model, which considers adversaries that have a bounded amount of space and unbounded computation time. There are many results known here [DM04, Vad04, AR99, CM97] and in particular, it is possible to construct Symmetric Key Encryption and Key Agreement protocols unconditionally in this model [CM97].

2 Preliminaries

In this section we establish notation that shall be used throughout the rest of the presentation and recall the notion of randomized encodings of functions. We state and prove some results about certain kinds of random matrices that turn out to be useful in Sect. 5. In Sects. 2.4 and 2.5, we present formal definitions of a general notion of adversaries with restricted computational power and also of several standard cryptographic primitives against such restricted adversaries (as opposed to the usual definitions, which are specific to probabilistic polynomial time adversaries).

2.1 Notation

For a distribution D, by $x \leftarrow D$ we denote x being sampled according to D. Abusing notation, we denote by $D(x)$ the probability mass of D on the element x. For a set S, by $x \leftarrow S$, we denote x being sampled uniformly from S. We also denote the uniform distribution over S by U_S, and the uniform distribution over $\{0,1\}^\lambda$ by U_λ. We use the notion of total variational distance between distributions, given by:

$$\Delta(D_1, D_2) = \frac{1}{2} \sum_x |D_1(x) - D_2(x)|$$

For distributions D_1 and D_2 over the same domain, by $D_1 \equiv D_2$ we mean that the distributions are the same, and by $D_1 \approx D_2$, we mean that $\Delta(D_1, D_2)$ is a negligible function of some parameter that will be clear from the context. Abusing notation, we also sometimes use random variables instead of their distributions in the above expressions.

For any $n \in \mathbb{N}$, we denote by $\lfloor n \rfloor_2$ the greatest power of 2 that is not more than n. For any n, k, and $d \le k$, we denote by $SpR_{k,d}$ the uniform distribution over the set of vectors in $\{0,1\}^k$ with exactly d non-zero entries, and by $SpM_{n,k,d}$ the distribution over the set of matrices in $\{0,1\}^{n \times k}$ where each row is distributed independently according to $SpR_{k,d}$.

We identify strings in $\{0,1\}^n$ with vectors in $\mathbb{F}_2^n$ in the natural manner. For a string (vector) x, $\|x\|$ denotes its Hamming weight. Finally, we note that all arithmetic computations (such as inner products, matrix products, etc.) in this work will be over $\mathbb{F}_2$, unless specified otherwise.

2.2   Constant-Depth Circuits

Here we state a few known results on the computational power of constant depth circuits that shall be useful in our constructions against AC^0 adversaries.

Theorem 2.1 (Hardness of Parity, [Hås14]). *For any circuit C with n inputs, size s and depth d,*

$$\Pr_{x \leftarrow \{0,1\}^n} [C(x) = \mathsf{PARITY}(x)] \le \frac{1}{2} + 2^{-\Omega(n/(\log s)^{d-1})}$$

Theorem 2.2 (Partial Independence, [Bra10, Tal14]). *Let D be a k-wise independent distribution over $\{0,1\}^n$. For any circuit C with n inputs, size s and depth d,*

$$\left| \Pr_{x \leftarrow D} C(x) = 1 - \Pr_{x \leftarrow \{0,1\}^n} C(x) = 1 \right| \le \frac{s}{2^{\Omega(k^{1/(3d+3)})}}$$

The following lemma is implied by theorems proven in [AB84, RW91] regarding the computability of polylog thresholds by constant-depth circuits.

Lemma 2.3 (Polylog Inner Products). *For any constant c and for any function $t : \mathbb{N} \to \mathbb{N}$ such that $t(\lambda) = O(\log^c \lambda)$, there is an AC^0 family $\mathcal{I}^t = \{ip_\lambda^t\}$ such that for any λ,*

- *ip_λ^t takes inputs from $\{0,1\}^\lambda \times \{0,1\}^\lambda$.*
- *For any $x, y \in \{0,1\}^\lambda$ such that $\min(\|x\|, \|y\|) \le t(\lambda)$, $ip_\lambda^t(x,y) = \langle x, y \rangle$.*

2.3   Sparse Matrices and Linear Codes

In this section we describe and prove some properties of a sampling procedure for random matrices. In interest of space, we will defer the proofs of the lemmas stated in this section to the full version.

We describe the following two sampling procedures that we shall use later. SRSamp and SMSamp abbreviate *Sparse Row Sampler* and *Sparse Matrix Sampler*, respectively. SRSamp(k, d, r) samples uniformly at random a vector from $\{0,1\}^k$ with exactly d non-zero entries, using r for randomness – it chooses a set of d distinct indices between 0 to $k-1$ (via rejection sampling) and outputs the vector in which the entries at those indices are 1 and the rest are 0. When we don't specifically need to argue about the randomness, we drop the explicitly written r. SMSamp(n,k,d) samples an $n \times k$ matrix whose rows are samples from SRSamp(k,d,r) using randomly and independently chosen r's.

Construction 2.1. Sparse row and matrix sampling.

SRSamp(k,d,r): Samples a vector with exactly d non-zero entries.

1. If r is not specified or $|r| < d^2 \lceil \log(k) \rceil$, sample $r \leftarrow \{0,1\}^{d^2 \lceil \log(k) \rceil}$ anew.
2. For $l \in [d]$ and $j \in [d]$, set $u_j^l = r_{((l-1)d+j-1)\lceil \log(k)\rceil+1} \cdots r_{((l-1)d+j)\lceil \log(k)\rceil}$.
3. If there is no l such that for all distinct $j_1, j_2 \in [d]$, $u_{j_1}^l \neq u_{j_2}^l$, output 0^k.
4. Else, let l_0 be the least such l.
5. For $i \in [k]$, set $v_i = 1$ if there is a $j \in [d]$ such that $u_j^{l_0} = i$ (when interpreted in binary), or $v_i = 0$ otherwise.
6. Output $v = (v_1, \ldots, v_k)$.

SMSamp(n,k,d): Samples a matrix where each row has d non-zero entries.

1. For $i \in [n]$, sample $r_i \leftarrow \{0,1\}^{d^2 \lceil \log(k)\rceil}$ and $a_i \leftarrow$ SRSamp(k,d,r_i).
2. Output the $n \times k$ matrix whose i-th row is a_i.

For any fixed k and $d < k$, note that the function $S_{k,d} : \{0,1\}^{d^2 \lceil \log(k) \rceil} \rightarrow \{0,1\}^k$ given by $S_{k,d}(x) = $ SRSamp(k,d,x) can be easily seen to be computed by a circuit of size $O((d^3+kd^2)\log(k))$ and depth 8. And so the family $\mathcal{S} = \{S_{\lambda, d(\lambda)}\}$ is in AC^0. When, in our constructions, we require computing SRSamp(k,d,x), this is to be understood as being performed by the circuit for $S_{k,d}$ that is given as input the prefix of x of length $d^2 \lceil \log(k) \rceil$. So if the rest of the construction is computed by polynomial-sized constant depth circuits, the calls to SRSamp do not break this property.

Recall that we denote by $SpR_{k,d}$ the uniform distribution over the set of vectors in $\{0,1\}^k$ with exactly d non-zero entries, and by $SpM_{n,k,d}$ the distribution over the set of matrices in $\{0,1\}^{n \times k}$ where each row is sampled independently according to $SpR_{k,d}$. The following lemma states that the above sampling procedures produce something close to these distributions.

Lemma 2.4 (Uniform Sparse Sampling). *For any n, and $d = \log^2(k)$, there is a negligible function ν such that for any k that is a power of two, when $r \leftarrow \{0,1\}^{\log^5(k)}$,*

1. $\Delta($SRSamp$(k,d,r), SpR_{k,d}) \leq \nu(k)$
2. $\Delta($SMSamp$(n,k,d), SpM_{n,k,d}) \leq n\nu(k)$

The following property of the sampling procedures above is easiest proven in terms of expansion properties of bipartite graphs represented by the matrices sampled. The analysis closely follows that of Gallager [Gal62] from his early work on Low-Density Parity Check codes.

Lemma 2.5 (Sampling Codes). *For any constant $c > 0$, set $n = k^c$, and $d = \log^2(k)$. For a matrix $\mathbf{H}$, let $\delta(\mathbf{H})$ denote the minimum distance of the code whose parity check matrix is $\mathbf{H}$. Then, there is a negligible function ν such that for any k that is a power of two,*

$$\Pr_{\mathbf{H} \leftarrow \mathsf{SMSamp}(n,k,d)} \delta(\mathbf{H}) \geq \frac{k}{\log^3(k)} \geq 1 - \nu(k)$$

Recall that a *δ-wise independent* distribution over n bits is a distribution whose marginal distribution on any set of δ bits is the uniform distribution.

Lemma 2.6 (Distance and Independence). *Let $\mathbf{H}$ (of dimension $n \times k$) be the parity check matrix of an $[n, (n-k)]_2$ linear code of minimum distance more than δ. Then, the distribution of $\mathbf{H}\mathbf{x}$ is δ-wise independent when $\mathbf{x}$ is chosen uniformly at random from $\{0,1\}^k$.*

The following is immediately implied by Lemmas 2.5, 2.6 and Theorem 2.2. It effectively says that AC^0 circuits cannot distinguish between $(\mathbf{A}, \mathbf{As})$ and $(\mathbf{A}, \mathbf{r})$ when $\mathbf{A}$ is sampled using SRSamp and $\mathbf{s}$ and $\mathbf{r}$ are chosen uniformly at random.

Lemma 2.7. *For any polynomial n, there is a negligible function ν such that for any Boolean family $\mathcal{G} = \{g_\lambda\} \in \mathsf{AC}^0$, and for any k that is a power of 2, when $\mathbf{A} \leftarrow \mathsf{SMSamp}(n(k), k, \log^2(k))$, $\mathbf{s} \leftarrow \{0,1\}^k$ and $\mathbf{r} \leftarrow \{0,1\}^{n(k)}$,*

$$|\Pr[g_\lambda(\mathbf{A}, \mathbf{As}) = 1] - \Pr[g_\lambda(\mathbf{A}, \mathbf{r}) = 1]| \leq \nu(\lambda)$$

2.4   Adversaries

Definition 2.8 (Function Family). *A function family is a family of (possibly randomized) functions $\mathcal{F} = \{f_\lambda\}_{\lambda \in \mathbb{N}}$, where for each λ, f_λ has domain D_λ^f and co-domain R_λ^f.*

In most of our considerations, D_λ^f and R_λ^f will be $\{0,1\}^{d_\lambda^f}$ and $\{0,1\}^{r_\lambda^f}$ for some sequences $\{d_\lambda^f\}_{\lambda \in \mathbb{N}}$ and $\{r_\lambda^f\}_{\lambda \in \mathbb{N}}$. Wherever function families are seen to act as adversaries to cryptographic objects, we shall use the terms *adversary* and *function family* interchangeably. The following are some examples of natural classes of function families.

Definition 2.9 (AC^0). *The class of (non-uniform) AC^0 function families is the set of all function families $\mathcal{F} = \{f_\lambda\}$ for which there is a polynomial p and constant d such that for each λ, f_λ can be computed by a (randomized) circuit of size $p(\lambda)$, depth d and unbounded fan-in using AND, OR and NOT gates.*

Definition 2.10 (NC^1). *The class of (non-uniform) NC^1 function families is the set of all function families $\mathcal{F} = \{f_\lambda\}$ for which there is a polynomial p and constant c such that for each λ, f_λ can be computed by a (randomized) circuit of size $p(\lambda)$, depth $c\log(\lambda)$ and fan-in 2 using AND, OR and NOT gates.*

2.5   Primitives Against Bounded Adversaries

In this section, we generalize the standard definitions of several standard crypto-graphic primitives to talk about security against different classes of adversaries. In the following definitions, C_1 and C_2 are two function classes, and $l, s : \mathbb{N} \to \mathbb{N}$ are some functions. Due to space constraints, we do not define all the primitives we talk about in the paper here, but the samples below illustrate how our definitions relate to the standard ones, and the rest are analogous. All definitions are present in the full version of the paper.

Implicit (and hence left unmentioned) in each definition are the following conditions:

- *Computability*, which says that the function families that are part of the primitive are in the class C_1. Additional restrictions are specified when they apply.
- *Non-triviality*, which says that the security condition in each definition is not vacuously satisfied – that there is at least one function family in C_2 whose input space corresponds to the output space of the appropriate function family in the primitive.

Definition 2.11 (One-Way Function). *Let $\mathcal{F} = \left\{ f_\lambda : \{0,1\}^\lambda \to \{0,1\}^{l(\lambda)} \right\}$ be a function family. $\mathcal{F}$ is a C_1-One-Way Function (OWF) against C_2 if:*

- ***Computability:*** *For each λ, f_λ is deterministic.*
- ***One-wayness:*** *For any $\mathcal{G} = \left\{ g_\lambda : \{0,1\}^{l(\lambda)} \to \{0,1\}^\lambda \right\} \in C_2$, there is a negligible function ν such that for any $\lambda \in \mathbb{N}$:*

$$\Pr_{x \leftarrow U_\lambda} f_\lambda(g_\lambda(y)) = y \mid y \leftarrow f_\lambda(x) \leq \nu(\lambda)$$

For a function class C, we sometimes refer to a C-OWF or an OWF against C. In both these cases, both C_1 and C_2 from the above definition are to be taken to be C. To be clear, this implies that there is a family $\mathcal{F} \in C$ that realizes the primitive and is secure against all $\mathcal{G} \in C$. We shall adopt this abbreviation also for other primitives defined in the above manner.

Definition 2.12 (Symmetric Key Encryption). *Consider function families $\mathcal{KeyGen} = \{\mathsf{KeyGen}_\lambda : \varnothing \to K_\lambda\}$, $\mathcal{Enc} = \{\mathsf{Enc}_\lambda : K_\lambda \times \{0,1\} \to C_\lambda\}$, and $\mathcal{Dec} = \{\mathsf{Dec}_\lambda : K_\lambda \times C_\lambda \to \{0,1\}\}$. $(\mathcal{KeyGen}, \mathcal{Enc}, \mathcal{Dec})$ is a C_1-Symmetric Key Encryption Scheme against C_2 if:*

- ***Correctness:*** *There is a negligible function ν such that for any $\lambda \in \mathbb{N}$ and any $b \in \{0,1\}$:*

$$\Pr \left[\mathsf{Dec}_\lambda(k,c) = b \; \middle| \; \begin{matrix} k \leftarrow \mathsf{KeyGen}_\lambda \\ c \leftarrow \mathsf{Enc}_\lambda(k,b) \end{matrix} \right] \geq 1 - \nu(\lambda)$$

- **Semantic Security:** *For any polynomials* $n_0, n_1 : \mathbb{N} \rightarrow \mathbb{N}$, *and any family* $\mathcal{G} = \left\{ g_\lambda : C_\lambda^{n_0(\lambda)+n_1(\lambda)+1} \rightarrow \{0,1\} \right\} \in \mathcal{C}_2$, *there is a negligible function* ν' *such that for any* $\lambda \in \mathbb{N}$:

$$\Pr \left[g_\lambda \left(\{c_i^0\}, \{c_i^1\}, c \right) = b \; \middle| \; \begin{array}{c} k \leftarrow \mathsf{KeyGen}_\lambda, b \leftarrow U_1 \\ c_1^0, \ldots, c_{n_0(\lambda)}^0 \leftarrow \mathsf{Enc}_\lambda(k,0) \\ c_1^1, \ldots, c_{n_1(\lambda)}^1 \leftarrow \mathsf{Enc}_\lambda(k,1) \\ c \leftarrow \mathsf{Enc}_\lambda(k,b) \end{array} \right] \leq \frac{1}{2} + \nu'(\lambda)$$

2.6   Randomized Encodings

The notion of randomized encodings of functions was introduced by Ishai and Kushilevitz [IK00] in the context of secure multi-party computation. Roughly, a randomized encoding of a deterministic function f is another deterministic function $\widehat{f}$ that is easier to compute by some measure, and is such that for any input x, the distribution of $\widehat{f}(x,r)$ (when r is chosen uniformly at random) reveals the value of $f(x)$ and nothing more. This reduces the computation of $f(x)$ to determining some property of the distribution of $\widehat{f}(x,r)$. Hence, randomized encodings offer a flavor of worst-to-average case reduction — from computing $f(x)$ from x to that of computing $f(x)$ from random samples of $\widehat{f}(x,r)$.

We work with the following definition of *Perfect Randomized Encodings* from [App14]. We note that constructions of such encodings for $\oplus\mathsf{L/poly}$ which are computable in NC^0 were presented in [IK00].

Definition 2.13 (Perfect Randomized Encodings). *Consider a deterministic function* $f : \{0,1\}^n \rightarrow \{0,1\}^t$. *We say that the deterministic function* $\widehat{f} : \{0,1\}^n \times \{0,1\}^m \rightarrow \{0,1\}^s$ *is a* Perfect Randomized Encoding (PRE) *of* f *if the following conditions are satisfied.*

- **Input independence:** *For every* $x, x' \in \{0,1\}^n$ *such that* $f(x) = f(x')$, *the random variables* $\widehat{f}(x, U_m)$ *and* $\widehat{f}(x', U_m)$ *are identically distributed.*
- **Output disjointness:** *For every* $x, x' \in \{0,1\}^n$ *such that* $f(x) \neq f(x')$, $Supp(\widehat{f}(x, U_m)) \cap Supp(\widehat{f}(x', U_m)) = \phi$.
- **Uniformity:** *For every* x, $\widehat{f}(x, U_m)$ *is uniform on its support.*
- **Balance:** *For every* $x, x' \in \{0,1\}^n$, $\left| Supp(\widehat{f}(x, U_m)) \right| = \left| Supp(\widehat{f}(x', U_m)) \right|$
- **Stretch preservation:** $s - (n+m) = t - n$

Additionally, the PRE is said to be surjective *if it also has the following property.*

- **Surjectivity:** *For every* $y \in \{0,1\}^s$, *there exist* x *and* r *such that* $\widehat{f}(x,r) = y$.

We naturally extend the definition of PREs to function families – a family $\widehat{\mathcal{F}} = \left\{ \widehat{f}_\lambda \right\}$ is a PRE of another family $\mathcal{F} = \{f_\lambda\}$ if for all large enough λ, $\widehat{f}_\lambda$ is a PRE of f_λ. Note that this notion only makes sense for deterministic functions, and the functions and families we assume or claim to have PREs are to be taken to be deterministic.

3   OWFs from Worst-Case Assumptions

In this section and in Sect. 4, we describe some constructions of crypto-graphic primitives against bounded adversaries starting from worst-case hardness assumptions. The existence of Perfect Randomized Encodings (PREs) can be leveraged to construct one-way functions and pseudo-random generators against bounded adversaries starting from a function that is hard in the worst-case for these adversaries. We describe this construction below.

Remark 3.1 (Infinitely Often Primitives). For a class $\mathcal{C}$, the statement $\mathcal{F} = \{f_\lambda\} \notin \mathcal{C}$ implies that for any family $\mathcal{G} = \{g_\lambda\}$ in $\mathcal{C}$, there are an infinite number of values of λ such that $f_\lambda \neq g_\lambda$. Using such a worst case assumption, we only know how to obtain primitives whose security holds for an infinite number of values of λ, as opposed to holding for all large enough λ. Such primitives are called *infinitely-often*, and all primitives constructed in this section and Sect. 4 are infinitely-often primitives.

On the other hand, if we assume that for every $\mathcal{G} \in \mathcal{C}$, there exists λ_0 such that for all $\lambda > \lambda_0$, $f_\lambda \neq g_\lambda$ we can achieve the regular stronger notion of security (that holds for all large enough security parameters) in each case by the same techniques.

Theorem 3.2 (OWFs, PRGs from PREs). *Let $\mathcal{C}_1$ and $\mathcal{C}_2$ be two function classes satisfying the following conditions:*

1. *Any function family in $\mathcal{C}_2$ has a surjective PRE computable in $\mathcal{C}_1$.*
2. *$\mathcal{C}_2 \nsubseteq \mathcal{C}_1$.*
3. *$\mathcal{C}_1$ is closed under a constant number of compositions.*
4. *$\mathcal{C}_1$ is non-uniform or randomized.*
5. *$\mathcal{C}_1$ can compute arbitrary thresholds.*

Then:

1. *There is a $\mathcal{C}_1$-OWF against $\mathcal{C}_1$.*
2. *There is a $\mathcal{C}_1$-PRG against $\mathcal{C}_1$ with non-zero additive stretch.*

Theorem 3.2 in effect shows that the existence of a language with PREs out-side $\mathcal{C}_1$ implies the existence of one way functions and pseudorandom generators computable in $\mathcal{C}_1$ secure against $\mathcal{C}_1$. Instances of classes that satisfy its hypoth-esis (apart from $\mathcal{C}_2 \nsubseteq \mathcal{C}_1$) include NC^1 and BPP. Note that this theorem does not provide constructions against AC^0 because AC^0 cannot compute arbitrary thresholds.

Proof Sketch. We start with a language in $\mathcal{C}_2 \setminus \mathcal{C}_1$ described by a function family $\mathcal{F} = \{f_\lambda\}$. Let $\widehat{\mathcal{F}} = \left\{ \widehat{f_\lambda} \right\}$ be its randomized encoding. Say f_λ takes inputs from $\{0,1\}^\lambda$. Then the PRG/OWF for parameter λ is the function $g_\lambda(r) = \widehat{f_\lambda}(0^\lambda, r)$.

Without loss of generality, say $f_\lambda(0^\lambda) = 0$ and $f_\lambda(z_1) = 1$ for some z_1. To show pseudorandomness, we first observe that, by the perfectness of the randomized encoding, the uniform distribution can be generated as an equal convex combination of $\widehat{f}_\lambda(0^\lambda, r)$ and $\widehat{f}_\lambda(z_1, r)$. The advantage in distinguishing $g_\lambda(r) = \widehat{f}_\lambda(0^\lambda, r)$ from uniform can hence be used to decide if a given input x is in the language because an equal convex combination of $\widehat{f}_\lambda(0^\lambda, r)$ and $\widehat{f}_\lambda(x, r)$ will be identical to $\widehat{f}_\lambda(0^\lambda, r)$ if $f_\lambda(x) = f_\lambda(0) = 0$, and otherwise will be identical to uniform.

We require the class to be closed under composition and to be able to compute thresholds in order to be able to amplify the success probability. The non-zero additive stretch comes from the fact that the PRE is stretch-preserving.

4   PKE Against NC^1 from Worst-Case Assumptions

In Theorem 3.2 we saw that we can construct one way functions and PRGs with a small stretch generically from Perfect Randomized Encodings (PREs) starting from worst-case hardness assumptions. We do not know how to construct Public Key Encryption (PKE) in a similar black-box fashion. In this section, we use certain algebraic properties of a specific construction of PREs for functions in $\oplus L/poly$ due to Ishai-Kushilevitz [IK00] to construct Public Key Encryption and Collision Resistant Hash Functions against NC^1 that are computable in $AC^0[2]$ under the assumption that $\oplus L/poly \not\subseteq NC^1$. We state the necessary implications of their work here. We start by describing sampling procedures for some relevant distributions in Construction 4.1.

In the randomized encodings of [IK00], the output of the encoding of a function f on input x is a matrix $\mathbf{M}$ sampled identically to $\mathbf{R}_1\mathbf{M}_0^\lambda\mathbf{R}_2$ when $f(x) = 0$ and identically to $\mathbf{R}_1\mathbf{M}_1^\lambda\mathbf{R}_2$ when $f(x) = 1$, where $\mathbf{R}_1 \leftarrow \mathsf{LSamp}(\lambda)$ and $\mathbf{R}_2 \leftarrow \mathsf{RSamp}(\lambda)$. Notice that $\mathbf{R}_1\mathbf{M}_1^\lambda\mathbf{R}_2$ is full rank, while $\mathbf{R}_1\mathbf{M}_0^\lambda\mathbf{R}_2$ has rank $(\lambda - 1)$. The public key in our encryption scheme is a sample $\mathbf{M}$ from $\mathbf{R}_1\mathbf{M}_0^\lambda\mathbf{R}_2$, and the secret key is a vector $\mathbf{k}$ in the kernel of $\mathbf{M}$. An encryption of 0 is a random vector in the row-span of $\mathbf{M}$ (whose inner product with $\mathbf{k}$ is hence 0), and an encryption of 1 is a random vector that is not in the row-span of $\mathbf{M}$ (whose inner product with $\mathbf{k}$ is non-zero). Decryption is simply inner product with $\mathbf{k}$. (This is very similar to the cryptosystem in [ABW10] albeit without the noise that is added there.)

Security follows from the fact that under our hardness assumption $\mathbf{M}$ is indistinguishable from $\mathbf{R}_1\mathbf{M}_1^\lambda\mathbf{R}_2$ (see Theorem 4.2), which has an empty kernel, and so when used as the public key results in identical distributions of encryptions of 0 and 1.

Theorem 4.1 (Public Key Encryption Against NC^1). *Assume $\oplus L/poly \not\subseteq NC^1$. Then, the tuple of families $(\mathcal{KeyGen}, \mathcal{Enc}, \mathcal{Dec})$ defined in Construction 4.2 is an $AC^0[2]$-Public Key Encryption Scheme against NC^1.*

Before beginning with the proof, we describe some properties of the construction. We first begin with two sampling procedures that correspond to sampling from $\widehat{f}(x, \cdot)$ when $f(x) = 0$ or $f(x) = 1$ as described earlier. We describe these again in Construction 4.3.

Construction 4.1. Sampling distributions from [IK00]

Let $\mathbf{M}_0^n$ and $\mathbf{M}_1^n$ be the following $n \times n$ matrices:

$$
\mathbf{M}_0 = \begin{pmatrix} 0 & & \cdots & 0 & 0 \\ 1 & 0 & & & 0 \\ 0 & 1 & \ddots & & \vdots \\ \vdots & & \ddots & \ddots & 0 \\ 0 & \cdots & & 0 & 1 & 0 \end{pmatrix}, \mathbf{M}_1 = \begin{pmatrix} 0 & & \cdots & 0 & 1 \\ 1 & 0 & & & 0 \\ 0 & 1 & \ddots & & \vdots \\ \vdots & & \ddots & \ddots & 0 \\ 0 & \cdots & & 0 & 1 & 0 \end{pmatrix}
$$

LSamp(n):

1. Output an $n \times n$ upper triangular matrix where all entries in the diagonal are 1 and all other entries in the upper triangular part are chosen at random.

RSamp(n):

1. Sample at random $\mathbf{r} \leftarrow \{0,1\}^{n-1}$.
2. Output the following $n \times n$ matrix:

$$
\begin{pmatrix} 1 & 0 & \cdots & 0 & | \\ 0 & 1 & \ddots & \vdots & r \\ \vdots & & \ddots & \ddots & 0 & | \\ 0 & \cdots & & 0 & 1 \\ 0 & \cdots & & 0 & 0 & 1 \end{pmatrix}
$$

Theorem 4.2 [IK00, AIK04]. *For any boolean function family $\mathcal{F} = \{f_\lambda\}$ in $\oplus L/\text{poly}$, there is a polynomial n such that for any λ, f_λ has a PRE $\widehat{f_\lambda}$ such that the distribution of $\widehat{f_\lambda}(x)$ is identical to $\mathsf{ZeroSamp}(n(\lambda))$ when $f_\lambda(x) = 0$ and is identical to $\mathsf{OneSamp}(n(\lambda))$ when $f_\lambda(x) = 1$.*

This implies that if some function in $\oplus L/\text{poly}$ is hard to compute on the worst-case then it is hard to distinguish between samples from $\mathsf{ZeroSamp}$ and $\mathsf{OneSamp}$. In particular, the following lemma follows immediately from the observation that $\mathsf{ZeroSamp}$ and $\mathsf{OneSamp}$ can be computed in NC^1.

Lemma 4.3. *If $\oplus L/\text{poly} \not\subseteq \mathsf{NC}^1$, then there is a polynomial n and a negligible function ν such that for any family $\mathcal{F} = \{f_\lambda\}$ in NC^1, for an infinite number of values of λ,*

$$
\left| \Pr_{\mathbf{M} \leftarrow \mathsf{ZeroSamp}(n(\lambda))} [f_\lambda(\mathbf{M}) = 1] - \Pr_{\mathbf{M} \leftarrow \mathsf{OneSamp}(n(\lambda))} [f_\lambda(\mathbf{M}) = 1] \right| \leq \nu(\lambda)
$$

Lemma 4.3 can now be used to prove Theorem 4.1 as described in Sect. 1.1. We defer the details to the full version.

Construction 4.2. Public Key Encryption

Let λ be the security parameter. Let $\mathbf{M}_0^\lambda$ be the $\lambda \times \lambda$ matrix described in Construction 4.1. Define the families $\mathcal{K}ey\mathcal{G}en = \{\mathsf{KeyGen}_\lambda\}$, $\mathcal{E}nc = \{\mathsf{Enc}_\lambda\}$, and $\mathcal{D}ec = \{\mathsf{Dec}_\lambda\}$ as follows.

KeyGen_λ:

1. Sample $\mathbf{R}_1 \leftarrow \mathsf{LSamp}(\lambda)$ and $\mathbf{R}_2 \leftarrow \mathsf{RSamp}(\lambda)$.
2. Let $\mathbf{k} = (\mathbf{r}\ 1)^T$ be the last column of $\mathbf{R}_2$.
3. Compute $\mathbf{M} = \mathbf{R}_1 \mathbf{M}_0^\lambda \mathbf{R}_2$.
4. Output $(\mathsf{pk} = \mathbf{M}, \mathsf{sk} = \mathbf{k})$.

$\mathsf{Enc}_\lambda(\mathsf{pk} = \mathbf{M}, b)$:

1. Sample $\mathbf{r} \in \{0, 1\}^\lambda$.
2. Let $\mathbf{t}^T = (0\ \dots\ 0\ 1)$, of length λ.
3. Output $\mathbf{c}^T = \mathbf{r}^T \mathbf{M} + b\mathbf{t}^T$.

$\mathsf{Dec}_\lambda(\mathsf{sk} = \mathbf{k}, \mathbf{c})$:

1. Output $\langle \mathbf{c}, \mathbf{k} \rangle$.

Construction 4.3. Sampling procedures

$\mathsf{ZeroSamp}(n)$: $\widehat{f}(x, r)$ where $f(x) = 0$

1. Sample $\mathbf{R}_1 \leftarrow \mathsf{LSamp}(n)$ and $\mathbf{R}_2 \leftarrow \mathsf{RSamp}(n)$.
2. Output $\mathbf{R}_1 \mathbf{M}_0 \mathbf{R}_2$.

$\mathsf{OneSamp}(n)$: $\widehat{f}(x, r)$ where $f(x) = 1$

1. Sample $\mathbf{R}_1 \leftarrow \mathsf{LSamp}(n)$ and $\mathbf{R}_2 \leftarrow \mathsf{RSamp}(n)$.
2. Output $\mathbf{R}_1 \mathbf{M}_1 \mathbf{R}_2$.

Remark 4.4. The computation of the PRE from [IK00] can be moved to NC^0 by techniques noted in [IK00] itself. Using similar techniques with Construction 4.2 gives us a Public Key Encryption scheme with encryption in NC^0 and decryption and key generation in $\mathsf{AC}^0[2]$. The impossibility of decryption in NC^0, as noted in [AIK04], continues to hold in our setting.

Remark 4.5. (This was pointed out to us by Abhishek Jain.) The above PKE scheme has what are called, in the terminology of [PVW08], "message-lossy" public keys – in this case, this is simply $\mathbf{M}$ when sampled from $\mathsf{OneSamp}$, as in the proof above. Such schemes may be used, again by results from [PVW08], to construct protocols for Oblivious Transfer where the honest parties are computable in NC^1 and which are secure against semi-honest NC^1 adversaries under the same assumptions (that $\oplus \mathsf{L/poly} \not\subseteq \mathsf{NC}^1$).

4.1 Collision Resistant Hashing

Note that again, due to the linearity of decryption, Construction 4.2 is additively homomorphic – if c_1 and c_2 are valid encryptions of m_1 and m_2, $(c_1 \oplus c_2)$ is a valid encryption of $(m_1 \oplus m_2)$. Furthermore, the size of ciphertexts does not increase when this operation is performed. Given these properties, we can use the generic transformation from additively homomorphic encryption to collision resistance due to [IKO05], along with the observation that all operations involved in the transformation can still be performed in $\mathsf{AC}^0[2]$, to get the following.

Theorem 4.6. *Assume $\oplus\mathsf{L}/\mathsf{poly} \not\subseteq \mathsf{NC}^1$. Then, for any constant $c < 1$ and function s such that $s(n) = O(n^c)$, there exists an $\mathsf{AC}^0[2]$-CRHF against NC^1 with compression s.*

5   Cryptography Without Assumptions

In this section, we present some constructions of primitives unconditionally secure against AC^0 adversaries that are computable in AC^0. This is almost the largest complexity class (after AC^0 with MOD gates) for which we can hope to get such unconditional results owing to a lack of better lower bounds. In this section, we present constructions of PRGs with arbitrary polynomial stretch, Weak PRFs, Symmetric Key Encryption, and Collision Resistant Hash Functions. We end with a candidate for Public Key Encryption against AC^0 that we are unable to prove secure, but also do not have an attack against.

5.1   High-Stretch Pseudo-Random Generators

We present here a construction of Pseudo-Random Generators against AC^0 with arbitrary polynomial stretch that can be computed in AC^0. In fact, the same techniques can be used to obtain constant stretch generators computable in NC^0.

The key idea behind the construction is the following: [Bra10] implies that for any constant ϵ, an n^ϵ-wise independent distribution will fool AC^0 circuits of arbitrary constant depth. So, being able to sample such distributions in AC^0 suffices to construct good PRGs. As shown in Sect. 2.3, if $\mathbf{H}$ is the parity-check matrix of a code with large distance d, then the distribution $\mathbf{Hx}$ is d-wise independent for $\mathbf{x}$ being a uniformly random vector (by Lemma 2.6). Further, as was also shown in Sect. 2.3, even for rather large d there are such matrices $\mathbf{H}$ that are sparse, allowing us to compute the product $\mathbf{Hx}$ in AC^0.

Theorem 5.1 (PRGs Against AC^0). *For any polynomial l, the family $\mathcal{F}^l$ from Construction 5.1 is an AC^0-PRG with multiplicative stretch $\left(\frac{l(\lambda)}{\lambda}\right)$.*

Construction 5.1. AC^0-PRG against AC^0

For any polynomial l, we define the family $\mathcal{F}^l = \left\{ f_\lambda^l : \{0,1\}^\lambda \to \{0,1\}^{l(\lambda)} \right\}$ as follows. Lemma 2.5 implies for large λ, there is an $[l(\lambda), (l(\lambda) - \lambda)]_2$ linear code with minimum distance at least $\frac{\lambda}{\log^3(\lambda)}$ whose parity check matrix has $\log^2(\lambda)$ non-zero entries in each row. Denote this parity check matrix by $\mathbf{H}_{l,\lambda}$. The dimensions of $\mathbf{H}_{l,\lambda}$ are $l(\lambda) \times \lambda$.

$$f_\lambda^l(\mathbf{x}) = \mathbf{H}_{l,\lambda}\mathbf{x}$$

Proof. For any l, the most that needs to be done to compute $f_\lambda^l(x)$ is computing the product $\mathbf{H}_{l,\lambda}\mathbf{x}$. We know that each row of $\mathbf{H}_{l,\lambda}$ contains at most $\log^2(\lambda)$ non-zero entries. Hence, by Lemma 2.3, $\mathcal{F}^l$ is in AC^0. The multiplicative stretch being $\left(\frac{l(\lambda)}{\lambda}\right)$ is also easily verified.

For pseudo-randomness, we observe that the product $\mathbf{H}_{l,\lambda}\mathbf{x}$ is $\Omega\left(\frac{\lambda}{\log^3(\lambda)}\right)$-wise independent, by Lemma 2.6. And hence, Theorem 2.2 implies that this distribution is pseudo-random to adversaries in AC^0.

5.2 Weak Pseudo-Random Functions

In this section, we describe our construction of Weak Pseudo-Random Functions against AC^0 computable in AC^0 (Construction 5.2). Roughly, we know that for a random sparse matrix $\mathbf{H}$, $(\mathbf{H}, \mathbf{Hk})$ is indistinguishable from $(\mathbf{H}, \mathbf{r})$ where $\mathbf{r}$ and $\mathbf{k}$ are chosen uniformly at random. We choose the key of the PRF to be a random vector $\mathbf{k}$. On an input $\mathbf{x}$, the strategy is to use the input $\mathbf{x}$ to generate a sparse vector $\mathbf{y}$ and then take the inner product $\langle \mathbf{y}, \mathbf{k} \rangle$.

Construction 5.2. AC^0-PRF against AC^0

Let $\mathcal{I}^t = \{ip_\lambda^t\}$ be the inner product family with threshold promise t described in Lemma 2.3. Define families $\mathcal{K}eyGen = \{\mathsf{KeyGen}_\lambda\}$ and $\mathcal{E}val = \{\mathsf{Eval}_\lambda\}$ as follows.

KeyGen_λ:

1. Output a random vector $\mathbf{k} \leftarrow \{0,1\}^{\lfloor\lambda\rfloor_2}$.

$\mathsf{Eval}_\lambda(\mathbf{k}, \mathbf{r})$:

1. Compute $\mathbf{y} \leftarrow \mathsf{SRSamp}(\lfloor\lambda\rfloor_2, \log^2(\lfloor\lambda\rfloor_2), \mathbf{r})$.
2. Output $ip_{\lfloor\lambda\rfloor_2}^{\log^2(\lambda)}(\mathbf{k}, \mathbf{y})$.

Theorem 5.2 (PRFs Against AC^0). *The pair of families $(\mathcal{K}eyGen, \mathcal{E}val)$ defined in Construction 5.2 is a Weak AC^0-PRF against AC^0.*

The intuitive reason one would think Construction 5.2 might be pseudo-random is that a collection of random function values from a randomly sampled key seems to contain the same information as $(\mathbf{H}, \mathbf{Hk})$ where $\mathbf{k}$ is sampled uniformly at random and $\mathbf{H}$ is sampled using SMSamp: a matrix with sparse rows. We know from Lemma 2.5 that except with negligible probability, $\mathbf{H}$ is going to be the parity check matrix of a code with large distance, and when it is, the arguments from Sect. 5.1 show that $(\mathbf{H}, \mathbf{Hk})$ is indistinguishable from $(\mathbf{H}, \mathbf{r})$, where $\mathbf{r}$ is sampled uniformly at random.

The only fact that prevents this from functioning as a proof is that what the adversary gets is not $(\mathbf{y}, \langle \mathbf{y}, \mathbf{k} \rangle)$ where $\mathbf{y}$ is an output of SRSamp, but rather

$(\mathbf{r}, \langle \mathbf{y}, \mathbf{k} \rangle)$, where $\mathbf{r}$ is randomness that when used in SRSamp gives $\mathbf{y}$. One way to show that this is still pseudo-random is to reduce the case where the input is $(\mathbf{y}, \langle \mathbf{y}, \mathbf{x} \rangle)$ to the case where the input is $(\mathbf{r}, \langle \mathbf{y}, \mathbf{x} \rangle)$ using an AC^0-reduction. To do this, one would need an AC^0 circuit that would, given $\mathbf{y}$, sample from a distribution close to the uniform distribution over $\mathbf{r}$'s that cause SRSamp to output $\mathbf{y}$ when used as randomness. We implement this proof strategy in the full version.

Construction 5.2 of Weak PRFs achieves only quasi-polynomial security — that is, there is no guarantee that some AC^0 adversary may not have an inverse quasi-polynomial advantage in distinguishing between the PRF and a random function. Due to the seminal work of Linial-Mansour-Nisan [LMN93] and subsequent improvements in [Tal14], we know that this barrier is inherent and we cannot hope for exponential security

5.3 Symmetric Key Encryption

In this section, we present a Symmetric Key Encryption scheme against AC^0 computable in AC^0, which is also additively homomorphic – a property that shall be useful in constructing Collision Resistant Hash Functions later on.

In Sect. 5.2, we saw a construction of Weak PRFs. And Weak PRFs give us Symmetric Key Encryption generically (where $\mathsf{Enc}(\mathbf{k}, b) = (\mathbf{r}, \mathsf{PRF}(\mathbf{k}, \mathbf{r}) \oplus b)$). For the Weak PRF construction from Sect. 5.2, this implied scheme also happens to be additively homomorphic. But it has the issue that the last bit of the ciphertext is an almost unbiased bit and hence it is not feasible to do more than $\mathsf{polylog}(\lambda)$ homomorphic evaluations on collections of ciphertexts in AC^0. So, we construct a different Symmetric Key Encryption scheme that does not suffer from this drawback and is still additively homomorphic. Then we will use this scheme to construct Collision Resistant Hash Functions. This scheme is described in Construction 5.3. In this scheme we choose the ciphertext by performing rejection sampling in parallel. For encrypting a bit b, we sample a ciphertext $\mathbf{c}$ such that $\mathbf{c}$ is sparse and $\langle \mathbf{c}, \mathbf{k} \rangle = b$. This ensures that the we have an additively homomorphic scheme where all the bits are sparse.

Theorem 5.3 (Symmetric Encryption Against AC^0). *The tuple of families $(\mathcal{KeyGen}, \mathcal{Enc}, \mathcal{Dec})$ defined in Construction 5.3 is an AC^0-Symmetric-Key Encryption Scheme against AC^0.*

The key idea behind the proof is showing that for most keys $\mathbf{k}$, the distribution of a uniformly random bit along with its encryption, that is,

$$D_1 = \{(b, \mathsf{Enc}_\lambda(\mathbf{k}, b)) \mid b \leftarrow \{0, 1\}\}$$

is statistically close to the distribution of a random sparse vector along with its inner product with $\mathbf{k}$, that is,

$$D_2 = \left\{ (\langle \mathbf{r}, \mathbf{k} \rangle, \mathbf{r}) \mid \mathbf{r} \leftarrow \mathsf{SRSamp}(\lambda, \log^2 \lambda) \right\}$$

Construction 5.3. AC^0-Symmetric Key Encryption against AC^0

Let $\mathcal{I}^t = \{ip_\lambda^t\}$ be the inner product family with threshold promise t described in Lemma 2.3. Define families $\mathcal{K}eyGen = \{\mathsf{KeyGen}_\lambda\}$, $\mathcal{E}nc = \{\mathsf{Enc}_\lambda\}$, and $\mathcal{D}ec = \{\mathsf{Dec}_\lambda\}$ as below.

KeyGen_λ:

1. Output $\mathbf{k} \leftarrow \{0,1\}^{\lfloor \lambda \rfloor_2}$.

$\mathsf{Enc}_\lambda(\mathbf{k}, b)$:

1. For $i \in [\lambda]$, sample $\mathbf{c}_i \leftarrow \mathsf{SRSamp}(\lfloor \lambda \rfloor_2, \log^2(\lfloor \lambda \rfloor_2))$.
2. Choose the first i such that $\langle \mathbf{c}_i, \mathbf{k} \rangle = b$.
3. If such an i exists, output $\mathbf{c}_i$, else output $0^{\lfloor \lambda \rfloor_2}$.

$\mathsf{Dec}_\lambda(\mathbf{k}, \mathbf{c})$:

1. Output $ip_{\lfloor \lambda \rfloor_2}^{\log^2(\lambda)}(\mathbf{k}, \mathbf{c})$.

The second distribution is similar to the one that came up in the security proof of the weak PRF construction earlier, where we effectively showed that we can replace $\langle \mathbf{r}, \mathbf{k} \rangle$ with an independent random bit without being caught by AC^0 adversaries. We defer the complete proof to the full version.

5.4 Collision Resistant Hash Functions

To construct Collision Resistant Hash Functions (CRHFs), we use the additive homomorphism of the Symmetric Key Encryption scheme constructed in Sect. 5.3. Each function in the family of hash functions is given by a matrix whose columns are ciphertexts from the encryption scheme, and evaluation is done by treating the input as a column vector and computing its product with this matrix (effectively computing a linear combination of ciphertexts). To find collisions, the adversary needs to come up with a vector in the kernel of this matrix. We show that constant depth circuits of polynomial size cannot do this for most such matrices. This is because the all-zero vector is a valid encryption of 0 in our encryption scheme, and as this scheme is additively homomorphic, finding a subset of ciphertexts that sum to zero is roughly the same as finding a subset of the corresponding messages that sum to 0, and this is a violation of semantic security.

Theorem 5.4 (CRHFs Against AC^0). *For any polylogarithmic function s, the pair of families $(\mathcal{K}eyGen^s, \mathcal{E}val^s)$, from Construction 5.4 is an AC^0-CRHF with compression s.*

We refer the reader to the sketch of the proof of Theorem 1.7 (an informal version of Theorem 5.4) towards the end of Sect. 1.1 and leave the proof of Theorem 5.4 to the full version.

Construction 5.4. AC^0-CRHFs against AC^0

Let $\mathcal{I}^t = \{ip_\lambda^t\}$ be the inner product family with threshold promise t described in Lemma 2.3. Let $(\mathcal{KeyGen}^{Enc}, \mathcal{Enc}^{Enc})$ be the SKE scheme from Construction 5.4. Let $l(\lambda) = \left\lfloor \frac{\lambda}{s(\lambda)} \right\rfloor_2$.

For any $s : \mathbb{N} \to \mathbb{N}$ such that $s(\lambda) = O(\log^c(\lambda))$ for some constant c, we define the families $\mathcal{KeyGen}^s = \{\mathsf{KeyGen}_\lambda^s\}$ and $\mathcal{Eval}^s = \{\mathsf{Eval}_\lambda^s\}$ as follows.

$\mathsf{KeyGen}_\lambda^s$:

1. Sample $\mathbf{k} \leftarrow \mathsf{KeyGen}_{l(\lambda)}^{Enc}$, and $b_1, \dots, b_\lambda \leftarrow \{0, 1\}$.
2. Output $\mathbf{M} = (\mathbf{m}_1, \dots, \mathbf{m}_\lambda)$, where $\mathbf{m}_i \leftarrow \mathsf{Enc}_{l(\lambda)}^{Enc}(\mathbf{k}, b_i)$.

$\mathsf{Eval}_\lambda^s(\mathbf{M}, \mathbf{x})$:

1. Note that $\mathbf{M} = (\mathbf{m}_1, \dots, \mathbf{m}_\lambda)$, where each $\mathbf{m}_i$ is of length $l(\lambda)$.
2. For $j \in [l(\lambda)]$, let $r_j = (\mathbf{m}_{1j}, \dots, \mathbf{m}_{\lambda j})$ (the jth bit of each $\mathbf{m}_i$).
3. Output $(h_1, \dots, h_{l(\lambda)})$, where $h_j = ip_\lambda^{4s(\lambda)\log^2(\lambda)}(r_j, \mathbf{x})$.

5.5   Candidate Public Key Encryption Scheme

In Lemma 2.7 we showed that the distribution $(\mathbf{A}, \mathbf{Ak})$ where $\mathbf{A}$ was sampled as a sparse matrix and $\mathbf{k}$ was a random vector is indistinguishable from $(\mathbf{A}, \mathbf{r})$ where $\mathbf{r}$ is also a random vector, for a wide range of parameters. We need at least one of the two $\mathbf{A}$ or $\mathbf{k}$ to be sparse to enable the computation of $\mathbf{Ak}$ in AC^0. If we make the analogous indistinguishability assumption with the key being sparse – that is, that $(\mathbf{A}, \mathbf{Ak})$ is indistinguishable from $(\mathbf{A}, \mathbf{r})$ where $\mathbf{A} \leftarrow \{0, 1\}^{\lambda \times \lambda}$, $\mathbf{k} \leftarrow \mathsf{SRSamp}(\lambda, \log^2 \lambda)$ and $\mathbf{r} \leftarrow \{0, 1\}^\lambda$ – this allows us to construct a Public Key Encryption scheme against AC^0 computable in AC^0.

This is presented in Construction 5.5, and is easily seen to be secure under Assumption 5.5. This candidate is very similar to the LPN based cryptosystem due to Alekhnovich [Ale03]. Note that while the correctness of decryption in Construction 5.5 is not very good, this may be easily amplified by repetition without losing security, as the error is one-sided.

Assumption 5.5. *Distributions* $D_1 = (\mathbf{A}, \mathbf{Ak})$ *where* $\mathbf{A} \leftarrow \{0, 1\}^{\lambda \times \lambda}$, $\mathbf{k} \leftarrow \mathsf{SRSamp}(\lambda, \log^2 \lambda)$ *and* $D_2 = (\mathbf{A}, \mathbf{r})$ *where* $\mathbf{r} \leftarrow \{0, 1\}^\lambda$ *are indistinguishable by* AC^0 *adversaries with non-negligible advantage.*

The most commonly used proof technique in this paper – showing k-wise independence for a large k – cannot be used to prove the security of this scheme because due to the sparsity of the key, the distribution $(\mathbf{A}, \mathbf{Ak})$ is not k-wise independent.

Conclusions and Open Questions. We construct various cryptographic primitives secure against parallel-time-bounded adversaries. Our constructions against AC^0 are unconditional whereas our constructions against NC^1 require the assumption

---

Construction 5.5. Public key encryption

Let $\mathcal{I}^t = \{ip_\lambda^t\}$ be the inner product family with threshold promise t described in Lemma 2.3. Define families $\mathcal{K}eyGen = \{\mathsf{KeyGen}_\lambda\}$, $\mathcal{E}nc = \{\mathsf{Enc}_\lambda\}$, and $\mathcal{D}ec = \{\mathsf{Dec}_\lambda\}$ as below.

KeyGen_λ:

1. Sample $\mathbf{A} \leftarrow \{0,1\}^{\lambda \times \lambda - 1}$, $\mathbf{k} \leftarrow \mathsf{SRSamp}(\lambda - 1, \log^2 \lambda)$
2. Output $(\mathsf{pk}, \mathsf{sk}) = ((\mathbf{A}, \mathbf{Ak}), (\mathbf{k} \circ 1))$.

$\mathsf{Enc}_\lambda(\mathsf{pk}, b)$:

1. If $b = 0$, sample $\mathbf{t} \leftarrow \mathsf{SRSamp}(\lambda, \log^2 \lambda)$ and output $\mathbf{t}^T \mathsf{pk}$
2. Else if $b = 1$, output $\mathbf{t} \leftarrow \{0,1\}^\lambda$

$\mathsf{Dec}_\lambda(\mathsf{sk}, \mathbf{c})$:

1. Output $ip_{\lfloor \lambda \rfloor_2}^{\log^2(\lambda)}(\mathsf{sk}, \mathbf{c})$.

---

that $\mathsf{NC}^1 \neq \oplus \mathsf{L/poly}$. Our constructions make use of circuit lower bounds [Bra10] and non-black-box use of randomized encodings for logspace classes [IK00].

There are several open questions that arise out of this work. Perhaps the most immediate are:

1. Unconditional lower-bounds are known for slightly larger classes like $\mathsf{AC}^0[p]$ when p is a prime power. Can we get cryptographic primitives from those lower-bounds?
2. Construct a public key encryption scheme secure against AC^0, either by proving the security of our candidate proposal (see Sect. 5.5) or by completely different means.

   Natural ways of doing this lead us to a fascinating question about the complexity of AC^0 circuits. Braverman [Bra10] shows that *any* n^ϵ-wise independent distribution fools all AC^0 circuits. Our candidate encryption, however, produces ciphertexts that come from a $\log^c(n)$-wise distribution for some constant c. This raises the following question: *Can we show some fixed poly-log-wise independent distribution (that is not n^ϵ-wise independent) that fools AC^0 circuits of arbitrary depth?* (This question came up during discussions with Li-Yang Tan.)
3. We relied on the assumption that $\oplus \mathsf{L/poly} \not\subseteq \mathsf{NC}^1$ to construct primitives secure against NC^1. It would be desirable to relax the assumption to $\mathsf{P} \not\subseteq \mathsf{NC}^1$.

A related extension of Merkle's work is to construct a public-key encryption scheme resistant against $O(n^c)$ time adversaries (for some $c > 2$) under worst-case hardness assumptions.

Acknowledgements. We thank Prabhanjan Ananth for several useful discussions towards the beginning of the project. We would also like to thank the anonymous reviewers for their careful comments.

References

[AB84] Ajtai, M., Ben-Or, M.: A theorem on probabilistic constant depth computations. In: Proceedings of the 16th Annual ACM Symposium on Theory of Computing, April 30–May 2 1984, Washington, DC, USA, pp. 471–474 (1984)

[ABW10] Applebaum, B., Barak, B., Wigderson, A.: Public-key cryptography from different assumptions. In: Proceedings of the Forty-Second ACM Symposium on Theory of Computing, pp. 171–180. ACM (2010)

[AGGM06] Akavia, A., Goldreich, O., Goldwasser, S., Moshkovitz, D.: On basing one-way functions on NP-hardness. In: Proceedings of the 38th Annual ACM Symposium on Theory of Computing, Seattle, WA, USA, May 21–23 2006, pp. 701–710 (2006)

[AGHP93] Alon, N., Goldreich, O., Håstad, J., Peralta, R.: Addendum to "simple construction of almost k-wise independent random variables". Random Struct. Algorithms $4(1)$, 119–120 (1993)

[AIK04] Applebaum, B., Ishai, Y., Kushilevitz, E.: Cryptography in NC^0. In: Proceedings of the 45th Annual IEEE Symposium on Foundations of Computer Science, FOCS 2004, 17–19 October 2004, Rome, Italy, p. 166. IEEE Computer Society Press (2004)

[Ajt83] Ajtai, M.: 11-formulae on finite structures. Ann. Pure Appl. Logic $24(1)$, 1–48 (1983)

[Ale03] Alekhnovich, M.: More on average case vs approximation complexity. In: Proceedings of the 44th Annual IEEE Symposium on Foundations of Computer Science, pp. 298–307. IEEE (2003)

[App14] Applebaum, B.: Cryptography in NC^0. In: Applebaum, B. (ed.) Cryptography in Constant Parallel Time, pp. 33–78. Springer, Heidelberg (2014)

[AR99] Aumann, Y., Rabin, M.O.: Information theoretically secure communication in the limited storage space model. In: Wiener, M. (ed.) CRYPTO 1999. LNCS, vol. 1666, pp. 65–79. Springer, Heidelberg (1999)

[AR15] Applebaum, B., Raykov, P.: On the relationship between statistical zero-knowledge and statistical randomized encodings. Electron. Colloq. Comput. Complex. (ECCC) 22, 186 (2015)

[Bar86] Mix Barrington, D.A.: Bounded-width polynomial-size branching programs recognize exactly those languages in NC^1. In: Proceedings of the 18th Annual ACM Symposium on Theory of Computing, 28–30 May 1986, Berkeley, California, USA, pp. 1–5 (1986)

[BB15] Bogdanov, A., Brzuska, C.: On basing size-verifiable one-way functions on NP-hardness. In: Dodis, Y., Nielsen, J.B. (eds.) TCC 2015, Part I. LNCS, vol. 9014, pp. 1–6. Springer, Heidelberg (2015)

[BGI+01] Barak, B., Goldreich, O., Impagliazzo, R., Rudich, S., Sahai, A., Vadhan, S.P., Yang, K.: On the (Im)possibility of obfuscating programs. In: Kilian, J. (ed.) CRYPTO 2001. LNCS, vol. 2139, p. 1. Springer, Heidelberg (2001)

[BGI08] Biham, E., Goren, Y.J., Ishai, Y.: Basing weak public-key cryptography on strong one-way functions. In: Canetti, R. (ed.) TCC 2008. LNCS, vol. 4948, pp. 55–72. Springer, Heidelberg (2008)

[BM84] Blum, M., Micali, S.: How to generate cryptographically strong sequences of pseudo-random bits. SIAM J. Comput. $13(4)$, 850–864 (1984)

[BR93]   Bellare, M., Rogaway, P.: Random oracles are practical: a paradigm for designing efficient protocols. In: Denning, D.E., Pyle, R., Ganesan, R., Sandhu, R.S., Ashby, V. (eds.) Proceedings of the 1st ACM Conference on Computer and Communications Security, CCS 1993, Fairfax, Virginia, USA, 3–5 November 1993, pp. 62–73. ACM (1993)

[Bra10] Braverman, M.: Polylogarithmic independence fools AC^0 circuits. J. ACM **57**(5), 28:1–28:10 (2010)

[BT03]   Bogdanov, A., Trevisan, L.: On worst-case to average-case reductions for NP problems. In: Proceedings of the 44th Symposium on Foundations of Computer Science (FOCS 2003), 11–14 October 2003, Cambridge, MA, USA, pp. 308–317. IEEE Computer Society (2003)

[BV11]   Brakerski, Z., Vaikuntanathan, V.: Efficient fully homomorphic encryption from (standard) LWE. In: Ostrovsky, R. (ed.) FOCS, pp. 97–106. IEEE (2011). Invited to SIAM Journal on Computing

[CM97]   Cachin, C., Maurer, U.M.: Unconditional security against memory-bounded adversaries. In: Kaliski, B.S. (ed.) CRYPTO 1997. LNCS, vol. 1294, pp. 292–306. Springer, Heidelberg (1997)

[DH76]   Diffie, W., Hellman, M.E.: New directions in cryptography. IEEE Trans. Inf. Theor. **22**(6), 644–654 (1976)

[DM04]   Dziembowski, S., Maurer, U.M.: On generating the initial key in the bounded-storage model. In: Cachin, C., Camenisch, J.L. (eds.) EUROCRYPT 2004. LNCS, vol. 3027, pp. 126–137. Springer, Heidelberg (2004)

[FSS84] Furst, M.L., Saxe, J.B., Sipser, M.: Parity, circuits, and the polynomial-time hierarchy. Math. Syst. Theor. **17**(1), 13–27 (1984)

[Gal62] Gallager, R.G.: Low-density parity-check codes. IRE Trans. Inf. Theor. **8**(1), 21–28 (1962)

[Gen09] Gentry, C.: Fully homomorphic encryption using ideal lattices. In: STOC, pp. 169–178 (2009)

[GGH+13] Garg, S., Gentry, C., Halevi, S., Raykova, M., Sahai, A., Waters, B.: Candidate indistinguishability obfuscation and functional encryption for all circuits. In: FOCS 2013, pp. 40–49 (2013)

[GGM86] Goldreich, O., Goldwasser, S., Micali, S.: How to construct random functions. J. ACM (JACM) **33**(4), 792–807 (1986)

[GM82]   Goldwasser, S., Micali, S.: Probabilistic encryption and how to play mental poker keeping secret all partial information. In: STOC 1982, pp. 365–377 (1982)

[GMR85] Goldwasser, S., Micali, S., Rackoff, C.: The knowledge complexity of interactive proof-systems (extended abstract). In: Proceedings of the 17th Annual ACM Symposium on Theory of Computing, 6–8 May 1985, Providence, Rhode Island, USA, pp. 291–304 (1985)

[GR12]   Goldwasser, S., Rothblum, G.N.: How to compute in the presence of leakage. In: 53rd Annual IEEE Symposium on Foundations of Computer Science, FOCS 2012, New Brunswick, NJ, USA, 20–23 October 2012, pp. 31–40 (2012)

[Hås86] Håstad, J.: Almost optimal lower bounds for small depth circuits. In: Proceedings of the 18th Annual ACM Symposium on Theory of Computing, 28–30 May 1986, Berkeley, California, USA, pp. 6–20 (1986)

[Has87] Hastad, J.: One-way permutations in NC^0. Inf. Process. Lett. **26**(3), 153–155 (1987)

[Hås14] Håstad, J.: On the correlation of parity and small-depth circuits. SIAM J. Comput. **43**(5), 1699–1708 (2014)

[HILL99] Håstad, J., Impagliazzo, R., Levin, L.A., Luby, M.: A pseudorandom generator from any one-way function. SIAM J. Comput. **28**(4), 1364–1396 (1999)

[IK00] Ishai, Y., Kushilevitz, E.: Randomizing polynomials: a new representation with applications to round-efficient secure computation. In: Proceedings of the 41st Annual Symposium on Foundations of Computer Science, pp. 294–304. IEEE (2000)

[IKO05] Ishai, Y., Kushilevitz, E., Ostrovsky, R.: Sufficient conditions for collision-resistant hashing. In: Kilian, J. (ed.) TCC 2005. LNCS, vol. 3378, pp. 445–456. Springer, Heidelberg (2005)

[IL89] Impagliazzo, R., Luby, M.: One-way functions are essential for complexity based cryptography (extended abstract). In: 30th Annual Symposium on Foundations of Computer Science, Research Triangle Park, North Carolina, USA, 30 October–1 November 1989, pp. 230–235. IEEE Computer Society (1989)

[IR88] Impagliazzo, R., Rudich, S.: Limits on the provable consequences of one-way permutations. In: Goldwasser, S. (ed.) CRYPTO 1988. LNCS, vol. 403, pp. 8–26. Springer, Heidelberg (1990)

[LMN93] Linial, N., Mansour, Y., Nisan, N.: Constant depth circuits, fourier transform, and learnability. J. ACM (JACM) **40**(3), 607–620 (1993)

[Mau92] Maurer, U.M.: Conditionally-perfect secrecy and a provably-secure randomized cipher. J. Cryptol. **5**(1), 53–66 (1992)

[Mer78] Merkle, R.C.: Secure communications over insecure channels. Commun. ACM **21**(4), 294–299 (1978)

[MP06] Micali, S., Pass, R.: Local zero knowledge. In: Proceedings of the 38th Annual ACM Symposium on Theory of Computing, Seattle, WA, USA, 21–23 May 2006, pp. 306–315 (2006)

[MST06] Mossel, E., Shpilka, A., Trevisan, L.: On epsilon-biased generators in NC^0. Random Struct. Algorithms **29**(1), 56–81 (2006)

[NW94] Nisan, N., Wigderson, A.: Hardness vs randomness. J. Comput. Syst. Sci. **49**(2), 149–167 (1994)

[PVW08] Peikert, C., Vaikuntanathan, V., Waters, B.: A framework for efficient and composable oblivious transfer. In: Wagner, D. (ed.) CRYPTO 2008. LNCS, vol. 5157, pp. 554–571. Springer, Heidelberg (2008)

[RAD78] Rivest, R., Adleman, L., Dertouzos, M.: On data banks and privacy homomorphisms. In: Foundations of Secure Computation, pp. 169–177. Academic Press (1978)

[Raz87] Razborov, A.A.: Lower bounds on the size of bounded depth circuits over a complete basis with logical addition. Math. Notes Acad. Sci. USSR **41**(4), 333–338 (1987)

[RSA78] Rivest, R.L., Shamir, A., Adleman, L.M.: A method for obtaining digital signatures and public-key cryptosystems. Commun. ACM **21**(2), 120–126 (1978)

[RST15] Rossman, B., Servedio, R.A., Tan, L.-Y.: An average-case depth hierarchy theorem for boolean circuits. Electron. Colloq. Comput. Complex. (ECCC) **22**, 65 (2015)

[RW91] Ragde, P., Wigderson, A.: Linear-size constant-depth polylog-treshold circuits. Inf. Process. Lett. **39**(3), 143–146 (1991)

[Smo87] Smolensky, R.: Algebraic methods in the theory of lower bounds for boolean circuit complexity. In: Proceedings of the 19th Annual ACM Symposium on Theory of Computing, New York, New York, USA, pp. 77–82 (1987)

[SW14] Sahai, A., Waters, B.: How to use indistinguishability obfuscation: deniable encryption, and more. In: Shmoys, D.B. (ed.) Symposium on Theory of Computing, STOC 2014, New York, NY, USA, May 31–June 03 2014, pp. 475–484. ACM (2014)

[Tal14] Tal, A.: Tight bounds on the fourier spectrum of AC^0. Electron. Colloq. Comput. Complex. (ECCC) **21**, 174 (2014)

[TX13] Trevisan, L., Xue, T.: A derandomized switching lemma and an improved derandomization of AC^0. In: Proceedings of the 28th Conference on Computational Complexity, CCC 2013, K.lo Alto, California, USA, 5–7 June 2013, pp. 242–247 (2013)

[Vad04] Vadhan, S.P.: Constructing locally computable extractors and cryptosystems in the bounded-storage model. J. Cryptol. **17**(1), 43–77 (2004)

[Vio12] Viola, E.: The complexity of distributions. SIAM J. Comput. **41**(1), 191–218 (2012)

[Yao82] Yao, A.C.: Theory and applications of trapdoor functions (extended abstract). In: 23rd Annual Symposium on Foundations of Computer Science, Chicago, Illinois, USA, 3–5 November 1982, pp. 80–91. IEEE Computer Society (1982)

TWORAM: Efficient Oblivious RAM in Two Rounds with Applications to Searchable Encryption

Sanjam Garg[1], Payman Mohassel[2], and Charalampos Papamanthou[3($\boxtimes$)]

[1] University of California, Berkeley, USA
[2] Visa Research, Foster City, USA
[3] University of Maryland, College Park, USA
cpap@umd.edu

Abstract. We present TWORAM, an asymptotically efficient oblivious RAM (ORAM) protocol providing oblivious access (read and write) of a memory index y in exactly *two* rounds: The client prepares an encrypted query encapsulating y and sends it to the server. The server accesses memory M obliviously and returns encrypted information containing the desired value M[y]. The cost of TWORAM is only a multiplicative factor of security parameter higher than the tree-based ORAM schemes such as the path ORAM scheme of Stefanov et al. [34].

TWORAM gives rise to interesting applications, and in particular to a 4-round symmetric searchable encryption scheme where search is sublinear in the worst case and the search pattern is not leaked—the access pattern can also be concealed assuming the documents are stored in the obliviously accessed memory M.

1 Introduction

Oblivious RAM (ORAM) is a cryptographic primitive for accessing a remote memory M of n entries in a way that memory accesses do not reveal anything about the accessed index $y \in \{1, \ldots, n\}$. Goldreich and Ostrovsky [16] were the first to show that ORAM can be built with $\mathsf{poly}(\log n)$ bandwidth overhead[1], and since then, there has been a fruitful line of research on substantially reducing this overhead [9,29,34,36], in part motivated by the tree ORAM framework proposed by Shi et al. [31]. However, *most* existing practical ORAM protocols are interactive, requiring the client to perform a "download-decrypt-compute-encrypt-upload" operation several times (typically $O(\log n)$ rounds are involved). This can be a bottleneck for applications where low latency is important.

In this paper, we consider the problem of building an efficient round-optimal ORAM scheme. In particular, we propose TWORAM, an ORAM scheme enabling a client to obliviously access a memory location M[y] in two rounds, where the client sends an encrypted message to the server that encapsulates y, the server

---

[1] We define *bandwidth overhead* as the number of bits transferred between the client and the server during a single memory access, including the data block.

© International Association for Cryptologic Research 2016
M. Robshaw and J. Katz (Eds.): CRYPTO 2016, Part III, LNCS 9816, pp. 563–592, 2016.
DOI: 10.1007/978-3-662-53015-3_20

performs the oblivious computation, and sends a message back to the client, from which the client can retrieve the desired value $M[y]$.

TWORAM's worst-case bandwidth overhead is $O(\kappa \cdot p)$ where p is the bandwidth overhead of a tree-based ORAM scheme and κ is the security parameter. For instance, in Path-ORAM [34], it is $p = \log^3 n$ for a block of size $O(\log n)$ bits. In other words, in order to obliviously read a data block of $O(\log n)$ bits using TWORAM, one needs to exchange, *in the worst case*, a $O(\kappa \cdot \log^3 n)$ bits with the server, just in two rounds.

1.1   Existing Round-Optimal ORAM Protocols

Williams and Sion [37] devised a round-optimal ORAM scheme based on a customized garbling scheme and Bloom filters. Lu and Ostrovsky also include an optimized construction for single-round oblivious RAM in their seminal garbled RAM paper [28]. Subsequent to our work, Fletcher et al. [10] also provide single-round ORAM by generalizing the approach of [37] to use a garbling scheme for branching programs. All aforementioned approaches are symmetric-key and are built on top of the hierarchical ORAM framework as introduced by Goldreich and Ostrovsky [16]. Our approach however is based on the tree-based ORAM framework as introduced by Shi et al. [31], yielding worst-case logarithmic costs by construction, thus avoiding involved deamortization procedures. Burst ORAM [21] is also round-optimal, yet it requires linear storage at the client side.

Other less efficient approaches to construct round-optimal ORAM schemes are generic constructions based on garbled RAM [11,12,14]. However, such generic approaches are prohibitively inefficient. For instance, for the non-black-box Garbled RAM approaches [12,14], the bandwidth overhead grows with $\mathsf{poly}(\log n, \kappa, |f|)$, where $|f|$ is the size of the circuit for computing the one-way function f and κ is the security parameter. This leads to inefficient constructions, that are only of theoretical interest. Also, for the black-box Garbled RAM approach [11] the bandwidth overhead grows with $\mathsf{poly}(\log n, \kappa)$, and is independent of $|f|$. However, the construction itself is asymptotically very inefficient. Specifically in [11] the authors do not provide details on how large the involved polynomials are, which will depend on the choice of various parameters. According to our back-of-the-envelope calculation, however, the polynomial is at least $\kappa^5 \cdot \log^7 n$. A key reason for this inefficiency is that they require certain expensive ORAM operations, specifically "eviction," to be performed inside a garbled circuit. We eliminate this source of inefficiency by moving these expensive ORAM operations outside of the garbled circuits.

1.2   TWORAM's Technical Highlights

Our construction is inspired by the ideas from the recent, black-box garbled RAM work by Garg et al. [11]. We specifically use those ideas on top of the tree ORAM algorithms [31]. Our new ideas help avoid certain inefficiencies involved in the original construction of [11], yielding an asymptotically better protocol.

Our first step is to abstract away certain details of eviction-based tree ORAM algorithms, such as Path-ORAM [34], circuit ORAM [36] and Onion ORAM [9]. These algorithms work as follows: The memory M that must be accessed obliviously is stored as a sequence of L trees $T_1, T_2, \ldots, T_L$. The actual data of M is stored encrypted in the tree T_L, while the other trees store *position map* information (also encrypted). Only T_1 is stored on the client side. Roughly speaking, to access an index y in M, the client accesses T_1 and sends a path index p_2 to the server. The server then, successively accesses paths $p_2, p_3, \ldots, p_L$ in $T_2, T_3, \ldots, T_L$. However the paths are accessed adaptively: in order to learn p_i, one needs to first access p_{i-1} in T_{i-1}, and have all the information (also known as buckets) stored in its nodes decrypted. This is where existing approaches require $O(L)$ rounds of interaction: decryption can only take place at the client side, which means all the information on the paths must be communicated back to the client.

TWORAM's Core Idea. In order to avoid the roundtrips described above, we do not use standard encryption. Instead, we hardcode the content of each bucket inside a garbled circuit [38]. In other words, after the trees $T_2, T_3, \ldots, T_L$ are produced, the client generates one garbled circuit per each internal node in each tree. The function of this garbled circuit is very simple: Informally, it takes as input an index x; loops through the blocks bucket[i] contained in the current bucket until it finds bucket[x], and returns the index $\pi =$ bucket[x] of the next path to be followed. Note that the index π is returned in form of a *garbled input* for the next garbled circuit, so that the execution can proceed by the server until T_L is reached, and the final desired value can be returned to the client (see Fig. 3 for a more formal description).

This simplified description ignores some technical hurdles. Firstly, security of the underlying ORAM scheme requires that the location where bucket[x] is found remains hidden. In particular, the garbled circuit which has the value bucket[x] inside should not be identifiable by the server. We resolve this issue as follows. For every bucket that the underlying ORAM needs to touch, all the corresponding garbled circuits are executed in a specific order and the value of interest is carried along the way and output only by the final evaluated circuit in that tree.

Secondly, the above approach only works well for a single memory access, since the garbled circuits can only be used once. Fortunately, as we show in the paper, only a logarithmic number of garbled circuits are touched for each access. These circuits can be downloaded by the client who decodes the hardcoded values, performs the eviction strategy locally (on plaintext data), and sends fresh garbled circuits back to the server. This step does not increase the number of rounds (from two to three), since sending the fresh garbled circuits to the server can be "piggybacked" onto the message the client prepares for the next memory access.

Finally, in order to ensure the desired efficiency, and to avoid a blowup of polynomial multiplicative factor in security parameter, we develop optimizations that help ensure that the sizes of the circuits garbled in our construction remain small and proportional to the underlying ORAM.

1.3   Application: 4-Round Searchable Encryption with No Search Pattern Leakage

An SSE scheme allows a client to outsource a database (defined as a set of document/keyword set pairs $DB = (d_i, W_i)_{i=1}^N$) to a server in an encrypted format, where a search query for w returns d_i where $w \in W_i$.

Several recent works [3,20,26,39] demonstrate attacks against property-preserving encryption schemes (which also enable search on encrypted data), by taking advantage of the leakage associated with these schemes. Thought these attacks do not lead to concrete attacks against existing SSE schemes, they underline the importance of examining the feasibility of solutions that avoid leakage. A natural building block for doing so is ORAM. We use TWORAM to obtain the first constant-round, and asymptotically efficient SSE that can hide search/access patterns.

Our construction combines TWORAM and a non-recursive Path-ORAM (whose position map of the first level is not outsourced) such that searching for w requires (i) a single access on TWORAM; (ii) $|DB(w)|$ parallel accesses to the non-recursive Path-ORAM (note that an access to a non-recursive ORAM requires only two rounds).

In particular, we use TWORAM to store pairs of the form $(w, (count_w, access_w))$, where w is a keyword, $count_w$ is the number of documents containing w and $access_w$ is the number of times w has been accessed so far. The keyword/document pairs $(w||i, d_i)$ (where d_i is the i-th document containing w) are then stored in the non-recursive Path-ORAM where their position in the Path-ORAM tree (namely the random path they are mapped to) is determined on the fly by using a PRF F as $F_k(w||i, access_w)$ (therefore there is no need to store the position map locally). To search for keyword w, we first access TWORAM to obtain $(count_w, access_w)$ (and increment $access_w$), and then generate all positions to look up in the Path-ORAM using the PRF F. These lookups can be parallelized and updating the paths can be piggybacked to the next search.

The above yields a construction with 4 rounds of interaction. Note that naively using ORAM for SSE would incur $|DB(w)|$ ORAM accesses which imply *at least* $|DB(w)|$ roundtrips (depending on the number of rounds of the underlying ORAM). As we said before, our construction *does not leak the search pattern*, by providing randomly generated tokens every time a search is performed. If we choose to store all documents in the obliviously-accessed memory, the access pattern can also be concealed.

1.4   Other Related Work

Oblivious RAM. ORAM protocols with a non-constant number of roundtrips can be categorized into *hierarchical* [17,18,24,27], motivated by the seminal work of Goldreich and Ostrovsky [16], and *tree-based* [9,29,34,36], motivated by the seminal work of Shi et al. [31]. We note however, that, by picking the data block size to be big (e.g., $\sqrt{n}$ bits), the number of rounds in tree-based ORAMs can be made constant, yet bandwidth increases beyond polylogarithmic, so such a parameter selection is not interesting.

Searchable Encryption. Song et al. [32] were the first to explore feasibility of searchable encryption. Since then, many follow-up works have designed new schemes for both static data [4,6,8] and dynamic data [5,15,22,23,33,35]. The security definitions also evolved over time and were eventually established in the work of [6,8]. Unlike our construction, *all these approaches use deterministic tokens*, an therefore leak the search patterns. The only proposed approaches that are constant-round and have randomized tokens (apart from constructing SSE through Garbled RAM) are the ones based on functional encryption [30]. However, such approaches incur a linear search overhead. We also note that one can obtain SSE with no search pattern leakage by using interactive ORAMs such as Path-ORAM [34], or other variants optimized for binary search [13].

Secure Computation for RAM Programs. A recent line of work studies efficient secure two-party computation of RAM programs based on garbled circuits [1,19]. These constructions can also be used to design SSE that hide the search pattern—yet these approaches do not lead to constant-round SSE schemes, requiring the client to perform computation proportional to the size of the search result.

2 Definitions for Garbled Circuits and Oblivious RAM

In this section, we recall definitions and describe building blocks we use in this paper. We use the notation $\langle C', S' \rangle \leftrightarrow \Pi \langle C, S \rangle$ to indicate that a protocol Π is executed between a client with input C and a server with input S. After the execution of the protocol the client receives C' and the server receives S'. For non-interactive protocols, we just use the left arrow notation ($\leftarrow$) instead.

2.1 Garbled Circuits

Garbled circuits were first constructed by Yao [38] (see Lindell and Pinkas [25] and Bellare et al. [2] for a detailed proof and further discussion). A *circuit garbling scheme* is a tuple of PPT algorithms (GCircuit, Eval), where GCircuit is the circuit garbling procedure and Eval the corresponding evaluation procedure. More formally:

- $(\tilde{C}, \mathsf{lab}) \leftarrow \mathsf{GCircuit}(1^\kappa, C)$: GCircuit takes as input a security parameter κ, and a Boolean circuit C. This procedure outputs a *garbled circuit* $\tilde{C}$ and *input labels* lab, which is a set of pairs of random strings. Each pair in lab corresponds to every input wire of C (and in particular each element in the pair represents either 0 or 1).
- $y \leftarrow \mathsf{Eval}(\tilde{C}, \mathsf{lab}_x)$: Given a garbled circuit $\tilde{C}$ and *garbled input* lab_x, Eval outputs $y = C(x)$.

Input Labels and Garbled Inputs. For a specific input x, we denote with lab_x the *garbled inputs*, a "projection" of x on the input labels. E.g., for a Boolean circuit of two input bits z and w, it is $\mathsf{lab} = \{(z_0, z_1), (w_0, w_1)\}$, $\mathsf{lab}_{00} = \{z_0, w_0\}$, $\mathsf{lab}_{01} = \{z_0, w_1\}$, etc.

Correctness. Let (GCircuit, Eval) be a circuit garbling scheme. For correctness, we require that for any circuit C and an input x for C, we have that that $C(x) = \mathsf{Eval}(\tilde{C}, \mathsf{lab}_x)$, where $(\tilde{C}, \mathsf{lab}) \leftarrow \mathsf{GCircuit}(1^\kappa, C)$.

Security. Let (GCircuit, Eval) be a circuit garbling scheme. For security, we require that for any PPT adversary A, there is a PPT simulator Sim such that the following distributions are computationally indistinguishable:

- $\mathsf{Real}_\mathsf{A}(\kappa)$: A chooses a circuit C. Experiment runs $(\tilde{C}, \mathsf{lab}) \leftarrow \mathsf{GCircuit}(1^\kappa, C)$ and sends $\tilde{C}$ to A. A then chooses an input x. The experiment uses lab and x to derive lab_x and sends lab_x to A. Then it outputs the output of the adversary.
- $\mathsf{Ideal}_\mathsf{A,Sim}(\kappa)$: A chooses a circuit C. Experiment runs $(\tilde{C}, \sigma) \leftarrow \mathsf{Sim}(1^\kappa, |C|)$ and sends $\tilde{C}$ to A. A then chooses an input x. The experiment runs $\mathsf{lab}_x \leftarrow \mathsf{Sim}(1^\kappa, \sigma)$ and sends lab_x to A. Then it outputs the output of the adversary.

The above definition guarantees adaptive security, since the adversary gets to choose input x after seeing the garbled circuit $\tilde{C}$. We only know how to instantiate garbling schemes with adaptive security in the random oracle model. In the standard model, existing garbling schemes achieve a weaker static variant of the above definition where the adversary chooses both C and input x at the same time and before receiving $\tilde{C}$.

Concerning complexity, we note that if the cleartext circuit C has $|C|$ gates, the respective garbled circuit has size $O(|C|\kappa)$. This is because every gate in the circuit is typically replaced with a table of four rows, each row storing encryptions of labels (each encryption has κ bits).

2.2 Oblivious RAM

We recall *Oblivious RAM* (ORAM), a notion introduced and first studied by Goldreich and Ostrovsky [16]. ORAM can be thought of as a compiler that encodes the memory into a special format such that accesses on the compiled memory do not reveal the underlying access patterns on the original memory. An ORAM scheme consists of protocols (SETUP, OBLIVIOUSACCESS).

- $\langle \sigma, \mathsf{EM} \rangle \leftrightarrow \text{SETUP}\langle(1^\kappa, \mathsf{M}), \perp \rangle$: SETUP takes as input the security parameter κ and a memory array M and outputs a secret state σ (for the client), and an encrypted memory EM (for the server).
- $\langle(\mathsf{M}[y], \sigma'), \mathsf{EM}' \rangle \leftrightarrow \text{OBLIVIOUSACCESS}\langle(\sigma, y, v), \mathsf{EM} \rangle$: OBLIVIOUSACCESS is a protocol between the client and the server, where the client's input is the secret state σ, an index y and a value v which is set to null in case the access is a read operation (and not a write). Server's input is the encrypted memory EM. Client's output is M[y] and an updated secret state σ' and the server's output is an updated encrypted memory EM' where M[y] = v, if $v \neq$ null.

Correctness. Consider the following correctness experiment. Adversary A chooses memory M_0. Consider EM_0 generated with $\langle\sigma_0, \mathsf{EM}_0\rangle \leftrightarrow$ SETUP$\langle(1^\kappa, \mathsf{M}_0), \perp\rangle$). The adversary then adaptively chooses memory locations to read and write. Denote the adversary's read/write queries by $(y_1, v_1), \ldots,$ (y_q, v_q) where $v_i =$ null for read operations. A wins in the correctness game if $\langle (M_{i-1}[y_i], \sigma_i), EM_i \rangle$ are not the final outputs of the protocol OBLIVIOUSACCESS$\langle (\sigma_{i-1}, y_i, v_i), EM_{i-1} \rangle$ for any $1 \leq i \leq q$, where M_i, EM_i, σ_i are the memory array, the encrypted memory array and the secret state, respectively, after the i-th access operation, and OBLIVIOUSACCESS is run between an honest client and server. The ORAM scheme is correct if the probability of A in winning the game is negligible in κ.

Security. An ORAM scheme is secure in the semi-honest model if for any PPT adversary A, there exists a PPT simulator Sim such that the following two distributions are computationally indistinguishable.

- Real$_A(\kappa)$: A chooses M_0. Experiment then runs $\langle \sigma_0, EM_0 \rangle \leftrightarrow$ SETUP $\langle (1^\kappa, M_0), \bot \rangle$. For $i = 1, \ldots, q$, A makes adaptive read/write queries (y_i, v_i) where $v_i =$ null on reads, for which the experiment runs the protocol

$$\langle (M_{i-1}[y_i], \sigma_i), EM_i \rangle \leftrightarrow \text{OBLIVIOUSACCESS} \langle (\sigma_{i-1}, y_i, v_i), EM_{i-1} \rangle.$$

  Denote the full transcript of the above protocol by t_i. Eventually, the experiment outputs $(EM_q, t_1, \ldots, t_q)$ where q is the total number of read/write queries.
- Ideal$_{A,Sim}(\kappa)$: The experiment outputs $(EM_q, t'_1, \ldots, t'_q) \leftrightarrow Sim(q, |M_0|, 1^\kappa)$.

3   TWORAM Construction

Our TWORAM construction uses an abstraction of tree-based ORAM schemes, e.g., Path-ORAM [34]. We start by describing this abstraction informally. Then we show how to turn the interactive Path-ORAM protocol (e.g., the one by Stefanov et al. [34]) into a two-round ORAM protocol, using the abstraction that we present below. We now give some necessary notation that we need for understanding our abstraction.

3.1   Notation

Let $n = 2^L$ be the size of the initial memory that we wish to access obliviously. This memory is denoted by $A_L[1], A_L[2], \ldots, A_L[n]$ where $A_L[i]$ is the i-th *block* of the memory. Given location y that we wish to access, let $y_L, y_{L-1}, \ldots, y_1$ be defined recursively as $y_L = y$ and $y_i = ceil(y_{i+1}/2)$, for all $i = L-1, L-2, \ldots, 1$. For example, for $L = 4$ and $y = 13$, we have

- $y_1 = ceil(ceil(ceil(y/2)/2)/2) = 2.$
- $y_2 = ceil(ceil(y/2)/2) = 4.$
- $y_3 = ceil(y/2) = 7.$
- $y_4 = 13.$

Also define $b_i = 1 - y_i \% 2$ to be a bit (namely b_i indicates if y_i is even or not). Finally, on input a value x of $2 \cdot L$ bits, select$(x, 0)$ selects the first L bits of x, while select$(x, 1)$ selects the last L bits of x. We note here that both y_i and b_i are functions of y, but we do not indicate this explicitly so that not to clutter notation.

3.2   Path-ORAM Abstraction

We start by describing our abstraction of Path-ORAM construction. In Appendix A we describe formally how this abstraction can be used to implement the interactive Path-ORAM algorithm [34] (with $\log n$ rounds of interaction). We note that the details in Appendix A are provided only for helping better understanding. Our construction can be understood based on just the abstraction defined below.

Roughly speaking, Path-ORAM algorithms encode the original memory A_L in the form of L memories

$$A_L, A_{L-1}, \ldots, A_1,$$

where A_L stores the original data and the remaining memories A_i store information required for accessing data in A_L *obliviously*. Each A_i has 2^i entries, each one storing blocks of $2 \cdot L$ bits (for ease of presentation we assume the block size is $\Theta(\log n)$ but our results apply with other block parameterizations as well). Memories $A_L, A_{L-1}, \ldots, A_2$ are stored in trees $T_L, T_{L-1}, \ldots, T_2$ respectively. The smallest memory A_1 is kept locally by the client. The invariant that is maintained is that any block $A_i[x]$ will reside in *some* leaf-to-root path of tree T_i, and specifically on the path that starts from leaf x_i in T_i. The value x_i itself can be retrieved by accessing A_{i-1}, as we detail in the following.

Reading a Value $A_L[y]$. To read a value $A_L[y]$, one first reads $A_1[y_1]$ from local storage and computes $x_2 \leftarrow$ select$(A_1[y_1], b_1)$ (recall definitions of y_1 and b_1 from Sect. 3.1). Then one traverses the path starting from leaf x_2 in T_2. This path is denoted with $T_2(x_2)$. Block $A_2[y_2]$ is guaranteed to be on $T_2(x_2)$. Then one computes $x_3 \leftarrow$ select$(A_2[y_2], b_2)$, and continues in this way. In the end, one will traverse path $T_L(x_L)$ and will eventually retrieve block $A_L[y]$. See Fig. 1.

Updating the Paths. Once the above process finishes, we need to make sure that we do not access the same leaf-to-root paths in case we access $A_L[y]$ again in the future—this would violate obliviousness. Thus, for $i = 2, \ldots, L$, we perform the following tasks:

1. We remove all blocks from $T_i(x_i)$ and copy them into a data structure C_i called *stash*. In our abstraction, stash C_i is viewed as an extension of the root of tree T_i;

2. In the stash C_{i-1}, we set select$(A_{i-1}[y_{i-1}], b_{i-1}) \leftarrow r_i$, where r_i is a fresh random number in $[1, 2^i]$ that replaces x_i from above. This effectively means that block $A_i[y_i]$ should be reinserted on path $T_i(r_i)$, when eviction from stash C_i takes place;

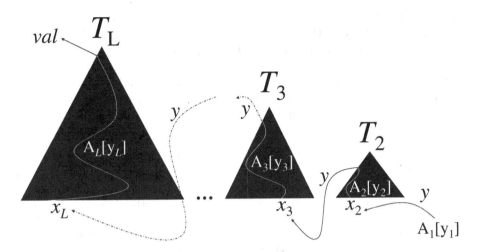

Fig. 1. Our Path-ORAM abstraction for reading a value $val = A_L[y]$. $A_1[y_1]$ is read from local storage and defines x_2. x_2 defines a path p_2 in T_2. By traversing p_2 the algorithm will retrieve $A_2[y_2]$, which will yield x_3, which defines a path p_3 in T_3. Repeating this process yields a path p_L in T_L, traversing which yields the final value $A_L[y_L] = A_L[y]$. Note that y is passed from tree T_{i-1} to tree T_i so that the index y_i (and the bit b_i) can be computed for searching for the right block on path p_i.

3. We evict blocks from stash C_i back to tree $T_i(x_i)$, respecting the new assignments made above.

Syntax. A *Path-ORAM* consists of three procedures (INITIALIZE, EXTRACT, UPDATE) with syntax:

- $\mathcal{T} \leftarrow$ INITIALIZE($1^\kappa, A_L$): Given a security parameter κ and memory A_L as input, SETUP outputs a set of $L - 1$ trees $\mathcal{T} = \{T_2, T_3, \ldots, T_L\}$ and an array of two entries A_1. A_1 is stored locally with the client and $T_2, \ldots, T_L$ are stored with the server.
- $x_{i+1} \leftarrow$ EXTRACT($i, y, T_i(x_i)$) for $i = 2, \ldots, L$. Given the tree number i, the final memory location of interest y and a leaf-to-root path $T_i(x_i)$ (that starts from leaf x_i) in tree T_i, EXTRACT outputs an index x_{i+1} to be read in the next tree T_{i+1}. The client can obtain x_2 from local storage as $x_2 \leftarrow$ select($A_1[y_1], b_1$). The obtained value x_2 is sent to the server in order for the server to continue execution. Finally, the server outputs x_{L+1}, which is the desired value $A_L[y]$.

EXTRACTBUCKET Algorithm. In Path-ORAM [34], internal nodes of the trees store more than one block $(z, A_i[z])$, in the form of *buckets*. We note that EXTRACT can be broken to work on individual buckets along a root-to-leaf path in a tree T_i. In particular, we can define the algorithm $\pi \leftarrow$ EXTRACTBUCKET(i, y, b) where i is the tree of interest, y is the memory location that needs to be accessed, and b is a bucket corresponding to a particular node on the leaf-to-root path. π will be found at one of the nodes on the leaf-to-root path. Note that the algorithm EXTRACT can be implemented by repeatedly calling EXTRACTBUCKET for every b on $T_i(x_i)$.

- $\{A_1, T_2(x_2), \ldots, T_L(x_L)\} \leftarrow$ UPDATE$(y, op, val, A_1, T_2(x_2), \ldots, T_L(x_L)$. Procedure UPDATE takes as input the leaf-to-root paths (and local storage A_1) that were traversed during the access and accordingly updates these paths (and local storage A_1). Additionally, UPDATE ensures the new value val is written to $A_L[y]$, if operation op is a "write" operation.

An implementation of the above abstractions, for Path-ORAM [34], is given in Algorithms 1, 2 and 3 in Appendix A.1. Note that the description of the UPDATE procedure [34] abstracts away the details of the eviction strategy. The SETUP and OBLIVIOUSACCESS protocols of the interactive Path-ORAM *using these abstractions* are given in Figs. 6 and 7 respectively in the Appendix A.2. It is easy to see that the OBLIVIOUSACCESS protocol has $\log n$ rounds of interactions. By the proof of Stefanov et al. [34], we get the following:

Corollary 1. *The protocols* SETUP *and* OBLIVIOUSACCESS *from* Figs. 6 and 7 *respectively in* Appendix A.2 *comprise a secure ORAM scheme (as defined in* Sect. 2.2*) with* $O(\log n)$ *rounds, assuming the encryption scheme used is CPA-secure.*

We recall that the bandwidth overhead for Path-ORAM [34] is $O(\log^3 n)$ bits and the client storage is $O(\log^2 n) \cdot \omega(1)$ bits, for a block size of $2 \cdot L = 2 \cdot \log n$ bits.

3.3   From $\log n$ Rounds to Two Rounds

Existing Path-ORAM protocols implementing our abstraction require $\log n$ rounds (see OBLIVIOUSACCESS protocol in Fig. 7). The main reason for that is the following: In order for the server to determine the index of leaf x_i from which the next path traversal begins, the server needs to access $A_{i-1}[y_{i-1}]$, which is stored *encrypted* at some node on the path starting from leaf x_{i-1} in tree T_{i-1}— see Fig. 1. Therefore the server has to return all encrypted nodes on $T_{i-1}(x_{i-1})$ to the client, who performs the decryption locally, searches for $A_{i-1}[y_{i-1}]$ (via the EXTRACTBUCKET procedure) and returns the value x_i to the server (see Line 10 of the OBLIVIOUSACCESS protocol in Fig. 7).

Our Approach. To overcome this difficulty, we do not encrypt the blocks in the buckets. Instead, for each bucket stored at a tree node u, we prepare a garbled circuit that hardcodes, among other things, the blocks that are contained in the bucket. Subsequently, this garbled circuit executes the EXTRACTBUCKET algorithm on the hardcoded blocks and outputs either $\perp$ or the next leaf index π, depending on whether the search performed by EXTRACTBUCKET was successful or not. The output, whatever that is, is fed as a garbled input to either the left child bucket or the right child bucket (depending on the currently traversed path) or the next root bucket (in case u is a leaf) of u. In this way, by the time the server has executed all the garbled circuits along the currently traversed

---

Circuit C[u, bucket, leftLabels, rightLabels, nextRootLabels](p, y, π)
Inputs: (p, y, π).
Outputs: Next node to be executed and garbled inputs for its bucket circuit.
Hardcoded parameters: [$u = (i, j, k)$, bucket, leftLabels, rightLabels, nextRootLabels].

1: **if** $\pi = \bot$ **then**
2:      Set $\pi \leftarrow$ EXTRACTBUCKET(i, y, bucket);   ▷ π will be the desired value x_{i+1}.
3: **end if**
4: **if** u is not a leaf **then**
5:      Based on p, **return** *either* (left(u), leftLabels$_{(p,y,\pi)}$);   ▷ Go to left child.
                          or (right(u), rightLabels$_{(p,y,\pi)}$);   ▷ Go to right child.
6: **else**
7:      **return** (nextRoot(u), nextRootLabels$_{(\pi,y,\bot)}$);   ▷ Go to next root.
8: **end if**

Fig. 2. Formal description of the naive bucket circuit. Notation: Given lab, the set of *input labels* for a garbled circuit, we let lab$_a$ denote the garbled input labels (i.e., the labels taken from lab) corresponding to the input value a.

path, he will be able to pass the index π to the next tree as a garbled input, and continue the execution in the same way without having to interact with the client. Therefore the client can obliviously retrieve his value $A_L[y]$ in only two rounds of communication.

Unfortunately, once these garbled circuits have been consumed, they cannot be used again since this would violate security of garbled circuits. To avoid this problem, the client downloads all the data that was accessed before, decrypts them, runs the UPDATE procedure locally, recomputes the garbled circuits that were consumed before, and stores the new garbled circuits locally. In the next access, these garbled circuits will be sent along with the query. Therefore the total number of communication rounds is equal to two (note that this approach requires permanent client storage—for transient storage, the client will have to send the garbled circuits immediately which would increase the rounds to three). We now continue with describing the bucket circuit that needs to be garbled for our construction.

Naive Bucket Circuit. To help the reader, in Fig. 2 we describe a naive version of our bucket circuit that leads to an inefficient construction. Then we give the full-fledged description of our bucket circuit in Fig. 3. The naive bucket circuit has *inputs*, *outputs* and *hardcoded parameters*, which we detail in the following.

Inputs. The input of the circuit is a triplet consisting of the following information:

1. The index of the leaf p from which the currently explored path begins;
2. The final location to be accessed y;
3. The output from previous bucket π (can be the actual value of the next index to be explored or $\bot$).

Outputs. The outputs of the circuit are the next node to be executed, along with its garbled inputs. For example, if the current node u is not a leaf (see Lines 4 and 5 in Fig. 2), the circuit outputs the garbled inputs of either the left or the right child, whereas if the current node is a leaf (see Lines 6–8 in Fig. 2), the circuit outputs the garbled inputs of the next root to be executed. Note that outputting the garbled inputs is easy, since the bucket circuit hardcodes the input labels of the required circuits. Finally we note that the EXTRACTBUCKET(i, y, bucket) algorithm used in Fig. 2 can be found in Appendix A.1—see Algorithm 2.

Hardcoded Parameters. The circuit for node u hardcodes:

1. The node identifier u that consists of a triplet (i, j, k) where
   - $i \in \{2, \ldots, L\}$ is the tree number where node u belongs to;
   - $j \in \{0, \ldots, 2^{i-1}\}$ is the depth of node u;
   - $k \in \{0, \ldots, 2^j - 1\}$ is the oder of node u in the specific level.

   For example, the root of tree T_3 will be denoted $(3, 0, 0)$, while its right child will be $(3, 1, 1)$.
2. The bucket information bucket (i.e., blocks $(x, A_i[x], r)$ contained in node u—recall r is the path index in T_i assigned to $A_i[x]$);
3. The *input labels* leftLabels, rightLabels and nextRootLabels that are used to compute the *garbled inputs* for the next circuit to be executed. Note that leftLabels and rightLabels are used to prepare the next garbled inputs when node u is an internal node (to go either to the left or the right child), while nextRootLabels are used when node u is a leaf (to go to the next root).

Final Bucket Circuit. In the naive circuit presented before, we hardcode the input labels of the root node root of every tree T_i into all the nodes/circuits of tree T_{i-1}. Unfortunately, in every oblivious access, the garbled circuits of all roots are consumed (and therefore root's circuit as well), hence *all* the garbled circuits of tree T_{i-1} will have to be recomputed from scratch. This cost is $O(n)$, thus very inefficient. We would like to mimimize the number of circuits in T_{i-1} that need to be recomputed and ideally make this cost proportional to $O(\log n)$.

To achieve that, we observe that, *instead of hardcoding input labels* nextRootLabels *in the garbled circuit of every node of tree* T_{i-1} , *we can just pass them as garbled inputs to the garbled circuit of every node of tree* T_{i-1}. The final circuit is given in Fig. 3. Note that the only difference of the new circuit from the naive circuit is in the computation of the garbled inputs

$$\text{leftNewLabels}_{(p,y,\pi,\text{nextRootLabels})}$$

and

$$\text{rightNewLabels}_{(p,y,\pi,\text{nextRootLabels})},$$

where nextRootLabels is added in the subscript (see Line 5 of both Figs. 3 and 2), to account for the new input of the new circuit. Note also that we indicate the change in the input format by using "leftNewLabels" instead of just "leftLabels" and "rightNewLabels" instead of just "rightLabels". nextRootLabels have the same meaning in both circuits.

Circuit $C[u, \mathsf{bucket}, \mathsf{leftNewLabels}, \mathsf{rightNewLabels}](p, y, \pi, \mathsf{nextRootLabels})$
Inputs: $p, y, \pi, \mathsf{nextRootLabels}$.
Outputs: Next node to be executed and garbled inputs for its bucket circuit.
Hardcoded parameters: $[u = (i, j, k), \mathsf{bucket}, \mathsf{leftNewLabels}, \mathsf{rightNewLabels}]$.

1: **if** $\pi = \bot$ **then**
2:      Set $\pi \leftarrow$ EXTRACTBUCKET(i, y, bucket);   ▷ π will be the desired value x_{i+1}.
3: **end if**
4: **if** u is not a leaf **then**
5:      Based on p, **return** *either* $(\mathsf{left}(u), \mathsf{leftNewLabels}_{(p,y,\pi,\mathsf{nextRootLabels})})$;   ▷ Go to left child.
                     or $(\mathsf{right}(u), \mathsf{rightNewLabels}_{(p,y,\pi,\mathsf{nextRootLabels})})$;   ▷ Go to right child.
6: **else**
7:      **return** $(\mathsf{nextRoot}(u), \mathsf{nextRootLabels}_{(\pi,y,\bot)})$;   ▷ Go to next root.
8: **end if**

Fig. 3. Formal description of the final bucket circuit.

3.4 Protocols SETUP and OBLIVIOUSACCESS of our construction

We now describe in detail the SETUP and OBLIVIOUSACCESS protocols of TWORAM.

SETUP. The SETUP protocol is described in Fig. 4. Just like the setup for the interactive ORAM protocol (see Fig. 6 in Appendix A.2), in TWORAM, the client does some computation locally in the beginning (using his secret key) and then outputs some "garbled information" that is being sent to the server. In particular:

1. After producing the trees $T_2, T_3, \ldots, T_L$ using algorithm INITIALIZE, the client prepares the garbled circuit of Fig. 3 for all the nodes $u \in T_i$, for all trees T_i. It is important this computation takes place from the leaves towards the root (that is why we write $j \in \{i-1, \ldots, 0\}$ in Line 2 of Fig. 4), since a garbled circuit of a node u hardcodes the input labels of the garbled circuits of its children—so these need to be readily available by the time u's garbled circuit is computed.

2. Apart from the garbled circuits, the client needs to prepare garbled inputs for the $\mathsf{nextRootLabels}$ inputs of all the roots of the trees T_i. These are essentially the β_i's computed in Line 4 of Fig. 4.

OBLIVIOUSACCESS. The OBLIVIOUSACCESS protocol of TWORAM is described in Fig. 5. The first step of the protocol is similar to that of the interactive scheme (see Fig. 7 in Appendix), where the client accesses local storage A_1 to compute the path index x_2 that must be traversed in T_2. However, the main difference is that, instead of sending x_2 directly, the client sends the *garbled input that corresponds to* x_2 for the root circuit of tree T_2, denoted with α in Fig. 5.

We note here that α is not enough for the first garbled circuit to start executing, and therefore the server complements this garbled input with β_2 (see Server Line 1), the other half that was sent by the client before and that represents the garbled inputs for the input labels of the next root. Subsequently, the server

---

Protocol $\langle \sigma, \mathsf{EM} \rangle \leftrightarrow \textsc{Setup}\langle (1^\kappa, \mathsf{M}), \bot \rangle$:

Client:

1: Pick a κ-bit secret key s. Run $\{A_1, T_2, \ldots, T_L\} \leftarrow \mathsf{Initialize}(1^\kappa, \mathsf{M})$;
2: For all $i \in 2, \ldots, L$, for all $j \in \{i-1, \ldots 0\}$, for all $k \in \{0, \ldots, 2^j - 1\}$, let $u = (i, j, k)$ be the (j, k)-th node in T_i and let bucket be its bucket. Compute:

$$(\tilde{\mathsf{C}}^u, \mathsf{lab}^u = (\mathsf{cState}^u, \mathsf{nState}^u)) \leftarrow \mathsf{GCircuit}(1^\kappa, \mathsf{C}[u, \mathsf{bucket}, \mathsf{lab}^{(i,j+1,2k)}, \mathsf{lab}^{(i,j+1,2k+1)}]),$$

$$\mathsf{X}^u \leftarrow \mathsf{Enc}_s(\mathsf{bucket}, \mathsf{lab}^{(i,j+1,2k)}, \mathsf{lab}^{(i,j+1,2k+1)}),$$

where
- C is defined in Figure 3 and (Enc, Dec) is a semantically-secure encryption scheme;
- cState^u are the input labels for the triplet (p, y, π) for node u;
- nState^u are the input labels for nextRootLabels for node u.

3: For all u, send to server $\tilde{\mathsf{C}}^u, \mathsf{X}^u$;
4: For all $i \in \{2, \ldots, L-1\}$ send to server $\beta_i = \mathsf{nState}_{\mathsf{cState}^{(i+1,0,0)}}^{(i,0,0)}$. Namely, β_i are garbled inputs for nextRootLabels for node $(i, 0, 0)$.
5: **return** $(s, A_1, \mathsf{cState}^{(2,0,0)})$ as σ;

Server:

1: **return** all data sent by the client from above as EM;

Fig. 4. SETUP protocol for TWORAM.

starts executing the garbled circuits one-by-one, using the outputs of the first circuit, as garbled inputs to the second one, and so on. Eventually, the clients reads and decrypts all paths $T_i(x_i)$, retrieving the desired value (see Client Line 2). Finally, the client runs the UPDATE, re-garbles the circuits that got consumed and waits until the next query to send them back. We can now state the main result of our paper.

Theorem 1. *The protocols* SETUP *and* OBLIVIOUSACCESS *from* Figs. 4 *and* 5 *respectively comprise a two-round secure ORAM scheme (as defined in* Sect. 2.2*), assuming the garbling scheme used is secure (as defined in* Sect. 2.1*) and the encryption scheme used is CPA-secure.*

The proof of the above theorem can be found in Appendix A.3. Concerning complexity, it is clear that the only overhead that we are adding on Path-ORAM [34] is a garbled circuit per bucket—this adds a multiplicative security parameter factor on all the complexity measures of Path-ORAM. E.g., the bandwidth overhead of our construction is $O(\kappa \cdot \log^3 n)$ bits (for blocks of $2 \log n$ bits).

3.5   Optimizations

Recall that in the garbling procedure of a circuit C, one has the following choices: (i) either to garble C in a way that during evaluation of the garbled circuit on x the output is the cleartext value $C(x)$; (ii) or to garble C in a way that during evaluation of the garbled circuit on x the output is the garbled labels corresponding to the value $C(x)$. We now describe an optimization for a specific circuit C that we will be using in our construction that uses the above observation.

Protocol $\langle (\mathsf{M}[y], \sigma'), \mathsf{EM}' \rangle \leftrightarrow \textsc{ObliviousAccess}\langle (\sigma, y, val), \mathsf{EM} \rangle$:

Client:

1: Compute $x_2 \leftarrow \mathsf{select}(A_1[y_1], b_1)$ and send to server $\alpha = \mathsf{cState}^{(2,0,0)}_{(x_2, y, \perp)}$;

Server:

1: Let $output = \alpha \| \beta_2$, where β_2 is defined in Line 4 of Protocol SETUP in Figure 4;
2: Set $i = 2$ and $j = 0$ and $k = 0$;
3: **while** $i \leq L$ **do**
4:     $(\mathsf{nextNode}, output) \leftarrow \mathsf{Eval}(\tilde{\mathsf{C}}^{(i,j,k)}, output)$;
5:     **if** $j = i - 1$ and $i < L$ **then** $\triangleright$ node (i, j, k) is a leaf of trees $T_2, T_3, \ldots, T_{L-1}$
6:         $i = i + 1$; $\triangleright$ go to next tree
7:         $output = output \| \beta_i$; $\triangleright$ prepare the garbled inputs for the root of the next tree
8:     **end if**
9:     **if** $\mathsf{nextNode} = \mathsf{left}(i, j, k)$ **then** $\triangleright$ node (i, j, k) is a not a leaf; decide whether to go next
10:         $j = j + 1; k = 2k$; $\triangleright$ go to left child
11:     **else**
12:         $j = j + 1; k = 2k + 1$; $\triangleright$ go to right child
13:     **end if**
14: **end while**
15: Let $x_2, \ldots, x_L$ be the indices of the paths that have been accessed above;
16: Send to client $output$ and all X^u's corresponding to nodes u on paths $T_2(x_2), T_3(x_3), \ldots, T_L(x_L)$;

Client:

1: Decrypt all information contained in X^u and reconstruct $T_2(x_2), T_3(x_3), \ldots, T_L(x_L)$;
2: Retrieve block $(y, A_L[y], x_L)$ from $T_L(x_L)$; Set $val = A_L[y]$;
3: Run $\{A_1, T_2(x_2), \ldots, T_L(x_L)\} \leftarrow \textsc{Update}(y, op, val, A_1, T_2(x_2), \ldots, T_L(x_L))$;
4: For all $i \in 2, \ldots, L$, for all $j \in \{i - 1, \ldots 0\}$, let $u = (i, j, k)$ be the (j, k)-th node in $T_i(x_i)$ and let bucket be its bucket. Compute:

$$(\tilde{\mathsf{C}}^u, \mathsf{lab}^u = (\mathsf{cState}^u, \mathsf{nState}^u)) \leftarrow \mathsf{GCircuit}(1^\kappa, \mathsf{C}[u, \mathsf{bucket}, \mathsf{lab}^{(i,j+1,2k)}, \mathsf{lab}^{(i,j+1,2k+1)}]) ,$$

$$\mathsf{X}^u \leftarrow \mathsf{Enc}_s(\mathsf{bucket}, \mathsf{lab}^{(i,j+1,2k)}, \mathsf{lab}^{(i,j+1,2k+1)}) ,$$

    where C is defined in Figure 3. Send to server $\tilde{\mathsf{C}}^u$ and X^u;
5: Store locally $(s, A_1, \mathsf{cState}^{(2,0,0)})$ as σ';
6: For all $i \in \{2, \ldots, L - 1\}$ send to server $\beta_i = \mathsf{nState}^{(i,0,0)}_{\mathsf{cState}^{(i+1,0,0)}}$;
7: **return** (val, σ').

Server:

1: **return** the data received by the client as EM'.

Fig. 5. ObliviousAccess protocol for TWORAM.

General Optimization. Consider a circuit that performs the following task: It hardcodes two k-bit strings s_0 and s_1, takes an input a bit b and outputs s_b. This cleartext circuit has size $O(k)$, so the garbled circuit for that will have size $O(k^2)$. To improve upon that we consider a circuit C' that takes as input bit b and outputs the same bit b! This cleartext circuit has size $O(1)$. However, to make sure that the output of the garbled version of C' is always s_b, we garble C' by outputting the garbled label corresponding to b, namely s_b (i.e., using (ii) from above). In particular, during the garbling procedure we use s_0 as the garbled label output for output $b = 0$ and we use s_1 as the garbled label output for the output $b = 1$. Note that the size of the new garbled circuit has size $O(k)$, yet it has exactly the same I/O behavior with the garbling of C, which has size $O(k^2)$.

- **Improving cState —Not Hard-Coding Input Labels Inside the Bucket Circuit.** In the construction we described, we include the input labels leftLabels, rightLabels in the circuit C[u, bucket, leftLabels, rightLabels]. Consequently, the size of the ungarbled version of this circuit grows with the size of leftLabels and rightLabels which is $\kappa \cdot |\text{cState}|$. We can easily use the general optimization described above, for each bit of $|\text{cState}|$, to make the size of the ungarbled version of our circuit only grow with $|\text{cState}|$.

- **Improving nState —Input Labels Passing.** In the construction described previously, for each tree, an input value nState is passed from the root to a leaf node in the tree. However this value is used only at the leaf node. Recall that the nState value passed from the root to a leaf garbled circuits in the tree T_i is exactly the value $\text{cState}^{i+1,0,0}$, the input labels of the root garbled circuit of the tree T_{i+1}. Since each ungarbled circuit gets this value as input, therefore each of one of them needs to grow with $\kappa \cdot |\text{cState}|$.[2] We will now describe an optimization such that the size of the garbled version, rather than the clear version, grows linearly in $\kappa \cdot |\text{cState}|$.

Note that in our construction the value $\text{cState}^{i+1,0,0}$ is not used at all in the intermediate circuits as it gets passed along the garbled circuits for tree T_i. In order to avoid this wastefulness, for all nodes $i \in \{1, \ldots, L\}, j \in [i], k \in [2^j]$ we sample a value $r^{(i,j,k)}$ of length $\kappa \cdot |\text{cState}|$ and hardcode the values $r^{(i,j,k)} \oplus r^{(i,j+1,2k)}$ and $r^{(i,j,k)} \oplus r^{(i,j+1,2k+1)}$ inside the garbled circuit $\tilde{C}^{i,j,k}$ which output the first of two values if the execution goes left and the second if the execution goes right. Note that a garbled circuits grows only additively in $\kappa \cdot |\text{cState}|$ because of this change. This follows by using the first optimization. Additionally, we include the value $\text{cState}^{i+1,0,0} \oplus r^{(i,0,0)}$ with the root node of the tree T_i. The leaf garbled circuit $(i, i-1, k)$ in tree T_i is constructed assuming $r^{(i,i-1,k)}$ is the sequence of input labels for the root garbled circuit of the tree T_{i+1}.[3] Let $\alpha_0, \ldots \alpha_{i-1}$ be the strings output during the root to a leaf traversal in tree T_i. Now observe that $\text{cState}^{i+1,0,0} \oplus r^{(i,0,0)} \oplus_{j \in [i]} \alpha_j$ is precisely $\text{cState}^{i+1,0,0} \oplus r^{(i,i-1,k)}$ where k is the leaf node in the traversed path. At this point it is easy to see that given the output of the leaf grabled circuit for tree T_i one can compute the required input labels for the root of tree T_{i+1}.

The update mechanism in our construction can be easily adapted to work with this change. Here note that we would now include the values $r^{(i,j,k)}, r^{(i,j+1,2k)}$ and $r^{(i,j+1,2k+1)}$ in the ciphertext $X^{(i,j,k)}$. Also note that we will use fresh $r^{(\cdot,\cdot,\cdot)}$ values whenever a fresh garbled circuit for a node is generated. The security argument now additionally uses the fact that the outputs generated by garbled circuits in two separate root to leaf traversals depend on completely independent $r^{(\cdot,\cdot,\cdot)}$ values.

Note that the above modification leaks what value is passed by the executed leaf garbled circuit in tree T_i to the root garbled circuit in tree T_{i+1}. This can be deduced based on what bit values of $\text{cState}^{i+1,0,0} \oplus r^{(i,0,0)}$ are revealed.

---

[2] This efficiency is achieved when the first optimization is used.

[3] Note that here the first optimization allows us to ensure that the size of the garbled leaf circuit, rather than the clear leaf circuit, grows with the length of $r^{(i,i-1,k)}$ as these hard-codings are performed.

This can be tackled by randomly permuting the labels in $\mathsf{cState}^{i+1,0,0}$ and passing the information on this permutations along with in the tree to leaf garbled circuits. Note that the size of this information is small.

Taken together these two optimizations reduce the size of each garbled circuit to $O(\kappa \cdot (|\mathsf{bucket}| + |\mathsf{cState}|))$. Since $|\mathsf{bucket}| > |\mathsf{cState}|$ this expression reduces to $O(\kappa \cdot |\mathsf{bucket}|)$. This implies that the overhead of our construction is just κ times the overhead of the underlying Path ORAM scheme.

4   Searchable Encryption Construction Using **TWORAM**

The natural way of designing an SSE scheme that does not leak the search and access patterns using an ORAM scheme is to first use a data structure for storing keyword-document pairs, setup the data structure in memory using an ORAM setup and then read/write from it using ORAM operations. Since ORAM hides the read/write access patterns, but it does not hide the number of memory accesses, one needs to ensure that the number of memory accesses for each operation is also data-independent. Fortunately, this can be achieved by not letting the key used for the hash table be the output of a pseudorandom function applied to the keyword w, and not the keyword w itself.

We start by giving some definitions and then describe constructions that can be instantiated using any ORAM scheme. We then show how to obtain a significantly more efficient instantiation using a combination of TWORAM and a non-recursive Path-ORAM scheme.

4.1   Hash Table Definition

A hash table is a data structure commonly used for mapping keys to values [7]. It often uses a hash function h that maps a key to an index (or a set of indices) in a memory array M where the value associated with the key may be found. In particular, h takes as input a keyword key and outputs a set of indices $i_1, \ldots, i_c$ where c is a parameter. The value associated with key is in one of the locations $\mathsf{M}[i_1], \ldots \mathsf{M}[i_c]$. The keyword is not in the table if it is not in one of those locations. Similarly, to write a new $(key, value)$ pair into the table, $(key, value)$ is written into the first empty location among $i_1, \ldots, i_c$. More formally, we define a hash table $H = (\mathsf{hsetup}, \mathsf{hlookup}, \mathsf{hwrite})$ using a tuple of algorithms and a parameter c denoting an upper bound on the number of locations to search.

- $(h, \mathsf{M}) \leftarrow \mathsf{hsetup}(S, size)$: hsetup takes as input an initial set S of keyword-value pairs and a maximum table size $size$ and outputs a hash function h and a memory array M.
- $value \leftarrow \mathsf{hlookup}(key)$: hlookup computes $\{i_1, \ldots, i_c\} \leftarrow h(key)$, looks for a key-value pair $(key, \cdot)$ in $\mathsf{M}[i_1], \ldots, \mathsf{M}[i_c]$. If such a pair is found it returns the second component of the pair (i.e., the value), else it returns $\perp$.
- $\mathsf{M} \leftarrow \mathsf{hwrite}(key, value)$: hwrite computes $i_1, \ldots, i_c \leftrightarrow h(key)$, if $(key, value)$ already exists in one of those indices in M it does nothing, else it stores $(key, value)$ in the first empty index.

4.2   Searchable Encryption Definition

A database D is a set of document/keyword-set pair

$$\mathsf{DB} = (d_i, W_i)_{i=1}^N.$$

Let $W = \cup_{i=1}^N W_i$ be the universe of keywords. A keyword search query for w should return all d_i where $w \in W_i$. We denote this subset of DB by $\mathsf{DB}(w)$. A searchable symmetric encryption scheme consists of protocols SSESETUP, SSESEARCH and SSEADD. The following formalization first appeared in [6,8].

- $\langle \sigma, \mathsf{EDB} \rangle \leftrightarrow \mathrm{SSESETUP}\langle (1^\kappa, \mathsf{DB}), \bot \rangle$: SSESETUP takes as client's input database DB and outputs a secret state σ (for the client), and an encrypted database EDB which is outsourced to the server.
- $\langle (\mathsf{DB}(w), \sigma'), \mathsf{EDB}' \rangle \leftrightarrow \mathrm{SSESEARCH}\langle (\sigma, w), \mathsf{EDB} \rangle$: SSESEARCH is a protocol between the client and the server, where client's input is the secret state σ and the keyword w he is searching for. Server's input is the encrypted database EDB. Client's output is the set of documents containing w, i.e. $\mathsf{DB}(w)$ as well an updated secret state σ' and the server obtains an updated encrypted database EDB$'$.
- $\langle \sigma', \mathsf{EDB}' \rangle \leftrightarrow \mathrm{SSEADD}\langle (\sigma, d), \mathsf{EDB} \rangle$: SSEADD is a protocol between the client and the server, where client's input is the secret state σ and a document d to be inserted into the database. Server's input is the encrypted database EDB. Client's output is an updated secret state σ' and the server's output is an updated encrypted database EDB$'$ which now contains the new document d.

Correctness. Consider the following correctness experiment. An adversary A chooses a database DB_0. Consider the encrypted database EDB_0 generated using SSESETUP (i.e., $\langle \sigma_0, \mathsf{EDB}_0 \rangle \leftrightarrow \mathrm{SSESETUP}\langle (1^\kappa, \mathsf{DB}_0), \bot \rangle$). The adversary then adaptively chooses keywords to search and documents to add to the database, and the respective protocols SSESEARCH and SSEADD are run between an honest client and server, outputting the updated EDB, DB and σ. Denote the operations chosen by the adversary with $w_1, \dots, w_q$. A wins in the correctness game if for some search query w_i it is

$$\langle (\mathsf{DB}_i(w_i), \sigma_i), \mathsf{EDB}_i \rangle \neq \mathrm{SSESEARCH}\langle (\sigma_{i-1}, w_i), \mathsf{EDB}_{i-1} \rangle,$$

where $\mathsf{DB}_i, \mathsf{EDB}_i$ are the database and encrypted database, respectively, after the i-th search. The SSE scheme is correct if the probability of A winning the game is negligible in κ.

Security. We discuss security in the semi-honest model. It is parametrized by a leakage function $\mathcal{L}$, which explains what the adversary (the server) learns about the database and the search and update queries, while interacting with a secure SSE scheme. A SSE scheme is $\mathcal{L}$-secure if for any PPT adversary A, there exist a simulator Sim such that the following two distributions are computationally indistinguishable.

– $\text{Real}_A(\kappa)$: A chooses DB_0. The experiment then runs

$$\langle \sigma_0, \mathsf{EDB}_0 \rangle \leftrightarrow \text{SSESETUP}\langle (1^\kappa, \mathsf{DB}_0), \bot \rangle.$$

A then adaptively makes search queries w_i, which the experiment answers by running the protocol $\langle \mathsf{DB}_{i-1}(w_i), \sigma_i \rangle \leftrightarrow \text{SSESEARCH}\langle (\sigma_{i-1}, w_i), \mathsf{EDB}_{i-1} \rangle$. Denote the full transcripts of the protocol by t_i and with EDB' the final encrypted database. Add queries are handled in a similar way. Eventually, the experiment outputs

$$(\mathsf{EDB}, t_1, \ldots, t_q),$$

where q is the total number of search/add queries made by A.
– $\text{Ideal}_{A,\text{Sim},\mathcal{L}}(\kappa)$: A chooses DB_0. The experiment runs

$$(st_0, \mathsf{EDB}_0) \leftrightarrow \text{Sim}(\mathcal{L}(\mathsf{DB}_0)),$$

where st_0 is the initial state of the simulator. On input any search query w_i from A, the experiment adds $(w_i, search)$ to the history H, and on an add query d_i it adds (d_i, add) to H. It then runs $(t_i, st_i) \leftrightarrow \text{Sim}(st_{i-1}, \mathcal{L}(\mathsf{DB}_{i-1}, H))$. Eventually, the experiment outputs $(\mathsf{EDB}', t_1, \ldots, t_q)$ where q is the total number of search/add queries made by A.

Leakage. The level of security one obtains from a SSE scheme depends on the leakage function $\mathcal{L}$. Ideally $\mathcal{L}$ should only output the total number $\sum_{w \in W} |\mathsf{DB}(w)|$ of (w, d) pairs, the total number of unique keywords $|W|$ and $|\mathsf{DB}(w)|$ for any searched keyword w. Achieving this level of security is only possible if the SSESEARCH operation outputs the documents themselves to the client. If instead (as is common for applications with large document sizes), it returns document identifiers which the client then uses to retrieve the actual documents, any SSE protocol would also leak the access pattern.

4.3 SSE from any ORAM

First Approach. The common way of storing a database of documents in a hash table is to insert a key-value pair (w, d) into the table for any keyword w in a document d. Searching for a document with keyword w then reduces to looking up w in the table. If there is more than one document containing a keyword w, a natural solution is to create a bucket B_w storing all the documents containing w and storing the bucket in position pt_w of an array A. One then inserts (w, pt_w) in a hash table. Now, to search for a keyword w, we first look up (w, pt_w), and then access $A[pt_w]$ to obtain the bucket B_w of all the desired documents. A subtle issue is that the distribution of bucket sizes would leak information about the database even before any keyword is searched. As a result, for this approach to be fully-secure, one needs to pad each bucket to an upperbound on the number of searchable documents per keyword.

Next we describe the SSE scheme more formally. Given a hash table $H = (\text{hsetup}, \text{hlookup}, \text{hwrite})$, and an ORAM scheme $ORAM = (\text{SETUP}, \text{OBLIVIOUSACCESS})$, we construct an SSE scheme (SSESETUP, SSESEARCH, SSEADD) as follows.

1. $\langle \sigma, \mathsf{EDB} \rangle \leftrightarrow \mathrm{SSESETUP}\langle (1^\kappa, max, \mathsf{DB}), \bot \rangle$: Given an initial set of documents DB, client lets S be the set of key-value pairs (w, pt_w) where pt_w is an index to an array of buckets A such that $\mathsf{A}[pt_w]$ stores the bucket of all documents in DB containing w. Each bucket is padded to the maximum size max with dummy documents.

   Client first runs $\mathsf{hsetup}(S, size)$ to obtain (h, M). $size$ is the maximum size of hash table H. Then client and server run $\langle \sigma_1, \mathsf{EM} \rangle \leftrightarrow \mathrm{SETUP}\langle (1^\kappa, \mathsf{M}), \bot \rangle$. Cleint and server also run $\langle \sigma_2, \mathsf{EA} \rangle \leftrightarrow \mathrm{SETUP}\langle (1^\kappa, \mathsf{A}), \bot \rangle$

   Note that server's output is $\mathsf{EDB} = (\mathsf{EM}, \mathsf{EA})$ and client's output is $\sigma = (\sigma_1, h, \sigma_2)$.

2. $\mathrm{SSESEARCH}\langle (\sigma, w), \mathsf{EDB} \rangle$: Client computes $i_1, \ldots, i_c \leftarrow h(w)$. Then, client and server run $\mathrm{OBLIVIOUSACCESS}\langle ((\sigma_1, i_j, \mathsf{null}), \mathsf{EM} \rangle$ for $j \in \{1, \ldots, c\}$ for client to obtain $\mathsf{M}[i_j]$. If client does not find (w, pt_w) in one of the retrieved locations it lets $pt_w = 0$, corresponding to a dummy access to the index 0 in A.

   Client and server then run $\mathrm{OBLIVIOUSACCESS}\langle (\sigma_2, pt_w, \mathsf{null}), \mathsf{EA}) \rangle$ for client to obtain the bucket B_w stored in $\mathsf{A}[pt_w]$. Client outputs all the non-dummy documents in B_w.

3. $\mathrm{SSEADD}\langle (\sigma, d), \mathsf{EDB} \rangle$: For every w in d, client computes $i_1, \ldots, i_c \leftarrow h(w)$ and client and server run $\mathrm{OBLIVIOUSACCESS}\langle (\sigma_1, i_j, \mathsf{null}), \mathsf{EM} \rangle$ for $j \in \{1, \ldots, c\}$ for client to obtain $\mathsf{M}[i_j]$. If (w, pt_w) is in the retrieved locations let i_j^* be the location it was found at. If not, let pt_w be the first empty location in A, and let $i*_j$ be the first empty location from the retrieved ones in M. Client and server run $\mathrm{OBLIVIOUSACCESS}\langle (\sigma_1, i_j^*, (w, pt_w)), \mathsf{EM} \rangle$.

   Client and server run $\mathrm{OBLIVIOUSACCESS}\langle (\sigma_2, pt_w, \mathsf{null}), \mathsf{EA} \rangle$ to retrieve $\mathsf{A}[pt_w]$. Let B_w be the retrieved bucket. Client inserts d in the first dummy entry of B_w, denoting the new bucket by B'_w. Client and server run

$$\mathrm{OBLIVIOUSACCESS}\langle (\sigma_2, pt_w, B'_w), \mathsf{EA} \rangle.$$

The main disadvantage of the above construction is that we need to anticipate an upper bound on the bucket sizes, and pad all buckets to that size. Given that in practice there are often keywords that appear in a large number of documents, and keywords that only appear in a few, the padding will lead to inefficiency. Our next solution addresses this issue but instead has a higher round complexity.

Second Approach. Instead of storing all documents matching a keyword w in one bucket, we store each of them separately in the hash table, using a different keyword. In particular, we can store the key-value pair $(w\|i, d)$ in the hash table for the ith document d containing w. This works fine except that it requires looking up $w\|count$ for an incremental counter $count$ until the keyword is no longer found in the table.

To make this approach cleaner and the write operations more efficient, we maintain two hash tables, one for storing the counter representing the number of documents containing the keyword, and one storing the incremental key-value pairs as described above. To lookup a keyword w, one first looks up the counter $count$ in the first table and then makes $count$ lookup queries to the second table.

We now describe the above SSE scheme in more detail. Given a hash table H = (hsetup, hlookup, hwrite) and a scheme $ORAM$ = (SETUP, OBLIVIOUSACCESS), we construct an SSE scheme (SSESETUP, SSESEARCH, SSEADD) as follows:

1. $\langle \sigma, \mathsf{EDB} \rangle \leftrightarrow$ SSESETUP$\langle (1^\kappa, \mathsf{DB}), \perp \rangle$: Given an initial set of documents DB. Let S_1 be the set of $(w, count_w)$ pairs and S_2 be the set of key-value pairs $(w||i, d_i)$ for $1 \leq i \leq count_w$ where $count_w$ is the number of documents containing w, and d_i denotes the ith document in DB containing w.

   Cleint runs hsetup$(S_i, size_i)$ to obtain (h_i, M_i). $size_i$ is the maximum size of the hash table H_i. Then client and server run $\langle \sigma_i, \mathsf{EM}_i \rangle \leftrightarrow$ SETUP$\langle (1^\kappa, \mathsf{M}_i), \perp \rangle$. Note that server's output is $\mathsf{EDB} = (\mathsf{EM}_1, \mathsf{EM}_2)$ and client's output is $\sigma = (\sigma_1, \sigma_2, h_1, h_2)$.

2. SSESEARCH$\langle (\sigma, w), \mathsf{EDB} \rangle$: Client computes $i_1, \ldots, i_c \leftarrow h_1(w)$ and client and server run OBLIVIOUSACCESS$\langle (\sigma_1, i_j, \mathsf{null}), \mathsf{EM}_1 \rangle$ for $j \in \{1, \ldots, c\}$ for client to obtain $(w, count_w)$ among the retrieved locations. If such a pair is not found, client lets $count_w = 0$.

   For $1 \leq k \leq count_w$, client computes $i_1^k, \ldots, i_c^k \leftarrow h_2(w||k)$ and client and server run OBLIVIOUSACCESS$\langle (\sigma_2, i_j^k, \mathsf{null}), \mathsf{EM}_2 \rangle)$ for $j \in \{1, \ldots, c\}$ for client to obtain $\mathsf{M}_2[i_j^k]$. Client outputs d for all d where $(w||k, d)$ is in the retrieved locations from M_2.

3. SSEADD$\langle (\sigma, d), \mathsf{EDB} \rangle$: For every w in d, client computes $i_1, \ldots, i_c \leftarrow h_1(w)$ and client and server run OBLIVIOUSACCESS$\langle (\sigma_1, i_j, \mathsf{null}), \mathsf{EM}_1 \rangle$ for $j \in \{1, \ldots, c\}$ for client to obtain $\mathsf{M}_1[i_j]$. If $(w, count_w)$ is in the retrieved locations let i_j^* be the location it was found at. If not, let $count_w = 0$ and let i_j^* be the first empty location from the retrieved ones. Client and server run OBLIVIOUSACCESS$\langle (\sigma_1, i_j^*, (w, count_w + 1)), \mathsf{EM}_1 \rangle$ to increase the counter by one.

   Client then computes $i_1', \ldots, i_c' \leftarrow h_2(w||count_w + 1)$ and client and server run OBLIVIOUSACCESS$\langle (\sigma_2, i_j', \mathsf{null}), \mathsf{EM}_2 \rangle$ to retrieve $\mathsf{M}_2[i_j']$ for $j \in \{1, \ldots, c\}$. Let i_k' be the first empty location among them. Client and server run

$$\text{OBLIVIOUSACCESS}\langle (\sigma_2, i_k', (w||count + 1)), \mathsf{EM}_2 \rangle.$$

The main disadvantage of our second approach is that for each search, it requires $count_w$ ORAM accesses to retrieve all matching documents. This means that the bandwidth/computation overhead of ORAM scheme is multiplied by $count_w$ which can be large for some keywords. More importantly, it would require $O(count_w)$ rounds since the ORAM accesses cannot be parallelized in our constant-round ORAM construction. In particular, note that each memory garbled circuit in the construction can only be used once and needs to be replaced before the next memory access. Finally, the constant-round ORAM needs to store a memory array that is proportional to the number of (w, d) tuples associated with the database, which is significantly larger than the number of unique keywords, increasing the storage overhead of the resulting SSE scheme.

Next, we address all these efficiency concerns, showing a construction that only requires a single ORAM access using our constant-round construction.

4.4   SSE from Path-ORAM

The idea is to not only store a per-keyword counter $count_w$ as before, but also to store a $access_w$ that represents the number of search/add queries performed on w so far. Similar to the previous approach, the tuple $(w, (count_w, access_w))$ is stored in a hash table that is implemented using our constant-round ORAM scheme TWORAM. The $count_w$ is incremented whenever a new document containing w is added and the $access_w$ is incremented after each search/add query for w.

The tuples $(w||i, d_i)$ for all d_i containing w are then stored in a one-level (non-recursive) Path-ORAM. In order to avoid storing a large client-side position map for this non-recursive Path-ORAM, we generate/update the positions pseudorandomly using a PRF $F_K(w||i||access_w)$. Since each document d_i has a different index and each search/add query for w will increment $access_w$, the pseudorandomness property of F ensures that this way of generating the position maps is indistinguishable from generating them at random. Now the client only needs to keep the secret key K. Note that since we are using a one-level Path-ORAM to store the documents, we can handle multiple parallel accesses without any problems, hence obtaining a constant-round search/add complexity. Furthermore, we only access TWORAM(which uses garbled circuits) once per keyword search to retrieve the tuple $(w, (count_w, access_w))$, so TWORAM's overhead is not multiplied by $count_w$ for each search/add query. Similarly, the storage overhead of TWORAMis only for a memory array of size $|W|$ (number of unique keywords in documents) which is significantly smaller than the number of keyword-document pairs needed in the general approach.

We need to make a few small modifications to the syntax of the abstraction for Path-ORAM here. First, since we generate the position map on the fly using a PRF, it is convenient to modify the syntax of the UPDATE procedure to take the new random position as input, instead of internally generating it in our original syntax. Also, since we are not extracting an index y from the Path-ORAM and instead are extracting a tuple of the form $(w||i, d_i)$, we will pass $w||i$ as input in place of y in the EXTRACT and UPDATE operations.

We now describe the SSE scheme. Given a hash table H = (hsetup, hlookup, hwrite), our constant-round ORAM scheme TWORAM = (SETUP, OBLIVIOUSACCESS), a single level Path-ORAM scheme with procedures (INITIALIZE, EXTRACT, UPDATE), and a PRF function F, we build an SSE scheme (SSESETUP, SSESEARCH, SSEADD) as follows:

1. $\langle \sigma, \text{EDB} \rangle \leftrightarrow \text{SSESETUP}\langle (1^\kappa, \text{DB}), \bot \rangle$: Given an initial set of documents DB, let S be the set of $(w, (count_w, access_w = 0))$ where $count_w$ is the number of documents containing w, and $access_w$ denotes the number of times the keyword w has been searched/added.

    Client runs hsetup($S, size$) to obtain (h, M). $size$ is the anticipated maximum size of the hash table H. Then client and server run $\langle \sigma_s, \text{EM} \rangle \leftrightarrow \text{SETUP}\langle (1^\kappa, \text{M}), \bot \rangle$.

    Let A_L be an initially empty memory array with a size that estimates an upper bound on total number of (w, d) pairs ind DB. Client runs $\mathcal{T} \leftarrow$

INITIALIZE($1^\kappa, A_L$), and only sends the tree T_L for the last level to server, and discards the rest.

Client generates a PRF key $K \leftarrow \{0,1\}^\kappa$.

For every item $(w, (count_w, access_w))$ in S, and for $1 \le i \le count_w$ (in parallel):

(a) Client lets $val_{w,i} = (w||i, d_i)$ where d_i denotes the ith document in DB containing w.

(b) Client lets $x_{w,i} = F_K(w||i||access_w)$ and sends $x_{w,i}$ to server who returns the encrypted buckets on path $T_L(x_{w,i})$ which client decrypts itself.

(c) Client runs $\{T_L(x_{w,i})\} \leftarrow$ UPDATE($w||i, write, val_{w,i}, T_L(x_{w,i}), x'_{w,i}$), where $x'_{w,i} = F_K(w||i||access_w + 1)$, to insert $val_{w,i}$ into the path along its new path $T_L(x'_{w,i})$. Client then encrypts the updated path $T_L(x_{w,i})$ and sends it to server who updates T_L.

Note that server's output is EDB $= (\text{EM}, T_L)$ and client's output is $\sigma = (\sigma_s, h, K)$.

2. SSESEARCH$\langle(\sigma, w), \text{EDB}\rangle$: Client computes $i_1, \ldots, i_c \leftarrow h(w)$ and client and server run OBLIVIOUSACCESS$\langle(\sigma_s, i_j, \text{null}), \text{EM}\rangle$ for $j \in \{1, \ldots, c\}$. If client finds $(w, (count_w, access_w))$ in one of the retrieved locations, let i_j^* be the location it was found at. If such a pair is not found the search ends here. Client and server run OBLIVIOUSACCESS$\langle(\sigma_s, i_j^*, (w, count_w, access_w + 1)), \text{EM}\rangle$ to increase the $access_w$ by one.

For $1 \le i \le count_w$ (in parallel):

(a) Client lets $x_{w,i} = F_K(w||i||access_w)$ and sends $x_{w,i}$ to server who returns $T_L(x_{w,i})$ which client decrypts.

(b) Client runs $(w||i, d_i) \leftarrow$ EXTRACT($L, w||i, T_L(x_{w,i})$), and outputs d_i. Client runs $\{T_L(x_{w,i})\} \leftarrow$ UPDATE($w||i, read, (w||i, d_i), T_L(x_{w,i}), x'_{w,i} = F_K(w||i||access_w + 1)$) to update the location of $(w||i, d_i)$ to $x'_{w,i}$. Client then encrypts the updated path and sends it to server to update T_L.

3. SSEADD$\langle(\sigma, d), \text{EDB}\rangle$:

For every w in d:

(a) Client computes $i_1, \ldots, i_c \leftarrow h(w)$ and client and server run

$$\text{OBLIVIOUSACCESS}\langle(\sigma_s, i_j, \text{null}), \text{EM}\rangle,$$

for $j \in \{1, \ldots, c\}$. If client finds $(w, (count_w, access_w))$ in one of the retrieved locations, let i_j^* be the location it was found at. Else, it lets i_j^* be the first empty location among the retrieved ones.

(b) Client and server run OBLIVIOUSACCESS$\langle(\sigma_s, i_j^*, (w, (count_w + 1, access_w + 1))), \text{EM}\rangle$ to increase $count_w$ and $access_w$ by one.

(c) Client lets $x_{w,count_w} = F_K(w||count_w||acess_w)$ and sends $x_{w,count_w}$ to server who returns encrypted $T_L(x_{w,count_w})$ back. Client decrypts the path.

(d) Client lets $x' = F_K(w||count_w + 1||access_w + 1)$ and runs $\{T_L(x_{w,count_w})\} \leftarrow$ UPDATE($w||i, write, (w||count_w + 1, d), T_L(x_{w,count_w}), x'$) to update the path. Client then encrypts the updated path and sends it to server to update T_L.

Before stating the security theorem for the above SSE scheme, we first need to make the leakage function associated with the scheme more precise. The leakage function $\mathcal{L}(\mathsf{DB}, H)$ for our scheme outputs the following (DB is the database and H is the search/add history): $|W|$, number unique keywords in all documents; $|\mathsf{DB}(w)|$ for every w searched; $\sum_{w \in W} |\mathsf{DB}(w)|$ i.e. the number of (w, d) pairs where w is in d. See Appendix A.4 for the proof.

Theorem 2. *The above SSE scheme is $\mathcal{L}$-secure (cf. Definition of Sect. 4), if* TWORAM *is secure (cf. Definition in Sect. 2.2), F is a PRF, and the encryption used in the one-level Path-ORAM is CPA-secure.*

Efficiency. The setup cost for our SSE scheme is the sum of the setup cost for TWORAM for a memory of size $|W|$, and the setup for a one-level Path-ORAM of size $n = \sum_{w \in W} |\mathsf{DB}(w)|$ which is $O(n \log n \log \log n)$.

The bandwidth cost for each search/add query w is the cost of one ORAM read in TWORAMplus $O(|\mathsf{DB}(w)| * (\log n \log \log n))$ for $n = \sum_{w \in W} |\mathsf{DB}(w)|$.

Acknowledgments. This work was done in part while the authors were visiting the Simons Institute for the Theory of Computing, supported by the Simons Foundation and by the DIMACS/Simons Collaboration in Cryptography through NSF grant #CNS-1523467. Sanjam Garg was supported in part from a DARPA/ARL SAFE-WARE award, AFOSR Award FA9550-15-1-0274, and NSF CRII Award 1464397. Charalampos Papamanthou was supported in part by NSF grants #1514261 and #1526950, by a NIST award, by a Google Faculty Research Award and by Yahoo! Labs through the Faculty Research Engagement Program (FREP). The views expressed are those of the authors and do not reflect the official policy or position of the Department of Defense, the National Science Foundation, or the U.S. Government.

A   More Details on Path ORAM

A.1   Path ORAM Abstraction Algorithms

Algorithm 1. Setting up path ORAM data structures.

1: **procedure** $\mathcal{T} \leftarrow$ INITIALIZE($1^\kappa, A_L$)
2:      Let π_L be a random permutation from $[n]$ to $[n]$;
3:      Store $(x, A_L[x], \pi_L(x))$ at leaf $\pi_L(x)$ of tree T_L;
4:      **for** $i = L$ down to 3 **do**
5:          Set $A_{i-1}[x] = \pi_i(2x-1)||\pi_i(2x)$ for $x = 1, \ldots, 2^{i-1}$;
6:          Let π_{i-1} be a random permutation from $[2^{i-1}]$ to $[2^{i-1}]$;
7:          Store $(x, A_{i-1}[x], \pi_{i-1}(x))$ at leaf $\pi_{i-1}(x)$ of tree T_{i-1};
8:      **end for**
9:      Let A_1 be an array of 2 entries such that $A_1[x] = \pi_2(2x-1)||\pi_2(2x)$ for $x = 1, 2$;
10:     **return** $\{A_1, T_2, \ldots, T_L\}$;
11: **end procedure**

Algorithm 2. Extraction algorithm for buckets.

```
1: procedure π ←EXTRACTBUCKET(i, y, b)
2:     Search bucket b to retrieve block (y_i, A_i[y_i], p);
3:     if found then
4:         return π ← select(A_i[y_i], b_i);     ▷ π is the index of the path to be explored
               in T_{i+1}.
5:     else
6:         return ⊥;
7:     end if
8: end procedure
```

Algorithm 3. Update algorithm. It takes as input $L-1$ paths and local storage A_1 and outputs new paths, based on the new assignments of positions.

```
1: procedure {A_1, T_2(x_2), ..., T_L(x_L)} ←UPDATE(y, val, A_1, T_2(x_2), ..., T_L(x_L))
2:     select(A_1[y_1], b_1) ← r_2;                    ▷ r_i is random in [1, 2^{i+1}].
3:     for i = 2 to L do
4:         T_i.root ← T_i.root ∪ readPath(T_i(x_i));     ▷ T_i.root serves as the stash C_i.
5:         Update block (y_i, A_i[y_i], x_i) to (y_i, A_i[y_i], r_i) in T_i.root;
6:         select(A_i[y_i], b_i) ← r_{i+1};     ▷ if i = L do if val ≠ null, A_L[y] ← val, else
               NOOP.
7:         [T_i.root, T_i(x_i)] ← evictPath(T_i.root);
8:     end for
9:     return A_1, T_2(x_2), T_3(x_3), ..., T_L(x_L);
10: end procedure
```

Protocol ⟨σ, EM⟩ ↔ SETUP⟨(1^κ, M), ⊥⟩:

Client:

1: Pick a κ-bit s; Run {A_1, T_2, ..., T_L} ← INITIALIZE(1^κ, M);
2: For all $i > 1$, for all $u \in T_i$, set B_u ← Enc_s(bucket_u), where Enc_s(.) is a CPA-secure encryption;
3: For all $i > 1$, for all $u \in T_i$, send to server data B_u;
4: **return** s and A_1 as σ;

Server:

1: **return** all data B_u sent by the client from above as EM;

Fig. 6. Formal description of the SETUP protocol for the interactive ORAM [34].

A.2   Path ORAM Protocols with $\log n$ Rounds of Interaction Using the Abstraction

A.3   Proof of Security for TWORAM

Now we prove TWORAM is a secure realization of an oblivious RAM scheme as described in Sect. 2.2. We start by arguing correctness. Note that the garbled circuits implement the exact same procedures as are required in our abstraction. Therefore the correctness of our scheme follows directly from the correctness of the underlying Path ORAM scheme and garbled circuits construction. Next we argue security. In other words we need to argue that for any adversary A, there exists a simulator Sim for which the following two distributions are computationally indistinguishable.

– $\text{Real}_A^\Pi(\kappa)$: A chooses M. The experiment then runs $\langle\sigma, \text{EM}\rangle \leftrightarrow$ $\text{SETUP}\langle(1^\kappa, \text{M}), \perp\rangle$. A then provides read/write queries (y_i, v) where $v = \text{null}$ on reads, for which the experiment runs the protocol

$$\langle(\text{M}[y_i], \sigma_i), \text{EM}_i\rangle \leftrightarrow \text{OBLIVIOUSACCESS}\langle(\sigma_{i-1}, y_i, v), \text{EM}_{i-1}\rangle.$$

Denote the full transcript of the protocol by t_i. Eventually, the experiment outputs $(\text{EM}, t_1, \ldots, t_q)$ where q is the total number of read/write queries.
– $\text{Ideal}_{\text{Sim}}^\Pi(\kappa)$: The experiment outputs $(\text{EM}, t'_1, \ldots, t'_q) \leftarrow \text{Sim}(q, |\text{M}|, 1^\kappa)$.

Our Simulator. Note that the simulator needs to provide to the server, for all u $\tilde{C}^u, X^u$ and for all $i \in \{2, \ldots, L\}$ $\beta_i := \text{nState}_{\text{cState}^{(i+1,0,0)}}^{(i,0,0)}$. Furthermore replacement circuits need to be provided as read/write queries are implemented. Our simulator Sim generates these as follows:

– For each $u = (i, j, k)$, let $(\tilde{C}^u, \text{lab}^u \leftarrow \text{GCircuit}(1^\kappa, P[u, b_u, \text{lab}_0^{(i,j+1,2k+b)}]))$, where b_u is random bit and P is a circuit that, if $j = i$ outputs $(\text{nextRoot}, \text{lab}_0^{(i+1,0,0)})$, else if $b = 0$ then it outputs $(\text{left}, \text{lab}_0^{(i,j+1,2k+b)})$ and $(\text{right}, \text{lab}_0^{(i,j+1,2k+b)})$ otherwise.
– Each X^u is generated as as encryption of a zero-string, namely $\text{Enc}_s(0)$. Similarly for all $i \in \{2, \ldots, L\}$ $\beta_i := \text{nState}_0^{(i,0,0)}$.

Note that as the provided garbled circuits are executed, replacements circuits need to be given and they are generated in the same manner as above.

Proof of Indistinguishability. The proof follows by a hybrid argument.

– H_0: This hybrid corresponds to the honest execution $\text{Real}_A^\Pi(\kappa)$ as done honestly.
– H_1: This hybrid is same as H_0 except that we now generate all the X^u values as encryptions of zero-strings of appropriate length. Specifically, for each u we set $X^u \leftarrow \text{Enc}_s(0)$.
  The indistinguishability between H_0 and H_1 follows from the security of the encryption scheme (Enc, Dec).
– H_2: In this hybrid the simulator maintains the entire Path ORAM tree internally but does not include it in the provided garbled circuits. In other words garbled circuits are generated as follows:
  • For each $u = (i, j, k)$, let $(\tilde{C}^u, \text{lab}^u \leftarrow \text{GCircuit}(1^\kappa, P[u, b_u, \text{lab}_0^{(i,j+1,2k+b)}]))$, where b_u is 0 or 1 depending on whether the execution as per ORAM would go left or right and P is a circuit that, if $j = i$ outputs $(\text{nextRoot}, \text{lab}_0^{(i+1,0,0)})$, else if $b = 0$ then it outputs $(\text{left}, \text{lab}_0^{(i,j+1,2k+b)})$ and $(\text{right}, \text{lab}_0^{(i,j+1,2k+b)})$ otherwise.
  • Each X^u is generated as as encryption of a zero-string, namely $\text{Enc}_s(0)$. Similarly for all $i \in \{2, \ldots, L\}$ $\beta_i := \text{nState}_0^{(i,0,0)}$.

Protocol $\langle (\mathsf{M}[y], \sigma'), \mathsf{EM}' \rangle \leftrightarrow \text{OBLIVIOUSACCESS}\langle (\sigma, y, val), \mathsf{EM} \rangle$:

Client(1):

1: Compute $x_2 \leftarrow \text{select}(A_1[y_1], b_1)$. Send to server index x_2; $\triangleright$ run **Server(2)**

Server(i):

1: For all $u \in T_i(x_i)$ send to client B_u; $\triangleright$ run **Client(i)**

Client(i):

1: $\pi = \bot$;
2: **for** $u \in T_i(x_i)$ **do**
3:      $\text{bucket}_u \leftarrow \text{Dec}_s(\mathsf{B}_u)$;
4:      **if** $\pi = \bot$ **then**
5:          $\pi \leftarrow \text{EXTRACTBUCKET}(i, y, \text{bucket}_u)$;
6:          $x_{i+1} = \pi$;
7:      **end if**
8: **end for**
9: **if** $i < L$ **then**
10:     Send to server new index x_{i+1}; $\triangleright$ run **Server($i+1$)**
11: **else**
12:     $\{A_1, T_2(x_2), \ldots, T_L(x_L)\} \leftarrow \text{UPDATE}(y, val, A_1, T_2(x_2), \ldots, T_L(x_L))$;
13:     For all $i > 1$, for all $u \in T_i(x_i)$, send to server $\mathsf{B}_u \leftarrow \text{Enc}_s(\text{bucket}_u)$; $\triangleright$ run **Server(L)**
14:     **return** x_{L+1} as $\mathsf{M}[y]$ and A_1 as σ';
15: **end if**

Server(L):

1: **return** the data received by the client as EM';

Fig. 7. Formal description of the OBLVIOUSACCESS protocol for the interactive ORAM [34].

The indistinguishability between H_1 and H_2 follows by a sequence of hybrids where each garbled circuit is replaced by a simulated garbled circuit. Here these hybrids must be performed in sequence in which garbled circuits are consumed. Note that for the unconsumed garbled circuits the input labels aren't provided (or hardcoded inside any other circuit) and hence they can also be simulated.

- H_3: Same as H_2, except that each b_u is now chosen uniformly random, independent of the Path ORAM execution. Note that this is same as the simulator. The indistinguishability between H_2 and H_3 from the security of the Path ORAM scheme.

This concludes the proof. $\square$

A.4   Proof of Security for the SSE scheme

We prove Theorem 2 on security of the SSE scheme next, Following the definition of Sect. 4, we first describe a simulator Sim who generates the transcripts for the ideal distribution $\text{Ideal}_{A,\text{Sim},\mathcal{L}}^{\Pi}(\kappa)$. Sim takes as input $\mathcal{L}(\mathsf{DB}, H)$, and does the following: To generate full transcripts of the constant round ORAM scheme for the adversary A, Sim runs Sim', the simulator that exists for that scheme due its security (see definition of Sect. 2.2). That is, he runs $(\mathsf{EM}, t_1, \ldots, t_q) \leftarrow \mathsf{Sim}'(q, |\mathsf{M}|, 1^\kappa)$, where he drives $|\mathsf{M}|$ from $|W|$. To simulate the transcripts of the path-ORAM component, it generates a one-level path ORAM tree T_L for a memory array of size $\sum_{w \in W} |\mathsf{DB}(w)|$ filled with all 0 values. For each read/add query, it replaces the PRF-genenerated paths by uniformly random paths, and generates freshly generated ciphertexts of 0 for updated paths. Sim knows the number of paths to retrieve/update for each query from the leakage function which outputs $|\mathsf{DB}(w)|$ for every query w. This completes the description of the simulator. We now need to show that $\mathsf{Ideal}_{A, \mathsf{Sim}, \mathcal{L}}^{\Pi}(\kappa)$ is indistinguishable from $\mathsf{Real}_A^{\Pi}(\kappa)$, which constitutes the first in the sequence of our Hybrids:

Proof of Indistinguishability. The proof follows by a hybrid argument.

- H_0: This hybrid corresponds to the honest execution $\mathsf{Real}_A^{\Pi}(\kappa)$ for the SSE scheme which we repeat here for completeness. A chooses DB. The experiment then runs $\langle \mathsf{EDB}, \sigma \rangle \leftrightarrow \mathsf{SSESETUP}\langle(1^\kappa, \mathsf{DB}), \perp\rangle$. A then adaptively makes search queries w_i, which the experiment answers by running the protocol $\langle \mathsf{DB}_{i-1}(w_i), \sigma_i \rangle \leftrightarrow \mathsf{SSESEARCH}\langle(\sigma_{i-1}, w_i), \mathsf{EDB}_{i-1}\rangle$. Denote the full transcripts of the protocol by t_i. Add queries are handled in a similar way. Eventually, the experiment outputs $(\mathsf{EDB}, t_1, \ldots, t_q)$ where q is the total number of search/add queries made by A.
- H_1: Similar to H_0, except that the portions of t_i's corresponding to the constant-round ORAM are instead generated by $\mathsf{Sim}'(q, |\mathsf{M}|, 1^\kappa)$ where Sim' is the simulator in the proof of the ORAM scheme.
  The indistinguishability of H_0 and H_1 follows from security of the ORAM scheme.
- H_2: Similar to H_1 except that all ciphertexts in the path ORAM tree are replaced by encryptions of 0, and all updated ciphertexts will be fresh encryption of 0.
  The indistinguishability of H_2 and H_1 follows from the semantic security of the encryption scheme used in the path ORAM.
- H_3: Similar to H_2 except that all PRF-generated positions are replaced by uniformly random positions. Note that H_3 is essentially $\mathsf{Ideal}_{A, \mathsf{Sim}, \mathcal{L}}^{\Pi}(\kappa)$.
  The indistinguishability of H_3 and H_2 follows from the pseudorandomness of the the the PRF.

This concludes the proof.   □

References

1. Afshar, A., Hu, Z., Mohassel, P., Rosulek, M.: How to efficiently evaluate RAM programs with malicious security. In: Oswald, E., Fischlin, M. (eds.) EUROCRYPT 2015. LNCS, vol. 9056, pp. 702–729. Springer, Heidelberg (2015)
2. Bellare, M., Hoang, V.T., Rogaway, P.: Foundations of garbled circuits. In: CCS, pp. 784–796 (2012)
3. Cash, D., Grubbs, P., Perry, J., Ristenpart, T.: Leakage-abuse attacks against searchable encryption. In: CCS, pp. 668–679 (2015)

4. Cash, D., Jarecki, S., Jutla, C., Krawczyk, H., Roşu, M.-C., Steiner, M.: Highly-scalable searchable symmetric encryption with support for boolean queries. In: Canetti, R., Garay, J.A. (eds.) CRYPTO 2013, Part I. LNCS, vol. 8042, pp. 353–373. Springer, Heidelberg (2013)

5. Chang, Y.-C., Mitzenmacher, M.: Privacy preserving keyword searches on remote encrypted data. In: Ioannidis, J., Keromytis, A.D., Yung, M. (eds.) ACNS 2005. LNCS, vol. 3531, pp. 442–455. Springer, Heidelberg (2005)

6. Chase, M., Kamara, S.: Structured encryption and controlled disclosure. In: Abe, M. (ed.) ASIACRYPT 2010. LNCS, vol. 6477, pp. 577–594. Springer, Heidelberg (2010)

7. Cormen, T.H., Stein, C., Rivest, R.L., Leiserson, C.E.: Introduction to Algorithms, 2nd edn. McGraw-Hill Higher Education, New York (2001)

8. Curtmola, R., Garay, J., Kamara, S., Ostrovsky, R.: Searchable symmetric encryption: improved definitions and efficient constructions. In: CCS, pp. 79–88 (2006)

9. Devadas, S., van Dijk, M., Fletcher, C.W., Ren, L., Shi, E., Wichs, D.: Onion ORAM: a constant bandwidth blowup oblivious RAM. In: TCC, pp. 145–174 (2016)

10. Fletcher, C., Naveed, M., Ren, L., Shi, E., Stefanov, E.: Bucket ORAM: single online roundtrip, constant bandwidth oblivious RAM. Cryptology ePrint Archive, Report 2015/1065 (2015). http://eprint.iacr.org/

11. Garg, S., Lu, S., Ostrovsky, R.: Black-box garbled RAM. In: FOCS, pp. 210–229 (2015)

12. Garg, S., Lu, S., Ostrovsky, R., Scafuro, A.: Garbled RAM from one-way functions. In: STOC, pp. 449–458 (2015)

13. Gentry, C., Goldman, K.A., Halevi, S., Julta, C., Raykova, M., Wichs, D.: Optimizing ORAM and using it efficiently for secure computation. In: De Cristofaro, E., Wright, M. (eds.) PETS 2013. LNCS, vol. 7981, pp. 1–18. Springer, Heidelberg (2013)

14. Gentry, C., Halevi, S., Lu, S., Ostrovsky, R., Raykova, M., Wichs, D.: Garbled RAM revisited. In: Nguyen, P.Q., Oswald, E. (eds.) EUROCRYPT 2014. LNCS, vol. 8441, pp. 405–422. Springer, Heidelberg (2014)

15. Goh, E.-J.: Secure indexes. Cryptology ePrint Archive, Report 2003/216 (2003). http://eprint.iacr.org/2003/216/

16. Goldreich, O., Ostrovsky, R.: Software protection and simulation on oblivious RAMs. J. ACM 43(3), 431–473 (1996)

17. Goodrich, M.T., Mitzenmacher, M.: Privacy-preserving access of outsourced data via oblivious RAM simulation. In: Aceto, L., Henzinger, M., Sgall, J. (eds.) ICALP 2011, Part II. LNCS, vol. 6756, pp. 576–587. Springer, Heidelberg (2011)

18. Goodrich, M.T., Mitzenmacher, M., Ohrimenko, O., Tamassia, R.: Privacy-preserving group data access via stateless oblivious RAM simulation. In: SODA, pp. 157–167 (2012)

19. Gordon, S.D., Katz, J., Kolesnikov, V., Krell, F., Malkin, T., Raykova, M., Vahlis, Y.: Secure two-party computation in sublinear (amortized) time. In: CCS, pp. 513–524 (2012)

20. Islam, M.S., Kuzu, M., Kantarcioglu, M.: Access pattern disclosure on searchable encryption ramification, attack and mitigation. In: NDSS (2012)

21. Dautrich Jr., J.L., Stefanov, E., Shi, E.: Burst ORAM: minimizing ORAM response times for bursty access patterns. In: Usenix Security, pp. 749–764 (2014)

22. Kamara, S., Papamanthou, C.: Parallel and dynamic searchable symmetric encryption. In: FC, pp. 258–274 (2013)

23. Kamara, S., Papamanthou, C., Roeder, T.: Dynamic searchable symmetric encryption. In: CCS, pp. 965–976 (2012)
24. Kushilevitz, E., Lu, S., Ostrovsky, R.: On the (in)security of hash-based oblivious RAM and a new balancing scheme. In: SODA, pp. 143–156 (2012)
25. Lindell, Y., Pinkas, B.: A proof of security of Yao's protocol for two-party computation. J. Cryptol. **22**(2), 161–188 (2009)
26. Liu, C., Zhu, L., Wang, M., Tan, Y.-A.: Search pattern leakage in searchable encryption: attacks and new construction. Inf. Sci. **265**, 176–188 (2014)
27. Lu, S., Ostrovsky, R.: Distributed oblivious RAM for secure two-party computation. In: Sahai, A. (ed.) TCC 2013. LNCS, vol. 7785, pp. 377–396. Springer, Heidelberg (2013)
28. Lu, S., Ostrovsky, R.: How to garble RAM programs? In: Johansson, T., Nguyen, P.Q. (eds.) EUROCRYPT 2013. LNCS, vol. 7881, pp. 719–734. Springer, Heidelberg (2013)
29. Moataz, T., Mayberry, T., Blass, E.-O.: Constant communication ORAM with small blocksize. In: CCS, pp. 862–873 (2015)
30. Shen, E., Shi, E., Waters, B.: Predicate privacy in encryption systems. In: Reingold, O. (ed.) TCC 2009. LNCS, vol. 5444, pp. 457–473. Springer, Heidelberg (2009)
31. Shi, E., Chan, T.-H.H., Stefanov, E., Li, M.: Oblivious RAM with $O((\log N)^3)$ worst-case cost. In: Lee, D.H., Wang, X. (eds.) ASIACRYPT 2011. LNCS, vol. 7073, pp. 197–214. Springer, Heidelberg (2011)
32. Song, D.X., Wagner, D., Perrig, A.: Practical techniques for searches on encrypted data. In: IEEE Symposium on Security and Privacy, pp. 44–55 (2000)
33. Stefanov, E., Papamanthou, C., Shi, E.: Practical dynamic searchable encryption with small leakage. In: NDSS (2014)
34. Stefanov, E., van Dijk, M., Shi, E., Fletcher, C.W., Ren, L., Xiangyao, Y., Devadas, S.: Path ORAM: an extremely simple oblivious RAM protocol. In: CCS, pp. 299–310 (2013)
35. van Liesdonk, P., Sedghi, S., Doumen, J., Hartel, P., Jonker, W.: Computationally efficient searchable symmetric encryption. In: Jonker, W., Petković, M. (eds.) SDM 2010. LNCS, vol. 6358, pp. 87–100. Springer, Heidelberg (2010)
36. Wang, X.S., Hubert Chan, T.-H., Shi, E., Circuit, O.: On tightness of the Goldreich-Ostrovsky lower bound. In: CCS, pp. 191–202 (2015)
37. Williams, P., Sion, R.: Single round access privacy on outsourced storage. In: CCS, pp. 293–304 (2012)
38. Yao, A.C.: Protocols for secure computations (extended abstract). In: FOCS (1982)
39. Zhang, Y., Katz, J., Papamanthou, C.: All your queries are belong to us: the power of file-injection attacks on searchable encryption. In: Usenix Security (2016)

Bounded Indistinguishability
and the Complexity of Recovering Secrets

Andrej Bogdanov[1]([✉]), Yuval Ishai[2,3], Emanuele Viola[4],
and Christopher Williamson[1]

[1] Chinese University of Hong Kong, Hong Kong, China
{andrejb,chris}@cse.cuhk.edu.hk
[2] Technion, Haifa, Israel
yuvali@cs.technion.ac.il
[3] UCLA, Los Angeles, USA
[4] Northeastern University, Boston, USA
viola@ccs.neu.edu

Abstract. Motivated by cryptographic applications, we study the notion of *bounded indistinguishability*, a natural relaxation of the well studied notion of bounded independence.

We say that two distributions μ and ν over Σ^n are *k-wise indistinguishable* if their projections to any k symbols are identical. We say that a function $f:\Sigma^n \to \{0,1\}$ is *ϵ-fooled by k-wise indistinguishability* if f cannot distinguish with advantage ϵ between any two k-wise indistinguishable distributions μ and ν over Σ^n.

We are interested in characterizing the class of functions that are fooled by k-wise indistinguishability. While the case of k-wise independence (corresponding to one of the distributions being uniform) is fairly well understood, the more general case remained unexplored.

When $\Sigma = \{0,1\}$, we observe that whether f is fooled is closely related to its approximate degree. For larger alphabets Σ, we obtain several positive and negative results. Our results imply the first efficient secret sharing schemes with a high secrecy threshold in which the secret can be reconstructed in AC^0. More concretely, we show that for every $0 < \sigma < \rho \leq 1$ it is possible to share a secret among n parties so that any set of fewer than σn parties can learn nothing about the secret, any set of at least ρn parties can reconstruct the secret, and where both the sharing and the reconstruction are done by constant-depth circuits of size poly(n). We present additional cryptographic applications of our results to low-complexity secret sharing, visual secret sharing, leakage-resilient cryptography, and eliminating "selective failure" attacks.

1 Introduction

For a finite alphabet Σ, a distribution μ over Σ^n is *k-wise independent* if its projection to every k coordinates is uniform. There is a large body of work

---

A full version of this paper appears in [8].

M. Robshaw and J. Katz (Eds.): CRYPTO 2016, Part III, LNCS 9816, pp. 593–618, 2016.
DOI: 10.1007/978-3-662-53015-3_21

studying bounded independence, namely, the conditions under which a given function $f: \Sigma^n \to \{0, 1\}$ cannot distinguish between any distribution on n bits that is k-wise independent and the uniform distribution with advantage ϵ, for various choices of ϵ and k. Classes of functions that are fooled by bounded independence include combinatorial rectangles [23], small-depth circuits [7,9,32, 40,45], and sign polynomials [19,20], to name a few.

In this work we consider a relaxation of bounded independence that we call *bounded indistinguishability*. Two distributions μ and ν over Σ^n are *k-wise indistinguishable* if for all subsets $S \subseteq [n]$ of size k, the projections $\mu|_S$ and $\nu|_S$ of μ and ν to the coordinates in S are identical. For instance, if μ (resp., ν) is uniform over n-bit strings whose parity is 0 (resp., 1), then μ and ν are both $(n-1)$-wise independent and hence are also $(n-1)$-wise indistinguishable. However, if we let $\mu' = \mu \circ \mu$ (i.e., a concatenation of two identical copies of μ) and similarly $\nu' = \nu \circ \nu$, then μ' and ν' are still $(n-1)$-wise indistinguishable but are not even 2-independent.

Bounded indistinguishability arises naturally in cryptographic applications that involve secret sharing or secure multiparty computation. We will be interested in the complexity of distinguishing between two k-wise indistinguishable distributions.

Definition 1. *For $\epsilon \in (0, 1)$, we say that a function $f: \Sigma^n \to \{0, 1\}$ is ϵ-fooled by k-wise indistinguishability if for any two k-wise indistinguishable distributions μ and ν over Σ^n, $|\Pr[f(\mu) = 1] - \Pr[f(\nu) = 1]| \leq \epsilon$.*

Our goal is to understand which functions f are fooled by k-wise indistinguishability. For instance, polylogarithmic independence fools all AC^0 circuits [9]. Is this also the case for polylogarithmic indistinguishability?

We start by observing that over the binary alphabet $\Sigma = \{0, 1\}$, whether f is fooled by k-wise indistinguishability is closely related to the *approximate degree* of f, a notion introduced in the seminal work of Nisan and Szegedy [35]. This connection is central to our work so we formalize it next. The ϵ-approximate degree of a function $f: \{0, 1\}^n \to \{0, 1\}$ is defined to be the smallest degree of a real-valued polynomial $p: \{0, 1\}^n \to \mathbb{R}$ such that $|f(x) - p(x)| \leq \epsilon$ for every $x \in \{0, 1\}^n$.

Theorem 1. *For every n, k, $\epsilon \in (0, 1)$, and $f: \{0, 1\}^n \to \{0, 1\}$, the following are equivalent:*

1. *f is not ϵ-fooled by k-wise indistinguishability.*
2. *The $\epsilon/2$-approximate degree of f is bigger than k.*

Proof. It follows from linear programming duality (see for example Sect. 3 in [42] or Theorem 1 in [11]) that 2. is equivalent to the following statement:

3. *There exists a function $g: \{0, 1\}^n \to \mathbb{R}$ such that (i) $\sum_{x \in \{0,1\}^n} g(x) f(x) > \epsilon/2$, (ii) $\sum_x |g(x)| = 1$, and (iii) $\sum_x g(x) \prod_{i \in S} x_i = 0$ for every set $S \subseteq [n]$ of size at most k (including the empty set).*

We now show that 1. and 3. are equivalent. To see that 1. implies 3., we assume without loss of generality that $\Pr[f(\mu) = 1] - \Pr[f(\nu) = 1] > \epsilon$ and set $g(x) = \frac{1}{2C}(\mu(x) - \nu(x))$, where C is the statistical distance between μ and ν. The first two requirements for g are immediate. The third requirement follows from k-wise indistinguishability of μ and ν.

To see that 3. implies 1., set $\mu(x) = 2\max\{g(x), 0\}$ and $\nu(x) = 2\max\{-g(x), 0\}$. Since $\sum g(x) = 0$ and $\sum |g(x)| = 1$, we have $\sum \mu(x) = \sum \nu(x) = 1$ and so μ and ν are probability distributions. Condition (i) implies that $\Pr[f(\mu) = 1] - \Pr[f(\nu) = 1] > \epsilon$. Finally, by linearity we have that condition (iii) implies that μ and ν are indistinguishable by k-juntas so they are k-wise indistinguishable. $\qquad\square$

As a corollary, we get a similar connection between being *non-trivially* fooled by bounded indistinguishability and *threshold degree*, a notion introduced in the classical work of Minsky and Papert [33]. Recall that the threshold degree of a function $f:\{0,1\}^n \rightarrow \{0,1\}$ is the smallest degree of a real-valued polynomial $p:\{0,1\}^n \rightarrow \mathbb{R}$ such that the sign of $p(x)$ corresponds to $f(x)$ for every $x \in \{0,1\}^n$.

Corollary 1. *For every n, k and $f:\{0,1\}^n \rightarrow \{0,1\}$, the following are equivalent:*

1. *There is a pair of k-wise indistinguishable distributions μ, ν that are perfectly distinguished by f, namely $|\Pr[f(\mu) = 1] - \Pr[f(\nu) = 1]| = 1$.*
2. *The threshold degree of f is bigger than k.*

Combining the above with known results on approximate degree, we conclude that bounded indistinguishability over $\Sigma = \{0,1\}$ behaves very differently from bounded independence. For example, $O(1)$-wise independence suffices to $1/3$-fool the OR function on n bits, but $\Omega(\sqrt{n})$-wise indistinguishability is required, due to the corresponding lower bound on the approximate degree of OR [35]. This answers the aforementioned question of whether polylogarithmic indistinguishability fools AC^0 in the negative. A separation of $\Omega(n)$ is achieved by the Majority function: $O(1)$-wise independence suffices to $1/3$-fool this function [19], but $\Omega(n)$-wise indistinguishability is required by Paturi's lower bound [38].

We turn to study the case of larger alphabets Σ. Here the equivalence with previously studied notions seems to break down. We restrict the attention to alphabets of the form $\Sigma = \{0,1\}^s$, viewing the function f as being computed by a circuit with sn input bits. This setting comes up naturally in cryptographic applications, as explained below. But first we remark that, over such larger alphabets, we construct "simple" functions f that are *not* fooled by k-wise indistinguishability for much larger values of k than what is known for $\Sigma = \{0,1\}$. For example, over $\Sigma = \{0,1\}^{\text{poly}(n)}$ we show that $(n - n/\text{poly}\log n)$-wise indistinguishability does not $(1 - 2^{-n})$-fool AC^0 (Theorem 2), and that $0.99n$-wise indistinguishability does not 0.99-fool DNF (Corollary 10). In contrast, over alphabet $\Sigma = \{0,1\}$ it is only known that $\tilde{\Omega}(n^{2/3})$-wise indistinguishability does not fool AC^0 (by work of Aaronson and Shi [2] and Theorem 1).

1.1 Secret Sharing Schemes

A secret sharing scheme allows a dealer to share a secret between n parties, so that any k parties learn nothing about the secret from their shares whereas any r parties can reconstruct the secret from their shares. Unlike the case of *threshold secret sharing*, where $r = k + 1$, we allow a bigger gap between r and k. Such secret sharing schemes are often referred to as *ramp schemes*.

We are interested in the computational complexity of sharing and (especially) reconstructing secrets. A simple secret sharing scheme for $k = n - 1$ and $r = n$ shares a bit s into n bits $s_1, \ldots, s_n$ that are random subject to the restriction that their parity is s. This scheme cannot be implemented by constant depth circuits (in the class AC^0) as reconstruction requires computing the parity of n bits. Other secret sharing schemes, such as Shamir's [41], employ linear functions over finite fields and suffer from the same limitation.

A pair of k-wise indistinguishable distributions (μ, ν), together with a function f that *can* tell the two distributions apart, can be viewed as a secret sharing scheme for a one-bit secret: Shares of 0 and 1 are samples of μ and ν, respectively, and f is the reconstruction algorithm. Applying this connection together with techniques for sampling by constant-depth circuits, we obtain the following secret sharing scheme in the class AC^0.

Theorem 2 (Secret sharing in AC^0). *Let d be a constant. For every n and δ there exist:*

- *Sharing in AC^0: circuits S_0, S_1 of constant depth and size $\mathrm{poly}(n, \log 1/\delta)$ that sample $(n - n/(\log n)^d)$-wise indistinguishable distributions μ, ν over Σ^n, $\Sigma = \{0,1\}^{\mathrm{poly}(n)}$,*
- *Reconstruction in AC^0: a circuit R of size $\mathrm{poly}(n)$ and depth $d + O(1)$ such that $\Pr[R(\mu) = 0] \geq 1 - \delta$ and $\Pr[R(\nu) = 1] \geq 1 - \delta$.*

Moreover, the circuits S_0, S_1, and R can be constructed deterministically in time polynomial in n and $\log 1/\delta$.

Theorem 2 gives an explicit construction, but requires that all n parties participate in reconstruction. If one does not insist on a fully explicit construction and settles for a probabilistic construction that fails with negligible probability, the secrecy-recovery gap can be moved to an arbitrary location: In Theorem 13 we obtain an AC^0 secret sharing scheme that provides secrecy against any σn parties and allows reconstruction by any ρn parties for any pair of constants $0 \leq \sigma < \rho \leq 1$ and sufficiently large n.

We obtain several other schemes with incomparable features. If we do not insist on sharing in AC^0 and only require that reconstruction be done in AC^0, then we can achieve similar results with *perfect reconstruction* ($\delta = 0$). This variant builds on Corollary 1 and known results on the threshold degree of DNF [33]. Alternatively, we can strengthen Theorem 2 by allowing an AC^0 sharing algorithm that indicates failure with probability δ, but otherwise supports perfect reconstruction. In Corollary 10, we improve the reconstruction function complexity to a polynomial-size DNF formula (with terms of size $O(\log n)$), at the cost of a small constant reconstruction error and a slightly worse secrecy threshold.

Finally, we complement the above positive results with some negative results, showing limitations of secret reconstruction by disjunctions of juntas (Theorem 17) or small decision trees (Theorem 19). In particular, the negative results imply that the positive result of Corollary 10 for DNF reconstruction does not hold if the secrecy threshold is much closer to n or if the DNF is restricted to have a polynomial-size decision tree.

Techniques. In Sect. 2 we rephrase known results on approximate degree in the language of secret sharing using the connection in Theorem 1. The resulting schemes have AC^0 reconstruction, but achieve somewhat poor secrecy ($k \leq n^{2/3}$) and do not come with algorithms for sampling the shares. In Sect. 2.1 we show that the distributions of the shares can be sampled in AC^0. Then, in Sect. 2.2 we give a reduction that trades alphabet size for secrecy, allowing us to derive our main positive results. This reduction makes use of unbalanced disperser graphs. Our negative results, presented in Sect. 2.4, are obtained by reducing the large alphabet to a binary alphabet using a suitable set system, and then using Fourier analysis for obtaining the negative result in the binary case.

Related work. The randomized encoding technique of Applebaum et al. [6] can transform any secret sharing scheme into one where the shares are sampled by circuits in which each output depends on a fixed number of random bits (i.e., in the class NC^0), but at the cost of further increasing the complexity of reconstruction. Druk and Ishai [21] and Cramer et al. [16] consider the question of minimizing the circuit *size* of secret sharing. They construct near-threshold schemes (i.e., with $r = (1 + \epsilon) \cdot k$) in which sharing and reconstruction can be performed by circuits of size $O(n)$; however, the depth of these circuits is logarithmic in n. The above results left open the existence of nontrivial secret sharing schemes in which reconstruction can be done by constant depth circuits or by other "simple" nonlinear functions, even when the computational complexity of sharing the secret is unbounded.

1.2 Visual Cryptography

Naor and Shamir [34] initiated the study of "visual cryptography" — a method for sharing secrets which allows for a physical implementation using transparencies. It can be phrased as a secret sharing scheme with ℓ-bit shares, where reconstruction proceeds by first applying bitwise-OR to the shares and then applying an approximate threshold function (with constant fractional threshold gap). The bitwise-OR is implemented by physically stacking transparencies, and the approximate threshold function is implemented by visually distinguishing between ℓ-tuples of bits (pixels) that have a low Hamming weight and those that have a high Hamming weight. The ratio between the threshold gap and ℓ is referred to as the *contrast*.

It is known that the optimal contrast of such visual schemes vanishes exponentially with the secrecy parameter k [30,34], assuming that one requires sharp threshold reconstruction by any subset of $r = k + 1$ parties. The latter assumption has been made in all works on visual cryptography we are aware of.

In Sect. 2.3 we give a visual "ramp scheme" that allows a quadratic gap between the secrecy and reconstruction thresholds:

Theorem 3 (Visual Secret Sharing). *For every n and r there exists a pair of distributions μ, ν over $\{0,1\}^n$ that are $\Omega(\sqrt{r})$-wise indistinguishable so that for every subset $S \subseteq [n]$ of size r,*

$$|\Pr[\mathrm{OR}(\mu|_S) = 1] - \Pr[\mathrm{OR}(\nu|_S) = 1]| \geq 0.2.$$

Moreover, μ and ν are samplable by explicit circuits S_0, S_1 of constant depth and size polynomial in n.

The benefits are a dramatic improvement in contrast, making it independent of k and visually noticeable even for large k, as well as shorter (1-bit) shares and simpler reconstruction. The latter two properties are also achieved by other probabilistic visual schemes from the literature [15,31]. However, this is the first visual scheme whose (probabilistic) contrast does not vanish exponentially with k. To give a better sense of the achievable parameters, in Appendix A we give some specific parameter choices along with an image demonstrating the level of contrast we achieve.

1.3 Additional Cryptographic Applications

The above positive results for secret sharing rely on functions f that are *not* fooled by bounded indistinguishability. Such functions can be used to recover a secret from its shares. We observe that when f *is* fooled by bounded indistinguishability, this has positive consequences for leakage-resilient cryptography. Concretely, in every implementation of a cryptographic primitive that guarantees *local secrecy,* in the sense that different values of the underlying secrets induce k-wise indistinguishable distributions of the internal state, leaking the output of f on the internal state does not compromise the secrets.

Therefore *all* secret sharing schemes with a sufficiently high secrecy parameter k protect the secret against global leakage functions that output few bits, where each output bit has a low approximate degree (significantly smaller than k). More concretely:

Theorem 4. *There exists a universal constant C such that the following holds. Let μ, ν be k-wise indistinguishable distributions over $\{0,1\}^n$. Let $L:\{0,1\}^n \rightarrow \{0,1\}^t$ be a leakage function such that the $1/3$-approximate degree of each of its t outputs is at most d. Then the statistical distance between $L(\mu)$ and $L(\nu)$ is bounded by δ, provided that $k \geq Cdt(t + \log \frac{1}{\delta})$.*

This theorem can be applied to leakage functions whose outputs are computed by small decision trees or disjunctions of small juntas. It can also be applied to establish leakage resilience of protocols for secure multiparty computation and the related object of "private circuits." See Sects. 3.1 and 3.2 for more details and concrete applications.

Eliminating Selective Failure Attacks. The above applications can be relevant to any $f: \Sigma^n \rightarrow \{0,1\}$ that is fooled by bounded indistinguishability. We show that the special case where $f = \mathrm{OR}$ can be useful for eliminating so-called "selective failure" attacks. A selective failure attack is an attack that makes a computation fail only if the input satisfies some predicate. Such attacks enable an adversary to tamper with the computation and learn a bit of information about the secret input even when the tampering is detected and the output is replaced by an indication of failure. Selective failure attacks arise in different areas of cryptography and are often difficult to protect against.

We propose the following natural methodology for protecting against such attacks. Suppose that the computation of $g(w)$ can be reduced to n sub-computations $g_1(w_1), \ldots, g_n(w_n)$, where each k of the w_i jointly hide w. The computation of g via this reduction fails if at least one of the sub-computations fails. Assume further that an adversary tampers with each sub-computation g_i by choosing an arbitrary function of $F_i(w_i)$ that determines whether this sub-computation fails. Then, a corollary of Theorem 4 (with $t = 1$ and $L = \mathrm{OR}$) is that if $k \gg \sqrt{n}$ (the approximate degree of OR), then no tampering strategy can significantly correlate the event of failure with w. In the full version [8] we describe a simple concrete application of this methodology to eliminating selective failure attacks in error-detecting coding schemes.

Organization. In Sect. 2 we present our results on secret sharing. In Sect. 2.4 we prove our negative results and in Sect. 3 we give the details of some of the additional cryptographic applications described above. In Appendix D we discuss an approximate notion of bounded indistinguishability.

2   Secret Sharing

In this section we prove our results on secret sharing. Our starting observation is that bounded indistinguishability is closely related to the complexity of secret sharing. Specifically, the distributions μ and ν over Σ^n capture the joint distributions of shares obtained by sharing the secrets 0 and 1, respectively. The k-wise indistinguishability of the distributions corresponds to the parties gaining no information from any k shares. However, if bounded indistinguishability does *not* fool some function $f: \Sigma^n \rightarrow \{0,1\}$ we can think of f as the reconstruction function that maps the shares back to the secret.

In this setting it is natural to think of the distinguishing advantage as being close to (and ideally equal to) one. We will be interested in the complexity of the function f as well as the complexity of sampling μ and ν.

A different connection between secret sharing and approximation theory is obtained in the visual cryptography literature [34] (see also [30] and the citations therein). However, it was confined to analyzing the so-called contrast of visual cryptography schemes.

We give next a formal definition of secret sharing for a one-bit secret.[1]

Definition 2. *An (n, k, r) bit secret sharing scheme with alphabet Σ, reconstruction function $f \colon \Sigma^r \to \{0, 1\}$ and reconstruction advantage α is a pair of k-wise indistinguishable distributions μ and ν over Σ^n such that μ and ν are k-wise indistinguishable but for every set S of size r we have $\Pr[f(\mu|_S) = 1] - \Pr[f(\nu|_S) = 1] \geq \alpha$. Here $\mu|_S$ is the projection of μ to the symbols in S, and similarly for ν. The secret sharing scheme has* perfect reconstruction *if $\alpha = 1$. The scheme is* explicit *if f is explicit and there are explicit algorithms to sample μ and ν.*

As mentioned earlier, the distributions μ and ν are the joint distributions of shares obtained by sharing the secret 0 and 1, respectively. We sometimes omit reference to the alphabet when $\Sigma = \{0, 1\}$ and omit r from the notation when $r = n$.

We note that Item 1. in Theorem 1 is equivalent to the assertion that there exists an (n, k) bit secret sharing scheme (with $r = n$ and one-bit shares) with reconstruction function f having reconstruction advantage ϵ. Item 1. in Corollary 1 is equivalent to the assertion that there exists a similar scheme with perfect reconstruction.

Theorem 1, combined with the body of works on approximate and threshold degree immediately gives the following consequences.

Corollary 2. *The following secret sharing schemes over $\Sigma = \{0, 1\}$ exist:*

1. *An $(n, \Omega(\sqrt{\delta n}))$ bit secret sharing scheme with reconstruction by OR with advantage $1 - \delta$, for any δ.*
2. *An $(n, \Omega(n))$ bit secret sharing scheme with reconstruction by majority with constant advantage.*
3. *An $(n, \Omega((n/\log n)^{2/3})$ bit secret sharing scheme with reconstruction by the element distinctness DNF and constant reconstruction advantage.*
4. *An $(n, \Omega(n^{1/3}))$ bit secret sharing scheme with perfect reconstruction by the DNF $AND_{n^{1/3}} \circ OR_{n^{2/3}}$.*
5. *An $(n, \Omega(\sqrt{n}))$ bit secret sharing scheme with perfect reconstruction by some AC^0 function.*

Proof. The schemes follows by Theorem 1 and the following works: 1. by Nisan and Szegedy [35] and refinements by Bun and Thaler [11] (Proposition 14); 2. by Paturi [38]; 3. by Aaronson and Shi [2]; 4. by Minsky and Papert [33]; and 5. by Sherstov [43].

---

[1] Restricting the attention to a one-bit secret is without loss of generality; an ℓ-bit secret can be shared by invoking a scheme for a one-bit secret ℓ times in parallel.

These results show that for an interesting range of parameters, the reconstruction procedure of a secret sharing scheme can be implemented by simple functions, and in particular by constant depth circuits.

Bounded Independence Versus Bounded Indistinguishability. In many secret sharing schemes (e.g., Shamir's scheme [41] over a field of characteristic 2), the distributions μ and ν are not only k-wise indistinguishable but also k-wise *independent*. Such distributions cannot be distinguished by AC^0 functions and sign polynomials of degree 2 unless k is at most polylogarithmic in n. In contrast, the above examples give examples of k-wise indistinguishable distributions that are distinguishable by such function even when k grows polynomially with n.

Remark 5. *Aaronson [1] considers a different relaxation of bounded independence that has a dramatic effect on distinguishability by AC^0 functions. He considers distributions where for any k bits the probability that those bits take any fixed value is within $\epsilon 2^{-k}$ of 2^{-k} and gives a family of depth 3 polynomial-size circuit that distinguishes such a distribution from a uniform one with constant advantage for any k and $\epsilon = k \cdot \mathrm{poly} \log(n)/n$.*

2.1 Sampling the Shares in AC^0

In this section we show the existence of secret sharing schemes in which sharing the secret can be performed by constant-depth circuits, i.e., in the class AC^0, and reconstructing the secret can be done by a "simple" function. (As discussed in Sect. 1.1, the problem of minimizing the complexity of sharing alone is much simpler and can be solved via the techniques of [6].)

We start by showing how to sample distributions that are exponentially close to the k-wise indistinguishable distributions corresponding to the schemes we described. In Appendix C we give a refinement that gives distributions that are (exactly) k-wise indistinguishable, i.e., we achieve perfect secrecy.

Theorem 6. *For schemes 1. to 4. in Corollary 2 there exist pairs of circuit families of constant depth and size polynomial in n and $\log(1/\epsilon)$ that sample distributions within statistical distance ϵ of μ and ν, respectively.*

We leave the existence of efficient samplers for scheme 5. as an open question.

Note that we can achieve statistical distance $\epsilon = 2^{-n^c}$ for any constant c with circuits of size $\mathrm{poly}(n)$. The reason for this loss in statistical distance is that our distributions over the shares have probability masses that may not be powers of two, and so if we want to sample them using random bits we have to incur some slight error.

We now give the proof of this theorem. Our analysis relies on known explicit constructions of "dual polynomials," i.e., of the function g in Item 3. in Theorem 1. This area of research has been quite active since Špalek [44] gave the first explicit dual polynomial for OR.

Let Γ be a group of permutations acting on $[n]$. Then Γ also acts on $\{0,1\}^n$ by permuting the coordinates. The next claim is immediate.

Claim 7. *Let Γ be a group of permutations on $[n]$. Assume $f(x) = f(\sigma x)$ for all $x \in \{0,1\}^n$ and all $\sigma \in \Gamma$. If (μ, ν) is an (n, k, r) bit secret sharing scheme with reconstruction function f and advantage α, then so is $(\overline{\mu}, \overline{\nu})$ where*

$$\overline{\mu}(x) = \mathrm{E}_{\sigma \sim \Gamma}[\mu(\sigma x)] \quad \text{and} \quad \overline{\nu}(x) = \mathrm{E}_{\sigma \sim \Gamma}[\nu(\sigma x)].$$

In particular, if f is symmetric under permutation of its input coordinates, then the distributions μ and ν can be assumed to assign the same probability to all strings of the same Hamming weight. These $n + 1$ probabilities can be found in polynomial time by solving a linear program.

Moreover, we argue that in such a case μ is AC^0-samplable; the same argument applies to ν. Let μ' be the distribution on Hamming weights induced by μ. To sample from μ, we first sample a weight $w \in \{0, \ldots, n\}$ from μ', then output a random permutation of the string $1^w 0^{n-w}$. Both of these steps can be implemented in AC^0; cf. [47].

Therefore secret sharing with reconstruction by OR and majority can both be implemented in AC^0.

A description of the bit sharing scheme for element distinctness can be extracted from the work of Bun and Thaler [12]. They first construct a bit secret sharing scheme for a partial function f whose inputs are strings of length N over an alphabet Σ of size $O(N)$. In the yes inputs of f all symbols are distinct, while in the no inputs all symbols occur exactly twice. Their distributions μ and ν are supported on strings where m/a symbols occur exactly a times and $(N - m)/b$ symbols occur exactly b times for various choices of m, a, b.

We can represent the input to f as a binary string $x_1 \ldots x_N \in (\{0, 1\}^\Sigma)^N$, where x_i is an indicator vector for the i-th input symbol of f. Under this representation, f is a partial boolean function from $\{0, 1\}^{|\Sigma| \cdot N}$ to $\{0, 1\}$. By Claim 7 we may assume μ and ν are invariant over both permutations of the alphabet and permutations of the input positions. Now μ and ν can be sampled by first sampling (m, a, b) from the marginal distribution, then writing down an arbitrary string with the correct counts, and applying random permutations to both the alphabet and the input positions. All of these steps can be implemented in AC^0. The bit secret sharing scheme for OR is obtained by projecting the entries of μ and ν on random subsets of size n, which can also be implemented by sampling a random permutation.

An explicit description of the bit sharing scheme for the Minsky-Papert function can be extracted from the work of O'Donnell and Servedio [37] (Appendix A). They first sample an integer t of magnitude at most $n^{1/3}$ (even for μ, odd for ν) then choose an independent random string of Hamming weight $(t - i)^2$ in the i-th block. Both steps can be implemented in AC^0.

2.2 Trading Alphabet Size for Secrecy

We now give a general method of composing secret sharing schemes. We will apply this method to improve the secrecy of the above schemes at the cost of an increase in alphabet size and a slight increase in depth of the reconstruction. Our construction makes use of disperser graphs.

Definition 3. *A $n \times m$ bipartite graph G with left degree d is a (k, ϵ) disperser if any subset of $[n]$ of size k has at least $(1 - \epsilon)n$ neighbors in $[m]$.*

The loss in reconstruction efficiency is related to the degree d of the disperser. So we obtain the best results with Zuckerman's construction of dispersers with degree linear in $\log n/\epsilon$.

Theorem 8 (Theorem 1.9 of [48] with $\alpha = 1/2$). *For every constant δ, and for every n and ϵ there is an explicit (n^δ, ϵ) disperser G with $d = O(\log n/\epsilon)$ and $m = \delta n/2$.*

We now show how to turn an (n, k) secret sharing scheme L over alphabet $\{0, 1\}$ into a $(m, m - \epsilon m)$ secret sharing scheme R over alphabet $\{0, 1\}^n$. The alphabet is actually $\{0, 1\}^{d'}$ where d' is the maximum right-hand side degree of the disperser graph. It is possible to obtain d' close to the average degree nd/m, but in our settings this will always be $n^{\Omega(1)}$ and so for simplicity we do not optimize this parameter.

The parties of L and R are associated to the left and right vertices of the bipartite graph respectively. To share a bit in R, first sample shares for L and label each left vertex $v \in [n]$ by its corresponding share $s(r) \in \{0, 1\}$. Now for each of the d edges $e_1, \ldots, e_d$ incident to r, choose a bit $s(e_i)$ at random conditioned on $s(e_1) \oplus \cdots \oplus s(e_d) = s(r)$. The share $s(w)$ of each right vertex $w \in [m]$ is the concatenation of the edge-shares $s(e)$ over all its $\leq n$ incident edges e.

To reconstruct, apply the process in reverse: First distribute $s(w)$ for $w \in [m]$ to its incident edges, then calculate $s(v), v \in [n]$ as $s(e_1) \oplus \cdots \oplus s(e_d)$ and output $f(s(1), \ldots, s(n))$, where f is the reconstruction function of L.

Lemma 1. *If G is a (k, ϵ) disperser graph and L is a (n, k) secret sharing scheme then R is a $(m, m - \epsilon m)$ secret sharing scheme with the same reconstruction advantage.*

Proof. It is easy to see that the reconstruction advantage is preserved. Next we argue secrecy.

For contradiction, assume that L is k-secret but R is not $(n - \epsilon n)$-secret. Then there exists a subset $S \subseteq [m]$ of size $\leq m - \epsilon m$ such that the parties in S can distinguish shares of 0 from shares of 1. Consider the joint distribution of the shares assigned to all the edges incident to S. If any vertex $v \in [n]$ has a neighbor outside S, then the edge-shares associated to v's neighbors inside S are uniformly random and independent of all the other edge-shares incident to S (even conditioned on all the values $s(v)$). Therefore, the two distributions must be distinguishable even when restricted to those edges whose right vertices have all their neighbors in S. Let T be the set of all such right vertices. Then the shares of S in L are determined by the shares of T in R. By the disperser property of G, T has size at most k, so the shares in T are indistinguishable, contradicting our assumption.

We note that Alon et al. [4] applied a similar construction to amplify the distance of linear error-correcting codes, while Damgård et al. [18] used it (in more general form) for improving the tolerance of multiparty computations. Both these applications make use of dispersers (in fact, expanders) G that are balanced (with $m = n$) and of constant degree d. In contrast, we apply it to unbalanced graphs whose left degree is logarithmic in the number of vertices.

If we set $k = n^\alpha$ for some constant $\alpha > 0$, we obtain the following consequence. Here $f \circ XOR_d$ denotes a function that can be computed by composing f by XORs over d inputs.

Theorem 9. *Let $\alpha > 0$ be a constant. Suppose that there exists a (n, n^α) secret sharing scheme with reconstruction function $f:\{0,1\}^n \to \{0,1\}$ over alphabet $\{0,1\}$. Then there exists a $(m, (1 - \epsilon)m)$ secret sharing scheme over alphabet $\{0,1\}^n$ with reconstruction function of the type $f \circ XOR_d$ with $d = O((\log n)/\epsilon)$ and $m = \Omega(n^\alpha)$.*

We now have all the pieces to prove Theorem 2.

Proof (of Theorem 2). Instantiate Theorem 9 with Item 4 in Corollary 2. The reconstruction function involves computing parities on poly $\log n$ bits which can be done in AC^0. To sample the shares efficiently use Theorem 24.

Several other schemes are possible. We highlight the following one in which reconstruction is done by a DNF, although it is not perfect.

Corollary 10. *For every constant $\epsilon > 0$, there is an explicit $(n, (1 - \epsilon)n)$-secret sharing scheme with reconstruction error ϵ over the alphabet $\{0,1\}^{\text{poly}(n)}$ with reconstruction by a $\text{poly}(n)$-size DNF with terms of size $O(\log n)$.*

Proof. Instantiate Theorem 9 with Item 1 in Corollary 2. The reconstruction function is an OR of $O((n/\log n)^2)$ XORs of size $O((\log n)/\epsilon)$, which can be computed by a polynomial-size DNF. The shares can be sampled in AC^0 by Theorem 6.

2.3   Reconstruction by a Subset of the Parties

In this section we give several secret sharing schemes that allow for reconstruction by a subset of the parties. Our starting point is the secret sharing scheme with reconstruction by the OR function.

Claim 11. *For every r, δ, and n there is an explicit $(n, \Omega(\sqrt{\delta n}), r)$ bit secret sharing scheme with reconstruction by OR with advantage at least $r/n - \delta$.*

Here, by OR we mean the class of OR functions on subsets of r input bits. We will need the following fact which is implicit in the proof of Theorem 1.

Remark 12. *Without loss of generality, the distributions μ and ν can be assumed to have disjoint support.*

Proof (of Claim 11). Let (μ, ν) be any (n, k) bit sharing scheme for OR with reconstruction advantage $1 - \delta$. By Remark 12 and Claim 7 we may assume μ and ν are disjoint and symmetric, so $\nu(0^n) = 1 - \delta$ and all strings in the support of μ have nonzero Hamming weight. For any subset of r parties, the probability that they jointly observe a nonzero entry under ν is then at most δ. By symmetry of μ, the probability that they observe nonzero entry under μ is at least r/n. Therefore $\Pr[f(\mu) = 1] - \Pr[f(\nu) = 1] \geq r/n - \delta$.

If we set $\delta = r/2n$ we obtain an $(n, \Omega(\sqrt{r}), r)$ bit secret sharing scheme with reconstruction by OR with advantage $\delta = r/2n$. In the next result we make this advantage constant.

We now prove Theorem 3, namely the existence of a $(n, \Omega(\sqrt{r}), r)$ bit secret sharing scheme with reconstruction by OR with constant advantage.

Proof (of Theorem 3). First we construct a scheme over alphabet $\{0, 1\}^{1/\delta}$ for $\delta = 2n/r$ which we assume to be an integer. To share a zero and a one respectively, sample $1/\delta$ independent shares using the scheme in Claim 11 and give the i-th party the i-th bit from each copy. By the proof of Claim 11 for any $\Omega(\sqrt{r})$ parties the OR of their i-th copies of their shares of one and zero evaluate to 1 with probability at least $1 - (1 - 2\delta)^{1/\delta}$ and at most $1 - (1 - \delta)^{1/\delta}$, respectively. The difference between these two numbers is always positive and tends to $1/e - 1/e^2$ as $1/\delta$ increases.

To reduce the alphabet to binary, we replace each party's share by the OR of its constituent bits.

If we allow for more complexity in reconstruction and larger shares, the gap between the secrecy and reconstruction parameters can be improved and the reconstruction error can be made negligible.

Theorem 13. *For every pair of constants $0 \leq \sigma < \rho \leq 1$ and sufficiently large m there exists a $(m, \sigma m, \rho m)$ bit secret sharing scheme with reconstruction by circuits of size polynomial in m and depth 4 and advantage $1 - 2^{-m^c}$ for any constant c over alphabet $\Sigma = \{0, 1\}^{\mathrm{poly}(m)}$.*

To prove Theorem 13, we apply the composition method from Sect. 2.2 using a bipartite graph with the following dispersion properties.

Claim 14. *For all constants $\delta > 0$ and $0 \leq \sigma < \rho \leq 1$, and every sufficiently large n there exist numbers $m = n^{\Omega(1)}$, $r \leq n$, and $d = O(\log n)$ and an $n \times m$ bipartite graph G with left degree d such that*

1. *For every subset $S \subseteq [m]$ of size at most σm, the set of vertices in $[n]$ all of whose neighbors are in S has size at most r^δ (i.e., G is a $(r^\delta, 1-\sigma)$-disperser), and*
2. *For every subset $R \subseteq [m]$ of size at least ρm, the set of vertices in $[n]$ all of whose neighbors are in R has size at least r.*

We then amplify the reconstruction error in Theorem 3 using the following claim.

Claim 15. *For every integer t, if there exists a (m, k, r) bit secret sharing scheme with reconstruction by size s and depth d circuits and constant advantage over alphabet Σ then there exists a (m, k, r) bit secret scheme with reconstruction by circuits of size $st + \mathrm{poly}(t)$ and depth $d + 2$ and advantage $1 - 2^{-\Omega(t)}$ over alphabet Σ^t.*

Proof (of Theorem 13). We apply the construction described in Sect. 2.2 to the $(n, \Omega(\sqrt{r}), r)$ scheme from Theorem 3 and the graph from Claim 14 with $\delta = 0.49$. Secrecy follows from Theorem 9. Reconstruction proceeds as in Sect. 2.2, except that only those parties in $[n]$ that have received all of their shares participate in the process. By property 2 of Claim 14, if at least ρm parties on the right participate in the reconstruction then at least r parties on the left receive all their share and the secret is reconstructed with constant advantage. By Claim 15 with $t = m^c$, the advantage can be amplified to $1 - 2^{-m^c}$ as desired.

Proof (of Claim 14). We show that a random graph has both properties with nonzero probability. Choose each of the d neighbors of each left vertex independently and uniformly at random. For a fixed set $S \subseteq [m]$ of size σm, the expected number of left vertices all of whose neighbors are in S equals $n\sigma^d$. By the multiplicative Chernoff-Hoeffding bound and a union bound, the probability that there exists a set S and a set of left vertices of size $2n\sigma^d$ all of whose neighbors are in S is at most $2^m \exp(-n\sigma^d/8)$. By a similar argument, the probability that there exists a set $R \subseteq [m]$ of size ρm such that fewer than $n\rho^d/2$ vertices have all their neighbors in R is at most $2^m \exp(-n\rho^d/3)$.

We set $d = \log_{\rho^\delta/\sigma}(2^{1+\delta}n^{1-\delta})$, $r = (\rho/\sigma)^{d/(1-\delta)}$, and $m = \lfloor r^\delta/2 \rfloor$. This choice of parameters ensures that $n\rho^d/2 = r$, $2n\sigma^d = r^\delta$, and $r, m = n^{\Omega(1)}$. Moreover, both probabilities of interest tend to zero at the rate $\exp(-\Omega(r^\delta)) = \exp(-n^{\Omega(1)})$ so a graph with the desired properties exists for sufficiently large n.

Proof (of Claim 15). For every pair of constants $0 \leq \ell < h \leq 1$, Ajtai [3] shows the existence of a Boolean function family $ApxMaj$ of depth 3 and size polynomial in its input such that $ApxMaj$ accepts all strings of relative Hamming weight at least h and rejects all strings of relative Hamming weight at most ℓ. These circuits are made explicit in [46].

Let S be the assumed secret sharing scheme. Choose h and ℓ so that the success probability of reconstructing a one from its shares in S bounds h strictly from above and the failure probability of reconstructing a zero in S bounds ℓ strictly from below. To share a bit, sample k independent copies of shares of S and give the i-th party the i-th bit of each copy. To perform the reconstruction, first apply the reconstruction algorithm for S for each copy, then apply $ApxMaj$ to all k reconstructed bits.

The secrecy of S is inherited by construction. We now analyze the probability of correct reconstruction by r parties. By the multiplicative Chernoff bound, the probability that fewer than hk copies of S reconstruct a one correctly, or that more than ℓk copies of S reconstruct a zero incorrectly, is $2^{-\Omega(k)}$. If this does not happen, $ApxMaj$ correctly recovers the secret bit.

2.4   Limitations

In this section we prove negative results on the existence of secret sharing schemes, or equivalently positive results on functions being fooled by bounded indistinguishability. Our main technical contribution consists of proving negative results that hold even over large alphabets Σ. However, we first start with the case $\Sigma = \{0,1\}$ because this provides motivation and is useful for larger Σ.

In the case $\Sigma = \{0,1\}$ we note an upper bound of $n(1 - 1/\text{poly} \log n)$ on the approximate-degree of AC^0. While it follows from standard Fourier-analytic techniques, we are not aware that it has been observed before. In terms of secret sharing schemes it shows that the secrecy is at most $n(1 - 1/\text{poly} \log n)$ if reconstruction is to be done in AC^0.

Claim 16. *Every function $f:\{0,1\}^n \to \{0,1\}$ that has a size s depth d circuit has $n^{-h/2}$-approximate degree $n - h$ for $h = \Omega_d(n/(\log s)^{d-1}(\log n))$.*

Proof. We will work with the function $F:\{-1,1\}^n \to \{-1,1\}$ given by $F(X) = 1-2f((1+X)/2)$. We construct a polynomial $P:\{-1,1\}^n \to \mathbb{R}$ that approximates F pointwise within $2n^{-h/2}$. Let

$$P(X) = \sum_{S \subseteq [n], |S| \leq n-h} \hat{F}(S) \prod_{i \in S} X_i,$$

where $\hat{F}(S) = E[F(X) \prod_{i \in S} X_i]$ are the Fourier coefficients of F, see e.g. O'Donnell's book [36] for background.

Håstad [24] shows that $|\hat{F}(S)| \leq 2^{-c|S|/(\log s)^{d-1}}$, where c is some constant that depends only on d. For every $X \in \{-1,1\}^n$,

$$|F(X) - P(X)| = \left| \sum_{S:|S|>n-h} \hat{f}(S) \prod_{i \in S} X_i \right|$$
$$\leq \sum_{S:|S|>n-h} |\hat{f}(S)| \leq n^h \cdot 2^{-c(n-h+1)/(\log s)^{d-1}},$$

which is at most $2n^{-h/2}$ for $h = \min\{n/2, cn/4(\log s)^{d-1}(\log n)\}$.

The following upper bound on the approximate degree of the OR function was obtained by Kahn et al. [29]. The special case $\delta = 1/3$ was first established by Nisan and Szegedy [35].

Lemma 2. *For every n and δ, the δ-approximate degree of OR on n bits is $O(\sqrt{n \log(1/\delta)})$.*

It follows from Theorem 1 that there does not exist a $(n, \omega(\sqrt{n \log(1/\delta)})$ secret sharing scheme over the alphabet $\{0,1\}$ with reconstruction by OR and advantage δ.

We now derive two negative consequences for secret sharing schemes with more complex reconstruction functions and over alphabets of arbitrary size.

Theorem 17. *For every Σ of the form $\{0,1\}^s$ and all n, m, d, h such that $h \leq n/(3 \ln n \cdot \exp(6\sqrt{\ln(2m) \cdot \ln d}))$ if $f \colon \Sigma^n \to \{0, 1\}$ is an OR of m functions each of which depends on at most d inputs then there is no $(n, n - h)$ secret sharing scheme with reconstruction function f and advantage $1/3$.*

In particular, Theorem 17 shows that if reconstruction is done by a DNF of size $m = \mathrm{poly}(n)$ and with terms of size $d = n^{o(1)}$ then the secrecy must be at most $n - h = n - n^{1-o(1)}$.

The proof of the theorem relies on the following combinatorial claim.

Claim 18. *For every N, M, n, m, d, and h such that $h \ln n, M \ln N + 1 \leq n/(3d^M(2m)^{M/N})$ and every collection $\mathcal{S}$ of m subsets of $[n]$, each of size d, there exists a collection $\mathcal{T}$ of N subsets of $[n]$ such that*

1. *for every set $S \in \mathcal{S}$ there is at least one set $T \in \mathcal{T}$ such that S is a subset of T, and*
2. *for every M sets $T_1, \ldots, T_M \in \mathcal{T}$, $|T_1 \cup \cdots \cup T_M| < n - h$.*

Proof (of Theorem 17). Suppose for contradiction that such a secret sharing scheme S exists. Let $S_i \subseteq [n]$ be the set of variables in the i-th term of f and $\mathcal{S} = \{S_1, \ldots, S_n\}$. For $N = \log_d(2m)$, $M = 2\sqrt{N}$, and sufficiently large n the set system $\mathcal{T} = \{T_1, \ldots, T_N\}$ given by Claim 18 exists. Assign to each term t of f a single set $T(t) \in \mathcal{T}$ that covers it as guaranteed by Property 1 of the Claim.

Consider the following N-party secret sharing scheme T for OR. To share, first run the secret sharing for S and evaluate each term t of f using the shares as inputs. Then assign each party i in T the OR of all the terms t such that $T(t) = T_i$. To reconstruct take the OR of all the shares of T. By construction, this equals f evaluated on the shares of S, so T has the same reconstruction advantage as S.

By Property 2 of Claim 18, each collection of M parties of T observes fewer than $n - h$ shares of S, so T is an (N, M) secret sharing scheme. By Lemma 2 T cannot have reconstruction advantage $1/3$, so neither can S.

Proof (of Claim 18). We choose the M sets of $\mathcal{T}$ at random such that each element in $[n]$ is included in each set in $\mathcal{T}$ independently with probability $1 - q$ for $q = (1/d)(1/2m)^{1/N}$. On the one hand, by a union bound, the probability that some set $S \in \mathcal{S}$ fails to be covered by any set of $\mathcal{T}$ is at most $m(qd)^N$, which is at most $1/2$ by our choice of q. On the other hand, by a union bound, the probability that property 2 is violated is at most

$$\binom{N}{M} \cdot \binom{n}{n-h} \cdot (1 - q^M)^{n-h} \leq \exp(M \ln N + h \ln n - (n-h)q^M)$$

$$\leq \exp(M \ln N + h \ln n - (2n/3)q^M)$$

$$\leq 1/e$$

by the assumed inequality. By a union bound, both desired properties are satisfied with probability at least $1 - 1/2 - 1/e > 0$.

Next we obtain a stronger negative result in the case in which the reconstruction is done by a decision tree.

Theorem 19. *Let $\Sigma = \{0,1\}^s$. If $f:\Sigma^n \to \{0,1\}$ has a binary decision tree with at most S leaves then there is no $(n, \omega(\sqrt{n \log(S/\epsilon)}))$-bit secret sharing scheme with reconstruction function f and advantage ϵ.*

In particular, a secret sharing scheme with constant advantage and whose reconstruction function is a polynomial-size decision tree can only be secure against coalitions of $O(\sqrt{n \log n})$ parties.

Proof. First assume f is an OR of a subset of literals. If a secret sharing scheme with reconstrcution function f, secrecy parameter $\omega(\sqrt{n(\log 1/\delta)})$, and advantage δ existed, then a scheme with the same parameters would exist for a binary alphabet as each party's shares can be replaced by the respective OR of the relevant literals, contradicting Lemma 2. By symmetry the same conclusion holds for ANDs of subsets of literals.

If f has a decision tree with $\leq S$ leaves, then we can write f as a sum of at most S ANDs of literals, one for each path in the decision tree that leads to a 1-leaf. This sum is over the reals yet it will always take a boolean value because at most one AND will evaluate to one. If there existed a secret sharing scheme with reconstruction function f, advantage ϵ and the desired properties, by a hybrid argument one of the constituent ANDs would have advantage ϵ/S in the same scheme. Setting $\delta = \epsilon/S$ yields the desired conclusion.

3   Additional Cryptographic Applications

In this section we present additional applications of our results on bounded indistinguishability in cryptography. These applications can be viewed as different instances of *leakage-resilient cryptography*.

The broad goal of leakage-resilient cryptography is to maintain the security of cryptographic primitives even if partial information about their secrets is leaked to an adversary. The type of information being leaked is typically captured by a *leakage function* $L:\{0,1\}^n \to \{0,1\}^t$ taken from a leakage class $\mathcal{L}$, where the input for L represents the internal (secret) state of the primitive and its output represents the partial information available to the adversary. For simplicity we will start by considering the case of single-bit leakage (i.e., $t = 1$) and later extend the results to the more general case.

Our motivating observation is that if two possible distributions of secret states are k-wise indistinguishable, and moreover k-wise indistinguishability implies $\mathcal{L}$-indistinguishability, then obtaining leakage-resilience against $\mathcal{L}$ reduces to obtaining resilience against k-*local* leakage, namely the class of all projection functions $P:\{0,1\}^n \to \{0,1\}^k$. Obtaining provable security against k-local leakage is typically much easier than obtaining provable security against bigger leakage classes, and can be achieved via standard techniques for secret sharing and secure multiparty computation (MPC).

The above observation may be relevant to any cryptographic scheme that maintains a sufficient level of local secrecy. We illustrate its usefulness by presenting applications in the contexts of secret sharing, error detecting codes, and private circuits.

3.1 Leakage-Resilience of Secret Sharing Schemes

The implication 1. $\implies$ 2. in Theorem 1 can be reformulated in the following equivalent way.

Claim 20. *Let* μ, ν *be* k-*wise indistinguishable distributions over* $\{0,1\}^n$. *Let* $L{:}\{0,1\}^n \to \{0,1\}$ *be a leakage function whose* ϵ-*approximate degree is at most* k. *Then*

$$|\Pr[L(\mu) = 1] - \Pr[L(\nu) = 1]| \leq \epsilon.$$

Claim 20 implies that *every* (m, k) bit secret sharing scheme over $\Sigma = \{0,1\}^\ell$ is resilient against leakage functions $L{:}\{0,1\}^{m\ell} \to \{0,1\}$ whose approximate degree is at most k. The same holds for secret sharing schemes with bigger secrets.

Many secret sharing schemes from the literature are in fact k-wise independent for a large value of k, in the sense that any k bits in μ and ν are uniformly distributed. This is the case, for instance, for Shamir's scheme [41] over fields of characteristic 2. In such a case one can appeal to stronger results about bounded independence. For instance, Braverman's theorem [9] implies resilience to every AC^0 leakage function L even when k is polylogarithmic in n, whereas the approximate degree of some AC^0 functions is known to be as big as $\Omega(n^{2/3})$. One could also apply similar results in the case of *biased* k-wise independence, namely μ and ν are k-wise indistinguishable and moreover every k bits are independently distributed (but may each have a different bias). See, e.g., Lemma 5.2 in [14] for the case of OR distinguishers.

However, there are cases in which it is undesirable or even impossible to guarantee a high level of independence. For instance, when considering secret sharing schemes with special properties, such as ones supporting multiplication, bounded independence may come at a significant price [13,39]. Alternatively, the shares of a k-wise independent secret sharing scheme may be subject to local encoding or to adversarial tampering, after which they are no longer k-wise independent but are still k-wise indistinguishable.

Finally, we extend Claim 20 to the case of a leakage function L with t output bits. For convenience, we restate Theorem 4 from the Introduction.

Theorem 21. *There exists a universal constant* C *such that the following holds. Let* μ, ν *be* k-*wise indistinguishable distributions over* $\{0,1\}^n$. *Let* $L{:}\{0,1\}^n \to \{0,1\}^t$ *be a leakage function such that the* $1/3$-*approximate degree of each of its* t *outputs is at most* d. *Then the statistical distance between* $L(\mu)$ *and* $L(\nu)$ *is bounded by* δ, *provided that* $k \geq Cdt(t + \log \frac{1}{\delta})$.

Proof. Using an indistinguishability variant of Vazirani's statistical XOR lemma (cf. [27, Lemma 1]), it suffices to prove that every $L':\{0,1\}^n \rightarrow \{0,1\}$ obtained by taking the parity of a subset of the outputs of L, we have $|\Pr[L'(\mu) = 1] - \Pr[L'(\nu) = 1]| \leq \delta'$ where $\delta' = \delta \cdot 2^{-t/2}$. Using Lemma 4, the 1/3-approximate degree of each such L' is $O(dt)$ and by Lemma 3 its approximate degree is $O(dt \log \frac{1}{\delta'})$. Applying Claim 20, $k = \Omega(dt(t + \log \frac{1}{\delta}))$ suffices to guarantee that the distinguishing advantage of L' is bounded by δ' as required.

3.2 Private Circuits

We now describe an application of Claim 20 to *private circuits*, a computational model for leakage-resilient cryptography. We consider the simpler stateless variant of private circuits with encoded inputs and outputs (see, e.g., [28, Sect. 3] and [25, Sect. 4.1]) and privacy with respect to a general leakage class $\mathcal{L}$. Informally, such a private circuit is a (possibly randomized) boolean circuit that transforms a randomly encoded input into a randomly encoded output while providing the guarantee that the output of any $\mathcal{L}$-leakage on the n circuit wires reveals essentially nothing about the input. More formally:

Definition 4 ($((\mathcal{L}, \epsilon)$-Private Circuit). *A private circuit for $g:\{0,1\}^{n_i} \rightarrow \{0,1\}^{n_o}$ is defined by a triple (I, C, O), where*

- *$I:\{0,1\}^{n_i} \rightarrow \{0,1\}^{\hat{n}_i}$ is a randomized input encoder;*
- *C is a deterministic or randomized boolean circuit with input $\hat{w} \in \{0,1\}^{\hat{n}_i}$, output $\hat{y} \in \{0,1\}^{\hat{n}_o}$, and n wires;*
- *$O:\{0,1\}^{\hat{n}_o} \rightarrow \{0,1\}^{n_o}$ is a deterministic output decoder.*

For a leakage function $L:\{0,1\}^n \rightarrow \{0,1\}^t$ and $\epsilon > 0$, we say that (I, C, O) is an (L, ϵ)-private implementation of g if the following requirements hold.

- *Correctness: For any input $w \in \{0,1\}^{n_i}$ we have $\Pr[O(C(I(w))) = g(w)] = 1$, where the probability is over the randomness of I and (possibly) C.*
- *Privacy: For any $w, w' \in \{0,1\}^{n_i}$, the statistical distance between $L(C[I(w)])$ and $L(C[I(w')])$ is at most ϵ, where $C[x]$ denotes the (randomized) values of the n wires of C on input x.*

For a class $\mathcal{L}$ of leakage functions, we say that (I, C, O) is an $(\mathcal{L}, \epsilon)$-private implementation of g if it is an (L, ϵ)-private implementation of g for every $L \in \mathcal{L}$, and that it is a k-private implementation of g if it is an $(\mathcal{L}, 0)$-private implementation of g for the class $\mathcal{L}$ of projection functions that output k bits of the input.

Without any requirements on I and O, the above definition can be satisfied by having I compute a leakage-resilient secret sharing of the input which is passed by C directly to the decoder. To rule out such a solution we require the encoder and the decoder to be *universal* (i.e., depend only on n_i, n_o and the circuit size of g and not on g itself). Furthermore, we would like the decoder size to be considerably smaller than the circuit size of g. These requirements effectively force C to perform the bulk of the computation in a leakage-resilient manner.

While there are asymptotically efficient constructions of k-private circuits obtained via MPC techniques [17,25,28], much less is known about defending against larger leakage classes. We use the connection between approximate degree and bounded indistinguishability to bootstrap from k-private circuits to $(\mathcal{L}, \epsilon)$-private circuits for larger classes $\mathcal{L}$. More accurately, we show that in many cases k-privacy automatically implies $(\mathcal{L}, \epsilon)$-privacy for a large $\mathcal{L}$ and negligible ϵ. A similar result for a special type of leakage called "noisy leakage" was obtained in [22]. The parameters of the leakage-resilient circuits we obtain via bounded indistinguishability are quite limited, since our approach requires the privacy threshold k to be rather close to the circuit size. An interesting research direction is to obtain better parameters by exploiting additional structural properties of the distributions induced by private circuit constructions.

Combining MPC-based constructions of k-private circuits with known bounds on approximate degree, we obtain the following corollary (see [8] for proof):

Corollary 22. *Any NC-function $g:\{0,1\}^{n_i} \to \{0,1\}^{n_o}$ of circuit size s admits an $(\mathcal{L}, 2^{-\sigma})$-private implementation (I, C, O), where $|I| = \tilde{O}(s)$, $|C| = \tilde{O}(s)$, and $|O| = \tilde{O}(n_o + k)$, for the following choices of $\mathcal{L}$, σ, and k:*

1. *$\mathcal{L}$ is the class of decision trees of size S, $k = \sigma\sqrt{s\log(S)}$, and $\sigma \le \sqrt{s/\log(S)}$.*
2. *$\mathcal{L}$ is the class of read-once DNF (or CNF) formulas, $k = \sigma s^{1/2}$, and $\sigma \le s^{1/2}$.*
3. *$\mathcal{L}$ is the entire class AC^0, $k = \sigma s^c$, and $\sigma \le s^{1-c}$, assuming that all AC^0 functions on n-bit inputs have a $1/3$-approximate degree of $O(n^c)$ for some constant $c < 1$.*

Extension to Multi-bit Leakage. The above corollary can be extended to leakage functions L with t bits of output by relying on Theorem 4 instead of Claim 20. The general form of the corollary can be obtained by replacing each occurrence of σ with σt^2.

The Case of Disjunctive Leakage. Private circuits that resist *disjunctive leakage*, namely an OR of an arbitrary subset of wires or their negations, have found applications to constant-round secure two-party computation [26]. While it was shown in [26] that every k-private circuit can be transformed into such a disjunction-resilient circuit with a constant multiplicative overhead to the circuit size, this transformation is nontrivial and has a significant concrete cost. We note that for the purpose of this application it is essential that the encoder be small, and thus Corollary 22 is not useful even for the case of NC circuits.

Instead, we rely on the following corollary of Claim 20 to show that the same k-private circuits to which the transformation from [26] was applied are in fact already resilient against disjunctive leakage.

Claim 23. *Let μ, ν be k-wise indistinguishable distributions over Σ^n for $\Sigma = \{0,1\}^{\ell}$. Let $L:\{0,1\}^{\ell n} \to \{0,1\}$ be a disjunctive leakage function. Then*

$$|\Pr[L(\mu) = 1] - \Pr[L(\nu) = 1]| \leq 2^{-\Omega(k/\sqrt{n})}.$$

Proof. By decomposing L into n disjunctive functions that operate separately on each ℓ-bit symbol, $L(\mu)$ and $L(\nu)$ can be written as $OR(\mu')$ and $OR(\nu')$ (respectively), where μ' and ν' are k-wise indistinguishable distributions over $\{0,1\}^n$. The claim then follows from Claim 20 and the approximate degree of OR. □

The k-private circuits employed in [26] are based on MPC protocols that resist a constant fraction of corrupted parties. As such, they have the property that their N wires can be partitioned into n "symbols" in $\Sigma = \{0,1\}^{N/n}$, such that the wire distributions on different inputs are k-wise indistinguishable over Σ for $k = \Omega(n)$. Thus, Claim 23 implies that these k-private circuits achieve a good level of disjunctive resilience without any modification.

Acknowledgements. We thank Daniel Wichs for helpful discussions. Andrej and Emanuele thank Chin Ho Lee for putting them in touch. Emanuele thanks Daniel Wichs for asking whether bounded indistinguishability fools AND, and Mark Bun and Justin Thaler for many discussions about the approximate degree literature.

The first and fourth authors were supported by RGC GRF grants CUHK410113 and CUHK14208215. The second author was supported by ERC starting grant 259426, ISF grant 1709/14, and BSF grant 2012378. Research done in part while visiting the Simons Institute for the Theory of Computing, supported by the Simons Foundation and by the DIMACS/Simons Collaboration in Cryptography through NSF grant #CNS-1523467. Research also supported from a DARPA/ARL SAFEWARE award, NSF Frontier Award 1413955, NSF grants 1228984, 1136174, 1118096, and 1065276. This material is based upon work supported by the Defense Advanced Research Projects Agency through the ARL under Contract W911NF-15-C-0205. The views expressed are those of the author and do not reflect the official policy or position of the Department of Defense, the National Science Foundation, or the U.S. Government. The third author was supported by NSF grant CCF-1319206. Work done in part while a visiting scholar at Harvard University, with support from Salil Vadhan's Simons Investigator grant, and in part while visiting the Simons Institute for the Theory of Computing, supported by the Simons Foundation.

A   Parameters for Visual Scheme

We demonstrate some specific parameter choices for our visual secret sharing scheme. For given k and α, the corresponding entry in the next table gives the minimum value of n for which an (n, k) bit secret sharing scheme for OR with distinguishing advantage (i.e., contrast) α exists. To compute these exact parameters we formulated the problem as a linear program and used the CVXOPT linear programming solver to perform the calculation. The images were recovered from instantiations of the scheme with parameter settings $k = 8, n = 21$ and $k = 8, n = 46$, respectively.

k	$\alpha = 0.1$	$\alpha = 0.3$
2	3	4
3	4	8
4	7	13
5	9	19
6	13	26
7	16	35
8	21	46
9	26	57

B   Useful Properties of Approximate Degree

We rely on the following two lemmas on approximate degree. The first lemma (cf. [19, Claim 3.8]) shows that approximation quality can be traded for degree.

Lemma 3. *Let $0 < \epsilon' < \epsilon \leq 1/3$. Suppose that the ϵ-approximate degree of f is k. Then the ϵ'-approximate degree of f is $O(k \cdot \log \frac{\epsilon}{\epsilon'})$.*

The second lemma relates the approximate degree of the parity of t functions to a bound on their approximate degree. It follows by composing the functions using a "robust" polynomial for parity of degree $O(t)$ [10]. A simpler bound, obtained by applying Lemma 3 and multiplying the t approximations, adds an additional $\log t$-factor to the degree.

Lemma 4. *Let $f_1, f_2, \ldots, f_t$ be boolean functions whose $1/3$-approximate degree is at most k. Then the $1/3$-approximate degree of $f = f_1 \oplus f_2 \oplus \cdots \oplus f_t$ is $O(kt)$.*

C   Sharing in AC^0 with Perfect Secrecy

In this section we describe ways to maintain perfect secrecy while still generating the shares in AC^0. Let p be a distribution over $\{0,1\}^n$. We say that a distribution q over $\{0,1\}^n \cup \{\perp\}$ is ϵ-near p if $\Pr[q = \perp] \leq \epsilon$ and p equals $q | q \neq \perp$, i.e., q conditioned on the event $q \neq \perp$. We think of '$\perp$' as failure and we generally use the word 'near' to indicate sampling with failure.

Theorem 24. *For schemes 1. to 4. in Corollary 2 the following holds. Let μ and ν be the distributions on $\{0,1\}^n$ of the shares of 0 and 1 respectively. Let c be an integer. There exists explicit AC^0 circuits of size polynomial in n that sample distributions $\mu_\perp$ and $\nu_\perp$ such that:*

1. *(Secrecy) If μ and ν are k-wise indistinguishable then so are $\mu_\perp$ and $\nu_\perp$.*
2. *(Reconstruction) $\mu_\perp$ and $\nu_\perp$ are ϵ-near μ and ν, respectively, for $\epsilon = 2^{-n^c}$.*

By Item 1. we achieve perfect secrecy, and Item 2. guarantees that reconstruction works up to a small error.

Proof (of Theorem 24). For simplicity let us consider the scheme for OR. As mentioned earlier, in this case μ and ν are symmetric. Let μ' and ν' be the corresponding distributions on Hamming weights. By inspection of the dual polynomial for OR, see [44], the probability mass functions of μ' and ν' is at any point a multiple of $1/m$, where m is an integer with poly(n) bits.

We now describe the near sampler for μ. First, pick a uniform number in $\{0,1\}^{n'}$ where $n' \geq m$. If the number is bigger than m then output $\bot$. Otherwise, use that to compute a sample i of μ'. This involves computing '$\leq$', which can be done in AC^0. Then the task is to output a uniform string of Hamming weight i. Because AC^0 can nearly sample the uniform distribution of permutations of $[n]$, cf. [47], this uniform string can indeed be sampled. The same process is applied to ν.

Conditioned on not failing, the process is sampling μ and ν as desired. What remains to be seen is that the probability of outputting $\bot$ does not depend on whether we are nearly sampling μ or ν. This holds by inspection. Indeed, in the first step we fail in either case if we obtain a number that is larger than m. In the second the failure probability is that of the sampler of a uniform permutation, which is independent of which distribution we are sampling.

D   Exact vs. Almost Bounded Indistinguishability

In this Appendix we show that k-wise indistinguishability is "robust to noise" in the following sense: Any pair of distributions that are "almost" k-wise indistinguishable is close to a pair of truly k-wise indistinguishable distributions. Alon, Goldreich, and Mansour proved an analogous statement for k-wise independence (Theorem 2.1 in [5]).

Theorem 25. *Let μ and ν be two distributions on $\{-1,1\}^n$. Suppose that no test $T:\{-1,1\}^k \to \{0,1\}$ on k bits can distinguish μ and ν with advantage bigger than ϵ. Then there exist two distributions μ^* and ν^* such that μ^* has statistical distance $\leq 2\epsilon n^k$ from μ, ν^* has statistical distance $\leq 2\epsilon n^k$ from ν, and μ^* and ν^* are k-wise indistinguishable.*

Proof. For a subset I of $[n]$ let $\chi_I:\{-1,1\}^n \to \{-1,1\}$ be $\chi_I(x) = \prod_{i \in I} x_i$. It suffices to prove the conclusion for the tests χ_I where $|I| \leq k$. This is because if $\sum_x (\mu'(x) - \nu'(x))T(x) \geq \alpha$, then writing T in Fourier expansion we have $\sum_I \hat{T}_I \sum_x (\mu'(x) - \nu'(x))\chi_I(x) \geq \alpha$, and so there exists a test χ_I giving advantage at least $\alpha/2^k$.

For a function $f:\{-1,1\}^n \to \mathbb{R}$ we write $[f, I]$ for $\sum_x f(x)\chi_I(x)$, and call it the I coefficient of f. We "adjust" the coefficients of μ and ν by repeating the following step. Let $I \subseteq [n]$ be a non-empty subset of size at most k. By hypothesis, $|[\mu - \nu, I]| = \alpha \leq \epsilon$. Without loss of generality let $[\mu, I] \leq [\nu, I]$. Set $\mu':=\mu + \alpha(\chi_I + 1)/2^n$, and $\nu':=\nu + \alpha/2^n$. Now we have $[\mu' - \nu', I] = 0$, while $[\mu'-\nu', J] = [\mu-\nu, J]$ for $J \neq I$. Moreover, note that $\sum_x |\mu'(x)| = \sum_x |\mu(x)|+\alpha$, $\sum_x |\mu(x) - \mu'(x)| = \alpha$, and that the same holds for ν.

Repeating the adjustment $\leq n^k$ times, we get two non-negative functions μ' and ν' such that $[\mu' - \nu', I] = 0$ for every I of size at most k, and $\sum_x |\mu(x) - \mu'(x)| \leq \epsilon n^k$, and the same for ν', and also $\sum_x |\mu'(x)| = \sum_x |\nu'(x)| = 1 + \sigma$, for some $0 \leq \sigma \leq \epsilon n^k$.

Finally, let $\mu^* = \mu/(1 + \sigma)$ and $\nu^* = \nu/(1 + \sigma)$. We have $[\mu^* - \nu^*, I] = 0$ for every I of size at most k. The distance of $\mu*$ from μ is $\leq (1 + \sigma)^{-1}(\sum_x |\mu(x) - \mu^*(x)| + \sigma \sum_x \mu(x)) \leq 2\epsilon n^k$, and the same for ν. $\qquad\square$

References

1. Aaronson, S.: A counterexample to the generalized Linial-Nisan conjecture. Electronic Colloquium on Computational Complexity, Technical report 109 (2010)
2. Aaronson, S., Shi, Y.: Quantum lower bounds for the collision and the element distinctness problems. J. ACM **51**(4), 595–605 (2004)
3. Ajtai, M.: Approximate counting with uniform constant-depth circuits. In: Advances in Computational Complexity Theory, pp. 1–20 (1993)
4. Alon, N., Bruck, J., Naor, J., Naor, M., Roth, R.M.: Construction of asymptotically good low-rate error-correcting codes through pseudo-random graphs. IEEE Trans. Inf. Theor. **38**(2), 509–516 (1992)
5. Alon, N., Goldreich, O., Mansour, Y.: Almost k-wise independence versus k-wise independence. Inf. Process. Lett. **88**(3), 107–110 (2003)
6. Applebaum, B., Ishai, Y., Kushilevitz, E.: Cryptography in NC⁰. SIAM J. Comput. **36**(4), 845–888 (2006)
7. Bazzi, L.M.J.: Polylogarithmic independence can fool DNF formulas. SIAM J. Comput. **38**(6), 2220–2272 (2009)
8. Bogdanov, A., Ishai, Y., Viola, E., Williamson, C.: Bounded indistinguishability, the complexity of recovering secrets. Electronic Colloquium on Computational Complexity (ECCC), vol. 22, p. 182 (2015)
9. Braverman, M.: Polylogarithmic independence fools AC⁰ circuits. J. ACM **57**(5), 1–6 (2010)
10. Buhrman, H., Newman, I., Röhrig, H., de Wolf, R.: Robust polynomials and quantum algorithms. Theor. Comput. Syst. **40**(4), 379–395 (2007)
11. Bun, M., Thaler, J.: Dual lower bounds for approximate degree and Markov-Bernstein inequalities. In: Fomin, F.V., Freivalds, R., Kwiatkowska, M., Peleg, D. (eds.) ICALP 2013, Part I. LNCS, vol. 7965, pp. 303–314. Springer, Heidelberg (2013)
12. Bun, M., Thaler, J.: Dual polynomials for collision and element distinctness (2015). www.eccc.uni-trier.de/
13. Cascudo, I., Chen, H., Cramer, R., Xing, C.: Asymptotically good ideal linear secret sharing with strong multiplication over *any* fixed finite field. In: Halevi, S. (ed.) CRYPTO 2009. LNCS, vol. 5677, pp. 466–486. Springer, Heidelberg (2009)
14. Chari, S., Rohatgi, P., Srinivasan, A.: Improved algorithms via approximations of probability distributions. J. Comput. Syst. Sci. **61**(1), 81–107 (2000)
15. Cimato, S., Prisco, R.D., Santis, A.D.: Probabilistic visual cryptography schemes. Comput. J. **49**, 97–107 (2006)
16. Cramer, R., Damgård, I.B., Döttling, N., Fehr, S., Spini, G.: Linear secret sharing schemes from error correcting codes and universal hash functions. In: Oswald, E., Fischlin, M. (eds.) EUROCRYPT 2015. LNCS, vol. 9057, pp. 313–336. Springer, Heidelberg (2015)

17. Damgård, I., Ishai, Y., Krøigaard, M.: Perfectly secure multiparty computation and the computational overhead of cryptography. In: Gilbert, H. (ed.) EUROCRYPT 2010. LNCS, vol. 6110, pp. 445–465. Springer, Heidelberg (2010)
18. Damgård, I., Ishai, Y., Krøigaard, M., Nielsen, J.B., Smith, A.: Scalable multiparty computation with nearly optimal work and resilience. In: Wagner, D. (ed.) CRYPTO 2008. LNCS, vol. 5157, pp. 241–261. Springer, Heidelberg (2008)
19. Diakonikolas, I., Gopalan, P., Jaiswal, R., Servedio, R.A., Viola, E.: Bounded independence fools halfspaces. SIAM J. Comput. **39**(8), 3441–3462 (2010)
20. Diakonikolas, I., Kane, D., Nelson, J.: Bounded independence fools degree-2 threshold functions. In: Proceedings of 51st FOCS (2010)
21. Druk, E., Ishai, Y.: Linear-time encodable codes meeting the Gilbert-Varshamov bound and their cryptographic applications. In: Proceedings of ITCS 2014, pp. 169–182 (2014)
22. Duc, A., Dziembowski, S., Faust, S.: Unifying leakage models: from probing attacks to noisy leakage. In: Nguyen, P.Q., Oswald, E. (eds.) EUROCRYPT 2014. LNCS, vol. 8441, pp. 423–440. Springer, Heidelberg (2014)
23. Even, G., Goldreich, O., Luby, M., Nisan, N., Velickovic, B.: Efficient approximation of product distributions. Random Struct. Algorithms **13**(1), 1–16 (1998)
24. Håstad, J.: On the correlation of parity and small-depth circuits. SIAM J. Comput. **43**(5), 1699–1708 (2014)
25. Ishai, Y., Kushilevitz, E., Li, X., Ostrovsky, R., Prabhakaran, M., Sahai, A., Zuckerman, D.: Robust pseudorandom generators. In: Fomin, F.V., Freivalds, R., Kwiatkowska, M., Peleg, D. (eds.) ICALP 2013, Part I. LNCS, vol. 7965, pp. 576–588. Springer, Heidelberg (2013)
26. Ishai, Y., Kushilevitz, E., Ostrovsky, R., Prabhakaran, M., Sahai, A.: Efficient non-interactive secure computation. In: Paterson, K.G. (ed.) EUROCRYPT 2011. LNCS, vol. 6632, pp. 406–425. Springer, Heidelberg (2011)
27. Ishai, Y., Sahai, A., Viderman, M., Weiss, M.: Zero knowledge LTCs and their applications. In: Raghavendra, P., Raskhodnikova, S., Jansen, K., Rolim, J.D.P. (eds.) RANDOM 2013 and APPROX 2013. LNCS, vol. 8096, pp. 607–622. Springer, Heidelberg (2013)
28. Ishai, Y., Sahai, A., Wagner, D.: Private circuits: securing hardware against probing attacks. In: Boneh, D. (ed.) CRYPTO 2003. LNCS, vol. 2729, pp. 463–481. Springer, Heidelberg (2003)
29. Kahn, J., Linial, N., Samorodnitsky, A.: Inclusion-exclusion: exact and approximate. Combinatorica **16**(4), 465–477 (1996)
30. Krause, M., Simon, H.: Determining the optimal contrast for secret sharing schemes in visual cryptography. Comb. Probab. Comput. **12**(3), 285–299 (2003)
31. Kuwakado, H., Tanaka, H.: Image size invariant visual cryptography. IEICE Trans. Fundam. Electron. Commun. Comput. Sci. **82**(10), 2172–2177 (1999)
32. Linial, N., Nisan, N.: Approximate inclusion-exclusion. Combinatorica **10**(4), 349–365 (1990)
33. Minsky, M., Papert, S.: Perceptrons. MIT Press, Cambridge (1969)
34. Naor, M., Shamir, A.: Visual cryptography. In: De Santis, A. (ed.) EUROCRYPT 1994. LNCS, vol. 950, pp. 1–12. Springer, Heidelberg (1995)
35. Nisan, N., Szegedy, M.: On the degree of Boolean functions as real polynomials. Comput. Complex. **4**, 301–313 (1994)
36. O'Donnell, R.: Analysis of Boolean Functions. Cambridge University Press, Cambridge (2014)
37. O'Donnell, R., Servedio, R.A.: New degree bounds for polynomial threshold functions. Combinatorica **30**(3), 327–358 (2010)

38. Paturi, R.: On the degree of polynomials that approximate symmetric boolean functions (preliminary version). In: Proceedings of STOC 1992, pp. 468–474 (1992)
39. Randriambololona, H.: Asymptotically good binary linear codes with asymptotically good self-intersection spans. IEEE Trans. Inf. Theor. **59**(5), 3038–3045 (2013)
40. Razborov, A.A.: A simple proof of Bazzi's theorem. ACM Trans. Comput. Theor. (TOCT) **1**(1), 1–4 (2009)
41. Shamir, A.: How to share a secret. Commun. ACM **22**(11), 612–613 (1979)
42. Sherstov, A.A.: The pattern matrix method. SIAM J. Comput. **40**(6), 1969–2000 (2011)
43. Sherstov, A.A.: The power of asymmetry in constant-depth circuits. In: Proceedings of FOCS 2015 (2015)
44. Spalek, R.: A dual polynomial for OR (2008). CoRR, abs/0803.4516
45. Tal, A.: Tight bounds on The Fourier Spectrum of AC^0. Electronic Colloquium on Computational Complexity, Technical report TR14-174 (2014). www.eccc.uni-trier.de/
46. Viola, E.: On approximate majority and probabilistic time. Comput. Complex. **18**(3), 337–375 (2009)
47. Viola, E.: The complexity of distributions. SIAM J. Comput. **41**(1), 191–218 (2012)
48. Zuckerman, D.: Linear degree extractors and the inapproximability of max clique and chromatic number. Theor. Comput. **3**(1), 103–128 (2007)

Two-Message, Oblivious Evaluation of Cryptographic Functionalities

Nico Döttling[1], Nils Fleischhacker[2]([✉]), Johannes Krupp[2], and Dominique Schröder[2,3]

[1] University of California, Berkeley, USA
[2] CISPA, Saarland University, Saarbrücken, Germany
fleischhacker@cs.uni-saarland.de
[3] Friedrich-Alexander-University, Nuremberg, Germany

Abstract. We study the problem of two round oblivious evaluation of cryptographic functionalities. In this setting, one party P_1 holds a private key sk for a provably secure instance of a cryptographic functionality $\mathcal{F}$ and the second party P_2 wishes to evaluate $\mathcal{F}_{sk}$ on a value x. Although it has been known for 22 years that *general* functionalities cannot be computed securely in the presence of malicious adversaries with only two rounds of communication, we show the existence of a round optimal protocol that obliviously evaluates *cryptographic* functionalities. Our protocol is provably secure against malicious receivers under standard assumptions and does not rely on heuristic (setup) assumptions. Our main technical contribution is a novel *non*black-box technique, which makes *non*black-box use of the security reduction of $\mathcal{F}_{sk}$. Specifically, our proof of malicious receiver security uses *the code* of the reduction, which reduces the security of $\mathcal{F}_{sk}$ to some hard problem, in order to break that problem directly. Instantiating our framework, we obtain the first two-round oblivious pseudorandom function that is secure in the standard model. This question was left open since the invention of OPRFs in 1997.

1 Introduction

An oblivious evaluation protocol of a cryptographic functionality $\mathcal{F}$, is a two-party protocol in which one party P_1, called the sender, holds a function $\mathcal{F}_{sk} \in \mathcal{F}$ and the second party P_2, called the receiver, wishes to evaluate $\mathcal{F}_{sk}$ on x. Sender security says that P_1 remains oblivious of x while receiver security guarantees that the security of $\mathcal{F}_{sk}$ is preserved, i.e., evaluating $\mathcal{F}_{sk}$ obliviously should be as secure as having direct access to $\mathcal{F}$, even if a malicious party deviates from the protocol arbitrarily. Although it has been known for 22 years that *general* functionalities cannot be computed securely in the presence of malicious adversaries with only two rounds (messages) of communication [29], we show the existence of a two message protocol that obliviously evaluates *cryptographic* functionalities. The functionalities covered by our framework have the following properties:

- There is a security experiment Exp that characterizes the security of $\mathcal{F}$.
- The experiment Exp gives the adversary access to an oracle $\mathcal{O}$.

M. Robshaw and J. Katz (Eds.): CRYPTO 2016, Part III, LNCS 9816, pp. 619–648, 2016.
DOI: 10.1007/978-3-662-53015-3_22

– There is a black-box reduction $\mathcal{B}$ with certain properties that reduces the security of $\mathcal{F}_{sk}$ to a hard problem π.

Our framework subsumes popular two-party protocols, such as blind signatures and oblivious pseudorandom functions (OPRF). In fact, our framework yields the first OPRF with only two rounds of communication in the standard model — a problem that has been open since their invention in 1997 [49].

Technical Contribution. Our main technical contribution is a *non*black-box proof technique, which is *non*black-box *in the reduction*. To explain what being *non*black-box means, consider an instance P of a cryptographic functionality $\mathcal{F}$. Assume further that this instance is provably secure, i.e., there is a reduction $\mathcal{B}$ that turns any adversary $\mathcal{A}$ breaking the security of P into an algorithm solving the underlying hard problem π. Our protocol then shows that P can be evaluated securely. The corresponding proof of malicious receiver security makes *non*black-box use of the underlying *code* of the reduction $\mathcal{B}$. This proof does *not* reduce the security to P but to the underlying hard problem exploiting the code of $\mathcal{B}$. To best of our knowledge, this is the first result that shows how to make *non*black-box use of the code of a given security reduction. We call this class of reductions *oblivious reductions*.

1.1 Impossibility of Malicious Security and Induced Game-Based Security

Ideally one would like to achieve the standard notion of simulation based security against malicious adversaries. This notion says that the malicious receiver and sender learn only $f(x)$ (except what can trivially be learned from $f(x)$) and that the private input of the other party remains hidden. Unfortunately, it is well known that standard simulation based security notions cannot be achieved for two-round secure function evaluation (SFE) [29]. In fact, if one uses black-box techniques only, then at least five rounds of communication are necessary [36].

Since there is no hope in achieving malicious simulation-based security, we propose an alternative definition of malicious security for the setting of secure evaluation of cryptographic primitives: On a high-level, our security notions for malicious receiver says that the security properties of the underlying cryptographic primitive is preserved even against malicious adversaries. More precisely, we consider the secure evaluation of cryptographic functionalities, which are equipped with a game-based security notion. In our formalization the adversary in the corresponding security experiment has black-box access to the primitive. Then, we define an induced security notion by replacing black-box calls to the primitive in the security game with instances of the two-round SFE protocol. I.e., instead of giving the adversary black-box access to the primitive, it acts as a malicious receiver in an SFE session with the sender. Achieving this notion and showing that the underlying security guarantees are preserved is non-trivial, because the adversary is *not* semi-honest and may not follow the protocol.

Regarding security against corrupted senders, we show that malicious sender security *and* induced game-based security against malicious receivers cannot be achieved under (standard) non-interactive assumptions. In fact, our result is more general as it rules out protocols with three moves. Our impossibility is constructive and shows that our notion captures the standard definition of blind signatures. But for blind signatures it is well known that a large class of three-move blind signture schemes cannot be proven secure in the standard model under non-interactive assumptions [16]. Since our blind signature scheme belongs to this class, it follows that achieving both notions of malicious security is impossible. Thus, we also need to weakening the security against malicious senders and we stick to the standard notion of semi-honest security.

1.2   Oblivious Reductions: A Nonblack-Box Proof Technique

We give a high-level overview of our protocol and proof strategy. Our starting point is an instance $\mathcal{F}_{sk}$ of some cryptographic functionality $\mathcal{F}$ (such as the pseudorandom function functionality). The corresponding security proof is a *black-box* reduction $\mathcal{B}$ to some underlying hard problem π. Our goal is to obliviously compute $\mathcal{F}_{sk}$ in a secure two-party protocol Π with only two rounds of communication. Our protocol is simple and uses a certain type of homomorphic encryption and works as follows: The receiver encrypts its input x using the homomorphic encryption scheme, it sends the ciphertext $c \leftarrow \mathsf{Enc}(x)$ to the sender. The sender evaluates the function $\mathcal{F}_{sk}$ on c computing $c' \leftarrow \mathsf{Eval}(c, \mathcal{F}_{sk})$ and returns c' to the receiver, who obtains $\mathcal{F}_{sk}(x)$ by simply decrypting c'. Using fully homomorphic encryption as the underlying encryption scheme, it is well known that this protocol is secure against *semi-honest* adversaries [23].

However, we are interested in achieving *malicious* security and we achieve our goal using a specific type of homomorphic encryption scheme in combination with our novel *non*black-box proof technique. We provide an efficient reduction from the security of the homomorphically evaluated primitive to the underlying problem π directly using the code of the reduction $\mathcal{B}$. Our proof technique works for a large and natural class of black-box reductions that we call *oblivious*. Loosely speaking, a reduction is oblivious, if it only knows an upper bound on the number of the oracle queries, but does neither learn the query nor the answer. We give several examples of known oblivious reductions in Sect. 2.2 and we sketch the basic ideas of this technique in the following.

In the first step of our proof (see Fig. 1), we run a security experiment where the malicious receiver $\mathcal{A}$ has oracle access to a homomorphically evaluated functionality $\mathsf{Eval}(c, \mathcal{F}_{sk})$. In the second step, the experiment is transformed in the following way. First, the adversary's oracle inputs are extracted via an unbounded extractor, the functionality is evaluated on the extracted input, and finally the output is encrypted (with the right distribution). Assuming that the homomorphic encryption is (statistically) circuit private, we show that this modification does not change the success probability of the adversary. While extracting an input x from a ciphertext c is not possible in polynomial-time, it does not change the success probability of $\mathcal{A}$.

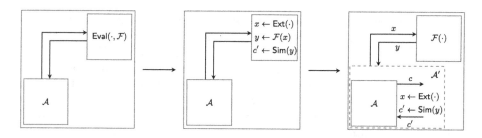

Fig. 1. Oblivious reduction part 1 of 2.

In the third step (see right picture of Fig. 1), we move the extraction and simulation procedures from the security experiment into the adversary itself, obtaining an unbounded adversary $\mathcal{A}'$. That is, the modified attacker $\mathcal{A}'$ runs $\mathcal{A}$ as a black-box. Whenever $\mathcal{A}$ sends c to its oracle, then $\mathcal{A}'$ extract x from c, invokes its own oracle obtaining $y \leftarrow F(x)$, and returns the encryption of y to $\mathcal{A}$. Obviously, the adversary $\mathcal{A}'$ does not run in polynomial-time, but this does not change its success probability, as we have only re-arranged the algorithms from one machine into another, but the overall composed algorithm remained the same.

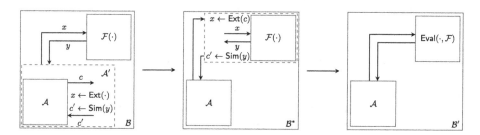

Fig. 2. Oblivious reduction part 2 of 2.

Consider the three steps shown in Fig. 2. In the first part, the unbounded adversary is plugged into the oblivious black-box reduction $\mathcal{B}$, which reduces the security of $\mathcal{F}$ to some hard problem π. This step is legitimate because the reduction only makes black-box use of the adversary. Observe that a black-box reduction cannot tell the difference between a polynomial-time adversary and an unbounded adversary, but only depends on the adversary's advantage in the security experiment. Thus, $\mathcal{B}^{\mathcal{A}'}$ is an inefficient adversary against the problem π. In our next modification we move the extraction and simulation algorithms from the adversary $\mathcal{A}'$ into the oracle-circuit. While this is just a bridging step, the inefficient algorithms for extraction and simulation are now part of the reduction. That is, whenever $\mathcal{A}$ queries c to its oracle, then the reduction $\mathcal{B}^*$ first extracts x from c and runs the simulation of $\mathcal{B}$ afterwards in order to compute the simulated answer $y \leftarrow \mathcal{F}_{sk}(x)$. Subsequently, $\mathcal{B}^*$ encrypts y as c' and sends this answer to $\mathcal{A}$. As a result, we obtain an inefficient reduction $\mathcal{B}^*$ that uses the code of the underlying reduction.

In the last step of our proof, we turn $\mathcal{B}^*$ into an *efficient* reduction $\mathcal{B}'$ against the underlying hard problem π (last picture in Fig. 2). Here, we again exploit the statistical circuit privacy of the homomorphic encryption scheme and replace the inefficient computation by the homomorphic evaluation of $\mathcal{F}$.

Running-Time of the Reduction. One may have the impression that we cheated in our proof by building a reduction that is not efficiently computable. This is not the case. A closer look at the formal proof reveals that the computationally inefficient steps are happening "inside" of the parts where we exploit the statistical circuit privacy. Thus, in some sense one may view this step as a game "in the head" while running an efficient reduction.

1.3   Our Contribution

The main contributions of this work are the following:

- We put forward the study of two-message secure evaluation of cryptographic functionalities.
- We propose a novel security model which says that the underlying security properties of the cryptographic functionality must be preserved, even if the malicious receiver does not follow the protocol.
- We show that security against malicious receivers with respect to our notion of induced game-based security and malicious senders cannot be achieved simultaneously in the standard model. In fact, our impossibility result is more general as it covers protocols with three moves.
- We suggest a protocol that is provably secure in this model under standard assumptions. The corresponding security proof relies on a novel *non*black-box technique that is *nonblack*-box in the reduction. We believe that this technique might be of independent interest.
- As an instance of our protocol, we present the first two-move oblivious pseudorandom function and solve a problem that was open since their invention in 1997.

1.4   Related Work

In this section, we discuss related works in the areas of secure two-party computation, round optimal oblivious PRFs and blind signatures.

Secure Two-Party Computation. The seminal works of Yao [58] and Goldreich et al. [28] show that any polynomial-time function can be securely computed in various settings. Recent works have shown protocols for secure two- and multi-party computation with practical complexity such as [7, 13, 44, 51]. A central measure of efficiency for interactive protocols is the round complexity. It was shown that secure two-party computation of arbitrary functionalities cannot be realized with only two rounds [29, 42, 43], and if the security proof uses black-box techniques only, then 5 rounds are needed [36]. On the other hand, several meaningful functionalities can be realized with only two (resp. less than five) rounds. Research in this area has gained much attention in the past and upper and lower bounds for many cryptographic protocols were discovered, such as for (concurrent) zero-knowledge proofs and arguments [5,15,26,27,29,56] and [10,39,54], blind signatures [16,19,20], as well as two- and multi-party computation [3,4,21,32,41,58] and [12,22,37].

Round Optimal Oblivious PRFs. Oblivious pseudorandom functions are in essence pseudorandom functions (PRFs) that are obliviously evaluated in a two-party protocol. This means that the sender S holds a key k of a PRF F and the receiver R a value x and wishes to learn $F(k, x)$. OPRFs have many applications, such as private key-word search [17], or secure computation of set intersection [34]. However, besides the popularity of this primitive, no scheme in the standard model is known with only two-rounds of communication. The first OPRF scheme was proposed by Naor and Reingold and it requires $\mathcal{O}(\lambda)$ rounds [49]. Freedman et al. [17] used previous work of Naor and Pinkas [46,47] to extend this to a constant round protocol assuming the hardness of DDH. Note that this protocol realizes a "weak PRF", which allows the receiver to learn information about the key k as long as this information does not change the pseudorandomness of future queries. Jarecki and Liu suggested the first round optimal OPRFs in the random oracle model [34].

Round Optimal Blind Signatures. A blind signature scheme [11] allows a signer to interactively issue signatures for a user such that the signer learns nothing about the message being signed (blindness) while the user cannot compute any additional signature without the help of the signer (unforgeability). Constructing round-optimal blind signature schemes in the standard model has been a long standing open question. Fischlin and Schröder showed that all previously known schemes having at most three rounds of communication, cannot be proven secure under non-interactive assumptions in the standard model via black-box reductions [16]. Subsequently, several works used a technique called "complexity leveraging" to circumvent this impossibility result [19,20] and recently, Fuchsbauer, Hanser, and Slamanig suggested a round optimal blind signature scheme that is secure in the generic group model [18]. In fact, it is still unknown if round optimal blind signatures, based on standard assumptions, exist in the standard model.

1.5   Outlook

Our work also shows that the "quality" of the proof has implication on the usability of the primitive in other contexts. In particular, having an oblivious black-box reduction, in contrast to a non-oblivious one, implies that the primitive can be securely evaluated in our framework while the underlying security is preserved. In fact, our results show a certain degree of composability of cryptographic functionalities and round optimal secure function evaluation.

Outline. We define our security model in Sect. 2. Our protocol is then given in Sect. 3. Section 4 shows how our result can be applied to achieve oblivious pseudorandom functions. The impossibility result is given in Sect. 4.

Notations. The security parameter is λ. By $y \leftarrow A(x; r)$ we refer to a (PPT) algorithm A that gets as input some value x and some randomness r and returns an output y. If X is a set, then $x \xleftarrow{\$} X$ means that x is chosen uniformly at random from X. The statistical distance $\Delta(A, B)$ of two probability distributions A and B is defined as $\Delta(A, B) = \frac{1}{2} \sum_v |Pr(A = v) - Pr(B = v)|$.

2   Secure Computation of Cryptographic Functionalities

In the following section, we formalize experiments, the corresponding notion of security of an experiment, oblivious black-box reduction, and our notion of secure computation of cryptographic primitives. Our formalization of experiments is similar to the one by Bellare and Rogaway [6], but our goal is to formalize oblivious reduction, i.e., reduction that only knows an upper number on the number of oracle queries made by an adversary and which does not see the actual queries to the oracle.

Please note that in the literature the term "round" has been used both to refer to a single message (either from A to B *or* from B to A) and to refer to two messages (one from A to B *and* one from B to A). Since none of the two seems to be favoured over the other, in this work we will stick to the former usage, i.e., a "round" refers a *single* message despite its direction.

2.1   Cryptographic Security Experiment

In this section, we formalize security experiments for cryptographic primitives P, where we view P as a collection of efficient algorithms. The basic idea of our notion is to define a framework, similar to the one of Bellare and Rogaway [6], for cryptographic experiments. Our framework provides some basic algorithm, such as initialization, an update mechanism, and a method to test if the adversary succeeds in the experiment. Moreover, it also define oracles that may be queried by the attacker. The most important aspect of our formalization is that the experiment is oblivious about the adversary's queries to its oracle. This means that the experiment may know an upper bound on the total number of queries, but does not learn the queries, or the corresponding answers.

Formally, the experiment consists of four algorithm. The first algorithm, Init, initializes the environment of the security experiment and computes publicly available informations pp and private informations st that may be hardcoded into the oracle that will be used by the attacker in the corresponding security notion. The algorithm Init receives a upper bound q on number of oracle queries as input. This is necessary because several security experiments, such as the one of blind signatures, require a concrete bound on the number of queries. This oracle, denoted by OA, obtains $(\mathsf{pp},\mathsf{st})$ and some query x, and it either returns an answer y, or $\perp$ to indicate failure. The update algorithm Update allows to re-program the oracle. The test algorithm Test checks the validity of some value out with respect to public and private informations pp and st, respectively.

Definition 1 (Security Experiment). *A security experiment for a cryptographic primitive* P *is a tuple of four algorithms defined as follows:*

Initialization. *The initialization algorithm* $\mathsf{Init}(1^\lambda, q)$ *gets as input the security parameter* 1^λ *and an upper bound* q *on the number queries. It outputs some public information* pp *together with some private information* st.

Oracle. *The oracle algorithm* $\mathsf{OA}(\mathsf{pp},\mathsf{st},x)$ *gets as input a string* pp, *state information* st, *and a query* x. *It answers with special symbol* $\perp$ *if the query is invalid, and otherwise with a value* y.

Update. *The stateful algorithm* $\mathsf{Update}(\mathsf{st},\mathsf{resp})$ *takes as input some state information* st *and a string* resp. *It outputs some updated information* st.

Testing. *The* $\mathsf{Test}(\mathsf{pp},\mathsf{st},\mathsf{out})$ *algorithm gets as input the input of the attacker* pp, *state information* st, *the output of the attacker* out, *and outputs a bit* b *signifying whether the attacker was successful.*

In almost all cases, the oracle OA embeds an algorithm from the primitive P, such as the signing algorithm in case of signature, or the encryption algorithm in case of the CPA (resp. CCA) security game. Given the formalization of a security experiment, we are ready to formalize the corresponding notion of security. Loosely speaking, a cryptographic primitive is secure, if the success probability of the adversary in this experiment is only negligible bigger than the guessing probability. Since our notions covers both computational and decisional cryptographic experiments, we follow the standard way of introducing a function ν that serves as a security threshold and which corresponds to the guessing probability. In our formalization, the adversary $\mathcal{A}$ is a stateful algorithm that runs r rounds of the security experiment. This algorithm is initially intitialized with an empty state $\mathsf{st}_{\mathcal{A}} := \emptyset$. Our formalization could also handle non-uniform adversaries by setting this initial state to some string.

Definition 2 (Security of a Cryptographic Primitive). *Let* $\mathsf{Exp} = (\mathsf{Init}, \mathcal{O}, \mathsf{Update}, \mathsf{Test})$ *be a security experiment for a cryptographic primitive* P, *and let* $\mathcal{A}$ *be an adversary having a state* $\mathsf{st}_{\mathcal{A}}$ *querying the oracle exactly once per invocation. Further let* $\nu : \mathbb{N} \to [0,1]$ *be a function. In abuse of notation, we denote by* $\mathsf{Exp}^\mathsf{P}(\mathcal{A})$ *the following cryptographic security experiment:*

> *Game* $\mathsf{Exp}^\mathsf{P}(\mathcal{A})$
> $\quad (\mathsf{pp},\mathsf{st}) \leftarrow \mathsf{Init}(1^\lambda, q)$
> $\quad \mathsf{st}_{\mathcal{A}} := \emptyset$
> $\quad \textit{for } i = 1 \textit{ to } q \textit{ do}$
> $\qquad (\mathsf{resp}_i, \mathsf{st}_{\mathcal{A}}) \leftarrow \mathcal{A}^{\mathcal{O}(\mathsf{pp},\mathsf{st},\cdot)}(\mathsf{pp}, \mathsf{st}_{\mathcal{A}})$
> $\qquad (\mathsf{pp},\mathsf{st}) \leftarrow \mathsf{Update}(\mathsf{st}, \mathsf{resp}_i)$
> $\quad \mathsf{out} := \mathsf{resp}_q$
> $\quad b \leftarrow \mathsf{Test}(\mathsf{pp},\mathsf{st},\mathsf{out})$
> $\quad \textit{Return } b$
>
> *Oracle* $\mathcal{O}(\mathsf{pp},\mathsf{st},x)$
> $\quad y \leftarrow \mathsf{OA}(\mathsf{pp},\mathsf{st},x)$
> $\quad \textit{Return } y$

We define the advantage of the adversary $\mathcal{A}$ as

$$\mathbf{Adv}^P(\mathcal{A}) := \left| \mathrm{Prob}\left[\mathsf{Exp}^P(\mathcal{A}) = 1 \right] - \nu(\lambda) \right|.$$

A cryptographic primitive is secure with respect to $\mathsf{Exp}^P(\mathcal{A})$ if the advantage $\mathbf{Adv}^P(\mathcal{A})$ is negligible (in λ).

Remark 1. Observe that in our formalization of a cryptographic security experiment, all algorithms, except for the adversary, are oblivious of the queries to the oracle. The reason is that the output of the oracle is returned to the adversary only and no other algorithm obtains this value. In particular, the update algorithm does not receive the output as an input and also the test algorithm, which determines if the attacker is successful, only receives pp, st, and out as an input and no input or output from OA.

The CCA Secure Encryption Experiment. Our formalization of cryptographic experiments covers standard security notions, such as CCA security for public-key encryption schemes (obviously, the adaption to CCA secure private-key encryption is trivial). Recall that a public-key encryption scheme HE = (Kg, Enc, Dec) consists of a key generation algorithm $(ek, dk) \leftarrow \mathsf{Kg}(1^\lambda)$, an encryption algorithm $c \leftarrow \mathsf{Enc}(ek, m)$, and a decryption algorithm $m \leftarrow \mathsf{Dec}(dk, c)$ and the corresponding security experiment of CCA is a two stage game. In the first stage, the attacker has access to a decryption oracle and may query this oracle on arbitrary values. Subsequently, the attacker outputs two messages of equal length and receives a challenge ciphertext that encrypts one of the messages depending on a randomly chosen bit b. In the second stage of the experiment, the attacker gets access to a modified decryption oracle that answers all queries, except for the challenge ciphertext. Eventually, the attacker outputs a bit b' trying to predict b and it wins the security experiment if its success probability is non-negligibly bigger than $1/2$.

In our formalization, the game of CCA security is a 2-round experiment. The initialization algorithm Init generates a key-pair (ek, dk) of a public-key encryption scheme, it chooses a random bit b, and sets $i = 1, r = 2$ and $c_b = \emptyset$. The public parameters pp contain (ek, i, r, c_b) and the private state is (dk, b). The input of the oracle OA is (pp, x), it parses pp as (ek, i, r, c_b) and behaves as follows: If $i = 1$, then it returns the decryption of x, i.e., it outputs $y = \mathsf{Dec}(dk, x)$. If $i = 2$, then OA outputs $\mathsf{Dec}(dk, x)$ if $x \neq c_b$, and $\perp$ otherwise. At some point, the adversary $\mathcal{A}$ outputs as its response $\mathsf{resp} = (m_0, m_1, \mathsf{st}_\mathcal{A})$ two challenges messages m_0, m_1 and some state information $\mathsf{st}_\mathcal{A}$. The update algorithm Update(st, resp, cnt) extracts b from st and updates the public parameters pp by replacing c_b with $c_b \leftarrow \mathsf{Enc}(ek, m_b)$ and by setting $i = 2$. Moreover, it stores the messages m_0 and m_1 in st. In the next stage of the experiment, the oracle OA returns $\perp$ when queried with c_b. Eventually, $\mathcal{A}$ outputs a bit b' as its response resp. The test algorithm Test extracts m_0, m_1, and b from st and b' from resp. It returns 0 if $|m_0| \neq |m_1|$ or if $b' \neq b$. Otherwise, it outputs 1.

The Unforgeability Experiment. The classical security experiment of existential unforgeability under chosen messages attacks for signature schemes is not covered by our formalization. The reason is that the testing algorithm outputs 1 if the forged message m^* is different from all queries $m_1, \ldots, m_i$ the attacker $\mathcal{A}$ queried to OA. Thus, the testing algorithm is clearly not oblivious of $\mathcal{A}$'s queries to OA. However, one can easily define a modified experiment that is implied by the classical experiment. Similar to the unforgeability notion of blind signatures, we let the attacker query the signing oracle q times and the attacker succeeds if it outputs $q + 1$ messages-signature pairs such that all messages are distinct and all signatures are valid. Clearly, giving a successful adversary against this modified game, one can easily break the classical notion by guessing which of the $q + 1$ pairs is the forgery.

2.2 Oblivious Black-Box Reductions

Hard Computational Problem. We recall the definition of hard computational problems due to Naor [45].

Definition 3 (Hard Problem). *A computational problem $\pi = (Ch, t)$ is defined by a machine Ch (the challenger) and a threshold function $t = t(\lambda)$. We say that an adversary $\mathcal{A}$ breaks the problem π with advantage ϵ, if*

$$\Pr[\langle Ch, \mathcal{A} \rangle = 1] \geq t(\lambda) + \epsilon(\lambda),$$

over the randomness of Ch and $\mathcal{A}$. If π is non-interactive, then the interaction between $\mathcal{A}$ consists of Ch providing an input instance to $\mathcal{A}$ and $\mathcal{A}$ providing an output to Ch. The problem π is hard if ϵ is negligible for all efficient adversaries $\mathcal{A}$.

All standard hardness assumptions used in cryptography can be modeled in this way, for instance the DDH assumption. The goal of a reduction is to show that the security of a cryptographic primitive P can be reduced to some underlying hard assumption. This is shown by contraposition assuming that the cryptographic primitive is insecure with respect to some security experiment. Then, the reduction gets as input an instance of the underlying hard problem, it runs a black-box simulation of the attacker and shows, via simulation of the security experiment, that it can use the adversary to solve the underlying hard problem. Since the problem is assumed to be hard, such an attacker cannot exist. A reduction is black-box if it treats the adversary as a black-box and does not look at the code of the attacker. A comprehensive discussion about the different types of black-box reductions and techniques is given in [55]. For our purposes we need a specific class of black-box reductions that we call *oblivious*. Loosely speaking, a black-box reduction is oblivious if it only knowns an upper bound on the number of oracle queries made by the attacker, but does neither know the query nor the answer. Intuitively, this motion allows the reduction to program the oracle once for each round of the security game.

Definition 4 (Oblivious Black-Box Reductions). *Let* P *be a cryptographic primitive with an associated security experiment* Exp. *Moreover, let* π *be a hard problem. Let* $\mathcal{B}$ *be an oracle algorithm with the following syntax.*

- $\mathcal{B}$ *is an adversary against the problem* π
- $\mathcal{B}$ *has restricted black-box access to a machine* $\mathcal{A}$, *which is an adversary for the security experiment* Exp
- $\mathcal{B}$ *gets as auxiliary input an upper bound* q *on the number of oracle queries* $\mathcal{A}$ *makes in each invocation.*

By restricted black-box access to $\mathcal{A}$ *we mean that* $\mathcal{B}$ *is allowed to program an oracle* $\mathcal{O}_{\mathcal{B}}$, *choose inputs* pp, st$_\mathcal{A}$ *and get the output* (resp, st$_\mathcal{A}$) $\leftarrow \mathcal{A}^{\mathcal{O}_{\mathcal{B}}(\cdot)}$(pp, st$_\mathcal{A}$). *As before, we assume that* $\mathcal{A}$ *queries its oracle exactly once per invocation (We stress that* $\mathcal{B}$ *does not see* $\mathcal{A}$'*s oracle queries).*

We say that $\mathcal{B}$ *is an oblivious black-box reduction from the security of* Exp *to* π *if it holds for every (possibly inefficient) adversary* $\mathcal{A}$ *against* Exp *that if* $\mathbf{Adv}_{\mathcal{A}}^{\mathsf{Exp}}(\lambda)$ *is non-negligible, then* $\mathbf{Adv}_{\mathcal{B}^{\mathcal{A}}}^{\pi}(\lambda)$ *is also non-negligible.*

2.3 Secure Function Evaluation for Cryptographic Primitives

In this section, we propose our security notions for two-round secure function evaluation of cryptographic primitives P. A two-round SFE protocol is a protocol between two parties, a sender $\mathcal{S}$ and a receiver $\mathcal{R}$. The sender provides as input a function f from a family $\mathcal{F}$ and the receiver an input x to the function. At the end of the protocol, the sender gets no output (except for a signal that the protocol is over), whereas the receiver's output is $f(x)$. The function that is realized by our SFE protocols is a function of the primitive P. Since we view P as a collection of algorithms, our SFE protocol evaluates the underlying functionality. For example, in the case of signature schemes this collection consists of a key generation, a signing, and a verification algorithm. Securely evaluating this primitive means to securely evaluate the signing algorithm.

In the following, we introduce our security definitions. Roughly speaking, receiver security says that the security of the underling cryptographic primitive is preserved. This property must hold even against malicious receivers. Moreover, our security notion for the sender holds with respect to semi-honest senders.

Induced Game-Based Malicious Receiver Security. Regarding security, ideally we would like to achieve that the receiver learns nothing but $f(x)$, which is usually modeled via standard simulation based security notions. However, it is well known that standard simulation based security notions fail in the regime of two-round secure function evaluation [29]. Thus, our goal is to achieve a weaker notion of security, which roughly says that the security of the underlying cryptographic primitive is preserved. More precisely, we consider the secure evaluation of cryptographic primitives, which are equipped with a game based security notion. In our formalization the adversary in the corresponding security experiment has black-box access to the primitive. Then, we define an induced security notion by replacing black-box calls to the primitive in the security game with instances of the two round SFE protocol. I.e., instead of giving the adversary black access to the primitive, it acts as a malicious receiver in an SFE session with the sender. Achieving this notion and showing that the underlying security guarantees are preserved is non-trivial, because the adversary is *not* semi-honest and may not follow the protocol.

Definition 5 (Induced Game-Based Malicious Receiver Security). *Let* $\mathsf{Exp} = (\mathsf{Init}, \mathcal{O}, \mathsf{Update}, \mathsf{Test})$ *be a cryptographic security experiment for a primitive* P. *Let* $\Pi = (\mathcal{S}, \mathcal{R})$ *be a two-round SFE protocol for a function F of* P. *The induced security experiment* Exp' *is defined by replacing* $\mathcal{O}$ *with instances of* Π, *where the adversary is allowed to act as a malicious receiver.*

In the following, we study the implications of our security notion with respect to the security of the underlying cryptographic primitive. It is not very difficult to see, that if a protocol is perfectly correct and securely realizes our notion of induced game-based security, then it immediately implies the security of the underlying cryptographic primitive. Second, one can also show that the converse is not true, by giving a counterexample. The basic idea of the counterexample is to build a two-round SFE protocol that completely leaks the circuit and thus the entire private input of the sender. The main result of our paper is a two-round SFE protocol that preserves the underlying security guarantees.

Semi-honest Sender Security. We define security against semi-honest senders via the standard simulation based definition [24].

Definition 6 (Semi-honest Sender Security). *Let* $\Pi = (\mathcal{S}, \mathcal{R})$ *be a two-party protocol for a functionality F. We say that* Π *is semi-honest sender secure, if there exists a PPT simulator* Sim *such that it holds for all receiver inputs x and all sender inputs f that*

$$(x, f, \mathsf{view}(\mathcal{S}), \langle \mathcal{S}, \mathcal{R}(x) \rangle) \stackrel{\text{COMP.}}{\approx} (x, f, \mathsf{Sim}(\mathsf{f}), f(x))$$

3   2-Round SFE via 1-Hop Homomorphic Encryption

In this section, we present our protocol and prove that it is induced game-based malicious receiver secure (Definitions 5) and semi-honest sender secure (Definition 6).

3.1   1-Hop Homomorphic Encryption

1-hop homomorphic encryption schemes are a special kind of homomorphic encryption schemes that allow a server to compute on encrypted data. Given a ciphertext c produced by the encryption algorithm Enc, the evaluation algorithm Eval can evaluate a circuit C from $\mathcal{C}$ on c. After this no further computation on the output ciphertext is supported. We recall the definition of 1-hop homomorphic encryption schemes and the corresponding notions of security [23].

Definition 7 (1-Hop Homomorphic Encryption). *Let* $C : \{0,1\}^n \rightarrow \{0,1\}^o$ *be a family of circuits. A 1-hop homomorphic encryption scheme* $\mathsf{HE} = (\mathsf{Kg}, \mathsf{Enc}, \mathsf{Dec}, \mathsf{Eval}, C_1, C_2)$ *for* C *consists of the following efficient algorithms:*

Key Generation. *The input of the key generation algorithm* $\mathsf{Kg}(1^\lambda)$ *is the security parameter* λ *and it returns an encryption key* ek *and a decryption key* dk.

Encryption. *The encryption algorithm* $\mathsf{Enc}(ek, m)$ *takes as input an encryption key* ek *and a message* $m \in \{0,1\}^n$ *and returns a ciphertext* $c \in C_1$.

Evaluation. *The evaluation algorithm* $\mathsf{Eval}(ek, c, C)$ *takes as input a public encryption key* ek, *a ciphertext* c *generated by* Enc *and a circuit* $C \in C$ *and returns a ciphertext* $c' \in C_2$.

Decryption. *The decryption algorithm* $\mathsf{Dec}(dk, c)$ *takes as input a private decryption key* dk *and a ciphertext* c' *generated by* Eval *and returns a message* $y \in \{0,1\}^o$.

We recall that the standard notions of completeness and compactness [23]. A homomorphic encryption scheme is complete if the probability of a decryption error is 0. It is compact if the size of the output ciphertext c' of the evaluation algorithm Eval is independent of the size of the circuit C. Moreover, we recall the standard notion of IND-CPA-security for homomorphic encryption schemes: Given a public key ek for the scheme, no PPT adversary succeeds to distinguish encryptions of two adversarially chosen messages m_0 and m_1.

For our purposes we need a homomorphic encryption scheme with malicious circuit privacy. This property says that even if both maliciously formed public key and ciphertext are used, encrypted outputs only reveal the evaluation of the circuit on some well-formed input x^*. We recall the definition in the following.

Definition 8 (Malicious Circuit Privacy). *A 1-hop homomorphic encryption scheme* $\mathsf{HE} = (\mathsf{Kg}, \mathsf{Enc}, \mathsf{Dec}, \mathsf{Eval}, C_1, C_2)$ *for a family* C *of circuits is (maliciously) circuit private if there exist unbounded algorithms* $\mathsf{Sim}_{\mathsf{HE}}(ek, c, y)$, *and deterministic* $\mathsf{Ext}_{\mathsf{HE}}(ek, c)$ *such that for all* λ, *and all* ek, *all* $c \in C_1$ *and all circuits* $C \in C$ *it holds that*

$$\mathsf{Sim}_{\mathsf{HE}}(ek, c, C(x)) \stackrel{\text{STAT.}}{\approx} \mathsf{Eval}(ek, C, c),$$

where $x = \mathsf{Ext}_{\mathsf{HE}}(ek, c)$.

Instantiations. We consider instantiations of maliciously circuit private 1-hop homomorphic encryption. Maliciously circuit private homomorphic encryption for logarithmic depth circuits can be achieved by combining information-theoretic garbled circuits (aka randomized encodings) [2,33,38] with two-message oblivious transfer [1,30,48].

Theorem 1 [1,2,30,33,38,48]. *Under numerous number-theoretic assumptions, there exist a non-compact maliciously circuit private homomorphic encryption scheme that support circuits of logarithmic depth.*

Ostrovsky et al. [52] provide a construction that bootstraps a maliciously circuit privacy scheme that supports only evaluation of logarithmic depth circuits into a scheme that supports all circuits (i.e., it is fully homomorphic).

Theorem 2 (Theorem 1 in [52]). *Assume there exists a compact semi-honest circuit private fully homomorphic encryption scheme* FHE *with decryption circuits of logarithmic depth and perfect completeness. Assume further that there exists a (non-compact) maliciously circuit private homomorphic encryption scheme for logarithmic depth circuits. Then there exists a maliciously circuit private fully homomorphic encryption scheme with perfect completeness.*

3.2   Construction

We can now state the two message SFE protocol. If f is a cryptographic functionality that takes input s from the sender, input x from the receiver and random coins r, we augment the functionality such that both parties contribute to the random coins. I.e., both parties also input random string r_S and r_R and the random coins for the functionality is set to $r_S \oplus r_R$.

Construction 1. *Let* HE *be a 1-hop homomorphic encryption scheme. The interactive protocol that realizes* $\mathcal{F} : (s, r_S, x, r_R) \to (\bot, f(s, r_S; x, r_R))$ *is shown in Fig. 3.*

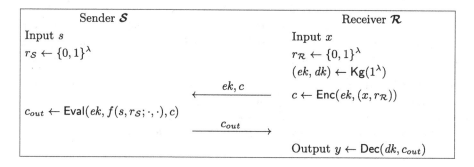

Fig. 3. Oblivious two-party protocol

The following theorem shows that security against malicious receivers with respect to our definition of induced game-based security.

Theorem 3. *Let* P *be a cryptographic primitive and* Exp *be the corresponding security experiment. If there exists an efficient oblivious black-box reduction* $\mathcal{B}$ *that reduces security of* P *to a hard problem* π, *then the protocol* Π *is secure with respect to* Exp'. *Formally, there exists an efficient reduction* $\mathcal{B}'$ *that reduces the security of* Π *to* π.

Proof. Assume there exists a PPT adversary $\mathcal{A}$ that has non-negligible advantage ϵ_1 in the security experiment Exp'.

Step 1. In the first step, we change the security experiment Exp' to an indistinguishable experiment Exp^*. In particular, we implement $\mathcal{A}$'s oracles differently. In Exp', the oracle gets a sender-input $(s, r_{\mathcal{S}})$ and a receiver-message (ek, c), computes $c' \leftarrow \mathsf{Eval}(ek, f(s, r_{\mathcal{S}}; \cdot, \cdot), c)$ and outputs c' to $\mathcal{A}$. In Exp^*, the oracle is implemented as follows. Given sender-input $(s, r_{\mathcal{S}})$ and a receiver-message (ek, c), the oracle first computes $(x, r_{\mathcal{R}}) \leftarrow \mathsf{Ext}_{\mathsf{HE}}(ek, c)$. Then it computes $y \leftarrow f(s, r_{\mathcal{S}}; x, r_{\mathcal{R}})$ and then $c' \leftarrow \mathsf{Sim}_{\mathsf{HE}}(ek, c, y)$ and finally outputs c' to $\mathcal{A}$

$\mathsf{OA}_1(ek, c)$
  $c' \leftarrow \mathsf{Eval}(ek, f(s, r_{\mathcal{S}}; \cdot, \cdot), c)$
  Return c'

$\mathsf{OA}_2(ek, c)$
  $(x, r_{\mathcal{R}}) \leftarrow \mathsf{Ext}_{\mathsf{HE}}(ek, c)$
  $y \leftarrow f(s, r_{\mathcal{S}}; x, r_{\mathcal{R}})$
  $c' \leftarrow \mathsf{Sim}_{\mathsf{HE}}(ek, c, y)$
  Return c'

We claim that $\epsilon_2 = \mathbf{Adv}_{\mathsf{Exp}^*}(\mathcal{A}) \geq \mathbf{Adv}_{\mathsf{Exp}'}(\mathcal{A}) - \mathsf{negl}(\lambda)$. We establish this via a hybrid argument. Assume that $\mathcal{A}$ makes at most $\ell = \mathsf{poly}(\lambda)$ oracle queries. Define $\ell + 1$ hybrid experiments $\mathcal{H}_0, \ldots, \mathcal{H}_\ell$. $\mathcal{H}_0$ simulates the oracle as in experiment $\mathsf{Exp}'(\mathcal{A})$, whereas $\mathcal{H}_\ell$ simulates it as in $\mathsf{Exp}^*(\mathcal{A})$. In $\mathcal{H}_i$ the first i oracle queries to $\mathcal{A}$ are answered as in $\mathsf{Exp}'(\mathcal{A})$, whereas the last $\ell - i$ oracle queries of $\mathcal{A}$ are answered as in $\mathsf{Exp}^*(\mathcal{A})$. It follows by the statistical circuit privacy of HE that the statistical distance between each $\mathcal{H}_i$ and $\mathcal{H}_{i+1}$ is at most ν for a negligible ν. Thus, by the triangle inequality the statistical distance between $\mathsf{Exp}'(\mathcal{A})$ and $\mathsf{Exp}^*(\mathcal{A})$ is at most $\ell \cdot \nu$, which is negligible. Note that the experiment Exp^* is not efficient anymore.

Step 2. The second step is a bridging step: We move both the extractor $\mathsf{Ext}_{\mathsf{HE}}$ and the simulator $\mathsf{Sim}_{\mathsf{HE}}$ into a new adversary $\mathcal{A}_2$, which internally simulates $\mathcal{A}$. The adversary $\mathcal{A}_2$ is an *unbounded* adversary against the experiment Exp with advantage ϵ_2. Adversary $\mathcal{A}_2$ works as follows. When adversary $\mathcal{A}$ sends an oracle query (ek, c), $\mathcal{A}_2$ computes $(x, r_{\mathcal{R}}) \leftarrow \mathsf{Ext}_{\mathsf{HE}}(ek, c)$ and sends x to its own oracle (in the Exp experiment). Once it receives an oracle output y, it computes $c' \leftarrow \mathsf{Sim}_{\mathsf{HE}}(ek, c, y)$ and forwards c' to $\mathcal{A}$.

Adversary $\mathcal{A}_2(\mathsf{pp}, \mathsf{st}_{\mathcal{A}})$
  Has access to oracle OA
  $(\mathsf{resp}, \mathsf{st}_{\mathcal{A}}) \leftarrow \mathcal{A}^{\mathsf{OA}'(\cdot)}(\mathsf{pp}, \mathsf{st}_{\mathcal{A}})$
  Return $(\mathsf{resp}, \mathsf{st}_{\mathcal{A}})$

Oracle $\mathsf{OA}'(ek, c)$
  $(x, r_{\mathcal{R}}) \leftarrow \mathsf{Ext}_{\mathsf{HE}}(ek, c)$
  $y \leftarrow \mathsf{OA}(x)$
  $c' \leftarrow \mathsf{Sim}_{\mathsf{HE}}(ek, c, y)$
  Return c'

We claim that $\mathsf{Exp}(\mathcal{A}_2)$ is identically distributed to $\mathsf{Exp}^*(\mathcal{A})$. To see this, note that we've just regrouped the algorithms $\mathsf{Ext}_{\mathsf{HE}}$ and $\mathsf{Sim}_{\mathsf{HE}}$ into $\mathcal{A}_2$ and removed the dependency of y from $r_{\mathcal{R}}$. However, since $f(s, r_{\mathcal{S}}; x, r_{\mathcal{R}})$ computes the function $F(s, x, r_{\mathcal{S}} \oplus r_{\mathcal{R}})$, the distribution of y does not depend on $r_{\mathcal{R}}$ (as $r_{\mathcal{S}}$ is chosen uniformly at random).

Step 3. In the third step, we combine the adversary $\mathcal{A}_2$ with the oblivious black-box reduction $\mathcal{B}$, which yields an (unbounded) adversary $\mathcal{B}^{\mathcal{A}_2}$ with non-negligible advantage ϵ_3 against the hard problem π (as $\epsilon_2 = \epsilon_1 - \mathsf{negl}(\lambda)$ is non-negligible).

Note at this stage that while the reduction $\mathcal{B}$ is efficient, the π-adversary $\mathcal{B}^{\mathcal{A}_2}$ is not efficient as $\mathcal{A}_2$ is not efficient.

Step 4. The fourth step is again a bridging step: We move the extractor $\mathsf{Ext}_{\mathsf{HE}}$ and the simulator $\mathsf{Sim}_{\mathsf{HE}}$ into the oracle simulated by $\mathcal{B}$, thus obtaining a new reduction $\mathcal{B}^*$. More precisely, $\mathcal{B}^*$ simulates $\mathcal{B}$, but when $\mathcal{B}$ invokes the adversary with input $(\mathsf{pp}, \mathsf{st}_\mathcal{A})$ and oracle circuit OA, $\mathcal{B}^*$ constructs the following new oracle OA^*. On input (ek, c), OA^* computes $(x, r_\mathcal{R}) \leftarrow \mathsf{Ext}_{\mathsf{HE}}(ek, c)$, $y \leftarrow \mathsf{OA}(x)$, $c' \leftarrow \mathsf{Sim}_{\mathsf{HE}}(ek, c, y)$ and outputs c'. We claim that $\langle Ch, \mathcal{B}^{\mathcal{A}_2} \rangle$ and $\langle Ch, \mathcal{B}^{*\mathcal{A}} \rangle$ are identically distributed (where Ch is the challenger for the hard problem π). In fact, we have merely rearranged the algorithms $\mathsf{Ext}_{\mathsf{HE}}$ and $\mathsf{Sim}_{\mathsf{HE}}$ from the adversary $\mathcal{A}_2$ into the oracle OA^*. Note that now the reduction $\mathcal{B}^*$ is inefficient, whereas the adversary $\mathcal{A}$ is efficient.

Step 5. In the fifth and final step, we change the way the reduction $\mathcal{B}^*$ implements its oracles, obtaining an efficient reduction $\mathcal{B}'$. We will use the circuit privacy of HE a second time to implement the oracles efficiently. Whereas $\mathcal{B}^*$ constructs oracle circuit OA^* from oracle circuit OA provided by $\mathcal{B}$, $\mathcal{B}'$ proceeds as follows. On input (ek, c), the oracle OA' computes $c' \leftarrow \mathsf{Eval}(ek, \overline{\mathsf{OA}}(\cdot), c)$ and outputs c'. Define the circuit $\overline{\mathsf{OA}}$ to compute the function $\overline{\mathsf{OA}}(x, r) = \mathsf{OA}(x)$, i.e., it just drops its second input. Using the malicious circuit privacy of HE, we can establish that $\langle Ch, \mathcal{B}^{*\mathcal{A}} \rangle$ and $\langle Ch, \mathcal{B}'^{\mathcal{A}} \rangle$ are statistically close in the same fashion as in step 1. Finally, note that both $\mathcal{B}'$ and $\mathcal{A}$ are efficient, therefore $\mathcal{B}'^{\mathcal{A}}$ is efficient. $\qquad\square$

The following theorem shows that our protocol is secure against semi-honest senders. Note that achieving security against malicious senders is not possible (under standard assumptions). The corresponding impossibility results is given in Sect. 5.

Theorem 4. *If* HE *is an IND-CPA secure 1-hop homomorphic encryption scheme, then* Π *is secure against semi-honest senders.*

Proof. We will first provide the simulator Sim. The main idea of Sim is to run the protocol Π between a simulated sender $\mathcal{S}$ and a simulated receiver $\mathcal{R}$, where the receivers input is 0^n. After the protocol terminates, Sim outputs the view of the sender $\mathcal{S}$.

Now assume there exists a PPT distinguisher $\mathcal{D}$ that distinguishes the distributions $(x, s, \mathsf{view}(\mathcal{S}), \langle \mathcal{S}, \mathcal{R}(x) \rangle)$ and $(x, f, \mathsf{Sim}(s), f(x))$ with non-negligible advantage ϵ for some inputs s and x. We will construct an adversary $\mathcal{A}$ that breaks the IND-CPA security of HE with advantage ϵ. Given a public key ek, $\mathcal{A}$ chooses a random $r_\mathcal{R}$ and $r'_\mathcal{R}$, sets $m_0 = (x, r_\mathcal{R})$ and $m_1 = (0^n, r'_\mathcal{R})$ and sends (m_0, m_1) to the IND-CPA experiment. Let c be the challenger ciphertext. $\mathcal{A}$ chooses random coins $r_\mathcal{S}$, and runs $\mathcal{S}$ with input s, $r_\mathcal{S}$ and receiver message c. Let $\mathsf{view}(\mathcal{S})$ be the simulated view of the sender. Next, $\mathcal{A}$ computes $y = f(s, r_\mathcal{S}; x, r_\mathcal{R})$. Finally, $\mathcal{A}$ runs $\mathcal{D}$ on input $(x, s, \mathsf{view}(\mathcal{S}), y)$ and outputs whatever $\mathcal{D}$ outputs.

We claim that $\mathcal{A}$ breaks the IND-CPA security of HE with advantage ϵ. First assume the IND-CPA challenge bit is 0. In this case the IND-CPA returns to $\mathcal{A}$ an encryption of $m_0 = (x, r_{\mathcal{R}})$. Thus, the view that $\mathcal{A}$ simulates is identically distributed to $\mathcal{S}$'s view in the real experiment, but also the output y has the same distribution as in the real experiment. On the other hand, if the IND-CPA choice bit is 1, then $\mathsf{view}(\mathcal{S})$ is identically distributed to $\mathsf{Sim}(s)$ and the output $y = f(s, r_{\mathcal{S}}; x, r_{\mathcal{R}})$ is independently distributed of $\mathsf{view}(\mathcal{S})$ (as $r_{\mathcal{R}}$ is independent of $\mathsf{view}(\mathcal{S})$). Thus we conclude

$$\mathbf{Adv}_{\text{IND-CPA}}(\mathcal{A}) = |\Pr[\text{IND-CPA}^0(\mathcal{A}) = 1] - \Pr[\Pr[\text{IND-CPA}^1(\mathcal{A}) = 1]|$$
$$= |\Pr[\mathcal{D}(x, s, \mathsf{view}(\mathcal{S}), \langle \mathcal{S}(s), \mathcal{R}(x)\rangle) = 1]$$
$$- \Pr[\mathcal{D}(x, s, \mathsf{Sim}(s), f(s; x)) = 1]| = \epsilon,$$

which concludes the proof. $\square$

4   Round-Optimal Oblivious Pseudorandom Functions

Our technique yields the first two message oblivious pseudorandom function in the standard model. Oblivious pseudorandom functions are in essence pseudorandom functions that are obliviously evaluated in a two-party protocol. This means that the sender $\mathcal{S}$ holds a key k of a PRF F and the receiver $\mathcal{R}$ a value x and wishes to learn $F(k, x)$. As already discussed in the introduction, OPRFs have many applications, such as private key-word search [17], or secure computation of set intersection [34]. However, despite the popularity of this primitive, no scheme in the standard model is known with only two rounds of communication. Regarding the security of OPRFs, we wish to express that the sender $\mathcal{S}$ does not learn anything about the value x, and the receiver $\mathcal{R}$ learns only the pseudorandom value $F(k, x)$. First recall the standard definition of pseudorandom functions.

Definition 9 (Pseudorandom Functions). *An efficiently computable two-argument function PRF is called pseudorandom function, if it holds for every PPT distinguisher $\mathcal{D}$ that*

$$\mathbf{Adv}(\mathcal{D}) = |\Pr[\mathcal{D}^{PRF(k,\cdot)} = 1] - \Pr[\mathcal{D}^{H(\cdot)} = 1]| \leq \mathsf{negl}(\lambda),$$

where k is a randomly chosen key F and H is a random function with the same domain and range as F.

We will now turn to the standard definition of oblivious pseudorandom functions. This notion require for an OPRF protocol Π two properties to be satisfied. First, we require that Π is a secure two-party protocol realizing a function $F(k, x)$, where k is the sender input and x is the receiver input. Second, we require that $F(k, \cdot)$ is a pseudorandom function. The first part of this definition captures the idea that the π allows the receiver to learn one function value of $F(k, \cdot)$ per invocation only. The second requirement ensures that such function values are pseudorandom.

While this definition is appealing due to its modularity, it is impossible in the two message setting, even if we only consider semi-honest senders[1]. To circumvent this impossibility, we propose a security notion which captures both intuitive requirements in a single definition. In this definition, a PPT distinguisher is given access to an oracle, which implements either an OPRF sender or an unbounded simulator Sim with access to a truly random function H. Since we are considering two-message OPRF protocols, the distinguisher's queries to its oracle are simply the first message of a malicious receiver. Since we are in the two-message setting, the simulator has a very simple structure: It extracts the receiver's queries by brute force, forwards them to the random function H, and then simulates a response by the sender using the random function's output. This definition contains a minor loophole: It does not rule out trivial simulators, i.e., it does not require the simulator to use the random function it is given access to at all. The simulator could do anything, even simulating the real protocol (which would give perfect indistinguishability between the two distributions), which would defeat the purpose of the definition. To fix this, we will give the distinguisher direct access to the random function H. In the real execution, this is mirrored by giving the distinguisher access to an oracle that implements an honest receiver interacting with the sender. Now, the distinguisher can actually cross check the answers of the simulator. This definition has some flavor of the universal composability framework [9] and Nielsen's definition of non-programmable random oracles [50]. Think of a complex scenario where multiple receivers interact with one OPRF sender (e.g. a server). We may think of the distinguisher in our definition as an environment in control of several malicious receivers over which it has full control, but it can also choose inputs and observe outputs of honest receivers. Then this definition requires that from the environments view the OPRF server *looks like* it actually implements a truly random function.

Definition 10 (Security Against Malicious Receiver for Oblivious Pseudorandom Functions). *Let $\Pi = (\mathcal{S}, \mathcal{R})$ be a two-message protocol. We say that π is a two-message oblivious pseudorandom function, if for every PPT distinguisher $\mathcal{D}$ there exists a (possibly unbounded) simulator* Sim, *such that*

$$\mathbf{Adv}(\mathcal{D}) = |\Pr[\mathcal{D}^{\langle \mathcal{S}(k), \cdot \rangle, \langle \mathcal{S}(k), \mathcal{R}(\cdot) \rangle} = 1] - \Pr[\mathcal{D}^{\mathsf{Sim}^H(\cdot), H(\cdot)} = 1]| \leq \mathsf{negl}(\lambda),$$

where k is a randomly chosen input for $\mathcal{S}$ and H is a random function (with appropriate domain and range). Here, $\langle \mathcal{S}(k), \cdot \rangle$ is a session of π where $\mathcal{D}$ can choose the first message of the receiver and receives the second message by the sender. In $\langle \mathcal{S}(k), \mathcal{R}(\cdot) \rangle$, $\mathcal{D}$ chooses the input for $\mathcal{R}$ and obtains the output of $\mathcal{R}$.

The security guarantee for the receiver is the standard simulation based security against semi-honest senders (Definition 6).

---

[1] The impossibility is analogous to the impossibility of simulation based two message oblivious transfer. Consider a malicious receiver that gets auxiliary input z, which the malicious receiver sends as its first message. An efficient simulator Sim for this malicious receiver must extract in input x given only z.

Remark 2. We remark several points. First, if a simulator Sim is non-trivial by construction, we can omit the second oracle of the distinguisher. Basically, the only property we need to ensure non-triviality is that if the simulator gets messages from an honest receiver, then this composed system actually implements in the random function H. Formally, this requirement can be written as $\langle \mathsf{Sim}^H, \mathcal{R}(\cdot) \rangle \equiv H(\cdot)$, i.e., if an honest receiver interacts with a simulator Sim with access to H, then this protocol implements H. If this is guaranteed, then the oracles $\langle \mathcal{S}(k), \cdot \rangle$ and $\mathsf{Sim}^H(\cdot)$ are sufficient: Given such an oracle OA (which is either of the two), the distinguisher $\mathcal{D}$ can simulate the *honest* oracle by $\langle \mathsf{OA}, \mathcal{R}(\cdot) \rangle$. In our construction the simulator Sim will be canonical: It extracts the first message, sends the extracted input to the random function H, and uses the output to simulate the senders message. This simulator is non-trivial by construction, and thus giving the distinguisher access to a single oracle will be sufficient. Moreover, while Definition 10 allows the simulator Sim to depend on the distinguisher $\mathcal{D}$, our canonic simulator will be universal in the sense that it works for any PPT distinguisher $\mathcal{D}$.

Pseudorandom Functions with Oblivious Black-Box Reductions. To apply the technique developed in Sect. 3, we require a pseudorandom function with an oblivious black-box reduction. Most constructions of PRFs in the literature do not possess such a reduction. In particular, most reductions need to *program* the distinguishers oracle adaptively depending on prior oracle inputs of the distinguisher. For example, the security reduction of the construction of Goldreich, Goldwasser and Micali [25], which reduces the security of the PRF on that of the underlying pseudorandom generator is based on a hybrid argument and needs to keep a list of the distinguisher's distinct oracle queries to be able to answer oracle queries consistently. This however contradicts our notion of obliviousness.

Fortunately, there are constructions of pseudorandom functions with oblivious black-box reductions to their underlying hard problems. One example of such a PRF is the Naor Reingold PRF [49]. While the security reduction provided in [49] is not oblivious, there is simple way of converting this reduction into an oblivious black-box reduction using q-wise independent functions (Appendix A). More generally, there is a recent line of work that aims at constructing large-domain pseudorandom functions from small-domain pseudorandom functions via oblivious black-box reductions [8,14]. The baseline of these results is that large domain PRFs can be constructed by combining several small-domain (i.e., poly-sized domain) PRFs in a suitable way. The pseudorandomness of large domain PRFs is established by replacing one of the small-domain PRFs (depending on the query bound of the adversary) with a random function in a single shot. Since the small-domain PRF has a domain of just polynomial size, the reduction can (non-adptively) query its oracle on all inputs and retrieve the entire function table. Thus, there is no need of adaptively programming the distinguishers oracle based on previous queries. In order to use the framework we developed in Sect. 3, it will be convenient to use an alternative definition of pseudorandom functions. In Definition 9, the distinguishers goal is to distinguish the PRF from a truly random function. However, if we do not know any bound on the distinguisher's number of queries in advance, the only (known) way to simulate a random function is by evaluating the random function lazily: Every time the distinguisher queries the random function on a new input, the simulation samples a random image and adds it into a table of input and output values. If a certain input has been queried before, it's image is retrieved from the table. However, such a simulation is necessarily stateful. To overcome this, we use an equivalent definition of pseudorandom functions which takes into account that a every PPT distinguisher has a polynomial upper bound on the number of its oracle queries. Once such a bound q is known, we can simulate a random function statelessly with an efficient q-wise independent function.

Definition 11 (q-Wise Independent Function). *Let F be an efficiently computable two argument function that takes a seed s and an input x. We say that F is a q-wise independent functions, if it holds for all pairwise distinct $x_1, \ldots, x_q$ that $F(s, x_1), \ldots, F(s, x_q)$ are distributed independently and uniformly random over the choice of the seed s.*

There are various constructions of efficient q-wise independent functions, such as the classical construction of Wegman and Carter [57] which is based on random degree q polynomials in large finite fields.

Definition 12 (Pseudorandom Functions, Equivalent Definition). *An efficiently computable two-argument function PRF is called pseudorandom function, if there exists a family $\{F_q\}_q$ of functions, where F_q is q-wise independent, such that the following holds. For every $q = \mathsf{poly}(\lambda)$ and every PPT distinguisher $\mathcal{D}$ that queries its oracle at most q times it holds that*

$$\mathbf{Adv}(\mathcal{D}) = |\Pr[\mathcal{D}^{PRF(k, \cdot)} = 1] - \Pr[\mathcal{D}^{F_q(s, \cdot)} = 1]| \leq \mathsf{negl}(\lambda),$$

where k is a randomly chosen key for PRF and s is a randomly chosen seed for F_q.

Theorem 5 [8,14,49]. *Under various standard hardness assumptions (pseudorandom generators, DDH, LWE) there exist pseudorandom functions with oblivious black-box reduction to their underlying hardness assumption.*

Construction. The construction is expectably simple. We combine Construction 1 with a pseudorandom function that possesses an oblivious black-box reduction to some hard problem π, which is provided by Theorem 5. For this instantiation, we need to instantiate Construction 1 with a maliciously circuit private *fully* homomorphic encryption scheme (such as provided by Theorem 2), as there is no a priori upper bound on the size of the circuits that implement q-wise independent functions. For convenience, we write down the protocol as follows. Let PRF be a pseudorandom function and HE be a fully homomorphic encryption scheme. The OPRF protocol Π is given as follows.

Protocol Π_{OPRF}
Setup
    $\mathcal{S}_0(1^\lambda)$: Choose a random key k for PRF
Query
    $\mathcal{R}_1(x)$
        $(ek, sk) \leftarrow \mathsf{Kg}(1^\lambda)$
        $c \leftarrow \mathsf{Enc}(ek, x)$
        Send (ek, c) to $\mathcal{S}$
    $\mathcal{S}(k, (ek, c))$:
        $c' \leftarrow \mathsf{Eval}(ek, PRF(k, \cdot), x)$
        Send c' to $\mathcal{R}$
    $\mathcal{R}_2(c')$:
        $y \leftarrow \mathsf{Dec}(sk, c')$
        Output y

We can now prove the main theorem of this section.

Theorem 6. *Let* HE *be an IND-CPA secure maliciously circuit private fully homomorphic encryption scheme with perfect completeness (as provided by Theorem 2) and PRF be a pseudorandom function with an oblivious black-box reduction to hard problem π. Then the protocol Π_{OPRF} is an OPRF protocol with security against semi-honest senders and malicious receivers.*

Proof. We begin with the proof of security against malicious receivers defining the universal simulator Sim. Let $\mathsf{Ext_{HE}}$ and $\mathsf{Sim_{HE}}$ be the extractor and simulator for the statistical circuit privacy of HE. Simulator Sim is given as follows.

Simulator $\mathsf{Sim}^H(ek, c)$
    Has oracle access to a function H
    $x \leftarrow \mathsf{Ext_{HE}}(ek, c)$
    $y \leftarrow H(x)$
    $c' \leftarrow \mathsf{Sim_{HE}}(ek, y, c)$
    return c'

Now, let $\mathcal{D}$ be a PPT distinguisher that makes at most $q = \mathsf{poly}(\lambda)$ oracle queries and has non-negligible advantage ϵ against the malicious receiver security experiment of Π_{OPRF}, i.e.,

$$|\Pr[\mathcal{D}^{\langle\mathcal{S}(k),\cdot\rangle,\langle\mathcal{S}(k),\mathcal{R}(\cdot)\rangle} = 1] - \Pr[\mathcal{D}^{\mathsf{Sim}^H(\cdot),H(\cdot)} = 1]| \geq \epsilon.$$

First of all, notice that since $\mathcal{D}$ makes at most q queries to its oracles, we can efficiently (and statelessly) simulate the random function H by an efficiently computable q-wise independent function F_q, i.e., we get

$$|\Pr[\mathcal{D}^{\langle\mathcal{S}(k),\cdot\rangle,\langle\mathcal{S}(k),\mathcal{R}(\cdot)\rangle} = 1] - \Pr[\mathcal{D}^{\mathsf{Sim}^{F_q(s,\cdot)}(\cdot),F_q(s,\cdot)} = 1]| \geq \epsilon.$$

Our proof strategy will now be as follows. We will use $\mathcal{D}$ to construct a distinguisher $\mathcal{D}'$ with advantage $\epsilon' = \epsilon - \mathsf{negl}(\lambda)$ against the *induced security experiment* for PRF under the homomorphic encryption HE (c.f. Definition 5). Recall that the pseudorandom function PRF possesses an oblivious black-box reduction $\mathcal{B}$ to some hard problem π. Thus, Theorem 3 yields an efficient reduction $\mathcal{B}'$ such that $\mathcal{B}'^{\mathcal{D}'}$ has non-negligible advantage against π, contradicting its hardness.

We will now consider the induced security experiment for PRF. Therefore, we will first define a sender algorithm $\mathcal{S}'$. Basically, $\mathcal{S}'$ homomorphically evaluates the q-wise independent function F_q.

$\mathcal{S}'(s, (ek, c))$
    $c' \leftarrow \mathsf{Eval}(ek, F_q(s, \cdot), c)$
    return c'

Thus, while $\mathcal{S}$ homomorphically evaluates the pseudorandom function PRF, $\mathcal{S}'$ homomorphically evaluates the q-wise independent function F_q. Thus, the induced security experiment of the experiment given in Definition 12 asks to distinguish the oracles $\langle \mathcal{S}(k), \cdot \rangle$ and $\langle \mathcal{S}'(s), \cdot \rangle$.

We will now construct a distinguisher $\mathcal{D}'$ against the induced security experiment of PRF using the distinguisher $\mathcal{D}$. $\mathcal{D}'$ is given as follows.

Distinguisher $\mathcal{D}'(1^\lambda)$	Oracle $\mathsf{OA}_2(x)$
Has access to oracle OA_1	$y \leftarrow \langle \mathsf{OA}_1, \mathcal{R}(x) \rangle$
out $\leftarrow \mathcal{D}^{\mathsf{OA}_1(\cdot), \mathsf{OA}_2(\cdot)}(1^\lambda)$	Return y
Return out	

We claim that

$$| \Pr[\mathcal{D}'^{\langle \mathcal{S}(k), \cdot \rangle} = 1] - \Pr[\mathcal{D}^{\langle \mathcal{S}'(s), \cdot \rangle} = 1]| \geq \epsilon - \mathsf{negl}(\lambda), \tag{1}$$

i.e., $\mathcal{D}'$ has non-negligible advantage $\epsilon - \mathsf{negl}(\lambda)$ against the induced security experiment of PRF.

We claim that if $\mathsf{OA}_1 = \langle \mathcal{S}(k), \cdot \rangle$, then the output of $\mathcal{D}'^{\langle \mathcal{S}(k), \cdot \rangle}(1^\lambda)$ is identically distributed to the output $\mathcal{D}^{\langle \mathcal{S}(k), \cdot \rangle, \langle \mathcal{S}(k), \mathcal{R}(\cdot) \rangle}(1^\lambda)$. To see this, note that the oracle OA_2 implemented by $\mathcal{D}'$ is precisely $\langle \mathcal{S}(k), \mathcal{R}(\cdot) \rangle$ in this case.

On the other hand, if $\mathsf{OA}_1 = \langle \mathcal{S}'(s), \cdot \rangle$, then we claim that the output of $\mathcal{D}'^{\langle \mathcal{S}'(s), \cdot \rangle}$ is distributed statistically close to the output of $\mathcal{D}^{\mathsf{Sim}^{F_q}(\cdot), F_q(\cdot)}(1^\lambda)$. To see this, note first that in this case the oracle OA_2 provided by $\mathcal{D}'$ to $\mathcal{D}$ can be expressed as follows.

$\mathsf{OA}_2(x)$
    $(ek, sk) \leftarrow \mathsf{Kg}(1^\lambda)$
    $c \leftarrow \mathsf{Enc}(ek, x)$
    $c' \leftarrow \mathsf{Eval}(ek, F_q(s, \cdot), c)$
    $y \leftarrow \mathsf{Dec}(sk, c')$
    return y

It follows immediately from the perfect completeness of HE that OA_2 implements exactly $F_q(s, \cdot)$. It remains to show that the oracles $\langle S'(s), \cdot \rangle$ and $\mathsf{Sim}^{F_q}(\cdot)$ are statistically close. However, as $S'(s)$ homomorphically evaluates F_q, it follows from the malicious circuit privacy of HE that both oracles produce distributions that are statistically close, even given F_q. Thus, we can use a standard q-step hybrid argument over the queries of $\mathcal{D}$ to establish that $\mathcal{D}'^{\langle S'(s), \cdot \rangle}$ and $\mathcal{D}^{\mathsf{Sim}^{F_q}(\cdot), F_q(\cdot)}(1^\lambda)$ are statistically close. Thus, (1) follows and we can apply Theorem 3 to arrive at a contradiction. Security against semi-honest senders follows directly from Theorem 4, which concludes the proof. $\qquad \square$

5 Impossibility of Malicious Sender Security

In this section, we show that malicious receiver security (w.r.t. our notion of induced game-based security) and malicious sender security cannot be achieved simultaneously. Our impossibility result is constructive in the sense that we show that our framework covers the standard security notion of blind signatures. However, Fischlin and Schröder showed that a large class of three-move blind signature schemes cannot be proven secure under standard assumptions [16]. Since our framework falls into this class, the impossibility result follows.

Blind Signatures. Blind signatures [11] implement a carbon copy envelope allowing a signer to issue signatures for messages such that the signer's signature on the envelope is imprinted onto the message in the sealed envelope. In particular, the signer remains oblivious about the message (blindness), but at the same time no additional signatures without the help of the signer can be created (unforgeability). Constructing round-optimal blind signature schemes in the standard model has been a long standing open question. Fischlin and Schröder showed that all previously known schemes having at most three rounds of communication, cannot be proven secure under non-interactive assumptions in the standard model via black-box reductions [16]. Subsequently, several works used a technique called "complexity leveraging" to circumvent this impossibility result [19,20] and recently, Fuchsbauer, Hanser, Slamanig suggested a round optimal blind signature scheme that is secure in the generic group model [18]. In fact, it is still unknown if round optimal blind signatures, based on standard assumptions, exist in the standard model.

By applying our technique to the oblivious computation of signatures, we obtain a round optimal blind signature scheme without complexity leveraging and whose security can be based on standard cryptographic assumptions. Since our scheme belongs to the class characterized by Fischlin and Schröder it is not possible to prove blindness w.r.t. malicious adversaries.

Security Definition for Blind Signatures. We recall the unforgeability definition of blind signatures [35,53] that can be expressed within our formalization of a cryptographic experiment.

Definition 13 (Unforgeability). *An interactive signature scheme* $\mathsf{BS} = (\mathsf{KG}, \langle \mathcal{S}, \mathcal{U} \rangle, \mathsf{Vf})$ *is called* unforgeable *if for any efficient algorithm* $\mathcal{A}$ *(the malicious user) the probability that experiment* $\mathsf{Forge}_{\mathcal{A}}^{\mathsf{BS}}(\lambda)$ *evaluates to* 1 *is negligible (as a function of* λ*) where*

Experiment $\mathsf{Forge}_{\mathcal{A}}^{\mathsf{BS}}(\lambda)$

> $(sk, pk) \leftarrow \mathsf{KG}(1^{\lambda})$
> $((m_1^*, \sigma_1^*), \ldots, (m_{k+1}^*, \sigma_{k+1}^*)) \leftarrow \mathcal{A}^{\langle \mathcal{S}(sk), \cdot \rangle^{\infty}}(pk)$
> *Return* 1 *iff*
>> $m_i^* \neq m_j^*$ *for all* i, j *with* $i \neq j$, *and*
>> $\mathsf{Vf}(pk, m_i^*, \sigma_i^*) = 1$ *for all* i, *and*
>> $\mathcal{S}$ *has returned* ok *in at most* k *interactions.*

The corresponding definition of blindness says that it should be infeasible for a malicious signer $\mathcal{S}^*$ to decide which of two messages m_0 and m_1 has been signed first in two executions with an honest user $\mathcal{U}$. If one of these executions has returned $\bot$ then the signer is not informed about the other signature (Otherwise the signer could trivially identify one session by making the other abort.). If one restricts this definition the semi-honest adversaries, then this definition is immediately implied by Definition 6.

Construction. Our construction instantiates our general framework as defined in Construction 1 with a signature scheme $\mathsf{DS} = (\mathsf{Kg}_{\mathsf{Sig}}, \mathsf{Sig}, \mathsf{Vf})$ that has an oblivious black-box reduction to some underlying hard problem π. For this instantiation, we need maliciously circuit private homomorphic encryption for logarithmic depth circuits that can be achieved by combining information-theoretic garbled circuits (aka randomized encodings) [2,33,38] with two-message oblivious transfer [1,30,48] as provided by Theorem 1. Moreover, we need a digital signature scheme that can computed via a logarithmic depth circuit. Such a signature scheme can be obtained by using the non-apaptively secure signature scheme by Applebaum et al. [2]. However, this scheme is only non-adaptively secure, which means the adversary has to commit to all messages before learning the public-key and the signature. Using the standard transformation based on chameleon hash functions [31,40] one can convert any non-adaptively secure signature scheme into one that is adaptively secure. Here we actually deal with two reductions. One that deals with adversaries that find collisions of the chameleon hash function and one that deals with adversaries that do not find hash collisions, but still manage to forge signatures. The first reduction is easily seen to be obliviously black-box, as the reduction possesses the signing key for the signature scheme an hash collisions can be easily recovered from the adversary's output. Here the signing circuit is the same as in the real experiment. The second reduction has the following structure. If q is the query bound of the adversary, the reduction computes chameleon hashes on q random values and has them (non-adaptively) signed by the signing oracle. Each time the adversary queries its signing oracle, the reduction uses up one of the precomputed signatures of the chameleon hashes by computing a hash collision with the adversary's query and returning the corresponding signature to the adversary. Note that since the reduction is allowed to reprogram the signing circuit after each query, we only need to hard-wire a single hash value and trapdoor at a time into the signing oracle circuit. Since chameleon hash functions can easily be obtained from the discrete logarithm assumption involving only two modular exponentiations and a multiplication [40], this transformation can also be computed by a circuit of logarithmic depth. Thus we obtain an oblivious black-box reduction to the non-adaptive unforgeability of the signature scheme where every circuit used by the reduction has a most an a priori known logarithmic depth. We obtain the following theorem.

Theorem 7. *Let* HE *be an IND-CPA secure maliciously circuit private homomorphic encryption scheme with perfect completeness for circuits of logarithmic depth and let* DS *be a signature scheme compute by a circuit of logarithmic depth and with an oblivious black-box reduction to hard problem* π. *Then the protocol* Π_{BS} *defined above is a blind signature protocol with security against semi-honest senders and malicious receivers.*

Given this theorem, we obtain our impossibility result in the following corollary.

Corollary 1 (Impossibility of Malicious Sender Security, Informal). *There exists no two-move secure evaluation protocol for cryptographic functionalities that is secure against malicious receivers and senders based on standard assumptions.*

Acknowledgement. Nico Döttling gratefully acknowledges support by the DAAD (German Academic Exchange Service) under the postdoctoral program (57243032). This work was in part supported by European Research Council Starting Grant 279447. Research supported in part from a DARPA/ARL SAFEWARE award, AFOSR Award FA9550-15-1-0274, and NSF CRII Award 1464397. The views expressed are those of the author and do not reflect the official policy or position of the Department of Defense, the National Science Foundation, or the U.S. Government. Nils Fleischhacker, Johannes Krupp and Dominique Schröder were supported by the German Federal Ministry of Education and Research (BMBF) through funding for the Center for IT-Security, Privacy and Accountability (CISPA – www.cispa-security.org) and the project PROMISE. Moreover, it was supported by the Initiative for Excellence of the German federal and state governments through funding for the Saarbrücken Graduate School of Computer Science and the DFG MMCI Cluster of Excellence. Part of this work was also supported by the German research foundation (DFG) through funding for the collaborative research center 1223 and by the DAAD PPP USA program (57129666). We would like to thank the anonymous reviewers of CRYPTO 2016 for their helpful comments.

A   An Oblivious Black-Box Reduction for Naor-Reingold PRF

Lemma 1. *The Naor-Reingold PRF is secure under the DDH assumption and the reduction is oblivious.*

Proof. Given an adversary $\mathcal{A}$ who can distinguish the Naor-Reingold PRF with non-negligible probability $\epsilon(\lambda)$ from a truly random function making at most q queries to its oracle, consider the following oblivious reduction $\mathcal{B}$ against DDH:

$\mathcal{B}$ gets as input a DDH instance $(g, g^a, g^b, g^{\tilde{c}})$, where either $\tilde{c} = a \cdot b$ or not. We restrict the reduction to the case where $a, b, \tilde{c} \neq 0$ (otherwise it is trivial to tell whether $\tilde{c} = a \cdot b$). $\mathcal{B}$ will choose a random $j \xleftarrow{\$} \{1, \dots, \lambda\}$ and pick a random q-wise independent function $F \xleftarrow{\$} \mathcal{F}^q$. It will then sample values $(a_{j+1}, \dots, a_\lambda) \xleftarrow{\$} \mathbb{Z}_p$ and program the oracle OA for $\mathcal{A}$ as follows:

OA(x):

$\qquad \overline{x}\, x_j \dots x_\lambda = x$, where $\overline{x}$ is the $(j-1)$-bit prefix of x

$\qquad \alpha = F(\overline{x})$

$\qquad$ If $x_j = 0$:

$$\qquad\qquad \text{Return } \left((g^b)^\alpha\right)^{\prod_{k=j+1}^{\lambda} a_k^{x_k}}$$

$\qquad$ else

$$\qquad\qquad \text{Return } \left((g^{\tilde{c}})^\alpha\right)^{\prod_{k=j+1}^{\lambda} a_k^{x_k}}$$

The reduction $\mathcal{B}$ will invoke $\mathcal{A}^{\mathsf{OA}}$ and output 1 exactly whenever $\mathcal{A}^{\mathsf{OA}}$ does.

If $\tilde{c} = a \cdot b$, then for $j = 1$ the oracle perfectly simulates the Naor-Reingold PRF $\mathsf{PRF}_{\vec{a}}$ with key $\vec{a} = (b\alpha, a, a_2, \dots, a_\lambda)$ (since $\overline{x}$ will be the empty string, α will be constant). Furthermore, if $\tilde{c} \neq a \cdot b$, then for $j = \lambda$ the oracle perfectly simulates a q-wise independent function f (observed as truly random by $\mathcal{A}$):

$$\mathrm{Prob}\left[\mathcal{B}^{\mathcal{A}}(g, g^a, g^b, g^{\tilde{c}}) = 1 \,\middle|\, \tilde{c} = a \cdot b \wedge j = 1\right] = \mathrm{Prob}\left[\mathcal{A}^{\mathsf{PRF}_{\vec{a}}}(1^\lambda) = 1\right]$$
$$\mathrm{Prob}\left[\mathcal{B}^{\mathcal{A}}(g, g^a, g^b, g^{\tilde{c}}) = 1 \,\middle|\, \tilde{c} \neq a \cdot b \wedge j = \lambda\right] = \mathrm{Prob}\left[\mathcal{A}^{f}(1^\lambda) = 1\right]$$

Since $g^{\tilde{c}}$ is independent of g^b in case of $\tilde{c} \neq a \cdot b$ it holds that

$$\mathrm{Prob}\left[\mathcal{B}^{\mathcal{A}}(g, g^a, g^b, g^{\tilde{c}}) = 1 \,\middle|\, \tilde{c} \neq a \cdot b \wedge j = i\right]$$
$$= \mathrm{Prob}\left[\mathcal{B}^{\mathcal{A}}(g, g^a, g^b, g^{\tilde{c}}) = 1 \,\middle|\, \tilde{c} = a \cdot b \wedge j = i+1\right]$$

And therefore

$$\left|\mathrm{Prob}\left[\mathcal{B}^{\mathcal{A}}(g, g^a, g^b, g^{\tilde{c}}) = 1 \,\middle|\, \tilde{c} = a \cdot b\right]\right.$$
$$\left. - \mathrm{Prob}\left[\mathcal{B}^{\mathcal{A}}(g, g^a, g^b, g^{\tilde{c}}) = 1 \,\middle|\, \tilde{c} \neq a \cdot b\right]\right|$$
$$= \left|\frac{1}{\lambda} \cdot \sum_{i=1}^{\lambda} \mathrm{Prob}\left[\mathcal{B}^{\mathcal{A}}(g, g^a, g^b, g^{\tilde{c}}) = 1 \,\middle|\, \tilde{c} = a \cdot b \wedge j = i\right]\right.$$
$$\left. - \frac{1}{\lambda} \cdot \sum_{i=1}^{\lambda} \mathrm{Prob}\left[\mathcal{B}^{\mathcal{A}}(g, g^a, g^b, g^{\tilde{c}}) = 1 \,\middle|\, \tilde{c} \neq a \cdot b \wedge j = i\right]\right|$$
$$= \frac{1}{\lambda}\left|\mathrm{Prob}\left[\mathcal{B}^{\mathcal{A}}(g, g^a, g^b, g^{\tilde{c}}) = 1 \,\middle|\, \tilde{c} = a \cdot b \wedge j = 1\right]\right.$$
$$\left. - \mathrm{Prob}\left[\mathcal{B}^{\mathcal{A}}(g, g^a, g^b, g^{\tilde{c}}) = 1 \,\middle|\, \tilde{c} \neq a \cdot b \wedge j = \lambda\right]\right|$$
$$= \frac{1}{\lambda}\left|\mathrm{Prob}\left[\mathcal{A}^{\mathsf{PRF}_{\vec{a}}}(1^\lambda) = 1\right] - \mathrm{Prob}\left[\mathcal{A}^{f}(1^\lambda) = 1\right]\right| \geq \frac{1}{\lambda}\epsilon(\lambda)$$

Thus this reduction will break the DDH assumption with non-negligible probability. As the reduction does not see the queries $\mathcal{A}$ makes to the oracle OA, it is oblivious according to Definition 4. This concludes the proof.

References

1. Aiello, W., Ishai, Y., Reingold, O.: Priced oblivious transfer: how to sell digital goods. In: Pfitzmann, B. (ed.) EUROCRYPT 2001. LNCS, vol. 2045, p. 119. Springer, Heidelberg (2001). 3.1, 1, 5
2. Applebaum, B., Ishai, Y., Kushilevitz, E.: Cryptography in NC^0. In: 45th Annual Symposium on Foundations of Computer Science, pp. 166–175. IEEE Computer Society Press, October 2004. 3.1, 1, 5
3. Bar-Ilan, J., Beaver, D.: Non-cryptographic fault-tolerant computing in constant number of rounds of interaction. In: Proceedings of the Eighth Annual ACM Symposium on Principles of Distributed Computing, pp. 201–209. ACM (1989). 1.4
4. Beaver, D., Micali, S., Rogaway, P.: The round complexity of secure protocols (extended abstract). In: 22nd Annual ACM Symposium on Theory of Computing, pp. 503–513. ACM Press, May 1990. 1.4
5. Bellare, M., Jakobsson, M., Yung, M.: Round-optimal zero-knowledge arguments based on any one-way function. In: Fumy, W. (ed.) EUROCRYPT 1997. LNCS, vol. 1233, pp. 280–305. Springer, Heidelberg (1997). 1.4
6. Bellare, M., Rogaway, P.: The security of triple encryption and a framework for code-based game-playing proofs. In: Vaudenay, S. (ed.) EUROCRYPT 2006. LNCS, vol. 4004, pp. 409–426. Springer, Heidelberg (2006). 2, 2.1
7. Bendlin, R., Damgård, I., Orlandi, C., Zakarias, S.: Semi-homomorphic encryption and multiparty computation. In: Paterson, K.G. (ed.) EUROCRYPT 2011. LNCS, vol. 6632, pp. 169–188. Springer, Heidelberg (2011). 1.4
8. Berman, I., Haitner, I.: From non-adaptive to adaptive pseudorandom functions. In: Cramer, R. (ed.) TCC 2012. LNCS, vol. 7194, pp. 357–368. Springer, Heidelberg (2012). 4, 5
9. Canetti, R.: Universally composable security: a new paradigm for cryptographic protocols. In: 42nd Annual Symposium on Foundations of Computer Science, pp. 136–145. IEEE Computer Society Press, October 2001. 4
10. Canetti, R., Kilian, J., Petrank, E., Rosen, A.: Black-box concurrent zero-knowledge requires omega (log n) rounds. In: 33rd Annual ACM Symposium on Theory of Computing, pp. 570–579. ACM Press, July 2001. 1.4
11. Chaum, D.: Blind signature system. In: Advances in Cryptology - CRYPTO 1983, p. 153. Plenum Press, New York (1983). 1.4, 5
12. Cramer, R., Damgård, I.B.: Secure distributed linear algebra in a constant number of rounds. In: Kilian, J. (ed.) CRYPTO 2001. LNCS, vol. 2139, p. 119. Springer, Heidelberg (2001). 1.4
13. Damgård, I., Pastro, V., Smart, N., Zakarias, S.: Multiparty computation from somewhat homomorphic encryption. In: Safavi-Naini, R., Canetti, R. (eds.) CRYPTO 2012. LNCS, vol. 7417, pp. 643–662. Springer, Heidelberg (2012). 1.4
14. Döttling, N., Schröder, D.: Efficient pseudorandom functions via on-the-fly adaptation. In: Gennaro, R., Robshaw, M.J.B. (eds.) CRYPTO 2015. LNCS, vol. 9215, pp. 329–350. Springer, Heidelberg (2015). 4, 5
15. Feige, U., Shamir, A.: Zero knowledge proofs of knowledge in two rounds. In: Brassard, G. (ed.) CRYPTO 1989. LNCS, vol. 435, pp. 526–544. Springer, Heidelberg (1990). 1.4

16. Fischlin, M., Schröder, D.: On the impossibility of three-move blind signature schemes. In: Gilbert, H. (ed.) EUROCRYPT 2010. LNCS, vol. 6110, pp. 197–215. Springer, Heidelberg (2010). 1.1, 1.4, 5, 5

17. Freedman, M.J., Ishai, Y., Pinkas, B., Reingold, O.: Keyword search and oblivious pseudorandom functions. In: Kilian, J. (ed.) TCC 2005. LNCS, vol. 3378, pp. 303–324. Springer, Heidelberg (2005). 1.4, 4

18. Garg, S., Rao, V., Sahai, A., Schröder, D., Unruh, D.: Round optimal blind signatures. In: Rogaway, P. (ed.) CRYPTO 2011. LNCS, vol. 6841, pp. 630–648. Springer, Heidelberg (2011). 1.4, 5

19. Garg, S., Gupta, D.: Efficient round optimal blind signatures. In: Nguyen, P.Q., Oswald, E. (eds.) EUROCRYPT 2014. LNCS, vol. 8441, pp. 477–495. Springer, Heidelberg (2014). 1.4, 5

20. Garg, S., Rao, V., Sahai, A., Schröder, D., Unruh, D.: Round optimal blind signatures. In: Rogaway, P. (ed.) CRYPTO 2011. LNCS, vol. 6841, pp. 630–648. Springer, Heidelberg (2011). 1.4, 5

21. Gennaro, R., Ishai, Y., Kushilevitz, E., Rabin, T.: The round complexity of verifiable secret sharing and secure multicast. In: 33rd Annual ACM Symposium on Theory of Computing, pp. 580–589. ACM Press, July 2001. 1.4

22. Gennaro, R., Ishai, Y., Kushilevitz, E., Rabin, T.: On 2-round secure multiparty computation. In: Yung, M. (ed.) CRYPTO 2002. LNCS, vol. 2442, p. 178. Springer, Heidelberg (2002). 1.4

23. Gentry, C.: Fully homomorphic encryption using ideal lattices. In: Mitzenmacher, M., (ed.) 41st Annual ACM Symposium on Theory of Computing, pp. 169–178. ACM Press, May/June 2009. 1.2, 3.1, 3.1

24. Goldreich, O.: Foundations of Cryptography: Volume 2, Basic Applications. Cambridge University Press, New York (2004). 2.3

25. Goldreich, O., Goldwasser, S., Micali, S.: How to construct random functions (extended abstract). In: 25th Annual Symposium on Foundations of Computer Science, pp. 464–479. IEEE Computer Society Press, October 1984. 4

26. Goldreich, O., Kahan, A.: How to construct constant-round zero-knowledge proof systems for NP. J. Cryptology 9(3), 167–190 (1996). 1.4

27. Goldreich, O., Krawczyk, H.: On the composition of zero-knowledge proof systems. SIAM J. Comput. 25(1), 169–192 (1996). 1.4

28. Goldreich, O., Micali, S., Wigderson, A.: How to play any mental game or a completeness theorem for protocols with honest majority. In: Aho, A. (ed.) 19th Annual ACM Symposium on Theory of Computing, pp. 218–229. ACM Press, May 1987. 1.4

29. Goldreich, O., Oren, Y.: Definitions and properties of zero-knowledge proof systems. J. Cryptol. 7(1), 1–32 (1994). 1, 1.1, 1.4, 2.3

30. Halevi, S., Kalai, Y.T.: Smooth projective hashing and two-message oblivious transfer. J. Cryptol 25(1), 158–193 (2012). 3.1, 1, 5

31. Hohenberger, S., Waters, B.: Short and stateless signatures from the RSA assumption. In: Halevi, S. (ed.) CRYPTO 2009. LNCS, vol. 5677, pp. 654–670. Springer, Heidelberg (2009). 5

32. Ishai, Y., Kushilevitz, E.: Randomizing polynomials: a new representation with applications to round-efficient secure computation. In: 41st Annual Symposium on Foundations of Computer Science, pp. 294–304. IEEE Computer Society Press, November 2000. 1.4

33. Ishai, Y., Paskin, A.: Evaluating branching programs on encrypted data. In: Vadhan, S.P. (ed.) TCC 2007. LNCS, vol. 4392, pp. 575–594. Springer, Heidelberg (2007). 3.1, 1, 5

34. Jarecki, S., Liu, X.: Efficient oblivious pseudorandom function with applications to adaptive OT and secure computation of set intersection. In: Reingold, O. (ed.) TCC 2009. LNCS, vol. 5444, pp. 577–594. Springer, Heidelberg (2009). 1.4, 4

35. Juels, A., Luby, M., Ostrovsky, R.: Security of blind digital signatures. In: Kaliski Jr., B.S. (ed.) CRYPTO 1997. LNCS, vol. 1294, pp. 150–164. Springer, Heidelberg (1997). 5

36. Katz, J., Ostrovsky, R.: Round-optimal secure two-party computation. In: Franklin, M. (ed.) CRYPTO 2004. LNCS, vol. 3152, pp. 335–354. Springer, Heidelberg (2004). 1.1, 1.4

37. Katz, J., Ostrovsky, R., Smith, A.: Round efficiency of multi-party computation with a dishonest majority. In: Biham, E. (ed.) EUROCRYPT 2003. LNCS, vol. 2656, pp. 578–595. Springer, Heidelberg (2003). 1.4

38. Kilian, J.: Founding cryptography on oblivious transfer. In: 20th Annual ACM Symposium on Theory of Computing, pp. 20–31. ACM Press, May 1988. 3.1, 1, 5

39. Kilian, J., Petrank, E.: Concurrent and resettable zero-knowledge in poly-loalgorithm rounds. In: 33rd Annual ACM Symposium on Theory of Computing, pp. 560–569. ACM Press, July 2001. 1.4

40. Krawczyk, H., Rabin, T.: Chameleon signatures. In: ISOC Network and Distributed System Security Symposium - NDSS 2000. The Internet Society, February 2000. 5

41. Lindell, Y.: Parallel coin-tossing and constant-round secure two-party computation. In: Kilian, J. (ed.) CRYPTO 2001. LNCS, vol. 2139, p. 171. Springer, Heidelberg (2001). 1.4

42. Lindell, Y.: Bounded-concurrent secure two-party computation without setup assumptions. In: 35th Annual ACM Symposium on Theory of Computing, pp. 683–692. ACM Press, June 2003. 1.4

43. Lindell, Y.: Lower bounds for concurrent self composition. In: Naor, M. (ed.) TCC 2004. LNCS, vol. 2951, pp. 203–222. Springer, Heidelberg (2004). 1.4

44. Lindell, Y., Pinkas, B.: Secure two-party computation via cut-and-choose oblivious transfer. J. Cryptology 25(4), 680–722 (2012). 1.4

45. Naor, M.: On cryptographic assumptions and challenges. In: Boneh, D. (ed.) CRYPTO 2003. LNCS, vol. 2729, pp. 96–109. Springer, Heidelberg (2003). 2.2

46. Naor, M., Pinkas, B.: Oblivious transfer and polynomial evaluation. In: 31st Annual ACM Symposium on Theory of Computing, pp. 245–254. ACM Press, May 1999. 1.4

47. Naor, M., Pinkas, B.: Oblivious transfer with adaptive queries. In: Wiener, M. (ed.) CRYPTO 1999. LNCS, vol. 1666, p. 573. Springer, Heidelberg (1999). 1.4

48. Naor, M., Pinkas, B.: Efficient oblivious transfer protocols. In: Kosaraju, S.R. (ed.) 12th Annual ACM-SIAM Symposium on Discrete Algorithms, pp. 448–457. ACM-SIAM, January 2001. 3.1, 1, 5

49. Naor, M., Reingold, O.: Number-theoretic constructions of efficient pseudo-random functions. In: 38th Annual Symposium on Foundations of Computer Science, pp. 458–467. IEEE Computer Society Press, October 1997. 1, 1.4, 4, 5

50. Nielsen, J.B.: Separating random oracle proofs from complexity theoretic proofs: the non-committing encryption case. In: Yung, M. (ed.) CRYPTO 2002. LNCS, vol. 2442, p. 111. Springer, Heidelberg (2002). 4

51. Nielsen, J.B., Nordholt, P.S., Orlandi, C., Burra, S.S.: A new approach to practical active-secure two-party computation. In: Safavi-Naini, R., Canetti, R. (eds.) CRYPTO 2012. LNCS, vol. 7417, pp. 681–700. Springer, Heidelberg (2012). 1.4

52. Ostrovsky, R., Paskin-Cherniavsky, A., Paskin-Cherniavsky, B.: Maliciously circuit-private FHE. In: Garay, J.A., Gennaro, R. (eds.) CRYPTO 2014, Part I. LNCS, vol. 8616, pp. 536–553. Springer, Heidelberg (2014). 3.1, 2

53. Pointcheval, D., Stern, J.: Security arguments for digital signatures and blind signatures. J. Cryptol. **13**(3), 361–396 (2000). 5
54. Prabhakaran, M., Rosen, A., Sahai, A.: Concurrent zero knowledge with logarithmic round-complexity. In: 43rd Annual Symposium on Foundations of Computer Science, pp. 366–375. IEEE Computer Society Press, November 2002. 1.4
55. Reingold, O., Trevisan, L., Vadhan, S.P.: Notions of reducibility between cryptographic primitives. In: Naor, M. (ed.) TCC 2004. LNCS, vol. 2951, pp. 1–20. Springer, Heidelberg (2004). 2.2
56. Richardson, R., Kilian, J.: On the concurrent composition of zero-knowledge proofs. In: Stern, J. (ed.) EUROCRYPT 1999. LNCS, vol. 1592, p. 415. Springer, Heidelberg (1999). 1.4
57. Wegman, M.N., Carter, L.: New classes and applications of hash functions. In: 20th Annual Symposium on Foundations of Computer Science, San Juan, Puerto Rico, 29–31 October 1979, pp. 175–182 (1979). 4
58. Yao, A.C.C.: Protocols for secure computations (extended abstract). In: 23rd Annual Symposium on Foundations of Computer Science, pp. 160–164. IEEE Computer Society Press, November 1982. 1.4

Author Index

Printed in the United States
By Bookmasters